INTERNATIONAL HUMAN RIGHTS IN CONTEXT

LAW, POLITICS, MORALS

INTERNATIONAL HUMAN RIGHTS IN CONTEXT

LAW, POLITICS, MORALS

Text and Materials

HENRY J. STEINER

Jeremiah Smith, Jr. Professor of Law and Director of Law School Human Rights Program, Harvard University

and

PHILIP ALSTON

Professor of International Law, European University Institute, Florence; Professor of Law, Australian National University

CLARENDON PRESS · OXFORD

Oxford University Press, Walton Street, Oxford OX2 6DP
Oxford New York
Athens Auckland Bangkok Bombay
Calcutta Cape Town Dar es Salaam Delhi
Florence Hong Kong Istanbul Karachi
Kuala Lumpur Madras Madrid Melbourne
Mexico City Nairobi Paris Singapore
Taipei Tokyo Toronto
and associated companies in
Berlin Ibadan

Oxford is a trade mark of Oxford University Press

Published in the United States by
Oxford University Press Inc., New York

First published 1996
Reprinted 1996

British Library Cataloguing in Publication Data
Data available

Library of Congress Cataloging in Publication Data
Data available
ISBN 0-19-825427-X
ISBN 0-19-825426-1 (pbk)

Printed and bound in Great Britain
on acid-free paper by
Biddles Ltd, Guildford and King's Lynn

Preface

As little as two decades ago, rare was the university whose curriculum included human rights studies. Much has changed, to the point where one no longer questions why such studies should be offered but rather how they could be ignored.

The human rights movement—the book uses the term to include governmental and intergovernmental as well as non-governmental developments in human rights since 1945—grew out of the disasters of World War II. A mere half century later human rights ideals deeply inform both the practice and theory of international law and politics. This coursebook critically examines the movement's achievements and prospects that now form an indelible part of our legal, political and moral landscape. The failures and dilemmas of efforts to realize human rights ideals are as instructive for an understanding of that landscape.

The university curriculum readily evidences this striking relevance of human rights to fields of study as diverse as law, government, international relations and institutions, moral theory, public health, world financial institutions, ecology, economic development, religion, education and anthropology. The frailties of the human rights movement are evident, and explored throughout this book. Nonetheless, the ideals animating that movement have become a part of modern consciousness, a lens through which to see the world, a universal discourse, a subject in their own right as well as a vital component of many others, a potent rhetoric and aspiration.

This coursebook presents a broad understanding of human rights ideals and the human rights system. It describes, analyzes, criticizes, proposes and provokes. It poses many questions that mean to engage the reader with the book's themes, thereby making the student an active participant with both fellow students and teacher in an ongoing inquiry about what human rights norms, processes and institutions have meant and have the potential to mean for the world.

The central premise to the book is that an introductory course on human rights should educate students to see the "big picture." Such a course should encourage students not only to master the basic history and doctrine and institutional structures, but also to reflect critically about international human rights as a whole. Such are the dual purposes of the authors' extensive text and questions throughout the book.

Specialized skills and knowledge—relevant, say, to domestic or international litigation, work in an international organization, investigative missions, or lobbying before UN or state organs—can readily be acquired when needed by the student of human rights who becomes a professional practitioner. The conceptual framework provided by this book will equip the student seriously engaging with it to work in various roles to advance human rights: legal advocate, activist in a non-governmental organization, policy analyst, scholar, and so on.

Other premises and assumptions of the coursebook follow.

1. The strongest focus is on legal materials, both international and domestic, hardly surprising in the light of a human rights movement that has struggled to assume so law-like a character. Thus the coursebook is attentive throughout to legal texts, to the lawmaking processes of custom and treaty, and to the work of institutions that have some "legal" output. Its readings and discussions often explore the relationships among law, politics and morals.

But what then are the appropriate boundaries of a book that has an important focus on "law"? Clearly the coursebook could not achieve its goals if it held to a formal or positivist conception of law. It could not illuminate the character of international human rights without consideration of some of the related fields of inquiry noted above.

Thus readings from different disciplines and formal legal materials like treaties or committee decisions engage each other, forming part of a larger system of interacting events and ideas. Such varied materials should be accessible to a broad range of university students, from academic backgrounds as diverse as law, government, political economy, philosophy, theology, business or public health. Each of the human rights courses that we have taught in law schools has benefited from the participation of students from such other university faculties. The coursebook can indeed be employed both within law schools and several other faculties.

2. A book emphasizing legal norms and institutions will characteristically devote much attention to the work of courts. In some states, particularly the Western democracies, courts play a leading role in resolving human rights controversies and developing human rights norms. But in most states, particularly authoritarian and politically repressive ones, courts play a reduced or insignificant role in this field. Relative to Western democracies, their competence to review executive or legislative action may be sharply reduced or eliminated, their jurisdiction limited, at times their decrees ignored and their judges subjected to political pressure and threat. The struggle for rights becomes fully a political struggle in which courts may at best be marginal actors.

Nor have international tribunals of a universal character played a major role in settling human rights disputes and developing human rights ideals. The coursebook includes opinions of international courts and arbitral tribunals, but they are rare. The prominence of the court in the European human rights system stands as the striking exception. Apart from that regional system (and, to a lesser extent, the Inter-American system) and from aspects of United States constitutional law that are specifically relevant to two chapters, the opinions of courts figure only rarely in this coursebook. In addition to our own text and questions, most materials therefore consist of secondary readings from a range of disciplines together with diverse primary source materials: treaties, resolutions and statutes; debates, reports, decisions and other acts of intergovernmental assemblies, commissions and committees as well as of non-governmental organizations; and so on.

3. Human rights are violated within states, not in outer space or on the high seas or (for most of the topics examined in this coursebook) during combat between states. It could therefore be argued that rights should be studied within the framework of different states—say, human rights *in* Kenya, *in* Pakistan, Peru, China, France, Israel, the United States. Such a book might consist of contextual studies of human rights issues—regulation of police, freedom of the press, religion and the state, discrimination, and so on—that would draw on different national histories and political cultures. It would have the character and high value of studies in comparative law, history and culture.

This book follows a different path. The distinctive aspect of the human rights movement of the last half century has been its invention and creation on the international level. Hence our stress is on the *international* human rights system, as well as on the vital relationships between that system and states' internal orders. Although many illustrations throughout the book draw on human rights violations within one or another state, most of the materials address international norms, processes and institutions, both in their own terms and in terms of their reciprocal relationships with internal orders including the trend in recent decades toward the spread of constitutionalism among states.

4. In vital respects, a book that concentrates on international human rights must count on students' rudimentary knowledge of international law. How else can a class discuss human rights issues involving treaties and custom, intergovernmental institutions and processes, and the internalization of treaty or customary norms within a state? Nonetheless, in our experience a substantial percentage of the students taking an introductory course on human rights lack that knowledge.

How to handle this situation? We saw no alternative to providing in the early chapters an introduction to basic conceptions, sources, processes and norms of international law. Of course such an introduction hardly achieves the depth and sophistication of a course devoted principally to that rich subject. But it suffices to enable students to grapple with the later chapters. The coursebook thus serves as a selective introduction to public international law as illustrated by international human rights.

5. Such an examination of international law must take account of the contemporary transformation of some of its basic characteristics that bear on human rights issues—for example, the changing nature of custom and of international pressures or sanctions. It must address phenomena such as the effects of UN resolutions and the distinctive functions and powers of international organizations. It must explore the effect of human rights norms on deeply rooted conceptions like state autonomy and sovereignty.

The book emphasizes one change in the character of international law brought about by the development of international human rights. Broadly speaking, treaties and custom, declarations and resolutions, and the many institutions dealing with human rights infuse international law with ideals of a novel depth and scope. Human rights norms could hardly influence state conduct if they merely restated states' actual behavior, or merely expressed the agreements among states about matters serving broadly shared, reciprocal interests. Human rights ideals make vast and distinctive demands on all states, and therefore have the inevitable effect of distancing international law to a greater degree than has traditionally been the case from existing state behavior. No state fully meets such olympian standards, and many states contemptuously ignore and seriously violate them.

Moreover, human rights norms concentrate on matters internal to a state, on how a state acts towards and governs its own citizens. As a consequence, states may lack concrete, material interest in protesting other states' violations of international human rights law that injure such other states' citizens and that may have little probable international effect.

These basic characteristics of international human rights require that a coursebook give heightened attention to problems of implementation and enforcement of the new ideal norms. The old techniques simply won't work. In such circumstances, experimentation in international implementation and enforcement becomes inevitable. Thus the materials stress the international political and legal institutions that are meant to develop international human rights and to persuade delinquent states to adjust their behavior to the human rights ideals. They demonstrate the importance of linking norms to institutions, indeed of thinking of international human rights in terms of complex reciprocal relationships among norms, processes and institutions. The book thereby engages the student in a critical analysis of the architecture and functions of international human rights organizations, both intergovernmental and non-governmental.

* * * * *

The book's organization stems from the premise that the task is to educate students to see this field as a whole.

After Part A's brief introduction, Part B starts by examining the historical background in the first half of this century to the growth of international human rights in the second half. At the same time, it develops basic ideas about international law that are illustrated through tribunals' decisions on human rights issues. Part B then turns to the normative structure of international human rights: the Universal Declaration, the basic covenants and conventions. It introduces several themes that continue throughout the coursebook: cultural relativism, the significance of organizing and expressing norms in terms of rights or duties, the relevance to human rights of the "public–private" distinction, and changing conceptions of statehood and sovereignty.

No one substantive human rights problem dominates these chapters on civil, political, economic and social rights. Rather the materials draw on many problems to illustrate their themes, including free speech and the right to housing, torture and special regimes for minorities, laws of war and the right to education, female circumcision and asserted differences in Asian and Western views of rights. No one state or region dominates the discussion. The chapters draw illustrations from many countries and cultures.

Part C explores the reasons for institutionalizing human rights norms in a variety of international organizations and organs. If Part B teaches a good deal about general international law normatively, this part probes the nature and functions of international institutions and processes in the distinctive field of human rights. Surely the human rights movement has been as creative in its institutional architecture as in its development of standards of conduct.

The materials concentrate on the universal United Nations system, both Charter-based and treaty-based. They also examine the three regional systems and non-governmental organizations. Like Part B, this part's chapters draw on diverse human rights problems, here to illustrate the functions and processes of institutions: disappearances and homosexuality, political participation and arbitrary detention, religious repression and democratization.

Part D completes the book's conceptual framework by examining how states do or could implement human rights norms. It looks at the internal processes of the state to explore the reciprocal influences between the international and national dimensions of human rights. Hence this part examines (primarily liberal) constitutionalism, methods of internalization of international human rights law, and actions by states (rather than international organizations) to enforce norms against other states by imposing human rights conditions on trade and aid. Part D illustrates its themes primarily through the internal law and foreign economic policy of the United States, but also through the experience of European states.

The remainder of the book draws on this completed framework to examine some substantive topics of high contemporary relevance. Part E consists of one extended case study, women's rights as a part of international human rights. That topic is particularly appropriate for a major case study, not simply because of its significance for human rights as a whole. Its issues lend themselves to a review and further development of basic themes in the entire book. Moreover, no part of the human rights movement expresses its dynamism more powerfully than the trend of the last decade toward recognition and development of women's rights.

Part F selects three human rights topics from among a great range of possibilities: self-determination and autonomy regimes, individual criminal responsibility for violations of international law and international criminal tribunals, and the relation of human rights to development. Teachers with particular interests not covered by the book can readily add materials on such topics as humanitarian intervention, human rights in relation to public health, religious freedom, rights of children, refugees and asylum, population control, or environmental protection.

Introductions to several of the coursebook's parts and chapters expand on this brief statement of structure and purposes. See particularly pages 24–6, 166, 330–31, 708–9, 886, and 970.

* * * * *

The coursebook grows primarily out of teaching materials for a general course on human rights at Harvard Law School that have been improved over a six-year period. Henry Steiner was responsible for Chapters 1–4, 6, 9, and 11–14. Philip Alston was responsible for Chapters 5, 7–8 and 16. The authors divided the sections of Chapters 10 and 15.

We have used the teaching materials leading to this coursebook for a one-semester course. Within that period, it has not been possible to cover all the chapters. We included in our courses the large majority of the materials in Parts A to E. We then selected some chapters in Part F or used additional, specially prepared materials for a given topic as time permitted and as our interests suggested. But other routes through the book are surely plausible. Although organized within the conceptual framework described above, the chapters in Parts B to E need not be taught in lock step. Most of them can stand independently; their order can be varied; and depending on the purposes of the teacher, not all of them need be included in a course.

Some practical detail. Rather than require students to purchase a separate booklet for documents, we included edited documents that are relevant to several portions of the book in the Annex on Documents. On the other hand, documents relevant only to a particular section appear in that section. The authors' text refers to frequently cited documents (treaties, declarations, statutes, and so on) only by their titles; an Annex on Citations sets forth official citations and refers to more readily available books or periodicals in which complete documents appear. An Annex on Bibliography lists periodicals and books in the field of human rights that carry the interested student well beyond the coursebook's scope.

We have sharply edited most of the primary and secondary materials so as to make the readings as compact as possible. Omissions (except for footnotes) are indicated by the conventional use of ellipses. Footnotes that we retained are renumbered within the consecutive footnote numbering of each chapter.

Henry J. Steiner
Philip Alston

August 1995

Summary of Contents

PART D. STATES AS PROTECTORS AND ENFORCERS OF HUMAN RIGHTS 707

PART E. AN ILLUSTRATIVE STUDY 885

PART F. CURRENT TOPICS 969

Contents

Acknowledgements

We gratefully acknowledge the permissions extended by the following publishers and authors to reprint excerpts from the indicated publications.

American Anthropological Association. Permission to reprint excerpts from American Anthropological Association's Statement on Human Rights, 49 Amer. Anthropologist, No. 4, 539 (October–December 1947). Not for further reproduction.

American Law Institute. Permission to reprint excerpts from Restatement (Third), Foreign Relations Law of the United States. Copyright 1987 by the American Law Institute.

American Society of International Law. Permission to reprint excerpts from American Society of International Law, Panel: Human Rights, Business and International Financial Institutions, Proceedings of 88th Annual Meeting 1994 (1995); Hilary Charlesworth, C. Chinkin and S. Wright, Feminist Approaches to International Law, 85 Am. J. Int. L. 613 (1991); Stephen Cohen, Conditioning U.S. Security Assistance on Human Rights Practices, 76 Am. J. Int. L. 246 (1982); James Crawford, The ILC Adopts a Statute for an International Criminal Court, 89 Am. J. Int. L. 404 (1995); Karen Engle, After the Collapse of the Public/Private Distinction: Strategizing Women's Rights, in Dorinda Dallmeyer (ed.), Reconceiving Reality: Women and International Law (1993); Thomas Franck, The Emerging Right to Democratic Governance, 86 Am. J. Int. L. 46 (1992); Leo Gross, The Peace of Westphalia 1648–1948, 42 Am. J. Int. L. 20 (1948); Frederic Kirgis, Jr., The Degrees of Self-Determination in the United Nations Era, 88 Am. J. Int. L. 304 (1994); James O'Brien, The International Tribunal for Violations of International Humanitarian Law in the Former Yugoslavia, 87 Am. J. Int. L 639 (1993); W. Michael Reisman, Amending the UN Charter: The Art of the Feasible, American Society of International Law, Proceedings of 88th Annual Meeting 1994 (1995); W. Michael Reisman, Sovereignty and Human Rights in Contemporary International Law, 84 Am. J. Int. L. 866 (1990); Oscar Schachter, Human Dignity as a Normative Concept, 77 Am. J. Int. L. (1983); Paul Szasz, The International Conference on the Former Yugoslavia and the War Crimes Issue, American Society of International Law, Proceedings of the 87th Annual Meeting 29 (1993). Copyright © 1948, 1982, 1983, 1990, 1991, 1992, 1993, 1994, 1995 by The American Society of International Law.

Amnesty International. Permission to reprint excerpts from Rape and Sexual Abuse: Torture and Ill Treatment of Woman in Detention (1992); Amnesty International Report 1994 (AI Index POL 10/02/94).

Berghahn Books. Permission to reprint excerpts from Pannikar, Is the Notion of Human Rights a Western Concept?, 120 Diogenes 75 (Winter 1982).

University of British Columbia Law Review. Permission to reprint excerpts from Karl Klare, Legal Theory and Democratic Reconstruction, 25 U. of Brit. Colum. L. Rev. 69 (1991).

British Institute of International and Comparative Law. Excerpts from Martti Koskenniemi, National Self-Determination Today: Problems of Legal Theory and Practice, 43 Int. & Comp. L. Q. 241 (1994). Reprinted with permission from the British Institute of International and Comparative Law, 17 Russell Square, London WC1B 5DR.

University of California Press and Regents of the University of California. Permission to reprint excerpts from Ernst Haas, When Knowledge is Power: Three Models of Change in International Organizations (1990). Copyright © 1989 The Regents of the University of California.

California Western School of Law. Permission to reprint excerpts from Jack Donnelly, In Search of the Unicorn: The Jurisprudence and Politics of the Right to Development, 15 Calif. Western Int. L. J. 473 (1985).

Cambridge University Press. Permission to reprint excerpts from Louis Sohn, Interpreting the Law, in Oscar Schachter and Christopher Joyner (eds.), United Nations Legal Order 169 (1995); John Murphy, International Crimes, in Oscar Schachter and Christopher Joyner (eds.), United Nations Legal Order 993 (Vol. 2, 1995). Reprinted with permission of Cambridge University Press.

Carleton University Press, Inc. Permission to reprint excerpts from Martha Jackman, Constitutional Rhetoric and Social Justice: Reflections on the Justiciability Debate, in Joel Bakan and D. Schneiderman (eds.), Social Justice and the Constitution: Perspectives on a Social Union for Canada 17 (1992).

Centre for International and Public Law, Faculty of Law, Australian National University. Permission to reprint excerpts from Yash Ghai, Human Rights and Governance: The Asia Debate, 15 Aust. Y'bk Int. L. 1 (1994).

Columbia Human Rights Law Review. Permission to reprint excerpts from Isabelle Gunning, Arrogant Perception, World Travelling and Multicultural Feminism, 23 Colum. Hum. Rts. L. Rev. 189 (1992); Michael Posner and Candy Whittome, The Status of Human Rights NGOS, 25 Colum. Hum. Rts. L. Rev. 269 (1994).

Columbia University Press. Permission to reprint excerpts from Elvin Hatch, Culture and Morality: The Relativity of Values in Anthropology (1983); Louis Henkin, Constitutionalism and Human Rights, in Louis Henkin and Albert Rosenthal (eds.), Constitutionalism and Rights: The Influence of the United States Constitution Abroad 383 (1990); Louis Henkin, The Age of Rights (1990). Copyright © 1983, 1990 by Columbia University Press.

Connecticut Law Review. Permission to reprint excerpts from Elizabeth Schneider, The Violence of Privacy, 23 Conn. L. Rev. 973 (1992).

Dag Hammarskjold Foundation. Permission to reprint excerpts from Erskine Childers and Brian Urquhart, We The Peoples, in Renewing The United Nations System 171 (1994).

Encyclopaedia Britanica, Inc. Permission to reprint excerpts from Burns Weston, Human Rights, 20 Encyclopaedia Britannica 656 (15th ed. 1992).

Reprinted with permission from *Encyclopaedia Britannica*, 15th edition, © 1992 by Encyclopaedia Britannica, Inc.

N. P. Engel Verlag. Permission to reprint excerpts from Andrew Drzemczewski and Meyer-Ladewig, Principal Characteristics of the New ECHR Control Mechanism, As Established by Protocol No. 11, 15 Hum. Rts. L. J. 81 (1994); Manfred Nowak, UN Covenant on Civil and Political Rights: CCPR Commentary (1993); Jorg Polakiewicz and Valérie Jacob-Foltzer, The European Human Rights Convention in Domestic Law, 12 Hum. Rts. L. J. 65 and 125 (1991); Christian Tomuschat, Quo Vadis Argentoratum? The Success Story of the European Convention on Human Rights—And A Few Dark Stains, 13 Hum. Rts. L. J. 401 (1992).

European Journal of International Law. Permission to reprint excerpts from Bernhard Graefrath, Universal Criminal Jurisdiction And An International Criminal Court, 1 Eur. J. Int. L. 67 (1990); Martti Koskenniemi, The Politics of International Law, 1 Eur. J. Int. L. 4 (1990).

Farrar, Strauss and Giroux, Inc. Permission to reprint excerpts from Ronald Dworkin, Introduction to Nunca Más, in Nunca Más (Report of the Argentine National Commission of the Disappeared) (1986). Introduction copyright © 1986 by Ronald Dworkin.

Foreign Affairs. Permission to reprint excerpts from Theodor Meron, The Case for War Crimes Trials in Yugoslavia, 72 For. Aff. 122 (1993). Copyright 1993 by the Council on Foreign Relations, Inc.

Foreign Policy. Permission to reprint excerpts from Roger Sullivan, Discarding the China Card, 86 For. Pol. 3 (Spring 1992); Bilhari Kausikan, Asia's Different Standard, 92 For. Pol. 24 (Fall 1993). Copyright 1992 and 1993 by the Carnegie Endowment for International Peace.

Foundation Press. Permission to reprint several judicial decisions in the edited form appearing in, and to reprint parts of several Notes appearing in, Henry Steiner and Detlev Vagts, Transnational Legal Problems (3d ed. 1986). Copyright by the Foundation Press, Inc.

Greenwood Publishing Group, Inc. Excerpts from Philip Alston, The Fortieth Anniversary of the Universal Declaration, in Jan Berting et al. (eds.), Human Rights in a Pluralist World: Individuals and Collectivities 1 (1990). Reprinted with permission of Greenwood Publishing Group, Inc., Westport, CT. Copyright © 1990.

HarperCollins College Publishers. Permission to reprint excerpts from Samuel Huntington, American Ideals Versus American Institutions, in G. John Ikenberry (ed.), American Foreign Policy: Theoretical Essays 223 (1989). Copyright © 1989 by G. John Ikenberry.

HarperCollins Publishers. Permission to reprint excerpts from Immanuel Kant, The Doctrine of Virtue, English translation copyright © 1964 by Mary J. Gregor.

Harvard Human Rights Journal (earlier, Harvard Human Rights Yearbook). Permission to reprint excerpts from Abdullahi Ahmed An Na'im, Human Rights in the Muslim World, 3 Harv. Hum. Rts. J. 13 (1990); Kenneth

Anderson, Nuremberg Sensibility: Telford Taylor's Memoir of the Nuremberg Trials, 7 Harv. Hum. Rts. J. 281 (1994); Demetrios Marantis, Human Rights, Democracy, and Development: The European Community Model, 7 Harv. Hum. Rts. J. 1 (1994); Wendy Patten and J. Andrew Ward, Empowering Women to Stop AIDS in Cote d'Ivoire and Uganda, 6 Harv. Hum. Rts. J. 210 (1993); Henry Steiner, Political Participation as a Human Right, 1 Harv. Hum. Rts. Y'Bk 77 (1988). Copyright © 1988, 1990, 1993, 1994, by the President and Fellows of Harvard College.

Harvard Law Review. Permission to reprint excerpts from Henry Steiner, The Youth of Rights, 104 Harv. L. Rev. 917 (1991). Copyright 1991 by the Harvard Law Review Association.

Harvard Law School Human Rights Program. Permission to reprint excerpts from Henry Steiner, Diverse Partners: Non-Governmental Organizations in the Human Rights Movement (1991); Jack Tobin and Jennifer Green, Guide to Human Rights Research (1994). Copyright © 1991 and 1994 by the President and Fellows of Harvard College.

Harvard School of Public Health. Permission to reprint excerpts from Nancy Birdsall, Pragmatism, Robin Hood, and Other Themes: Good Government and Social Well-Being in Developing Countries, in Lincoln Chen, Arthur Kleinman, and Norma Ware (eds.), Health and Social Change in International Perspective (1994). © 1994, Lincoln Chen and Harvard School of Public Health, Boston, Massachusetts.

Harvard Women's Law Journal. Permission to reprint excerpts from Kay Boulware-Miller, Female Circumcision: Challenges to the Practice as a Human Rights Violation, 8 Harv. Women's L. J. 155 (1985). Copyright © 1985 by the President and Fellows of Harvard College.

Hastings Law Journal. Permission to reprint excerpts from José Zalaquett, Balancing Ethical Imperatives and Political Constraints: The Dilemma of New Democracies Confronting Past Human Rights Violations, 43 Hastings L. J. 1425 (1992). © 1992 by University of California, Hastings College of Law.

Human Rights Watch. Permission to reprint excerpts from Women's Rights Project and Americas Watch, Criminal Injustice: Violence Against Women in Brazil (1991); Human Rights Watch World Report 1993 (1994); Human Rights Watch World Report 1994 (1995).

Indian Law Institute. Permission to reprint excerpts from Upendra Baxi, Judicial Discourse: Dialectics of the Face and the Mask, 35 J. of Ind. L. Instit. 1 (1993).

The Jewish Publication Society. Permission to reprint excerpts from David Sidorsky, Contemporary Reinterpretations of the Concept of Human Rights, in Sidorsky (ed.), Essays on Human Rights 88 (1979).

Johns Hopkins University Press. Permission to reprint excerpts from Philip Alston and Gerard Quinn, The Nature and Scope of States Parties' Obligations under the ICESCR, 9 Hum. Rts. Q. 156 (1987); David Kramer and David Weissbrodt, The 1980 UN Commission on Human Rights and the Disappeared, 3 Hum. Rts. Q. 18 (1981); Tom Farer, Looking at Looking at

Nicaragua: The Problematique of Impartiality in Human Rights Inquiries, 10 Hum. Rts. Q. 141 (1988); Cecilia Medina, The Inter-American Commission on Human Rights and the Inter-American Court of Human Rights: Reflections on a Joint Venture, 12 Hum. Rts. Q. 439 (1990); Robert Drinan and Teresa Kuo, The 1991 Battle for Human Rights in China, 14 Hum. Rts. Q. 21 (1992).

Journal of Law and Religion. Permission to reprint excerpts from Robert Cover, Obligation: A Jewish Jurisprudence of the Social Order, 5 J. of Law and Relig. 65 (1987). Permission granted by Professor Robert Destro, The Catholic University of America.

Kluwer Academic Publishers. Permission to reprint excerpts from; Mohammed Bedjaoui, The Right to Development, in M. Bedjaoui (ed.), International Law: Achievements and Prospects 1177 (1991); Rudolph Bernhardt, The Convention and Domestic Law, in R. St. J. MacDonald, F. Matscher and H. Petzold (eds.), The European System for the Protection of Human Rights 25 (1993); Antonio Cassese, A. Clapham and J. Weiler, 1992—What Are Our Rights?: Agenda for a Human Rights Action Plan, in A. Cassese et al. (eds.), Human Rights and the European Community: Methods of Protection 1 (1991); Yoram Dinstein, International Criminal Law, 5 Israel Y'bk on Hum. Rts. 55 (1975); Louis Henkin, International Law: Politics, Values and Functions, in 216 Collected Courses of the Hague Academy of International Law 13 (Vol. IV 1989); Rosalyn Higgins, Comments, in Catherine Brolman, R. Lefeber and M. Zieck (eds.), Peoples and Minorities in International Law 29 (1993); Virginia Leary, International Labour Conventions and National Law (1982); Catherine MacKinnon, On Torture: A Feminist Perspective on Human Rights, in Kathleen Mahoney and P. Mahoney (eds.), Human Rights in the Twenty-First Century 21 (1993); Oscar Schachter, Intrnational Law in Theory and Practice (1991); Christian Tomuschat, Self-Determination in a Post-Colonial World, in Tomuschat (ed.), Modern Law of Self-Determination 1 (1993); Nicholas Valticos, Foreword to B. G. Ramcharan (ed.), International Law and Fact-Finding in the Field of Human Rights (1982); P. van Dijk and G. J. H. Hoof, Theory and Practice of the European Convention on Human Rights (2nd ed. 1990); G. J. H. van Hoof, The Legal Nature of Economic, Social and Cultural Rights: A Rebuttal of Some Traditional Views, in Philip Alston and Katarina Tomasevski (eds.), The Right to Food 97 (1984). Kluwer Law and Taxation Publishers: Peter Baehr, Amnesty International and its Self-Imposed Mandate, 12 Neth. Q. H. R. 5 (1994). Sitjhoff and Noordhoff Publishers: Georges Abi-Saab, The Legal Formulation of a Right to Development, in Hague Academy of International Law, The Right to Developlment at the International Level 159 (1979); Henry Schermers, International Institutional Law (1980). Reprinted by permission of Kluwer Academic Publishers.

Lawyers Committee for Human Rights. Permission to reprint excerpts from Makau Mutua and Peter Rosenblum, Zaire: Repression as Policy (1990); Linking Security Assistance and Human Rights (1989).

Lokayan. Permission to reprint excerpts from Rajni Kothari, Human Rights—A Movement in Search of a Theory, in Smitu Kothari and Harsh Sethni, Rethinking Human Rights: Challenges for Theory and Action (1989).

Longman Group. Permission to reprint excerpts from Robert Jennings and Arthur Watts (eds.), Oppenheim's International Law, Vol. 1 (9th ed. 1992).

Manchester University Press. Permission to reprint excerpts from J. G. Merrills, The Development of International Law by the European Court of Human Rights (2nd ed. 1993); A. H. Robertson, and J. G. Merrills, Human Rights in Europe: A Study of the European Convention on Human Rights (3rd ed. 1993).

Melbourne University Law Review. Permission to reprint excerpts from Philip Alston, Revitalising United Nations Work on Human Rights and Development, 18 Melb. U. L. Rev. 216 (1991).

MIT Press. Permission to reprint excerpts from Jack Donnelly, International Human Rights: A Regime analysis, 40 Int. Org. 599 (1986). © 1986 by The World Peace Foundation and the Massachusetts Institute of Technology.

Michigan Journal of International Law. Permission to reprint excerpts from Gregory Fox, Self-Determination in the Post-Cold War Era: A New Internal Focus?, 16 Mich. J. Int. L. 733 (1995).

Minority Rights Group International. Permission to reprint excerpts from Report 923/3, Female Genital Mutilation, Proposals for Change.

The New Republic. Permission to reprint excerpts from Amartya Sen, Freedoms and Needs, The New Republic 31 (Jan. 10 and 17, 1994). © 1994, The New Republic, Inc.

The New York Review of Books. Permission to reprint excerpts from Aryeh Neier, What Should be Done About the Guilty?, February 1, 1990; Amartya Sen, More Than 100 Million Women Are Missing, December 20, 1990. Copyright © 1990, Nyrev., Inc.

The New York Times. Permission to reprint excerpts from For Many Brides in India, A Dowry Buys Death, 12/30/93, p. 4; U.S. Prods Indonesia on Rights, 1/19/94, p. D1; China Rejects Call from Christopher for Rights Gains, 3/13/94, p. A1; Gauging the Consequences of Spurning China, 3/21/94, p. D1; U.S. Signals China It May End Annual Trade-Rights Battles, 3/24/94, p. A1; Many in U.S. Back Singapore's Plan to Flog Youth, 4/5/94, p. A6; U.S. Is To Maintain Trade Privileges for China's Goods, 5/27/94, p. A1; Rights Issues Aside, Asia Deals Rise, 8/1/94, p. D1; Israel Resumes Sealing of Houses as Punishment, 12/4/94, p. 3; President Imposes Trade Sanctions on Chinese Goods, 2/5/95, p. A1; China Warns of New Peril to U.S. Ties, 2/23/95, p. A9; U.N. Investigator Tells of 'Heinous' Rights Violations in Iraq, 2/28/95, p. A6; U.N. Rights Panel Declines to Censure China, 3/9/95, p. A5; Argentine Tells of Dumping 'Dirty War' Captives into Sea, 3/13/95, p. 1; A Prophet Tests the Honor of His Own Country, 3/14/95, p. A4; Clinton to Urge a Rights Code for Business Dealing Abroad, 3/27/95, p. D1; Kenya Crackdown is Aimed at

Opposition and the Press, 4/3/95, p. 3; Gap in Wealth in U.S. Called Widest in West, 4/17/95, p. 1. Copyright © 1993/94/95 by The New York Times Company.

Northwestern Journal of International Law and Business. Permission to reprint excerpts from Diane Orentlicher and Timothy Gelatt, Public Law, Private Actors: The Impact of Human Rights on Business Investors in China, 14 Nw. J. of Int. L. and Bus. 66–69, 80–85, 96–100, 108–110 (1993). Reprinted by special permission of Northwestern University School of Law.

Notre Dame Law Review. Permission to reprint excerpts from Henry Steiner, Ideals and Counter-Ideals in the Struggle Over Autonomy Regimes for Minorities, 66 Notre Dame L. Rev. 1539 (1991). © by Notre Dame Law Review, University of Notre Dame.

Oxford University Press. Permission to reprint excerpts from Philip Alston, Appraising the United Nations Human Rights Regime, also The Commission on Human Rights, in Alston (ed.), The United Nations and Human Rights: A Critical Appraisal 1, 126 (1992); Ian Brownlie, Principles of Public International Law (4th ed. 1990); Andrew Clapham, Human Rights in the Private Sphere (1993); Jean Drèze and Amartya Sen, Hunger and Public Action (1989); Louis Henkin, International Human Rights and Rights in the United States, in Theodor Meron (ed.), Human Rights in International Law 25 (1984); Rosalyn Higgins, Problems and Process: International Law and How We Use It (1994); Eugene Kamenka, Human Rights, Peoples' Rights, in James Crawford (ed.), The Rights of Peoples (1988); Dominic McGoldrick, The Human Rights Committee (1994); Theodor Meron, Human Rights Law-Making in the United Nations (1986); Terkel Opsahl, The Human Rights Committee, in Philip Alston (ed.), The United Nations and Human Rights: A Critical Appraisal 369 (1992).

Oxford University Press, Inc. Permission to reprint excerpts from Radhika Coomaraswamy, Uses and Usurpation of Constitutional Ideology; Yash Ghai, The Theory of the State in the Third World and the Problematics of Constitutionalism; and Andras Sajo and Vera Losonci, Rule by Law in East Central Europe: Is the Emperor's New Suit a Straitjacket?, all appearing in Douglas Greenberg et al. (eds.), Constitutionalism and Democracy: Transitions in the Contemporary World (1993), at respectively 159, 186 and 321 (1993). Published in New York.

University of Pennsylvania Press. Permission to reprint excerpts from Rhoda Howard, Dignity, Community, and Human Rights, in Abdullahi Ahmed An-Na'im (ed.), Human Rights in Cross-Cultural Perspectives 81 (1992).

Princeton University Press. Permission to reprint excerpts from Myron Weiner, The Child and the State in India: Child Labor and Education Policy in Comparative Perspective (1991). © 1991 by Princeton University Press.

Random House, Inc. Permission to reprint excerpts from Inis L. Claude, Jr., Swords Into Plowshares (4th ed. 1984). Copyright © 1974 by Inis L. Claude, Jr.

Reconstruction. Permission to reprint excerpts from Janet Fleischman, The Liberian Tragedy, 2 Reconstruction 47 (1992).

Reed Books. Permission to reprint excerpts from Jomo Kenyatta, Facing Mount Kenya: The Tribal Life of the Gikuyu (1965).

Lynne Rienner Publishers. Permission to reprint excerpts from (pp. 1–6) R. B. J. Walker and Saul Mendlovitz (eds.), Contending Sovereignties: Redefining Political Community (1990). © 1990 by Lynne Rienner Publishers, Inc.

Roosevelt Study Center. Permission to reprint excerpts from Philip Alston, The Fortieth Anniversary of the Universal Declaration, in J. Berting et al. (eds.), Human Rights in a Pluralist World, Individuals and Collectivities 1 (1990).

Routledge & Kegan Paul. Permission to reprint excerpts from Tom Campbell, The Left and Rights: A Conceptual Analysis of the Idea of Socialist Rights (1983). Reprinted with permission of the author.

Routledge Publishing Company. Permission to reprint excerpts from Steven Lukes, Is There an Alternative to Market Utopianism?, in Christopher G. A. Bryant and Edmund Mokrzycki (eds.), The New Great Transformation? Change and Continuity in East-Central Europe (1994); Hayter, Female Circumcision, Is There a Legal Solution?, J. of Soc. Welf. L. (UK) 323 (Nov. 1984).

Transaction Publishers. Permission to reprint excerpts from Hans Morgenthau, Human Rights and Foreign Policy, in Kenneth Thompson (ed.), Moral Dimensions of American Foreign Policy 341 (1984). Copyright © 1984 by Transaction, Rutgers–The State University of New Jersey.

UNESCO Publishing. Permission to reprint excerpts from Theo van Boven, Distinguishing Criteria of Human Rights, in Karel Vasak and Philip Alston (eds.), The International Dimensions of Human Rights, Vol. 1 (1982); Charles Taylor, Human Rights: The Legal Culture, in UNESCO, Philosophical Foundations of Human Rights 49 (1986).

Vanderbilt Journal of Transnational Law. Permission to reprint excerpts from Thomas Buergenthal, The United Nations Truth Commission for El Salvador, 27 Vanderbilt J. Trans. L. 497 (1994). Copyright by Vanderbilt Journal of Transnational Law.

Virginia Journal of International Law. Permission to reprint excerpts from Hurst Hannum. Rethinking Self-Detrmination, 34 Va. J. Int. L. 1 (1993). Copyright by Virginia Journal of International Law.

Wisconsin International Law Journal. Permission to reprint excerpts from David Kennedy, A New Stream of International Law Scholarship, 7 Wisc. Int. L. J. 1 (1988).

World Health Organization. Permission to reprint excerpts from A Traditional Practice that Threatens Health—Female Circumcision, in 40 WHO Chronicle 31 (1986).

Yale Journal of International Law. Permission to reprint excerpts from Andrew Byrnes, The 'Other' Human Rights Treaty Body: The Work of the

Committee on the Elimination of Discrimination Against Women, 14 Yale J. of Int'l L. 1 (1989).

Yale Law Journal. Permission to reprint excerpts from Diane Orentlicher, Settling Accounts: The Duty to Prosecute Human Rights Violations of a Prior Regime, 100 Yale L. J. 2537, 2568–76 (1991). Reprinted by permission of the Yale Law Journal Company and Fred B. Rothman & Company.

The authors express their appreciation for the contributions made to the research for this book and to its preparation for publication by Madelaine Chiam, Meredeth Hoffmann, Deborah Isser, Kerry Rittich, James Vázquez-Azpiri, and Norma Wasser.

PART A

INTRODUCTORY NOTIONS

This coursebook examines the world of contemporary human rights—legal norms, international institutions and processes, national and international actors, rhetoric, political bases and context, moral ideals, and so on. Over a mere half century, the human rights movement that grew out of World War II has become an indelible part of our legal, political and moral landscape. The book uses this term 'movement' to include governmental and intergovernmental as well as non-governmental developments since 1945, unlike some contemporary usage that restricts it to non-governmental activities.

The three dimensions of law, politics and morals are interrelated, indeed inseparable whether one thinks of the movement in theoretical or instrumental terms. The political and moral aspects of human rights are self-evident; it is the international legal aspect that is novel. The rules and standards of contemporary human rights are expressed not only through states' constitutions, laws and practices, but also through international custom and treaties, and the work products (decisions about action, forms of adjudication, studies, investigative reports, recommendations) of a range of international institutions and organs. Of course a movement of such breadth and complexity involves many aspects of life and thought, so that the book necessarily includes materials from a range of disciplines that bear on its themes.

Throughout, the materials underscore the youth of this movement, and the task of students committed to its ideals to see themselves not as novices within an established, even frozen, framework of ideas and institutions, but rather as moulders and architects of the movement's ongoing development. Put another way, the book's goal is not only to train the student to work effectively within the present structure and boundaries of the human rights movement. The materials also seek to impart a broad and critical understanding of that movement, as well as ideas about the directions in which it may or ought to be heading.

The Preface sets forth the book's pedagogical goals, conceptual structure and formal organization. Students may wish to read it now.

1

Global Snapshots

This introductory chapter assumes no special knowledge about the foundations or content of international human rights. Rather it is meant to spur thoughts about many issues that the later chapters examine.

The following illustrative excerpts from reports of newspapers and organizations plunge you into the diverse human rights problems that plague the world. When reading these materials, consider some of the following questions:

How do we know when and where human rights violations occur and what precisely happened? Who or what institution 'finds the facts,' and through what methods?

Are intergovernmental and non-governmental organizations hostile towards or cooperative with each other, or sometimes one and sometimes the other?

What is the source of the rules or standards under which officials and non-governmental organizations evaluate and criticize a state?

Do the states being criticized appear to accept the legitimacy of those rules and standards? The legitimacy of being criticized about internal matters by intergovernmental organizations or other states?

What kind of relationships, if any, do you see among the types of violations described or problems raised in these excerpts? For example, the excerpts clearly consider physical abuse to be a violation of human rights. What else?

Are you clear in each story that (if the reported facts are true) there has been a violation of human rights? How would you identify the violation in each story?

Are (international) human rights violations committed only by states, or are non-governmental forces and individuals also accused of such violations? Do all the stories involve governments as the direct violators of rights?

What steps if any seem to be taken to influence or force a state to end violations?

STORIES ABOUT HUMAN RIGHTS

Four Cited for Human Rights Work

United Press International, Dec. 8, 1993

Reebok Corp. Wednesday presented its 1993 Human Rights awards to four people, including a California minister targeted for death by a white supremacist group.

In ceremonies in Boston, the recipients of the sixth annual awards by the sportswear manufacturer received $25,000 each for human rights organizations of their choice.

. . .

Botte was cited for her work to save Asian children from prostitution. Reebok noted she has rescued hundreds of young women from brothels in Thailand.

Mubarak faced numerous personal threats, Reebok said, while documenting human rights violations of citizens detained by Egyptian authorities.

Kashinawa was honored for his work to defend the economic rights and cultural integrity of the indigenous people of the Brazilian rainforest.

U.N. Investigator Tells of 'Heinous' Rights Violations in Iraq

by Barbara Crosette, *New York Times*, Feb. 28, 1995, p. A6

A former Foreign Minister of the Netherlands who has been monitoring human rights in Iraq told a United Nations Commission today that over the last year President Saddam Hussein made an already grave situation 'heinous' for Iraqis through officially sanctioned amputations, maiming and branding of dissidents and military deserters.

'It is simply shocking that near the end of the 20th century, any state should so publicly and unashamedly incorporate heinous practices into its law,' said the former Dutch Foreign Minister, Max van der Stoel. He said that such practices, which threaten many thousands of Iraqis arrested for both petty crimes and political opposition are 'absolutely outlawed' under international conventions. The series of laws mandating physical punishment was enacted in the summer and fall of 1994.

. . .

Mr. van der Stoel has not been permitted to visit Iraq to write his reports, the most recent of which was submitted to the Human Rights Commission less than two weeks ago. He relies on material published in official Iraqi gazettes and information collected from Iraqis who have fled or who have managed to smuggle out of the country video tapes purporting to document abuses.

One tape pictured an army deserter with his ears cut off and blood still flowing on his face. Others show people whose foreheads had been branded with an X to indicate guilt, most often in cases where other mutilation had been done. Mr. van der Stoel said he believed 'tens of thousands' of military deserters or draft evaders lived in fear of torture and amputation under laws that can be applied retroactively.

He said that doctors who complain about brutality are often persecuted for their outspokenness. Occasionally there are reports of children being maimed along with their parents.

. . .

U.S. Prods Indonesia on Rights

by Thomas Friedman, *New York Times*, Jan. 19, 1994, p. D1

Treasury Secretary Lloyd Bentsen told President Suharto of Indonesia today that while the United States believed his Government was making progress on workers' rights it should do more if it wanted to retain its preferential trade privileges with Washington.

Mr. Bentsen was visiting Indonesia as part of an Asian tour intended to advance the Clinton Administration's efforts to get American businesses to focus more on the rapidly expanding economies of Indonesia, Thailand and China.

. . .

Given Indonesia's key role in the forum, and the fact that its economy is expected to spend more than $100 billion on infrastructure projects by the year 2000, one reason for Mr. Bentsen's visit here is to try to defuse a potential spat between Jakarta and Washington about the possible revocation of American trade benefits because of Indonesian human rights abuses. This is a smaller version of the problem Mr. Bentsen will face when he arrives in China Tuesday, after his stop in Bangkok.

The issue is this: Indonesia is a beneficiary of a trade provision called the Generalized System of Preferences. Under the system, developing countries are allowed to send a variety of goods duty-free to the United States. To maintain the benefits, the law states that a country has to be constantly demonstrating that it is making progress toward widening the rights of its workers.

In June, the Clinton Administration said it was reviewing Indonesia's trade status because of the Indonesian Government's continued failure to permit basic workers' rights, most notably the right to organize. Indonesia permits only one Government-directed union.

The Administration gave Indonesia until Feb. 15 to show more progress on workers' rights, particularly the right to strike, or otherwise face possible revocation of its trade privileges. American officials said the Indonesians had been taking steps to loosen up on workers' rights to organize. In recent months, Indonesia has authorized the formation of 14 new trade unions but has insisted that they remain under the supervision of the state-run labor federation, known by its Indonesian acronym S.P.S.I.

Since Indonesia is now leading the Asia-Pacific Economic Cooperation forum in the Non-Aligned Movement and represents an enormous market with 180 million consumers, the Clinton Administration is loath to impose any trade sanctions on Jakarta.

. . .

A Prophet Tests the Honor of His Own Country

by John Darnton, *New York Times*, March 14, 1995, p. A4

In physical stature Yasar Kemal, Turkey's best known and best-loved novelist, is decidedly Hemingwayesque—a big, barrel-chested man who looks as if he's ready to wrestle with the devil. And that, a lot of people feel, is exactly what he's doing.

. . .

Turkey has charged the author under Article 8 of the Anti-Terror Law. That is a catch-all provision about advocating separatism, and it is being applied because of an article he wrote for the Jan. 10 issue of Der Spiegel, the German newsmagazine, describing the oppression of fellow Kurds in his country.

Then when he published the essay in a collection of writings by Turkish authors here, titled 'Freedom of Expression and Turkey,' he was charged again under Article 8, along with the publisher. The book was banned and confiscated.

There was another charge in between—something about causing divisions in society, he said.

The writer faces two to five years in jail on two of the charges, three to six on the third. His trial is to start on May 5. It consumes him and has totally intruded on his writing.

. . .

His appearances for hearings in court so far have tended to put the spotlight on Article 8. The Human Rights Association, a nationwide monitoring group, says 118 people are currently jailed under it. Another 2,139 have been convicted but are appealing their sentences, and 5,600 others have been charged and are awaiting trial.

. . .

The article in Der Spiegel, 'Campaign of Lies,' accused the Government of systematically oppressing Turkey's 10 million to 15 million Kurds, especially in the southeast. In 10 years of insurrection there, human rights groups have documented widespread executions, disappearances in the custody of security forces, torture and the burning of villages.

'A tragedy of the human race is going on,' he said in an interview, and nobody—neither the United States nor Europe—pays attention. It is generally thought that about 14,000 have died in the conflict.

Mr. Kemal says in the article that he does not advocate a separate state for Kurds. But he suggests that he can understand, given their treatment, why some Kurds do. That is close to heresy in Turkey, where the 1923 republic founded by Ataturk was based on the principle that ethnic groups must suppress their identities to build a strong nation based on a common sense of Turkishness.

President Suleyman Demirel, asked about Mr. Kemal in an interview in his palace in Ankara, gave a figurative shrug of helplessness. He said: 'Yasar Kemal is a very famous author. I like him very much. He has done beautiful things. He uses Turkish. But he is coming from a Kurdish family—that's all right.

'But what he did in his article, that's unfortunate. I don't think the whole thing is bad. People in Turkey were divided—many said he shouldn't have done it. He did it. The prosecutor cannot do anything else.'

. . .

Argentine Tells of Dumping 'Dirty War' Captives into Sea

by Calvin Sims, *New York Times*, March 13, 1995, p. A1

Many of the victims were so weak from torture and detention that they had to be helped aboard the plane. Once in flight, they were injected with a sedative by an Argentine Navy doctor before two officers stripped them and shoved them to their deaths.

Now, one of those officers has acknowledged that he pushed 30 prisoners out of planes flying over the Atlantic Ocean during the right-wing military Government's violent crackdown in the 1970's.

The former officer, Adolfo Francisco Scilingo, 48, a retired navy commander, became the first Argentine military man to provide details of how the military dictatorship then in power disposed of hundreds of kidnapping and torture victims of what was known as the dirty war by dumping them, unconscious but alive, into the ocean from planes.

In his account, which was published this month in the Argentine newspaper Página 12, Mr. Scilingo said that he took part in two of the 'death flights' in 1977 and that most other officers at the Navy School of Mechanics in Buenos Aires, where he served, were also involved in such flights. He estimated that the navy conducted the flights every Wednesday for two years, 1977 and 1978, and that 1,500 to 2,000 people were killed.

'I am responsible for killing 30 people with my own hands,' Mr. Scilingo said in an interview after his account was published.

'But I would be a hypocrite if I said that I am repentant for what I did. I don't repent because I am convinced that I was acting under orders and that we were fighting a war.'

Mr. Scilingo's disclosure has reopened a bitter debate here over the so-called dirty war, in which more than 4,000 people were killed and 10,000 others disappeared during the military juntas from 1976 to 1983, according to an official Government inquiry.

Mr. Scilingo said he was motivated to tell his story because of what he called the navy's indifference to the plight of the rank and file who carried out orders to torture and kill prisoners. He said he was so tormented by the memory of his two death flights that he could not sleep at night without taking sleeping pills or drinking heavily.

'I'm not confessing to clear my conscience,' Mr. Scilingo said. 'I'm talking because I feel like the navy had abandoned us, left us to the wolves, the very ones who were loyal and followed orders.'

He said that after his first flight, in which he slipped and almost fell through the portal from which he was throwing bodies, he became so distraught that he confessed his actions to a military priest, who absolved him, saying the killings 'had to be done to separate the wheat from the chaff.'

'At first it didn't bother me that I was dumping these bodies into the ocean because as far as I was concerned they were war prisoners,' Mr. Scilingo said in the interview.

'There were men and women, and I had no idea who they were or what they had done. I was following orders. I did not get too close to the prisoners, and they had no idea what was going to happen to them.'

But he said he had a slight change of heart during the first mission, after a noncommissioned officer, who had not been informed of what the mission entailed, began to express reservations about dumping people into the ocean.

'I reached over to try to comfort him, and I slipped and nearly fell through the door,' Mr. Scilingo said. 'That's when it first hit me exactly what we were doing. We were killing human beings. But still we continued.'

He went on: 'When we finished dumping the bodies, we closed the door to the plane, it was quiet, and all that was left was the clothing which was taken back and thrown away. I went home that night and had two glasses of whiskey and went to sleep.'

Asked to describe the second mission, in which he said he dumped 17 people into the ocean, Mr. Scilingo said he could no longer discuss the details because he was about to break down.

'I have spent many nights sleeping in the plazas of Buenos Aires with a bottle of wine, trying to forget,' he said. 'I have ruined my life. I have to have the radio or television on at all times or something to distract me. Sometimes I am afraid to be alone with my thoughts.'

He said senior military officers had told participants in the flights that the church hierarchy sanctioned the missions as 'a Christian form of death.'

Outrage over Mr. Scilingo's disclosures was so strong here that the Roman Catholic Church, which in the past has been reluctant to talk about the dirty war, publicly denounced the torture and killings of that era.

Speaking on the behalf of Catholic bishops, Bishop Emilio Bianchi di Cárcano said no Christian could condone killings committed by Argentina's former military rulers. He denied that the church had ever been consulted over 'death flights.'

Bishop Bianchi di Cárcano said that the bishops had written to the military asking for information about the fate of political prisoners, but that the generals had never offered a clear reply.

President Carlos Saúl Menem, who granted broad pardons to military officers and others accused of human rights abuses, called Mr. Scilingo a 'criminal' and ordered the navy to strip the officer of his rank as a result of a conviction for fraud in a car-theft case in 1991.

Speaking to reporters, Mr. Menem, a former dissident who was imprisoned for five years by the military, defended his decision to issue the pardons, saying it was necessary for the country to move forward and to stop the military discontent that led to three barracks uprisings in the 1980's and in 1990.

But human rights groups and families of victims criticized Mr. Menem, saying that for political reasons the President was belittling the first detailed confirmation of what had long been charged: that the military had disposed of victims at sea and that the Catholic Church had sanctioned its actions.

. . .

Capt. Héctor Césari, a spokesman for the navy, said no interviews would be given about Mr. Scilingo because he was no longer associated with the military after being stripped of his rank.

'We don't know if he is motivated by vengeance or money,' Captain Césari said. 'But that is not our issue because Scilingo's statements are his own responsibility; they are his problem.'

Mr. Scilingo said that he began writing letters to navy officials, urging the military to disclose what happened during the dictatorship, but that their only response was to offer him money to keep silent and then to threaten to take away his military medical and social insurance.

Many Argentines, especially those whose relatives or friends were not killed or tortured during the dictatorship, say it is futile to continue to nurture old hostilities from the dirty war. Indeed, some, like Carmen Herrera, who watched the Mothers of the Plaza de Mayo protest on Thursday, argue that the military crackdown was justified to defeat a leftist guerrilla insurgency.

But others, like Horacio Verbitsky, a reporter for Página 12 to whom Mr. Scilingo gave his account, say that for the country to move ahead, all sectors of Argentine society, including the military and the church, must acknowledge their role in the crackdown.

'For a wound to heal and scar properly, you first have to clean it thoroughly and not leave infection inside,' Mr. Verbitsky said.

Many in U.S. Back Singapore's Plan to Flog Youth

New York Times, April 5, 1994, p. A6

Spare the rod and spoil the child may no longer be an axiom of American parenting, but many Americans are surprisingly unsympathetic to the plight of an Ohio youth who was convicted of vandalizing cars in Singapore and now faces a flogging in that country.

Michael P. Fay, 18 years old, was sentenced to four months in prison, a $2,215 fine and six lashes on his bare buttocks with a rattan cane by a martial arts expert after pleading guilty last October to vandalizing cars with spray paint and eggs and tearing down traffic signs in Singapore, where he lived with his mother and stepfather.

Mr. Fay and his parents say that he is innocent and that he agreed to a plea bargain because they thought he would be sentenced for a lesser crime than vandalism, which carries a mandatory sentence of flogging.

President Clinton and Representative Tony P. Hall, an Ohio Democrat, have appealed to the Singapore Government, arguing that the flogging, which may cause permanent scars, is excessive punishment for the youth.

A judge turned down Mr. Fay's appeal of his sentence on Thursday. His lawyer, Theodore Simon, and Mr. Fay's family are expected to make a final appeal for clemency to the Singapore President, Ong Teng Cheong. Amnesty International sees the Fay case as one more reason to refocus international attention on the inhumaneness of flogging.

. . .

Chin Hock Seng, first secretary at the Singapore Embassy, told The Associated Press that more than 100 letters and 200 phone calls had been received from Americans in recent weeks. 'The vast majority express very strong support for Singapore,' Mr. Chin said.

A spokesman for The Dayton Daily News, Mr. Fay's hometown newspaper, also said its mail was running against the youth.

And on the streets of Washington, there was little compassion for the young man. Women were slightly more sympathetic, worrying about the physical damage done by caning.

'If you've ever had your antenna ripped off your car, you can sympathize with the Government of Singapore,' said Ben Johnson, 26, a media relations director of Ein Communications Inc., a public relations firm. 'Lash him. Vandalism is a cowardly and insubordinate act.'

Margaret Henderson, a retired employee at the Department of Defense, who was window shopping on Connecticut Avenue, thought Mr. Fay deserved punishment, 'but certainly not as severe as that.'

The enthusiasm for such measures reflects American frustration with crime and young miscreants, some experts say.

. . .

It is Singapore's flogging law that disturbs Amnesty International. Estrellita Jones, coordinator of Asian affairs for the United States section of Amnesty International, said that the law, which her organization has protested for years, violates international laws and is 'unusually cruel and degrading' punishment.

'This is not the paddle or the whip that some of us got in public school,' she said. 'They tie you down like an animal. This is a very traumatic experience both mentally and physically.'

Ms. Jones said for 1,218 floggings in Singapore reported to Amnesty International in 1987 and 1988, Singaporeans accounted for 984 and foreigners for 234.

For Many Brides in India, A Dowry Buys Death

by Eduard Gargan *New York Times*, Dec. 30, 1993, p. A4

A few days ago, the police arrested Nagavani's husband. It came none too soon: she's still alive.

'My husband wanted a house in his name,' she began, whispering. 'He wanted a 30,000-rupee scooter,' worth about $1,000. 'He said if I did not give him this, he would take me to the top of a building and push me down.'

'He beat me,' she continued, a quiet matter-of-factness cushioning her words. 'He hit me on the back. He used to poke me with a needle on my back. He kept saying, I am an engineer and we must have lots of things. Last night they arrested my husband for dowry harassment.'

The experience of Miss Nagavani, who has only one name, is increasingly common. Despite a 32-year-old statute banning dowries—the money and gifts given by a bride's parents to the groom—the practice has now spread both among untouchables, who never traditionally gave dowries, and, with a vengeance, among the growing middle classes.

And with the spread of the practice has come a rapid rise in the killing of women for not providing dowries that are opulent enough, that are in the eyes of the husband and his family too meager for their status and needs.

Here in India's Silicon Valley, a growing city of high tech, computer enthusiasts and a newly entrenched middle class, dowry abuse has reached epidemic proportions. In the first two months of 1993, the months for which records are available, 161 cases of dowry abuse, including death, were turned over to the city's detectives.

In 1992, 4,785 women were killed by their husbands for not having provided adequate dowries, according to Government statistics. By the last day of 1992, 146 men were awaiting trial in the Delhi High Court for killing or abusing their wives in dowry-related cases, according to the Indian Department of Justice.

. . .

Mohandas K. Gandhi condemned the giving and receiving of dowries as he led India toward independence. For Gandhi, the quest for political freedom was bound up in what he saw as the need to transform the Indian soul. Part of that struggle in Gandhi's eyes included abolishing the dowry.

'A strong public opinion should be created in condemnation of the degrading practice of dowry, and young men who soil their fingers with such ill-gotten gold should be excommunicated from society,' Gandhi wrote. 'The system has to go. Marriage must cease to be a matter of arrangement made by parents for money.'

After Gandhi's admonitions, the new Indian Government passed the Dowry Prohibition Act, which was intended to do just that but which has been ignored by rich and poor alike, by Government ministers and street sweepers, lawyers and computer engineers. In the marriage pages of The Times of India on Sundays, pages of classified advertisements by families looking for husbands for their daughters or wives for their sons, the code words that indicate the need for a dowry are 'decent marriage.' And most of the advertisements are for a decent marriage.

. . .

'Men feel they have a right to strike women,' Ms. Fernandes said. 'The law hardly works, the police hardly work. In India, I think the kind of inequalities in the relationship between men and women are more visible and more deep than elsewhere. Almost every day a woman is killed.'

. . .

Kenya Crackdown Is Aimed at Opposition and the Press

by Donatella Lorch, *New York Times*, April 3, 1995, p. A3

Recent actions by Kenya's Government have been perceived by Western diplomats as the most serious crackdown on the opposition and the press since multiparty elections in 1992.

In the last month, President Daniel arap Moi has ordered the arrest of anyone who insulted him and he declared that a guerrilla movement was poised to invade from neighboring Uganda. At the same time, the governing party's newspaper has published reports about the February 18th Movement, described as a Communist rebel group led by an exile named John Odongo.

The Government has also accused the opposition of colluding with the rebels.

'Nobody takes it seriously,' one diplomat said about the threat of a guerrilla invasion. 'What does this mean for Kenya? The image is of a Kenya increasingly perceived as politically oppressive. This could be a step back for democracy.'

. . .

Four opposition members of Parliament, among others, have been arrested on charges of maligning the Office of the President, one of them accused of calling the Government satanic.

Other apparent opposition targets have included a legal aid center that was firebombed and a magazine whose office was torched after it published a report implicating a presidential adviser in a murder. The Government has denied involvement in both attacks.

Reasons for the crackdown vary, but it is widely seen as a tactic to divert attention from longstanding problems like unemployment, poverty and corruption. It has also come at a time of tensions in the President's governing party, the Kenya African National Union, with hardliners and moderates at odds over a range of issues like economic policy and political reform.

The Government has also accused Time, Newsweek and The Washington Post of 'outrageous lies and deliberate distortion of events in this country,' after they published reports about the crackdown.

'Any attempts to derail law and order by anyone including the media will not be condoned as the security of this nation is not negotiable,' said the Minister of Information, Johnstone Makau, who demanded an apology and raised the possibility of deporting the publications' correspondents.

. . .

Gap in Wealth in U.S. Called Widest in West

by Keith Bradsher, *New York Times*, April 17, 1995, p. A1

New studies on the growing concentration of American wealth and income challenge a cherished part of the country's self-image: They show that rather than being an egalitarian society, the United States has become the most economically stratified of industrial nations.

Even class societies like Britain, which inherited large differences in income and wealth over centuries going back to their feudal pasts, now have greater economic equality than the United States, according to the latest economic and statistical research, much of which is to be published soon.

Economic inequality has been on the rise in the United States since the 1970's. Since 1992, when Bill Clinton charged that Republican tax cuts in the 1980's had broadened the gap between the rich and the middle class, it has become more sharply focused as a political issue.

Many of the new studies are based on the data available then, but provide new analyses that coincide with a vigorous debate in Congress over provisions in the Republican Contract With America.

Indeed, the drive by Republicans to reduce Federal welfare programs and cut taxes is expected, at least in the short term, to widen disparities between rich and poor.

Federal Reserve figures from 1989, the most recent available, show that the wealthiest 1 percent of American households—with net worth of at least $2.3 million each—owns nearly 40 percent of the nation's wealth. By contrast, the wealthiest 1 percent of the British population owns about 18 percent of the wealth there—down from 59 percent in the early 1920's.

Further down the scale, the top 20 percent of Americans—households worth $180,000 or more—have more than 80 percent of the country's wealth, a figure higher than in other industrial nations.

Income statistics are similarly skewed. At the bottom end of the scale, the lowest-earning 20 percent of Americans earn only 5.7 percent of all the after-tax income paid to individuals in the United States each year.

. . .

Liberal social scientists worry about poor people's shrinking share of the nation's resources, and the consequences in terms of economic performance and social tension.

Margaret Weir, a senior fellow in government studies at the Brookings Institution, called the higher concentration of incomes and wealth 'quite divisive,' especially in a country where the political system requires so much campaign money.

'It tilts the political system toward those who have more resources,' she said, adding that financial extremes also undermined the 'sense of community and commonality of purpose.'

Robert Greenstein, executive director of the Center on Budget and Policy Priorities, a Washington research group, observed, 'When you have a child poverty rate that is four times the average of Western European countries that are our principal industrial competitors, and when those children are a significant part of our future work force, you have to worry about the competitive effects as well as the social-fabric effects.'

Conservatives have tended to pay less attention to rising inequality, and some express skepticism about the statistics or their significance.

. . .

Murray L. Weidenbaum, professor of economics at Washington University in St. Louis and chairman of the Council of Economic Advisers under President Ronald Reagan in 1981–1982, said he thought the measures tended to overstate the gap by overlooking Government programs like food stamps or Medicaid.

Still, he said he was uncomfortable with greater concentration of wealth 'unless there's a rapid turnover' in which 'this year's losers will be next year's winners.'

He noted that many wealthy people have bad years and that a lot of middle-class people, like graduate students, briefly look statistically as if they are starving. The United States does have 'very substantial mobility,' he added.

. . .

Britain Out of Step with E.U. Neighbours

Guardian, Jan. 12, 1994, p. 6

The age of consent for sex between gay men is higher in the UK—21—than anywhere else in the European Union, according to research at the University of Utrecht for the European Commission.

Other member states have ruled that the minimum legal age for homosexual and heterosexual activity should be the same—usually 16, or below.

There are still a few small pockets where homosexuality remains illegal: Bosnia, Serbia, Macedonia, Romania and Bielarus. But the European Court of Human Rights has ruled that such prohibition infringes the right to private life.

Ireland last year abolished gross indecency and buggery and established 17 for homosexual and heterosexual activity. Cyprus has promised to lift its ban. This leaves Britain as the member of the Council of Europe with the most restrictive laws.

. . .

Commission officials said they did not have competence under the EU treaties to combat discrimination. The European Court of Human Rights has outlawed national legislation banning homosexuality but has refused to get involved in ages of consent.

In Denmark same-sex couples can enter 'registered partnerships' which have most of the legal effects of marriage for the purposes of maintenance . . .

Israel Resumes Sealing of Houses as Punishment

by Clyde Haberman, *New York Times*, Dec. 4, 1994, Sec. 1, p. 3

In a courtyard a few hundred yards from El Bireh City Hall, Omar al-Asmar's relatives have lived in tents for more than four years, barely protected from rain and cold now that an early winter slashes at the canvas sheets.

The spare four-room house they rented was sealed shut by the Israeli Army in 1990 after Mr. Asmar was sent to prison for acts that included throwing gasoline bombs at Israeli soldiers. Shutters were welded and doors boarded up.

. . .

Even after Mr. Asmar, 28, was released from prison in May and started working in a pastry shop around the corner, the army insisted that the house remain shut despite family attempts to reopen it.

'They punished my whole family, not only my son,' said Halima al-Asmar, the mother of three sons and three daughters. 'We have no property here. We had no income after he was arrested. So we had no choice but to move into the tents.'

For years, such cases formed the core of the Israeli policy of demolishing or sealing houses in the West Bank and Gaza Strip to combat Palestinian resistance to Israel's military occupation. Hundreds of homes were shut or destroyed, most in the first four years of the uprising that began in December 1987.

Israel has defended the practice as a deterrent against Arab attacks, calling it legal under emergency regulations imposed in 1945 by the British Mandate of Palestine then in existence.

But Palestinians say, and human-rights groups agree, that the policy violates international law, amounting to collective punishment while doing nothing to stop the uprising. If anything, they say, resentment bred by such practices produces even more rock-throwers and killers.

The debate lost a good deal of steam after Yitzhak Rabin became Prime Minister in July 1992. Under the Rabin Government, house demolitions and sealings slowed to a trickle, and then stopped after September 1993 with the agreement giving Palestinians self-rule in Gaza and the West Bank town of Jericho.

But suddenly the issue is back.

Last week, with the blessing of the Israeli High Court of Justice, the army blew up most of the family home of a West Bank Palestinian who carried out the suicide bombing on a Tel Aviv bus that killed 22 passengers and himself in October.

This week, the authorities sealed shut the Jerusalem houses of three men involved in the kidnapping and killing of an Israeli soldier, Nahshon Waxman, also in October. Two of the three were themselves killed in a failed army rescue mission.

After a recent spurt of suicide attacks by Islamic fundamentalists, Israel decided that it had to revive the policy. Government lawyers argued, and the court agreed, that the only thing that might stop a suicide bomber was the knowledge that his family would suffer for his actions.

'This is one of the few weapons of deterrence that we have to deal with a situation that's becoming intolerable,' an army official said.

As in the past, Israeli and Palestinian human rights groups challenged the deterrence claim, and denounced the practice as punishing the wrong people because most of the culprits in these cases were dead. Besides, they noted, Israel had not touched the Qiryat Arba apartment of Dr. Baruch Goldstein, the Israeli settler who machine-gunned 29 Palestinians to death in Hebron in February and was himself killed.

. . .

'What's unfair is that the doctor who killed people, this Jewish criminal, did not also have his house sealed,' said Taysir al-Natshe, whose son, Hassan, was one of the slain Waxman kidnappers. Mr. Natshe, who owns a clothing store in Jerusalem's Old City, watched as windows and doors were welded shut at his home this week.

'This is my house that they've sealed,' he said. 'What's my crime? I did nothing wrong.'

I Have No Words

Ottawa Citizen, Feb. 4, 1993, p. B1

They sat in a row, eyes down, four survivors of Serb-run concentration camps waiting to be questioned about the nightmare they endured.

Of the 24 Bosnian Muslims who arrived in Ottawa last week, the three men and one woman were the only ones who felt up to talking about their eight-month ordeal.

. . .

Faudin revealed a hell on earth. Like the other prisoners, he spent his days with his head down, hands behind his back. A sideways glance was enough to bring a beating. So was not eating a meal in the allotted 45 seconds. He was forced to drink five litres of motor oil, destroying his intestines. He lost 85 pounds.

The prisoners were made to watch all manner of cruelty: fathers forced to have oral sex with their sons, women raped in front of husbands and sons, people quartered after their limbs were tied to cars that drove off in different directions.

Many had their fourth and fifth fingers cut off at the knuckles, leaving the other fingers fixed in the two-fingered salute to Serbian nationalism.

. . .

'One man had to bite off his brother's testicles and then beat him up,' recalled Faudin, head down, arms crossed. 'In Manjaca camp, 100 people died daily. Those shot with bullets were the lucky ones.'

'I have no words to describe the methods the Serbian villains used to humiliate us,' said Avdo Nalic, formerly of Sanski Most. 'All the scenes of horror of the Second World War were used on us. They are destroying the Muslim people.'

The eight refugee families in Ottawa are among 5,000 prisoners released from Serb camps since the United Nations High Commissioner for Refugees and the International Red Cross appealed to the Bosnian Serb forces in November. They were let go on the condition they not return to Bosnia. Canada has accepted 500 of them.

. . .

The refugees have mixed feeling about being in Canada. They are grateful for the haven, but fear the international community is unintentionally helping achieve the Serbian goal of ethnic cleansing by agreeing to the no-return condition.

Mobutu Says Amnesty Aims to Destabilize Zaire

Reuter Library Report, Sept. 17, 1993

Zaire's President Mobutu Sese Seko has accused Amnesty International of trying to destabilise his country with a report accusing security forces of appalling human rights abuses.

A statement from the presidency read on state-run television on Thursday night also said an Amnesty International report accusing Zaire of killing hundreds and even thousands of civilians could wreck talks between the government and the opposition.

'Was it the best moment for Amnesty International to pour oil onto the flames by publishing a report which makes allegations which are clearly tendentious, and which aim only at sabotaging the ongoing political consultations and at influencing certain international circles with the aim of destabilising the Republic of Zaire?' the statement said.

'The Presidency of the Republic would like to remind those who dream of a "Somalicisation" of Zaire that they would do better to look elsewhere,' said the statement, monitored by the British Broadcasting Corporation.

'Indeed, the important thing for now is to do everything to ensure the success of the current negotiations in order to secure the foundations of the irreversible option of pluralist democracy, of the respect of human rights, of equality and transparency in the management of public affairs.'

The London-based human rights group said government security forces had killed thousands of civilians and members of the political opposition since 1990 and urged the international community to intervene.

'Zaire is sliding inexorably towards a total breakdown of law and order and the government is using the country's worsening political and economic situation as an excuse for appalling human rights violations,' the report released on Thursday said.

. . .

U.S. State Department, Country Reports on Human Rights Practices for 1994
1995, p. 934

Russia
The Russian Constitution establishes democratic governance through three branches of power with checks and balances: the Presidency and the Government, headed by the Prime Minister, a bicameral legislature, or Federal Assembly, consisting of the State Duma and Federation Council; and the courts. The Constitution also lays a comprehensive groundwork for observance and enforcement of human rights set forth in the Universal Declaration of Human Rights.

The constitutional referendum of December 1993, which was concurrent with parliamentary elections, strengthened President Boris Yeltsin's position vis-a-vis the legislature. He continued to proclaim his commitment to political reform and the transition to a modern market economy. Nevertheless, the process of institutionalizing democracy and a modern market economy lagged due, in part, to the slow enactment of laws and development of regulatory institutions, widespread unfamiliarity with democratic and market principles, and a reaction against 'democrats' and free market advocates because of social dislocation.

. . .

The overall human rights record in Russia in 1994 remained uneven. The concept of the rule of law has yet to be institutionalized and implemented. In his desire to combat rapidly increasing crime, President Yeltsin signed two decrees in June which contradict constitutional rights to protection against arbitrary arrest and illegal search, seizure, and detention. The Constitutional Court, the only body having the authority to settle constitutional disputes, remained inoperative because of governmental inability to fill vacancies. Courts of general jurisdiction, with several notable exceptions, remained timid in asserting their authority. Moreover, judges generally feel uncomfortable with the idea of having the duty and responsibility to declare actions taken by the executive branch and regional authorities as unconstitutional.

In August Sergey Kovalev, the Human Rights Commissioner, (a position provided for in the Constitution), published a highly critical report on the human rights situation in 1993, noting that law enforcement officials reportedly beat and physically abused detainees; military officers failed to discipline those who engaged in 'dedovshchina,' the violent hazing of conscripts, which led to numerous deaths and injuries; and prison officials appeared unable to correct life-threatening situations in pretrial detention centers.

Claiming the need to prevent Chechnya's secession from Russia, Russian troops crossed into the Russian Federation's Republic of Chechnya on December 11. Beginning in late December following major Chechen resistance, there was massive aerial and artillery bombardment of Chechnya's capital, Groznyy, resulting in a heavy loss of civilian life and hundreds of thousands of internally displaced persons. These actions were in conflict with a number of Russia's international obligations, including those concerning the protection of civilian noncombatants and notification of troop movements. In late December, Human Rights Commissioner Kovalev accused Russian troops of violating human rights on a 'massive scale' in Chechnya. The international community and human rights nongovernmental organizations (NGO's) called upon both parties to respect international humanitarian law and the human rights of noncombatant civilians.

Although freedom of speech and press is widely respected, there were physical attacks on journalists by unknown persons and at least one killing. Journalists reporting on the conflict in Chechnya were harassed and threatened. Other human rights abuses include official and societal discrimination against people from the Caucasus, some continuing societal discrimination against Jews, violence against women, lack of attention to the welfare of children and the handicapped, and bureaucratic obstacles to the development of independent labor unions.

. . .

Soldiers Kill Thousands in Rwanda Camp

Boston Globe, April 23, 1995, p. 2

Government soldiers yesterday killed thousands of people who were trying to flee Rwanda's largest refugee camp, relief workers and military officials said.

UN troops found more than 4,000 people, most of them members of Rwanda's Hutu majority, dead and 650 wounded in Kibeho camp in southwest Rwanda in a search of only half the site, Reuters reported early this morning.

Maj. Mark Mackay, spokesman for the Integrated Operations Center, a UN-affiliated organization, said reports from the Kibeho camp indicated that the victims had been killed by soldiers or died in stampedes. 'There were bodies all over the place,' Mackay said.

. . .

The killings occurred on the fifth day of a campaign by Rwanda's Tutsi-led army to close the camp at Kibeho, 80 miles southwest of Kigali, and force its residents to return to their homes. Kibeho is one of nine camps inside Rwanda that hold a total of 250,000 people.

'What happened today was pure stupidity on the part of the soldiers,' a relief worker said.

The afternoon's shootings began after a group of camp residents tried to break away from the cordon of soldiers around the area, prompting hundreds of others to try to flee. 'The floodgates opened,' said Shaharyar Khan, special representative of the UN secretary general in Kigali.

Witnesses said shooting lasted two hours, as soldiers reportedly sprayed crowds with automatic-weapons fire and launched at least one rocket-propelled grenade into the camp.

. . .

Most of the camps' residents are Hutus who fled their homes in fear of revenge killings by the Tutsi minority, an estimated 500,000 of whom were massacred by Hutu extremists last spring.

The Tutsi-led Rwandan Patriotic Front ousted the hard-line Hutu regime that fostered the massacres, and Rwanda's new government has been attempting to close the camps for months, saying the areas have become hiding places for Hutus who took part in last year's genocide. Soldiers encircled Kibeho to screen its inhabitants before sending them out of the camp.

. . .

PART B
HISTORICAL DEVELOPMENT AND NORMATIVE FRAMEWORK OF INTERNATIONAL HUMAN RIGHTS

It would be possible to study human rights issues not at the international level but in the detailed contexts of different states' histories, socio-economic and political structures, legal systems, religions, cultures and so on. With respect to its legal dimension, a human rights course that was so organized would stress foreign and comparative law, in the sense of a contextual and comparative analysis of law and practice in selected foreign states. It could devote its full attention to the analysis of states like Zaire, China, Saudi Arabia, Italy, the United States, or Guatemala. It could stress different internal traditions and distinctive internal problems, the trend in many states toward (at least as a formal matter) liberal constitutionalism, and so on. International law could play a peripheral role, relevant only when it exerted some clear influence on the national scene.

Contrast many other subjects where international law necessarily occupies a central position. Imagine, for example, that our interest were not human rights but the humanitarian law of war (rules for the conduct of war), or the regulation of fisheries, or immunities of diplomats from arrest, or the regulation of trade barriers like tariffs. Each of those fields is inherently, intrinsically, *international* in character. Each involves relations *between* states or between citizens of one state and other states. We could not meaningfully examine any one of them without examining international custom and treaties, international institutions and processes.

Violations of human rights are different. They are rooted within states, and need not on their surface involve any international consequences whatsoever. Police torture defendants to extract confessions; the attorney general shuts the opposition press as elections approach; prisoners are raped; courts decide cases according to executive command; women are barred from education. Each of these events could profitably be studied entirely within a state's (or region's, culture's) internal framework, just as law students in many countries traditionally devote most of their energy to studying their internal legal–political system, including that system's provision for civil liberties and human rights. All these considerations point toward the foreign–comparative law approach to the study of human rights.

Nonetheless, it would be profoundly misleading to develop a framework for the study of human rights without including as a major ingredient the international legal and political aspects of the field: laws, processes, and institutions. We have stressed that human rights is in vital respects a distinctive field of international law. But it is characteristically imagined in the modern world as an international movement whose study necessarily involves international law and institutions, as well as a movement within and among states. Internal developments in many states have been too influenced by that law and those institutions, as well as by pressures from other states trying to enforce international law, for students to ignore the international dimensions. These internal and international aspects of human rights are now rarely 'split,' but rather complexly intertwined and reciprocally influential with respect to the causes and effects of violations of human rights, the reactions and sanctions of intergovernmental bodies or other states, the transformations of internal orders, and so on.

From another perspective as well it would be impossible to grasp the character of the human rights movement without a basic knowledge about international law and its contributions to the human rights movement. The aspirations of that movement to universal validity are necessarily rooted in that body of law. Many of the distinctive organizations intended to help to realize those aspirations are creations of international law.

For such reasons, this coursebook does not concentrate on the internal law and politics of one or a few states. It does however draw on internal matters in a large number of states from all regions for brief illustrative examples of one or another problem, or for lengthier case studies. Throughout it emphasizes the international law aspect of the human rights movement, the distinctive development of the last half century, particularly the growth of customary law and treaties on human rights and the creation of organizations, both intergovernmental and non-governmental, involved with human rights issues.

2

International Law Concepts and Doctrinal Background Relevant to the Human Rights Movement

This chapter gives the student some familiarity with the traditional conceptions and processes of international law. That familiarity must include an understanding of the modes of argument or argumentative strategies about international law, of custom or treaties or general principles, of ways of dealing with the tension between national sovereignty and international regulation, and so on. International human rights is only one branch of international law. Nonetheless, the arguments of advocates, diplomats, and legal or other scholars and activists about that body of law necessarily have the same general structure and invoke the same phenomena as arguments relevant to other fields of international law.

Section A provides readings about international law: processes, sources, some basic political and jurisprudential assumptions. It is brief and selective in providing background for later portions of the book that develop its themes. These readings serve as review for students with an academic background in international law, and as introduction to fundamental conceptions for those first entering the field. Section B illustrates and probes the processes and concepts introduced in Section A through examination of decisions of tribunals during the first half of the twentieth century that involve international law doctrines forming an important part of the background to the contemporary human rights movement.

A. BASIC CONCEPTIONS AND PROCESSES

How, by what means or methods, have the international aspects of the human rights movement developed? By what processes are international legal rules made or elaborated or applied, and international institutions created and administered? Section A introduces these issues through brief and general observations about international law that should be sufficient for grasping the international law aspects of the human rights materials that follow. The readings are compact and abstract, and shy on illustrations. The case studies in Section B offer those illustrations and will strengthen your grasp of the present materials.

Article 38 of the Statute of the International Court of Justice, the judicial organ of the United Nations that is provided for in the UN Charter, now constitutes a traditional point of departure for examining such issues. The question to which that article responds is put from a court's perspective: to what 'sources'—or, some would say, 'evidence'—of law does an international court look, to what 'processes' of lawmaking should it be attentive, if its mission is to decide disputes submitted to it—in the words of Article 38—'in accordance with international law'? One might ask why the perspective of a court should be relevant to a field like international human rights where courts have not played a vital role in the development of the universal human rights system. It is relevant for our purposes because an international court's perspective provides a useful point of departure for examining basic characteristics of international law, whatever the institutions or processes through which that law develops.

Article 38 answers the question noted above in a formal and positivist manner. It defines the task of the court in terms of the court's *application* of an identifiable body of international law. Its skeletal list, one that omits important contemporary processes of international lawmaking, expresses a conception of the judicial function that is radically different from that of, say, a legal realist. Article 38(1) provides:

> 1. The Court, whose function is to decide in accordance with international law such disputes as are submitted to it, shall apply:
>
> a. international conventions, whether general or particular, establishing rules expressly recognized by the contesting states;
>
> b. international custom, as evidence of a general practice accepted as law;
>
> c. the general principles of law recognized by civilized nations;
>
> d. subject to the provisions of Article 59 [stating that decisions of the Court have no binding force except between the parties to the case], judicial decisions and the teachings of the most highly qualified publicists of the various nations, as subsidiary means for the determination of rules of law.

COMMENT ON THE ROLE OF CUSTOM

We start the inquiry with custom, which has been referred to as the oldest and original source of international law. Customary law remains indispensable to an adequate understanding of human rights law. It figures in many fora, from the broad debates about human rights within the United Nations to the arguments of counsel before an international or national tribunal.

By customary law, we refer to conduct, or the conscious abstention from certain conduct, of states that becomes in some measure a part of international legal order. By virtue of a developing custom, particular conduct may be considered to be permitted or obligatory in legal terms, or abstention from particular conduct may come to be considered a legal duty.

Consider the 1950 statement of a noted scholar describing the character of the state practice that can build a customary rule of international law: (1) 'concordant practice' by a number of states relating to a particular situation; (2) continuation of that practice 'over a considerable period of time'; (3) a conception that the practice is required by or consistent with international law; and (4) general acquiescence in that practice by other states.[1] Other scholars have contested some of these observations, and today many authorities would contend that custom has long been a less rigid, more flexible and dynamic force in lawmaking.

Clause (b) of Article 38(1) states that the I.C.J. shall apply 'international custom, as evidence of a general practice accepted as law.' Contemporary formulations of custom have overcome some of the difficulties in understanding this terse and confusing phrase, but three of the terms there used remain contested and vexing: 'general,' 'practice,' and 'accepted as law.'

Section 102 of the *Restatement (Third), Foreign Relations Law of the United States*, presents a contemporary and clearer formulation of customary law that draws broadly on scholarly, judicial, and diplomatic sources. Many authorities on international law, surely in the developed world and to varying degrees in the states of the Third World, could accept that formulation as an accurate description and guide. After including custom as one of the sources of international law, it provides in clause (2): 'Customary international law results from a general and consistent practice of states followed by them from a sense of legal obligation.'

Each of these terms—'general,' 'consistent,' 'practice,' 'followed,' and 'sense of legal obligation'—is defined in a particular way. For example, the *Restatement*'s comments on Section 102 say: state practice includes diplomatic acts and instructions, public measures, and official statements, whether unilateral or in combination with other states in international organizations; inaction may constitute state practice as when a state acquiesces in another state's conduct that affects its legal rights; the state practice necessary may be of 'comparatively short duration;' a practice can be general even if not universally followed; there is no 'precise formula to indicate how widespread a practice must be, but it should reflect wide acceptance among the states particularly involved in the relevant activity.'

Other *Restatement* comments address the question of the sense of legal obligation, or *opinio juris* in the conventional Latin phrase. For example, to form a customary rule, 'it must appear that the states follow the practice from a sense of legal obligation (*opinio juris sive necessitatis*); hence a practice gen-

[1] M. Hudson, Working Paper on Article 24 of the Statute of the International Law Commission, UN Doc. A/CN.4/16, March 3, 1950, p. 5.

erally followed 'but which states feel legally free to disregard' cannot form such a rule; *opinio juris* need not be verbal or in some other way explicit, but may be inferred from acts or omissions. The comments also note that a state that is created after a practice has ripened into a rule of international law 'is bound by that rule.'

The *Restatement* (in the Reporter's Notes to Section 102) notes some of the perplexities in the concept of customary law:

> Each element in attempted definitions has raised difficulties. There have been philosophical debates about the very basis of the definition: how can practice build law? Most troublesome conceptually has been the circularity in the suggestion that law is built by practice based on a sense of legal obligation: how, it is asked, can there be a sense of legal obligation before the law from which the legal obligation derives has matured? Such conceptual difficulties, however, have not prevented acceptance of customary law essentially as here defined.

The following discussion, and later readings in Chapter 2, address a few dilemmas of customary law that are particularly relevant to understanding its role in human rights law.

Consider the need to evaluate state practice with respect to (1) *opinio juris* and (2) the reaction of other states to a given state's conduct. Suppose that a state's 'abstention' is primarily at issue—for example, state X neither arrests nor asserts judicial jurisdiction over a foreign ambassador, which is one aspect of the law of diplomatic immunities that developed as customary law long before it was subjected to treaty regulation. During the period when this customary law was being developed, it would have been relevant to inquire why states generally did not arrest or prosecute foreign ambassadors. For example, assume that X asserted that it was not legally barred from such conduct but merely exercised its discretion, as a matter of expediency, not to arrest or prosecute. Abstention by X coupled with such an explanation would not as readily have contributed to the formation of a customary legal rule. On the other hand, assume that a decision by the executive or courts of X not to arrest or assert judicial jurisdiction over the ambassador rested explicitly on the belief that international law required such abstention. Such practice of X would then constitute classic evidence of *opinio juris*.

Consider a polar illustration, where X acts in a way that immediately and adversely affects the interests of other states rather than abstains from conduct. Suppose that X imprisons without trial the ambassador from state Y, or imprisons many local residents who are citizens of Y. Surely it has not acted out of a sense of an international law *duty*. If it considered international law to be relevant at all, it may have concluded that its conduct was not prohibited by customary law. Or it may have decided that even if imprisonment was prohibited, it would nonetheless violate international law.

In this type of situation, the conception of *opinio juris* is less relevant, indeed irrelevant to the state's conduct. What does appear central to a determination of the legality of X's conduct is the *reaction* of other states—in this instance, particularly Y. That reaction of Y might be one of tacit acquiescence, thus tending to support the legality of X's conduct, or, more likely on the facts here given, Y might make a diplomatic protest (perhaps accompanied by economic pressures) that would criticize X's action as a violation of international law. Action and reaction, acts by a state perhaps accompanied by claims of the act's legality, followed by reaction-responses by other states affected by those acts, here constitute the critical components of the growth of a customary rule.

These simplified illustrations suggest some of the typical dynamics of traditional customary international law. What is common to both illustrations—abstention from arrest, and arrest—is that the interests of at least two states were directly involved. Of course states other than Y may well have taken an interest in X's action; after all, those states also have ambassadors and citizens in foreign countries. All of these possibilities are relevant to understanding *The Paquete Habana*, p. 60, *infra*.

COMMENT ON TREATIES

In Article 38(1) of the Statute of the International Court of Justice, the Court is instructed to apply 'a. international conventions, whether general or particular, establishing rules expressly recognized by the contesting states.' Treaties, that is, appear first on the list. They have become the primary expression of international law and, particularly when multilateral, the most effective if not the only path toward international regulation of many contemporary problems. Treaties, for example, have been the principal means for development of the human rights movement. Only treaties, not custom or general principles, can create international institutions in which state parties participate and to which they may owe duties.

The terminology for this voluminous and diverse body of international law varies. International agreements are referred to as pacts, protocols (generally supplemental to another agreement), covenants, conventions, charters, exchanges of notes and concordats (agreements between a nation and the Holy See), as well as treaties—terms that are more or less interchangeable in legal significance. Within the internal law of some countries such as the United States, the term 'treaty' (as contrasted, say, with international executive agreement) has a particular constitutional significance. See p. 742, *infra*.

Consider the different purposes which treaties serve. Some reaching critical national interests have a basic political character: alliances, peace settlements, atom bomb test ban. Others, while less politically charged, also involve relationships between governments or government agencies and affect private parties only indirectly: agreements on foreign aid, cooperation in the provision of governmental services such as weather forecasting and the mails. But treaties often have a direct and specific impact upon private parties. Tariff accords, income tax conventions, and treaties of friendship, commerce and navigation have for many decades determined some of the conditions under which the nationals or residents of one signatory can export to, or engage in business activities within, the other signatory's territory. Most significant for this book's purposes, human rights conventions have sought to extend protection to all persons against governmental abuse.

Domestic analogies to the treaty help to portray its distinctive character. Some treaties settling particular disputes resemble a private accord and satisfaction: an agreement over boundaries, an agreement to pay a stated sum as compensation for injury to the receiving nation or its nationals. Others are closer in character to private contracts of continuing significance or to domestic legislation, for they regulate recurrent problems by defining rights and obligations of the parties and their nationals: agreements over rules of navigation, income taxation, or the enforcement of foreign-country judgments. Indeed, the term 'international legislation' has gained some currency, particularly with respect to multilateral treaties such as those on human rights that impose rules on states intended to regulate their conduct.

Nonetheless, domestic legislation differs in several critical respects from the characteristic treaty. A statute is generally enacted by the majority of a legislature and binds all members of the relevant society. Even changes in a constitution, which usually require approval by the legislature and other institutions or groups, can be accomplished over substantial dissent. The ordinary treaty, on the other hand, is a consensual arrangement. With few exceptions, such as Article 2(6) of the UN Charter, it purports to bind or benefit only parties. Alteration of its terms by one party generally requires the consent of all.

The other principal domestic analogy is the contract. Like the treaty, a contract can be said to make or create law between the parties—for within the facilitative framework of governing law but subject to that law's mandatory norms and constraints, courts recognize and enforce contract-created duties. The treaty shares a contract's consensual basis, but treaty law lacks the breadth and relative inclusiveness of a national body of contract law. It has preserved a certain Roman law flavor ('*pacta sunt servanda*', '*rebus sic stantibus*') acquired during the long period from the Renaissance to the nineteenth century, when continental European scholars dominated the field. But treaty law often reflects the diversity of approaches to contract law that lawyers in different countries bring to the topic, a diversity that is particularly striking on issues of treaty interpretation.

Duties Imposed by Treaty Law

Whatever its purpose or character, the international agreement is generally recognized from the perspective of international law as an authoritative starting point for legal reasoning about any dispute to which it is relevant. The maxim *pacta sunt servanda* is at the core of treaty law. It embodies a widespread recognition that commitments publicly, formally and (more or less) voluntarily made by a nation should be honored. As stated in Article 26 of the Vienna Convention on the Law of Treaties, referred to below: 'Every treaty in force is binding upon the parties to it and must be performed by them in good faith.'

Whatever the jurisprudential or philosophical bases for this norm, one can readily perceive the practical reasons for and the national interests served by adherence to the principle that *pacta sunt servanda.* The treaty represents one of the most effective means for bringing some order to relationships among states or their nationals, and for the systematic development of new principles responsive to the changing needs of the international community. It is the prime legal form through which that community can realize some degree of stability and predictability, and seek to institutionalize ideals like peaceful settlement of disputes and protection of human rights.

Acceptance of the primary role of the treaty does not, however, mean that a problem between two countries is adequately solved from the perspective of legal ordering simply by execution of a treaty with satisfactory provisions. A body of law has necessarily developed to deal with questions analogous to those addressed by domestic contract law—for example, formation of a treaty, its interpretation and performance, remedies for breach, and amendment or termination. But that body of law is often fragmentary and vague, reflecting the scarcity of decisions of international tribunals and the political tensions of some aspects of treaty law.

There have been recurrent efforts to remedy this situation through more or less creative codification of the law of treaties. The contemporary authoritative text grows out of a United Nations Conference on the Law of Treaties that adopted in 1969 the Vienna Convention on the Law of Treaties, UN Doc, A/Conf., 39/27, 63 A.J.I.L. 875 (1969). That Convention became effective in 1980 and (as of 1993) had been ratified by about 90 states from all regions of the world. For reasons stemming largely from tensions between the Executive and the Congress over authority over different types of international agreements, the United States has not ratified the Vienna Convention. Nonetheless, in its provisions on international agreements the *Restatement (Third), Foreign Relations Law of the United States* (1987) 'accepts the Vienna Convention as, in general, constituting a codification of the customary international law governing international agreements, and therefore as foreign relations law of the United States. . . . ' Vol. I, p. 145.

Treaty Formation

A treaty is formed by the express consent of its parties. Although there are no precise requirements for execution or form, certain procedures have become standard. By choice of the parties, or in order to comply with the internal rules of a signatory country that are considered at pp. 726–50 *infra*, it may be necessary to postpone the effectiveness of the agreement until a national legislative body has approved it and national executive authorities have ratified it. Instruments of ratification for bilateral agreements are then exchanged. In the case of multilateral treaties, such instruments are deposited with the national government or international organization that has been designated as the custodian of the authentic text and of all other instruments relating to the treaty, including subsequent adhesions by nations that were not among the original signatories. Thereafter a treaty will generally be proclaimed or promulgated by the executive in each country.

Consent

Given the established principle that treaties are consensual, what rules prevail as to the character of that consent? Do domestic-law contract principles about the effect of duress carry over to the international field? A leader came to power in Germany with a platform stressing the proposition that the Versailles Treaty was dictated by force and therefore was not binding. The newly emergent countries have made claims that treaties under which the colonial powers withdrew were imposed upon them by the threat of force or as a condition to withdrawal, and therefore are not valid. Sometimes analogous claims are made by states of the Third World when they are accused of violating human rights conventions to which they are parties.

In a domestic legal system, a party cannot enforce a contract which was signed by a defendant at gunpoint. One could argue that victorious nations cannot assert rights under a peace treaty obtained by a whole army. It is not surprising that the large powers are reluctant to recognize that such forms of duress can invalidate a treaty. If duress were a defense, it would be critical to define its contours, for many treaties result from various forms of military, political or economic pressure. The paucity of and doubts about international institutions with authority to develop answers to such questions underscore the reluctance to open treaties to challenge on these grounds.

The different approaches of the victorious Western powers towards the defeated nations in World War I and World War II reflect the growing awareness that treaties, whatever their 'legal' character, will survive only insofar as they bring satisfactory solutions to developing political, economic, and social problems. Even from a legal perspective, the advent of the UN Charter after World War II, with its explicit rejection of war as a permitted instrument of national policy, may herald some evolution of legal doctrine in this field. Article 52 of the Vienna Convention states: 'A treaty is void if its conclusion has been procured by the threat or use of force in violation of the principles of international law embodied in the Charter of the United Nations.' Attempts at Vienna to broaden the scope of coercion to include economic duress failed, although they resulted in a declaration condemning the use of such practices.

Reservations

Problems of consent that have no precise parallel in national contract law arise in connection with reservations to treaties, i.e., unilateral statements made by a state accepting a treaty 'whereby it purports to exclude, or vary the legal effect of certain provisions of the treaty in their application to a state.' (Article 2(1)(d) of the Vienna Convention). With bilateral treaties, no conceptual difficulties arise: ratification with reservations amounts to a counteroffer; the other state may explicitly accept (or reject) that ratification or may be held to have tacitly accepted it by proceeding with its ratification process or with compliance with the treaty. With multilateral treaties the problems may be quite complex. The traditional rule said that acceptance by all parties was required. The expanding number of states has required more flexibility.[2]

A treaty's text may explicitly permit some reservations, thus deliberately sacrificing uniformity of obligation for the sake of more widespread adherence. Such a listing of permissible reservations may by implication prohibit others or there may be explicit limitations. When neither such circumstance applies, Article 19 of the Vienna Convention provides that a state can formulate a reservation unless it is 'incompatible with the object and purpose of the treaty.' A state faced with unpalatable reservations by another state must, under the Vienna Convention, make its objection formally known if it is to prevent the reservation from becoming effective as to it. A series of reservations and objections can produce a complex set of relations among different states.

Given the increased number of reservations, some of great significance, that many states are attaching to their ratifications of basic human rights treaties, questions about those reservations' validity under general treaty law or under the terms of a specific treaty have become matters of high concern within the human rights movement. See pp. 756–58 and 767–76, *infra*.

Violations of and Changes in Treaties

Violation of a treaty may lead to diplomatic protests and a claim before an international tribunal. But primarily because of the limited and qualified consents of nations to the jurisdiction of international tribunals, the offended party will usually resort to other measures. In a national system of contract law, well developed rules govern such measures. They may distinguish between a minor breach not authorizing the injured party to terminate its own performance, and a material breach providing justification for such a move. Article 60 of the Vienna Convention provides that a material breach (as defined) of

[2] This is particularly true with such agreements as human rights conventions where widespread adherence is thought to be more important in advancing the rights of individuals than is exact correspondence among the commitments of the consenting states. See Advisory Opinion on Reservations to the Genocide Convention, [1951] I.C.J.Rep. 15, p. 757, *infra*.

a bilateral treaty entitles the other party to terminate the treaty or suspend its performance thereunder. These rules necessarily grow more complex for multilateral treaties. For reasons that Chapters 2 and 3 describe, they may have little relevance for the ways that states respond to violations of human rights treaty norms by other states.

Amendments raise additional problems. The treaty's contractual aspect suggests that the consent of all parties is necessary. Parties may however agree in advance (see Article 108 of the UN Charter) to be bound with respect to certain matters by the vote of a specified number. Such provisions in a multilateral treaty bring it closer in character to national legislation. They may be limited to changes which do not impose new obligations upon a dissenting party, although a state antagonistic to an amendment could of course withdraw. Absent some such provisions, a treaty may aggravate rather than resolve a fundamental problem of international law, how to achieve in a peaceful manner the changes in existing arrangements that are needed to adapt them to developing political, social or economic conditions.

One of the most contentious issues in treaty law is whether the emergence of conditions that were unforeseeable or unforeseen at the time of the treaty's conclusion terminates or modifies a party's obligation to perform. This problem borders the subject of treaty interpretation, considered below, since it is often described as a question whether an implied condition or an escape clause, called the *clausula rebus sic stantibus*, should be read into a treaty. Mature municipal legal systems have developed rules for handling situations wherein the performance of one party is rendered impossible or useless by intervening conditions. 'Impossibility,' 'frustration,' 'force majeure' and 'implied conditions' are the concepts used in Anglo-American law.

At the international level, possibilities of changes in conditions that upset assumptions underlying an agreement are enhanced by the long duration of many treaties, the difficulty in amending them and the rapid political, economic and social vicissitudes in the twentieth century. Thus nations, including the United States, have occasionally used *rebus sic stantibus* as the basis for declaring treaties no longer effective. International tribunals have shed no illumination on the field. Article 62 of the Vienna Convention states that a 'fundamental change of circumstances' which was not foreseen by the parties may not be invoked as a ground for terminating a treaty unless '(a) the existence of those circumstances constituted an essential basis of the consent of the parties . . . and (b) the effect of the change is radically to transform the extent of obligations still to be performed under the treaty.'

Treaty Interpretation

There is no shortcut to a reliable sense of how a given treaty will be construed. Even immersion in a mass of diplomatic correspondence and cases would not develop such a skill. In view of the variety of treaties and of approaches to their interpretation, such learning would more likely shed light on the possibilities than provide a certain answer to any given question.

One obstacle to helpful generalization about treaty interpretation is the variety of purposes which treaties serve. Different approaches are advisable for treaties that lay down rules for a long or indefinite period, in contrast with those settling past disputes. The long-term treaty must benefit from a certain flexibility and room for development if it is to survive changes in circumstances and relations between the parties. Changes in conditions like those that make *rebus sic stantibus* an attractive doctrine may lead a court or executive official to interpret a treaty liberally so as to give it a sensible application to new circumstances.

The very style of a treaty will influence the approach of an official charged with interpreting it. Certain categories, such as income tax conventions, lend themselves to a detailed draftsmanship that will often be impractical and undesirable in a constitutional document such as the UN Charter. Conventions such as those relating to human rights necessarily use broad and abstract terms and standards like fairness or *ordre public*.

In a national legal system, lawyers and courts can seek to give specific content to general statutory standards by resort to a common-law background or to a constitutional tradition, indeed by reference to the entire legal culture and society within which these standards become operative. Interpretation can reach towards generally understood practices, customs or purposes—even if the lawyer or judge may encounter choice and contradiction in practices and purposes rather than consensus. It may be far more difficult to interpret treaty standards of comparable generality embracing an international community with diverse national traditions. The problem becomes acute in multilateral treaties among nations from different regions, for one method of securing agreement to a treaty in the first instance may be to conceal rather than resolve differences through resort to general standards. In addition, the difficulty in achieving agreement over amendments to long-continuing multilateral treaties may encourage their draftsmen to express a 'consensus' through norms of a general character, which have a better chance of surviving and carrying their broad purposes through changed conditions among the signatories.[3]

Maxims similar to those found in domestic fields exist for treaties as well. The Vienna Convention contains several. Article 31 provides that a 'treaty shall be interpreted in good faith in accordance with the ordinary meaning to be given to the terms of the treaty in their context and in the light of its object and purpose.' Article 32 goes on to add that recourse may be had to supplementary

[3] The distinction suggested between interpretation of certain categories of international agreements and of domestic instruments becomes less tenable when one considers the range of domestic law expressed in open-textured norms of an abstract and general character. This is most evident in constitutional litigation. Decisions of the United States Supreme Court under the Due Process Clause and Equal Protection Clause of the Fourteenth Amendment to the U.S. Constitution offer apt illustrations. Consider also the critical and often fresh or innovative terms in regulatory legislation such as antitrust or civil rights statutes that raise just as vexing and contested issues of interpretation.

means if interpretation pursuant to Article 31 produces a meaning that is 'ambiguous or obscure' or an outcome 'manifestly absurd or unreasonable.'

One way to build a framework for construing treaties is to consider the continuum which lies between 'strict' interpretation according to the 'plain meaning' of the treaty, and interpretation according to the interpreter's view of the best means of implementing the purposes or realizing the principles expressed by the treaty. Of course both extremes of the spectrum are untenanted. One cannot wholly ignore the treaty's words, nor can one always find an unambiguous and relevant text that resolves the immediate issue.

Part of the difficulty is that treaties may be drafted in several languages. If domestic courts deem it unwise to 'make a fortress out of the dictionary,' it would seem particularly unwise when dictionaries in several languages (and in different legal systems according different meanings to linguistically similar terms) must be resorted to. Sometimes corresponding words in the different versions may shed more light on the intended meaning; at other times, they are plainly inconsistent.

Reliance upon literal construction or 'strict' interpretation, may however be an attractive method or technique to an international tribunal that is sensitive to its weak political foundation. It may be tempted to take refuge in the position that its decision is the ineluctable outcome of the drafter's intention expressed in clear text, and not a choice arrived at on the basis of the tribunal's understanding of policy considerations or relevant principles that may resolve a dispute over interpretation. Reliance on legislative history or *travaux préparatoires* can achieve the same result of placing responsibility on the drafters. The charge of 'judicial legislation' evokes strong reactions in the United States; it inevitably influences judges of international tribunals and heightens the temptation to take refuge in the dictionary.

Relationships between Treaties and Custom

Thus far we have considered custom and treaties in separate compartments, although as later materials show these two 'sources' or law-making processes of international law are complexly interrelated. For example, the question often arises of the extent to which a treaty should be read in the light of preexisting custom. A treaty norm of great generality may naturally be interpreted against the background of relevant state practice or policies. In such contexts, the question whether the treaty is intended to be 'declaratory' of preexisting customary law or to change that law may become relevant.

Moreover, treaties may give birth to rules of customary law. Assume a succession of bilateral treaties among many states, each containing a provision giving indigent aliens who are citizens of the other state party the right to counsel at the government's expense in a criminal prosecution. The question may arise whether these bilateral treaties create a custom that would bind a state not party to any of them. Polar arguments will likely be developed by parties to such a dispute, for example: (1) The non-party state cannot be bound by those treaties since it has not consented. The series of bilateral treaties simply constitutes special exceptions to the traditional customary law that leaves the state's discretion unimpaired on this matter. Indeed, the necessity that many states saw for treaties underscores that no obligation existed under customary law. (2) A solution worked out among many states should be considered relevant or persuasive for the development of a customary law setting standards for all countries. Similarly, the network of treaties may have become dense enough and state practice consistent with the treaty may have become general enough to build a customary norm binding all states. Article 38 of the Vienna Convention signals rather than resolves this issue by stating that nothing in its prior articles providing generally that a treaty does not create obligations for a third state precludes a rule set forth in a treaty from becoming binding on a third State 'as a customary rule of international law, recognized as such.'

In contemporary international law, broadly ratified multilateral treaties are more likely than a series of bilateral treaties to generate the argument that treaty rules have become customary law binding non-parties. Some of the principal human rights treaties, for example, have about 120 state parties from all parts of the world. Of course, one must distinguish between substantive norms in multilateral treaties that are alleged to constitute custom binding non-parties, and institutional arrangements created by the treaties in which parties have agreed, for example, to submit reports or disputes to a treaty organ.

Treaties and International Organizations

In addition to setting forth specific rules which are to govern the conduct of the parties, a treaty may establish machinery for the development of further norms. This applies particularly to multilateral agreements, which may be specialized or general in subject matter, regional or worldwide in scope. At its simplest, such a treaty may provide for periodic meetings at which the signatories' representatives will exchange views. From such discussions the representatives may go on to negotiate new agreements, for the presence of delegates from several countries makes possible the adjustment at one time of interlocking problems that affect each of them. At the next level, the treaty may authorize the parties' delegates to pass advisory or recommendatory resolutions. Such meetings can produce draft conventions which will be submitted to the members for consideration and possible ratification.

In more advanced arrangements the structure created by treaty will include organs or agencies exercising stated powers. Sometimes they are authorized to mediate or put pressure on disputants to arbitrate. Sometimes their authority extends to issuing non-binding declarations on relevant issues and recommendatory resolutions. Such powers can go further and include competence to issue binding interpretations of the treaty as well as regulations, directives or resolutions binding upon the parties (a limited legislative function) or to apply the treaty to specific situations in an authoritative way (a limited judicial function). Finally, the treaty may give a stated majority of the members the power to amend the agreement with respect to some or all provisions so as to bind all parties.

At some point in this progression we find that the treaty has created an international 'organization'. The growth of a permanent staff maintaining a continuous interim activity and the acquisition of a budget and buildings signal the emergence of a distinct entity with some life of its own. The members may endow this entity with a juridical personality and empower it to make contracts or treaties with private parties or governments and be a party to lawsuits. They may also confer upon the organization and its officials various immunities.

Such international organizations' concerns and functions now include diverse fields like peacekeeping, trade and monetary matters, fisheries and other regimes of the high seas, environmental protection, the regulation of basic commodities, and protection of human rights. In all such cases, the issues to be addressed do not lend themselves to adequate resolution through the development of customary rules, through a network of bilateral treaties, or through multilateral treaties that contain only substantive rules without institutional mechanisms for their promotion, implementation or enforcement. Such organizations may be universal or regional. Issues of human rights, for example, figure pervasively in the work of the United Nations and fully occupy some organs within the UN such as the Commission on Human Rights (Chapters 3, 5). There are also regional human rights organizations in Europe, the Americas and Africa (Chapter 10).

NOTE

The following three readings introduce debates about international law—concerning state sovereignty, the foundations and justifications for international law, and the structure of international legal argument—that figure throughout this coursebook. The three authors-scholars differ widely in their method and approach toward the subject, as evidenced even in the slender excerpts that follow. Those differences inform later materials and debates about human rights law itself.

The excerpts from Brownlie's treatise expand and probe the notions of custom set forth in the earlier Comment. Those from Schachter's book explore fundamental questions about the nature of international law, such as its relation to power, realist and positivist perspectives on international law, the relation of custom or general principles to state sovereignty, and so on. The article by Koskenniemi (a condensed version of his 1989 book, *From Apology to Utopia*) asks different questions in its examination of dilemmas and contradictions built into the structure of and argument developing international law.

For the novice, the readings will be difficult to grasp fully at this stage. Bear in mind that the balance of Chapter 2 as well as Chapters 3–5 refer on several occasions to the ideas in these readings, drawing on and illustrating them.

IAN BROWNLIE, PRINCIPLES OF PUBLIC INTERNATIONAL LAW

(4th ed. 1990), at 3.

[These excerpts start with the author's comments on Article 38 of the Statute of the International Court of Justice.]

These provisions are expressed in terms of the function of the Court, but they represent the previous practice of arbitral tribunals, and Article 38 is generally regarded as a complete statement of the sources of international law. Yet the article itself does not refer to 'sources' and, if looked at closely, cannot be regarded as a straightforward enumeration of the sources. The first question which arises is whether paragraph 1 creates a hierarchy of sources. They are not stated to represent a hierarchy, but the draftsmen intended to give an order and in one draft the word 'successively' appeared. In practice the Court may be expected to observe the order in which they appear: (a) and (b) are obviously the important sources, and the priority of (a) is explicable by the fact that this refers to a source of mutual obligations of the parties. . . .

. . . Moreover, it is probably unwise to think in terms of hierarchy dictated by the order (a) to (d) in all cases. Source (a) relates to *obligations* in any case; and presumably a treaty contrary to a custom or to a general principle part of the *jus cogens* would be void or voidable. Again, the interpretation of a treaty may involve resort to general principles of law or of international law. A treaty may be displaced or amended by a subsequent custom, where such effects are recognized by the subsequent conduct of the parties.

. . .

[The following observations are taken from a section entitled 'International Custom'.]

Evidence. The material sources of custom are very numerous and include the following: diplomatic correspondence, policy statements, press releases, the opinions of official legal advisers, official manuals on legal questions, e.g. manuals of military law, executive decisions and practices, orders to naval forces etc., comments by governments on drafts produced by the International Law Commission, state legislation, international and national judicial decisions, recitals in treaties and other international instruments, a pattern of treaties in the same form, the practice of international organs, and resolutions relating to legal questions in the United Nations General Assembly. Obviously the value of these sources varies and much depends on the circumstances.

The elements of custom

(a) Duration. Provided the consistency and generality of a practice are proved, no particular duration is required: the passage of time will of course be a part of the evidence of generality and consistency. A long (and, much less, an immemorial) practice is not necessary, and rules relating to airspace and the continental shelf have emerged from fairly quick maturing of practice. The International Court does not emphasize the time element as such in its practice.

(b) Uniformity, consistency of the practice. This is very much a matter of appreciation and a tribunal will have considerable freedom of determination in many cases. Complete uniformity is not required, but substantial uniformity is . . .

. . .

(c) Generality of the practice. This is an aspect which complements that of consistency. Certainly universality is not required, but the real problem is to determine the value of abstention from protest by a substantial number of states in face of a practice followed by some others. Silence may denote either tacit agreement or a simple lack of interest in the issue. . . .

(d) Opinio juris et necessitatis. The Statute of the International Court refers to 'a general practice *accepted as law*'. Brierly speaks of recognition by states of a certain practice 'as obligatory', and Hudson requires a 'conception that the practice is required by, or consistent with, prevailing international law'. Some writers do not consider this psychological element to be a requirement for the formation of custom, but it is in fact a necessary ingredient. The sense of legal obligation, as opposed to motives of courtesy, fairness, or morality, is real enough, and the practice of states recognizes a distinction between obligation and usage. The essential problem is surely one of proof, and especially the incidence of the burden of proof.

In terms of the practice of the International Court of Justice—which provides a general guide to the nature of the problem—there are two methods of approach. In many cases the Court is willing to assume the existence of an *opinio juris* on the basis of evidence of a general practice, or a consensus in the literature, or the previous determinations of the Court or other international tribunals. However, in a significant minority of cases the Court has adopted a more rigorous approach and has called for more positive evidence of the recognition of the validity of the rules in question in the practice of states. The choice of approach appears to depend upon the nature of the issues (that is, the state of the law may be a primary point in contention), and the discretion of the Court.

. . .

The persistent objector. The way in which, as a matter of practice, custom resolves itself into a question of special relations is illustrated further by the rule that a state may contract out of a custom in the process of formation. Evidence of objection must be clear and there is probably a presumption of acceptance which is to be rebutted. Whatever the theoretical underpinnings of the principle, it is well recognized by international tribunals, and in the practice of states. Given the majoritarian tendency of international relations the principle is likely to have increased prominence.

. . .

OSCAR SCHACHTER, INTERNATIONAL LAW IN THEORY AND PRACTICE

(1991), at 5.

Ch. 1. The Nature and Reality of International Law

. . .

The Uses of Law and the Role of Power

In discussing the complexity of international law and its relative autonomy, we have more or less assumed its 'reality'. We did so while recognizing that it is a product of political and social forces, that it is dependent on behaviour and that it is an instrument to meet changing ends and values. All these aspects give rise to questions as to the reality of international law. These questions are epitomized in the not uncommon view that we cannot have genuine and effective law in a society of sovereign States dominated by power and self-interest. . . .

One way of approaching the issues is to consider the views of the sceptics of international law—those who doubt, for one reason or another, that international law can contribute significantly to international order. Four kinds of sceptical positions merit our attention. The first emphasizes the dominance of power over law. The second asserts the dependence of international law on the will of national States. A third position points to the deep divisions of belief, aims and culture in international society and questions whether a common authoritative legal system is realizable. The fourth sceptical position lays stress on the fragility of a legal system that lacks centralized institutions to determine authoritatively what the law is and to enforce it. I will discuss the reasoning behind these positions and comment on their validity.

The thesis that law is subordinate to power is commonly referred to as a 'realist' (or 'realpolitik') point of view. Power, in this context, refers to the ability of a State to impose its will on others or, more broadly, to control outcomes contested by others. The components of power are military, economic, political and psychological; international society exhibits, in striking degree, an unequal distribution of these components of power. The unequal distribution of power is a pervasive and dominating element in the relations of States. States strive to augment their power, perceiving power sometimes as an end in itself and more commonly as a means to attain more freedom of action and other objectives. For the realist, the crucial importance of power and the pursuit of self-interest by States make it virtually inevitable that the law—both in its creation and application—should yield to those determinants. Law may play a subordinate role as an aid to stability, a 'gentle civilizer of events' but it cannot be relied on 'to suppress the chaotic and dangerous aspirations of governments'. Political theorists reach similar conclusions in their analysis of the State system. From a 'structural' standpoint (as they sometimes put it), a system based on the sovereignty of States is a system of 'co-ordinate relations', in which formal authority is decentralized. While this fundamental condition does not rule out the use of a legal system to provide a required degree of order and predictability, the individual States in the last analysis are not subordinated to any superior authority. Hence the effective limits on their action derive from their own perception of national interest and the countervailing power of others. In consequence, as Raymond Aron has put it, international society is 'an anarchical order of power in which might makes right'.

These general theoretical conclusions accord with the widespread popular belief that in international relations power rather than law governs. . . . The fact that legal arguments are almost always made by the alleged violators only tends to add to the cynicism about law since such self-serving legal arguments are not submitted to adjudication or other third-party determinations. The absence of compulsory or generally accepted judicial settlement of international disputes is taken as compelling evidence that the law is not taken seriously and hence that 'power politics' prevails.

Since we cannot deny the crucial role of power in the relations of States, we should seek to understand its specific impact on the international legal system. Plainly, international law is not an ideal construct, created and given effect solely in terms of its internal logic. Nor can it be understood only as an instrument to serve human needs and aims (though it is that too). International law must also be seen as the product of historical experience in which power and the 'relation of forces' are determinants. Those States with power (i.e., the ability to control the outcomes contested by others) will have a disproportionate and often decisive influence in determining the content of rules and their application in practice. Because this is the case, international law, in a broad sense, both reflects and sustains the existing political order and distribution of power.

. . . A popular view, probably prevalent all over the world, is that powerful States, like powerful individuals in many countries, can often flout the law and get away with it. . . .

We need to consider not only that States break the rules but also that they generally conform to them even against their immediate interest. Nobody denies that States, powerful and not so powerful, observe international law most of the time. There are various reasons why they do so. Much compliance can be attributed to institutionalized habit; officials follow the rules as a matter of practice, and in countless decisions they look to treaty obligations, to precedents that evidence custom and to general principles of law expounded in treatises and manuals. Many of the decisions involve no apparent clash of law and self-interest but numerous cases arise in which a government refrains from action (or non-action) that it would otherwise take if there were no legal grounds limiting its discretion. This is most evident when specific treaties apply—say, of commerce, navigation, reciprocal exchanges—but it is also true of the many cases covered by the unwritten customary rules applicable to many areas of inter-State relations. We must remember that the law is not applied as if it were governed by a computer programmed to respond to every contingency. In actuality, a good part of law observance takes place because the officials concerned do not even consider the option of violation; they have, so to speak, 'internalized' the rules so that possibilities of action contrary to the law do not even rise to conscious decision-making. It would, for example, be virtually certain today that a State would not consider sending its troops in to another country to collect revenues to pay a debt (as the United States once did in Haiti). Nor would it consider asserting jurisdiction over foreign flag vessels on the high seas. We often tend to forget how much of the basic law of nations is so thoroughly embedded in the minds and habits of officials that it is given effect without conscious decision making.

It is of course also true that cases arise in which officials do have to consider whether the law should be applied when it appears to be in the immediate interest of the government not to do so. The responses to such situations depend on a variety of considerations. Most obviously, governments will weigh a possible breach by them against their interest in reciprocal observance by the other party. They will also consider the likelihood of retaliation and other self-help measures by that party. Nor would they ignore the negative consequences of a reputation for repudiating their obligations. In many countries, officials would be sensitive to anticipated criticism by influential domestic leaders or groups who place high value on the country's reputation for legality generally or on observance of the particular obligations involved. The possibility of judicial enforcement in domestic tribunals may in some cases serve as a deterrent to non-compliance. Remedies by the aggrieved State may also be available under some circumstances in international mechanisms, perhaps through arbitral or judicial means or through loss of benefits under treaty régimes. . . . Violations, in short, are rarely cost-free even to powerful States.

. . .

The Dependency of International Law on the Will of States

. . . If States have no superior authority and their relations are 'co-ordinate', rather than hierarchical, does it not follow that a State is bound only by the legal rules to which it agrees to submit? . . .

These questions are as old as international law itself and have given rise to considerable theoretical writing. The idea that the will of States is the basis of international law and hence that the law is dependent on the consent of States is referred to in international law theory as 'voluntarism' or 'consensualism'. Voluntarism is not only a theory held by academic scholars. It is also an expression of the strongly held conception of State sovereignty dominant in most governments. . . .

The general idea that international law rests on the will of the States has been applied in various ways, with quite different significance. It has been applied to international law as a whole (particularly customary law) and to particular rules of international law. In regard to the latter, the requirement of consent has been directed both to the creation of new rules of law and to their use in particular cases. We will sort out the questions raised in each category by considering the following five propositions:

(1) international law as a general system is accepted by all States and hence is an expression of their will;

. . .

(3) the creation of a new rule or repeal of an old rule of customary law requires the consent of States;
(4) a State which has not consented to a customary law rule is free at any time to reject its application to that State;
(5) any State is free to exercise its sovereign right to reject the application of a customary law rule on the ground that it is not in accordance with that State's will.

(I should note that the term customary law rule as used above is intended to refer not only to rules in the narrow sense of that term but also to principles, standards, practices, concepts and procedures that are considered as legal grounds for asserting rights and obligations.)

(1) The first proposition asserts an empirical fact amply supported by practice. A possible objection is that the acceptance of international law cannot be attributed to the 'will' of the States since they have no choice in the matter. The very claim to be a State with authority over a given territory and population involves recognition of the basic international law rules. We shall consider this further in regard to the second proposition. However, even if it is true that 'membership' in any society presupposes adherence to its basic rules (*ubi societas, ibi ius*), it is not inconsistent with the fact that States accept the particular system of international law now in force. At least in that sense, the system rests on their consent just as a domestic law system may be said to rest basically on the consent of the people. Without their general consent, there could be no durable operative system of law. Although voluntarism in this form may seem to be a 'weak' version, it is important to recognize that the system of law has in general been accepted by the community of States.

Acceptance of the system is in itself a plausible basis for the obligation to abide by the particular rules valid in that system.

. . .

(3) The third proposition is that the creation of a new rule or repeal of an existing rule requires the consent of States. Inasmuch as customary law arises through uniformities of State conduct accompanied by the belief of States that they are conforming to what amounts to a legal obligation, the States that participated in such conduct and recognize the obligation created by it can reasonably be considered to have consented to the rule thus established. . . .

How many States are required to establish 'general' practice and how frequent, numerous and consistent the practice must be are questions which cannot be answered in categorical propositions. Generality, frequency, density, consistency, duration are in principle required but whether they are met in regard to a specific rule depends on the circumstances of the case. This seems to leave the emergence of customary law rather mysterious. It is not easy to generalize about the factors that should help in deciding how many States are enough, how 'dense' the practice need be, how long in time and so on. . . . For our present purpose, it is enough to note the broad agreement among authorities that general custom does not require universal consent of all States. . . .

(4) The most significant test of voluntarist-consensualist theory is raised by the fourth proposition—namely that a State is not bound by a rule to which it has not consented. The proposition requires critical analysis. Non-consenting States divide into two categories: (1) States that have manifested neither acceptance nor objection; (2) States which have openly objected to the rule in general or to its specific application to the objecting State. If we accept the principle that general custom does not require universal consent it follows that the assent of a particular State is not necessary for a general rule to come into being and to bind all States. . . .

A special problem is presented by the second category of non-consenting States—those that have openly manifested their dissent to a customary rule. One might ask why a dissenting State should be able to avoid a general customary rule if universal consent is not considered necessary and non-assenting States are considered bound. . . . Even if this is accepted as law, it is subject to some limitations. . . .

. . . [I]t may be questioned whether the exception for a dissenting State would apply to a new principle of customary law regarded as fundamental or of major importance. . . . If . . . principles are regarded as fundamental and of major importance to the generality of States with respect to an area beyond any State's jurisdiction, a good case can be made for denying the dissenting State the right to avoid the obligations that all other States incur as a consequence of the acceptance of the new principles. The issue cannot reasonably be decided solely by reference to voluntarist theory. It would be germane to consider a variety of factors including the circumstances of adoption of the new principle, the reasons for its importance to the generality of States, the grounds for dissent, and the relevant position of the dissenting States. . . .

(5) The fifth proposition constitutes the 'strongest' use of voluntarist theory. It would allow a State to reject the application of a customary rule to it simply on the ground that it was contrary to the State's present will. No State, to my knowledge, has openly espoused that position. It would amount to a denial of customary law and it is most unlikely that any State would be prepared to take that position. However, a State may seek to avoid submitting to a rule by adopting a more moderate form of consensualism. . . . A strong preference in favour of State sovereignty and voluntarism on the part of members of a tribunal (or other decision-makers) will tend to raise the threshold of practice and *opinio juris* considered necessary to establish a universal rule.

Voluntarism in a somewhat different guise can also be found when a State objects to a rule on the ground that it is incompatible with a vital interest of the State. In such case, the recalcitrant State rests not simply on its will per se but on a superior norm of self-interest that is said to prevail over law. . . .

Significantly, no State appears to claim a right of this kind although some have interpreted self-defence broadly (as we shall see) or have turned to general notions of sovereignty and independence to justify departing from rules that are deemed against their vital interest.

. . .

Ch. IV. General Principles and Equity

. . .

The Broad Expanse of General Principles of Law

We can distinguish five categories of general principles that have been invoked and applied in international law discourse and cases. Each has a different basis for its authority and validity as law. They are

(1) The principles of municipal law 'recognized by civilized nations'.
(2) General principles of law 'derived from the specific nature of the international community'.
(3) Principles 'intrinsic to the idea of law and basic to all legal systems'.
(4) Principles 'valid through all kinds of societies in relationships of hierarchy and co-ordination'.
(5) Principles of justice founded on 'the very nature of man as a rational and social being'.

Although these five categories are analytically distinct, it is not unusual for a particular general principle to fall into more than one of the categories. For example, the principle that no one shall be a judge in his own cause or that a victim of a legal wrong is entitled to reparation are considered part of most if not all, systems of municipal law and as intrinsic to the basic idea of law.

Our first category, general principles of municipal law, has given rise to a considerable body of writing and much controversy. Article 38 (1) *(c)* of the Statute of the Court does not expressly refer to principles of national law but rather general principles 'recognized by civilized nations'. . . . Elihu Root, the American member of the drafting committee, prepared the text finally adopted and it seemed clear that his amendment was intended to refer to principles 'actually recognized and applied in national legal systems'. The fact that the subparagraph was distinct from those on treaty and custom indicated an intent to treat general principles as an independent source of law, and not as a subsidiary source. As an independent source, it did not appear to require any separate proof that such principles of national law had been 'received' into international law.

However, a significant minority of jurists holds that national law principles, even if generally found in most legal systems, cannot *ipso facto* be international law. One view is that they must receive the *imprimatur* of State consent through custom or treaty in order to become international law. The strict positivist school adheres to that view. A somewhat modified version is adopted by others to the effect that rules of municipal law cannot be considered as recognized by civilized nations unless there is evidence of the concurrence of States on their status as international law. Such concurrence may occur through treaty, custom or other evidence of recognition. This would allow for some principles, such as *res judicata*, which are not customary law but are generally accepted in international law. . . .

. . .

. . . The most important limitation on the use of municipal law principles arises from the requirement that the principle be appropriate for application on the international level. Thus, the universally accepted common crimes—murder, theft, assault, incest—that apply to individuals are not crimes under international law by virtue of their ubiquity. . . .

At the same time, I would suggest a somewhat more positive approach for the emergent international law concerned with the individual, business companies, environmental dangers and shared resources. Inasmuch as these areas have become the concern of international law, national law principles will often be suitable for international application. This does not mean importing municipal rules 'lock, stock and barrel', but it suggests that domestic law rules applicable to such matters as individual rights, contractual remedies, liability for extra-hazardous activities, or restraints on use of common property, have now become pertinent for recruitment into international law. In these areas, we may look to representative legal systems not only for the highly abstract principles of the kind referred to earlier but to more specific rules that are sufficiently widespread as to be considered 'recognized by civilized nations'. . . .

The second category of general principles included in our list comprises principles derived from the specific character of the international community. The most obvious candidates for this category of principles are . . . the necessary principles of co-existence. They include the principles of *pacta sunt servanda*, non-intervention, territorial integrity, self-defence and the legal equality of States. Some of these principles are in the United Nations Charter and therefore part of treaty law, but others might appropriately be treated as principles required by the specific character of a society of sovereign independent members.

. . .

The foregoing comments are also pertinent to the next two categories of general principles. The idea of principles '*jus rationale*' 'valid through all kinds of human societies'. . . is associated with traditional natural law doctrine. At the present time its theological links are mainly historical as far as international law is concerned, but its principal justification does not depart too far from the classic natural law emphasis on the nature of 'man', that is, on the human person as a rational and social creature.

The universalist implication of this theory—the idea of the unity of the human species—has had a powerful impetus in the present era. This is evidenced in at least three significant political and legal developments. The first is the global movements against discrimination on grounds of race, colour and sex. The second is the move toward general acceptance of human rights. The third is the increased fear of nuclear annihilation. These three developments strongly reinforce the universalistic values inherent in natural law doctrine. They have found expression in numerous international and constitutional law instruments as well as in popular movements throughout the world directed to humanitarian ends. Clearly, they are a 'material source' of much of the new international law manifested in treaties and customary rules.

In so far as they are recognized as general principles of law, many tend to fall within our fifth category—the principles of natural justice. This concept is well known in many municipal law systems (although identified in diverse ways). 'Natural justice' in its international legal manifestation has two aspects. One refers to the minimal standards of decency and respect for the individual human being that are largely spelled out in the human rights instruments. We can say that in this aspect, 'natural justice' has been largely subsumed as a source of general principles by the human rights instruments . . .

. . .

NOTE

Compare with Schachter's views about the relationship between law and force the following observations of Stanley Hoffmann, in *The Study of International Law and the Theory of International Relations*, 1963 Proc. Am. Soc. Int. L. 26:

> It is however essential for the social scientist to understand that law is not merely a policy among others in the hands of statesmen, but that it is a

> tool with very special characteristics and roles. . . . Most important is the fact that law has a distinct solemnity of effects: it is a normative instrument that creates rights and duties. Consequently it has a function that is both symbolic and conservative; it enshrines, elevates, consecrates the interests or ideas it embodies. We understand, thus, why law is an important stake in the contests of nations. What makes international law so special a tool for states is this solemnity of effects, rather than the fact that its norms express common interests; for this is far too simple: some legal instruments such as peace treaties reflect merely the temporary, forced convergence of deeply antagonistic policies. A situation of dependence or of superiority that is just a fact of life can be reversed through political action, but once it is solemnly cast in legal form, the risks of action designed to change the situation are much higher: law is a form of policy that changes the stakes, and often 'escalates' the intensity, of political contests; it is a constraint comparable to force in its effects.

MARTTI KOSKENNIEMI, THE POLITICS OF INTERNATIONAL LAW

1 Eur. J. Int. L. 4 (1990), at 6

[This theoretical essay examines the character and politics of international law and the structure of international legal argument. It both develops earlier readings' conceptions of international law and custom, and provides an illuminating framework for understanding the particular character and effects of the human rights movement within the larger field of international law. The excerpts below stress the author's discussion of (i) the conceptions of apology and utopia that are central to his theory; (ii) dilemmas of state sovereignty and of consensual and justice-oriented theories of obligation in international law; and (iii) complexities in argument based on customary international law.]

The vision of a Rule of Law between states . . . is yet another reformulation of the liberal impulse to escape politics. . . .

In this article . . . I shall attempt to show that our inherited ideal of a World Order based on the Rule of Law thinly hides from sight the fact that social conflict must still be solved by political means and that even though there may exist a common legal rhetoric among international lawyers, that rhetoric must, *for reasons internal to the ideal itself*, rely on essentially contested—political—principles to justify outcomes to international disputes.

II. The Content of the Rule of Law: Concreteness and Normativity

Organizing society through legal rules is premised on the assumption that these rules are objective in some sense that political ideas, views, or preferences are not. To show that international law is objective—that is, independent from international politics—the legal mind fights a battle on two fronts.

On the one hand, it aims to ensure the *concreteness* of the law by distancing it from theories of natural justice. On the other hand, it aims to guarantee the *normativity* of the law by creating distance between it and actual state behaviour, will, or interest. Law enjoys independence from politics only if both of these conditions are simultaneously present.

. . . To avoid political subjectivism and illegitimate constraint, we must base law on something concrete—on the actual (verifiable) behaviour, will and interest of the members of society-states. The modern view is a *social conception of law*. For it, law is not a natural but an artificial creation, a reflexion of social circumstances.

According to the requirement of normativity, law should be applied regardless of the political preferences of legal subjects. In particular, it should be applicable even against a state which opposes its application to itself. As international lawyers have had the occasion to point out, legal rules whose content or application depends on the will of the legal subject for whom they are valid are not proper legal rules at all but apologies for the legal subject's political interest.

. . .

This argumentative structure, however, which forces jurists to prove that their law is valid because concrete and normative in the above sense, both creates and destroys itself. For it is impossible to prove that a rule, principle or doctrine (in short, an argument) is both concrete and normative simultaneously. The two requirements *cancel each other*. An argument about concreteness is an argument about the closeness of a particular rule, principle or doctrine to state practice. But the closer to state practice an argument is, the less normative and the more political it seems. The more it seems just another apology for existing power. An argument about normativity, on the other hand, is an argument which intends to demonstrate the rule's distance from state will and practice. The more normative a rule, the more political it seems because the less it is possible to argue it by reference to social context. It seems utopian and—like theories of natural justice—manipulable at will.

The dynamics of international legal argument are provided by the constant effort of lawyers to show that their law is either concrete or normative and their becoming thus vulnerable to the charge that such law is in fact political because apologist or utopian. Different doctrinal and practical controversies turn on transformations of this dilemma. It lies behind such dichotomies as 'positivism'/'naturalism', 'consent'/'justice', 'autonomy'/'community', process'/ 'rule', etc., and explains why these and other oppositions keep recurring and do not seem soluble in a permanent way. They recur because it seems possible to defend one's legal argument only by showing either its closeness to, or its distance from, state practice. They seem insoluble because both argumentative strategies are vulnerable to what appear like valid criticisms, compelled by the system itself.

This provides an argumentative structure which is capable of providing a valid criticism of each substantive position but which itself cannot justify any.

The fact that positions are constantly taken and solutions justified by lawyers, demonstrates that the structure does not possess the kind of distance from politics for which the Rule of Law once seemed necessary. It seems possible to adopt a position only by a political choice: a choice which must ultimately defend itself in terms of a conception of justice.

III. Doctrinal Structures

Two criticisms are often advanced against international law. One group of critics has accused international law of being too political in the sense of being too dependent on states' political power. Another group has argued that the law is too political because founded on speculative utopias. The standard point about the non-existence of legislative machineries, compulsory adjudication and enforcement procedures captures both criticisms. From one perspective, this criticism highlights the infinite flexibility of international law, its character as a manipulable facade for power politics. From another perspective, the criticism stresses the moralistic character of international law, its distance from the realities of power politics. According to the former criticism, international law is too *apologetic* to be taken seriously in the construction of international order. According to the latter, it is too *utopian* to the identical effect.

International lawyers have had difficulty answering these criticisms. The more reconstructive doctrines have attempted to prove the normativity of the law, its autonomy from politics, the more they have become vulnerable to the charge of utopianism. The more they have insisted on the close connexion between international law and state behaviour, the less normative their doctrines have appeared. Let me outline the four positions which modern international lawyers have taken to prove the relevance of their norms and doctrines. These are mutually exclusive and logically exhaustive positions and account for a full explanation of the possibilities of doctrinal argument.

Many of the doctrines which emerged from the ashes of legal scholarship at the close of the First World War explained the failure of pre-war international doctrines by reference to their apologist character. Particular objects of criticism were 'absolutist' doctrines of sovereignty, expressed in particular in . . . doctrines stressing the legal significance of the balance of power or delimiting the legal functions to matters which were unrelated to questions of 'honour' or 'vital interest.' . . . By associating the failure of those doctrines with their excessive closeness to state policy and national interest and by advocating the autonomy of international legal rules, these jurists led the way to the establishment of what could be called a *rule approach* to international law, stressing the law's normativity, its capacity to oppose state policy as the key to its constraining relevance.

This approach insists on an objective, format test of pedigree (sources) which will tell which standards qualify as legal rules and which do not. If a rule meets this test, then it is binding. Though there is disagreement between rule approach lawyers over what constitutes the proper test, there is no dispute about its importance. . . .

The second major position in contemporary scholarship uses these criticisms to establish itself. A major continental interpretation of the mistakes of 19th-century lawyers and diplomats explains them as a result of naive utopianism: an unwarranted belief in the viability of the [1815 Congress of Vienna system], with its ideas of legality and collective intervention. It failed because it had not been able to keep up with the politics of emergent nationalism and the increasing pace of social and technological change. Lawyers such as Nicolas Politis or Georges Scelle stressed the need to link international law much more closely to the social—even biological—necessities of international life. Roscoe Pound's programmatic writings laid the basis for the contemporary formulation of this approach by criticizing the attempt to think of international law in terms of abstract rules. It was, rather, to be thought of "in terms of social ends."

According to this approach—the *policy approach*—international law can only be relevant if it is firmly based in the social context of international policy. Rules are only trends of past decision which may or may not correspond to social necessities. 'Binding force' is a juristic illusion. Standards are, in fact more or less effective and it is their effectiveness—their capacity to further social goals—which is the relevant question, not their formal 'validity.'

But this approach is just as vulnerable to well-founded criticisms as the rule approach. By emphasizing the law's concreteness, it will ultimately do away with its constraining force altogether. If law is only what is effective, then by definition, it becomes an apology for the interests of the powerful. If, as Myres McDougal does, this consequence is avoided by postulating some 'goal values' whose legal importance is independent of considerations of effectiveness, then the (reformed) policy approach becomes vulnerable to criticisms which it originally voiced against the rule approach. In particular, it appears to assume an illegitimate naturalism which . . . is in constant danger of becoming just an apology of some states' policies.

The rule and the policy approaches are two contrasting ways of trying to establish the relevance of international law in the face of what appear as well-founded criticisms. The former does this by stressing the law's normativity, but fails to be convincing because it lacks concreteness. The latter builds upon the concreteness of international law, but loses the normativity, the binding force of its law. It is hardly surprising, then, that some lawyers have occupied the two remaining positions: they have either assumed that international law can neither be seen as normatively controlling nor widely applied in practice (the sceptical position), or have continued writing as if both the law's binding force as well as its correspondence with developments in international practice were a matter of course (idealist position). The former ends in cynicism, the latter in contradiction.

. . .

. . . On the one hand, the 'idealist' illusion is preserved that law can and does play a role in the organization of social life among states. On the other, the 'realist' criticisms have been accepted and the law is seen as distinctly secondary to power and politics. Modern doctrine, as Philip Allott has shown, uses a mixture of positivistic and naturalistic, consensualistic and non-consensualistic, teleological, practical, political, logical and factual arguments in happy confusion, unaware of its internal contradictions. . . .

. . .

[The following excerpts, about consensualism and justice-based explanations of international law and their relevance to argument about customary international law, appear in a section entitled 'Sources'.]

Though there is no major disagreement among international lawyers about the correct enumeration of sources (treaties, custom, general principles), the rhetorical force of sources ('binding force') is explained from contrasting perspectives. Their importance is sometimes linked with their capacity to reflect state will (consensualism). At other times, such binding force is linked with the relationship of sources arguments with what is 'just', 'reasonable', 'in accordance with good faith', or some other non-consensual metaphor.

Standard disputes about the content or application of international legal norms use the contradiction between consent and justice based explanations. One party argues in terms of consent, the other in terms of what is just (reasonable, etc.). But neither argument is fully justifiable alone. A purely consensual argument cannot ultimately justify the application of a norm against non-consenting states (apologism). An argument relying only on a notion of justice violates the principle of the subjectivity of value (utopianism). Therefore, they must rely on each other. Arguments about consent must explain the relevance and content of consent in terms of what seems just. Arguments about justice must demonstrate their correctness by reference to what states have consented to. Because these movements (consent to justice; justice to consent) make the originally opposing positions look the same, no solution can be made by simply choosing one. A solution now seems possible only by either deciding what is it that states 'really' will or what the content of justice 'really' is. Neither question, however, is answerable on the premises of the Rule of Law.

For the modern lawyer, it is very difficult to envisage, let alone to justify, a law which would divorce itself from what states think or will to be the law. The apparent necessity of consensualism seems grounded in the very criticism of natural norms as superstition. Yet, the criticisms against full consensualism—its logical circularity, its distance from experience, its inherent apologism—are well-known. Consensualism cannot justify the application of a norm against a state which opposes such application unless it creates distance between the norm and the relevant state's momentary will. It has been explained, for example, that though law emerges from consent, it does not need every state's consent all the time, that a general agreement . . . is sufficient to apply the norm.

But these explanations violate the principle of sovereign equality—they fail to explain why a state should be bound by what another state wills. This can, of course, be explained from some concept of social necessity. But in such case we have already moved away from pure consensualism and face the difficulty of explaining the legal status of the assumed necessity and why it should support one norm instead of another.

A more common strategy is to explain that the state has originally consented (by means of recognition, acquiescence, by not protesting or by 'tacitly' agreeing) although it now denies it has. Such an argument is extremely important in liberal legitimation discourse. It allows defending social constraint in a consensual fashion while allowing the application of constraint against a state which denies it consent. But even this argument fails to be convincing because it must ultimately explain itself either in a fully consensual or fully non-consensual way and thereby become vulnerable to the objections about apologism or utopianism.

Why should a state be bound by an argument according to which it has consented, albeit 'tacitly'? If the reason is stated in terms of respecting its own consent, then we have to explain why we can know better than the state itself what it has consented to. Even consensualists usually concede that such knowledge is not open to external observers . . .

Tacit consent theorists usually explain that the question is not of 'real' but of 'presumed' will. But what then allows the application of the presumption against a state denying that it had ever consented to anything like it? At this point the tacit consent lawyer must move from consensualism to non-consensualism. Tacit consent— or the presumption of consent—binds because it is 'just' or in accordance with reasonableness or good faith, or it protects legitimate expectations or the like. . . .

. . .

The structure, importance and weaknesses of tacit consent is nowhere more visible that in the orthodox argument about customary international law.

According to this argument, binding custom exists if there is a material practice of states to that effect and that practice is motivated by the belief that it is obligatory. . . .

. . .

Modern lawyers have rejected fully materialistic explanations of custom as apologist, incapable of distinguishing between factual constraint and law. If the possibility is excluded that this distinction can be made by the justice of the relevant behaviour, then it can only be made by reference to the psychological element, the *opinio juris*. But, as many students of the [International Court of Justice] jurisprudence have shown, there are no independently applicable criteria for ascertaining the presence of the *opinio juris*. The ICJ has simply inferred its presence or absence from the extent and intensity of the material practice it has studied. . . .

. . .

Customary law doctrine remains indeterminate because it is circular. It assumes behaviour to be evidence of states' intentions (*opinio juris*) and the latter to be evidence of what behaviour is relevant as custom. To avoid apologism (relying on the state's present will), it looks at the psychological element from the perspective of the material: to avoid utopianism (making the distinction between binding and non-binding usages by reference to what is just), it looks at the material element from the perspective of the psychological. It can occupy neither position in a permanent way without becoming vulnerable to criticism compelled by the other. . . .

. . .

Many have been dissatisfied with the modern strategy of arguing every imaginable non-written standard as 'custom.' Sir Robert Jennings, among others, has noted that what we tend to call custom: 'is not only not customary law: it does not even faintly resemble a customary law.' But if a non-written standard is not arguable in terms of material practices or beliefs relating to such practices then it can only exist as natural law—being defensible only by reference to the political importance of its content. . . .

. . .

Conclusion

Theorists of the present often explain our post-modern condition as a result of a tragedy of losses. For international lawyers, the Enlightenment signified loss of faith in a natural order among peoples, nations and sovereigns. To contain political subjectivism, 19th and 20th-century jurists put their faith variably on logic and texts, history and power to find a secure, objective foothold. Each attempt led to disappointment. One's use of logic depended on what political axioms were inserted as the premises. Texts, facts and history were capable of being interpreted in the most varied ways. In making his interpretations the jurist was always forced to rely on conceptual matrices which could no longer be defended by the texts, facts or histories to which they provided meaning. They were—and are—arenas of political struggle.

. . .

Social theorists have documented a recent modern turn in national societies away from the *Rechtstaat* [Rule of Law state] into a society in which social conflict is increasingly met with flexible, contextually determined standards and compromises. The turn away from general principles and formal rules into contextually determined equity may reflect a similar turn in the development of international legal thought and practice. There is every reason to take this turn seriously—though this may mean that lawyers have to re-think their professional self-image. For issues of contextual justice cannot be solved by the application of ready-made rules or principles. Their solution requires venturing into fields such as politics, social and economic casuistry which were formally delimited beyond the point at which legal argument was supposed to stop in order to remain 'legal.' To be sure, we shall remain uncertain. Resolutions based on political acceptability cannot be made with the kind of certainty post-Enlightenment lawyers once hoped to attain. And yet, it is only by their remaining so which will prevent their use as apologies for tyranny.

NOTE

The ideas developed by Koskenniemi provide useful insights into the dilemmas of contemporary argument about international human rights. A fundamental theme is the relation between 'law' and 'politics' in argument about international law, and between the notion of an international order based on the Rule of Law or based on political views and preferences (including theories of justice or social utility). The questions below explore these themes.

QUESTIONS

1. The two columns below composed of terms used in the preceding excerpts attempt to organize some of Koskenniemi's themes. In his discussion of international-law norms and argument, Koskenniemi employs sets of terms expressed as dichotomies or oppositions (or at least expressed as alternative methods or approaches toward understanding international law). These sets are set forth on the same lines in the two columns below—i.e., concrete is paired with normative, autonomy is paired with community, and so on. Koskenniemi also suggests relationships within each column—i.e., relationships in the first column among concrete, apology and consent, relationships in the second column among normative, utopia and justice.

concrete	normative
close to state practice	distant from state practice
apology	utopia
positivism	naturalism
consent	justice
autonomy	community
policy	rule

a. Can you describe the relationships that the author suggests both (i) between the paired terms (like apology and utopia) and (ii) among the terms in each column (like concrete, apology, and consent)?

b. Suppose that the rule that a state cannot arrest another state's ambassador were expressed only in the form of customary international law—that is, it is not expressed in a treaty. State X arrests the ambassador to it from state Y, which protests the arrest. Develop two arguments for Y to the effect that the arrest violates international law. The first argument could be based on the terms in the first column, the second argument on the terms in the second column.

c. Suppose that state X tortures its political prisoners to extract information. Other states criticize X, arguing that its conduct violates customary international law. Try to construct the two types of arguments sketched in paragraph (b). Are the arguments different in structure from those developed for paragraph (b)?

d. Koskenniemi contends that whichever argument you make in either case, it would contain elements of the other column's structure of argument. Does this appear to be the case?

2. Compare Schachter's use of the terms 'realism' and 'voluntarism'. Do either or both of these terms, as defined by Schachter, fit comfortably within Koskenniemi's scheme?

3. Compare the views of Schachter and Koskenniemi about the role of consent in the formation and application of customary international law. What are the differences?

ADDITIONAL READING

Note: Sections entitled 'Additional Reading' appear throughout the book. Readers should be aware of the general list of books concerned with human rights that appears in the Annex on Bibliography.

Contemporary scholarship on international law in general includes the following books and treatises, all of which have sections on human rights: Rudolph Bernhardt (ed.), *Encyclopedia of Public International Law*, 1981– , vols. 1–12; Ian Brownlie, *Principles of Public International Law* (4th ed. 1990); Martti Koskenniemi, *Apology and Utopia* (1989); Louis Henkin, *International Law: Politics, Values and Functions*, 216 Collected Courses of the Hague Academy of International Law 13 (Vol. IV, 1989); Robert Jennings and Arthur Watts, *Oppenheim's International Law* (9th ed. 1992); Oscar Schachter, *International Law in Theory and Practice* (1991).

B. HISTORICAL ANTECEDENTS TO CONTEMPORARY HUMAN RIGHTS

Section B serves two functions simultaneously. Through decisions of courts and other types of tribunals, it illustrates problems in applying the ideas about international law—sources, evidence, processes, structure of argument—that were discussed in the readings of Section A. It renders those readings more concrete by examining, for example, how jurists grasp and apply ideas about customary or treaty law, how they conceive of international law or international legal argument in general, and how such conceptions influence their decisions.

This section uses as illustrations four opinions of tribunals dating from the start of the twentieth century to the eve of the development of the human rights movement at the end of World War II. The four tribunals have radically different characters. In sequence, the opinions are those of the United States Supreme Court (sitting as a court of prize), of an international arbitral tribunal formed under a U.S.–Mexican convention, of the Permanent Court of International Justice formed under the League of Nations (the predecessor to the present International Court of Justice formed under the United Nations Charter), and of the International Military Tribunal formed by the allied powers after World War II for the Nuremberg trials of war criminals.

The second function of Section B is to describe some vital antecedents to the growth of the human rights movement. Each opinion involves a different field of international law that formed part of the relevant background to, and nourished, that movement—in the same sequence, the laws of (the conduct of) war, the protection of aliens under the international law of state responsibility, the regimes protecting minorities that were established by treaty in some European states after World War I, and the imposition of personal criminal responsibility for war crimes and crimes against humanity in the Nuremberg trials of Nazi leaders.

These four topics, while central to an understanding of the antecedents to the contemporary human rights movement, do not exhaust those antecedents. Other developments from the early part of the nineteenth century are sketched in the concluding reading of this chapter.

The selection of four tribunals' opinions to present some background to the human rights movement might suggest to the reader that tribunals have played a major role in resolving disputes over international human rights and in simultaneously developing that body of law. An analogous suggestion with respect to the role and contribution of courts in a domestic legal system like that of the United States would be accurate. But that observation would be seriously misleading in an international context, at least outside the distinctive regional arrangement of the European human rights system in which a court does play a major role.

For the most part, courts and other tribunals have played peripheral roles in the development of what we now term universal human rights law. In the following chapters, their opinions figure only rarely in discussions that concentrate on different institutions and processes. Nonetheless, the four opinions provide excellent points of departure for exploring the questions now before us. They provide much information about the conceptions of international law and the substantive doctrines of earlier periods, as well as about matters such as the processes by which international law develops, the relation between international law and state sovereignty, and the relevance of international law to individual rights.

NOTE

The following decision in *The Paquete Habana* deals with an earlier period in the development of the laws of war, here naval warfare, and with a theme that became central in the later treaty development of this field—the protection of noncombatant civilians and their property (here, civilian fishing vessels) against the ravages of war. Within the framework of the laws of war, this case involves *jus in bello*, the ways in which war ought to be waged, rather than the related but distinct *jus ad bellum*, the determination of those conditions (if any) in which a *just* or justified war can be waged, in which war is legal.

In its analysis of the question before it, the U.S. Supreme Court here illustrates a classical understanding of international customary law, thus advancing our inquiry into custom that began with the readings in Section A.

THE PAQUETE HABANA

Supreme Court of the United States, 1900.
175 U.S. 677, 20 S.Ct. 290.

Mr. Justice Gray delivered the opinion of the court:

These are two appeals from decrees of the district court of the United States for the southern district of Florida condemning two fishing vessels and their cargoes as prize of war.

Each vessel was a fishing smack, running in and out of Havana, and regularly engaged in fishing on the coast of Cuba; sailed under the Spanish flag; was owned by a Spanish subject of Cuban birth, living in the city of Havana; was commanded by a subject of Spain also residing in Havana; and her master and crew had no interest in the vessel, but were entitled to shares, amounting in all to two thirds, of her catch, the other third belonging to her owner. Her cargo consisted of fresh fish, caught by her crew from the sea, put on board as they were caught, and kept and sold alive. Until stopped by the blockading squadron she had no knowledge of the existence of the war or of any blockade. She had no arms or ammunition on board, and made no attempt to run the blockade after she knew of its existence, nor any resistance at the time of the capture.

. . .

Both the fishing vessels were brought by their captors into Key West. A libel for the condemnation of each vessel and her cargo as prize of war was there filed on April 27, 1898; a claim was interposed by her master on behalf of himself and the other members of the crew, and of her owner; evidence was taken, showing the facts above stated; and on May 30, 1898, a final decree of condemnation and sale was entered, 'the court not being satisfied that as a matter of law, without any ordinance, treaty, or proclamation, fishing vessels of this class are exempt from seizure'.

Each vessel was thereupon sold by auction; the Paquete Habana for the sum of $490; and the Lola for the sum of $800. There was no other evidence in the record of the value of either vessel or of her cargo. . . .

. . .

We are then brought to the consideration of the question whether, upon the facts appearing in these records, the fishing smacks were subject to capture by the armed vessels of the United States during the recent war with Spain.

By an ancient usage among civilized nations, beginning centuries ago, and gradually ripening into a rule of international law, coast fishing vessels, pursuing their vocation of catching and bringing in fresh fish, have been recognized as exempt, with their cargoes and crews, from capture as prize of war.

This doctrine, however, has been earnestly contested at the bar; and no complete collection of the instances illustrating it is to be found, so far as we are aware, in a single published work, although many are referred to and discussed by the writers on international law, notable in 2 Ortolan, Règles Internationales et Diplomatie de la Mer (4th ed.) lib. 3, chap. 2, pp. 51–56; in 4 Calvo, Droit International (5th ed.) §§ 2367– 2373; in De Boeck, Propriété Privée Ennemie sous Pavillon Ennemìe, §§ 191–196; and in Hall, International Law (4th ed.) § 148. It is therefore worth the while to trace the history of the rule, from the earliest accessible sources, through the increasing recognition of it with occasional setbacks, to what we may now justly consider as its final establishment in our own country and generally throughout the civilized world.

The earliest acts of any government on the subject, mentioned in the books, either emanated from, or were approved by, a King of England.

In 1403 and 1406 Henry IV issued orders to his admirals and other officers, entitled 'Concerning Safety for Fishermen—*De Securitate pro Piscatoribus.*' By an order of October 26, 1403, reciting that it was made pursuant to a treaty between himself and the King of France; and for the greater safety of the fishermen of either country, and so that they could be, and carry on their industry, the more safely on the sea, and deal with each other in peace; and that the French King had consented that English fishermen should be treated likewise,—it was ordained that French fishermen might, during the then pending season for the herring fishery, safely fish for herrings and all other fish, from the harbor of Gravelines and the island of Thanet to the mouth of the Seine and the harbor of Hautoune. . . .

. . .

The same custom would seem to have prevailed in France until towards the end of the seventeenth century. For example, in 1675, Louis XIV. and the States General of Holland by mutual agreement granted to Dutch and French fishermen the liberty, undisturbed by their vessels of war, of fishing along the coasts of France, Holland, and England. . . .

The doctrine which exempts coast fishermen, with their vessels and cargoes, from capture as prize of war, has been familiar to the United States from the time of the War of Independence.

. . .

In the treaty of 1785 between the United States and Prussia, article 23 . . . provided that, if war should arise between the contracting parties, 'all women and children, scholars of every faculty, cultivators of the earth, artisans, manufacturers, and fishermen, unarmed and inhabiting unfortified towns, villages, or places, and in general all others whose occupations are for the common subsistence and benefit of mankind, shall be allowed to continue their respective employments, and shall not be molested in their persons, nor shall their houses or goods be burnt or otherwise destroyed, nor their fields wasted by the armed force of the enemy, into whose power, by the events of war, they may happen to fall; but if anything is necessary to be taken from them for the use of such armed force, the same shall be paid for at a reasonable price.' . . .

Since the United States became a nation, the only serious interruptions, so far as we are informed, of the general recognition of the exemption of coast fishing vessels from hostile capture, arose out of the mutual suspicions and recriminations of England and France during the wars of the French Revolution.

. . .

On January 24, 1798, the English government by express order instructed the commanders of its ships to seize French and Dutch fishermen with their boats. . . . After the promulgation of that order, Lord Stowell (then Sir William Scott) in the High Court of Admiralty of England condemned small Dutch fishing vessels as prize of war. In one case the capture was in April, 1798, and the decree was made November 13, 1798. *The Young Jacob and Johanna*, 1 C.Rob. 20. . . .

. . .

On March 16, 1801, the Addington Ministry, having come into power in England, revoked the orders of its predecessors against the French fishermen; maintaining, however, that 'the freedom of fishing was nowise founded upon an agreement, but upon a simple concession;' that 'this concession would be always subordinate to the convenience of the moment,' and that 'it was never extended to the great fishery, or to commerce in oysters or in fish.' And the freedom of the coast fisheries was again allowed on both sides. . . .

Lord Stowell's judgment in *The Young Jacob and Johanna*, 1 C.Rob. 20, above cited, was much relied on by the counsel for the United States, and deserves careful consideration.

The vessel there condemned is described in the report as 'a small Dutch fishing vessel taken April, 1798, on her return from the Dogger bank to Holland;' and Lord Stowell, in delivering judgment, said: 'In former wars it has not been usual to make captures of these small fishing vessels; but this rule was a rule of comity only, and not of legal decision; it has prevailed from views of mutual accommodation between neighbouring countries, and from tenderness to a poor and industrious order of people. In the present war there has, I presume, been sufficient reason for changing this mode of treatment; and as they are brought before me for my judgment they must be referred to the general principles of this court; they fall under the character and description of the last class of cases; that is, of ships constantly and exclusively employed in the enemy's trade.' And he added: 'it is a further satisfaction to me, in giving this judgment, to observe that the facts also bear strong marks of a false and fraudulent transaction.'

Both the capture and the condemnation were within a year after the order of the English government of January 24, 1798, instructing the commanders of its ships to seize French and Dutch fishing vessels, and before any revocation of that order. Lord Stowell's judgment shows that his decision was based upon the order of 1798, as well as upon strong evidence of fraud. Nothing more was adjudged in the case.

But some expressions in his opinion have been given so much weight by English writers that it may be well to examine them particularly. The opinion begins by admitting the known custom in former wars not to capture such vessels; adding, however, 'but this was a rule of comity only, and not of legal decision.' Assuming the phrase 'legal decision' to have been there used, in the sense in which courts are accustomed to use it, as equivalent to 'judicial decision,' it is true that so far as appears, there had been no such decision on the point in England. The word 'comity' was apparently used by Lord Stowell as synonymous with courtesy or goodwill. But the period of a hundred years which has since elapsed is amply sufficient to have enabled what originally may have rested in custom or comity, courtesy or concession, to grow, by the general assent of civilized nations, into a settled rule of international law. . . .

The French prize tribunals, both before and after Lord Stowell's decision, took a wholly different view of the general question. . . .

The English government [by orders in council of 1806 and 1810] unqualifiedly prohibited the molestation of fishing vessels employed in catching and bringing to market fresh fish. . . .

Wheaton, in his Digest of the Law of Maritime Captures and Prizes, published in 1815, wrote: 'It has been usual in maritime wars to exempt from capture fishing boats and their cargoes, both from views of mutual accommodation between neighboring countries, and from tenderness to a poor and industrious order of people. This custom, so honorable to the humanity of civilized nations, has fallen into disuse; and it is remarkable that both France and England mutually reproach each other with that breach of good faith which has finally abolished it.' Wheaton, Captures, chap. 2, § 18.

This statement clearly exhibits Wheaton's opinion that the custom had been a general one, as well as that it ought to remain so. His assumption that it had been abolished by the differences between France and England at the close of the last century was hardly justified by the state of things when he wrote, and has not since been borne out.

. . .

In the war with Mexico, in 1846, the United States recognized the exemption of coast fishing boats from capture. . . .

. . .

In the treaty of peace between the United States and Mexico, in 1848, were inserted the very words of the earlier treaties with Prussia, already quoted, forbidding the hostile molestation or seizure in time of war of the persons, occupations, houses, or goods of fishermen. 9 Stat. at L. 939, 940.

. . .

France in the Crimean war in 1854, and in her wars with Italy in 1859 and with Germany in 1870, by general orders, forbade her cruisers to trouble the coast fisheries, or to seize any vessel or boat engaged therein, unless naval or military operations should make it necessary.

. . .

Since the English orders in council of 1806 and 1810 . . . in favor of fishing vessels employed in catching and bringing to market fresh fish, no instance has been found in which the exemption from capture of private coast fishing vessels honestly pursuing their peaceful industry has been denied by England or by any other nation. And the Empire of Japan (the last state admitted into the rank of civilized nations), by an ordinance promulgated at the beginning of its war with China in August, 1894, established prize courts, and ordained that 'the following enemy's vessels are exempt from detention,' including in the exemption 'boats engaged in coast fisheries,' as well as 'ships engaged exclusively on a voyage of scientific discovery, philanthrophy, or religious mission.' Takahashi, International Law, 11, 178.

International law is part of our law, and must be ascertained and administered by the courts of justice of appropriate jurisdiction as often as questions of right depending upon it are duly presented for their determination. For this purpose, where there is no treaty and no controlling executive or legislative act or judicial decision, resort must be had to the customs and usages of civilized nations, and, as evidence of these, to the works of jurists and commentators who by years of labor, research, and experience have made themselves peculiarly well acquainted with the subjects of which they treat. Such works are resorted to by judicial tribunals, not for the speculations of their authors concerning what the law ought to be, but for trustworthy evidence of what the law really is. Hilton v. Guyot, 159 U.S. 113, 163, 164, 214, 215, 40 L.Ed. 95, 108, 125, 126, 16 Sup.Ct.Rep. 139.

. . .

Chancellor Kent says: 'In the absence of higher and more authoritative sanctions, the ordinances of foreign states, the opinions of eminent statesmen, and the writings of distinguished jurists, are regarded as of great consideration on questions not settled by conventional law. In cases where the principal jurists agree, the presumption will be very great in favor of the solidity of their maxims; and no civilized nation that does not arrogantly set all ordinary law and justice at defiance will venture to disregard the uniform sense of the established writers on international law.' 1 Kent, Com. 18.

It will be convenient, in the first place, to refer to some leading French treatises on international law, which deal with the question now before us, not as one of the law of France only, but as one determined by the general consent of civilized nations. . . . [Discussion of French treatises omitted.]

. . .

No international jurist of the present day has a wider or more deserved reputation than Calvo, who, though writing in French, is a citizen of the Argentine Republic, employed in its diplomatic service abroad. In the fifth edition of his great work on international law, published in 1896, he observes, in § 2366, that the international authority of decisions in particular cases by the prize courts of France, of England, and of the United States is lessened by the fact that the principles on which they are based are largely derived from the internal legislation of each country; and yet the peculiar character of maritime wars, with other considerations, gives to prize jurisprudence a force and importance reaching beyond the limits of the country in which it has prevailed. He therefore proposes here to group together a number of particular cases proper to serve as precedents for the solution of grave questions of maritime law in regard to the capture of private property as prize of war. Immediately, in § 2367, he goes on to say: 'Notwithstanding the hardships to which maritime wars subject private property, notwithstanding the extent of the recognized rights of belligerents, there are generally exempted, from seizure and capture, fishing vessels.' . . .

The modern German books on international law, cited by the counsel for the appellants, treat the custom by which the vessels and implements of coast fishermen are exempt from seizure and capture as well established by the practice of nations. Heffter, § 137; 2 Kalterborn, § 237, p. 480; Bluntschli, § 667; Perels, § 37, p. 217.

. . .

Two recent English text-writers cited at the bar (influenced by what Lord Stowell said a century since) hesitate to recognize that the exemption of coast fishing vessels from capture has now become a settled rule of international law. Yet they both admit that there is little real difference in the views, or in the practice, of England and of other maritime nations; and that no civilized nation at the present day would molest coast fishing vessels so long as they were peaceably pursuing their calling and there was no danger that they or their crews might be of military use to the enemy. . . .

But there are writers of various maritime countries, not yet cited, too important to be passed by without notice. . . .

[The opinion quotes from writing from the Netherlands, Spain, Austria, Portugal and Italy.]

This review of the precedents and authorities on the subject appears to us abundantly to demonstrate that at the present day, by the general consent of the civilized nations of the world, and independently of any express treaty or other public act, it is an established rule of international law, founded on considerations of humanity to a poor and industrious order of men, and of the mutual convenience of belligerent states, that coast fishing vessels, with their implements and supplies, cargoes and crews, unarmed and honestly pursuing their peaceful calling of catching and bringing in fresh fish, are exempt from capture as prize of war.

. . .

This rule of international law is one which prize courts administering the law of nations are bound to take judicial notice of, and to give effect to, in the absence of any treaty or other public act of their own government in relation to the matter.

. . .

To this subject in more than one aspect are singularly applicable the words uttered by Mr. Justice Strong, speaking for this court: 'Undoubtedly no single nation can change the law of the sea. The law is of universal obligation and no statute of one or two nations can create obligations for the world. Like all the laws of nations, it rests upon the common consent of civilized communities. It is of force, not because it was prescribed by any superior power, but because it has been generally accepted as a rule of conduct. Whatever may have been its origin, whether in the usages of navigation, or in the ordinances of maritime states, or in both, it has become the law of the sea only by the concurrent sanction of those nations who may be said to constitute the commercial world. . . . [Of these facts] we may take judicial notice. Foreign municipal laws must indeed be proved as facts, but it is not so with the law of nations.' The Scotia, 14 Wall. 170, 187, 188, sub nom. Sears v. The Scotia, 20 L.Ed. 822, 825, 826.

The position taken by the United States during the recent war with Spain was quite in accord with the rule of international law, now generally recognized by civilized nations, in regard to coast fishing vessels.

On April 21, 1898, the Secretary of the Navy gave instructions to Admiral Sampson, commanding the North Atlantic Squadron, to 'immediately institute a blockade of the north coast of Cuba, extending from Cardenas on the east to Bahia Honda on the west.' Bureau of Navigation Report of 1898, appx. 175. The blockade was immediately instituted accordingly. On April 22 the President issued a proclamation declaring that the United States had instituted and would maintain that blockade, 'in pursuance of the laws of the United States, and the law of nations applicable to such cases.' 30 Stat. at L. 1769. And by the act of Congress of April 25, 1898, chap. 189, it was declared that the war between the United States and Spain existed on that day, and had existed since and including April 21. 30 Stat. at L. 364.

On April 26, 1898, the President issued another proclamation which, after reciting the existence of the war as declared by Congress, contained this further recital: 'It being desirable that such war should be conducted upon principles in harmony with the present views of nations and sanctioned by their recent practice.' This recital was followed by specific declarations of certain rules for the conduct of the war by sea, making no mention of fishing vessels. 30 Stat. at L. 1770. But the proclamation clearly manifests the general policy of the government to conduct the war in accordance with the principles of international law sanctioned by the recent practice of nations.

. . .

Upon the facts proved in either case, it is the duty of this court, sitting as the highest prize court of the United States, and administering the law of nations, to declare and adjudge that the capture was unlawful and without probable cause; and it is therefore, in each case,—

Ordered, that the decree of the District Court be reversed, and the proceeds of the sale of the vessel, together with the proceeds of any sale of her cargo, be restored to the claimant, with damages and costs.

Mr. Chief Justice Fuller, with whom concurred *Mr. Justice Harlan* and *Mr. Justice Mckenna*, dissenting:

The district court held these vessels and their cargoes liable because not 'satisfied that as a matter of law, without any ordinance, treaty, or proclamation, fishing vessels of this class are exempt from seizure.'

This court holds otherwise, not because such exemption is to be found in any treaty, legislation, proclamation, or instruction granting it, but on the ground that the vessels were exempt by reason of an established rule of international law applicable to them, which it is the duty of the court to enforce.

I am unable to conclude that there is any such established international rule, or that this court can properly revise action which must be treated as having been taken in the ordinary exercise of discretion in the conduct of war.

. . .

This case involves the capture of enemy's property on the sea, and executive action, and if the position that the alleged rule *proprio vigore* limits the sovereign power in war be rejected, then I understand the contention to be that, by reason of the existence of the rule, the proclamation of April 26 must be read as if it contained the exemption in terms, or the exemption must be allowed because the capture of fishing vessels of this class was not specifically authorized.

The preamble to the proclamation stated, it is true, that it was desirable that the war 'should be conducted upon principles in harmony with the present views of nations and sanctioned by their recent practice,' but the reference was to the intention of the government 'not to resort to privateering, but to adhere to the rules of the Declaration of Paris;' and the proclamation spoke for itself. The language of the preamble did not carry the exemption in terms, and the real question is whether it must be allowed because not affirmatively withheld, or, in other words, because such captures were not in terms directed.

. . .

It is impossible to concede that the Admiral ratified these captures in disregard of established international law and the proclamation, or that the President, if he had been of opinion that there was any infraction of law or proclamation, would not have intervened prior to condemnation.

. . .

In truth, the exemption of fishing craft is essentially an act of grace, and not a matter of right, and it is extended or denied as the exigency is believed to demand.

It is, said Sir William Scott, 'a rule of comity only, and not of legal decision.'

. . .

It is difficult to conceive of a law of the sea of universal obligation to which Great Britain has not acceded. And I am not aware of adequate foundation for imputing to this country the adoption of any other than the English rule.

. . .

It is needless to review the speculations and repetitions of the writers on international law. Ortolan, De Boeck, and others admit that the custom relied on as consecrating the immunity is not so general as to create an absolute international rule; Heffter, Calvo, and others are to the contrary. Their lucubrations may be persuasive, but not authoritative.

In my judgment, the rule is that exemption from the rigors of war is in the control of the Executive. He is bound by no immutable rule on the subject. It is for him to apply, or to modify, or to deny altogether such immunity as may have been usually extended.

. . .

COMMENT ON THE HUMANITARIAN LAW OF WAR

The opinion in *The Paquete Habana* has the aura of a humane world in which, if war occurs, the fighting should be as compassionate in spirit as possible. It rests the rule of exemption of coastal fishing vessels 'on considerations of humanity to a poor and industrious order of men, and [on] the mutual convenience of fishing vessels.' The opinion seems more than a mere fourteen years distant from the savagery of World War I, and remote from that war's successors during this century with their massive atrocities and civilian casualties in engagements of close to total war by one or both sides.

The intricate body of international law considered by the Supreme Court grew out of centuries of primarily customary law, although custom was supplemented and informed centuries ago by selective bilateral treaties. Custom remains essential to argument about the laws of war to this day, indeed to the jurisdictional provisions of the International Criminal Tribunal for the Former Yugoslavia discussed in Ch. 15. But this field is increasingly dominated by multilateral treaties that have both codified customary standards and rules and developed new ones in numerous international conferences. Multilateral declarations and treaties started to achieve prominence in the second half of the nineteenth century. The treaties now include the Hague Conventions concluded around the turn of the century, the four Geneva Conventions of 1949 (as well as two significant Protocols of 1977 to those conventions), and several discrete treaties since World War II on matters like bans on particular weapons and cultural property.

The basic Geneva Conventions (185 statesparties as of January 1995) and the two Protocols (Protocol No. 1, 136 parties; Protocol No. 2, 126 parties) cover a vast range of problems stemming from land, air or naval warfare, including the protection of wounded combatants and prisoners of war, of civilian populations and civilian objects affected by military operations or present in occupied territories, and of medical and religious personnel and buildings. As suggested by this list, the provisions of the four Conventions and two Protocols constitute the principal regulation of *jus in bello*.

This entire corpus of custom and treaties has come to be know as the 'international humanitarian law of war,' the broad purpose being—in the words of the landmark St. Petersburg Declaration of 1868—'alleviating as much as possible the calamities of war.' Here lies the tension, even contradiction, within this body of law. Putting aside the question of a war's legality (an issue central to the Judgment of the International Military Tribunal at Nuremberg, p. 102, *infra*, and today governed by the UN Charter), a war fought in compliance with the standards and rules of the laws of war permits massive intentional killing or wounding and massive other destruction that, absent a war, would violate fundamental human rights norms.

Hence all these standards and rules stand at some perilous and problematic divide between brutality and destruction (i) that is permitted or privileged and (ii) that is illegal and subject to sanction. Principles like 'proportionality' in choosing military means or like the avoidance of 'unnecessary suffering' to the civilian population are employed to help to draw the line. The powerful ideal of reducing human suffering that animates this humanitarian law of war thus is countered by the goal of state parties to a war—indeed, in the eyes of states, the paramount goal—of gaining military objectives and victory while reducing as much as possible the losses to one's own armed forces.

The generous mood of *The Paquete Habana* toward the civilian population and its food-gathering needs was reflected in the various Hague Conventions regulating land and naval warfare that were adopted during the ensuing decade. Note Article 3 of the Hague Convention of 1907 on Certain Restrictions with Regard to the Exercise of the Right to Capture in Naval War, 36 Stat. 2396, T.S. No. 544, which proclaimed in 1910: 'Vessels used exclusively for fishing along the coast . .. are exempt from capture . . . .'

The efforts to protect civilian populations and their property took on renewed vigor after World War II through the Geneva Conventions of 1949. Consider Protocol No. 1 to the Geneva Conventions, adopted in 1977 and ratified (as of January 1995) by 136 states. Article 48 enjoins the parties to a conflict to 'distinguish between the civilian population and combatants and between civilian objects and military objectives.' Their operations are to be directed 'only against military objectives.' Article 52 defines military objectives to be 'objects which, by their nature, location, purposes or use make an effective contribution to military action and whose total or partial destruction, capture or neutralization, in the circumstances ruling at the time, offers a definite military advantage.' Article 54 is entitled, 'Protection of Objects Indispensable to the Survival of the Civilian Population.' It states that '[s]tarvation of civilians as a method of warfare is prohibited.' Specifically, parties are prohibited from attacking or removing 'objects indispensable to the survival of the civilian population, such as foodstuffs . . ., for the specific purpose of denying them for their sustenance value to the civilian population or to the adverse Party. . . .' An exception is made for objects used by an adverse party as sustenance 'solely' for its armed forces or 'in direct support of military action.'

Consider some special characteristics of *The Paquete Habana.* (1) Note the emphasis on the fact that the Supreme Court here sat as a *prize court* administering the law of nations, and note its references to the international character of the law maritime. Indeed, the Court almost assumed the role of an international tribunal, a consideration stressed in the excerpts from the scholar Calvo. Nonetheless, the Court's statement that 'international law is part of our law' and must be 'ascertained and administered by courts of justice' as often as 'questions of right' depending on it are presented for determination, has been drawn on in many later judicial decisions in the United States involving unrelated international-law issues.

(2) An antiquarian aspect of the decision and period is that the naval personnel who captured the fishing vessels participated in the judicial proceedings, for at the time of the war captors were entitled to share in the proceeds of the sale of lawful prizes. That practice has ended and proceeds are now paid into the Treasury. 70A Stat. 475 (1956), 10 U.S.C.A. §§7651–81.

(3) The Court looked to a relatively small number of countries for evidence of state practice, dominantly in Western Europe. It referred to Japan as 'the last state admitted into the rank of civilized nations.' Even at the start of the twentieth century, the world community creating international law was a small and relatively cohesive one. Contrast the multinational and multicultural character of an assembly of states today drafting a convention on the laws of war or a human rights convention, and imagine the range of states to which references might be made in a contemporary judicial opinion considering the customary law of international human rights.

QUESTIONS

1. Suppose that an international tribunal rather than U.S. courts had heard this controversy, and had sought to decide it within the framework of Article 38 of the Statute of the International Court of Justice. Assuming that this tribunal came to the same conclusion, would any observations in the Supreme Court's opinion be likely to have been omitted or changed by such an international tribunal? Which observations? Suppose, for example, that the historical record was identical with that reported by the Supreme Court except for the fact that the United States had consistently objected to this rule of exemption and had often refused to follow it.

2. Does the method of the Court in 'ascertaining' the customary rule appear consistent with some of the observations about the nature of custom and the processes for its development in the readings in Section A? Consider, for example, how the Supreme Court directly or implicitly deals with (i) the issue of *opinio juris,* (ii) the relevance of treaties, and (iii) the departure from the rule of exemption during the Napoleonic wars. With which of these aspects of the opinion does the dissenting opinion differ?

COMMENT ON THE LAW OF STATE RESPONSIBILITY AND THE *CHATTIN* CASE

The *Chattin* case that follows was decided under a 1923 General Claims Convention between the United States and Mexico, 43 Stat. 1730, T.S. No. 678. That treaty provided that designated claims against Mexico of U.S. citizens (and vice versa) for losses or damages suffered by persons or by their properties that (in the case of the U.S. citizens) had been presented to the U.S. government for interposition with Mexico and that had remained unsettled 'shall be submitted to a Commission consisting of three members for decision in accordance with the principles of international law, justice and equity.' Each state was to appoint one member, and the presiding third commissioner was to be selected by mutual agreement (and by stipulated procedures failing agreement).

These arbitrations grew out of and further developed the law of state responsibility for injuries to aliens, a branch of international law that was among the important predecessors to contemporary human rights law. That body of law addressed only certain kinds of conflicts—not including, for example, conflicts originating in the first instance in a dispute between a claimant state (X) and a respondent state (Y). Thus it did not cover a dispute, say, based on a claim by X that Y had violated international law by its invasion of X's territory or by its imprisonment of X's ambassador.

Rather, the claims between states that were addressed by the law of state responsibility for injuries to aliens grew out of disputes arising in the first instance between a citizen-national of X and the government of Y. For example, respondent state Y allegedly imprisoned a citizen of claimant state X without hearing or trial, or seized property belonging to citizens of X—allegations which, if true, could show violations of international law. Note that these illustrations involve action leading to injury of X's citizens by governmental officials or organs (executive, legislative, judicial) of Y. The law of state responsibility required that the conduct complained of be that of the state or, in less clear and more complex situations, be ultimately attributable to the state.

In the normal case, the citizen of X would seek a remedy within Y, probably through its judiciary—release from jail, return of the seized property or compensation for it. Indeed, before invoking the aid of his own government, the citizen of X would generally be required under the relevant treaty to pursue such a path, to 'exhaust local remedies.' But that path may prove to be fruitless, because of lack of recourse to Y's judiciary, because that judiciary is corrupt, or because of Y's law adverse to the citizen of X that would be applied by its judiciary. In such circumstances, the injured person may turn to his own government X for diplomatic protection.

The 1924 decision of the Permanent Court of International Justice in the *Mavrommatis Palestine Concessions (Jurisdiction)* case, P.C.I.J., Ser. A, No. 2, gave classic expression to such diplomatic protection. It pointed out that when a state took up the cause of one of its subjects (citizens-nationals) in a dispute originating between that subject and respondent state, the dispute

> entered upon a new phase; it entered the domain of international law, and became a dispute between two States . . . . It is an elementary principle of international law that a State is entitled to protect its subjects, when injured by acts contrary to international law committed by another State, from whom they have been unable to obtain satisfaction through the ordinary channels. By taking up the case of one of its subjects and by resorting to diplomatic action or international judicial proceedings on his behalf, a State is in reality asserting its own rights—its right to ensure, in the person of its subjects, respect for the rules of international law.

Precisely what action to take, what form of diplomatic protection to extend, lay within the discretion of the claimant state. If it decided to intervene and thereby make the claim its own, it might espouse the claim through informal discussions with the respondent state, or make a formal diplomatic protest, or exert various economic and political pressures to encourage a settlement (extending at times to military intervention), or, if these strategies failed, have recourse to international tribunals. Such recourse was infrequent. International tribunals to whose jurisdiction states had consented for the resolution of disputes between them were rare. Moreover, states were reluctant to raise controversies between their citizens and foreign states to the level of interstate conflict before an international tribunal except where a clear national interest gave reason to do so.

An arbitral tribunal to which the claimant state turned may have been created by agreement between the disputing states to submit to it designated types of disputes. That agreement may have been part of a general arbitration treaty (which after World War II found scant use) covering a broad range of potential disputes between the two parties. Or it may have been a so-called compromissory clause (*compromis*) in a treaty dealing with a specific subject that bound the parties to submit to arbitration disputes that might arise under that treaty. Of course, two states could always agree to submit specified disputes to arbitration, as in the 1923 General Claims Convention between the United States and Mexico under which *Chattin* was decided.

In 1921, *ad hoc* arbitral tribunals were first supplemented by an international court, the Permanent Court of International Justice provided for in the Covenant of the League of Nations. Again, problems of states' consent to jurisdiction and states' reluctance to start interstate litigation limited the role of that court (and indeed the role of its successor, the International Court of Justice created under the Charter of the United Nations) in developing the law of state responsibility (or, today, in developing the international law of human rights).

The growth in the nineteenth and twentieth centuries of the law of state responsibility for injury to aliens was the product of and evidenced by a range of state interactions— diplomatic protests and responses, negotiated settlements, arbitral decisions—and the writings of scholars. Before World War II, there was little attempt at formal codification or creative development of this body of law through treaties—that is, treaties spelling out the content of what international law required of a state in its treatment of aliens.

As it developed, the international law of state responsibility reflected the more intense identification of the individual with his state (or later, the identification of the corporation with the state of its incorporation, or of most of its shareholders) that accompanied the nationalistic trends of that era. This body of law would not have developed so vigorously but for Western colonialism and economic imperialism that reached their zenith during this period. Transnational business operations centered in Europe, and later in the United States as well, penetrated those regions now known as the Third World or developing countries.

In such circumstances, given the obvious links between the success and wealth of corporations in their foreign ventures and national wealth and power, the security of the person and property of a national or corporation operating in a foreign part of the world became a concern of his or its government. That concern manifested itself in the vigorous assertion of diplomatic protection and in the enhanced activity of arbitral tribunals. In the late nineteenth and early twentieth centuries, some such arbitrations occurred under the pressure of actual or threatened military force by the claimant states, particularly against Latin American governments.

A statement in an arbitral proceeding in 1924 by Max Huber, a Judge of the Permanent Court of International Justice, cogently expressed some basic principles of that era's consensus (among states of the developed world) about the law of state responsibility:[4]

> . . . It is true that the large majority of writers have a marked tendency to limit the responsibility of the State. But their theories often have political inspiration and represent a natural reaction against unjustified interventions in the affairs of certain nations. . . .
>
> . . . The conflicting interest with respect to the problem of compensation of aliens are, on the one hand, the interest of a State in exercising its public power in its own territory without interference or control of any nature by foreign States and, on the other hand, the interest of the State in seeing the rights of its nationals established in foreign countries respected and well protected.
>
> Three principles are hardly debatable:
>
> . . .

[4] Judge Huber delivered these remarks in his role as a Reporter (in effect, arbitrator) in a dispute between Great Britain and Spain involving claims of British subjects against Spanish authorities for alleged mistreatment. British Claims in the Spanish Zone of Morocco, 2 U.N.R.I.A.A. 615, 639 (1924).

(2) In general, a person established in a foreign country is subject to the territorial legislation for the protection of his person and his property, under the same conditions as nationals of that country.

(3) A State whose national established in another State is deprived of his rights has a right to intervene *if the injury constitutes a violation of international law*. . . .

. . . The territorial character of sovereignty is so essential a trait of contemporary public law that foreign intervention in relationships between a territorial State and individuals subject to its sovereignty can be allowed only in extraordinary cases. . . .

. . . This right of intervention has been claimed by all States; only its limits are under discussion. By denying this right, one would arrive at intolerable results: international law would become helpless in the face of injustices tantamount to the negation of human personality, for that is the subject which every denial of justice touches.

. . . No police or other administration of justice is perfect, and it is doubtless necessary to accept, even in the best administered countries, a considerable margin of tolerance. However, the restrictions thus placed on the right of a State to intervene to protect its nationals assume that the general security in the country of residence does not fall below a certain standard . . . .

How was it determined whether, in Huber's words, an 'injury' to an alien 'constitutes a violation of international law,' or whether the administration of justice in a given country fell below 'a certain standard?' To what materials would, for example, an arbitral tribunal turn for help in defining the content of that standard? What types of argument and justifications would inform the development of this body of international law? Consider these questions as you read the *Chattin* arbitration.

UNITED STATES OF AMERICA (B.E. CHATTIN) v. UNITED MEXICAN STATES

United States-Mexican Claims Commission, 1927.
Opinions of Commissioners under the Convention Concluded September 8, 1923 between the United States and Mexico, 1926–27. at 422.
4 U.N.R.I.A.A. 282.

[Chattin, a United States citizen, was a conductor on a railroad in Mexico from 1908 to 1910, when he was arrested for embezzlement of fares. Chattin's trial was consolidated with those of several other Americans and Mexicans who had been arrested on similar charges. In February 1911 he was convicted and sentenced to two years' imprisonment. His appeal was rejected in July 1911. In the meantime the inhabitants of Mazatlán, during a political uprising, threw open the doors of the jail and Chattin escaped to the United States. In asserting Chattin's claims, the United States argued that the arrest was illegal, that Chattin was mistreated while in prison, that his trial was unreasonably delayed, and that there were irregularities in the trial. It claimed that Chattin suffered injuries worth $50,000 in compensation.

The Claims Commission had three members, including one from the United States (Nielsen) and from Mexico (MacGregor). The opinion of Van Vollenhoven, the Presiding Commissioner, distinguished between the legal bases for responsibility of a state because of acts or failures to act by its *judicial* and *non-judicial* branches of government. 'Acts of the *judiciary* . . . are not considered insufficient unless the wrong committed amounts to an outrage, bad faith, wilful neglect of duty, or insufficiency of action apparent to any unbiased man.' Acts of the legislative or executive branches were held to this same standard with respect to *indirect* liability (for example, a lack of appropriate action by government in response to conduct of private persons who initially caused the loss to the victim), but in cases of *direct* liability (for example, unwarranted arrest, reckless shooting by police), the standard was more demanding on the state and included forms of negligence. Van Vollenhoven, quoting from precedent, noted that 'as far as acts of the judiciary are involved . . . "it is a matter of the greatest political and international delicacy for one country to disacknowledge the judicial decision of a court of another country."'

The following excerpts from Van Vollenhoven's opinion deal with three of the detailed complaints about the conduct of the trial, a number of other complaints having been rejected as irrelevant or non-prejudicial.]

17. The allegation (e) that the accused has not been duly informed regarding the charge brought against him is proven by the record, and to a painful extent. The real complainant in this case was the railroad company, acting through its general manager; this manager, an American, not only was allowed to make full statements to the Court on August 2, 3, and 26, 1910, without ever being confronted with the accused and his colleagues, but he was even allowed to submit to the Court a series of anonymous written accusations, the anonymity of which reports could not be removed (for reasons which he explained); these documents created the real atmosphere of the trial. Were they made known to the conductors? Were the accused given an opportunity to controvert them? There is no trace of it in the record, nor was it ever alleged by Mexico. . . . The court record only shows that on January 13, and 16, 1911, the conductors and one of their lawyers were aware of the existence, not that they knew the contents, of these documents. . . . It is not shown that the confrontation between Chattin and his accusers amounted to anything like an effort on the Judge's part to find out the truth. Only after November 22, 1910, and only at the request of the Prosecuting Attorney, was Chattin confronted with some of the persons who, between July 13 and 21, inclusive, had testified of his being well acquainted with Ramírez. It is regrettable, on the other hand, that the accused misrepresents the wrong done him in this respect.

He had not been left altogether in the dark. According to a letter signed by himself and two other conductors dated August 31, 1910, he was perfectly aware even of the details of the investigations made against him; so was the American vice-consul on July 26, 1910. . . .

Owing to the strict seclusion to which the conductors contend to have been submitted, it is impossible they could be so well-informed if the charges and the investigations were kept hidden from them.

. . .

19. The allegation (h) that the witnesses were not sworn is irrelevant, as Mexican law does not require an 'oath' (it is satisfied with a solemn promise, *protesta*, to tell the truth), nor do international standards of civilization.

. . .

21. The allegation (j) that the hearings in open court lasted only some five minutes is proven by the record. This trial in open court was held on January 27, 1911. It was a pure formality, in which only confirmations were made of written documents, and in which not even the lawyer of the accused conductors took the trouble to say more than a word or two.

22. The whole of the proceedings discloses a most astonishing lack of seriousness on the part of the Court. There is no trace of an effort to have the two foremost pieces of evidence explained (paragraphs 14 and 17 above). There is no trace of an effort to find one Manuel Virgen, who, according to the investigations of July 21, 1910, might have been mixed in Chattin's dealings, nor to examine one Carl or Carrol Collins, a dismissed clerk of the railroad company concerned, who was repeatedly mentioned as forging tickets and passes and as having been discharged for that very reason. One of the Mexican brakemen, Batriz, stated on August 8, 1910, in court that 'it is true that the American conductors have among themselves schemes to defraud in that manner the company, the deponent not knowing it for sure'; but again no steps were taken to have this statement verified or this brakeman confronted with the accused Americans. . . . No investigation was made as to why Delgado and Sarabia felt quite certain that June 29 was the date of their trip, a date upon the correctness of which the weight of their testimony wholly depended. No search of the houses of these conductors is mentioned. Nothing is revealed as to a search of their persons on the days of their arrest; when the lawyer of the other conductors, Haley and Englehart, insisted upon such an inquiry, a letter was sent to the Judge at Culiacán, but was allowed to remain unanswered. Neither during the investigations nor during the hearings in open court was any such thing as an oral examination or cross-examination of any importance attempted. It seems highly improbable that the accused have been given a real opportunity during the hearings in open court, freely to speak for themselves. It is not for the Commission to endeavor to reach from the record any conviction as to the innocence or guilt of Chattin and his colleagues; but even in case they were guilty, the Commission would render a bad service to

the Government of Mexico if it failed to place the stamp of its disapproval and even indignation on a criminal procedure so far below international standards of civilization as the present one. If the wholesome rule of international law as to respect for the judiciary of another country . . . shall stand, it would seem of the utmost necessity that appellate tribunals when, in exceptional cases, discovering proceedings of this type should take against them the strongest measures possible under constitution and laws, in order to safeguard their country's reputation.

. . .

24. In Mexican law, as in that of other countries, an accused cannot be convicted unless the Judge is convinced of his guilt and has acquired this view from legal evidence. An international tribunal never can replace the important first element, that of the Judge's being convinced of the accused's guilt; it can only in extreme cases, and then with great reserve, look into the second element, the legality and sufficiency of the evidence.

. . .

26. From the record there is not convincing evidence that the proof against Chattin, scanty and weak though it may have been, was not such as to warrant a conviction. Under the article deemed applicable the medium penalty fixed by law was imposed, and deduction made of the seven months Chattin had passed in detention from July, 1910, till February, 1911. It is difficult to understand the sentence unless it be assumed that the Court, for some reason or other, wished to punish him severely. . . . The allegation that the Court in this matter was biased against American citizens would seem to be contradicted by the fact that, together with the four Americans, five Mexicans were indicted as well, four of whom had been caught and have subsequently been convicted—that one of these Mexicans was punished as severely as the Americans were—and that the lower penalties imposed on the three others are explained by motives which, even if not shared, would seem reasonable. . . . If Chattin's guilt was sufficiently proven, the small amount of the embezzlement (four pesos) need not in itself have prevented the Court from imposing a severe penalty.

. . .

29. Bringing the proceedings of Mexican authorities against Chattin to the test of international standards . . . there can be no doubt of their being highly insufficient. Inquiring whether there is convincing evidence of these unjust proceedings . . . the answer must be in the affirmative. Since this is a case of alleged responsibility of Mexico for injustice committed by its judiciary, it is necessary to inquire whether the treatment of Chattin amounts even to an outrage, to bad faith, to wilful neglect of duty, or to an insufficiency of governmental action recognizable by every unbiased man . . . and the answer here

again can only be in the affirmative.

30. An illegal arrest of Chattin is not proven. Irregularity of court proceedings is proven with reference to absence of proper investigations, insufficiency of confrontations, withholding from the accused the opportunity to know all of the charges brought against him, undue delay of the proceedings, making the hearings in open court a mere formality, and a continued absence of seriousness on the part of the Court. Insufficiency of the evidence against Chattin is not convincingly proven; intentional severity of the punishment is proven, without its being shown that the explanation is to be found in unfairmindedness of the Judge. Mistreatment in prison is not proven. Taking into consideration, on the one hand, that this is a case of direct governmental responsibility, and, on the other hand, that Chattin, because of his escape, has stayed in jail for eleven months instead of for two years, it would seem proper to allow in behalf of this claimant damages in the sum of $5,000.00, without interest.

NIELSEN, Commissioner, concurring:

. . .

So far as concerns methods of procedure prescribed by Mexican law, conclusions with respect to their propriety or impropriety may be reached in the light of comparisons with legal systems of other countries. And comparisons pertinent and useful in the instant case must be made with the systems obtaining in countries which like Mexico are governed by the principles of the civil law, since the administration of criminal jurisprudence in those countries differs so very radically from the procedure in criminal cases in countries in which the principles of Anglo-Saxon law obtain. This point is important in considering the arguments of counsel for the United States regarding irrelevant evidence and hearsay evidence appearing in the record of proceedings against the accused. From the standpoint of the rules governing Mexican criminal procedure conclusions respecting objections relative to these matters must be grounded not on the fact that a judge received evidence of this kind but on the use he made of it.

Counsel for Mexico discussed in some detail two periods of the proceedings under Mexican law in a criminal case. The procedure under the Mexican code of criminal procedure apparently is somewhat similar to that employed in the early stages of the Roman law and similar in some respects to the procedure generally obtaining in European countries at the present time. Counsel for Mexico pointed out that during the period of investigation a Mexican judge is at liberty to receive and take cognizance of anything placed before him, even matters that have no relation to the offense with which the accused is tried. The nature of some of the things incorporated into the record, including anonymous accusations against the character of the accused, is shown in the Presiding Commissioner's opinion. Undoubtedly in European countries a similar measure of latitude is permitted to a judge, but there seems to be an essential difference between procedure in those countries and that obtaining in the

Mexican courts, in that after a preliminary examination before a judge of investigation, a case passes on to a judge who conducts a trial. The French system, which was described by counsel for Mexico as being more severe toward the accused than is Mexican procedure, may be mentioned for purposes of comparison. Apparently under French law the preliminary examination does not serve as a foundation for the verdict of the judge who decided as to the guilt of the accused. The examination allows the examining judge to determine whether there is ground for formal charge, and in case there is, to decide upon the jurisdiction. The accused is not immediately brought before the court which is to pass upon his guilt or innocence. His appearance in court is deferred until the accusation rests upon substantial grounds. This trial is before a judge whose functions are of a more judicial character than those of a judge of investigation employing inquisitorial methods in the nature of those used by a prosecutor. When the period of investigation was completed in the cases of Chattin and the others with whom his case was consolidated, the entire proceedings so far as the Government was concerned were substantially finished, and after a hearing lasting perhaps five minutes, the same judge who collected evidence against the accused sentenced them.

. . .

International law requires that in the administration of penal laws an alien must be accorded certain rights. There must be some grounds for his arrest; he is entitled to be informed of the charge against him; and he must be given opportunity to defend himself.

. . .

. . . Positive conclusions as to the existence of some irregularities in a trial of a case obviously do not necessarily justify a pronouncement of a denial of justice. I do not find myself able fully to concur in the general trend of the argument of counsel for the United States that the record of the trial abounds in irregularities which reveal a purpose on the part of the judge at Mazatlán to convict the accused even in the absence of convincing proof of guilt. . . . I am of the opinion that the conclusions of the Commission must be grounded upon the record of the proceedings instituted against the accused. Having in mind the principles asserted by the Commission from time to time as to the necessity for basing pecuniary awards on convincing evidence of a pronounced degree of improper governmental administration, and having further in mind the peculiarly delicate character of an examination of judicial proceedings by an international tribunal, as well as the practical difficulties inherent in such examination, I limit myself to a rigid application of those principles in the instant case by concluding that the Commission should render an award, small in comparison to that claimed, which should be grounded on the mistreatment of the claimant during the period of investigation of his case. While deeply impressed with the importance of a strict application of the principles applicable to a case of this character, such application does not, in my opinion, preclude a full appreciation of human rights which it was contended in argument were grossly violated, and which it is clearly shown were in a measure disregarded with resultant injury to a man who languished in prison for seven months and was severely sentenced on scanty evidence for the alleged embezzlement of four pesos.

MacGREGOR, Commissioner, dissenting:

[MacGregor rejected the assertion 'that the accused was ignorant of a single one of the charges made against him, for the simple reason that the records formed in a criminal process are not secret, according to Mexican law, and are, from the time of their commencement at the disposal of the defendants or their counsel . . . .' He continued:]

8. It has been alleged that the trial proper (meaning by trial that part of the proceedings in which the defendants and witnesses as well as the Prosecuting Attorney and counsel appear personally before the Judge for the purpose of discussing the circumstances of the case) lasted five minutes at the most, for which reason it was a mere formality, implying thereby that there was really no trial and that Chattin was convicted without being heard. I believe that this is an erroneous criticism which arises from the difference between Anglo-Saxon procedure and that of other countries. Counsel for Mexico explained during the hearing of this case that Mexican criminal procedure is composed of two parts: Preliminary proceedings (sumario) and plenary proceedings (plenario). In the former all the information and evidence on the case are adduced; the corpus delicti is established; visits are made to the residences of persons concerned; commissions are performed by experts appointed by the Court; testimony is received and the Judge can cross-examine the culprits, counsel for the defense having also the right of cross-examination; public or private documents are received, etc. When the Judge considers that he has sufficient facts on which to establish a case, he declares the instruction closed and places the record in the hands of the parties (the defendant and his counsel on the one side evidence has been received are the parties in the cause requested to and the Prosecuting Attorney on the other [sic]), in order that they may state whether they desire any new evidence filed, and only when such file their respective final pleas. This being done, the public hearing is held, in which the parties very often do not have anything further to allege, because everything concerning their interests has already been done and stated. In such a case, the hearing is limited to the Prosecuting Attorney's ratification of his accusation, previously filed, and the defendants and their counsel also rely on the allegations previously made by them, these two facts being entered in the record, whereupon the Judge declares the case closed and it becomes ready to be decided. This is what happened in the criminal proceedings which have given rise to

this claim, and they show, further, that the defendants, including Chattin, refused to speak at the hearing in question or to adduce any kind of argument or evidence. In view of the foregoing explanation, I believe that it becomes evident that the charge, that there was no trial proper, can not subsist, for, in Mexican procedure, it is not a question of a trial in the sense of Anglo-Saxon law, which requires that the case be always heard in plenary proceedings, before a jury, adducing all the circumstances and evidence of the cause, examining and cross-examining all the witnesses, and allowing the prosecuting attorney and counsel for the defense to make their respective allegations. International law insures that a defendant be judged openly and that he be permitted to defend himself, but in no manner does it oblige these things to be done in any fixed way, as they are matters of internal regulation and belong to the sovereignty of States.

. . .

10. I admit that [other deficiencies] exist and that they show that the Judge could have carried out the investigation in a more efficient manner, but the fact that it was not done does not mean any violation of international law.

. . .

19. I consider that this is one of the most delicate cases that has come before the Commission and that its nature is such that it puts to a test the application of principles of international law. It is hardly of any use to proclaim in theory respect for the judiciary of a nation, if, in practice, it is attempted to call the judiciary to account for its minor acts. It is true that sometimes it is difficult to determine when a judicial act is internationally improper and when it is so from a domestic standpoint only. In my opinion the test which consists in ascertaining if the act implies damage, wilful neglect, or palpable deviation from the established customs becomes clearer by having in mind the damage which the claimant could have suffered. There are certain defects in procedure that can never cause damage which may be estimated separately, and that are blotted out or disappear, to put it thus, if the final decision is just. There are other defects which make it impossible for such decision to be just. The former, as a rule, do not engender international liability; the latter do so, since such liability arises from the decision which is iniquitous because of such defects. To prevent an accused from defending himself, either by refusing to inform him as to the facts imputed to him or by denying him a hearing and the use of remedies; to sentence him without evidence, or to impose on him disproportionate or unusual penalties, to treat him with cruelty and discrimination; are all acts which per se cause damage due to their rendering a just decision impossible. But to delay the proceedings somewhat, to lay aside some evidence, there existing other clear proofs, to fail to comply with the adjective law in its secondary provisions and other deficiencies of this kind, do not cause damage nor violate international law. Counsel for Mexico justly

stated that to submit the decisions of a nation to revision in this respect was tantamount to submitting her to a régime of capitulations. All the criticism which has been made of these proceedings, I regret to say, appears to arise from lack of knowledge of the judicial system and practice of Mexico, and, what is more dangerous, from the application thereto of tests belonging to foreign systems of law. For example, in some of the latter the investigation of a crime is made only by the police magistrates and the trial proper is conducted by the Judge. Hence the reluctance in accepting that one same judge may have the two functions and that, therefore, he may have to receive in the preliminary investigation (instrucción) of the case all kinds of data, with the obligation, of course, of not taking them into account at the time of judgment, if they have no probative weight. It is certain that the secret report, so much discussed in this case, would have been received by the police of the countries which place the investigation exclusively in the hands of such branch. This same police would have been free to follow all the clues or to abandon them at its discretion; but the Judge is criticized here because he did not follow up completely the clue given by Ramirez with respect to Chattin. The same domestic test—to call it such—is used to understand what is a trial or open trial imagining at the same time that it must have the sacred forms of common-law and without remembering that the same goal is reached by many roads. And the same can be said when speaking of the manner of taking testimony of witnesses, of cross-examination, of holding confrontations, etc.

20. In view of the above considerations, I am of the opinion that this claim should be disallowed.

COMMENT ON PROBLEMS IN DEVELOPING AN INTERNATIONAL MINIMUM STANDARD IN CRIMINAL PROCEDURE

As the opinions in *Chattin* indicate, fundamental differences exist between common-law and civil-law criminal procedure. Moreover, there are important differences within each group—for example, between Mexico and France. Consider some aspects of the more-or-less contemporary French judicial process that appear applicable in a case similar to *Chattin*—a case that would likely be regarded as a *délit* (misdemeanor) rather than *crime* (felony) in view of the conduct and small sum involved.

Official proceedings would likely start with a preparatory investigation by the *juge d'instruction* (examining magistrate), who may visit the scene of the crime, conduct searches, call witnesses and experts and so on. Witnesses are heard under oath without the accused being present, and their statements are put in writing. The accused is informed of the charge, and told that he need make no statement and has the right to counsel. Indeed, absent an express waiver, the accused cannot be examined or confronted unless counsel is present. Throughout these pretrial proceedings, the *dossier* (official file) grows, and counsel may consult it. Counsel may pose questions during proceedings only with the authorization of the examining magistrate.

After the investigation and pre-trial proceedings are completed, the magistrate determines if the evidence is sufficient to justify further proceedings, and if so, determines the appropriate court (in view of the nature of the charge) to which the case should be sent for trial. The *dossier* is transmitted to that court. The magistrate plays no further role. The case in handled (for *délits*) by a three-judge court. Trial hearings are public; there are no formal rules of procedure in the Anglo-American sense. The court's president interrogates the accused and poses questions to witnesses, including as useful the questions proposed by the parties.

Among the stark contrasts between the Mexican or French procedures and criminal trials in the United States is the traditional and constitutional requirement of trial by jury. Another pervasive difference is the significance in the Anglo-American jury trial of what has been loosely termed an accusatorial or prosecutorial rather than an inquisitorial approach. The court in the Anglo-American world is relatively quiescent. Initiative lies generally with the parties, who develop facts and seek evidence through personal or police inquiry, summon witnesses whom they examine, and generally give direction to the legal or factual inquiries pursued during pre-trial proceedings or at trial. The court aids them through the issuance of its official process, as necessary.

A critical ingredient of this approach is the emphasis which Anglo-American legal systems place upon direct confrontation and cross-examination. Compare, for example, the decision in *Pointer* v. *Texas*, 380 U.S. 400 (1965), with the Mexican and French procedures. The Supreme Court there held that the Due Process Clause of the Fourteenth Amendment forbade the prosecutor in a state criminal trial to use a transcript of the witness's testimony which was given at a preliminary hearing at which the defendant, present without counsel, did not cross-examine. The witness himself was absent from the trial. The Court stated that 'the Sixth Amendment's right of an accused to confront the witnesses against him is . . . a fundamental right and is made obligatory on the States by the Fourteenth Amendment. . . . There are few subjects, perhaps, upon which this Court and other courts have been more nearly unanimous than in their expressions of belief that the right of confrontation and cross-examination is an essential and fundamental requirement for the kind of fair trial which is this country's constitutional goal.' The transcript, concluded the Court, had not been taken under circumstances giving the defendant 'through counsel an adequate opportunity to cross-examine . . . .'

The *Chattin* opinions underscore the methodological problems in developing a minimum international standard of criminal procedure out of such diverse materials—a diversity there restricted to Europe and Latin America, hence far less perplexing than the worldwide diversity of legal cultures and criminal processes among the states of the contemporary world. A treaty was relevant to *Chattin*, but as indicated above, it addressed the scope and structure of the arbitration between the United States and Mexico rather than the international norms of criminal procedure to be applied. The only reference of the General Claims Convention to applicable norms was the terse provision in Article 1 that claims should be submitted to the tripartite Commission 'for decision in accordance with the principles of international law, justice and equity.'

Today a dispute like that in *Chattin* could draw on a human rights treaty, the International Covenant on Civil and Political Rights to be discussed in Chapter 3 that (as of September 1995) had 131 states parties. Article 14 of that Covenant dealing with criminal trials provides in relevant part:

> 1. All persons shall be equal before the courts . . . . [E]veryone shall be entitled to a fair and public hearing by [an] impartial tribunal . . . .
>
> 2. Everyone . . . shall have the right to be presumed innocent until proved guilty according to law.
>
> 3. . . . [E]veryone shall be entitled to the following minimum guarantees . . . .
>
> (d) To be tried in his presence and to defend himself in person or through legal assistance of his own choosing . . . ;
>
> (e) To examine, or have examined, the witnesses against him and to obtain the attendance and examination of witnesses on his behalf . . . .

QUESTIONS

1. How would you identify the most serious issues in the judicial process leading to Chattin's conviction—say, on the basis of a comparison between the Mexican and French systems? Does the challenge to Chattin's conviction implicate the Mexican system of criminal justice structurally—that is, do the two commissioners finding the conviction to violate an international standard indict the structure of the Mexican system of criminal justice in general? Or rather do they find an aberration from that system in this particular case?

2. How do the commissioners approach the task of identifying an 'international standard of civilization' (or, within the terms of the 1923 Convention, the relevant 'international law, justice and equity') against which they test the legality of the conviction and imprisonment of Chattin? Is this decision an illustration of a tribunal's defining and applying customary international law, or is some other 'source' or 'process' of international law also involved? Does any other provision of Article 38(1) of the I.C.J. Statute or does the discussion by Schachter at p. 47, *supra*, illuminate this last question?

3. What threshold problems do you see in the working out—by a tribunal, by a multilateral conference drafting a treaty, by a scholar—of a general international standard governing criminal process?

4. Would the tribunal's task have been much simpler if there had been a treaty between the U.S. and Mexico regulating treatment of aliens that included Article 14 of the ICCPR? Would Article 14 have resolved on its face all the issues?

5. Issues similar to those in *Chattin*, particularly the significance of having the same judge preside over pretrial proceedings and the actual trial, have been the subject of several decisions of the European Court of Human Rights under the European Convention for the Protection of Human Rights and Fundamental Freedoms. Do such issues appear easier to resolve in a regional context like Europe than in the Universal settin gof the ICCPR?

ADDITIONAL READING

Richard Baxter, *Reflections on Codification in Light of the International Law of State Responsibility for Injuries to Aliens*, 16 Syracuse L. Rev. 745 (1965); Richard Lillich, *The Current Status of the Law of State Responsibility*, in Lillich (ed.) *International Law of State Responsibility for Injuries to Aliens* (1983); Guha Roy, *Is the Law of Responsibility of States for Injuries to Aliens a Part of Universal International Law?*, 55 Am. J. Int. L. 863 (1961).

COMMENT ON THE MINORITIES REGIME AFTER WORLD WAR I

The *Minority Schools in Albania* opinion that follows illustrates treaties as a source and major expression of international law, and introduces another field of international law that influenced the growth of the human rights movement. This Comment provides some background to the opinion.

Treaties and other special regimes to protect minorities have a long history in international law dating from the emergence in the seventeenth century of the modern form of the political state, sovereign within its territorial boundaries. Within Europe religious issues became a strong concern since states often included more than one religious denomination, and abuse by a state of a religious minority could lead to intervention by other states where that religion was dominant. Hence peace treaties sometimes included provisions on religious minorities. In later centuries, the precarious situation of Christian minorities within the Ottoman Empire and of religious minorities in newly independent East European or Balkan states led to outbreaks of violence and to sporadic treaty regulation.

World War I ushered in an era of heightened attention to problems of racial, religious or linguistic minorities. The collapse of the great Austro-Hungarian and Ottoman multinational empires, and the chaos as the Russian empire of the Romanoffs was succeeded by the Soviet Union, led to much redrawing of maps and the creation of new states. President Wilson's Fourteen Points, however compromised they became in the Versailles Treaty and later arrangements, nonetheless exerted influence on the postwar settlements. In it and other messages, Wilson stressed the ideals of the freeing of minorities and the related 'self-determination' of peoples or nationalities. That concept of self-determination, so politically powerful and open to such diverse interpretations, continues to this day to be much disputed and to have profound consequences. It not only appears in the UN Charter but is given a position of high prominence in the two principal human rights covenants. Its current significance is examined in Chapter 4.

From concepts like self-determination and out of the legacy of nineteenth century liberal nationalism that saw the development of nation-states like Germany and Italy, the principle of nationalities took on a new force. Here was another ambiguous and disputed concept—the 'nation' or 'nationality' as distinct from the political state, the nation (often identified with a 'people') defined in cultural or historical terms, often defined more concretely in racial, linguistic and religious terms. One goal in displacing the old empires with new or redrawn states was to identify the nation with the state—ideally, to give each 'nation' its own state. Membership in a 'nation' would ideally be equivalent to membership in a 'state' consisting only or principally of that nation.

Within the pure realization of this ideal, all 'Poles,' for example, would be situated in Poland; there would be no 'Polish' minority in other states, and other 'nationalities' would not be resident in Poland. However, even the detaching of Poland after World War I from the empires and states that had absorbed different parts of it failed to acheive a strict congruence between the 'nation' and 'state'. There were polar moves, for example the creation of Yugoslavia as a multiethnic state that after seventy years has had such tragic consequences.

Of course the goal of total identification of state with nation—a goal itself disputed and in contradiction with other conceptions of the political state that did not emphasize cultural homogeneity or ethnic purity—could not be realized. Life and history were and remain too various and complex for such precise correlations. The nineteenth century examples of Germany and Italy, for example, were far from unitary; each had its national, ethnic, linguistic, and religious minorities. National or ethnic homogeneity could be achieved in the vast majority of the world's states only by the compulsory and massive migrations of minority groups, migrations far more systematic and coercive than were some of the population movements and exchanges after World War I. A 'nation' defined, say, in linguistic–religious terms would generally transcend national boundaries and be located in the territories of two or several sovereign states in the new world created by the postwar settlements. A Greek-speaking Christian minority would, for example, be present in the reconfigured Muslim Albania.

Bear in mind another confusing linguistic usage. The term 'national' is generally used in international law to signify the subjects or citizens of a state. Hence members of the 'German' nation (in the sense of a 'people' and 'culture') living in Poland could be Polish 'nationals' in the sense of being citizens of Poland. Or they could possess only German citizenship and be alien residents in Poland. In the *Minority Schools in Albania* case that follows, members of the

Greek-speaking Christian minority (part of a 'nation' in the cultural or ethnic sense) in Muslim Albania were 'nationals' (citizens) of Albania. One can imagine the ambiguity attending the frequent usage of the term 'national minorities,' which could mean at least (1) a group in a state belonging in the cultural or ethnic sense to a 'nation' that constituted a minority in that state, or (2) all minorities in a state who were 'nationals' (citizens) of that state.

After World War I, the victorious powers and the new League of Nations sought to address this situation. They confronted the impossibility, even if it were desirable, of creating ethnically homogeneous states. Hence they had to deal with the continuing presence in states of minorities which had frequently been abused in ways ranging from economic discrimination to pogroms and other violence that could implicate other states, spill across international boundaries and lead to war. The immediate trigger for the outbreak of World War I in the tormented Balkans was fresh in memory.

President Wilson had proposed that the Covenant of the League of Nations include norms governing the protection of minorities that would have embraced all members of the League. The other major powers rejected this approach, preferring discrete international arrangements to handle discrete problems of minorities in particular states of Central-East Europe and the Balkans rather than a universal treaty system. This compromise led to the regime of the so-called Minorities Treaties that were imposed on the new or reconfigured states of Central-East Europe and the Balkans.

For some states like Austria and Hungary, provisions for minority protection were included in the peace treaties. Other states like Poland or Greece signed minority protection treaties with the allied and associated powers. Some states like Albania and Lithuania made minority protection declarations as a condition for their membership in the League of Nations. There were also bilateral treaties protecting minorities such as one between Germany and Poland. Note that one of the features of this new regime was to insulate the victorious powers from international regulation of their treatment of their own citizens belonging to minorities.

Although there were significant variations among these treaties and declarations, many provisions were common. The 1919 Minorities Treaty between the Principal Allied and Associated Powers and Poland served as a model for later treaties and declarations. It provided for protection of life and liberty and religious freedom for all 'inhabitants of Poland.' All Polish nationals (citizens) were guaranteed equality before the law and the right to use their own language in private life and judicial proceedings. Members of racial, religious or linguistic minorities were guaranteed 'the same treatment and security in law and in fact' as other Polish nationals, and the right to establish and control at their expense their own religious, social and educational institutions. In areas of Poland where a 'considerable proportion' of Polish nationals belonged to minorities, an 'equitable share' of public funds would go to such minority groups for educational or religious purposes. In view of the particular history of oppression and violence, there were specific guarantees for Jews.

Like other minority treaties and declarations, the Polish treaty's provisions were placed under the guarantee of the League of Nations to the extent that 'they affect persons belonging to' minority groups. The League developed procedures to implement its duties, including a right of petition to it by beleaguered minorities claiming that a treaty regime or declaration had been violated, and including a minorities committees given the task of seeking negotiated solutions to such disputes. As shown by the *Minority Schools in Albania* case, the Council of the League could invoke in accordance with its usual procedures the advisory opinion jurisdiction of the Permanent Court of International Justice (P.C.I.J.), the first international court (supplementing *ad hoc* arbitral tribunals as in the *Chattin* case). The Court was created by the League in 1921, became dormant in World War II, and was then succeeded by the International Court of Justice created under the UN Charter.

MINORITY SCHOOLS IN ALBANIA

Advisory Opinion, Permanent Court of International Justice, 1935. Series A/B–No. 64.

[In 1920, the Assembly of the League of Nations adopted a recommendation requesting that if Albania were admitted into the League, it 'should take the necessary measures to enforce the principles of the Minorities Treaties' and to arrange the 'details required to carry this object into effect' with the Council of the League. Albania was admitted to membership a few days later. In 1921 the Council included on its agenda the question of protection of minorities in Albania.

The Greek government, in view of the presence of a substantial Christian minority of Greek origin in (dominantly Muslim) Albania, communicated to the League proposals for provisions going beyond the Minorities Treaties that were related to Christian worship and to education in the Greek language. The Council commissioned a report, and the reporter submitted to it a draft Declaration to be signed by Albania and formally communicated to the Council. The Declaration was signed by Albania and submitted to the Council in 1921, with basic similarities to but some differences from the typical clauses of the Minorities Treaties. The Council decided that the stipulations in the Declaration about minorities should be placed under the guarantee of the League from the date of the Declaration's ratification by Albania, which took place in 1922.

The first paragraph of Article 5 of the Declaration, at the core of the dispute that later developed, provided as follows:

> Albanian nationals who belong to racial, linguistic or religious minorities, will enjoy the same treatment and security in law and in fact as other

> Albanian nationals. In particular, they shall have an equal right to maintain, manage and control at their own expense or to establish in the future, charitable, religious and social institutions, schools and other educational establishments, with the right to use their own language and to exercise their religion freely therein.

Over the years, numerous changes in the laws and practices of the Albanian government led to questions about compliance with the Declaration. In 1933, the Albanian National Assembly modified Articles 206 and 207 of the Constitution, which had provided that 'Albanian subjects may found private schools' subject to government regulation, to state:

> The instruction and education of Albanian subjects are reserved to the State and will be given in State schools. Primary education is compulsory for all Albanian nationals and will be given free of charge. Private schools of all categories at present in operation will be closed.

The new provisions affecting Greek-language and other private schools led to petitions and complaints to the League from groups including the Greek minority in Albania. Acting within its regular powers, the Council requested the Permanent Court of International Justice in 1935 to give an advisory opinion whether, in light of the 1921 Declaration as a whole, Albania was justified in its position that it had acted in conformity with 'the letter and the spirit' of Article 5 because (as Albania argued) its abolition of private schools was a general measure applicable to all Albanian nationals, whether members of an ethnic majority or minority.

There follow excerpts from the opinion for the P.C.I.J. and from a dissenting opinion. For present purposes, the Albanian Declaration can be understood as tantamount to a treaty. The opinions draw no relevant distinction between the two, and refer frequently to the Minorities Treaties to inform their interpretation of the Declaration.]

The contention of the Albanian Government is that the above-mentioned clause imposed no other obligation upon it, in educational matters, than to grant to its nationals belonging to racial, religious, or linguistic minorities a right equal to that possessed by other Albanian nationals. Once the latter have ceased to be entitled to have private schools, the former cannot claim to have them either. This conclusion, which is alleged to follow quite naturally from the wording of paragraph I of Article 5, would, it is contended, be in complete conformity with the meaning and spirit of the treaties for the protection of minorities, an essential characteristic of which is the full and complete equality of all nationals of the State, whether belonging to the majority or to the minority. On the other hand, it is argued, any interpretation which would compel Albania to respect the private minority schools would create a privilege in favour of the minority and run counter to the essential idea of the law governing minorities. Moreover, as the minority régime is an extraordinary régime constituting a derogation from the ordinary law, the text in question should, in case of doubt, be construed in the manner most favourable to the sovereignty of the Albanian State.

According to the explanations furnished to the Court by the Greek Government, the fundamental idea of Article 5 of the Declaration was on the contrary to guarantee freedom of education to the minorities by granting them the right to retain their existing schools and to establish others, if they desired; equality of treatment is, in the Greek Government's opinion, merely an adjunct to that right, and cannot impede the purpose in view, which is to ensure full and effectual liberty in matters of education. Moreover, the application of the same régime to a majority as to a minority, whose needs are quite different, would only create an apparent equality, whereas the Albanian Declaration, consistently with ordinary minority law, was designed to ensure a genuine and effective equality, not merely a formal equality.

. . .

As the Declaration of October 2nd, 1921, was designed to apply to Albania the general principles of the treaties for the protection of minorities, this is the point of view which, in the Court's opinion, must be adopted in construing paragraph I of Article 5 of the said Declaration.

*

The idea underlying the treaties for the protection of minorities is to secure for certain elements incorporated in a State, the population of which differs from them in race, language or religion, the possibility of living peaceably alongside that population and co-operating amicably with it, while at the same time preserving the characteristics which distinguish them from the majority, and satisfying the ensuing special needs.

In order to attain this object, two things were regarded as particularly necessary, and have formed the subject of provisions in these treaties.

The first is to ensure that nationals belonging to racial, religious or linguistic minorities shall be placed in every respect on a footing of perfect equality with the other nationals of the State.

The second is to ensure for the minority elements suitable means for the preservation of their racial peculiarities, their traditions and their national characteristics.

These two requirements are indeed closely interlocked, for there would be no true equality between a majority and a minority if the latter were deprived of its own institutions, and were consequently compelled to renounce that which constitutes the very essence of its being as a minority.

In common with the other treaties for the protection of minorities, and in particular with the Polish Treaty of June 28th, 1919, the text of which it follows, so far as concerns the question before the Court, very closely and almost literally, the Declaration of October 2nd, 1921, begins by laying down that no person shall be placed, in his relations with the Albanian authorities, in a position of inferiority by reason of his language, race or religion. . . .

. . .

In all these cases, the Declaration provides for a régime of legal equality for all persons mentioned in the clause; in fact no standard of comparison was indicated, and none was necessary, for at the same time that it provides for equality of treatment the Declaration specifies the rights which are to be enjoyed equally by all.

. . .

It has already been remarked that paragraph I of Article 5 consists of two sentences, the second of which is linked to the first by the words *in particular*: for a right apprehension of the second part, it is therefore first necessary to determine the meaning and the scope of the first sentence.

This sentence is worded as follows:

> Albanian nationals who belong to racial, linguistic or religious minorities, will enjoy the same treatment and security in law and in fact as other Albanian nationals.

The question that arises is what is meant by the *same treatment and security in law and in fact*.

It must be noted to begin with that the equality of all Albanian nationals before the law has already been stipulated in the widest terms in Article 4. As it is difficult to admit that Article 5 set out to repeat in different words what had already been said in Article 4, one is led to the conclusion that 'the same treatment and security in law and in fact' which is provided for in Article 5 is not the same notion as the equality before the law which is provided for in Article 4.

. . .

This special conception finds expression in the idea of an equality in fact which in Article 5 supplements equality in law. All Albanian nationals enjoy the equality in law stipulated in Article 4; on the other hand, the equality between members of the majority and of the minority must, according to the terms of Article 5, be an equality in law and in fact.

It is perhaps not easy to define the distinction between the notions of equality in fact and equality in law; nevertheless, it may be said that the former notion excludes the idea of a merely formal equality; that is indeed what the Court laid down in its Advisory Opinion of September 10th, 1923, concerning the case of the German settlers in Poland (Opinion No. 6), in which it said that:

> There must be equality in fact as well as ostensible legal equality in the sense of the absence of discrimination in the words of the law.

Equality in law precludes discrimination of any kind; whereas equality in fact may involve the necessity of different treatment in order to attain a result which establishes an equilibrium between different situations.

It is easy to imagine cases in which equality of treatment of the majority and of the minority, whose situation and requirements are different, would result in inequality in fact; treatment of this description would run counter to the first sentence of paragraph I of Article 5. The equality between members of the majority and of the minority must be an effective, genuine equality; that is the meaning of this provision.

The second sentence of this paragraph provides as follows:

> In particular they shall have an equal right to maintain, manage and control at their own expense or to establish in the future, charitable, religious and social institutions, schools and other educational establishments, with the right to use their own language and to exercise their religion freely therein.

This sentence of the paragraph being linked to the first by the words 'in particular', it is natural to conclude that it envisages a particularly important illustration of the application of the principle of identical treatment in law and in fact that is stipulated in the first sentence of the paragraph. For the institutions mentioned in the second sentence are indispensable to enable the minority to enjoy the same treatment as the majority, not only in law but also in fact. The abolition of these institutions, which alone can satisfy the special requirements of the minority groups, and their replacement by government institutions, would destroy this equality of treatment, for its effect would be to deprive the minority of the institutions appropriate to its needs, whereas the majority would continue to have them supplied in the institutions created by the State.

Far from creating a privilege in favour of the minority, as the Albanian Government avers, this stipulation ensures that the majority shall not be given a privileged situation as compared with the minority.

It may further be observed that, even disregarding the link between the two parts of paragraph I of Article 5, it seems difficult to maintain that the adjective 'equal', which qualifies the word 'right', has the effect of empowering the State to abolish the right, and thus to render the clause in question illusory; for, if so, the stipulation which confers so important a right on the members of the minority would not only add nothing to what has already been provided in Article 4, but it would become a weapon by which the State could deprive the minority régime of a great part of its practical value. It should be observed that in its Advisory Opinion of September 15th, 1923, concerning the question of the acquisition of Polish nationality (Opinion No. 7), the Court referred to the opinion which it had already expressed in Advisory Opinion

No. 6 to the effect that 'an interpretation which would deprive the Minorities Treaty of a great part of its value is inadmissible'.

. . . The idea embodied in the expression 'equal right' is that the right thus conferred on the members of the minority cannot in any case be inferior to the corresponding right of other Albanian nationals. In other words, the members of the minority must always enjoy the right stipulated in the Declaration, and, in addition, any more extensive rights which the State may accord to other nationals. . . .

The construction which the Court places on paragraph 1 of Article 5 is confirmed by the history of this provision.

[Analysis of the proposals of the Greek Government and replies of the Albanian Government during the period of drafting of the Declaration omitted.]

The Court, having thus established that paragraph I of Article 5 of the Declaration, both according to its letter and its spirit, confers on Albanian nationals of racial, religious or linguistic minorities the right that is stipulated in the second sentence of that paragraph, finds it unnecessary to examine the subsidiary argument adduced by the Albanian Government to the effect that the text in question should in case of doubt be interpreted in the sense that is most favourable to the sovereignty of the State.

. . .

For these reasons,

The Court is of opinion,

by eight votes to three,

that the plea of the Albanian Government that, as the abolition of private schools in Albania constitutes a general measure applicable to the majority as well as to the minority, it is in conformity with the letter and spirit of the stipulations laid down in Article 5, first paragraph, of the Declaration of October 2nd, 1921, is not well founded.

. . .

DISSENTING OPINION BY SIR CECIL HURST, COUNT ROSTWOROWSKI AND M. NEGULESCO.

The undermentioned are unable to concur in the opinion rendered by the Court. They can see no adequate reason for holding that the suppression of the private schools effected in Albania in virtue of Articles 206 and 207 of the Constitution of 1933 is not in conformity with the Albanian Declaration of October 2nd, 1921.

. . .

The construction of the paragraph is clear and simple. The first sentence stipulates for the treatment and the security being the same for the members of the minority as for the other Albanian nationals. The second provides that as regards certain specified matters the members of the minority shall have an equal right. The two sentences are linked together by the words 'In particular' (*notamment*). These words show that the second sentence is a particular application of the principle enunciated in the first. If the rights of the two categories under the first sentence are to be the same, the equal right provided for in the second sentence must indicate equality between the same two categories, viz. the members of the minority and the other Albanian nationals. The second sentence is added because the general principle laid down in the first sentence mentions only 'treatment and security in law and in fact'—a phrase so indefinite that without further words of precision it would be doubtful whether it covered the right to establish and maintain charitable, religious and social institutions and schools and other educational establishments, but the particular application of the general principle of identity of treatment and security remains governed by the dominating element of equality as between the two categories.

The word 'equal' implies that the right so enjoyed must be equal in measure to the right enjoyed by somebody else. '*They shall have an equal right*' means that the right to be enjoyed by the people in question is to be equal in measure to that enjoyed by some other group. A right which is unconditional and independent of that enjoyed by other people cannot with accuracy be described as an 'equal right'. 'Equality' necessarily implies the existence of some extraneous criterion by reference to which the content is to be determined.

If the text of the first paragraph of Article 5 is considered alone, it does not seem that there could be any doubt as to its interpretation. It is, however, laid down in the Opinion from which the undersigned dissent that if the general purpose of the minority treaties is borne in mind and also the contents of the Albanian Declaration taken as a whole, it will be found that the 'equal right' provided for in the first paragraph of Article 5 cannot mean a right of which the extent is measured by that enjoyed by other Albanian nationals, and that it must imply an unconditional right, a right of which the members of the minority cannot be deprived.

. . .

As the opinion of the Court is based on the general purpose which the minorities treaties are presumed to have had in view and not on the text of Article 5, paragraph I, of the Albanian Declaration, it involves to some extent a departure from the principles hitherto adopted by this Court in the interpretation of international instruments, that in presence of a clause which is reasonably clear the Court is bound to apply it as it stands without considering whether other provisions might with advantage have been added to it or substituted for it, and this even if the results following from it may in some particular hypothesis seem unsatisfactory.

. . .

Furthermore, the suppression of the private schools—even if it may prejudice to some appreciable extent the interests of a minority—does not oblige them to abandon an essential part of the characteristic life of a minority. In interpreting Article 5, the question whether the possession of particular institutions may or may not be *important* to the minority cannot constitute the decisive consideration. There is another consideration entitled to equal weight. That is the extent to which the monopoly of education may be of importance to the State. The two considerations cannot be weighed one against the other: Neither of them—in the absence of a clear stipulation to that effect—can provide an objective standard for determining which of them is to prevail.

International justice must proceed upon the footing of applying treaty stipulations impartially to the rights of the State and to the rights of the minority, and the method of doing so is to adhere to the terms of the treaty—as representing the common will of the parties—as closely as possible.

. . .

If the intention of the second sentence: 'In particular they [the minority] shall have an equal right . . .', had been that the right so given should be universal and unconditional, there is no reason why the draftsman should not have dealt with the right to establish institutions and schools in the earlier articles [of the Declaration that set up fixed and universal standards for all Albanians on matters like protection of life and free exercise of religion]. The draftsman should have dealt with the liberty to maintain schools and other institutions on lines similar to those governing the right to the free exercise of religion, which undoubtedly is conferred as a universal and unconditional right. Instead of doing so the right conferred upon the minority is an 'equal' right . . .

. . .

COMMENT ON FURTHER ASPECTS OF THE MINORITY TREATIES

The *Minority Schools in Albania* opinions address many current issues that remain vexing. The discussions about the nature of 'equality' and assurances thereof, in particular about equality 'in law' and 'in fact,' inform contemporary human rights law as well as constitutional and legislative debates in many states with respect to issues like equal protection and affirmative action. The question whether the Declaration and the Court's opinion recognized only the rights of individual members of a minority, or also the right of the minority itself as a collective or group, remains one that vexes the discussion of minority rights. Protection aiming at the cultural survival of minorities continues to raise the troubling issue of which types of minorities merit such protection, and whether assurance of equal protection (with the majority) is sufficient for the purpose. See Chapter 14.

Although the issues debated within the minorities regime remain vital, the regime itself disappeared. Over the next two decades, its norms were roundly violated. Its international machinery within the League of Nations proved to be ineffectual, partly for the same lack of political will that led to other disastrous events in the interwar period. The failure of the regime was tragic in its consequences. Its noble purposes were distorted or blunted or ignored as Europe of the 1930s moved toward the horrors of World War II, the Holocaust and the brutalization and slaughter of so many other minorities. The settlements, norms and institutions after World War II designed to prevent further savagery against minorities stressed different principles and created radically different institutions, principally within the universal human rights system built in and around the United Nations.

Nonetheless, it is important to recognize the distinctive dilemmas and advances as well as the shortcomings of this minorities regime. Sovereignty in the sense of a state's (absolute) internal control over its own citizens was to some extent eroded. Treaties–declarations subjected aspects of the state's treatment of its own citizens to international law and and processes—that is, citizens who were members of a racial, religious or linguistic minority. Although the norms were expressed in bilateral treaties or declarations, the regime took on a multilateral aspect through its incorporation into the League as well as through the large number of nearly simultaneous treaties and declarations. The whole scheme was informed by multilateral planning, in contrast with the centuries-old examples of sporadic bilateral treaties protecting (usually religious) minorities. Minorities became a matter of formal international concern, the treaties-declarations fragmented the state into different sections of its citizens, and international law reached beyond the law of state responsibility to protect some of a state's own citizens.

The precise issue of the *Minority Schools in Albania* case is now addressed in the 1960 UNESCO Convention against Discrimination in Education. Article 5(1)(c) recognizes the 'right of members of national minorities to carry on their own educational activities, including the maintenance of schools and, depending on the educational policy of each State, the use or the teaching of their own language . . . .' The article subjects this right to several provisos. For example, its exercise should not prevent minorities from understanding the culture and language of the larger community as well, or prejudice national sovereignty.

QUESTIONS

1. The types of protections or assurances given by treaty to a distinctive group within a larger polity can be categorized in various ways, including the following. The assurance can be *absolute* (fixed, unconditional) or *contingent* (dependent on some reference group). For example, treaties of commerce between two states may reciprocally grant to citizens of each state the right to reside (for business purposes) and do business (as aliens) in the other state. Some assurances in such treaties will be absolute—for example, citizens of each state are given the right to buy or lease real property for residential purposes in the other state. Other assurances will be contingent—for example, citizens of each state are given the right to organize a corporation and qualify to do business in the other state on the same terms as citizens of that other state (so called 'national treatment'). Within this framework, how would you characterize the rights given to members of a designated minority by the Declaration? Do the majority and dissenting opinions differ about how to characterize them?

2. If you were a member of the Greek-speaking Christian minority, would you have been content with a Declaration that contained no more than a general equal protection clause? If not, why not? How would you justify your argument for more protection?

3. Why do the opinions refer to this minorities regime as 'extraordinary'? In what respects did it depart from classical conceptions of international law, or differ from the law of state responsibility?

4. Does a treaty necessarily solve the problems of the method to be employed in 'identifying' and 'applying' international law that were present in *The Paquete Habana* and *Chattin*? How would you characterize the methodologies or conceptions of interpretation that the majority and dissenting opinions reveal? How do those approaches to interpretation differ with respect to their basic assumptions about the relation between the minorities regime and general international law?

5. Consider how close to or distant from the minorities regime Article 27 of the International Covenant on Civil and Political Rights—the most invoked and influential human rights treaty today with (as of September 1995) 131 states parties—appears on its face to be. It provides:

> In those States in which ethnic, religious or linguistic minorities exist, persons belonging to such minorities shall not be denied the right, in community with the other members of their group, to enjoy their own culture, to profess and practise their own religion, or to use their own language.

ADDITIONAL READING

Nathaniel Berman, *'But the Alternative is Despair': European Nationalism and the Modernist Renewal of International Law*, 106 Harv. L. Rev. 1792 (1993); Danilo Türk, *Minority Issues in the UN: Norms and Institutions*, in Henry Steiner (ed.), *Ethnic Conflict and the UN Human Rights System* (forthcoming 1995).

COMMENT ON THE NUREMBERG TRIALS

The trial at Nuremberg in 1945–46 of major war criminals among the Axis powers, dominantly Nazi party leaders and military officials, gave the nascent human rights movement a powerful impulse. The UN Charter that became effective in 1945 included a few broad human rights provisions. But they were more programmatic than operational, more a program to be realized by states over time than legal rules to be applied immediately to states. Nuremberg, on the other hand, was concrete and applied: prosecutions, convictions, punishment. The prosecution and the Judgment of the International Military Tribunal were based on concepts and norms, some of which had deep roots in international law and some of which represented a significant development of that law that underlay the later formulation of major human rights norms.

The striking aspect of Nuremberg was that the trial and Judgment applied international law doctrines and concepts to impose criminal punishment on individuals for their commission of any of the three types of crimes under international law that are described below. The notion of crimes against the law of nations for which violators bore an individual criminal responsibility was itself an older one, but it had operated in a restricted field. As customary international law developed from the time of Grotius, certain conduct came to be considered a violation of the law of nations—in effect, a universal crime. Piracy on the high seas was long the classic example of this limited category of crimes. Given the common interest of all nations in protecting navigation against interference on the high seas outside the territory of any state, it was considered appropriate for the state apprehending a pirate to prosecute in its own courts. Since there was no international criminal tribunal, prosecution in a state court was the only means of judicial enforcement. To the extent that the state courts sought to apply the customary international law defining the crime of piracy, either directly or as it had become absorbed into national legislation, the choice of forum became less significant, for state courts everywhere, at least in theory, were applying the same law.

One specialized field, the humanitarian laws of war, had long included rules regulating the conduct of war, the so-called *jus in bello*. This body of law imposed sanctions against combatants who committed serious violations of the restrictive rules. Such application of the laws of war, and its foundation in customary norms and in treaties, figure in the Judgment below. But the concept of individual criminal responsibility was not systematically developed. It achieved a new prominence and a clearer definition after the Nuremberg Judgment, and the treaties and protocols noted at p. 69 *supra*. Gradually other types of conduct have been added to this small list of crimes under international law—for example, slave trading prior to Nuremberg and genocide thereafter.

As World War II came to an end, the Allied powers held several conferences to determine what policies they should follow towards the Germans responsible for the war and the massive, systematic barbarity and destruction of the period. These conferences culminated in the (U.S., U.S.S.R., Britain, France) London Agreement of August 8, 1945, 59 Stat. 1544, E.A.S. No. 472, in which the parties determined to constitute 'an International Military Tribunal for the trial of war criminals.' The Charter annexed to the Agreement provided for the composition and basic procedures of the Tribunal and stated in its three critical articles:

Article 6.

The Tribunal established by the Agreement referred to in Article 1 hereof for the trial and punishment of the major war criminals of the European Axis countries shall have the power to try and punish persons who, acting in the interests of the European Axis countries, whether as individuals or as members of organizations, committed any of the following crimes.

The following acts, or any of them, are crimes coming within the jurisdiction of the Tribunal for which there shall be individual responsibility:

(a) CRIMES AGAINST PEACE: namely, planning, preparation, initiation or waging of a war of aggression, or a war in violation of international treaties, agreements or assurances, or participation in a common plan or conspiracy for the accomplishment of any of the foregoing;

(b) WAR CRIMES: namely, violations of the laws or customs of war. Such violations shall include, but not be limited to, murder, ill-treatment or deportation to slave labor or for any other purpose of civilian population of or in occupied territory, murder or ill-treatment of prisoners of war or persons on the seas, killing of hostages, plunder of public or private property, wanton destruction of cities, towns or villages, or devastation not justified by military necessity;

(c) CRIMES AGAINST HUMANITY: namely, murder, extermination, enslavement, deportation, and other inhumane acts committed against any civilian population, before or during the war, or persecutions on political, racial or religious grounds in execution of or in connection with any crime within the jurisdiction of the Tribunal, whether or not in violation of the domestic law of the country where perpetrated.

Leaders, organizers, instigators and accomplices participating in the formulation or execution of a common plan or conspiracy to commit any of the foregoing crimes are responsible for all acts performed by any persons in execution of such plan.

Article 7.

The official position of defendants, whether as Heads of State or responsible officials in Government Departments, shall not be considered as freeing them from responsibility or mitigating punishment.

Article 8.

The fact that the Defendant acted pursuant to order of his Government or of a superior shall not free him from responsibility, but may be considered in mitigation of punishment if the Tribunal determines that justice so requires.

Note the innovative character of these provisions. The Tribunal of four judges (one from each of the major Allied Powers) was international in formation and composition and, although restricted to the four victorious powers creating The Tribunal, was radically different from the national military courts before which the laws of war had to that time generally been enforced. At the core of the Charter lay the concept of international crimes for which there would be 'individual responsibility,' a sharp departure from the then-existing customary law or conventions which gave prominence to the duties of (and sometimes to sanctions against) nations. Moreover, in defining crimes within the Tribunal's jurisdiction, the Charter went beyond the traditional 'war crimes' (paragraph (b) of Article 6) in two ways.

First, the Charter included the war-related 'crimes against peace'—so-called *jus ad bellum*, in contrast with the category of war crimes or *jus in bello*. International law had for a long time been innocent of such a concept. After a slow departure during the post-Reformation period from earlier distinctions of philosophers, theologians, and writers on international law between 'just' and 'unjust' wars, the European nations moved towards a conception of war as an instrument of national policy, much like any other, to be legally regulated only as to the manner of its conduct. The Covenant of the League of Nations did not frontally challenge this principle, although it attempted to control aggression through collective decisions of the League. The interwar period witnessed some fortification of the principles later articulated in the Nuremberg Charter, primarily through the Kellogg–Briand pact of 1927 referred to in the Judgment. Today the United Nations Charter requires members (Article 2(4)) to 'refrain in their international relations from the threat or use of force' against other states, while providing (Article 51) that nothing shall impair 'the inherent right of individual or collective self-defense if an armed attack occurs against a Member . . .' When viewed in conjunction with the Nuremberg Charter, those provisions suggest the contemporary effort to distinguish not between 'just' and 'unjust' wars but between 'self-defense' and 'aggression'—the word used in defining 'crimes against peace' in Article 6(a) of that Charter.

Second, Article 6(c) represented an important innovation. There were few precedents for use of the phrase 'crimes against humanity' as part of a description of international law, and its content was correspondingly indeterminate. On its face, paragraph (c) might have been read to include the entire program of the Nazi government to exterminate Jews and other civilian groups, in and outside Germany, whether 'before or during the war,' and thus to include the planning for and early persecution of Jews and other groups preceding the Holocaust as well as the Holocaust itself. Moreover, that paragraph included the persecution or annihilation by Germany of Jews who were German nationals as well as those who were aliens. The advance on the international law of state responsibility to aliens as described in the materials on the *Chattin* case, p. 75, *supra* is evident. Note how the Judgment of the Tribunal interpreted paragraph (c) with respect to these observations.

In other respects as well, the concept of 'crimes against humanity,' even in this early formulation, differed from earlier international law. War crimes were directed to combatants; crimes against humanity could be committed by civilians as well. War crimes could cover discrete as well as systematic action by a combatant—an isolated murder of a civilian by a combatant as well a systematic policy of wanton desruction of towns. Crimes against humanity were directed primarily to planned conduct, to systematic conduct, to massive destruction.

In defining the charges against the major Nazi leaders tried at Nuremberg and its successor tribunals, the Allied powers took care to exclude those types of conduct which had not been understood to violate existing custom or conventions and in which they themselves had engaged—for example, the massive bombing of cities with predictably high tolls of civilians.

JUDGMENT OF NUREMBERG TRIBUNAL

International Military Tribunal (Nuremberg), 1946.
41 Am. J. Int. L. 172 (1947).

. . .

The Law of the Charter

The jurisdiction of the Tribunal is defined in the [London] Agreement and Charter, and the crimes coming within the jurisdiction of the Tribunal, for which there shall be individual responsibility, are set out in Article 6. The law of the Charter is decisive, and binding upon the Tribunal.

The making of the Charter was the exercise of the sovereign legislative power by the countries to which the German Reich unconditionally surrendered; and the undoubted right of these countries to legislate for the occupied territories has been recognized by the civilized world. The Charter is not an arbitrary exercise of power on the part of the victorious Nations, but in the view of the Tribunal, as will be shown, it is the expression of international law existing at the time of its creation; and to that extent is itself a contribution to international law.

. . . With regard to the constitution of the Court, all that the defendants are entitled to ask is to receive a fair trial on the facts and law.

The Charter makes the planning or waging of a war of aggression or a war in violation of international treaties a crime; and it is therefore not strictly necessary to consider whether and to what extent aggressive war was a crime before the execution of the London Agreement. But in view of the great importance of the questions of law involved, the Tribunal has heard full argument from the Prosecution and the Defense, and will express its view on the matter.

It was urged on behalf of the defendants that a fundamental principle of all law—international and domestic—is that there can be no punishment of crime without a pre-existing law. '*Nullum crimen sine lege, nulla poena sine lege.*' It was submitted that *ex post facto* punishment is abhorrent to the law of all civilized nations, that no sovereign power had made aggressive war a crime at the time that the alleged criminal acts were committed, that no statute had defined aggressive war, that no penalty had been fixed for its commission, and no court had been created to try and punish offenders.

In the first place, it is to be observed that the maxim *nullum crimen sine lege* is not a limitation of sovereignty, but is in general a principle of justice. To assert that it is unjust to punish those who in defiance of treaties and assurances have attacked neighboring states without warning is obviously untrue, for in such circumstances the attacker must know that he is doing wrong, and so far from it being unjust to punish him, it would be unjust if his wrong were allowed to go unpunished. . . .

This view is strongly reinforced by a consideration of the state of international law in 1939, so far as aggressive war is concerned. The General Treaty for the Renunciation of War of 27 August 1928, more generally known as the Pact of Paris or the Kellogg–Briand Pact, was binding on 63 nations, including Germany, Italy and Japan at the outbreak of war in 1939. . . .

. . . The nations who signed the Pact or adhered to it unconditionally condemned recourse to war for the future as an instrument of policy, and expressly renounced it. After the signing of the Pact, any nation resorting to war as an instrument of national policy breaks the Pact. In the opinion of the Tribunal, the solemn renunciation of war as an instrument of national policy necessarily involves the proposition that such a war is illegal in international law; and that those who plan and wage such a war, with its inevitable and terrible consequences, are committing a crime in so doing. War for the solution of international controversies undertaken as an instrument of national policy certainly includes a war of aggression, and such a war is therefore outlawed by the Pact. . . .

. . . The Hague Convention of 1907 prohibited resort to certain methods of waging war. These included the inhumane treatment of prisoners, the employment of poisoned weapons, the improper use of flags of true, and similar matters. Many of these prohibitions had been enforced long before the date of the Convention; but since 1907 they have certainly been crimes, punishable as offenses against the law of war; yet the Hague Convention nowhere designates such practices as criminal, nor is any sentence prescribed, nor any mention made of a court to try and punish offenders. For many years past, however, military tribunals have tried and punished individuals guilty of violating the rules of land warfare laid down by this Convention. In the opinion of the Tribunal, those who wage aggressive war are doing that which is equally illegal, and of much greater moment than a breach of one of the rules of the Hague Convention. . . . The law of war is to be found not only in treaties, but in the customs and practices of states which gradually obtained universal recognition, and from the general principles of justice applied by jurists and practised by military courts. This law is not static, but by continual adaptation follows the needs of a changing world. Indeed, in many cases treaties do no more than express and define for more accurate reference the principles of law already existing.

. . .

All these expressions of opinion, and others that could be cited, so solemnly made, reinforce the construction which the Tribunal placed upon the Pact of Paris, that resort to a war of aggression is not merely illegal, but is criminal. The prohibition of aggressive war demanded by the conscience of the world, finds its expression in the series of pacts and treaties to which the Tribunal has just referred.

. . .

. . . That international law imposes duties and liabilities upon individuals as well as upon States has long been recognized. . . . Crimes against international law are committed by men, not by abstract entities, and only by punishing individuals who commit such crimes can the provisions of international law be enforced.

. . .

. . . The authors of these acts cannot shelter themselves behind their official position in order to be freed from punishment in appropriate proceedings. Article 7 of the Charter expressly declares:

> The official position of Defendants, whether as heads of State, or responsible officials in Government departments, shall not be considered as freeing them from responsibility, or mitigating punishment.

On the other hand the very essence of the Charter is that individuals have international duties which transcend the national obligations of obedience imposed by the individual state. He who violates the laws of war cannot obtain immunity while acting in pursuance of the authority of the state if the state in authorizing action moves outside its competence under international law.

It was also submitted on behalf of most of these defendants that in doing what they did they were acting under the orders of Hitler, and therefore cannot be held responsible for the acts committed by them in carrying out these orders. The Charter specifically provides in Article 8:

> The fact that the Defendant acted pursuant to order of his Government or of a superior shall not free him from responsibility, but may be considered in mitigation of punishment.

The provisions of this article are in conformity with the law of all nations. That a soldier was ordered to kill or torture in violation of the international law of war has never been recognized as a defense to such acts of brutality, though, as the Charter here provides, the order may be urged in mitigation of the punishment. The true test, which is found in varying degrees in the criminal law of most nations, is not the existence of the order, but whether moral choice was in fact possible.

. . .

War Crimes and Crimes against Humanity

. . . War Crimes were committed on a vast scale, never before seen in the history of war. They were perpetrated in all the countries occupied by Germany, and on the High Seas, and were attended by every conceivable circumstance of cruelty and horror. There can be no doubt that the majority of them arose from the Nazi conception of 'total war,' with which the aggressive wars were waged. For in this conception of 'total war,' the moral ideas underlying the conventions which seek to make war more humane are no longer regarded as having force or validity. Everything is made subordinate to the overmastering dictates of war. Rules, regulations, assurances, and treaties all alike are of no moment; and so, freed from the restraining influence of international law, the aggressive war is conducted by the Nazi leaders in the most barbaric way. Accordingly, War Crimes were committed when and wherever the Führer and his close associates thought them to be advantageous. They were for the most part the result of cold and criminal calculation.

. . .

. . . Prisoners of war were ill-treated and tortured and murdered, not only in defiance of the well-established rules of international law, but in complete disregard of the elementary dictates of humanity. Civilian populations in occupied territories suffered the same fate. Whole populations were deported to Germany for the purposes of slave labor upon defense works, armament production, and similar tasks connected with the war effort. Hostages were taken in very large numbers from the civilian populations in all the occupied countries, and were shot as suited the German purposes. Public and private property was systematically plundered and pillaged in order to enlarge the resources of Germany at the expense of the rest of Europe. Cities and towns and villages were wantonly destroyed without military justification or necessity.

. . .

Murder and Ill-treatment of Civilian Population

Article 6 (b) of the Charter provides that 'ill-treatment . . . of civilian population of or in occupied territory . . . killing of hostages . . . wanton destruction of cities, towns, or villages' shall be a war crime. In the main, these provisions are merely declaratory of the existing laws of war as expressed by the Hague Convention, Article 46. . . .

. . .

One of the most notorious means of terrorizing the people in occupied territories was the use of concentration camps . . . [which] became places of organized and systematic murder, where millions of people were destroyed.

In the administration of the occupied territories the concentration camps were used to destroy all opposition groups. . . .

A certain number of the concentration camps were equipped with gas chambers for the wholesale destruction of the inmates, and with furnaces for the burning of the bodies. Some of them were in fact used for the extermination of Jews as part of the 'final solution' of the Jewish problem. . . .

. . .

Slave Labor Policy

Article 6 (b) of the Charter provides that the 'ill-treatment or deportation to slave labor or for any other purpose, of civilian population of or in occupied territory' shall be a War Crime. The laws relating to forced labor by the inhabitants of occupied territories are found in Article 52 of the Hague Convention. . . . The policy of the German occupation authorities was in flagrant violation of the terms of this convention. . . . [T]he German occupation authorities did succeed in forcing many of the inhabitants of the occupied territories to work for the German war effort, and in deporting at least 5,000,000 persons to Germany to serve German industry and agriculture.

. . .

Persecution of the Jews

The persecution of the Jews at the hands of the Nazi Government has been proved in the greatest detail before the Tribunal. It is a record of consistent and systematic inhumanity on the greatest scale. Ohlendorf, Chief of Amt III in the RSHA from 1939 to 1943, and who was in command of one of the Einsatz groups in the campaign against the Soviet Union testified as to the methods employed in the extermination of the Jews. . . .

When the witness Bach Zelewski was asked how Ohlendorf could admit the murder of 90,000 people, he replied: 'I am of the opinion that when, for years, for decades, the doctrine is preached that the Slav race is an inferior race, and Jews not even human, then such an outcome is inevitable.'

. . .

. . . The Nazi Party preached these doctrines throughout its history, *Der Stürmer* and other publications were allowed to disseminate hatred of the Jews, and in the speeches and public declarations of the Nazi leaders, the Jews were held up to public ridicule and contempt.

. . . By the autumn of 1938, the Nazi policy towards the Jews had reached the stage where it was directed towards the complete exclusion of Jews from German life. Pogroms were organized, which included the burning and demolishing of synagogues, the looting of Jewish businesses, and the arrest of prominent Jewish business men. . . .

It was contended for the Prosecution that certain aspects of this anti-Semitic policy were connected with the plans for aggressive war. The violent measures taken against the Jews in November 1938 were nominally in retaliation for the killing of an official of the German Embassy in Paris. But the decision to seize Austria and Czechoslovakia had been made a year before. The imposition of a fine of one billion marks was made, and the confiscation of the financial holdings of the Jews was decreed, at a time when German armament expenditure had put the German treasury in difficulties, and when the reduction of expenditure on armaments was being considered. . . .

It was further said that the connection of the anti-Semitic policy with aggressive war was not limited to economic matters. . . .

The Nazi persecution of Jews in Germany before the war, severe and repressive as it was, cannot compare, however, with the policy pursued during the war in the occupied territories. . . . In the summer of 1941, however, plans were made for the 'final solution' of the Jewish question in Europe. This 'final solution' meant the extermination of the Jews. . . .

The plan for exterminating the Jews was developed shortly after the attack on the Soviet Union. . . .

. . .

. . . Adolf Eichmann, who had been put in charge of this program by Hitler, has estimated that the policy pursued resulted in the killing of 6 million Jews, of which 4 million were killed in the extermination institutions.

The Law Relating to War Crimes and Crimes against Humanity

. . .

The Tribunal is of course bound by the Charter, in the definition which it gives both of War Crimes and Crimes against Humanity. With respect to War

Crimes, however, as has already been pointed out, the crimes defined by Article 6, Section (b), of the Charter were already recognized as War Crimes under international law. They were covered by Articles 46, 50, 52, and 56 of the Hague Convention of 1907, and Articles 2, 3, 4, 46, and 51 of the Geneva Convention of 1929. That violation of these provisions constituted crimes for which the guilty individuals were punishable is too well settled to admit of argument.

But it is argued that the Hague Convention does not apply in this case, because of the 'general participation' clause in Article 2 of the Hague Convention of 1907. That clause provided:

> The provisions contained in the regulations (Rules of Land Warfare) referred to in Article 1 as well as in the present Convention do not apply except between contracting powers, and then only if all the belligerents are parties to the Convention.

Several of the belligerents in the recent war were not parties to this Convention.

In the opinion of the Tribunal it is not necessary to decide this question. The rules of land warfare expressed in the Convention undoubtedly represented an advance over existing international law at the time of their adoption. But the convention expressly stated that it was an attempt 'to revise the general laws and customs of war,' which it thus recognized to be then existing, but by 1939 these rules laid down in the Convention were recognized by all civilized nations, and were regarded as being declaratory of the laws and customs of war which are referred to in Article 6 (b) of the Charter.

. . .

With regard to Crimes against Humanity there is no doubt whatever that political opponents were murdered in Germany before the war, and that many of them were kept in concentration camps in circumstances of great horror and cruelty. The policy of terror was certainly carried out on a vast scale, and in many cases was organized and systematic. The policy of persecution, repression, and murder of civilians in Germany before the war of 1939, who were likely to be hostile to the Government, was most ruthlessly carried out. The persecution of Jews during the same period is established beyond all doubt. To constitute Crimes against Humanity, the acts relied on before the outbreak of war must have been in execution of, or in connection with, any crime within the jurisdiction of the Tribunal. The Tribunal is of the opinion that revolting and horrible as many of these crimes were, it has not been satisfactorily proved that they were done in execution of, or in connection with, any such crime. The Tribunal therefore cannot make a general declaration that the acts before 1939 were Crimes against Humanity within the meaning of the Charter, but from the beginning of the war in 1939 War Crimes were committed on a vast scale, which were also Crimes against Humanity; and insofar as the inhumane acts charged in the Indictment, and committed after the beginning of the war, did not constitute War Crimes, they were all committed in execution of, or in connection with, the aggressive war, and therefore constituted Crimes against Humanity.

[The opinion considered individually each of the 22 defendants at this first trial of alleged war criminals. It found 19 of the defendants guilty of one or more counts of the indictment. It imposed 12 death sentences. Most convictions were for War Crimes and Crimes against Humanity, the majority of those convicted being found guilty of both crimes.]

NOTE

Note the following statement in Ian Brownlie, *Principles of Public International Law* (4th ed. 1990), at 562:

> But whatever the state of the law in 1945, Article 6 of the Nuremberg Charter has since come to represent general international law. The Agreement to which the Charter was annexed was signed by the United States, United Kingdom, France, and USSR, and nineteen other state subsequently adhered to it. In a resolution adopted unanimously on 11 December 1946, the General Assembly affirmed 'the principles of international law recognized by the Charter of the Nuremberg Tribunal and the judgment of the Tribunal'.

There has been considerable expansion in the definitions of two of the crimes defined in Article 6. The field of individual criminal responsibility for war crimes has been both expanded and clarified, through provisions of the Geneva Conventions of 1949 and later instruments. The concept of crimes against humanity has both expanded in coverage and shed some limitations placed on it by the Judgment of the Tribunal. Such developments are described in the materials dealing with the current International Criminal Tribunal for the Former Yugoslavia (and later, for Rwanda as well) in Chapter 15. The notion of 'crimes against peace,' however, has fallen into relative disuse.

Compare with the Nuremberg Judgment the following provisions of the Convention on the Prevention and Punishment of the Crime of Genocide (116 parties as of September 1995) bearing on personal responsibility. The treaty parties 'confirm' in Art. 1 that genocide 'is a crime under international law which they undertake to prevent and to punish.' Persons committing acts of genocide (as defined) 'shall be punished, whether they are constitutionally responsible rulers, public officials or private individuals.' (Art. IV). The parties agree (Art. V) to enact the necessary legislation to give effect to the Convention and 'to provide effective penalties for persons guilty of genocide.' Under Art. VI, persons charged with genocide are to be tried by a tribunal 'of the State in the territory of which the act was committed, or by such international penal tribunal as may have jurisdiction with respect to those

Contracting Parties which shall have accepted its jurisdiction.' No international penal tribunal of general jurisdiction has been created.

VIEWS OF COMMENTATORS

There follow a number of authors' observations about the Judgment and the principles underlying the Nuremberg trials.

(1) In a review of a book by Sheldon Glueck entitled *The Nuremberg Trial and Aggressive War* (1946), the reviewer George Finch, 47 Am. J. Int. L. 334 (1947), makes the following arguments:

> As the title indicates, this book deals with the charges at Nuremberg based upon the planning and waging of aggressive war. The author has written it because in his previous volume he expressed the view that he did not think such acts could be regarded as 'international crimes.' He has now changed his mind and believes 'that for the purpose of conceiving aggressive war to be an international crime, the Pact of Paris may, together with other treaties and resolutions, be regarded as evidence of a sufficiently developed *custom* to be accepted as international law' (pp. 4–5). . . .
>
> The reviewer fully agrees with the author in regard to the place of custom in the development of international law. He regards as untenable, however, the argument not only of the author but of the prosecutors and judges at Nuremberg that custom can be judicially established by placing interpretations upon the words of treaties which are refuted by the acts of the signatories in practice, by citing unratified protocols or public and private resolutions of no legal effect, and by ignoring flagrant and repeated violations of non-aggression pacts by one of the prosecuting governments which, if properly weighed in the evidence, would nullify any judicial holding that a custom outlawing aggressive war had been accepted in international law. . . .

(2) In his article *The Nurnberg Trial*, 33 Va. L. Rev. 679 (1947), at 694, Francis Biddle, the American judge on the Tribunal, commented on the definition of 'crimes against humanity' in Article 6(c) of the Charter:

> . . . The authors of the Charter evidently realized that the crimes enumerated were essentially domestic and hardly subject to the incidence of international law, unless partaking of the nature of war crimes. Their purpose was evidently to reach the terrible persecution of the Jews and liberals within Germany before the war. But the Tribunal held that 'revolting and horrible as many of these crimes were,' it had not been established that they were done 'in execution of, or in connection with' any crime within its jurisdiction. After the beginning of the war, however, these inhumane acts were held to have been committed in execution of the war, and were therefore crimes against humanity.
>
> . . .

> Crimes against humanity constitute a somewhat nebulous conception, although the expression is not unknown to the language of international law. . . . With one possible exception . . . crimes against humanity were held [in the Judgment of the Tribunal] to have been committed only where the proof also fully established the commission of war crimes. Mr. Stimson suggested [that the Tribunal eliminate from its jurisdiction matters related to pre-war persecution in Germany], which involved 'a reduction of the meaning of crimes against humanity to a point where they became practically synonymous with war crimes.' I agree. And I believe that this inelastic construction is justified by the language of the Charter and by the consideration that such a rigid interpretation is highly desirable in this stage of the development of international law.

(3) Professor Hans Kelsen, in *Will the Judgment in the Nuremberg Trial Constitute a Precedent in International Law?*, 1 Int. L. Q. 153 (1947) at 164, was critical of several aspects of the London Agreement and the Judgment. But with respect to the question of retroactivity of criminal punishment, he wrote:

> The objection most frequently put forward—although not the weightiest one—is that the law applied by the judgment of Nuremberg is an ex post facto law. There can be little doubt that the London Agreement provides individual punishment for acts which, at the time they were performed were not punishable, either under international law or under any national law. . . . However, this rule [against retroactive legislation] is not valid at all within international law, and is valid within national law only with important exceptions. [Kelsen notes several exceptions, including the rule's irrelevance to 'customary law and to law created by a precedent, for such law is necessarily retroactive in respect to the first case to which it is applied. . . .']
>
> A retroactive law providing individual punishment for acts which were illegal though not criminal at the time they were committed, seems also to be an exception to the rule against ex post facto laws. The London Agreement is such a law. It is retroactive only in so far as it established individual criminal responsibility for acts which at the time they were committed constituted violations of existing international law, but for which this law has provided only collective responsibility. . . . Since the internationally illegal acts for which the London Agreement established individual criminal responsibility were certainly also morally most objectionable, and the persons who committed these acts were certainly aware of their immoral character, the retroactivity of the law applied to them can hardly be considered as absolutely incompatible with justice.

(4) In his biography entitled *Harlan Fiske Stone: Pillar of the Law* (1956), Alpheus Thomas Mason discussed Chief Justice Stone's views about the involvement of Justices of the U.S. Supreme Court in extrajudicial assignments and, in particular, Stone's views about President Truman's appointment of Justice Robert Jackson to be American Prosecutor at the trials. The following excerpts (at p. 715) are all incorporations by Mason in his book of quotations of Chief Justice Stone's remarks.

So far as the Nuremberg trial is an attempt to justify the application of the power of the victor to the vanquished because the vanquished made aggressive war, . . . I dislike extremely to see it dressed up with a false facade of legality. The best that can be said for it is that it is a political act of the victorious States which may be morally right . . . . It would not disturb me greatly . . . if that power were openly and frankly used to punish the German leaders for being a bad lot, but it disturbs me some to have it dressed up in the habiliments of the common law and the Constitutional safeguards to those charged with crime.

Jackson is away conducting his high-grade lynching party in Nuremberg . . . . I don't mind what he does to the Nazis, but I hate to see the pretense that he is running a court and proceeding according to common law. This is a little too sanctimonious a fraud to meet my old-fashioned ideas.

(5) Professor Herbert Wechsler, in *The Issues of the Nuremberg Trial*, 62 Pol. Sci. Q. 11 (1947), at 23, observed:

. . . [M]ost of those who mount the attack [on the Judgment on contentions including *ex post facto* law] hasten to assure us that their plea is not one of immunity for the defendants; they argue only that they should have been disposed of politically, that is, dispatched out of hand. This is a curious position indeed. A punitive enterprise launched on the basis of general rules, administered in an adversary proceeding under a separation of prosecutive and adjudicative powers is, in the name of law and justice, asserted to be less desirable than an *ex parte* execution list or a drumhead court-martial constituted in the immediate aftermath of the war. . . . Those who choose to do so may view the Nuremberg proceeding as 'political' rather than 'legal'—a program calling for the judicial application of principles of liability politically defined. They cannot view it as less civilized an institution than a program of organized violence against prisoners, whether directed from the respective capitals or by military commanders in the field.

QUESTIONS

1. Recall clause (c) of Article 38(1) of the Statute of the I.C.J., p. 27, *supra*. Could the Tribunal have relied on that clause to respond to charges of *ex post facto* application of Article 6(c) to individuals who were responsible for the murder of groups of Germans or aliens?

2. Do you agree with the Tribunal's restrictive interpretation of Article 6(c)? Consider the commentary of Francis Biddle.

3. How do you react to the criticism by Finch of the Tribunal's use of treaties in deciding whether customary international law included a given norm? Recall the comments about the growth of customary law by Schachter and Koskenniemi.

ADDITIONAL READING

In addition to the sources cited in the text, see three books of Telford Taylor: *Nuremberg Trials: War Crimes and International Law* (1949); *Nuremberg and Vietnam: An American Tragedy* (1978); and *The Anatomy of the Nuremberg Trials: A Personal Memoir* (1992). See also Memorandum Submitted by the Secretary-General, The Charter and Judgment of the Nürnberg Tribunal: History and Analysis, U.N. Doc. A/CN.4/5 (1949); and Egon Schwelb, *Crimes against Humanity*, 23 Brit. Ybk. Int. L. 178 (1946).

NOTE

The following comments of Louis Henkin review some themes developed in this chapter and describe other antecedents in international law to contemporary human rights.

LOUIS HENKIN, INTERNATIONAL LAW: POLITICS, VALUES AND FUNCTIONS

216 Collected Courses of Hague Academy of International Law 13 (Vol. IV, 1989), at 208.

CHAPTER X
STATE VALUES AND OTHER VALUES: HUMAN RIGHTS

. . .

That until recently international law took no note of individual human beings may be surprising. Both international law and domestic legal norms in the Christian world had roots in an accepted morality and in natural law, and had common intellectual progenitors (including Grotius, Locke, Vattel). But for hundreds of years international law and the law governing individual life did not come together. International law, true to its name, was law only between States, governing only relations between States on the State level. What a State did inside its borders in relation to its own nationals remained its own affair, an element of its autonomy, a matter of its 'domestic jurisdiction'.

Antecedents of the International Law of Human Rights

In fact, neither the international political system nor international law ever closed out totally what went on inside a State and what happened to individuals within a State. Early, international law began to attend to internal matters that held special interest for other States, and those sometimes included concern for individual human beings, or at least redounded to the benefit of individual human beings. But what was in fact of interest to other States, and what was accepted as being of legitimate interest to other States (and therefore to the system and to law), were limited *a priori* by the character of the State system and its values. Of course, every State was legitimately concerned with what happened to its diplomats, to its diplomatic mission and to its property in the territory of another State. States were concerned, and the system developed norms to assure, that their nationals (and the property of their nationals) in the territory of another State be treated reasonably, 'fairly', and the system and the law early identified an international standard of justice by which a State must abide in its treatment of foreign nationals. States also entered into agreements, usually on a reciprocal basis, promising protection or privilege—freedom to reside, to conduct business, to worship—to persons with whom the other State party to the treaty identified because of common religion or ethnicity.

Concern for individual human welfare seeped into the international system in the eighteenth and nineteenth centuries in other discrete, specific respects. In the nineteenth century, European (and American) States abolished slavery and slave trade. Later, States began to pursue agreements to make war less inhumane, to outlaw some cruel weapons, to safeguard prisoners of war, the wounded, civilian populations. It is noteworthy that, in these instances, even less-than-democratic States began to attend to human values, though humanitarian limitations on the conduct of war may have brought significant cost to the State's military interests.

Following the First World War, concern for individual human beings was reflected in several League of Nations programmes. Building on earlier precedents in the nineteenth century, the dominant States pressed selected other States to adhere to 'minorities treaties' guaranteed by the League, in which States Parties assumed obligations to respect rights of identified ethnic, national or religious minorities among their inhabitants. . . . The years following the First World War also saw a major development in international concern for individual welfare, a development that is often overlooked and commonly underestimated: the International Labour Office (now the International Labour Organisation (ILO) was established and it launched a variety of programmes including a series of conventions setting minimum standards for working conditions and related matters.

In general, the principles of customary international law that developed, and the special agreements that were concluded, addressed only what happened to *some* people inside a State, only in respects with which other States were in fact concerned, and only where such concern was considered their proper business in a system of autonomous States. One can only speculate as to why States accepted these norms and agreements, but it may be reasonable to doubt whether those developments authentically reflected sensitivity to human rights generally. States attended to what occurred inside another State when such happenings impinged on their political-economic interests. States were concerned, and were deemed legitimately concerned, for the freedoms, privileges, and immunities of their diplomats because an affront to the diplomat affronted his prince (or his State), and because interference with a diplomat interfered with his functions and disturbed orderly, friendly relations. Injury to a foreign national or to his or her property was also an affront to the State of his or her nationality, and powerful States exporting people, goods, and capital to other countries in the age of growing mercantilism insisted on law that would protect the State interests that these represented.

. . .

Humanitarian developments in the law of war reflected some concern by States to reduce the horrors of war for their own people and a willingness in exchange to reduce them for others. Powerful States promoted minorities treaties because mistreatment of minorities with which other States identified threatened international peace. Those treaties were imposed selectively, principally on nations defeated in war and on newly created or enlarged States; they did not establish general norms requiring respect for minorities by the big and the powerful as well; they did not require respect for individuals who were not members of identified minorities, or for members of the majority. . . .

Even the ILO conventions, perhaps, served some less-than-altruistic purposes. Improvement in the conditions of labour was capitalism's defence against the spectre of spreading socialism which had just established itself in the largest country in Europe. States, moreover, had a direct interest in the conditions of labour in countries with which they competed in a common international market: a State impelled to improve labour and social conditions at home could not readily do so unless other States did so, lest the increase in its costs of production render its products non-competitive.

I have stressed the possibly political-economic (rather than humanitarian) motivations for early norms and agreements, identifying a State's concern for the welfare of some of its nationals as an extension of its Statehood and perhaps reflecting principally concern for State interests and values. If some norms and agreements in fact were motivated by concern for a State's own people generally, they did not reflect interest in the welfare of those in other countries, or of human beings generally. State interests rather than individual human interests, or at best the interests of a State's own people rather than general human concerns, also inspired voluntary inter-State co-operation to promote reciprocal economic interests. Occasional assistance to other States, even if for the benefit of the people of that State (as in flood or earthquake relief), was voluntary, out of friendship and generosity (rather than legal obligation), and was provided to the receiving State and penetrated its society only lightly and only with its consent.

I would not underestimate the influence of ideas of rights and constitutionalism in the seventeenth and eighteenth centuries, and of a growing and spreading enlightenment generally: Locke, Montesquieu, other Encyclopedists, Rousseau; the example of the Glorious Revolution in England and the establishment of constitutionalism in the United States; the influence of the

French Declaration of the Rights of Man and of the Citizen. Such ideas and examples have influenced developments inside countries, but they did not easily enter the international political and legal system. Concern by one country for the welfare of individual human beings inside another country met many obstacles, not least the conception and implications of Statehood in a State system. The human condition in other countries and the treatment of individuals by other Governments were not commonly known abroad since they were not included in the information sources of the time. Information (and concern) were filtered through the State system and through diplomatic sources, and human values as such were not the business of diplomacy. For those reasons, and for other reasons flowing from the State system, other States took little note and expressed little concern for what a Government did to its own citizens. In general, the veil of Statehood was impermeable. If occasionally something particularly horrendous happened—a massacre, a pogrom—and was communicated and made known by the available media of communication, it evoked from other States more-or-less polite diplomatic expressions of regret, not on grounds of law but of *noblesse oblige* or of common princely morality wrapped in Christian charity (whose violation gave princes and Christianity a bad name).

Even if the implications of Statehood had not been an obstacle, as regards any but the grossest violations of what we now call human rights, few if any States had moral sensitivity and moral standing to intercede. When a State invoked an international standard of justice on behalf of one of its nationals abroad, it may have been invoking a standard unknown and unheeded at home. Few States had constitutional protections and not many had effective legislative or common-law protections for individual rights. Torture and police brutality, denials of due process, arbitrary detention, perversions of law, were not wildly abnormal. Surely, few States recognized political freedoms—freedom of speech, association and assembly, universal suffrage. Many States denied religious freedom to some, and few States granted complete religious toleration; full equality to members of other than the dominant religion was slow in coming anywhere. Women were subject to rampant and deep-rooted inequalities and domination, often to abuse and oppression. Even today such violations are not the stuff of dramatic television programmes and do not arouse international revulsion and reaction; in earlier times, surely, violations of what are today recognized as civil and political rights caused little stir outside the country. A State's failure to provide for the economic and social welfare of its inhabitants was wholly beyond the ken of other States. There were no alert media of information and few civil rights or other non-governmental organizations to sensitize and activate people and Governments.

3

Evolution and Norms of the Universal Human Rights System

Chapter 3 introduces basic instruments of the human rights movement. Those instruments are at the core of the *universal* human rights system—universal in that it is based on treaties that aim at worldwide, universal membership. The chapter then examines the continuing significance for human rights of customary international law, extending the examination of that topic that Chapter 2 began. Its last section explores the effect of the human rights movement on the related conceptions of state sovereignty and domestic jurisdiction.

The core of the universal system consists of the United Nations Charter and related instruments. Three such instruments, composing the so-called International Bill of Rights, stand out in significance: the Universal Declaration of Human Rights of 1948, and two principal covenants that became effective in 1976: the International Covenant on Civil and Political Rights (ICCPR), and the International Covenant on Economic, Social and Cultural Rights (ICESCR). (Chapter 5 examines this second covenant.) As of September 1995, the ICCPR had 131 states parties, the ICESCR 132 parties.

Chapter 10 explores three *regional* human rights systems—regional in that they are based on treaties whose membership is restricted to states in a particular region: the European Convention for the Protection of Human Rights and Fundamental Freedoms (known as the European Convention on Human Rights) (30 states parties as of September 1995), the American Convention on Human Rights (25 parties), and the African Charter on Human and Peoples' Rights (49 parties).

A. FOUNDATIONS OF THE SYSTEM

COMMENT ON THE UNITED NATIONS CHARTER AND THE ORIGINS OF THE HUMAN RIGHTS MOVEMENT

The human rights movement is not simply a matter of norms—rules, standards, principles. To the contrary, those norms are imbedded in institutions, some of them state and some international. In particular, it is impossible to grasp this movement adequately without an appreciation of its close relation to and reliance on international organizations. Both the universal and regional human rights systems have vital links to such organizations. For example, the basic instruments of the universal system were drafted within the different organs of the United Nations and adopted by its General Assembly, before (in the case of treaties) being submitted to states for ratification.

Moreover, the United Nations Charter itself first gave formal and authoritative expression to the human rights movement that began at the end of World War II. Since its birth in 1945, the UN has served as a vital institutional spur to the development of the movement. The purpose of the present comments is to call attention to aspects of the UN and its Charter that bear on the following examination of basic rights instruments. You should become familiar now with the provisions (in the Documents Annex) of the Charter and the ICCPR that are referred to below. Other Charter provisions figure in later chapters.

Charter Provisions

Consider first the Charter's radical transformation of the branch of the laws of war concerning *jus ad bellum* (p. 101, *supra*). Recall that for several centuries that body of law had addressed almost exclusively *jus in bello*, the rules regulating the conduct of warfare rather than the justice or legality of the waging of war. The International Military Tribunal at Nuremberg was empowered to adjudicate 'crimes against peace,' part of *jus ad bellum* and the most disputed element of that Tribunal's mandate.

The Charter builds on the precedents to which the Nuremberg Judgment (p. 102, *supra*) refers and states the UN's basic purpose of securing and maintaining peace. It does so by providing in Article 2(4) that UN members 'shall refrain in their international relations from the threat or use of force against the territorial integrity or political independence of any state,' a rule qualified by Article 51's provision that nothing in the Charter 'shall impair the inherent right of individual or collective self-defence if an armed attack occurs' against a member.

The Charter has little to say directly about human rights. Its references to human rights are scattered, terse, even cryptic. The term 'human rights' appears infrequently, although in vital contexts. Note its occurrence in the following provisions: second paragraph of the Preamble, Article 1(3), Article 13(1)(b), Articles 55 and 56, Article 62(2) and Article 68.

Several striking characteristics of these provisions emerge. Many have a promotional or programmatic character, for they refer principally to the purposes or goals of the UN or to the competences of different UN organs: 'encouraging respect for human rights,' 'assisting in the realization of human rights,' 'promote . . . universal respect for, and observance of, human rights.' Not even a provision such as Article 56, which refers to undertakings of the member states rather than of the UN, contains clear language of obligation. It notes only that states 'pledge themselves' to action 'for the achievement' of purposes including the promotion of observance of human rights. Note also the prominence in these provisions of the notion of equal protection.

The UN and the Universal Declaration

Despite proposals to the contrary, the Charter stopped shy of incorporating a bill of rights. Instead, there were proposals for developing one through the work of a special commission that would give separate attention to the issue. That commission was contemplated by Charter Article 68, which provides that one of the UN organs, the Economic and Social Council (ECOSOC), 'shall set up commissions in economic and social fields and for the promotion of human rights . . . .' In 1946, ECOSOC established the Commission on Human Rights (sometimes referred to in this book as the UN Commission), which has evolved over the decades to become the world's single most important human rights organ. Chapter 7 examines its work. At this earlier time, the new Commission was charged primarily with submitting reports and proposals on an international bill of rights.

The UN Commission first met in its present form early in 1947, its members (representatives of the state members of the Commission) including such distinguished founders of the human rights movement as René Cassin of France, Charles Malik of the Lebanon, and Eleanor Roosevelt of the United States. Some representatives urged that the draft bill of rights under preparation should take the form of a declaration—that is, a recommendation by the General Assembly to member states (see Charter Article 13) that would exert a moral and political influence on states rather than constitute a legally binding instrument. Other representatives urged the Commission to prepare a draft convention containing a bill of rights that would, after adoption by the General Assembly, be submitted to states for their ratification.

The first path was followed. In 1948, the UN Commission adopted a draft Declaration, which in turn was adopted by the General Assembly that year as the Universal Declaration of Human Rights (UDHR), with 48 states voting in favor and eight abstaining—Saudi Arabia, South Africa, and the Soviet Union together with four East European states and a Soviet republic whose votes it controlled. (It is something of a jolt to realize today, in a decolonized and fragmented world of 185 member states, that UN membership in 1948 stood at 56 states.)

The Universal Declaration was meant to precede more detailed and comprehensive provisions in a single convention that would be approved by the General Assembly and submitted to states for ratification. After all, within the prevailing concepts of human rights at that time, the UDHR seemed to cover most of the field, including economic and social rights (see Articles 22–26) as well as civil and political rights. But during the years of drafting—years in which the Cold War took harsher and more rigid form, and in which the United States strongly qualified the nature of its commitment to the universal human rights movement—these matters became more contentious. The human rights movement was buffeted by ideological conflict and the formal differences of approach in a polarized world. One consequence was the decision in 1952 to build on the UDHR by dividing its provisions between two treaties, one on civil and political rights, the other on economic, social and cultural rights.

The plan to use the Universal Declaration as a springboard to treaties triumphed, but not as quickly as anticipated. The two principal treaties—the International Covenant on Civil and Political Rights (ICCPR) and the International Covenant on Economic, Social and Cultural Rights (ICESCR)—made their ways through the drafting and amendment processes in the Commission, the Third Committee and the General Assembly, where they were approved only in 1966. Another decade passed before the two Covenants achieved the number of ratifications necessary to enter into force.

During the 28 years between 1948 and 1976, specialized human rights treaties such as the Genocide Convention entered into force. But not until the two principal Covenants became effective did treaties achieve as broad coverage of human rights topics as the Universal Declaration. It was partly for this reason that the UDHR became so broadly known and frequently invoked. During these intervening years, it was the only broad-based human rights instrument available. To this day, it retains its symbolism, rhetorical force and significance in the human rights movement. It is the parent document, the initial burst of idealism and enthusiasm, terser, more general and grander than the treaties, in some sense the constitution of the entire movement. It remains the single most invoked human rights instrument.

Other UN Organs Related to Human Rights

Together with the UN Commission, other UN organs have played major roles in developing universal human rights. Their full significance becomes apparent through readings in this chapter and through the topics of later chapters. A brief description follows.

Chapter IV of the Charter sets forth the composition and powers of the General Assembly. Those powers are described in Articles 10–14 in terms such as 'initiate studies,' 'recommend,' 'promote,' 'encourage' and 'discuss.'

Particularly relevant are Articles 10 and 13. Article 10 authorizes the General Assembly to 'discuss any questions or any matters within the scope of the present Charter [and] . . . make recommendations to the Members of the United Nations . . . on any such questions or matters.' Article 13 authorizes the General Assembly to 'make recommendations' for the purpose of, *inter alia*, 'assisting in the realization of human rights.'

Contrast the stronger and more closely defined powers of the Security Council under Chapter VII. Those powers range from making recommendations to states parties about ending a dispute, to the power to take military action 'to maintain or restore international peace and security' (Article 42) after the Council 'determines the existence of any threat to the peace, breach of the peace, or act of aggression' (Article 39). Under Article 25, member states 'agree to accept and carry out' the Security Council's decisions on these and other matters. No such formal obligation of states attaches to recommendations or resolutions of the General Assembly.

Two of the six Main Committees of the General Assembly—committees of the whole, for all UN members are entitled to be represented on them—have also participated in the drafting or other processes affecting human rights. The Social, Humanitarian and Cultural Committee (Third Committee) and the Legal Committee (Sixth Committee) have reviewed drafts of proposed declarations or conventions and often added their comments to the document submitted to the plenary General Assembly for its ultimate approval.

Historical Sequence and Typology of Instruments

That part of the universal human rights movement consisting of intergovernmental instruments—that is, excluding for present purposes both national laws and non-governmental institutions forming part of the movement—can be imagined as a four-tiered normative edifice, the tiers described generally in the order of their chronological appearance.

(1) We have seen that the UN Charter, at the pinnacle of the human rights system, has relatively little to say about the subject. But what it does say has been accorded great significance. Through interpretation and extrapolation, the sparse text has constituted a point of departure for inventive development of the entire movement.

(2) The Universal Declaration of Human Rights, viewed by some as a further elaboration of the brief references to human rights in the Charter, occupies in some ways the primary position of constitution of the entire movement.

(3) The two principal covenants, which alone among the universal treaties have broad coverage of human rights topics, develop in more detail the basic categories of rights—civil and political; and economic, social and cultural—that figure in the Universal Declaration, and include additional rights as well. These covenants together with the UDHR form what is generally referred to as the International Bill of Rights.

(4) A host of multilateral human rights treaties (usually termed 'conventions,' for there are only the two basic 'covenants'), as well as resolutions or declarations with a more limited or focused subject than the comprehensive International Bill of Rights, have grown out of the United Nations (drafting by UN organs, approval by the General Assembly) and (in the case of treaties) have been ratified by large numbers of states. They develop further the content of rights that are more tersely described in the two covenants or, in some cases, that escape mention in them. This fourth tier consists of a network of treaties, most but not all of which became effective after the two Covenants, including: the Convention on the Prevention and Punishment of the Crime of Genocide, adopted 1948 (116 states parties as of September 1995), the International Convention on the Elimination of all Forms of Racial Discrimination, 1965 (143 parties), the Convention on the Elimination of all Forms of Discrimination against Women, 1979 (144 parties), the Convention against Torture and other Cruel, Inhuman or Degrading Treatment or Punishment, 1984 (90 parties), and the Convention on the Rights of the Child, 1989 (176 parties). Later chapters examine several of these treaty regimes.

NOTE

Consider the following observations in Louis Henkin, *International Law: Politics, Values and Functions*, 216 Collected Courses of the Hague Academy of International Law (Vol. IV, 1989), at 215.

> The United Nations Charter, a vehicle of radical political-legal change in several respects, did not claim authority for the new human rights commitment it projected other than in the present consent of States. Unlike the international standard of justice for foreign nationals, which derived from the age of natural law and clearly reflected common acceptance of some natural rights, the Charter is a 'positivist' instrument. It does not invoke natural rights or any other philosophical basis for human rights. (The principal Powers could not have agreed on any such basis.) The Charter Preamble links human rights with human dignity but treats that value as self-evident, without need for justification. Nor does the Charter define either term or give other guidance as to the human rights that human dignity requires. In fact, to help justify the radical penetration of the State monolith, the Charter in effect justifies human rights as a State value by linking it to peace and security.
>
> Perhaps because we now wish to, we tend to exaggerate what the Charter did for human rights. The Charter made the promotion of human rights a purpose of the United Nations; perhaps without full appreciation of the extent of the penetration of Statehood that was involved, it thereby recognized and established that relations between a State and its own inhabitants were a matter of international concern. But the Charter did not erode State autonomy and the requirement of State consent to new human rights law. . . .

> . . . Surely, the Charter did not provide, clearly and explicitly, that every State party to the Charter assumes legal obligations not to violate the human rights, or some human rights, of persons subject to its jurisdiction.
>
> In 1945, the principal Powers were not prepared to derogate from the established character of the international system by establishing law and legal obligation that would penetrate Statehood in that radical way; clearly, they themselves were not ready to submit to such law. Small Powers and non-governmental organizations indeed proposed the addition to the Charter of an international bill of rights, but it was not done. . . .
>
> . . .

QUESTIONS

Compare the Charter's human rights provisions with those of the UDHR. How many of the concrete rights declared in the UDHR appear in the Charter? Note how frequently one such right, equality of worth and treatment, appears in the Charter. Why would you imagine that it alone was given such prominence?

COMMENT ON RELATIONSHIPS BETWEEN THE UNIVERSAL DECLARATION AND THE ICCPR

You should now read the Universal Declaration and the substantive part (Articles 1–27) of the International Covenant on Civil and Political Rights. The comparisons below between the UDHR and the ICCPR assume a familiarity with these provisions.

(1) Under international law, approval by the General Assembly of a declaration like the UDHR has different consequences from a treaty that has become effective through the required number of ratifications. Of course the declaration will have solemn effects as the formal act of a deliberative body of global importance. Its subject matter, like that of the UDHR, may be of the greatest significance. But when approved or adopted, it is hortatory and aspirational, recommendatory rather than, in a formal sense, binding.

The Covenant, on the other hand, binds the states parties in accordance with its terms, subject to such formal matters as reservations (see p. 34, *supra*). Of course this statement of international law doctrine and its premise, *pacta sunt servanda*, do not end discussion. Disputes over interpretation may likely arise; some states will disagree with others as to what even basic provisions of the Covenant mean and require. What indeed is the 'commitment'? In absence of a consensus, which state party or which international institution can provide an authoritative answer, decision, interpretation? Even if there is a widespread consensus about meaning, we must confront the question of whether states will honor this 'binding' commitment and, if not, whether the UN or some member states will apply pressure against violators sufficient to persuade them to comply. Does or should the probability of enforcement against violators have any bearing on the legally binding character of an international agreement? Such are the questions underlying much of the discussion in Part C of this book.

We must take account of these and other qualifications to the apparently clear contrast between 'binding' and 'hortatory' instruments. In the case of the UDHR, the years have further blurred that contrast as arguments have developed for viewing all or part of the Declaration as legally binding, either as a matter of customary international law or as an authoritative interpretation of the UN Charter. See Section B of this chapter.

(2) A resolution of the General Assembly with the formal status of a recommendation can hardly create an international institution or an organ particular to the resolution. A treaty can. The Covenant creates an ongoing institution, a treaty organ: the Human Rights Committee. That organ gives institutional support to the Covenant's norms, for the Covenant imposes on states parties formal obligations (such as the submission of periodic reports) to the Committee. This Committee (referred to in this book as the ICCPR Committee), examined in Chapter 9, *infra*, is charged with the performance of the tasks defined both in the Covenant and in its Optional Protocol effective in 1976 (79 states parties as of January 1995).

(3) Both the UDHR and ICCPR are terse about their derivations or foundations in moral and political thought. Such statements as are made that have the character of justifications appear in the preambles. But clearly these instruments differ radically from, say, a tax treaty that expresses a compromise and temporary convergence of interests among the states parties. They speak to matters deep, lasting, purportedly universal. What then are the intuitions that shape them, their sources in intellectual history?

(4) Many rights declared in the Covenant closely resemble the provisions of the Universal Declaration, although they are stated in considerably greater detail. Compare, for example, the requirements for criminal trials in Articles 10 and 11 of the Universal Declaration with the analogous provisions in Articles 14 and 15 of the Covenant.

(5) Group or collective rights are either asserted or hinted at in the Covenant, most directly in Articles 1 and 27. They are absent from the Universal Declaration, which contains no equivalent to the powerful, historically influential first article in the Covenant, 'All peoples have the right of self-determination.' (Note that the International Covenant on Economic, Social and Cultural Rights contains an identical Article 1.)

(6) In both instruments the idea of 'rights' dominates. Article 29(1) of the Declaration does state that everyone 'has duties to the community in which alone the free and full development of his personality is possible.' The Covenant has no article referring to individuals' *duties*, though its Preamble has such a clause.

(7) Article 17 of the Declaration on the 'right to own property' and protection against arbitrary deprivation thereof does not figure among the rights declared in the Covenant.

(8) Note the structure of the ICCPR, which goes beyond the 'declaration' of rights in the UDHR to require the state to provide a remedial system in the event of violations of rights. In Article 2, states parties agree to 'ensure' to all persons within their territory the rights recognized by the Covenant, and to adopt such legislative or other measures as may be necessary to achieve that goal. Moreover, the parties agree to 'ensure' that any person whose rights are violated 'shall have an effective remedy,' and that 'the competent authorities shall enforce such remedies when granted.'

Hence the Covenant goes well beyond imposing a duty on states not to interfere in stated ways with individuals. States must develop and enforce a legal system adequate to respond to claims of violation. At its very start, the Covenant makes us aware that the duties of state parties are (to use a conventional contrast whose utility will be questioned in later chapters) both *negative/hands-off* (don't torture) and *positive/affirmative* (provide a legal system to which individuals can have recourse in order to seek remedies for violations).

(9) The rights declared in the Covenant are not by their terms restricted to rights against *governmental* interference. That is, interference by non-governmental, private actors (the rapist, say) could as destructively impair the right to 'security of person' (Article 9). The state's duty to provide effective remedies can then be read to attach to conduct (rape) that was initially non-governmental.

(10) Two types of provisions in the Covenant limit states' obligations thereunder. (*a*) Article 4 dealing with a public emergency permits under closely stated conditions a temporary *derogation*—that is, deviation—from many of the rights declared by the Covenant. Thus states may consciously, purposively, explicitly depart from such rights—for example, from rights in Article 9 relating to arrest and detention. Note that under paragraph (2) certain rights are non-derogable. (*b*) A number of articles include *limitation clauses*—that is, provisions indicating that a given right cannot be absolute but must be adapted to meet a state's interest in protecting public safety, order, health or morals, or national security. See, for example, Articles 18 and 19. In Articles 21 and 22, the limitation clause is phrased in terms of permitting those restrictions on a right 'which are necessary in a democratic society.' Compare the broad provision of Article 29(2) of the UDHR, which is not linked to a specific right.

(11) Article 5 of the UDHR bans 'cruel, inhuman or degrading' punishment, but that instrument is not explicit about capital punishment. See Article 6, para. 2 of the ICCPR. The Second Optional Protocol to the ICCPR, Aiming at the Abolition of the Death Penalty, became effective in 1991 (28 states parties as of September 1995). Article 1 provides that 'No one within the jurisdiction of a State Party to the present Protocol shall be executed. . . . Each State Party shall take all necessary measures to abolish the death penalty within its jurisdiction.'

One can organize or classify the rights declared in the Covenant in various ways, depending on the purpose of the typology. Consider the adequacy of the following scheme that embraces most of the Covenant's rights, although it excludes such distinctive provisions as Article 1 on the self-determination of peoples and Article 27 on the enjoyment by minorities of their own cultures:

(*a*) Protection of the individual's physical integrity, as in provisions on torture, arbitrary arrest, arbitrary deprivation of life;
(*b*) procedural fairness when government deprives an individual of liberty, as in provisions on arrest, trial procedure and conditions of imprisonment;
(*c*) equal protection norms defined in racial, religious, gender and other terms;
(*d*) freedoms of belief, speech and association, such as provisions on the practice of religion, press freedom, and the right to hold assembly and form associations; and
(*e*) the right to political participation.

These five categories of rights can be seen to lie on a spectrum, moving from those such as torture at one extreme over which there exists a broad formal–verbal consensus among states (whatever the degree of violation by many states and the susceptibility of the rights to varying interpretations), to those at the other extreme whose purposes, basic meanings and even validity are formally disputed. For example, few if any states (even those that practice it) formally justify torture. A good number of states, however, may justify some form of religious or gender discrimination that should (such states argue) be viewed as permitted by other norms or goals of the human rights movement, or may justify different forms of political participation.

NOTE

In a report on its Fiftieth Session in 1994, the Human Rights Committee created by the ICCPR noted a proposal submitted to it for a possible draft optional protocol to that instrument that would add Article 9 paras. 3 and 4 and Article 14 to the list of non-derogable provisions under Article 4(2). The Committee's reaction was negative. It 'was satisfied that States parties generally understood that the right to habeas corpus and *amparo* should not be limited in situations of emergency,' and that the remedies provided in the relevant parts of Article 9, 'read in conjunction with article 2 were inherent in the Covenant as a whole.' There was a 'considerable risk' that such an optional protocol 'might implicitly invite States parties to feel free to derogate from the provisions of article 9 of the Covenant during states of emergency if they do not ratify the proposed optional protocol.' Finally, the Committee observed that 'it would simply not be reasonable to expect that all provisions of article

14 can remain fully in force in any kind of emergency. Thus the inclusion of article 14 as such in the list of non-derogable provisions would not be appropriate.' GAOR 49th Sess., Supp. No. 40 (A/49/40), paras. 22–5.

Note the following observations about the progressive or immediate character of the obligations under the ICCPR, in Dominic McGoldrick, *The Human Rights Committee* (1991), at 12:

> There were marked differences of opinion during the drafting on the matter of the obligations that would be incurred by a State party to the ICCPR. Some representatives argued that the obligations under the ICCPR were absolute and immediate and that, therefore, a State could only become a party to the ICCPR after, or simultaneously with, its taking the necessary measures to secure those rights. If there were disparities between the Covenant and national law they could best be met by reservations. . . .
>
> Against this view it was argued that the prior adoption of the necessary measures in domestic law was not required by international law. . . .
>
> . . .
>
> Proposals to provide that the necessary measures be taken within a specified time limit or within a reasonable time were rejected as was a suggestion that each State fix its own time limit in its instrument of ratification. The only clear intentions of the [Commission on Human Rights] that emerged were those of avoiding excessive delays in the full implementation of the Covenant and of not introducing the general notion of progressiveness that was a feature of the obligations under the then draft [International Covenant on Economic, Social and Cultural Rights].
>
> The objections to the draft article 2(2) were again voiced in the Third Committee but the provision remained unchanged. The Committee's report stated that,
>
>> It represented the minimum compromise formula, the need for which, particularly in new States building up their body of legislation, was manifest. The notion of implementation at the earliest possible moment was implicit in article 2 as a whole. Moreover, the reporting requirement in article 49 (later article 40) would indeed serve as an effective curb on undue delay.

QUESTIONS

1. Relying only on the preambles and texts of the UDHR and the ICCPR, how would you identify the reasons for those instruments, their justifications in moral and political thought, the moral and political traditions from which they derive? Why do you suppose there was such a sparse statement of reasons or justifications in these instruments?

2. Do you see in either of these instruments any departure from 'universal' premises, rights, and related obligations of states? That is, are there concessions in any provisions to different cultures or regions that would allow those cultures or regions to privilege their own traditions rather than follow these instruments' rules—for example, by inflicting certain severe modes of criminal punishment, or governing by theocracy or inherited rule?

3. Note the breadth and the susceptibility to varying understandings of the 'limitations' on rights in the ICCPR Articles referred to above. Compare the different approach of the United States Constitution, whose Bill of Rights (the first ten amendments) contains no formal equivalent of such limitation clauses. It has been primarily the task of the United States Supreme Court to develop comparable notions through its decisional law resolving conflicts—for example, between free speech and national security or pornography. In general, does one or the other approach seem to you preferable, to be perhaps more protective of individual rights?

ADDITIONAL READING

For analysis of the drafting process for vital provisions of the ICCPR, see the essays in Louis Henkin (ed.), *The International Bill of Rights* (1981).

NOTE

It is frequently stated that all rights declared in the ICCPR are 'equal and interdependent.' Within that formulation, the right of an indigent person to assigned legal assistance in a criminal case in Article 14(3) (c) is of the same rank, interdependent with, the right not to be tortured in Article 7. The following readings explore this issue of equality or hierarchy.

THEO VAN BOVEN, DISTINGUISHING CRITERIA OF HUMAN RIGHTS

In Karel Vasak and P. Alston (eds.), The International Dimensions of Human Rights, Vol. 1 (1982), at 43.

. . .

There is another argument against making a distinction between fundamental human rights and other human rights. Such a distinction might imply that there is a hierarchy between various human rights according to their fundamental character. However, in modern human rights thinking the indivisibility of human rights and fundamental freedoms is prevalent. This idea of indivisibility presupposes that human rights form, so to speak, a single package and that they cannot rank one above the other on a hierarchical scale.

This may all be true, but there still remain weighty arguments which militate in favour of distinguishing fundamental human rights from other human rights. Such fundamental rights can also be called elementary rights or supra-positive rights, i.e. rights whose validity is not dependent on their acceptance by the subjects of law but which are at the foundation of the international community. . . .

. . .

. . . The intensity of the prevailing sentiments against racism and racial discrimination, the awareness of urgency and the political climate have made the principle of racial non-discrimination one of the foundations of the international community as represented in the UN. Members of this community are bound by this principle on the basis of the UN Charter, even if they do not adhere to the various international instruments specifically aimed at the elimination of racial discrimination and apartheid. . . .

There is also a great deal of law in humanitarian conventions and in international human rights instruments supporting the existence of very fundamental human rights. This is that part of human rights law which does not permit any derogation even in time of armed conflict or in other public emergency situations threatening the life of the nation. The common article 3 of the four Geneva Conventions of 1949, setting out a number of minimum humanitarian standards which are to be respected in cases of conflict which are not of an international character, enumerates certain acts which 'are and shall remain prohibited at any time and in any place whatsoever.' The following acts are mentioned: '(a) violence to life and person, in particular murder of all kinds, mutilation, cruel treatment and torture; (b) taking of hostages; (c) outrages upon personal dignity, in particular, humiliating and degrading treatment; (d) the passing of sentences and the carrying out of executions without previous judgment pronounced by a regularly constituted court, affording all the judicial guarantees which are recognized as indispensable by civilized nations.' The universal validity of these fundamental prescriptions is underlined by the words 'at any time and in any place whatsoever' in this common article 3 of the four 1949 Geneva Conventions.

The International Covenant on Civil and Political Rights enumerates in article 4, para. 2, the rights from which no derogation is allowed in time of public emergency, viz. the right to life (article 6), the right not to be subjected to torture or to cruel, inhuman or degrading treatment or punishment (article 7), the right not to be held in slavery or servitude (article 8, paras. 1 and 2), the right not to be imprisoned merely on the ground of inability to fulfil a contractual obligation (article 11), the prohibition of retroactive application of criminal law (article 15), the right to recognition everywhere as a person before the law (article 16) and the right to freedom of thought, conscience and religion (article 18). Regional human rights conventions contain a similar clause enumerating provisions from which no derogation may be made.

The fact that in a number of comprehensive human rights instruments at the worldwide and the regional level, certain rights are specifically safeguarded and are intended to retain their full strength and validity notably in serious emergency situations, is a strong argument in favour of the contention that there is at least a minimum catalogue of fundamental or elementary human rights.

. . .

NOTE

Compare with van Boven's observations those in Theodor Meron, *On a Hierarchy of International Human Rights*, 80 Am. J. Int. L. 1 (1986), at 21:

> . . . Hierarchical terms constitute a warning sign that the international community will not accept any breach of those rights. Historically, the notions of 'basic rights of the human person' and 'fundamental rights' have helped establish the *erga omnes* principle, which is so crucial to ensuring respect for human rights. Eventually, they may contribute to the crystallization of some rights, through custom or treaties, into hierarchically superior norms, as in the more developed national legal systems.
>
> Yet the balance of pros and cons does not necessarily weigh clearly on the side of the pros. Resort to hierarchical terms has not been matched by careful consideration of their legal significance. Few criteria for distinguishing between ordinary rights and higher rights have been agreed upon. There is no accepted system by which higher rights can be identified and their content determined. Nor are the consequences of the distinction between higher and ordinary rights clear. Rights not accorded quality labels, i.e., the majority of human rights, are relegated to inferior, second-class, status. Moreover, rather than grapple with the harder questions of rationalizing human rights lawmaking and distinguishing between rights and claims, some commentators are resorting increasingly to superior rights in the hope that no state will dare—politically, morally and perhaps even legally—to ignore them. In these ways, hierarchical terms contribute to the unnecessary mystification of human rights, rather than to their greater clarity.
>
> Caution should therefore be exercised in resorting to a hierarchical terminology. Too liberal an invocation of superior rights such as 'fundamental rights' and 'basic rights,' as well as *jus cogens*, may adversely affect the credibility of human rights as a legal discipline.

QUESTIONS

1. Suppose that you are a director of an international non-governmental human rights organization, like Human Rights Watch or the International Commission of Jurists. You must vote on an agenda for that organization indicating what types of violations it should investigate over the next year. From this perspective, how do you react to the views of van Boven and Meron about equality or hierarchy among human rights norms?

2. Suppose that you are part of a new, reforming and popularly elected government of State X that has just displaced a decade-long repressive, brutal regime forced from power by popular uprisings. Of course your political and material resources are now limited. How would you start to think about agendas and priorities for the new X with respect to realization of human rights norms?

NOTE

Among the intergovernmental organs or institutions referred to in this chapter, some such as the UN Commission are created pursuant to the UN Charter, others such as the Human Rights Committee under the ICCPR pursuant to distinct treaties. As a matter of convenience, the first set is often referred to as 'Charter-based' organs/institutions and the second set as 'treaty-based' organs/institutions.

Despite this distinction, bear in mind that the entire universal human rights regime is related to the United Nations. The human rights treaties are distinct from the Charter only up to a point. Thus the ICCPR and the other treaties noted in category (4) on p. 122, *supra*, all grew within the UN, from the time that they were first drafted in an organ like the UN Commission to their final approval by the General Assembly and submission to states for ratification. Typically of such other treaties, the ICCPR provides for a number of ongoing links to the UN. Articles 40 and 45 indicate that reports or 'general comments' adopted by the ICCPR Committee should or may be submitted by that Committee to the UN Secretary-General. Note also the provisions for amendment of the ICCPR in Article 51. Moreover, each of these separate treaty regimes like the ICCPR depends for funding on the regular biennial budget adopted by the General Assembly.

B. THE CONTINUING ROLE OF CUSTOM AND THE SIGNIFICANCE OF UN RESOLUTIONS

Given the multitude of human rights treaties, one might conclude that resort to customary international law and to general principles as components of argument about international law is no longer necessary. Since treaties occupy the field, recourse thereto should be sufficient. At least with respect to human rights, para. (a) of Article 38(1) of the Statute of the International Court of Justice, p. 27, *supra*, reigns supreme.

That conclusion would be wrong. As components of argument about international human rights, custom and general principles retain great importance, although less in their classical methods and forms as sketched in Chapter 2, than in their contemporary transformed character. To anticipate later discussions, custom occupies a central role in argument about matters as diverse as litigation under the Alien Tort Statute in the United States (p. 779, *infra*) and interpretation of the source of the International Criminal Tribunal for the Former Yugoslavia (p. 1050, *infra*).

This Section B traces the influence over the last half century of both older and contemporary notions of international customary law on the development of the human rights movement. It also explores the significance of such prominent new ingredients of argument about international human rights as resolutions and declarations voted by the UN General Assembly.

OPPENHEIM'S INTERNATIONAL LAW (ROBERT JENNINGS AND ARTHUR WATTS, EDS.)

Vol. 1 (9th ed. 1992), at 4.

That part of international law that is binding on all states, as is far the greater part of customary law, may be called *universal* international law, in contradistinction to *particular* international law which is binding on two or a few states only. *General* international law is that which is binding upon a great many states. General international law, such as provisions of certain treaties which are widely, but not universally, binding and which establish rules appropriate for universal application, has a tendency to become universal international law.

One can also distinguish between those rules of international law which, even though they may be of universal application, do not in any particular situation give rise to rights and obligations *erga omnes*, and those which do. Thus, although all states are under certain obligations as regards the treatment of aliens, those obligations (generally speaking) can only be invoked by the state whose nationality the alien possesses: on the other hand, obligations deriving from the outlawing of acts of aggression, and of genocide, and from the principles and rules concerning the basic rights of the human person, including protection from slavery and racial discrimination, are such that all states have an interest in the protection of the rights involved.[1] Rights and obligations *erga omnes* may even be created by the actions of a limited number of states. There is, however, no agreed enumeration of rights and obligations *erga omnes*, and the law in this area is still developing, as it is in the connected matter of a state's ability, by analogy with the *actio popularis* (or *actio communis*) known to some national legal systems, to institute proceedings to vindicate an interest as a member of the international community as distinct from an interest vested more particularly in itself. . . .

. . .

States may, by and within the limits of agreement between themselves, vary or even dispense altogether with most rules of international law. There are, however, a few rules from which no derogation is permissible. The latter—rules of *ius cogens*, or peremptory norms of general international law—have been defined in Article 53 of the Vienna Convention on the Law of Treaties 1969 (and for the purpose of that Convention) as norms 'accepted and recognised by the international community of states as a whole as a norm from which no derogation is permitted and which can be modified only by a subsequent norm of general international law having the same character'; and Article 64 contemplates the emergence of new rules of *ius cogens* in the future.

Such a category of rules of *ius cogens* is a comparatively recent development and there is no general agreement as to which rules have this character. The International Law Commission regarded the law of the Charter concerning the prohibition of the use of force as a conspicuous example of such a rule. Although the Commission refrained from giving in its draft Articles on the Law of Treaties any examples of rules of *ius cogens*, it did record that in this context mention had additionally been made of the prohibition of criminal acts under international law, and of acts such as trade in slaves, piracy or genocide, in the suppression of which every state is called upon to cooperate; the observance of human rights, the equality of states and the principle of

[1] [Eds.] The authors here refer in a footnote to the Case Concerning The Barcelona Traction, Light and Power Company, Limited (New Application 1962) (Belgium v. Spain), [1970] I.C.J. Rep. 4. The relevant portion of that opinion of the International Court of Justice reads, at paras 33–4: '. . . [A]n essential distinction should be drawn between the obligations of a State towards the international community as a whole, and those arising vis-à-vis another State in the field of diplomatic protection. By their very nature, the former are the concern of all States. In view of the importance of the rights involved, all States can be held to have a legal interest in their protection; they are obligations *erga omnes*. Such obligations derive, for example, in contemporary international law, from the outlawing of acts of aggression, and of genocide, as also from the principles and rules concerning the basic rights of the human person, including protection from slavery and racial discrimination. Some of the corresponding rights of protection have entered into the body of general international law . . . ; others are conferred by international instruments of a universal or quasi-universal character.'

self-determination. The full content of the category of *ius cogens* remains to be worked out in the practice of states and in the jurisprudence of international tribunals. . . .

The operation and effect of rules of *ius cogens* in areas other than that of treaties are similarly unclear. Presumably no act done contrary to such a rule can be legitimated by means of consent. . . .

OSCAR SCHACHTER, INTERNATIONAL LAW IN THEORY AND PRACTICE (1991), at 85.

Ch. VI. Resolutions and Political Texts

. . .

Few issues of international law theory have aroused as much controversy as that engendered by resolutions and declarations of the General Assembly which appear to express principles and rules of law. Their adoption by large majorities through voting or consensus procedures has been seen by many as attempts to impose obligatory norms on dissenting minorities and to change radically the way in which international law is made.

It is, of course, true that such resolutions are not a formal source of law within the explicit categories of Article 38 (1) of the Statute of the International Court of Justice. . . .

. . . As the central global forum for the international community, with the competence to discuss all questions of international concern, with institutional continuity and a constitutional framework of agreed purposes and principles, the Assembly has become a major instrument of States for articulating their national interests, and seeking general support for them. The conception of Assembly resolutions as expressions of common interests and the 'general will' of the international community has been a natural consequence. It also has naturally followed that in many cases the effort is made to transform the 'general will' thus expressed into law. One obvious way of accomplishing that transformation is to use a resolution as a basis for the preparation of a treaty by the Assembly itself or by a diplomatic conference convened by it. The treaties are then open for adherence by member States and other States. . . .

Legal uncertainty has, however, been created when the Assembly adopted resolutions which purported to assert legal norms without recourse to the treaty process. Such resolutions 'declared the law' either in general terms or as applied to a particular case. Neither in form nor in intent were they recommendatory. Surprising as it may seem, the authority of the General Assembly to adopt such declaratory resolutions was accepted from the very beginning. At its first session, in 1946, the Assembly considered the Nuremberg Principles and they 'affirmed' them in a unanimous resolution. In another resolution adopted at the same session, genocide was declared a crime under international law. This, too, was unanimous. No one questioned the Assembly's competence to adopt such resolutions despite the absence of explicit Charter authority to do so. The Assembly also interpreted and applied the Charter in particular cases, characterizing certain conduct as illegal. The resolutions condemning South Africa for apartheid and for its administration of South West Africa fall into this category. The competence of the Assembly to do this—that is, to designate conduct as illegal under the Charter and to assert obligations and rights applicable in particular cases—was not questioned.

What was, however, in question was the legal force of the declarations of law, whether general or particular. Could they be considered 'binding' when the Assembly lacked constitutional authority to adopt mandatory decisions concerning the subjects dealt with? If not binding, were they authoritative in some other sense? Was unanimity or near-unanimity a requirement for their authority? If nearly all States agreed on what is the law, was there a sufficient reason to deny effect to that determination? These and related questions gave rise to official perplexity and a considerable body of legal analysis. . . .

Lawyers are accustomed to pouring new wine into old bottles and keeping the old labels. Thus, the law-declaring resolutions that construed and 'concretized' the principles of the Charter—whether as general rules or in regard to particular cases—may be regarded as authentic interpretation by the parties of their existing treaty obligations. To the extent that they were interpretation, and agreed by all the member States, they fitted comfortably into an established source of law. . . .

But as we know the line between interpretation and new law is often blurred. Whenever a general rule is construed to apply to a new set of facts, an element of novelty is introduced; in effect, new content is added to the existing rule. This is even clearer when an authoritative body re-defines and makes more precise an existing rule or principle. In any such case, the question of degree can be raised: how far does the 'new' rule go beyond the agreed meaning of the old rule? Is the Charter being amended or simply interpreted? The answer is provided, not by logical analysis, but by the responses of those deemed competent to decide—namely, the States parties to the treaty. If they all agree in a formal resolution that the Charter means what the resolution says it does, that will be regarded as 'authentic' (that is, authoritative) interpretation. . . .

We come now to the declaratory resolution that purports to state the law independently of any Charter rule. . . . When all the States in the United Nations declare that a . . . norm is legally binding, it is difficult to dismiss that determination as ultra vires—or reduce it to a recommendation—because it was made in the General Assembly. The fact is that the declaration purports to express the *opinio juris communis*, not a recommendation. . . .

The . . . question is whether the assertion in good faith by all States that a norm is legally binding is sufficient to validate the norm as law even though

State practice is negligible or inconclusive. International lawyers differ on the answer. . . .

. . .

. . . Much of the debate has focused on the choice between two polar categories: 'binding' and 'hortatory' (i.e., without legal force *et al*). That categorization—however clear it may appear—seems much less appropriate than treating the law-declaring resolutions as evidence for the asserted proposition of law.

. . .

Ch. XV. International Human Rights

. . .

Human Rights as Customary International Law

. . .

. . . [A] juristic debate has taken place for some years on whether human rights in whole or in part has become part of general customary international law. . . . If we have so extensive a network of treaty obligations, as suggested earlier, how important is it to determine the extent of customary law? Two answers may be given. First despite the many treaties and the considerable numbers of States parties to most of them, a significant number of States have not adhered to many of the treaties. They are therefore neither bound by the treaty obligations nor entitled to invoke those obligations against the parties. It is therefore of some consequence to determine their obligations and rights under customary law. A second reason is that the recognition of human rights in customary law allows not only the treaty non-parties, but also the parties to have recourse to international law remedies not provided for in the treaties. . . .

Whether human rights obligations have become customary law cannot readily be answered on the basis of the usual process of customary law formation. States do not usually make claims on other States or protest violations that do not affect their nationals. In that sense, one can find scant State practice accompanied by *opinio juris*. Arbitral awards and international judicial decisions are also rare except in tribunals based on treaties such as the European and Inter-American courts of human rights. The arguments advanced in support of a finding that rights are a part of customary law rely on different kinds of evidence. They include the following:

— the incorporation of human rights provisions in many national constitutions and laws;
— frequent references in United Nations resolutions and declarations to the 'duty' of all States to observe faithfully the Universal Declaration of Human Rights;

— resolutions of the United Nations and other international bodies condemning specific human rights violations as violative of international law;
— statements by national officials criticizing other States for serious human rights violations;
— a dictum of the International Court of Justice that obligations erga omnes in international law include those derived 'from the principles and rules concerning the basic rights of the human person' (*Barcelona Traction* Judgment, 1970);
— some decisions in various national courts that refer to the Universal Declaration as a source of standards for judicial decision.

None of the foregoing items of 'evidence' of custom conform to the traditional criteria. General statements by international bodies (such as the United Nations General Assembly or the Tehran Conference on Human Rights) that the 'Universal Declaration constitutes an obligation for the members of the international community' are not without significance, but their weight as evidence of custom cannot be assessed without considering actual practice. National constitutions and legislation similarly require a measure of confirmation in actual behaviour. One can readily think of numerous constitutions that have incorporated many of the provisions of the Universal Declaration or other versions of human rights norms, but these provisions are far from realization in practice. Constitutions with human rights provisions that are little more than window-dressing can hardly be cited as significant evidence of practice or 'general principles' of law.

Should we then reject all of the affirmations of human rights principles as obligatory on the ground that infringements are widespread, often gross and generally tolerated by the international community? Or should we minimize the negative practice and treat the verbal affirmations as persuasive evidence that the Universal Declaration has now become customary law? Some international lawyers answering this second question in the affirmative have asserted that the Declaration 'is now part of the customary law of nations and therefore binding on all States'.

Although only a few legal scholars have taken this position, they are often cited by human rights advocates in national tribunals and in publications. The argument for treating the Declaration as law is also bolstered by noting that its principles have been included in many national constitutions and laws and consequently may be reasonably regarded as 'general principles of law accepted by civilized nations', a source of general international law under Article 38 of the Statue of the International Court of Justice. . . . A third theory for attributing legal force to the Declaration is premised on the provision of the United Nations Charter (Article 56) that pledges members to take action to achieve certain ends of the Charter, including human rights. It is suggested that the Declaration by authoritatively spelling out the recognized human rights gives specific content to the obligation.

These three lines of legal argument are generally linked by their proponents in human rights advocacy. It is not inconceivable that in time they will carry the day for the Declaration to be treated as obligatory. However, for the present, their reach exceeds their grasp. Neither governments nor courts have accepted the Universal Declaration as an instrument with obligatory force. Many have, of course, lauded its principles as standards to be achieved and in specific instances have rhetorically relied on the Declaration as a touchstone of legality. . . . But these particular references fall short of recognizing the Declaration as obligatory in law. It remains difficult to do so in the face of the clear intention of the governments to consider it as non-binding. . . .

This conclusion, however, does not dispose of claims that some important human rights included in the Declaration have become customary law (and/or general principles of law) and therefore binding on all States. The evidence for this must, of course, focus on the specific rights in question. As noted earlier, such evidence is rarely to be found in the traditional patterns of State practice involving claims and counter-claims between two States. Instead, one must look for 'practice' and *opinio juris* mainly in the international forums where human rights issues are actually discussed, debated and sometimes resolved by general consensus. These are principally organs of the United Nations and of regional bodies. In those settings, governments take positions on a general and specific level: they censure, condemn, or condone particular conduct. An evaluation of those actions and their effects on State conduct provides a basis for judgments on whether a particular right or principle has become customary international law. Such inquiries may have to be broadened to include pronouncements by national leaders, legislative enactments, judicial opinions, and scholarly studies. No single event will provide the answer. One essential test is whether there is a general conviction that particular conduct is internationally unlawful. Occasional violations do not nullify a rule that is widely observed. The depth and intensity of condemnation are significant indicators of State practice in this context. The extent of agreement across geographical and political divisions is also pertinent. Applying these indicators on a global scale is obviously not an easy task, nor is it a one-time effort. Attitudes, practices and expectations are in flux and judgments may often change. Nonetheless there is little doubt that some human rights are recognized as mandatory for all countries, irrespective of treaty. The most obvious are the prohibitions against slavery, genocide, torture and other cruel, inhuman and degrading treatment. No government would contend that these prohibitions apply only to parties to the treaties that outlaw them. The list does not stop there. The ALI Restatement (Third) of 1987[2] adds the following actions as unlawful 'for a state to practice, encourage or condone':

— The murder or causing the disappearance of individuals;
— Prolonged arbitrary detention;
— Systematic racial discrimination.

[2] [Eds.] Restatement (Third), Foreign Relations Law of the United States (1987). Section 702, to which the text refers, appears at p. 145, *infra*.

The Restatement's list also includes a more general category as violative of customary law, to wit: 'consistent patterns of gross violations of internationally recognized human rights'. . . .

The Restatement's enumeration of customary law human rights is well-founded as far as it goes, but developments affecting human rights in the past decade indicate that the list of customary law rights may have significantly increased. Studies carried out for the United Nations Commission on Human Rights which have examined national laws on a global scale as well as governmental and scholarly statements reveal that several rights have been widely invoked as principles of general international law. The examples include:

— The right to self-determination of peoples;
— The individual right to leave and return to one's country;
— The principle of non-refoulement for refugees threatened by persecution.

There is a well-intentioned tendency among human rights lawyers to add to the list of customary law rights, especially 'due process' rights. Theodor Meron, for example, has suggested that a number of basic rights of the accused stipulated in the Covenant on Political and Civil Rights (Article 14) are also customary law. His indicators are the inclusion of these rights in national law generally and references to them in treaties and other international instruments. However, evidence that these rights are secured in most countries is lacking. . . . Even where they are on the books, they are often honored in the breach, not the observance. . . .

However, recent developments in various parts of the world indicate that certain human rights have penetrated deeply into the consciousness of peoples in many countries. Violations are more and more resented in places where previously they had been ignored or seen as unavoidable. Most striking in this respect are the changes in Eastern Europe and in the Soviet Union in recent years (especially 1987–1990). Individual human rights were emphasized in the popular demands and given effect in the new political arrangements. The rights were asserted as fundamental entitlements recognized by the international community. Protection against arbitrary arrest, against political trials and against lack of procedural rights were emphasized. Freedom of speech and of peaceful assembly were demanded as basic rights. Political participation through genuine elections, ensured by secret ballot, was another salient demand. It is true that these rights are in the International Covenant on Civil and Political Rights to which the USSR and other States in Eastern Europe have long been parties though not in substantial compliance. But the treaty compliance issue was subsidiary; the main point was that the rights were now demanded as basic and essential whether or not in treaties. The dramatic reversal of the long prevalent political statist ideology had worldwide repercussions, foremost of which was an intensified awareness of the importance of basic rights.

. . .

Present tendencies also suggest that other human rights may be on their way to acceptance as general international law, especially in virtue of their widespread inclusion in national law plus general recognition of their international significance. Several economic and social rights may well meet that dual test—in particular, the right to basic sustenance, and to public assistance in matters of health, welfare and basic education. ILO practice indicates that trade union rights, including freedom of association, are widely accepted as 'international common law'. Also significant is the widespread recognition of the rights of women to full equality and to protection against discrimination. These rights have been affirmed and emphasized in numerous declarations as well as in some international conventions, though in many respects the gap between the proclaimed rights and actual conditions remains great. A positive trend is that many countries have enacted legislation that prohibits gender discrimination by State action. The prevalence of these laws in conjunction with the United Nations Charter's prohibition of discrimination based on sex provides a strong argument for holding that such discrimination is an international delict.

. . .

Whatever the doctrinal theory, the political dynamics that mark the demands for human rights make it almost certain that the international law of human rights will continue to have a deeper and broader basis than the treaties alone. The powers that govern States are not immune from pressures based on a social consciousness of the limits of State authority. . . .

. . .

ROSALYN HIGGINS, PROBLEMS AND PROCESS: INTERNATIONAL LAW AND HOW WE USE IT

(1994), at 19.

One of the special characteristics of international law is that violations of law can lead to the formation of new law. Of course, this characteristic is more troublesome for those who regard law as rules, and less troublesome for those who regard law as process. But whether one believes that international law consists of rules that have been derived from consent or natural law; or whether one believes international law is a process of decision-making, with appropriate reliance on past trends of decision-making in the light of current context and desired outcomes, there still remains the question of how the 'rules' or the 'trend of decision' change through time. And, in so far as these rules or trends of decisions are based on custom, then there is the related question of what legal significance is to be given to practice that is inconsistent with the perceived rules or trends of decision.

. . .

If a customary rule loses its normative quality when it is widely ignored, over a significant period of time, does this not lead to a relativist view of the substantive content of international law, with disturbing implications? . . .

A second example: all states agree that international law prohibits genocide (and that this total prohibition is today rooted in customary international law and not just in treaty obligations). So what if some states from time to time engage in genocide? Here we may safely answer that genocide, while it sometimes occurs and while its very nature makes *all* norm compliance shocking, is certainly not the majority practice. The customary law that prohibits genocide remains intact, notwithstanding appalling examples of non-compliance. Let us look at a third, more difficult example. No one doubts that there exists a norm prohibiting torture. No state denies the existence of such a norm; and, indeed, it is widely recognized as a customary rule of international law by national courts. But it is equally clear from, for example, the reports of Amnesty International, that *the great majority* of states systematically engage in torture. If one takes the view that non-compliance is relevant to the retention of normative quality, are we to conclude that there is not really any prohibition of torture under customary international law? . . .

. . .

New norms require both practice and *opinio juris* before they can be said to represent customary international law. And so it is with the gradual death of existing norms and their replacement by others. The reason that the prohibition on torture continues to be a requirement of customary international law, even though widely abused, is not because it has a higher normative status that allows us to ignore the abuse, but because *opinio juris* as to its normative status continues to exist. No state, not even a state that tortures, believes that the international law prohibition is undesirable and that it is not bound by the prohibition. A new norm cannot emerge without both practice and *opinio juris*; and an existing norm does not die without the great majority of states engaging in both a contrary practice and withdrawing their *opinio juris*.

VIEWS OF SCHOLARS

Compare the views of the following scholars with Schachter's remarks about (a) contemporary conceptions of customary law in relation to acts of international organizations and (b) the related issue of the status of the UDHR.

Louis Henkin[3]

Henkin refers to a number of fields of international law that initially developed by custom, with its requirement of 'consistent general practice plus

[3] International Law: Politics, Values and Functions, 216 Collected Courses of Hague Academy of International Law, Receuil des Cours 13 (Vol. IV, 1989), at 54.

opinio juris.' His examples include aspects of diplomatic immunities, many principles of the law governing treaties, state responsibility to foreign nationals, some laws of war, much of the law of the sea, and 'substantial human rights law.' He states:

> Unlike law made by treaty, this authentic customary law ordinarily has not been made intentionally, purposefully, by a 'conspiracy' of States. Nor is it the product of deliberate exercises of 'will' by States acting separately, aiming to develop new law. Customary law was not *made*, it *resulted*, from an accretion of practices, though often the practice of individual States was intended to conform to what others had done, and often it was thought to be required by law.

In explaining why the general principle of consent as a basis for obligation in international law doesn't apply to states created after a customary norm has been established (so that such states are bound by that norm), Henkin says:

> That result can be explained (and even justified) in that, unlike treaties, customary law is not *created* but *results*; that it is therefore not a product of the will of States but a 'systemic creation', reflecting the 'consent' of the international system, not the consent of individual States.

Henkin later comments on contemporary customary law and efforts by states to 'create new customary law by purposeful activity,' such as General Assembly resolutions.

> The purposive creation of custom is a radical innovation, and indeed reflects a radical conception. Whereas law was *made* by treaty but *grew* by custom, now there is some tendency to treat custom as a means, alternative to treaty-making, for deliberate legislation. Using the concept of custom for that purpose brings with it the traditional definition, but now practice sometimes means activity designed to create the norm rather than to reflect it.

Bruno Simma and Philip Alston[4]

The authors describe reasons for the effort to locate human rights norms in custom, including several of the reasons noted in the preceding excerpts from Schachter. They point to an additional factor. States may be parties to a human rights treaty like the ICCPR that is relevant to litigation in domestic courts—relevant, say, to a potential claim by a defendant in a criminal proceeding that the process to which he was subjected by a state party violates Article 14 of the ICCPR. However, the courts in that state may not be able to rule on the applicability of the ICCPR to the defendant's case, because that

[4] The Sources of Human Rights Law: Custom, Jus Cogens, and General Principle, 12 Australian Y. B. Int. L. 82 (1992).

treaty is neither automatically incorporated into the state's legal system nor has it been enacted as domestic law by the state's legislature. See pp. 726–50, *infra*. In such circumstances where a *treaty* does not constitute *internal* law that state courts can apply, international custom may nonetheless constitute a source of human rights law which such courts can invoke.

The traditional understanding of custom, the authors argue, emphasized state practice; 'deeds were what counted, not just words.' The rules thereby emerging may have been limited in scope, but they have 'several undoubted advantages. They are hard and solid; they have been carefully hammered out on the anvil of actual, tangible interaction among States; and they allow reasonably reliable predictions as to future State behavior.' Through institutions such as the General Assembly, the notion of 'practice' itself has changed into 'paper practice: the words, texts, votes and excuses themselves.'

Thomas Buergenthal[5]

Buergenthal states that 'few international lawyers would deny that the [Universal] Declaration is a normative instrument that creates legal obligations for the Member States of the UN'. Such dispute as exists about its 'legal character' concerns whether all or only some of the rights that it declares are 'binding as such and under what circumstances.' He observes:

> The process leading to the transformation of the Universal Declaration from a non-binding recommendation to an instrument having a normative character was set in motion, in part at least, because the effort to draft and adopt the Covenants remained stalled in the UN for almost two decades. During that time the need for authoritative standards defining the human rights obligations of UN Member States became ever more urgent. As time went on, the Declaration came to be utilized with ever greater frequency for that purpose.

Buergenthal cautions that a 'careful analysis of the relevant state practice' suggests that not all rights proclaimed in the Declaration have to date acquired the status of customary international law. He then quotes the views of a 'distinguished author,' Louis Sohn, who wrote:

> The Declaration . . . is now considered to be an authoritative interpretation of the U.N. Charter, spelling out in considerable detail the meaning of the phrase 'human rights and fundamental freedoms,' which Members States agreed in the Charter to promote and observe. The Universal Declaration has joined the Charter . . . as part of the constitutional structure of the world community. The Declaration, as an authoritative listing of human rights, has become a basic component of international

[5] International Human Rights in a Nutshell (1988). The quotation by Buergenthal of Louis Sohn that appears in the following text is taken from Sohn, The New International Law: Protection of the Rights of Individuals Rather Than States, 32 Am. U. L. Rev. 1, 16 (1982).

> customary law, binding all states, not only members of the United Nations.

Robert Jennings and Arthur Watts[6]

The authors discuss the 'considerable' impact on the traditional sources of international law of changes since World War II in the international community, particularly the growth of international organizations. The vital fact is that states

> have in a short space of time developed new procedures through which they can act collectively. While at present this can be regarded as merely providing a different forum for giving rise to rules whose legal force derives from the traditional sources of international law, there may come a time when the collective actions of the international community within the framework provided by international organisations will acquire the character of a separate source of law.

Through such organizations, the international community can express a general consensus on a particular matter, contributing to the development of custom 'in a way never before possible.' Collective action often takes the form of adoption of resolutions, which may serve functions as diverse as 'declaring' what is believed to be existing law, authoritatively interpreting a legal instrument such as the organization's own constitutive document, and articulating new norms. The authors emphasize

> the dual capacity in which states now act within international organisations, as individual states and as part of the collectivity of the membership of the organisation. It is the change in international organisations from being merely a gathering of individual states to a collective institution of the international community which has contributed most of the changing nature of international organisations in relation to the sources of international law.

Nonetheless, the authors caution, in assessing the legal as opposed to political character or significance of such resolutions, it is important to keep in mind that even a unanimously adopted resolution of the General Assembly 'does not necessarily reflect an *opinio juris* or give rise forthwith to a new customary rule,' and it is important to be attentive to 'the facts relating to the practice to which the resolution relates. . . .'

[6] Oppenheim's International Law (Jennings and Watts, eds.), Vol. 1 (9th ed. 1992), at 45.

QUESTION

Consider the following analysis: 'Far from becoming outmoded and even irrelevant because of the many widely ratified human rights treaties, custom has achieved heightened significance. Once viewed by positivists as "tacit consent" and thereby assimilated to the consensual regime of treaties, custom now has become the treaty itself in shallow disguise, resting apparently not on "tacit" but on "explicit" consent. A resolution voted unanimously or by a large majority in the General Assembly is "law" as much as a treaty. It is indeed superior to the treaty in that it binds all states including those still to be born. Given the new postwar structure of international organizations where states are almost continuously in immediate touch with each other, and thus where multilateral instruments can be decided on promptly rather than go through the awkward procedures of an earlier age, this development may have been inevitable. Probably these new modes of communication with their attendant convenience and speed best explain the heightened significance of custom in the field of human rights.' Do you agree?

RESTATEMENT (THIRD) THE FOREIGN RELATIONS LAW OF THE UNITED STATES

American Law Institute, 1987.

§702. Customary International Law of Human Rights

A state violates international law if, as a matter of state policy, it practices, encourages, or condones

(a) genocide,
(b) slavery or slave trade,
(c) the murder or causing the disappearance of individuals,
(d) torture or other cruel, inhuman, or degrading treatment or punishment,
(e) prolonged arbitrary detention,
(f) systematic racial discrimination, or
(g) a consistent pattern of gross violations of internationally recognized human rights.

Comment:

a. Scope of customary law of human rights. This section includes as customary law only those human rights whose status as customary law is generally accepted (as of 1987) and whose scope and content are generally agreed. The list is not necessarily complete, and is not closed: human rights not listed in this section may have achieved the status of customary law, and some rights might achieve that status in the future.

b. State policy as violation of customary law. In general, a state is responsible for acts of officials or official bodies, national or local, even if the acts were not authorized by or known to the responsible national authorities, indeed even if expressly forbidden by law, decree or instruction. The violations of human rights cited in this section, however, are violations of customary international law only if practiced, encouraged, or condoned by the government of a state as official policy. . . .
A government may be presumed to have encouraged or condoned acts prohibited by this section if such acts, especially by its officials, have been repeated or notorious and no steps have been taken to prevent them or to punish the perpetrators. . . .
Even when a state is not responsible under this section because a violation is not state policy, the state may be responsible under some international agreement that requires the state to prevent the violation. . . .

. . .

l. Gender discrimination. The United Nations Charter (Article 1(3)) and the Universal Declaration of Human Rights (Article 2) prohibit discrimination in respect of human rights on various grounds, including sex. Discrimination on the basis of sex in respect of recognized rights is prohibited by a number of international agreements, including the Covenant on Civil and Political Rights, the Covenant on Economic, Social and Cultural Rights, and more generally by the Convention on the Elimination of All Forms of Discrimination Against Women, which, as of 1987, had been ratified by 91 states and signed by a number of others. The United States had signed the Convention but had not yet ratified it. The domestic laws of a number of states, including those of the United States, mandate equality for, or prohibit discrimination against, women generally or in various respects. Gender-based discrimination is still practiced in many states in varying degrees, but freedom from gender discrimination as state policy, in many matters, may already be a principle of customary international law. . . .

. . .

m. Consistent pattern of gross violations of human rights. The acts enumerated in clauses (a) to (f) are violations of customary law even if the practice is not consistent, or not part of a 'pattern,' and those acts are inherently 'gross' violations of human rights. Clause (g) includes other infringements of recognized human rights that are not violations of customary law when committed singly or sporadically (although they may be forbidden to states parties to the International Covenants or other particular agreements); they become violations of customary law if the state is guilty of a 'consistent pattern of gross violations' as state policy. A violation is gross if it is particularly shocking because of the importance of the right or the gravity of the violation. All the rights proclaimed in the Universal Declaration and protected by the principal International Covenants are internationally recognized human rights, but some rights are fundamental and intrinsic to human dignity. Consistent patterns of violation of such rights as state policy may be deemed 'gross' *ipso facto*. These include, for example, systematic harassment, invasions of the privacy of the home, arbitrary arrest and detention (even if not prolonged); denial of fair trial in criminal cases; grossly disproportionate punishment; denial of freedom to leave a country; denial of the right to return to one's country; mass uprooting of a country's population; denial of freedom of conscience and religion; denial of personality before the law; denial of basic privacy such as the right to marry and raise a family; and invidious racial or religious discrimination. A state party to the Covenant on Civil and Political Rights is responsible even for a single, isolated violation of any of these rights; any state is liable under customary law for a consistent pattern of violations of any such right as state policy.

n. Customary law of human rights and jus cogens. Not all human rights norms are peremptory norms (*jus cogens*), but those in clauses (a) to (f) of this section are, and an international agreement that violates them is void.

o. Responsibility to all states (erga omnes). Violations of the rules stated in this section are violations of obligations to all other states and any state may invoke the ordinary remedies available to a state when its rights under customary law are violated.

NOTE

In the United States, the *Restatements of Law* represent an important reference for many legal issues, although they have no official, legal status. These *Restatements* are adopted and promulgated by the American Law Institute, a private organization not affiliated with the United States Government, whose membership consists of judges, legal academicians, and lawyers involved in private practice and in government. The drafts presented to the Institute for its approval and adoption are generally prepared by leading academicians, who may consult advisory committees with a broader membership.

Such was the case with the *Restatement (Third) of the Foreign Relations Law of the United States*, other portions of which appear in later chapters. The introduction to that *Restatement* states (at p. ix) that it is 'in no sense an official document of the United States.' It notes that in some particulars its rules 'are at variance' with positions taken by the United States Government. Nonetheless, despite this independence and non-official status, it is inevitable that a *Restatement* dealing with international law will in general reflect the broad positions taken by the United States rather than, say, inconsistent or polar positions taken by other, perhaps hostile states.

QUESTIONS

1. Suppose that persons unconnected to the government—secret groups or well-known gangs of private citizens that hate a given minority—attack members of that minority and their property. The state apparatus—police, prosecutors—gives much less attention to such attacks and the related harm than it does to crimes against other members of society. What provision(s) of the ICCPR would be relevant to a claim (a) that the members of the groups or gangs have violated that instrument, or (b) that the state itself has violated it? Does §702 of the *Restatement* affect your answer?

2. Does either the ICCPR or §702 support (a) a civil action for monetary compensation, based on international law and brought by the victim against members of the gang referred to in the preceding question, or (b) a criminal action based on international law brought by the state against the group members who assaulted the victim?

C. HUMAN RIGHTS CHALLENGE TO CONCEPTIONS OF SOVEREIGNTY AND DOMESTIC JURISDICTION

We are now familiar with some implications of the human rights movement's radical premise that a state's treatment of its own citizens, its internal governance on many significant matters, is subject to the norms of international human rights. How can that premise coexist with the classical conception of the state? Are there fixed conceptions of the state, of what it 'is' and 'must be,' and of what related conceptions like autonomy, sovereignty or domestic jurisdiction 'are' or 'must be,' that continue to stand in bold opposition to the challenge of the human rights movement? Or has the nature of the state, and therefore of these allied conceptions, undergone similar radical change? Such are the themes of this Section C.

LEO GROSS, THE PEACE OF WESTPHALIA, 1648–1948

42 Am. J. Int. L. 20 (1948).

[The Peace of Westphalia of 1648 provided a general settlement bringing to an end the Thirty Years' War among several European powers, the Holy Roman Empire, and parts thereof. The Peace, with its adverse consequences for the Empire, is broadly understood as the inauguration of the European state system. In this article, Gross considers the implications of this settlement and its underlying principles for the new international order and international law. The brief excerpts that follow comment on the waning and waxing of different conceptions of the state and of international law that were also examined in the writings of Schachter, p. 42, *supra*, and Koskenniemi, p. 50, *supra*—for example, positivism (voluntarism, consent, related to strict notions of state sovereignty) and naturalism (natural law, the sense of an objective order of nature and law standing above the states, related to qualified notions of sovereignty). Gross comments in these excerpts on the writings of some major theorists of international law who influenced the changing conceptions of the relation between the state and that law: Victoria (1486–1546), Suarez (1548–1617), Grotius (1583–1645), and Vattel (1714–67).]

The acceptance of the United Nations Charter by the overwhelming majority of the members of the family of nations brings to mind the first great European or world charter, the Peace of Westphalia. To it is traditionally attributed the importance and dignity of being the first of several attempts to establish something resembling world unity on the basis of states exercising untrammeled sovereignty over certain territories and subordinated to no earthly authority.

. . .

. . . It would seem possible to distinguish . . . trends of thought on the subject of the binding force of international law prior to 1648. In Victoria one might discern the attempt to base international law on an objective foundation irrespective of the will of the states and to conceive international law as a law above states. In Suarez the objective foundation is at least overshadowed if not replaced by a subjective foundation in the will of the states. Suarez presented the *jus gentium* as a law between states. . . .

. . .

Grotius and several subsequent writers still maintain natural or divine law alongside of customary law as a source of international law. It would seem, however, that with Grotius the accent begins to be transferred from the Law of Nature or divine law to that branch of human law which 'has received its obligatory force from the will of nations, or of many nations.'. . . . [I]t may be useful to conclude this brief survey with a few remarks about Vattel. Vattel, regarded as a Grotian, still maintains the distinction between different types of branches of the Law of Nations. Within the positive law of nations, based on the agreement of nations, he differentiates three divisions: the voluntary, the conventional, and the customary law. The voluntary law proceeds from their presumed consent, the conventional law from their express consent, and the customary law from their tacit consent. Positive international law is distinguished from the natural or necessary law of nations which Vattel undertakes to treat separately. . . .

. . . [Vattel] declares that while the necessary law is at all times obligatory upon the conscience, and that a nation must never lose sight of it when deliberating on the course it must pursue in order to fulfill its duty, it must consult the voluntary law 'when there is question of what it can demand from other states.' It many not be unreasonable to conclude that according to Vattel only those rules of the law of nations which proceed from and are based on the consent of states are enforceable in international relations. . . .

. . .

. . . From the 18th century and in particular, from Vattel onward, however, there can be no doubt as to the trend of the development. It was predominately positivist and consensual. The will of states seems to explain both the contents and the binding force of international law. The concept of the Family of Nations recedes in the background. To have paved the way for this development by liquidating, with a degree of apparent finality, the idea of the Middle Ages of an objective order of things personified by the Emperor in the secular realm, would seem to be one of the more vital aspects of the consequences of the Peace of Westphalia and of its place in the evolution of international relations. Viewed in this light the answer to the question formulated above cannot be doubtful. Instead of heralding the era of a genuine international community of nations subordinated to the rule of the law of nations, it led to the era of absolutist states, jealous of their territorial sovereignty to a point where the idea of an international community became an almost empty phrase and where international law came to depend upon the will of states more concerned with the preservation and expansion of their power than with the establishment of a rule of law. . . .

It may be said, by way of summary, that on the threshold of the modern era of international relations there were two doctrines with respect to the binding force of international law and the existence of an international community of states. The doctrine of Victoria is characterized by an objective approach to the problem of the binding force of international law and by an organic conception of the international community of states. The other doctrine . . . breaks to the fore in the work of Vattel who, emphasizing the independence rather than the interdependence of states, wrote the international law of political liberty. The growth of the voluntaristic conception of international law is accompanied by a weakening of the notion that all states form and are part of an international community. . . . In this era the liberty of states becomes increasingly incompatible with the concept of the international community, governed by international law independent of the will of states. On the contrary this era may be said to be characterized by the reign of positivism in international law. This positivism could not admit the existence of a society of states for the simple reason that it was unable to find a treaty or custom, proceeding from the will of states, which could be interpreted as the legal foundation of a community of states. In the nineteenth century, after the Napoleonic wars, there may be discerned in the Congress [of Vienna, 1814–15] and Concert [of Europe system growing out of the Congress] the beginning of a conscious effort to establish a community of states based on the will of all states or at least on the will of the Great Powers. The Hague Peace Conferences, the League of Nations and, we may confidently assert, the United Nations are further stages in this development cognizable by positive international law.

This reaction against the unrestrained liberty of states, recognized as self-destructive in its ultimate implications was accentuated by a reaction against the prevailing voluntaristic conception of international law. . . . It is common to all these schools of thought that they strive to vindicate for international law a binding force, independent of the will of the states and to substitute for the doctrine that international law is a law of coordination, the old-new doctrine, that international law is, and, if it is to be law, must be, a law of subordination, that is, a law above states.

An international law thus conceived could be interpreted as a law of an international community constituting a legal order for the existing states. It would seem doubtful, however, whether this result can be achieved without the creation of some new institutions or the strengthening of existing institutions. . . .

. . .

R. B. J. WALKER AND SAUL MENDLOVITZ, INTERROGATING STATE SOVEREIGNTY

in Walker and Mendlovitz (eds.), Contending Sovereignties: Redefining Political Community (1990), at 1.

The claim that profound structural transformations are undermining the principle of state sovereignty has been advanced by many analysts from quite different theoretical traditions. No longer, it is said, can states pretend to be autonomous or to exercise a monopoly on the legitimate use of violence in a specific territory. The most important forces that affect people's lives are global in scale and consequence. Even the most powerful states recognize serious global constraints on their capacity to affirm their own national interest above all else. In view of capacities for nuclear destruction, the global mobility of capital, and a new awareness of the fragility of the planetary ecology, the organization of political life within a fragmented system of states appears to be increasingly inconsistent with emerging realities.

. . .

. . . Despite persistent attempts to reify the state as an eternal presence in human affairs, and despite continuing appeals to national identity and the principle of nonintervention, questions about both the meaning and significance of state sovereignty are again firmly on political and scholarly agendas.

. . .

. . . Despite evidence about how states have changed historically, and about how different forms of the state have emerged in various parts of the world and in relation to different locations within geopolitical and economic structures, the state itself has been treated as either the only possible arena in which serious political life can take place or as a merely transient prelude to a universal community of humankind. Eternally present or imminently absent: these options are deeply embedded in our understanding of what it means to be realistic—or utopian—in political life. . . .
Between these options lies a litany of well-known though notoriously imprecise concepts: we may refer to processes of internationalization or globalization, for example, or speak of interdependence and world politics. . . .

. . .

What, for example, is it that is supposedly interdependent (or even dependent)? It could be states, in which case the term is little more than new packaging for a very old theme. The very notion of a state system implies that states are already interdependent in some sense. Much of the history of the modern state system can be interpreted as a series of responses to new forms of interdependence. The Treaty of Westphalia (1648), for example, affirmed a mutual recognition of the need to prevent religious warfare from overwhelming fragile accommodations between secular states. The formalizations of international law and tacit agreements about the rules of the game; the supposedly responsible behavior of the great powers and the institutionalization of international organizations; structural arrangements such as the balance of power; regional alliances; spheres of influence; diplomatic conventions: these are all responses to the interdependence of states. Yet they are usually understood as the necessary conditions of the interstate order that make state sovereignty possible, rather than as fundamental challenges to state sovereignty. Contemporary forms of interdependence may require more complex forms of cooperation between states, but these will only seem radically novel to those who confuse the history of the state system with abstract models of anarchy or a Hobbesian state of nature.

The term *interdependence* has also come to refer to much more than relations between states. It is used in conjunction with various accounts of economic, technological, social, and cultural interaction, of global networks of communication and exchange that refuse to privilege the territorial boundaries of sovereign states. Yet if it is not only states that are interdependent, it is not at all clear what the other subjects of interdependence might be. One might be referring, for example, to multinational corporations or professional organizations, to money traders, computer networks, social movements, or even individuals.

In this case the claim to novelty is more plausible, as well as disconcerting.

For here we quickly run up against the conventional limits of our understanding of what political life can be. What does it mean to speak of all individuals as in some sense interdependent, as part of a common system or even of a global society? What does it mean to speak of the global organization of production, distribution, and exchange, when people's political lives and identities are framed and articulated within particular states, each still jealous of its autonomy and national identity? . . .

. . .

The possibility of anything that might be called world politics rather than interstate relations is in fact denied explicitly by the principle of state sovereignty. This principle affirms the priority of particular peoples—the citizens of particular states—over any universalizing claims to humanity as such. The implication of this affirmation is already apparent from the infamy that still clings to the name of Machiavelli. Recognizing the incongruity between the demands of statemanship or civic *virtù* and the universalizing claims of Christian virtue, Machiavelli had the temerity to privilege the former over the latter. Machiavelli's reputation still lingers, but his radicalism has become formalized as the taken-for-granted foundation of modern political life.

. . .

The principle of state sovereignty formalizes a specific answer to questions about who we are as political beings that were posed in early modern Europe. As an answer, it poses new questions—and suggests appropriate answers to them—in turn. Once we affirm that we are citizens first and humans second—and that when push comes to shove, the claims of citizenship (nationalism, national interest, national security, and so on) must take priority over the claims of humanity in general (universal ethics, universal human rights), some way must be found to resolve the contradiction.

. . .

Yet while there has been considerable interest in the questions posed by the principle of state sovereignty, there has been much less reflection on the questions to which state sovereignty is itself an answer. Who are 'we'? What is the political community within which we ought to be thinking about principles of freedom and obligation, justice and democracy? How ought we to understand the relationship between specific communities and other communities, and between specific communities and humanity in general? How ought we to understand the apparent contradiction between the cultural parochialism of state sovereignty as a product of specifically Western experiences and the embrace of state sovereignty everywhere, not least in connection with the mobilization of nationalist resistances to Western hegemonies? How do states manage to sustain their claims to autonomous authority in the face of competing claims to cultural identity, economic interest, and local commitments? What, in fact, is the political status of 'humanity'?

VIEWS OF SCHOLARS

Consider the following views about the meaning of sovereignty in contemporary international law and argument. Brownlie[7] states that the

> sovereignty and equality of states represent the basic constitutional doctrine of the law of nations, which governs a community consisting primarily of states having a uniform legal personality. If international law exists, then the dynamics of state sovereignty can be expressed in terms of law, and, as states are equal and have legal personality, sovereignty is in a major aspect a relation to other states (and to organizations of states) defined by law.

He describes the principal corollaries of states' sovereignty and equality as

> (1) a jurisdiction, prima facie exclusive, over a territory and the permanent population living there; (2) a duty of non-intervention in the area of exclusive jurisdiction of other states; and (3) the dependence of obligations arising from customary law and treaties on the consent of the obligor.

Koskenniemi[8] observes that it is 'notoriously difficult to pin down the meaning of sovereignty,' but that nonetheless the literature characteristically starts with a definition. Usually the concept is connected with ideas of independence (external sovereignty) and self-determination (internal sovereignty). He quotes a classic definition in an arbitral decision to the effect that sovereignty 'in the relations between States signifies independence; independence in regard to a portion of the globe is the right to exercise therein, to the exclusion of any other States, the functions of a State.' Sovereignty thus implies freedom of action by a state.

If, argues Koskenniemi, this or any agreed-on definition of sovereignty had a clear, ascertainable meaning, then 'whether an act falls within the State's legitimate sphere of action could always be solved by simply applying [that definition] to the case.' But '[t]here simply is no fixed meaning, no natural extent to sovereignty at all.' Thus in disputes between two states, each may base its argument on its own sovereignty. Assuming that 'sovereignty had a fixed content would entail accepting that there is an antecedent material rule which determines the boundaries of State liberty regardless of the subjective will or interest of any particular State.' Such material boundaries not stemming from the free choice of the state 'will appear as unjustified coercion.' It is indeed 'impossible to define "sovereignty" in such a manner as to contain our present perception of the State's full subjective freedom and that of its objective submission to restraints to such freedom.'

[7] Ian Brownlie, Principles of Public International Law (4th ed. 1990), Ch. XIII, 287.

[8] Martti Koskenniemi, From Apology to Utopia: The Structure of International Legal Argument (1989), Ch. 4.

Antonio Cassese, in *Human Rights in a Changing World* (1990) at 13, identifies several characteristics of the prior international community. It was dominated by 'individualistic relationships among [state] members of this anarchic society,' such as the principle of bilateral reciprocity with respect to obligations and remedies. 'The international community was truly a juxtaposition of subjects, each concerned only with its own well-being and its freedom of manoeuvre, each pursuing only its own economic, political and military interests . . . .' Peoples and individuals did not 'count' but were 'absorbed and overshadowed by the "princes": the sovereign states, the only real actors on the world stage.' He comments on the postwar system:

> These two great doctrines [human rights and the self-determination of peoples] have had an extremely important role in international life. They have subverted the very foundations of the world community, by introducing changes, adjustments and realignments to many political and legal institutions. . . . To be sure, they have not changed the actual structure of that community or the main rules of the game. Sovereign states have remained the true holders of power; authority continues to be distributed among various powers; there are still no centralized agencies charged with stating, ascertaining and enforcing the law; each powerful state continues in the main to deal with national interests, with little weight being attributed to collective needs going beyond the requirement to harmonize aspirations and necessities of several states. Nevertheless, the two doctrines have introduced seeds of subversion into this framework, destined sooner or later to undermine and erode the traditional structures and institutions, and gradually—over a very broad space of time, the course of which cannot yet be foreseen—to revolutionize those structures and institutions.

NOTE

The next two readings stress those human rights provisions that point to popular political participation and sovereignty, and that challenge the traditional association of a state with its government, whatever that government's character. Human rights ideals affect the basic political structure of the state, and thus the legitimacy of the sovereign as well as the abstract idea of sovereignty. At least for certain purposes, should recognition of a state's sovereignty depend on that state's compliance with norms of popular participation that rest government on the consent of the people?

HENRY STEINER, THE YOUTH OF RIGHTS

104 Harv. L. Rev. 917 (1991), at 929.

. . . Unlike many components of classical international law, the human rights movement was not meant to work out matters of reciprocal convenience among states—for example, sovereign or diplomatic immunities—or to aim only at regulating areas of historical conflict among states—for example, uses of the sea or airspace, or treatment by a state of its alien population. Rather it reached broad areas of everyday life within states that are vital to the internal rather than international distribution of political power. As international law's aspirations grew, as that law became more critical of and hence more distanced from states' behavior, the potential for conflict between human rights advocates within a state and that state's controlling elites escalated.

Even the most consensual of rights, the right not to be tortured, has a subversive potential. If, as [an] Amnesty International report suggests, torture amounts to the price of dissent because it is 'most often used as an integral part of a government's security strategy,' abolishing torture lowers that price. Oppressive regimes prefer to keep the price high.

Other rights included in the *Universal Declaration* and the *Civil–Political Rights Covenant* influence the structure of government more directly. Abolishing discrimination on grounds of race, ethnicity, religion, or gender can radically alter economic and social arrangements and redirect political power. Protecting rights of speech, expression, and association will give citizens not only security against arbitrary state action, but also the chance to develop a diverse and vibrant civil society that can influence the directions of the state as effectively as governmental policies influence it. Entrenched structures of domination—landholding patterns, power over rural labor, virtual enslavement of children or women or given minorities—may become open to effective challenge.

The stakes for power rise as we move further along the spectrum of human rights. The major human rights instruments empower citizens to 'take part' in government and to vote in secrecy in genuine, periodic, and nondiscriminatory elections. In given circumstances, an authoritarian government can stop torturing and arresting without surrendering its monopoly of power. As events in Eastern Europe illustrate, however, such a government cannot grant the right to political participation without signing its death warrant. 'Throw out the rascals' speaks the more dramatically after decades of unchosen and oppressive regimes.

. . .

Particular clusters of civil-political rights thus challenge many of the world's governments in unavoidable, implacable ways. To some extent the range of human rights that I have mentioned respond to a 'disaster' dimension of the human rights movement. From that perspective, the movement seeks primarily to avert or terminate catastrophes stemming from gross abuses of power—at the extreme another genocide, at less draconian levels the systematic violations of rights to physical integrity in many repressive states. The premise to its goal of protection is captured in the familiar maxim, 'Power tends to corrupt and absolute power corrupts absolutely.' Its modern roots lie in the rule of law tradition, in the notion of law as fences protecting us from each other and, most important, from the state.

But the aspirations of the human rights movement reach beyond the goal of preventing disasters. The movement also has a 'utopian' dimension that envisions a vibrant and broadly based political community. Such a vision underscores the potential of the human rights movement for conflict with regimes all over the world. A society honoring the full range of contemporary human rights would be hospitable to many types of pluralism [and unwilling to embrace or impose] one final truth, at least to the point of allowing and protecting difference. It would not stop at the protection of negative rights but would encourage citizens to exercise their right to political participation, one path toward enabling peoples to realize the right to self-determination. It would ensure room for dissent and alternative visions of social and political life by keeping open and protecting access to the roads toward change.

Violations of human rights, particularly those of a systematic character, are then never gratuitous. Correctly or incorrectly, those holding power understand abuse and terror as instrumental to their keeping it. Sadism and cruelty will be all too evident, but they are harnessed to a politically purposeful scheme rather than fostered or permitted by states for their own sake. [The sorry condition of human rights] in many states should generally be traced to a recurrent structural phenomenon related to power and ideology rather than to individual pathologies.

The fight over rights therefore becomes the fight over the redistribution of power—sometimes direct and unadorned, sometimes imbedded in ideological struggles or in complex ethnic conflicts. How else could South African elites view the insistence of political opponents on one-person-one-vote? How else could the former communist leaders of Eastern Europe have viewed demands for political pluralism? Moreover, some leaders succeeding to power in societies undergoing transformation will surely experience the desire to retain it, denouncing the rights that they now proclaim. The struggle is ongoing, relentless.

. . .

W. MICHAEL REISMAN, SOVEREIGNTY AND HUMAN RIGHTS IN CONTEMPORARY INTERNATIONAL LAW

84 Am. J. Int. L. 866 (1990), at 869.

[Reisman contrasts an earlier conception of state sovereignty ('insulating from legal scrutiny and competence a broad category of events' within a state, later captured in the concept of 'domestic jurisdiction') with the conception inaugurated with the eighteenth century revolutions of the popular will as the source of political authority. Government authority became based on the consent of the people within the state. He refers to Article 21(3) of the UDHR with its reference to the 'will of the people' as the 'basis of the authority of government.']

Although the venerable term 'sovereignty' continues to be used in international legal practice, its referent in modern international law is quite different. International law still protects sovereignty, but—not surprisingly—it is the people's sovereignty rather than the sovereign's sovereignty. Under the old concept, even scrutiny of international human rights without the permission of the sovereign could arguably constitute a violation of sovereignty by its 'invasion' of the sovereign's *domaine réservé*. The United Nations Charter replicates the 'domestic jurisdiction–international concern' dichotomy, but no serious scholar still supports the contention that internal human rights are 'essentially within the domestic jurisdiction of any state' and hence insulated from international law.

This contemporary change in content of the term 'sovereignty' also changes the cast of characters who can violate that sovereignty. Of course, popular sovereignty is violated when an outside force invades and imposes its will on the people. One thinks of the invasion of Afghanistan in 1979 or of Kuwait in 1990. But what happens to sovereignty, in its modern sense, when it is not an outsider but some home-grown specialist in violence who seizes and purports to wield the authority of the government against the wishes of the people, by naked power, by putsch or by coup, by the usurpation of an election or by those systematic corruptions of the electoral process in which almost 100 percent of the electorate purportedly votes for the incumbent's list (often the only choice)? Is such a seizer of power entitled to invoke the international legal term 'national sovereignty' to establish or reinforce his own position in international politics?

Under the old international law, the internal usurper was so entitled, for the standard was de facto control: the only test was the effective power of the claimant. . . . [But that test] stands in stark contradiction to the new constitutive, human rights-based conception of popular sovereignty. . . .

. . .

In many countries, the internal political situation is murky and constitutional procedures for the orderly transfer of power are nonexistent or ineffective. In a flurry of coups and putsches, both outsiders and insiders may be unable to ascertain the popular will, especially if the disorder or tyranny has prevented it from being consulted or expressed. Even in the absence of elections—indeed, even when there are 'supervised' elections—it is often clear that the vast majority of the people detest those who have assumed power and characterize themselves as the government. It is more difficult, however, to say who the people would wish in their stead. . . .

But in circumstances in which free elections are internationally supervised and the results are internationally endorsed as free and fair and the people's choice is clear, the world community does not need to speculate on what constitutes popular sovereignty in that country.

When those confirmed wishes are ignored by a local caudillo who either takes power himself or assigns it to a subordinate he controls, a jurist rooted in the late twentieth century can hardly say that an invasion by outside forces to remove the caudillo and install the elected government is a violation of national sovereignty. Cross-border military actions should certainly never be extolled, for they are necessarily brutal and destructive of life and property. They may well be unlawful for a variety of other reasons. But if they displace the usurper and emplace the people who were freely elected, they can be characterized, in this particular regard, as a violation of sovereignty only if one uses the term anachronistically. . . .

This is not to say that every externally motivated action to remove an unpopular government is now permitted, or that officer corps that feel obsolescence hard upon them can claim a new raison d'être and start scouring the globe for opportunities for 'democratizing' interventions. Authoritative conclusions about the lawfulness of the unilateral use of force, no less than about any other unilateral action, turn on many contextual factors: e.g., the contingencies allegedly justifying the unilateral use, the availability of feasible persuasive alternatives, the means of coercion selected, the level of coercion used (the classic test of necessity and proportionality), whether the objectives of the intervener include internationally illicit aims, the aggregate consequences of inaction, and the aggregate consequences of action. But it is to say that the suppression of popular sovereignty may be a justifying factor, not a justification *per se* but a *conditio sine qua non*. And it is to say that the word 'sovereignty' can no longer be used to shield the actual suppression of popular sovereignty from external rebuke and remedy.

International law is still concerned with the protection of sovereignty, but, in its modern sense, the object of protection is not the power base of the tyrant who rules directly by naked power or through the apparatus of a totalitarian political order, but the continuing capacity of a population freely to express and effect choices about the identities and policies of its governors. . . . President Marcos violated Philippine sovereignty, General Noriega violated Panamanian sovereignty, and the Soviet blockade of Lithuania violated its sovereignty. Fidel Castro violates Cuban sovereignty by mock elections that insult the people whose fundamental human rights are being denied, no less than the intelligence of the rest of the human race. . . .

. . .

The international human rights program is more than a piecemeal addition to the traditional corpus of international law, more than another chapter sandwiched into traditional textbooks of international law. By shifting the fulcrum of the system from the protection of sovereigns to the protection of people, it works qualitative changes in virtually every component. . . .

. . .

Unambiguous situations, however, may be exceptions. When the internationally supervised elections result in an absence of consensus on who should govern, *or* the integrity of the elections is doubtful, *or* there have been no elections, *or* a civil insurrection has left diverse groups vying for power, no one can be sure that the unilateral intervener from the outside is implementing popular wishes. To varying extents, the intervener will be shaping them. In some circumstances, the banner of popular sovereignty can become a fig leaf for its suppression by foreign intervention, especially when governments bent on intervention maintain stables of alternative local leaders who can be brought forward to authorize an invasion at the appropriate time. In practice, therefore, there may be a factual 'gray' area between unequivocal expressions of popular will through internationally supervised, observed or validated elections, on the one hand, and the atrocities that warrant humanitarian intervention, on the other. Situations falling into the gray area will simply not lend themselves to unilateral action.

. . .

ADDITIONAL READING

Helmut Steinberger, *Sovereignty*, in R. Bernhardt (ed.), *Encyclopedia of Public International Law*, Vol. 10 (1987) at 397.

NOTE

The following decision of the Permanent Court of International Justice is among the best known judicial opinions on the question of domestic jurisdiction. Consider what relation the notion of domestic jurisdiction as used in this opinion bears to the ideas of sovereignty just discussed.

TUNIS–MOROCCO NATIONALITY DECREES

Permanent Court of International Justice, 1923.
P.C.I.J., Ser. B, No. 4.

[Article 15 of the Covenant of the League of Nations provided, in effect, that any dispute between members of the League likely to lead to a rupture that was not submitted to arbitration under another Covenant provision should be submitted to the Council of the League. The Council could hear the facts and make recommendations. Paragraph 8 of Article 15 stated:

> If the dispute between the parties is claimed by one of them, and is found by the Council, to arise out of a matter which by international law is solely within the domestic jurisdiction of that party, the Council should so report, and shall make no recommendation as to its settlement.

In 1921, nationality decrees were promulgated in Tunis and Morocco, both French Protectorates. They provided that persons born in Tunis or Morocco of parents at least one of whom was also born there had French (or Tunisian or Moroccan) nationality. Such nationality apparently implied a liability to military service. Concerned with enforcement of the decrees against British subjects, the British Government referred the dispute to the Council under Article 15. The French Government invoked Paragraph 8 of that Article. Pursuant to Article 14 of the Covenant, the Council then requested the P.C.I.J. for an advisory opinion on the question 'whether the dispute . . . is or is not by international law solely a matter of domestic jurisdiction . . . . ' Excerpts from the opinion of the Court follow.]

From one point of view, it might well be said that the jurisdiction of a State is *exclusive* within the limits fixed by international law—using this expression in its wider sense, that is to say, embracing both customary law and general as well as particular treaty law. But a careful scrutiny of paragraph 8 of Article 15 shows that it is not in this sense that exclusive jurisdiction is referred to in that paragraph.

The words 'solely within the domestic jurisdiction' seem rather to contemplate certain matters which, though they may very closely concern the interests of more than one State, are not, in principle, regulated by international law. As regards such matters, each State is sole judge.

The question whether a certain matter is or is not solely within the jurisdiction of a State is an essentially relative question; it depends upon the development of international relations. Thus, in the present state of international law, questions of nationality are, in the opinion of the Court, in principle within this reserved domain.

For the purpose of the present opinion, it is enough to observe that it may well happen that, in a matter which, like that of nationality, is not, in principle, regulated by international law, the right of a State to use its discretion is nevertheless restricted by obligations which it may have undertaken towards other States. In such a case, jurisdiction which, in principle, belongs solely to the State, is limited by rules of international law. Article 15, paragraph 8, then ceases to apply as regards those States which are entitled to invoke such rules, and the dispute as to the question whether a State has or has not the right to take certain measures becomes in these circumstances a dispute of an international character and falls outside the scope of the exception contained in this paragraph. . . .

. . .

. . . Under the terms of paragraph 8, the League's interest in being able to make such recommendations as are deemed just and proper in the circumstances with a view to the maintenance of peace must, at a given point, give way to the equally essential interest of the individual State to maintain intact its independence in matters which international law recognizes to be solely within its jurisdiction.

. . .

. . . [T]he mere fact that one of the parties appeals to engagements of an international character in order to contest the exclusive jurisdiction of the other is not enough to render paragraph 8 inapplicable. But when once it appears that the legal grounds (*titres*) relied on are such as to justify the provisional conclusion that they are of juridical importance for the dispute submitted to the Council, and that the question whether it is competent for one State to take certain measures is subordinated to the formation of an opinion with regard to the validity and construction of these legal grounds (*titres*), the provisions contained in paragraph 8 of Article 15 cease to apply and the matter, ceasing to be one solely within the domestic jurisdiction of the State, enters the domain governed by international law.

. . . [T]he Court holds, contrary to the final conclusions of the French Government, that it is only called upon to consider the arguments and legal grounds (*titres*) advanced by the interested Governments in so far as is necessary in order to form an opinion upon the nature of the dispute. . . .

[The opinion observed that whether the general competence of a State to enact nationality legislation for its own national territory extended to protected territory here depended on a number of treaties involving France, its Protectorates, and third countries including Britain, which claimed that treaties between it and Tunis and Morocco gave British subjects a measure of extraterritoriality incompatible with imposition of another nationality. The French Government claimed that such treaties had lapsed after Tunis and Morocco became Protectorates. The Court noted that in such circumstances, the question whether France had exclusive jurisdiction depended on 'an examination of the whole situation as it appears from the standpoint of international law. The question therefore is no longer solely one of domestic jurisdiction as defined above.' The Court would have to consider international-law rules about the duration and validity of treaties. Britain also relied on clauses in treaties between it and France. The Court stated that questions of interpretation of those clauses were not within the domestic jurisdiction of the French Government. For these reasons, the Court replied to the question submitted to it by the Council in the negative.]

COMMENT ON DOMESTIC JURISDICTION

As Brierly observed in 1925, the doctrine of domestic jurisdiction had become a catchword 'capable proving as great a hindrance to the orderly development of [international law] as the somewhat battered ideas of sovereignty, State equality, and the like have been in the present.'[9] As Brownlie puts it: 'The corollary of the independence and equality of states is the duty on the part of states to refrain from intervention in the internal or external affairs of other states.'[10]

[9] Brierly, Matters of Domestic Jurisdiction, VI Brit. Y. Bk. Int. L. 8 (1925).

[10] Ian Brownlie, Principles of Public International Law, 291 (4th ed., 1990).

In the *Tunis–Morocco* case, the notion of domestic jurisdiction was relevant to the competence of the Council of the League. It is relevant in other contexts as well, including many involving human rights issues. In all such settings, the states parties to a dispute or an international organ or a third-party decision-maker are required to identify for one or another purpose the critical boundary line between matters regulated to some degree by international law and matters recognized to be within the unfettered discretion of national governments (i.e., recognized to be a matter of domestic jurisdiction). The opinion in *Tunis–Morocco Nationality Decrees* tells us that the question of where to draw that line 'is an essentially relative question; it depends upon the development of international relations.'

The definition, or indeed the very notion, of a boundary line raises some vexing issues. The line could be described in several ways. For example, matters within a state's domestic jurisdiction could be identified by stating the sum of all relevant rules of international law that impose restraints or affirmative duties on states. A state's rules or conduct not within that sum could then be viewed as within its domestic jurisdiction. Alternatively, one could follow an 'essentialist' theory and attempt affirmatively to describe those matters recognized—now, forever?—to be within a state's domestic jurisdiction. Brownlie, *op. cit* at 292, notes that the concept 'is mysterious only because many have failed to see that it really stands for a tautology.'

We here describe the relevance of domestic jurisdiction in the context of the competence of political organs of the United Nations. Article 2(7) of the Charter provides:

> Nothing contained in the present Charter shall authorize the United Nations to intervene in matters which are essentially within the domestic jurisdiction of any State or shall require the Members to submit such matters to settlement under the present Charter; but this principle shall not prejudice the application of enforcement measures under Chapter VII.

Note the difference in wording between Article 15(8) of the Covenant of the League and this Article 2(7). As in other contexts in which the issue of domestic jurisdiction arises, Article 2(7) involves the fundamental tension in international relations between state sovereignty and external restraints on that sovereignty. All treaties, as well indeed as all norms of customary international law, implicity address this tension and resolve it in their different ways. In the case of the Charter, note how Articles 1 and 2 reflect the tension; compare the 'internationalist' spirit of Articles 1(1) and 2(6) with the traditional themes of state sovereignty in paras. (1), (4) and (7) of Article 2.

Within the United Nations, states have frequently invoked Article 2(7) to support their argument that the General Assembly is acting beyond its powers by debating certain issues, whether or not the debate in question terminates with concrete action by the Assembly such as the voting of a recommendation or resolution. States' arguments resisting the Assembly's consideration of a matter also involve the broad mandates for the General

Assembly of Article 10, authorizing it to 'discuss any questions or any matters within the scope of the present Charter' as well as to 'make recommendations' to member states on such questions or matters, and of Article 13, authorizing it to initiate studies and make recommendations for the purpose of 'assisting in the realization of human rights and fundamental freedoms.'

The issues leading to tensions between the General Assembly and a member state have frequently involved human rights issues, ranging from colonialism to the apartheid regime of South Africa. As early as 1946 the Assembly condemned discrimination in South Africa as in violation of the Charter despite that state's claim that how it treated its own citizens was a matter of domestic jurisdiction. In these situations, the Assembly has interpreted Article 2(7) restrictively, in order to permit debate and resolutions on issues that generally have a clear international impact, even if there is no consensus over the degree to which international law does or should regulate the matter.

The International Court of Justice has never been asked for an advisory opinion whether the Assembly's action with respect to a given human rights issue (debate alone, recommendation, general or specific resolution, declaration) amounted to a forbidden intervention into domestic jurisdiction. In the meantime, the continuing practice of the political organs of the UN in overriding the claim of domestic jurisdiction, particularly but not exclusively with respect to human rights issues, may well have created new law on the question, at least with respect to its meaning in Article 2(7). Given the breadth of the UN's concerns and its universal membership, that meaning is likely to erode the significance of claims of domestic jurisdiction in international law generally.

NOTE

Consider the following observations in Oscar Schachter, *International Law in Theory and Practice* (1991) at 332. Schachter notes the views of many international lawyers that 'domestic jurisdiction is only the negative of international obligation, it has no positive content of its own.' But, he argues, this logic

> does not entirely suffice to solve the problem of domestic jurisdiction. The reason for this is that States do in fact recognize that even though a 'matter' may be governed by an international obligation, it may under some circumstances be regarded as appropriate for domestic action alone. For example, a right to a fair trial is clearly internationally recognized; however an isolated or minor infringement of that right would not be regarded as of international concern. The same would hold for cases of discrimination on grounds of race or sex. The particular cases fall within domestic jurisdiction even if the relevant obligation applies to the class of infringements. It may be argued that the examples only show that States do not

take a violation seriously unless it is 'gross' or 'systematic' and that this reflects judgments made on political or prudential grounds. Even if that is true, it is not inconsistent with recognizing that States also have an obligation to recognize that the 'domestic jurisdiction' principle merits consideration in determining whether an infringement of a right should be taken up by others. . . .

QUESTIONS

1. Compare the preceding comment by Schachter with §702 of the Restatement, p. 145, *supra*. Are they to the same effect?

2. Does the *Tunis–Morocco* opinion leave any room for some eternal, 'essentialist' conception of domestic jurisdiction or state sovereignty, such that matters within it are forever immune from international regulation? How would you determine the content of such a conception—through what method, under what assumptions—and what difficulties would you encounter in doing so?

3. Consider the following thoughts: 'Did the human rights movement erode sovereignty and domestic jurisdiction? The answer is "yes, of course" and "no, of course not," depending on what you're referring to. The "yes" answer is almost tautological. First, human rights ideas are distanced from state behavior, way above it in terms of moral norms, hence very critical, utopian. Second, they are meant to control conduct within a state involving relations between a government and citizens. Put those two notions together, and sovereignty is no longer what it was. But have human rights norms seriously affected state behavior? That's where the "no" answer comes in.' Comment.

ADDITIONAL READING

Anthony D'Amato, *Domestic Jurisdiction*, in R. Bernhardt (ed.), *Encyclopedia of Public International Law*, Vol. 10 (1987), at 132.

4

What are Rights, Are They Everywhere, and Everywhere the Same?: Cultural Relativism

Thus far the materials have described but barely commented on the fundamental characteristic of the UDHR and ICCPR, their foundation in the rhetoric and concept of *rights*. Many view that rhetoric as unproblematic, as the central and inevitable component of a discourse about human dignity and about humane treatment of individuals by governments. Others, to the contrary, view a discourse about rights as alien and harmful, disruptive of traditional social structures. Consider the following queries:

(1) Why do we have a human rights rather than human duties movement? Why does the language of rights dominate the texts of the declarations and treaties as well as, in many states, the slogans and polemics of political debate?
(2) Is that language intrinsically superior to other possible ones. Is it essential to the values and goals of the human rights movement? Or is the currency of international human rights a matter of historical contingency, in that the postwar movement in fact found its roots in liberal rights-oriented political cultures?
(3) Do any particular characteristics or substantive content necessarily attach to the language of rights? Or are rights empty receptacles open to many different types of values and ideas—for example, a right to social stability and to ongoing tradition as opposed, say, to a right to free speech and political advocacy?
(4) Are the language of rights and the content of rights in the UDHR or ICCPR universal? Or are the values that are incorporated in these instruments particular (relative) to given cultures or states? If so, are there ways of bridging the differences among cultures or states so that all are bound to recognize the same rights?

Such are the questions that we encounter in Chapter 4. The first two sets of questions are addressed in Section A, while Sections B, C and D explore the last two sets under the broad rubric of cultural relativism.

A. THE NOTION OF 'RIGHTS': ORIGINS AND RELATION TO 'DUTIES'

We here consider different understandings, historical and contemporary, of the notion of 'rights' and inquire whether rights have inherent implications for a society's moral, political and socio-economic order. For example, does rights rhetoric in a constitution and statutes, or in a dominant moral and political theory, point to an individualistic or communitarian or other type of society? Does it necessarily assume certain institutional arrangements for government, such as a constitutional separation of powers and particularly an independent judiciary?

Exploration of such questions provides background to the discussion of cultural relativism in Section B. The readings begin with descriptions of the evolution from earlier concepts of natural law and natural rights to contemporary notions of right.

BURNS WESTON, HUMAN RIGHTS

20 New Encyclopaedia Britannica (15th ed. 1992) at 656.

. . .

The expression 'human rights' is relatively new, having come into everyday parlance only since World War II and the founding of the United Nations in 1945. It replaces the phrase 'natural rights,' which fell into disfavour in part because the concept of natural law (to which it was intimately linked) had become a matter of great controversy, and the later phrase 'the rights of Man' . . .

. . .

It was primarily for the 17th and 18th centuries, however, to elaborate upon this modernist conception of natural law as meaning or implying natural rights. The scientific and intellectual achievements of the 17th century . . . encouraged a belief in natural law and universal order; and during the 18th century, the so-called Age of Enlightenment, a growing confidence in human reason and in the perfectability of human affairs led to its more comprehensive expression. Particularly to be noted are the writings of the 17th-century English philosopher John Locke—arguably the most important natural law theorist of modern times—and the works of the 18th-century Philosophes centred mainly in Paris, including Montesquieu, Voltaire, and Jean-Jacques Rousseau. Locke argued in detail, mainly in writings associated with the Revolution of 1688 (the Glorious Revolution), that certain rights self-evidently pertain to individuals as human beings (because they existed in 'the state of nature' before humankind entered civil society); that chief among them are the rights to life, liberty (freedom from arbitrary rule), and property; that, upon entering civil society (pursuant to a 'social contract'), humankind surrendered to the state only the right to enforce these natural rights, not the rights themselves; and that the state's failure to secure these reserved natural rights (the state itself being under contract to safeguard the interests of its members) gives rise to a right to responsible, popular revolution. The Philosophes, building on Locke and others and embracing many and varied currents of thought with a common supreme faith in reason, vigorously attacked religious and scientific dogmatism, intolerance, censorship, and social-economic restraints. They sought to discover and act upon universally valid principles harmoniously governing nature, humanity, and society, including the theory of the inalienable 'rights of Man' that became their fundamental ethical and social gospel.

All this liberal intellectual ferment had, not surprisingly, great influence on the Western world of the late 18th and early 19th centuries. Together with the practical example of England's Revolution of 1688 and the resulting Bill of Rights, it provided the rationale for the wave of revolutionary agitation that then swept the West, most notably in North America and France. Thomas Jefferson, who had studied Locke and Montesquieu and who asserted that his countrymen were a 'free people claiming their rights as derived from the laws of nature and not as the gift of their Chief Magistrate,' gave poetic eloquence to the plain prose of the 17th century in the Declaration of Independence proclaimed by the 13 American Colonies on July 4, 1776: 'We hold these truths to be self-evident, that all men are created equal, that they are endowed by their Creator with certain unalienable Rights, that among these are Life, Liberty and the Pursuit of Happiness.' Similarly, the Marquis de Lafayette . . . imitated the pronouncements of the English and American revolutions in the [French] Declaration of the Rights of Man and of the Citizen of August 26, 1789. Insisting that 'men are born and remain free and equal in rights,' the declaration proclaims that 'the aim of every political association is the preservation of the natural and imprescriptible rights of man,' identifies these rights as 'Liberty, Property, Safety and Resistance to Oppression,' and defines 'liberty' so as to include the right to free speech, freedom of association, religious freedom, and freedom from arbitrary arrest and confinement (as if anticipating the Bill of Rights added in 1791 to the Constitution of the United States of 1787).

In sum, the idea of human rights, called by another name, played a key role in the late 18th- and early 19th-century struggles against political absolutism. It was, indeed, the failure of rulers to respect the principles of freedom and equality, which had been central to natural law philosophy almost from the beginning, that was responsible for this development. . . .

. . . [B]ecause they were conceived in essentially absolutist—'inalienable,' 'unalterable,' 'eternal'—terms, natural rights were found increasingly to come into conflict with one another. Most importantly, the doctrine of natural rights came under powerful philosophical and political attack from both the right and the left.

In England, for example, conservatives Edmund Burke and David Hume united with liberal Jeremy Bentham in condemning the doctrine, the former out of fear that public affirmation of natural rights would lead to social upheaval, the latter out of concern lest declarations and proclamations of natural rights substitute for effective legislation. In his *Reflections on the Revolution in France* (1790), Burke, a believer in natural law who nonetheless denied that the 'rights of Man' could be derived from it, criticized the drafters of the Declaration of the Rights of Man and of the Citizen for proclaiming the 'monstrous fiction' of human equality, which, he argued, serves but to inspire 'false ideas and vain expectations in men destined to travel in the obscure walk of laborious life.' Bentham, one of the founders of Utilitarianism and a nonbeliever, was no less scornful. 'Rights,' he wrote, 'is the child of law; from real law come real rights; but from imaginary laws, from 'law of nature,' come imaginary rights. . . . Natural rights is simple nonsense; natural and imprescriptible rights (an American phrase), rhetorical nonsense, nonsense upon stilts.' Hume agreed with Bentham: natural law and natural rights, he insisted, are unreal metaphysical phenomena.

This assault upon natural law and natural rights, thus begun during the late 18th century, both intensified and broadened during the 19th and early 20th centuries. John Stuart Mill, despite his vigorous defense of liberty, proclaimed that rights ultimately are founded on utility. The German jurist Friedrich Karl von Savigny, England's Sir Henry Maine, and other historicalists emphasized that rights are a function of cultural and environmental variables unique to particular communities. And the jurist John Austin and the philosopher Ludwig Wittgenstein insisted, respectively, that the only law is 'the command of the sovereign' (a phrase of Thomas Hobbes) and that the only truth is that which can be established by verifiable experience. By World War I, there were scarcely any theorists who would or could defend the 'rights of Man' along the lines of natural law. . . .

Yet, though the heyday of natural rights proved short, the idea of human rights nonetheless endured in one form or another. The abolition of slavery, factory legislation, popular education, trade unionism, the universal suffrage movement—these and other examples of 19th-century reformist impulse afford ample evidence that the idea was not to be extinguished even if its transempirical derivation had become a matter of general skepticism. But it was not until the rise and fall of Nazi Germany that the idea of rights—human rights—came truly into its own. . . .

. . .

To say that there is widespread acceptance of the principle of human rights on the domestic and international planes is not to say that there is complete agreement about the nature of such rights or their substantive scope—which is to say, their definition. Some of the most basic questions have yet to receive conclusive answers. Whether human rights are to be viewed as divine, moral, or legal entitlements; whether they are to be validated by intuition, custom, social contract theory, principles of distributive justice, or as prerequisites for happiness; whether they are to be understood as irrevocable or partially revocable; whether they are to be broad or limited in number and content—these and kindred issues are matters of ongoing debate and likely will remain so as long as there exist contending approaches to public order and scarcities among resources.

. . .

DAVID SIDORSKY, CONTEMPORARY REINTERPRETATIONS OF THE CONCEPT OF HUMAN RIGHTS

in Sidorsky (ed.), Essays on Human Rights 88 (1979), at 89.

. . .

In a context in which escalating rhetorical support for human rights as a cause célèbre coexists with systematic abuse of many of these rights, it is not surprising that the very meaning of human rights has become contested. Radically different definitions and interpretations of human rights have been proposed, each of which claims the banner of human rights. . . .

. . .

In turning to the contemporary interpretations and the current uses of the concept of human rights, even before examining the ways in which the idea is contested, an appropriate starting point is the recognition that the term seems to be fulfilling two different, although consistent, functions. On the one hand, the phrase *universal human rights* is used to assert that universal norms or standards are applicable to all human societies. This assertion has its roots in ancient ideas of universal justice and in medieval notions of natural law. . . .

On the other hand, the idea of human rights is used to affirm that all individuals, solely by virtue of being human, have moral rights which no society or state should deny. This idea has its classic source in seventeenth- and eighteenth-century theories of natural rights. . . .

. . .

The Theory of Natural Rights

The theory of natural rights had a major influence in the development of the political self-consciousness of modern Western society. The idea of natural rights as the sole justification for any political society was a challenge to all established political authority. From the perspective of the theory of natural rights, all the recognized theories of legitimacy—the divine rights of kings, the pragmatic necessity of stable political rule, conformity to divine or natural law or rootedness in historical and institutional traditions—were inadequate. A political regime was justified only if it satisfied the natural rights of its citizens.

The current function of the theory of human rights, unlike the doctrine of natural rights, is not primarily that of serving as a principle of legitimacy within a particular national state. It has become part of an effort to develop standards of achievement with respect to citizens' rights within an international community. Yet it is significant in this context to recognize the continuity between the traditional theory of natural rights and recent formulations of human rights. Six elements of that continuity merit special examination.

First, it was characteristic of theorists of natural rights to develop a list of specific rights. Although these rights allegedly derived from the universal and evident desires of all men, the content of various bills or declarations of rights differed. The appearance or omission of a specified right, like the right to property, was an index of the importance given in social policy to the defense or realization of that right. This tradition has been adopted in the theory of human rights and has resulted, for example, in the articulation of the more extended list of thirty rights that mark the Universal Declaration of Human Rights.

Second, in all traditional theories of natural rights, such rights were ascribed only to human beings. . . .

The further implication is that any human being, solely by virtue of his potential ability to exercise rational choice, had rights. It was in this sense that the theory of natural rights proposed the equality of all men. Since having natural rights was intrinsically connected to being a human being, there was a basis for the later transition from the phraseology of *natural rights* to that of *human rights*.

Third, a major characterization of natural rights derived from this belief that rights are the properties of persons capable of exercising rational choice. For, when men asserted their natural rights they were expressing their autonomy as individuals. Hence, the model or pattern for the exercise of natural rights became the protection of the sphere of the autonomous individual from arbitrary incursion by the state or other coercive association. The listing of the right to life, for example, did not involve a commitment to the extension or universalization of health care or to actions for shaping a safer environment but to a rule of law that would restrain arbitrary acts of violence, especially those of governmental authorities, against individuals. Similarly, the natural right to liberty did not refer to support of policies that would enhance self-realization through the universalization of education, but it did require the legal protection of individuals against arbitrary imprisonment.

Since the natural rights of men were bound intrinsically to their capacity to exercise rational choice as autonomous beings, the list of natural rights comprised what have been termed negative freedoms, rather than positive liberty, that is, the freedoms that protect the individual *from* the invasion of his domain of selfhood or privacy rather than the freedom of the individual or group *to* achieve its purposes or ideals. This stress is evident in the many detailed lists of the declarations or bills or rights that proliferated in the late eighteenth and early nineteenth centuries. The inclusion of a number of human rights that relate to social and economic development is a point of difference between the classic theory of natural rights and the theory that led to the Universal Declaration of Human Rights. That inclusion, and the priority to be assigned to social and economic rights, has become the single most contested item in discussions of contemporary theory of human rights.

. . .

Fifth, natural rights, as the adjective shows, derive from the order of *nature* or from the nature of 'natural man' but not from society or history. Indeed, as truths of nature they were held to be rationally self-evident: that is, if the meaning of the term is understood, then all rational beings could intuitively know that men had natural rights. In this view, the recognition that the theory of natural rights had its genesis in the rise to power of the middle class no more shows the relativity of natural rights than the fact that the calculus was discovered in seventeenth-century Germany or England makes its truth relative to that place and time. While rational intuition is no longer relevant for the contemporary views of human rights, the belief that rights are universal, and not relative, to particular social or historical culture has become, if anything, even more important in their use as international norms.

. . .

NOTE

The following three readings describe their authors' understandings of basic characteristics of rights in contemporary legal and political discourse. Two of the authors compare these characteristics with the tradition of natural law and natural rights; each draws different distinctions between those traditions and contemporary usage. All three authors stress the priority attaching to claims based on rights, and each notes the problem of conflicts among rights. Klare explores the issue of conflict, as well as the related problem of the indeterminacy of rights discourse. Taylor stresses that recent decades have moved us from an initial orthodoxy about the individualistic nature of rights to conceptions of collective rights and positive obligations of states.

EUGENE KAMENKA, HUMAN RIGHTS, PEOPLES' RIGHTS

in James Crawford (ed.), The Rights of Peoples (1988), at 127.

Rights are claims that have achieved a special kind of endorsement or success: legal rights by a legal system; human rights by widespread sentiment or an international order. All rights arise in specific historical circumstances. They are claims made, conceded or granted by people who are themselves historically and socially shaped. They are asserted by people on their own behalf or as perceived and endorsed implications of specific historical traditions, institutions and arrangements or of a historically conditioned theory of human needs and human aspirations, or of a human conception of a Divine plan and purpose. In objective fact as opposed to (some) subjective feeling, they are neither eternal nor inalienable, neither prior to society or societies nor independent of them. Some such rights can be singled out, and they often are singled out, as social ideals, as goals to strive toward. But even as such, they cannot be divorced from social content and context.

Claims presented as rights are claims that are often, perhaps usually, presented as having a special kind of importance, urgency, universality, or endorsement that makes them more than disparate or simply subjective demands. Their success is dependent on such endorsement—by a government or a legal system that has power to grant and protect such rights, by a tradition or institution whose authority is accepted in those circles that recognize these claims as rights, by widespread social sentiment, regionally, nationally, or internationally.

Claims, whether presented as rights or not, conflict. So do the traditions, institutions and authorities that endorse the claim as a right. They conflict both with each other and, often, in their internal structure, implications and working out. . . .

The concept of human rights is no longer tied to belief in God or natural law in its classical sense. But it still seeks or claims a form of endorsement that transcends or pretends to transcend specific historical institutions and traditions, legal systems, governments, or national and even regional communities. Like moral claims more generally, it asserts in its own behalf moral and sometimes even logical priority—connection with the very concept (treated as morally loaded) of what it means to be a human being or a person, or of what it means to behave morally. These are questions on which moral philosophers do have a certain expertise, at least in seeing where the difficulties lie, and on which they, like ordinary people throughout the world, have long disagreed and continue to disagree.

CHARLES TAYLOR, HUMAN RIGHTS: THE LEGAL CULTURE

in UNESCO, Philosophical Foundations of Human Rights (1986), at 49.

. . .

Our legal tradition, at least in the West, accords a special position to what are called 'personal' rights (*droits subjectifs*). By recognizing certain rights, we give individuals or groups within society the power to set limits on the actions of that society. . . . As a subject of law, I have the power under law to assert my right, thus rendering the decision or legislation in question null and void. . . .

The American philosopher of law, Ronald Dworkin, borrows an image from the game of bridge to explain this principle: it is as though the individual were given a 'trump' card that enables him to invalidate the results of the normal social decision-making process whenever it encroaches on his protected sphere. . . . The point is that if any attempt is made to inhibit my freedom of speech, I can play my 'trump' card and have the decision revoked in a court of law.

This legal system corresponds to certain conceptions of man and society. In order to grasp what makes it special, it may be compared with a system where political decisions are also restricted by a fundamental law but where there are no personal rights. Thus, the fundamental law might prohibit others from killing me, imprisoning me, silencing me, etc., without giving me the right to life, liberty, free speech, etc. in the modern sense. At first glance, it might seem that the same thing is being said in a different way in the two cases. In point of fact, however, there is a major difference between them.

In a system of personal rights, the law does not confine itself to proscribing action X or Y: the individual is also empowered to annul the measures causing X or Y. The limits imposed by the law are regarded as privileges enjoyed by individuals, as a legal power at their disposal.

. . .

The system of personal rights thus has two effects: (a) it places limits on the actions of governments and on collective decisions by offering a measure of protection to individuals and specific groups; (b) it offers individuals and specific groups the right to seek redress and gives them a margin of liberty in the imposition of these limits.

It shares characteristic (a) with every system of fundamental law, including pre-modern natural law. Characteristic (b), however, is peculiar to it.

We have seen, therefore, how deeply the system of personal rights is rooted in the political traditions and ways of thinking of the West today. It goes hand in hand with the assertion of the status of the individual, of his liberty and of his precedence over society. . . .

. . . This is not to say that the performance of Western societies in the field of human rights has been particularly brilliant. One has only to think of the dramatic interruptions (such as the Nazi regime, the Latin American dictatorships, etc.) or the systematic injustices practised throughout the course of Western history (the age-long violation of the rights of certain social classes and of women) or the external behaviour of the same societies—colonialism, imperialism—reflecting contempt for the rights of other peoples.

But the main question which arises is whether the system of personal rights can have equal relevance in political cultures which do not attach the same fundamental importance to the freedom of the individual or of specific groups. This requires not only (a) a conception of human well-being going beyond the individual's welfare as a member of society, but also (b) a conception of such well-being as including the power to determine, on the basis of this broader definition and within certain limits, the individual's relationship with society and with its forms and functions.

. . .

To sum up, we can identify two conceptions of rights: 'A', the original Western conception, that of personal rights, and 'B', the conception of rights as fundamental social objectives that cannot be abandoned. The two approaches logically tend towards different enumerations of rights: 'A' concentrates mainly on the individual's ability to determine the way society behaves towards him and hence is concerned with life, freedom of speech, of association, opinion and religion; the right to a trial that fulfils certain requirements if he is charged with an offence, etc. Conception 'B' naturally tends towards a 'social' bill of rights: full employment, guaranteed income, education, etc., which may be achieved by social action.

However, there is a basic ambiguity. Approaches 'A' and 'B' cannot be clearly distinguished as two independently valid points of view, because most of the supporters of 'B' would also claim to be fulfilling the conditions of 'A'. This confusion is not likely to be dispelled overnight.

. . .

We have been speaking so far in terms of individual rights. But certain rights may also be claimed on behalf of communities. Natural law originated in the seventeenth century in the context of a strictly atomistic view of society. It has developed since then and with it the designation of the rights that we claim. This has happened in two main ways. First, Romanticism drew attention to man's national dimension. It made us see man as a cultural being who develops his humanity through a language and through the body of knowledge that has already been expressed in a culture, whether it be in the form of art, music, literature, or family and political traditions, etc.

The result is that if gaps or deficiencies appear in the language or if the expressions of the culture become inaccessible for whatever reason, the very development of the individual's personality is threatened.

But languages and cultures are the property of certain communities. Isolated individuals cannot create or preserve them; only societies can. If there is a right to be claimed—that of preserving one's language and culture—it will have to be attributed to the community. In this case it is the right of the nation or of the cultural group that is at issue.

If this cultural view of man is accepted, it is natural to claim, as the complement of individual liberty, the right of peoples to self-determination. For if man is a cultural being, and there are significant differences between cultures, and if cultures are also reflected in differing political traditions, each people must be able to determine its own political destiny in the interests of full self-expression.

. . .

The second great change since the seventeenth century has been the gradual abandonment of atomistic premises. If the individual is regarded as self-sufficient, the right to life requires only that others refrain from attacking him. But when one considers the extent to which the very conditions of life—especially those of a fully human existence—are assured by society, it must be conceded that the rights of the individual may demand not only the non-interference of others but their active assistance.

. . .

The result is that today we proclaim such rights as that to education which can only be assured by collective action. The right to education is observed only if society assumes the task of promoting it. Modern enumerations of rights therefore entail positive action. And this is what facilitates the transition from conception 'A' to conception 'B' referred to above.

These two changes, concerning collective rights and positive obligations, have radically transformed our conception of rights. They have made it far more complex and difficult to apply. The requirements deriving from different kinds of rights, both individual and collective and imposing both negative and positive obligations, come into conflict with each other. It has become much more difficult to draw up a clear and coherent list of rights that may be applied without conflict in practical situations. Moreover, the conflict between different kinds of rights is often related to a worsening social conflict. The language of rights lends itself easily to extravagance and intransigence.

Nevertheless, it seems difficult, if not impossible, to abandon this language, for it corresponds to something essential in our modern culture. It is the reflection in the political and legal fields of our fundamental conception of the human person and human dignity. It would indeed be difficult for us to renounce it.

QUESTIONS

Taylor relates the notion of rights as powers of individuals to set limits on the actions of a society, to the notion of rights as 'trump' cards defeating state action 'whenever it encroaches on [the individual's] protected sphere.' He seems to relate this idea to 'Conception A' rights, while 'Conception B' rights are associated with rights as 'fundamental social objectives' that may be expressed in a 'social' bill of rights and that require 'positive action' by the state.

Do you find this distinction helpful? A 'right to food' surely falls into 'Conception B,' but what of the right to vote in elections, or to a fair trial, or to life? Might those too be seen as 'fundamental social objectives' and require 'positive action' by the state? What kind of action?

KARL KLARE, LEGAL THEORY AND DEMOCRATIC RECONSTRUCTION

25 U. of Brit. Colum. L. Rev. 69, (1991), at 97.

[The author here discusses the appropriate place of 'rights' in the formal legal structures and guarantees and in the legal–political discourse of the postcommunist states of Central-East Europe. He concludes that 'it seems obvious that postcommunist law should be founded upon an explicit charter of human rights guarantees. How could there be any doubt of the central place of rights in democratic legal reconstruction?' Nonetheless, Klare notes, recent debates among western legal scholars have developed a serious 'critique of rights,' and he undertakes to summarize 'some of the major lines of criticism advanced by the rights skeptics.' The following excerpts deal with two aspects of this critique and Klare's responses thereto.]

A second branch of rights skepticism concerns the efficacy and limitations of the rights tradition in relationship to social change. . . . [T]he skeptics call attention to certain self-imposed limitations internal to rights discourse stemming from its embrace of the public/private distinction. Rights thinking has predominantly concerned the relationship between the individual and the state. As traditionally understood, the human rights project is to erect barriers between the individual and the state, so as to protect human autonomy and self-determination from being violated or crushed by governmental power.

Unquestionably, a just society requires such protections, but human freedom can also be invaded or denied by nongovernmental forms of power, by domination in the so-called 'private sphere'. Human dignity is denied by *de jure* racial segregation, but it is also denied by employers who discriminate on the basis of race, Laws barring adult homosexuals from privately and consensually expressing their sexuality deny freedom and autonomy, but so, too, do homophobic social practices such as housing discrimination and gay bashing. The expression of dissent can be inhibited by the cost of media access as well as by abuses of state power. Rights charters almost invariably concern restrictions on state power and therefore leave intact many forms of 'private' domination, including hierarchies of class, race, gender and sexual preference. The skeptics argue that the vision of freedom embodied in the rights tradition is for this reason partial and incomplete.

Given the injustices committed by the Stalinist regimes, it is understandable that the first priority of postcommunist lawyers is to guard against the abuse of state power. . . . Granting this, the argument goes, to realize freedom in all aspects of life, to establish arrangements in all social contexts that will be committed to human dignity, self-realization and equality, requires a deep transformation, in both East and West, not only of governmental but also of non-state institutions and practices that are left untouched by conventional human rights doctrine. A strong version of rights skepticism suggests that the fixation on the individual/state relationship in the rights tradition actually diverts intellectual and political resources from other, needed approaches to social justice.

Here again, it is conceivable that rights discourse can be transformed to accommodate these criticisms; that we can articulate a panoply of self-determination rights in social and economic life. . . .

This brings us to a third aspect of contemporary rights skepticism, the so-called 'indeterminacy critique'. In its strongest versions, rights discourse purports to supply a political criteria for evaluating institutions and practices and for resolving social conflict. The very power of a claim of right is that it is founded upon universal values, that it transcends all particular understandings of appropriate social organization. A successful rights claimant trumps majoritarian sentiment regarding the good life. . . .

. . .

. . . The initial problem is that many of the most important rights concepts are formulated at an exceedingly high level of abstraction. Because human rights concepts tend to be very elastic and open-ended, they are capable of being given a wide range of meanings, including inconsistent meanings. Take freedom of speech, for example. One meaning is the right to dissent and to criticize the powers that be. Yet the right to free speech can also be given quite a different meaning, as, e.g., in the American cases barring government from trying to prevent the distortion of the electoral process by corporate campaign contributions. In the former interpretation, free speech permits individuals to unfreeze hierarchy and open up political debate, whereas in the latter case, the right to free speech is mobilized to reinforce domination by entrenched power. Or, take the right to privacy, the right to be left alone by government with regard to certain intimate matters. For most feminists, this connotes a right to choice about reproduction and abortion. But the right to privacy regarding intimate personal matters has long had a less savory invocation as a justification for why courts should not intervene to prevent or punish domestic violence. An interesting aspect of rights-fixated political cultures, such as we have in the United States, is that anybody and everybody can and does formulate their political claims in rights terms.

Thus, rights concepts are sufficiently elastic so that they can mean different things to different people. People who seek to reinforce hierarchy and perpetuate domination can speak the language of rights, often with sincerity. But there is an even deeper problem. Even those who would consistently invoke rights in the service of self-determination, autonomy and equality find that rights concepts are internally contradictory. That is because, like all of legal discourse, rights theory is an arena of conflicting conceptions of justice and human freedom. For example, democratic thinking, particularly within the liberal tradition, contains conceptions of rights as freedom of action and also of rights as guarantors of security. It contains conceptions of rights as protection from state power and also of rights to invoke state power to protect the individual from powerful private groups. Proponents of democracy have advanced conceptions of rights to freedom of association and also conceptions of rights of excluded minorities to insist on membership in important groups. Rights theories contain conceptions of equality as identical treatment of those similarly situated and also theories of equality as protection for those not similarly situated. Human rights discourse holds that its claims are universal yet also embodies a belief in the right of all peoples to cultural autonomy and self-determination.

Thus, choices must be made in elaborating any structure of human rights guarantees, just as in the course of specifying market structures, and the choices bear socially and politically significant consequences.

The problem is that rights discourse itself does not provide neutral decision procedures with which to make such choices.

. . .

. . . My point here is that, by itself, rights discourse does not and probably cannot provide us with the criteria for deciding between conflicting claims of right. In order to resolve rights conflicts, it is necessary to step outside the discourse. One must appeal to more concrete and therefore more controversial analyses of the relevant social and institutional contexts than rights discourse offers; and one must develop and elaborate conceptions of and intuitions about human freedom and self-determination by reference to which one seeks to assess rights claims and resolve rights conflicts.

If the processes of concretizing rights concepts and of resolving rights conflicts extend beyond the traditional discourse of rights onto the terrain of social theory and political philosophy, it follows that rights rhetoric must be politicized in order to serve as a foundation for legal reconstruction. . . . Surely it is insufficient to think of human rights practice in terms of obtaining the correct list of rights and then enacting them into a code. Rather, postcommunist lawyers must think of rights discourse and rights charters as relatively open media in which to advance visions of socially desirable institutions and practices. That is, the rights foundation of legal reconstruction is an invitation to make political philosophy, not only in promulgating the initial charters, but at every step along the way of articulating and interpreting rights concepts and filling them with concrete legal and institutional meaning. But this revised, 'politicized' conception of human rights discourse and practice in postcommunist legal reconstruction sits uneasily with the idea of an autonomous rule of law that is ostensibly the basis of the whole enterprise.

. . .

NOTE

Klare describes as one branch of 'rights skepticism' the belief that 'human freedom can also be invaded or denied by non-governmental forms of action, by domination in the so-called "private sphere".' Human rights, however, are principally associated with checks on the public sphere of state action. This theme of the distinction between the private and public recurs throughout the coursebook. It is developed more systematically in Chapter 13 on women's rights, and figures in the Note and Questions at the end of Section D of this chapter.

A number of different understandings about rights—their nature, content and consequences—appear in the readings in Chapters 3 and 4. Consider the following lists, setting forth in the left column assertions or understandings about rights that typify one important tradition of rights discourse within liberal societies against a list in the right column of polar understandings. These lists are relevant to the analysis of issues of cultural relativism in Sections B to D of this chapter. Consider these polar characterizations of 'rights' as:

inalienable	socially constructed, given and taken
absolute	contingent
universal	particular, culturally specific
eternal, ahistorical	time bound, historicist
rights based on equal human dignity	rights based on utility, power

QUESTION

Klare stresses that 'choices must be made in elaborating any structure of human rights guarantees . . . and the choices bear socially and politically significant consequences. The problem is that rights discourse itself does not provide neutral decision procedures with which to make such choices. . . . In order to resolve rights conflicts, it is necessary to step outside the discourse.'

Apply this observation to a conception as basic as the 'right to life.' What 'choices' about the meaning of this conception are before legislatures, courts or advocates elaborating this right, and through what methods or processes can differences among those choices be resolved? What does Klare mean by the necessity to 'step outside the discourse'?

NOTE

The following two readings comment on duty-oriented rather than rights-oriented ordering through law and cultural tradition. Cover comments on the legal culture of Judaism with its stress on obligations imposed by God rather than on rights. He suggests historical reasons why Western states and Judaism developed in these different ways. Kenyatta describes aspects of the education of the young in the Gikuyu people in Kenya, particularly the inculcation of elements of social obligations and duty. Although the cultural and religious contexts and the content of the duties referred to are radically different in these two readings, they both suggest important consequences of an orientation towards duty/obligation.

Note that the duties/obligations referred to in these readings are *not* the same as duties within a *scheme of rights* that are understood as correlative to the described rights. For example, your basic right to be free from torture imposes a correlative (that is, corresponding) duty on the state not to torture. The following readings talk of duties imposed on *individuals* rather than on the state or other collective entity (and also distinct from the characteristic duties of individuals to respect the rights of other individuals, as expressed in state tort or criminal law). A distant analogy to such imposition of duties on individuals is provided by Article 29(1) of the UDHR. But the language of individual duty in the universal human rights system is rare. A closer analogy to the present readings, particularly the excerpts from Kenyatta, is provided by the African Charter on Human and Peoples' Rights, examined at p. 689, *infra*.

ROBERT COVER, OBLIGATION: A JEWISH JURISPRUDENCE OF THE SOCIAL ORDER

5 J. of Law and Relig. 65 (1987).

I. Fundamental Words

Every legal culture has its fundamental words. When we define our subject this weekend as human rights, we also locate ourselves in a normative universe at a particular place. The word 'rights' is a highly evocative one for those of us who have grown up in the post-enlightenment secular society of the West. . . .

Judaism is, itself, a legal culture of great antiquity. It has hardly led a wholly autonomous existence these past three millennia. Yet, I suppose it can lay as much claim as any of the other great legal cultures to have an integrity to its basic categories. When I am asked to reflect upon Judaism and human rights, therefore, the first thought that comes to mind is that the categories are wrong. I do not mean, of course, that basic ideas of human dignity and worth are not powerfully expressed in the Jewish legal and literary traditions. Rather, I mean that because it is a legal tradition Judaism has its own categories for expressing through law the worth and dignity of each human being. And the categories are not closely analogous to 'human rights.' The principal word in Jewish law, which occupies a place equivalent in evocative force to the American legal system's 'rights', is the word 'mitzvah' which literally means commandment but has a general meaning closer to 'incumbent obligation.'

Before I begin an analysis of the differing implications of these two rather different key words, I should like to put the two words in a context—the contexts of their respective myths. For both of us these words are connected to fundamental stories and receive their force from those stories as much as from the denotative meaning of the words themselves. The story behind the term 'rights' is the story of social contract. The myth postulates free and independent if highly vulnerable beings who voluntarily trade a portion of their autonomy for a measure of collective security. The myth makes the collective arrangement the product of individual choice and thus secondary to the individual. 'Rights' are the fundamental category because it is the normative category which most nearly approximates that which is the source of the legitimacy of everything else. Rights are traded for collective security. But some rights are retained and, in some theories, some rights are inalienable. In any event the first and fundamental unit is the individual and 'rights' locate him as an individual separate and apart from every other individual.

I must stress that I do not mean to suggest that all or even most theories that are founded upon rights are 'individualistic' or 'atomistic.' Nor would I suggest for a moment that with a starting point of 'rights' and social contract one must get to a certain end. Hobbes as well as Locke is part of this tradition. And, of course, so is Rousseau. Collective solutions as well as individualistic ones are possible but, it is the case that even the collective solutions are solutions which arrive at their destination by way of a theory which derives the authority of the collective from the individual. . . .

The basic word of Judaism is obligation or mitzvah. It, too, is intrinsically bound up in a myth—the myth of Sinai. Just as the myth of social contract is essentially a myth of autonomy, so the myth of Sinai is essentially a myth of heteronomy. Sinai is a collective—indeed, a corporate—experience. The experience at Sinai is not chosen. The event gives forth the words which are commandments. In all Rabbinic and post Rabbinic embellishment upon the Biblical account of Sinai this event is the Code for all Law. All law was given at Sinai and therefore all law is related back to the ultimate heteronomous event in which we were chosen-passive voice.

. . .

What have these stories to do with the ways in which the law languages of these respective legal cultures are spoken? Social movements in the United States organize around rights. When there is some urgently felt need to change the law or keep it in one way or another a 'Rights' movement is started. Civil rights, the right to life, welfare rights, etc. The premium that is to be put upon an entitlement is so coded. When we 'take rights seriously' we understand them to be trumps in the legal game. In Jewish law, an entitlement without an obligation is a sad, almost pathetic thing. . . .

Indeed, to be one who acts out of obligation is the closest thing there is to a Jewish definition of completion as a person within the community. A child does not become emancipated or 'free' when he or she reaches maturity. Nor does she/he become *sui juris.* No, the child becomes bar or bat mitzvah, literally one who is of the obligations. Traditionally, the parent at that time says a blessing. Blessed is He that has exonerated me from the punishment of this child. The primary legal distinction between Jew and non-Jew is that the non-Jew is only obligated to the 7 Noachide commandments. . . .

THE USES OF RIGHTS AND OBLIGATIONS

The Jewish legal system has evolved for the past 1900 years without a state and largely without much in the way of coercive powers to be exercised upon the adherents of the faith. I do not mean to idealize the situation. The Jewish communities over the millennia have wielded power. Communal sanctions of banning and shunning have been regularly and occasionally cruelly imposed on individuals or groups. Less frequently, but frequently enough, Jewish communities granted quasi-autonomy by gentile rulers, have used the power of the gentile state to discipline dissidents and deviants. Nonetheless, there remains a difference between wielding a power which draws on but also depends on pre-existing social solidarity, and, wielding one which depends on violence. . . .

In a situation in which there is no centralized power and little in the way of coercive violence, it is critical that the mythic center of the Law reinforce the bonds of solidarity. Common, mutual, reciprocal obligation is necessary. The myth of divine commandment creates that web. . . . It was a myth that created legitimacy for a radically diffuse and coordinate system of authority. But while it created room for the diffusion of authority it did not have a place for individualism. One might have independent and divergent understandings of the obligations imposed by God through his chosen people, but one could not have a world view which denied the obligations.

The jurisprudence of rights, on the other hand, has gained ascendance in the Western world together with the rise of the national state with its almost unique mastery of violence over extensive territories. Certainly, it may be argued, it has been essential to counterbalance the development of the state with a myth which a) establishes the State as legitimate only in so far as it can be derived from the autonomous creatures who trade in their rights for security—i.e., one must tell a story about the States's utility or service to us, and b) potentially justifies individual and communal resistance to the Behemoth. It may be true as Bentham so aptly pointed out that natural rights may be used either apologetically or in revolutionary fashion, and there is nothing in the concept powerful enough analytically to constrain which use it shall be put to. Nevertheless, it is the case that natural right apologies are of a sort that in their articulation they limit the most far-reaching claims of the State, and the revolutionary ideology that can be generated is also of a sort which is particularly effective in countering organic statist claims.

Thus, there is a sense in which the ideology of rights has been a useful counter to the centrifugal forces of the western nation state while the ideology of mitzvoth or obligation has been equally useful as a counter to the centripetal forces that have beset Judaism over the centuries.

. . .

. . . [T]he Maimonides system contrasts the normative world of mitzvoth with the world of vanity—hebel. It seems that Maimonides, in this respect, as in so many others has hit the mark. A world centered upon obligation is not, really cannot be, an empty or vain world. Rights, as an organizing principle, are indifferent to the vanity of varying ends. But mitzvoths because they so strongly bind and locate the individual must make a strong claim for the substantive content of that which they dictate. The system, if it's content be vain, can hardly claim to be a system. The rights system is indifferent to ends and in its indifference can claim systemic coherence without making any strong claims about the fullness or vanity of the ends it permits.

. . .

JOMO KENYATTA, FACING MOUNT KENYA: THE TRIBAL LIFE OF THE GIKUYU

(1965), at 109.

[These excerpts are taken from a description by Kenyatta, who later became the first post-colonial president of Kenya, of the Gikuyu people (often rendered in English as 'Kikuyu') in that country. The excerpts stress elements of duty inculcated in Gikuyu children, and appear in Chapter 5, 'System of Education.']

[The children] are also taught definitely at circumcision the theory, as it were, of respect to their parents and kinsfolk. Under all circumstances they must stay with them and share in their joys and sorrows. It will never do to leave them and go off to see the world whenever they take the notion, especially when their parents are in their old age. They must give them clothes, look after their garden, herd their cattle, sheep and goats, build their grain stores and houses. It thus becomes a part of their outlook on life that their parents shall not suffer want nor continue to labour strenuously in their old age while their children can lend a hand and do things to give them comfort. This respect and duty to parents is further emphasised by the fact that the youth or girl cannot advance from one stage to another without the parent's will and active assistance. The satisfaction of all a boy's longings and ambitions depends on the father's and family's consent. . . .

. . .

The teaching of social obligations is again emphasised by the classification of age-groups to which we have already referred. This binds together those of the same status in ties of closest loyalty and devotion. Men circumcised at the same time stand in the very closest relationship to each other. When a man of the same age-group injures another it is a serious magico–religious offence. They are like blood brothers; they must not do any wrong to each other. It ranks with an injury done to a member of one's own family. The age-group (*riika*) is thus a powerful instrument for securing conformity with tribal usage. The selfish or reckless youth is taught by the opinion of his gang that it does not pay to incur displeasure. He will not be called to eat with the others when food is going. He may be put out of their dances, fined, or even ostracised for a time. If he does not change his ways he will find his old companions have deserted him.

. . . The age-groups do more than bind men of equal standing together. They further emphasise the social grades of junior and senior, inferior and superior. We see the same principle in evidence all through the various grades. . . .

Owing to the strength and numbers of the social ties existing between members of the same family, clan and age-group, and between different families and clans through which the tribe is unified and solidified as one organic whole, the community can be mobilised very easily for corporate activity. House-building, cultivation, harvesting, digging trap-pits, putting up fences around cultivated fields, and building bridges, are usually done by the group; hence the Gikuyu saying: '*Kamoinge koyaga ndere*,' which means collective activities make heavy tasks easier. In the old days sacrifices were offered and wars were waged by the tribe as a whole or by the clan. Marriage contracts and ceremonies are the affairs of families and not of individuals. Sometimes even cattle are bought by joint effort. Thus the individual boy or girl soon learns to work with and for other people. An old man who has no children of his own is helped by his neighbour's children in almost everything. His hut is built, his garden dug, firewood is cut and water is fetched for him. If his cattle, sheep or goats are lost or in difficulties the children of his neighbour will help to bring them back, at great pains and often at considerable risk. The old man reciprocates by treating the children as though they were his own. Children learn this habit of communal work like others, not by verbal exhortations so much as by joining with older people in such social services. . . . All help given in this way is voluntary, and kinsfolk are proud to help one another. There is no payment or expectation of payment. They are well feasted, of course. This is not regarded as payment, but as hospitality. The whole thing rests on the principle of reciprocal obligations. It is taken for granted that the neighbour whom you assist in difficulty or whose house you help to build will do the same for you when in similar need. Those who do not reciprocate these sentiments of neighbourliness are not in favour. . . .

. . .

The selfish or self-regarding man has no name or reputation in the Gikuyu community. An individualist is looked upon with suspicion and is given a nickname of *mwebongia*, one who works only for himself and is likely to end up as a wizard. He may lack assistance when he needs it. . . .

In the Gikuyu community there is no really individual affair, for every thing has a moral and social reference. The habit of corporate effort is but the other side of corporate ownership; and corporate responsibility is illustrated in corporate work no less than in corporate sacrifice and prayer.

In spite of the foreign elements which work against many of the Gikuyu institutions and the desire to implant the system of wholesale Westernisation, this system of mutual help and the tribal solidarity in social services, political and economic activities are still maintained by the large majority of the Gikuyu people. It is less practised among those Gikuyu who have been Europeanised or detribalised. The rest of the community look upon these people as mischief-makers and breakers of the tribal traditions, and the general disgusted cry is heard: '*Mothongo ne athogonjire borori*,' i.e. the white man had spoiled and disgraced our country.

. . .

The striking thing in the Gikuyu system of education, and the feature which most sharply distinguishes it from the European system of education, is the primary place given to personal relations. Each official statement of educational policy repeats this well-worn declaration that the aim of education must be the building of character and not the mere acquisition of knowledge. . . .

. . .

QUESTIONS

1. 'A duty-based social order seems inherently less subject to universalization (with respect to the duties imposed on individuals) than a rights-based social order (with respect to the rights attributed to individuals). That is, the content of duties (obligations toward elders, toward the community, toward God) seem to be very particular and bound to a given context, a product of a given religion or political or social culture, whereas individuals' rights seem to be more divorced from a particular context and can therefore be stated more abstractly.' Do you agree? Any examples?

2. '"Individual rights" necessarily imply equality among all rights holders, which is to say among all members of society. This in fact is what the contemporary human rights instruments declare. To the contrary, duties can be (and frequently are) defined so as to impose hierarchy, status, and discrimination in a given social order.' Do you agree? Any examples?

3. 'Different from a regime of rights, a regime of duties intrinsically exerts an inward, centripetal force. It draws individual duty-bearers into the society, connects them intricately with other individuals and the community in a variety of ways, blurs the separate identity of the individual from society, and leads to a more communal and collective structure of life.' Do you agree?

4. Note that Article 2(3) of the ICCPR requires states to provide all persons whose rights have been violated with 'an effective remedy,' and to develop particularly the possibilities of 'judicial remedy.' Do rights imply a preference for or even require individual (judicial or other) remedies (cf. Taylor, p. 173, *supra*)

against the state, whereas a regime of individual duties is less likely to provide such remedies?

5. Can you imagine instead of a human rights movement a 'human duties movement' that would be informed by similar values and goals? How would its constitutional foundation, a Universal Declaration of Human Duties, read?

NOTE

As several prior readings underscore, rights as a fundamental language of law, politics and morals grew within and are associated with the Western tradition of liberalism. This is not to say that the values and goals expressed through rights language, particularly through the kinds of rights that we have seen in the UDHR and the ICCPR, are exclusive to that tradition. Many of them, may be expressed through other languages as well.

As background for the discussion of cultural relativism in Section B, it will be helpful to have in mind some characteristics of liberal political thought and the liberal state. The following Comment provides that background.

COMMENT ON SOME CHARACTERISTICS OF THE LIBERAL POLITICAL TRADITION

Observers from different regions and cultures can agree that the human rights movement, with respect to its language of rights and the civil and political rights that it declares, stems principally from the liberal tradition of Western thought. That observation lies at the core of argument by states from non-Western parts of the world that some basic provisions in instruments like the UDHR or ICCPR are inappropriate and inapplicable to their circumstances. Those instruments, the argument goes, purport to give a genuinely universal expression to certain tenets of liberal political culture. But those tenets stem from and should be applied only to states within this Western political tradition. Thus liberal thought and practices inform much contemporary debate about the meaning and relevance of cultural relativism.

For the purpose of facilitating some comparisons between liberalism and the human rights movement, this Comment sketches basic characteristics that observers would associate with the different expressions of the liberal tradition during this century. The Comment has a limited historical scope, not reaching back to the origins of liberal thought in earlier centuries and the Age of Enlightenment, or to the significant changes in that body of thought in the nineteenth century.

The liberal political tradition has never been and surely is not today a monolithic body of thought requiring one and only one form of government. It is not then surprising that the very term 'liberal' has assumed different meanings, from the liberal economics associated with the *laissez faire* school of the nineteenth century to contemporary associations of liberalism in a country like the United States with a more active and engaged state concerned with the general welfare of the population and with regulation of the market and non-governmental actors—the modern regulatory and welfare state so familiar to Western states.

The contemporary expressions of liberal thought by theorists like Dworkin or Rawls depart significantly from the writings of the classical theorists influencing its development, like Bentham, Kant, Locke, Mill, Rousseau and Tocqueville. The differences among such classical writers are reflected in the distinct versions of liberal ideology and the varied structures and practices of self-styled liberal democracies. This variety and ongoing transformation suggest caution in making inclusive and dogmatic comparisons between, say, liberalism and the human rights movement, which has during this last half century generated its own internal conflicts and has undergone significant change.

No characteristic of the liberal tradition is more striking that its emphasis on the individual. Liberal political theory and the constitutive instruments of many liberal states frequently employ basic concepts or premises like the dignity and autonomy of the individual, and respect that is due the individual. The vital concept of equality informs these terms: the equal dignity of all human beings, the equal respect to which individuals are entitled, the equal right of all men and women for self-realization. It is not then surprising that equal protection and equal opportunities without repressive discrimination constitute so cardinal a value of contemporary liberalism. In general, the protection of minorities against invidious discrimination continues to be a central concern for the liberal state.

Such stress on the individual informs basic justifications for the state. The liberal state rests on, its legitimacy stems from, the consent of the people within it. Within liberal theory, that consent is both hypothetical, as in the notion of a social contract among the inhabitants of a state of nature to create the political state, and institutionalized through typical practices such as periodic elections. Such ideas are explicit in the basic human rights instruments. Note Article 21 of the UDHR ('The will of the people shall be the basis of the authority of government') and Article 25 of the ICCPR (the importance of elections 'guaranteeing the free expression of the will of the electors').

From the start, liberal theory has been attentive to the risk of abuse of the individual by the state. The rights language that is found in constitutional bills of rights, statutory provisions for basic rights, political traditions not expressed in positive law, and writings of theorists and advocates responds to this need for protection against the state. The rights with which the individual is endowed limit governmental power—the right not to be tortured, not to be discriminated against on stated grounds. Historically the protection of the property right against interference by the state and others played a major role in liberal theory.

Sometimes such rights are referred to as 'negative': the hands-off or non-interference rights (don't touch), or the right to be interfered with (as by arrest, imprisonment) only pursuant to stated processes. It is partly the prominence of the rights related to notions of individual liberty, autonomy and choice and of the right related to property protection that produces the sharp division in much liberal thought between the state and individual, between government and non-governmental sectors, between what are often referred to as the public and private realms or spheres of action.

This conception of negative rights, and of negative freedom as the absence of external constraints, together with the historical alliance of political liberalism with conceptions of a free market and *laissez faire*, led to liberalism's early emphasis on sharply limited government. The tension between that early ideology and background, and the growing emphasis over more than a century on the welfare and regulatory functions of the modern liberal state, remains central to much political and moral debate today.

That debate is related to an opposition that has developed in liberal thought between *negative* rights or negative liberty (freedom), and *positive or affirmative* rights or liberty (freedom). Those terms have acquired different meanings, to be explored in this and later chapters. For example, 'positive rights' have been described as entitlements of individuals that the state not simply respect the 'private' sphere of inviolability of the individual (the negative rights), but also 'act' in particular ways to benefit the individual, perhaps by providing education or health care. In this sense, the 'positive rights' of individuals such as the right to education or health care impose duties on the state to provide the necessary institutions or resources.

In a different and more ample sense, *positive liberty* has been described as 'liberty to' as opposed to 'liberty from'—for example, the liberty to realize oneself, to satisfy one's real interests, to achieve individual self-determination. One form of such positive liberty facilitated by the state would be governmental policies and institutions fostering the active political participation of citizens in electoral and other processes that help to determine the allocation ways in which public power is executed. Through such positive liberty, the individual can participate in the creation and recreation of self and state. The state readily and naturally becomes involved in this search by individuals for positive liberty, characteristically by creating the conditions that make the individual quest more likely to succeed, but at the dangerous authoritarian extreme by attempting to define the content of genuine self-realization and by coercing individuals to achieve it.

What an individual should seek in life, what idea of the good in life that individual holds, how the individual seeks self-realization, remain in the liberal state matters of individual choice to which both negative and positive conceptions of rights and freedom are relevant. That state must be open to a variety of ends, a variety of conceptions of the good, that individuals will express. The liberal state must then be a pluralist state. Its structure of rights, going beyond the rights to personal security and equal protection to include rights of conscience and speech and association, facilitates and protects the many types of diversity within pluralism, as well as ongoing argument in the public arena about the forms and goals of social and political life.

Precisely what governmental structures best realize such liberal principles is among the disputed features of the liberal tradition. The liberal state is closely associated with the ideal of the rule of law, hence with some minimum of separation of government powers such as an independent judiciary that can protect individual rights against executive abuse. The fear of tyranny of the majority lies at the foundation of the argument for restrictions on governmental power through a constitutional bill of rights limiting or putting conditions on what government can do. How to enforce that bill of rights against the executive and legislature has never achieved a consensus among liberal states. They vary in the degree to which they subject legislative action to judicial review, hence in the degree to which governmental power, even if supported by a freely voting majority of the population, can abridge or transform or abolish rights.

The liberal tradition continues to be subjected to deep challenges from within and without, and thus continues its process of evolutionary change. During and particularly after the cold war, its interaction with states of the developing world posed complex issues in relation to efforts of some of those states to develop new forms of government and economy. For example, the relationships in the former Communist states of Central and East Europe among liberalism, privatization and property rights, markets and regulation thereof, and the provision of welfare remain ambiguous and in flux. More generally, questions of the relationship between liberalism and a market economy, or liberalism and ethnic nationalism, have assumed heightened prominence. In a Western country such as the United States, liberalism responds to challenges from diverse perspectives such as communitarian ideas, civic republicanism, and multiculturalism (cultural particularism).

Some of these contemporary challenges underscore a continuing debate within liberalism, the two sides to which can lead to significantly different political and social orders: *individual* or *group* identity as primary. The group may be—to use the conventional porous and overlapping terms—national, linguistic, religious, cultural, ethnic. At the extreme, it is not compatible with the liberal creed for a governing order to subordinate individuals to the demands of such kinds of groups. With respect to the core values of liberalism, individual rights remain lexically prior to the demands of a culture or group for collective identity or group solidarity.

Nonetheless, the liberal state is hardly hostile to groups as such. It is not blind to the influence of groups (religious, cultural, ethnic) or of group and cultural identity in shaping the individual. Indeed, the political life of modern liberal democracies is largely constituted by the interaction, lobbying and other political participation of groups, some of which are natural in their defining characteristic (race, sex, elderly citizens), some formed out of shared interests (labor unions, business associations, environmental groups). The liberal state, by definition committed to pluralism, must accomodate different types of groups, and maintain the framework of rights in which they can struggle for recognition, power and survival.

Such issues indicate how much is open and debated within liberalism about the significance of the priority of the individual in the contemporary liberal state—or different types of liberal states. Should we, for example, understand the 'individual' *abstractly*, as similar in vital respects everywhere, both within the same state and universally? Or do we understand the individual *contextually*, as influenced or even determined by ethnic, cultural, national, religious and other traditions and communities? Should we even phrase the question in such dramatic contrasts, or should we rather assume that the answers are too complex for any clear choice between them?

Since the birth of the human rights movement, and particularly since the collapse of the Soviet Union, such issues about the individual and the collective have taken on great pungency in the contradictions bred, on the one hand, by the spread of both liberal ideology with its emphasis on the individual and of market ideology with its stress on private initiative, and on the other hand, by the often savage bursts of ethnic nationalism in many parts of the world with their stress on collective rather than individual identity. See Chapter 14, *infra*.

The emphasis in both liberalism and the human rights movement on individual 'rights' leads to one final observation related to several of this chapter's readings. Rights are no more determinate in meaning, no less susceptible to varying interpretations and disputes among states, than any other moral, political or legal conception—for example, 'property,' or 'sovereignty,' or 'consent,' or 'national security.' Consider, for example, the different values and aspirations that have been attributed to one conventional phrase in many human rights instruments, the 'right to life.' Within liberal states, different institutional solutions have been brought to the question of who should determine and develop the content of rights, who should resolve the many and puzzling conflicts among rights. In the international arena, this problem becomes all the more complex. What mechanisms, what institutional framework, what allocation or separation of powers, what blend of overtly political and judicial resolution of these issues, will we find in the international human rights movement? Such issues are examined in Part C.

ADDITIONAL READING

Two histories of liberalism are Harold Laski, *The Rise of Liberalism* (1936), and Guido de Ruggiero, *The History of European Liberalism* (1927). Two contemporary collections of articles on the liberal tradition by leading theorists writing from a range of perspectives are Nancy Rosenblum (ed.), *Liberalism and the Moral Life* (1989), and Michael Sandel (ed.), *Liberalism and its Critics* (1984). The latter book includes the prominent essay of Isaiah Berlin, *Two Concepts of Liberty*, which in turn is examined in Charles Taylor, *What's Wrong with Negative Liberty*, in Alan Ryan (ed.), *The Idea of Freedom* 175 (1979). Leading theoretical works developing the liberal tradition with respect to concepts of rights and justice include Ronald Dworkin, *Taking Rights Seriously* (1977), and *A Matter of Principle* (1985); and John Rawls, *A Theory of Justice* (1971), and *The Law of Peoples*, 20 Critical Inquiry 3b (1993). A defense of liberalism from a modern philosophical perspective is found in Richard Rorty, *Contingency, Irony, and Solidarity* (1989), particularly chapters 3 and 4. Criticism of inherited forms of liberal democracy and a program for its development are set forth in Roberto Unger, *Democratic Experimentalism* (forthcoming 1995).

B. UNIVERSALISM AND CULTURAL RELATIVISM

COMMENT ON THE UNIVERSALIST–RELATIVIST DEBATE

One of the intense debates in the human rights movement involves the 'universal' or 'relative' character, related to the 'absolute' or 'contingent' character, of the rights declared. The contest between the universal–relative positions and between the absolute–contingent positions is an old one. It took on renewed vigor in light of the human rights movement's erosion of notions of sovereignty, domestic jurisdiction, and cultural autonomy that in an earlier period had enjoyed greater strength. Indeed, the advocates of cultural relativism often employ the ideas clustered around those notions to sustain stronger, more traditional ideas of state autonomy.

To put it simply, the partisans of universality claim that international human rights like equal protection or physical security or free speech, religion and association are and must be the same everywhere. This applies at least as to the rights' substance, for universalists must concede that many basic rights (such as the right to a fair criminal trial) allow for culturally influenced forms of implementation or realization (i.e., not all states are required to use the jury in its Anglo-American form).

The advocates of cultural relativism claim that rights and rules about morality (most, some, a few) are encoded in and thus depend on cultural context, the term 'culture' often being used in a broad and diffuse way that may go beyond indigenous traditions and customary practices to include political and religious ideologies and institutional structures. Hence notions of right (and wrong) and moral rules necessarily differ throughout the world because the cultures in which they inhere themselves differ. Thus this relativist position can be understood simply to assert as an empirical matter that the world contains an impressive diversity in views about right and wrong that is linked to the diverse underlying cultures.

But the strong relativist position goes beyond arguing that there is—as a matter of fact, empirically—an impressive diversity. It attaches an important consequence to this diversity: that no transcendent or trans-cultural ideas of right can be found or agreed on, and hence that no culture (whether or not in the guise of enforcing international human rights) is justified in attempting to impose on others what must be understood as its own ideas. In this strong form, cultural relativism necessarily contradicts a basic premise of the human rights movement.

On their face, human rights instruments are surely on the 'universalist' side of this debate. The landmark instrument is the *Universal* Declaration of Human Rights. The two Covenants, with their numerous states parties from all the world's regions, speak in universal terms: 'everyone' has the right to liberty, 'all persons' are entitled to equal protection, 'no one' shall be subjected to torture, 'everyone' has the right to an adequate standard of living. Even the vital limitations of certain rights on grounds of 'public health or order' or 'national security' are cast in universal terms, although their interpretation will surely differ among cultures and states. The text of these basic instruments makes no explicit concession to cultural variation. (The regional instruments examined in Chapter 10, particularly the African Charter on Human and Peoples' Rights, do express cultural variation.)

To the relativist, these instruments and their pretension to universality may suggest nothing so strongly as the arrogance or 'cultural imperialism' of the West, given the West's traditional urge—expressed for example in political ideology (liberalism) and in religious faith (Christianity)—to view its own forms and beliefs as universal, and to attempt to universalize them. Moreover, the push to universalization of norms is said by some relativists to destroy diversity of cultures and hence amounts to another form of homogenization in the modern world. But the debate between these two positions follows no simple route. It is open to a range of views and strategies that the materials in Sections B and C explore.

During the cold war, such debates (sometimes no more than highly politicized accusations, routine polemics) were dominantly between the Communist world (and its sympathizers) and the Western democracies. The Western democracies charged the Communist world with violating many basic rights, particularly those of a civil and political character. That world replied both by charging the West with violations of the more important economic and social rights, and by asserting that the political and ideological structures of Communist states pointed toward a different understanding of rights.

That debate died more-or-less together with the Soviet Union. Today it continues in different form, often in North–South (or West–East) framework, or in a religious (West–Islam) framework, or more broadly between developing (Third World) and developed (Western–Northern) countries. It also includes non-state actors such as indigenous peoples.

NOTE

The two introductory readings come out of the rich anthropological literature on cultural relativism. Anthropologists have long had to wrestle with these issues, in the context of their ethnographic writings about diverse cultures with practices and values often departing radically from the West and subject to extreme censure from the perspective of Western moral thought. Such writings seek to describe, explain, understand the alien culture, within the framework of one or another theoretical perspective or methodology. Hence the role of the anthropologist has traditionally been very different from that of the human rights investigator and advocate who assesses a state's conduct against international standards.

Nonetheless, what should be the stance of the anthropologist toward practices and values that are offensive from a Western viewpoint? Ought she to be critical of them, or to the contrary be tolerant and accepting, or simply be distant and neutral while in the role of observer and explainer? If critical, under what standards would she criticize?

Most of the anthropological debate on such issues developed before the international human rights movement became prominent in the 1970s, and makes no reference to that movement. The Statement on Human Rights, p.198, *infra*, is an exception to this observation. Moreover, bear in mind in the following readings that the human rights movement addresses primarily states—individual rights against the state—whereas ethnographic writings involve peoples or tribes or societies, all non-state (often sub-state) entities. Hence these rich and suggestive anthropological writings are relevant to cultural relativism in human rights settings more by analogy than direct application.

ELVIN HATCH, CULTURE AND MORALITY: THE RELATIVITY OF VALUES IN ANTHROPOLOGY

(1983), at 8.

. . . Herskovits wrote that cultural relativism developed because of

> the problem of finding valid cross-cultural norms. In every case where criteria to evaluate the ways of different peoples have been proposed, in no matter what aspect of culture, the question has at once posed itself: 'Whose standards?' . . . [T]he need for a cultural relativistic point of view has become apparent because of the realization that there is no way to play this game of making judgments across cultures except with loaded dice.

Ethical relativism is generally conceived as standing at the opposite pole from absolutism, which is the position that there is a set of moral principles that are universally valid as standards of judgment. One absolutist ethical theory is the traditional Christian view that right and wrong are God-given, and that all people may be judged according to Christian values. A wide range of purely secular ethical theories have also developed. . . .

It is the *content* of moral principles, not their existence, that is variable among human beings. It seems that all societies have some form of moral system, for people everywhere evaluate the actions of kinsmen, neighbors, and acquaintances as virtuous, estimable, praiseworthy, and honorable, or as unworthy, shameful, and despicable. These evaluations take objective form as sanctions, such as open praise or rebuke; and in extreme cases, violence and execution. The ubiquity of the moral evaluation of behavior apparently is a feature which sets humanity apart from other organisms. . . .

. . .

Chapter 4. The Call for Tolerance

. . . By and large ethical relativists have been anthropologists and not philosophers, and it is chiefly in the anthropological literature that we find arguments in its favor. Two people in particular have stood out as its proponents in the United States, Melville Herskovits and Ruth Benedict, both of whom were students of Boas. Almost without exception, the philosophers are disapproving, for usually they mention ethical relativism only to criticize it while in the course of arguing some other ethical theory.

At least two very different versions of ethical relativism have been advanced by anthropologists, and these need to be distinguished since they have their own faults and virtues. The first is sometimes classified (erroneously, as we shall see) as a form of skepticism, and I will call it the Boasian version of ethical relativism. . . . Skepticism in ethics is the view that nothing is really either right or wrong, or that there are no moral principles with a reasonable claim to legitimacy. It has been suggested that the Boasian position differs from this on one main point: Boasian relativism implies that principles of right and wrong do have some validity, but a very limited one, for they are legitimate only for the members of the society in which they are found. The values of the American middle class are valid for middle-class Americans, but not for the Trobriand Islanders, and vice versa.

Philosophers have presented a wide range of arguments against Boasian ethical relativism. . . . According to this argument, Boasian relativism is in essence a moral theory that gives a central place to one particular value. . . . It contains a more or less implicit value judgment in its call for tolerance: it asserts that we *ought* to respect other ways of life. . . . Herskovits wrote: 'The very core of cultural relativism is the social discipline that comes of respect for differences—of mutual respect. Emphasis on the worth of many ways of life, not one, is an affirmation of the values in each culture.

. . .

. . . The call for tolerance was an appeal to the liberal philosophy regarding human rights and self-determinism. It expressed the principle that others ought to be able to conduct their affairs as they see fit, which includes living their lives according to the cultural values and beliefs of their society. Put simply, what was at issue was human freedom.

The call for tolerance (or for the freedom of foreign peoples to live as they choose) was a matter of immediate, practical importance in light of the pattern of Western expansion. As Western Europeans established colonies and assumed power over more and more of the globe, they typically wanted both to Christianize and civilize the indigenous peoples. Christian rituals were fostered or imposed, and 'pagan' practices were prohibited, sometimes with force. The practice of plural marriage was condemned as a barbaric custom, and Western standards of modesty were enforced in an attempt to improve morals by covering the body. In the Southwest of North America, Indians who traditionally had lived in scattered encampments were made to settle in proper villages like 'civilized' people. The treatment of non-Western societies by the expanding nations of the West is a very large blot on our history, and had the Boasian call for tolerance—and for the freedom of others to define 'civilization' for themselves—been heard two or three centuries earlier, this blot might not loom so large today.

. . .

To develop a moral theory around the principle of tolerance raises the need to justify that principle: what reasons or grounds can be given to make the case that cultural differences ought to be respected? . . . The relativists make the error of deriving an 'ought' statement from an 'is' statement. To say that values vary from culture to culture is to describe (accurately or not) an empirical state of affairs in the real world, whereas the call for tolerance is a value judgment of what ought to be, and it is logically impossible to derive the one from the other. The fact of moral diversity no more compels our approval of other ways of life than the existence of cancer compels us to value ill-health.

Let us assume that we do find a set of values shared by all cultures. Would the relativist want to claim that these moral principles are legitimate ones for the world to embrace? What if the universal standard we discover is that all people are intolerant of other cultures—which is not very far-fetched? Clearly the ethical relativists would not throw aside their value of tolerance, but they would be forced to recognize that it is an error to think that the presence or absence of universal values among human cultures is a suitable base on which to build a moral philosophy.

. . .

Chapter 5. The Limits of Tolerance

The Boasian version of ethical relativism is subject to even harsher criticism . . . in its commitment to the status quo. The approval it enjoins seems to be absolute, leaving no room for judgment. . . .

. . .

The moral principle of tolerance that is proposed by Boasian relativism carries the obligation that one cannot be indifferent toward other ways of life—it obligates us to approve what others do. So if missionaries or government officials were to interfere in Yanomamo affairs for the purpose of reducing violence, the relativist would be obligated to oppose these moves in word if not action. Similarly, by the strict logic of relativism, Chagnon was wrong to insist that the mother feed her emaciated child. The Boasian relativist is placed in the morally awkward position of endorsing the infant's starvation, the rape of abducted women, the massacre of whole villages. . . .

Chapter 6. A Growing Disaffection

. . .

We can now understand why ethical relativism has fallen on such hard times in spite of the resurgence of pessimism during the 1960s and later. First, it has been the experience of most anthropologists that non-Western peoples (and especially Third World nations) want change, at least to some extent: second, it is clear that they are often disadvantaged if it does not come; third, anthropologists by and large have altered their thinking about the relativity of material interests and improvement: most today consider these to be general values that can be applied throughout the world.

Not only has relativism fallen on hard times, it has become the subject of angry criticism, much of it from the Third World, which tends to conceive anthropologists as conservative in their attitudes toward change and therefore as promoting the subservience of the underdeveloped nations. . . .

. . .

Whatever the cause, according to the radical critique, relativism has played directly into the hands of the oppressors throughout the world by its tacit support of the status quo. The relativists have not recognized that the exotic cultures to which they grant equal validity are poverty-stricken, powerless, and oppressed. William Willis comments that the relativist 'avoids the distress and misery' of foreign peoples who are 'cringing and cursing at the aggressive cruelty' of the Western nations (p. 126). This avoidance of the matter of oppression 'helps explain the lack of outrage that has prevailed in anthropology until recent years.' Willis writes: 'Since relativism is applied only to "aboriginal" customs, it advises colored peoples to preserve those customs that contributed to initial defeat and subsequent exploitation. . . . Hence, relativism defines the good life for colored peoples differently than for white people, and the good colored man is the man of the bush' (p. 144). Instead of leaving cultures as they are, as museum pieces, we should help to bring about change—or, better, we should help the oppressed to bring about change.

. . .

AMERICAN ANTHROPOLOGICAL ASSOCIATION, STATEMENT ON HUMAN RIGHTS

49 Amer. Anthropologist No. 4, 539 (1947).

[In 1947, the Commission on Human Rights created under the UN Charter was considering proposals for a declaration on basic human rights. Ultimately the instrument took the form of the Universal Declaration of Human Rights voted by the UN General Assembly in 1948. The Statement from which the following excerpts are taken was submitted to the Commission in 1947 by the Executive Board of the American Anthropological Association. It uses several terms or designations to refer to the pending document that became the UDHR.]

The problem faced by the Commission on Human Rights of the United Nations in preparing its Declaration on the Rights of Man must be approached from two points of view. The first, in terms of which the Declaration is ordinarily conceived, concerns the respect for the personality of the individual as such and his right to its fullest development as a member of his society. In a world order, however, respect for the cultures of differing human groups is equally important.

These are two facets of the same problem, since it is a truism that groups are composed of individuals, and human beings do not function outside the societies of which they form a part. The problem is thus to formulate a statement of human rights that will do more than just phrase respect for the individual as an individual. It must also take into full account the individual as a member of the social group of which he is a part, whose sanctioned modes of life shape his behavior, and with whose fate his own is thus inextricably bound.

. . . How can the proposed Declaration be applicable to all human beings and not be a statement of rights conceived only in terms of the values prevalent in the countries of Western Europe and America?

. . .

If we begin, as we must, with the individual, we find that from the moment of his birth not only his behavior, but his very thought, his hopes, aspirations, the moral values which direct his action and justify and give meaning to his life in his own eyes and those of his fellows, are shaped by the body of custom of the group of which he becomes a member. The process by means of which this is accomplished is so subtle, and its effects are so far-reaching, that only after considerable training are we conscious of it. Yet if the essence of the Declaration is to be, as it must, a statement in which the right of the individual to develop his personality to the fullest is to be stressed, then this must be based on a recognition of the fact that the personality of the individual can develop only in terms of the culture of his society.

. . .

. . . Doctrines of the 'white man's burden' have been employed to implement economic exploitation and to deny the right to control their own affairs to millions of peoples over the world, where the expansion of Europe and America has not meant the literal extermination of whole populations. Rationalized in terms of ascribing cultural inferiority to these peoples, or in conceptions of their backwardness in development of their 'primitive mentality,' that justified their being held in the tutelage of their superiors, the history of the expansion of the western world has been marked by demoralization of human personality and the disintegration of human rights among the peoples over whom hegemony has been established.

The values of the ways of life of these peoples have been consistently misunderstood and decried. Religious beliefs that for untold ages have carried conviction and permitted adjustment to the Universe have been attacked as superstitious, immoral, untrue. And, since power carries its own conviction, this has furthered the process of demoralization begun by economic exploitation and the loss of political autonomy. . . .

We thus come to the first proposition that the study of human psychology and culture dictates as essential in drawing up a Bill of Human Rights in terms of existing knowledge:

1. *The individual realizes his personality through his culture, hence respect for individual differences entails a respect for cultural differences.*

There can be no individual freedom, that is, when the group with which the individual identifies himself is not free. There can be no full development of the individual personality as long as the individual is told, by men who have the power to enforce their commands, that the way of life of his group is inferior to that of those who wield the power.

. . .

2. *Respect for differences between cultures is validated by the scientific fact that no technique of qualitatively evaluating cultures has been discovered.*

This principle leads us to a further one, namely that the aims that guide the life of every people are self-evident in their significance to that people. . . .

3. *Standards and values are relative to the culture from which they derive so that any attempt to formulate postulates that grow out of the beliefs or moral codes of one culture must to that extent detract from the applicability of any Declaration of Human Rights to mankind as a whole.*

Ideas of right and wrong, good and evil, are found in all societies, though they differ in their expression among different peoples. What is held to be a human right in one society may be regarded as anti-social by another people, or by the same people in a different period of their history. The saint of one epoch would at a later time be confined as a man not fitted to cope with reality. Even the nature of the physical world, the colors we see, the sounds we hear, are conditioned by the language we speak, which is part of the culture into which we are born.

The problem of drawing up a Declaration of Human Rights was relatively simple in the eighteenth century, because it was not a matter of *human* rights, but of the rights of men within the framework of the sanctions laid by a single society. . . .

Today the problem is complicated by the fact that the Declaration must be of worldwide applicability. It must embrace and recognize the validity of many different ways of life. It will not be convincing to the Indonesian, the African, the Indian, the Chinese, if it lies on the same plane as like documents of an earlier period. . . .

. . .

QUESTIONS

1. From the perspective of the 1990s, how would you respond to (a) the charge that the UDHR must be a 'statement of rights conceived in terms of the values prevalent in the countries of Western Europe and America,' and (b) the charge that an attempt to impose different values on millions of peoples spread throughout the world is but another example of a Western colonialism that systematically destroyed for its own gain the cultures that it conquered and ruled?

2. Consider whether one should distinguish between arguments of cultural relativism (a) applied to indigenous peoples who are relatively isolated, often tribal, and long settled in a given territory, and who seem to be the subject of the Statement on Human Rights, and (b) applied to the contemporary state.

NOTE

Consider:

BRITANNUS (*shocked*):
Caesar, this is not proper.

THEODOTUS (*outraged*):
How?

CAESAR (*recovering his self-possession*):
Pardon him Theodotus: he is a barbarian, and thinks that the customs of his tribe and island are the laws of nature.

Caesar and Cleopatra, Act II
George Bernard Shaw

The next two articles examine cultural relativism from the perspective of two major religions and related national cultures: Hinduism and Islam. They express two very different approaches to this issue. Pannikar concentrates on India and on the 'traditional Hindu, Jain and Buddhist conceptions of reality.' He looks for equivalents of international human rights in a given non-Western culture—in his case, in Hindu society. 'There are no trans-cultural values, for the simple reason that a value exists as such only in a given cultural context. But there may be cross-cultural values, and a cross-cultural critique is indeed possible.'

An-Na'im examines the 'Muslim world.' Committed to international human rights and of the Islamic faith, he argues that 'human rights advocates in the Muslim world must work within the framework of Islam to be effective . . . [and] should struggle to have their interpretations of the relevant [Islamic] texts adopted as the new Islamic scriptural imperatives for the contemporary world.' Those interpretations would be broadly consistent with the norms of international human rights. An-Na'im's emphasis then is on the relation between the international system and a given religious tradition, and on the need for reconciliation through reinterpretation of the tradition, rather than on the identification of cross-cultural values among different systems.

PANNIKAR, IS THE NOTION OF HUMAN RIGHTS A WESTERN CONCEPT?

120 Diogenes 75 (1982).

We should approach this topic with great fear and respect. It is not a merely 'academic' issue. Human rights are trampled upon in the East as in the West, in the North as in the South of our planet. Granting the part of human greed and sheer evil in this universal transgression, could it not also be that Human Rights are not observed because in their present form they do not represent a universal symbol powerful enough to elicit understanding and agreement?

. . .

I. The Method of Inquiry

1. *Diatopical Hermeneutics*

It is claimed that Human Rights are universal. This alone entails a major philosophical query. Does it make sense to ask about conditions of universality when the very question about conditions of universality is far from universal? Philosophy can no longer ignore this inter-cultural problematic. Can we extrapolate the concept of Human Rights, from the context of the culture and history in which it was conceived, into a globally valid notion? Could it at least *become* a universal symbol? Or is it only one particular way of expressing—and saving—the *humanum?*

Although the question posed in the title is a legitimate one, there is something disturbing in this formulation as it was given to me. At least at first glance, it would seem to offer only one alternative: either the notion of

Universal Human Rights is a Western notion, or it is not. If it is, besides being a tacit indictment against those who do not possess such a valuable concept, its introduction into other cultures, even if necessary, would appear as a plain imposition from outside. It would appear, once again, as a continuation of the colonial syndrome, namely the belief that the constructs of one particular culture (God, Church, Empire, Western civilization, science, modern technology, etc.) have, if not the monopoly, at least the privilege of possessing a universal value which entitles them to be spread over all the Earth. If not, that is, if the concept of Universal Human Rights is not exclusively a Western concept, it would be difficult to deny that many a culture has let it slumber, thus again giving rise to an impression of the indisputable superiority of Western culture. . . .

Our question is a case in point of *diatopical hermeneutics*: the problem is how, from the [locus or context] of one culture, to understand the constructs of another.[1] It is wrong-headed methodology to begin by asking: Does another culture also have the notion of Human Rights?—assuming that such a notion is absolutely indispensable to guarantee human dignity. No question is neutral, for every question conditions its possible answers.

2. *The homeomorphic equivalent*

I was once asked to give the Sanskrit equivalents of the twenty-five key Latin words supposed to be emblematic of Western culture. I declined, on the grounds that that which is the foundation of one culture need not be the foundation for another. Meanings are not transferable here. Translations are more delicate than heart transplants. So what must we do? We must dig down to where a homogeneous soil or a similar problematic appears: we must search out the *homeomorphic equivalent*—to the concept of Human Rights in this case . . . .[2]

Thus we are not seeking merely to transliterate Human Rights into other cultural languages, nor should we be looking for mere analogies; we try instead to find the homeomorphic equivalent. If, for instance, Human Rights are considered to be the basis for the exercise of and respect for human dignity, we should investigate how another culture satisfies the equivalent need—and this can be done only once a common ground (a mutually understandable language) has been worked out between the two cultures. Or perhaps we should ask how the idea of a just social and political order could be formulated within a certain culture, and investigate whether the concept of Human Rights is a particularly appropriate way of expressing this order. A traditional Confucian might see this problem of order and rights as a question of 'good

[1] [Eds.] The author means by this term 'diatopical hermeneutics' (a) the fact that the contexts or loci of different cultures that are historically unrelated 'make it problematic to understand one tradition with the tools of another,' and (b) the related attempt through hermeneutics (the method of interpretation) to bridge that gulf.

[2] The two words Brahman and God, for instance, are neither analogous nor merely equivocal (nor univocal, of course). They are not exactly equivalent either. They are homeomorphic. They perform a certain type of respectively corresponding function in the two different traditions where these words are alive.

manners' or in terms of his profoundly ceremonial or ritual conception of human intercourse, in terms of *li*. A Hindu might see it another way, and so on.

. . .

. . . Human Rights are one window through which one particular culture envisages a just human order for its individuals. But those who live in that culture do not see the window. For this they need the help of another culture which sees through another window. Now I assume that the human landscape as seen through the one window is both similar to and different from the vision of the other. If this is the case, should we smash the windows and make of the many portals a single gaping aperture—with the consequent danger of structural collapse—or should we enlarge the viewpoints as much as possible and, most of all, make people aware that there are—and have to be—a plurality of windows? This latter option would be the one in favor of a healthy pluralism. This is much more than a merely academic question. There can be no serious talk about cultural pluralism without a genuine socio-economic political pluralism. . . .

II. Assumptions and Implications of the Western Concept

. . .

1. At the basis of the discourse on Human Rights there is the assumption of a *universal human nature* common to all peoples. Otherwise, a Universal Declaration could not logically have been proclaimed. This idea in its turn is connected with the old notion of a Natural Law.

But the contemporary Declaration of Human Rights further *implies*:

a) that this human nature must be *knowable*. . . .
b) that this human nature is known by means of an equally universal organ of knowledge, generally called *reason*. Otherwise, if its knowledge should depend on a special intuition, revelation, faith, decree of a prophet or the like, Human Rights could not be taken as *natural* rights—inherent in Man. . . .
c) . . . Man is the master of himself and the universe. He is the supreme legislator on Earth—the question of whether a Supreme Being exists or not remains open, but ineffective.

2. The second assumption is that of the *dignity of the individual*. Each individual is, in a certain sense, absolute, irreducible to another. This is probably the major thrust of the Modern question of Human Rights. Human Rights defend the dignity of the individual *vis-à-vis* Society at large, and the State in particular.

But this in turn implies:

a) not only the distinction but also the *separation* between individual and society. In this view the human being is fundamentally the individual.

Society is a kind of superstructure, which can easily become a menace and also an alienating factor for the individual. Human Rights are there primarily to protect the individual;

b) the *autonomy* of humankind *vis-à-vis* and often versus the Cosmos. . . . The individual stands in between Society and World Human Rights defend the autonomy of the human individual;

c) resonances of the idea of Man as *microcosmos* and reverberations of the conviction that Man is *imago dei*, and at the same time the relative independence of this conviction from ontological and theological formulations. The individual has an inalienable dignity because he is an end in himself and a kind of absolute. You can cut off a finger for the sake of the entire body, but can you kill one person to save another?

3. The third assumption is that of a *democratic social order*. Society is assumed to be not a hierarchical order founded on a divine will or law or mythical origin, but a sum of 'free' individuals organized to achieve otherwise unreachable goals. Human Rights, once again, serve mainly to protect the individual. Society here is not seen as a family or a protection, but as something unavoidable which can easily abuse the power conferred on it (precisely by the assent of the sum of its individuals). This Society crystallizes in the State, which theoretically expresses the will of the people, or at least of the majority. . . .

This implies:

a) that each individual is seen as equally important and thus equally responsible for the welfare of society. . . .

b) that Society is nothing but the sum total of the individuals whose wills are sovereign and ultimately decisive. . . .

c) that the rights and freedoms of the individual can be limited only when they impinge upon the rights and freedoms of other individuals, and in this way majority rule is rationally justified. . . .

. . . [The Universal] Declaration clearly was articulated along the lines of the historical trends of the Western world during the last three centuries, and in tune with a certain philosophical anthropology or individualistic humanism which helped justify them.

III. Cross-Cultural Reflections

1. *Is the Concept of Human Rights a Universal Concept?*

The answer is a plain *no*. . . .

No concept as such is universal. Each concept is valid primarily where it was conceived. If we want to extend its validity beyond its own context we shall have to justify the extrapolation. . . . To accept the fact that the concept of Human Rights *is* not universal does not yet mean that it *should* not *become* so. Now in order for a concept to become universally valid it should fulfill at least two conditions. . . . [I]t should be the universal point of reference for any problematic regarding human dignity. In other words, it should displace all other homeomorphic equivalents and be the pivotal center of a just social order. To put it another way, the culture which has given birth to the concept of Human Rights should also be called upon to become a universal culture. This may well be one of the causes of a certain uneasiness one senses in non-Western thinkers who study the question of Human Rights. They fear for the identity of their own cultures.

. . .

The following parallelism may be instructive. To assume that without the explicit recognition of Human Rights life would be chaotic and have no meaning belongs to the same order of ideas as to think that without the belief in one God as understood in the Abrahamic tradition human life would dissolve itself in total anarchy. This line of thinking leads to the belief that Atheists, Buddhists and Animists, for instance, should be considered as human aberrations. In the same vein: either Human Rights, or chaos. This attitude does not belong exclusively to Western culture. To call the stranger a barbarian is all too common an attitude among the peoples of the world. . . .

2. *Cross-Cultural Critique*

There are no trans-cultural values, for the simple reason that a value exists as such only in a given cultural context. But there may be cross-cultural values, and a cross-cultural critique is indeed possible. The latter does not consist in evaluating one cultural construct with the categories of another, but in trying to understand and criticize one particular human problem with the tools of understanding of the different cultures concerned, at the same time taking thematically into consideration that the very awareness and, much more, the formulation of the problem is already culturally bound. Our question is then to examine the possible cross-cultural value of the issue of Human Rights, an effort which begins by delimiting the cultural boundaries of the concept. The dangers of cultural westocentrism are only too patent today.

a) We have already mentioned the particular historical origins of the Declaration of Human Rights. To claim universal validity for Human Rights in the formulated sense implies the belief that most of the peoples of the world today are engaged in much the same way as the Western nations in a process of transition from more or less mythical *Gemeinschaften* (feudal principalities, self-governing cities, guilds, local communities, tribal institutions . . .) to a 'rationally' and 'contractually' organized 'modernity' as known to the Western industrialized world. This is a questionable assumption. No one can predict the evolution (or eventual disintegration) of those traditional societies which have started from different material and cultural bases and whose reaction to modern Western civilization may therefore follow hitherto unknown lines.

. . .

b) We may now briefly reconsider the three assumptions mentioned above. They may pass muster, insofar as they express an authentically valid human issue from one particular context. But the very context may be susceptible to a legitimate critique from the perspective of other cultures. To do this systematically would require that we choose one culture after another and examine the assumptions of the Declaration in the light of each culture chosen. We shall limit ourselves here to token reflections under the very broad umbrella of a pre-Modern, non-Western state of mind.

i) There is certainly a *universal human nature* but, first of all, this nature does not need to be segregated and fundamentally distinct from the nature of all living beings and/or the entire reality. Thus exclusively *Human* Rights would be seen as a violation of 'Cosmic Rights' and an example of selfdefeating anthropocentrism, a novel kind of apartheid. To retort that 'Cosmic Rights' is a meaningless expression would only betray the underlying cosmology of the objection, for which the phrase makes no sense. But the existence of a different cosmology is precisely what is at stake here. We speak of the laws of nature; why not also of her rights?[3]

Secondly, the interpretation of this 'universal human nature,' i.e. Man's self-understanding, belongs equally to this human nature. Thus to single out one particular interpretation of it may be valid, but it is not universal and may not apply to the entirety of human nature.

Thirdly, to proclaim the undoubtedly positive concept of Human Rights may turn out to be a Trojan horse, surreptitiously introduced into other civilizations which will then all but be obliged to accept those ways of living, thinking and feeling for which Human Rights is the proper solution in cases of conflict. . . .

ii) Nothing could be more important than to underscore and defend the *dignity of the human person*. But the person should be distinguished from the individual. The individual is just an abstraction, i.e. a selection of a few aspects of the person for practical purposes. My *person*, on the other hand, is also in 'my' parents, children, friends, foes, ancestors and successors. 'My' person is also in 'my' ideas and feelings and in 'my' belongings. If you hurt 'me,' you are equally damaging my whole clan, and possibly yourself as well. Rights cannot be individualized in this way. Is it the right of the mother, or of the child?—in the case of abortion. Or perhaps of the father and relatives as well? Rights cannot be abstracted from duties; the two are correlated. The dignity of the human person may equally be violated by your language, or by your desecrating a place I consider holy, even though it does not 'belong' to me in the sense of individualized private property. You may have 'bought' it for a sum of money, while it belongs to me by virtue of another order altogether. An individual is an isolated knot; a person is the entire fabric around that

[3] [Eds.] The author here uses a number of related concepts. *Cosmos* refers broadly to a universe regarded as an orderly and harmonious whole. *Cosmology* refers generally to a branch of philosophy concerned with the origin and structure of the universe. An *anthropocentric* view refers to one in which man is viewed as a central fact of the universe, and reality is interpreted in terms of human values and experiences. *Anthropomorphism* refers to the attribution of human characteristics to inanimate or natural phenomena.

knot, woven from the total fabric of the real. The limits to a person are not fixed, they depend utterly on his or her personality. Certainly without the knots the net would collapse; but without the net, the knots would not even exist.

To aggressively defend my individual rights, for instance, may have negative, i.e. unjust, repercussions on others and perhaps even on myself. The need for consensus in many traditions—instead of majority opinion—is based precisely on the corporate nature of human rights.

. . .

iii) *Democracy* is also a great value and infinitely better than any dictatorship. But it amounts to tyranny to put the peoples of the world under the alternative of choosing either democracy or dictatorship. Human Rights are tied to democracy. Individuals need to be protected when the structure which is above them (Society, the State or the Dictator—by whatever name) is not qualitatively superior to them, i.e. when it does not belong to a higher order. Human rights is a legal device for the protection of smaller numbers of people (the minority or the individual) faced with the power of greater numbers. . . . In a hierarchical conception of reality, the particular human being cannot defend his or her rights by demanding or exacting them independently of the whole. The wounded order has to be set straight again, or it has to change altogether. Other traditional societies have different means to more or less successfully restore the order. The rāja may fail in his duty to protect the people, but will a Declaration of Human Rights be a corrective unless it also has the power to constrain the rāja? Can a democracy be imposed and remain democratic?

. . .

In short, the cross-cultural critique does not invalidate the Declaration of Human Rights, but offers new perspectives for an internal criticism and sets the limits of validity of Human Rights, offering at the same time both possibilities for enlarging its realm, if the context changes, and of a mutual fecundation with other conceptions of Man and Reality.

. . .

IV. An Indian Reflection

The word 'Indian' here has no political connotations. It does not refer to the 'nation' with the third largest Islamic population in the world, but to the traditional Hindu, Jain and Buddhist conceptions of reality.

. . .

The starting point here is not the individual, but the whole complex concatenation of the Real. In order to protect the world, for the sake of the protection of this universe, says Manu, He, Svayambhū, the Self-existent, arranged the castes and their duties. Dharma is the order of the entire reality, that which keeps the world together. The individual's duty is to maintain his 'rights;' it is to find one's place in relation to Society, to the Cosmos, and to the transcendent world.

It is obvious from these brief paragraphs that here the discourse on 'Human Rights' would take on an altogether different character. It would distract us from the purpose of this article to look now for the homeomorphic equivalent of Human Rights in a culture pervaded with the conception of dharma. We adduce this Indian example only to be able to elaborate in a fuller way the question of our title.

. . .

3. Human Rights are not Rights only. They are also duties and both are interdependent. Humankind has the 'right' to survive only insofar as it performs the duty of maintaining the world (*lokasamgraha*). We have the 'right' to eat only inasmuch as we fulfill the duty of allowing ourselves to be eaten by a hierarchically higher agency. Our right is only a participation in the entire metabolic function of the universe.

We should have, if anything, a Declaration of Universal Rights and Duties in which the whole of Reality would be encompassed. Obviously, this demands not only a different anthropology but also a different cosmology and an altogether different theology—beginning with its name. . . .

. . .

6. Both systems (the Western and the Hindu) make sense from and within a given and accepted myth. Both systems imply a certain kind of consensus. When that consensus is challenged, a new myth must be found. The broken myth is the situation in India today, as it is in the world at large. That the rights of individuals be conditioned only by their position in the net of Reality can no longer be admitted by the contemporary mentality. Nor does it seem to be admissible that the rights of individuals be so absolute as not to depend at all on the particular situation of the individual.

In short, there is at present no endogenous theory capable of unifying contemporary societies and no imposed or imported ideology can be simply substituted for it. A mutual fecundation of cultures is a human imperative of our times.

. . .

V. By Way of Conclusion

Is the concept of Human Rights a Western conception?

Yes.

Should the world then renounce declaring or enforcing Human Rights?

. . .

No.

Three qualifications, however, are necessary:

1. For an authentic human life to be possible within the *megamachine* of the modern technological world, Human Rights are imperative. This is because the development of the notion of Human Rights is bound up with and given its meaning by the slow development of that megamachine. How far individuals or groups or nations should collaborate with this present-day system is another question altogether. But in the contemporary political arena as defined by current socio-economic and ideological trends, the defense of Human Rights is a sacred duty. Yet it should be remembered that to introduce Human Rights (in the definite Western sense, of course) into other cultures before the introduction of *techniculture* would amount not only to putting the cart before the horse, but also to preparing the way for the technological invasion—as if by a Trojan horse, as we have already said. And yet a technological civilization without Human Rights amounts to the most inhuman situation imaginable. The dilemma is excruciating. This makes the two following points all the more important and urgent.

2. Room should be made for other traditions to develop and formulate their own homeomorphic views corresponding to or opposing Western 'rights.' Or rather, these other world traditions should make room for themselves, since no one else is likely to make it for them. This is an urgent task; otherwise it will be impossible for non-Western cultures to survive, let alone to offer viable alternatives or even a sensible complement. Here the role of a cross-cultural philosophical approach is paramount. The need for human pluralism is often recognized in principle, but not often practiced, not only because of the dynamism which drives the paneconomic ideology, linked with the megamachine, to expand all over the world, but also because viable alternatives are not yet theoretically worked out.

3. An intermediary space should be found for mutual criticism that strives for mutual fecundation and enrichment. Perhaps such an interchange may help bring forth a new myth and eventually a more humane civilization. The dialogical dialogue appears as the unavoidable method.

. . .

If many traditional cultures are centered on God, and some other cultures basically cosmocentric, the culture which has come up with the notion of Human Rights is decisively anthropocentric. Perhaps we may now be prepared for a cosmotheandric vision of reality in which the Divine, the Human and the Cosmic are integrated into a whole, more or less harmonious according to the performance of our truly human rights.

ABDULLAH AHMED AN-NA'IM, HUMAN RIGHTS IN THE MUSLIM WORLD

3 Harv. Hum. Rts. J. 13 (1990).

INTRODUCTION

Historical formulations of Islamic religious law, commonly known as Shari'a, include a universal system of law and ethics and purport to regulate every aspect of public and private life. The power of Shari'a to regulate the behavior of Muslims derives from its moral and religious authority as well as the formal enforcement of its legal norms. As such, Shari'a influences individual and collective behavior in Muslim countries through its role in the socialization processes of such nations regardless of its status in their formal legal systems. For example, the status and rights of women in the Muslim world have always been significantly influenced by Shari'a, regardless of the degree of Islamization in public life. Of course, Shari'a is not the sole determinant of human behavior nor the only formative force behind social and political institutions in Muslim countries.

. . .

I conclude that human rights advocates in the Muslim world must work within the framework of Islam to be effective. They need not be confined, however, to the particular historical interpretations of Islam known as Shari'a. Muslims are obliged, as a matter of faith, to conduct their private and public affairs in accordance with the dictates of Islam, but there is room for legitimate disagreement over the precise nature of these dictates in the modern context. Religious texts, like all other texts, are open to a variety of interpretations. Human rights advocates in the Muslim world should struggle to have their interpretations of the relevant texts adopted as the new Islamic scriptural imperatives for the contemporary world.

A. Cultural Legitimacy for Human Rights

The basic premise of my approach is that human rights violations reflect the lack or weakness of cultural legitimacy of international standards in a society. Insofar as these standards are perceived to be alien to or at variance with the values and institutions of a people, they are unlikely to elicit commitment or compliance. While cultural legitimacy may not be the sole or even primary determinant of compliance with human rights standards, it is, in my view, an extremely significant one. Thus, the underlying causes of any lack or weakness of legitimacy of human rights standards must be addressed in order to enhance the promotion and protection of human rights in that society.

. . . This cultural illegitimacy, it is argued, derives from the historical conditions surrounding the creation of the particular human rights instruments. Most African and Asian countries did not participate in the formulation of the Universal Declaration of Human Rights because, as victims of colonization, they were not members of the United Nations. When they did participate in the formulation of subsequent instruments, they did so on the basis of an established framework and philosophical assumptions adopted in their absence. For example, the preexisting framework and assumptions favored individual civil and political rights over collective solidarity rights, such as a right to development, an outcome which remains problematic today. Some authors have gone so far as to argue that inherent differences exist between the Western notion of human rights as reflected in the international instruments and non-Western notions of human dignity. In the Muslim world, for instance, there are obvious conflicts between Shari'a and certain human rights, especially of women and non-Muslims.

. . . In this discussion, I focus on the principles of legal equality and nondiscrimination contained in many human rights instruments. These principles relating to gender and religion are particularly problematic in the Muslim world.

. . .

II. ISLAM, SHARI'A AND HUMAN RIGHTS

. . .

A. The Development and Current Application of Shari'a

To the over nine hundred million Muslims of the world, the Qur'an is the literal and final word of God and Muhammad is the final Prophet. During his mission, from 610 A.D. to his death in 632 A.D., the Prophet elaborated on the meaning of the Qur'an and supplemented its rulings through his statements and actions. This body of information came to be known as Sunna. He also established the first Islamic state in Medina around 622 A.D. which emerged later as the ideal model of an Islamic state. . . .

While the Qur'an was collected and recorded soon after the Prophet Muhammad's death, it took almost two centuries to collect, verify, and record the Sunna. Because it remained an oral tradition for a long time during a period of exceptional turmoil in Muslim history, some Sunna reports are still controversial in terms of both their authenticity and relationship to the Qur'an.

Because Shari'a is derived from Sunna as well as the Qur'an, its development as a comprehensive legal and ethical system had to await the collection and authentication of Sunna. Shari'a was not developed until the second and third centuries of Islam. . . .

. . .

Shari'a is not a formally enacted legal code. It consists of a vast body of jurisprudence in which individual jurists express their views on the meaning of the Qur'an and Sunna and the legal implications of those views. Although most Muslims believe Shari'a to be a single logical whole, there is significant diversity of opinion not only among the various schools of thought, but also among the different jurists of a particular school. . . .

Furthermore, Muslim jurists were primarily concerned with the formulation of principles of Shari'a in terms of moral duties sanctioned by religious consequences rather than with legal obligations and rights and specific temporal remedies. They categorized all fields of human activity as permissible or impermissible and recommended or reprehensible. In other words, Shari'a addresses the conscience of the individual Muslim, whether in a private, or public and official, capacity, and not the institutions and corporate entities of society and the state.

. . .

Whatever may have been the historical status of Shari'a as the legal system of Muslim countries, the scope of its application in the public domain has diminished significantly since the middle of the nineteenth century. Due to both internal factors and external influence, Shari'a principles had been replaced by European law governing commercial, criminal, and constitutional matters in almost all Muslim countries. Only family law and inheritance continued to be governed by Shari'a. . . .

Recently, many Muslims have challenged the gradual weakening of Shari'a as the basis for their formal legal systems. Most Muslim countries have experienced mounting demands for the immediate application of Shari'a as the sole, or at least primary, legal system of the land. These movements have either succeeded in gaining complete control, as in Iran, or achieved significant success in having aspects of Shari'a introduced into the legal system, as in Pakistan and the Sudan. Governments of Muslim countries generally find it difficult to resist these demands out of fear of being condemned by their own populations as anti-Islamic. Therefore, it is likely that this so-called Islamic fundamentalism will achieve further successes in other Muslim countries.

The possibility of further Islamization may convince more people of the urgency of understanding and discussing the relationship between Shari'a and human rights, because Shari'a would have a direct impact on a wider range of human rights issues if it became the formal legal system of any country. . . .

I believe that a modern version of Islamic law can and should be developed. Such a modern 'Shari'a' could be, in my view, entirely consistent with current standards of human rights. These views, however, are appreciated by only a tiny minority of contemporary Muslims. To the overwhelming majority of Muslims today, Shari'a is the sole valid interpretation of Islam, and as such *ought* to prevail over any human law or policy.

B. Shari'a and Human Rights

In this part, I illustrate with specific examples how Shari'a conflicts with international human rights standards. . . .

. . .

The second example is the Shari'a law of apostasy. According to Shari'a, a Muslim who repudiates his faith in Islam, whether directly or indirectly, is guilty of a capital offense punishable by death. This aspect of Shari'a is in complete conflict with the fundamental human right of freedom of religion and conscience. The apostasy of a Muslim may be inferred by the court from the person's views or actions deemed by the court to contravene the basic tenets of Islam and therefore be tantamount to apostasy, regardless of the accused's personal belief that he or she is a Muslim.

The Shari'a law of apostasy can be used to restrict other human rights such as freedom of expression. A person may be liable to the death penalty for expressing views held by the authorities to contravene the official view of the tenets of Islam. Far from being an historical practice or a purely theoretical danger, this interpretation of the law of apostasy was applied in the Sudan as recently as 1985, when a Sudanese Muslim reformer was executed because the authorities deemed his views to be contrary to Islam.[4]

A third and final example of conflict between Shari'a and human rights relates to the status and rights of non-Muslims. Shari'a classifies the subjects of an Islamic state in terms of their religious beliefs: Muslims, *ahl al-Kitab* or believers in a divinely revealed scripture (mainly Christian and Jews), and unbelievers. In modern terms, Muslims are the only full citizens of an Islamic state, enjoying all the rights and freedoms granted by Shari'a and subject only to the limitations and restrictions imposed on women. *Ahl al-Kitab* are entitled to the status of *dhimma*, a special compact with the Muslim state which guarantees them security of persons and property and a degree of communal autonomy to practice their own religion and conduct their private affairs in accordance with their customs and laws. In exchange for these limited rights, *dhimmis* undertake to pay *jizya* or poll tax and submit to Muslim sovereignty and authority in all public affairs. . . .

According to this scheme, non-Muslim subjects of an Islamic state can aspire only to the status of *dhimma*, under which they would suffer serious violations of their human rights. *Dhimmis* are not entitled to equality with Muslims. [Economic and family law illustrations omitted.]

. . .

[4] . . . The Salman Rushdie affair illustrates the serious negative implications of the law of apostary to literary and artistic expression. Mr Rushdie, a British national of Muslim background, published a novel entitled, The Satanic Verses, in which irreverent reference is made to the Prophet of Islam, his wives, and leading companions. Many Muslim governments banned the book because their populations found the author's style and connotations extremely offensive. The late Imam Khomeini of Iran sentenced Rushdie to death *in absentia* without charge or trial. . . .

IV. A CASE STUDY: THE ISLAMIC DIMENSION OF THE STATUS OF WOMEN

. . .

The present focus on Muslim violations of the human rights of women does not mean that these are peculiar to the Muslim world.[5] As a Muslim, however, I am particularly concerned with the situation in the Muslim world and wish to contribute to its improvement.

The following discussion is organized in terms of the status and rights of Muslim women in the private sphere, particularly within the family, and in public fora, in relation to access to work and participation in public affairs. This classification is recommended for the Muslim context because the personal law aspects of Shari'a, family law and inheritance, have been applied much more consistently than the public law doctrines.[6] The status and rights of women in private life have always been significantly influenced by Shari'a regardless of the extent of Islamization of the public debate.

A. Shari'a and the Human Rights of Women

. . . The most important general principle of Shari'a influencing the status and rights of women is the notion of *qawama*. *Qawama* has its origin in verse 4:34 of the Qur'an: 'Men have *qawama* [guardianship and authority] over women because of the advantage they [men] have over them [women] and because they [men] spend their property in supporting them [women].' According to Shari'a interpretations of this verse, men as a group are the guardians of and superior to women as a group, and the men of a particular family are the guardians of and superior to the women of that family.

. . . For example, Shari'a provides that women are disqualified from holding general public office, which involves the exercise of authority over men, because, in keeping with the verse 4:34 of the Qur'an, men are entitled to exercise authority over women and not the reverse.

Another general principle of Shari'a that has broad implications for the status and rights of Muslim women is the notion of *al-hijab*, the veil. This means more than requiring women to cover their bodies and faces in public.

[5] It is difficult to distinguish between Islamic, or rather Shari'a, factors and extra-Shari'a factors affecting the status and rights of women. The fact that women's human rights are violated in all parts of the world suggests that there are universal social, economic, and political factors contributing to the persistence of this state of affairs. Nevertheless, the articulation and operation of these factors varies from one culture or context to the next. In particular, the rationalization of discrimination against and denial of equality for women is based on the values and customs of the particular society. In the Muslim world, these values and customs are supposed to be Islamic or at least consistent with the dictates of Islam. It is therefore useful to discuss the Islamic dimension of the status and rights of women.

[6] The private/public dichotomy, however, is an artificial distinction. The two spheres of life overlap and interact. The socialization and treatment of both men and women at home affect their role in public life and vice versa. While this classification can be used for analysis in the Muslim context, its limitations should be noted. It is advisable to look for both the private and public dimensions of a given Shari'a principle or rule rather than assume that it has only private or public implications.

According to Shari'a interpretations of verses 24:31, 33:33,[7] 33:53, and 33:59[8] of the Qur'an, women are supposed to stay at home and not leave it except when required to by urgent necessity. When they are permitted to venture beyond the home, they must do so with their bodies and faces covered. *Al-hijab* tends to reinforce women's inability to hold public office and restricts their access to public life. They are not supposed to participate in public life, because they must not mix with men even in public places.

. . . In family law for example, men have the right to marry up to four wives and the power to exercise complete control over them during marriage, to the extent of punishing them for disobedience if the men deem that to be necessary.[9] In contrast, the co-wives are supposed to submit to their husband's will and endure his punishments. While a husband is entitled to divorce any of his wives at will, a wife is not entitled to a divorce, except by judicial order on very specific and limited grounds. Another private law feature of discrimination is found in the law of inheritance, where the general rule is that women are entitled to half the share of men.

In addition to their general inferiority under the principle of *qawama* and lack of access to public life as a consequence of the notion of *al-hijab*, women are subjected to further specific limitations in the public domain. For instance, in the administration of justice, Shari'a holds women to be incompetent witnesses in serious criminal cases, regardless of their individual character and knowledge of the facts. In civil cases where a woman's testimony is accepted, it takes two women to make a single witness. *Diya*, monetary compensation to be paid to victims of violent crimes or to their surviving kin, is less for female victims than it is for male victims.

. . . These overlapping and interacting principles and rules play an extremely significant role in the socialization of both women and men. Notions of women's inferiority are deeply embedded in the character and attitudes of both women and men from early childhood.

. . .

C. Muslim Women in Public Life

A similar and perhaps more drastic conflict exists between reformist and conservative trends in relation to the status and rights of women in the public domain. Unlike personal law matters, where Shari'a was never displaced by secular law, in most Muslim countries, constitutional, criminal, and other public law matters have come to be based on secular, mainly Western, legal

[7] [O Consorts of the Prophet . . .] And stay quietly in your houses, and make not a dazzling display, like that of the former Times of Ignorance; and establish regular prayer, and give regular charity; and obey God and His Apostle. And God only wishes to remove all abomination from you, ye Members of the Family, and to make you pure and spotless.

[8] O Prophet! Tell thy wives and daughters, and the believing women, that they should cast their outer garments over their persons (when abroad): that is most convenient, that they should be known (as such) and not molested. And God is Oft-Forgiving, Most Merciful.

[9] Polygamy is based on verse 4:3 of the Qur'an. The husband's power to chastise his wife to the extent of beating her is based on verse 4:34 of the Qur'an.

concepts and institutions. Consequently, the struggle over Islamization of public law has been concerned with the reestablishment of Shari'a where it has been absent for decades, or at least since the creation of the modern Muslim nation states in the first half of the twentieth century. In terms of women's rights, the struggle shall determine whether women can keep the degree of equality and rights in public life they have achieved under secular constitutions and laws.

. . .

. . . Educated women and other modernist segments of society may not be able to articulate their vision of an Islamic state in terms of Shari'a, because aspects of Shari'a are incompatible with certain concepts and institutions which these groups take for granted, including the protection of all human rights. To the extent that efforts for the protection and promotion of human rights in the Muslim world must take into account the Islamic dimension of the political and sociological situation in Muslim countries, a modernist conception of Islam is needed.

V. ISLAMIC REFORM AND HUMAN RIGHTS

. . .

Islamic reform needs must be based on the Qur'an and Sunna, the primary sources of Islam. Although Muslims believe that the Qur'an is the literal and final word of God, and Sunna are the traditions of his final Prophet, they also appreciate that these sources have to be understood and applied through human interpretation and action. . . .

A. An Adequate Reform Methodology

. . . The basic premise of my position, based on the work of the late Sudanese Muslim reformer *Ustadh* Mahmoud Mohamed Taha, is that the Shari'a reflects a historically-conditioned interpretation of Islamic scriptures in the sense that the founding jurists had to understand those sources in accordance with their own social, economic, and political circumstances. In relation to the status and rights of women, for example, equality between men and women in the eight and ninth centuries in the Middle East, or anywhere else at the time, would have been inconceivable and impracticable. It was therefore natural and indeed inevitable that Muslim jurists would understand the relevant texts of the Qur'an and Sunna as confirming rather than repudiating the realities of the day.

In interpreting the primary sources of Islam in their historical context, the founding jurists of Shari'a tended not only to understand the Qur'an and Sunna as confirming existing social attitudes and institutions, but also to emphasize certain texts and 'enact' them into Shari'a while de-emphasizing other texts or interpreting them in ways consistent with what they believed to be the intent and purpose of the sources. Working with the same primary sources, modern Muslim jurists might shift emphasis from one class of texts to the other, and interpret the previously enacted texts in ways consistent with a new understanding of what is believed to be the intent and purpose of the sources. This new understanding would be informed by contemporary social, economic, and political circumstances in the same way that the 'old' understanding on which Shari'a jurists acted was informed by the then prevailing circumstances. The new understanding would qualify for Islamic legitimacy, in my view, if it is based on specific texts in opposing the application of other texts, and can be shown to be in accordance with the Qur'an and Sunna as a whole.

For example, the general principle of *qawama*, the guardianship and authority of men over women under Shari'a, is based on verse 4:34 of the Qur'an. . . . This verse presents *qawama* as a consequence of two conditions: men's advantage over and financial support of women. The fact that men are generally physically stronger than most women is not relevant in modern times where the rule of law prevails over physical might. Moreover, modern circumstances are making the economic independence of women from men more readily realized and appreciated. In other words, neither of the conditions—advantages of physical might or earning power—set by verse 4:34 as the justification for the *qawama* of men over women is tenable today.

The fundamental position of the modern human rights movement is that all human beings are equal in worth and dignity, regardless of gender, religion, or race. This position can be substantiated by the Qur'an and other Islamic sources as understood under the radically transformed circumstances of today. For example, in numerous verses the Qur'an speaks of honor and dignity for 'humankind' and 'children of Adam,' without distinction as to race, color, gender, or religion. By drawing on those sources and being willing to set aside archaic and dated interpretations of other sources, such as the one previously given to verse 4:34 of the Qur'an, we can provide Islamic legitimacy for the full range of human rights for women.

Similarly, numerous verses of the Qur'an provide for freedom of choice and non-compulsion in religious belief and conscience.[10] These verses have been either de-emphasized as having been 'overruled' by other verses which were understood to legitimize coercion, or 'interpreted' in ways which permitted such coercion. For example, verse 9:29 of the Qur'an was taken as the foundation of the whole system of *dhimma*, and its consequent discrimination against non-Muslims. Relying on those verses which extoll freedom of religion rather than those that legitimize religious coercion, one can argue now that the *dhimma* system should no longer be part of Islamic law and that complete equality should be assured regardless of religion or belief. The same argument

[10] See, for example, verse 2:256 of the Qur'an which provides: 'Let there be no compulsion in religion: Truth stands out clear from error . . .' In verse 18:29 God instructs the Prophet: 'Say, the Truth is from your Lord. Let him who will, believe, and let him who will, reject [it]'.

can be used to abolish all negative legal consequences of apostasy as inconsistent with the Islamic principle of freedom of religion.

[Discussion omitted of mechanisms and methods within Islam for development and reform.]

. . . The ultimate test of legitimacy and efficacy is, of course, acceptance and implementation by Muslims throughout the world.

B. Prospects for Acceptance and Likely Impact of the Proposed Reform

. . .

. . . Governments of Muslim countries, like many other governments, formally subscribe to international human rights instruments because, in my view, they find the human rights idea an important legitimizing force both at home and abroad. . . .

Nevertheless, the proposed reform will probably be resisted because it challenges the vested interests of powerful forces in the Muslim world and may upset male-dominated traditional political and social institutions. These forces probably will try to restrict opportunities for a genuine consideration of this reform methodology. . . .

Consequently, the acceptance and implementation of this reform methodology will involve a political struggle within Muslim nations as part of a larger general struggle for human rights. I would recommend this proposal to participants in that struggle who champion the cause of justice and equality for women and non-Muslims, and freedom of belief and expression in the Muslim world. Given the extreme importance of Islamic legitimacy in Muslim societies, I urge human rights advocates to claim the Islamic platform and not concede it to the traditionalist and fundamentalist forces in their societies. I would also invite outside supporters of Muslim human rights advocates to express their support with due sensitivity and genuine concern for Islamic legitimacy in the Muslim world.

. . .

NOTE

Iran is a party to the International Covenant on Civil and Political Rights. Its obligation under Article 40 of the Covenant is to submit periodic reports to a body created by the Covenant, the Human Rights Committee. The report is meant to describe, and to evaluate in terms of consistency with the provisions of the Covenant, the protection of civil and political rights in Iran.

Iran submitted such a report to the Committee in 1982.[11] Its representative

[11] Report Submitted by Iran under Article 40 of International Covenant on Civil and Political Rights, CCPR/C/SR. 364, 365, 366, 368, July 19–21, 1982.

presenting and defending the report before the Committee pointed out at the start that 'the dynamic doctrine of Islam was the ideological foundation of the Islamic Revolution in Iran.' The Summary Records of the discussion with Committee members notes that the representative 'felt bound to emphasize that although many articles of the Covenant were in conformity with the teachings of Islam, there could be no doubt that the tenets of Islam would prevail whenever the two sets of laws were in conflict.'

In the course of the discussion, several Committee members asked for clarification about reports they had received about torture, mass executions, denial of equal protection to non-Muslims, and other reported or alleged human rights violations. The Iranian representative 'regretted that some members had repeated lies and accusations originating from imperialist quarters,' and felt that the Committee's 'claim to fairness and impartiality' had been called into question.

Serious and systemic human rights abuses in Iran, ranging from physical violence to censorship, continue, as described in the report of a rapporteur of the UN Commission on Human Rights at p. 407, *infra*. With respect to cultural matters, consider a report in the *New York Times*, Dec. 27, 1994, p. 2 noting that the Iranian Parliament attempted to curb the influence of foreign television programs by banning the use of satellite dishes. The measure still required approval by a Guardian Council, a body of twelve Islamic jurors. About 200,000 Iranians received television programs by satellite, including numbers of Western TV's most popular shows. A Parliamentary deputy stated: 'the crackdown on satellite dishes should be enforced like the fight against drugs.'

> Faced with growing fears about what senior clerics have characterized as a cultural invasion threatening traditional values, the Interior Ministry and Secret Service agencies have been ordered to prevent the import, distribution and use of satellite dishes 'with all the necessary means.'

* * * * *

Consider the following observation in Rosalyn Higgins, *Problems and Process: International Law and How We Use It* (1994), at 96:

> It is sometimes suggested that there can be no fully universal concept of human rights, for it is necessary to take into account the diverse cultures and political systems of the world. In my view this is a point advanced mostly by states, and by liberal scholars anxious not to impose the Western view of things on others. It is rarely advanced by the oppressed, who are only too anxious to benefit from perceived universal standards. The non-universal, relativist view of human rights is in fact a very state-centred view and loses sight of the fact that human rights are *human* rights and not dependent on the fact that states, or groupings of states, may behave differently from each other so far as their politics, economic

policy, and culture are concerned. I believe, profoundly, in the universality of the human spirit. Individuals everywhere want the same essential things: to have sufficient food and shelter; to be able to speak freely; to practise their own religion or to abstain from religious belief; to feel that their person is not threatened by the state; to know that they will not be tortured, or detained without charge, and that, if charged, they will have a fair trial. I believe there is nothing in these aspirations that is dependent upon culture, or religion, or stage of development. They are as keenly felt by the African tribesman as by the European city-dweller, by the inhabitant of a Latin American shanty-town as by the resident of a Manhattan apartment.

QUESTIONS

1. After challenging the notion of universal human rights, Pannikar concludes that human rights are imperative 'within the megamachine of the modern technological world.' Compare Cover's references (p. 183, *supra*) to the 'almost unique mastery of violence' of the modern state and of 'communal resistance to the Behemoth.' How do you understand these observations? Do they amount to a distinctive modern justification for a human rights system, whatever its sharp differences from the religion examined? What would that justification be?

2. 'One persuasive argument for cultural relativism is the preservation of difference, surely among the important values expressed by the human rights movement. Of course there must be limits to that argument so that in certain circumstances international human rights prevail—at the extreme, over the "differences" expressed by Nazis or Stalinists. In more typical divergences from international norms, such as forms of gender discrimination enforced by a state, perhaps the international community should defer to that state when, say, people in it including those apparently harmed by the norm or practice in question prefer that way of life. Is this not simply a matter of allowing people to choose for themselves by engaging in self-determination, another deep human rights value?'

How do you react to this proposal? Whom would you consult to determine what people preferred in, say, a state characterized by a rigid caste system, or by ethnic or gender discrimination? In a state like Iran that executes people convicted of consensual homosexual acts? How indeed would you determine which groups of people in the state were 'harmed' by the norm or practice? Might the very conceptions of the 'harm' suffered by such a group, and of the idea of putting a choice to that group, be a Western conception? How, for example, might an Islamic or Hindu believer respond to a proposal for consultation?

NOTE

Compare the following writings of Howard and Schachter with respect to their conceptions of human dignity, and the relationship of those conceptions to human rights. Consider how these conceptions differ, and what the consequences of that difference might be for a human rights regime.

RHODA HOWARD, DIGNITY, COMMUNITY, AND HUMAN RIGHTS

In Abdullahi An-Na'im (ed.), Human Rights in Cross-Cultural Perspectives 81 (1992).

. . .

. . . [Most] known human societies did not and do not have conceptions of human rights. Human rights are a moral good that one can accept—on an ethical basis—and that everyone ought to have in the modern state-centric world. To seek an anthropologically based consensus on rights by surveying all known human cultures, however, is to confuse the concepts of rights, dignity, and justice. One can find affinities, analogues, and precedents for the actual content of internationally accepted human rights in many religious and cultural (geographic and national) traditions; but the actual concept of *human* rights, as will be seen, is particular and modern, representing a radical rupture from the many status-based, nonegalitarian, and hierarchical societies of the past and present. . . .

Human rights are a modern concept now universally applicable in principle because of the social evolution of the entire world toward state societies. The concept of human rights springs from modern human thought about the nature of justice; it does not spring from an anthropologically based consensus about the values, needs, or desires of human beings. As Jack Donnelly puts it, the concept of human rights is best interpreted by constructivist theory:

> Human rights aim to establish and guarantee the conditions necessary for the development of the human person envisioned in . . . [one particular] underlying moral theory of human nature, thereby bringing into being that type of person. . . . The evolution of particular conceptions or lists of human rights is seen in the constructivist theory as the result of the *reciprocal interactions of moral conceptions and material conditions of life*, mediated through social institutions such as rights.

Human rights tend to be particularly characteristic of liberal and/or social democratic societies. . . .

Human rights adhere to the human being *by virtue of being human, and for no other reason*. . . .

This means that the human being who holds rights holds them not only against the state, but also against 'society,' that is, against his or her community or even family. This orientation is a radical departure from the way most human societies in the past—and many in the present—have been or are organized. For most human societies, insofar as 'rights' might be considered to be applicable at all, collective or communal rights would be preferred to individual human rights. . . . Collective or community rights imply permissible inegalitarian ranking of members in the interests of preservation of 'tradition.'

I define human dignity as *the particular cultural understandings of the inner moral worth of the human person and his or her proper political relations with society.* Dignity is not a claim that an individual asserts against a society; it is not, for example, the claim that one is worthy of respect merely because one is a human being. Rather, dignity is something that is granted at birth or on incorporation into the community as a concomitant of one's particular ascribed status, or that accumulates and is earned during the life of an adult who adheres to his or her society's values, customs, and norms: the adult, that is, who accepts normative cultural constraints on his or her particular behavior. . . .

Many indigenous groups (that is, the remnants of precapitalist societies destroyed—physically, culturally, or both—during the process of European conquest and/or settlement) now make claims for the recognition of their collective or communal rights. When they do so they are not primarily interested in the human rights of the individual members of their collectivities. Rather, they are interested in the recognition of their *collective dignity,* in the acknowledgment of the value of their collective way of life as opposed to the way of life of the dominant society into which they are unequally 'integrated.' . . .

. . .

Thus in most known past or present societies, human dignity is not private, individual, or autonomous. It is public, collective, and prescribed by social norms. The idea that an individual can enhance his or her 'dignity' by asserting his or her human rights violates many societies' most fundamental beliefs about the way social life should be ordered. Part of the dignity of a human being consists of the quiet endurance and acceptance of what a human rights approach to the world would consider injustice or inequality. . . .

. . .

What then is a human being? For many societies, the human being is the person who has learned and obeys the community's rules. A nonsocial atomized individual is not human; he or she is a species of 'other'—perhaps equivalent to a (presocialized) child, a stranger, a slave, or even an animal. There is very little room in most societies for Mead's 'I'—the individual, self-reflective being—to emerge over the 'me,' that part of a being that absorbs his or her community's culture and faithfully follows the rules and customs expected of a person of his or her station. The human group takes precedence over the human person.

. .

This does not mean that human rights are not relevant, in the late twentieth century, to those societies in the world that retain precapitalist, nonindividualist notions of human dignity, honor, and the social order. The rise of the centralized state makes human rights relevant the world over. It does mean that to look for universalistic 'roots' of human rights in different social areas of the world . . . or in different religious traditions, is to abstract those societies and religions from culture and history. One can find, in Judaism and Christianity for example, strong moral analogues to the content—although not the concept—of contemporary human rights. But one can also find moral precepts justifying inequality and denial of what are now considered fundamental human rights. . . .

. . .

. . . All societies do have underlying conceptions of human dignity and social justice. These conceptions can be identified; and certain commonalities of belief, for example, in the social value of work, can also be located on a transcultural basis. But in most known human societies, dignity and justice are not based on any idea of the *inalienable right* of the *physical,* socially *equal* human being against the claims of family, community, or the state. They are based on just the opposite, that is, the *alienable privileges* of *socially unequal* beings, considered to embody gradations of humanness according to socially defined status categories entitled to different degrees of respect.

While all societies have underlying concepts of dignity and justice, few have concepts of rights. Human rights, then, are a particular expression of human dignity. In most societies, dignity does not imply human rights. There is very little cultural—let alone universal—foundation for the concept, as opposed to the content, of human rights. The society that actively protects rights both in law and in practice is a radical departure for most known human societies. . . .

. . .

OSCAR SCHACHTER, HUMAN DIGNITY AS A NORMATIVE CONCEPT

77 Am. J. Int. L. 848 (1983).

The 'dignity of the human person' and 'human dignity' are phrases that have come to be used as an expression of a basic value accepted in a broad sense by all peoples.

Human dignity appears in the Preamble of the Charter of the United Nations as an ideal that 'we the peoples of the United Nations' are 'determined' to achieve. . . .

The term dignity is also included in Article 1 of the Universal Declaration of Human Rights. . . .

. . .

The Helsinki Accords in Principle VII affirm that the participating states will promote the effective exercise of human rights and freedoms, 'all of which derive from the inherent dignity of the human person.'

References to human dignity are to be found in various resolutions and declarations of international bodies. National constitutions and proclamations, especially those recently adopted, include the ideal or goal of human dignity in their references to human rights. Political leaders, jurists and philosophers have increasingly alluded to the dignity of the human person. . . . No other ideal seems so clearly accepted as a universal social good.

We do not find an explicit definition of the expression 'dignity of the human person' in international instruments or (as far as I know) in national law. Its intrinsic meaning has been left to intuitive understanding, conditioned in large measure by cultural factors. . . .

. . .

An analysis of dignity may begin with its etymological root, the Latin 'dignitas' translated as worth (in French, 'valeur'). One lexical meaning of dignity is 'intrinsic worth.' Thus, when the UN Charter refers to the 'dignity and worth' of the human person, it uses two synonyms for the same concept. The other instruments speak of 'inherent dignity,' an expression that is close to 'intrinsic worth.'

What is meant by 'respect' for 'intrinsic worth' or 'inherent dignity' of a person? 'Respect' has several nuanced meanings: 'esteem,' 'deference,' 'a proper regard for,' 'recognition of.' These terms have both a subjective aspect (how one feels or thinks about another) and an objective aspect (how one treats another). Both are relevant to our question, but it seems more useful to focus on the latter aspect for purposes of practical measures.

One general answer to our question is suggested by the Kantian injunction to treat every human being as an end, not as a means. Respect for the intrinsic worth of every person should mean that individuals are not to be perceived or treated merely as instruments or objects of the will of others. This proposition will probably be generally acceptable as an ideal. There may be more question about its implications. I shall suggest such implications as corollaries of the general proposition.

The first is that a high priority should be accorded in political, social and legal arrangements to individual choices in such matters as beliefs, way of life, attitudes and the conduct of public affairs. Note that this is stated as a 'high priority,' not an absolute rule. We may give it more specific content by applying it to political and psychological situations. In the political context, respect for the dignity and worth of all persons, and for their individual choices, leads, broadly speaking, to a strong emphasis on the will and consent of the governed. . . .

. . .

. . . [W]e believe that the idea of human dignity involves a complex notion of the individual. It includes recognition of a distinct personal identity, reflecting individual autonomy and responsibility. It also embraces a recognition that the individual self is a part of larger collectivities and that they, too, must be considered in the meaning of the inherent dignity of the person. . . .

We are led more deeply into the analysis of human dignity when we consider its relation to the material needs of human beings and to the ideal of distributive justice. Few will dispute that a person in abject condition, deprived of adequate means of subsistence, or denied the opportunity to work, suffers a profound affront to his sense of dignity and intrinsic worth. Economic and social arrangements cannot therefore be excluded from a consideration of the demands of dignity. At the least, it requires recognition of a minimal concept of distributive justice that would require satisfaction of the essential needs of everyone.

. . .

. . . The general idea that human rights are derived from the dignity of the person is neither truistic nor neutral. It has two corollaries that challenge conceptions prevalent in some societies and ideologies. The first corollary is the idea that basic rights are not given by authority and therefore may not be taken away; the second is that they are rights of the person, every person. It is not unrealistic to assume that ideas of this kind will have a role in challenging existing attitudes. When they are found in official declarations, they become part of the instruments of change, sometimes loudly proclaimed, at other times almost imperceptibly affecting ideas of legitimacy.

. . .

. . . Respect for human dignity may be realized in other ways than by asserting claims of right. In many cases, the application of a 'rights approach' to affronts to dignity would raise questions involving existing basic rights such as free speech. In other cases, respect for dignity may be more appropriately and effectively attained through social processes such as education, material benefits, political leadership and the like. . . .

These observations indicate that the central idea of human dignity has a wide range of applications outside of the sphere of human rights. It is therefore of some importance to treat it as a distinct subject and to consider ways that respect for dignity can be fostered through public and private agencies.

QUESTION

Is the view of Howard or of Schachter about human dignity more sympathetic to cultural relativism? Do you find it helpful to consider the issue of relativism and rights from the perspective of this more fundamental notion of human dignity?

C. CONTEMPORARY DEBATE BETWEEN THE WEST AND SOME ASIAN STATES

During the Cold War, the primary context for argument over universalism and relativism was the conflict between the West and Communist states, particularly between the U.S. and the U.S.S.R. Apart from the important issue of indigenous peoples (see p. 1006, *infra*), the primary context today would be the argument between the West and (a) Islamic states and (b) a number of East Asian states intent on economic development. We concentrate here on this second category, involving particularly arguments between the U.S. and the People's Republic of China. The Government of Singapore and various spokespersons for it have made a sustained and serious contribution to the debate.

BILHARI KAUSIKAN, ASIA'S DIFFERENT STANDARD

92 Foreign Policy 24 (1993).

East and Southeast Asia must respond to a new phenomenon: Human rights have become a legitimate issue in interstate relations. How a country treats its citizens is no longer a matter for its own exclusive determination. Others can and do legitimately claim a concern. There is an emerging global culture of human rights. . . .

In response, East and Southeast Asia are reexamining their own human rights standards.

. . . But there is a more general acceptance of many international human rights norms, even among states that have not acceded to the two covenants or are accused by the West of human rights abuses.

The human rights situation in the region, whether measured by the standard of civil and political rights or by social, cultural, and economic rights, has improved greatly over the last 20 years. As countries in the region become more prosperous, secure, and self-confident, they are moving beyond a purely defensive attitude to a more active approach to human rights. All the countries of the region are party to the U.N. Charter. None has rejected the Universal Declaration. There are references to human rights in the constitutions of many of the countries in the region. Countries like China, Indonesia, and even Burma have not just brushed aside Western criticism of their human rights records but have tried to respond seriously, asserting or trying to demonstrate that they too adhere to international human rights norms. They tend to interpret rather than reject such norms when there are disagreements. They discuss human rights with Western delegations. They have released political prisoners; and Indonesia, for instance, has even held commissions of inquiry on alleged abuses and punished some officials found guilty.

Abuses and inconsistencies continue. But it is too simplistic to dismiss what has been achieved as mere gestures intended to appease Western critics. Such inclinations may well be an element in the overall calculation of interests. And Western pressure undeniably plays a role. But in themselves, self-interest and pressure are insufficient and condescendingly ethnocentric Western explanations. They do less than justice to the states concerned, most of which have their own traditions in which the rulers have a duty to govern in a way consonant with the human dignity of their subjects, even if there is no clear concept of 'rights' as has evolved in the West. China today, for all its imperfections, is a vast improvement over the China of the Cultural Revolution. So too has the situation in Taiwan, South Korea, and the Association of Southeast Asian Nations (ASEAN) improved. Western critics who deny the improvements lose credibility.

. . .

The diversity of cultural traditions, political structures, and levels of development will make it difficult, if not impossible, to define a single distinctive and coherent human rights regime that can encompass the vast region from Japan to Burma, with its Confucianist, Buddhist, Islamic, and Hindu traditions. Nonetheless, the movement toward such a goal is likely to continue. What is clear is that there is a general discontent throughout the region with a purely Western interpretation of human rights. The further development of human rights there will be shaped primarily by internal developments, but pressure will continue to come from the United States and Europe.

Human rights did not evolve in a vacuum. During the Cold War, the Western promotion of human rights was shaped by and deployed as an ideological instrument of the East-West struggle. The post-Cold War human rights dialogue between the West and Asia will be influenced by the power structure and dynamics of a more regionalized world, built around the United States, Europe, and Asia, which is replacing Cold War alliances and superpower competition. Trade and security will, as always, be foremost on the international agenda, and human rights will not be an issue of the first order. But human rights touch upon extraordinarily delicate matters of culture and values. And human rights issues are likely to become more prominent.

. . .

Meanwhile, relations among the United States, Europe, and Asia may lead the West to use human rights as an instrument of economic competition. . . . The lengthening catalogue of rights and freedoms in international human rights law now encompasses such matters as pay, work conditions, trade unions, standard of living, rest and leisure, welfare and social security, women's and children's rights, and the environment. The pressures and temptation to link economic concerns with human rights will certainly rise if economic strains increase. That is not to say that the West is insincere in its commitment to human rights. But policy motivations are rarely simple; and it is difficult to believe that economic considerations do not to some degree influence Western attitudes toward such issues as, say, the prison labor component of Chinese exports, child labor in Thailand, or some of the AFL-CIO complaints against Malaysian labor practices. President Bill Clinton's declared intention to press for human rights in China in return for continuing to grant most-favored-nation trading status to Beijing makes the linkage explicit.

But efforts to promote human rights in Asia must also reckon with the altered distribution of power in the post-Cold War world. Power, especially economic power, has been diffused. For the last two decades, most of East and Southeast Asia has experienced strong economic growth and will probably keep growing faster than other regions well into the next century. . . .

Of course, the economic growth has not been even. The United States and Europe are still major markets for almost all East and Southeast Asian countries, many of whom remain aid recipients. But Western leverage over East and Southeast Asia has been greatly reduced. The countries in the region are reacting accordingly. . . .

East and Southeast Asia are now significant actors in the world economy. There is far less scope for conditionality and sanctions to force compliance with human rights. The region is an expanding market for the West. . . .

. . .

. . . [T]he trend, already evident under Reagan and Bush in the attendant strains between the United States and many Asian countries, is away from rights as relatively precisely defined in international law, toward the promotion of hazier notions of 'freedom' and 'democracy.' The human rights apparatus is now even more open to manipulation by competing legislative, executive, judicial, media, and special interests devoted to such transcendent American values.

. . .

. . . Distance makes it easier to be virtuous; proximity makes for prudence. If, for instance, tough sanctions break the grip of the State Law and Order Restoration Council in Burma or the Communist party in China, the results could be violent. If disorder breaks out in Burma or China, it is not the United States or Europe that will pay the immediate price. Is the West prepared to intervene and remain engaged, perhaps for decades, to restore order? China will be a formidable political and economic force by the turn of the century. That does not mean human rights abuses in China must be overlooked. But if the promotion of human rights ignores Chinese realities and interests, expect China to find ways to exert countervailing pressures. And it will have the wherewithal to try to reshape any international order it sees as threatening.

For the first time since the Universal Declaration was adopted in 1948, countries not thoroughly steeped in the Judeo-Christian and natural law traditions are in the first rank: That unprecedented situation will define the new international politics of human rights. It will also multiply the occasions for conflict. In the process, will the human rights dialogue between the West and East and Southeast Asia become a dialogue of the deaf, with each side proclaiming its superior virtue without advancing the common interests of humanity? Or can it be a genuine and fruitful dialogue, expanding and deepening consensus? The latter outcome will require finding a balance between a pretentious and unrealistic universalism and a paralyzing cultural relativism. The myth of the universality of all human rights is harmful if it masks the real gap that exists between Asian and Western perceptions of human rights. The gap will not be bridged if it is denied.

The June 1993 Vienna U.N. conference on human rights did not even attempt to do so. The West went to Vienna accusing Asia of trying to undermine the ideal of universality, and determined to blame Asia if the conference failed. Inevitably, Asia resisted. The result after weeks of wrangling was a predictable diplomatic compromise ambiguous enough so all could live with it, but that settled very few things. There was no real dialogue between Asia and the West, no genuine attempt to address the issues or forge a meeting of the minds. If anything, the Vienna conference may only have hardened attitudes on both sides and increased the deep skepticism with which many Asian countries regard Western posturing on human rights.

. . .

For many in the West, the end of the Cold War was not just the defeat or collapse of communist regimes, but the supreme triumph and vindication of Western systems and values. It has become the lens through which they view developments in other regions. There has been a tendency since 1989 to draw parallels between developments in the Third World and those in Eastern Europe and the former USSR, measuring all states by the advance of what the West regards as 'democracy.' That is a value-laden term, itself susceptible to multiple interpretations, but usually understood by Western human rights activists and the media as the establishment of political institutions and practices akin to those existing in the United States and Europe.

There is good reason to doubt whether the countries of the former USSR and Eastern Europe will really evolve into 'democracies' anytime soon, however this term is defined, or even whether such a transformation would necessarily always be for the better, given the ethnic hatreds in the region. But the Western approach is ideological, not empirical. The West needs its myths; missionary zeal to whip the heathen along the path of righteousness and remake the world in its own image is deeply ingrained in Western (especially American) political culture. It is entirely understandable that Western human rights advocates choose to interpret reality in the way they believe helps their cause most.

. . .

The hard core of rights that are truly universal is smaller than many in the West are wont to pretend. Forty-five years after the Universal Declaration was adopted, many of its 30 articles are still subject to debate over interpretation and application—not just between Asia and the West, but within the West itself. Not every one of the 50 states of the United States would apply the provisions of the Universal Declaration in the same way. It is not only pretentious but wrong to insist that everything has been settled once and forever. The Universal Declaration is not a tablet Moses brought down from the mountain. It was drafted by mortals. All international norms must evolve through continuing debate among different points of view if consensus is to be maintained.

Most East and Southeast Asian governments are uneasy with the propensity of many American and some European human rights activists to place more emphasis on civil and political rights than on economic, social, and cultural rights. They would probably not be convinced, for instance, by a September 1992 report issued by Human Rights Watch entitled *Indivisible Human Rights: The Relationship of Political and Civil Rights to Survival, Subsistence and Poverty*. They would find the report's argument that 'political and civil rights, especially those related to democratic accountability,' are basic to survival and 'not luxuries to be enjoyed only after a certain level of economic development has been reached' to be grossly overstated. Such an argument does not accord with their own historical experience. That experience sees order and stability as preconditions for economic growth, and growth as the necessary foundation of any political order that claims to advance human dignity.

The Asian record of economic success is a powerful claim that cannot be easily dismissed. Both the West and Asia can agree that values and institutions are important determinants of development. But what institutions and which values? The individualistic ethos of the West or the communitarian traditions of Asia? The consensus-seeking approach of East and Southeast Asia or the adversarial institutions of the West? . . .

. . .

One explanation of the contradictions in Asian attitudes is that popular pressures against East and Southeast Asian governments may not be so much for 'human rights' or 'democracy' but for good government: effective, efficient, and honest administrations able to provide security and basic needs with good opportunities for an improved standard of living. To be sure, good government, human rights, and democracy are overlapping concepts. Good government requires the protection of human dignity and accountability through periodic fair and free elections. But they are not always the same thing; it cannot be blithely assumed, as many in the West have, that more democracy and human rights will inevitably lead to good government, as the many lost opportunities of the Aquino government demonstrated. The apparent contradictions mirror a complex reality: Good government may well require, among other things, detention without trial to deal with military rebels or religious and other extremists; curbs on press freedoms to avoid fanning racial tensions or exacerbating social divisions; and draconian laws to break the power of entrenched interests in order to, for instance, establish land reforms.

Those are the realities of exercising authority in heterogeneous, unevenly modernized, and imperfectly integrated societies with large rural populations and shallow Western-style civic traditions. . . .

. . .

Future Western approaches on human rights will have to be formulated with greater nuance and precision. It makes a great deal of difference if the West insists on humane standards of behavior by vigorously protesting genocide, murder, torture, or slavery. Here there is a clear consensus on a core of international law that does not admit of derogation on any grounds. The West has a legitimate right and moral duty to promote those core human rights, even if it is tempered by limited influence. But if the West objects to, say, capital punishment, detention without trial, or curbs on press freedoms, it should recognize that it does so in a context where the international law is less definitive and more open to interpretation and where there is room for further elaboration through debate. The West will have to accept that no universal consensus may be possible and that states can legitimately agree to disagree without being guilty of sinister designs or bad faith. Trying to impose pet Western definitions of 'freedom' and 'democracy' is an incitement to destructive conflict, best foregone in the interest of promoting real human rights.

The international law on human rights provides a useful, relatively precise, and common framework for the human rights dialogue between West and East. It helps prevent 'human rights' from becoming a mere catchphrase for whatever actions the West finds contrary to its preferences or too alien to comprehend. But the implementation, interpretation, and elaboration of the international law on human rights is unavoidably political. It must reflect changing global power structures and political circumstances. It will require the West to make complex political distinctions, perhaps refraining from taking a position on some human rights issues, irrespective of their merits, in order to press others where the prospects for consensus are better.

. . .

Yet it is only through such thickets of compromise, contradiction, and ambiguity that further progress on human rights can be made. Those in the West concerned about human rights in East and Southeast Asia, therefore, must be asked a simple question: Do you ultimately want to do good, or merely posture to make yourselves feel good?

NOTE

Consider the relevance to Kausikan's remarks of the following information:

(1) In 1987 the Singapore Government detained various social workers and activists. Several U.S. Congressmen wrote to complain about the detentions. A statement (in reply) of the Minister for Home Affairs stressed the fragility and heterogeneity of Singapore.

> We are vulnerable to powerful centrifugal forces and volatile emotional tides. Like many other developing countries, Singapore's major problem of nationhood is simply to stay united as one viable nation. . . . Singapore has repeatedly encountered subversive threats from within and without. . . . The very secrecy of covert operations precludes garnering evidence to meet the standards of the criminal law for conviction. In many cases of racial agitation, the process of trial itself will provide further opportunity for inflammatory rabble rousing. . . . Preventive detention is not a blemish marring our record; it is a necessary power underpinning our freedom.[12]

(2) The *New York Times*, Jan. 18, 1995, p. A6, reported that a Singapore court found a U.S. scholar and U.S.-owned newspaper (the *International Herald Tribune*) guilty of contempt of court over a published opinion article that was critical of what it called 'intolerant regimes' in Asia that use 'a compliant' judiciary to drive opposition politicians into bankruptcy through defamation and other suits. Heavy fines were imposed. The Justice noted that the article's clear reference was to Singapore, and that the description of compliant courts had 'scandalized the Singapore judiciary.' The U.S. State Department had protested this contempt proceeding, noting that 'people have a right to freedom of expression.'

Responding to similar critical allegations in the past about suits brought by members of the governing party, the Singapore government had insisted that the cases were decided on the merits without pressure on the judiciary. A few months earlier, a judicial decision found that the same newspaper had defamed the former Prime Minister (Lee Kuan Yew) through its reference to 'dynastic politics' in Singapore and other Asian states. It was public knowledge that Lee desired to see his son eventually become Prime Minister, but Lee claimed that the article amounted to an accusation of nepotism.

(3) A speech by Prime Minister Goh in which he 'delivered a wide-ranging denunciation of liberal ideas' was reported in *The Australian*, Aug. 23, 1994, p. 10. Stating that Singapore could not accept unmarried single-parent families, Goh noted that 'unmarried mothers would no longer be allowed to buy

[12] Quoted in Yash Ghai, Human Rights and Governance: The Asia Debate. 15 Aust. Y. B. Int. L. 1, 9 (1994). Ghai comments on another aspect of bypassing the formal legal system not discussed in the Minister's statement: coerced and doctored 'confessions' extracted from detainees and shown on national television.

government-subsidised flats direct from the Housing Development Board.' Goh also reaffirmed the Government's refusal to allow female civil servants the same medical benefits as their male counterparts, explaining that 'changing the rule would alter the balance of responsibility between men and women.' He further said that

> the government policy of stopping aid to families when they broke up was harsh but right. It was the Government's underlying philosophy to 'channel rights, benefits and privileges through the head of the family so that he can enforce the obligations and responsibilities of family members.' . . . To promote family togetherness, Mr. Goh announced a government grant of $S30,000 ($27,000) for people who want to buy a flat near their parents.

The Prime Minister acknowledged some problems, noting that Singaporeans were 'more preoccupied with materialism and individual rewards,' and that divorce rates were rising.

HUMAN RIGHTS IN CHINA

Information Office of the State Council, Beijing, 1991.

[This White Paper represents the most important official statement made by the government of the Peoples' Republic of China on human rights issues. The excerpts are taken from Section X, 'Active Participation in International Human Rights Activities.']

China pays close attention to the issue of the right to development. China believes that as history develops, the concept and connotation of human rights also develop constantly. . . . To the people in the developing countries, the most urgent human rights are still the right to subsistence and the right to economic, social and cultural development. Therefore, attention should first be given to the right to development. . . .

Over a long period in the UN activities in the human rights field, China has firmly opposed to any country making use of the issue of human rights to sell its own values, ideology, political standards and mode of development, and to any country interfering in the internal affairs of other countries on the pretext of human rights, the internal affairs of developing countries in particular, and so hurting the sovereignty and dignity of many developing countries. Together with other developing countries, China has waged a resolute struggle against all such acts of interference, and upheld justice by speaking out from a sense of fairness. China has always maintained that human rights are essentially matters within the domestic jurisdiction of a country. Respect for each country's sovereignty and non-interference in internal affairs are universally recognized principles of international law, which are applicable to all fields of international relations, and of course applicable to the field of human rights as well. Section 7 of Article 2 of the Charter of the United Nations stipulates that 'Nothing contained in the present Charter shall authorize the United Nations to intervene in matters which are essentially within the domestic jurisdiction of any state'. . . These provisions of international instruments reflect the will of the overwhelming majority of countries to safeguard the fundamental principles of international law and maintain a normal relationship between states. They are basic principles that must be followed in international human rights activities. The argument that the principle of non-interference in internal affairs does not apply to the issue of human rights is, in essence, a demand that sovereign states give up their state sovereignty in the field of human rights, a demand that is contrary to international law. Using the human rights issue for the political purpose of imposing the ideology of one country on another is no longer a question of human rights, but a manifestation of power politics in the form of interference in the internal affairs of other countries. Such abnormal practice in international human rights activities must be eliminated.

China is in favor of strengthening international cooperation in the realm of human rights on the basis of mutual understanding and seeking a common ground while reserving differences. However, no country in its effort to realize and protect human rights can take a route that is divorced from its history and its economic, political and cultural realities. . . . It is neither proper nor feasible for any country to judge other countries by the yardstick of its own mode or to impose its own mode on others. Therefore, the purpose of international protection of human rights and related activities should be to promote normal cooperation in the international field of human rights and international harmony, mutual understanding and mutual respect. Consideration should be given to the differing views on human rights held by countries with different political, economic and social systems, as well as different historical, religious and cultural backgrounds. International human rights activities should be carried on in the spirit of seeking common ground while reserving differences, mutual respect, and the promotion of understanding and cooperation.

China has always held that to effect international protection of human rights, the international community should interfere with and stop acts that endanger world peace and security, such as gross human rights violations caused by colonialism, racism, foreign aggression and occupation, as well as apartheid, racial discrimination, genocide, slave trade and serious violation of human rights by international terrorist organizations. . . .

. . . Interference in other countries' internal affairs and the pushing of power politics on the pretext of human rights are obstructing the realization of human rights and fundamental freedoms. . . .

NOTE

The Second World Conference on Human Rights, held in Vienna in June 1993, was preceded by regional preparatory meetings in four parts of the world. A group of Asian countries held such a regional meeting in Bangkok earlier that year. It culminated in the Bangkok Governmental Declaration.[13]

In that Declaration, the participating states emphasized (para. 5) 'principles of respect for national sovereignty and territorial integrity as well as non-interference in the internal affairs of States, and the non-use of human rights as an instrument of political pressure.' Although para. 7 stressed the universality of human rights and that no violation could be justified, para. 8 asserted that human rights 'must be considered in the context of a dynamic and evolving process of international norm-setting, bearing in mind the significance of national and regional particularities and various historical, cultural and religious backgrounds.'

Human rights non-governmental organizations (NGOs) from the Asia–Pacific region also gathered at Bangkok to transact their business immediately before the intergovernmental meeting began: The Bangkok NGO Declaration on Human Rights provided in para. 1:

> *Universality*. We can learn from different cultures in a pluralistic perspective . . . . Universal human rights are rooted in many cultures. We affirm the basis of universality of human rights which afford protection to all of humanity. . . . While advocating cultural pluralism, those cultural practices which derogate from universally accepted human rights, including women's right, must not be tolerated. *As human rights are of universal concern and are universal in value, the advocacy of human rights cannot be considered to be an encroachment upon national sovereignty.*

The Vienna Declaration[14] adopted by states at the Second World Conference on Human Rights, provides in Sec. I, para. 5 that

> [a]ll human rights are universal, indivisible, and interdependent and interrelated. . . . While the significance of national and regional particularities and various historical, cultural and religious backgrounds must be borne in mind, it is the duty of States, regardless of their political, economic and cultural systems, to promote and protect all human rights and fundamental freedoms.

QUESTIONS

1. 'It is the last resort of a state labelled a violator of human rights to contend (as does Kausikan, a governmental official in Singapore) that Western advocacy or international human rights is to some important degree a guise for supporting

[13] Set forth in 14 Hum. Rts. L. J. 370 (1993).

[14] Set forth in 14 Hum. Rts. L. J. 352 (1993).

the material interests of Western states in protecting their own markets, in opening foreign markets to Western goods, and generally in spreading market economics.' Comment.

2. 'Kausikan's article is not about cultural relativism at all. It never refers to the culture of the people in Singapore that the West is allegedly threatening. It is about the power of a one-party government, and the effort to insulate that government and its conduct from international human rights scrutiny. What do Kausikan's general arguments, or his precise examples of where Singapore justifiably deviates from human rights norms, have to do with popular, religious or other culture in Singapore? I thought that international human rights were meant to benefit individuals, not governmental elites.' Comment. How would you distinguish between a 'genuine' resort to cultural relativism and argument based on a one-party government's effort to blunt international criticism of its rule?

3. Are Kausikan's article and the P.R.C. White Paper both open to the interpretation that states making economic progress believe human rights to endanger that process? How would you best state that argument? What responses would you make to it?

4. In its White Paper, the P.R.C. invokes the traditional rhetoric of domestic jurisdiction. What relationship do you see between cultural relativism and domestic jurisdiction?

5. The White Paper gives several illustrations at the end of these excerpts of situations where the P.R.C. believes in international enforcement of human rights, even to the point of accepting enforcement that would interfere with and arrest a state's acts. How would you characterize those situations? Do they have any common themes? Do any of them, from a Western perspective, implicate the P.R.C.?

NOTE

The two following writings by Ghai and Donnelly consider cultural relativism in general, but bear on the preceding questions.

YASH GHAI, HUMAN RIGHTS AND GOVERNANCE: THE ASIA DEBATE

15 Austral. Y. Bk. Int. L. 1 (1994), at 5.

... It is easy to believe that there is a distinct Asian approach to human rights because some government leaders speak as if they represent the whole continent when they make their pronouncements on human rights. This view is reinforced because they claim that their views are based on perspectives which emerge from the Asian culture or religion or Asian realities. The gist of their position is that human rights as propounded in the West are founded on individualism and therefore have no relevance to Asia which is based on the primacy of the community. It is also sometimes argued that economic underdevelopment renders most of the political and civil rights (emphasised in the West) irrelevant in Asia. Indeed, it is sometimes alleged that such rights are dangerous in view of fragmented nationalism and fragile Statehood.

It would be surprising if there were indeed one Asian perspective, since neither Asian culture nor Asian realities are homogenous throughout the continent. All the world's major religions are represented in Asia, and are in one place or another State religions (or enjoy a comparable status: Christianity in the Philippines, Islam in Malaysia, Hinduism in Nepal and Buddhism in Sri Lanka and Thailand). To this list we may add political ideologies like socialism, democracy or feudalism which animate peoples and governments of the region. Even apart from religious differences, there are other factors which have produced a rich diversity of cultures. A culture, moreover, is not static and many accounts given of Asian culture are probably true of an age long ago. Nor are the economic circumstances of all the Asian countries similar. Japan, Singapore and Hong Kong are among the world's most prosperous countries, while there is grinding poverty in Bangladesh, India and the Philippines. The economic and political systems in Asia likewise show a remarkable diversity, ranging from semi-feudal kingdoms in Kuwait and Saudi Arabia, through military dictatorships in Burma and formerly Cambodia, effectively one party regimes in Singapore and Indonesia, communist regimes in China and Vietnam, ambiguous democracies in Malaysia and Sri Lanka, to well established democracies like India. There are similarly differences in their economic systems, ranging from tribal subsistence economies in parts of Indonesia through highly developed market economies of Singapore, Hong Kong and Taiwan and the mixed economy model of India to the planned economies of China and Vietnam. Perceptions of human rights are undoubtedly reflective of these conditions, and suggest that they would vary from country to country.

Perceptions of human rights are reflective of social and class positions in society. What conveys an apparent picture of a uniform Asian perspective on human rights is that it is the perspective of a particular group, that of the ruling elites, which gets international attention. What unites these elites is their notion of governance and the expediency of their rule. For the most part, the political systems they represent are not open or democratic, and their publicly expressed views on human rights are an emanation of these systems, of the need to justify authoritarianism and occasional repression. It is their views which are given wide publicity domestically and internationally.

. . .

. . . [S]ome Asian governments claim that their societies place a higher value on the community than in the West, that individuals find fulfilment in their participation in communal life and community tasks, and that this factor constitutes a primary distinction in the approach to human rights. . . . This argument is advanced as an instance of the general proposition that rights are culture specific.

The 'communitarian' argument is Janus-faced. It is used against the claim of universal human rights to distinguish the allegedly Western, individual-oriented approaches to rights from the community centred values of the East. Yet it is also used to deny the claims and assertions of communities in the name of 'national unity and stability'. It suffers from at least two further weaknesses. First, it overstates the 'individualism' of Western society and traditions of thought. . . .

Secondly, Asian governments (notwithstanding the attempt in the Singapore Paper to distinguish the 'nation' and the community) fall into the easy but wrong assumption that they or the State are the 'community'. . . . Nothing can be more destructive of the community than this conflation. The community and State are different institutions and to some extent in a contrary juxtaposition. The community, for the most part, depends on popular norms developed through forms of consensus and enforced through mediation and persuasion. The State is an imposition on society, and unless humanised and democratised (as it has not been in most of Asia), it relies on edicts, the military, coercion and sanctions. It is the tension between them which has elsewhere underpinned human rights. In the name of the community, most Asian governments have stifled social and political initiatives of private groups. . . . Governments have destroyed many communities in the name of development or State stability. . . .

Another attack on the community comes from the economic, market oriented policies of the governments. Although Asian capitalism appears to rely on the family and clan associations, there is little doubt that it weakens the community and its cohesion. The organising matrix of the market is not the same as that of the community. Nor are its values or methods particularly 'communitarian'. The moving frontier of the market, seeking new resources, has been particularly disruptive of communities which have managed to preserve intact a great deal of their culture and organisation during the colonial and post-colonial periods. The emphasis on the market, and with it individual rights of property are also at odds with communal organisation and enjoyment of property. . .

A final point is the contradiction between claims of a consensus and harmonious society, and the extensive arming of the state apparatus. The pervasive use of draconian legislation like administrative detention, disestablishment of societies, press censorship, and sedition, belies claims to respect alternative views, promote a dialogue, and seek consensus. The contemporary State intolerance of opposition is inconsistent with traditional communal values and processes. . . .

JACK DONNELLY, UNIVERSAL HUMAN RIGHTS IN THEORY AND PRACTICE

(1989), at 118.

. . .

4. CULTURE AND RELATIVISM

The cultural basis of cultural relativism must be considered too, especially because numerous contemporary arguments against universal human rights standards strive for the cachet of cultural relativism but actually are entirely without cultural basis.

Standard arguments for cultural relativism rely on such examples as the precolonial African village, Native American tribes, and traditional Islamic social systems, but we have seen that human rights are foreign to such communities, which employed other mechanisms to protect and realize human dignity. . . . [W]here there is a thriving indigenous cultural tradition and community, arguments of cultural relativism offer a strong defense against outside interference—including disruptions that might be caused by introducing 'universal' human rights.

Such communities, however, are increasingly the exception rather than the rule. They are not, for example, the communities of the teeming slums that hold an ever-growing proportion of the population of most Third World states. Even most rural areas of the Third World have been substantially penetrated, and the local culture 'corrupted,' by foreign practices and institutions, including the modern state, the money economy, and 'Western' values, products, and practices. In the Third World today we see most often not the persistence of traditional culture in the face of modern intrusions, or even the development of syncretic cultures and values, but rather a disruptive 'Westernization,' cultural confusion, or the enthusiastic embrace of 'modern' practices and values. In other words, the traditional culture advanced to justify cultural relativism far too often no longer exists. But communitarian defenses of traditional practices usually cannot be extended to modern nation-states and contemporary nationalist regimes.

Therefore, while recognizing the legitimate claims of self-determination and cultural relativism, we must be alert to cynical manipulations of a dying, lost, or even mythical cultural past. We must not be misled by complaints of the inappropriateness of 'Western' human rights made by repressive regimes whose practices have at best only the most tenuous connection to the indigenous culture; communitarian rhetoric too often cloaks the depredations of corrupt and often Westernized or deracinated elites.

Arguments of cultural relativism are far too often made by economic and political elites that have long since left traditional culture behind. While this may represent a fundamentally admirable effort to retain or recapture cherished traditional values, it is at least ironic to see largely Westernized elites warning against the values and practices they have adopted.

. . .

. . . Leaders sing the praises of traditional communities—while they wield arbitrary power antithetical to traditional values, pursue development policies that systematically undermine traditional communities, and replace traditional leaders with corrupt cronies and party hacks. Such cynical manipulation of tradition occurs everywhere.

. . .

The cynicism of many claims of cultural relativism can also be seen in the fact that far too often they are for foreign consumption only. The same elites that raise culture as a defense against external criticisms based on universal human rights often ruthlessly suppress inconvenient local customs, whether of the majority or of a minority. National unification certainly will require substantial sacrifices of local customs, but the lack of *local* cultural sensitivity shown by many national elites that strongly advocate an international cultural relativism suggests a very high degree of self-interest.

. . .

D. ILLUSTRATION: THE DEBATE OVER FEMALE CIRCUMCISION

Readings in Sections B and C suggested the important degree to which human rights debates over universalism and cultural relativism had as a subject a related set of issues deeply imbedded in most cultures: gender discrimination, sexuality and family. From the perspective of the human rights movement, those issues raise particularly acute problems in many developing countries. In recent decades, those countries experienced strong external and internal pressures to rethink and revise, sometimes radically, their traditional beliefs and practices. The relentless assault of the developed world on other cultures, the penetration of those cultures by trade, investment, tourism and media, the universalization of ideas and values like human rights, have inevitably launched transformative processes that are often referred to as modernization.

This Section D presents a case study involving two of these issues, sexuality and family, and implicitly involving gender discrimination as well. The focus is on practices that are variously referred to in the following readings as (with strikingly different political and moral innuendos and sometimes agendas) female circumcision *or* female genital mutilation. The issue of the consistency of this culturally imbedded practice with human rights norms has become a matter of intense, often angry public debate, both in the West and in the portions of Africa where the practice is now centered. Women's groups and health institutions around the world have become parties to that debate.

It is not only the currency and importance of these issues that justify their selection for this case study. The questions posed by challenges to this cultural practice and the responses by Africans and others raise troubling issues about cultural relativism, as well as about the reach of human rights norms to non-governmental practices. We can here examine in a concrete setting some of the broad concepts in Section B's readings.

The following readings draw on the views of a number of organizations and individual commentators. When reading them, consider (a) what the main lines of argument in the discussion (or debate, or campaign) are, and how that discussion is understood from the different perspectives presented; (b) what provisions in the UDHR or ICCPR are most relevant to a criticism of this practice; (c) what changes those criticisms point towards; and (d) the degree to which differences in views are explainable as fundamentally different (cultural, historical) perceptions of the practice *or* as questions of strategy about how to address and change the practice.

VIEWS OF COMMENTATORS ABOUT FEMALE CIRCUMCISION

A Traditional Practice that Threatens Health—Female Circumcision

40 World Health Organization Chronicle 31 (1986).

. . .

The traditional practices of a society are closely linked with the living conditions of the people and with their beliefs and priorities. In societies where women's needs have been subordinated to those of men, traditional practices often serve to reinforce their disadvantage, with direct and indirect effects on their health.

Traditionally, the reproductive role of women is surrounded by myths and taboos that underpin practices pertaining to menstruation, pregnancy and childbirth. Some of these practices are health-promoting but many are dangerous, even life-threatening. . . .

. . .

One traditional practice that has attracted much attention in the last decade is female circumcision. Although not the most lethal of the practices affecting women's health, its adverse effects are undeniable. Seventy million women are estimated to be circumcised, with several thousand new operations performed each day. It is a custom that is still widespread only in Africa north of the equator, though mild forms of female circumcision are reported from some countries in Asia too. However, history reveals that female circumcision of some kind has been practised at one time or another on every continent. . . .

. . .

There are three main types of female circumcision.

1. *Circumcision proper*, known in Muslim countries as *sunna* (which means 'traditional'), is the mildest but also the rarest form. It involves the removal only of the clitoral prepuce.
2. *Excision* involves the amputation of the whole of the clitoris and all or part of the labia minora.
3. *Infibulation*, also known as *Pharaonic circumcision*, involves the amputation of the clitoris, the whole of the labia minora, and at least the anterior two-thirds and often the whole of the medial part of the labia majora. The two sides of the vulva are then stitched together with silk, catgut or thorns, and a tiny sliver of wood or a reed is inserted to preserve an opening for urine and menstrual blood. The girl's legs are usually bound together from ankle to knee until the wound has healed, which may take anything up to 40 days.

Initial circumcision is carried out before a girl reaches puberty, the age range being anywhere from one week to 14 years. The operation is generally the responsibility of the traditional midwife, who rarely uses even a local anaesthetic. She is assisted by a number of women to hold the child down, and these frequently include the child's own relatives. . . .

. . .

Most of the adverse health consequences are associated with Pharaonic circumcision. Haemorrhage and shock from the acute pain are immediate dangers of the operation, and, because it is usually performed in unhygienic circumstances, the risks of infection and tetanus are considerable. . . .

Implantation dermoid cysts are a very common complication; these often grow to the size of a grapefruit. . . .

Infections of the vagina, urinary tract and pelvis occur frequently. Pelvic infection can result in sterility. Between 20% and 25% of infertility in Sudan has been attributed to Pharaonic circumcision.

Not surprisingly, a woman who has been infibulated suffers great difficulty and pain during sexual intercourse, which can be excruciating if a neuroma has formed at the point of section of the dorsal nerve of the clitoris. Consummation of marriage often necessitates the opening up of the scar by the husband using his fingers, a razor or a knife. Very little research has been done on the sexual experience of circumcised women, a subject surrounded by taboos and personal inhibition in most societies. However, the operation is known to destroy much or all of the vulval nerve and pressure endings, and seems likely to delay arousal and impair orgasm. One study among infibulated women found that few even knew of the existence of orgasm.

During childbirth infibulation causes a variety of serious problems including prolonged labour and obstructed delivery, with increased risk of fetal brain damage and fetal loss. . . .

Though some observers believe female circumcision was originally a means of suppressing female sexuality and attempting to ensure chaste or monogamous behaviour, others believe that it was started long ago among herders as a protection against rape for the young girls who took the animals out to pasture. In fact its origins have proved impossible to trace. Not surprisingly, a variety of reasons are advanced by its adherents for continuing to support the practice today. As the word 'sunna' suggests, some Muslim people believe it is religiously ordained. However, there is no support for female circumcision in the Koran, nor is it practised in Saudi Arabia, the cradle of Islam. Other adherents believe that intact female genitalia are 'unclean'; that an uncircumcised woman is likely to be promiscuous; even that the operation improves the life chances of a woman's offspring. Some say it is a ritual initiation into womanhood.

None of the reasons given bears close scrutiny. They are, in fact, rationalizations for a practice that has woven itself into the fabric of some societies so completely that 'reasons' are no longer particularly relevant, since invalidating them does not stop the practice.

Significantly, female circumcision is usually associated with poverty, illiteracy and low status of women—with communities in which people face hunger, ill-health, overwork, lack of clean water. In such settings an uncircumcised woman is stigmatized and not sought in marriage, which helps explain the paradox that the victims of the practice are also its strongest proponents. They can scarcely afford not to be. In the best of circumstances, people are reluctant to question tradition or take an independent line lest they lose social approval. In poverty-stricken communities struggling to survive, social acceptance and support may mean the difference between life and death.

However, the signs are that education and a widening range of choices for women are slowly but surely undermining the practice. And men, too, from societies that customarily circumcise their women are beginning to express their own ambivalence or outright dislike of the custom. . . .

. . .

Wherever a colonial administration of the past or a government of today has tried to ban it outright, it has simply been practised with greater secrecy, and those suffering health complications have been inhibited from seeking professional help. Such an approach ignores the fact that those practising female circumcision believe in it, and that deeply entrenched attitudes of and towards circumcised women cannot be changed overnight. It ignores the need to *replace* the practice and not merely repress it: girls and women need to find other forms and types of social status, approval, and respectability. It ignores, too, the fact that the operation is a principal source of income to traditional birth attendants and even midwives, who cannot afford to relinquish it unless an alternative living is available.

Roman Catholic missionaries in Ethiopia in the sixteenth century tried to stop the practice among their converts, but when men refused to marry the girls a reversal of the policy had to be demanded urgently from Rome. Today abolitionists who attempt to move fast similarly come up against the brick wall of a conservative society feeling itself threatened, or, if the campaigners are outsiders, against suspicion that they are meddling in cultural affairs that are none of their business. Many of these lessons have been learned, and the present approach is through national or local organizations, using as far as possible the skills and experience of those whose work is among villagers normally, such as teachers, social workers and health personnel.

. . .

In August 1982 WHO made a formal statement of its position to the United Nations Commission on Human Rights. This statement endorsed the recommendations of the Khartoum seminar, namely:

— that governments should adopt clear national policies to abolish the practice, and to inform and educate the public about its harmfulness;
— that programmes to combat it should recognize its association with extremely adverse social and economic conditions, and should respond sensitively to women's needs and problems;
— that the involvement of women's organizations at the local level should be encouraged, since it is with them that awareness and commitment to change must begin.

In the same statement the Organization expressed its unequivocal opposition to any medicalization of the operation, advising that under no circumstances should it ever be performed by health professionals or in health establishments.

. . .

Female Genital Mutiliation: Proposals for Change

Minority Rights Group International, Report 92/3 (1992) at 11.

. . .

Female genital mutilation is a complex issue, for it involves deep-seated cultural practices which affect millions of people. However, it can be divided into (at least) four distinct issues.

1 *Rights of women.* Female genital mutilation is an extreme example of the general subjugation of women, sufficiently extreme and horrifying to make

women and men question the basis of what is done to women, what women have accepted and why, in the name of society and tradition.

The burning of Indian widows and the binding of the feet of Chinese girl children are other striking examples, sharp enough and strange enough to throw a spotlight on other less obvious ways in which women the world over submit to oppression. . . .

. . .

2 *Rights of children.* An adult is quite free to submit her or himself to a ritual or tradition, but a child, having no formed judgement, does not consent but simply undergoes the operation (which in this case is irrevocable) while she is totally vulnerable. The descriptions available of the reactions of children—panic and shock from extreme pain, biting through the tongue, convulsions, necessity for six adults to hold down an eight-year-old, and death—indicate a practice comparable to torture.

Many countries signatory to Article 5 of the Universal Declaration of Human Rights (which provides that no one shall be subjected to torture, or to cruel, inhuman or degrading treatment) violate that clause. Those violations are discussed and sometimes condemned by various UN commissions. . . .

. . .

In September 1990, the United Nations Convention on the Rights of the Child went into force. It became part of International Human Rights Law. Under Article 24(3) it states that:

> States Parties shall take all effective and appropriate measures with a view to abolishing traditional practices prejudicial to the health of children.

. . .

3 *The right to good health.* There is no medical reputable practitioner who insists that mutilation is good for the physical or mental health of girls and women. . . .

. . .

LEGISLATION

In Africa

Formal legislation forbidding genital mutilation, or more precisely infibulation, exists in the **Sudan**. A law first enacted in 1946 allows for a term of imprisonment up to five years and/or a fine. However, it is not an offence (under Article 284 of the Sudan Penal Code for 1974) 'merely to remove the free and projecting part of the clitoris'.

. . .

In September 1982, President Arap Moi took steps to ban the practices in **Kenya**, following reports of the deaths of 14 children after excision. A traditional practitioner found to be carrying out this operation can be arrested by the Chiefs Act and brought before the law.

. . .

In Western countries

A law prohibiting female excision, whether consent has been given or not, came into force in **Sweden** in July 1982, carrying a two-year sentence. In **Norway**, in 1985, all hospitals were alerted to the practice. **Belgium** has incorporated a ban on the practice. Several states in the **USA** have incorporated female genital mutilation into their criminal code.

In the **UK**, specific legislation prohibiting female circumcision came into force at the end of 1985. A person found guilty of an offence is liable to up to five years' imprisonment or to a fine. . .

. . .

Kay Boulware-Miller, Female Circumcision: Challenges to the Practice as a Human Rights Violation

8 Harv. Womens L. J. 155 (1985), at 165.

. . .

A. The Rights of the Child

The Declaration of the Rights of the Child, adopted by the UN General Assembly in 1959, asserts that children must be guaranteed the opportunity to develop physically in a healthy and normal way.

. . .

First, to challenge female circumcision as a violation of the rights of the child suggests that women who permit the operation are incompetent and abusive mothers who, in some ways, do not love their children. The success of this approach therefore depends in part on how it is implemented; if African women are offended by the implication that they are poor mothers, they will likely reject the children's rights argument altogether.

The second problem with the rights of the child approach is that it conflicts with parents' desires to rear children independently and their notions of what is in their children's best interests. While women may not wish to see their daughters harmed, they may also feel strongly that they should be able to rear their children according to their own cultural norms and traditions. Besides, if mothers value the economic, social, and cultural benefits of the operation, they are unlikely to be persuaded that it should not be performed on their daughters. Moreover, the strong social and cultural pressures to continue the practice work against parents who would prefer not to submit their daughters to the operation. . . .

The third problem with this approach is that it almost exclusively focuses on the physical harm done to a child when she is circumcised and does not address the positive feelings she may have as a circumcised woman. In African communities with strong cultural and traditional ties, the perceived need to be circumcised mitigates the hellish remembrances of the event. Little girls who are initially hurt, betrayed, and degraded by the operation later come to feel socially and morally acceptable because they have been circumcised. As the girls grow into women they may forget the pain and argue that the practice need not be banned. Furthermore, it is difficult to attack a practice as harmful to children when it later gives them both social and economic benefits.

A final problem with approaching this issue from the rights of the child perspective is that many young girls believe that they want to be circumcised. The stigma associated with not being circumcised attaches early, virtually compelling a choice to undergo the operation. . .

To argue that the girls themselves are opposed to the operation is therefore difficult; indeed, recent studies indicate that adolescent girls 'voluntarily' undergo the operation.

In some cases, however, African girls recognize how horrifying the practice can be. One reason given for the large number of Ethiopian girls in the Eritrean People's Liberation Front Army is that they were running away from forced marriages and the 'knife.' Others who have been circumcised feel outraged and betrayed by the *excieuse* (the man or woman who circumcised them), their fathers, their aunts, but most of all by their mothers.

. . .

C. The Right to Health

The right to health argument, which attacks female circumcision for producing menacing health problems, is likely to be successful. Women and governments in Africa accept this approach because it can be integrated into preexisting values and social and economic priorities, and does not require a reformulation of rights and policies. Combatting female circumcision as a violation of the right to health may therefore help unify the campaign to eradicate it.

African women accept the right to health argument for a number of reasons. First, as sexuality is not an openly discussed topic among most African women, they can more easily discuss the operation in terms of health effects. Further, since African countries face numerous health problems that demand immediate attention, framing the issue in terms of a right to health is seen as more legitimate than arguing for sexual or corporal integrity.

. . .

Sabelle Gunning, Arrogant Perception, World Travelling and Multicultural Feminism: The Case of Female Genital Surgeries
23 Colum. Hum. Rts. L. Rev. 189 (1991–92), at 238.

. . .

Arguably, most of the activity reviewed and criticized by the human rights system is not culturally based. In cases of torture or forced disappearances, accused governments generally deny the fact or any knowledge thereof. With a cultural practice, the condemned act is acknowledged and defended: the practice is viewed 'as conduct which has evolved for a specific purpose within a culture and is endorsed as a legitimate expression of that purpose.' However governments may not be actually involved in the practice, because private citizens willingly nurture their cultural norms.

One problem therefore is whether human rights which, like the rest of international law, is aimed at public or government actions can be used to alter the behavior of private parties. Feminists have argued persuasively that the public-private distinction is a false one and that the real question is not whether law, in this case human rights law, should apply to the private as well as the public, but rather 'what types of private acts are and are not protected.' If one can decide that a particular act is a violation, even if performed by private citizens, one can hold governments responsible. For example, when one reviews the international definition of torture one sees that it is not only active or direct government participation which is prohibited, but also government 'consent or acquiescence.'

It may be argued that that language is designed to hold accountable governments that are believed to be responsible for torturous acts but who have created sufficient 'plausible deniability' to make it difficult to prove complicity. Still, it reflects a willingness to pressure governments to do something about 'private' acts. The practical problem is that if governments really do not have control over private actions, then the primary tool of human rights enforcement, governmental embarrassment, will not be nearly as effective. This is particularly true with a practice like female genital surgeries, where the governments involved may either refuse to be embarrassed or become angry at the attack on the culture; thus they reject the interference. Moreover, even if a government is embarrassed, the cost of implementing an eradication law, as has been explained, could be enormously socially disruptive and ineffective.

. . .

. . . One is not stuck between choosing 'universal standards' and 'everything is relative.' It is not that there are 'universals' out there waiting to be discovered. But through dialogue, shared values can become universal and be safeguarded. The process by which these universal standards are created is important. A dialogue, with a tone that respects cultural diversity, is essential.

From that dialogue a consensus may be reached, understanding that as people and cultures interact they do change and learn from each other.

CEDAW, Female Circumcision

General Recommendation. No. 14, 9th Sess., 1990
UN Doc. A/45/38, 1 Int. Hum. Rts. Rep. 21 (No. 1, 1994)

[The Committee on the Elimination of Discrimination against Women is created by the Convention to Eliminate all Forms of Discrimination against Women, examined in Chapter 13, *infra*, and is charged with specified tasks in the implementation of that Convention. The Committee is authorized to make general recommendations based on reports that it receives from the states parties.]

Recommends that States parties:

(a) Take appropriate and effective measures with a view to eradicating the practice of female circumcision. Such measures could include:

(i) The collection and dissemination by universities, medical or nursing associations, national women's organizations or other bodies of basic data about such traditional practices;

(ii) The support of women's organizations at the national and local levels working for the elimination of female circumcision and other practices harmful to women;

(iii) The encouragement of politicians, professionals, religious and community leaders at all levels, including the media and the arts, to cooperate in influencing attitudes towards the eradication of female circumcision;

(iv) The introduction of appropriate educational and training programmes and seminars based on research findings about the problems arising from female circumcision;

(b) Include in their national health policies appropriate strategies aimed at eradicating female circumcision in public health care. Such strategies could include the special responsibility of health personnel, including traditional birth attendants, to explain the harmful effects of female circumcision;

(c) Invite assistance, information and advice from the appropriate organizations of the United Nations system to support and assist efforts being deployed to eliminate harmful traditional practices;

(d) Include in their reports to the Committee under articles 10 and 12 of the Convention on the Elimination of All Forms of Discrimination against Women information about measures taken to eliminate female circumcision.

K. Hayter, Female Circumcision—Is There a Legal Solution?

J. of Soc. Welf. L. (U.K.) 323 (Nov. 1984), at 325.

[These remarks concerned a pending bill in Parliament to prohibit female circumcision in the U.K.]

Clearly the effects of female circumcision and the enforced suppression of female sexuality is to be abhorred. The overall response of members of the House of Lords reflects this view, as summarised in the speech of Baroness Gaitskell's where she states, 'The primitive attitude to female circumcision rests not only on tradition, but on the male desire for the female to be pure for him . . . That is not only the most cruel, but also . . . the most primitive, and the most important aspect of the matter which we should reject.' But is this moral indignation sufficient to justify legal intervention to prohibit consensual acts performed on women over 16 in accordance with the cultural requirements of a minority group? Support for the view that it is can be seen in the statement of Lord Devlin, in his Maccabean Lecture in 1959 entitled *The Enforcement of Morals*. Here it is argued that legal intervention is justified where in the collective moral judgment of a society a practice cannot be tolerated. But unless moral repugnance can be founded on the broader principle that all acts amounting to sexual oppression of women are not to be tolerated, legislation on this issue could be interpreted as purely discriminatory against the minority groups concerned. It is interesting to note here that clitoradectomies were openly performed on children and women in England and the United States as late as 1945 as a 'cure' for masturbation and 'promiscuity.' The demise of this practice in recent years may indicate a general change in attitudes towards female sexuality. If this is correct then, it is suggested, the legality of certain western practicès will also require review. Purely elective cosmetic surgery is an obvious case where the right of the individual to consent to treatment is not seriously questioned. Breast reduction, for example, is an unnecessary and mutilating operation involving considerable pain and scarring to the patient. If justification for its performance were called for, medical evidence of anxiety and depression brought on by the woman's dissatisfaction with her body would undoubtedly be sufficient to outweigh the injury inherent in the treatment. Indeed, the Government's proposed amendment to the Bill which would safeguard the right of western women to undergo surgery on mental health grounds reinforces this view. Precisely the same justification would be pleaded in support of the legality of female circumcision and should, by analogy, in the absence of further justification for its prohibition, be sufficient. In both cases the women's perception of themselves reflects the demands of the social group to which they belong. This justification is the greater in the case of female circumcision where its necessity extends beyond mere aesthetic appeal, being crucial to the women's status within the group.

Additionally, the imposition of the moral values of the majority onto minority groups would seem inappropriate in a multiracial society in which the current trend is towards tolerance of others' cultural practices. An analogy can be drawn here between female circumcision and the circumcision of Jewish males, which does receive social and legal tolerance. Clearly nice distinctions can be drawn between a mere custom in the case of female circumcision and a strict religious requirement in the latter case. But is this really the criteria to be used to limit the bounds of toleration? The essential element in both appears to be the unquestioned and entrenched nature of the practices which are part of the social fabric of the groups concerned. Arguably both should, prima facie, be tolerated on this basis alone. A valid distinction between the two practices, however, is the degree of injury involved in female circumcision which is not associated with male circumcision.

. . . Legal intervention is, however, justified to protect persons from what is offensive or injurious, particularly where the individual is young, weak in body or mind or in a state of particular physical or economic dependence. Arguably the practice of female circumcision bears characteristics which bring it within these exceptions thus justifying legal intervention, which are not present in other forms of elective surgery. Clearly it is applicable to the circumcision of female children and this approach has been taken to prohibit indigenous practices, notably the tattooing of minors. To subject women over 16 to the same degree of legal paternalism appears, prima facie, to be a denial of their right to self-determination and a slight on the intellectual capacity of the women members of these groups. This issue underlies objections to legal limitations on a woman's right to elect for abortion. It is possible, however, that the cloistered lifestyle and acute state of economic dependence in which the women practising female circumcision find themselves may provide some justification for a paternalistic approach here. Access to research findings and wider views which refute the necessity for female circumcision are denied to them and the traditional view of the practice is enforced within the closed environment. They are not, therefore, in a position to form a balanced judgment in their own best interests. By criminalising female circumcision the law may assist in freeing women who are powerless to help themselves by reducing the social pressure to conform.

. . .

AAWORD, A Statement on Genital Mutilation

in Miranda Davies (ed.), *Third World—Second Sex: Women's Struggles and National Liberation* (1983), at 217.

The Association of African Women for Research and Development (AAWORD) is a group of African women researchers dedicated to doing women's research from an African perspective. They are based in Dakar, Senegal where their first official meeting was held in December 1977.

. . .

This new crusade of the West has been led out of the moral and cultural prejudices of Judaeo-Christian Western society: aggressiveness, ignorance or even contempt, paternalism and activism are the elements which have infuriated and then shocked many people of good will. In trying to reach their own public, the new crusaders have fallen back on sensationalism, and have become insensitive to the dignity of the very women they want to 'save'. They are totally unconscious of the latent racism which such a campaign evokes in countries where ethnocentric prejudice is so deep-rooted. And in their conviction that this is a 'just cause', they have forgotten that these women from a different race and a different culture are also *human beings*, and that solidarity can only exist alongside self-affirmation and mutual respect.

. . .

AAWORD, whose aim is to carry out research which leads to the liberation of African people and women in particular, *firmly condemns* genital mutilation and all other practices—traditional or modern—which oppress women and justify exploiting them economically or socially, as a serious violation of the fundamental rights of women.

. . .

However, as far as AAWORD is concerned, the fight against genital mutilation, although necessary, should not take on such proportions that the wood cannot be seen for the trees. . . .

. . . [T]o fight against genital mutiliation without placing it in the context of ignorance, obscurantism, exploitation, poverty, etc., without questioning the structures and social relations which perpetuate this situation, is like 'refusing to see the sun in the middle of the day'. This, however, is precisely the approach taken by many Westerners, and is highly suspect, especially since Westerners necessarily profit from the exploitation of the peoples and women of Africa, whether directly or indirectly.

Feminists from developed countries—at least those who are sincerely concerned about this situation rather than those who use it only for their personal prestige—should understand this other aspect of the problem. They must accept that it is a problem for *African women,* and that no change is possible without the conscious participation of African women. . . .

. . .

Hope Lewis, Between 'Irua' and 'Female Genital Mutilation'

8 Harv. Hum. Rts. J. 1 (1995), at 31.

A primary concern expressed in African feminist texts is the tendency among Western human rights activists to essentialize the motivations for practicing FGS [Female Genital Surgery] as rooted either in superstition or in the passive acceptance of patriarchal domination. In rejecting these characterizations, African feminists seek to recapture and control the representation of their own cultural heritage.

. . .

The African feminist literature on FGS emphasizes the importance of the cultural context in which FGS occurs and the complexity of the justifications for its continued practice. It contends that western feminist discourse fails to ask the questions that would help place its patriarchal aspects in a broader context: Are boys initiated at the same time as girls? Are the risk and consequences of initiation rituals for boys as life-threatening and long-lasting as they are for girls? What socioeconomic purposes does FGS serve for a particular group? Are there alternative ways of fulfilling those purposes or challenging their necessity? If so, how should domestic and international actors identify and support those alternatives? Finally, do domestic and international actors contribute to the continuation of harmful traditional practices?

Merwine, Letter to Editor

New York Times, Nov. 24, 1993, at A24

To the Editor:

A. M. Rosenthal condemns female circumcision, a traditional practice common to many African and Arabic peoples, as 'female mutilation' . . . From the Western liberal tradition, and certainly from a feminist perspective, Mr. Rosenthal is correct.

However, from the African viewpoint the practice can serve as an affirmation of the value of woman in traditional society.

This tradition has long been a source of conflict between Western and African values.

. . .

The operation completed, a fee was provided by the young women, usually in the form of a cooked meal, to their moruithia. At this point, they became full members of the Kikuyu and were no longer considered girls.

The importance of the ceremony among traditional Kikuyu cannot be understated, for each girl showed by her act of courage that she was ready to be married. Of equal importance, she now became a member of an age-set. An age-set is a group of people of similar age who tend to act together in their society for the rest of their lives. To the Kikuyu, female circumcision is much more than a mere physical act.

. . .

The sentiments expressed long ago in Kenya are almost certainly shared by the peoples who practice the custom today. To demand, as Mr. Rosenthal does, that economic aid be used to force a change in a tradition central to many Africans and Arabs is the height of ethnocentrism.

A better approach would be for Western peoples to try to understand the importance of these traditions to those who practice them. The West could encourage Africans to have the surgical part of the ceremony performed by competent medical practitioners. That would eliminate potential infection and restrict the extent of excision. This is being done in many African states. Such a policy would allow the West to uphold its values while avoiding the appearance of arrogance.

NOTE

As the comments of Gunning in the preceding readings make clear, the practice of female circumcision raises the distinctive question of who (if anyone), which party or actor, is violating international human rights. We have several times observed that the human rights movement was primarily intended to prohibit or require conduct of states. They are the ratifiers of the instruments, the makers of customary law, the defenders of the rights declared in the different instruments. Here lies the problem. Apparently no state enforces the practice of female circumcision, or instructs or advocates through its affiliated religious or educational institutions that the practice be continued.

Female circumcision therefore introduces a theme that runs through this book, the degree to which the human rights movement regulates directly or indirectly the conduct of non-governmental—and in this sense, private—actors. What mandate does one find in the basic texts, the UDHR and ICCPR, or in related customary law (say, as summarized by Restatement §702, p. 145, *supra*) to bring such actors, directly or indirectly, within the prohibitions of human rights law? We have thus far seen one instance, the dramatic instance of the Nuremberg trials, p. 102, *supra*, in which norms of international law were applied to, indeed used to impose *criminal* punishment on, individuals who included both government officials and (in the above sense) 'private' persons.

The broad problem here presented is then the reach of the human rights movement to 'private' actors and actions that cannot be attributed directly to the 'public' state. This theme, the public–private divide and its implications for the character and growth of international human rights, recurs in different forms in later chapters. Some questions that follow explore it now with respect to the practice of female circumcision.

QUESTIONS

1. African state X is a party to the ICCPR. Its government takes no formal, legal position on female circumcision, which is undergone by a substantial number of girls in X in the different forms described in the readings. No law, no subsidy, no official policy, requires or facilitates or prohibits the practice. Suppose that you are a member of a non-governmental human rights organization within X challenging this widespread practice on the ground that it violates the ICCPR. Of course you must anticipate the defenses to your charge.

a. Whom would you charge with committing the violation?

b. On what provisions of the Covenant would you rely for that charge, and what arguments would you make based on those provisions' text?

c. What provisions of the Covenant could be drawn on to refute your charge—say, provisions allegedly protecting the practice? For example, do you find among the rights declared by the ICCPR some that are relevant to challenging the practice and some relevant to defending it? In your answer, would you distinguish between circumcision at age three and age sixteen?

2. Does the practice of female circumcision raise issues of cultural relativism? That is, do you believe that the practice should be permitted in some states and prohibited in others? Or do you believe that it should be permitted or prohibited equally everywhere? If the first, what are the distinctions that lead you to defend the practice in, say, state X but not state Y?

3. 'It is no wonder that challenges to female circumcision have generated so much controversy in states where it is practiced. Could the line-up be worse from the perspective of getting things done? It's West vs. the rest, the uneducated and backward rest. It's whites vs. non-whites. It's science vs. culture.' Comment.

4. Given the resentment indicated by the prior readings against Western criticism of female circumcision, what strategy should an opponent of the practice develop to lessen or eliminate it? With whom would she work, what message would she deliver, what goals would she establish?

5. What do you view as the strongest justifications of the practice, and how as a critic of the practice would you respond to them?

5

ECONOMIC AND SOCIAL RIGHTS

A. OVERVIEW AND HISTORICAL BACKGROUND

COMMENT ON HISTORICAL ORIGINS

The Universal Declaration of Human Rights (UDHR) recognizes two sets of human rights: the 'traditional' civil and political rights, as well as economic, social and cultural rights. In transforming the Declaration's provisions into legally binding obligations, the United Nations adopted two separate International Covenants which, taken together, constitute the bedrock of the international normative regime in relation to human rights.

The 'official' position, dating back to the Universal Declaration and reaffirmed in innumerable resolutions since that time, is that the two sets of rights are, in the words adopted by the second World Conference on Human Rights in Vienna, 'universal, indivisible and interdependent and interrelated' (Vienna Declaration, para. 5). But this formal consensus masks a deep and enduring disagreement over the proper status of economic, social and cultural rights. At one extreme lies the view that these rights are superior to civil and political rights both in terms of an appropriate value hierarchy and in chronological terms. At the other extreme we find the view that economic and social rights do not constitute rights (as properly understood) at all and that treating them as rights will inevitably undermine the enjoyment of individual freedom, justify large-scale state interventionism and provide an excuse to downgrade the importance of civil and political rights.

Although variations on these extremes have dominated both diplomatic and academic discourse, the great majority of governments has actually taken some sort of intermediate position. For the most part that position has involved (a) support for the equal status and importance of economic and social rights (as of September 1995, 132 states were parties to the International Covenant on Economic, Social and Cultural Rights—hereafter, the ICESCR), together with (b) failure to take particular steps to entrench those rights constitutionally, to adopt any legislative or administrative provisions based explicitly on the recognition of specific economic and social rights as human rights, or to provide effective means of redress to individuals or groups alleging violations of those rights.

Even before the final adoption of the UDHR, the debate over the relationship between the two sets of rights had become a casualty of the Cold War: the Communist countries abstained from voting on its adoption in the General Assembly on the grounds that the economic and social rights provisions were inadequate. Moreover, at least since the 1970s, it has taken on an important North–South dimension. As a result, the debate carries a lot of ideological baggage. It is diffuse, often not well thought through and inextricably linked to some of the most basic political choices confronting any society. Nevertheless, with the rejection of communism, the widespread embrace of free-market economic solutions, and increasing global economic and social integration, economic and social rights are certain to remain at the center of controversy in the years ahead. Moreover, the issues which they raise have important implications for other aspects of human rights law.

The historical origins of the recognition of economic and social rights are diffuse. Those rights have drawn strength for example, from the injunctions reflected in different religious traditions to care for those in need and those who cannot look after themselves. In Catholicism, papal encyclicals have long promoted the importance of the right to subsistence with dignity, while 'liberation theology' has sought to build upon this 'preferential option for the poor.' Virtually all of the major religions manifest comparable concern for the poor and oppressed.[1] Other sources include philosophical analyses as diverse as those of Thomas Paine, Karl Marx, Immanuel Kant and John Rawls; the political programs of the nineteenth century Fabian socialists in Britain, Chancellor Bismarck in Germany (who introduced social insurance schemes in the 1880s), and the New Dealers in the U.S.; and constitutional precedents such as the Mexican Constitution of 1917, the first and subsequent Soviet Constitutions, and the 1919 Constitution of the Weimar Republic (embodying the *Wohlfahrstaat* concept).

For present purposes, however, our focus is on the evolution of international human rights law. The most appropriate starting point is the International Labour Organisation (ILO).

> Established by the Treaty of Versailles in 1919 to abolish the 'injustice, hardship and privation' which workers suffered and to guarantee 'fair and humane conditions of labour', it was conceived as the response of Western countries to the ideologies of Bolshevism and Socialism arising out of the Russian Revolution.[2]

[1] See Martin Shupack, The Churches and Human Rights: Catholic and Protestant Human Rights Views as Reflected in Church Statements, 6 Harv. Hum. Rts. J. 127 (1993).

[2] Virginia Leary, Lessons from the Experience of the International Labour Organisation, in Philip Alston (ed.), The United Nations and Human Rights: A Critical Appraisal, 580, 582 (1992).

In the inter-war years, the ILO adopted international minimum standards in relation to a wide range of matters which now fall under the rubric of economic and social rights. They included, *inter alia*, conventions dealing with freedom of association and the right to organize trade unions, forced labor, minimum working age, hours of work, weekly rest, sickness protection, accident, invalidity and old-age insurance, and freedom from discrimination in employment. The Great Depression of the early 1930s served to emphasize the need for forms of social protection of those who were unemployed and gave a strong impetus to a push for full employment policies such as those advocated by Keynes in his *General Theory of Employment, Interest and Money* (1936).

Partly as a result of these developments, various proposals were made during the drafting of the UN Charter for the inclusion of provisions enshrining the maintenance of 'full employment' as a commitment to be undertaken by Member States. The strongest version, known after its principal proponents as the 'Australian Pledge,' read:

> All members of the United Nations pledge themselves to take action both national and international for the purpose of securing for all peoples, including their own, improved labour standards, economic advancement, social security, and employment for all who seek it: and as part of that pledge they agree to take appropriate action through the General Assembly, the Economic and Social Council, the [ILO]. . .[3]

Despite significant support for the proposal, the United States was strongly opposed on the grounds that any such undertaking would involve interference in the domestic, economic and political affairs of states. Ultimately agreement was reached on Article 55(a) of the Charter, which calls on Member States to promote 'higher standards of living, full employment, and conditions of economic and social progress and development' but does not call for any specific follow-up at the international level.

But U.S. opposition in this context did not signify the rejection of economic and social rights *per se*. Indeed, in 1941 President Roosevelt had nominated 'freedom from want' as one of the four freedoms that should characterize the future world order. He spelled out this vision in his 1944 State of the Union address[4]:

> We have come to a clear realization of the fact that true individual freedom cannot exist without economic security and independence. 'Necessitous men are not free men.' People who are out of a job are the stuff of which dictatorships are made.
>
> In our day these economic truths have become accepted as self-evident.

[3] See generally Ruth Russell and Jean Muther, A History of the United Nations Charter: The Role of the United States 1940–1945 (1958) 786.

[4] Eleventh Annual Message to Congress (Jan. 11, 1944), in J. Israel (ed.), The State of the Union Messages of the Presidents (1966), Vol. 3, 2875, 2881.

We have accepted, so to speak, a second bill of rights, under which a new basis of security and prosperity can be established for all—regardless of station, race, or creed.

Among these are:

The right to a useful and remunerative job in the industries, or shops, or farms, or mines of the Nation;

The right to earn enough to provide adequate food and clothing and recreation;

The right of every farmer to raise and sell his products at a return which will give him and his family a decent living;

The right of every businessman, large and small, to trade in an atmosphere of freedom from unfair competition and domination by monopolies at home or abroad;

The right of every family to a decent home;

The right to adequate medical care and the opportunity to achieve and enjoy good health;

The right to adequate protection from the economic fears of old age, sickness, accident, and unemployment;

The right to a good education.

All of these rights spell security. And after this war is won we must be prepared to move forward, in the implementation of these rights, to new goals of human happiness and well-being.

. . .

This approach was subsequently reflected in a draft international Bill of Rights, completed in 1944, by a Committee appointed by the American Law Institute (ALI). In addition to listing the rights contained in the U.S. Bill of Rights (the first ten amendments to the Constitution), the Institute's proposal advocated international recognition of a range of rights and acceptance of the correlative duties in relation to: (1) education ('The state has a duty to require that every child within its jurisdiction receive education of primary standard; to maintain or insure that there are maintained facilities for such education which are adequate and free; and to promote the development of facilities for further education which are adequate and effectively available to all its residents'); (2) work ('The state has a duty to take such measures as may be necessary to insure that all its residents have an opportunity for useful work'); (3) reasonable conditions of work ('The state has a duty to take such measures as may be necessary to insure reasonable wages, hours, and other conditions of work'); (4) adequate food and housing ('The state has a duty to take such measures as may be necessary to insure that all its residents have an opportunity to obtain these essentials'); and (5) social security ('The state has a duty to maintain or insure that there are maintained comprehensive arrangements for the promotion of health, for the prevention of sickness and accident, and for the provision of medical care and of compensation for loss of livelihood'). See Statement of Essential Human Rights, UN Doc. A/148 (1947), Arts. 11–15.

In relation to each of the proposed rights, a Comment by the Committee drew attention to the fact that it had already been recognized in the 'current or recent constitutions' of many countries; e.g. 40 countries in the case of the right to education; 9 for the right to work; 11 for the right to adequate housing; 27 for the right to social security.

Although these proposals were never formally endorsed by the ALI, they were submitted directly to the United Nations and were to prove highly influential in the preparation of the first draft of the Universal Declaration in 1947. In the drafting of the relevant provisions (Articles 22–28), strong support for the inclusion of economic and social rights came from the United States (a delegation led by Eleanor Roosevelt), Egypt, several Latin American countries (particularly Chile) and from the (Communist) countries of Eastern Europe. Australia and the United Kingdom opposed their inclusion,[5] as did South Africa which objected firstly that 'a condition of existence does not constitute a fundamental human right merely because it is eminently desirable for the fullest realisation of all human potentialities' and secondly that if the proposed economic rights were to be taken seriously it would be 'necessary to resort to more or less totalitarian control of the economic life of the country.'[6]

After the adoption of the Universal Declaration in 1948, the next step was to translate the rights it recognized in Articles 22–28 into binding treaty obligations. This process took from 1949 to 1966. The delay was due to reasons including the Cold War, U.S. opposition to the principle of international human rights treaties, and the scope and complexity of the proposed obligations. By 1955, the main lines of what was later to become the International Covenant on Economic, Social and Cultural Rights were agreed, although some of the formulations were amended between 1963 and 1965. The following analysis of the drafting process, prepared by the UN, captures the main dilemmas and controversies relating to the inclusion of economic, social and cultural rights.

ANNOTATIONS ON THE TEXT OF THE DRAFT INTERNATIONAL COVENANTS ON HUMAN RIGHTS

UN Doc. A/2929 (1955), at 7.

General Problems Relating to the Draft Covenants

. . .

One Covenant or Two

. . .

[5] See B. Andreassen, Article 22, and Asbjørn Eide, Article 25, in Asbjørn Eide *et al.* (eds.), The Universal Declaration of Human Rights: A Commentary (1992)

[6] UN Doc. E/CN.4/82/Add.4 (1948) 11, 13.

[Between 1949 and 1951 the Commission on Human Rights worked on a single draft covenant dealing with both of the categories of rights. But in 1951 the General Assembly, under pressure from the Western-dominated Commission, agreed to draft two separate covenants] . . . to contain 'as many similar provisions as possible' and to be approved and opened for signature simultaneously, in order to emphasize the unity of purpose,

7. It was clear that the opinion of United Nations Members was divided as to whether there should be one or two covenants. It should be noted, however, that those in favour of having two covenants as well as those in favour of a single covenant were generally agreed that 'the enjoyment of civil and political freedoms and of economic, social and cultural rights are interconnected and interdependent' and that 'when deprived of economic, social and cultural rights, man does not represent the human person whom the Universal Declaration regards as the ideal of the free man'. The divergence of opinion appeared to arise from a difference of approach rather than of purpose.

8. Those who were in favour of drafting a single covenant maintained that human rights could not be clearly divided into different categories, nor could they be so classified as to represent a hierarchy of values. All rights should be promoted and protected at the same time. Without economic, social and cultural rights, civil and political rights might be purely nominal in character; without civil and political rights, economic, social and cultural rights could not be long ensured. There should, therefore, be a single covenant which would embrace all human rights and by which States would solemnly undertake to promote and guarantee them all.

9. Those in favour of drafting two separate covenants argued that civil and political rights were enforceable, or justiciable, or of an 'absolute' character, while economic, social and cultural rights were not or might not be; that the former were immediately applicable, while the latter were to be progressively implemented; and that, generally speaking, the former were rights of the individual 'against' the State, that is, against unlawful and unjust action of the State, while the latter were rights which the State would have to take positive action to promote. Since the nature of civil and political rights and that of economic, social and cultural rights, and the obligations of the State in respect thereof, were different, it was desirable that two separate instruments should be prepared.

10. The question of drafting one or two covenants was intimately related to the question of implementation. If no measures of implementation were to be formulated, it would make little difference whether one or two covenants were to be drafted. Generally speaking, civil and political rights were thought to be 'legal' rights and could best be implemented by the creation of a good offices committee, while economic, social and cultural rights were thought to be 'programme' rights and could best be implemented by the establishment of a system of periodic reports. Since the rights could be divided into two broad categories, which should be subject to different procedures of implementation, it would be both logical and convenient to formulate two separate covenants.

11. However, it was argued that not in all countries and territories were all civil and political rights 'legal' rights, nor all economic, social and cultural rights 'programme' rights. A civil or political right might well be a 'programme' right under one régime, an economic, social or cultural right a 'legal' right under another. A covenant could be drafted in such a manner as would enable States, upon ratification or accession, to announce, each in so far as it was concerned, which civil, political, economic, social and cultural rights were 'legal' rights, and which 'programme' rights, and by which procedures the rights would be implemented.

12. [It was also said] that there should be only one covenant, on civil and political rights, and that economic, social and cultural rights, which could only be promoted progressively, should not be embodied in a legal instrument at all.

BRIEF CLAUSES OR ELABORATE PROVISIONS

13. There were two schools of thought regarding the manner in which articles on substantive rights should be drafted. One school held that each article should be a brief clause of a general character; another school was of the opinion that each right, its scope and substance, its limitations, as well as the obligations of the State in respect thereof, should be drafted with the greatest possible precision.

14. The first school maintained that, in general instruments of such a comprehensive character as the covenants, it was impossible to set forth the scope and substance of each right in great detail. While there were concepts of rights which might be generally acceptable there were also concepts which varied a great deal from one legal system to another and might not be universally applicable. It would be better to provide that 'no-one shall be held in slavery or in servitude' than to define exactly what slavery or servitude was. It would be better to provide that 'the States Parties to the Covenant recognize the right of everyone to social security' than to attempt to define the precise content of that right. The covenants could only contain general provisions, and the precise scope and substance of each right should be left to national legislation.

. . .

16. As to the obligations of States, according to this school of thought, the covenants could provide in a general manner that the States parties should guarantee civil and political rights in accordance with law, and should recognize and progressively promote economic, social and cultural rights. To enumerate the specific acts that States might perform in respect of civil or political rights or to determine in advance the particular measures they should take in respect of economic, social or cultural rights would be going far beyond the scope of the covenants. Furthermore, no directory of specific obligations could be exhaustive.

17. Finally, the covenants were not the only or the final instruments on

human rights. The rights set forth in the covenants could be elaborated—individually or severally—in a series of international conventions, should the community of nations so desire. . . .

18. The other school held the view that the covenants on human rights should not be a second edition of the Universal Declaration. . . .

19. [The] scope and substance of each right should be precisely defined. It was not sufficient to declare that 'everyone shall be entitled to a fair and public hearing'; it was far more important to specify minimum guarantees under which that right could be fully protected. It was not sufficient to declare that everyone shall have the 'right to education'; it was far more important to set forth the legal standards in respect of each level of education. To declare the existence of a right, without indicating its content, would leave much to be desired.

COMMENT ON THE ICESCR

This Covenant was adopted by the General Assembly in res. 2200A (XXI) of 16 December 1966 and entered into force on 3 January 1976. It is divided into five 'Parts'. Part I recognizes the right to self-determination; Part II defines the general nature of states parties' obligations; Part III enumerates the specific substantive rights; Part IV deals with international implementation; and Part V contains typical final provisions of a legal nature. Part III recognizes the rights to: work, just and favorable conditions of work; rest and leisure; form and join trade unions and to strike; social security; special protection for the family, mothers and children; an adequate standard of living, including food, clothing and housing; physical and mental health; education; and scientific and cultural life. The right to property, although recognized in the Universal Declaration, is not included. Its omission is due to the inability of governments to agree on a formulation governing social takings and the compensation therefor.

To grasp the following materials, it is important to become familiar now with Parts II and III of the Covenant.

The interdependence of the civil and political rights, and the economic, social and cultural rights, has always been part of UN doctrine. This is reflected in the Preamble to the ICESCR which states, in terms mirroring those used in the ICCPR, that 'in accordance with the Universal Declaration . . . , the ideal of free human beings enjoying freedom from fear and want can only be achieved if conditions are created whereby everyone may enjoy his economic, social and cultural rights, as well as his civil and political rights.'

The interdependence principle was accepted partly as a necessary political compromise between the two principal competing visions. But it also reflects the fact that the two sets of rights can neither logically nor practically be separated in entirely watertight compartments. Thus, for example, the right to form trade unions is contained in the ICESCR while the right to freedom of association is recognized in the ICCPR. The ICESCR also recognizes various 'liberties' and 'freedoms' in relation to schooling, scientific research and creative activity. Similarly, while the right to education and the parental liberty to choose a child's school are dealt with in the former (Art. 13), the liberty of parents to choose their child's religious and moral education is recognized in the latter (Art. 18).

Moreover, the prohibition of discrimination in relation to the provision of, and access to, educational facilities and opportunities can be derived from both Art. 2 of the ICESCR and Art. 26 of the ICCPR. The European Convention, which is generally considered to cover only civil and political rights issues, states (in Art. 2 of Protocol 1) that 'no person shall be denied the right to education.' While the supervisory organs have tended to adopt a restrictive, essentially non-discrimination based, approach to the interpretation of this provision, it is nevertheless generally considered to encompass an obligation on the part of the state 'to provide the existence and maintenance of a minimum of educational facilities.'[7] There is reason to expect that, over time, interpretations of this formulation will contribute to a further blurring of the supposed distinctions between the two sets of rights.

Although the title of the ICESCR expressly refers to 'cultural rights' and Art. 15 (1) recognizes 'the right of everyone . . . to take part in cultural life,' these rights have attracted relatively little attention in this context. Rather, they have tended to be dealt with in relation to the ICCPR, whether under its non-discrimination clause (Art. 2(1)), the minorities provision (Art. 27), or specific rights such as freedoms of expression, religion, and association and the right to 'take part in the conduct of public affairs.' Nevertheless, it is clear that this neglect of the specifically economic and social rights dimensions of cultural rights is unjustified and should be remedied.

The principal UN body concerned with economic, social and cultural rights is the Committee on Economic, Social and Cultural Rights, established in 1987 to monitor the compliance of states parties with their obligations under the Covenant (sometimes referred to as the ICESCR Committee). An initial report by each state party is due within two years, and subsequent reports are required at five-year intervals. The Committee consists of eighteen independent experts, elected by the Economic and Social Council for four-year terms and reflecting an equitable geographic distribution. Its principal activities are the adoption of 'general comments' and the examination of states parties' reports leading to the adoption by the Committee of 'concluding observations' thereon. An overview of its activities is provided at p. 316, *infra*.

[7] Manfred Nowak, The Right to Education, in Asbjørn Eide *et al* (eds.), Economic, Social and Cultural Rights: A Textbook 189, 204 (1994).

NOTE

Since 1990 the United Nations Development Programme (UNDP) has produced an annual Human Development Report which has done much to stimulate and re-orient the debate over development priorities. The following excerpt from the 1991 report gives an indication of the magnitude of the challenge. It is followed by part of the statement made by the ICESCR Committee to the Second World Human Rights Conference at Vienna.

UNITED NATIONS DEVELOPMENT PROGRAMME, HUMAN DEVELOPMENT REPORT

(1991), at 2.

[The world's population in 1990 was estimated at 5.3 billion.]

. . .

Poverty—Over one billion people live in absolute poverty.
Nutrition—Some 180 million children, one in three, suffer from serious malnutrition.
Health—One and a half billion people are deprived of primary health care. Nearly three million children die each year from immunizable diseases. About half a million women die each year from causes related to pregnancy and childbirth.
Education—About a billion adults cannot read or write. Well over 100 million children of primary school age are not in school.
Gender—Disparities between men and women remain wide, with female literacy still only two-thirds that of males. Girls' primary enrolment rates are a little over half that of boys', and much of women's work still remains underpaid and undervalued.

People in all developing regions share these problems, but the most urgent problems tend to differ. In Latin America, South Asia and the Arab States, poverty is reinforced by the very unequal distribution of assets.

. . .

. . . In Africa, almost two-thirds of the people lack access to safe water, and fewer than half the children attend primary school. The problem of absolute poverty is increasingly concentrated in Africa. Even in East and South-East Asia, where overall economic growth has been fast, half the people still lack access to safe water and basic health care.

In the industrial countries of the North, average income is much higher than in the South, and almost everyone has access to basic social services. But human deprivation and distress have not disappeared. Indeed, the analyses for

the industrial countries and the developing countries show many points of similarity, although the extent and character of deprivation are different:
Poverty—Over 100 million people live below the poverty line in the industrial market economies. If the USSR and Eastern Europe are included, the number is at least 200 million.
Unemployment—In ten industrial countries, the rate is between 6% and 10%, and in another three it is beyond 10%.
Gender—Female wages are, on average, only two-thirds those of men, and women's parliamentary representation is but a seventh that of men.
Social fabric—In many industrial countries, the social fabric continues to unravel fast—old cultural and social norms are disappearing, with nothing cohesive to take their place. The all-too-frequent result is isolation and alienation. There is evidence of high rates of drug addiction, homelessness, suicide, divorce and single-parent homes.

STATEMENT TO THE WORLD CONFERENCE ON HUMAN RIGHTS ON BEHALF OF THE COMMITTEE ON ECONOMIC, SOCIAL AND CULTURAL RIGHTS

UN Doc. E/1993/22, Annex III.

2. [The principle of the equality of the two sets of rights] has often been more honoured in the breach than in the observance.

. . .

5. The shocking reality . . . is that States and the international community as a whole continue to tolerate all too often breaches of economic, social and cultural rights which, if they occurred in relation to civil and political rights, would provoke expressions of horror and outrage and would lead to concerted calls for immediate remedial action. In effect, despite the rhetoric, violations of civil and political rights continue to be treated as though they were far more serious, and more patently intolerable, than massive and direct denials of economic, social and cultural rights.

. . .

7. Statistical indicators of the extent of deprivation, or breaches, of economic, social and cultural rights have been cited so often that they have tended to lose their impact. The magnitude, severity and constancy of that deprivation have provoked attitudes of resignation, feelings of helplessness and compassion fatigue. Such muted responses are facilitated by a reluctance to characterize the problems that exist as gross and massive denials of economic, social and cultural rights. Yet it is difficult to understand how the situation can realistically be portrayed in any other way.

8. The fact that one fifth of the world's population is afflicted by poverty,

hunger, disease, illiteracy and insecurity is sufficient grounds for concluding that the economic, social and cultural rights of those persons are being denied on a massive scale. Yet there continue to be staunch human rights proponents—individuals, groups and Governments—who completely exclude these phenomena from their concerns. Such an approach to human rights is inhumane, distorted and incompatible with international standards. It is, in addition, ultimately self-defeating.

9. Democracy, stability and peace cannot long survive in conditions of chronic poverty, dispossession and neglect. Political freedom, free markets and pluralism have been embraced with enthusiasm by an ever-increasing number of peoples in recent years, in part because they have seen them as the best prospect of achieving basic economic, social and cultural rights. If that quest proves to be futile the pressures in many societies to revert to authoritarian alternatives will be immense. Moreover, such failures will generate renewed large-scale movements of peoples involving additional flows of refugees, migrants and so-called 'economic refugees', with all of their attendant tragedy and problems. . . .

B. THE CHALLENGE OF ECONOMIC AND SOCIAL RIGHTS

COMMENT ON OBJECTIONS TO ECONOMIC AND SOCIAL RIGHTS

Economic and social rights have been challenged on many grounds. In political terms, no group of states has consistently followed up its rhetorical support for these rights at the international level with practical and sustained programs of implementation. Nevertheless, formal support for economic, social and cultural rights has been near universal. The principal exception has been the United States whose attitude has varied considerably from one administration to another. Under President Johnson the U.S. voted in the General Assembly in 1966 in favor of adopting the Covenant. Although neither the Nixon nor Ford Administrations were opposed to these rights, neither actively promoted them. The Carter Administration adopted a different approach epitomized by Secretary of State Vance's 'Law Day Speech' at the University of Georgia, in which he defined human rights as including:

> First, . . . the right to be free from governmental violation of the integrity of the person. . . . Second, . . . the right to the fulfilment of such vital needs as food, shelter, health care and education. . . . Third, . . . the right to enjoy civil and political liberties. . . . (76 *Dept. of State Bulletin* 505 (1977)).

In 1978, President Carter signed the Covenant and sent it to the Senate for its advice and consent to ratification. No action was taken on the Covenant, even in Committee.

This approach was reversed by the Reagan and Bush Administrations which opposed the concept of economic and social rights on the grounds that while

> the urgency and moral seriousness of the need to eliminate starvation and poverty from the world are unquestionable . . . the idea of economic and social rights is easily abused by repressive governments which claim that they promote human rights even though they deny their citizens the basic . . . civil and political rights.[8]

Subsequently, in international forums, the U.S. opposed measures designed to promote economic and social rights. In 1993, Secretary of State Christopher indicated that the Clinton Administration would press for ratification of the Covenant, although no timetable was set.

However, at least in formal terms for the purposes of international law, debate over whether economic rights are 'really' rights was settled long ago. Whatever the disputes over their nature and consequences, they are an integral part of the Universal Declaration. Some contend that the UDHR as a whole should be considered part of international customary law (see p. 143, *supra*); such a position necessarily includes the provisions dealing with economic and social rights (Arts. 22–28).[9]

Moreover, these rights are an integral part of most of the core universal human rights treaties. For example, Article 5 of the Convention on the Elimination of All Forms of Racial Discrimination requires states parties 'to guarantee . . . equality before the law' in the enjoyment of economic, social and cultural rights. Similarly, both the Convention on the Elimination of All Forms of Discrimination against Women and the Convention on the Rights of the Child contain extensive obligations relating to those rights. Since the latter has, as of March 1995, 170 states parties, there are only some 15 member states of the UN which have not assumed formal treaty obligations relating to some economic, social and cultural rights in at least some contexts. Nevertheless, the philosophical and ideological dimensions of the debate continue to be important.

The greatest challenge is to identify effective approaches to implementation—i.e. to the means by which economic, social and cultural rights can be given effect and governments can be held accountable to fulfill their obligations. The Covenant says only that governments must use 'all appropriate means' to work towards the stated ends. Such means may be universally valid or relevant or may be quite specific to a particular culture or legal system. The Covenant gives no further pointers, beyond noting that legislative measures

[8] Introduction, U.S. Dept of State, Country Reports on Human Rights Practices for 1992, 5.

[9] See generally Bruno Simma and P. Alston, The Sources of Human Rights Law: Custom, Jus Cogens and General Principles, 12 Aust. Y. B. Int'l L. 82–108 (1992).

are likely to be important. It is clear, however, that neither legislation nor effective remedies of a primarily judicial nature, which are both central to the domestic implementation framework contained in the ICCPR, will *per se* be sufficient in relation to the ICESCR.

We must therefore consider such issues as whether 'traditional' legal approaches, including justiciability, are important; whether constitutional or legislative entrenchment is indispensable or even desirable; what the implications are of the limitation of a state party's obligations to use the maximum of its 'available resources'; and whether a 'programmatic' approach is viable and, from both legal and practical points of view, sufficient.

The principal reasons for the relative under-development of economic and social rights by comparison with their civil and political rights counterparts include (1) the ambivalence of most governments, but particularly those from the Third World; (2) the demonstrated reluctance of non-governmental organizations (NGOs) to focus specifically on economic and social rights; and (3) the lack of innovative legal and other approaches to implementation by those governments that clearly do support the concept.

In relation to the first point it has often been argued in international fora, especially by Third World governments, that economic and social rights have been neglected by the international community. Thus, for example, at its 1992 Jakarta Summit meeting the Non-Aligned Movement 'expressed concern over a tendency to address aspects of human rights selectively, . . . and to neglect economic, social and cultural rights . . . .' UN Doc. A/47/675 (1992), p. 47, para. 75. Despite occasional rejections of this criticism, it is clearly warranted.[10] Ironically, however, even the strongest proponents of economic and social rights have rarely put forward concrete proposals for their systematic implementation at either the national or international levels. The only exception in this regard relates to the right to development, discussed in Chapter 16.

In relation to the second point, development NGOs active at the international level have tended to assume that the language of rights would not be effective in persuading potential donor states to contribute funds, for those states might not wish to see their benevolence merely as an obligation to help meet the rights of other states. Moreover, the donee states might see the invocation of human rights language as constituting an interference in their internal affairs. Moreover, development specialists are often justifiably sceptical of legal (or at least legalistic) approaches. They assume that such approaches invariably favor the rich and powerful rather than provide tools for the use of disadvantaged members of society.

International human rights NGOs have tended to adopt a primarily civil and political rights-driven agenda. Consider the following statement by the then Executive Director of Human Rights Watch:[11]

[10] Cf. Morris Abram, Human Rights and the United Nations: Past as Prologue, 4 Harv. Hum. Rts. J. 69 (1991); and Philip Alston, Revitalising United Nations Work on Human Rights and Development, 18 Melb. U. L. Rev. 216 (1991).

[11] Aryeh Neier, Human Rights, in J. Krieger (ed.), The Oxford Companion to Politics of the World 403 (1993).

> The view that economic and social questions should also be thought of in terms of rights was not solely confined to the Soviet bloc. It is reflected in several provisions of the Universal Declaration of Human Rights and in an International Covenant on Economic, Social and Cultural Rights. . . . Moreover, human rights activists in a number of Third World countries, especially in Asia, have long held the view that both kinds of concerns are rights. Their argument has not proved persuasive in the West, however, and none of the leading international nongovernmental groups concerned with human rights has become an advocate of economic and social rights.

Although the accuracy of this statement is open to question on several counts, the fact remains that many leading international NGOs, including Human Rights Watch and Amnesty International, have not taken specific measures to promote awareness of economic and social rights issues, or to take direct account of them in their principal activities.

In relation to the third point, consider the following excerpts.

JEAN DRÈZE AND AMARTYA SEN, HUNGER AND PUBLIC ACTION

(1989), at 20.

2.1. Deprivation and the Law

When millions of people die in a famine, it is hard to avoid the thought that something terribly criminal is going on. The law, which defines and protects our rights as citizens, must somehow be compromised by these dreadful events. Unfortunately, the gap between law and ethics can be a big one. The economic system that yields a famine may be foul and the political system that tolerates it perfectly revolting, but nevertheless there may be no violation of our lawfully recognized rights in the failure of large sections of the population to acquire enough food to survive.

The point is not so much that there is no law against dying of hunger. That is, of course, true and obvious. It is more that the legally guaranteed rights of ownership, exchange and transaction delineate economic systems that can go hand in hand with some people failing to acquire enough food for survival. In a private ownership economy, command over food can be established by either growing food oneself and having property rights over what is grown, or selling other commodities and buying food with the proceeds. There is no guarantee that either process would yield enough for the survival of any particular person or a family in a particular social and economic situation. The third alternative, other than relying on private charity, is to receive free food or supplementary income from the state. These transfers rarely have the status of legal rights, and furthermore they are also, as things stand now, rather rare and limited.

For a large part of humanity, about the only substantial asset that a person owns is his or her ability to work, i.e. labour power. If a person fails to secure employment, then that means of acquiring food (e.g. by getting a job, earning a wage, and buying food with this income) fails. If, in addition to that, the laws of the land do not provide any social security arrangements, e.g. unemployment insurance, the person will, under these circumstances, fail to secure the means of subsistence. And that can result in serious deprivation—possibly even starvation death. In seeking a remedy to this problem of terrible vulnerability, it is natural to turn towards a reform of the legal system, so that rights of social security can be made to stand as guarantees of minimal protection and survival.

LOUIS HENKIN, INTERNATIONAL HUMAN RIGHTS AND RIGHTS IN THE UNITED STATES

in Theodor Meron (ed.), Human Rights in International Law (1984), at 33.

Like American rights, international human rights inevitably implicate the purposes for which government is created. But—unlike American rights originally—international rights surely do *not* reflect a commitment to government-for-limited-purposes-only. On the contrary, born after various socialisms were established and spreading, and after commitment to welfare economics and the welfare state was nearly universal, international human rights imply rather a conception of government as designed for many purposes and seasons. The rights deemed to be fundamental include not only freedoms which government must not invade, but also rights to what is essential for human well-being, which government must actively provide or promote. They imply a government that is activist, intervening, committed to economic-social planning for the society, so as to satisfy economic-social rights of the individual.

. . .

Different conceptions of equality may underlie the major difference between American and international human rights. While international human rights are as consistent with varieties of capitalism as with various socialisms, the Universal Declaration and the International Covenant on Economic, Social and Cultural Rights advance a few small, important steps toward an equality of enjoyment, declaring that all individuals are entitled to have society supply their basic human needs and other economic–social benefits if the individual cannot do so. That Covenant adds a galaxy of rights, most of them unknown to American constitutionalism. . . .

. . . Although states adhering to the Economic Covenant undertake only to realize these rights 'progressively' and 'to the maximum of [their] available resources', the Covenant uses the language of right, not merely of hope; of undertaking and commitment by governments, not merely of aspiration and goal. Some have asked whether it is meaningful to call what is promised there 'rights', since the undertakings are vague and long-term; they are unenforceable, if only because they require major governmental planning and programs and are conditioned upon availability of resources. But in international law and rhetoric, they are legal rights, and in many societies, including the United States, the language of rights is increasingly used and the sense of entitlement to such benefits is becoming pervasive.

. . .

Let there be no doubt. The United States is now a welfare state. But the United States is not a welfare state by constitutional compulsion. Indeed, it became a welfare state in the face of powerful constitutional resistance: federalism and, ironically, notions of individual rights—economic liberty and freedom of contract—held the welfare state back for half a century; and a constitutional amendment was required to permit the progressive income tax which was essential to make the welfare state possible. Jurisprudentially, the United States is a welfare state by grace of Congress and of the states. . . .

. . . Surely the United States has moved far from 'negative' government, from thinking that the poor are a special and natural category of people and are not the responsibility of society and government but only of church and charity. In theory, Congress could probably abolish the welfare system at will, and the states could probably end public education. But that is a theoretical theory. The welfare system and other rights granted by legislation (for example, laws against private racial discrimination) are so deeply imbedded as to have near-constitutional sturdiness. . . . And Americans have begun to think and speak of social security and other benefits as matters of entitlement and right.

QUESTIONS

1. Writing a decade after the passage above was written, Henkin concluded that '[r]ights in the United States remain in essence eighteenth century freedoms. . . . [W]elfare rights are legislative, not constitutional, and are subject to political, ideological and budgetary restraints.'[12] Do you agree with his characterization of the U.S. as a 'welfare state' and, if so, does this imply acceptance by the U.S. of the principle of economic and social rights?

2. Despite Henkin's optimism about the durability of U.S. welfare entitlements, the *New York Times* reported on Feb. 10, 1995 (p.1) that: 'House [of Representatives] Republican leaders moved today to undo more than a half-century of social welfare policy by proposing to eliminate the right of poor women and children to receive cash assistance from the Government. . . . Benefits would no longer be automatically available to people who met certain eligibility crite-

[12] Louis Henkin, Economic Rights Under the United States Constitution, 32 Colum. J. Transnat'l L. 97, 128 (1994).

ria.' Does the apparent vulnerability of welfare rights to political and ideological tides negate any claim they may have had to human rights status in the U.S.? Is the situation in this respect fundamentally different from those Western European social welfare states which do not have constitutionally entrenched economic and social rights?

3. Henkin has also observed that 'the contemporary international category of economic and social rights . . . treats economic rights principally as "welfare rights." It includes a right to property, but says nothing about freedom of economic enterprise or of a right to participate in a market economy.'[13] Should these rights have been reflected in the ICESCR, or in the ICCPR? Would it be appropriate to include them in a future protocol to either or both Covenants, or are the key principles now explicit or at least implicit when those treaties are read together? Is criticism of these omissions an argument for reverting to a single integrated Covenant that eliminates the distinctions between the two sets of rights?

NOTE

Consider the following remarks taken from a discussion organized by the Harvard Law School Human Rights Program and the François-Xavier Bagnoud Center for Health and Human Rights, and published as *Economic and Social Rights and the Right to Health* 13 (1995). The speaker was Albie Sachs, then a professor in South Africa, now a member of South Africa's newly created Constitutional Court.

> My faith in rights may appear terribly naïve here in the United States, where so many movements have come and gone, but it is a burgeoning period in my country. We must accept the imagery and language and symbolism that is most appropriate to the occasion. To begin with, the language of rights negates the core principles of apartheid: it says, *You have the right to be who you are*. Beyond that, it encourages pluralism: it says, *You have the right to be different*. Women are empowered, the disabled are empowered; gay and lesbian activists reflect the broader liberation movement. And then it establishes a framework for the allocation of resources that is very empowering for the poor. The most compelling health needs, such as child immunizations and clean drinking water, are a matter of right, and do not depend on some remote notion of efficacy.
>
> At this stage of nation-building, when we are sitting down with our oppressors, the rights rhetoric is very helpful. It helps to allay their fears: we will not lock them up or kick them out or boot them into the country; we want to escape this cycle of domination, subordination, resistance, and revolution, which never ends. As an internationally accepted aspiration, it

[13] Ibid., pp. 97–8.

appeals to the best in all of us. They have the right to their freedoms, we have the right to forgive.

C. AN OBLIGATIONS-BASED APPROACH

We here consider the principal issues and approaches that have emerged in relation to the implementation of economic and social rights.

COMMENT ON DIFFERENCES BETWEEN THE TWO COVENANTS

Three differences between the two major Covenants should be noted. First, the terminology used in the ICESCR to describe the obligations of states varies in significant respects from that used in the other Covenant. Whereas the ICCPR contains terms such as 'everyone has the right to . . .' or 'no one shall be . . . ,' the ICESCR employs the formula 'States Parties . . . recognize the right of everyone to . . . .' The principal exceptions concern Art. 3 (equal rights of men and women) and Art. 8 (trade union-related rights), under each of which the states 'undertake to ensure' the relevant rights; and Art. 2(2) (non-discrimination) where the undertaking is 'to guarantee.'

Second, the obligation of states parties stated in Art. 2(1) is recognized to be subject to the availability of resources. And third, the obligation is one of progressive realization. The key provisions are Arts. 2(1) and 2(2):

> 1. Each State Party to the present Covenant undertakes to take steps, individually and through international assistance and cooperation, especially economic and technical, to the maximum of its available resources, with a view to achieving progressively the full realization of the rights recognized in the present Covenant by all appropriate means, including particularly the adoption of legislative measures.
>
> 2.The States Parties to the present Covenant undertake to guarantee that the rights enunciated in the present Covenant will be exercised without discrimination of any kind as to race, colour, sex, language, religion, political or other opinion, national or social origin, property, birth or other status.

This language has been subject to conflicting critiques. On the one hand, it is often suggested that the nature of the obligation is so onerous that virtually no government will be able to comply. Developing countries, in particular, are seen to be confronting an impossible challenge. On the other hand, it is argued that the relative open-endedness of the concept of progressive realization, particularly in light of the qualification related to the availability of resources, renders the obligation devoid of meaningful content. That in turn suggests that governments can present themselves as defenders of economic and social rights, without their policies and behavior being constrained in any way. A related criticism is that the Covenant imposes only 'programmatic' obligations upon governments—that is, obligations to be fulfilled incrementally through the ongoing execution of a program. It therefore becomes difficult if not impossible to determine when those obligations have been met.

NOTE

The next two readings examine the importance of conceptions of duty and obligation in general philosophical terms, while the two other readings that follow do so in relation to the Covenant. While the Covenant contains no provision for a binding interpretation of the nature and scope of the states' obligations in question, the ICESCR Committee is in a position to adopt more or less authoritative statements of interpretation, especially in the form of its 'general comments.' The Committee's third such comment (the fourth reading below) is directed explicitly at the issue of obligations.

IMMANUEL KANT, THE DOCTRINE OF VIRTUE,
in The Metaphysics of Morals, 1797 (M. J. Gregor, trans. 1964), at 116.

§24.

When we are speaking of laws of duty (not laws of nature) and, among these, of laws governing men's external relations with one another, we are considering a moral (intelligible) world where, by analogy with the physical world, *attraction* and *repulsion* bind together rational beings (on earth). The principle of *mutual love* admonishes men constantly to *come nearer* to each other; that of the *respect* which they owe each other, to keep themselves at a *distance* from one another. And should one of these great moral forces fail, 'then nothingness (immorality), with gaping throat, would drink the whole kingdom of (moral) beings like a drop of water'. . . .

§25.

In this context, however, *love* is not to be taken as a *feeling* (aesthetic love), *i.e.* a pleasure in the perfection of other men; it does not mean *emotional* love (for others cannot oblige us to have feelings). It must rather be taken as a maxim of *benevolence* (practical love), which has beneficence as its consequence.

The same holds true of the *respect* to be shown to others: it is not to be taken merely as the *feeling* that comes from comparing one's own *worth* with another's (such as mere habit causes a child to feel toward his parents, a pupil toward his teacher, a subordinate in general toward his superior). Respect is rather to be taken in a practical sense (*observantia aliis praestanda*), as a *maxim* of limiting our self-esteem by the dignity of humanity in another person.

Moreover, the duty of free respect to others is really only a negative one (of not exalting oneself above others) and is thus analogous to the juridical duty of not encroaching on another's possessions. Hence, although respect is a mere duty of virtue, it is considered *narrow* in comparison with a duty of love, and it is the duty of love that is considered *wide*.

The duty of love for one's neighbour can also be expressed as the duty of making others' *ends* my own (in so far as these ends are only not immoral), The duty of respect for my neighbour is contained in the maxim of not abasing any other man to a mere means to my end (not demanding that the other degrade himself in order to slave for my end).

By the fact that I fulfill a duty of love to someone I obligate the other as well: I make him indebted to me. But in fulfilling a duty of respect I obligate only myself, contain myself within certain limits in order to detract nothing from the worth that the other, as a man, is entitled to posit in himself.

. . .

§30.

It is every man's duty to be beneficent—that is, to promote, according to his means, the happiness of others who are in need, and this without hope of gaining anything by it.

For every man who finds himself in need wishes to be helped by other men. But if he lets his maxim of not willing to help others in turn when they are in need become public, *i.e.* makes this a universal permissive law, then everyone would likewise deny him assistance when he needs it, or at least would be entitled to. Hence the maxim of self-interest contradicts itself when it is made universal law—that is, it is contrary to duty. Consequently the maxim of common interest—of beneficence toward the needy—is a universal duty of men, and indeed for this reason: that men are to be considered fellow-men—that is, rational beings with needs, united by nature in one dwelling place for the purpose of helping one another.

. . .

Casuistical Questions

. . .

The ability to practice beneficence, which depends on property, follows largely from the injustice of the government, which favours certain men and so introduces an inequality of wealth that makes others need help. This being the case, does the rich man's help to the needy, on which he so readily prides himself as something meritorious, really deserve to be called beneficence at all?

. . .

§*38*.

Every man has a rightful claim to *respect* from his fellow-men and is *reciprocally* obligated to show respect for every other man.

Humanity itself is a dignity; for man cannot be used merely as a means by any man (either by others or even by himself) but must always be treated at the same time as an end. . . .

TOM CAMPBELL, THE LEFT AND RIGHTS: A CONCEPTUAL ANALYSIS OF THE IDEA OF SOCIALIST RIGHTS

(1993), at 142.

. . .

Summarising the previous section, we may say that the characteristic justificatory principle of socialist organisation is the equal satisfaction of need at the highest level of fulfilment. . . .

. . .

Interpreting the need principle as having primarily to do with the allocation of pre-existing resources for the satisfaction of human needs, the characteristic type of right that it justifies is one which places obligations on those in a position to provide the wherewithal for the satisfaction of the needs of others. Such rights will be positive or affirmative rights in that they correlate with obligations that others take positive steps to meet the needs of right-holders. This assumes the rejection of the liberal assumption that, while there is a general moral requirement not to harm others, there is no equivalent requirement to assist them, even when in need. . . .

. . .

Consequently, the relationships between right-holders and obligation-bearers is not the simple one-to-one correlation which holds in the case of such typical liberal rights as the right to freedom from bodily injury which are primarily rights which all members of a society have against all other members. The fact that meeting needs requires the co-operative effort of many individuals, involving public or collective procedures, means that the fulfilment of typically socialist duties must be mediated through social procedures. In this sense the individual's rights in a socialist society are rights against society and the correlative obligations can also be seen as obligations to society since their fulfilment does not depend on the individual who has the obligation identifying and acting directly towards the individuals who have the correlative rights. In this formulation 'society' is simply a way of referring to the indispensable system of rule-governed co-operation which forms the intermediary between rights and duties.

This model of societally mediated rights and duties has, from the socialist point of view, the defect that it seems to assume the activity of B (the obligation-bearer) and the passivity of A (the right-holder), whereas the socialist ideal is of a society in which individuals are fulfilled primarily through the active co-operative creative ventures. In part this objection is met by pointing out that any particular person is both an A and a B, both a receiver of what he needs and a contributor to the social mechanisms designed to see that he and others receive what they need. It can therefore be argued that it is to obligations, not rights, that we should look for the instantiation of active, productive socialist man.

. . .

The social function of the socialist concept of rights would not, however, be centred on courts of law. The chief purpose of societal rules designed to further the concerns of individuals would be within the normal administrative operations of whatever institutional arrangements were thought necessary to organise communal life so as to meet approved human needs. It is an essential corollary of the dominance within a socialist society of positive or affirmative rights that those entrusted with the tasks of societal organisation would be the primary initiators in the protection of individual rights and themselves the bearers of significant power-rights to enable them to carry out their functions. Socialist rights are more typically directives and enablements than claims. They signify the proper ends and capacities of organised co-operative activity rather than the ultimate recourse of aggrieved individuals. This places them at the conscious centre of any human group organised to satisfy the needs of socialised man.

. . .

QUESTIONS

1. 'If we were to make a rough analogy, Kant's duty of respect recalls the ICCPR, while the duty of beneficence recalls the ICESCR.' Do you agree?

2. Under the ICCPR, say in relation to torture, the state must itself refrain from torture and must use reasonable efforts to enforce that obligation against any private, non-governmental actor who tortures. In that particular sense, the duties (correlative to individual rights) under the ICCPR affect individuals as well

as the state; individuals too would violate a right declared by the Covenant by torturing. Is the same true of the ICESCR? On its face, it imposes duties on the state. But does it also impose on private, non-governmental actors something akin to a duty of beneficence, such that one is under an obligation to assist a starving neighbor? Who would enforce that obligation?

G. J. H. VAN HOOF, THE LEGAL NATURE OF ECONOMIC, SOCIAL AND CULTURAL RIGHTS: A REBUTTAL OF SOME TRADITIONAL VIEWS

in Philip Alston and K. Tomasevski (eds.), The Right to Food (1984), at 97.

. . . A considerable number of people, particularly in the West, and including eminent and influential international lawyers, still hold views attributing a second-rate status to economic, social and cultural rights. In their extreme form such views may be summed up as follows: either the position is taken which expressly denies a legally binding character to economic, social and cultural rights, or those rights are alleged to differ from civil and political rights in such fundamental respects that it becomes impossible to escape the conclusion that the former are inferior from a legal point of view.

. . .

In the following sections we will outline our own position on the basis of a discussion of the studies engaged in by two scholars, whose work is broadly representative of the prevailing school of thought concerning the legal nature of economic, social and cultural human rights. These authors are E. Vierdag and M. Bossuyt.[14] . . .

. . .

. . . Arguments that economic, social and cultural rights differ in fundamental respects from civil and political rights may vary in their implications. They may amount, or at least come close, to the proposition that economic, social and cultural rights are not really law. Alternatively, without denying the legally binding character of economic, social and cultural rights, it may be alleged that the differences between them and civil and political rights may be so great that the former are in fact considered to be second-rate human rights.

Bossuyt's approach would seem to be representative of the latter school of

[14] [Eds.] The works referred to by the author are E. Vierdag, The Legal Nature of the Rights Granted by the International Covenant on Economic, Social and Cultural Rights, 9 Neths. Ybk. Int.l L. 69 (1978) and Marc Bossuyt, La distinction entre les droits civils et politiques et les droits economiques, sociaux et culturels, 8 Hum. Rts. J. 783 (1975).

thought. The differences he construes between civil and political rights on the one hand, and economic, social and cultural rights on the other are very far-reaching. The ultimate reason for these differences is, according to Bossuyt, to be found in the fact that the realization of the latter set of rights requires a financial effort on the part of the State, while in respect of the former this is not the case. The fact that the implementation of at least a number of civil and political rights, such as the right to a fair trial and the right concerning periodic elections, may cost the State money, is dismissed by Bossuyt on the ground that the expenditure involved is very modest and, at any rate, does not go beyond the minimum required to ensure the very existence of the State. It is submitted, however, that in relation to the resources which at least some developing countries have available, the expenditure involved in, for instance, the holding of free and secret elections or the setting up of an adequate judiciary and legal aid system may be quite considerable.

Another, legally more tangible, argument used by Bossuyt is that civil and political rights require non-interference on the part of the State, whereas the implementation of economic, social and cultural rights requires active intervention by the State. The former are, therefore, said to create negative obligations, whereas the latter create positive obligations. In this rigid form the distinction put forward by Bossuyt is, in my view, difficult to uphold. There are well known examples of civil and political rights which in fact demand active intervention on the part of the State, such as the right to a fair trial. Similarly, not all the rights categorized as economic, social and cultural, fit the mould of State-intervention. Freedom to form trade unions constitutes the classic example, although much depends on how exactly this freedom is defined.

Despite these difficulties in the non-interference intervention dichotomy, Bossuyt deduces from it a difference in both the content and the character of the rights. As to the first aspect, Bossuyt argues that civil and political rights are of necessity invariable in their content, as minimum rights cannot vary from one country to another. Economic, social and cultural rights on the other hand are said to have a variable content depending upon the level of economic development of the State concerned. But the case law concerning the European Convention for the Protection of Human Rights and Fundamental Freedoms sheds a very different light on the allegedly invariable content of civil and political rights. Although the Convention applies to a region consisting of a comparatively homogeneous group of States, it has nevertheless been interpreted by the European Commission and the European Court of Human Rights so as to deny the existence of a uniform European standard applicable in all cases. In other words, the fact that the content given to a particular right or freedom protected by the Convention may deviate markedly from one Contracting State to the other has not been judged to be incompatible with the Convention.

As to the difference in character, Bossuyt asserts the existence of a fundamental distinction between the two categories of rights. To the civil and political rights he assigns an absolute character: 'In recognizing civil rights, positive law can only protect those things that man already possesses'. Conversely, the economic, social and cultural rights can, according to Bossuyt, only be enjoyed 'to the extent that these rights have become subjective rights'. It is difficult to rebut views ascribing an absolute character to one or another type of human rights, when they are, as seems to be the case here, based on some concept of Natural Law. For such concepts ultimately prove not to be scientifically verifiable to those who are not already convinced of their validity. Nevertheless, I know of no convincing argument to support the contention that, for instance, the right to life is absolute because it emanates from human dignity, while, for instance, the right to food does not have the same absolute character.

. . .

One of the most important characteristics of the international law of human rights is its dynamic nature. This partly reflects the fact that this area of international law, or at least the considerably increased attention paid to it, is comparatively young. As a result, important parts of the field covered by it have not yet been sufficiently mapped out, which has been responsible for uncertainties and ambiguities. Secondly, and more importantly in this respect, the law of human rights is not only comprehensive, but also concerns the very foundations of society. Even more than other fields of law it is therefore influenced by developments in that society. Particularly in present times which witness far-reaching changes, the law of human rights finds itself in a state of more or less permanent development.

. . .

Parallel to and in interaction with these material changes, psychological processes have occurred which can be described as a general emancipatory movement. This movement is characterized in particular by the phenomenon that with respect to various aspects of life people have come to take on a less submissive, or rather more assertive attitude vis-à-vis the State. This has sometimes resulted in a more or less paradoxical situation. On the one hand more far-reaching demands are made upon the State in providing for the well-being and well-fare of its population; on the other hand such a deeper intervention on the part of the State is warded off on the basis of the argument that it encroaches upon traditional freedoms. . . .

As far as the international law of human rights is concerned, the upshot of the developments condensed in this brief survey is that traditional schemes for analysis have become obsolete in a number of respects.

In these circumstances the quest for innovative approaches to the promotion of human rights assumes particular importance. With respect to economic, social and cultural rights, it may be productive to approach the problem of implementation from the angle of obligations. In this regard four 'layers' of State obligations may be discerned which, for lack of better terms, may be called: an obligation to respect, an obligation to protect, an obligation to ensure, and an obligation to promote.

. . .

This model obviously needs to be elaborated and refined. Nevertheless, even in the form outlined above, it would seem to offer a number of possibilities to develop a more fruitful approach to the international law of human rights. First and foremost, it stresses the unity between civil and political rights, and economic, social and cultural rights, as long as it is recognized that the various 'layers' of obligations can be found in each separate right or freedom. In other words, civil and political rights do not always consist only of obligations to respect and obligations to protect. Under the impact of the above-mentioned processes of increased interdependence and emancipation this type of right now sometimes involves obligations to ensure and even obligations to promote as well. An example is freedom of expression, which, at least in some countries, has come to include, apart from the prohibition of censorship, an obligation to create conditions favourable to the freedom to demonstrate (police-escort, police-protection, etc.) and to pluralism in the press and the media in general.

Conversely, economic, social and cultural rights are not always made up only of obligations to ensure and/or obligations to promote. They may also involve elements of obligations to respect and obligations to protect. For instance, the obligation to respect and protect the right to adequate housing, as laid down in Article 11 of the Covenant, would in my view be violated, if the government's policy, even in the least developed countries, allowed the hovels of poor people to be torn down and replaced by luxury housing which the original inhabitants could not afford and without providing them with access to alternative housing on reasonable terms. It bears reiterating therefore that obligations to fulfil and to promote cannot from a legal point of view too readily be assumed to be empty shells.

The foregoing may be illustrated a little further by briefly applying the model of the various 'layers' of obligations to the right to food. The following examples of the different types of obligations can thus be discerned. The obligation to respect the right to food implies that a government may not expropriate land from people for whom access to control over that land constitutes the only or main asset by which they satisfy their food needs, unless appropriate alternative measures are taken.

Similarly, the obligation to protect the right to food includes the duty on the part of the State to prevent others from depriving people in one way or another, for instance by force or economic dominance, from their main resource base to satisfy their food needs, such as access to land, water, markets, or jobs.

The obligation to ensure the right to food requires a State to take steps in case members of its population prove incapable of providing themselves with food of a sufficient quantity and quality. In emergency situations a State must, to the maximum of its resources, make available the necessary food stuffs. Alternatively or in addition it must seek the assistance of other States to cope with the hunger problem. This raises the question as to the existence of an obligation on the part of other States to provide the required assistance to the best of their ability, as in my view can be deduced from article 11(2). In a more structural way the obligation to ensure the right to food may entail the duty of a State to initiate land reforms in order to improve the production and distribution of food. Although it is difficult *in abstracto* to indicate what measures a State has to take, in concrete situations particular measures can be said to be pertinent for the State concerned in order to meet its obligation to ensure the right to food. At any rate, a State violates this obligation when, in the face of food shortage, it does nothing.

The same holds, *mutatis mutandis*, for the obligation to promote the right to food. As was pointed out this type of obligation encompasses measures aimed at long term goals, and in the case of the right to food it may therefore consist of, for instance, the duty on the part of the government to set training programmes for farmers in an effort to improve methods of production and thus raise productivity in agriculture.

Apart from the fact that the model of various 'layers' of obligations allows for a more integrated approach to the issue of human rights it would seem to offer an additional advantage in that it constitutes a more promising point of departure for dealing with the issue of enforcement or implementation of human rights. From the point of view of effectiveness it may be expected that different types of obligations require different forms of enforcement or implementation. The model of 'layers' opens the possibility to tailor the system of enforcement or implementation to the various types of obligations.

COMMITTEE ON ECONOMIC, SOCIAL AND CULTURAL RIGHTS, GENERAL COMMENT No.3 (1990)

UN Doc. E/1991/23, Annex III.

The nature of States parties obligations (article 2, paragraph 1).

1. Article 2 is of particular importance to a full understanding of the Covenant and must be seen as being a dynamic relationship with all of the other provisions of the Covenant. It describes the nature of the general legal obligations undertaken by States parties to the Covenant. Those obligations include both what may be termed (following the work of the International Law Commission) obligations of conduct and obligations of result. While great emphasis has sometimes been placed on the difference between the formulations used in this provision and that contained in the equivalent article 2 of the Covenant on Civil and Political Rights, it is not always recognized that there are also significant similarities. In particular, while the Covenant provides for progressive realization and acknowledges the constraints due to the

limits of available resources, it also imposes various obligations which are of immediate effect. Of these, two are of particular importance in understanding the precise nature of States parties obligations. One of these, . . . is the 'undertaking to guarantee' that relevant rights 'will be exercised without discrimination . . .'.

2. The other is the undertaking in article 2(1) 'to take steps', which in itself, is not qualified or limited by other considerations. . . . [W]hile the full realization of the relevant rights may be achieved progressively, steps towards that goal must be taken within a reasonably short time after the Covenant's entry into force for the States concerned. Such steps should be deliberate, concrete and targeted as clearly as possible towards meeting the obligations recognized in the Covenant.

3. The means which should be used in order to satisfy the obligation to take steps are stated in article 2(1) to be 'all appropriate means, including particularly the adoption of legislative measures'. The Committee recognizes that in many instances legislation is highly desirable and in some cases may even be indispensable. For example, it may be difficult to combat discrimination effectively in the absence of a sound legislative foundation for the necessary measures. In fields such as health, the protection of children and mothers, and education, as well as in respect of the matters dealt with in articles 6 to 9, legislation may also be an indispensable element for many purposes.

4. . . . [H]owever, the adoption of legislative measures, as specifically foreseen by the Covenant, is by no means exhaustive of the obligations of States parties. Rather, the phrase 'by all appropriate means' must be given its full and natural meaning. . . . [T]he ultimate determination as to whether all appropriate measures have been taken remains for the Committee to make.

. . .

7. Other measures which may also be considered 'appropriate' for the purposes of article 2(1) include, but are not limited to, administrative, financial, educational and social measures.

8. The Committee notes that the undertaking 'to take steps . . . by all appropriate means including particularly the adoption of legislative measures' neither requires nor precludes any particular form of government or economic system being used as the vehicle for the steps in question, provided only that it is democratic and that all human rights are thereby respected. Thus, in terms of political and economic systems the Covenant is neutral and its principles cannot accurately be described as being predicated exclusively upon the need for, or the desirability of a, socialist or capitalist system, or a mixed, centrally planned, or laisser-faire economy, or upon any other particular approach. . . .

9. . . . The concept of progressive realization constitutes a recognition of the fact that full realization of all economic, social and cultural rights will generally not be able to be achieved in a short period of time. In this sense the obligation differs significantly from that contained in article 2 of the Covenant on Civil and Political Rights which embodies an immediate obligation to respect

and ensure all of the relevant rights. Nevertheless, the fact that realization over time, or in other words progressively, is foreseen under the Covenant should not be misinterpreted as depriving the obligation of all meaningful content. It is on the one hand a necessary flexibility device, reflecting the realities of the real world and the difficulties involved for any country in ensuring full realization of economic, social and cultural rights. On the other hand, the phrase must be read in the light of the overall objective, indeed the raison d'être of the Covenant which is to establish clear obligations for States parties in respect of the full realization of the rights in question. It thus imposes an obligation to move as expeditiously and effectively as possible towards that goal. Moreover, any deliberately retrogressive measures in that regard would require the most careful consideration and would need to be fully justified. . . .

10. . . . [T]he Committee is of the view that a minimum core obligation to ensure the satisfaction of, at the very least, minimum essential levels of each of the rights is incumbent upon every State party. Thus, for example, a State party in which any significant number of individuals is deprived of essential foodstuffs, of essential primary health care, of basic shelter and housing, or of the most basic forms of education is, *prima facie*, failing to discharge its obligations under the Covenant. If the Covenant were to be read in such a way as not to establish such a minimum core obligation, it would be largely deprived of its raison d'être. By the same token, it must be noted that any assessment as to whether a State has discharged its minimum core obligation must also take account of resource constraints applying within the country concerned. Article 2(1) obligates each State party to take the necessary steps 'to the maximum of its available resources'. In order for a State party to be able to attribute its failure to meet at least its minimum core obligations to a lack of available resources it must demonstrate that every effort has been made to use all resources that are at its disposition in an effort to satisfy, as a matter of priority, those minimum obligations.

11. . . . [T]he obligations to monitor the extent of the realization, or more especially of the non-realization, of economic, social and cultural rights, and to devise strategies and programmes for their promotion, are not in any way eliminated as a result of resource constraints. . . .

12. Similarly, the Committee underlines the fact that even in times of severe resources constraints whether caused by a process of adjustment, of economic recession, or by other factors the vulnerable members of society can and indeed must be protected by the adoption of relatively low-cost targeted programmes.

. . .

QUESTIONS

1. 'With respect to economic rights such as food, housing and health, the notion of duty seems to be more appropriate for legal instruments than the concept of individual rights.' Do you agree?

2. Consider the analysis of economic and social rights in terms of the four levels of obligation—respect, protect, ensure, promote—originally suggested by the author Shue[15] and applied above to the right to food by van Hoof.

a. Is it a convincing analytical framework? For example, does it have the potential to illuminate the specific obligations of states parties to the Covenant in relation to other rights such as those to health, education or housing?

b. Does the framework also have utility for understanding the implications of civil and political rights? Can you apply the scheme to, say, the right to freedom from torture or the right to political participation through voting?

3. Do you consider the Committee's analysis to be too demanding, or insufficiently so? How does it relate to the approach outlined by van Hoof? Do you agree with the Committee's statement in para. 10 concerning minimum core obligations?

4. How do you react to the following ways of characterizing civil and political rights, and economic and social rights, by stating bold differences between the two sets?

Civil and Political Rights	*Economic and Social Rights*
law	politics, policy
rights	needs, wants, political claims
binding	hortatory, at best directive
no resource excuse for failure	resource contingent, legitimate restraint
immediate full application	progressive realization, programmatic
determinate	open-textured
judicial remedies, enforceable	non-justicable, non-enforceable
negative, hands off	positive, interventionist

Would an advocate of economic and social rights necessarily deny some or all of the asserted contrasts?

ADDITIONAL READING

Philip Alston and Gerard Quinn, The Nature and Scope of States Parties' Obligations Under the International Covenant on Economic, Social and Cultural Rights, 9 Hum. Rts Q. 156 (1987).

[15] Henry Shue, Basic Rights (1980).

D. THE RELEVANCE OF RESOURCE CONSTRAINTS

NOTE

Although some critics of the concept of economic and social rights would acknowledge the validity of the Committee's analysis in General Comment No. 3, *supra*, in relation to industrialized countries, they would argue that the majority of developing countries has no chance of meeting any 'minimum core obligation' in respect of all the rights recognized in the Covenant. For example, Maurice Cranston has written that: '[f]or a government to provide social security . . . it has to have access to great capital wealth . . . . The government of India, for example, simply cannot command the resources that would guarantee each' Indian an adequate standard of living.[16] For many critics it follows that, in the absence of large-scale international aid or of rapid domestic economic growth (or both), the government's hands are tied and little can be expected of it in response to its obligations under the Covenant. (Note that the issue of international aid is considered in Chapter 16, *infra*.)

In the post-Cold War world, in which oppressive and corrupt regimes are less likely to be propped up by 'powerful allies,' it has become common to speak of the phenomenon of 'failed states' (those whose governments are perceived to be incapable of performing the most elementary functions of governance). This has tended to reinforce the opinion that many Third World countries, perhaps even the majority, cannot be expected to meet the Covenant's demands. The following readings address that opinion.

MYRON WEINER, THE CHILD AND THE STATE IN INDIA: CHILD LABOR AND EDUCATION POLICY IN COMPARATIVE PERSPECTIVE

(1991), at 3.

The governments of all developed countries and many developing countries have removed children from the labor force and required that they attend school. They believe that employers should not be permitted to employ child labor and that parents, no matter how poor, should not be allowed to keep their children out of school. Modern states regard education as a legal duty, not merely a right: parents are required to send their children to school, children are required to attend school, and the state is obligated to enforce compulsory education. Compulsory primary education is the policy instrument by which the state effectively removes children from the labor force. The state

[16] Human Rights: Real and Supposed, in D.D.Raphael (ed.), Political Theory and the Rights of Man 43, 51 (1967).

thus stands as the ultimate guardian of children, protecting them against both parents and would-be employers.

This is not the view held in India. Primary education in India is not compulsory, nor is child labor illegal. The result is that less than half of India's children between ages six and fourteen—82.2 million—are not in school. They stay at home to care for cattle, tend younger children, collect firewood, and work in the fields. They find employment in cottage industries, tea stalls, restaurants, or as household workers in middle-class homes. They become prostitutes or live as street children, begging or picking rags and bottles from trash for resale. Many are bonded laborers, tending cattle and working as agricultural laborers for local landowners.

. . .

Most children who start school drop out. Of those who enter first grade, only four out of ten complete four years of school. Depending upon how one defines 'work' (employment for wages, or full-time work whether or not for wages), child laborers in India number from 13.6 million to 44 million, or more.

Indian law prohibits the employment of children in factories, but not in cottage industries, family households, restaurants, or in agriculture. Indeed, government officials do not regard the employment of children in cottage industries as child labor, though working conditions in these shops are often inferior to those of the large factories. . . .

India is a significant exception to the global trend. . . .

. . . Poverty has not prevented governments of other developing countries from expanding mass education or making primary education compulsory. Many countries of Africa with income levels lower than India have expanded mass education with impressive increases in literacy. . . . Between 1961 and 1981 the total number of adult illiterates in India increased by 5 million per year, from 333 million to 437 million. India is the largest single producer of the world's illiterates.

. . .

. . .Why is the Indian state unable—or unwilling—to deal with the high and increasing illiteracy, low school enrollments, high dropout rates, and rampant child labor? . . . Why has the state not taken legislative action when the Indian Constitution calls for a ban on child labor and for compulsory primary-school education. . . .

The central proposition of this study is that India's low per capita income and economic situation is less relevant as an explanation than the belief systems of the state bureaucracy, a set of beliefs that are widely shared by educators, social activists, trade unionists, academic researchers, and, more broadly, by members of the Indian middle class. These beliefs are held by those outside as well as those within government, by observant Hindus and by those who regard themselves as secular, and by leftists as well as by centrists and rightists.

At the core of these beliefs are the Indian view of the social order, notions concerning the respective roles of upper and lower social strata, the role of education as a means of maintaining differentiations among social classes, and concerns that 'excessive' and 'inappropriate' education for the poor would disrupt existing social arrangements.

Indians reject compulsory education, arguing that primary schools do not properly train the children of the poor to work, that the children of the poor should work rather than attend schools that prepare them for 'service' or white-collar occupations, that the education of the poor would lead to increased unemployment and social and political disorder, that the children of the lower classes should learn to work with their hands rather than with their heads (skills more readily acquired by early entry into the labor force than by attending schools), that school dropouts and child labor are a consequence, not a cause, of poverty, and that parents, not the state, should be the ultimate guardians of children. Rhetoric notwithstanding, India's policy makers have not regarded mass education as essential to India's modernization. They have instead put resources into elite government schools, state-aided private schools, and higher education in an effort to create an educated class that is equal to educated classes in the West and that is capable of creating and managing a modern enclave economy.

The Indian position rests on deeply held beliefs that there is a division between people who work with their minds and rule and people who work with their hands and are ruled, and that education should reinforce rather than break down this division. These beliefs are closely tied to religious notions and to the premises that underlie India's hierarchical caste system. . . .

. . . [T]here is historical and comparative evidence to suggest that the major obstacles to the achievement of universal primary education and the abolition of child labor are not the level of industrialization, per capita income and the socioeconomic conditions of families, the level of overall government expenditures in education, nor the demographic consequences of a rapid expansion in the number of school age children, four widely suggested explanations. India has made less of an effort . . . than many other countries not for economic or demographic reasons but because of the attitudes of government officials, politicians, trade union leaders, workers in voluntary agencies, religious figures, intellectuals, and the influential middle class toward child labor and compulsory primary-school education. Of particular importance . . . are the attitudes of officialdom itself. . . .

UNITED NATIONS DEVELOPMENT PROGRAMME, HUMAN DEVELOPMENT REPORT (1990), at 4.

7. Developing countries are not too poor to pay for human development *and* take care of economic growth.

The view that human development can be promoted only at the expense of economic growth poses a false tradeoff. It misstates the purpose of development and underestimates the returns on investment in health and education. These returns can be high, indeed. Private returns to primary education are as high as 43% in Africa, 31% in Asia and 32% in Latin America. Social returns from female literacy are even higher—in terms of reduced fertility, reduced infant mortality, lower school dropout rates, improved family nutrition and lower population growth.

Most budgets can, moreover, accommodate additional spending on human development by reorienting national priorities. In many instances, more than half the spending is swallowed by the military, debt repayments, inefficient parastatals, unnecessary government controls and mistargetted social subsidies. Since other resource possibilities remain limited, restructuring budget priorities to balance economic and social spending should move to the top of the policy agenda for development in the 1990s.

Special attention should go to reducing military spending in the Third World—it has risen three times as fast as that in the industrial nations in the last 30 years, and is now approaching $200 billion a year. Developing countries as a group spend more on the military (5.5% of their combined GNP) than on education and health (5.3%). In many developing countries, current military spending is sometimes two or three times greater than spending on education and health. There are eight times more soldiers than physicians in the Third World.

Governments can also do much to improve the efficiency of social spending by creating a policy and budgetary framework that would achieve a more desirable mix between various social expenditures, particularly by reallocating resources:

— from curative medical facilities to primary health care programmes,
— from highly trained doctors to paramedical personnel,
— from urban to rural services,
— from general to vocational education,
— from subsidising tertiary education to subsidising primary and secondary education,
— from expensive housing for the privileged groups to sites and services projects for the poor,
— from subsidies for vocal and powerful groups to subsidies for inarticulate and weaker groups and

— from the formal sector to the informal sector and the programmes for the unemployed and the underemployed.

Such a restructuring of budget priorities will require tremendous political courage. But the alternatives are limited, and the payoffs can be enormous.

PAKISTAN'S REAL POVERTY

The Economist, March 5, 1994, at 32.

. . .

It seems possible that Pakistan is beginning to redress the worst of its social wrongs: a disastrous educational record. Its failure in this area has undoubtedly damaged its economy as well as the hopes of its people. As a report last year by the World Bank analysing East Asia's success pointed out, one of the few policies successful countries had in common was heavy investment in education. Pakistan shows up badly not just against those countries, but also against its poorer neighbours, such as India (see table).

Some explain Pakistan's particular failure through its religion, which has discouraged the education of women. Some put it down to social structure—feudalstyle landlords and tribal chiefs who have no interest in giving the lower orders a chance to escape the slot they were born into. Others still blame the former military dictator, Zia ul Haq, under whom the educational system suffered especially badly.

Aid donors, who in Paris on February 25th agreed to provide $2.5 billion-worth of aid for Pakistan in 1994–95, have been badgering Pakistan for years to spend more on schooling. Pressure from them, combined with a growing realisation among Pakistani bureaucrats and businessmen that the country would never grow fast without a healthy, well-educated population, seems at last to have paid off. This year, after three years' planning and negotiation, an expensive national programme has got underway to try to improve Pakistan's record, not just in education, but also in health and population planning.

The Social Action Programme, a combined effort by the government and the donors, is worth $8 billion over five years. Pakistan will pay $6 billion, mostly by reallocating money from other sectors rather than by increasing public expenditure. Money will be shifted from universities (where state funding mostly benefits the already-prosperous) mostly to primary schools.

Already, the primary education budget has increased by two-thirds between 1991–92 and 1993–94. Politicians' discretion on education spending has been curtailed, so an MP can no longer get two schools built in a marginal village and leave the surrounding area with none. A 'community school' project is designed to allocate money as efficiently as possible, by providing funding only for those villages whose people prove there is a demand for education.

. . .

None of this is irrevocable, of course. The SAP can still fall victim to the current fiscal squeeze or to the tendency of politicians to use money on vote-buying handouts rather than on investment in the future. . . .

A lesson for Pakistan

	GDP per head 1991 PPP$*	Years of schooling 1990 average	Primary and secondary enrolment rate 1987–90, %
Indonesia	2,730	3.9	81
Pakistan	1,970	1.9	29
China	1,680	4.8	88
India	1,150	2.4	68

Source: UNDP *Purchasing-power parity

NANCY BIRDSALL, PRAGMATISM, ROBIN HOOD, AND OTHER THEMES: GOOD GOVERNMENT AND SOCIAL WELL-BEING IN DEVELOPING COUNTRIES

in Lincoln Chen, A. Kleinman and N. Ware (eds.), Health and Social Change in International Perspective 375 (1994) at 381.

[The author notes that, both to promote fairness (irrespective of ability to pay) and to compensate for market failure, governmental involvement in social programs has been characteristic of the twentieth century. In developing countries there are both pros and cons. The former include the high proportion of poor people, the relatively high return for social/public investment in fields like education or health, and the greater likelihood of capital market failures. The latter include the scarcity of administrative skills and the relative inefficiency of government. In low-income developing countries, the share of central government spending on social programs is less than 10%, compared to some 50% in developed countries.

Nevertheless, she points to 'substantial evidence that in the developing countries, government involvement in the social sectors . . . has been a marked success,' so that, for example, gross primary school enrolment in Africa has more than doubled since 1960 (36% to 75%) while adult literacy has gone from 9% to 42%. By the same token, social spending in these countries has declined significantly, due mainly to debt servicing requirements. 'Between the early 1970s and 1985, the share of central government budgets in all developing countries going to health fell from 7 to 4 percent; the share going to education fell from 14 to 10 percent.']

. . . Recent World Bank studies have emphasized in particular the extent to which the pattern of government spending in health, education, housing and social security is inequitable, generally favoring the rich and the middle-class over the poor. In health and education, the problem is often referred to as the 'resource allocation' problem. In health, it is one of substantial spending of public resources on generally high-cost, curative hospital care that tends to benefit the rich, compared with low-cost primarily preventive services that tend to benefit the poor. In education, it is one of substantial spending of public resources on high-cost university education, again benefiting primarily the rich, compared with low-cost primary and secondary education. Many governments thus fail to play Robin Hood, i.e., to assure that through social programs some national resources are transferred from the rich to the poor.

This failure means much government spending is ineffective as well as inequitable. Because it is among the poor that health and education indicators are still low, spending that does not reach the poor has limited impact on average indicators for nations as a whole. The situation is particularly ironic because the unit costs of many programs that would reach the poor (primary health care, basic education) are relatively low.

. . .

A Pragmatic Framework for Assessing Government's Role

Social Programs: Definition and Society's Objectives

. . .

Most social services, along with other goods and services such as roads, sanitation systems and communications networks, are 'quasi-public' goods; they are neither pure public goods (such as clean air or national defense), for which all benefits of use are captured by all members of society, nor pure private goods (such as apples or clothing), for which all benefits of use are captured by the individual who consumes the good. . . .

Among quasi-public goods, social services are those for which society has, in addition, a particular concern about fairness. Fairness dictates that certain 'merit goods' be available to all members of society, irrespective of ability to pay. What these goods and services are (and hence which should be the subject of public policy) will vary across societies, depending on income, stage of development, and values. Access to basic education and health care are almost universally included; some societies, particularly in the industrial countries, include unemployment, disability and old age insurance, decent housing, and child care. Note that difficult issues of definition (and fairness) arise not only around what services should be included, but around how much of certain services (vocational and university education as well as primary and secondary education? coronary bypass operations as well as prenatal care?), and of what quality, should be included.

. . .

One way to think about the role of government in the social sectors is therefore to be concerned with the appropriate mix of inputs of government, given not only society's objectives and fiscal constraints, but also given initial conditions and potential opportunities regarding the private alternative (is there a private sector, and whom does it serve?).

. . .

Possible combinations of inputs by government, in descending order of government involvement, are:

1. Public provision of services, with full financing (through taxes) or partial financing (through a combination of taxes and user charges) by:
— central or local governments, as in the case of fully funded public schools and health clinics;
— quasi-independent public agencies (e.g., a publicly financed hospital or university with an independent board of trustees);

2. Public financing, full or partial, of the provision of such services by the private sector, through:
— subsidies to individuals, including vouchers directly to potential consumers (allowing consumers to purchase directly from alternative public or private providers), and reductions of personal income taxes associated with private spending on health, education or other social services;
— subsidies to private providers of services, including grants (e.g., to nonprofit organizations to run particular programs), reimbursement to providers of some costs by government (e.g., through public health insurance programs), and exemption from taxes;
3. Public regulation of private providers, including restriction of provision of services to nonprofit (vs. profit making) organizations; accreditation or other licensing of providers; control over inputs (licensing of teachers and doctors); price controls over outputs;
4. Public provision of information to consumers regarding particular services, products or providers (leaving decisions to consumers) in the public and private sector.

. . .

. . . [W]here *provision* of services is concerned, there is no simple 'correct' division between the public and private sectors. It may, for example, be efficient for the public sector to provide a full range of services (as is the case with the National Health Service in England, or with the state of California's university system), even including provision of some purely 'private' goods, to all income groups in a population.

There are at least three reasons that can justify public provision of apparently private goods, even to the nonpoor.

— First, joint production and consumption of public goods with private goods may be the most cost-effective approach, particularly in relatively small and relatively poor countries.

. . .

— Second, quality of services is likely to suffer if the government takes responsibility only for providing services to the poor. Or it may be more costly to government to find and target the poor than to simply provide a service to everyone.

. . .

— Third, and more generally, various market failures mean government, if it does not provide a service directly, will need to monitor provision by the private sector. In some circumstances, the social costs of monitoring and regulation may exceed the social costs of direct provision.

QUESTIONS

1. Weiner's analysis rejects resource constraints as a significant impediment to the realization of the right to universal primary education. If you accept his analysis, is there any reason not to find the Government of India in violation of its obligations under the ICESCR (which it ratified in 1979)? Recall the discussion of cultural relativism in Chapter 4(B). Is this a case in which cultural factors (religious, caste-based, and gender-based) could be cited to justify governmental inaction in response to claims of economic and social rights? Or is it one that highlights the need to insist upon respect for those human rights?

2. What conclusions would you draw from the change of education policy in Pakistan, as described by *The Economist*?

3. The ICESCR Committee states in para. 8 of General Comment No. 3, *supra*, that the Covenant 'neither requires nor precludes any particular form of government or economic system . . . '. By contrast, many commentators seem to assume that it is a recipe for socialism or at least for governmental 'welfarism.' Henkin, for example, in the excerpt *supra*, characterizes the Covenant as implying 'a government that is activist, intervening, committed to economic–social planning for the society . . . .' Similarly, Nowak argues that the right to education 'is based on the socialist philosophy which holds that human rights can only be guaranteed by positive state action.'[17]

Does this mean that economic and social rights are no longer viable, indeed are anachronistic, in an era characterized by widespread deregulation; the systematic privatization of activities ranging from the provision of essential services

[17] Manfred Nowak, The Right to Education, in A. Eide *et al* (eds.), Economic, Social and Cultural Rights: A Textbook 189, 196 (1994).

such as electricity, gas, water and sewerage to the running of prisons and postal services; and seemingly irresistible pressure to reduce both the scope and level of government spending?

4. Consider the following comment on the implications of the right to education set forth in an Article in the 1944 ALI Statement of Essential Rights (see other excerpts at p. 259 *supra*), which laid much of the groundwork for the inclusion of economic and social rights in the Universal Declaration.

> . . . The age limits within which the individual is to be considered a 'child' are left to reasonable interpretation in the light of local physiological and other conditions. . . . The Article does not make attendance at school compulsory although the great majority of children will be able to meet the requirements only by attendance at a public or private school.
>
> . . . The expression 'adequate and free' does not prohibit private schools from charging tuition or other fees. It does, however, impose upon the state the duty of insuring that there are maintained schools at which each child has the opportunity to receive a primary education free.
>
> The Article does not prescribe the extent to which schools and other educational facilities for 'further education' may be provided by the state or by churches, endowed institutions, or other voluntary bodies; the nature of the public control, if any, exercised over privately provided schools; the conditions under which privately provided schools may receive financial assistance from the state; or the status of universities. It does impose on the state the responsibility of insuring that adequate educational facilities are provided by either public or private action, a responsibility which would include the duty of providing such facilities itself whenever they are not effectively provided in some other manner. Thus the Article, while affirming the responsibility of the state, allows unlimited variety in the means by which the responsibility is discharged.
>
> The phrase 'to promote the development of facilities' recognizes the inevitability of gradualness in the implementation of the right to education; the interpretation of the phrase 'adequate and effectively available' will vary with local conditions from either a quantitative or qualitative standpoint. Facilities adequate at one stage of social and economic development will cease to be adequate as further progress becomes possible.

Is this an adequate formulation, albeit at a very general level, of the obligations flowing from the right to education? If so, does it point the way to an approach which does not depend inexorably on 'big government'? If not, does it prove the point made by Henkin and Nowak?

5. Birdsall does not use the language of human rights. Is her analysis fully compatible with the principles underlying economic and social rights? Consider, for example, her relativist approach in defining 'merit goods' according to various characteristics of the society.

NOTE

The Vienna World Conference on Human Rights called for a greater effort to be made to develop indicators in relation to the realization of economic and social rights. Consider the following analysis by Robertson.[18]

> [F]inancial, natural, human, technological, and informational resources are the most important resources in achieving ICESCR rights. State direction of these resources, to their maximum extent, toward the goal of promoting economic, social, and cultural well-being would signify . . . substantial if not total compliance with Article 2 . . . .
>
> . . .
>
> *V. How Can Compliance Be Measured?*
>
> [In relation to financial resources, it] would not seem possible to arrive at any one indicator. Rather, what seems reasonable are comparisons between certain countries, and between expenditures on ICESCR rights and other items. For example, if developed countries with comparable economies are spending different amounts on realizing ICESCR rights, then that is indicative, in the case of the low-spender, of non-compliance with Article 2. The same would be true of developing countries similarly situated. This is not to say that the high spenders are in compliance. It simply means that by one indicator the low spender is not. Similarly, comparisons could be drawn between ICESCR and other expenditures. Military expenditures provide a good example. In 1991, developed countries spent 3.6 percent of their GNP on military expenditures compared with health expenditures of 5.3 percent. Developing countries spent only 1.6 percent on health and more than twice as much on the military, 3.8 percent. Some individual developing countries had horrible military to health ratios: Sudan 20 to 1; Syria 34 to 1; Pakistan 38 to 1. Other developing countries were much closer to having a balance. Indicators could be developed which illustrate the trade-offs which countries have made. For example, spending more on the military than on health, education, housing, or social assistance would be indicative of non-compliance. Similar ratios could be developed which compare ICESCR expenditures to other expenditures which clearly do not claim the priority status of ICESCR rights. The greater the number of indicators that can be developed on domestic spending, the greater the insight will be into whether a state has accorded due priority to the obligation to devote the maximum available resources to ICESCR rights.

[18] Robert Robertson, Measuring State Compliance with the Obligation to Devote the 'Maximum Available Resources' to Realizing Economic, Social, and Cultural Rights, 16 Hum. Rts. Q. 693, 697 (1994).

QUESTION

How satisfactory is such an approach? Does it suggest that a system of indicators might be able to be developed for formal use by the ICESCR Committee in assessing compliance?

E. THE QUESTION OF JUSTICIABILITY

If one single issue has dominated the debate over economic and social rights, it is the question of whether those rights are justiciable at the national level. In the view of many observers, justiciability need not be seen as the indispensable characteristic of a human right (as opposed to a legal right in Hohfeldian terms). Nevertheless, programmatic approaches that depend solely upon legislatures and/or administrative agencies for the implementation of economic and social rights have left many observers, lawyers in particular, uneasy.

The question of justiciability (i.e. whether the courts can, and at least sometimes will, provide a remedy for aggrieved individuals claiming a violation of those rights) is seen by some as a 'red herring' or distraction from the real issues. Others, in contrast, view it as a *conditio sine qua non* for an entitlement to warrant classification as a human right. The ICESCR Committee has sought to respond to this issue, but has stopped short of characterizing justiciability as an indispensable element. Consider the following formulation.

COMMITTEE ON ECONOMIC, SOCIAL AND CULTURAL RIGHTS, GENERAL COMMENT NO. 3 (1990)

UN Doc. E/1991/23, Annex III.

. . .

5. Among the measures which might be considered appropriate, in addition to legislation, is the provision of judicial remedies with respect to rights which may, in accordance with the national legal system, be considered justiciable. The Committee notes, for example, that the enjoyment of the rights recognized, *without discrimination*, will often be appropriately promoted, in part, through the provision of judicial or other effective remedies. Indeed, those States parties which are also parties to the International Covenant on Civil and Political Rights are already obligated (by virtue of articles 2(1), 2(3) 3 and 26 of that Covenant) to ensure that any person whose rights or freedoms (including the right to equality and non-discrimination) recognized in that

Covenant are violated, 'shall have an effective remedy' (article 2(3)(a)). In addition, there are a number of other provisions, including articles 3, 7(a)(i), 8, 10(3), 13(2)(a), 13(3), 13(4) and 15(3) which would seem to be capable of immediate application by judicial and other organs in many national legal systems. Any suggestion that the provisions indicated are inherently non-self-executing would seem to be difficult to sustain.

6. Where specific policies aimed directly at the realization of the rights recognized in the Covenant have been adopted in legislative form, the Committee would wish to be informed, *inter alia*, as to whether such laws create any right of action on behalf of individuals or groups who feel that their rights are not being fully realized. In cases where constitutional recognition has been accorded to specific economic, social and cultural rights, or where the provisions of the Covenant have been incorporated directly into national law, the Committee would wish to receive information as to the extent to which these rights are considered to be justiciable (i.e. able to be invoked before the courts).

NOTE

The rights referred to by the Committee in the preceding General Comment No. 3 are equal rights of men and women (Art. 3), equal pay for equal work (Art. 7(a)(i)), the right to form and join trade unions and the right to strike (Art. 8), the right of children to special protection (Art. 10(3)), the right to free, compulsory, primary education (Art. 13 (2)(a)), the liberty to choose a non-public school (Art. 13(3)), the liberty to establish schools (Art. 13(4)), and the freedom for scientific research and creative activity (Art. 15(3)). The extent to which these rights are actually justiciable varies considerably from one country to another. Nevertheless, that is sufficient to demonstrate that the rights in question are not intrinsically non-justiciable.

In relation to the remaining rights in the ICESCR, such as the rights to work (Art. 6), health (Art. 12), food, clothing, housing (Art. 11) and education (Art. 13), Vierdag concludes a lengthy critique of the Covenant in the following terms:[19]

> What are laid down in provisions such as Articles 6, 11 and 13 of the ICESCR are consequently not rights of individuals, but broadly formulated *programmes* for governmental policies in the economic, social and cultural fields.
>
> It is suggested that it is misleading to adopt an instrument that by its very title and by the wording of its relevant provisions purports to grant 'rights' to individuals but in fact appears not to do so, or to do so only marginally. It is also regrettable that, in this way, a notion of 'right' is

[19] E. Vierdag. The Legal Nature of the Rights Granted by the International Covenant on Economic, Social and Cultural Rights, 9 Neths. Ybk. Int. L. 69, 103 (1978).

introduced into international law that is utterly different from the concept of 'right of an individual' as it is traditionally understood in international law and employed in practice. . . .

QUESTIONS

The first two paragraphs of the preceding Note each lists a group of rights declared in the ICESCR. Should the two lists necessarily be treated similarly with respect to the issue of justiciability, i.e. either all amenable to adjudication or none amenable? Or could one draw a distinction between the two lists? What distinction, and what consequences might it suggest?

NOTE

In the following readings, bear in mind several aspects of the question of justiciability. (1) Are the rights as recognized in the ICESCR formulated in a manner that is sufficiently precise to enable judges to apply them in concrete cases? (2) To the extent that such cases will involve decisions about public spending priorities, should such decisions remain the exclusive domain of the executive and legislature? (3) Are judges well-suited in terms of their expertise, social and political background and the facilities available to them, to make such decisions? (4) Does the justiciability test need to be applied in a narrow, traditional manner, or are there more creative approaches, perhaps still involving the courts in some way, which would satisfy those demanding formal, institutionalized measures of implementation for economic and social rights?

In the early 1990s the question of the justiciability of economic and social rights was extensively debated in both Canada and South Africa in the context of constitutional reform. The debate in both countries also served to explode assumptions that social progressives would automatically favor the entrenchment of economic rights. In Canada the proposal was for a social charter rather than for the recognition of rights enforceable in the courts.

In the following article, Jackman reviews the principal arguments that were used against the latter approach and responds to some of them. In South Africa the principal focus of debate was on the possible inclusion of a set of 'directive principles' in the post-Apartheid constitution.

MARTHA JACKMAN, CONSTITUTIONAL RHETORIC AND SOCIAL JUSTICE: REFLECTIONS ON THE JUSTICIABILITY DEBATE

in Joel Bakan and D. Schneiderman (eds.), Social Justice and the Constitution: Perspectives on a Social Union for Canada 17 (1992), at 18.

The Ontario government discussion paper on the social charter canvassed a number of options. These included strengthening existing constitutional commitments to equalization and the reduction of interregional disparities; expanding the welfare component of the interprovincial mobility rights clause under section 6 of the *Canadian Charter of Rights and Freedoms*; entrenching a general statement of principles which would inform social and economic policy making; and adding certain basic and justiciable social rights to the constitution. The official proposals released by the Ontario government in February 1992 were, however, disappointingly modest. The government opted for an expanded version of the non-justiciable principles set out in section 36 of the *Constitution Act, 1982*, abandoning any mention of justiciable individual rights to specific programs or services.

In its proposals for a new 'social covenant,' released at the end of February 1992, the Beaudoin–Dobbie Committee also rejected the idea of a justiciable social charter in favour of a largely declaratory one. The Committee recommended that a new section 36.1 be added to the constitution, one 'which would commit governments to fostering' a number of 'social commitments,' including comprehensive, universal, portable, publicly administered and accessible health care; adequate social services and social benefits; high quality education; the rights of workers to organize and bargain collectively; and the integrity of the environment. The Committee explained its preference for a social covenant framed in terms of constitutional commitments, rather than justiciable social rights, on the grounds that:

> While these commitments are in many ways as important to Canadians as their legal rights and freedoms, they are different. These commitments express goals, not rights and they embrace responsibilities of enormous scope. Therefore, while these are appropriate subjects for constitutional recognition, elected governments should retain the authority to decide how they can best be fulfilled. We believe that the matters addressed in the Social Covenant are best resolved through democratic means.

The Committee acknowledged, however, that in order for the commitments contained in the covenant to 'be more than merely words,' governmental compliance must be subject to some form of public review. The Committee recommended that this review be carried out by a specialized commission, which would hold public hearings and table periodic reports in Parliament and in provincial and territorial legislatures. . . .

The Case against Justiciability

The unwillingness reflected in [these proposals] to support a social charter containing justiciable social rights is shared by many constitutional scholars and others on the left. Criticism is based on three main arguments.

The first argument against the entrenchment of justiciable social rights is based on perceived distinctions between classical and social rights, and their appropriateness as a matter for judicial review. Classical, or 'negative' rights, it is argued, reflect natural and inherent traits in human beings. In order for these traits to become rights, the state need only recognize their existence and refrain from interfering with them. Social or 'positive' rights, on the other hand, do not come into existence automatically upon their recognition. Rather, the state must act affirmatively to create them or to ensure the conditions necessary for their enjoyment. And, while the content of classical rights is seen as relatively universal and precise, social rights are considered highly subjective and imprecise in character.

These differences have significant implications for the role of the judiciary. . . . To vindicate a claim for social rights, however, the court may have to order the state to act affirmatively, in some more or less specific way. Such judicial remedies will often require substantial and long-term state intervention and expenditure. Because of the contingent nature of social rights, and the lack of objective standards in this area, many fear that judges deciding social rights claims will substitute their own values for those of democratically elected, and accountable, legislatures. In enforcing social rights, courts will effectively engage in social policy making, forcing governments to expend public funds in ways they did not plan or choose. In addition to eroding public confidence in the independence and integrity of the judiciary, it is claimed that this usurpation of policy making by the courts will lead to an abdication of responsibility by governments, and to frustration and eventual apathy in voters. . . .

Social rights and judicial competence

A second set of arguments against justiciable social rights suggests that courts are institutionally incompetent to deal with the complex social, political and resource allocation issues which would inevitably be raised by individual social rights claims. For a number of reasons, including the complexity of judicial procedures, the incomprehensibility of judicial language, the formality and adversarial nature of judicial process, and the prohibitive cost of litigation, courts are virtually inaccessible to the poor.

. . . In remedying social rights violations, courts will potentially be required to tell governments what benefits or services they must provide, and in what quality or quantity. Such determinations require a thorough grasp of social and economic conditions in the society, as well as knowledge of public perceptions of the community's needs and means. Legislatures, and not courts, it is argued, are in the best position to make the complex judgements which these questions require.

The pursuit of rights as a mechanism for social change

A final objection to justiciable social rights reflects broader concerns relating to rights-seeking as a form of social action and as a mechanism for social change. In one variant of the critique, the pursuit of rights is seen as a new and damaging social trend which pits individualistic American-style conceptions of citizenship against traditional Canadian collectivist visions of the relationship between individual, community, and the state. In this conception, the individual as bearer of rights is seen as an incomplete and flawed construction of personhood, or as one commentator describes it, 'an emaciated, juridical conception of ourselves.'

More prominent in legal critiques of rights and rights-seeking is the argument that the constitutional entrenchment of rights represents a 'liberal lie,' promoting a false expectation in disadvantaged individuals and groups that the pursuit of legal rights through the courts can effect lasting social change, or that formal recognition of rights can have any real transformative impact on underlying social institutions, values, and conditions. Rights, it is claimed, operate instead to perpetuate existing power structures in society, and to channel potentially radical demands for change into legal claims which, by definition, will not be disruptive of the social and economic *status quo*.

An Assessment of the Social Rights Critique

. . . The distinction between negative and positive rights is . . . open to serious challenge. Once one moves beyond the most personal of human rights—such as the right to freedom of conscience or belief—the arguments that classical rights reflect natural and inherent traits in human beings, that they are absolute in character, or that their recognition imposes negligible costs on the state, are hard to sustain.

. . .

It is important to realize that traditional distinctions between classical or negative rights, and social or positive rights, and the willingness to provide for judicial enforcement of one but not the other, operate in fact to discriminate against the poor. To be in a position to complain about state interferences with rights, one has to exercise and enjoy them. But without access to adequate food, clothing, income, education, housing and medical care, it is impossible to benefit from most traditional human rights guarantees. . . .

From the perspective of the disadvantaged, arguments relating to democracy and judicial legitimacy must also be reassessed. The insistence that courts are incompetent to engage in social policy making ignores their historic and continuing intervention in this area. Courts create social policy not only in their interpretation of the *Canadian Charter of Rights and Freedoms*, but also in interpreting federal and provincial economic, labour, environmental and other legislation, as well as in their resolution of private law disputes. . . . The view that justiciable social rights are inimical to democracy fails to recognize the role of the courts as an avenue for challenges to the antidemocratic behaviour of Parliament and its delegates. It also fails to consider the extent to which social rights can, by expanding the number of participants and the quality of participation, enhance democratic decision making by elected governments and other public institutions.

Critiques of rights-seeking must also be re-examined. The argument that meaningful social change cannot be achieved through the pursuit of legal rights suggests that the judiciary is somehow distinct from ordinary politics and government, and that legal institutions, legal reasoning, and litigation are not legitimate arenas for social action and debate. By failing to accept judicial challenges and rights-seeking as legitimate strategies for reform, the anti-rights critique recreates the dichotomy, challenged by progressive lawyers since the 1920s, between politics and law.

While the socially and economically disadvantaged may not be adequately represented by the courts, neither are they well represented by other branches of government. If judicial representation and access are a real concern, reforms of the judicial appointments process and judicial procedure are a more effective response than excluding the claims of the disadvantaged from the courts.

. . .

Conclusion

. . .

[A] constitution is more than a legal document. It is a highly symbolic and ideologically significant one—reflecting both who we are as a society, and who we would like to be. Inclusion of certain rights and principles in the constitution says a great deal about their stature and importance; omission of others has the same effect. . . .

QUESTIONS

1. Assess Jackman's analysis in light of the following arguments against a social charter for Canada, made by another commentator, Bakan.

(a) [A] government might argue that its decision to lower or refuse to raise social assistance rates or minimum wages is justified because the current or new rate is equal to or higher than the minimum rate required by the social charter (as determined by an enforcement agency, or the government's own legal advisors). Ironically, along these lines, a social charter might actually *prompt* governments to cut spending and lower regulatory standards where the minimum levels determined to be acceptable under a social right are lower than the existing ones. . . .

(b) By necessity, an effective social charter will make some social goals priorities over others, and these priorities, not to mention the trade-offs they will require, may be undesirable from the perspective of many on the left and in social movements. Not all would agree, for example, that spending on health or education, each explicitly protected by a social right, is more important than spending on shelters for battered women, First Nations media and advocacy groups, foreign aid, or public broadcasting, to take just a few programs that might not come to be within the scope of social charter protection. . . .

(c) [For many] supporters, . . . a social charter would be a symbol of citizenship for economically disadvantaged and disempowered people; its absence would symbolize a denial of that citizenship. As well, they point out, the symbolism of social rights would inspire and mobilize people to demand from governments fulfillment of social obligations. I find little comfort in these arguments and I am concerned that the symbolism of social rights might backfire badly. One of the symbolic effects of a social charter, for example, might be to obscure and go some way towards legitimating a reality of social wrongs. . . . My concern is that social rights themselves may be understood as remedies for social inequality, and this will obfuscate the continuing need for social change. . . .[20]

2. Does Bakan's third proposition also challenge arguments to the effect that rights rhetoric is useful in the social field as a mobilizing agent and as a means of elevating the moral status of the claim?

3. Consider the justiciability of the claims made by the plaintiff in the following situations. Assume that State X has ratified both Covenants. (a) State X has legislated a plan administered by a government agency to provide food stamps for people below a certain income level. The plaintiff sues the agency because he has been denied stamps on the grounds that his higher income (as calculated by the agency pursuant to its formula interpreting the legislation) disqualifies him from the plan. (b) The plaintiff sues the agency on the ground that the plan disqualifies and thus illegally discriminates against people who have not lived in X for the past year. (c) X has failed to develop a national food strategy; it has no legislation providing food for the poor. The plaintiff argues that X is in violation of the ICESCR and requests the court to order the X legislature to draw up and enact such legislation at once.

4. 'It's not only with respect to economic and social rights that the role of courts has been overstated by human rights advocates. Only in a few states, and most of them concentrated in the West, have courts played vital roles in a state's development of civil and political rights. For the most part, respect by government for such rights has grown out of political battles, and sometimes violent battle.' Comment.

[20] Joel Bakan, What's Wrong with Social Rights, in J. Bakan and D. Schneiderman (eds.), Social Justice and the Constitution: Perspectives on a Social Union for Canada 85 (1992).

5. 'At least the courts in democratic states can work out of coherent traditions in giving content to and developing basic civil and political rights, like fair process in criminal trials, or equal protection, or free speech. But what guides are available to courts when they work with such nebulous conceptions in the ICESCR as a right to food, or health? Where do they turn—unlike a legislature, which turns to the full political process of interest groups, debate and elections?' Comment.

NOTE

Another technique for promoting economic and social rights, additional to making them justiciable, involves the inclusion of 'directive principles' in a constitution. Such principles are considered to be distinct from, and usually inferior in status to, rights that appear in the constitution without the qualification 'directive'. These principles appear in different forms in diverse constitutions including those of Ireland, Papua New Guinea and Nigeria.

It is the Indian experience, however, that holds the greatest interest for our purposes. With the fall of *apartheid* and the first democratic elections in South Africa in the early 1990s, considerable attention was given to the desirability of that country's following the Indian model by including social rights as directive principles in a new constitution.[21] The Indian Constitution of 1950 contains one chapter dealing with 'fundamental rights' (Part III) which consists largely of civil and political rights, and another (Part IV) dealing with 'directive principles of state policy.'

THE CONSTITUTION OF INDIA

in Albert Blaustein and G. Flanz, Constitutions of the Countries of the World, Release 94–7 (1994), at 77.

PART IV

. . .

39. The state shall, in particular, direct its policy towards securing—

(a) that the citizens, men and women equally, have the right to an adequate means of livelihood;

[21] See Patrick Macklem and C. Scott, Constitutional Ropes of Sand or Justiciable Guarantees?: Social Rights in a New South African Constitution, 141 U. Penn. L. Rev. 1 (1992): Nicholas Haysom, Constitutionalism, Majoritarian Democracy and Socio-Economic Rights, 8 S. Af. J. Hum. Rts. 451 (1992); Etienne Mureinik, Beyond a Charter of Luxuries: Economic Rights in the Constitution, *ibid.*, 464. See also Frederic Fourie, The Namibian Constitution and Economic Rights, 6 S. Af. J. Hum. Rts. 363 (1990).

(b) that the ownership and control of the material resources of the community are so distributed as best to subserve the common good;
(c) that the operation of the economic system does not result in the concentration of wealth and means of production to the common detriment;
(d) that there is equal pay for equal work for both men and women;
(e) that the health and strength of workers, men and women, and the tender age of children are not abused and that citizens are not forced by economic necessity to enter avocations unsuited to their age or strength;
(f) that children are given opportunities and facilities to develop in a healthy manner and in conditions of freedom and dignity and that childhood and youth are protected against exploitation and against moral and material abandonment.

. . .

41. The State shall, within the limits of its economic capacity and development, make effective provision for securing the right to work, to education and to public assistance in cases of unemployment, old age, sickness and disablement, and in other cases of undeserved want.

42. The State shall make provision for securing just and humane conditions of work and for maternity relief.

. . .

46. The State shall promote with special care the educational and economic interests of the weaker sections of the people, and, in particular, of the Scheduled Castes and the Scheduled Tribes, and shall protect them from social injustice and all forms of exploitation.

47. The State shall regard the raising of the level of nutrition and the standard of living of its people and the improvement of public health as among its primary duties . . . .

NOTE

Over the years the Indian courts as well as the legislature have redefined the relationship between fundamental rights and directive principles. The important element in the present context is the extent to which the directive principles have gone from being clearly non-justiciable to providing the basis of a right of action at first instance before the Indian Supreme Court. The relationship between the two has been summarized as follows:[22]

. . .

[22] B. De Villiers, The Socio-Economic Consequences of Directive Principles of State Policy: Limitations on Fundamental Rights, 8 S. Af. J. Hum. Rts. 188, 198 (1992).

(i) The Constituent Assembly, after thorough consideration, decided to provide for two chapters in the Constitution dealing with fundamental rights. The fundamental rights are justiciable and provide for a wide range of civil, political, cultural, economic and social rights. The directive principles of state policy are not justiciable, but as the soul of the Constitution provide the framework according to which governments of the future should act. In more than one sense the directive principles constitute a social contract to which political parties have to conform.

(ii) The fundamental rights are central to individual rights and freedoms and form the basis of a democratic state. The directive principles are the backbone of state action and planning, and emphasize the socio-economic responsibility of the state towards its citizens.

(iii) The Constituent Assembly realized the importance of these two sets of principles being used in tandem in government and administration. Fundamental rights were regarded as essential to democratic processes, but they would be of no avail to a poverty-stricken society.

(iv) The Supreme Court has gone through various phases in interpreting the relationship between fundamental rights and directive principles. Initially there was a firm adherence to the supremacy of fundamental rights. After several constitutional amendments, public debate and disputes over court decisions, the Supreme Court has adopted a more balanced and integrated approach in order to interpret harmoniously the two chapters.

(v) The current attitude of the Supreme Court is that fundamental rights should be understood within the framework of directive principles. Legislation which may limit fundamental rights is upheld if it is reasonable, in the public interest and shows a clear nexus with the directive principles.

. . .

UPENDRA BAXI, JUDICIAL DISCOURSE: THE DIALECTICS OF THE FACE AND THE MASK

35 J. of the Indian Law Institute 1 (1993), at 7.

[In this reading Baxi describes the way in which the directive principles have been used in the context of what is sometimes called 'public interest litigation,' or what he prefers to term 'social action litigation' (SAL).]

. . . The procedure is that of epistolary jurisdiction, where letters written by ordinary citizens to courts get converted into writ petitions. And these letters do not allege violation of fundamental rights of their authors; the authors allege such violation of the rights of the impoverished groups of Indian society—be they people in custody, victims of police violence, forced, bonded, migrant, contract labour, child workers, rickshaw pullers, hawkers, self-employed people, pensioners, pavement dwellers, slum dwellers, fishermen or Sri Lankan Tamils used as bonded labourers. The law of standing, that is persons who can bring complaints of rights-violation, has been thus revolutionised; and access to constitutional justice has been fully democratised.

The second major procedural innovation brought in by SAL is the collection of social data and legal evidence concerning the plight of these impoverished groups. Courts now appoint socio-legal commissions of public citizens, social scientists and others to examine the conditions alleged to be violative of people's rights; and the reports of the commission, constituted at state expense, provide the material for doing justice. Increasingly, universities and research institutes are directed by the court to function as commissions.

. . .

The third major aspect of the SAL is in the area of relief. Compensation and rehabilitation for victims deprived of their fundamental rights now constitute a constitutional right; the Supreme Court undertook a detailed monitoring of the rehabilitation of the blinded of Bhagalpur and since then has fashioned many a measure of compensation and rehabilitation; it has ordered the administration of theosuplhate [sic] injections to the Bhopal victims and upheld the constitutional validity of the Bhopal Act by reading into it the obligation to provide monthly interim relief . . . ; it has provided elaborate directives for treatment of prisoners and undertrials in jails; it has given specific directives for humane and just conditions of work for migrant workers and forced labourers.

The fourth salient aspect of the SAL is the development of constitutional jurisprudence itself. The right to compensation for violation of fundamental rights is now fully emergent. Indians have now a right to speedy trial though nowhere explicitly formulated by the Constitution. The custodial inmates have a right to dignity and immunity from cruel, unusual or degrading treatment. . . . Above all, the jurisprudence of SAL insists on a simple postulate of civilised jurisprudence: administration shall act in accordance with the law and the Constitution. This is a monumental achievement of SAL jurisprudence as this insistence has been vividly and memorably, in concrete contexts of the growing lawlessness of the state [sic].

Fifth, without being exhaustive, the SAL processes have at times resulted in a mini-takeover of the administrative regime of certain institutions of administration, which have displayed a congenital inability to work in accordance with the law and the Constitution. The most conspicuous illustration of this, perhaps, is the Agra Protective Home for Women, virtually run by the judiciary for well over ten years. The Supreme Court is doing its best to enable, by constant invigilation, the State of Bihar to ensure proper prison administration, to the point of maintenance of the record of undertrial and convict populations in the various jails of Bihar. In environmental cases, the Supreme Court has ordered the closure or monitored the pollution potential of private industries. The State High Courts have not lagged behind in these and related areas.

. . .

But when, at the end of the day, judicial orders arrive, the executive is left with a series of rather painful choices. To implement them is to engage in tasks of renovation of power which the executive loathes in the first place: *e.g.*, appointment of vigilance committees under the bonded labour law, prosecution of officers allegedly guilty of blindings, torture and tyrannies, expansion of the minimum wage of factories inspectorate, avoidance of sex-based discrimination in public works, including famine relief, running of remand homes for women and juvenile institutions in accordance with its own declared laws and policies, and reformation of jails. The SAL outcomes are not perceived as opportunities to reshape power but rather as obstacles in the exercise of *real* power.

. . .

. . . Constitutional adjudication in SAL becomes, then, a dialogue between judiciary and executive on the nature of public power and its public purposes. The executive has yet to accept the role of the pupil, although the courts have all too eagerly donned the didactic robes of pedagogues for democracy and constitutionalism.

QUESTIONS

1. Does the 'access-to-justice' approach provide a persuasive response to those who argue that the inappropriateness of adjudication renders economic and social rights unfit to be characterized as human rights at all?[23]

2. In light of the Indian SAL experience as described by Baxi, is it desirable for 'constitutional adjudication' to have become 'a dialogue between judiciary and executive on the nature of public power and its public purposes'? From the perspective of promoting human rights, is it desirable for the judiciary to have 'donned the didactic roles of pedagogues for democracy and constitutionalism'?

3. Consider the arguments listed by Jackman against a justiciable social charter in light of the experience with SAL in India.

[23] For a more detailed exploration of these issues see M. Cappelletti, The Judicial Process in Comparative Perspective (1989).

F. DEVELOPMENT BY STATES OF A PROGRAMMATIC APPROACH

It has often been observed that, for the promotion of economic and social rights, the principal alternative to reliance on judicial remedies is a programmatic approach. Although there is no single structure for such an approach, it generally implies the formulation of a program of action aimed at ensuring the realization of the right in question, with the government playing a more or less central role. The program might also provide for participation in decision-making by those affected.

One formulation of a programmatic approach has been called 'access to justice'. The following reading, which is taken from an analysis prepared in the context of a review of new options for promoting human rights within the framework of the European Union (known then as the Community), explains what such an approach implies.

The second reading addresses the issue of whether a programmatic approach can be effective in practice. Drèze and Sen focus on what they see as the indispensable role of public action in relation to the elimination of hunger.

CASSESE, CLAPHAM AND WEILER, 1992—WHAT ARE OUR RIGHTS?: AGENDA FOR A HUMAN RIGHTS ACTION PLAN

in A. Cassese et al., Human Rights and the European Community: Methods of Protection 1 (1991), at 57.

[The authors argue that the 'normative–judicial model,' premised on judicial review, has shortcomings which are generally overlooked. They include: (1) various societal groups—for cultural, linguistic, socio-economic and other reasons—will be unaware of the possibility of their seeking judicial review; (2) for reasons of cost or standing, marginally-affected individuals will rarely take an issue to court; (3) an individual might not even be aware that her rights (e.g. to privacy) have been violated; and (4) the reluctance of many judges to address issues with major implications for public expenditure.]

. . .

The normative–judicial model must be complemented by an approach which insists on effective vindication over and above the normative content and judicial enforcement.

Again, for the sake of convenience only, we would like to refer to the complementary approach as the 'Access-to-Justice' approach—denoting the range of procedural devices developed to make rights truly effective.

. . .

2. *Differentiation*

Clearly the key weakness of the normative–judicial model is that it treats, for most purposes, all rights in the same way in terms of their method of vindication. . . . Clearly, in the field, say, of data protection, an administrative provision such as a Commissioner for Data Protection and a requirement that public authorities report on any plan to open a new data base which would include personal information on individuals (such as we find in, for example, the British and the German systems) would be as important and maybe even more important than a simple judicial remedy for a violation of privacy. Clearly in fields such as consumer protection or environmental protection, where interests and rights are diffuse and fragmented, consideration of different rules of standing may be called for . . . .

. . .

3. *Information*

. . . [E]ach right, depending on its addressees—whether private individuals or public authorities—calls for a different approach to disseminating the content of the right and the means for its vindication. Sometimes legislation is enough. Other times one would have to make a concerted effort to cross cultural, socio-economic and other barriers in order to inform potential victims of their rights. In practically all Member States, large sums of social benefits remain each year uncollected by persons with entitlements—i.e. persons whom society has deemed it right should enjoy these entitlements. The usual reason is simply ignorance.

4. *Impact and means*

[I]t is necessary when discussing [economic and social rights] to consider the financial resources necessary for their vindication, and the impact that consecration of such a right may have on the state.

The converse . . . could be the requirement in fields such as environmental protection that new legislation and new public policies in diverse areas should be accompanied by impact studies on, say, the ecological balance of the environment. In the long run, such a requirement, by rendering the policy formation more transparent, may be as beneficial as an individual right which is to be protected by the judges.

5. *Proceduralization*

Under this heading one would have to investigate the different legal procedural devices which should accompany each right or cluster of rights. Should the right be available only against violations by public authorities (vertical vindication) or also against private bodies (horizontal vindication)? What time limitations, if any, should be imposed? What rules of standing should apply? What kind of legal aid should be available and from whom, etc.?

6. *Institutionalization*

In the same vein, it is clear that different rights, or different clusters of rights, may require different institutional arrangements for their vindication. Here one may consider such centralized institutions as a Community Commissioner on Human rights whose task it would be to monitor and generally supervise the status, content, and vindication of human rights in the Community. But it is more likely that decentralized institutional arrangements such as advice bureaux may eventually be more effective.

. . .

NOTE

Discussions of the right to food often involve the assumption that if there is a lack of available food in a given situation, it is impossible for the government concerned to remedy the problem. But Amartya Sen has shown that 'food availability decline' has often been absent in famine situations.

> Starvation is the characteristic of some people not *having* enough food to eat. It is not the characteristic of there *being* not enough food to eat. While the latter can be a cause of the former, it is but one of many *possible* causes. Whether and how starvation relates to food supply is a matter for factual investigation. (*Poverty and Famines: An Essay on Entitlement and Deprivation* (1981) at 1.)

He shows, for example, that in the Great Bengal Famine of 1943, which claimed between 1.5 and 3 million victims, food may not have been plentiful but there was no less of it than in previous and subsequent years (in which there were no famines). The disaster was therefore due to factors other than a lack of food. They included: a war-related boom; speculation and panic hoarding; administrative ineptitude and chaos; a decline in the relative purchasing power of agricultural laborers; and the fact that, for political reasons, a famine was never declared. That omission resulted in a failure to trigger preplanned policy responses appropriate to such situations (*ibid*, pp. 52–83).

In a later analysis, Drèze and Sen argue that there is an essential role for public action in eliminating both famines and persistent food deprivation.

DRÈZE AND SEN, HUNGER AND PUBLIC ACTION (1989), at 257.

. . .

In the field of famine prevention, the decisive role of public action is illustrated not only by the elimination of famines in India since independence, but also by the unsung and underappreciated achievements of many African countries. These experiences firmly demonstrate how easy it is to exterminate famines if public support (e.g., in the form of employment creation) is well planned on a regular basis to protect the entitlements of vulnerable groups. Ensuring that the concerned governments take early and effective steps to prevent a threatening famine is itself a matter of public action. It is also clear that the eradication of famines need not *await* a major breakthrough in raising the per-capita availability of food, or in radically reducing its variance (even though these goals are important in themselves and can be—and must be—promoted in the long run by well-organized public policy). Public action can decisively eliminate famines *now*, without waiting for some distant future.

Regarding the elimination of regular, persistent deprivation (as opposed to the eradication of intermittent famines), the analysis presented here has indicated the positive contribution that can be made by public provisioning (especially of education and health services) and more generally by public support (including such different policies as epidemiological control, employment generation and income support for the vulnerable). Expectations based on general reasoning are, in fact, confirmed by the empirical experiences of different countries.

Public support in these different forms has played a major part in combating endemic deprivation not only in economies that are commonly seen as 'interventionist' (e.g., China, Costa Rica, Jamaica, Sri Lanka), but also in the market-oriented economies with high growth (e.g., Hong Kong, Singapore, South Korea); . . . . Indeed, the contrast between what was called 'growth-mediated security' (as in, say, South Korea) and 'unaimed opulence' (as in, say, Brazil) relates closely to the extensive and well-planned use of public support in the former cases, in contrast with the latter . . .

When it comes to enhancing basic human capabilities and, in particular, beating persistent hunger and deprivation, the role played by public support—including public delivery of health care and basic education—is hard to replace.

The crucial role of public support in diverse economic environments is well illustrated by the intertemporal variations in the experience of China. The radical transformation in the health and nutritional status of the Chinese population (visible *inter alia* in a sharp increase in life expectancy, a dramatic decline of infectious and parasitic diseases, and improved anthropometric indicators) took place *before* the reforms of 1979, at a time of relatively moderate growth of GNP but enormously effective public involvement in the promotion of living conditions. The post-reform period has seen an impressive acceleration in the growth of GNP and private incomes, but also a crisis of public provisioning (especially of health services), and an *increase* in mortality. Much more is involved in increasing human capabilities—and in preventing their decline—than the stimulation of economic growth through revamping private incentives and market profits.

We have also discussed how the crucial role of public support in removing endemic deprivation is visible not only in the achievements and failures of developing countries today, but also in the historical experiences of the rich and industrialized countries. This is well illustrated by the sharp increases in longevity in Britain during the decades of the world wars, which were periods of rapid expansion of public support in the form of public food distribution, employment generation and health care provisioning (not unconnected with the war efforts). There is nothing particularly *ad hoc* in the findings regarding the contribution of public support to human lives in the developing countries today.

Public action is not, of course, just a question of public delivery and state initiative. It is also, in a very big way, a matter of participation by the public in the process of social change. As we have discussed, public participation can have powerful positive roles in both 'collaborative' and 'adversarial' ways *vis-à-vis* governmental policy. The collaboration of the public is an indispensable ingredient of public health campaigns, literacy drives, land reforms, famine relief operations, and other endeavours that call for cooperative efforts for their successful completion. On the other hand, for the initiation of these endeavours and for the government to act appropriately, adversarial pressures from the public *demanding* such action can be quite crucial. For this adversarial function, major contributions can be made by political activism, journalistic pressures and informed public criticism. Both types of public participation—collaborative and adversarial—are important for the conquest of famines and endemic deprivation.

To emphasize the vital role of public action in eliminating hunger in the modern world must not be taken as a general denial of the importance of incentives, nor indeed of the particular role played by the specific incentives provided by the market mechanism. Incentives are, in fact, central to the logic of public action. But the incentives that must be considered are not only those that offer profits in the market, but also those that motivate governments to implement well-planned public policies, induce families to reject intrahousehold discrimination, encourage political parties and the news media to make reasoned demands, and inspire the public at large to cooperate, criticize and coordinate. This complex set of social incentives can hardly be reduced to the narrow—though often important—role of markets and profits.

. . .

QUESTION

Drèze and Sen see a decisive role for public action. Does this necessarily involve a comprehensive programmatic approach? Vierdag (see p. 279, supra) has argued that the term 'programmatic' automatically implies that the 'so-called right' is not enforceable by the individual 'in a court of law or similar body.' Is the public participation, both collaborative and adversarial, of which Drèze and Sen speak sufficient in itself to provide the necessary element of governmental accountability?

G. INTERNATIONAL SUPERVISION BY THE ICESCR COMMITTEE: CASE STUDY OF HOUSING

Neither the UN Commission nor its Sub-Commission has been very active in relation to the implementation of economic, social and cultural rights. The principal responsibility has thus fallen to the ICESCR Committee, established in 1987 to supervise compliance by states parties with their obligations under the ICESCR.

In many respects, the Committee functions in the same way as the ICCPR Human Rights Committee, examined in Chapter 9, *infra*. On the basis of regular reports submitted by states parties in accordance with the Committee's 'reporting guidelines', and of its own deliberations during 'days of general discussion' and in the drafting of 'general comments,' the ICESCR Committee seeks to achieve three principal objectives: (1) development of the normative content of the rights recognized in the Covenant; (2) acting as a catalyst to state action in developing national 'benchmarks' and devising appropriate mechanisms for establishing accountability, and providing means of vindication to aggrieved individuals and groups at the national level; and (3) holding states accountable at the international level through the examination of reports.

In the following case study of the right to adequate housing we consider the type of reports that the Committee has sought to elicit through its guidelines, the use it has made of 'general comments' to develop the content of the right, and the outcome of the examination of a report on the Dominican Republic. The 'concluding observations' on that report that are excerpted constitute the final stage in a process involving: the submission of the governmental report, receipt by the Committee of detailed NGO information, a detailed discussion of the relevant issues by the Committee and representatives of the Dominican Government, and drafting by the Committee in closed session of its observations. Those observations are then made public at the same time as they are provided to the government concerned.

COMMITTEE ON ECONOMIC, SOCIAL AND CULTURAL RIGHTS, REPORTING GUIDELINES

UN Doc. E/1991/23, Annex IV.

[These guidelines, adopted by the Committee in 1991 and subject to revision over time, regulate the form and contents of the reports that states parties to the ICESCR are required to submit. An initial report is required within two years of ratification and periodic reports are due every five years thereafter.]

The right to adequate housing

(a) Please furnish detailed statistical information about the housing situation in your country.

(b) Please provide detailed information about those groups within your society that are vulnerable and disadvantaged with regard to housing. Indicate, in particular:

(i) The number of homeless individuals and families;

(ii) The number of individuals and families currently inadequately housed and without ready access to basic amenities . . . ;

(iii) The number of persons currently classified as living in 'illegal' settlements or housing;

(iv) The number of persons evicted within the last five years . . . .

. . .

(c) Please provide information on the existence of any laws affecting the realization of the right to housing.

(d) Please provide information on all other measures taken to fulfil the right to housing, including:

(i) Measures taken to encourage 'enabling strategies' whereby local community-based organizations and the 'informal sector' can build housing and related services. Are such organizations free to operate? Do they receive Government funding?

(ii) Measures taken by the State to build housing units and to increase other construction of affordable, rental housing;

(iii) Measures taken to release unutilized, under-utilized or mis-utilized land;

(iv) Financial measures taken by the State including details of the budget of the Ministry of Housing or other relevant Ministry as a percentage of the national budget;

. . .

4. Please give details on any difficulties or shortcomings encountered in the fulfilment of the rights enshrined in article 11 and on the measures taken to remedy these situations (if not already described in the present report).

COMMITTEE ON ECONOMIC, SOCIAL AND CULTURAL RIGHTS, GENERAL COMMENT No. 4 (1991)

UN Doc. E/1992/23, Annex III.

The right to adequate housing (art. 11 (1) of the Covenant)

1. Pursuant to article 11 (1) of the Covenant, States parties 'recognize the right of everyone to an adequate standard of living for himself and his family, including adequate food, clothing and housing, and to the continuous improvement of living conditions'. . . .

. . .

7. In the Committee's view, the right to housing should not be interpreted in a narrow or restrictive sense which equates it with, for example, the shelter provided by merely having a roof over one's head or views shelter exclusively as a commodity. Rather it should be seen as the right to live somewhere in security, peace and dignity. . . .

8. . . . While adequacy is determined in part by social, economic, cultural, climatic, ecological and other factors, the Committee believes that it is nevertheless possible to identify certain aspects of the right that must be taken into account for this purpose in any particular context. They include the following:

(a) *Legal security of tenure*. Tenure takes a variety of forms, including rental (public and private) accommodation, cooperative housing, lease, owner-occupation, emergency housing and informal settlements, including occupation of land or property. Notwithstanding the type of tenure, all persons should possess a degree of security of tenure which guarantees legal protection against forced eviction, harassment and other threats. . . .
(b) *Availability of services, materials, facilities and infrastructure*. An adequate house must contain certain facilities essential for health, security, comfort and nutrition. All beneficiaries of the right to adequate housing should have sustainable access to natural and common resources, safe drinking water, energy for cooking, heating and lighting, sanitation and washing facilities, means of food storage, refuse disposal, site drainage and emergency services;

(c) *Affordability*. Personal or household financial costs associated with housing should be at such a level that the attainment and satisfaction of other basic needs are not threatened or compromised. Steps should be taken by States parties to ensure that the percentage of housing-related costs is, in general, commensurate with income levels. . . .

(d) *Habitability*. Adequate housing must be habitable, in terms of providing the inhabitants with adequate space and protecting them from cold, damp, heat, rain, wind or other threats to health, structural hazards, and disease vectors. . . .

(e) *Accessibility*. Adequate housing must be accessible to those entitled to it. Disadvantaged groups must be accorded full and sustainable access to adequate housing resources. . . .

(f) *Location*. Adequate housing must be in a location which allows access to employment options, health-care services, schools, child-care centres and other social facilities. This is true both in large cities and in rural areas where the temporal and financial costs of getting to and from the place of work can place excessive demands upon the budgets of poor households. Similarly, housing should not be built on polluted sites nor in immediate proximity to pollution sources that threaten the right to health of the inhabitants;

(g) *Cultural adequacy*. The way housing is constructed, the building materials used and the policies supporting these must appropriately enable the expression of cultural identity and diversity of housing.

. . .

10. Regardless of the state of development of any country, there are certain steps which must be taken immediately. . . . [M]any of the measures required to promote the right to housing would only require the abstention by the Government from certain practices and a commitment to facilitating 'self-help' by affected groups. To the extent that any such steps are considered to be beyond the maximum resources available to a State party, it is appropriate that a request be made as soon as possible for international cooperation in accordance with articles 11 (1), 22 and 23 of the Covenant, and that the Committee be informed thereof.

. . .

12. While the most appropriate means of achieving the full realization of the right to adequate housing will inevitably vary significantly from one State party to another, the Covenant clearly requires that each State party take whatever steps are necessary for that purpose. This will almost invariably require the adoption of a national housing strategy. . . . [S]uch a strategy should reflect extensive genuine consultation with, and participation by, all of those affected, including the homeless, the inadequately housed and their representatives. . . .

13. Effective monitoring of the situation with respect to housing is another obligation of immediate effect. For a State party to satisfy its obligations

under article 11 (1) it must demonstrate, *inter alia*, that it has taken whatever steps are necessary, either alone or on the basis of international cooperation, to ascertain the full extent of homelessness and inadequate housing within its jurisdiction. . . .

COMMITTEE ON ECONOMIC, SOCIAL AND CULTURAL RIGHTS, CONCLUDING OBSERVATIONS ON THE REPORT OF THE DOMINICAN REPUBLIC

UN Doc. E/C.12/1994/15, at 3.

[The ICESCR Committee notes that 'detailed and precise information' that it has received over several years from NGO sources indicates, *inter alia*, that: 30,000 families in the Zona Norte are threatened with forced eviction; thousands of families have already been evicted from Faro a Colon and from other specified cities; 3,000 relocated families received neither compensation nor relocation allowances; and the housing conditions of some 750 families relocated after Hurricane David in 1979 are grossly inadequate. That information was relayed to the Government by the Committee.]

10. While the Government presented the Committee with information as to the achievements and shortcomings of its various policies in relation to housing, the Committee did not receive any information which would lead it to conclude that these problems do not exist or have been adequately addressed.

11. It therefore expresses its serious concern at the nature and magnitude of the problems relating to forced evictions and calls upon the Government of the Dominican Republic to take urgent measures to promote full respect for the right to adequate housing. In this regard, the Committee notes that whenever an inhabited dwelling is either demolished or its inhabitants evicted, the Government is under an obligation to ensure that adequate alternative housing is provided. . . .

. . .

13. . . . [T]he Committee was also informed that less than 17 per cent of Government-built housing units are provided to the poorest sectors of society.

14. On the basis of the detailed information available to it the Committee also wishes to emphasize its concern at the 'militarization' of La Cienaga–Los Guandules, the long-standing prohibition on improving or upgrading existing dwellings for the more than 60,000 residents of the area, and the inadequate and heavily polluted living conditions. The situation is especially problematic given that these communities were originally established as relocation areas for evictees in the 1950s. Since that time the Government has failed to confer legal security of tenure on residents or to provide basic civic services.

. . .

16. The Committee is also concerned at the effects Presidential decrees can and do have upon the enjoyment of the rights recognized in the Covenant. It wishes to emphasize in this regard the importance of establishing judicial remedies which can be invoked, including in relation to Presidential decrees, in order to seek redress for housing rights violations. The Committee is not aware of any housing rights matters that have been considered by the Supreme Court in relation to article 8 (15) (b) of the Constitution. In so far as this might be taken to indicate that the provision has not so far been subject to judicial review, the Committee expresses the hope that greater reliance will be placed upon it in future as a means by which to defend the right to adequate housing.

. . .

19. All persons residing in extremely precarious conditions such as those residing under bridges, on cliff sides, in homes dangerously close to rivers, ravine dwellers, residents of Barrancones and Puente Duarte, and the more than 3,000 families evicted between 1986–1994 who have yet to receive relocation sites . . . should all be ensured, in a rapid manner, the provision of adequate housing in full conformity with the provisions of the Covenant.

20. The Government should confer security of tenure on all dwellers lacking such protection at present, with particular reference to areas threatened with forced eviction.

. . .

23. The Committee requests the Government to apply existing housing rights provisions in the Constitution and for that purpose to take measures to facilitate and promote their application. Such measures could include: (a) adoption of comprehensive housing rights legislation; (b) legal recognition of the right of affected communities to information concerning any governmental plans actually or potentially affecting their rights; (c) adoption of urban reform legislation which recognizes the contribution of civil society in implementing the Covenant and addresses questions of security of tenure, regularization of land-ownership arrangements, etc.

24. In order to achieve progressively the right to housing, the Government is requested to undertake, to the maximum of available resources, the provision of basic services (water, electricity, drainage, sanitation, refuse disposal, etc.) to dwellings and ensure that public housing is provided to those groups of society with the greatest need. It should also seek to ensure that such measures are undertaken with full respect for the law.

25. In order to overcome the existing problems recognized by the Government in its dialogue with the Committee, the Government is urged to give consideration to initiatives designed to promote the participation of those affected in the design and implementation of housing policies. Such initiatives could include: (a) a formal commitment to facilitating popular participation in the urban development process; (b) legal recognition of community-based

organizations; (c) the establishment of a system of community housing finance designed to open more lines of credit for poorer social sectors; (d) enhancing the role of municipal authorities in the housing sector; (e) improving coordination between the various governmental institutions responsible for housing and considering the creation of a single governmental housing agency.

. . .

QUESTIONS

1. The ICESCR's key provision makes reference to resource constraints and progressive realization. In light of the wording of Art. 2(1), how do you assess the way the three preceding documents interpret and apply these provisions?

2. 'The housing standards in General Comment No. 4 are high, so high that they are way beyond the capacity of most states. To this extent they are not only unrealistic but ineffective. The Committee should instead have identified priorities indicating what all states must or must not do. Such standards could more readily be made operational.' Do you agree or disagree? What priorities would you suggest?

3. How do you evaluate the significance of the Committee's 'concluding observations'? Suppose the Dominican Republic does nothing in response. What recourse does the Committee have under the ICESCR? Could it have been more specific about what was to be done when, and demanded compliance reports?

ADDITIONAL READING

Matthew Craven, *The International Covenant on Economic, Social and Cultural Rights: A Perspective on its Development* (1995); Philip Alston, *The Committee on Economic, Social and Cultural Rights*, in P. Alston (ed.), *The United Nations and Human Rights: A Critical Appraisal* 473 (1992); Bruno Simma, *The Implementation of the International Covenant on Economic, Social and Cultural Rights*, in F. Matscher (ed.), *The Implementation of Economic and Social Rights* 75 (1991); Scott Leckie, *Towards an International Convention on Housing Rights: Options at Habitat II*, (1994); and *An Overview and Appraisal of the Fifth Session of the UN Committee on Economic, Social and Cultural Rights*, 13 Hum. Rts Q. 545 (1991).

H. INTERNATIONAL ENFORCEMENT: IDENTIFYING VIOLATORS

The challenge is to see whether the various approaches that have been explored above are workable in practice and provide an adequate basis upon which the international community can proceed to evaluate the performance of states in relation to their obligations regarding economic and social rights. The two studies that follow provide an opportunity to test the utility of these approaches in the context of a variety of issues arising in different societies.

CHRISTIAN TOMUSCHAT, THE RIGHT TO HEALTH IN GUATEMALA

UN Doc. E/CN.4/1993/10, at 60.

[The Commission on Human Rights (Res. 1992/78) requested the UN Secretary-General to appoint an independent expert to examine the human rights situation in Guatemala. The expert, a German international law professor, submitted this report in 1993.]

223. . . . The Ministry of Public Health covers only 25 per cent of the population; the Guatemalan Social Security Institute (IGSS) covers 15 per cent, and the private sector 14 per cent. The total cover is thus 54 per cent of the population, concentrated exclusively in urban areas. This means that the remaining 46 per cent (i.e. 4.5 million people) is not provided with any kind of health service, not even a defective one. Most of these people are rural and indigenous. In addition, . . . the installations in 70 per cent of the country's 35 main hospitals are in a state of disrepair; many of them have already reached the end of their useful life. Basic equipment such as boilers, laundries, electricity plant, lifts, etc., are in a state of deterioration and in 45 per cent of the hospitals the support services such as radiology, laboratories, blood banks and sterilization facilities are described as '60 per cent in poor condition'.

224. In his written opinion of 9 April 1992 concerning the operating conditions in the Escuintla regional hospital, the Human Rights Procurator stated that 'the investigations carried out have shown that this hospital is in a very poor state which makes it difficult to operate normally. . . '.

225. In contrast, . . . the new departmental hospital of Huehuetenango . . . is of modern design and will have a capacity of 172 beds; its final cost, including equipment, is estimated at about 20 million United States dollars. However, the hospital will have to cater to the needs of a very large population (about 80,000 in the town and 600,000 in the department).

226. All these factors explain why the infant mortality rate remains high. Current data indicate that about 100 children die every day in Guatemala,

most of them from hunger and causes connected with malnutrition, and that 58 per cent of Guatemalan children aged between 3 and 36 months suffer from some degree of malnutrition; this figure rises to 72 per cent in rural and poor areas.

. . .

228. [An NGO, the] Human Rights Commission of Guatemala [has] reported . . . about the defects in health services in rural areas, where there is only one doctor for every 25,000 inhabitants [and] 80 per cent of the rural population does not have access to medical services. It adds that 68 per cent of health personnel works in hospitals, 25 per cent in health centres, and 7 per cent in rural communities; 45 per cent of the doctors live in the capital, where 20 per cent of the total population is concentrated.

NOTE

The Tomuschat report dealt primarily with abuses of civil and political rights but also, as this excerpt demonstrates, with issues of economic and social rights. In responding to the report, however, the Commission on Human Rights (Res. 1993/88) made no mention of the latter rights. It did invite the Government 'to give priority also to economic and social development programmes' (para. 13).

QUESTION

Guatemala ratified the ICESCR in 1988. Does the evidence contained in the report demonstrate any clear violations by the Government of its obligations under the Covenant? Under which provisions? How might Guatemala respond to a charge that it had violated the Covenant?

VAN DER STOEL, SITUATION OF HUMAN RIGHTS IN IRAQ

UN Doc. A/48/600 (1993).

[In the wake of Iraq's invasion of Kuwait in 1990 and its ouster in 1991, the Commission on Human Rights appointed a Special Rapporteur (a former Dutch Foreign Minister, Max van der Stoel) to examine alleged violations by the Iraqi Government. The violations related in particular to the Kurdish population in the north of Iraq, the Shi'ite population in the south and those living in the area of the southern marshes (the 'Marsh Arabs'). Iraq is a party to the ICESCR. The following excerpt relates specifically to the 'Marsh Arabs'.

In previous reports the Special Rapporteur had concluded that the Government of Iraq was intent on de-populating the entire Marsh area because of its inhabitants' suspected opposition to the regime and the difficulties of policing such an area. It sought to achieve this goal through the re-settlement in 'secure' urban camps of all of the inhabitants of the area. This was to be achieved through instructions to re-locate as directed, backed up by military and police operations, the imposition of a total economic blockade and the drainage and destruction of the marshes.]

II. The Situation in the Southern Marsh Area

. . .

C. Violations of economic rights

. . .

34. In view of the Government of Iraq's obligations to take steps to provide for adequate food, clothing and housing and to achieve the highest standards of physical and mental health, the existence of prohibitions, restrictions and administrative requirements interfering with access constitute violations. Reports and testimonies indicate that such interferences exist and impact the Marsh Arabs in severe ways. For example, access to food rations is extremely limited for them because the system of food rationing employs the use of ration cards, which change periodically and for which recipients must be duly registered. However, this system assumes the prior possession of identification cards which . . . Marsh Arabs find extremely difficult to acquire in a considerable number of cases. Thus, the peculiar situation of the Marsh Arabs coupled with the registration process acts as an interference with their access to basic foodstuffs. In place of food rations, Marsh Arabs are left to acquire their provisions on the open market, which has been inflicted with hyper-inflation. Yet, even should they be able to afford provisions, testimonies allege that Government forces often confiscate acquired goods. . . .

35. Aside from the difficulties associated with the normal system of rationing, testimonies allege that, for those who previously had access, the Government stopped distributing rations within the marsh villages after the Gulf war of 1991. Instead, the Army is said to have established checkpoints which have the effect of stopping people from procuring food and other supplies in the towns. Moreover, other conditions on access to food rations within the towns have been reported: one witness claimed that he was refused his rations because he would not join the Army.

36. Access to health care is also said to be extremely limited for the Marsh Arab people. As in the case of access to food rations, the problem of identification cards also interferes with access to health care available through

the urban medical centres. Since Government doctors reportedly no longer come to the marsh villages, the population is left to its own means. Clandestine and illegal medical services are reportedly offered by individuals and groups operating on a humanitarian basis. . . .

37. In terms of needs, reports and testimonies indicate that malnutrition and disease are widespread within the marsh area. Waterborne diseases such as bilharzia are said to be rampant. . . .

. . .

39. Without a doubt, a major contributing cause to the increasing needs of the Marsh Arabs is the loss of their natural resources as the marshes have dried. While attacks on marsh settlements have resulted in the death of livestock, the greater cause of death of fish and animals is attributable to the depletion of water and the deterioration of that which remains. Reports, testimonies and video recordings indicate that the previously clean waters of the marshes have become toxic. This has also resulted in the loss of the Marsh Arabs' source of clean drinking water.

40. The precise cause and nature of the toxins in the water is unknown. Reports and testimonies allege that the Government has purposefully poisoned the waters through the introduction of chemicals and other poisons. Some witnesses claim to have observed the dumping of chemicals, while others recount how governmental authorities encouraged fishermen to pour bottled chemicals into the water, supposedly to improve their catches. However, the Special Rapporteur is much more of the opinion that the reduced waters have become stagnant and also extremely polluted by the huge amounts of industrial and agricultural waste, together with raw sewage, which have been running into the marshes subsequent to the destruction of urban water treatment facilities. In this connection, the Special Rapporteur notes the submission of the Government of Iraq in which the Government recounts damage caused to the environment through the discharge into surface water and soil of substantial amounts of 'various poisonous and dangerous chemical substances'—discharge resulting, according to the Government, from the destruction of civil installations by Allied bombing during the 1991 Gulf war (A/CONF.157/4, paras. 35–40 and annex I). Whatever the correctness of this explanation, the Special Rapporteur stresses that the Government of Iraq remains obliged to take steps to limit this discharge; . . .

41. . . . [T]he Special Rapporteur has also received reports, testimonies and photographs indicating that the agricultural region south-east of Amara has experienced significant flooding. Specifically, it has been reported that the farms and date orchards in [certain areas] were flooded in mid-June of 1993 just at the time of harvest; wheat, barley and rice crops were lost, and date orchards were destroyed.

42. . . . [T]he Government has apparently failed to respond to the factual situation irrespective of causes. For example, there are said to have been no efforts to ameliorate the situation of the Marsh Arabs. Vaccinations have

apparently also not been carried out, although the Government felt it appropriate to ask most humanitarian agencies operating in the south last year to leave. Moreover, testimonies allege that artillery bombardments of marsh villages are not followed by any efforts to assist the injured, innocent or not. Similarly, witnesses allege that there are no efforts to assist victims of mines which have been laid in the waters of the marshes and which may be hit by Marsh Arabs in their daily movements.

QUESTIONS

1. How would you describe, in legal terms, the precise obligations that the Special Rapporteur considers to have been violated?

2. Could these violations be dealt with as effectively under the rubric of civil and political rights? Which provisions of the ICCPR, to which Iraq is a party, apply?

3. Are the problems involved in proving violations of economic and social rights in this particular context such as to distinguish them, in some fundamental ways, from civil and political rights? Suppose the charge was that Iraq had failed to build enough hospitals or to provide enough food for the population. Would the distinction from civil and political rights be greater or lesser than in this case?

4. How would you assess a claim of the Iraqi Government that the absence of 'available resources' nullifies any obligation that it might otherwise have in relation to some, or all, of the issues raised by the Special Rapporteur?

5. Under what circumstances might it be acceptable for the Iraqi Government to drain the marshes? The Government claims that its objectives are: '(a) to wash away salt-encrusted soils; (b) to reclaim land for cultivation; and (c) to increase water available for irrigation' (UN Doc. A/48/600 (1993), para. 46). Does the Covenant oblige a government to desist from such activities even if undertaken in the name of national development?

6. Many of those affected in this case are indigenous peoples (the Ma'dan). Does this element make a fundamental difference in the principles you would endorse? To what extent do your principles draw upon notions of civil and political, rather than economic and social, rights?

ADDITIONAL READING

Asbjørn Eide, C. Krause and A. Rosas (eds.), *Economic, Social and Cultural Rights: A Textbook* (1994); Franz Matscher (ed.), *The Implementation of Economic and Social Rights* (1991); Ralph Beddard and D. Hill (eds.), *Economic, Social and Cultural Rights: Progress and Achievement* (1992); Ellen Frankel Paul *et al.* (eds.), *Economic Rights* (1992); Partha Dasgupta, *An Inquiry into Well-Being and Destitution* (1993); Martha Nussbaum and A. Sen (eds.), *The Quality of Life* (1993); Joseph Wronka, *Human Rights and Social Policy in the 21st Century* (1992); A. Glenn Mower Jr., *International Cooperation for Social Justice: Global and Regional Protection of Economic/Social Rights* (1985); Craig Scott and P. Macklem, *Constitutional Ropes of Sand or Justiciable Guarantees: Social Rights in a New South African Constitution?*, 141 U. Penn. L. Rev. 1 (1992); Lammy Betten (ed.), *The Future of European Social Policy* (1989); Asbjørn Eide *et al.* (eds.) *Food as a Human Right* (1984); Krzysztof Drzewicki *et al.* (eds.), *Social Rights as Human Rights: A European Challenge* (1994); Amartya Sen, *Inequality Reexamined* (1992); David Harris, *The European Social Charter* (1984); Philip Alston, *U.S. Ratification of the Covenant on Economic, Social and Cultural Rights: The Need for an Entirely New Strategy*, 84 Am. J. Int'l L. 365 (1990); Barbara Starck, *Economic Rights in the United States and International Human Rights Law: Toward an 'Entirely New Strategy'*, 44 Hastings L. J. 79 (1992); Danilo Türk, *The Realization of Economic, Social and Cultural Rights: Final Report submitted by the Special Rapporteur*, UN Doc. E/CN.4/Sub.2/1992/16; Human Rights Program, Harvard Law School and the François-Xavier Bagnoud Center for Health and Human Rights, *Economic and Social Rights and the Right to Health* (1995); Audrey R. Chapman (ed.), *Health Care Reform: A Human Rights Approach* (1994).

PART C

HUMAN RIGHTS INSTITUTIONS AND PROCESSES

Part C turns from the norms stressed in Part B to an examination of the international institutions embodying, developing, monitoring and enforcing those norms. The invention of institutions by the human rights movement has been among its most striking and important features. None of the institutions discussed in Part C existed before the end of World War II.

Neither Part B nor Part C can fully exclude the other and become an airtight compartment. Norms and institutions are too intricately and fundamentally connected. For example, several UN organs necessarily figured in the discussion in Chapter 3 of the development of the UDHR and ICCPR, as well as in the discussion in Chapter 5 of the ICESCR. In general, standard setting, the grandest achievement of the human rights movement's first half century, began within institutional processes. Similarly, the materials in Part C necessarily involve law making by international institutions in such varied settings as comments by committees on treaty texts or decisions by international courts.

Nonetheless our concerns here sharply change. They involve institutions as such, their constitutional structure, processes, functions, powers, and so on. Above all, we are here concerned with their role in 'enforcement' rather than standard setting.

Part C examines intergovernmental and non-governmental organizations, both universal and regional in scope, whose work basically involves international human rights. Its materials inevitably raise broader issues about international *organizations* or *institutions* as such.[1] An analogy to Part B is pertinent. There we gave some attention to characteristics of international law, as essential background to the study of international human rights norms. Here too we give some attention to characteristics of international organizations, again brief but sufficient for this book's purposes.

Chapter 6 introduces such broader characteristics of international institutions. The remaining chapters in Part C start with the UN human rights organs (or bodies), continue with non-governmental organizations, turn to the ICCPR Human Rights Committee to illustrate a treaty organ, and conclude with aspects of regional human rights organizations in Europe, the Americas and Africa.

[1] A word about terminology. The authors' text in this book uses the terms '*organization*' and '*institution*' more or less interchangeably. Excerpts from writings by other authors included in this book may, however, attach particular meanings to one or the other term. For example, some scholars of international 'organizations' view that designation as indicating possession of specific legal attributes, such as juridical personality and capacity to sue. From this perspective, the UN is the paradigmatic 'organization'. On the other hand the Human Rights Committee created under the ICCPR, a treaty *organ* or *body* (interchangeable terms), would not readily fit within such a meaning of the term. Some scholars view the term 'institution' as suggesting a richer and more complex and autonomous entity that goes beyond the form of a rudimentary organization that merely organizes states within a loose cooperative framework.

6

The Need for Institutions: Introductory Ideas

Except for non-governmental organizations, the international institutions and organs examined in Part C were created by multilateral treaties or derive their authority and legitimacy from such treaties. Customary law does not create institutions. These materials explore the more developed and complex institutional arrangements, going well beyond the kind of rudimentary organization created by a multilateral treaty that, say, provides only for periodic meetings of states parties to exchange views on a matter or draft conventions to be submitted to states for ratification.

Each of the five treaties here relevant—the Charter, the ICCPR, and the treaties creating the three regional human rights systems—creates a broad institution or one or more treaty organs that perform certain functions by exercising stated powers. To one or another extent, that institution or organ takes on a life of its own. It will have its own officials, staff and budget. Despite its pervasive links with the states parties (who are directly represented in the membership of many treaty organs such as the UN General Assembly or UN Commission on Human Rights), the institution or organ can be meaningfully considered separately from those parties (unlike, say, the 'rudimentary' form of organization described above). If in some major respects it depends on the states parties' 'will,' in other major respects it possesses autonomy. Thus an institution or organ may become a significant participant in international relations, adding to the traditional system of sovereign states and qualitatively changing the nature of international life.

The institutions here considered vary enormously. Even with respect to its human rights activities, the UN includes a panoply of organs of diverse constitution, competence and powers. On the other hand, the ICCPR creates only one organ, a committee of a radically different character from any UN organ. Despite some common attributes, each institution and organ considered in Part C has its distinctive character.

There follow excerpts from the Table of Contents of a leading treatise on the subject that suggest one framework for thinking about the constitution and operations of international institutions. They indicate the variety referred to above, how much choice states have, how many possibilities are before them when many states negotiate about the original design or the reform of these institutions. The excerpts include the entire section of the Table of Contents on voting since a reading below draws on voting arrangements to illustrate its argument.

HENRY SCHERMERS, INTERNATIONAL INSTITUTIONAL LAW

(1980), at ix.

Table of Contents

D. Implied Powers

. . .

Chapter Five. ***Supervisory Organs***

I. Parliamentary Organs

. . .

D. Tasks

1. Control over the executive
2. Control over the budget

. . .

II. Judicial Organs

. . .

Chapter Six. ***Decision-Making Process***

. . .

V. Voting

A. Unanimity

1. Consensus and unanimity
2. Organizations requiring unanimity
3. Exceptions to unanimity

B. Voting Power

1. Equality of voting power
2. Inequality of voting power
   a. Permanent seats and weighted representation
   b. Weighted voting
      (i) Desirability
      (ii) Existing systems
   c. Veto

C. Required Majority

1. Kinds of majorities
2. Calculation of majorities
   a. Majority of membership
   b. Majority of the votes
   c. Abstention
   d. Absence
   e. Invalid vote
3. Unqualified majority
   a. Voting between two alternatives
   b. Voting between several alternatives
   c. Multiple elections
4. Two-thirds majority
5. Other qualified majorities
6. Qualified minorities
7. Factors influencing the majority to be preferred
   a. The need for a decision
   b. The effect of the decision
   c. Structure and procedure of the decision-making organ

D. Methods of Voting

1. Simultaneous open voting

2. Roll-call or recorded vote
3. Secret vote
4. Vote by correspondence
5. Alteration of votes cast

E. Conditional Voting

Chapter Seven. ***Financing***

. . .

Chapter Eight. ***Legal Order***

. . .

II. Constitution

. . .

C. Amendment of the Constitution

. . .

III. Decisions of the Organization

A. Internal Rules

1. Rules on the functioning of the organization

. . .

B. External Rules

1. Recommendations

. . .

2. Declarations

. . .

3. Conventions

. . .

Chapter Ten. ***Enforcement***

. . .

II. Supervision of the Implementation of Rules

A. Supervision by Other Members Acting on their Own Account

B. Supervision by or on behalf of the Organization

. . .

III. Official Recognition of Violations

. . .

V. Sanctions

A. Sanctions by the Organization

. . .

3. Suspension of representation

. . .

7. Expulsion from the organization
8. Sanctions through other organizations
9. Military enforcement

. . .

B. Sanctions by the Other Members

C. Sanctions against Individuals

D. Enforcement within the National Legal Order

. . .

COMMENT ON RELATIONS BETWEEN NORMS AND INSTITUTIONS

Imagine a human rights system consisting solely of the rules or principles growing out of treaties and customary law. That is, we 'assume away' the UN organs dealing with human rights issues, the organs under treaties like the ICCPR, and regional human rights organs. We do away entirely with the 'institutionalization' of norms in these different entities.

Simply imagining a human rights system shorn of its intergovernmental institutions and processes makes us realize how profoundly different the human rights movement would be. Only state governments and state institutions would be available to meet the need for development, monitoring and enforcement of norms. Of course one hopes that most states would take their obligations seriously, internally enforcing their treaty obligations through their own equivalent constitutional norms or through specific internalization of treaty norms (see Chapter 11). One also knows, however, that many states will not so act, and that some among them will engage in gross and systematic abuses. From the international perspective, human rights norms would be freely floating rather than anchored in any international regime, dependent for their effectiveness on the willingness of treaty parties to apply pressures to delinquent states.

Whatever the inadequacies of existing international human rights institutions, a state party to the basic treaties that commits serious violations would have much less to fear in such an imagined world. The state may have ratified the ICCPR and ICESCR without the intention of complying with essential norms of either one, thereby gaining some international respect and legitimacy from its formal participation in the human rights system. The ratification may have been seen as relatively costless, for, even within the existing system, the state may view the probability of serious enforcement as slender. In this new imagined world, that probability almost disappears.

We have previously noted a distinctive characteristic of human rights as a topic of treaty law. States parties to such a treaty lack the usual material incentives (as exist in, say, a reciprocal tax exemption treaty) to act against a violator state. The violation consists in that state's abuse of *its own* citizens. Why should state X invest its energies in trying to persuade state Y to stop jailing its own political dissidents? It would be foolish to assume that sustained inquiry, let alone serious pressures and sanctions, would originate in other states parties, which at most might suspend economic or military aid to the delinquent state (see Chapter 12).

It should then be no surprise that the traditional defenses by states of sovereignty and domestic jurisdiction are applied today more strongly against the enforcement powers or actions of international institutions than against standard setting. Through its collective acts and decisions, ranging from the holding of a debate to investigative missions or the ordering of sanctions, the institution can hurt the delinquent state. It becomes the bridge between states (the very creators of international norms) and the norms themselves. It can make those norms 'real'.

All the institutions and organs examined in Part C reflect this tension or contradiction. States create institutions that, to one or another degree, are meant to discipline them. Much depends on the institution's architecture—its constitution and processes, its modes of decision-making such as majority vote or consensus, its powers. Here lie some of the most acerbic fights and deepest divisions in the human rights movement.

NOTE

The following readings raise in a preliminary and abstract way some basic questions about international institutions that recur in the later specialized chapters. They start with excerpts from a book of Inis Claude that refers to both the constitutional (internal) and substantive (external) problems of international organizations.

INIS CLAUDE, SWORDS INTO PLOWSHARES

(4th ed., 1984), at 6.

. . . To understand that international agencies are products not of the aspirations of idealists standing outside of and above international politics, but of the necessities felt by statesmen operating within the arena of international politics, is to sense the fact that international organization is a functional response to the complexities of the modern state system, an organic development rooted in the realities of the system rather than an optional experiment fastened upon it. For one who grasps this fact, the issue of whether we should have international organization is no more meaningful than the issue of whether urbanization should result in the provision of more extensive public services and the imposition of more elaborate governmental regulations. One recognizes international organization as a distinctive modern aspect of world politics, a relatively recent growth, but an established trend. Particular organizations may come and go, but international organization as a generic phenomenon is here to stay. The collapse of the League of Nations led almost automatically to consideration of the nature of its replacement, and similar failure by the United Nations might be expected to produce the same reaction. A sense of history provides the basis for the understanding that international organization has become a necessary part of the system for dealing with international problems, and that 'to organize or not to organize' is no longer an open question for statesmen or a useful one for students of international relations.

. . .

. . .The problems confronted by international organizations may be divided into two categories: constitutional problems—the problems *of* international organizations, and substantive problems—the problems *with which* the organizations are designed to grapple. The first group consists of internal matters, related to the management and functioning of the organizations, while the second includes external issues requiring solution. Constitutional problems are occasioned *by* the establishment of international organizations; substantive problems are the occasions *for* the establishment of such agencies.

However definite the dividing line between these two classes of problems may be in logic, it is not so in practice. The nature and intensity of world problems determine the nature and scope of organizational efforts, and thereby define the constitutional problems which emerge. Decisions concerning the internal development of international agencies are inevitably influenced by external political considerations, and, conversely, the solution of substantive political problems is affected by the degree of constitutional development achieved by international organizations. The two problem areas cannot be divorced.

One of the major tasks of twentieth-century statesmanship is to strike a balance between obsessive concern with institutional problems—which makes international organization an end in itself, and exclusive concentration upon substantive issues of current world politics—which neglects the building of an adequate institutional apparatus for international relations. . . .

. . . It is useful for statesmen to be reminded that they cannot expect international organizations to serve them well, now or in the future, if their urge to exploit these institutions for immediate political advantage overrides all consideration of the requirements of sound constitutional development. International organizations cannot become effective means to the ends that states envisage unless they are treated, to some degree, as ends in themselves.

. . .

NOTE

The excerpts from Ernst Haas present the issue of control by one or a few states of the processes and powers of an international organization. That issue permeates the materials in Chapter 7 on the UN General Assembly and Commission on Human Rights.

ERNST HAAS, WHEN KNOWLEDGE IS POWER (1990), at 57.

All international organizations have a heterogeneous membership. Their members differ in size, military power, population, resource endowment, and degree of industrialization. The members also differ greatly from one another in the extent to which they are permeable—that is, subject to being 'penetrated' economically, culturally, and politically by their stronger neighbors.

Most organizations have their own 'superpower' capable of playing a hegemonic role if it chose to do so. All organizations (except those of Eastern and Western Europe) count democratic, totalitarian, and authoritarian governments among their members. Even organizations that consist almost entirely of economically less developed countries display significant differences in the degree of development among their members. All universal organizations include members with capitalist, socialist, and mixed economies.

All organizations are characterized by major inequalities in power, however defined, among their members. Consequently, they are subject to rule by hegemonic states or hegemonic coalitions of states. The hegemony need not be expressed in the direct imposition of the preferences of the stronger on the weaker. It usually takes the form of higher financial contributions and disproportionate roles for the nationals of the stronger members in organizational secretariats. At the extreme, this kind of hegemony is illustrated by the role of the United States in U.N. agencies. On the one hand, a decision to reduce American financial contributions from 25 percent of the budget to 20 percent threatened to ruin organizational programs. On the other hand, the United States had been signally unsuccessful in translating its superpower status into consistent influence over the content of programs, having lost many programmatic battles for almost two decades except in organizations in which greater power is recognized in the form of weighted voting.

In addition to disproportionate influence due to financial prowess or voting privilege, the principle of sovereign equality is also contradicted in practice by the tendency of the more powerful states to constitute themselves into an inner elite that is consulted far more consistently by the formal heads of secretariats, commissions, and councils than are representatives of less important states. Membership in this elite differs with topic and issue. It almost always includes the delegates of the superpowers and of Japan, Britain, France, West Germany, India, and, increasingly, Brazil in organizations to which these states belong. But it may also include the delegates of smaller states if the country in question happens to be salient to the issue at stake. Sweden, Singapore, and Tanzania have played inner-elite roles on some occasions in the United Nations.

. . .

International organizations share a certain marginality with respect to the core activities in international politics. Few foreign policy initiatives depend on international organizations for their success. States risk little by investing symbolically in the programs of such organizations; the core of one's foreign policy remains intact even if little concrete help is provided by the organizations. In most instances, any result from a symbolic investment in organizational action will not be experienced until much later. Foreign policy relies on nonorganizational means to a far greater extent than on institutionalized multi-lateral efforts. There are, of course, exceptions for large and small states. Immediate benefits can accrue from a successful peacekeeping operation for the losing side in a war; ambitious economic development and technical assistance projects are sources of prestige and employment for one's nationals; a country beset with refugees benefits immediately from multilateral aid; even the superpowers may benefit from a successful mediation to prevent crisis escalation. The great lines of foreign policy, however, are only marginally and gradually influenced by what goes on in international organizations.

NOTE

The following article by David Kennedy talks of the 'move' toward institutions as a key to understanding modern international law. These excerpts stress the complex, dual, even contradictory positions of states with respect to institutions. States parties make the treaty that creates the institution in which they become members. They lead lives 'within and without the institution.' They form part of its internal processes, participating in decisions about its external action. At the same time they remain sovereign, autonomous states that can become the objects of such action. Through his discussion of different voting arrangements, Kennedy illustrates some of these dilemmas.

DAVID KENNEDY, A NEW STREAM OF INTERNATIONAL LAW SCHOLARSHIP

7 Wisc. Int. L. J. 1 (1988), at 39.

. . .

. . . [F]ew areas of public international law doctrine today remain free of the network of institutions understood to have been set in motion in 1918. The corpus of modern doctrine . . ., is relentlessly procedural, harnessing each substantive aspiration into the policy objective of some institutional regime. Seen either from history or doctrine, then, the move to institutions is the key to modern international law.

From this perspective, institutions—and the discipline of international institutions—are different from doctrines . . . Public international law turns to institutions, turns into institutions, as a turn to practice, to engagement with sovereign society, as a move to realism and the politics of regime management.

We see then, in the relationship between our two disciplines—public international law and international institutions—a familiar division of labor. The one handles issues of independent legal judgment, the other problems of sovereign engagement. Of course this image is an oversimplification—we saw the repetition throughout public international law of a shrewd equivocation about the independence and normative nature of international law doctrine, and we are likely to find the discipline of international institutions riddled with doctrinal independence, procedural channels, consensual covenants, and the like.

But still, between them they handle international law's more general aspiration to both remain independent and connect with sovereign power. Perhaps this can be done ever so much better the more the division is blurred, or the division of labor proliferated throughout both disciplines. At least we never need face either sovereign autonomy or legal dominion in their pure form. They exist only as rather unstable and hesitant invocations and reference points. . . .

. . .

Questions of constitutional structure are normally considered in relationship to a constituting text—be it the League Covenant, the U.N. Charter or the U.N. Convention of the Law of the Sea. The texts establishing international institutions are remarkably similar in basic structure. In broad outline, all set out the membership, decisionmaking procedures, and respective competences of legislative and administrative organs. Sometimes provisions for reference to an independently established or integrated dispute settlement procedure is added. . . .

Leaving dispute resolution aside for a moment, no document seems complete, seems fully to have established a plenary [legislative plenary organ] if it does not indicate who will participate, how they will decide and how their collective being will be known. This pattern repeats the temporal logic of establishment: signatories are transformed into members, the interactions of members are structured, and the organ which they constitute is named. Membership marks a break between life within and without the institution. The organ is the name given the object established. Voting inserts a text between these two moments—both reminiscent of the particularity of members and generative of the constituted organ. The basic historical narrative could hardly be more familiar—a move from politics through text to institutional action.

. . .

. . . Focus for a moment on voting. People writing about institutional design in our discipline have devoted a great deal of energy to voting structure—the allocation of votes among members or the voting configurations required for action. On the one hand, the voting mechanism seems completely internal to the organization, a mere procedure for translating membership into organ activity. Such a sense focuses reformist energy on a technical procedure which might easily be changed, even if it seems too removed from context to provide a fully convincing account of the institution's practices. On the other hand, the voting mechanism draws a connection between the original members and the activities of their institution—connecting the preinstitutional context to the actions of the organization. The result is a double position—within and without the institution.

As such, voting exists uneasily between membership—itself the break between the institution and its creators—and organs—themselves the link between the institution and its context and object. Voting reaches back to members, defining them, and forward to organs, reminding them of their past. The problem of voting is to translate membership into action, orchestrating a smooth movement from constitution as members (frozen in the intentions of the establishing document) to institutional action within the competences of the organ in question. Voting thus both marks the inside of the plenary and asserts a relationship with both a preinstitutional constituency and an implementing organization, thereby linking two constituted beings—states as members with institutions as actors. . . .

. . .

This central relationship demands much of voting. It must accommodate both the authority of sovereign members and the cooperative activity of the institution. . . . Voting must move from sovereign autonomy to cooperation. . . .

. . .

. . . [Voting] must both ratify and express a particular distribution of power merely promised states by membership and be the mechanism by which the community makes up its collective mind and expresses itself vis-a-vis specific state powers—a relationship posited by the instrumental posture of organs. If the institution must be open and closed, voting must be deferential to and expressive of state power and yet also control, channel and ultimately reapportion that power as the voice and mind of the international institution. . . .

. . .

. . . Over the past sixty-five years, scholars considering voting in international institutions have advocated plenary decisionmaking by unanimity, majority vote and consensus. They have expressed their enthusiasm for and disillusionment with each scheme in remarkably similar terms. Each, in turn, has been credited with an ability simultaneously to defer to sovereign authority and express sovereign cooperation. As each decisionmaking scheme fell out of favor, it was criticized for permitting or encouraging either the anarchy of organizational collapse or the tyranny of institutional capture.

During the Hague period, and into the first days of the League [of Nations], scholars defended unanimity voting as a move from sovereign decentralization, in which international law could grow only through the relatively cumbersome mechanism of treaty drafting or the quite lengthy process of customary accretion, to institutionalization. At the turn of the century, unanimity symbolized the achievement of an institutional life among states, for it permitted autonomous sovereigns to sit in standing plenaries without forswearing their sovereign prerogative.

By the mid-1930's, unanimity no longer seemed so attractive. Scholars began to suggest that the League either need not, as a matter of law, or did not, as a matter of practice, continue to abide by a rigid unanimity rule. These texts advanced arguments against unanimity and in favor of some alternative voting scheme (usually majority voting) to those which had been advanced in support of unanimity during the preceding period. Unanimity, as a matter of theory and practice, could neither respect sovereign autonomy nor generate sovereign cooperation. It permits states to be held hostage by one bad actor, both preventing international action and centralizing international authority so as to override sovereign authority. By reducing international cooperation to the lowest common denominator of sovereign accord, unanimity emasculates the institution and sabatoges cooperation. In short, unanimity slows the momentum of institutional life and permits backsliding to anarchy.

Majority voting seemed much better. It would decentralize international authority, allowing states to defend their interests without waiting for the go ahead from one recalcitrant sovereign. At the same time, majority voting allows for more powerful and decisive institutional action, rendering the international institution persuasive by keeping it in touch with the greater part of the community. A strong international institution, in turn, fosters community. These arguments prevailed in 1945, and the post-war institutions exhibit a veritable cornucopia of majoritarian and weighted voting formulas.

By the mid-1960's, however, the luster was off majoritarianism. Weighted and majority voting—and particularly the veto—seemed a step backward, away from organization toward anarchy or irrelevance. On the one hand, majority voting produces a tyranny of the majority, allowing international organizations to be far too assertive, thereby threatening the sovereign authority of the minority. On the other hand, majority voting is the enemy of international cooperation. By encouraging rash decisions which reflect passing fads, majority voting leaves the institution powerless in the face of sovereign autonomy. By ignoring the interests of the minority, it debases the currency of international institutional outputs and causes the institution to lose respect. Majority voting fails the cooperative sovereign as much as the autonomous one.

By 1975, the fashionable international institution made up its mind by concensus. By exactly translating political reality into institutional action, concensus keeps the institution in step with all states. The minority feels attended to, included, respected: neither the big powers nor the blocs are able to control the majority any more. Consensus is the perfect form of institutional deference. Moreover, consensus permits the institution to make powerful decisions and ensures compliance with such decisions as are taken. The very experience of coming to consensus builds community. Finally, as we might expect, by 1980, the bloom was beginning to be off consensus—and the reasons were familiar. The institution was hostage to one hold out autonomous state—and the individual sovereign felt bullied into agreement by a powerful consensus building plenary practice.

This rather fickle rotation among voting procedures repeats the same arguments in each generation. Good procedures instantiate both autonomy and cooperation among sovereigns. Bad procedures fail to banish the threats of anarchy and tyranny. The move from one to another—from unanimity to majoritarianism to consensus—also marks a certain maturity. Although the arguments for and against consensus sound similar to those advanced for and against unanimity, these procedures are quite dissimilar. In many ways, consensus is the very opposite of a voting mechanism, producing no actual record of inter-sovereign accord, it seems to presume the accord behind the institutional output.

We might say that unanimity positions voting close to membership, consensus close to organs. Seen this way, the move among voting mechanisms is simply a repetition of the more general institutionalizing move from membership through voting to organs. Unanimity suggests an immature plenary, constantly recapitulating the moment of establishment. Consensus suggests a mature organizational voice finally released from its members. Majority voting seems a middle ground, a half-way house of trust, in which formalization of minority rights is still necessary to shackle the organ to members.

. . .

NOTE

The attraction of international organizations as a means of solving international problems was perhaps at its height after World War II, an attraction and enthusiasm captured in the creation of the United Nations. Over the decades, enthusiasm waned as it became clear that institutions could simply incorporate the polarities and conflicts of the 'outside' world, become lethargic administrators through inertia and stale bureaucracy, and experience manipulation and corruption. They could share many vices of states themselves.

With the end of the Cold War, fresh hopes looked less to the creation of new institutions than to the revivification of existing ones, particularly the UN. Those hopes too have faded. Much current discussion looks toward reforms of major institutions like the UN to permit them to play a more significant and effective role.

Human rights institutions are fully part of this cycle of enthusiasm and despair. Each of the later chapters involves evaluation of existing institutions and inquires into the kinds of changes that are necessary, desirable, feasible.

The excerpts from Michael Reisman present one of many possible perspectives on necessary reforms. Reisman's approach is not radical, it does not propose restructuring. He prefers to work within the existing treaty structures to improve them, indeed if possible without returning to states parties for ratification of treaty amendments.

The ideas below underscore an issue that returns in the following chapters. Should international organizations continue to have so exclusively 'statist' a character excluding the presence and participation of a range of nongovernmental international actors—human rights groups, religious organizations, environmental or women's groups, business organizations, trade unions, transnational political parties, and so on? How do we assure that international institutions embrace a popular component? Consider Reisman's proposals.

W. MICHAEL REISMAN, AMENDING THE UN CHARTER: THE ART OF THE FEASIBLE

American Society of International Law, Proceedings of 88th Annual Meeting 1994 (1995), at 108.

Reality is both an important element in constitutional design and an important corrective in constitutional dreams. The point has special relevance to discussions about changing the United Nations. No reality-based discussion of Charter revision can ignore Article 108, which allows for amendments by two-thirds of the General Assembly members but requires ratification by all the Security Council permanent members in order for those amendments to come into force. This does not mean that reform is unattainable, but rather that international constitutional amendment, in its conventional understanding, is the least likely to succeed and to be satisfactory.

...

... [T]he Charter is not a bad piece of drafting, in the sense that is has served and serves the interests of those who created it. Dissatisfactions with the Charter are found not within the Charter, as it were, but within the world constitutive process. New forces have gathered with different conceptions of what type of order the world requires. Most of the 180-some governments now in the United Nations are not Permanent Members of the Security Council and have no expectation of ever reaching that position. They are concerned about certain aspects of its operations. Less appreciated, but far more important, the nongovernmental entities that are now increasingly critical international actors pursue a set of goals quite different from those implicit in the state-centric Charter. Thanks largely to these non-state entities, issues such as human rights and environmental protection have been pushed to the forefront of international legal concerns. Non-state entities have different agendas and different priorities than those of the elite states within the United Nations.

Discussions of Charter reform frequently restrict their attention to the Charter, without relating it to these larger processes, which it is intended to influence and whose dynamics will determine its success. Reform proposals target a few dramatic but unrealistic or essentially marginal changes that confirm the state-centric structure of the Charter while ignoring a range of constitutional options that, though less dramatic, are more feasible and could yield substantial improvements in the overall performance of the Organization's functions. Change is directed at the Charter itself, which is viewed from within its four corners. Viewed from *outside*, however, the Charter's chief defect is that it is exclusively an interstate organization; this cannot be appreciated if one looks only at the Charter, for the whole Charter is premised on *étatism*. The sorts of changes the Charter itself contemplates can only redistribute power *within* that framework. But if one places the Charter in the largest world context, where many transnational and national nongovernmental entities now play increasingly critical roles in many decisions, the restriction on participation in the United Nations becomes glaringly apparent.

. . . If one of the purposes of Charter reform is the enhancement of programs such as human rights and the environment, then an instrumental goal should be a larger role for individuals and groups not affiliated with states. The 'deadest letters' in the UN Charter are in its very first words: 'We the peoples of the United Nations.' The peoples created the Charter, but were then summarily and totally dismissed from it; the Charter allows them no role whatsoever.

. . .

[The author considers proposals to expand the permanent membership of the Security Council, as well as suggestions that the International Court of Justice develop procedures for reviewing Security Council and General Assembly action under the Charter or international law. In the end, he advocates more effective control of the Security Council through the combined votes of its non-permanent membership, and through techniques permitting the General Assembly to follow more closely the Council's potential responses to crises and to make its views about those responses known to the Council. Such 'constitutional control' would not require amendment and would strengthen the voice of the Assembly.]

Making the General Assembly Credible and Effective

The real source of constitutional dissatisfaction in the United Nations is not the way the Security Council is structured, for in the final analysis there must be some sort of executive instrument with restricted membership and real power. The dissatisfaction relates instead to the reduced role of the General Assembly—the only arena in which the great majority of states can operate—and the inefficient way the Assembly performs its assigned functions. If, as I have suggested, UN constitutional reform would misfire by weakening or diluting the Security Council, it would certainly benefit by making the General Assembly more efficient. Happily, many important constitutive changes in this area are within the province of the Assembly and cannot be derailed by opposition from the Security Council or by a withholding of ratification by all or some of the Permanent Members.

. . .

As a modern parliamentary body, the General Assembly is incredibly inefficiently organized. Its six plenary committees are too large, and the functions of intelligence gathering and processing, of promotion and of prescription are ineffectively performed. Confirmation of the low esteem of the committees even among the membership is reflected in low attendance records and the manifest somnolence of so many of those who must attend. The Assembly could increase its efficiency by discarding the plenary committees entirely and creating, in their place, smaller, representative, functional committees and subcommittees with no more than twenty-five members, to hold hearings, take testimony, draft resolutions or even more ambitious instruments, and oversee those parts of the international administration that pertain to their specific functions. The rules of these committees and subcommittees would have to be revised to allow for direct testimony by interested individuals and NGOs as well as by state representatives. Once such committees were formed and began to function, NGOs, performing the functions of lobbyists, would fill the vacuum and could begin to direct their attention to the committees. In this more refined focus, their energies would be magnified. Hopefully, some of these committees would be of enough interest to warrant media coverage through C-Span and similar operations, thus reaching a wider audience. Since the new committees would be performing important functions, attendance would revive.

. . .

The improvement of the output of committees would make the action of the plenary General Assembly more meaningful. . . .

7

Intergovernmental Enforcement of Human Rights Norms: The UN System

This chapter takes a thorough, systematic look at the UN human rights system insofar as it bears on enforcement. Hence it complements Chapter 3, which looked at that system with respect to the generation of norms. The stress here is on the UN as an organization, and on the roles and effectiveness of its different organs or bodies concerned with human rights issues. The materials trace the historical evolution of these organs over several decades, examine the UN's current and significant procedures or processes in this field, suggest criteria for evaluation of the UN's record, and explore possible reforms.

A. DEVELOPMENT OF THE UN HUMAN RIGHTS SYSTEM

COMMENT ON ENFORCEMENT AND THE UN ORGANS INVOLVED

Conceptions of Enforcement

For individuals whose human rights are being violated, and for the groups that seek to defend them, the effectiveness of the United Nations' human rights system depends to an important degree upon its ability to 'enforce' respect for the legal norms that originated within it. But the very concept of such international 'enforcement' is controversial and resisted by a significant number of governments (a few of which do so overtly, while many others use more subtle methods). As suggested by Chapter 6, it is therefore not surprising that, although the UN has been very effective in setting standards (often consensually) in many human rights fields, enormous energy has been invested in hotly contested efforts to establish institutions and procedures capable of securing enforcement.

An evaluation of the UN's performance will be strongly influenced by one's starting point or perspective on world order. Do we assume that the 'globalization' of issues such as human rights is desirable, or perhaps even unavoidable, so that a nation's treatment of its own nationals is a legitimate concern of all others (an *erga omnes* approach in terms of international law)? Or do we hold to a more traditional image of the sovereign state that emphasizes the inviolability of national boundaries for human rights as well as other purposes? Even if the former, do we envisage a world in which an effective multilateral organization (which might or might not be the UN) should be able to act against the will of the government(s) concerned to enforce universal norms? Or do we believe that although some degree of globalization is inevitable, say with respect to standard setting, the actual implementation by individual governments of those standards, each in its own way, remains the most effective, desirable or realistic approach?

Another important consideration in evaluating the UN's performance involves reciprocity. Are we prepared to accept that the measures that we would happily support against another country might, in a different context, be applied against our own? Do we assume that international enforcement actions must be applied equally to powerful nations and to smaller states, so that we should only adopt policies that can be applied across the board, consistently?

The answers to all such questions will depend partly on the definition of enforcement. Do we mean by it only police actions, or should the notion extend to the other extreme of UN action, a debate or recommendatory resolution of the General Assembly? The only use of the term 'enforcement' in the UN Charter occurs in relation to the enforcement under Chapter VII of decisions of the Security Council (Article 45). This has led some international lawyers to equate enforcement with the use of, or threat to use, economic or other sanctions or armed force. It is true that most dictionary definitions of enforcement include an element of compulsion, although it may be moral as well as physical. It is also true that the possible use of force for human rights purposes has won increasing support in recent years, but this is surely not what is meant by calls for the UN to 'enforce', routinely, universal human rights norms.

At the other extreme, enforcement has been defined as 'comprising all measures intended and proper to induce respect for human rights.'[1] But that definition is so open-ended that it provides no criteria against which to evaluate the UN's performance. It puts the emphasis on intentions rather than on results achieved. It suggests that we might be content, for example, if the UN confined its efforts to the adoption of resolutions and other such hortatory activities.

In the materials that follow, three strong emphases or biases are to be observed. First, the concern of most UN organs—surely of the Commission

[1] Rudolf Bernhardt, General Report, in Bernhardt and Jolowicz (eds.), International Enforcement of Human Rights 5 (1985).

on Human Rights to which we give primary attention—is overwhelmingly with civil and political rights rather than with the economic and social rights that were examined in Chapter 5. Second, the emphasis is on responding to relatively discrete but gross and noticeable violations—large numbers of disappearances or killings of political opponents, violent ethnic conflict, a declaration of martial law. The emphasis is not on what might be termed persistent, endemic, and commonplace violations that are often ignored by other states—including such fundamental and serious violations as systematically entrenched discrimination against women or particular minority groups. Third, little attention is here given to consciousness-raising through education or promotional activities that many observers would identify as indispensable components of an effective UN program.

Such choice of themes can be said to reflect accurately the practice of the United Nations as a whole. Indeed, perhaps the principal criterion used by the vast majority of governments and commentators in assessing the UN's performance is the extent to which it reacts effectively to gross violations. But this criterion should not be accepted uncritically. It should be noted that a significant amount of the UN's important work concentrates on the longer-term, structural dimensions of human rights issues. Such activities include standard-setting, the promotion of greater awareness of those standards both within (e.g. in peace-keeping, or in the work of the World Bank and UN Development Programme) and outside the UN system, and the provision of advice and assistance (known in the UN as 'advisory services').[2]

UN Organs: Coverage of this Chapter

In the following reading, Louis Henkin discusses notions of compliance and enforcement and points to the differences in this regard between international law in general and international human rights law in particular. He employs the notion of a 'two-track' approach in relation to the UN's enforcement machinery. That is, he refers to:

(1) the *Charter-based organs* whose creation is directly mandated by the UN Charter (such as the General Assembly, the Economic and Social Council and the Commission on Human Rights) or has been authorized by one those bodies (such as the Sub-Commission on Prevention of Discrimination and Protection of Minorities, and the Commission on the Status of Women); and

(2) the *treaty-based organs* (such as the Human Rights Committee formed under the ICCPR, referred to in this book as the 'ICCPR

[2] More details of these programs can be obtained from the annual Report of the Secretary-General on the Work of the Organization (issued in 1995 as UN Doc. A/50/1) and the annual Report of the United Nations High Commissioner for Human Rights (issued in 1995 as E/CN.4/1995/98).

Committee' in order to reduce confusion with the UN Commission on Human Rights that is identified as the 'UN Commission') that have been created by six other human rights treaties originating in UN processes and that are intended to monitor compliance by states with their obligations under those treaties.

Except for some comments by Henkin and Alston in the introductory readings, the treaty-based organs (or bodies) are dealt with separately in Chapter 9.

Among the Charter-based organs, this chapter concentrates on the work of the UN Commission. Of course the debates, declarations, resolutions and recommendations of the General Assembly, the vital plenary organ, play an essential role. A number of readings draw on the work of the Assembly when it is particularly relevant to the problem under discussion.

The materials refer less frequently to the work of the Security Council that has only rarely directly addressed human rights violations as such. They include (p. 365, *infra*) the most significant historical action of the Council in this regard, relating to the *apartheid* issue. Since the end of the cold war, the Council has assumed a new importance in this field; some of its recent involvement is described in Chapter 15 on international criminal tribunals.

LOUIS HENKIN, INTERNATIONAL LAW: POLITICS, VALUES AND FUNCTIONS

in 216 Collected Courses of the Hague Academy of International Law 13 (Vol. IV, 1989), at 251.

Compliance with international law as to civil and political rights . . . takes place within a State and depends on its legal system, on its courts and other official bodies. But, as with other international obligations, the international system can exert influence on the State to comply. . . . States observe international law from developed habit and from commitment to order generally; because States have an interest in maintaining norms which they themselves made or to which they consented and in which all have a common interest (or which represent agreed compromises and accommodations); because the system has developed a (less-than-perfect) culture of compliance; because of the availability of 'horizontal enforcement' so that a would-be violator is deterred by the anticipated response of the victim State or of others which would visit undesirable consequences upon a violator. These inducements do not work in the same way for human rights law, precisely because that law promotes human values rather than State values. Because a State's commitments under human rights law are directed towards its own people and may require major internal readjustments, the habit of compliance with international standards may not have developed; indeed, ingrained attitudes and habits inconsistent with those standards may have to be broken, and established practices abandoned. Even after 40 years under the Universal Declaration, even years after States were persuaded to make legal commitments to human rights, many States have not yet 'internalized' their international human rights undertakings, have not developed strong commitment to constitutionalism, rights and the rule of law, have not established institutions to nurture that commitment.

The general culture of compliance with international law also is less effective for human rights law. The international human rights system is still 'settling in'. States have not yet wholly shed the idea that conditions inside a State, including how a State treats its own inhabitants, are no one else's business. . . . States have not yet wholly assimilated the fact that they have an *international obligation* to respect the rights of their citizens, that an act of torture or other inhuman treatment, for example, is a violation of an international law.

Compliance with international human rights obligations—i.e., respect for human rights at home—is more responsive to domestic forces, to the domestic constitutional culture, than to any international culture pressing for compliance with international human rights norms. The causes of human rights violations are cultural, political, internal, close to home—an underdeveloped commitment to constitutionalism, to the rule of law, to the idea of individual rights, to limitations on government; political-social-economic underdevelopment and instability; evil or stupid State leaders, fostering a culture that tolerates brutality and repression; inefficient administration. In such circumstances, external inducements to comply with international human rights law are remote and not readily felt.

Horizontal enforcement, the principal inducement for international compliance generally, also works differently and less effectively for human rights. Like other international obligations, human rights undertakings run to other States, to the other parties to a covenant or convention, or to all States when the obligation is under customary law and *erga omnes*. . . . In principle, human rights obligations, like other international obligations, create rights in the promisees and afford them remedies. But while State promisees are entitled to pursue such remedies, they have not been sufficiently motivated to do so and do not in fact do so. For the real beneficiaries are not the State promisees but the inhabitants of the promisor State, and, in general, States—even if they have adhered to international agreements—do not have a strong interest in human rights generally, and are not yet politically acclimated and habituated to responding to violations of rights of persons abroad other than their own nationals. Many States, themselves still lacking an entrenched human rights culture, themselves vulnerable to charges of violation, are reluctant to respond (surely to respond unilaterally) to a violation by another friendly State of the human rights of the State's own inhabitants. What is more, the principal element of horizontal deterrence is missing. A State promisee cannot respond to a violation by retaliation or the threat of retaliation; such retaliation would itself violate human rights. And the threat that 'if you violate the human rights of your inhabitants, we will violate the human rights of our inhabitants' hardly serves as a deterrent. The result is that the temptation to violate human rights law is stronger than for other international law while fear of reaction by other States or of other adverse consequences is weaker.

In all, then, there has been a basic disappointment of original expectations about the enforcement of international human rights law.

. . .

Some of the weaknesses in the international enforcement of human rights law were clear from the beginning of the Human Rights Movement; others have become clearer during the intervening decades. In any event, for these (and other) reasons, the International Human Rights Movement has developed special 'enforcement machinery'.

Special enforcement machinery has followed two principal tracks. Some has been established by particular human rights agreements, such as the Human Rights Committee under the Covenant on Civil and Political Rights and the commissions and courts under the European and American conventions. A second track of enforcement consists of United Nations bodies—the General Assembly, the Economic and Social Council (ECOSOC), and especially the Human Rights Commission and its subsidiary units. Their activities are sometimes seen as politics, not law, but these bodies invoke norms and are properly seen as part of the enforcement system.

It is difficult to assess which of these 'tracks' has been more successful; surely they have both contributed to compliance. But they work differently. A monitoring body created by a human rights covenant or convention addresses only compliance by States parties to that agreement and only with the norms established by the agreement. The mandate, authority and procedures of the monitoring body are defined by the agreement. United Nations bodies, on the other hand, often address human rights issues as part of their general mandate as defined by the United Nations Charter and by General Assembly resolutions. They are not themselves monitoring bodies, but have sometimes created *ad hoc* monitors and have sometimes condemned violations. In principle, they might address human rights violations by virtually any State, since nearly all States are parties to the United Nations Charter; in fact, political bodies are likely to address only selected, dramatic human rights violations by selected countries.

. . .

Surely, horizontal enforcement is available to enforce the customary law of human rights. Obligations of customary law in respect of human rights are *erga omnes* and all States can act (peacefully) to induce compliance. They can protest, make claims, and even bring suit if the parties had consented to the compulsory jurisdiction of the International Court of Justice or to some relevant system of arbitration. [See the discussion of obligations *erga omnes* at p. 132, *supra*.]

. . .

International human rights law benefits significantly from enforcement also by political bodies [that are within the UN system, as well as by bodies created by the human rights treaties]. In general, international political bodies have attended only to the enforcement of norms of extraordinary political significance such as the law of the Charter on the use of force, but political bodies have devoted extraordinary efforts to promoting law on human rights and for that and other reasons they have not avoided the demands of enforcement of—inducing compliance with—that law.[3]

If law is politics, enforcement of law in the inter-State system is also heavily political. Political influence brought to bear in the organs and suborgans of the United Nations determined the enforcement machinery that found its way into covenants and conventions. (Political forces, I have suggested, have influenced also how that machinery has worked.) But United Nations bodies themselves have also been an arena for charges of human rights violations, sometimes evoking resolutions of condemnation.

One cannot appraise these activities with precision or with confidence, but clearly they have served as some inducement to terminate or mitigate violations, perhaps even as some deterrent. Political bodies, however, are subject to their own political laws. The larger bodies—notably the United Nations General Assembly—are more visible, more newsworthy, therefore more 'politicized', therefore less likely to apply human rights norms judicially, impartially. In such bodies, human rights are more susceptible to being subordinated to non-human rights considerations. There, voting, including 'bloc-voting', has led to 'selective targeting' of some States, sometimes exaggerating their violations, and overlooking those of other States, including some that are guilty of gross violations. Smaller political bodies, such as the Human Rights Commission, are also inhabited by government representatives concerned for State values and friendly relations, but increasingly they are able to be somewhat less 'political', more evenhanded, as well as more activist in the cause of human rights.

In monitoring human rights as in other matters, the influence of the United Nations bodies has reflected the transformation of the United Nations by the influx of new Members and the dominance of the Organization by the Third World. In general, Third World States have been committed to non-alignment and were therefore reluctant to support an active United Nations role in human rights monitoring, which, in the past at least, Communist States (and others particularly committed to State values) resisted. . . .

. . . On the other hand, African States provided a particular impetus to human rights activism when they led the United Nations to take a strong stance against apartheid, make it a perennial issue, adopt perennial resolutions of condemnation and exhortations to sanctions. On two occasions, in 1963 and in 1970, they induced the Security Council to call on States not to provide South Africa with weapons and in 1977 the Council imposed a

[3] [Eds.] See generally Sydney Bailey, *The UN Security Council and Human Rights* (1994).

mandatory arms embargo. It was the desire to attack apartheid by every means available that led to the development of 'working group' procedures under ECOSOC, originally designed for complaints of apartheid, then extended to other 'consistent patterns of gross violations of human rights and fundamental freedoms'. Votes on apartheid in United Nations bodies, adherence to the Convention on Racial Discrimination, and later to the Convention on the Suppression and Punishment of the Crime of *Apartheid*, became elements in friendly relations with African States. The influence of African and other Third World States has rendered racial discrimination the most serious international crime, and helps explain why the Convention on Racial Discrimination has the largest number of State adherences, with machinery, we saw, reaching farther than that provided in the Covenant on Civil and Political Rights.

PHILIP ALSTON, APPRAISING THE UNITED NATIONS HUMAN RIGHTS REGIME

in Alston (ed.), The United Nations and Human Rights: A Critical Appraisal 1 (1992), at 4.

. . . [T]he essential role of each of the treaty bodies is to monitor and encourage compliance with a specific treaty regime, while the political organs have a much broader mandate to promote awareness, to foster respect, and to respond to violations of human rights standards. Each treaty-based organ has been established either pursuant to the terms of a specific treaty or for the specific purpose of monitoring compliance with such a treaty. The Charter-based organs on the other hand derive their legitimacy and their mandate, in the broadest sense, from the human rights-related provisions of the Charter.

Within that overall framework the treaty-based organs are distinguished by: a limited clientele, consisting only of States Parties to the treaty in question; a clearly delineated set of concerns reflecting the terms of the treaty; a particular concern with developing the normative understanding of the relevant rights; a limited range of procedural options for dealing with matters of concern; caution in terms of setting precedents; consensus-based decision-making to the greatest extent possible; and a non-adversarial relationship with States Parties based on the concept of a 'constructive dialogue'.

By contrast, the political organs generally: focus on a diverse range of issues; insist that every state is an actual or potential client (or respondent), regardless of its specific treaty obligations; work on the basis of a constantly expanding mandate, which should be capable of responding to crises as they emerge; engage, as a last resort, in adversarial actions *vis-à-vis* states; rely more heavily upon NGO inputs and public opinion generally to ensure the effectiveness of their work; take decisions by often strongly contested majority voting; pay comparatively little attention to normative issues; and are very wary about establishing specific procedural frameworks within which to work, preferring a more *ad hoc* approach in most situations.

It is, of course, easy to overstate the differences between the two types of organ and to underestimate the ability of one type to emulate certain characteristics of the other. Thus a Charter-based organ might choose to play down its political character and devote some of its efforts to a systematic clarification of the normative content of a specific right while a treaty-based organ might play down its constructive dialogue approach in order to indicate its strong disapproval of a state's behaviour. Nevertheless, the differences of mandate, content, and style between the two types of organs are sufficiently clear and consistent as to justify using this as the principal distinction for purposes of the present analysis.

3. *EVOLUTION OF THE CHARTER-BASED ORGANS*

. . .

The remaining principal organs form part of a definite hierarchy with the Security Council at the apex. The problem, however, is that the Council has a long history of refusing to consider itself as an organ for the promotion of respect for human rights, except in so far as a given situation constitutes a threat to international peace and security. Although there are many examples of the Council taking up human rights issues . . . its reluctance to alter its stance still persists . . . . [The] Council's position thus remains (artificially) detached from that of the remaining principal organs.

[In addition] the formal hierarchy consists of the General Assembly and ECOSOC as principal organs and, underneath them as 'functional commissions', the Commission on Human Rights and the Commission on the Status of Women. This is as far as the system goes in terms of specific Charter authorization. However, each of these organs is entitled to create whatever subsidiary mechanisms it considers necessary to enable it to carry out its own responsibilities and functions under the Charter. It is at this point that the proliferation begins. . . . The only such body to which specific attention need be drawn in this context is the Sub-Commission on Prevention of Discrimination and Protection of Minorities. [In contrast to the Commission, which composed entirely of governmental representatives, the Sub-Commission consists of 26 independent experts, elected by the Commission upon the nomination of governments. It meets for four weeks annually in Geneva in August, although its session is generally preceded by various Working Groups dealing with the rights of indigenous populations, contemporary forms of slavery, and communications. The degree of independence of its members varies radically.] It has long stood out because of its relative independence, its flexible agenda and working methods, its preparedness to act as a pressure group *vis-à-vis* its parent body (the Commission), and its ambiguous and often antagonistic relationship with that parent.

While in principle, the lines of authority are reasonably clear-cut and obvious, in practice they are much less so. The ECOSOC, which once played a major role as an intermediary between the Assembly and the Commission, is now little more than a rubber stamp. The Assembly, on the other hand, has come to play an important initiating role in a number of areas while at the same time deferring (in substance, although not in form) to the Commission in many respects. Similarly, any analysis that portrayed the Sub-Commission as little more than a subsidiary organ to advise the Commission and do its bidding (and no more) might be an accurate reflection of the initial design as seen by some of its creators, but would otherwise be singularly out of touch with today's reality. The result of these various evolutionary trends is a much more complex interrelationship among the group of bodies than any diagram or flow-chart could ever convey.

NOTE

The most important UN body in the human rights field is the Commission on Human Rights. This UN Commission, established in 1946, currently consists of 53 member governments elected for three-year terms by the ECOSOC. To a significant extent its working methods reflect the regional composition of the Commission (Asia, Africa, Eastern Europe, Latin America, and Western Europe and Others—the last category including Canada, Australia and New Zealand, and, in practice, the U.S.). Working groups consisting of five member governments are commonly established to include one member from each group. The position of Chairperson rotates annually among the groups. The regional groups also caucus regularly during the Commission's annual six week session in Geneva starting in April, as from 1996. Since 1992 there has also been provision for emergency sessions of the Commission, three of which have been held as of June 1995 (two relating to the former Yugoslavia and one to Rwanda).

The Commission reports to the Economic and Social Council (ECOSOC). Any resolution or decision with financial consequences requires the latter's approval, unless an earlier authorization can be invoked. It is extremely rare for ECOSOC to refuse to endorse anything the Commission has decided.

Because of the extent to which the Commission's role has changed over the years, a sense of its historical evolution is critical to an understanding of how it operates today. The following excerpts describe that evolution.

PHILIP ALSTON, THE COMMISSION ON HUMAN RIGHTS

in Alston (ed.), The United Nations and Human Rights: A Critical Appraisal 126 (1992).

. . . The key issue with respect to the Commission's terms of reference was whether its mandate was to be *limited* and thus restricted to specified objectives or *general* and therefore open-ended.

Partly due to the determined lobbying efforts of non-governmental organizations, the San Francisco Conference in 1945 included a series of human rights provisions in the UN Charter. One concern of a number of delegations at the Conference was to secure the adoption of a declaration of rights.

. . . These efforts were unsuccessful but there emerged from the Conference a clear expectation, as noted by President Truman in his closing speech, that the major task of the Commission on Human Rights, the establishment of which was provided for in the Charter, would be the drafting of an international Bill of Rights. . . .

Subsequently, the Preparatory Commission which met in London in late 1945 to make provisional arrangements for the different UN organs, suggested that the work of the Commission should be directed towards:

(*a*) formulation of an international bill of rights;
(*b*) formulation of recommendations for an international declaration or convention on such matters as civil liberties, status of women, freedom of information;
(*c*) protection of minorities;
(*d*) prevention of discrimination on grounds of race, sex, language, or religion; and
(*e*) any matters within the field of human rights considered likely to impair the general welfare or friendly relations among nations.

Through the inclusion of (*e*) in the proposed terms of reference the Preparatory Commission clearly envisaged a relatively open-ended, and thus politically active, mandate for the Commission. . . . [But] the paragraph was deleted by the Council at its first session, thus providing the preparatory session of the Commission (the so-called 'nuclear' session) with only a limited mandate.

However, when the Nuclear Commission of nine members met at Hunter College, New York, in April–May 1946, with Eleanor Roosevelt as Chairman, it expressed a definite preference for an expansive approach. In its report to ECOSOC it advocated that the Commission be composed of independent experts and that it be given an open-ended mandate including a general brief not only to assist ECOSOC and the Assembly in their work but also to 'aid the Security Council in the task entrusted to it by Article 39 of the Charter, by pointing to cases where a violation of human rights committed in one country may, by its gravity, its frequency, or its systematic nature, constitute a threat to the peace'. . . . But this conception proved to be anathema to the governments represented in ECOSOC which, at it second session, in June 1946, promptly scrapped the proposals for an independent membership and for a role in dealing with violations constituting a threat to peace. However, perhaps by way of compensation, it agreed not to limit the mandate to the specific issues already listed, but to add a new version of paragraph (*e*), giving the Commission a role with respect to 'any other matter concerning human rights not covered by items (*a*) (*b*) (*c*) and (*d*). Thus, after a lengthy battle, the Commission was endowed with a *general* mandate. In fact, the new provision was even more open-ended and susceptible to a broad and flexible interpretation than was the original proposal.

. . .

THE COMMISSION'S INTERPRETATION OF ITS MANDATE

For the first twenty years of its existence the Commission struggled valiantly and successfully to avoid becoming an overtly political organ. Despite the significant human rights dimensions of the Cold War, the decolonization debate and many other matters being brought before the Assembly and the Security Council, the Commission managed to confine its efforts to standard-setting with a variety of other technical pursuits thrown in for good measure. In 1947 the Commission adopted its oft-criticized statement that it had 'no power to take any action in regard to any complaints concerning human rights', thereby effectively removing itself from the front line of the vast majority of human rights-related political battles. . . .

THE STANDARD-SETTING ROLE OF THE COMMISSION

The drafting of international instruments in the field of human rights has been a continuous occupation of the Commission and its achievements in this regard are by no means confined to the first decade of its existence. Indeed, until the early 1970s standard-setting continued to be by far its most productive and enduring activity. Even in the early 1990s, it is apparent that the development of new standards in a range of specialized fields will continue to occupy much of the Commission's agenda for the foreseeable future. . . .

In the space of only two sessions, in 1947–8, the Commission completed a comprehensive draft of the Universal Declaration, which was proclaimed by the Assembly less than six months later. By 1954, after a remarkably sustained and intensive drafting effort, the Commission had completed its drafts of the two Covenants and forwarded them on to the Assembly. . . .

Starting in 1977 the Commission's standard-setting role entered a new, and ultimately very productive, phase. In that year the Assembly, which had chosen to ignore it in drafting the Declaration against Torture, chose the Commission as the preferred forum in which to draft a Convention on the same subject. Also, in 1978, the Polish government selected the Commission as the place to launch an initiative designed to produce a convention on childrens' rights. These initiatives resulted in the adoption of Conventions of major importance in 1984 and 1989 respectively. In each case virtually all of the drafting work was done by the Commission. Another major achievement during this period was the 1986 Declaration on the Right to Development.

. . .

RESPONDING TO VIOLATIONS

All too frequently the sole criterion which is used for judging the performance and achievements of the Commission is the extent to which it has succeeded in responding to specific violations of human rights and providing a degree of protection to actual and potential victims. While there are strong grounds for insisting that the Commission's performance be judged on the basis of a much broader range of considerations, the reality is that its effectiveness in responding to violations remains the benchmark which the great majority of governments and other observers continue to use. . . .

(a) The Historical Evolution: An Overview

. . . [T]here have been three very distinct phases: (1) 1946–66, during which time the Commission was not prepared to address the issue of specific violations at all; (2) 1967–78, when the Commission struggled to evolve procedures which were initially designed to respond only to problems associated with racism and colonialism; and (3) 1979 to the present, when the procedures developed earlier have been applied in an increasingly creative and tailored fashion to an ever-widening range of countries and types of violations. Thus any meaningful evaluation must at least recognize these distinct phases and acknowledge that it is only in the present (third) phase that the member States of the United Nations have made any serious effort to respond to violations in a manner that at all purports to be objective and even-handed. . . .

(i) 1946–1966: Abdication of Responsibility For twenty years, until 1967, the official and oft-repeated position of the Commission was that it had 'no power to take any action in regard to any complaints concerning human rights'. . . .

. . . [The] Commission was simply informed, on a confidential basis and in the most telegraphic fashion, of the complaints that were being received, without any information as to the identity of the authors involved. No action of any description was taken as a result and the sole justification for the ensuing charade was that Commission members could gain a better idea of the sort of concrete problems existing in the world at any given time.

. . .

(ii) 1967–1978: A Gradual Assumption of Responsibility [By the mid-1960s there had been a dramatic change in the composition of the major UN organs as a reflection of the influx of new members, mainly newly independent

African and Asian states. Membership of the Commission went from 18 in 1960 to 32 in 1967 (20 of which were from the Third World.]

. . .

Another important development was the adoption of the Convention on the Elimination of All Forms of Racial Discrimination (CERD) in 1965 which included provision, in Article 14, for the submission of complaints by individuals or groups against States which accepted the procedure. This development was intimately linked with the emergence of the new Third World majority and the preparedness, indeed determination, of that majority to make racism a major concern and to develop whatever procedures might be necessary in order to combat it. In any such endeavours they were virtually assured the support of the Eastern Europeans who were happy to encourage attention to what were assumed to be quintessentially Western sins. By the mid-1960s they also had a significant (but by no means unqualified) degree of support from an American government which had committed itself to the implementation of civil rights domestically and had nominated black Americans to both ECOSOC (Clyde Ferguson) and the Sub-Commission (Beverly Carter). The willingness of the Third World majority to develop new procedures applied even though there was a significant 'risk' that the scope or reach of the procedure might eventually be extended to address problems other than racism.

The adoption of the CERD complaints procedure cleared the way for an Optional Protocol to the Covenant on Civil and Political Rights to be adopted the following year. It too enabled complaints to be brought against States Parties. These two procedures, even though they were not to bear much fruit for another couple of decades and would in any event apply only to the small number of ratifying states, constituted very important breakthroughs in terms of the principles involved, particularly that of accountability.

The most significant development, however, was a decision by the Third World countries, strongly supported by the Eastern Europeans, that a general, non-treaty-based communications-type procedure would be useful as an additional means by which to pursue the struggle against racist and colonialist policies, particularly in southern Africa. . . . It was on the initiative of the Special Committee on Decolonization, in 1965, that the attention of the Commission was drawn to petitions alleging human rights violations in southern Africa. The following year ECOSOC accordingly invited the Commission to consider urgently the question of the violation of human rights, including policies of racial discrimination and segregation and of apartheid in all countries, with particular reference to colonial and other dependent countries and territories, and to submit its recommendations on measures to halt those violations. In the Commission the view prevailed that an artificial restriction of such a mandate to only a single category of countries was not justified. As a result, the ambiguity of the Commission's original mandate was resolved by a General Assembly request that it give urgent consideration to ways and means of improving the capacity of the United Nations to put a stop to violations of human rights *wherever they may occur*. This latter phrase was to prove to be of vital importance, despite the clear desire of most of the resolution's key proponents to confine the focus to racist and colonialist situations.

These developments resulted in the adoption of what eventually turned out to be two separate procedures. The procedure established under ECOSOC Resolution 1235 (XLII) established the principle that violations could be examined and responded to, and provided the necessary authorization for the Commission to engage in public debate on the issue each year. The procedure under ECOSOC Resolution 1503 (XLVIII) provided a carefully and deliberately constrained procedure by which situations which appear to reveal 'a consistent pattern of gross and reliably attested violations of human rights' could be pursued with the governments concerned, but in private. [Both the 1235 and 1503 procedures are examined below in this chapter.]

. . . [B]oth procedures generated considerable controversy and engendered high hopes while actually accomplishing relatively little during their first decade in operation. The inevitable teething problems involved in taking procedures which had been established on the basis of a hotchpotch of competing aspirations and transforming them into flexible and effective tools for inquiry and assessment, backed by at least a minimum of political will, were to take up almost all of the 1970s.

(iii) 1979 to the Present: Evolution Towards an Effective Response It was not until more than a decade had elapsed after the adoption of the 1235 procedure that a convergence of factors facilitated the next quantum leap in United Nations practice in response to violations. This took the form of: a more active and (albeit minimally) more public approach to the 1503 confidential procedures; increasing acceptance that confidential consideration did not preclude public consideration with respect to any given State; and a substantially expanded range of fact-finding activities. Unlike 1967 when the impetus had come primarily from a change in the composition of the Commission and ECOSOC, the developments in the late 1970s were environmental. The Commission had excelled itself in 1976–7 by a failure to act publicly with regard to horrendous violations in Pol Pot's Democratic Kampuchea, Amin's Uganda, Bokassa's Central African Empire, Macias' Equatorial Guinea, the military's Argentina and Uruguay, and several other situations. Thus a 1976 investigation by the (London) *Sunday Times* concluded that the Commission worked 'in almost total secrecy' and had deliberately constructed 'a bureaucratic and procedural maze' as a result of which 'delay has been institutionalized and the aim has not been to protect the victims but the oppressors'.

But the developing consciousness of public opinion (almost by definition among the élites in the West and a limited number of Third World countries), combined with the higher profile given to human rights issues by the Carter Administration and several of its allies contributed to a climate in which task expansion was almost an imperative. That the West was prepared to take the initiative, despite the fact that many of its 'allies' were among the worst

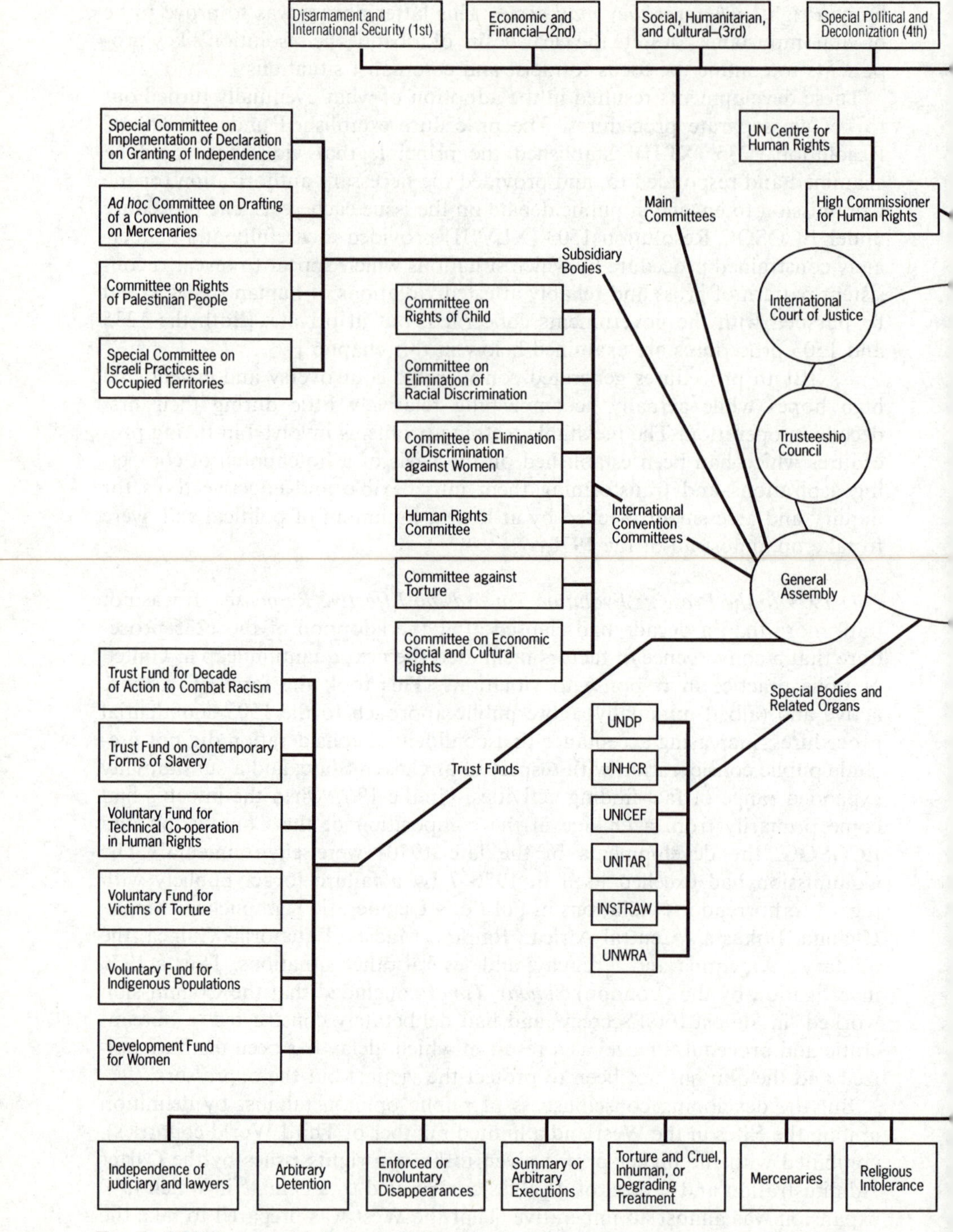

FIG. 1. UN Organs with Responsibilities in the Human Rights Area [as of July 1995]

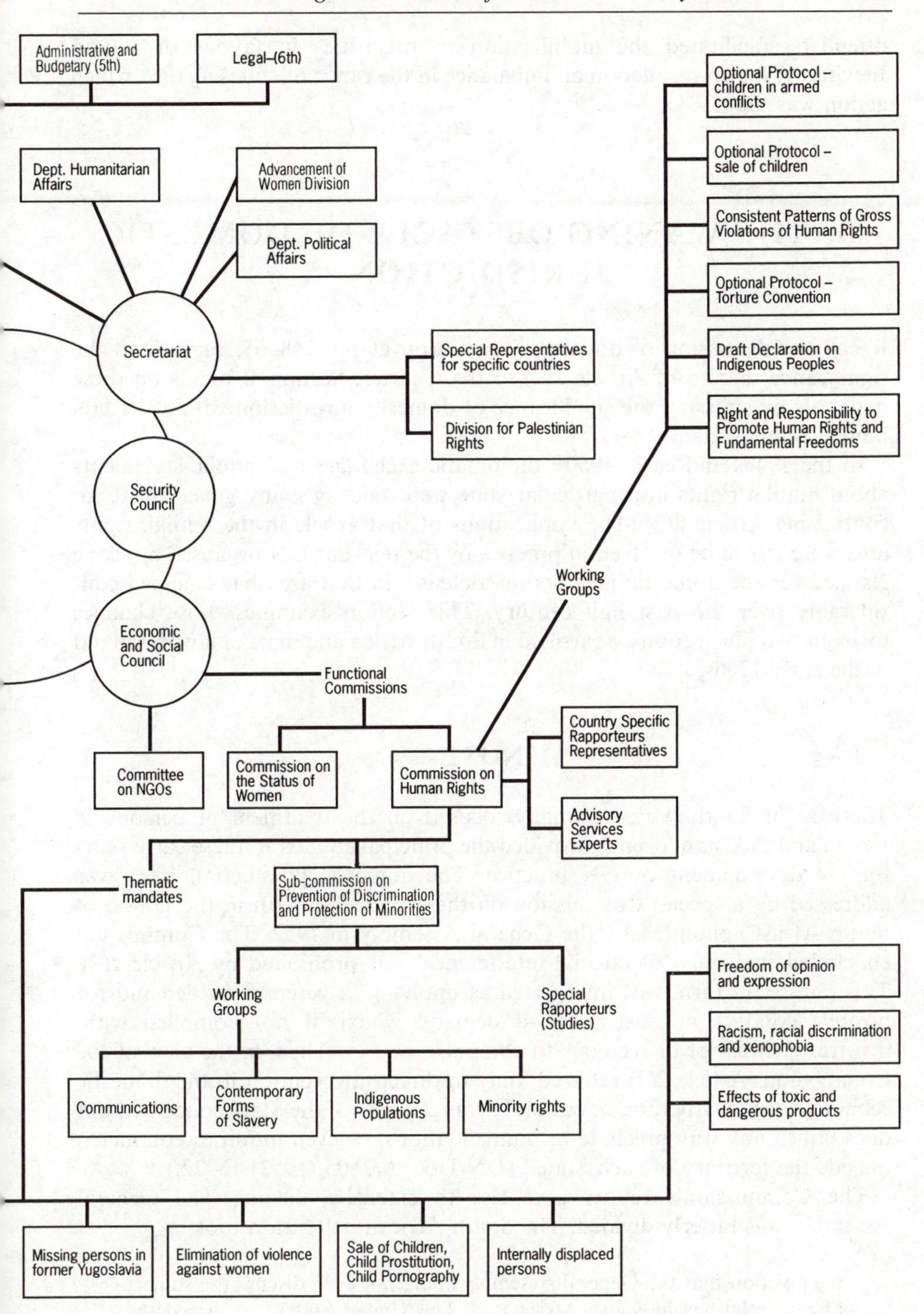
Administrative and Budgetary (5th)
Legal–(6th)
Dept. Humanitarian Affairs
Advancement of Women Division
Dept. Political Affairs
Secretariat
Special Representatives for specific countries
Division for Palestinian Rights
Security Council
Economic and Social Council
Functional Commissions
Committee on NGOs
Commission on the Status of Women
Commission on Human Rights
Country Specific Rapporteurs Representatives
Advisory Services Experts
Working Groups
Optional Protocol – children in armed conflicts
Optional Protocol – sale of children
Consistent Patterns of Gross Violations of Human Rights
Optional Protocol – Torture Convention
Draft Declaration on Indigenous Peoples
Right and Responsibility to Promote Human Rights and Fundamental Freedoms
Thematic mandates
Sub-commission on Prevention of Discrimination and Protection of Minorities
Working Groups
Special Rapporteurs (Studies)
Communications
Contemporary Forms of Slavery
Indigenous Populations
Minority rights
Freedom of opinion and expression
Racism, racial discrimination and xenophobia
Effects of toxic and dangerous products
Missing persons in former Yugoslavia
Elimination of violence against women
Sale of Children, Child Prostitution, Child Pornography
Internally displaced persons

offenders, facilitated the mobilization of majorities in favour of action. Inevitably, it also resulted in an imbalance in the range of states against which action was taken. . . .

B. THE WANING OBSTACLE OF DOMESTIC JURISDICTION

Recall the discussion of domestic jurisdiction at pp. 148–65, *supra*, and the significance thereto of Article 2(7) of the Charter. Section B builds on these materials by stressing the significance of domestic jurisdiction within the UN human rights system.

In the 1940s and early 1950s, diplomatic exchanges and public statements about human rights in a particular state were said by many governments to contravene Article 2(7). The implications of that article in the human rights area were left to be resolved in practice by the relevant UN organs. The scope claimed for the domestic jurisdiction 'defense' in that area has changed considerably over the past half century. This section examines those changes through two illustrations: apartheid in South Africa and martial law in Poland in the early 1980s.

NOTE

The case of South Africa (initially focused on the treatment of persons of Indian and Pakistani origin) provided the principal context in these early years for the development of UN practice. The domestic jurisdiction issue was addressed by a special Commission on the Racial Situation in the Union of South Africa appointed by the General Assembly in 1952. The Commission concluded that only 'dictatorial interference' was prohibited by Article 2(7). This phrase, in turn, was interpreted as implying 'a peremptory demand for positive conduct or abstention—a demand which, if not complied with, involves a threat of or recourse to compulsion . . . .' Thus, in the view of the Commission, Article 2(7) referred 'only to direct intervention in the domestic economy, social structure, or cultural arrangements of the State concerned but does not in any way preclude recommendations, or even inquiries conducted outside the territory of such State.' (UN Doc. A/2505 (1953) 16–22.)

The Commission's report gave rise to extensive debate. The General Assembly was bitterly divided. The South African delegation took

> the position that the General Assembly could not even discuss the subject of race relations in South Africa. . . . The United States . . . urged the Assembly to confine itself to action that might have some practical results.

> . . . [It] declared it had observed with increasing concern the tendency of the Assembly to place on its agenda subjects the international character of which was doubtful. . . . [It] stated its opposition to addressing a recommendation to only one Member and to expressing regret over domestic legislation, and it abstained in the vote [on the proposed resolution under discussion].[4]

Apartheid in South Africa (and to a lesser extent the situation in Rhodesia) continued to provide the primary context within which the issue of domestic jurisdiction was debated within UN organs. The following excerpts from an article of Louis Sohn pick up the narrative in the early 1960s and describe some high points in the slow evolution of ideas about domestic jurisdiction. This reading also serves to illustrate the vital, decades-long process of elaboration or interpretation of the Charter by UN organs through which effectively new understandings about the UN's role and powers were reached. Moreover, it bears directly on this chapter's discussion of modes of enforcement of human rights norms, by emphasizing the role of the General Assembly in giving publicity to a situation and stimulating action responsive to it, and the (rare) role of the Security Council in ordering economic and other sanctions against a delinquent state.

Before reading these excerpts, you should become familiar with the provisions in Chapters VI and VII of the UN Charter, particularly Articles 36–41 and 48.

LOUIS SOHN, INTERPRETING THE LAW

in Oscar Schachter and Christopher Joyner (eds.), United Nations Legal Order 169 (1995), at 211.

. . .

A 1962 General Assembly resolution finally crossed the borderline between exhortatory recommendations and enforcement. It pointed out that 'the actions of some Member States indirectly provide encouragement to the Government of South Africa to perpetuate its policy of racial segregation, which had been rejected by the majority of that country's population' and requested Member States to take the following measures, separately or collectively, in conformity with the Charter, to bring about the abandonment of these policies:

(a) breaking off diplomatic relations with the Government of the Republic of South Africa or refraining from establishing such relations;

(b) closing their ports to all vessels flying the South African flag;

[4] James Green, The United Nations and Human Rights 785 (1956).

(c) enacting legislation prohibiting their ships from entering South African ports;

(d) boycotting all South African goods and refraining from exporting goods, including all arms and ammunition, to South Africa;

(e) refusing landing and passage facilities to all aircraft belonging to the Government of South Africa and companies registered under the laws of South Africa.

In addition, the Assembly decided to establish a Special Committee to 'keep the racial policies of the Government of South Africa under review,' and to 'report either to the General Assembly or to the Security Council, or to both, as may be appropriate.'

Another line had thus been crossed. The General Assembly recommended that Member States take the specified measures against Sough Africa, and encouraged them to take them immediately, separately or collectively, without waiting for a decision of the Security Council.

The Assembly recognized the Council's primary jurisdiction to enact binding sanctions, and requested it to enact them in order to ensure compliance with both organs' resolutions. A year later, Security Council Resolution 181 expressed its conviction that 'the situation in South Africa is seriously disturbing international peace and security'; . . . and called upon all states 'to cease forthwith the sale and shipment of arms, ammunition of all types and military vehicles to South Africa.'

. . .

The 1963 developments clarified the interpretation of Article 2(7) of the Charter by the Assembly and the Council. The Security Council joined the General Assembly in holding implicitly that human rights provisions of the Charter prevail over the prohibition in Article 2(7) of the Charter. . . .

From 1965 to 1979, the cases of South Africa and Southern Rhodesia became intertwined. Over this period the Security Council and the General Assembly applied to Rhodesia a variety of measures that were also subsequently applied to South Africa. . . .

An important next step was taken by the General Assembly in 1966, when it condemned the policies of apartheid of the South African Government as 'a crime against humanity' and suggested to the Security Council that 'universally applied mandatory economic sanctions are the only means of achieving a peaceful solution.'

. . .

In 1970, the Security Council . . . adopted a resolution which recognized . . . that 'the situation resulting from the continued application of the policies of apartheid and the constant build-up of the South African military and police forces . . . constitutes a potential threat to international peace.' It then called upon all states to strengthen the arms embargo:

> (a) by implementing fully the arms embargo against South Africa unconditionally and without reservations whatsoever;
>
> . . .
>
> (e) by prohibiting investment in, or technical assistance for, the manufacture of arms and ammunition, aircraft, naval craft, or other military vehicles;
>
> (f) by ceasing provision of military training for members of the South African armed forces and all other forms of military co-operation with South Africa;
>
> . . .

. . . Nevertheless, the Special Committee on the Policies of Apartheid immediately complained that this resolution was not mandatory, and that the arms embargo was being violated. . . .

. . .

In 1973, another step was taken to make condemnation of apartheid part of international law, when the General Assembly approved the draft Convention on the Suppression and Punishment of the Crime of Apartheid, prepared by the Commission on Human Rights, and opened it for signature and ratification. . . .

. . .

In its 1976 resolution on South Africa, the Assembly recognized 'that the consistent defiance by the racist régime of South Africa of United Nations resolutions on apartheid and the continued brutal repression, including indiscriminate mass killings, by that régime leave no alternative to the oppressed people of South Africa but to resort to armed struggle to achieve their legitimate rights'; and declared that the situation in South Africa constituted 'a grave threat to the peace, requiring action under Chapter VII of the Charter of the United Nations'; and appealed to all states and organizations 'to provide all assistance required by the oppressed people of South Africa and their national liberation movements during their legitimate struggle.' By this resolution, the Assembly not only encouraged a civil war, but also authorized foreign assistance to the 'freedom fighters.'

In 1977 another barrier to sanctions against South Africa fell. . . . Security Council Resolution 418 in 1977 concerned the mandatory arms embargo. The Council recognized that 'the military build-up by South Africa and its persistent acts of aggression against the neighbouring States seriously disturb the security of those States.' The Council then decided that, in view of its conviction that 'a mandatory arms embargo needs to be universally applied against South Africa in the first instance,' it is obliged to act 'under Chapter VII of the Charter of the United Nations.' It determined that, 'having regard to the policies and acts of the South African Government, the acquisition by South Africa of arms and related *matérial* constitutes a threat to the maintenance of international peace and security,' and decided that 'all states shall cease forthwith any provision to South Africa of arms and related *matérial* of all types, including the sale or transfer of weapons and ammunition, military vehicles and equipment, paramilitary police equipment, and spare parts of the aforementioned.' Toward this end, the Council called 'upon all States, including States non-members of the United Nations, to act strictly in accordance with the provisions of the present resolution.' In reaction to this resolution, the Secretary-General issued a statement in which he observed that 'this was the first time in the history of the United Nations that mandatory sanctions have been imposed against a Member State.'

. . .

In 1984, the Security Council declared that the so-called 'new constitution' adopted by 'the exclusively white electorate' was contrary to the principles of the Charter of the United Nations and was therefore 'null and void.' It also rejected the so-called 'negotiated settlement' based on the Bantustan structure.

In 1989, the General Assembly adopted a record twelve resolutions and an ambitious Declaration on Apartheid and Its Destructive Consequences in Southern Africa. . . .

The Declaration contains a set of principles for a new constitutional order of South Africa, a set of requirements for creating an appropriate climate for negotiations, guidelines for the process of negotiations, and a program of action. According to the Declaration, the new constitutional order should be based on the Charter of the United Nations and the Universal Declaration of Human Rights. . . .

This declaration represented a turning point in the relations between the United Nations and South Africa. Its main emphasis was on substituting meaningful negotiations and achievable goals for hostile rhetoric. The door was opened for the Secretary-General to assist in achieving the new goals, rather than directing the strangling of the South African economy. For the first time, the United Nations presented to the South African Government a detailed plan on how the transition to new South Africa can be accomplished, in which all, including the white minority, will be able to live normal lives in peace.

. . .

After more than forty years of confrontation between South Africa and the United Nations over the right of the United Nations 'to intervene in matters which are essentially within the domestic jurisdiction' of South Africa, the view finally prevailed that the issue of establishing in South Africa a non-racial and democratic constitution that would guarantee equal rights to all citizens was no longer a matter essentially of domestic jurisdiction. . . .

. . .

. . . While a 'threat to the peace' was first narrowly interpreted, it was considerably broadened over the years, far beyond the early tentative approach to include 'potential threats.' Apartheid in South Africa became transformed through interpretations of United Nations law from a social evil, to a repugnant practice, to a crime under international law, to a threat to the peace that must not be tolerated by the international community and which warranted the imposition of mandatory economic sanctions against the deviant government.

NOTE

As the General Assembly increasingly exercised its Charter-based authority to 'discuss any questions or any matters within the scope of the . . . Charter' (Art. 10) and to 'initiate studies and make recommendations for the purpose of . . . [*inter alia*] assisting in the realization of human rights' (Art. 13), governments turned their objections to more intrusive UN-related activities such as requests by UN organs to states to admit human rights monitors of one type or another, concerted campaigns to mobilize world public opinion, and the imposition of unilateral or multilateral sanctions.

Today, the Article 2(7) defence is invoked primarily in response to threats of forceful intervention. In recent years a coalition of Third World states, as noted below, has sought to resist this development.

In *Inter-State Accountability for Violations of Human Rights* (1992), Menno Kamminga traces the evolution of the domestic jurisdiction defense in relation to cases such as those of Vietnam in the early 1960s, Israel since the late 1960s, Chile in the 1970s and countries such as Iran and Afghanistan in the 1980s. A wide range of governments have proved to be unable to maintain a strong defensive interpretation of Article 2(7) that is consistent. Although they might wish to invoke the principle in their own defense or the defense of one of their allies, they have been prepared to insist on measures against certain states, thereby creating precedents that have been difficult to resist or distinguish in subsequent cases.

The only significant exception in this regard has been the People's Republic of China, which has abstained from condemning human rights violations in other states of which it clearly did not approve. But even China supported resolutions condemning human rights violations in Afghanistan, southern Africa and the Israeli Occupied Territories. Recall China's contemporary views about human rights, domestic jurisdiction and intervention that appear in the excerpts from the 1991 White Paper on Human Rights in China, at p. 233, *supra*.

In the early 1980s the Communist countries of Eastern Europe consistently invoked Article 2(7) in response to attempts to undertake a study of the situation under martial law in Poland. Martial law was declared in 1981 under the Polish Constitution partly in response to challenges by Solidarity and other trade union groups to the powers of the central government. In accordance with its obligations under Article 4 of the ICCPR, the Polish Government informed the UN Secretary-General 'of the temporary derogations from or limitations [to] some of the Covenant's provisions, to the extent strictly required by the situation, along with the reasons therefor' (UN Doc. E/CN.4/1983/SR.40/Add.1, para. 42). A resolution by the UN Commission condemning the declaration of martial law provoked the following response.

STATEMENT BY THE REPRESENTATIVE OF POLAND

UN Commission on Human Rights, 1983
UN Doc. E/CN.4/1983/SR.40/Add.1.

41. *MR SOKALSKI* (Poland) . . . [T]he Government of Poland had rejected [the resolution] as unlawful, null and void, politically harmful and morally two-faced. Poland had also publicly declared that it would not participate in any way in the implementation of the resolution, which represented a miscarriage of international justice on the part of some members of the Commission. . . .

. . .

47. Martial law was obviously a painful and by no means normal state of affairs, but not nearly so abnormal as the situation that had prevailed previously. It had represented neither a military coup nor an attempt to install a military dictatorship. The Polish people had sufficient strength, wisdom and national zeal to institute and live under an efficient democratic system of socialist government. Martial law was not an end in itself, not an attack, but an instinctive defence aimed at reversing an exceptionally serious public emergency that had threatened the life of the nation. . . . Developments in Poland had clearly fallen within the sphere of domestic jurisdiction and, in accordance with Article 2, paragraph 7, of the Charter, the United Nations was precluded from discussing them, a further reason why Poland had rejected resolution 1982/26 as a serious violation of the Charter principles.

. . .

52. The proclamation of martial law in Poland had complied with all the requirements of domestic and international law, had been throughout a sovereign Polish decision, and had in no way endangered any other State. . . .

. . .

54. The Polish constitutional authorities were alone entitled to pass judgement on the necessity for and justifiability of the action taken. . . .

55. In September 1982, the Minister for Foreign Affairs of Poland had had discussions with the Secretary-General . . . The detailed memorandum which the Secretary-General had received had spelled out clearly that, along with the intensive efforts for an early lifting of the remaining restrictions, continued efforts were being made to create conditions favourable to national accord, economic reconstruction and revitalization, and ensure that the policy of reforms and social renewal was irreversible. Poland would take every opportunity for greater co-operation with the Secretary-General on all matters of mutual interest, and in his capacity as the depositary of the International Covenant on Civil and Political Rights. It would do nothing under duress or under the unacceptable terms of illegal resolutions.

56. Pursuant to General Assembly and Economic and Social Council resolutions, United Nations organs, and notably the Commission on Human Rights, were entitled to consider human rights questions concerning Member States only if the following criteria were met: firstly, that a particular situation represented a gross, massive and flagrant violation of human rights and fundamental freedoms; secondly, that the situation represented a consistent pattern of such violations; thirdly, that the situation endangered international peace and security; and lastly, that consideration of the situation was without prejudice to the functions and powers of organs already in existence . . . .

NOTE

The end of the Cold War brought a dramatic change of policy with respect to domestic jurisdiction. Consider the statement adopted in October 1991 by the Moscow Meeting of the Conference on the Human Dimension of the Conference on Security and Co-operation in Europe (the CSCE, or Helsinki process, described at p.577. *infra*):

> The participating States emphasize that issues relating to human rights, fundamental freedoms, democracy and the rule of law are of international concern, as respect for these rights and freedoms constitutes one of the foundations of the international order. They categorically and irrevocably declare that the commitments undertaken in the field of the human dimension of the CSCE are matters of direct and legitimate concern to all participating States and do not belong exclusively to the internal affairs of the State concerned.[5]

Despite this clear erosion of the domestic jurisdiction defense, many commentators consider that the concept has not become meaningless. Recall the comments of Oscar Schachter, at p. 164, *supra*. Felix Ermacora's conclusion

[5] 30 Int'l Leg. Mat. 1670, 1672 (1991).

is that only matters relating to 'a state's governmental system' are essentially within domestic jurisdiction. Nonetheless,

> [e]ven this only holds true in so far as the state can claim legitimacy in terms of the right to self-determination . . . by virtue of a representative government, and by respect for human rights. No particular governmental system is dictated by the Charter or required by general international law; no UN resolutions prescribe a specific governmental system. A government must only exhibit the criteria mentioned above.[6]

A clear indication of the continuing attempt to grapple with these issues is contained in two recent statements. The first is a 1994 General Assembly resolution sponsored by a coalition of Third World states and rejected by the West. It was adopted by 110 votes in favor, 35 against and 24 abstentions. General Assembly Resolution 49/186 (1994) reads in part:

> The General Assembly,
>
> . . .
>
> 6. *Considers it necessary* for all Member States to promote international cooperation on the basis of respect for the independence, sovereignty and territorial integrity of each State, including the right of every people to choose freely its own socio-economic and political system, with a view to solving international economic, social and humanitarian problems; . . .

The second statement is a resolution on 'the principle of non-intervention in internal affairs' adopted in 1990 by the highly respected Institute of International Law (UN Doc. E/CN.4/1990/NGO/55). It provides, *inter alia*:

> [2] . . . Without prejudice to the [UN Charter] . . . States, acting individually or collectively, are entitled to take diplomatic, economic and other measures towards any other State which has violated the obligation [to ensure the observance of human rights], provided such measures are permitted under international law and do not involve the use of armed force in violation of the Charter . . . .
>
> . . .
>
> [3] Diplomatic representations as well as purely verbal expressions of concern or disapproval regarding any violation of human rights are lawful in all circumstances.

[6] Ermacora, Article 2(7), in Bruno Simma *et al* (eds.), The Charter of the United Nations: A Commentary 139, 152 (1994).

QUESTIONS

1. What conclusions, if any, can be drawn from the manner in which the approach of states and of the principal UN organs to the issue of domestic jurisdiction has evolved since 1945?

2. What would be the implications of accepting either the 1982 Polish or the 1991 Chinese views for determining those situations in which the international community could legitimately concern itself with a state's human rights violations?

3. Can you suggest guidelines for a contemporary conception of domestic jurisdiction that would recognize its historical, essential purpose without unduly constraining the international community's ability to respond to human rights violations?

4. Is para. 6 of GA Res. 49/186 above compatible with the emerging approach to domestic jurisdiction that is reflected in the Moscow statement of the CSCE?

5. Consider the following observations of Martti Koskenniemi:

> The values of the international system as expressed in public international law are those of liberal individualism transposed to the interstate level. The system denies the existence of an external standard of judgment that is valid regardless of whether it is held by states themselves. Its law, Louis Henkin observed recently, 'is designed to further each state's realization of its own notion of the Good.' The 'system' exists only in the formal sense of a shared vocabulary and a set of institutional practices that states use for cooperation or conflict. It is an artificial creation, a contrived synthesis of power and ideas. In short, it is not an organism that embodies some autonomous ideal of authentic communal life.
>
> This is the understanding of the present state system shared and challenged by each of the discourses discussed above: human rights, nationalism, and world order. Each posits an ideal of *authenticity* outside the system that contrasts with the formal and artificial character of statehood. These ideals, however, cannot be sustained within liberal political rhetoric. To the contrary, their enforcement might appear as precisely the kind of authoritarianism against which the state system was created.
>
> . . .
>
> By establishing and consenting to human rights limitations on their own sovereignty, states actually define, delimit, and contain those rights, thereby domesticating their use and affirming the authority of the state as the source from which such rights spring. . . .
>
> . . .
>
> The most universally acclaimed principles of international law continue to be those of sovereignty, self-determination, territorial integrity, non-intervention, and consent, each of which is a reaffirmation of statehood as the law's creative center. . . .

. . .

Statehood survives and should continue to survive for the foreseeable future because its formal-bureaucratic rationality provides a safeguard against the totalitarianism inherent in a commitment to substantive values, which forces those values on people not sharing them.

. . .

These reflections . . . do not require an unreserved commitment to statehood . . . [It] remains a second best—defensible but only to the extent that there can be no general agreement about the authentic purpose of social life. . . .[7]

a. Human rights, says the author, both 'share and challenge' the understanding of the state system set forth in the first paragraph. How does it do each of these things?

b. What implications do these views have for the continuing vitality of the concept of domestic jurisdiction?

c. Can the human rights movement be viewed as what the author refers to as a 'commitment to substantive values' that has an inherent 'totalitarianism' in it? Does that movement tend to destroy cultural and other diversity associated with the existing state system and with related concepts like domestic jurisdiction?[7]

C. THE UN COMMISSION'S MAIN PROCEDURES FOR RESPONDING TO VIOLATIONS

The cumulative result of post-1967 developments in the UN Commission is that there are now three different procedures used by the Commission to respond to violations. They are: (1) confidential consideration of a situation under the 1503 procedure; (2) public debate under the 1235 procedure, which may lead to the appointment of a Special Rapporteur of the Commission, a Special Representative of the Secretary-General, or some other designated individual or group to investigate a situation; and (3) the designation of a 'thematic' rapporteur or Working Group to consider violations anywhere relating to a specific theme (such as torture, disappearances, or arbitrary detention).

In principle, each of these procedures is relatively distinct from the others in terms of its origins, the nature of its mandate, the steps to be followed and the types of outcome available. In practice, there is considerable overlap. It is conceivable that different aspects of a particular situation would be under review by all three procedures at the same time. Moreover, the same situation might be considered simultaneously by one or more of the treaty bodies (such as the ICCPR Human Rights Committee or the Committee Against Torture)

[7] Martti Koskenniemi, The Future of Statehood, 32 Harvard Int. L. J. 397, 404 (1991).

to be considered in Chapter 9 (and as well by one of the regional organizations described in Chapter 10).

For example, before the invasion of Kuwait, the human rights situation in Iraq was under consideration through the 1503 procedure and by several of the thematic mechanisms. Following Iraq's expulsion from Kuwait, its situation was taken up under the 1235 procedure and the thematic mechanisms as well as by several treaty bodies. Had there been a functioning regional organization in existence, consideration might have been extended to it as well.

This duplication raises questions which we briefly note. The first is: why have the relevant procedures, as well as the institutional arrangements, been permitted to develop in this way? One explanation follows.

> . . . The system has grown 'like Topsy' and the boundaries between the different organs are often only poorly delineated. For the most part, this pattern has hardly been accidental. Rather, it is the inevitable result of a variety of actors seeking to achieve diverse, and perhaps sometimes even irreconcilable, objectives within the same overall institutional framework. If an existing body could not do a particular job, whether because of some intrinsic defects, sheer incompetence, or, more likely, political intransigence, the preferred response was to set up yet another. In a very short space of time States and individual actors would develop a vested interest in the maintenance of the new body in the same form. This pattern simply repeated itself when a new policy agenda, to which none of the existing bodies was sufficiently responsive, emerged.
>
> In general then, the evolution of the regime has reflected specific political developments. Its expansion has depended upon the effective exploitation of the opportunities which have arisen in any given situation from the prevailing mix of public pressures, the cohesiveness or disarray of the key geopolitical blocks, the power and number of the offending state(s) and the international standing of their current governments, and a variety of other, often rather specific and ephemeral, factors. . . .[8]

A second question is whether it is desirable and feasible to rationalize the various procedures in order to reduce overlapping institutional competences. The 1993 World Conference on Human Rights at Vienna gave impetus to reform proposals in this area, but it remains necessary to overcome one of the principal obstacles to reform. That obstacle is the suspicion that the 'real' agenda of some governments advocating reform is the elimination of the most effective (and intrusive) procedures while retaining those that are largely cosmetic and ineffectual. Genuine reformers have thus tended to assume that the maintenance of duplicative arrangements is a necessary price to pay for ensuring the continuing existence of better procedures.

Before being in a position to consider possible reforms, we need to

[8] Philip Alston, Appraising The United Nations Human Rights Regime, in P. Alston (ed.), The United Nations and Human Rights: A Critical Appraisal 1, 2 (1992).

understand the existing procedures and the ways in which they have functioned. We will consider each procedure in turn.

1. THE 1503 PROCEDURE: PROS AND CONS OF CONFIDENTIALITY

The 1503 procedure involving communications (complaints) to the UN stems from Economic and Social Council Resolution 1503 (XLVIII)(1970). It was adopted after, and built upon, resolution 1235 (XLII)(1967). However, as a mechanism for investigating specific situations involving violations, it developed more rapidly than the 1235 procedure and has often been used as a precursor to action under the latter. Thus we consider it first.

The UN Commission had previously (under Resolutions 75 (V)(1947) and 728 F (XXVIII)(1959)) used communications only as a means of identifying general trends, thus providing no response whatsoever to the particular violations at issue. The adoption of the 1503 procedure involved a typical horse-trading exercise in which governments with competing objectives sought to reconcile their goals through the use of open-ended and flexible language. Both the resolution itself and the subsequent Sub-Commission resolution that laid down the admissibility criteria for communications are perfect case studies in ambiguity.

ECOSOC RESOLUTION 1503 (XLVIII)(1970)

The Economic and Social Council,

. . .

1. *Authorizes* the Sub-Commission on Prevention of Discrimination and Protection of Minorities to appoint a working group consisting of not more than five of its members, with due regard to geographical distribution, to meet once a year in private meetings for a period not exceeding ten days immediately before the sessions of the Sub-Commission to consider all communications, including replies of Governments thereon, received by the Secretary-General under Council resolution 728 F (XXVIII) of 30 July 1959 with a view to bringing to the attention of the Sub-Commission those communications, together with replies of Governments, if any, which appear to reveal a consistent pattern of gross and reliably attested violations of human rights and fundamental freedoms within the terms of reference of the Sub-Commission;

. . .

4. *Further requests* the Secretary-General:

(*a*) To furnish to the members of the Sub-Commission every month a list of communications prepared by him in accordance with Council resolution 728 F (XXVIII) and a brief description of them, together with the text of any replies received from Governments;
(*b*) To make available to the members of the working group at their meetings the originals of such communications listed as they may request. . . ;

. . .

5. *Requests* the Sub-Commission on Prevention of Discrimination and Protection of Minorities to consider in private meetings, in accordance with paragraph 1 above, the communications brought before it in accordance with the decision of a majority of the members of the working group and any replies of Governments relating thereto and other relevant information, with a view to determining whether to refer to the Commission on Human Rights particular situations which appear to reveal a consistent pattern of gross and reliably attested violations of human rights requiring consideration by the Commission;
6. *Requests* the Commission on Human Rights after it has examined any situation referred to it by the Sub-Commission to determine:

(*a*) Whether it requires a thorough study by the Commission and a report and recommendations thereon to the Council in accordance with paragraph 3 of Council resolution 1235 (XLII);

[Eds. The other alternative is for the Commission to appoint an '*ad hoc* committee' to investigate the situation. But that requires 'the express consent of the State concerned and shall be conducted in constant co-operation with that State and under conditions determined by agreement with it.' All 'available means at the national level' must have been exhausted and the situation should not relate to a matter being dealt with under other international procedures. Members of the committee are appointed by the Commission but subject to the consent of the government concerned. Its work is to be entirely confidential and it should 'strive for friendly solutions' at all times. It then reports to the Commission 'with such observations and suggestions as it may deem appropriate.' This procedure has never been used.]

. . .

8. *Decides* that all actions envisaged in the implementation of the present resolution by the Sub-Commission on Prevention of Discrimination and Protection of Minorities or the Commission on Human Rights shall remain confidential until such time as the Commission may decide to make recommendations to the Economic and Social Council;

. . .

10. *Decides* that the procedure set out in the present resolution for dealing with communications relating to violations of human rights and fundamental freedoms should be reviewed if any new organ entitled to deal with such communications should be established within the United Nations or by international agreement.

SUB-COMMISSION RESOLUTION 1 (XXIV)(1971)

The Sub-Commission on Prevention of Discrimination and Protection of Minorities,

. . .

Adopts the following provisional procedures for dealing with the question of admissibility of communications referred to above:

(1) Standards and criteria

(a) The object of the communication must not be inconsistent with the relevant principles of the Charter, of the Universal Declaration of Human Rights and of the other applicable instruments in the field of human rights.

(b) Communications shall be admissible only if, after consideration thereof, together with the replies of any of the Governments concerned, there are reasonable grounds to believe that they may reveal a consistent pattern of gross and reliably attested violations of human rights and fundamental freedoms, including policies of racial discrimination and segregation and of *apartheid*, in any country, including colonial and other dependent countries and peoples.

(2) Source of communications

(a) Admissible communications may originate from a person or group of persons who, it can be reasonably presumed, are victims of the violations referred to in subparagraph (1)(b) above, any person or group of persons who have direct and reliable knowledge of those violations, or non-governmental organizations acting in good faith in accordance with recognized principles of human rights, not resorting to politically motivated stands contrary to the provisions of the Charter of the United Nations and having direct and reliable knowledge of such violations.

(b) Anonymous communications shall be inadmissible; . . . ;

(c) Communications shall not be inadmissible solely because the knowledge of the individual authors is second-hand, provided that they are accompanied by clear evidence.

(3) Contents of communications and nature of allegations

(a) The communication must contain a description of the facts and must indicate the purpose of the petition and the rights that have been violated.

(b) Communications shall be inadmissible if their language is essentially abusive and in particular if they contain insulting references to the State against which the complaint is directed. Such communications may be considered if they meet the other criteria for admissibility after deletion of the abusive language.

(c) A communication shall be inadmissible if it has manifestly political motivations and its subject is contrary to the provisions of the Charter of the United Nations.

(d) A communication shall be inadmissible if it appears that it is based exclusively on reports disseminated by mass media.

(4) Existence of other remedies

(a) Communications shall be inadmissible if their admission would prejudice the functions of the specialized agencies of the United Nations system.

(b) Communications shall be inadmissible if domestic remedies have not been exhausted, unless it appears that such remedies would be ineffective or unreasonably prolonged. Any failure to exhaust remedies should be satisfactorily established.

(c) Communications relating to cases which have been settled by the State concerned in accordance with the principles set forth in the Universal Declaration of Human Rights and other applicable documents in the field of human rights will not be considered.

(5) Timeliness

A communication shall be inadmissible if it is not submitted to the United Nations within a reasonable time after the exhaustion of the domestic remedies as provided above.

QUESTION

The best way to get a sense of the loopholes in the 1503 procedure that governments have tried to exploit over the years is to take a hypothetical case and formulate as many arguments as possible to the effect that the matter does not fit within the relevant guidelines, and therefore should not be considered.

Consider this case. The government you represent is responding to a communication alleging that 100 members of a group have been arbitrarily detained for a period of six months. The communication has been brought by a small NGO based in your country. The group's secretary is also the leader of a political party opposed to the government. The complaint, based upon newspaper reports and accounts provided by relatives of eleven detained persons, mentions that the government is widely considered to be both oppressive and exploitative. The communication notes that the state's courts have always done the government's bidding and that it would be a waste of time and resources to ask the courts to order the release of the detainees.

What arguments would you make against admissibility?

NOTE

Until the mid-1980s the UN received, on average, 25,000 complaints per year. Recently this number has ballooned to around 300,000, but many of these complaints are identical as a result of letter-writing campaigns by groups with a large and active membership. Because the procedure was carefully designed to ensure that governments would not lightly be accused of violations and because it is concerned not with individual cases, but with 'situations,' the processes involved are tedious and time-consuming. Consider the following.

PHILIP ALSTON, THE COMMISSION ON HUMAN RIGHTS

in Alston (ed.), The United Nations and Human Rights: A Critical Appraisal 126 (1992), at 146.

. . .

[T]he procedure currently consists of the following four steps. Firstly, the Communications Working Group of the Sub-Commission begins the procedure by sorting through the various complaints that have been received in the preceding year. The group consists of five Sub-Commission members and each takes primary responsibility for complaints dealing with a specific cluster of rights. . . . The Group requires an affirmative majority vote before any 'situation' can be forwarded on to the Sub-Commission in plenary.

In recent years the Sub-Commission's Working Group has sent its parent a list of about eight to ten countries a year on average. It might also identify a few situations for specific consideration the following year, thus implicitly serving notice on the governments concerned. At the second stage, when the full Sub-Commission considers each of its Working Group's 'nominees', it opts by a simple majority vote to send the country on to the Commission, to drop it, or to reconsider it the following year. While the complainant is not informed at this stage, the government concerned is invited to submit written observations and to defend itself before the Commission, thus creating a clear inequality of opportunities to participate in the proceedings. As the third stage of the procedure the Commission also establishes a Communications Working Group. Its task is to draft recommendations as to the action which the Commission might wish to take on each country situation that is put before it. At the fourth stage the Commission itself devotes several days at each of its annual sessions to a consideration of all the relevant material. At the end of its deliberations its Chairman announces the names of the countries that have been considered and the names of those that have been let off the hook. He or she does not, however, provide any details as to the nature of the allegations or the specific action taken by the Commission. As to the latter, several options are available—in addition to dropping the case altogether. The first is to keep it 'under review' which means that more evidence will be admitted next year and the government concerned will be called to account again. It also provides an incentive for at least cosmetic measures to be adopted and for strong lobbying to be undertaken in the meantime. The second option is to send an envoy to seek further information on the spot and report back to the Commission. The third is to appoint an *ad hoc* committee which, with the approval of the state concerned, can conduct a confidential investigation aimed at finding a 'friendly solution'. The Committee then reports to the Commission 'with such observations and suggestions as it may deem appropriate'. This procedure has, as of the end of the 1991 session, never been invoked. The fourth and final option is for the Commission to transfer the case to the 1235 procedure, thereby going public and permitting 'a thorough study by the Commission and a report and recommendations thereon to the Council'. The role of the Council in this respect has so far proved to be absolutely minimal.

Thus, in the absence of a decision to go public, the entire procedure is shrouded in secrecy, with each of its stages being accomplished in confidential sessions by the bodies concerned. Nevertheless, the details have invariably been leaked to the media for one reason or another and the complete documentation on several country situations has been made available as a result of Commission decisions to release all the relevant documentation in cases concerning Equatorial Guinea, Uruguay, Argentina, and the Philippines.

In statistical terms, the 1503 procedure has 'touched' an impressive number of countries. On the basis of unofficial sources it seems that at least forty-five countries have been reported to the Commission under the procedure since 1972. More authoritatively, however, it can be said that since the Commission began to officially identify the situations before it, in 1978, thirty-nine States have been subject to scrutiny as of the end of the 1991 session. Of these, twelve were in Africa, twelve in Asia, twelve in Latin America, two in Eastern Europe, and one in Western Europe.

NOTE

The release of the documentation on Uruguay, which was under 1503 review from 1978 to 1985, provides insights into some of the ways in which the procedure might be rendered ineffectual. Human rights NGOs had reported to the Commission (UN Doc. E/CN.4/R.27(1977)) thousands of cases of arbitrary arrest and torture, a significant number of disappearances and deaths and the 'destruction of the rule of law.' In response, the Commission requested the then Secretary-General (Kurt Waldheim) to undertake a confidential mission to make 'direct contacts with the Government . . . with a view to keeping the Commission informed and to improving the human rights situation.' He appointed Under-Secretary-General Javier Pérez de Cuéllar (later to become Secretary-General) as his envoy. His report read, in part, as follows:

REPORT OF THE SECRETARY-GENERAL ON URUGUAY PURSUANT TO RESOLUTION 15 (XXXIV) OF THE COMMISSION ON HUMAN RIGHTS

UN Doc. E/CN.4/R.50/Add.2 (1980)(Confidential), at 2.

. . .

The discussions held in the Commission . . . indicate that the Commission wanted 'direct contacts' to be established with the Governments of certain countries, in order to improve the human rights situation and keep abreast of developments. It is clear from the discussions that these "direct contacts" should meet the following criteria:

(a) Exclusive dealings with the authorities;
(b) Respect for the sovereignty of the State;
(c) A cautious and flexible approach to the problem;
(d) Confidentiality of the contacts.

. . .

2. *Direct contacts with the Uruguayan Government*

Mr. Pérez de Cuéllar arrived at Montevideo on 10 December 1979 and was met by . . . Ambassador Edmundo Narancio, and by a representative of the Commander in Chief of the Army, Colonel Regino Brugeno. Throughout the visit the latter was most helpful in establishing liaison with the various Uruguayan authorities . . . .

Five days after his arrival, Mr. Pérez de Cuéllar was received by the President of the Republic, Dr. Aparicio Méndez, who was accompanied by the Commanders-in-Chief of the three armed forces; by the President of the Council of State, Dr. Hamlet Reyes and the members of the Council, who were also the members of a 'Commission on Individual Rights' of the Council; by the Ministers for Foreign Affairs, Defense, Interior, Justice and Labour; by the Commander-in-Chief of the Army, and by the President of the Military Tribunal.

Mr. Pérez de Cuéllar was not only treated with courtesy at all times but he was also given every facility for the accomplishment of his mission.

. . .

First of all the authorities interviewed stated—as a basic premise—that the Uruguayan Constitution of 1967 (article 168, paragraph 17) authorizes the President of the Republic to take 'prompt security measures in serious or unforeseen cases of foreign attack or internal disorder'.

Secondly, they gave a detailed and documented account of the political, economic and social crisis suffered by Uruguay from 1967 onwards and of the way in which various clandestine Marxist groups helped to turn this crisis into a veritable challenge to the authority of the State and reduced the country to a state of internal chaos through their terrorist activities, which constituted outright subversion or intense disorder.

. . .

On 27 June 1973, considering that the Legislature was impeding the course of justice by refusing to suspend the immunity from prosecution of the members of Parliament involved in verified seditious activities, and was failing to perform or misrepresenting its constitutional functions, the President of the Republic dissolved the two Houses and ordered that their powers should be exercised by a Council of State consisting of 20 members, whose functions would be the specific functions of the Parliament, 'to monitor the actions of the Executive Branch in connexion with respect for the individual rights of the human person and the compliance of that branch with the Constitutional and legal provisions' and to 'draft a Constitutional Reform reaffirming the basic principles of democratic and representative government, to be adopted by the electorate at a future plebiscite'.

In June 1976 the Executive began to issue 'Constitutional decrees or Institutional Acts' introducing transitional amendments to the 1967 Constitution, pending the formulation of definitive Constitutional norms. The amendments included the establishment of the National Council, consisting of the members of the Council of State, and the members of the Council of General Officers (Junta de Oficiales Generales), with the power to appoint the President of the Republic and other heads of Government branches and to hear charges made against those dignitaries.

In September 1977 the Executive announced its intention of promulgating a new Constitution in 1980 and holding general elections in 1981.

3. *General situation of human rights in Uruguay*

The Uruguayan authorities with whom Mr. Pérez de Cuéllar discussed human rights in Uruguay most strongly re-emphasized that those rights not only form part of Uruguay's legal tradition and were recognized by the Constitution and the Constitutional Decree of 20 October 1976, but had been continually respected and were still being respected in the struggle against subversion in that country.

First of all the authorities most emphatically stated that neither in the past nor at present had there been, or were there, in Uruguay any prisoners arrested or detained for political reasons. . . .

Secondly, they stated that the 'State Security Act', making acts detrimental to the nation an offence, was a law enacted in 1972 by the Parliament elected in 1971, and therefore constitutionally unobjectionable. They maintained that the detentions under article 168, paragraph 17, of the Constitution were exceptions. At present there are only 20 such cases.

. . .

As for the trials of detainees, the authorities gave their assurance that they always acted within the constitutional and legal framework prescribed by Parliament, and particularly the Security Act. The procedure provided for in that Act, which continues to be applied, requires that the arrested person be brought before the competent Examining Military Judge within 10 days. The Government therefore rejects the charge that the prisoners are held incommunicado for months without being brought before a judge. It explained that if they had not been brought before a judge they would have invoked the right of habeas corpus under the Constitution (articles 12, 17, 23 and 30 of the Constitution). It stated that only three out of a thousand detainees had invoked that right. The constitutional principle of the right of defence was also respected, according to the authorities.

. . .

The authorities most strongly denied that any ill treatment or torture of prisoners, either before the trial or during their stay in prison, had been tolerated.

With regard to prison conditions and the treatment of prisoners, Mr. Pérez de Cuéllar was able to pay a lengthy visit to the . . . houses of detention. . . . [T]hose prisons did not house any common criminals who had not committed offences against State security. Mr. Pérez de Cuéllar was able to observe that the prison conditions were extremely reasonable as regards hygiene (cleanliness, lighting and ventilation), neatness, food and medical services. He was able to see that normal arrangements were made for recreation and outdoor activities. He visited special reading rooms, music rooms and workshops for the teaching or practice of handicrafts, such as carpentry, printing, weaving, upholstering, mechanics and bookbinding. The prisoners can hear radio programmes, national and international news reports, and information on and broadcasts of sports events. The guards are armed only with wooden truncheons for purposes of self defence. He was told that visits to prisoners were normal and regular, and he himself was able to observe the presence of relatives waiting to pay such visits. As regards correspondence, including parcels and foodstuffs, he was assured that they were of the kind allowed in any detention centre.

As for the physical condition of the detainees, Mr. Pérez de Cuéllar found that they all appeared healthy. Among the many prisoners whom he was able to meet, were Messrs. Altessor, Estrella, Massera and Turcanski, whose cases are known internationally.

. . .

4. The authorities interviewed made repeated references to the campaign being waged outside the country about the trial conditions and pressure on prisoners in Uruguay. They feel that this campaign is being engineered by exiled or fugitive opposition politicians, with the support of foreign political groups interested in projecting an image of inhuman repression in Uruguay.

QUESTION

This report remains controversial. (a) How would you characterize its strengths and weaknesses? (b) How would you describe the purposes of this report from the point of view of its author? How do you assess those purposes? (c) Does the element of confidentiality lend itself inevitably to the approach of this report? If not, what measures could be taken, within the constraints imposed by the procedure, to ensure a more satisfactory outcome?

PHILIP ALSTON, THE COMMISSION ON HUMAN RIGHTS

in Alston (ed.), The United Nations and Human Rights: A Critical Appraisal 126 (1992), at 151.

. . .

The experience gained with the 1503 procedure over almost twenty years is sufficient to enable some general observations to be made about it. With respect to the range of situations taken up it appears that: (*a*) alleged violations of economic, social, and cultural rights have never been examined seriously, despite a resolution by the Commission urging that they should be; (*b*) the Commission has responded to violations of only a limited range of civil and political rights, which in turn has ensured that while Third World countries are disproportionately represented on the 1503 blacklist, developed countries (both West and East) have only very rarely been called to account, (*c*) there is probably a *de facto* limit of 6 to 8 on the number of countries that the Commission is prepared to take on at any one time, but since a similar principle applies to the public country-specific procedures 1503 effectively doubles the number of countries that can be 'named'; (*d*) despite the focus on 'situations' some individuals have been directly assisted under the procedure; and (*e*) many of the situations dealt with under the Commission's public procedures have been raised earlier in the 1503 context.

With respect to the actual functioning of the procedure it is clear that some of the early wrinkles were effectively ironed out in the course of the first decade. In particular, the relationship between the public and private procedures, [i.e. the public 1235 procedure discussed below and the private 1503 procedure], which was often invoked in an effort to confine discussions to the latter domain in the 1970s, was resolved in such a way as to make it clear that only the proceedings themselves and the details of decisions taken were confidential. There is thus no bar to focusing on the same country in both procedures at the same time. Nevertheless, the procedure is designed to move extremely slowly and deliberations are, as a result, too often based on outdated information. . . .

Another criticism of the functioning of the procedure concerns the Commission's apparent failure to structure its working methods under 1503. Consequently, the response in one situation will often bear little relationship to that adopted in another generally comparable one. While no two situations are identical and attempts to draw comparisons are often flawed, there is still room to seek some consistency in responses, to seek a more uniform structure for reports prepared, and to question States representatives in a more systematic fashion. From the records that are available, the 'dialogue' that takes place is usually unadulterated political horse-trading rather than a probing inquiry into the facts and a quest for the most effective potential response. Thus, those who have observed the procedure in practice have noted that the level of debate in the closed sessions is sometimes abysmally low and reveals that delegations have done little, if any, serious preparation.

Providing a general evaluation of the 1503 procedure is difficult but necessary. The question is whether the procedure provides adequate returns in terms of the investments of faith, time, energy, and media attention that have been made in it; whether, on balance, it succeeds in putting enough pressure on enough countries in ways that could not more effectively be achieved by other means. Some of those means might already be in existence, others might more easily be created if the 1503 procedure were to be eliminated.

The first point to be made is that the historical value of the 1503 procedure cannot be doubted. In many ways it laid the groundwork for the development of the potentially effective public (i.e., resolution 1235-based) response to violations which began to come into its own after 1979. It put paid once and for all to the domestic jurisdiction *canard*; it accustomed States to the need to defend themselves and gave them practice in examining (and prosecuting) the performance of others; it galvanized some of the NGOs at a time when some of the other procedures offered even lower rates of return; and it exposed the Commission and Sub-Commission to the real world of violations more effectively than any earlier exercise had. But a valuable historical role does not of itself justify the procedure's retention in the 1990s. Among commentators, the procedure has had both its defenders and its critics. Among the latter was Amnesty International which, in the mid-1970s, characterized the confidentiality of 1503 as 'an undisguised stratagem for using the United Nations, not as an instrument for promoting and protecting and exposing large-scale violations of human rights, but rather for concealing their occurrence'. In a similar vein the then Director of the UN's human rights Secretariat, Theo van Boven, asked in a thinly veiled allusion to the procedure in his opening statement to the Commission in 1980:

> Is it satisfactory to place so much emphasis on the consideration of situations in confidential procedures thereby shutting out the international community and oppressed peoples? Are certain procedures in danger of becoming, in effect, screens of confidentiality to prevent cases discussed thereunder from being aired in public? While there is probably no alternative to trying to co-operate with the Governments concerned, should we

> allow this to result in the passage of several years while the victims continue to suffer and nothing meaningful is really done?

Writing a decade later, in 1990, Iain Guest concluded that 'confidentiality has not persuaded governments to cooperate with the United Nations' and that '1503 has become truly dangerous to human rights—and that it offers a useful refuge to repressive regimes'.

While some of the procedure's proponents would disagree strongly with such assessments, their enthusiasm is nevertheless usually rather restrained. Writing in 1980, Tardu was optimistic about the procedure's future while conceding that its past performance had left much to be desired. In 1982 Zuijdwijk, after an exhaustive study of the procedure, damned it with faint praise by concluding that it is 'worth sending a petition to the United Nations under resolution 1503 (XLVIII) but petitioners should limit their expectations'. Tolley is equally sparing in his praise for 1503 although he attaches importance to the fact that most target governments feel threatened by it and that 'only four states have failed to have observers at two consecutive [Commission] sessions when their government was under Resolution 1503 review'. An experienced Sub-Commission member, Marc Bossuyt, has strongly defended the procedure on the grounds that: (1) 1503 review facilitates subsequent consideration of a country under the public procedures—a hypothesis that would appear to be at least questionable; and (2) that it enables attention to be paid to situations that would otherwise be ignored. His views are predicated on the assumptions that a government's reputation suffers from remaining under 1503 review and that a continuing dialogue is the key to success in dealing with human rights violations. . . .

In general, it would seem . . . that the shortcomings of the procedure are so considerable, its tangible achievements so scarce, the justifications offered in its favour so modest, and the need for an effective and universally applicable petition procedure so great, that it is time to re-evaluate its future. The principal options are major reform or abolition. . . .

FRANK NEWMAN AND DAVID WEISSBRODT,

in International Human Rights 109 (1990), at 122.

. . .

The 1503 process may encourage governments to engage in an exchange of views and possibly to improve the situation without the glare of substantial publicity. Many governments, including the U.S., regularly respond to 1503 communications. They may not provide a substantive answer to public criticism heard at the Commission or its Sub-Commission, because they have insufficient opportunity during the busy debate (under resolution 1235) to research and submit a good response. Also, unless there is a substantial consensus for action under 1235, governments may realize the public criticism can safely be ignored. In addition, a few governments are so offended by public criticism that they stubbornly refuse to take action to improve the situation. Hence, the 1503 procedure may sometimes afford a better opportunity for constructive dialogue than the public 1235 process. Until 1989 Iran was an example of a country which was intransigent in the face of public actions under 1235. It is, though, doubtful whether Iran would have been more responsive under 1503.

While the 1503 process is painfully slow, complex, secret, and vulnerable to political influence at many junctures, it does afford an incremental technique for placing gradually increasing pressure on offending governments. Nevertheless, the confidentiality of the process can be used as a barrier to effective U.N. action in the case of governments that do not respond to incremental pressure and continue to engage, over several years in grave and widespread violations of human rights. If the objective is to obtain prompt publicity or public action for serious human rights violations, the 1503 process is inappropriate.

QUESTIONS

1. How would you evaluate the 1503 procedure? What criteria for effectiveness would you use? What reforms, if any, would you suggest in order to make the procedure more effective?

2. You are a lawyer working with an NGO in state X, ruled by a repressive military government. Your client, a journalist hostile to the regime, is one of a number of people—exact figures are not available, but an informed guess would be around 800—in this country of 10 million people that have recently been imprisoned because of their advocacy for a democratic regime. You have succeeded in seeing her only after 38 days of detention, during which she claims to have been tortured. Your client asks whether there are international procedures available for seeking relief. Assuming that a communication under the 1503 procedure would be held admissible, would you recommend that she request you to proceed?

2. THE 1235 PROCEDURE: PROCESSES AND PARTICIPANTS

a. *The Process and its Outcomes*

ECOSOC Resolution 1235 (XLII)(1967) established the procedure on the basis of which the Commission holds an annual public debate focusing on gross violations in a number of states. The pertinent parts of the resolution are as follows.

ECOSOC RESOLUTION 1235 (XLII)(1967)

The Economic and Social Council,

. . .

1. *Welcomes* the decision of the Commission on Human Rights to give annual consideration to the item entitled 'Question of the violation of human rights and fundamental freedoms, including policies of racial discrimination and segregation and of *apartheid*, in all countries, with particular reference to colonial and other dependent countries and territories,' without prejudice to the functions and powers of organs already in existence or which may be established within the framework of measures of implementation included in international covenants and conventions on the protection of human rights and fundamental freedoms; . . .

2. *Authorizes* the Commission on Human Rights and the Sub-Commission on Prevention of Discrimination and Protection of Minorities, . . . to examine information relevant to gross violations of human rights and fundamental freedoms, as exemplified by the policy of *apartheid* as practised in the Republic of South Africa and in the Territory of South West Africa under the direct responsibility of the United Nations and now illegally occupied by the Government of the Republic of South Africa, and to racial discrimination as practised notably in Southern Rhodesia. . . .

3. *Decides* that the Commission on Human Rights may, in appropriate cases, and after careful consideration of the information thus made available to it, in conformity with the provisions of paragraph 1 above, make a thorough study of situations which reveal a consistent pattern of violations of human rights, as exemplified by the policy of *apartheid* as practised in the Republic of South Africa and in the Territory of South West Africa under the direct responsibility of the United Nations and now illegally occupied by the Government of the Republic of South Africa, and racial discrimination as practised notably in Southern Rhodesia, and report, with recommendations thereon, to the Economic and Social Council;

4. *Decides* to review the provisions of paragraphs 2 and 3 of the present resolution after the entry into force of the International Covenants on Human Rights;

. . .

COMMENT ON THE 1235 PROCEDURE AND ITS POSSIBLE OUTCOMES

These provisions provide a vivid illustration of the extent to which the Commission's mandate has evolved. The ways in which violations are now dealt with by the Commission, still under the rubric of the 1235 procedure, bear only a passing resemblance to the procedure formally authorized by the resolution.

The procedure now operates to legitimate two types of activity of the Commission. The first, in accordance with the mandate, involves the holding of an annual public debate in which governments and NGOs are given an opportunity to identify publicly those country-specific situations that they consider to be particularly deserving of the Commission's attention. The second involves studying and investigating particular situations (or individual cases) through the use of whatever techniques the Commission deems appropriate.

Although this section of Chapter 7 focuses primarily on fact-finding problems and the issues raised by reporting about governments' conduct, it should be noted that such investigative activity is only authorized in relation to a small proportion of the situations identified during the annual debate. This does not mean that the remaining situations are entirely neglected. The Commission can contribute in other ways to the pressures felt by a government accused of violations. In theory, there is a broad range of outcomes that might follow the identification of a serious country situation by another government or by an NGO within the framework of the 1235 debate. They include:

- the mere mention of a situation in the debate might embarrass a country (sometimes referred to as the sanction of 'shaming'), generate media coverage, or influence another country's foreign policy;
- an NGO might use the occasion to pressure other governments to take up the issue on a bilateral or multilateral basis;
- a draft resolution might be circulated, and then withdrawn, perhaps after a strong lobbying effort by the government concerned or in response to concessions offered by that government;
- the Chairman of the Commission might issue a statement of exhortation, with the (de facto, if not formal) approval of the Commission;
- the Commission might adopt a resolution calling for all available information to be submitted to it with a view to considering the matter at a later session;
- the Commission might call upon the government to respond to the allegations in detail and in writing before its next session;

- the Commission might adopt a resolution criticizing the government (for which purpose, language ranging from the diplomatic to the highly critical might be used) and calling upon it to take specific measures;

- the Commission might appoint a Special Rapporteur or other individual or group to examine the situation and submit a report to the Commission on the basis of a visit to the country; or

- the Commission might call upon the Security Council to take up the issue, with a view to considering the adoption of sanctions or some other punitive measure.

As later materials suggest, the impact of any of these measures will vary greatly depending on factors such as: the nature of the violations and especially the extent to which their continuation is central to the government's strategy for retaining power; the relative influence of domestic pressure groups; the degree of support for the measure in the Commission; the openness of the country concerned to external influences; the vulnerability of the country to trade, aid or other pressures; and the attitude taken by the country's allies and its regional neighbours.

Before examining the practice of the Commission under the 1235 procedure of appointing a Special Rapporteur or other individual or group to examine the situation, it is necessary to consider the assumptions and procedures on the basis of which international fact-finding has developed.

b. The Quest for Fact-finding Guidelines

In the context of this Chapter, the procedures and guidelines set forth in the following materials bear directly on investigations and reporting within the UN system. But they are also relevant to fact-finding by non-governmental organizations (see. p. 473, *infra*) and by regional organs such as the Inter-American Commission on Human Rights (p. 640, *infra*).

NOTE

The notion that the international community will seek to 'find facts' that may not accord with, or even flatly contradict, those provided officially by a sovereign government would have been virtually unthinkable not so many years ago. For example, when the 1907 Hague Convention relating to international commissions of inquiry was adopted, its scope was carefully limited so as to cover only 'disputes involving neither honour nor essential interests.' Today, international fact-finding is an accepted and relatively common activity. It is carried out not only by a large number of international organizations but also by individual states.

Fact-finding depends for its credibility and potential impact upon the extent to which it is perceived to have been thorough, politically objective and procedurally fair. For that reason attempts have been made to draw up rules or guidelines for fact-finders. In the following materials, Nicholas Valticos introduces some general issues. An excerpt from the State Department's annual Country Report on Human Rights Practices emphasizes the difficulty of getting to the truth of the matter.

NICHOLAS VALTICOS, FOREWORD

in B.G. Ramcharan (ed.), International Law and Fact-finding in the Field of Human Rights (1982), at vii.

. . . Today fact-finding is . . . frequently called for as part of the action taken by the international community to secure respect for human rights.

. . . It is no longer a matter of ascertaining the facts in cases which merely involve the interests of two States. As was pointed out by an ILO commission of inquiry in 1962, the purpose of inquiry is to gather 'thorough and objective information' on issues of 'public importance'. . . . [I]ssues of major importance to both the international community and the State concerned are often at stake. What type of action can be taken in such cases to meet the requirements of the international community while taking account of the susceptibilities of the State involved?

The problem is not confined to international action in the field of human rights. Fact-finding may be needed, in varying measure, in all procedures through which it is sought to check the conformity of particular situations with international standards of commitments. . . .

These preoccupations have many features in common, yet fact-finding in the field of human rights has a special importance, and also encounters special difficulties, both because of the subject-matter and because of the importance attached to it by public opinion, which regards it as the acid test of the effectiveness of international organisations. Fact-finding in respect of human rights is however all the more difficult, because it frequently concerns the action and essential interests, if not indeed the very structure, of the States involved, who are therefore less inclined to accept international intervention in such matters. The issues often have political aspects and are the subject of discussion in political bodies, a factor which necessarily complicates their examination. Lastly, . . . fact-finding on questions concerning human rights has been undertaken by various organisations and bodies in differing contexts, and the methods used have not always been similar.

. . .

Having myself taken part in such procedures, I should like to set out some general reflections on the problem.

In the first place, as we are on the frequently unstable terrain of international law, it is necessary, as has previously been recognised, not to confine oneself within unduly rigid categories or rules. In international law, functions intertwine—at times, indeed, too much— and judicial aspects cannot always be distinguished clearly from non-judicial ones. It is therefore not always possible, in international fact-finding, to transpose internal judicial procedures in full. Nor is it always possible—or even desirable—to establish unduly detailed rules which may turn out not to be applicable in practice. If procedures are too formal and judicial and rules too detailed, they may prove not to be adapted to the great variety of situations, to the susceptibilities and objections of the States concerned, or to practical needs.

One conclusion to be drawn from this is that it is necessary to have available a variety of procedures suited to different situations, ranging from quasi-judicial inquiries to methods involving a minimum of formality such as 'direct contacts'. . . .

Does this mean a purely empirical approach, with a different method in each case, marked by compromise and bargaining? Obviously not. On the contrary, even though fact-finding should not be constricted within unduly rigid *a priori* rules, it must be based on certain rules and respect certain principles, which are highlighted in the present volume.

The principles must be such that, having regard to the procedure followed and the persons entrusted with it, the fact-finding process enjoys the confidence of the international community as well as of the State concerned. It thus becomes possible more readily to obtain the co-operation of the latter, while not leaving the international community in any doubt about the integrity and reliability of the findings.

These principles must naturally be based on the principal concepts of *due process of law* in domestic procedures (such as the age-old rule 'auditur et altera pars'), but they must also make allowance for the special features of this kind of international action. Thus, in the event of on-the-spot visits, it will not normally be possible for a representative of the complainant to be present, nor will it be appropriate for a representative of the party complained against to take part in interviews with private individuals. The latter party should, however, be given an opportunity to comment on allegations received in the course of such visits. Similarly, precautions have sometimes to be taken to ensure the safety of witnesses and to protect them against intimidation or reprisals (or the mere fear of reprisals). More generally, one must bear in mind that the proceedings are aimed at ensuring observance of standards adopted by the world community (or by a regional organisation): they should therefore be investigatory in nature, particularly as regards the gathering of evidence, and should seek to obtain the fullest possible information on the matters at issue.

A process as difficult as human rights fact-finding calls not only for procedural safeguards. In a divided and distrustful world, and on questions where there exist profound differences of views, fact-finding itself and the conclusions and recommendations emanating from it are more likely to find acceptance if it is entrusted to independent and impartial persons. Not only logic, but also several decades of experience lead to that conclusion.

U.S. DEPARTMENT OF STATE, COUNTRY REPORTS ON HUMAN RIGHTS PRACTICES FOR 1994

(1995), Appendix A.

We base the annual Country Reports on Human Rights Practices on information available from all sources, including American and foreign government officials, victims of human rights abuse, academic and congressional studies, and reports from the press, international organizations, and nongovernmental organizations (NGO's) concerned with human rights. We find particularly helpful, and make reference in most reports to, the role of NGO's, ranging from groups in a single country to those that concern themselves with human rights worldwide. . . .

. . .

We have attempted to make these country reports as comprehensive as space will allow, while taking care to make them objective and as uniform as possible in both scope and quality of coverage. . . .

It is often difficult to evaluate the credibility of reports of human rights abuses. With the exception of some terrorist organizations, most opposition groups and certainly most governments deny that they commit human rights abuses and often go to great lengths to conceal any evidence of such acts. There are often few eye-witnesses to specific abuses, and they frequently are intimidated or otherwise prevented from reporting what they know. On the other hand, individuals and groups opposed to a particular government sometimes have powerful incentives to exaggerate or fabricate abuses, and some governments similarly distort or exaggerate abuses attributed to opposition groups. . . . Many governments that profess to oppose human rights abuses in fact secretly order or tacitly condone them or simply lack the will or the ability to control those responsible for them. Consequently, in judging a government's policy, it is important to look beyond statements of policy or intent in order to examine what in fact a government has done to prevent human rights abuses, including the extent to which it investigates, tries, and appropriately punishes those who commit such abuses. We continue to make every effort to do that in these reports.

NOTE

The following readings involve attempts to codify international practice. The first excerpt is from a detailed set of rules drawn up by the United Nations for a precedent-setting fact-finding mission to Chile in 1978. Despite attempts to do so, the UN has been unable to reach agreement on a set of model rules. Hence the Memorandum of Understanding agreed upon with the Government of Chile remains an important benchmark. Then follows a set of Rules drawn up by the International Law Association, an NGO whose recommendations carry no formal weight but whose professional standing ensures that these Rules are seriously considered by the international community.[9]

MEMORANDUM OF UNDERSTANDING WITH THE GOVERNMENT OF CHILE

UN Doc. A/33/331 (1978), Annex VII.

1. This memorandum reflects the exchanges between the *Ad Hoc* Working Group and representatives of the Government of Chile at the meetings held in New York from 22 to 26 May 1978.

A. Visit of the Ad Hoc Working Group to Chile

2. The representatives of the Government of Chile informed the Group that conditions now existed which enabled the Government to agree to a visit by the Group to Chile in fulfilment of the Group's mandate. The Group recognized the unprecedented nature of the visit it would undertake to Chile and the Group expressed its determination to fulfil its mandate in an objective and impartial manner and, within the terms of its mandate, to take steps to that end in co-operation with the Government of Chile. Taking into consideration the Government's wish that the visit take place in the near future and the need for adequate preparation it was agreed that the visit would begin on or about 12 July 1978 for an effective duration of two weeks, which the Group determined to be the minimum time necessary to carrying out adequately the visit as part of its mandate.

B. Facilities to be enjoyed by the Group during the visit

3. It was agreed that the Group would enjoy the following facilities during the visit which are necessary to the carrying out of its mandate:

(a) *Freedom of movement*

The members of the Group and the accompanying Secretariat staff will enjoy freedom of movement throughout the country.

[9] Mention should also be made of a Declaration on Fact-finding by the United Nations in the Field of the Maintenance of International Peace and Security, adopted by General Assembly Resolution 46/59 in 1991. It is addressed not to human rights matters but to the field of international peace and security. Nevertheless, some of its principles are potentially as applicable to human rights inquiries.

(b) *Freedom of investigation*

The Group, its members and the accompanying Secretariat staff will have access to prisons, places of detention and interrogation centres, will be able to interview freely and privately persons, groups and representatives of entities and institutions and will have access to pertinent files and other documents or materials which it considers necessary to its inquiry. The Government will provide members of the Group and the Secretariat staff with official identity documents stipulating the above. The representatives of the Government of Chile noted that access to places, persons and files, documents or other materials under the jurisdiction of the judicial authorities is subject to the authorization of the competent officials and that access to places connected with national security is similarly subject to the authorization of the competent officials. The representatives of the Government of Chile undertook to make the necessary arrangements with the appropriate authorities prior to and during the visit to ensure the Group's freedom of investigation. Visits or interviews with private persons or institutions would take place with due respect for the normal rights of said persons or institutions.

C. Assurances given by the Government of Chile in connexion with the visit

4. The representatives of the Government of Chile assured the Group that no person who had been in contact with the Group would for that reason be subjected to coercion, sanctions, punishment or judicial proceedings. The Group attaches particular importance to these guarantees.

5. The representatives of the Government of Chile assured the Group that the required measures will be taken to ensure the privacy and unimpaired conduct of the activities of the Group and the security of the members of the Group, the Secretariat staff and the Group's records and documents while in Chile.

6. The freedoms and assurances mentioned in paragraphs 3, 4 and 5 would be officially communicated to the Group in writing by the Government of Chile.

D. Basic rules of procedure

7. The Group reiterated that it could not diminish or depart from its terms of reference as determined by the Commission on Human Rights and the General Assembly nor could it delegate or renounce its sole responsibility for the interpretation of its mandate. Paragraphs 8 through 14 reflect the Group's understanding of certain aspects of its mandate.

8. The Group was of the opinion that its future reports to the Commission on Human Rights and the General Assembly should cover the situation of human rights in Chile from the most recent extension of its mandate, it being understood that the substance and conclusions of prior reports would in no way be affected, in part or in whole. Cases and

situations already mentioned in previous reports which continued to exist could be studied and reported on by the Group in fulfilment of its mandate.

9. Meetings and hearings of witnesses would, as provided for in rules 5 and 16 of the Group's rules of procedure, be held *in camera.*

E. Exchange of information between the Group and the Government

10. In order to facilitate collaboration between the Group and the Government of Chile the Group would communicate to the representatives of the Government of Chile in the measure possible the Group's views on areas of its concern relating to the situation of human rights in that country. The Group would also communicate to the Government of Chile information on testimonies, individual cases or events of concern to the Group relating to the situation of human rights in Chile insofar as such transmission of information is consistent with the Group's mandate and the Group's obligations to persons supplying information or mentioned therein. This would be done with a view to enabling the Government to submit information and views on these matters.

11. These exchanges would take place during and after the Group's visit and for this purpose a special meeting for two days between the Group and the Government would take place after the end of the visit. During the visit contacts with the liaison officer(s) in this regard would be maintained as appropriate. Information or views from the Government of Chile will be taken into consideration by the Group in the preparation of its report and included therein as appropriate.

12. In connexion with the contents of the substantive areas of the Group's report on which the Government of Chile had not previously had the occasion to submit its information or its views, the Group would make known to the Government the substantive contents of such areas before the final adoption of its report. Information and views of the Government in this regard would be taken into consideration by the Group and, as appropriate, included in its report. The Group agreed that it would annex to its report the observations of the Government, provided they were received prior to the end of the Group's meetings at which it adopts its report. Should they be received after such meetings they would be brought out in an addendum to the document containing the Group's report.

F. Information and evidence

13. The Group will continue to carefully weigh the evidentiary value of all the information it receives taking into consideration, *inter alia*, the character of the source of the information, its direct and reliable nature, the potential motivations of the source and the concordant nature of other information. The Group is aware that in taking these factors into consideration certain information or evidence, in some cases that from

official sources, both national and international, may, in given circumstances, have greater probative value than other information or evidence and that such be reflected in the findings of the Group. The Group is also aware that in relation to certain matters, for example economic, social and cultural rights, official documents, reports and studies may, if on the point, be relevant but not to the exclusion of other evidence. The Group will take into consideration the relevant information and conclusions arrived at by the specialized agencies and other international bodies on matters within their respective competence.

14. The Group wishes to point out that the nature of its mandate and the task it is required to perform requires that it be responsible for the final decisions of the probative value of information and evidence in light of all the relevant circumstances.

15. In connexion with its visit to Chile the Group will transmit to the Government of Chile an indicative but not exhaustive list of persons and representatives of institutions the Group may wish to interview and the places and institutions it may wish to visit to enable the Government to take those steps necessary to facilitate the visit. In the exercise of its freedom of movement and investigation the Group will itself definitively decide on its programme and the persons it will interview and the places it will visit.

. . .

H. Privileges and immunities of the members of the Group and Secretariat staff

17. The Government of Chile agreed that the members of the Group and the Secretariat staff would enjoy, in addition to the privileges and immunities to which they are entitled by virtue of the Convention on the Privileges and Immunities of the United Nations, full diplomatic privileges and immunities. This will be confirmed in writing by the Government of Chile.

. . .

INTERNATIONAL LAW ASSOCIATION, THE BELGRADE MINIMAL RULES OF PROCEDURE FOR INTERNATIONAL HUMAN RIGHTS FACT-FINDING MISSIONS

75 Am. J. Int. L. 163 (1981).

I. *Terms of Reference (Mandate)*

1. The organ of an organization establishing a fact-finding mission should set forth objective terms of reference which do not prejudge the issues to be

investigated. These terms should accord with the instrument establishing the organization.

2. The resolution authorizing the mission should not prejudge the missions's work and findings.

3. While terms of reference should not unduly restrict the mission in the investigation of the subject and its context, they should be so specific as to indicate the nature of the subject to be investigated.

II. Selection of Fact-Finders

4. The fact-finding mission should be composed of persons who are respected for their integrity, impartiality competence and objectivity and who are serving in their personal capacities.

5. Where the mandate of the mission concerns one or several specific states, in order to facilitate the task of the mission, the government or governments concerned, whenever possible, should be consulted in regard to the composition of the mission.

6. Any person appointed a member of the fact-finding mission should not be removed from membership except for reasons of incapacity or gross misbehaviour.

7. The chairman and the rapporteur of the fact-finding mission should not be replaced during the term of the mission except for reasons of incapacity or gross misbehaviour.

8. Once a fact-finding mission has been established and its chairman and members appointed, no persons should be added to the mission as members except to fill a vacancy in the mission.

III. Collection of Evidence

9. At the commencement of the mission, all material relevant to the purpose of the mission should be made available to it, with the assistance of the organization concerned.

10. Fact-finding missions should operate with staff sufficient to permit the independent collection of data and should be assisted by such independent experts as the mission may deem necessary.

11. Fact-finding missions may invite the submission of evidence that is in writing and contains specific statements of fact that are in their nature verifiable.

12. The state concerned should have an opportunity to comment in writing on data referred to in paragraph 10 and statements referred to in paragraph 11.

13. Both the petitioners, such as states, non-governmental organizations, or groups of individuals, and the states concerned may present lists of witnesses to the fact-finding mission. The fact-finding mission should make its own determination as to which witnesses it will hear.

14. Petitioners ought ordinarily to be heard by the fact-finding mission in public session with an opportunity for questioning by the states concerned.

15. The fact-finding mission shall in advance require the state concerned to provide adequate guarantee of non-retaliation against individual petitioners, witnesses and their relatives.

16. In case a guarantee, as referred to in paragraph 15, is provided to the satisfaction of the fact-finding mission, the latter should, on hearing witnesses, either provide an opportunity for the state concerned to be present and to question witnesses, or make available to the state concerned a record of the witnesses' testimony for comment.

17. The fact-finding mission may withhold information which, in its judgement, may jeopardize the safety or well-being of those giving testimony, or of third parties, or which in its opinion is likely to reveal sources.

18. On the basis of data generated by its staff, written statements, and testimony of witnesses, the fact-finding mission should make its own determination as to whether it needs to conduct an on-site inspection.

IV. The On-Site Investigation

19. The fact-finding mission should draw up its programme of work, including the list of witnesses it wishes to interview at the site of the investigation, places it wishes to visit, and the sequence, timing and location of its activities on the site.

20. The fact-finding mission may operate as a whole or in smaller groups assigned to conduct specific parts of the investigation.

21. The fact-finding mission should insist on interviewing any persons it deems necessary, even if incarcerated.

V. Final Stage

22. After conclusion of the on-site investigation, members of the fact-finding mission should draw up a set of preliminary findings and submit these, together with supplementary questions where appropriate, to the state concerned, giving it an opportunity, within a reasonable time, to present comments and/or to rectify the matter investigated.

23. A final report shall be prepared by the chairman reflecting the consensus of the fact-finding mission. In the absence of a consensus, the mission's report should contain the findings of the majority as well as any views of dissenting members.

24. In case a decision is made to publish the report, it should be published in its entirety.

25. The organization establishing the fact-finding mission should keep under review the compliance of states with their undertaking regarding non-reprisal against petitioners, witnesses, their relatives and associates.

NOTE

Any consideration of fact finding with respect to human rights (or a great many other subjects) must keep in mind one pervasive, ongoing and transformative feature of the modern world. Electronic communication through e-mail or the Internet, as well as the now familiar transmission of messages by facsimile, make possible the instant diffusion of information to a vast audience. These technologies have made it almost impossible to contain knowledge of new, serious human rights violations within any part of the world, however remote and isolated.

This recent revolution in communications puts the contemporary fact-finder in a radically different situation from predecessors. As recently as the early decades of this century, horrific events within isolated parts of Europe such as the rural Balkans (during the 1912–13 wars preceding World War I) became known only through difficult expeditions by investigators who brought their findings out with them.[10]

The contemporary fact-finder has a further powerful advantage over predecessors. The human rights researcher into a given country or problem has access to ever expanding and more efficient data bases that apply search technologies to organized data and thereby yield detailed information about a vast range of matters from all parts of the globe.

Of course this vast expansion in 'information' available to a broad public gives no assurance of its accuracy. The problems of verification of allegations of human rights violations may be no less serious. But at a minimum the allegations are known, a large number of unrelated sources may have made them, and those receiving the information are able to make their decisions about how to proceed.

QUESTIONS

1. Are these instruments about standards or methods of fact-finding of help when an investigator interviewing varied people within a state hears conflicting accounts of events? Suppose, for example, that X has disappeared. Relatives claim that X, a known opponent of the government, was murdered and his body disposed of. The government claims that X left home to fight with guerillas in the mountains.

2. If you were an investigator writing a report about violations, what criteria would you apply in deciding whether a violation had occurred? Would you follow a standard like 'more likely than not'? Would it be relevant whether the government, which had information about the relevant events, remained silent, or said that it had no knowledge, or offered an account of what had occurred that did not amount to a violation?

[10] See the account of the difficulties in getting a fact-finding Commission underway in George Kennan, The Balkan Crises: 1913 and 1993, in The Other Balkan War 3, 6–9. (A Carnegie Endowment Book 1993).

c. Case Studies of Fact-finding Under the 1235 Procedure

The following excerpt provides an overview of the UN Commission's experience in relation to fact-finding between 1979 and 1992, as background for the case studies that follow.

PHILIP ALSTON, THE COMMISSION ON HUMAN RIGHTS

in Alston (ed.), The United Nations and Human Rights: A Critical Appraisal 126 (1992), at 163.

. . .

(iv) *The Selection of Countries Targeted* [Although an enormous number of country situations have been discussed by the Commission under the 1235 procedure, only 23 special procedures (i.e. Special Rapporteurs, Rapporteurs, Special Representatives, Working Groups) were set up between 1979 and 1994. In relation to those countries that have been targeted, a number of imbalances exist. Eastern Europe was only subject to two investigations prior to 1992, remarkably few African countries have been identified and, in the Middle East, only Israel and Iraq have been singled out. Western Europe has not been targeted at all.]

Without seeking to justify these apparent imbalances, a number of factors help to explain them. In the first place, the great powers (such as the United States, the USSR, China, the United Kingdom, and France) and the major regional powers (such as India and Brazil) have all enjoyed a degree of immunity resulting from their political and economic clout. The case of China in 1990, when no State was prepared to introduce a much discussed and very watered-down draft resolution, best illustrates this regrettable phenomenon. The close allies of the great powers (e.g., El Salvador, Guatemala, Cuba, Poland, and Afghanistan) have not, however, avoided the spotlight. Secondly, regional solidarity has been an important means of preventing African and Arab nations from being scrutinized. It has been less effective in Asia and Latin America. Thirdly, it is clear that the Commission was only moved to action in cases involving widespread, documented repression of a serious kind. The 'mere' suppression of democracy, the violation of economic and social rights, the denial of the cultural and other rights of minorities and indigenees, and comparable violations were not deemed sufficient to warrant the creation of a special procedure. As a general rule it seems that blood needed to be spilled, and in large quantities. The main exceptions (such as Cuba and Poland) resulted from the investment of massive political capital (especially by the United States).

Finally, the disproportionately strong focus on Latin America may be explained by: the strength of (the myth surrounding) the democratic/rights tradition in that region, the brutality of the violations, the close involvement of the United States and thus the special concern of activist American NGOs, the detailed and powerful reporting of the Inter-American Commission on Human Rights in the late 1970s and early 1980s, and the very active role of NGOs and the churches in the countries concerned. But, notwithstanding the plausibility of such explanatory factors, the focus on Latin America during the 1980s provoked persistent accusations that the Commission applied double standards. American support for many of the governments that the Commission had singled out as 'gross violators' led the US Representative to the United Nations, Jeanne Kirkpatrick, to tell the General Assembly in 1981 that: '[H]uman rights has become a bludgeon to be wielded by the strong against the weak, by the majority against the isolated, by the blocs against the unorganized . . . The activities of the United Nations with respect to Latin America offer a particularly egregious example of moral hypocrisy.' . . . Nevertheless, while allegations of egregious hypocrisy can no longer be sustained, it remains true that double standards have prevailed and that many countries which have been thoroughly deserving of scrutiny have been intentionally overlooked. Only time and a political preparedness on the part of states that are members of the Commission to be critical of their allies (when such criticism is warranted) can cure this defect. By the same token, the Commission's achievements in relation to the countries that have been targeted should not be under estimated simply on the grounds that other States should also have been selected.

. . .

(vi) *Designation and Selection of Rapporteurs* The implementation of special procedures has been entrusted to a wide range of entities. While 'working groups' and 'special rapporteurs' were initially the favoured means of fact-finding, various other designations have been added over the years. They include: 'rapporteurs', 'envoys', 'special representatives', 'experts', 'independent experts', 'delegations', etc. The different terminology was originally intended to reflect an unstated hierarchy according to the gravity of the response. But the Commission's creativity, combined with its inconsistency in this regard, has served to blur the significance of these distinctions, at least in the minds of all but those diplomats who continue to fight with such vigour and enthusiasm to secure one designation rather than another. . . .

The selection of rapporteurs (using the term generically) has been a quintessentially political process, the mechanics of which do little to ensure that expertise and competence will be the principal qualities sought. . . .

. . . The task of identifying individuals to undertake the special procedures has generally been left to the Chairman of the Commission who is expected to 'consult' with the regional groups before making an appointment. The target country has sometimes been closely involved in the selection process, an approach which has not always been welcomed by human rights activists.

When the task of reporting on the situation in Poland was entrusted to a representative of the Secretary-General the resulting reports were brief, superficial, and unduly solicitous of the government's viewpoint. In light of a similarly unsatisfactory experience under the confidential procedures there would appear to be very good grounds for concluding that the task of objective fact-finding is inherently incompatible with the 'neutrality' which Secretaries-General seem to assume they must demonstrate, even in human rights matters.

There is no doubt that there has been great unevenness in the quality of the reports produced by different rapporteurs. Some have been accused of being 'in the pockets' of the government being investigated while the efforts of others have been deemed 'spineless' and 'needlessly apologetic'. One proposed solution is that they be 'appointed by a body that is insulated from political pressure and whose members serve for sufficient time so that they can be called to account for their choices over a period of years'. But the feasibility of such a system in the United Nations' inevitably politicized milieu is at best questionable. The present system at least enables some pressure to be brought to bear on the Chairman not to appoint an individual who is patently unqualified. A better approach would seem to be that proposed in a UN report in 1970 according to which violations would be investigated by a standing body which would be 'scrupulously non-political' and 'should strive to offer all guarantees of impartiality, efficiency and rectitude'. Such a system would also have its potential defects but it is difficult to accept that they would outweigh those inherent in the existing approach.

Another valid criticism is that the range of people from which the Commission generally draws its rapporteurs is unduly narrow. [The first woman ever appointed under the public procedures was Radhika Coomaraswamy, Special Rapporteur on Violence against Women, appointed in 1994.] The nominees are usually diplomats, and expertise in human rights law or in the politics and culture of the region concerned is often not prominent among their attributes. . . .

(vii) *The Mandate: Prosecutors, Solution Seekers or Fact-finders?* The terms used to describe the formal mandates given to country rapporteurs have varied considerably. But whether they have been asked to 'study', 'inquire into', 'investigate', or 'examine', most rapporteurs have tended to assume considerable flexibility and to approach each situation as they see fit. The Commission, for its part, has generally not sought to impose any procedural straitjackets and has been reluctant to criticize the approach adopted by individual rapporteurs. Not surprisingly, this lack of structure has resulted in enormous disparities of style, methodology, content, and focus from one report to another. It has also enabled individual rapporteurs to assume that they have a *carte blanche* in determining the nature of their reports.

For analytical purposes, three principal approaches to country-reporting may be discerned. The first emphasizes the *fact-finding and documentation function*. In this view the function of reporting is to record the facts, to provide a reliable historical record, and to provide the necessary raw material against the background of which the political organs can determine the best strategy under the circumstances. In this approach, facts and their substantiation are the key. The report prepared by the Special Commission appointed in Argentina in 1984 is an excellent example in this regard. The second approach assumes that the *prosecutorial/publicity function* is paramount. Thus the rapporteur's role is not to establish whether violations have occurred but to marshal as much evidence as possible to support a condemnation that, in many instances, will already have been made. Thus Bailey has described the Special Committee on southern Africa as a fact-finding body 'only in the sense that it collects and collates facts in pursuit of a predetermined political aim'. The goal in such cases is to mobilize world public opinion and to provide the basis on which the earlier conviction can be justified. The third approach is to emphasize the *conciliation function*. The rapporteur's role is not to confront the violators but to seek solutions which will improve, even if not necessarily resolve, the situation. The perceived challenge is to steer a middle course between the positions of the accused and their accusers and the emphasis is on dialogue and co-operation between the Commission and the government. The high (or some would say low) point of this approach was the report on Guatemala in 1984 in which the Rapporteur stated his 'belief that the Commission . . . wishes to encourage as well as condemn'. He concluded by urging the Commission to adopt 'a constructive approach'. In a subsequent report, the same person, albeit redesignated as a Special Representative, concluded (again to the dismay of most observers) that the relevant government's efforts to guarantee human rights, 'even if not yet perfect or complete, should receive the support of the international community'.

Although each of these models has had its adherents among the various rapporteurs the Commission itself has never acknowledged that different situations appear to call for different approaches. It would seem wholly inappropriate, and even unfair, for such an important decision to be left to the predisposition of the rapporteur. It is inevitably unseemly and most probably unproductive for a rapporteur to engage in haggling with a government over whether his report will be 'constructive' and mild, or 'condemnatory' and uncompromising. An exercise which purports to be engaged in 'fact-finding', but then denies or plays down the seriousness of the violations in order to maintain a conciliatory stance, insults the victims, misleads the public and the Commission, and discredits the United Nations. It is the Commission itself which should determine what type of report is called for and it would be possible to draw up some general (perhaps informal) guidelines for that purpose. . . .

(viii) *Procedures and Fact-finding Methodology* It has been persuasively argued that 'if fact-finding is to become more than another chimera, the sponsoring institutions must develop universally applicable minimal standards of due process to control both the way the facts are established and what is done with them afterwards'. But in these respects the Commission's procedures got off to a bad start and have yet to recover fully. The first two situations dealt with (apartheid and Israeli occupation) were hardly conducive to dispassionate and objective reporting and the resulting precedents were strongly criticized. As a result, the 1968 International Conference on Human Rights, held in Teheran, recognized 'the importance of well defined rules of procedure for the orderly and efficient discharge of their functions' by UN human rights bodies and called for 'model rules' to be prepared. A Working Group of the Commission subsequently took five years to come up with an incomplete draft which was never adopted but which ECOSOC requested be brought to the attention of all relevant bodies. In principle, therefore, the model rules are still relevant but, in practice, they have long since been forgotten. The only other comparable attempts to devise procedures designed to ensure both fairness and effectiveness were the rules drawn up by a fact-finding mission to South Vietnam in 1963 and the arrangements contained in a memorandum between the Working Group on Chile and the Chilean government in connection with the Group's on-site visit in 1978.

Despite the lack of any formal guidelines it has been suggested that the various rapporteurs do in fact generally follow certain 'basic rules and procedures'. But insofar as this may be the case, the rules in question are so rudimentary that they do not go anywhere near ensuring the degree of integrity of process that most commentators would agree to be essential if consistency, credibility, effectiveness, and fairness are to be achieved. Thus, for example, one rapporteur could happily observe that he was 'much indebted to the Army . . . for their safe-keeping and their transport, with many other kindnesses and facilities', without thereby running foul of any procedural rules for independent fact-finding. In brief, the existing methodology for fact-finding by UN rapporteurs is *ad hoc*, inconsistent, and often unsatisfactory. Perhaps the clearest demonstration of their inferiority is to compare the reports produced by UN fact-finders with those produced by their Inter-American counterparts. In the case of Guatemala and, albeit to a slightly lesser extent, Chile, the difference in quality and thoroughness is enormous. Similarly invidious comparisons may be made with respect to the ILO, with the case of Poland being especially telling. Yet in spite of these patent shortcomings, the Commission has tended to operate on the assumption that any attempt to move towards a more carefully structured and credible process would only be hijacked by those opposed to effective fact-finding. But it is now time to challenge that assumption and to move towards a methodology which would ensure that UN reports consistently pass muster in terms of their procedural probity. . . .

(ix) *The Preparedness of Governments to Co-operate* The picture in this regard is a very mixed one. Neither South Africa nor Israel have ever formally co-operated with the Commission's investigative organs. Chile permitted the Working Group to visit on only one occasion and almost certainly considered that to have been a grave mistake given the Group's uncompromising report on its findings. At a later stage, the Chilean government relented after a new rapporteur released what a leading NGO termed an 'informal, summary and highly selective' report in 1985. As a result the rapporteur was able to conduct on-site investigations each year from 1986 to 1989. Poland was singularly uncooperative although a visit by the Secretary-General's representative was eventually tolerated. Afghanistan was entirely negative at first and then changed its tune dramatically to permit a visit by the rapporteur in 1987. Iran took a similar approach while Guatemala, Bolivia, and El Salvador have all permitted relatively unrestricted access to the respective rapporteurs. Cuba and Romania (after the fall of President Ceaucescu) have also been prepared to admit the Commission's representatives.

Overall, the factors that induce governmental co-operation seem to be rather diverse and to depend very much on a calculation, taking account of all of the relevant circumstances, as to the relative costs of co-operation versus non co-operation. The costs of the latter are being steadily increased but they are still far from being consistently prohibitive. But whether or not an on-site visit is permitted, the great majority of governments have sought to defend themselves systematically and vigorously within the Commission. Thus detailed rebuttals of country-specific reports are now very much the norm rather than the exception.

NOTE

The following readings are reports prepared for the Commission by special rapporteurs. The purpose of examining them is not to focus on the situation in the countries concerned, but rather to assess the adequacy of the reports in the light of the procedural standards noted above, as well as to get a sense of some difficulties involved in such fact-finding.

REPORT OF THE SPECIAL RAPPORTEUR ON CUBA

UN Doc. E/CN.4/1993/39.

[At the request of the Commission, the Special Rapporteur (Mr Carl-Johan Groth) reported in 1993 on a wide range of civil and political rights-related issues. The Government did not permit him to make an on-site visit. The following excerpts deal only with the rights to freedom of opinion, assembly and association; some objections raised by the Cuban Government; and the question of the United States' economic embargo against Cuba.]

. . .

IV. RIGHTS TO FREEDOM OF OPINION, ASSEMBLY AND ASSOCIATION

[The Special Rapporteur reviewed the relevant provisions of both the Universal Declaration and the Cuban Constitution in relation to these rights.]

. . .

29. The information received by the Special Rapporteur from individuals and non-governmental groups suggests that the framework of protection established by these provisions and the manner in which they are applied do not duly conform to the principle laid down in the Universal Declaration of Human Rights, because persons associated with groups whose purpose is to denounce violations of human rights, or groups of differing tendencies which are critical of the current political system, are harassed, even though their actions are perfectly peaceful and they address the authorities in a respectful manner. Furthermore, these groups that have emerged in recent years have not been able to obtain legalization since the numerous applications to the Ministry of Justice for legalization under the Associations Act have gone unanswered.

. . .

31. As regards the scale of the human rights violations referred to in the present section, the Special Rapporteur considers it worth mentioning, as an example, a document prepared by the organization 'Americas Watch', dated 30 September 1992 and updated in January 1993, in which it is stated that more than 250 people associated with human rights groups have been detained since 1989. At least 50 of them have apparently been serving sentences of up to 10 years for their activities in this connection. Others are being held in custody pending trial. At least half of the total have been in custody since September 1991. Many others have remained in custody for short terms in police or State Security Department premises.

. . .

A. Trial and sentencing

33. An Amnesty International report of December 1992 indicates that it is difficult to estimate the number of persons sentenced for political reasons since the authorities do not provide information on them, the activities of the national groups which try to follow up these cases are severely restricted and, for some time now, the international human rights organizations have not been allowed access to the country. However, Amnesty International estimates that there are at least 300 to 500 of them at present and that possibly half of them have been sentenced for trying to leave the country illegally.

. . .

IX. CONCLUSIONS AND RECOMMENDATIONS

80. The Government of Cuba has so far refused to cooperate at all with the Special Rapporteur, but nevertheless the Special Rapporteur continues to hope that, in the not too distant future, there will be an opportunity to open a dialogue and to make working visits to Cuba.

81. The Cuban Government's position is based on the arguments set out below. . . .

82. Secondly, the Government maintains that there are no human rights violations in Cuba, at least not on a scale comparable to the situations of mass violations existing in many countries being monitored by the Commission on Human Rights under agenda item 12 [relating to the 1235 procedure]. Accordingly, the Government maintains that the study of the human rights situation in Cuba in this context is unjustified and is entirely politically motivated.

83. In this connection, the Special Rapporteur wishes to make a number of observations. His mandate does indeed have a political origin, in that it derives from a resolution of the Commission on Human Rights, which is a politically constituted body. Regardless of this, however, it is the obligation of the Special Rapporteur, as of all rapporteurs and experts appointed by the Commission on Human Rights, to discharge his duties with impartiality, independence and objectivity, taking as points of reference the values embodied in international human rights instruments, particularly the Universal Declaration. While these instruments may also be said to have a political content, in that they are the result of a specific conception of the individual, society and the State, they have been universally recognized and now represent a minimum standard, regardless of the social or ideological situation prevailing in a given country.

. . .

85. This report and its recommendations focus on such basic rights as freedom of opinion, assembly and association, as well as trade union freedom and religious freedom. It also describes the practices followed by the Government in regard to freedom of movement and draws attention to the situation of the prison population, particularly of persons serving sentences for offences with political connotations. The report does not mention the right to life because, at present, violations in this area, such as summary or arbitrary executions or enforced or involuntary disappearances, are not typical of the Cuban situation. Economic, social and cultural rights have also been omitted.

86. There can be no doubt that the Cuban political system and Cuban society have special characteristics and, as a result, so do the human rights situation and the systematic violations committed. However, this in no way justifies such violations or makes them acceptable. One of the characteristics mentioned in the report which deserves special consideration is the fact that dissidence within the society is expressed in an entirely peaceful manner, in an attempt to avoid any confrontation or violence.

87. In the light of the human rights situation in Cuba as described in this report, the Special Rapporteur, through the Commission on Human Rights, recommends that the Government of Cuba should adopt the following measures with a view to improving the observance of fundamental rights:

(a) Ratify the principal human rights instruments to which Cuba is not a party, in particular, the International Covenant on Civil and Political

Rights and its Optional Protocols and the International Covenant on Economic, Social and Cultural Rights;

(b) Cease persecuting and punishing citizens for reasons relating to the freedom of peaceful expression and association;

(c) Permit legalization of independent groups, especially those seeking to carry out human rights or trade-union activities, and allow them to act within the law, but independently;

(d) Respect the guarantees of due process, in accordance with the provisions set forth in international instruments;

(e) Ensure greater transparency and guarantees in the prison system, so as to avoid incidents of excessive violence exercised against prisoners. In this connection, it would be a major achievement to renew the agreement with the International Committee of the Red Cross and to allow independent national groups access to prisons;

(f) Review sentences imposed for offences with political connotations and for trying to leave the country illegally;

(g) Expedite and make more transparent the procedure for applying for permission to leave and enter the country, while at the same time avoiding measures of retaliation against the applicants. Family reunification cases should be given priority attention. On this subject, the Special Rapporteur is aware of the need for persons wishing to travel to have visas for entry into other countries.

88. It would be desirable for Cuba's policy on travel abroad to be less strict and, for other countries, in their turn and as a counterpart, to relax the restrictions imposed on their own citizens regarding travel to and communication with Cuba. This, in addition to resolving difficulties from a strictly humanitarian aspect, would help to end the artificial and painful isolation in which the Cuban people live.

89. While not overlooking the urgent need for specific measures, as proposed above, the Special Rapporteur nevertheless wishes to point out that any analysis concerning the situation and implementation of human rights in Cuba must, as a point of departure, accept the fact that the Government is, and has for a long time been, surrounded by an international climate extremely hostile to many of its policies and, in some cases, even to its very existence. This hostile international climate does not seem to have been affected by the vast political, military and economic changes that have taken place in the world in the last few years. Similarly, the changes which have occurred in the previously socialist European countries, as also in the policies of many third world countries, seem to have so far had no impact on Cuba's internal policy. On the other hand, the abrupt breakdown in the flow of aid previously received from abroad, as well as the almost total exclusion of Cuba as a beneficiary of the multilateral financing and technical assistance agencies, have not given the Government much room for manoeuvre in this field. In the opinion of the

Special Rapporteur, a policy *vis à vis* Cuba based on economic sanctions and other measures designed to isolate the island constitute, at the present stage, the surest way of prolonging an untenable internal situation, as the only remedy that would be left for not capitulating to external pressure would be to continue desperate efforts to stay anchored in the past. International sanctions, especially if accompanied by conditions implying the adoption of specific measures, be they political or economic, are totally counterproductive if it is the international community's intention to improve the human rights situation and, at the same time, to create conditions for a peaceful and gradual transition to a genuinely pluralist and civil society. Any suggestion along the lines that the future sovereignty of the Cuban people could be contingent on external powers or forces would, in the collective memory of the Cuban people, evoke traumatic experiences of their not-very-distant history and their fight for independence, and would be a very effective obstacle to the achievement of changes which could be very welcome in other circumstances.

. . .

QUESTIONS

1. The extent to which the Special Rapporteur must depend upon NGO sources is apparent from these brief extracts. What are the implications for the report, if any, of that dependence?

2. In the case of Cuba, implementation of the Special Rapporteur's recommendations would be tantamount to a radical change in the country's system of government. Is it appropriate for this implication to be neither recognized nor addressed? On what basis—that is, on what human rights norms or other principles—has the Special Rapporteur made this recommendation?

3. Given that the U.S. embargo is not mentioned in the Commission's resolution, is it an appropriate matter for the Special Rapporteur to raise? To what extent should underlying structural influences be addressed in such 'fact-finding' reports? In other words, how discrete (limited to the precise human rights violations at issue) or how broad-ranging (looking to a country's total situation) should UN inquiry be?

REPORT OF THE SPECIAL REPRESENTATIVE ON IRAN

UN Doc. E/CN.4/1991/35.

[The Commission on Human Rights appointed a Special Representative to investigate and report on the human rights situation in Iran. The Special Representative's mandate was contained in Commission Resolution 1984/54 and was several times extended. In 1990, the Commission requested the Special Representative, Reynaldo Pohl from El Salvador, to submit to it a final report at its 47th (1991) Session. There follow excerpts from that report.]

. . .

B. Written communications concerning allegations received by the Special Representative and transmitted to the Government

. . .

12. On 17 December 1990 the Special Representative wrote the following letter to the Permanent Representative of the Islamic Republic of Iran to the United Nations Office at Geneva:

. . .

> 'Since I intend to finalize my report by mid-January 1991, I would be most grateful if any further replies your Government might wish to submit could be communicated to the Centre for Human Rights not later than 10 January 1991, so that they may be included in my report and be taken into consideration in the conclusions and recommendations I intend to place before the Commission.'

14. On 10 January 1991, the Permanent Representative of the Islamic Republic of Iran to the United Nations Office at Geneva addressed to the Special Representative the following letter:

> '. . . [I] would like to draw your attention to the fact that the deadline for submission of replies set forth for 10 January 1991 is absolutely impractical to meet as investigations of this nature are time consuming. The time, therefore, needs to be extended to a reasonable date.'

. . .

15. By letter dated 11 January 1991, the Permanent Representative of the Islamic Republic of Iran to the United Nations Office at Geneva transmitted replies to the letters dated 27 June and 9 July 1990 addressed to him by the Special Representative, as follows:

> '. . . I have the pleasure to inform you that since 10 December 1990, the following persons have been released: Reza Sadr, Ezatollah Sahabi, Farhad Behbahani, Abbas Ghaem Al-Sabahi, Mahmoud Naimpoor, Nour Ali Tabandeh and Hossein Shah-Hosseini.
>
> 'With regard to the other detainees, as soon as the judicial proceedings are concluded, you will be informed of the final results.'

16. By letter dated 14 January 1991, the Special Representative replied to the letter of the Permanent Representative dated 10 January 1991 as follows:

'. . .

> While I appreciate the difficulty of investigating fully the most recent allegations transmitted by letter dated 28 December 1990, I have to remind you that the majority of the allegations received in 1990 were communicated to you by letters dated 20 August and 8 October 1990 and for certain individual cases even earlier. In addition, replies are still outstanding for allegations sent to you in previous years.'

. . .

II. Information Received by the Special Representative

29. The following paragraphs contain allegations of human rights violations received by the Special Representative and transmitted to the Government of the Islamic Republic of Iran by memoranda dated 20 August, 8 October, 28 December 1990 and 29 January 1991. Replies received from the Government with regard to the alleged incidents and cases have also been reflected in this section. . . .

A. Right to life

30. According to a report by the daily *Abrar*, a man condemned for fornication was publicly executed in Mashad in early 1990. Agence France Presse reported on 16 January that a 31-year-old woman convicted of prostitution had been stoned to death in Bandar Anzali. On 31 January *Jomhouri Islami* published a declaration of the Komiteh Commander of the Province of West Azerbaijan, according to which five persons engaged in prostitution and corruption had been stoned to death. According to a report by *Ressalat* on 15 February 1990, Gholam Reza Masouri was hanged in Arak for pederasty.

31. *Jomhouri Islami* reported on 17 February 1990 that Bolouch Ismalel Zehi had been executed for drug-trafficking. On 10 January Radio Tehran announced that 31 persons convicted of drug-trafficking had been executed, 23 of them in Tehran, 3 at Shiraz, 3 at Sabzevar and 2 at Saveh. According to *Ressalat* of 11 February, a married couple accused of drug-trafficking was sentenced to death in Saveh.

32. Dailies from various countries published the statement of Mitra Moazez (21), claiming that she had been forced to witness the death by burning of a 37-year-old woman and two 18-year-old men in an Iranian prison. . . .

. . .

36. On 28 April 1990, the newspaper *Kayhan* published a report by the Islamic Republic News Agency to the effect that the Prosecutor General of the Tehran Revolutionary Court had announced that 10 persons accused of

espionage would be executed in the next few days. Other sources reported directly to the Special Representative that Mr. Jamshead Amiry Bigvand, former Director of the Marodasht Shiraz Petrochemical Laboratory, and 13 other persons had allegedly been convicted on the charge of espionage for the United States of America, an offence for which capital punishment might be applied. Reportedly these persons had been held for months in solitary confinement at Evin prison, and had not been allowed to avail themselves of legal assistance of their own choosing. It was further alleged that confessions had been extracted under torture and that some of them had been compelled to make extrajudicial confessions which were broadcast by Iranian television. The Special Representative requested the Government, by a letter dated 8 May 1990, to enable all 14 persons to benefit from all the procedural safeguards provided for in articles 6 and 14 of the International Covenant on Civil and Political Rights. . . .

38. By a letter dated 5 June 1990, the Permanent Representative of the Islamic Republic of Iran to the United Nations Office at Geneva forwarded to the Special Representative the following response of the judicial authorities of the Islamic Republic of Iran:

> 'According to the article 37 of the Constitution of the Islamic Republic of Iran, and as contained in the second paragraph of article 14 of the International Covenant on Civil and Political Rights, no person shall be considered guilty by law unless the accusation against him is proved by a competent court and the courts are naturally obliged to act accordingly;
>
> 'In the light of information received by the Islamic Revolutionary Court, those people were arrested and tried in accordance with the law. In addition, they were entitled to appoint a legal counsel and they duly and freely defended themselves during the trial;'

. . .

42. It has further been reported that on 14 February 1990 a judicial panel sent to Hamadan on behalf of the Head of the Judiciary issued the following sentences:

(a) Gholamhossein Golzar, 27 years old, discharged employee of the Agricultural Bank of Hamadan: 74 lashes for committing robbery; 92 lashes for participation in a forbidden act, and decapitation by the just sword of the Imam Ali;

. . .

(c) Reza Khanian, 23 years old, fruit and vegetable centre clerk: 74 lashes for committing robbery; 50 lashes for participation in a forbidden act; amputation of hand for committing assault and battery and hanging by scaffold.

43. The newspaper *Kayhan* announced on 3 January 1990 that Khodakaram Zamani, given a retributory death sentence for the murder of Morad-Ali Rezai, was executed in the main square of Khorramabad.

. . .

53. It has also been widely reported that the Iranian Government has endorsed the death sentence against the British author Salman Rushdie. On 5 June 1990, the Leader of the Islamic Republic of Iran reportedly stated that the *fatwa* (religious verdict) of the late Imam Khomeini concerning the author was based on divine rulings and remained irrevocable. On 26 December 1990 the Leader of the Islamic Republic of Iran reiterated that the *fatwa* cannot be revised or repealed by anyone at any time.

54. By a letter of 22 January 1991, the Permanent Representative of the Islamic Republic of Iran also referred to the Rushdie case, as follows:

> '. . . It should be pointed out that as a result of the criminal act of Mr. Rushdie which was a direct insult to the most sacred values of Muslims, tens of people lost their lives in different parts of the world. It is very surprising to note that the Special Representative addressed this political issue under the humanitarian mandate without making any reference whatsoever to those whose blood was spilled in protest to this criminal act. The Special Representative is expected to show as much sensitivity to the right to life of those Muslims who lost their lives as that which he extends to the culprit.'

. . .

59. It has been reported that as of January 1990 persons have been executed in the Islamic Republic of Iran for their homosexual or lesbian tendency.

60. In its reply of 22 January 1991, the Government of the Islamic Republic of Iran stated that 'according to the Islamic Shariat, homosexuals who confess to their acts and insist on that are condemned to death'.

. . .

B. Enforced or involuntary disappearances

73. The Special Representative wishes to refer to the report of the Working Group on Enforced or Involuntary Disappearances (E/CN.4/1991/20), which has transmitted to the Government of the Islamic Republic of Iran 451 cases of missing persons, 7 of which were reported to have occurred in 1990. So far only one case has been clarified by information received from non-governmental sources.

C. Right to freedom from torture or cruel, inhuman or degrading treatment or punishment

74. Reports on torture and ill-treatment during imprisonment have continued to be received since the first visit of the Special Representative to the Islamic

Republic of Iran. It was also alleged that mutilations and corporal punishment are being applied. In this context, it is pertinent to note an Agence France Presse report that, according to *Kayhan*, a person convicted of robbery suffered the amputation of four fingers on his right hand in Ghasr prison at Tehran. . . .

75. In the letter of 22 January 1991, the Government of the Islamic Republic of Iran stated that 'any physical abuse or torture under any name is denied in the prisons and they are totally baseless. But in regard to amputations, it is worth mentioning that the divine religion of Islam permits certain punishments for certain crimes and in the field of *Qesas* which entails amputations'.

. . .

IV. Conclusions and Recommendations

. . .

464. The Special Representative has expressed the view, which is also his belief, that, on the basis of existing international principles and known facts, a critical analysis should be made in order to arrive at some recommendations and conclusions in the light of criteria of probability. The aim is, of course, to base conclusions and recommendations not on findings similar to those of a court of justice, but, rather, on probability and reasonable belief. . . .

465. This exercise is particularly complex and complicated because of the specific circumstances which characterize the monitoring of human rights in the Islamic Republic of Iran. It is obvious that there are persons who are making every effort, and even going to any lengths, to politicize this exercise, politicization being taken to mean a departure from the objective consideration of the facts and their consequences and the use of human rights as an instrument in the struggle for political power. Reports on international monitoring procedures may be used for political purposes and such use is beyond the scope of the competent United Nations bodies, but the aim of the exercise is not to support or destabilize Governments, but, rather, to encourage the fulfilment of international obligations relating to human rights. The purpose of international monitoring is to bring about fulfilment of international human rights obligations by gathering information and evaluating, analysing and criticizing it.

466. The mandate relating to Iran is one of the most controversial of all the mandates on which international monitoring has focused in particular countries in recent years. This is probably the result of the radical polarization of political forces, the conflict between opinions that have turned into pre-established, inflexible, intransigent credos and the struggle between national and international political interests. The situation is being followed attentively and even passionately in and outside the United Nations and in the media throughout the world.

. . .

468. The Special Representative has concentrated on objective consideration of the facts and the preparation of observations and recommendations on the basis of critria of probability without giving in to the pressure and Manichaeism of persons or groups not connected with international monitoring. However, every voice which has expressed an opinion or criticism and drawn attention to procedures has been listened to. This task is one plagued by doubts, questions, conflicts of conscience and contradictory requests and it is a difficult one because the aim is to guarantee objective and independent international monitoring and compliance with international instruments, without complacency or fear.

. . .

471. During the second visit, information was received on the categories of acts which have been considered in previous reports, such as executions, the lack of a defence lawyer, failure to notify detainees of the charges against them immediately following arrest, difficulties in making trials public and ill-treatment and torture. Consistent information was also received on restrictions on freedom of the press, the publication of books and artistic creation, and delays and difficulties in exercising the right to freedom of association, including the right to form political parties. . . .

. . .

473. It should be noted that the laws governing political parties in particular and associations in general, contain provisions which in theory refer to the safeguarding of the Constitution and moral and religious principles, but whose practical effect is that associations whose objective is political propaganda and participation in electoral activities or the protection of human rights are not legally recognized. . . .

474. The positive measures adopted by the Government include: (a) the replies to many allegations which were communicated to it and which are reproduced *in extenso* in the present report; (b) the favourable outcome in a number of cases submitted by the Special Representative for consideration on humanitarian grounds; (c) the periodic adoption of clemency measures which benefit both ordinary prisoners and political prisoners; (d) the release of seven of the signers of the so-called 'Letter by the 90'; and (e) the decree of 31 December 1990, which requires a defence lawyer to be present at all stages in criminal proceedings. . . .

. . .

476. The Government acknowledged 113 executions between the first and second visits. Calculations based on the gathering of information broadcast on the official radio indicate that about 500 persons were executed between January and October 1990. The second visit had barely ended when Iranian radio reported further executions. The media reportedly announced several

dozen executions in January 1991. It is also known that some missing persons were in some way linked with executed persons and their disappearance is thus a matter of concern to their families and friends.

477. According to the information received, most of the executions concern persons accused of drug trafficking and the others are for ordinary offences of various kinds and political offences. The study of Iranian legislation clearly indicates that, because there are no gradations in the penalties for various types of criminal involvement and as a result of very general wording, the death penalty tends to be applied on a large scale. . . .

. . .

478. Executions in the Islamic Republic of Iran continue to go beyond the narrow limits within which the International Covenant on Civil and Political Rights allows the application of capital punishment.

. . .

483. The situation of followers of the Baha'i faith continues to be uncertain, given the unequal treatment they receive in different provinces and cities, depending on the ideas and temperament of individual officials. No reports of executions have been received in recent months. There do not appear to be any exception to the prohibition on admission to universities and there have been very few exceptions to refusals to grant legal recognition to inheritance rights . . . The General Assembly requested the Government to ensure that all individuals within its territory and subject to its jurisdiction, including religious groups, enjoy the rights recognized in international instruments (resolution 45/173).

484. During his second visit to the country, the Special Representative talked in private homes to people unconnected with the Government who lead a normal life and have no judicial or police problems. These people agreed to be interviewed after taking precautions to preserve their anonymity since they feared reprisals if it became known that they had given information about the human rights situation prevailing in the country. They said they feared mainly the activities of irregular groups and of Komiteh and Pasdarán agents who use intimidatory tactics. Other people interviewed at the United Nations Development Programme offices and the Esteghlal Hotel voiced the same fear.

. . .

492. The Special Representative is of the opinion and belief that there are arguments in favour of extending the international monitoring of the situation of human rights in the Islamic Republic of Iran with the objective of assuring, through co-operation, criticism, international public opinion and measures adopted by the Iranian authorities, that Iranian legislation, administration and practice are brought fully into line with the international instruments in force.

493. On the basis of the aforementioned information, the Special Representative, as in previous reports, ventures to make a number of recommendations. . . . The Special Representative states opinions and cannot act as a substitute for the Government, the General Assembly or the Commission on Human Rights, whose respective attributions include decision-making powers. It is for them to consider the recommendations and decide on them.

494. The Special Representative wishes to state that, in his view, it would be appropriate to adopt the following measures:

(a) The Government should take immediate action to reduce drastically the application of the death penalty, and, while technical reforms are being introduced into penal legislation, clemency and the right of pardon should be exercised broadly;

(b) Just as the penalty of flogging is being gradually replaced by a fine or imprisonment, consideration should be given to replacing the penalties regarded by the international organizations as forms of torture, including stoning and amputation;

(c) The Government should be urged to initiate forthwith or to speed up the pace of legislative and administrative reform to make national institutions compatible with the international human rights instruments, beginning with the introduction of technical reforms to penal legislation, as well as to introduce remedies to make moral and economic redress effective and to assign responsibility for abuses or excesses of power;

(d) The Government should carefully supervise the enjoyment of equal rights and equal treatment for all citizens, regardless of their political opinions or their religious beliefs;

(e) The Government should be urged to take, immediately and urgently, effective measures to establish a climate of confidence and legal certainty in institutions to enable citizens to express themselves without fear or intimidation;

(f) The Government should take care to apply the rules of due process of law . . . ;

(g) A specific agreement should be concluded soon with the International Committee of the Red Cross so that prison visits may be carried out regularly and without exceptions;

(h) The legal functioning of independent organizations should be authorized, including political organizations and organizations that seek to defend human rights;

(i) The prior examination of books and forms of artistic creation in general should end;

(j) Measures should be adopted to guarantee genuine freedom for the media and journalists should enjoy full guarantees for their professional activities;

(k) Compensation should be granted to persons affected by violations of human rights or to members of their families.

. . .

QUESTIONS

1. On what human rights instruments or bodies of norms does the rapporteur draw in criticizing Iran's actions? What norms, indeed, ought the UN Commission invoke in assessing state conduct? Is there 'one' decisive instrument? Is it relevant whether the state involved has ratified a particular treaty?

2. Are any concessions made in this report to Iran's distinctive situation as an Islamic state, to cultural relativism?

3. Given the rapporteur's inability to enter Iran, how does he 'find' the facts? Are there identified procedures for resolving conflicting assertions? What standard or burden of proof does the rapporteur appear to employ to determine if allegations are proven?

4. What functions are the recommendations meant to serve? To whom are they realistically addressed?

3. THE EVOLUTION OF TECHNIQUES FOR PROMOTING COMPLIANCE: 'THEMATIC' MECHANISMS AND BEYOND

COMMENT ON RANGE OF THEMATIC MECHANISMS

The 1503 procedure is extremely time-consuming and, in principle at least, deals only with overall 'situations' rather than with individual cases. The 1235 procedure, by contrast, is in principle capable of addressing any type of case. However, until such time as a specific fact-finding procedure (such as a rapporteur) is established for the relevant country, the procedure is limited to general debate within the many constraints of the Commission's annual session.

It is then obvious that these two procedures leave a serious gap in terms of the Commission's ability to respond to the plight of individuals or groups in countries that are not under specific 1235 investigations. The 'thematic' mechanisms that have been developed by the Commission since 1980 go a long way towards filling that gap. Each mechanism focuses on a particular 'theme' or genre of violations rather than on a particular country. By the end of the

Commission's 1995 session the following 'thematic' mandates were in existence:

> **Working Groups:** *Enforced or involuntary disappearances*—5 members; *Arbitrary detention*—5 members.
>
> **Special Rapporteurs:** *Extrajudicial, summary or arbitrary executions*—Mr Ndiaye, Senegal; *Freedom of opinion and expression*—Mr Hussain, India; *Racism, racial discrimination and xenophobia*—Mr Glele-Ahanhanzo, Benin; *Torture and other cruel, inhuman or degrading treatment*—Mr Rodley, UK; *Religious intolerance*—Mr Amor, Tunisia; *Use of mercenaries*—Mr Bernales-Ballesteros, Peru; *Sale of children, child prostitution and child pornography*—Ms. Calcetas-Santos, Philippines; *Internally displaced persons*—Mr Deng, Sudan; *Independence and impartiality of the judiciary*—Mr Cumaraswamy, Malaysia; *Violence against women*—Ms Coomaraswamy, Sri Lanka; and *Effects of Toxic and Dangerous Products on Human Rights*.

In their quest for effectiveness, the thematic mechanisms have developed certain techniques that include: requests to governments for information on specific cases; an urgent action procedure involving a request that a government take immediate action to rectify or clarify a case; and on-site visits for a more intensive and enduring examination of a series of cases. The manner in which these techniques are used, if at all, varies enormously from one mechanism to another. The mechanisms also vary in the extent to which they have sought to explore the doctrinal aspects of the issues, to develop the applicable normative framework, and to apply political pressure to individual governments.

The following materials provide an overview of the different approaches adopted by three of the mechanisms—those dealing with disappearances, arbitrary detention, and violence against women. Our objectives are to understand the techniques used; to consider the reasons and pros and cons for such diversity; to evaluate the overall system that has developed in a piecemeal and *ad hoc* manner; and to consider what changes might be made in these approaches.

One last consideration is vital. Given the range of UN responses—the 1503 procedure, the 1235 procedure and the different methods of implementing it through country-specific investigations, and the thematic mechanisms—is it now appropriate in evaluating the UN human rights system to consider any one of these responses separately, independently of its relation to the others within this larger evolving system?

COMMENT ON THE DISAPPEARANCES WORKING GROUP

We turn to the first of these mechanisms to be established—the Working Group on Disappearances. It provides a classic and unusually well-documented case study of the legal, political and other factors that have enabled, or compelled, the Commission to act. In addition, it raises important doctrinal questions relating to state responsibility for human rights violations.

In the late 1970s it proved extremely difficult to mobilize sufficient political support within the United Nations to mount an investigation of the disappearances that were taking place in Argentina during its 'dirty war'. Representatives of the military junta then in control were skilful in their manipulation of the relevant international fora. Many governments were ambivalent about 'naming' Argentina, for a variety of reasons ranging from trade interests to fear that they might be next on the list. The means used by the UN to get around this array of obstacles was to avoid a country-specific inquiry by establishing the first 'thematic' mechanism. Argentina was optimistic that the thematic approach would avoid singling out any one country, demonstrate that not just Argentina but many countries had problems, and give a significant number of governments a strong incentive to ensure that the new mechanism would be kept under careful political control so as not to become effective.

Since that time the Disappearances Working Group has played an important role in developing certain dimensions of the international human rights system. It has also served as a model for the creation of a growing range of thematic mechanisms.

Before exploring the techniques used by these groups, one should obtain a sense of the circumstances that made this breakthrough possible. Although it is tempting in this field to attribute most successes and failures to political factors, legal issues are often either of genuine concern or are sufficiently compelling that they can be used as shields by governments seeking to avoid scrutiny. One such issue of major importance in the 1970s was whether a government could legally, and should politically, be held accountable for violations for which it was not directly responsible—in other words for acts that were not committed directly by government officials or their properly delegated agents. The practice of disappearances that became common in some Latin American countries in the 1970s raised this issue. The disappearances were said to have been the responsibility of vigilante groups or death squads that were not linked to the government. Could the Commission then hold the government accountable?

The issue first arose in relation to Chile. An Expert appointed by the Commission, Mr Dieye of Senegal, tackled the issue directly in a 1979 report to the General Assembly.[11] Although his analysis of the issue of state responsibility for violations of human rights is less comprehensive than that of the Inter-American Court of Human Rights in 1988 in the *Velásquez-Rodríguez* Case (see p. 650, *infra*), it laid the groundwork for some important international legal developments required for effective action in response to gross violations. The Expert concluded that a state is responsible in international law

[11] UN Doc. A/34/583/Add.1 (1979), paras. 168–72.

for a range of acts or omissions if a person disappears subsequent to being detained by a state authority; if the state authorities do not react promptly to reliable reports of disappearances; if the relevant legal remedies are ineffective or non-existent; if, in the face of reliable evidence, the state does not act to clarify the situation; or if it takes no action to establish individual responsibility within the national framework.

The Expert concluded that the situation in Chile satisfied all the necessary conditions for establishing governmental responsibility. He relied on various facts, including: (1) the cases of at least 600 persons arrested by state authorities had not been clarified; (2) in making arrests, the secret police (the DINA) were not subject to restrictions compatible with international law; (3) available remedies were ineffective; (4) the Government had done nothing to strengthen the investigative system; (5) the Government had not cooperated fully with the UN and other international organizations.

The conclusions reached by the Expert were accepted by the Assembly and the Commission. Their implications were not confined to the situation in Chile, and they played a crucial role in setting the scene for action against Argentina.

ANTONIO CASSESE, HUMAN RIGHTS IN A CHANGING WORLD

(1990), at 128.

. . .

The first documented reports on the forced disappearances and tortures in Argentina reached the UN Sub-Commission in August 1976, in the course of its annual session. Some representatives of Argentinian human rights groups consigned voluminous documentation to various members of the Sub-Commission, trusting that they would take some initiative. I was then a member of the Sub-Commission, and along with the French expert, Nicole Questiaux (later a Minister in the first year of Mitterrand's presidency), we decided to check the reports and, should they prove well-founded, draw up a draft resolution expressing the UN's concern at what was happening in Argentina. A rather mild initiative; nevertheless, it aroused a great deal of alarm in the Argentine Embassy to the UN. The Ambassador immediately tried to speak to us, and even invited us to lunch. His invitation was flatly refused, but he continued to do everything he could to dissuade us from pursuing the initiative. His arguments, accusing the authors of the documentation distributed to us of being known terrorists, certainly did nothing to sway us. The following day, each of us was called before our respective Ambassadors in Geneva, who had received strongly-worded messages from their capitals: Buenos Aires had telephoned the capitals of our two countries, to have our Ambassadors to Geneva direct us to stop what we were doing. Fortunately, both the Ambassadors took note of our status as independent experts and simply conveyed to us the steps the Argentinian authorities had taken. The Argentinian Ambassador to Geneva, having heard of our decision not to retreat from our initial position, called another meeting, where he quickly moved from polite 'recommendations' to the most open threats (he was then chairman of the 'Group of 77', i.e. of the Third World states, a position that gave him considerable influence as regards our possible re-election and our 'diplomatic' future within the UN). These Argentinian protests and threats, far from discouraging us, only served to strengthen our resolve to persist. . . .

NOTE

The Sub-Commission subsequently adopted a resolution (Res. 2A (XXIX) of 1976) expressing deep concern that human rights appeared to be in jeopardy in Argentina and 'expressed the hope that international standards' would be respected. In 1978 the Inter-American Commission on Human Rights issued a damning indictment of the situation and set the scene for a more energetic UN response. The report by the Expert on Chile, described above, provided the legal basis on which UN organs could hold the Government of Argentina accountable for the disappearances which it persisted in attributing to unknown, private agents with no governmental connections. The UN General Assembly expressed concern in December 1978 and called for action by the Commission.

Clever manœuvering by Argentina, however, prevented the Commission from adopting a resolution in March 1979 and it returned to the matter in 1980. In the meantime, significant pressure for action had been generated by a campaign by Argentinean and international NGOs. The following excerpt describes the relevant events in the UN Commission in 1980.

DAVID KRAMER AND DAVID WEISSBRODT, THE 1980 UN COMMISSION ON HUMAN RIGHTS AND THE DISAPPEARED

3 Hum. Rts. Q. 18 (1981), at 20.

. . .

[S]trong interest in the issue of disappearances among all the Western delegates caused difficulties as the delegates squabbled among themselves to determine who would take a leadership role in formulating and sponsoring a resolution on disappearances. France, having come to the Commission with a prepared draft resolution, took the initiative. The French proposed that a group of three experts acting in their individual capacities would examine all reports of disappearances in any part of the world. The experts would be empowered to seek information from the governments and families concerned and to take appropriate action, in consultation with the governments concerned. . . . The Western bloc reached agreement on all the provisions contained in the draft except for paragraph 6(a), which defined disappeared persons. To several of the Western countries—the United States, Canada, Australia, and the Netherlands—broadly defining disappeared persons as those who cannot be located immediately or following a brief investigation once an abduction was reported to the government was an open invitation to other countries to find the resolution imprecise and unworkable. The definition seemed to cover all *missing* persons, instead of all *disappeared* persons. It did not exclude a voluntary disappearance by someone who, for personal reasons, might wish to hide from the authorities or perhaps from their family. Nor did it exclude those cases in which it was thought that a person had voluntarily left the country. Yet the French insisted on their language. . . .

The indecision of the Western bloc created a vacuum which Argentina tried to fill with a resolution of its own. . . .

[That] resolution urged governments to inform the Secretary-General of the measures they were adopting to cope with the problem of disappearances. It called upon governments to express their opinion as to what procedures might be appropriate to deal with disappeared persons without encroaching upon the sovereignty of any nation. A working group of five persons would then meet a week before the 1981 session of the Human Rights Commission to evaluate the submissions of the government and to make appropriate recommendations to the Commission. The Argentine proposal, essentially postponing any action for at least a year, was circulated just before the beginning of public debate.

On 22 February 1980 at the end of the third week of the Commission session, the public debate on disappearances began. The Western nations had still not agreed among themselves on the text of a resolution. The nonaligned nations were still reacting to the Argentine proposal. The socialist bloc was waiting to see what the nonaligned countries would do. As a consequence, no nation was prepared to speak.

The floor then unexpectedly went to the NGOs. The representative of Amnesty International (AI) spoke first. He defined the nature of the problem and described AI's efforts in collecting and submitting to the United Nations thousands of names of disappeared persons in Argentina, Afghanistan, Democratic Kampuchea, Ethiopia, Nicaragua, and Uganda. He then proceeded to discuss the case of two Argentines, who had been disappeared, tortured, and imprisoned in a secret camp, but who had escaped to tell of their experiences. Argentina harshly interrupted the speech and demanded to know from the Chairman what right an NGO had to attack a government in front of the Commission. . . . [Uruguay and Ethiopia both supported Argentina's position, while Canada and the U.S. supported Amnesty.] The Chair . . . ruled that although NGOs may not attack particular countries, they may provide the Commission with information about particular countries.

This initial public debate of 22 February 1980 revealed several things about

how the issue of disappearance was to fare before the Human Rights Commission. First, the debate exemplified just how critical the issue was for countries—such as Argentina—accused of disappearing persons. Second, the public debate disclosed a strategy that might defeat any action to establish a U.N. mechanism for handling disappearances. It was clear that the NGOs, which seemed to be primarily Western in orientation, advocated an effective resolution. It was also clear that they had the full support of the Western nations. If the initiative surrounding the issue of disappearances was viewed as Western, there was a significant danger that, for that reason alone, Third World countries might not support it. Clearly, no resolution could pass the Commission without the support of a good number of Third World nations.
. . .

[A proposal drawn up by Iraq, with the help of Cuba, was eventually accepted by the Western countries with several amendments. The key issue was whether the Working Group to be established would be authorized to respond to specific 'cases' or only to urgent 'situations'. The final compromise authorized the Group to 'examine questions relevant to' disappearances. After much drama, the draft was adopted without a vote, thus preserving a shaky consensus and not requiring individual states to vote for or against it.]

The battle over the meaning of the proposal immediately began. In explaining their position after the adoption of the resolution, the USSR, Argentina, and Ethiopia urged a restrictive reading of the resolution. The USSR, for instance, pointedly referred to the fact that the working group had been created for one year only. It presumed that the group would work by strict consensus and that the duration of any meetings would be limited. It saw no need for the group to meet at all until two to three weeks before the 1981 session of the Human Rights Commission.

A reading that would allow the working group to consider individual cases of disappearances was urged by the U.S., Australia, Cyprus, Netherlands, and Canada. The United States specifically pointed out that the mandate of the working group of experts would allow it to consider the thousands of individual cases in existence.

. . .

NOTE

The debates that took place in the Commission in 1980 led to the establishment of the Working Group with the mandate (or terms of reference) contained in Resolution 20 (XXXVI)(1980). In implementing such a mandate, much depends on the interpretation given by the members of the Working Group to terms that are vague or open-ended, and in turn on their perception of the political climate in the Commission.

The mandate is currently renewed every three years. In 1993, a fairly average year, the Group met three times for a total of less than four weeks. It met with governments, NGOs and others and transmitted 3,162 new cases to governments. Of these '523 had been received in 1993, while the rest were part of the . . . backlog; 122 . . . were reported to have occurred in 1993; 151 were transmitted under the urgent action procedure, of which 18 were clarified during the year' (UN Doc. E/CN.4/1994/26, para. 30). On average the Group undertakes one or two on-site visits each year. It sends a single report annually to the UN Commission.

COMMISSION RESOLUTION 20 (XXXVI)(1980)

. . .

1. *Decides* to establish for a period of one year a working group consisting of five of its members, to serve as experts in their individual capacities, to examine questions relevant to enforced or involuntary disappearances of persons;
2. *Requests* the Chairman of the Commission to appoint the members of the group;
3. *Decides* that the working group, in carrying out its mandate, shall seek and receive information from Governments, intergovernmental organizations, humanitarian organizations and other reliable sources;
4. *Requests* the Secretary-General to appeal to all Governments to co-operate with and assist the working group in the performance of its tasks and to furnish all information required;
5. *Further requests* the Secretary-General to provide the working group with all necessary assistance, in particular staff and resources they require in order to perform their functions in an effective and expeditious manner;
6. *Invites* the working group, in establishing its working methods, to bear in mind the need to be able to respond effectively to information that comes before it and to carry out its work with discretion;
7. *Requests* the working group to submit to the Commission at its thirty-seventh session a report on its activities, together with its conclusions and recommendations;

. . .

COMMISSION RESOLUTION 1992/30

. . .

2. *Takes note* of the report of the Working Group and thanks the Working Group for continuing to improve its methods of work and for recalling the humanitarian spirit underlying its mandate;
3. *Decides* to extend for three years the mandate of the Working Group . . .

4. . . . reminds the Working Group of the obligation to discharge its mandate in a discreet and conscientious manner;

. . .

7. *Reminds* the Working Group of the need to observe, in its humanitarian task, United Nations standards and practices regarding the receipt of communications, their consideration, their evaluation, their transmittal to Governments and the consideration of Government replies;
8. *Notes with concern* that some Governments have never provided substantive replies concerning disappearances alleged to have occurred in their countries;
9. *Deplores* the fact that, as the Working Group points out in its report, some Governments have not acted on the recommendations contained in the Working Group's reports concerning them nor replied to its requests for information on those matters, and requests the Working Group to continue to provide the Commission with all information on action taken further to its recommendations;
10. *Urges* the Governments concerned, particularly those which have not yet responded to communications transmitted to them by the Working Group, to cooperate with and assist the Working Group so that it may carry out its mandate effectively, and in particular to reply expeditiously to its requests for information;

. . .

14. *Encourages* the Governments concerned to give serious consideration to inviting the Working Group to visit their countries so as to enable the Working Group to fulfil its mandate even more effectively;

. . .

METHODS OF WORK

Report of the Working Group on Disappearances UN Doc. E/CN.4/1988/19, paras. 16–30.

16. The Working Group's . . . main objective . . . is to assist families in determining the fate and whereabouts of their missing relatives. . . . To this end, the Working Group endeavours to establish a channel of communication between the families and the Governments concerned, with a view to ensuring that sufficiently documented and clearly identified individual cases which the families, directly or indirectly, have brought to the Group's attention, are investigated and the whereabouts of the missing person clarified. The Group's role ends when the fate and whereabouts of the missing person have been clearly established as a result of investigations by the Government or the search by the family, irrespective of whether that person is alive or dead. The

Group's approach is strictly non-accusatory. It does not concern itself with the question of determining responsibility for specific cases of disappearance or for other human rights violations which may have occurred in the course of disappearances. In sum, the Group's activity is humanitarian in nature.

. . .

19. In transmitting cases of disappearances, the Working Group deals exclusively with Governments, basing itself on the principle that Governments must assume responsibility for any violation of human rights on their territory. If, however, disappearances are attributed to terrorist or insurgent movements fighting the Government on its own territory, the Working Group has refrained from processing them. The Group considers that, as a matter of principle, such groups may not be approached with a view to investigating or clarifying disappearances for which they are held responsible.

. . .

21. In order to enable Governments to carry out meaningful investigations, the Working Group provides them with information containing at least a minimum of basic data. In addition, the Working Group constantly urges the sources of reports to furnish as many details as possible on the identity of the missing person (if available, identity card numbers) and the circumstances of the disappearance. The Group requires the following minimum elements:

(a) Full name of the missing person;

(b) Date of disappearance, i.e., day, month and year of arrest or abduction or day, month and year when the missing person was last seen. When the missing person was last seen in a detention centre, an approximate indication is sufficient (i.e. March or spring 1980);

(c) Place of arrest or abduction or where the missing person was last seen (at least indication of town or village);

(d) Parties presumed to have carried out the arrest or abduction or to hold the missing person in unacknowledged detention;

(e) Steps taken to determine the fate or whereabouts of the missing person or at least an indication that efforts to resort to domestic remedies were frustrated or have otherwise been inconclusive.

22. Reported cases of disappearances are placed before the Working Group for detailed examination during its sessions. Those which fulfil the requirements as outlined above are transmitted, upon the Group's specific authorization, to the Governments concerned requesting them to carry out investigations and to inform the Group about their results. . . .

. . .

24. At least once a year the Working Group reminds every Government concerned of the cases which have not yet been clarified. . . .

25. All replies received from Governments on reports of disappearances are examined by the Working Group and summarized in the Group's annual report to the Commission . . . . Any information given on specific cases is forwarded to the sources of those reports who are invited to make observations thereon or to provide additional details on the cases.

26. If the reply clearly indicates where the missing person is (whether alive or dead) and if that information is sufficiently definite for the family to be reasonably expected to accept it, the Working Group considers the case clarified. . . .

. . .

30. The Working Group retains cases on its files as long as the exact whereabouts of the missing persons have not been determined. . . . This principle is not affected by changes of Government in a given country. However, the Working Group accepts the closure of a case on its files when the competent authority specified in the relevant national law pronounces, with the concurrence of the relatives and other interested parties, on the presumption of death of a person reported missing.

THE QUESTION OF DISAPPEARANCES IN THE FORMER YUGOSLAVIA

Report of the Working Group on Disappearances
UN Doc. E/CN.4/1993/25, paras. 38–44.

. . .

38. From the very early years of its existence, the Working Group has consistently taken the view that cases occurring in the context of an international armed conflict should not be taken up by the Group. That position was occasioned by the Iran–Iraq war. The Group argued at the time that taking up all cases of disappearance occurring in international armed conflicts, including the disappearance of combatants, would be a task far surpassing the resources of the Group. It also argued that, in any event, there already existed an international agency, namely the International Committee of the Red Cross, entrusted with the duty of tracing disappeared persons in such circumstances. . . .

39. As regards the situation in the former Yugoslavia, the Working Group is not aware of any authoritative position within the United Nations system which might give it guidance as to whether the armed conflict in that area is of an international or an internal character, nor as from what date it assumed such a character, nor whether the conflict might be characterized differently

for different parts of the area at any given time. The Security Council consistently refers to 'the armed conflict' and avoids qualifying it as either international or internal. Legal advisers differ on the subject. The Working Group has no independent means of establishing the character of the conflict and acting accordingly.

40. . . . It is obvious that if the Group were asked to involve itself in the situation in the former Yugoslavia, its resources would be totally inadequate to meet an influx of such magnitude. Even at present, due to the scant human resources at the Centre for Human Rights, the Working Group is trying to cope with an existing backlog of over 8,000 cases of disappearance waiting for transmission to the Governments concerned.

41. Apart from the question of resources, the methods of work of the Working Group . . . are not really geared to handling situations of the size and nature of the one in the former Yugoslavia. The Group's approach has consistently been to consider cases on an individual basis; this would, of course, become an illusion if attempted in a situation where the disappearances are on a very large scale, an experience the Group already suffered in the case of Iraq regarding disappearances that occurred after the end of the war with Iran.

. . .

43. The world is looking at one of the most dramatic episodes of humanitarian crisis and large-scale violation of human rights since the Second World War. The United Nations is bound to concern itself eventually with all aspects of the situation and can hardly turn a blind eye on one particular aspect, such as the occurrence of thousands of disappearances. Relatives, interested parties and the public at large would fail to understand the absence of significant action on the part of the United Nations. On the other hand, when the United Nations does take steps in the matter, its action should be commensurate to the situation addressed. Action which failed to meet the minimum standards of effectiveness, and therefore failed to contribute significantly towards resolving the problem of disappearances, might equally be harmful to the image of the world organization. If the Working Group were to assume the responsibility itself, its involvement in the matter would amount, at best, to a bookkeeping exercise, which would hardly do justice to the proportions of the problem.

NOTE

Although the Working Group sought the Commission's advice on this matter, it was left to find its own solution. Thus, in its 1994 report (UN Doc. E/CN.4/1994/26, para. 43), it proposed the following approach which the Commission subsequently endorsed:

. . .

(a) All cases of missing persons in any part of the former Yugoslavia should be considered under the same special procedure, geared to the exigencies of the situation;

(b) That special procedure should be implemented as a joint mandate by one member of the Working Group on Enforced or Involuntary Disappearances and the Special Rapporteur on the situation of human rights in the former Yugoslavia, resulting in joint reports to be submitted to the Commission on Human Rights;

(c) The Secretary-General should provide sufficient financial and personnel resources to the special procedure in order to guarantee its effective functioning.

A CASE STUDY

Report of the Working Group on Disappearances, UN Doc. E/CN.4/1994/26, paras. 237–54.

India

. . .

238. During the period under review, the Working Group transmitted 45 newly reported cases of disappearance to the Government of India, of which 14 were reported to have occurred in 1993. Twenty of these cases were transmitted under the urgent action procedure.

. . .

242. The newly reported cases of disappearance were submitted by Amnesty International, Human Rights Trust, the Sikh Human Rights Group and the International Human Rights Organization. These organizations reported that during 1993 most reported cases of disappearance took place in the Punjab region.

243. . . . The forces named as being responsible are, primarily, the army and the police. The missing persons include persons suspected of belonging to separatist groups, members of trade unions, lawyers, judges, journalists and human rights workers. Many other allegations of disappearances reported to have occurred in Kashmir and Jammu were also received, but owing to the methods of work of the Working Group, which require that all of the essential elements of the case be provided, these cases were not transmitted to the Government. It was reported to the Working Group that the situation in Kashmir and Jammu did not allow for a comprehensive compilation of information on cases, and that relatives of disappeared persons and human rights workers were often concerned about their own physical safety, in part because of harassment, threats or attacks often directed against them. For example, it

was alleged that one human rights worker who had frequently represented the families of disappeared persons before the Jammu and Kashmir High Court, was extrajudicially executed in Srinagar at the end of 1992.

244. Two laws which have allowed for preventive detention were particularly cited as contributing to the conditions in which disappearances were likely to take place: the Terrorist and Disruptive Activities Act (TADA) and the Public Security Act (PSA). In addition to allowing for preventive detention, these laws allow for prolonged detention without the many other normal safeguards available under the criminal codes. . . . Widespread torture, alleged to have occurred during periods of prolonged and incommunicado detention, was also reported to be an important element contributing to the phenomenon of disappearance. It was further noted that during these periods of detention women were particularly vulnerable to rape.

245. In regard to responsibility for human rights violations and in particular disappearances, it was reported to the Working Group that the police and other authorities acted with almost total impunity. Official investigations were reportedly rare and the Working Group was informed that the trial and conviction of authorities held responsible for such violations has occurred in approximately 1 per cent of all reported cases. Compensation to the victim or to the victim's family, most often without criminal prosecution, was reported to have been awarded in some cases.

Information and views received from the Government

246. By letter dated 7 January 1993, the Government of India informed the Working Group that in one case of disappearance the authorities had not detained the person concerned.

. . .

248. By letter dated 17 November 1993, the Government transmitted information to the Working Group concerning the newly constituted National Human Rights Commission. . . .

249. By letters dated 25, 26 and 30 November 1993, the Permanent Mission of India to the United Nations Office at Geneva provided information on 36 cases of disappearances. In six cases the Working Group decided to apply the six-month rule, while the information provided on the 30 other cases was considered by the Working Group insufficient to constitute a clarification.

250. By letter dated 30 November 1993, the Permanent Mission of India to the United Nations Office at Geneva responded to the general allegations contained in the Working Group's letter of 20 October 1993. It was stressed that the Indian Constitution established all the relevant institutions to safeguard democracy, namely an independent judiciary, a parliamentary form of government, a free press and commitment to the rule of law. All actions of State officials were subject to judicial review. In particular, a magisterial inquiry was mandatory for deaths in custody, and by means of a 'public interest litigation'

any individual or group could bring instances of violations of human rights to the attention of the High Court and the Supreme Court.

251. This commitment to pluralist democracy and the rule of law was, however, confronted with terrorism. According to the Government, in the past decade, terrorist violence had taken a toll of about 12,000 lives in Punjab and 4,000 lives in Jammu and Kashmir, including nearly 2,000 policemen and security forces personnel. In addition, systematic religion-based extremism had resulted in an exodus of 250,000 persons from the Kashmir valley to other parts of India.

252. Against this background, special legislation such as the . . . TADA . . . had had to be enacted in so-called 'disturbed' areas.

253. Nevertheless, the right to habeas corpus remained in force in all circumstances, and detainees always remained under judicial custody. There was thus no provision guaranteeing any form of impunity to security forces, and in Jammu and Kashmir alone, disciplinary action had been taken against 170 officers and men of the army and security forces. Custodial rape, if proven, could carry a life sentence.

254. According to the Government, the extrajudicial execution of a human rights worker, Mr. H. N. Wanchoo, in Srinagar on 5 December 1992, had been carried out by persons belonging to the terrorist organization Jamait-UV-Mujahideen, who had distorted these events to blame the authorities.

NOTE

Statistics in the report indicate that, of 213 cases transmitted to the Government over a five year period, specific responses had been received in relation to 66. As a result, 19 cases had been clarified by the Government and one by a NGO.

ABUSE OF HUMAN RIGHTS COMMISSION

The Economist, Feb. 4, 1995, p. 24.

. . .

To coincide with the start of the annual meeting of the United Nations Commission on Human Rights, Amnesty International has published yet another report on brutality in Kashmir. It contains details of 715 cases in which, it claims, people have been shot or tortured to death in Kashmir since the current insurgency started there in 1989.

These reports infuriate the Indian government. In the past, it has responded either by denying their truth, or by insisting that it was taking action to improve matters. The denials do not wash. . . . [T]here are enough detailed dossiers, including evidence from doctors and policemen, to convince any impartial observer that many terrible things have taken place.

The government has tried to show that it is making amends. In July 1993, it said that allegations of deaths in custody would be investigated. By May 1994, it announced that action had been taken in 174 cases, but refused to say what offences had been involved, or to name those who had been punished.

In October 1993, it appointed a National Human Rights Commission, an independent body with a brief to investigate brutality. But the commission is not allowed to investigate allegations against the army and the paramilitary forces, which are involved in almost all the cases at issue in Kashmir.

Given the Indian government's sensitivity to public discussion of the issue, the United Nations Commission on Human Rights could play an influential role. The UN's Special Rapporteur . . . has produced some carefully worded documents about the allegations, expressing 'serious concerns' and asking for an invitation to investigate them. None has been forthcoming. A resolution from the commission criticising Indian behaviour in Kashmir, which would probably sting the government more sharply than anything else, has not been forthcoming either.

At last year's meeting of the commission, Pakistan's prime minister, Benazir Bhutto, tried to table a resolution. The many countries that were either friends of India's, or had something to hide, made it clear that they would vote against it. China and Iran (friends of Pakistan's, with much to hide) persuaded Pakistan to 'defer' the resolution. In India, the government's avoidance of censure was splattered across the front pages of newspapers as a triumph.

Because the UN's Commission on Human Rights has failed to condemn India's brutality in Kashmir, it has, by default, sanctioned it.

REPORT OF THE WORKING GROUP ON DISAPPEARANCES

UN Doc. E/CN.4/1993/25, Paras. 74–88.

. . .

74. On 18 December 1992 the General Assembly, in resolution 47/133, adopted the United Nations Declaration on the Protection of all Persons from Enforced Disappearance. The Working Group, which participated actively in the elaboration of this Declaration, welcomes it as a milestone in the united efforts to combat the practice of disappearance and considers that it constitutes an important basis for its own future work. Many proposals and recommendations which the Working Group has adopted over the years and published in its annual reports have been reflected in the Declaration. In accordance with the Declaration, the systematic practice of disappearance is of the nature of a crime against humanity and constitutes a violation of the right to recognition as a person before the law, the right to liberty and security of the person, and the right not to be subjected to torture; it also violates or constitutes a grave threat to the right to life. States are under an

obligation to take effective legislative, administrative, judicial or other measures to prevent and terminate acts of enforced disappearance, in particular to make them continuing offences under criminal law and to establish civil liability.

75. The Declaration also refers to the right to a prompt and effective judicial remedy as well as unhampered access of national authorities to all places of detention, the right to habeas corpus, the maintenance of centralized registers of all places of detention, the duty to investigate fully all alleged cases of disappearance, the duty to try alleged perpetrators of disappearances before ordinary (not military) courts, the exemption of the criminal offence of acts of enforced disappearance from statutes of limitations, and special amnesty laws and similar measures leading to impunity.

. . .

87. . . . The Working Group . . . recommends that the Commission on Human Rights establish a reporting system by which all Governments are requested to submit to the Secretary-General periodic reports on all measures they have adopted to implement the Declaration and on all difficulties affecting its implementation.

88. It further recommends that the Commission on Human Rights entrust the Working Group with the task of examining these reports and transmitting comments and recommendations to the Governments concerned. Whenever requested by the Working Group, Governments should send representatives for the examination of their reports.

NOTE

Note the following excerpts from Commission on Human Rights Resolution 1994/39:

> 17. *Requests* the Working Group, in the exercise of its mandate, to take into account the provisions of the Declaration on the Protection of All Persons from Enforced Disappearance, and to modify its working methods if necessary;
>
> . . .
>
> 19. *Encourages* States, as some have already done, to provide concrete information on measures taken to give effect to the Declaration, as well as obstacles encountered;

COMMENT ON ON-SITE VISITS: THE CASE OF EAST TIMOR

Beginning in 1985, the Working Group on Disappearances pioneered the practice of conducting on-site visits to enable it to examine cases in greater depth, to galvanize public interest in its work and to apply greater pressure, both domestically and internationally, to governments. Most of the other thematic mechanisms subsequently adopted the same practice. Governments are not obliged to accept such missions, although a refusal might be taken as a lack of good faith and provide justification for criticism. No established procedures govern the initiation, organization or methodology of such visits. Nonetheless, the standards developed in relation to fact-finding under Resolution 1235 are generally applicable.

A visit to Indonesia and East Timor in July 1994 by the Special Rapporteur on Extrajudicial, Summary or Arbitrary Executions, Mr. Bacre Waly Ndiaye, is reasonably typical of such undertakings.[12] The Special Rapporteur had sought to visit other regions of Indonesia as well as East Timor, but the Government restricted his visit to the latter. The visit was initiated by a request from the Rapporteur in November 1993, based upon the killings that had occurred at the Santa Cruz Cemetry in Dili, the capital of East Timor, in November 1991. The killings had been filmed by a British photographer and the footage got widespread coverage in the West. The Rapporteur based his mission on human rights treaty provisions; the Principles on the Effective Prevention and Investigation of Extra-Legal, Arbitrary and Summary Executions, adopted by the ECOSOC in 1989 (Res. 1989/65); and the Declaration on the Protection of All Persons from Enforced Disappearance (see p. 435, *supra*).

The Rapporteur spent ten days in Indonesia, including four and a half days in East Timor. He met with a range of Government Ministers, the most senior military officers, NGO and other Timorese leaders, church officials and the Ambassadors of the United States and the Netherlands. He noted that his only purpose was 'to examine respect for the right to life,' irrespective of the political status or level of economic development of East Timor. He observed that at least three other of the UN's 'thematic mechanisms' had focused on the Santa Cruz killings and called upon the Government to take various measures.

The Indonesian authorities had reported 19 persons killed in the incident, while a National Commission of Inquiry had estimated the number at about 50. The Rapporteur observes that according to testimonies he gathered, the total was 'between 150 and 270, although some estimated it to be around 400.' He noted that the 'Indonesian Government and military authorities expressed regret for the Santa Cruz killings, which they consider as a tragic accident that

[12] UN Doc. E/CN.4/1995/61/Add.1 (1995).

arose out of a provocative action by anti-integration [of East Timor into Indonesia] elements.'

The Rapporteur recommended that, in the future, all cases of executions and disappearances should be investigated in a thorough, prompt and impartial manner, according to international standards. He called for the creation of 'a civilian police force as a matter of urgency,' and recommended the creation of a new commission of inquiry composed of independent and impartial experts, including some who are 'internationally recognized,' with full powers of investigation, an adequate budget, and powers to protect witnesses and their families. In addition to ascertaining the facts the new commission should determine 'the chain of command and the identity of all the perpetrators and their superiors, and their individual responsibility [for] the human rights violations.' He also recommended that, in order to restore popular confidence in the government and to end the impunity enjoyed by the Indonesian armed forces, the civil courts should deal with all relevant matters, the independence of the judiciary should be 'improved and guaranteed,' and equitable compensation should be granted to victims and their families.

In attaching particular importance to the role of NGOs, he made the following recommendation:

> 84. The Special Rapporteur believes that the involvement of non-governmental organizations in all questions relating to human rights in East Timor—e.g. investigation, monitoring, legal assistance, information and training—should be allowed and encouraged by the Indonesian authorities:
>
> (a) Independent NGOs should be created in East Timor and allowed to operate freely throughout the territory. At this stage, the Special Rapporteur feels that the involvement of the Catholic clergy (which at the moment is the only institution whose involvement with human rights questions is tolerated by the Indonesian authorities) in such organizations would be essential;
>
> (b) Indonesian and international human rights NGOs should be granted full access to East Timor.
>
> 85. The Special Rapporteur believes that the National Human Rights Commission is not the most appropriate mechanism to deal with human rights violations in East Timor. Its mandate, the means of action at its disposal and its methods of work are insufficient. Furthermore, it is not trusted by the population of East Timor. . . .

COMMENT ON THE WORKING GROUP ON ARBITRARY DETENTION

We now examine the mandates and the 1993 report of one of the more recently created thematic mechanisms—the Working Group on Arbitrary

Detention. The Group initially considered adopting the same working methods as the Disappearances Working Group, but its 1993 report indicates a radically different approach. We look at the Group's mandate, its own explanation of its working methods, a case study involving the United States and the Group's conclusions and recommendations.

The Group's original mandate was stated in the following terms by the UN Commission (Res. 1991/42):

> . . .
>
> 2. *Decides* to create, for a three-year period, a working group composed of five independent experts, with the task of investigating cases of detention imposed arbitrarily or otherwise inconsistently with the relevant international standards set forth in the Universal Declaration of Human Rights or in the relevant international legal instruments accepted by the States concerned;
> 3. *Decides* that the working group, in carrying out its mandate, shall seek and receive information from Governments and intergovernmental and non-governmental organizations, and shall receive information from the individuals concerned, their families or their representatives;
> 4. *Invites* the working group to take account, in fulfilling its mandate, of the need to carry out its task with discretion, objectivity and independence; . . .

By 1993 the Commission had endorsed the principal elements contained in the strategy developed by the Working Group. Note the rather casual terms in which the Commission (in Res. 1993/36) provides its imprimatur and considers whether, on the basis of the readings that follow, the Working Group has succeeded in balancing the various directives it has been given.

> 1. *Expresses its appreciation* to the Working Group on Arbitrary Detention for the way in which it carries out its task, more particularly for the importance that it attaches to respect for the adversarial procedure in its dialogue with States, and to seeking the cooperation of all those concerned by the cases submitted to it for consideration;
>
> . . .
>
> 4. *Considers* that the Working Group, within the framework of its mandate, and aiming still at objectivity, could take up cases on its own initiative;
>
> . . .
>
> 6. *Takes note* of the 'deliberations' adopted by the Working Group . . . ;
>
> . . .
>
> 9. *Requests* Governments concerned to give the necessary attention to the 'urgent appeals' addressed to them by the Working Group on a strictly

humanitarian basis and without prejudging its final decision on the character of the detention;

10. *Calls upon* Governments concerned to pay due heed to the Working Group's decisions and, where necessary, to take appropriate steps and inform the Working Group, within a reasonable period of time, of the follow-up to the Group's recommendations so that it can report thereon to the Commission;

The Working Group's methods of work are, in form, very similar to those of the Disappearances Working Group. Thus, it meets three times a year for a total of up to four weeks. It bases its legal analysis upon applicable treaty provisions and various declarations, in particular the Body of Principles for the Protection of All Persons under Any Form of Detention or Imprisonment, adopted by the UN General Assembly in 1988. In determining the range of cases it will consider, the Working Group has identified the following three categories:

I. Cases in which the deprivation of freedom is arbitrary, as it manifestly cannot be linked to any legal basis (such as continued detention beyond the execution of the sentence or despite an amnesty act, etc.); or

II. Cases of deprivation of freedom when the facts giving rise to the prosecution or conviction concern the exercise of the rights and freedoms protected by articles 7, 13, 14, 18, 19, 20 and 21 of the Universal Declaration of Human Rights and articles 12, 18, 19, 21, 22, 25, 26 and 27 of the International Covenant on Civil and Political Rights; or

III. Cases in which non-observance of all or part of the international provisions relating to the right to a fair trial is such that it confers on the deprivation of freedom, of whatever kind, an arbitrary character.

The Working Group receives information from all sources. It requires a minimum amount of information for admissibility of a case, and transmits cases to the government concerned for a reply. During 1993, the Group considered 335 cases, of which 231 were declared to be arbitrary and 59 were pending. It undertakes urgent action procedures (1) where there are 'sufficiently reliable allegations' that a person's arbitrary detention constitutes a 'serious danger to that person's health or even life;' and (2) 'where the particular circumstances of the situation warrant' it. The Group undertakes on-site visits where invited, and it reports annually to the Commission.

In practice, however, it has developed an approach which is radically different from those of other thematic mechanisms.[13] The Group considers that its 'investigation [of cases] should be of an adversarial nature so as to assist it in obtaining the cooperation of the State concerned.' Each case is characterised as a 'communication.' At least 90 days after transmitting the case to the government, the Group issues a 'decision' which is presented in a quasi-

[13] See generally 'Report of the Working Group on Arbitrary Detention', UN Doc. E/CN.4/1994/27, Annex I.

judicial fashion. Even if the person concerned is subsequently released 'the Working Group reserves the right to decide, on a case-by-case basis, whether or not the deprivation of liberty was arbitrary' and, if so, it 'shall make recommendations to the Government' which will also be included in the annual report. In addition, the Group adopts formal 'deliberations', designed to develop the doctrinal underpinnings of it work. The Group will generally not deal with situations of international armed conflict.

The Working Group has explained the function of its 'deliberations' in the following terms (UN Doc. E/CN.4/1993/24, p. 9):

. . .

> [In its reports, the Working Group has] identified certain legal situations which deserved particular attention. . . . [T]he Working Group felt that it would then consider whether these legal situations could characterize a detention as arbitrary. The Working Group deemed that this would make it possible for the Governments concerned to appreciate, not in the abstract but with reference to the identity of the legal situations prevailing in their respective jurisdictions, why detentions in the context of these legal situations were declared arbitrary. In addition, the Working Group felt that consideration of these situations would help formalize certain principles which might hitherto not have been considered relevant for the purposes of declaring a particular detention arbitrary.

DELIBERATION 02

Report of the Working Group on Arbitrary Detention, UN Doc. E/CN.4/1993/24, p. 9.

[In 1991 the Cuban Government challenged the working methods of the Group on several grounds. The most significant legal challenge was directed toward reliance by the Group on 'documents of a merely declaratory nature' (i.e. the 1988 Body of Principles) and on the provisions of treaties such as the ICCPR that Cuba had not ratified. The Working Group responded as follows.]

. . .

15. The Working Group would point out that resolution 1991/42, which lays down its mandate, refers expressly to 'the . . . international legal instruments accepted by the States concerned' as an international reference standard for the Working Group, in addition to the Universal Declaration of Human Rights. . . .

(a) Legal definition of 'instrument'

16. As interpreted in legal writings generally, the term 'legal instruments' covers all legal texts, whether they are conventional, that is to say binding,

instruments, such as conventions, covenants, protocols and other treaties or such forms of agreement as resolutions or gentlemen's agreements. . . .

. . .

18. The use of the word 'instruments' without further qualification in paragraph 2 of resolution 1991/42 therefore shows that it was not the intention of the Commission on Human Rights to confine the reference standards of the Working Group to treaties and other similar instruments but that it also wished to include in it acts of agreement, such as resolutions.

(b) 'Declaratory' nature

. . .

20. The Body of Principles is an instrument declaratory of pre-existing rights, inasmuch as the main purpose of many of its provisions is to set forth, and sometimes develop, principles already recognized under customary law.

21. It should be noted that, in the case of mere acts of agreement (and this applies to General Assembly resolutions), legal writers draw a distinction between those which are declaratory of pre-existing rights (as in the abovementioned example of most of the provisions of the Body of Principles or the Declaration on Territorial Asylum or the Declaration on Torture, etc.) and those—purely declaratory—instruments whose purpose is not to produce such an effect (for example, resolutions which take note of a report of a working group, or which institute a decade on a given theme).

22. The Working Group also wishes to point out in this connection that, according to legal writers, in the case of a non-party State, the same applies to any convention. . . . [T]o take the case of the Covenant again, it has a binding effect with respect to States parties and a declaratory effect with respect to non-party States.

23. In the light of the foregoing, the Working Group considers that, when it takes a decision on whether a case of detention is arbitrary, it is justified in referring, in categories I, II and III which it established in connection with its methods of work both to:

> the International Covenant on Civil and Political Rights, even if the Working Group has before it a case concerning a non-party State, in view of the tenacity of the declaratory effect of the quasi-totality of its provisions;
>
> and the Body of Principles, again on account of the declaratory effect of its substantive provisions.

(c) The concept of 'accepted' instrument

24. When it comes not to treaty instruments having binding force but to acts of agreement, the question is whether they can still be regarded as having been 'accepted'. . . .

DECISION NO. 48/1993 (UNITED STATES OF AMERICA),

Report of the Working Group on Arbitrary Detention, UN Doc. E/CN.4/1994/27, p. 135.

Communication addressed to the Government of the United States of America on 6 November 1992.

Concerning: Humberto Alvarez Machaín, on the one hand, and the United States of America, on the other.

. . .

2. The Working Group notes with appreciation the information forwarded by the Government concerned in respect of the case in question, received with slight delay. . . .

. . .

5. The Working Group considers that:

(a) Regarding the facts, there are no substantial differences. . . between the complainant's version and the version supplied by the Government. Accordingly, the Group holds it true that Dr. Humberto Alvarez Machaín, a doctor of Mexican nationality living in Mexico, was abducted (the expression used by the United States Government and in the ruling of the United States Supreme Court) on 2 April 1990 . . . at his medical office in Guadalajara, Mexico. . . . According to the complainant, the persons who seized him were 'paid agents of the DEA' (Drug Enforcement Administration, a United States Government agency to investigate and suppress drug trafficking). . . . According to the complaint, after being held incommunicado for over 20 hours and being physically and psychologically abused—something the Government denies—he was taken by private plane to the border town of El Paso, Texas, where he was arrested by DEA officials.
(b) Nor is there any controversy about the grounds invoked for the deprivation of freedom: on 31 January 1990 a United States Federal Grand Jury charged Dr Alvarez Machaín with taking part in the kidnapping and murder of DEA Special Agent Enrique Camarena Salazar in Mexico. Alvarez Machaín is said to have administered drugs to Camarena to facilitate his continued torture and interrogation. In the opinion of the Grand Jury, these acts constitute [various] crimes . . . covered by United States federal laws.
(c) When he was brought before . . . the District Court for the Central District of California . . . Alvarez Machaín said that his abduction . . . had been a violation of the 1978 Extradition Treaty between [the US and Mexico]. This allegation was admitted by the Court . . . [The Ninth Circuit Court of Appeals rejected an appeal by the US Government, but in June

1992 the Supreme Court, by a 6–3 majority] reversed the decisions of the lower courts and held that 'forcible abduction does not prohibit . . . trial in a United States court for violation of criminal law.'

. . .

(e) . . . Dr Alvarez Machaín was tried . . . and . . . acquitted on all counts on 14 December 1992 and released. . . .
(g) In view of the importance of . . . principles presented by this case, the Working Group deems it advisable to declare whether or not the deprivation of freedom . . . was arbitrary.
(h) [For that purpose] the Working Group must basically weigh up the following issues:
(1) Whether international treaty law governing relations between the United States of America and Mexico permits or prohibits the abduction of one person from the territory of one country to the territory of another, in order for him to be tried;
(2) If the matter is not resolved in treaty law, whether customary international law permits or prohibits abduction of this kind.

. . .

(n) [I]t may be maintained that the Extradition Treaty does not explicitly prohibit abduction, just as it does not prohibit someone being held under an extradition application from being tortured or executed by the requested country. However, it is obvious that this is implicitly prohibited when the subject matter—cooperation in the struggle against crime by surrendering offenders—is regulated in all dimensions by the treaty in question.
Abduction is the opposite of surrender. . . .

. . .

. . . [T]he object and purpose of the Treaty, and an analysis of the context, lead to the unquestionable conclusion that abduction . . . is a breach of the 1978 Treaty.
(o) Furthermore, both Mexico and the United States are also parties to the [OAS] Convention on Extradition . . . [of] 1933. . . .
This . . . is a comprehensive legal text which regulates the grounds and the procedures for surrendering wanted persons and it details cases in which extradition can be denied. Obviously, abduction is prohibited.
The deprivation of freedom, as a consequence of the arrest, is therefore arbitrary.
(p) The foregoing conclusion makes it pointless to analyse the second issue mentioned in paragraph (h) of this decision. Nevertheless, the importance of the matter is such that it needs to be resolved.
[C]ustomary international law . . . is unquestionably part of the internal law of the [USA] . . . .
Another basic principle of international law and of international rela-

tions is respect for the territorial sovereignty of States, a principle which, in addition to prohibiting the use of force and intervention by one State in the affairs of another—includes refraining from committing acts of sovereignty in the territory of another State, particularly acts of coercion or judicial investigation.

[Relying upon international judicial decisions as well as resolution 138 (1960) of the Security Council in the Eichmann case, the Working Group concluded that the deprivation of Alvarez Machaín's freedom] is not justified in customary international law.

Accordingly, with all the more reason it must be inferred that the deprivation of the freedom of Humberto Alvarez Machaín is not justified in customary international law.

(q) There are further considerations. First, the United States never tried to request the extradition of Alvarez Machaín or of any of the other participants. . . .

Nor did the United States have grounds for doubting the courts in Mexico. Indeed, everything indicates that Mexico scrupulously tried, in its courts, the persons responsible for the death of DEA agent Enrique Camarena. . . .

(r) In [this case] no legal basis whatsoever can be found to justify the deprivation of freedom . . . since [it] took place without the orders of any authority whatsoever and, indeed, both the District Court and the Court of Appeals declared it unlawful. In the circumstances, the deprivation of freedom is a breach of article 9 of the Universal Declaration of Human Rights and article 9 of the International Covenant on Civil and Political Rights, and principle 2 of the Body of Principles for the Protection of All Persons under Any Form of Detention or Imprisonment. Accordingly, the detention is arbitrary, falling within category I of the principles applicable in the consideration of the cases submitted to the Working Group.

7. [T]he Working Group requests the Government of the United States of America to take the necessary steps to remedy the situation. . . .

NOTE

In 1994 the Commission of Human Rights, responding to a decision of the Vienna World Human Rights Conference and concerted lobbying by women's and other human rights groups, appointed a Special Rapporteur on Violence against Women. Questions relating to that issue are further considered in Chapter 13, *infra*, within the framework of the Convention on Elimination of All Forms of Discrimination against Women.

COMMISSION ON HUMAN RIGHTS, RESOLUTION 1994/45

. . .

6. *Decides* to appoint, for a three-year period, a special rapporteur on violence against women, including its causes and its consequences, who will report to the Commission on an annual basis beginning at its fifty-first session;
7. *Invites* the Special Rapporteur, in carrying out this mandate, and within the framework of the Universal Declaration of Human Rights and all other international human rights instruments, including the Convention on the Elimination of All Forms of Discrimination against Women and the Declaration on the Elimination of Violence against Women, to:

(a) Seek and receive information on violence against women, its causes and its consequences from Governments, treaty bodies, specialized agencies, other special rapporteurs responsible for various human rights questions and intergovernmental and non-governmental organizations, including women's organizations, and to respond effectively to such information;
(b) Recommend measures, ways and means, at the national, regional and international levels, to eliminate violence against women and its causes, and to remedy its consequences;
(c) Work closely with other special rapporteurs, special representatives, working groups and independent experts of the Commission on Human Rights and the Sub-Commission on Prevention of Discrimination and Protection of Minorities and with the treaty bodies, taking into account the Commission's request that they regularly and systematically include in their reports available information on human rights violations affecting women, and cooperate closely with the Commission on the Status of Women in the discharge of its functions.

PRELIMINARY REPORT SUBMITTED BY THE SPECIAL RAPPORTEUR ON VIOLENCE AGAINST WOMEN, ITS CAUSES AND CONSEQUENCES

UN Doc. E/CN.4/1995/42.

[Excerpts from the report of Ms. Radhika Coomaraswamy follow:]

7. The different forms of violence against women include, as spelled out in the above resolution, all violations of the human rights of women in situations of armed conflict, and in particular, murder, systematic rape, sexual slavery

and forced pregnancy, as well as all forms of sexual harassment, exploitation and trafficking in women, the elimination of gender bias in the administration of justice and the eradication of the harmful effects of certain traditional or customary practices, cultural prejudice and religious extremism.

8. The Special Rapporteur has understood her mandate to contain two components. The first consists of setting out the elements of the problem before her, the international legal standards and a general survey of incidents and issues as they relate to the many problem areas. The second component consists of identifying and investigating factual situations, as well as allegations which may be forwarded to the Special Rapporteur by concerned parties.

9. With regard to the second component, the Special Rapporteur deems it useful to take a more specific approach by endeavouring to identify more precisely situations of violence against women. For this purpose the Special Rapporteur, in a spirit of dialogue, will approach concerned Governments and request clarifications on allegations regarding violence against women that she may have received. . . .

10. Taking into consideration the alarming situation of violence against women throughout the world, the Special Rapporteur intends to establish dialogue with Governments concerning allegations and prospective field missions with a view to assisting the Governments concerned to find durable solutions for the elimination of violence against women in their societies.

11. In addition . . . the Special Rapporteur is planning to undertake a number of field missions . . .

. . .

317. Finally, the Special Rapporteur encourages the formulation of an optional protocol to the Convention on the Elimination of All Forms of Discrimination against Women allowing for an individual right of petition once local remedies are exhausted. This will ensure that women victims of violence will have a final recourse under an international human rights instrument to have their rights established and vindicated.

QUESTIONS

1. What accounts for the different approaches adopted under these three procedures within the broad category of thematic mechanisms? Are they required by the mandates, do they follow logically from the subject matter, are they a reflection of the different levels of sensitivity of governments in relation to these different issues? Can you offer other explanations?

2. Which of the approaches do you favor and why? What changes would you propose in the procedures followed by one or another of the Groups?

3. What are the differences between the 1503 procedure and the 'communications' approach adopted by the Working Group on arbitrary detention? Does the latter initiative put the 1503 procedure in a different light?

4. The Commission asked the Disappearances Working Group 'to carry out its work with discretion' (1980) and 'in a discrete and conscientious manner' (1992). It also recalled the 'humanitarian spirit underlying' the Group's mandate. Similar terms were used with respect to the Working Group on Arbitrary Detention. What is the reason for and significance of such terminology? In what ways does the Arbitrary Detention Working Group follow an 'adversarial' as opposed to an 'accusatorial' approach 'in its dialogue with States'?

5. How do you assess the report and recommendations on East Timor, from the perspective of both the international community and the Indonesian Government?

6. Is it a good thing for so many mechanisms to focus on the same case, as illustrated by the East Timor case, or is there a saturation point? Consider a situation involving abuses which comes under the purview of several different thematic mechanisms, each looking at the situation from its own mandate. From the perspective of effective and efficient enforcement of human rights norms, what do you see as the advantages and disadvantages of having so many different thematic mechanisms?

7. Do you agree with an observation in *The Economist* that, when the UN Commission fails to condemn a situation, it endorses it by default?

D. OVERVIEW AND EVALUATION

PHILIP ALSTON, APPRAISING THE UNITED NATIONS HUMAN RIGHTS REGIME,

in Alston (ed.), The United Nations and Human Rights: A Critical Appraisal 1 (1992), at 14.

. . .

What then are the key questions that need to be asked in evaluating the effectiveness of the UN regime. It is generally accepted in the academic literature that there are five such questions: When? Where?, For Whom?, What?, and Why? *When?* raises the issue of an appropriate time-frame. The 1993 World Conference, for example, has been asked 'to review and assess progress . . . since the adoption of the Universal Declaration' in 1948. It is indisputable that by almost any measure immense progress has been achieved since that time. But other time-frames could equally well have been chosen. Since the first World Conference in 1968, for example, or over the last decade, the results might look significantly different, although much would still depend on how the other questions were framed. . . .

Where? refers to the scope and focus of the evaluation. Donnelly, for example, in his analysis of the human rights regime looks specifically only at the Human Rights Committee and the Commission on Human Rights. From a feminist perspective this immediately biases the outcome by excluding the Commission on the Status of Women and CEDAW. Excluding the other treaty bodies as well as the Assembly and the Sub-Commission also ensures the presentation of only a partial picture. On the other hand, . . . attempting to cover the entire regime provides a potentially unmanageable focus.

The question *for whom?* poses an immediate problem in terms of the human rights regime. The outcome of the evaluation will be radically different depending on whether the standpoint adopted is primarily that of the victims of violations, human rights activists, governments, UN officials, or the press. The answer is by no means self-evident when the UN is asked to undertake the evaluation, since its approach in general seeks to balance the concerns of its different constituencies, while at the same time playing down any suggestion that trade-offs are being made. For this reason, another question is appropriately added in the present context: *by whom?* On occasion, the UN uses independent experts, but more often than not such people will be 'insiders' of one kind or another with strong vested interests in maintenance of the status quo. Should members of the expert committees, for example, be asked to evaluate their own activities . . . ? How objective, or outspoken, can the Secretariat be in such a context? What weight should be placed on NGO evaluations? Presumably, evaluators from a range of backgrounds are required but then the frame of reference of each can still differ radically from those of others, thereby diminishing the comparability of the results.

What? adds another level of questions after *where?* Having determined, for example, that the focus will be on the Human Rights Committee it must still be decided whether its interstate complaints jurisdiction is best ignored as being certainly unproductive or is carefully examined in the hope of breathing life into a potentially important procedure. Similarly, should the task of developing a sophisticated jurisprudence based on the Covenant be treated as being of primary or only secondary importance? A focus on the Commission raises an even more complex array of issues, particularly the question of the importance attached to its function of responding to violations.

Finally, the question *why?* will be answered rather differently by an academic, an activist, a government minister, and an international official. The answers might range from a general quest to gain a better understanding of modes of international co-operation, or a desire to increase efficiency defined in managerial terms, or a mandate 'to formulate concrete recommendations for improving [overall] effectiveness' (as in the case of the World Conference), through to a desire to ensure an immediate and productive response to all future reports of alleged violations.

A more formal or legalistic approach to evaluation, and one more consonant with the past practice of the UN, is to ignore most of these questions (implicitly dismissing them as practically or politically unanswerable) and to seek to compare existing practice with the stated objectives of the system. The problem, of course, is that the 'system' *per se* does not exist in such terms. Thus, the evaluator might turn either to the mandate provided in the UN Charter or to the more specific terms of reference outlined in the constituent instruments relating to each of the organs in question. But in the case of the principal Charter-based organs, the terms of the Charter provisions give little practical guidance for evaluation purposes and the constituent instruments may not be a great deal more helpful. In the case of the Commission, for example, its standard-setting mandate refers only to three specific issues, with respect to one of which (freedom of information) it has been able to achieve virtually nothing. . . . Similarly, the word 'protection' is only mentioned in relation to 'minorities', but it is unthinkable that the Commission's success in providing 'protection' could be evaluated solely in that regard. In reality, most of the Commission's activities are justified by reference to the catch-all provision in its mandate referring to 'any other matter concerning human rights'. That hardly constitutes a meaningful yardstick for evaluation purposes, however.

JACK DONNELLY, INTERNATIONAL HUMAN RIGHTS: A REGIME ANALYSIS

40 Int'l Org. 599 (1986), at 613.

4. Political foundations of the international human rights regime

The international human rights regime is a relatively strong promotional regime, composed of widely accepted substantive norms, largely internationalized standard-setting procedures, some general promotional activity, but very limited international implementation, which rarely goes beyond information exchange and voluntarily accepted international assistance for the national implementation of international norms. There is no international enforcement. Such normative strength and procedural weakness, however, is the result of conscious political decisions.

Regimes are political creations to overcome perceived problems arising from inadequately regulated or insufficiently coordinated national action. Robert O. Keohane offers a useful market analogy: regimes arise when sufficient international 'demand' is met by a state (or group of states) willing and able to 'supply' international norms and decision-making procedures. The shape and strength of an international regime reflect who wants it, who opposes it, and why—and how the conflicting objectives, interests, and capabilities of the parties have been resolved. As Krasner puts it, in each issue-area there are makers, breakers, and takers of (potential) international regimes; understanding the structure of a regime (or its absence) requires that we know who has played which roles, when and why, and what agreements they reached. In this section I shall examine the interaction of supply and demand which has led to the international human rights regime. . . .

. . .

A cynic might suggest, with some basis, that these postwar 'achievements' simply reflect the minimal international constraints and very low costs of a declaratory regime: decision making under the Universal Declaration remained entirely national, and it would be more than twenty years until resolution 1503 and nearly thirty years before even the rudimentary promotion and monitoring procedures of the Covenants came into effect. Yet prior to the war even a declaratory regime had rarely been contemplated. In the late 1940s, human rights became, for the first time, a recognized international issue-area.

Moving much beyond a declaratory regime, however, has proved difficult. As we have seen, procedural innovations have been modest. Even the legal elaboration of substantive norms has been slow and laborious: for example, it took nine years to move from a declaration to a convention on torture: work on stronger, more precise norms on religious liberty is now in its third decade. It is in this relative constancy of the regime—critics and frustrated optimists are likely to say stagnation—that the weakness of the demand is most evident.

To the extent—probably considerable—that the international human rights regime arose from postwar frustration, guilt, or unease, the very proclamation ('supply') of the Declaration, along with the adoption of the Genocide Convention, seems to have satisfied the demand. To the extent—again probably considerable—that it rested on an emotional reaction to the horrors of Hitler and the war, time sadly but predictably blunted the emotion. Time also revealed both the superficial, merely verbal commitment of many states and substantive disagreements over particular rights, causing enthusiasm to wane further. And with the cold war heating up, not only was the desire to move on to other issues strong, but East–West rivalry itself soon came to infect and distort the discussion of human rights.

The most important problem, however, was and remains the fact that a stronger international human rights regime does not rest on any perceived material interest of a state or coalition willing and able to supply it. In the absence of a power capable of compelling compliance, states participate in or increase their commitment to international regimes more or less voluntarily. Barring extraordinary circumstances, states participate in an international regime only to achieve *national* objectives in an environment of perceived international interdependence, to address national problems caused by the existing international state of affairs.

Both theory and practice suggest that states will relinquish authority only to obtain a significant benefit beyond the reach of separate national action or to avoid bearing a major burden. Furthermore, relinquishing sovereign authority must appear 'safe' to states who are notoriously jealous of their sovereign prerogatives. A stronger international human rights regime simply does not present a safe prospect of obtaining otherwise unattainable national benefits.

Moral interests such as human rights may be no less 'real' than material interests. They are, however, less tangible, and policy, for better or worse, tends to be made in response to relatively tangible national objectives. . . .

Furthermore, the extreme sensitivity of human rights practices makes the very subject intensely threatening to most states. National human rights practices often would be a matter for considerable embarrassment should they be subject to full international scrutiny, and compliance with international human rights standards in numerous countries would mean the removal of those in power.

In addition, and perhaps most important, human rights are ultimately a profoundly *national*—not international—issue. States are the principal violators of human rights and the principal actors governed by the regime's norms; international human rights are concerned primarily with how a government treats inhabitants of its own country. . . .

Human rights are also a national matter from the perspective of practical political action. . . . Who is to prevent a government from succumbing to the temptations and arrogance of position and power? Who can *force* a government to respect human rights? The only plausible candidates are the people whose rights are at stake.

Foreign actors may overthrow a repressive government. With luck and skill, foreign actors may even be able to place good people in charge of finely crafted institutions based on the best of principles. They may provide tutelage, supervision, and monitoring: moral and material support: and protection against enemies. This scenario, however, is extremely unlikely, especially if we do not impute unrealistically pure motives and unbelievable skill and dedication to external powers, for whom 'humanitarian intervention' usually amounts to little more than a convenient cover for partisan politics. And in any case, a regime's ultimate success—its persistence in respecting, implementing, and enforcing human rights—depends on *internal* political factors.

A government that respects human rights is almost always the legacy of persistent national political struggles against human rights violations. Most governments that respect human rights have been created not from the top down, but from the bottom up. Domestically, paternalistic solutions, in which human rights are given rather than taken, are likely to be unstable. Internationally, paternalism is no more likely to be successful.

But if international regimes arise primarily because of international interdependence—the inability to achieve perceived national objectives by independent national action—how can we account for the creation and even modest growth of the international human rights regime? First and foremost, the 'moral' concerns that brought the regime into being in the first place persist. Butchers such as Pol Pot and Idi Amin still shock the conscience of mankind and provoke a desire to reject them as not merely reprehensible but prohibited by clear and public, authoritative international norms; even regimes with dismal human rights records seem to feel impelled to join in condemning the abuses of such rulers, and lesser despots as well.

Although cynics might interpret such uses of the language of human rights as merely craven abuse of the rhetoric of human rights, it can just as easily be seen as an implicit, submerged, or deflected expression of a sense of *moral* interdependence. Although states—not only governments but often the public as well—often are unwilling to translate this perceived moral interdependence into action or into an international regime with strong decision-making powers, they also are unwilling (or at least politically unable) to return to treating national human rights practices as properly beyond all international norms and procedures.

. . .

States also may miscalculate or get carried away by the moment, and procedures may evolve beyond what the regime's participants originally intended. For example, ECOSOC resolution 1235, which provides the principal basis for the Commission on Human Rights' public study and discussions of human rights situations in individual countries, was explicitly established in 1967 to focus principally on the pariah regimes in Southern Africa, but it has evolved into a procedure with universal application (or at least, one that may be applied to any country that a majority of members decide to consider). Although procedures seldom expand to such an extent, the possibility should not be overlooked.

The current international human rights regime thus represents a politically acceptable international mechanism for the collective resolution of principally national problems. Because perception of the problem rests on a politically weak sense of *moral* interdependence, however, there is no powerful demand for a stronger regime: even policy coordination seems too demanding, and there is little reason for states to accept international monitoring, let alone authoritative international decision making.

. . .

This is not to belittle the importance of international procedures—the more effective the monitoring and enforcement procedures, the stronger the regime and the more likely it is to achieve its objectives—but, rather, to stress the fact that regime procedures largely reflect underlying political perceptions of interest and interdependence. Compliance with regime norms rests primarily on authority and acceptance, not force or even enforcement.

. . .

PHILIP ALLOTT, EUNOMIA: NEW ORDER FOR A NEW WORLD

(1990), at 287.

. . .

15.66. But, as so often in human social experience, the installation of human rights in the international constitution after 1945 has been paradoxical. The idea of human rights quickly became perverted by the self-misconceiving of international society. Human rights were quickly appropriated by governments, embodied in treaties, made part of the stuff of primitive international relations, swept up into the maw of an international bureaucracy. The reality of the idea of human rights has been degraded. From being a source of ultimate anxiety for usurping holders of public social power, they were turned into bureaucratic small-change. Human rights, a reservoir of unlimited power in all the self-creating of society, became a plaything of governments and lawyers. The game of human rights has been played in international statal organizations by diplomats and bureaucrats, and their appointees, in the setting and the ethos of traditional international relations.

15.67. The result has been that the potential energy of the idea has been dissipated. Alienation, corruption, tyranny, and oppression have continued wholesale in many societies all over the world. And in all societies governments have been reassured in their arrogance by the idea that, if they are not proved actually to be violating the substance of particularized human rights, if they can bring their willing and acting within the wording of this or that formula with its lawyerly qualifications and exceptions, then they are doing well enough. The idea of human rights should intimidate governments or it is worth nothing. If the idea of human rights reassures governments it is worse than nothing.

15.68. But, once again, there is room for optimism, on two grounds. (1) The idea of human rights having been thought, it cannot be unthought. It will not be replaced, unless by some idea which contains and surpasses it. (2) There are tenacious individuals and non-statal societies whose activity on behalf of the idea of human rights is not part of international relations but is part of a new process of international reality-forming.

. . .

NOTE

In April 1994, the first UN High Commissioner for Human Rights (HCHR), José Ayala Lasso of Ecuador, took office. The General Assembly (Res. 48/141) created the post while providing it with inadequate resources. The key provision instructs the HCHR to play 'an active role . . . in preventing the continuation of human rights violations throughout the world . . . .' As of mid-1995, the role that the HCHR would play in practice had not yet been defined, although the first HCHR has indicated that his role is to supplement and not replace the work of the Commission's Special Rapporteurs and thematic mechanisms. The principal challenges for the new post include defining its relationship to the rest of the UN system including the General Assembly and Commission on Human Rights, receiving adequate resources for staff and active work, and implementing the mandate in the face of governmental resistance to an active role. See generally Helena Cook, *The Role of the High Commissioner for Human Rights: One Step Forward or Two Steps Back?*, appearing in 89 Proc. Ann. Mtg. Am. Soc. Int. L. (1995); and UN Doc. E/CN.4/1995/98.

QUESTIONS

1. Why do governments participate in the universal human rights system? Do they indeed have a choice?

2. How would you identify what you believe to be the major weaknesses in the institutional structures and processes for enforcement of human rights norms? Which of them do you believe can realistically be improved, and what such realistic improvements would you propose?

3. Should the criteria for effectiveness be the same in relation to all the procedures and mechanisms considered in this chapter? Or do they aim at different goals that imply different measures of effectiveness?

ADDITIONAL READING

For an extensive bibliography, see Philip Alston (ed.), *The United Nations and Human Rights: A Critical Appraisal* (1992) 677–748; Asbjørn Eide, *The Sub-Commission on Prevention of Discrimination and Protection of Minorities*, in *ibid*, 311–64; Iain Guest, *Behind the Disappearances: Argentina's Dirty War Against Human Rights and the United Nations* (1990); Howard Tolley Jr., *The U.N. Commission on Human Rights* (1987); Theodor Meron, *Human Rights Law-Making in the United Nations: A Critique of Instruments and Processes* (1986); Nigel Rodley, *The Treatment of Prisoners Under International Law* (1987); Menno Kamminga, *Inter-State Accountability for Violations of Human Rights* (1992); Thomas Franck and H. Scott Fairley, *Procedural Due Process in Human Rights Fact-finding by International Agencies*, 74 Am. J. Int'l L. (1980) 308; B. G. Ramcharan (ed.), *International Law and Fact-Finding in the Field of Human Rights* (1982); Kurt Herndl, *Recent Developments Concerning United Nations Fact-finding in the Field of Human Rights*, in Manfred Nowak *et al* (eds.), *Progress in the Spirit of Human Rights* (1988) 1–36; Rudolf Bernhardt and J. A. Jolowicz (eds.), *International Enforcement of Human Rights* (1987); Richard Bilder, *Rethinking Human Rights: Some Basic Questions*, [1969] Wis. L. Rev. 171.

8

The Role of Non-Governmental Organizations

Most of Part C examines intergovernmental human rights organizations—universal ones in Chapters 7 and 9, regional ones in Chapter 10. References to non-governmental organizations (NGOs) inevitably appear in each of those chapters, for NGOs pervade and are a vital part of the broader human rights movement. Above all, human rights NGOs bring out the facts. They also contribute to standard setting as well as to the promotion, implementation and enforcement of human rights norms. They provoke and energize. Decentralized and diverse, they proceed with a speed and decisiveness and range of concerns impossible to imagine for most of the work of bureaucratic and politically cautious intergovernmental organizations.

NGOs operate subject to differing mandates, each responding to its own priorities and methods of action, bringing a range of viewpoints to the human rights movement. It is inconceivable that this movement, whatever its weaknesses, could have achieved as much in its first half century without the spur and inventiveness of NGOs. This chapter explores their character, constitution, functions, methods of work and effects.

Within individual states, it is often the human rights NGOs that call governments to account and compel reconsideration of policies and programs that have been designed in disregard or violation of human rights norms. It is not only the domestic policies of a state that figure in NGO reporting and advocacy. The development in a few states of a foreign economic policy that takes into account human rights violations in other states owes much to the information provided and pressures exerted by NGOs.

At the international level, and particularly in the United Nations context, the frequent reluctance of governmental actors to criticize their counterparts from other countries and the limited supply of independent sources of information have contributed to making NGOs the lynchpins of the system as a whole. In situations in which NGO information is not available or where the NGOs are either unable or unwilling to generate political pressures upon the governments concerned, the chances of a weak response by the international community, or of none at all, are radically increased. A high proportion of the most significant initiatives to draft new international instruments, to establish new procedures and machinery, and to identify specific governments as violators have come as a result of concerted NGO campaigns designed to mobilize public opinion and lobby governmental support.

Of course the term non-governmental organization, standing alone, reaches too broadly for this chapter. It includes myriad types of organizations that are (more-or-less, depending on the state) independent of the state: economic interest groups like labor unions, consumer unions or industrial associations; racial, gender and religious groups; issue-oriented groups like environmental or educational organizations; groups representing the elderly or the young; public interest groups that are anti-corruption or pro-universal health care; and so on—the many types of associations of citizens and of non-state institutions that some theorists view as the essential ingredients of so-called civil society.

This chapter is concerned with a subset of NGOs that are involved with human rights issues—a category that, as we shall see, can have porous boundaries. The chapter first explores the range of work of NGOs before concentrating on large international NGOs and their relationships with the UN human rights system.

A. NGOs: DIVERSITY IN CHARACTER AND WORK

Section A both explores the diverse character of NGOs, and the complex relationships between domestic NGOs and international ones (INGOs).

HENRY STEINER, DIVERSE PARTNERS: NON-GOVERNMENTAL ORGANIZATIONS IN THE HUMAN RIGHTS MOVEMENT

(1991), at 5.

[These excerpts are taken from the report of a retreat for human rights NGO activists from around the world, held in Crete in 1989.]

. . .

A. What Makes a Group a 'Human Rights' NGO?

Participants distinguished between 'national' NGOs limiting their activities to their home country, and 'international' NGOs (INGOs) that act in two or more countries. . . .

Our discussions about threshold definitions underscored how diverse were the participants' understandings of the human rights movement. Certain questions recurred. For a public interest group to qualify as a 'human rights' NGO, must it base its criticism of state conduct on international human rights law—the Universal Declaration of Human Rights (UDHR), the Civil–Political Rights Covenant, and so on? Most participants thought not. They were more persuaded by the nature of the claims made and the goals advanced by a group than by the formal source of the norms that it invoked to criticize state conduct.

In fact, some participants noted, the choice between relying on a national legal system or on international law norms to support advocacy before national institutions often amounts to a question of strategy. Employing domestic standards rather than international law might be politically expedient—more political clout, less risk that an NGO will be viewed as inspired by alien doctrine. On the other hand, international law has strategic advantages in countries whose domestic legal norms are of little assistance. . . .

Participants suggested other reasons for national NGOs to rely on domestic law. In countries that have ratified few covenants, NGOs arguing before national courts can criticize domestic law or conduct as violating international norms only by invoking the less determinate customary international law. . . . [P]ublic interest groups in the U.S. that vindicate civil and political rights—for example, the American Civil Liberties Union (ACLU), or the NAACP Legal Defense and Education Fund (LDF), an NGO committed to ending discrimination against blacks—base their advocacy on domestic constitutional law and only rarely refer to international human rights. Nonetheless, from an international perspective, these groups are as much 'human rights' NGOs as, say, Tutela Legal in El Salvador, which invokes primarily international standards and has close links with INGOs.

Public interest groups in fields like consumer or environmental protection or workers' safety regulation fall within more ambiguous categories. Whether they are classified as 'human rights' organizations does not, however, appear to have operational significance.

. . .

Many participants concluded that self-perception and self-definition by NGOs constitute the only sensible method of identifying human rights organizations. It would be impractical and unwise to maintain a protective boundary around some core or traditional preserve of human rights work, such as the protection of individuals against violence or discrimination. Who would define and monitor such a boundary, and what sanctions could be imposed on organizations crossing it but still claiming to be human rights NGOs? An attempt at authoritative definition could block a natural and important growth of the human rights movement, such as its earlier evolution toward economic and social rights, or its present initiatives toward linking human rights concerns with developmental and environmental issues. Other participants, however, stressed that to be effective, it was important for NGOs to hold to clearly defined mandates based on consensual legal norms.

Like the human rights movement itself, NGOs are in a state of flux. The costs of such change and uncertainty—a threat to the human rights movement's core identity, a blurring of fields, disagreements about whether employing the rhetoric of 'human rights' for certain goals will strengthen or hurt the movement as a whole—inhere in a dynamic, decentralized, multicultural, universal movement.

ARYEH NEIER, NOT ALL HUMAN RIGHTS GROUPS ARE EQUAL

Letter to the Editor, New York Times, May 27, 1989, p. 22.

. . .

A phenomenon that has concerned the human rights movement is the proliferation of groups claiming to speak in the name of the human rights cause but actually engaged in efforts to promote one or another side in a civil conflict. Nowhere has this been more of a factor than in Nicaragua. On both sides, the human rights issue has been a weapon to use against the enemy.

It may be useful, accordingly, to suggest a few questions to raise in distinguishing partisan efforts from genuine efforts to promote human rights. These include:

• Is the organization funded by or otherwise linked to any party to the conflict?

• Is it impartial? If it is an international group, does it regularly criticize abuses by governments of all political persuasions and geopolitical alignments? In situations of sustained armed conflict, does it criticize violations of the laws of war by both sides according to the same criteria?

• Does it engage in systematic field research and does it avoid sweeping comments, except to the degree that these are sustained by its detailed findings through field research?

• Does it exercise care in the use of language? For example, does it refer to 'torture' when the word 'mistreatment' would be more appropriate? When does it allege 'atrocities' as a 'strategy'?

• Does it acknowledge contradictory evidence, such as a government's prosecutions of its own personnel for abuses, or the statements of one witness that cast doubt on the statements of another?

Above all, of course, it is the record that an organization has compiled over time that indicates whether it deserves credence when it reports on human rights. . . .

RAJNI KOTHARI, HUMAN RIGHTS—A MOVEMENT IN SEARCH OF A THEORY

in Smitu Kothari and H. Sethni, Rethinking Human Rights: Challenges for Theory and Action 19 (1989).

[Rajni Kothari is a leading Indian intellectual, long associated with the Indian civil rights movement. This analysis is directed specifically to the Indian situation.]

. . .

An essential paradox afflicts the human rights scene. On the one hand, everyone associated with this movement or that—women, ecologists, trade unionists, peasantry, tribals, even exponents of state's right's—claims to be struggling for civil liberties and democratic rights. On the other hand, when it comes to the growth of the human rights 'movement', most of these tend to hold back, keen on retaining their specialized identities and afraid of being swamped by a generalised platform or body. In insisting on their diverse specializations, they have pushed human rights activists towards playing a specialized role. Many of the human rights activists have themselves contributed to such an image by insisting that a human rights body should confine itself to fighting against atrocities committed by the state, not in dealing with the sources of these atrocities in the structure of the state and of civil society. While individual activists may involve themselves in political activities, including in party politics, it is not the role of human rights bodies to get so entangled. So both by the reckoning of others and by their own, human rights activists are a breed apart. And, I may add, a breed that is shrinking in number, influence and, to an extent, even legitimacy.

. . .

Two important reasons lie behind [the failure of human rights NGOs to have a greater impact] and the general state of institutional decay and paralysis of will, each of which also brings out the conceptual flaws in the human rights movement as it has developed in India. First, it is quite clear that the movement has entered a state of exhaustion. The struggle against the state is becoming an unequal one, constantly frustrating and causing depletion in ranks, in turn leaving the field to an ageing leadership that is still living in the past as far as its grasp of the human rights problematique is concerned. State violence and state terror are on the increase, the judiciary is proving insensitive to real suffering, and middle class opinion (including the bulk of the national Press) is strongly conditioned by the spectre of terrorism and the slogan of national unity in danger. Human rights activists in the various regions have been found to fight increasingly losing battles. . . .

The 'flaw' behind this state of affairs lies in a conception of human rights that is essentially state-centred. The very state that the human rights movement is trying to struggle against defines the terms of the latter's thinking and action, and in fact receives further endorsement from it. Organisations of civil liberties and democratic rights have done little to reduce the all-embracing and monopolistic role and character of the modern state. They do expose its misdemeanours and tendentious nature; they bring out its malfunctioning; they occasionally lay bare the interests that shape and influence its behaviour; . . . . But they have done precious little in bringing out either the nature of the state's undermining of civil society, or the extent to which the state has become an instrument of national chauvinism on the one hand and arrogant majoritarianism on the other; even less the manner in which the state has been used for speeding the neurosis about 'national interest' and 'national security' and in the process weakened the deeper roots of national cohesion and cultural integrity. . . .

The second major reason that lies behind the decline and disarray in the human rights movement is in many ways conceptual, and can be traced to the very genesis and historic antecedents of the idea of human rights. Some of the other social movements that also employ the phraseology and rhetoric of 'rights', such as the women's movement, also suffer from the same historic legacy. This is the legacy of universal human rights grounded in the western conception of rights, that is at once grounded in an individualistic ethic and takes as a philosophical given that social formations are homogeneous and hence amenable to universal formulations of individual and community, liberty and democracy. This, in my view, is a basic flaw when it comes to highly diverse, plural and continental size societies like ours that also happen to be predominantly rural. Even the distinction between legal and political rights and social and economic rights the U.N. Charter talks about, fails to get over this negative legacy from the West. The tradition of civil liberties and even of democratic rights is too grounded in this conception of 'universal man', whereas the more powerful assertions of 'rights' that have occurred have been either community based or class based. There is therefore, a hiatus between theory and practice. And yet many of the 'demands' that continue to be voiced are couched in either citizen versus state terms, or in terms of one class of citizens versus another class (not 'class' as a collective category but rather a statistical one).

. . .

. . . [I]n India a 'civil society' has not yet emerged (or is confined to the narrow elite strata in a few metropoles), and the only organized form in which castes, communities and tribes are seen to be brought together is the state. The result is that in absence of a vibrant and self conscious civil society, the state is found to impinge directly on diverse communities and imposes, a centralised and homogeneous structure and culture, forcing in turn the latter to express their voices of protest and dissent in separate, regional and often sectarian and communal forms, both fomented by the agents of the state and provoking a backlash from it on behalf of an imagined majority called the 'Hindus'. . . .

[The author then argues that if the non-governmental human rights movement in India is to transcend its marginality and become the 'catalyst of a renewed democratic political struggle'], it will need to give up both its state-centredness and its Western pedigree, become relevant to the fundamental task of constructing a civil society that it is rooted in the unique social and ecological constellation of India, and with this in view, develop a comprehensive strategy. But first it must develop a relevant political theory of democratic transformation that can provide the basis for a movement in a radically different social setting than one that obtains in the West. The imported theory of human rights is already proving counter-productive.

. . .

HENRY STEINER, DIVERSE PARTNERS: NON-GOVERNMENTAL ORGANIZATIONS IN THE HUMAN RIGHTS MOVEMENT

(1991), at 19.

[These excerpts are taken from the report of a retreat for human rights NGO activists from around the world, held in Crete in 1989.]

. . .

Criticisms of First World NGOs

Most criticisms of the mandates and activities of First World NGOs came from Third World participants. . . .

Individualistic orientation and inattention to structural factors

Critics stressed the tendency of First World NGOs to concentrate on individual cases involving governments' violations of identifiable persons' rights to personal security. This individualistic orientation characterizes most well-known INGOs. . . .

It was a shared perception that emphasis on individual cases has a humane foundation and appeal. It grounds an NGO's activities in graphic facts or events—a murdered or tortured or jailed victim, 'human rights with a human face.' Others can empathize with that victim, and support for the victim's cause can more readily be organized. Not only is the victim graspable, as a person rather than a statistic, but the problem itself seems manageable. The task is not to save society but to save a victim or punish a victimizer. . . .

A main point of the critics was that stress on individual cases, even as those cases are aggregated into statistical data, may blur the big picture, the systemic and structural issues that underlie and in some sense explain violations: land-holding patterns, rooted forms of control through intimidation of workers and rural labor, ethnic and class discrimination, unrepresentative political formations, maldistribution of resources and power. . . .

In situations of mass violations such as those a few decades ago in Indonesia or Uganda, stress by NGOs on the individual victims may even be counter-productive in that it can deflect attention from systematic abuse. The critics agreed that statistics and social analysis may be less persuasive than flesh-and-blood victims in mobilizing opposition to a government's actions. They insisted, however, that only through such analysis could one gain the understanding of violations' causes that was necessary for basic change.

For many First World NGOs, seeking that understanding lies outside the formal and legitimate scope of human rights work. It belongs to the realm of political, economic and social theory and analysis. A prominent example is Amnesty International, whose reports inquiring into human-rights conditions in many countries suffer from a failure to probe the contexts for violations in any among them.

Too narrow a mandate

It was argued that Western-derived norms stressing individual rights stem from a historical experience and from social formations that are alien to the Third World. Most of that world lacks a civil society that is both vital and relatively autonomous from the state, and that has the resources enabling it to influence and contain state policies. The conditions are lacking for the political life assumed by Western constitutionalism—namely, voluntary formation of political and other associations, interest groups giving a voice to many constituencies, and widespread citizen participation through elections.

For the First World, the assumption may well be valid that rights of speech, association and participation will bring about a vibrant and democratic political process. NGOs can efficaciously direct their resources to the protection of such rights. Those same protections would not have the same effects in many Third World countries where fundamental violations of rights can coexist with apparent freedoms of speech. The victims of violations may remain outside the scope of debate and political contest.

. . .

In particular, the critics continued, economic and social rights—and in the view of a few participants, group rights—must figure as parts of an integrated view of needs and rights in Third World societies. Properly understood, civil and political rights are necessary but insufficient conditions for progress toward the ideals expressed in the full range of the postwar human rights instruments.

B. Third World NGOs

. . . Many Third World NGOs may speak a different language from the West to describe the character of the human rights movement and the explanations for violations. Thus the participants from Tutela Legal in El Salvador, the Vicaria de la Solidaridad in Chile, and the Servicio Paz y Justicia (SERPAJ) in Uruguay, held a range of views about which factors best explained the failure of governments to respect human rights. For example, one of these Latin American participants described the underlying human rights problem as one of North-South economic relations and stressed that without their reform no significant progress was likely in observing human rights norms.

Nonetheless, such NGOs often act in ways comparable to their Western counterparts. During the periods of extreme repression in these three states, the principal activity of each NGO involved the protection of human life. Exigent circumstances, and the notion of first things first, led to their exclusive attention to the same gross matters that occupied First World INGOs doing investigative work in those countries. As severe repression ended in, say, Uruguay, differences between Third World NGOs and INGOs surfaced as SERPAJ sought new directions in social and economic work.

. . .

Some participants urged a more active, reform-oriented conception of NGOs—not an exclusive conception, for the critics agreed that functions now performed by NGOs should be continued. A larger number of human rights organizations should address needs not now met by the NGO community.

. . . As noted by a Latin American participant, 'When you look at the wider situation, protests about individual cases don't lead to an independent judiciary. How do we create effective democratic reforms? The international community is not as strong on these issues. This requires some form of mass mobilization, some economic leverage.'

. . .

Economic, political and social theory applied to social analysis and prescription could lead in many directions. For example, a participant suggested that an NGO believing that much governmental repression was traceable to conditions stemming from an export-led model of economic development should face that problem directly, rather than stop at the defense of particular victims or at urging a government to ratify a convention prohibiting torture. Such legal aid and standard setting are helpful only so far as they go, palliative rather than cure. Another illustration involved NGOs that bring court actions to require a government to enforce existing laws protecting rural labor or a low caste. If government officials and agencies consistently ignore these laws because of the dominant position of elites hostile to their enforcement, such NGO activity becomes futile and other avenues toward structural change must be pursued. A further example involving violence against leaders of rural labor led to the claim that NGOs must examine patterns of land control that perpetuate master-slave relationships.

Participants addressed the violence stemming from economic development plans that unsettle peoples long rooted in the affected land, destroying their economy or even property. NGOs believing that such development plans violate the group rights of minorities or indigenous peoples, or even nondiscrimination rights, should challenge those plans directly rather than stop at protesting identifiable violence.

...

... The Western practice of invoking judicial processes to challenge violations rests on deep political traditions. ...

In the Third World, however, not many genuine solutions will be found through the judiciary. Many judiciaries will respond from fear or tradition to executive pressure, or will observe self-limitations that insulate them from troubling issues with political consequences. The exceptional judiciaries that address basic issues of human rights are apt to have their decrees in favor of human rights plaintiffs ignored to the extent that those decrees require structural reform. Nor will significant reforms be achieved through proceedings instituted by individuals or NGOs before organs of IGOs. Third World NGOs must follow other routes.

Criticisms of Positions of Third World NGOs

Reporting of facts and analysis of social structure. Several observations ... responded to the criticism of a 'case-by-case' approach. Continuing reporting of individual violations reveals a pattern, one with human faces rather than only statistics. As evidenced by the work of groups like the Madres de la Plaza de Mayo in Argentina, persistent attention to such violations can create the political space in which to challenge the system of oppression as a whole. ...

Nor was it accurate to draw so sharp a contrast between attention to violations and analysis of their causes. Reports of INGOs, such as those of the LCHR [Lawyers Committee for Human Rights] or Human Rights Watch, briefly refer to such factors as military control over civilian government, unequal power among ethnic groups, and concentration of economic resources in a small elite. To be sure, social or political analysis is not the primary thrust of these reports, which are meant to bring violations to public light. In addition, INGOs do work in advocating sanctions and educating the public. The principal INGOs are more than the statisticians of violence that some participants seemed to suggest.

Moreover, there is the problem of 'first things first.' Who will report facts if most NGOs, national and international, did not? Facts are the point of departure, the essential primary information, for any serious human rights work. ...

Intergovernmental human rights organs, caught up in global or regional power struggles, are not nearly as effective as NGOs in investigating and reporting facts. . . .

Another aspect of 'first things first' is important. Social analysis and reform are long run necessities. But the immediate necessity for human rights groups is to stop atrocities, to give support to victims subjected to abuse by their own governments while an apathetic world goes about its business. Which groups would perform this function if NGOs deserted their missions and reports for deep think? Did not the performance of such functions give courage to actual and potential human rights fighters in many countries? If not exactly a protective shield, NGOs' investigative missions and widely distributed reports give some measure of international protection to dissidents.

Goals of NGOs.

. . .

Even though conceptions of right are surely relevant to broad-ranging advocacy for social change, proposals that NGOs challenge basic political and economic structures of a society differ dramatically from the NGO work of two decades. Where, it was asked, could a mandate for so enveloping a task be found within the corpus of human rights law? Was not the strength and success of the human rights movement attributable partly to the fact that norms were relatively discrete, set only basic ground rules, expressed no final vision of the good or just society, and rested largely on consensus?

. . .

To the extent that NGOs base their prescriptions for society not solely on a body of human rights norms but on broader social analysis, how are they to be distinguished from other institutions in the vast and controverted world of social analysis—think tanks, academics, government policy makers? As one commentator among many on political choices, an NGO might repel or enlist our sympathies. But it would shed the objectivity that traditional human rights groups claim in their reporting of facts and judgments about violations. A claim of 'accurate' analysis, of accurate diagnosis and prognosis, could not be supported in the way that, say, Amnesty International could support its account of violations and its condemnation of governmental action under a widely acknowledged norm prohibiting torture. The public interest sector of the human rights movement could lose its character as a defender of legality and readily merge into the broad political process.

NOTE

Consider the following observations in Issa Shivji, *The Concept of Human Rights in Africa* (1989), at 59:

> Western NGOs and scholarship, no doubt, concentrate a lot of time and energy on exposing human rights violations in Africa by African states. In itself this is a useful activity, for publicity may hopefully, to whatever small extent, act as a deterrent to potential violators. But, since this is not done in the context of imperialism, this activity, we submit, objectively reproduces imperialist ideology of human rights.

[The author then quotes from Noam Chomsky and Edward Herman, who criticize the failure of Western intellectuals to focus on the violations committed by their own governments.]

> . . . [T]he same 'hypocrisy and opportunism' is reproduced within the African scholarship and activity on human rights. The few African NGOs, funded as they are by their Western counter-parts or other Western funding agencies, rarely touch on the role of imperialism. They do not even expose the crimes of their own states. Instead much time is spent on refining legal concepts of human rights and the machinery for implementation.

MICHAEL POSNER AND CANDY WHITTOME, THE STATUS OF HUMAN RIGHTS NGOs

25 Colum. Hum. Rts L. Rev. 269 (1994), at 272.

II. Obstacles Confronting Human Rights NGOs

A. Government Reaction to NGOs

In spite of the critical contribution which NGOs make to the strengthening of civil society and to upholding the rule of law, governments often seek to limit or prevent their effective operation. This is particularly true for local human rights groups. . . . Some governments, especially those with poor human rights records, perceive 'non-governmental' organizations to be 'anti-government' organizations which are a threat to security and stability. Government with poor human rights records are particularly sensitive to criticism from independent groups in their own society. While criticism from abroad can be easily dismissed as propaganda or as influenced by governments hostile to the country concerned, public denouncement of a government's acts by its own citizens is harder to dismiss. In such situations, NGOs routinely face government-created obstacles in their efforts to operate freely, including restrictions on their rights to raise funds from abroad, on their ability to disseminate materials freely and on their ability to gain permission to operate as a legal entity. Even a government that is publicly committed to allowing NGOs to operate freely may react with hostility when national NGOs draw attention to their abuses of human rights.

The reaction of many other governments is less subtle. State-sponsored physical attacks on and persecution of human rights advocates are well-documented and widespread. In 1993, the Lawyers Committee for Human Rights documented over 250 cases of attacks on lawyers and judges, involving some 450 people in fifty countries.

. . .

A further frustration for many local NGOs is their inability to achieve accountability for human rights violations through domestic means of redress. In many cases, they are turning to intergovernmental fora to aid them in finding effective remedies.

B. Current Law: Examples of Restrictions On NGOs

1. The People's Republic of China

. . .

Current PRC law condemns 'counter-revolutionaries' who would 'sabotage' the 'people's democratic dictatorship' or 'disclose state secrets.' These terms are not defined. The 1993 National Security Law specifically includes 'stealing, prying into or illegally transmitting state secrets' in its definition of 'behavior that endangers national security.' The government has deemed independent human rights reporting to constitute counterrevolutionary propagandizing and disclosure of important state secrets. Even mere reporting of events unfavorable to the government is considered to be disclosure of state secrets.

2. Singapore

Independent human rights advocacy groups are forbidden by law in Singapore. Organizations such as the Singapore Law Society, which have attempted to address human rights issues, have often been the target of harassment and intimidation by the Singaporan government.

. . .

3. Egypt

. . . Egyptian Law 32 of 1964 requires that all non-governmental organizations seek official permission if they wish to operate freely. Even if an association has been granted this permission, the Ministry of Social Affairs has intrusive powers that enable it to prohibit the organization from affiliating with groups outside Egypt, to change the composition of its board of directors, and to require organizations to merge against their will.

Some human rights organizations have been denied permission to register under the law.

. . .

6. Tunisia

. . . [I]n 1991, after the Tunisian League for Human Rights (LTDH) had issued a number of public statements critical of the government's human rights policies, government-owned newspapers discontinued their coverage of LTDH activities. With the government enjoying a virtual monopoly over mass media outlets, this severely impeded the LTDH's ability to disseminate its findings.

. . .

QUESTIONS

1. Is Kothari's diagnosis of the dilemma in which the human rights movement finds itself valid only in relation to the Indian situation or is it transferable to most Third World countries, and perhaps even beyond?

2. The report by Steiner expresses the strongly held views of many Western commentators who emphasize the importance of objectivity, accuracy and measurability for assuring the credibility of the reporting of a human rights NGO. Does the choice of such criteria beg the question as to the 'proper' focus of the non-governmental side of the human rights movement?

3. 'Human rights NGOs, local or international, should do what they were set up to do, and nothing more. They find the facts. For most of history, horrific events in one part of the world were not at that time known about elsewhere. At least that has changed today, both because of the investigations and reporting of NGOs and the revolutionary developments in media. Perhaps states and international organizations do too little, often nothing, to arrest the massive violations that can no longer be hidden. But at least they and their populations *know*. That knowledge is the predicate to debate, publicity, and possible action. Let NGOs stay with their essential mission and not weaken themselves and confuse the world by attempting more.' Comment.

4. 'International NGOs are but another method for imposing Western concepts upon the Third World. Originating in the developed world, they inevitably remain alien in such other settings. Their agendas are dictated by external assumptions, and they ignore or suppress vital issues like exploitation by their home states in the Third World and the responsibility of those states for human rights violations abroad.' How do you respond to such an analysis?

5. 'INGOs, transient in a foreign state because their representatives arrive only to gather facts and then leave, are incapable of bringing about basic change in that state, such as a change in political structure. At best their activity can lead

to modest, often short-lived improvement. Only the local NGOs can achieve the basic political change that is essential for long-run respect for human rights. That is where the energy and funds ought to go.' Do you agree?

NOTE

One of the recurrent themes in analyses of the prospects of democratization in the post-Cold War era concerns the actual or potential role of 'civil society'. That term has been widely used to describe the role played by the non-state actors that have emerged to fill the space left by the demise of the Communist Party in Eastern and Central Europe.

Most readings in Section A of this chapter contain at least a passing reference to the term. Some of them seem to use it as though it were synonymous with non-governmental organizations as a collective group, while others apparently mean it to encompass 'social movements' in the broadest sense as well as NGOs in particular, without necessarily specifying the differences between the two. An example of the latter approach is the 1995 report of the Commission on Global Governance, which indeed expanded its conception of civil society to the transnational stage. It asserted that global governance,

> once viewed primarily as concerned with intergovernmental relationships, now involves not only governments and intergovernmental institutions but also non-governmental organizations (NGOs), citizens' movements, transnational corporations, academia, and the mass media. The emergence of a global civil society, with many movements reinforcing a sense of human solidarity, reflects a large increase in the capacity and will of people to take control of their own lives.[1]

The concept of civil society is vigorously debated in the theoretical literature: its historical origins, its relation to liberal or conservative thought, its role in democratization, and so on.[2] It has been described so broadly as to encompass all non-state sectors—some would say, 'society' as opposed to the formal structure of the 'state'. It has been portrayed as mediating between the state and its citizens. It has been associated with the more limited sphere of the market through which individuals seek to maximize their private interests. It has been described as the voluntary, self-generated collective action of citizens in a public sphere to articulate their ideas and interests, monitor and

[1] Our Global Neighborhood: The Report of the Commission on Global Governance 335 (1995).

[2] See, e.g., Jean Cohen and Andrew Arato, Civil Society and Political Theory (1992); Ernest Gellner, Conditions of Liberty: Civil Society and Its Rivals (1994); Adam Seligman, The Idea of Civil Society (1992).

influence the state, and participate in different forms of political life. Exploration of this complex concept is beyond our present intention; we mean only to call attention to ways in which it has been used in the literature about NGOs in general and human rights NGOs in particular.

QUESTIONS

1. The Commission on Global Governance puts immense faith in the concept of 'global civil society' which it considers to be 'best expressed in the global non-governmental movement' (p. 254). Consider the large variety of institutions that would fall within the notion of a 'non-governmental movement.' Are there risks in placing immense faith in such a movement as part of a trend toward global governance? What risks? Are human rights NGOs part of those risks?

2. The Commission also calls upon the international community 'to create the public–private partnerships that enable and encourage non-state actors to offer their contributions to effective global governance' (p. 255). One commentator, however, has called for caution as local NGOs 'proliferate under encouragement of or incentives from foreign donors, operate increasingly as consulting firms, and lose touch with, or accountability to, their local constituencies.'[3] Can the local human rights NGO community be 'encouraged', supported and brought by outside funders directly into the processes of 'effective global governance' (as paid monitors, as participants in humanitarian assistance projects, and so on) without losing some essential characteristics for a human rights NGO?

B. INGOs AND THE UNITED NATIONS

Part B examines the work of the large international NGOs and their relations with the United Nations. Amnesty International is the pre-eminent example of an INGO on which these readings concentrate.

COMMENT ON INGOs

Although the following readings use Amnesty International (AI) as a case study, there is a wide range of INGOs, the majority of which do not follow the AI model of being a membership-based organization. The spectrum of major human rights INGOs includes:

- the International Committee of the Red Cross, a predominantly Swiss organization with close links to the Swiss Government;

[3] Yash Ghai, Human Rights And Governance: The Asia Debate, 15 Aust. Y.B. Int'l L. 1, 32 (1994).

• the International Commission of Jurists, based in Geneva with national affiliates but dependent primarily upon grants, including from governments;
• the Lawyers Committee for Human Rights, based in New York and drawing financial support from foundations and law firms; and
• the Anti-Slavery International, based in London, originally established in the 1840s as the Anti-Slavery Society.

For the most part, INGOs' activities focus on the preparation of reports on country situations throughout the world as well as on generic problems involving many countries like censorship or prison conditions. INGOs distribute these reports, provide the information in them to the media, and use that information to engage in lobbying or other forms of advocacy before national executive officials or legislatures and international organizations. Sometimes they initiate or join (as *amicus curiae*) in litigation. Their other activities include the drafting of proposed legislation and of international standards, human rights education, conferences and other forms of promotion, protests and demonstrations, letter-writing and urgent action campaigns, monitoring and critiquing the work of governmental and intergovernmental human rights agencies, and working with and strengthening their local affiliates.

The largest of the INGOs based in the United States is Human Rights Watch (HRW). In its Human Rights Watch World Report 1994 (1995), HRW characterizes itself in the following way (at vii):

> Human Rights Watch conducts regular, systematic investigations of human rights abuses in some seventy countries around the world. It addresses the human rights practices of governments of all political stripes, of all geopolitical alignments, and of all ethnic and religious persuasions. In internal wars it documents violations by both governments and rebel groups. Human Rights Watch defends freedom of thought and expression, due process and equal protection of the law; it documents and denounces murders, disappearances, torture, arbitrary imprisonment, exile, censorship and other abuses of internationally recognized human rights.
>
> Human Rights Watch began in 1978 with the founding of its Helsinki division. Today, it includes five divisions covering Africa, the Americas, Asia, the Middle East, as well as the signatories of the Helsinki accords. It also includes five collaborative projects on arms transfers, children's rights, free expression, prison conditions, and women's rights.

HRW's advocacy has traditionally been centred on the executive branch of the U.S. Government and the Congress, as well as on the U.S. media. In recent years, it has enhanced its international profile through the establishment of offices in London and Brussels (to monitor and influence the human rights policies of the governments of the European Union), in Moscow (both to investigate and influence Russian human rights policy), in Hong Kong (primarily focused on China), and in Belgrade, Zagreb, Rio de Janeiro and Dushanbe (Tajikistan).

NOTE

The following readings start with reflections on problems of fact-finding growing out of an investigative mission in Nicaragua of Human Rights Watch/America. Of course such missions raise the problems in fact-finding that were examined in Chapter 7 with respect to the UN human rights system—that is, problems in investigations and reports by rapporteurs appointed by the UN Commission or by working groups of that Commission. Recall (pp. 394–401, *supra*) the different suggested criteria for fact-finding, and the tension between an ideal of objectivity and the reality of political contexts and possibilities.

The readings continue with a general description of INGOs' work before turning to a closer look at Amnesty International.

TOM FARER, LOOKING AT LOOKING AT NICARAGUA: THE PROBLEMATIQUE OF IMPARTIALITY IN HUMAN RIGHTS INQUIRIES

10 Hum. Rts. Q. 141 (1988), at 142.

. . .

THE PROBLEMATICS OF HUMAN RIGHTS ASSESSMENT

Particularly in the context of domestic conflict, balanced assessment is difficult enough when vehement political interests do not intrude. The various human rights treaties allow governments temporarily to suspend such important rights as free speech, press, and association in order to maintain public order. Moreover, respected intergovernmental enforcement agencies like the Inter-American Commission on Human Rights and the European Court of Human Rights have conceded to governments a considerable margin of discretion in deciding whether conditions require suspension. In any event, the possibility of lawful suspension removes what would otherwise be a helpful, bright-line distinction between permissible and delinquent behavior by governments.

Certain rights can never be suspended—in legal argot they are 'non-derogable.' They include the right to life, the right to be protected from torture, and the right not to be convicted or punished without due process. Despite the categorical character of these rights, their violation may nevertheless leave a fact-finding and assessing body with uncertainties that can only be resolved with the aid of assumptions about credibility, about the meaning and implications of evidence, that are immanent in the fact finder's value system and symbiotic with his or her political orientation.

To indict the regime, it is not enough that people die violently or disappear. In addition, you must conclude that senior officials either have ordered the delinquencies or tolerate them. Some or all parts of the security forces may be out of control. In that event, the central question is whom to treat as the government—occupants of what the constitution labels the country's highest offices or occupants of the headquarters of the armed forces. Moreover, particularly in the course of an armed conflict, even fairly effective governments may find it difficult to enforce relatively humane combat and interrogation guidelines on its troops and security operatives.

. . .

A third obstacle to accurate portrayal of human rights in a given country is the sheer difficulty of getting the basic facts straight. The most bestial governments deem the costs of exposure sufficiently high to justify measures of concealment. Torture can be carried out on a large-scale without benefit of elaborate and readily identifiable plant and equipment. Military schools, abandoned warehouses, and all sorts of other innocuous venues are readily adaptable. Use the existing electrical outlets, add a few hooks, attach a few extra tubs to the plumbing system and, without any prodigies of improvisation, you have a perfectly adequate and easily dismantled interrogation center. In developing countries, many alleged violations occur in the countryside. Rudimentary systems of communication and transportation delay word of alleged violations from reaching fact finders and impede access once a charge has been filed. The trails may be very cold and, in the short time available to fact finders, it may be possible to follow only a few trails and they may be the wrong ones.

A fourth difficult feature of human rights assessment that makes it terribly vulnerable to ideological and political pressure is the universality of imperfection. Look closely at any country in the world and you will find violations of human rights. The comparatively handsome performance of capitalist democracies in Western Europe has not deprived the region's human rights Commission and Court of sufficient business to keep them busy. . . .

One nevertheless requires the aid neither of an elaborate brief nor of preternatural insight to appreciate that the condition of human rights in the United States circa 1987 is qualitatively better than in South Africa or that Malaysia was paradise compared to Pol Pot's Cambodia. But a great many nations do, of course, offer less stark contrasts.

Virtually all the respective private, as well as public, human rights monitoring bodies have rejected interstate comparisons. In doing so, they have attempted to avoid exposing themselves to charges of political bias and to disputes that would shift attention from the bedrock finding of human rights violations. Moreover, by eschewing comparison, they have sought in good faith to neutralize their own predispositions. They have, in addition, argued that the existence of still worse conditions elsewhere is simply irrelevant. Violations are violations wherever and however frequently they occur; the international community should do what it can to terminate them.

The considerable virtues of this principled refusal to compare must be weighed against a considerable vice. Exposure alone is a weapon of limited effect, particularly if its use does not signal the coming deployment of the more tangible forms of coercion available only to states. Since the application of coercive measures is not cost-free, and since even the most morally sensitive of governments has many other foreign policy interests, national action on behalf of human rights is necessarily selective. The comparative severity of present or prospective violations being one factor influencing choice, comparisons will be made with or without the aid of human rights monitoring bodies.

. . .

Except in the fairly rare cases of mass murder and torture (by any definition of the term)—Cambodia under the Khmer Rouge, Uganda under Idi Amin, the Soviet Union under Stalin, Europe during the Nazi occupation—comparisons require the exercise of problematic etiologies and value judgments about the relative importance of different rights and the mitigating effect (if any) of the domestic and international context. How, for instance, do you compare governments that kill with guns to governments that allow people to die of starvation and malnutrition? Or governments that maintain themselves by suppressing free speech and free association with those that use death squads to that end? Should you compare governments without reference to each country's political culture and institutional heritage? One reason Pinochet's night-and-fog regime in Chile evokes more active hostility than fetid conservative governments in places like Zaire and Paraguay is the country's tradition of democratic rule and its evident potential for the renewal of democracy.

. . .

FRAMING THE NICARAGUAN CASE

When I visited Nicaragua in December 1986 as part of a four-member fact-finding group dispatched by Americas Watch, a private human rights monitoring agency, I had behind me eight years of fact-finding experience as a member and the first gringo president of the Inter-American Commission on Human Rights. In that capacity I had already visited Nicaragua three times, twice since the fall of Somoza. From the outset of this fourth visit, I felt a peculiarly intense self-consciousness about the assessment process itself. This strong feeling of bifurcation into observer and participant did not stem exclusively from my awareness of those problematic features of fact-finding and assessment I have sketched here. They had haunted me in the past. The peculiar sensitivity I felt on this occasion testified both to the tarnished hopes for the country I had shared with so many other Americans at the time of Somoza's fall and to the confidence-shaking intellectual offensive of the anti-Sandinista coalition in the United States.

. . .

In the case of Third-World countries like Nicaragua that on the basis of the raw facts, would seem to stand somewhere among the broad middle ranks of human rights performers, ideologically determined frames for organizing the data invariably influence, and not infrequently govern, overall assessments of government behavior. (I speak of persons making a good faith effort at objectivity. Ideology—by which I mean a cluster of ideas, ordered by some normative vision, that purport to explain the way the world works and how people should live—may also induce observors to lie about the raw facts in the unexpressed name of higher purposes.) These middling cases give fullest rein to ideological influence over assessment precisely because the facts, by themselves, do not confirm the goals of the regime's dominant personalities and the means they are prepared to use for their attainment. Nor do they speak clearly concerning the direction the society is likely to take irrespective of its leaders' intentions. They are ambiguous even with respect to the propriety of a good deal of current behavior.

[Farer points out the alternative explanations that are available for certain policies or actions of a government under investigation, such as (in the case of the Nicaraguan Sandinista government) closing down an opposition newspaper or expelling some church officials. One's judgment about such acts may depend on one's worldview, ideological orientation, or general belief or disbelief in the good faith of a government. He concludes that, in the case of Nicaragua, numbers of governmental acts about which he was required to make a judgment in his report for Americas Watch involved 'genuinely tough issues to resolve, particularly since . . . governments enjoy a margin of discretion in choosing means for dealing with threats to public security.']

NIGEL RODLEY, THE WORK OF NON-GOVERNMENTAL ORGANIZATIONS IN THE WORLD-WIDE PROMOTION AND PROTECTION OF HUMAN RIGHTS

90/1 U.N. Bulletin of Human Rights 84 (1991), at 85.

Organization

There are probably as many organizational structures as there are human rights organizations. However, three general models may be identified: (*a*) a group of individuals form and control the organization, with provision for self-renewal; (*b*) a number of organizations federate to combine their activities; (*c*) a centralized 'democratic' structure is created, with the constituent units both controlling the organization and carrying out its activities. The national/international mix may vary. Most will have permanent secretariats.

Thus, the ICRC [International Committee of the Red Cross] and the Anti-Slavery Society fall into the first category, with their controlling membership coming from nationals of one country—Switzerland and the United Kingdom respectively. The International Federation of Human Rights falls into the second category, since it is composed of national leagues working to defend human rights in their various countries and electing an international board. Amnesty International (AI) falls into the third category, having national sections which elect an international executive and work against human rights violations in any country but their own. . . .

What all these organizations have in common is their work to defend human rights internationally. Their financing will also generally reflect their non-governmental nature, though some will take substantial sums of Government money, especially if, like the ICRC, they have governmentally assigned responsibilities, while others, such as AI, will not accept government donations.

The NGOs also have to determine the area of specialization that they will undertake, since the broad field of human rights covers the pursuit of justice in many forms and requires priority-setting. Most NGOs, including all those just mentioned, specialize in the area of civil and political rights. . . .

This orientation is understandable: it is easier to identify and prescribe a remedy for a violation of civil rights than economic rights.. Nevertheless, most of the organizations recognize the interrelationship of politico-civil and socio-economic rights. Several, such as the ICJ and the ILHR, are increasingly addressing issues of socio-economic rights, by such means as seminars, segments of country reports and so on. . . . A new organization, Rights and Humanity, deals with economic and social rights as a central aspect of its mandate.

Information

By far the most important product of human rights NGOs is information, particularly information about human rights violations. . . .

. . .

Action

Most NGO action is aimed at disseminating information and is determined by strategies followed by the particular NGO, which may be general or country-specific. For example, the International Committee of the Red Cross operates under a principle of confidentiality. Its reports, generally the product of visits to prisoners within its mandate, are addressed to the Government in question, not to the public. The Government is free to publish the reports, but if it does so selectively, the ICRC retains the right to publish them in full. Even NGOs that tend to publish as a normal technique may well first send a text intended for publication to the authorities concerned. . . .

Most NGOs see publication of their documentation as the principal means of putting an end to the human rights violations they record. The form of such publication will be determined by two main factors: one, the nature of the material, and two, the perceived efficacy of the available modes of publication. For urgent situations a news release or press conference may be used. For specific country studies, a book or monograph-style format can be appropriate . . .

Much work is done by issuing brief documents on individual cases. Some NGOs use their own membership and the networks of other NGOs to make urgent appeals. . . .

. . . Some times it may be appropriate to address the judiciary, although it is important to avoid appearing to bring extraneous influence to bear on a specific court considering a specific case (this assumes that the judiciary is independent, often a politically necessary assumption, however far from reality it may be).

Some activities may also be directed at the legislature. . . .

Increasingly, NGOs are becoming interested in short- and long-term preventive work. For instance, they may seek to oppose the sending of people from one country to another where there is the risk of a human rights violation occurring in the receiving country. They may also promote human rights awareness in the general population, in formal education, or in the education of particular professional sectors, such as lawyers, judges, the police and the armed forces.

Nevertheless, the executive arm of government will remain the main addressee of NGO activities, for the executive is the branch of government most involved in human rights violations and the one most likely to be sensitive to public opinion. And public opinion, both national and international, is the main forum in which human rights violations may be judged. . . .

. . . NGOs may well target their information on one Government towards other Governments. International NGOs are particularly well placed to do this: their national affiliates are a means of communication to their own Governments, and often the very fact that they represent part of the domestic constituency may oblige those Governments to take the information in question into account. This does not mean that NGOs necessarily advocate 'sanctions' against Governments accused of violating human rights: some do, usually selectively; others do not. Nevertheless, the more excessive a Government's resort to human rights violations, the harder it becomes for other Governments to maintain the 'atmosphere' necessary for business-as-usual.

NGOs and the UN

. . .

NGOs have been able to be actively involved in . . . norm-creating exercises. Indeed, it has sometimes been their public activities that have demonstrated the need for the United Nations to take action. . . . [T]his was manifestly the case with regard to the problem of torture, the United Nations first having taken up the issue in response to a 1973 NGO campaign aimed at abolishing the practice. Furthermore their consultative status has permitted NGOs to participate, without vote, in the drafting discussions. They can propose ideas for inclusion and can explain, on the basis of their experience, why those ideas need to be incorporated in the text. They can also approach the drafting

Governments at the national level with a view to influencing the instructions given to their representatives in the drafting body. Prominent recent examples of such NGO participation are that of Amnesty International, the International Commission of Jurists and the International Association of Penal Law in the drafting of the Convention against Torture and Other Cruel, Inhuman or Degrading Treatment or Punishment, and that of a NGO group led by Defence for Children International in the drafting of the Convention on the Rights of the Child.

Another facet of United Nations preventive work is its programme of advisory services and technical assistance, which in recent years has undergone major expansion, especially since the establishment of a voluntary fund to finance it. The fund can, among other things, help Governments acquire the necessary training and resources to assist them to respect human rights. This can be useful to NGOs which are often asked by Governments for such assistance, or at least advice in obtaining it. At the same time, NGOs have pointed to the need to ensure that the advice and assistance programme is not exploited by Governments engaged in extensive human rights violations. . . .

The area of United Nations human rights activity of most concern to NGOs has come to be that of protection or implementation. Until some 20 years ago, such activity was virtually non-existent. The United Nations did not take up individual cases, ignored most situations of human rights violations, however serious, and did not even permit NGOs at United Nations meetings to document human rights violations, either orally or in writing. . . .

Successive heads of the Human Rights Centre . . . have acknowledged the preponderant role played by NGOs in furnishing the information on human rights violations that constitute the subject-matter of the various procedures. . . .

QUESTIONS

1. Consider some of the problems in investigative missions by INGOs that Farer identifies. (a) Does the question of political assumptions and ideological beliefs arise only when a NGO report must go beyond 'stating what happened' (say, a newspaper was closed by the government during a proclaimed 'public emergency' under Article 4 of the ICCPR) to reach the 'legal' conclusion whether or not such action amounted to a human rights violation (say, whether that closure was 'strictly required by the exigencies of the situation' under Article 4)? (b) Does this illustration merely suggest that INGOs frequently and inevitably interpret treaty provisions (like Article 4 of the ICCPR) in order to write their reports, and that such traditional interpretation/application should not prejudice the claim of an INGO to apolitical objectivity in its country report?

2. Do you agree with Rodley that the orientation of most INGOs towards civil and political rights 'is understandable'? In what ways does an INGO face

different tasks in reporting about a country with respect to (a) torture and disappearances, and (b) the homeless and the right to housing? Does the second task seem more complex, less manageable? Why?

NOTE

The mandate is among the most significant aspects of a NGO. It states the purposes of the organization, the subject matter or region of special concentration, and so on. Mandates differ significantly; the mandate of AI that appears below as part of its Statute is in important respects distinctive. Of course mandates need not remain constant over time. They change with an organization's growth and shifting purposes, as well as in response to the evolving global context for its activities. The following readings trace the manner in which the AI mandate has been interpreted and expanded since AI was founded in 1961.

Some background information about AI will be helpful. At the beginning of 1994 it had over 1,100,000 members in over 150 countries, with 4,349 local groups based in over 80 countries. In 1993 it recorded 922 cases involving the release of prisoners of conscience and initiated 551 new 'urgent action' procedures and repeated 318 such actions. Its central budget in London was close to £13 million (approximately U.S. $22 million). In total, four times that amount was raised by national sections. *Amnesty International Report 1994*, at 352.

STATUTE OF AMNESTY INTERNATIONAL

Amnesty International Report 1994, at 332–3.

Object and Mandate

1. The object of Amnesty International is to contribute to the observance throughout the world of human rights as set out in the Universal Declaration of Human Rights.

In pursuance of this object, and recognizing the obligation on each person to extend to others rights and freedoms equal to his or her own, Amnesty International adopts as its mandate:

To promote awareness of and adherence to the Universal Declaration of Human Rights and other internationally recognized human rights instruments, the values enshrined in them, and the indivisibility and interdependence of all human rights and freedoms;

To oppose grave violations of the rights of every person freely to hold and to express his or her convictions and to be free from discrimination by reason of ethnic origin, sex, colour or language, and of the right of every person to physical and mental integrity, and, in particular, to oppose by all appropriate means irrespective of political considerations:

(a) the imprisonment, detention or other physical restrictions imposed on any person by reason of his or her political, religious or other conscientiously held beliefs or by reason of his or her ethnic origin, sex, colour or language, provided that he or she has not used or advocated violence (hereinafter referred to as 'prisoners of conscience'); Amnesty International shall work towards the release of and shall provide assistance to prisoners of conscience;

(b) the detention of any political prisoner without fair trial within a reasonable time or any trial procedures relating to such prisoners that do not conform to internationally recognized norms;

(c) the death penalty, and the torture or other cruel, inhuman or degrading treatment or punishment of prisoners or other detained or restricted persons, whether or not the persons affected have used or advocated violence;

(d) the extrajudicial execution of persons whether or not imprisoned, detained or restricted, and 'disappearances', whether or not the persons affected have used or advocated violence.

Methods

2. In order to achieve the aforesaid object and mandate, Amnesty International shall:

(a) at all times make clear its impartiality as regards countries adhering to the different world political ideologies and groupings;

(b) promote as appears appropriate the adoption of constitutions, conventions, treaties and other measures . . . ;

(c) support and publicize the activities of and cooperate with international organizations and agencies which work for the implementation of the aforesaid provisions;

(d) take all necessary steps to establish an effective organization of sections, affiliated groups and individual members;

(e) secure the adoption by groups of members or supporters of individual prisoners of conscience . . . ;

(f) provide financial and other relief to prisoners of conscience and their dependants and to persons who have lately been prisoners of conscience . . . ;

(g) provide legal aid, where necessary and possible, to prisoners of conscience . . . ;

. . .

(i) investigate and publicize the 'disappearance' of persons where there is reason to believe that they may be victims of violations of the rights set out in Article 1 hereof;

(j) oppose the sending of persons from one country to another where they can reasonably be expected to become prisoners of conscience or to face torture or the death penalty;

(k) send investigators, where appropriate, to investigate allegations that the rights of individuals under the aforesaid provisions have been violated or threatened;

(l) make representations to international organizations and to governments whenever it appears that an individual is a prisoner of conscience or has otherwise been subjected to disabilities in violation of the aforesaid provisions;

(m) promote and support the granting of general amnesties of which the beneficiaries will include prisoners of conscience;

(n) adopt any other appropriate methods for the securing of its object and mandate.

PETER BAEHR, AMNESTY INTERNATIONAL AND ITS SELF-IMPOSED LIMITED MANDATE

12 Neths Q.H.R. 5 (1994).

Amnesty International is today the largest international non-governmental organization in the field of human rights in the world. . . .

. . .

. . . The Statute may be amended by the International Council, which represents the international membership and which nowadays meets once every two years, by a majority of two thirds of the votes cast. The Council can also, by ordinary majority, adopt resolutions to further specify or interpret the Statute.

. . .

2. Meaning of the Mandate

. . . [Since 1991] a distinction has been made between the notions of *object*, *mandate*, and *methods*. [See Statute, *supra*.]

. . .

Mandate issues—though generally considered of great importance—are notoriously difficult. The average Amnesty-member usually lacks the experience and the knowledge of decisions taken in the past which may have created precedents. Only very few members and certain individuals in the International Secretariat have the necessary command over the intricacies of the mandate to give knowledgeable interpretations of the existing mandate and develop options that are sufficiently clear for the membership to act upon.

. . .

3. Criteria

The mandate should be sufficiently flexible to develop with changing circumstances in the world. . . . At the same time, such changes should not be too radical in order to reach the necessary consensus among the membership.

. . .

Political impartiality is seen as contributing to Amnesty's credibility and thus to its effectiveness. The need for maintaining its political impartiality as well as the wish to avoid even the *appearance* of such partiality, was the main reason why Amnesty has not found it possible to condemn the best known legally based system of human rights violations in the world: Apartheid in South Africa. It wanted to avoid having to take up all kinds of discriminatory legislation in other parts of the world as well and limited itself to '. . . condemning and opposing those laws and practises of Apartheid which permit the imprisonment of people on grounds of conscience or race; the denial of fair trial to political prisoners; torture; or the death penalty.' This compromise text was arrived at with considerable difficulty. It has been challenged at regular intervals at meetings of the membership. Especially African members of the organization, but also members of other sections, have found it difficult to accept that Amnesty did not outright condemn Apartheid.

. . .

5. Three Illustrations

Below, three illustrations are offered of mandate discussions within Amnesty International. . . .

5.1. The Violence Clause

From the beginning, Amnesty's mandate has included what has come to be known as the 'violence clause', which means that the organization will not call for the unconditional release of any persons who have used or advocated violence. . . . Although the exact reasons for adopting the clause are difficult to retrieve, the following are usually given for the retention of the clause;

- *character of the membership*: Amnesty's membership is composed of people of different persuasions, pacifists as well as non-pacifists. . . . The non-violence clause enables people of such different persuasion to work together for the common goal: achieving respect for human rights within Amnesty's mandate.

- *political impartiality*: Amnesty's influence with regard to Governments depends on the fact that they tend on the whole to accept its political impartiality with regard to intergovernmental relations and internal relations between Governments and opposition groups. If Amnesty would demand the release of persons engaged in violent opposition to the government, it could become identified with the opposition and thereby risk losing its credibility and its influence.
- *distinguishing political from criminal actions*: it is sometimes difficult . . . to distinguish political from purely criminal actions and Amnesty wants to avoid getting into a position where it has to sort out the types of violence that are and that are not legitimate.
- *ethical considerations*: in many circumstances the use of violence may cause the loss of life or limb, which is seen as morally repugnant by many, if not most people.

Amnesty makes an effort to point out that the above does not mean that it is opposed to the use of violence as such. It has 'no views' on such use, but it refuses to work for the release of persons who have used or advocated violence. Indeed this limitation is not considered when it is asking for a prompt and fair trial for *all* political prisoners and in its fight against the death penalty (which is in fact usually applied to persons who have used violence) and torture or other cruel, inhuman or degrading treatment or punishment.

. . .

5.2. Political Non-Governmental Entities

[In this omitted section, the author examines AI's changing attitude toward investigating and reporting about the actions (killings, torture) of non-governmental entities with a political character, such as a group in a civil war claiming partial political control, or an armed opposition to a government that exercises some degree of control over a territory and its population. The author's illustrations include Hezbollah in Lebanon, the FMLN in El Salvador, Sendero Luminoso in Peru and the IRA in Ireland, and thus includes groups whose authority was often 'transient in character.' AI gradually moved from the position that international human rights norms were binding only on states and began to deal with abuses by non-governmental entities. 'The 1991 International Council decided . . . that henceforth Amnesty would deal with abuses by political non-governmental entities, whether or not the entity has the attributes of a government. . . .']

. . .

5.3. Homosexuality

For many years the issue was debated, whether Amnesty should work for the release of persons imprisoned or detained for their homosexual identity or orientation or for homosexual acts committed in private and between consenting adults. This issue even threatened to split up the organization along multicultural lines. While members in Western Europe and North America tended to consider it rather obvious that Amnesty should work for such persons, many people in Asian, African as well as some Latin American countries thought otherwise.

Agreement existed only to work for (a) persons imprisoned or detained for advocating equality for homosexuals, (b) persons charged with homosexuality as a pretext, while the real reason for their imprisonment was the expression of their political, religious or other conscientiously held beliefs, (c) persons subjected to medical treatment while in prison with the aim of modifying their homosexual orientation without their agreement.

The issue was hotly debated for a number of years, resulting in various voluminous studies of the matter. The differences of view can be roughly summarized as follows. Those in favour of working for imprisoned homosexuals argued that '(homo)sexual orientation' and '(homo)sexual identity' are attributes similar or subordinate to the category of 'sex', which has for many years been in the mandate and that it would only be logical to add this category of persons to the mandate. The opponents felt that in societies where homosexuality is considered a physical ailment or a reflection of socially deviant behaviour, activities on behalf of such individuals would be seen as not related to human rights and make Amnesty look ridiculous; moreover, if Amnesty were to work for imprisoned homosexuals, this would risk to involve it in having to deal with all sorts of other sexual practices.

. . . The International Council adopted by consensus a simple resolution deciding 'to consider for adoption as prisoners of conscience persons who are imprisoned solely because of their homosexuality, including the practice of homosexual acts in private between consenting adults'. The resolution expressed the realization that this decision would increase the difficulty of the development of Amnesty in many parts of the world and instructed the International Executive Committee to draft guidelines regarding action on behalf of imprisoned homosexuals, 'taking into consideration the cultural background of various areas where we have sections and groups or countries in which AI is proposing development.'

Thus ended the major internal debate on the issue of homosexuality. The result had certain features of a political compromise. It was not explicitly decided how the position taken related to the text of the Statute. There is nothing in the Statute, except for a broad interpretation of the word 'sex,' which would cover such activities. The issue of whether the consequence of the decision on homosexuality would indeed open the debate on other sexual practices, was never faced. Nor was a fundamental debate held on the issue of whether Amnesty's new stand on homosexuality meant a departure from its emphasis on universal human rights values. Politically, it was also a matter of attracting one constituency (the gay community) at the risk of possibly losing another (potential Amnesty International-members in the Third World).

. . .

6. Conclusion

. . .

[As one expert on Amnesty International's mandate] has rightly pointed out, there exists an inherent tension between logic of the Amnesty movement and the logic of human rights. While the logic of the movement is its specificity, the logic of human rights consists of their inseparability from each other. He has also made the point that Amnesty International could not become a general human rights organization—promoting all human rights equally—without destroying its identity, losing its membership and abandoning its techniques.

NOTE

In its 1994 Report, Amnesty International observed (p. 4):

> The patterns of human rights abuses that Amnesty International seeks to combat have become more and more complex in recent years. . . . A spate of local wars, often accompanied by the virtual disintegration of state authority, have spread turmoil and terror.
>
> . . .
>
> Nationalist, ethnic and religious conflict, famine and repression have led to massive movements of refugees. More and more countries, especially wealthy ones, have closed their doors to people who need and deserve sanctuary.
>
> These horrific events illuminate the interdependence and indivisibility of human rights more powerfully than any abstract argument. In this increasingly volatile human rights environment, Amnesty International offers a structure for people from all continents to campaign for human rights and influence events worldwide as well as in their own regions. Amnesty International is determined to adapt to the evolving realities by adjusting its responses. It must play its role within the broader human rights movement. And it must transcend its Western roots and continue to develop as a truly international, multicultural human rights movement.
>
> . . .
>
> Top of the agenda for Amnesty International's members and supporters in the coming decade is a worldwide campaign against political killings and 'disappearances'. . . .

PHILIP ALSTON, THE FORTIETH ANNIVERSARY OF THE UNIVERSAL DECLARATION

in J. Berting et al. (eds.) Human Rights in a Pluralist World, Individuals and Collectivities 1 (1990), at 12.

. . . Amnesty . . . has been extraordinarily effective . . . . Its success has been linked inextricably to its narrowly defined mandate resulting in a focus on a very specific range of civil and political rights. . . .

. . .

. . . [It] is my contention that the result, albeit not the intention, of Amnesty's efforts is the widespread dissemination of a conception of human rights which is partial (in the sense of being incomplete) and is not a faithful reflection of the Universal Declaration and the assumptions underpinning that document (from which the Declaration derives its strength and standing). I would suggest that the great majority of Amnesty members and activists would, if pressed, provide a rather distorted list of basic human rights which would reflect the list of core 'mandate' issues pursued by Amnesty and little more. This distortion is essentially a consequence of the enormous success of the organization's determination not to be diverted from the core issues as it sees them. If that consequence could be characterized as intentional then Amnesty should not claim fidelity to the Universal Declaration and should concede that its concerns, far from being a reflection of universal minorities, mirror more closely values associated with the Western liberal tradition. In that case, any pretense to ideological blindness or neutrality should be abandoned. I believe, however, that for the most part that consequence is entirely unintended. If that is the case, it is time for the organization squarely and openly to address the issue of how it can best combat the risk of being perceived to endorse an unduly selective conception of human rights, while at the same time maintaining its core focus which can be justified in terms of manageability, legal specificity and operational potential.

I would not suggest that this will be an easy task, and there will be some among its membership who will resist it fiercely (for reasons of both effectiveness and ideology). Nevertheless, there are some relatively straightforward and painless first steps which could be taken by way of an expanded public information and education campaign . . . . [A] very widely distributed booklet entitled *What Does Amnesty International Do?* notes that Amnesty's 'work is based on principles set forth in the . . . Universal Declaration.' It states that while the organization's focus on prisoner's rights leads it to concentrate 'on a specific program in the human rights field' this 'does not imply that the rights it does not deal with are less important.' But at no point in the twenty-five-page booklet is any indication given as to what those other rights are. It is hard to believe that a brief educational reference to those other rights would in any way impair Amnesty's need to maintain a limited range of concerns.

The principal retort to my objection is likely to be that Amnesty, like all other human rights NGOs, is entitled to be specific in its focus and that other groups can better fill the gaps that Amnesty leaves. But this argument overlooks the fact that Amnesty, whether it likes it or not, is the single dominant force in the entire field. It is bigger, richer, better organized, more representative and more influential than most of the other groups put together. As a result, it is precluded from taking refuge in a justification which, when proffered by smaller NGOs, must (reluctantly) be accepted. As the great powers themselves often need to be reminded, with power and influence comes responsibility. Much of that responsibility may be unwanted, but it cannot simply be shrugged off.

QUESTIONS

1. Amnesty's Statute and the development of its mandate are fraught with underlying complexity. (a) What other solutions might AI have adopted in relation to the question of *apartheid*? (b) What practical difficulties do you see in relation to the application of the 'violence clause'? (c) Do you consider the policy on homosexuality to be a sell-out or an appropriate compromise? Or do you believe that it is wrong since Amnesty, in order to achieve its goals, must draw only on accepted universal human rights standards and hence should stay out of this field?[4]

2. There was strong resistance to the extension of Amnesty's mandate to cover non-governmental entities. Some major NGOs continue to resist this trend (although Human Rights Watch was a pioneer in embracing the broader focus). In 1995 the Commission on Global Governance urged recognition of the reality that 'governments are only one source of threats to human rights and [that often] government action alone will not be sufficient to protect many human rights. This means that all citizens . . . should accept the obligation to recognize and help protect the rights of others' (p. 56). What are the implications of such a development of the conceptual framework of international human rights law?

3. What are some of the practical consequences for its mandate and operations that might flow from Amnesty's commitment to 'transcend its western roots and continue to develop as a truly international, multicultural human rights movement'?

[4] The issue of the consistency of legislation barring homosexual acts (even between consenting adults) with human rights norms arises in the following chapters. See the opinion of the ICCPR Committee in the Toonen case, p. 545, *infra*, and the decision of the European Court of Human Rights in the Norris case, p. 618, *infra*.

NOTE

The following readings examine the attitude, or perhaps contradictory attitudes, of the UN to INGOs. On the one hand, Article 71 of the UN Charter acknowledges the important role of NGOs. Moreover, the UN mechanisms for responding to violations rely heavily on INGOs to provide alternative sources of information, as well as to heighten the external pressures on states that are indispensable to those mechanisms' effectiveness. On the other hand, NGOs have only grudgingly been accorded speaking and other lobbying 'rights,' and their status before UN bodies is regularly challenged by hostile governmental forces. At the same time, the existing 'consultative arrangements' within the UN system are overloaded and unsustainable. We therefore turn, at the end of this chapter, to the challenge of devising appropriate reforms.

ARRANGEMENTS FOR CONSULTATION WITH NON-GOVERNMENTAL ORGANIZATIONS

UN ECOSOC Resolution 1296 (XLIV) (1968).

The following principles shall be applied in establishing consultative relations with non-governmental organizations:

1. The organization shall be concerned with matters falling within the competence of the Economic and Social Council with respect to . . . human rights.
2. The aims and purposes of the organization shall be in conformity with the spirit, purposes and principles of the Charter of the United Nations.
3. The organization shall undertake to support the work of the United Nations and to promote knowledge of its principles and activities . . . .
4. The organization shall be of representative character and of recognized international standing; it shall represent a substantial proportion, and express the views of major sections, of the population or of the organized persons within the particular field of its competence, covering, where possible, a substantial number of countries in different regions of the world. Where there exist a number of organizations with similar objectives, interests and basic views in a given field, they shall, for the purposes of consultation with the Council, form a joint committee or other body authorized to carry on such consultation for the group as a whole. It is understood that when a minority opinion develops on a particular point within such a committee, it shall be presented along with the opinion of the majority.
5. The organization shall have an established headquarters, [and] a democratically adopted constitution . . . .

. . .

7. Subject to paragraph 9 below, the organization shall be international in its structure. . . .

Any international organization which is not established by inter-governmental agreement shall be considered as a non-governmental organization for the purpose of these arrangements, including organizations which accept members designated by governmental authorities, provided that such membership does not interfere with the free expression of views of the organization.

8. The basic resources of the international organization shall be derived in the main part from contributions of the national affiliates or other components or from individual members. Where voluntary contributions have been received, their amounts and donors shall be faithfully revealed to the Council Committee on Non-Governmental Organizations. . . . Any financial contribution or other support, direct or indirect, from a Government to the international organization shall be openly declared. . . .

9. National organizations shall normally present their views through international non-governmental organizations to which they belong. . . . National organizations, however, may be admitted after consultation with the Member State concerned in order to help achieve a balanced and effective representation of non-governmental organizations reflecting major interests of all regions and areas of the world, or where they have special experience upon which the Council may wish to draw.

. . .

36. The consultative status . . . shall be suspended up to three years or withdrawn in the following cases:

(a) If there exists substantiated evidence of secret governmental financial influence to induce an organization to undertake acts contrary to the purposes and principles of the Charter of the United Nations;

(b) If the organization clearly abuses its consultative status by systematically engaging in unsubstantiated or politically motivated acts against States Members of the United Nations contrary to and incompatible with the principles of the Charter;

(c) If, within the preceding three years, an organization had not made any positive or effective contribution to the work of the Council or its commissions or other subsidiary organs.

. . .

MICHAEL POSNER AND CANDY WHITTOME, THE STATUS OF HUMAN RIGHTS NGOs

25 Colum. Hum. Rts. L.Rev. 269 (1994), at 283.

Perhaps as part of a backlash by governments hostile towards the idea of more open participation by NGOs within the United Nations system, there have been . . . recent moves by some governments to diminish the protection offered by international law to human rights defenders.

In June 1993, during the Second World Conference on Human Rights in Vienna (Vienna Conference), much time was spent on drafting the passage which relates to the contribution of NGOs. From the viewpoint of NGOs, the final text represents a significant setback for human rights advocates. It states that

> [n]on-governmental organizations *genuinely* involved in the field of human rights should enjoy the rights and freedoms recognized in the Universal Declaration of Human Rights and the protection of the national law. . . . Non-governmental organizations should be free to carry out their human rights activities, without interference, within the framework of national law and the Universal Declaration of Human Rights.

The reference to national law undermines a fundamental premise of international human rights law: international standards are necessary precisely because national law so often offers inadequate protection.

. . .

To be effective participants in the international debate on human rights, NGOs—local and international—need effective access to the key players in the U.N. system. Yet such access is still hard, if not almost impossible, to come by for most local NGOs.

. . .

The implications of the current arrangements for formal NGO relations with the United Nations are significant. Two questions, in particular, arise: (1) does consultative status provide NGOs effective access to the United Nations system? and (2) how can the NGOs who do not have, or are not eligible for, consultative status, obtain effective access? Those with consultative status need to assess whether it grants them effective and meaningful participation in United Nations fora. For example, NGOs were excluded from the drafting sessions at the Vienna Conference where the final document, the Vienna Declaration and Programme of Action, was negotiated. Moreover, many of the 'rights' granted are, in practice, less useful than they might at first appear. The right to make oral statements at the United Nations Commission on Human Rights, for example, is often rendered virtually useless because of the fact that the debate is so crowded, with so many governments and NGOs wanting to speak, that NGOs are frequently allotted the least popular time, late at night, when there are few government delegates to hear or respond to them.

Even this situation compares favorably with that of the majority of local NGOs that are unable to obtain consultative status at all. In most cases it is only if a local NGO is affiliated with an NGO with consultative status that is willing to let the local group speak or circulate documents under its name can a local group obtain even informal access to meetings such as those of the United Nations Commission on Human Rights. Attending such meetings simply to lobby delegates is not an option unless a national NGO can persuade an international one to give it accreditation.

On the other hand, NGOs do not need consultative status to participate in many of the other human rights activities of the United Nations. They may, for example, submit information to various thematic rapporteurs and treaty-monitoring bodies. . . .

The question of access to the United Nations is becoming increasingly important and complex as the number of local human rights groups continues to grow. At present there are over 900 NGOs in consultative status . . . and the great majority of these are international NGOs. Up until the early 1980s, when there were few local NGOs, the system could cope with the numbers. In recent years, the number of NGOs has increased dramatically—over 1500 NGOs were represented at the Vienna Conference alone. If all these groups were to be awarded consultative status and given the same rights and privileges as those with such status now enjoy, it is arguable that the system would collapse . . . How can the situation be remedied, with a greater number of NGOs [being] granted more effective access, without overloading the system to the breaking point?

The question of access for NGOs is currently being addressed within a new intergovernmental working group within the United Nations. In July 1993, ECOSOC established an open-ended working group consisting of representatives of all interested states to undertake a 'general review' with 'a view to updating, if necessary, Council Resolution 1296 (XLIV), as well as introducing coherence in the rules governing the participation of non-governmental organizations in international conferences convened by the United Nations.'

. . .

REPORT OF THE OPEN-ENDED WORKING GROUP ON THE REVIEW OF ARRANGEMENTS FOR CONSULTATION WITH NON-GOVERNMENTAL ORGANIZATIONS

UN Doc. A/49/215 (1994), at 12.

. . .

55. Many representatives of Member States noted that the review of arrangements for consultations with non-governmental organizations was timely and necessary, especially in view of recent developments, in order to reflect the current needs and realities. Some were of the view that Council resolution 1296 (XLIV) needed updating to enable an increasing number of non-governmental organizations to participate in United Nations activities while retaining a reasonable filter to exclude those non-governmental organizations whose objectives were incompatible with those of the Charter of the United Nations.
56. One delegation stressed that non-governmental organizations considered for consultative status should strictly adhere to the principles and purposes of the Charter of the United Nations, and that organizations indulging in terrorism to destabilize legitimate Governments should have no place whatsoever in the United Nations system. It suggested that a set of rights and responsibilities and a code of conduct for non-governmental organizations should be considered.

. . .

63. . . . Concern was expressed at the recent practice of asking non-governmental organizations to form 'coalitions' and 'constituencies' and to speak through a spokesperson. Such forced consensus would result in destroying the diversity of opinions. It was suggested that those participatory rights, warranted by Council resolution 1296 (XLIV), be fully reinstated. . . .

. . .

66. A number of representatives of Member States stressed the need for a broader and more diverse participation of non-governmental organizations from developing countries in the consultative relationship. In that regard, several Member States suggested the establishment of a fund to assist the participation of non-governmental organizations from developing countries in United Nations meetings, in particular United Nations conferences and their preparatory meetings. Others, while agreeing that there was a need to facilitate the participation of non-governmental organizations from developing countries, expressed doubts about establishing a trust fund.
67. Several delegations opposed the establishment of a trust fund based on compulsory financial contributions from non-governmental organizations.

. . .

69. Some delegations suggested that in order to increase the representation of non-governmental organizations from developing countries so as to reflect the principle of equitable geographical distribution, the Committee on Non-Governmental Organizations, when considering applications for consultative status, should give priority to those emanating from non-governmental organizations in developing countries.

. . .

71. One delegation suggested that measures should be taken to ensure that a certain percentage of non-governmental organizations participating in United Nations conferences were from developing countries. However, several delegations expressed their opposition to the idea of establishing quotas for non-governmental organizations.

. . .

75. One non-governmental organization called for the creation of 'umbrella national networks' at the regional and national levels, which would include representatives of the poor, disadvantaged or marginalized sectors, particularly from developing countries. Those 'umbrella networks' should be eligible for consultative status.

. . .

84. Many representatives of Member States suggested that consultative arrangements with non-governmental organizations be extended beyond the Economic and Social Council and its bodies in order to encompass the General Assembly and its Main Committees and subsidiary bodies. Several delegations suggested that the Security Council and other bodies dealing with peace, security and disarmament should be encompassed.

. . .

89. Another delegation stated that the participation of non-governmental organizations in conferences should be decided on a case-by-case basis, and that it was essential to get a 'no objection' clearance for this from the States concerned.
90. Many speakers stressed the importance of including non-governmental organizations in national delegations and expressed the hope that the practice would be widely followed.

ERSKINE CHILDERS AND B. URQUHART, WE, THE PEOPLES

in Renewing the United Nations System 171 (1994), at 172.

Until quite recently the hauteur of most secretariat and diplomatic officials about NGOs was matched by the disinterest and disdain of large portions of the NGO community for the UN as merely 'another bureaucracy'. The very UN phrase for them—'non-governmental'—has been seen as two-edged: negative, distancing on the part of an intergovernmental body, but perhaps usefully protective against officious outsiders.

. . .

A turning-point

Quite suddenly, however, there is a new dynamic between the UN system and 'the peoples of the United Nations' reflected in various popular movements. It is stimulated by six main forces:

• The thousands (at least 25,000) of NGOs working on Environment in its now enlarged dimension have perceived the indispensable role of multilateral machinery. They have become impatient that the United Nations has been kept so marginalized in global economic issues, which they are not prepared to entrust to the Bretton Woods institutions (or the G-7).

• After many years of holding 'the UN' at arms' length, the constantly expanding human rights NGO community has decided to assert itself in all relevant UN bodies. The 1993 UN Human Rights Conference in Vienna was perhaps the decisive illustration of this significant trend.

• A huge number of citizen supporters of humanitarian relief have seen their NGOs working alongside the UN in more and more crises, not always in happy circumstances. Humanitarian NGO leaders want to get involved in UN policy issues. Senior UN officials and UN diplomats have increasingly acknowledged that this NGO community is indispensable to the UN's humanitarian and even perhaps to its peacekeeping work.

• NGOs of the international women's movement have mobilized the interest of hundreds of thousands of women in a wide range of international political, economic and social issues. They have perceived the centrality of the UN system in these issues. Their leaders are equally determined to take a hand in the future of the UN.

• The world parliamentary community is increasingly engaged. Parliamentarians for Global Action, with some 700 member-MPs from 45 countries, has discussed some form of parallel assembly. Other groups are also interested. . . .

• Underlying these developments are the general transnational expansion of the democratic idea, ever larger popular concern for human rights and UN promotion of them, and the constantly increasing sense of solidarity among young people through transnational media and music.

There are two main reasons why these developments must be actively nourished and built upon.

A wider support base for the UN

. . .

The United Nations has frequently suffered severely from inadequate public knowledge and awareness of its work, especially at critical moments when member-governments have misused the organization or impugned its activities.

Governments are, for the most part, represented in the UN system by national civil servants, making an additional layer between citizens and the world institution. The UN is therefore very remote from its ultimate constituents, and far too dependent on how (and whether) a small number of national officials and media commentators interpret it to them.

. . .

Citizens have as much right to be informed, and to be heard, about their governments' policies in the UN system as about any other policies.

. . .

Enfranchisement by necessity

. . .

The second reason . . . is that governments can no longer make effective progress without citizen involvement.

In many cases the citizenry are so poorly informed that when large additional funds are needed for UN work, the public support base for such decisions is perilously small. The political confusion in many countries over the need for larger contributions of military and other personnel for UN operations has recently also exposed the effects of inadequate public knowledge and debate.

Governments are already faced with the need to ratify far-reaching global ecology treaties. If citizens and their NGOs had not educated themselves on these issues, their governments would have little or no public support for such accords. In the next two decades governments, if they wish to face up to the future, will also have to adopt genuinely global macro-economic policies through the UN, as national commitments. Their citizens will have to be far better informed and engaged in these global issues, if they are to support such policies.

. . .

The NGO dilemma

By the end of 1993 their recent successes had confronted the NGO community with serious new problems. Several hundred additional environmental NGOs had been accredited to the UN. Other 'issue-oriented' international NGOs were clamouring for greater access.

On one hand, even the most sympathetic governments were having to point out that it would hardly be manageable for UN organs and subsidiary bodies, with their own membership from a few dozen to a maximum of 184, to hear spoken representations from hundreds of NGOs; and that their delegations could not even cope with the volume of official documents, leave alone every NGO's individual proposal-paper.

On the other hand, the older-accredited NGOs were beginning to feel overwhelmed and even 'sidelined' by the energetic and often very well-prepared newcomers from the fields of environment, women's rights, humanitarian relief and other high-profile issue areas. Governments still not particularly well-disposed towards NGOs could find ample ammunition from within these tensions and disarrays. . . .

There is thus a considerable danger that the major gains the NGO community has recently made could be seriously dissipated if it is unable to agree on efficient proposals for improved access and representation.

The essence of the change facing the NGO community may be summarized as requiring the best possible answers to a number of strategic questions:

. . .

a. The optimum *places* of NGO influence on governmental decision-making in the system, as between pressure and advocacy within member-countries, lobbying key delegations at UN organs or other bodies, and representation to the total intergovernmental community including at UN world conferences;

b. How NGOs working in a given issue-area can best consolidate their analyses and proposals in order to mount an effective representation, given that they outnumber by many thousands the number of Delegations at any UN body;

c. What facilities of the Secretariat itself must NGOs, in their now greatly increased numbers, minimally seek, and how they can propose these in the most cogent manner;

d. How the NGO community can more generally help to build bridges of communication between the citizens whom they variously but not electorally represent, and the member-governments in the UN system.

QUESTIONS

1. What reforms would you propose to the Resolution 1296 arrangements governing the role of NGOs *vis-à-vis* UN bodies? What about setting aside a percentage of seats (at least 10%, perhaps as much as 40%) in all UN bodies to be occupied by NGO representatives?

2. 'The position of NGOs in relation to the UN human rights system is guaranteed to be a no-win situation. The UN can do very little without the mass, popular support that only the NGOs can provide, but if the latter are given any serious influence, governments will simply withdraw their support from the system. Either the UN itself or the NGOs must be marginalized; governments will never tolerate the two of them working together in tandem.' Discuss with particular reference to human rights NGOs.

3. Various authors have noted the important role played by NGOs in the drafting of international standards. An assessment of the role of NGOs in drafting environmental standards reached the following conclusion:

> One lesson for the future is that although a greater participation of NGOs in regime-building processes in the areas of environment and development should be welcomed and encouraged, the terms of access for NGOs need to be carefully assessed. It is necessary to consider that the character of a multilateral negotiation usually changes drastically from its initiation to the conclusion of a final agreement that terminates the process. For example, the demand for new knowledge and information is much more limited at the end of the negotiation, when concessions are exchanged and deals are made, than when the agenda is set and issues are clarified in the earlier stages of the process. Some NGOs may be capable of performing a constructive role from the beginning to the end of the process. However, it may be more reasonable to constrain or circumscribe the activities of certain NGOs in the regime-building process or negotiation. For instance, some scientific organizations that are in a position to supply significant information during the process stage of issue clarification may have little to contribute when hard bargaining and the exchange of concessions prevail. In contrast, other NGOs may have but a layman's knowledge about the technical issues, but can be in a position to communicate the agreements reached in the negotiation to the general public or to certain social strata. Maybe distinctions such as these should be taken into consideration when access rules are formulated for NGOs.[5]

To what extent, if any, might a similar analysis be warranted in relation to human rights standard-setting?

ADDITIONAL READING

D. Shelton, *The Participation of Nongovernmental Organizations in International Judicial Proceedings*, 88 Am. J.Int'l L. 611 (1994); *Put Our World To Rights: Towards a Commonwealth Human Rights Policy* (1991); Jerome Shestack, *Sisyphus*

[5] Gunnar Sjöstedt et al, in B. Spector et al (eds.), *Negotiating International Regimes: Lessons Learned from the United Nations Conference on Environment and Development* 233, 238 (1994).

Endures: The International Human Rights NGO, 24 N.Y.L.S.L.Rev. 89–124 (1978); R. Lagoni, *Article 71*, in B. Simma *et al* (eds.), *The Charter of the United Nations: A Commentary* 902–15 (1994); P. Chiang, *Non-Governmental Organizations at the United Nations* (1981); Howard Tolley Jr., *The International Commission of Jurists: Global Advocates for Human Rights* (1994); Manfred Nowak, *World Conference on Human Rights: The Contribution of NGOs, Reports and Documents* (1994); Lowell Livezey, *Non-Governmental Organizations and the Idea of Human Rights* (1988); Arthur Blaser, *Human Rights in the Third World and Development of International Nongovernmental Organizations*, in Shepherd and Nanda (eds.), *Human Rights and Third World Development* (1985); D. Weissbrodt and J. McCarthy, *Fact-Finding by International Nongovernmental Human Rights Organizations*, 22 Va. J. Int'l L. 1–89 (1981); Theo van Boven, *The Role of Non-Governmental Organizations in International Human Rights Standard-Setting: A Prerequisite of Democracy*, 20 Calif. Western Int'l L.J. 207–25 (1990); Amnesty International, *'Disappearances' and Political Killings, Human Rights Crisis of the 1990s: A Manual for Action (1994)*; *idem, Human Rights Watch World Report 1995* (1994); Fund for Peace and Jacob Blaustein Institute for the Advancement of Human Rights, *Human Rights Institution-Building* (1994).

9

Treaty Organs: The ICCPR Human Rights Committee

Chapter 9 continues the inquiry begun in Chapter 7 into the structure, roles, functions and processes of international human rights bodies. We continue to emphasize the relationships among human rights norms, institutions and processes, as well as the reasons and techniques for 'institutionalization' of norms.

The Commission on Human Rights (UN Commission) created under the UN Charter (hence a 'Charter' organ) and examined in Chapter 7 remains the most complex and politically interesting of the specifically human rights organs with universal reach. It differs markedly in organization, functions and powers, as well as notoriety, from the six organs (or bodies) related to six universal human rights treaties on civil and political rights, economic and social rights, racial discrimination, gender discrimination, torture and children's rights. Each such body is distinctive in some respects; each has functions only with respect to the treaty creating it or for which it was created; each such treaty regime is now to some extent 'administered' or 'implemented' or 'developed' by that body.

Chapter 9 selects one such 'treaty' organ, the Human Rights Committee created by and functioning within one of the world's two principal human rights treaties, the International Covenant on Civil and Political Rights. We continue to use the abbreviation 'ICCPR Committee.'

Similar questions to those in Chapter 7 arise. Why has the ICCPR Committee assumed the character, structure and functions that are here described and analyzed? What purposes are served by the present arrangements? Are there other purposes that should be served? In which respects, and why, does the ICCPR Committee differ from the UN Commission? Should the structure and functions of the ICCPR Committee now be reconsidered, in the light of a half century of experience of the human rights movement and two decades of experience with this Committee?

In thinking about such questions, two thoughts should be kept in mind. (1) The ICCPR Committee forms part of a complex of Charter and treaty organs. Should it be understood and evaluated as an isolated organ functioning within one treaty, or as part of this larger complex? If the latter, should one ask whether the Committee's performance and functions should be evaluated as complementary to those of other organs—that is, be best understood as part of a universal system of institutions? (2) Even within the ICCPR, can we understand each of the Committee's powers or functions discussed below in isolation from the others, or should all the functions be seen as part of a treaty system, hence as complementary and as occupying different parts of a scheme within the treaty?

Section A examines the ICCPR Committee. Section B develops comparisons with the five other universal human rights treaties that also create special organs.

A. POWERS, FUNCTIONS, AND PERFORMANCE OF THE ICCPR COMMITTEE

1. INTRODUCTION

COMMENT ON THE FORMAL ORGANIZATION OF THE ICCPR COMMITTEE

You should now read the provisions bearing on the organization and functions of the ICCPR Committee, set forth in Articles 28–45 of the Covenant and in its First Optional Protocol. The following discussion incorporates some of those provisions.

Bear in mind the three dominant functions of the Committee set forth in these instruments. (1) Article 40 requires states parties to 'submit reports' on measures taken to 'give effect' to the undertakings of the Covenant and on progress in the enjoyment of rights declared by the Covenant. The Committee is to 'study' these reports. (2) The same article instructs the Committee to transmit 'such general comments as it may consider appropriate, to these states parties. (3) The Optional Protocol—a distinct agreement requiring separate ratification—authorizes the Committee to receive and consider 'communications' from individuals claiming to be victims of violations by states parties of the Covenant, and to forward its 'views' about communications to the relevant individuals and states. The materials in this chapter consider each of these principal activities of the Committee.

Articles 28–31 of the Covenant provide the crucial information about the Committee's membership. The 18 members are to have 'high moral character and recognized competence in the field of human rights.' Consideration is to be given to the utility of including 'some persons having legal experience.' In fact, all members of the Committee have had such experience in some capacity: private practice, the academy, public interest work, diplomacy, judicial offices, or government. Note the characteristic provisions of Article 31(2) that consideration in elections 'shall be given to equitable geographical distribution of membership and to the representation of the different forms of civilization and of the principal legal systems.'

Under Article 28(3), all members are to be 'elected and shall serve in their personal capacity.' The UN term for such members is 'experts,' as opposed to the 'representatives' of states who sit on the UN Commission on Human Rights. If we link the references to 'personal capacity' to references to 'high moral character' and 'recognized competence' of members with an emphasis on their legal experience, the compelling inference is that Committee members are to act independently of the governments of their states, not under orders of their government—as does, for example, a state's representative (often with a rank of ambassador) on the UN Commission.

Generally this aspiration appears to have been realized, but in many contexts, 'independence' in the sense identified has been a relative rather than absolute concept. Consider members who are nationals of (and originally nominated for election by) states of an authoritarian character directed, say, by a single party, a military clique, or a personal dictator. Moreover, since membership on the Committee is a part-time business, members have continued to hold government (diplomatic and other) posts, again qualifying the degree of possible independence from their governments' positions on given issues.

The Committee meets for three sessions annually, each three weeks long, at UN headquarter in Geneva (twice) and New York. Working groups referred to below meet for one week prior to the start of each session. Since emoluments (effectively salaries) paid by the UN are low and the work is part-time, members hold jobs, closer to full time, and must fit the Committee's work into already busy schedules. Most meetings (the dominant exceptions being meetings considering 'communications' under the Optional Protocol or considering drafts of General Comments) are public, are poorly attended by outsiders and gather little press coverage. The ICCPR Committee has never enjoyed or indeed sought the publicity and notoriety of the UN Commission.

Decisions of the Committee should formally be by majority vote pursuant to Article 39(2). In fact, all decisions to date have been taken by consensus, although as a formal matter any member could demand a vote on any issue. This unbroken practice of reaching decisions by consensus (say, on decisions about the Committee's observations on a state's report, p. 505, *infra*, or about the text of a General Comment, p. 522, *infra*) meets with varying reactions from Committee members. Its advantages in avoiding the factional battles that have dominated much of the life of the UN Commission and in permitting the Committee to move ahead as a unit are obvious. Its undoubted if indeterminate historical effects on the action taken by the Committee are as obvious: compromise, the blunting of positions, the failure to take the bolder step.

Committee members have said that the practice has had the general effect of not permitting an individual member to hold out for a different position from the large majority, but also has generated a lot of give-and-take while encouraging members holding minority views to go along with a clear trend or dominant opinion. In one activity, the writing of 'views' about communications discussed at p. 535, *infra*, Committee practice has allowed individual members to write dissenting opinions.

Like the UN Commission but not over so long a period (the Covenant entered into force only in 1976), the ICCPR Committee has witnessed vast changes in global politics. The ideological disputes and at times sharp discord that characterized the years of drafting of the Covenant and the Committee's first decade during the full sway of the Cold War have now become blunted or have disappeared. Nor has North–South controversy erupted as forcefully on the Committee as within the UN Commission with its less precise, more extensive mandate and broadly political character. Nonetheless, the earlier disputes and compromises, particularly those during the period of drafting of the ICCPR and Optional Protocol when the Committee's basic structure and functions were determined, have left a strong imprint on the Committee today. History's traces are indeed everywhere in the Committee's activities.

NOTE

Consider the following brief summaries by two authors of the nature of the earlier disputes and their continuing influence. The first is taken from Dominic McGoldrick, *The Human Rights Committee* (1991) at 13:

> 1.18 There was general agreement during the drafting that the primary obligation under the ICCPR would be implementation at the national level by States. There was continuing disagreement, however, on the question whether there should also be international measures of implementation. A minority of States, principally the Soviet bloc, insisted that there should be provisions to ensure implementation but that there should be no international measures of implementation. It was argued that such measures were a system of international pressure intended to force States to take particular steps connected with the execution of obligations under the Covenant. They were, therefore, contrary to the principle of domestic jurisdiction in article 2(7) of the United Nations Charter, would undermine the sovereignty and independence of States and would upset the balance of powers established by the UN Charter. Moreover, the establishment of petitions systems would transform complaints into international disputes with consequent effects upon peaceful international relations.
>
> 1.19 Against these views it was argued that the undertaking of international measures of implementation was an exercise of domestic jurisdiction and not an interference with it. International measures were essential to the effective observance of human rights, which were matters of international concern. However, even within those States that agreed that international measures were essential, there were significant differences of

> opinion as to the appropriate types of measures. The proposals included an International Court of Human Rights empowered to settle disputes concerning the Covenant; settlement by diplomatic negotiation and, in default, by *ad hoc* fact-finding Committees; the establishment of an Office of High Commissioner (or Attorney-General) for Human Rights; the establishment of reporting procedures covering some or all of the provisions in the Covenant; empowering the proposed Human Rights Committee to collect information on all matters relevant to the observance and enforcement of human rights and to initiate an inquiry if it thought one necessary.
>
> . . .
>
> 1.21 The lengthy drafting process of the ICCPR largely coincided with the depths of cold war confrontation, the explosive development of notions of self-determination and independence, the accompanying political tensions of large scale decolonization, and the consequential effects of a rapidly altering balance of diplomatic power within the United Nations. In retrospect then it must be acknowledged that it was much more difficult to agree on the text of a Covenant containing binding legal obligations and limited measures of international implementation than it had been to agree upon the statement of political principles in the Universal Declaration in 1948. . . .

The second summary is by Torkel Opsahl, The Human Rights Committee, in Philip Alston (ed.), The United Nations and Human Rights (1992) at 371:

> . . . The draft Covenant prepared in 1954 by the Commission envisioned a quasi-judicial Human Rights Committee quite different in its powers and functions from that which actually came into existence. It was another twelve years before the General Assembly's Third Committee debated the proposed implementation provisions, at which time they were drastically altered. The majority was opposed to making obligatory the procedure for interstate communications. . . .
>
> All of the various positions, except that of dispensing with the Committee altogether, were taken into account by a formula worked out by the Afro-Asian group. According to this version, the Committee's only compulsory role would be to study and comment generally upon the reports of States Parties, a function originally intended for the Commission on Human Rights. Many of the details of this proposal were amended, which later caused doubts and disagreements about the proper role of the Committee in the reporting system. The functions relating to communications were made entirely optional, and arrangements providing for the consideration of individual complaints of violations were separated from the Covenant and put in the Optional Protocol. In other words, the result was a compromise between those States which favoured strong international measures and those which emphasized the primacy of national sovereignty and responsibility. As is inevitably the case with such compromises, many specific issues were left unresolved, perhaps inten-

tionally. As a result the subsequent evolution of the arrangements has had to be shaped by a continuing give-and-take within the Committee over many years.

QUESTION

'The lesson to be learned from the drafting of the ICCPR provisions bearing on the Committee's functions is straightforward. Those terse, inevitably ambiguous provisions are a product of international power politics—the cold war between East and West, the major powers resisting serious enforcement of the ICCPR, other states wanting a more significant international regime, and so on. Power shifts, issues shift, possibilities change. The Committee ought to interpret these provisions of the ICCPR in the light of today's context of politics, power and ideology, not in the light of the transient context at the time of drafting.'

Do you agree? Why, or why not? Keep the implications of this position in mind as you read the later materials in this chapter.

2. STATE REPORTING

COMMENT ON PERIODIC REPORTS OF STATES

Submission by states of reports to a treaty organ about their internal implementation of human rights obligations has now become a familiar requirement. But consider how revolutionary a practice this must have appeared at the time of the first proposals. As little as 60 years ago, it would have seemed nearly inconceivable that most of the world's states would periodically submit a report to an international body about their internal matters involving many aspects of relations between government and citizens, and then participate in a discussion about that report with members of an international body drawn from all over the world.

The critical provision is Article 40, requiring states to 'submit reports on the measures they have adopted which give effect to the rights recognized herein and on the progress made in the enjoyment of those rights.' Reports shall 'indicate the factors and difficulties, if any, affecting the implementation of the present Covenant.' The ICCPR Committee is to 'study' the reports and transmit its 'general comments' to the states parties.

Discussions of reports are public proceedings, though attended by few persons other than Committee members. The proceedings amount less to a systematic 'study' (to use the term of Article 40) than to an examination of the report with members speaking individually, making comments and posing questions. The representative of the state responds to comments and questions. Reports are now presented by a state party every five years, though the Committee has exercised its authority to request reports at shorter intervals in cases such as national emergencies.

The requirements for reports and the processes for discussing them depend on the purposes that the reports and the discussions thereof are meant to serve. One can imagine two polar purposes: (1) confront the state with facts about it derived from other sources than the report, condemn the report if it fails to describe and account for violations, and possibly condemn the state as well, particularly for ongoing violations; *or* (2) view the report as an occasion for one of a series of ongoing discussions between the state and the Committee looking toward improvement of the human rights situation—as an occasion for 'constructive dialogue,' to use the term employed by a number of Committee members. There are many formulations in between, some of which are suggested in the two reports of Iraq and Austria that follow.

The General Comments at p. 522, *infra*, indicate some of the many problems that the Committee has encountered in the reports submitted to it: incomplete coverage, abstraction and formality that lead states to stress their unenforced constitutional or statutory provisions rather than to offer a realistic description of practices; great delays in filing reports. Delays have been particularly troublesome. As of May 1993, 65 states had failed to submit initial or periodic reports that had become due. The situation has continued to worsen.

One of the vexing problems for the Committee has been how to acquire knowledge about the reporting country independently of the report of that country. Otherwise, questions would be guesswork rather than informed. Committee members and the Committee as a whole lack a Secretariat and staff that might be helpful in acquiring such knowledge. They rely to an increasing extent on groups about which the Covenant is silent—principally NGOs that provide members with recent investigative reports and similar data. NGOs may attend the open discussions about reports but, unlike NGOs with consultative status before the UN Commission on Human Rights, they have no right of intervention in those proceedings and hence can't formally address the Committee members as a group.

The Covenant makes no provision about the way in which a state should prepare a report, but the Committee has issued guidelines. The reporting process, from preparation through Committee proceedings, gets little publicity. In fact, few states include groups of their citizens—interest groups, particular lobbies, ethnic or gender groups or indigenous peoples, human rights NGOs and so on—in the process of preparation. Little effort is made in a state to give publicity to the Committee's consideration of that state's report. The entire process is poorly understood by the informed public, and, absent a large scandal or exceptional mobilization of interest, public apathy may be unshakable.

The following excerpts from the discussion of periodic reports to the Committee of Iraq and Austria illustrate many characteristics of this process.

Be attentive to: (1) a comparison between the two discussions, with respect to attitudes of members of the Committee and the state representatives; (2) the types of issues raised by the discussions, from the relatively minute and specific to the global; (3) the degree to which the reports and discussions consider human rights instruments, institutions and regimes other than the ICCPR; and (4) the intended purposes and likely effects of these discussions.

REPORT OF IRAQ TO THE HUMAN RIGHTS COMMITTEE

Third Periodic Report under Article 40 of Covenant
UN Doc. A/46/40 (1991), Supp. No. 40, at 150
UN Doc. A/47/40 (1994), Supp. No 40, at 41.

[These excerpts from the Committee's consideration of Iraq's third periodic Report under Article 40 of the ICCPR cover two sessions of the Committee. Consideration began in the Committee's 42nd Session in 1991, and was completed in its 43rd Session later that year.]

620. The report was introduced by the representative of Iraq, who stressed his Government's willingness to pursue a frank and constructive dialogue with all United Nations bodies concerned with human rights, and especially with the Committee in its efforts to enhance the implementation of the Covenant.

. . .

622. Referring to the Committee's special request for information on the application of articles 6, 7, 9 and 27 of the Covenant, the representative pointed out that the recent Kuwait crisis had been the subject of several Security Council resolutions, which Iraq had accepted and would be implementing responsibly and with good will. Matters that were still pending before the Security Council could not be regarded as falling within the Committee's competence.

623. Members of the Committee, for their part, observed that by ratifying or acceding to the Covenant, States parties accepted the Committee's competence and could not evade their obligations under that instrument. The Committee had competence to monitor the implementation of the Covenant independently of any other obligations arising from Security Council recommendations and decisions or international instruments other than the Covenant. The Committee was well aware that the situation of Iraq was difficult. However, the root cause of those difficulties was the Iraqi intervention in Kuwait on 2 August 1990 and not the counter-action undertaken by the international community.

. . .

625. With reference to [the right to life], members of the Committee . . . wished to know whether Revolutionary Command Council Decree No. 840 of 1986, prescribing severe penalties for offences against the President, was still in force; how often and for what offences the death penalty had been imposed since the consideration of Iraq's second periodic report; how often the death penalty had been carried out, in particular with respect to minors; what legal remedies were available to persons sentenced to death; whether there had been any violations of the rules and regulations governing the use of firearms by the police and security forces and, if so, what measures had been taken to prevent their recurrence and what disciplinary and other measures had been taken against those found guilty; whether any investigations had been carried out in respect of alleged disappearances of individuals and killings of persons in the course of military operations by the Iraqi armed forces and, if so, with what results; and what compensation was being made available in respect of casualties and disappearances in Kuwait following the events of 2 August 1990 and for damages resulting from the deliberate setting on fire of oil wells.
626. Recalling also the Committee's concerns about events occurring in Iraq before the Gulf war, which constituted serious violations of the Covenant, particularly of its articles 6 and 14, members of the Committee requested information regarding the reported manufacture of nuclear weapons in Iraq and the alleged use of chemical weapons by the army in 1987 against the population of Halabja and about the current status of Mr. Jan Richtes, a foreigner who had been tried in Iraq in 1987. It was also observed that while the report referred to Iraq's full cooperation with the Special Rapporteur of the Commission on Human Rights on summary or arbitrary executions, it said nothing about the measures the Government had taken to prevent the practice of arbitrary and extrajudicial executions to which the Special Rapporteur had drawn attention. The report was also largely silent about measures adopted to ensure protection of the right to life in connection with the recent 'riots', although it was clear that the taking of hostages, the killing of hundreds of civilians in the Kirkuk region or the massive aerial bombardments in the Kurdish sector could not be considered as actions appropriate to dealing with riots. In the foregoing connection, information was requested on disciplinary and judicial measures that had been taken against those responsible for such acts. Concerning the large number of cases identified by the Working Group of the Commission on Human Rights on Enforced or Involuntary Disappearances that had not yet been elucidated, it was also asked whether appropriate investigations were under way.

. . .

628. In his reply, the representative of the State party said that Revolutionary Command Council Decree No. 840 of 1986 was still in force but was undergoing review by a high-level committee. Criminal courts were obliged to report to the Public Prosecutor all cases in which the death penalty had been imposed for automatic transmission to the Court of Appeal. Prisoners under sentence

could also appeal directly. Death sentences could not be implemented without the issuance of a Decree of the Republic, and sentenced persons also had the right of appeal to the President of the Republic. In the uncertain situation following the end of the Gulf war, Iraq had been obliged to use the armed forces to put down insurrections and maintain the sovereignty of the State. Disappearances of individuals and killings of persons were mainly the work of rioters. Some persons who had been reported to have disappeared, had in fact fled the country.

. . .

631. With regard to the cases of involuntary disappearance that had not been cleared up, the representative said that that matter concerned primarily a tribe in northern Iraq composed of more than 2,300 persons who had collaborated with Iran during its occupation of Iraqi territory and who had left the territory with the occupation forces.

632. The representative further denied that the Iraqi armed forces had used chemical weapons against civilians. . . .

. . .

636. Members of the Committee also referred to the four judgements concerning acts of torture mentioned in the third periodic report of Iraq and to detailed information on the practice of torture in Iraq furnished by the Special Rapporteur on Torture of the Commission on Human Rights, by Amnesty International and by other international organizations. Such allegations could not be refuted by the Iraqi authorities; the members asked whether all the complaints concerning acts of torture had really been investigated and, if so, with what results. They also asked how many Iraqi soldiers had been tried for rape during the occupation of Kuwait; whether Iraq applied the United Nations Standard Minimum Rules for the Treatment of Prisoners; whether representatives of governmental and non-governmental organizations had been permitted to visit detention centres; and whether any persons had died following torture. They also asked what specific measures had been taken to prevent maltreatment in places of detention; whether Iraqi legislation included any provision enabling the State to take action *ex officio* in cases of torture; and whether the Iraqi Government would be prepared to conduct impartial investigations with the assistance of international experts.

637. In his reply, the representative of the State party referred to the provisions of the Iraqi Constitution and Penal Code designed to prohibit and punish any act of torture and to the criminal and civil procedures laid down to enable victims of torture to claim moral or material compensation. Investigations were conducted by the courts, which received complaints of torture and which took the necessary steps within their competence against the offenders. The Attorney-General played an essential role: it was his responsibility to institute proceedings on any information he received concerning acts of torture and to follow up the matter until judgement was passed. . . .

638. In addition, directive No. 4 of 1988 required the Department of Public Prosecution to investigate prison conditions in order to verify that they conformed to the regulations. The Department's representative saw to it that physicians visited detention centres. He received complaints from detainees, whom he met in private and he instituted criminal proceedings against those responsible for ill-treatment or torture.

. . .

[Questions of members about liberty and security of the person have been omitted.]

643. In his reply, the representative of the State party referred to the provisions of the Iraqi Constitution and Penal Code concerning the conditions for lawful arrest and the penalties for unlawful arrest. As soon as he was arrested, a person was entitled to contact relatives and his counsel. The maximum period of detention in custody was 24 hours for offences punishable by three years' imprisonment or less; a detainee could be released on bail or surety. The period of detention in custody could be extended by the court. Bail was not granted where crimes which carried the death penalty were involved. Appeals against all judicial decisions relating to arrest lay with the competent regional criminal court.

. . .

648. Referring to Shiites currently in the marshes, who had been bombed and prevented by brutal means from obtaining assistance, members asked whether the United Nations and Amnesty International would be allowed to have access to them and to assist them. They also wished to know how many members of the Executive Council, established in 1989, were Kurdish and how many belonged to other groups; to what extent the Council was independent in governing the Autonomous Region; what positive measures had been taken to protect the fundamental rights of Kurdish, Shiite and Assyrian minorities; and what the situation was with regard to holy places in the towns that had been subjected to heavy bombardment.

. . .

650. The representative further stated that the State supported the right of persons belonging to minorities to enjoy their own culture by publishing books and by broadcasting on radio and television in the local languages. The Kurdish language was the official language in the Autonomous Region of Kurdistan, and a major university existed in the region. According to the Iraqi Constitution, Kurds were considered not as a minority but as a people on an equal footing with the Arab people. Members of the Executive and Legislative Councils were elected by free and secret ballot. Negotiations between Kurdish representatives and the authorities in Baghdad were proceeding well and

would reach a successful conclusion. Kurds in Iraq had political and cultural rights that did not exist for Kurds in other countries. The Iraqi Constitution enshrined the principle of non-discrimination, and the principle applied to religious matters.

651. Members of the Committee said that while they had hoped that a constructive dialogue between the Committee and Iraq would be possible, unfortunately that had not proven to be the case. Rather, the representative of the State party had engaged in a kind of monologue or 'stonewalling' and had sought constantly to evade certain issues and to avoid responding to the legitimate questions posed by members of the Committee. In the latter connection, they referred to questions they had raised regarding such important issues as disappearances, unlawful executions, including the execution of minors, torture and the existence of political prisoners, which had not received clear replies or had remained unanswered.

652. . . . Members also disagreed with the implication in the report that the difficult situation concerning human rights in the country was due primarily to the Gulf war and to the sanctions that had been adopted against Iraq by the international community, noting in that connection the existence in Iraq of reliably attested human rights violations, including summary executions and arbitrary detention, well before the invasion of Kuwait on 2 August 1990.

653. Members of the Committee also expressed deep concern with regard to the existence in Iraq of special courts, as well as death sentences without any possibility of appeal; the lack of protection of freedom of expression; the situation of the Shiites in the country; and the repressive action of the Government, particularly against the Kurds and the Shiites. Indeed, it was their overall impression that a situation of serious human rights violations that had already been very disturbing in 1987 had persisted and worsened throughout the intervening period.

. . .

655. The representative of the State party reaffirmed his Government's desire to cooperate as fully as possible with the Committee even though he could not accept the criticism that inadequate replies had been given to the questions concerning protection of the right to life.

. . .

[The following excerpts are taken from the completion of consideration by the Committee of Iraq's third periodic report in the following 43rd session of the Committee.]

183. In his introductory statement, the representative of the State party drew the Committee's attention to a number of important developments in the field of human rights that had occurred in his country since the consideration of

the first part of the report. Much of the legislation objected to by the Committee had been repealed. Decree No. 416 of the Revolutionary Command Council had thus been suspended, the Revolutionary Court had been abolished and a decree had been adopted granting amnesty to persons convicted of political crimes, from which 187 persons had benefited. Furthermore, a law on political parties had come into effect on 16 September 1991. A Code of Human Rights, setting out provisions of international human rights instruments as well as those of Iraqi legislation, was in preparation, which would serve as a basis for incorporating such international standards into domestic law. Lastly, there was a continuous dialogue between the Government and the Kurds to seek an improved formula for greater autonomy for Iraqi Kurdistan.

. . .

185. Observing that economic, social and cultural rights and civil and political rights were closely interrelated, the representative said that the current blockade of Iraq was posing a danger to the right of people, particularly children, the elderly and the sick, to health, food and other basic needs. Furthermore, the shortage of medicines and pesticides had increased the incidence of disease. Cases of typhoid, hepatitis and cholera had sharply increased and infant mortality had risen from 5 to 21 per 1,000 between August 1990 and August 1991. Those circumstances had to be taken into account by the Committee in analysing the situation in Iraq. Since it was impossible to enjoy civil and political rights while being denied economic, social and cultural rights, the economic blockade should be lifted so that the Iraqi people could enjoy all their human rights.

. . .

187. In addition, members wished to know what concrete measures had been taken in order to attain the Government's objectives of reconstruction, the establishment of democracy and a multiparty system, freedom of association, freedom of the press and the supremacy of law; whether the Covenant had specifically been taken into consideration in drafting the new Constitution and the law on political parties; whether the Covenant had been incorporated into Iraqi law and could be invoked before the courts; and what the remaining restrictions were under the state of emergency. . . .

. . .

189. In his reply, the representative of the State party explained that, since the submission of the report, the law on political parties had been adopted and that the proposed new Constitution was to be submitted for approval in a referendum once the National Assembly had completed its discussion on it. Under a general rule embodied in a law, international instruments were considered as an integral part of domestic legislation. The purpose of drafting a

code of human rights was precisely to clarify that point for those who applied the law and to remedy shortcomings in national legislation that might be inconsistent with international instruments. The Covenant was now considered to be part of Iraqi legislation and its provisions could be invoked by private individuals before the courts. . . .

. . .

[Questions of Committee members about political participation and parties have been omitted.]

212. In his reply, the representative of the State party explained that Law No. 30/1991 on political parties had been promulgated and was now in force. That law guaranteed the equality of all parties, which had full freedom to establish themselves and to publish their literature, and would lead to an increase in the number of political parties and hence to broader participation by Iraqi citizens in public life. The new Constitution, when promulgated, would certainly be in line with the principles of that Act and would provide an appropriate framework for the encouragement of a multiparty system and consequently a diversity of ideas and opinions. Although nothing was yet known as to the place that was to be attributed under the new Constitution to the party in power, it would be inconceivable for the Constitution to make a distinction between the various political movements. . . .

Concluding observations by individual members

213. Members of the Committee expressed their appreciation to the representative of the State party for his cooperation in presenting the third periodic report of Iraq and for having engaged in an open dialogue with the Committee. Although the report had been somewhat overdue, great efforts had been made in difficult circumstances to submit it on time. Information had been updated as requested and efforts had been made to provide the Committee with answers to its questions. Furthermore, a certain degree of progress in the implementation of the Covenant had been noted, including drafting a code of human rights, abolishing the Revolutionary Court, moves towards permitting the establishment of political parties, formulating a new Constitution and adopting an amnesty law. Iraq was thus making an endeavour to bring its domestic law into line with the Covenant and was taking some steps towards pluralism and democracy.

214. While welcoming those measures, members regretted that many of their questions had not received satisfactory replies and felt that the rights specified in the Covenant were neither adequately protected nor properly implemented. Serious concern was expressed, in particular, regarding the treatment of Kurds in northern Iraq and of Shiites in the south. . . .

REPORT OF AUSTRIA TO HUMAN RIGHTS COMMITTEE

Second Periodic Report under Article 40 of Covenant
UN Doc. A/47/40 (1994), Supp. No. 40, at 26.

[The following excerpts are taken from consideration by the Human Rights Committee at its 43rd Session in 1991 of the second periodic report by Austria.]

81. The report was introduced by the representative of the State party, who drew members' attention, in particular, to the fact that the Second Optional Protocol aiming at the abolition of the death penalty was currently before the Austrian parliament, with ratification expected in early 1992.
82. With reference to that issue, members asked what measures Austria had taken to give effect to the rights recognized in the Covenant and whether there were any difficulties in that regard. Members also inquired about the remedies available to individuals whose rights under the Covenant had been violated. Concerning the promotion of human rights, they wished to know, in particular, whether a commission, ombudsman or similar institution would be established, as well as about measures taken to increase public awareness of the Covenant and the Optional Protocol.
83. Members were concerned about the status of the Covenant, given that Austria had incorporated into its domestic law the European Convention on Human Rights, but not the Covenant. They wondered whether those parts of the Covenant that were not reflected in the European Convention, if not the Covenant in its entirety, could be incorporated into Austria's domestic law. In addition, members wished to know whether there was any governmental machinery for monitoring legislation to ensure its compatibility with Austria's international obligations under the Covenant; how complaints would be handled in the light of the provisions of the Optional Protocol;
84. In his reply, the representative of the State party said that the Covenant, though not an integral part of the domestic law, was recognized as an instrument prescribing obligations under international public law. Fundamental human rights in Austria had been guaranteed since the enactment of the Basic Law in 1867 and the ratification of the European Convention on Human Rights, which in 1964 was made part of domestic constitutional law. Notwithstanding the fact that neither a judge nor an administrative authority was required to apply the provisions of the Covenant directly, there were no difficulties in giving effect to the rights recognized in it. . . .

. . .

86. The Government had no intention to set up a commission on human rights or a special agency to promote human rights. However, the Office of the Ombudsman had been in existence since 1976 and all government institutions were ready to provide information on human rights upon request. While the

public was less aware of the Covenant than of the European Convention on Human Rights, it was generally aware of its provisions and of those of the Optional Protocol. The Austrian Government believed that the provisions of the Basic Law and of the European Convention on Human Rights, as amended by subsequent protocols, would ensure compliance with the provisions of the Covenant. Furthermore, the text of every statute or decree was scrutinized in the light of the fundamental rights and freedoms provided for in the Covenant, the European Convention and domestic law. To ensure that any person whose rights or freedoms were violated would have effective remedies, Austria was prepared to change its domestic legislation to provide for new remedies or to allow the use of existing remedies, if regarded by the Human Rights Committee as suitable, in the same manner as it had done in respect of the decisions of the European Court on earlier occasions.

. . .

88. In connection with that issue, members wished to know how the Austrian Constitution guaranteed the rights provided for in article 2 (1) of the Covenant; whether women received equal pay and what measures had been taken to promote women's participation in the various sectors of society; what was the proportion between the sexes in educational institutions; how the special Federal Constitutional Act against racial discrimination had been applied in practice; . . .

89. In his reply, the representative said that discrimination was prohibited by the Constitution. Since virtually all the rights contained in the Covenant were also embodied in the European Convention, which had become a part of constitutional law, Austrian law necessarily contained provisions similar to those of the Covenant.

. . .

91. The figures for 1988–1989 showed that about 50 per cent of children attending day-care centres and primary and secondary schools were female, and that one third of all university students were female. A report on the measures taken to promote the participation of women in the life of the country would be brought to the attention of the Committee as soon as it was completed. On the issue of equal pay, the representative acknowledged that such equality had not been guaranteed in Austria. The Government had therefore established an Equal Pay Committee and it was expected that matters would improve slowly.

. . .

[Questions of Committee members about the right to life have been omitted.]

95. In his reply, the representative said . . . [that 'a]ctive' euthanasia was regarded as illegal and contrary to medical ethics. Attempted suicide had

ceased to be an offence, but helping a person to commit suicide was still punishable. Termination of pregnancy during the first three months was not punishable by law. The Code of Criminal Procedure had been amended in 1987 with a view to imposing greater penalties for pollution-related offences, but discussion of that sensitive issue was still continuing. A system under which cases of AIDS were registered had been introduced and the principle of anonymity was strictly respected. Homosexual prostitution had also been decriminalized to enable the application of preventive measures and to fight AIDS more effectively. It was also planned to provide drug addicts with substitute products.

96. With reference to that issue, members of the Committee wished to know whether there had been any allegations of violations of obligations under article 7 of the Covenant and whether statistics regarding ill-treatment of detainees were available. Noting that torture had been practised in Austria, members requested information on the measures that had been taken to prevent ill-treatment, on the competent investigative authorities and complaints procedures, and concerning the main problems faced by the prison commissions and how such problems had been addressed. . . .

97. In addition, members wished to know whether there was any procedure providing for the review of compulsory confinement decisions; whether the European Committee for the Prevention of Torture had detected any cases of violations of the provisions of the European Convention for the Prevention of Torture and Inhuman or Degrading Treatment or Punishment; how soon after arrest a detainee was allowed to contact counsel; . . .

98. Replying to the questions raised by members, the representative confirmed that there had been allegations of violations of obligations under article 7 of the Covenant. In 1989 a decree was issued, ordering that justifiable allegations should be the subject not only of a police inquiry, but also of a thorough investigation by an independent examining magistrate. Whilst it was too early to provide statistics on the results of the inquiries (a decree stating that statistics should be kept of allegations of ill-treatment during detention was issued in May 1991), the representative referred to a report published by Amnesty International in 1990 in which 14 individual cases had been mentioned. Even before the publication of that report, a decree containing strict orders concerning the rights of detainees to communicate with counsel and the possibility of warning third parties of the arrest had been given. . . .

. . .

100. On the issue of admissibility of evidence obtained through ill-treatment, the representative explained that the Austrian parliament's declaration, made when the Convention against Torture and Other Cruel, Inhuman or Degrading Treatment or Punishment was ratified, had in effect prohibited the use of statements obtained by torture. The Austrian Government was aware that the Code of Criminal Procedure would have to be amended in the light of article 15 of the European Convention on Human Rights so as to provide

explicitly for the prohibition of the use as evidence of confessions obtained through ill-treatment. In the view of the Austrian authorities, article 15 of the European Convention was even more restrictive than article 7 of the Covenant.

. . .

103. In response to members' queries as to whether the maximum period of pretrial detention could be reduced from five days to three days, the representative said that the Government intended to amend the Code of Criminal Procedure in order to reduce the maximum from five to four days, thus bringing the practice into line with the decisions of the European Court of Human Rights. . . .

. . .

Concluding observations by individual members

120. Members of the Committee expressed warm appreciation for the high quality of the report which was informative and straightforward. They also welcomed the candor and competence of the State party representatives in answering the Committee's questions, which had made for a useful and constructive dialogue.

121. While recognizing Austria's traditions and the Government's efforts to promote respect for human rights, members expressed continuing concern about a number of areas where, in their view, further improvements were needed. One such concern related to the status of the Covenant in relation to Austrian law. . . .

122. Other concerns raised by members related to such matters as the independence of the administrative courts; the inadequacy of protection extended to detainees at the interrogation stage; the impartiality of the mechanisms for investigating cases involving alleged torture and ill-treatment by the police; . . .

123. The representative of the State party said the dialogue had been extremely interesting and thanked the Committee for the warm welcome it had accorded to his delegation.

. . .

COMMENT ON CHANGING PRACTICE OF THE COMMITTEE FOR REPORTS

Consistent with the ICCPR Committee's long-standing practice of not expressing a collective review about a report, the reports of Iraq and Austria conclude with a section entitled 'Concluding observations by individual members.' Those observations are then individual. In 1992, the Committee reconsidered this issue and decided that in future discussions of reports

> comments would be adopted reflecting the views of the Committee as a whole at the end of the consideration of each State Party report. That would be in addition to, and would not replace, comments made by members. . . . Such comments . . . were to provide a general evaluation of the State Party report and of the dialogue with the delegation and to underline positive developments . . . , factors and difficulties affecting the implementation of the Covenant, as well as specific issues of concern. . . . Comments were also to include suggestions and recommendations formulated by the Committee. . . . UN Doc. A/47/40 (1994), at 18, para. 45.

Consider an early instance of this important change in practice in the Committee's consideration of the second periodic report of Peru, UN Doc. A/47/40 (1994), at 69. It concludes with a section entitled 'Comments of the Committee.' The introduction to that section states that the Committee 'regrets that its concerns have not been adequately addressed' and 'notes with disappointment' that promised answers had not been given. After stating that it 'welcomes' certain developments, the Committee observes that the assumption of power by the Peruvian military and the declaration of a state of emergency have 'rendered ineffective the implementation' of certain rights under the Covenant.

> The Committee expresses its deep concern about the terrorism that appears to be part of daily life in Peru. The Committee condemns the atrocities perpetrated by insurgent groups. . . . Nonetheless, the Committee also censures excessive force and violence used by the military, the paramilitary, the police and armed civilian groups. . . . The Committee considers that combating terrorism with arbitrary and excessive State violence cannot be justified under any circumstances.
>
> [With respect to the detention of opposition political and labor leaders and journalists], the Committee does not find the reasons for such detentions convincing. Nor can the unavailability of certain rights to those and other persons . . . be legally justified. The Committee also observes with concern that many people including women and children are held for prolonged periods before trial in police cells. That is not compatible with the rights guaranteed under article 9 of the covenant. . . .

These collective comments, drafted by a rapporteur and then adopted by the Committee as a whole, conclude with the hope 'that the democratic system will be re-established as soon as possible.'

The Committee's collegial comments on Japan's third periodic report, CCPR/C/79/Add.28, 5 Nov. 1993, included several interesting observations. Para. 3 complimented the Japanese Government for giving 'wide publicity' to its report, thus enabling many NGOs 'to become aware of the contents of the report and to make known their particular concerns.' In para. 4, the Committee took note of difficulties experienced by the Government in implementing the ICCPR due to social factors like 'the traditional concept of the different roles of the sexes, the unique relationship between individuals and the group they belong to, and the unconscious particularities due to the homogeneity of the population.' It went on in para. 10 to express 'concern' at continuing gender discrimination in employment. The comments gave strong attention in para. 11 to 'the discriminatory legal provisions concerning children born out of wedlock.' In a section of the comments entitled 'Suggestions and Recommendations,' the Committee 'recommends that Japan takes measures toward the abolition of the death penalty.' (In fact, spurred by the provisions of the Second Optional Protocol to the ICCPR, effective in 1991 and with 25 states parties as of January 1995, the Committee has shown increasing interest in states' use of the death penalty.)

In its comments on Iran's second periodic report, CCPR/C/79/Add.25, 3 August 1993, the Committee in para. 9 'condemned' the death sentence pronounced without trial against the foreign author Salman Rushdie for having written a book viewed as blasphemous toward Islam. It spoke specifically to various forms of discrimination against women:

> 13. The Committee observes that the persistence and extent of discrimination against women is incompatible with the provisions of article 3 of the Covenant and refers, in particular, to the punishment and harassment of women who do not conform with a strict dress code; the need for women to obtain their husband's permission to leave home; their exclusion from the magistracy; discriminatory treatment in respect of the payment of compensation to the families of murder victims, depending on the victim's gender and in respect of the inheritance rights of women; prohibition against the practice of sports in public; and segregation from men in public transportation.

The comments on the second periodic report of Togo, CCPR/C/79/Add.36, 10 August 1994, included the following observations under the heading 'Suggestions and Recommendations':

> 18. The Committee deems it necessary that specific measures be taken to ensure that human rights are respected by the military and security forces. Vigorous action should be taken to ensure that persons closely associated with human rights abuses do not re-enter the police, army or security forces. Urgent steps should be undertaken to ensure that the composition of the army equitably represents various ethnic groups of the Togolese population, including currently under-represented minority groups and that the army remains subject to the control of the elected civil government.

Other new developments about reports

The ICCPR Committee submitted to the 1993 World Conference on Human Rights in Vienna several documents on its work and plans. One such document dealt with reports.[1] It reaffirmed that the purpose of the meetings in which reports were discussed with state representatives was to 'establish a constructive dialogue between the Committee and the State party,' and summarized several recent developments in the Committee's approach to reports.

1. The Committee listed the several times it had requested states parties to submit reports urgently about a new situation such as declarations of emergency or widespread violence, generally within three months of the request. The states included Iraq, Peru and parts of the former Yugoslavia.
2. 'The Committee has also given its support to the Secretary-General's proposal that ways should be explored of empowering human rights bodies to bring massive human rights violations to the attention of the Security Council.'
3. In cases where it was unable to obtain required information, and to learn what had happened with respect to its recommendations in its concluding comments, the Committee was considering requesting the state party concerned 'to agree to receive a mission, consisting of one or two members of the Committee, with a view to collecting information the Committee needs to carry out its functions under the Covenant. Such a decision would only be taken after the Committee had satisfied itself that no adequate alternative approach was available . . . .'

NOTE

The Committee's proposal in para. 2 above to bring 'massive human rights violations to the attention of the Security Council' is relevant to the discussion in the Comment on p. 1042, *infra* on of recent decisions by the Security Council about conflicts involving such massive violations.

QUESTIONS

1. Based on these illustrations, how do you evaluate the Committee's purpose of achieving 'constructive dialogue' with the reporting state? With what types of states is it most likely to realize this goal? Indeed, how would one know whether the discussion of the state report had been 'constructive'?

[1] Work of the Human Rights Committee under article 40 of the Covenant on Civil and Political Rights, UN Doc. A/48/40 (1993), Part I, Annex X at 218.

2. If you were a Committee member reviewing a state report, questioning the state representative, and then helping to prepare the Committee's collective comments about the report, how would you decide what the state and others had done, what the facts were? In the discussion of the Iraq report, there were stark differences between the state representative's account of events and what Committee members believed. How did the members 'know'? Did they look to investigations by other institutions? What standard would they have applied in the event of controverted facts—a preponderance of the evidence? What evidence? In para. 652, Committee members referred to 'reliably attested' violations. Attested by whom, by what institutions? Compare in these respects how the UN Commission on Human Rights might gather facts necessary to support, say, a condemnatory resolution.

3. How do you react to the Committee's thoughts (in para. 3 at the end of the preceding Comment) about sending a mission?

4. Para. 651 in the Iraq report is an atypically strong, direct, critical expression by Committee members of their reactions to a reporting state. Why do you think such strong language appeared with respect to Iraq? How do you understand the more typical, milder language of paras. 213–14 at the end of the next discussion session?

5. Consider the different tone of the discussion about the Austrian report, including the members' concluding observations. From the perspective of advancing observance of human rights, which of the reports and discussion about Iraq or Austria is the more helpful, the more 'constructive'? What does your answer suggest about the likelihood that the processes of treaty organs like the ICCPR can bring about structural change in states that commit serious and systematic human rights violations?

6. Do you think that the Committee's new procedure of formulating comments of the Committee as a whole is a useful change? Will it promote 'constructive dialogue'?

7. The Committee comments addressed to Japan, Iran and Togo include a range of matters. Is the Committee going beyond strict 'application' of the ICCPR to these state's report and conduct? Is it developing, elaborating, creatively interpreting the terms of the ICCPR, and viewing its own role as a treaty organ more spaciously? Illustrate your answers.

3. GENERAL COMMENTS

EXCERPTS FROM THE COMMITTEE'S GENERAL COMMENTS

The text of the ICCPR is terse about what is intended. It invites to diverse interpretations. Article 40, after setting forth the undertaking of states to submit periodic reports to the Committee, provides in para. 4 that the Committee 'shall study the reports' submitted by states and 'shall transmit its reports, and such *general comments* as it may consider appropriate' to the states (emphasis added). Under para. 5, states 'may submit to the Committee observations on any comments that may be made' pursuant to para. 4.

So many possibilities existed for the elaboration of this opaque text, depending on how the Committee answered questions like: Are the general comments to be directed only to states' reports? Are they to vary with the report, addressing concretely this or that problem of this or that state? Alternatively, are they to remain truly 'general' in the sense that they do not pertain exclusively to one state but rather address issues of general relevance to all or many states? Are they to deal only with the processes of reporting or also with substantive provisions of the Covenant? Are they to elaborate those substantive provisions?

Within what framework would the Committee seek to answer such questions—clear meaning of the text, *travaux préparatoires* of the Covenant, major purposes of the Covenant, evolving consensus among members of the Committee or among states parties to the Covenant? Partial answers to these questions—of course, the answers are never 'final' but always subject to change over time—have emerged gradually, some of them of critical importance to the entire exercise. For example, it was early decided that a 'general comment' could not be addressed to a specific state but had to speak with a certain generality.

This Comment reviews General Comments (GCs) of the Committee from several perspectives: their functions, what they reveal about the Committee's understanding of the Covenant, the significance of GCs for the elaboration of the Covenant and for the human rights movement in general. It would be helpful to reread the articles of the Covenant that are referred to when you read the excerpts below from several GCs.

The GCs up to 1992 are set forth in UN Doc. HRI/GEN/1, 4 Sept. 1992, 'Compilation of General Comments and General Recommendations Adopted by Human Rights Treaty Bodies,' at 1–34. Page references below are to this compilation. The GCs up to 1993 appear in 1 Int. Hum. Rts. Reports (No. 2, 1994). Later GCs on the question of reservations to ratifications of the ICCPR and on Article 27 are set forth at pp. 774 and 992, *infra*.

1. Aid to states in filing reports under Article 40

The most prominent theme since the first General Comment (GC) of 1981 has been instructions to states about the reports to be filed under Article 40. The introduction to UN Doc. CCPR/C/21/Rev.1, 19 May 1989, explains the purposes of GCs as follows:

> The Committee wishes to reiterate its desire to assist States parties in fulfilling their reporting obligations. These general comments draw attention to some aspects of this matter but do not purport to be limitative or to attribute any priority between different aspects of the implementation of the Covenant. . . .
>
> The Committee so far has examined 77 initial reports, 34 second periodic reports and, in some cases, additional information and supplementary reports. . . .
>
> The purpose of these general comments is to make this experience available for the benefit of all States parties in order to promote their further implementation of the Covenant; to draw their attention to insufficiencies disclosed by a large number of reports; to suggest improvements in the reporting procedure and to stimulate the activities of these States and international organizations in the promotion and protection of human rights. . . .

* * * * *

GC No. 2. 'Reporting Guidelines' (1981) (p. 2) provides in para. 3:

> 3. The Committee considers that the reporting obligation embraces not only the relevant laws and other norms relating to the obligations under the Covenant but also the practices and decisions of courts and other organs of the State party as well as further relevant facts which are likely to show the degree of the actual implementation and enjoyment of the rights recognized in the Covenant, the progress achieved and factors and difficulties in implementing the obligations under the Covenant.

* * * * *

A number of GCs underscore the Committee's frustration with the inadequacy of many reports. For example, *GC No. 13, 'Article 14'* (1984) (p. 13), concerned with procedures for a fair trial, states in para. 3:

> 3. The Committee would find it useful if, in their future reports, States parties could provide more detailed information on the steps taken to ensure that equality before the courts, including equal access to courts, fair and public hearings and competence, impartiality and independence of the judiciary are established by law and guaranteed in practice. In particular, States parties should specify the relevant constitutional and legislative

texts which provide for the establishment of the courts and ensure that they are independent, impartial and competent, in particular with regard to the manner in which judges are appointed, the qualifications for appointment, and the duration of their terms of office; the condition governing promotion, transfer and cessation of their functions and the actual independence of the judiciary from the executive branch and the legislative.

* * * * *

GC No. 16, 'Article 17' (1988) (p. 20), concerned with privacy, states in part:

> 2. In this connection, the Committee wishes to point out that in the reports of States parties to the Covenant the necessary attention is not being given to information concerning the manner in which respect for this right is guaranteed by legislative, administrative or judicial authorities, and in general by the competent organs established in the State. In particular, insufficient attention is paid to the fact that article 17 of the Covenant deals with protection against both unlawful and arbitrary interference. That means that it is precisely in State legislation above all that provision must be made for the protection of the right set forth in that article. At present the reports either say nothing about such legislation or provide insufficient information on the subject.
>
> . . .
>
> 6. The Committee considers that the reports should include information on the authorities and organs set up within the legal system of the State which are competent to authorize interference allowed by the law. It is also indispensable to have information on the authorities which are entitled to exercise control over such interference with strict regard for the law, and to know in what manner and through which organs persons concerned may complain of a violation of the right provided for in article 17 of the Covenant. States should in their reports make clear the extent to which actual practice conforms to the law. State party reports should also contain information on complaints lodged in respect of arbitrary or unlawful interference, and the number of any findings in that regard, as well as the remedies provided in such cases.

* * * * *

At times, particularly with respect to provisions of the Covenant that are open-textured and that have been further developed in later human rights declarations or treaties, the GCs develop and clarify the substantive content of an article by specifying the particular information that states are to provide about its implementation. Consider, for example, *GC No. 17, 'Article 24'* (1989) (p. 22), dealing with rights of children. Paras. 5 and 6 make particular requests:

5. . . . Reports by States parties should indicate how legislation and practice ensure that measures of protection are aimed at removing all discrimination in every field, including inheritance, particularly as between children who are nationals and children who are aliens or as between legitimate children and children born out of wedlock.
6. . . . However, since it is quite common for the father and mother to be gainfully employed outside the home, reports by States parties should indicate how society, social institutions and the State are discharging their responsibility to assist the family in ensuring the protection of the child. . . . The Committee considers it useful that reports . . . provide information on the special measures of protection adopted to protect children who are abandoned or deprived of their family environment in order to enable them to develop in conditions that most closely resemble those characterizing the family environment.

* * * * *

In a similar vein, note the range of information required by para. 9 of GC *No. 18, 'Non-discrimination'* (1989) (p. 25):

9. Reports of many States parties contain information regarding legislative as well as administrative measures and court decisions which relate to protection against discrimination in law, but they very often lack information which would reveal discrimination in fact. When reporting on articles 2(1), 3 and 26 of the Covenant, States parties usually cite provisions of their constitution or equal opportunity laws with respect to equality of persons. While such information is of course useful, the Committee wishes to know if there remain any problems of discrimination in fact, which may be practised either by public authorities, by the community, or by private persons or bodies. The Committee wishes to be informed about legal provisions and administrative measures directed at diminishing or eliminating such discrimination.

* * * * *

Finally, note the degree of detail about a system of implementation of a norm that is requested in para. 6 of *GC No. 21, 'Article 10,'* (1992) (p. 32), on conditions of detention.

States parties should include in their reports information concerning the system for supervising penitentiary establishments, the specific measures to prevent torture and cruel, inhuman and degrading treatment, and how impartial supervision is ensured.

2. Restatement, interpretation and elaboration of provisions of the Covenant

Consider the purposes served by the following excerpts from GCs. Compare those excerpts from the (frequently lengthy) GCs with the terse text of the articles subjected to commentary in order to consider whether the GCs merely restate the articles or, on the other hand, whether they elaborate or interpret the relevant provision in a significant way.

GC No. 4, 'Article 3' (1981) (p. 3) deals with gender equality. Note that the ICCPR as a whole—unlike, say, the Convention on the Elimination of all Forms of Discrimination against Women, Chapter 13, *infra*—makes no reference to affirmative action. Note also, in relation to para. 3, the phrase in Article 3, 'all civil and political rights set forth in the present Covenant.' There is a similar phrase in Article 2.

> 1. Article 3 of the Covenant requiring, as it does, States parties to ensure the equal right of men and women to the enjoyment of all civil and political rights provided for in the Covenant, has been insufficiently dealt with in a considerable number of States reports and has raised a number of concerns, two of which may be highlighted.
> 2. Firstly, article 3, as articles 2(1) and 26 in so far as those articles primarily deal with the prevention of discrimination on a number of grounds, among which sex is one, requires not only measures of protection but also affirmative action designed to ensure the positive enjoyment of rights. This cannot be done simply by enacting laws. Hence, more information has generally been required regarding the role of women in practice with a view to ascertaining what measures, in addition to purely legislative measures of protection, have been or are being taken to give effect to the precise and positive obligations under article 3 and to ascertain what progress is being made or what factors or difficulties are being met in this regard.
> 3. Secondly, the positive obligation undertaken by States parties under that article may itself have an inevitable impact on legislation or administrative measures specifically designed to regulate matters other than those dealt with in the Covenant but which may adversely affect rights recognized in the Covenant. One example, among others, is the degree to which immigration laws which distinguish between a male and a female citizen may or may not adversely affect the scope of the right of the woman to marriage to non-citizens or to hold public office.

* * * * *

GC No. 6, 'Article 6' (1982) (p. 5) and *GC No. 14, 'Article 6'* (1984) (p. 17) both address the right to life, a phrase as central to the human rights corpus as it is elliptic and open to a range of interpretations. Consider the following excerpt from *GC No. 6:*

> 1. The right to life enunciated in article 6 of the Covenant has been dealt with in all State reports. It is the supreme right from which no derogation

is permitted even in time of public emergency which threatens the life of the nation (art. 4). However, the Committee has noted that quite often the information given concerning article 6 was limited to only one or other aspect of this right. It is a right which should not be interpreted narrowly.

. . .

5. Moreover, the Committee has noted that the right to life has been too often narrowly interpreted. The expression 'inherent right to life' cannot properly be understood in a restrictive manner, and the protection of this right requires that States adopt positive measures. In this connection, the Committee considers that it would be desirable for States parties to take all possible measures to reduce infant mortality and to increase life expectancy, especially in adopting measures to eliminate malnutrition and epidemics.

* * * * *

Compare the preceding provisions from the earlier GC with *GC No. 14*, issued two years later:

3. While remaining deeply concerned by the toll of human life taken by conventional weapons in armed conflicts, the Committee has noted that, during successive sessions of the General Assembly, representatives from all geographical regions have expressed their growing concern at the development and proliferation of increasingly awesome weapons of mass destruction, which not only threaten human life but also absorb resources that could otherwise be used for vital economic and social purposes, particularly for the benefit of developing countries, and thereby for promoting and securing the enjoyment of human rights for all.
4. The Committee associates itself with this concern. It is evident that the designing, testing, manufacture, possession and deployment of nuclear weapons are among the greatest threats to the right to life which confront mankind today. This threat is compounded by the danger that the actual use of such weapons may be brought about, not only in the event of war, but even through human or mechanical error or failure.

. . .

6. The production, testing, possession, deployment and use of nuclear weapons should be prohibited and recognized as crimes against humanity.
7. The Committee accordingly, in the interest of mankind, calls upon all States, whether Parties to the Covenant or not, to take urgent steps, unilaterally and by agreement, to rid the world of this menace.

* * * * *

GC No. 12. 'Article 1' (1984) (p. 11) examines one of the most influential, debated and contested provisions of this Covenant, the right of 'peoples' to 'self-determination'. A central question is the degree to which Article 1 addresses the issue of so-called 'external' self-determination, the right of a 'people' to independent statehood, as in the process of decolonization. Does it also address more general questions such as the form of government to which the population of a state is entitled in order to achieve 'internal' self-determination? Most of this GC tracks and restates the provisions of the article. Consider what light the following excerpts might shed on the questions noted.

> 1. In accordance with the purposes and principles of the Charter of the United Nations, article 1 of the International Covenant on Civil and Political Rights recognizes that all peoples have the right of self-determination. The right of self-determination is of particular importance because its realization is an essential condition for the effective guarantee and observance of individual human rights and for the promotion and strengthening of those rights. It is for that reason that States set forth the right of self-determination in a provision of positive law in both Covenants and placed this provision as article 1 apart from and before all of the other rights in the two Covenants.
>
> . . .
>
> 4. With regard to paragraph 1 of article 1, States parties should describe the constitutional and political processes which in practice allow the exercise of this right.

* * * * *

GC No. 13. 'Article 14' (1984) (p. 13) examines numerous provisions of Article 14 on questions of criminal process in particular.

> 7. The Committee has noted a lack of information regarding article 14, paragraph 2 and, in some cases, has even observed that the presumption of innocence, which is fundamental to the protection of human rights, is expressed in very ambiguous terms or entails conditions which render it ineffective. By reason of the presumption of innocence, the burden of proof of the charge is on the prosecution and the accused has the benefit of doubt. No guilt can be presumed until the charge has been proved beyond reasonable doubt. Further, the presumption of innocence implies a right to be treated in accordance with this principle. It is, therefore, a duty for all public authorities to refrain from prejudging the outcome of a trial.
> 8. . . . Article 14 (3) (a) applies to all cases of criminal charges, including those of persons not in detention. The Committee notes further that the right to be informed of the charge 'promptly' requires that information is given in the manner described as soon as the charge is first made by a competent authority. In the opinion of the Committee this right must arise when in the course of an investigation a court or an authority of the prosecution decides to take procedural steps against a person suspected of a crime or publicly names him as such. . . .

* * * * *

GC No. 16. 'Article 17' (1988) (p. 20) elaborates this brief article on interference with privacy. One illustration follows:

> 10. The gathering and holding of personal information on computers, databanks and other devices, whether by public authorities or private individuals or bodies, must be regulated by law. Effective measures have to be taken by States to ensure that information concerning a person's private life does not reach the hands of persons who are not authorized by law to receive, process and use it, and is never used for purposes incompatible with the Covenant. In order to have the most effective protection of his private life, every individual should have the right to ascertain in an intelligible form, whether, and if so, what personal data is stored in automatic data files, and for what purposes. Every individual should also be able to ascertain which public authorities or private individuals or bodies control or may control their files. If such files contain incorrect personal data or have been collected or processed contrary to the provisions of the law, every individual should have the right to request rectification or elimination.

* * * * *

GC No. 17. 'Article 24' (1989) (p. 22) concerns that article's brief and general provision on children's rights. Note that a treaty devoted to such issues was later drafted and is now in effect—the Convention on the Rights of the Child. Consider the following comment:

> 3. In most cases, however, the measures to be adopted are not specified in the Covenant and it is for each State to determine them in the light of the protection needs of children in its territory and within its jurisdiction. The Committee notes in this regard that such measures, although intended primarily to ensure that children fully enjoy the other rights enunciated in the Covenant, may also be economic, social and cultural. For example, every possible economic and social measure should be taken to reduce infant mortality and to eradicate malnutrition among children and to prevent them from being subjected to acts of violence and cruel and inhuman treatment or from being exploited by means of forced labour or prostitution, or by their use in the illicit trafficking of narcotic drugs, or by any other means. In the cultural field, every possible measure should be taken to foster the development of their personality and to provide them with a level of education that will enable them to enjoy the rights recognized in the Covenant, particularly the right to freedom of opinion and expression. . . .

* * * * *

GC No. 18. 'Non-discrimination' (1989) (p. 25) deals with several provisions of the Covenant—Articles 2, 3 and 26 among others—that state the principle of non-discrimination. Note also the references to other human rights treaties.

6. The Committee notes that the Covenant neither defines the term 'discrimination' nor indicates what constitutes discrimination. However, article 1 of the International Convention on the Elimination of All Forms of Racial Discrimination provides that the term 'racial discrimination' shall mean any distinction, exclusion, restriction or preference based on race, colour, descent, or national or ethnic origin which has the purpose or effect of nullifying or impairing the recognition, enjoyment or exercise, on an equal footing, of human rights and fundamental freedoms in the political, economic, social, cultural or any other field of public life. Similarly, article 1 of the Convention on the Elimination of All Forms of Discrimination against Women provides that 'discrimination against women' shall mean any distinction, exclusion or restriction made on the basis of sex which has the effect or purpose of impairing or nullifying the recognition, enjoyment or exercise by women, irrespective of their marital status, on a basis of equality of men and women, of human rights and fundamental freedoms in the political, economic, social, cultural, civil or any other field.

7. While these conventions deal only with cases of discrimination on specific grounds, the Committee believes that the term 'discrimination' as used in the Covenant should be understood to imply any distinction, exclusion, restriction or preference which is based on any ground such as race, colour, sex, language, religion, political or other opinion, national or social origin, property, birth or other status, and which has the purpose or effect of nullifying or impairing the recognition, enjoyment or exercise by all persons, on an equal footing, of all rights and freedoms.

8. The enjoyment of rights and freedoms on an equal footing, however, does not mean identical treatment in every instance. In this connection, the provisions of the Covenant are explicit. For example, article 6, paragraph 5, prohibits the death sentence from being imposed on persons below 18 years of age. The same paragraph prohibits that sentence from being carried out on pregnant women. . . . Furthermore, article 25 guarantees certain political rights, differentiating on grounds of citizenship.

. . .

10. The Committee also wishes to point out that the principle of equality sometimes requires States parties to take affirmative action in order to diminish or eliminate conditions which cause or help to perpetuate discrimination prohibited by the Covenant. For example, in a State where the general conditions of a certain part of the population prevent or impair their enjoyment of human rights, the State should take specific action to correct those conditions. Such action may involve granting for a time to the part of the population concerned certain preferential treatment in specific matters as compared with the rest of the population. However, as long as such action is needed to correct discrimination in fact, it is a case of legitimate differentiation under the Covenant.

. . .

12. . . . In the view of the Committee, article 26 does not merely duplicate the guarantee already provided for in article 2 but provides in itself an autonomous right. It prohibits discrimination in law or in fact in any field

regulated and protected by public authorities. Article 26 is therefore concerned with the obligations imposed on States parties in regard to their legislation and the application thereof. Thus, when legislation is adopted by a State party, it must comply with the requirement of article 26 that its content should not be discriminatory. In other words, the application of the principle of non-discrimination contained in article 26 is not limited to those rights which are provided for in the Covenant.
13. Finally, the Committee observes that not every differentiation of treatment will constitute discrimination, if the criteria for such differentiation are reasonable and objective and if the aim is to achieve a purpose which is legitimate under the Covenant.

* * * * *

GC No. 19. 'Article 23' (1990) (p. 27) concerns the protection of the family. Note the limited discussion of family planning. What are its implications, for example, for the right to practice contraception or to abort?

4. . . . The Covenant does not establish a specific marriageable age either for men or for women, but that age should be such as to enable each of the intending spouses to give his or her free and full personal consent in a form and under conditions prescribed by law. In this connection, the Committee wishes to note that such legal provisions must be compatible with the full exercise of the other rights guaranteed by the Covenant; thus, for instance, the right to freedom of thought, conscience and religion implies that the legislation of each State should provide for the possibility of both religious and civil marriages. In the Committee's view, however, for a State to require that a marriage, which is celebrated in accordance with religious rites, be conducted, affirmed or registered also under civil law is not incompatible with the Covenant. . . .
5. The right to found a family implies, in principle, the possibility to procreate and live together. When States parties adopt family planning policies, they should be compatible with the provisions of the Covenant and should, in particular, not be discriminatory or compulsory. . . .
6. . . . During marriage, the spouses should have equal rights and responsibilities in the family. This equality extends to all matters arising from their relationship, such as choice of residence, running of the household, education of the children and administration of assets. Such equality continues to be applicable to arrangements regarding legal separation or dissolution of the marriage.

* * * * *

GC No. 20. 'Article 7' (1992) (p. 29) concerns torture and cruel or degrading treatment or punishment.

2. The aim of the provisions of article 7 of the International Covenant on Civil and Political Rights is to protect both the dignity and the physical and mental integrity of the individual. It is the duty of the State party

to afford everyone protection through legislative and other measures as may be necessary against the acts prohibited by article 7, whether inflicted by people acting in their official capacity, outside their official capacity or in a private capacity. . . .

. . .

5. The prohibition in article 7 relates not only to acts that cause physical pain but also to acts that cause mental suffering to the victim. In the Committee's view, moreover, the prohibition must extend to corporal punishment, including excessive chastisement ordered as punishment for a crime or as an educative or disciplinary measure. . . .

. . .

11. . . . To guarantee the effective protection of detained persons, provisions should be made for detainees to be held in places officially recognized as places of detention and for their names and places of detention, as well as for the names of persons responsible for their detention, to be kept in registers readily available and accessible to those concerned, including relatives and friends. To the same effect, the time and place of all interrogations should be recorded, together with the names of all those present and this information should also be available for purposes of judicial or administrative proceedings. Provisions should also be made against incommunicado detention. In that connection, States parties should ensure that any places of detention be free from any equipment liable to be used for inflicting torture or ill-treatment. The protection of the detainee also requires that prompt and regular access be given to doctors and lawyers and, under appropriate supervision when the investigation so requires, to family members.

12. It is important for the discouragement of violations under article 7 that the law must prohibit the use of admissibility in judicial proceedings of statements or confessions obtained through torture or other prohibited treatment.

. . .

15. The Committee has noted that some States have granted amnesty in respect of acts of torture. Amnesties are generally incompatible with the duty of States to investigate such acts; to guarantee freedom from such acts within their jurisdiction; and to ensure that they do not occur in the future. States may not deprive individuals of the right to an effective remedy, including compensation and such full rehabilitation as may be possible.

* * * * *

GC No. 21. 'Article 10' (1992) (p. 32) applies to detained person. Note again the references to other instruments.

4. Treating all persons deprived of their liberty with humanity and with respect for their dignity is a fundamental and universally applicable rule. Consequently, the application of this rule, as a minimum, cannot be

dependent on the material resources available in the State party. This rule must be applied without distinction of any kind, such as race, colour, sex, language, religion, political or other opinion, national or social origin, property, birth or other status.
5. States parties are invited to indicate in their reports to what extent they are applying the relevant United Nations standards applicable to the treatment of prisoners: the Standard Minimum Rules for the Treatment of Prisoners (1957), the Body of Principles for the Protection of All Persons under Any Form of Detention or Imprisonment (1988), the Code of Conduct for Law Enforcement Officials (1978) and the Principles of Medical Ethics relevant to the Role of Health Personnel, particularly Physicians, in the Protection of Prisoners and Detainees against Torture and Other Cruel, Inhuman or Degrading Treatment or Punishment (1982).

. . .

10. As to article 10, paragraph 3, which concerns convicted persons, the Committee wishes to have detailed information on the operation of the penitentiary system of the State party. No penitentiary system should be only retributory; it should essentially seek the reformation and social rehabilitation of the prisoner. . . .

. . .

13. . . . Lastly, under article 10, paragraph 3, juvenile offenders shall be segregated from adults and be accorded treatment appropriate to their age and legal status in so far as conditions of detention are concerned, such as shorter working hours and contact with relatives, with the aim of furthering their reformation and rehabilitation. Article 10 does not indicate any limits of juvenile age. While this is to be determined by each State party in the light of relevant social, cultural and other conditions, the Committee is of the opinion that article 6, paragraph 5, suggests that all persons under the age of 18 should be treated as juveniles, at least in matters relating to criminal justice. States should give relevant information about the age groups of persons treated as juveniles. In that regard, States parties are invited to indicate whether they are applying the United Nations Standard Minimum Rules for the Administration of Juvenile Justice, known as the Beijing Rules (1987).

NOTE

Consider the following observations of the Committee in its 1994 Report, vol. 1, on its activities, GAOR 49th Sess., Supp. No. 40 (A/49/40), para. 50:

General comments . . . do not purport to . . . attribute any priority among the different aspects [of the Covenant] in terms of implementation. They are intended to make the Committee's experience available for the benefit of all States parties, so as to promote more effective implementation of the Covenant; to draw their attention to insufficiencies disclosed by a large number of reports; to suggest improvements in the reporting procedure to clarify the requirements of the Covenant; and to stimulate the activities of States parties and international organizations in the promotion and protection of human rights. General comments are also intended to . . . strengthen cooperation among States in the universal promotion and protection of human rights.

QUESTIONS

1. 'In its General Comments, the Committee has exercised a power of its own creation. It is understandable that the Committee developed formal rules and processes for the submission of reports; such matters affect relations between the Committee and states parties to the ICCPR. It is a different matter for the Committee to express its collective understanding of the content of the ICCPR's substantive provisions regulating what states may do internally—what privacy amounts to, what due process requires, and so on. Modest interpretation of ICCPR articles is perhaps inevitable, but numbers of the General Comments amount to a bold elaboration, an emphatic development of ideas in the Covenant itself, to "legislation by Committee."'

a. Do you agree or disagree? What illustrations from the preceding selection of excerpts from General Comments best support this statement?

b. Some General Comments clearly go beyond the literal text of the ICCPR. Using any illustrations that you choose, how would you describe what the Committee is here doing? Interpretation or elaboration based on the literal meaning of the ICCPR's text, or on the larger internal context of all provisions of the ICCPR and interrelations among them? Interpretation or elaboration in light of the broad object and purpose of the ICCPR as a whole? Creative development in the light of changing world conditions, conflicts and ideologies? Working out the common understanding of representative states parties to the ICCPR of the meaning of various provisions?

c. What arguments could the Committee make to justify this trend toward more significant General Comments that go beyond merely restating the ICCPR's text?

2. (a) What authority do you believe that these General Comments should have before other institutions—say, a General Comment giving a precise interpretation of an ICCPR article? For example, a question of interpretation of that article might arise before the UN Commission on Human Rights in a debate about whether a country was in violation of the ICCPR, before a national court applying the ICCPR to a case before it, before the International Court of Justice in a case where one state accuses another of violating that article. Should the General Comment be treated as decisive? (b) Should the answer depend on which of the descriptions in Question 1(b) of what the Committee is doing seems

most accurate? (c) Does any other national or international institution have such a power of authoritative interpretation of the ICCPR? Should there be one such institution?

3. General Comments can be long debated within the Committee, in closed proceedings, and then be issued in final form—though, as we have seen, subject to revision by a later General Comment on the same ICCPR article. Should the process be changed, so that the Committee issues to states parties, IGOs and NGOs a preliminary version of a proposed General Comment and solicits reactions thereto? The Committee would then take those reactions into account as it thought appropriate when preparing the final form of the General Comment. What advantages and disadvantages do you see in such a process?

4. INDIVIDUAL COMMUNICATIONS

COMMENT ON COMMUNICATIONS UNDER THE OPTIONAL PROTOCOL

As of September 1995, 84 of the 131 states parties to the ICCPR were also parties to this First Optional Protocol. Note some of its critical provisions. The communications must be 'from individuals . . . who claim to be victims of a violation' by a state party to the Protocol 'of any of the rights set forth in the Covenant.' After being notified of the communication, the state party shall 'submit to the Committee written explanations or statements clarifying the matter . . . .' The Committee considers communications 'in the light of all written information made available to it by the individual and by the State Party concerned.' It will not consider a communication before ascertaining that the matter is 'not being examined under another procedure of international investigation or settlement.' Examination of the communications takes place at 'closed meetings.' The Committee is to forward 'its views' to the individual and state concerned.

Note some of the consequences or negative implications of these terse and ambiguous provisions, supplemented by the Committee's own procedural rules:

(1) The proceedings are in no sense a continuation of or appeal from judicial proceedings (if there were any) in the state in which the dispute originated. They are fresh, distinct proceedings that may involve the same two parties to prior proceedings or different parties.
(2) Unlike the requirements for procedures 1235 and 1503 of the UN Commission on Human Rights, the communication to the ICCPR

Committee need not allege that the violation complained of is systemic—that is, involves a consistent pattern in or practice of the state or some recurrent phenomenon. In theory, an isolated, atypical violation suffices to found the communication.

(3) There are no provisions for oral hearings, let alone confrontation between the parties, or for independent fact-finding (such as examination of the parties or witnesses or independent experts, or on-site visits) by the Committee. The Optional Protocol refers only to written proceedings.

(4) There are no hearings or debates of Committee members about how to deal with the communication that are open to the public.

(5) There is no provision setting forth the precise legal effect of 'views' or setting forth what follow-up should take place if the Committee's views indicating that a state should take particular action (such as payment of compensation or release of a prisoner) are ignored by that state. In fact, there have been many instances of states continuing to ignore the Committees 'views,' even in cases of the most serious violations. Recent changes in the Committee's procedure have led it to request the state to advise it within 90 days of what action has been taken with respect to views, and to appoint a Special Rapporteur to monitor compliance by states with the Committee's views and report to the Committee thereon.

The procedure developed by the Committee to handle communications involves the formation of a Working Group of five Committee members who meet a week before each three-week session to consider the admissibility of the communication. Only the Committee can decide that a communication is inadmissible.

The Optional Protocol is one of only three human rights treaties of universal reach, the two others being the conventions on racial discrimination and torture, that provide (some form of) an individual remedy at the international level. As of July 1994, 587 communications had been registered, of which 193 led to the adoption of views by the Committee. In 142 such cases, the Committee found violations of the ICCPR. The Committee had declared 201 communications inadmissible.[2]

Note that the optional provision in Article 41 of the ICCPR for the filing of a communication by one state party against another has never been utilized, although as of January 1995, 44 states had made the necessary declaration under that article.

[2] The figures in this paragraph are taken from the 1994 Report of the Human Rights Committee, Vol. I, GAOR 49th Sess., Supp. No. 40 (A/49/40), at paras. 377 and 459.

TORKEL OPSAHL, THE HUMAN RIGHTS COMMITTEE

in Philip Alston (ed.), The United Nations and Human Rights 369 (1992), at 421.

More ambitious proposals for the handling of such complaints were discarded at an early stage. No private complaints system, to a Court or otherwise, could be made obligatory. Not even in an optional form was any such procedure allowed to become part of the Covenant itself. But a majority of the General Assembly decided at the eleventh hour to create a separate document, the Optional Protocol, allowing for 'communications from individuals claiming to be victims of violations of any of the rights set forth in the Covenant'. . . .

. . .

On some points, this system resembles those of the Council of Europe and the Organization of American States, but it is much simpler. The Committee is the only competent organ and its procedure is not very elaborate. Its 'views' are not to be understood as strictly binding in law and cannot be enforced. The Protocol itself seems to consider a case closed as soon as the views have been forwarded to the parties. There is no follow-up procedure leading, as in the European system, from the report of a Commission to binding decisions by other competent organs (such as the European Committee of Ministers or Court of Human Rights).

The consideration of communications takes place in closed meetings. The files and summary records of the Committee's deliberations remain confidential, but the texts of its final decisions, called 'views', have been made public from the beginning. The decision to publish the views was taken without any express authority from the Optional Protocol and did not correspond to any explicit rule in the then Provisional Rules of Procedure. Thus it was a bold and important precedent. . . .

. . .

. . . [D]espite the confidentiality of the proceedings under the Optional Protocol, ample material is now available. However, as a source of case law, the material is less developed than in, for instance, the European system, because the Committee does not go to the same lengths in its published reasoning.

Lack of more widespread publicity may be one of the reasons why individual communications have not arrived in greater numbers, compared to the European system, and which hardly amount to a representative selection of problems in the countries concerned. In the first fifteen years only 468 complaints had arisen from thirty-six States Parties, and their distribution was very uneven. The absence of cases against, for instance, the Central African Republic and the Congo might simply reflect lack of knowledge about the procedure. In other cases, such as the States Parties that are also bound by the

European Convention on Human Rights, the more effective regional procedure may be preferred by most people who wish to complain.

. . .

At the same time, the Committee has not, or not yet, displayed much initiative or allowed for much development of its procedures at the merits stage. Besides recalling the time-limit of six months for the State Party, and transmitting any replies to the author for possible further information or observations, the Committee's decisions reveal that it has remained largely passive. The paucity of the relevant rules on the subject is evidence that the Committee has not taken such an active role as an investigative body might be expected to do. Some members have expressed the opinion that the Optional Protocol does not permit the Committee to take such steps as carrying out a more independent inquiry by specifying its requests, resorting to active fact-finding by sending delegates, arranging for oral hearings of the parties and witnesses, or taking steps towards mediation or settlement of the matter. It is true that the Protocol only refers to the duty of the Committee to consider 'all *written* material made available to it' by the *parties*. But this clause only says what the Committee shall consider, not what, in addition, it *may* consider or do. In the absence of consensus on the Committee's powers and policy as regards a more active role, a minimalist line has again prevailed in its practice.

In its handling of complaints, however, the Committee has been able to reach substantive conclusions on a large number of issues, often unfavourable to States Parties. And a flexible approach may, once the practical need for it is demonstrated, outweigh the cautious interpretation which holds that the Committee cannot do anything not expressly foreseen under the Protocol.

The Committee has in many cases not only expressed its 'view' that there have been violations of specific rights, but also added what in its opinion should follow, stating as a separate conclusion a duty to provide individual reparation and take preventive measures for the future. This is now a settled interpretation of its role. The Committee could go further in seeking solutions in co-operation with the parties. Certainly it might, with their consent, develop its fact-finding role and allow oral pleadings.

. . .

The other side of this approach concerns monitoring *violations* of these rights through complaints procedures. This should not be more than a secondary aim. Universality in such procedures is less necessary and more difficult to apply. It is true that complaints procedures, more than abstract inquiries, may throw specific light even over matters of general interest. Case law can also be more specific than general comments on how provisions ought to be understood. Nevertheless, the Committee will never be able to control violations in all parts of the world through complaints procedures. It would not be practical to develop its machinery to the extent which this control would require, since this would include hearings, fact-finding, visits, investigations, use of many languages, etc. Theoretically the procedures could become permanent and deal annually with thousands of cases rather than dozens. But there is a better alternative; building and strengthening regional complaints systems, like the European, American, and African ones, offers several advantages in the areas of logistics, local trust, and homogeneity. States Parties from the regions which have such systems in place have already shown their preference for them through reservations.

. . .

HERRERA RUBIO v. COLOMBIA

Communication No. 161/1983, Human Rights Committee Views of Committee, November 2, 1987.
UN Doc. A/43/40 (1988), Supp. No. 40, at 190.

[Herrera Rubio, a Colombian citizen and resident, submitted this communication on his own behalf and in respect of his deceased parents. He alleged arrest by members of the armed forces, imprisonment in a military camp, and torture in an attempt to extract information on a guerrilla movement about which he had no knowledge. He described the tortures in horrific detail, and said that he was threatened with his parents' death. His parents were abducted by individuals in military uniform identified as members of the 'counter guerrilla.' Their bodies (the father's decapitated) were later found. Herrera Rubio was released in 1982 pursuant to an amnesty law of that year concerning political detainees.

In this communication, the author asserted that all incidents complained of occurred 'in a region under military control where violations of the rights of the civilian population have allegedly become general practice.' He claimed violations of Articles 6, 7, 9, 10 and 17 of the ICCPR.

In accordance with its usual procedures, a Working Group of the Committee first considered the communication and requested specific information from the state party. No reply was received from the state. The Committee then concluded, on the basis of the information before it, that the communication was admissible.

The state party did reply to further requests for information, alleging that the deaths of the parents 'were duly investigated and that no evidence was found to support charges against military personnel.' Hence the dossier was closed. Colombia also provided the text of the decision of a penal court that had made a judicial investigation concluding that the killings 'had been perpetrated by armed persons, without, however, being able to determine to which group they belonged.'

With respect to the author's allegations that he had been tortured, the Attorney-General Delegate for the Armed Forces decided not to open a formal investigation. That decision, noting the difficulty of obtaining evidence about events five years old, stated in part:

> [Herrera Rubio's] statement on the alleged acts of torture are not credible in view of the fact that three months elapsed from the time of the alleged ill-treatment before the complainant reported it to the Court. On witnessing his statement as an accused person . . . , this office put on record that 'the accused appeared normal physically and mentally . . . .'

The author responded to some of the state party's arguments by claiming that (1) the Government's allegations that a guerrilla group might have killed the parents were 'at odds with the facts of the case,' (2) the Attorney-General had not explained why the author's petitions to government officials while in prison went unanswered, and (3) the delay in filing a complaint about torture was explainable by obvious psychological pressures on prisoners and by strategic considerations.

In light of the 'conflicting statements by the parties,' the Working Group of the Committee requested additional information from Colombia about its investigations of the author's charges. The reply was general and unresponsive to specific requests.

The Committee then decided 'to base its views on the following facts and considerations' as set out below.]

10.3 Whereas the Committee considers that there is reason to believe, in the light of the author's allegations, that Colombian military persons bear responsibility for the deaths of José Herrera and Emma Rubio de Herrera, no conclusive evidence has been produced to establish the identity of the murderers. In this connection the Committee refers to its general comment No. 6 (16) concerning article 6 of the Covenant, which provides, *inter alia*, that States parties should take specific and effective measures to prevent the disappearance of individuals and establish effective facilities and procedures to investigate thoroughly, by an appropriate impartial body, cases of missing and disappeared persons in circumstances which may involve a violation of the right to life. The Committee has duly noted the State party's submissions concerning the investigations carried out in this case, which, however, appear to have been inadequate in the light of the State party's obligations under article 2 of the Covenant.

10.4 With regard to the author's allegations of torture, the Committee notes that the author has given a very detailed description of the ill-treatment to which he was subjected and has provided the names of members of the armed forces allegedly responsible. In this connection, the Committee observes that the initial investigations conducted by the State Party may have been concluded prematurely and that further investigations were called for in the light of the author's submission of 4 October 1986 and the Working Group's request of 18 December 1986 for more precise information.

10.5 With regard to the burden of proof, the Committee has already established in other cases (for example, Nos. 30/1978 and 85/1981) that this cannot rest alone on the author of the communications, especially considering that the author and the State-party do not always have equal access to the evidence and that frequently the State party alone has access to relevant information. In the circumstances, due weight must be given to the authors' allegations. It is implicit in article 4, paragraph 2, of the Optional Protocol that the State party has the duty to investigate in good faith all allegations of violation of the Covenant made against it and its authorities, and to furnish to the Committee the information available to it. In no circumstances should a State party fail to investigate fully allegations of ill-treatment when the person or persons allegedly responsible for the ill-treatment are identified by the author of a communication. The State party has in this matter provided no precise information and reports, *inter alia*, on the questioning of military officials accused of maltreatment of prisoners, or on the questioning of their superiors.

11. The Human Rights Committee, acting under article 5, paragraph 4, of the Optional Protocol to the International Covenant of Civil and Political Rights, is of the view that the facts as found by the Committee disclose violations of the Covenant with respect to:

> Article 6, because the State party failed to take appropriate measures to prevent the disappearance and subsequent killings of José Herrera and Emma Rubio de Herrera and to investigate effectively the responsibility for their murders; and
> Article 7 and article 10, paragraph 1, because Joaquín Herrera Rubio was subjected to torture and ill-treatment during his detention.

12. The Committee, accordingly, is of the view that the State party is under an obligation, in accordance with the provisions of article 2 of the Covenant, to take effective measures to remedy the violations that Mr. Herrera Rubio has suffered and further to investigate said violations, to take action thereon as appropriate and to take steps to ensure that similar violations do not occur in the future.

BWALYA V. ZAMBIA

Communication No. 314/1988, Human Rights Committee
Views of Committee, July 14, 1993.
UN Doc. 4/48/40 (1994), Supp. 40, at 52; 1 Int. Hum. Rts. Reports 84 (No. 2, 1994).

[Bwalya, a Zambian citizen, ran in 1983 for a parliamentary seat on a program committed to changing the Government's policy toward particularly the homeless and unemployed. He claimed that in retaliation for his activism, the authorities threatened him. He was dismissed from employment and with his family expelled from his home. He was later arrested and detained for 31 months on charges of belonging to an organization considered illegal under Zambia's one-party Constitution and of conspiring to overthrow the Government of then President Kaunda.

In his communication to the Committee, Bwalya alleged violations by Zambia of several provisions of the Covenant. He denied participating in a conspiracy against the Government and claimed that he continued to be denied freedom of movement.

The Committee found the communication admissible with respect to claims of violations of Articles 9, 12, 19, 25 and 26 of the ICCPR. It noted 'with concern that, with the exception of a brief note informing the Committee of the author's release, the State party has failed to cooperate on the matter under consideration'—for example, by failing to examine allegations against it or to provide the Committee with relevant information at its disposal. 'In the circumstances, due weight must be given to the author's allegations, to the extent that they have been substantiated.' The Committee continued:]

6.2 In respect of issues under article 19, the Committee considers that the uncontested response of the authorities to the attempts of the author to express his opinions freely and to disseminate the political tenets of his party constitute a violation of his rights under article 19.

6.3 The Committee has noted that when the communication was placed before it for consideration, Mr. Bwalya had been detained for a total of 31 months, a claim that has not been contested by the State party. It notes that the author was held solely on charges of belonging to a political party considered illegal under the country's (then) one-party constitution and that, on the basis of the information before the Committee, Mr. Bwalya was not brought promptly before a judge or other officer authorized by law to exercise judicial power to determine the lawfulness of his detention. This, in the Committee's opinion, constitutes a violation of the author's right under article 9, paragraph 3, of the Covenant.

. . .

6.5 The author has claimed, and the State party has not denied, that he continues to suffer restrictions on his freedom of movement, and that the authorities have refused to issue a passport to him. This, in the Committee's opinion, amounts to a violation of article 12, paragraph 1, of the Covenant.

6.6 As to the alleged violation of article 25 of the Covenant, the Committee notes that the author, a leading figure of a political party in opposition to the former President, has been prevented from participating in a general election campaign as well as from preparing his candidacy for this party. This amounts to an unreasonable restriction on the author's right to 'take part in the conduct of public affairs' which the State party has failed to explain or justify. In particular, it has failed to explain the requisite conditions for participation in the elections. Accordingly, it must be assumed that Mr. Bwalya was detained and denied the right to run for a parliamentary seat in the Constituency of Chifubu merely on account of his membership in a political party other than that officially recognized; in this context, the Committee observes that restrictions on political activity outside the only recognized political party amount to an unreasonable restriction of the right to participate in the conduct of public affairs.

6.7 Finally, on the basis of the information before it, the Committee concludes that the author has been discriminated against in his employment because of his political opinions, contrary to article 26 of the Covenant.

7. The Human Rights Committee, acting under article 5, paragraph 4, of the Optional Protocol to the International Covenant on Civil and Political Rights, is of the view that the facts as found by the Committee disclose violations of articles 9, paragraphs 1 and 3, 12, 19, paragraph 1, 25(a) and 26 of the Covenant.

8. Pursuant to article 2 of the Covenant, the State party is under an obligation to provide Mr. Bwalya with an appropriate remedy. The Committee urges the State party to grant appropriate compensation to the author. The State party is under an obligation to ensure that similar violations do not occur in the future.

9. The Committee would wish to receive information, within 90 days, on any relevant measures taken by the State party in respect of the Committee's Views.

PAUGER V. AUSTRIA

Communication No. 415/1990, Human Rights Committee
Views of Committee, March 26, 1992.
UN Doc. A/47/40, Supp. No. 40, at 325.

[Pauger, an Austrian citizen and resident, was a university professor whose wife died in 1984. She had been a civil servant. Pauger submitted a claim under the Pension Act of 1965, which granted preferential treatment to widows who could receive a pension regardless of their income, whereas widowers (surviving husbands) could receive a pension only if they lacked other income. His claim was therefore rejected, a decision upheld by the Constitutional Court of Austria. The legislation was subsequently amended to provide a pension to widowers, retroactively, but in amounts below those received by widows. Pauger again sought a full pension, but the rejection of his claim was again upheld by the Constitutional Court.

In filing his communication with the Human Rights Committee, Pauger alleged a violation of Article 26 of the Covenant, claiming that 'the differentiation between widows and widowers is arbitrary and cannot be said to be based on reasonable and objective criteria.' The Committee declared the communication admissible in March 1991. Excerpts follow from its views under Article 5(4) of the Optional Protocol.]

The State party's explanations and the author's comments thereon

5.1 In its submission, dated 8 October 1991, the State party argues that the former Austrian pension legislation was based on the fact that in the overwhelming majority of cases only the husband was gainfully employed, and therefore only he was able to acquire an entitlement to a pension from which his wife might benefit. It submits that, in response to changed social conditions, it amended both family legislation and the Pension Act; equality of the husband's position under pension law is to be accomplished in a number of successive stages, the last of which will be completed on 1 January 1995.

5.2 The State party further submits that new legislation, designed to change old social traditions, cannot be translated into reality from one day to the other. It states that the gradual change in the legal position of men with regard to their pension benefits was necessary in the light of the actual social conditions, and hence does not entail any discrimination. In this context, the State party points out that the equal treatment of men and women for purposes of civil service pensions has financial repercussions in other areas, as the pensions will have to be financed by the civil servants, from whom pension contributions are levied.

6.1 In his reply to the State party's submission, the author argues that pursuant to amendments in family law, equal rights and duties have existed for both spouses since 1 January 1976, in particular with regard to their income and their mutual maintenance. He further submits that in the public sector men and women receive equal payment for equal services and have also to pay equal pension fund contributions. The author states that there is no convincing reason as to why a period of nearly two decades since the emancipation of men and women in family law should be necessary for the legal emancipation in pension law to take place.

6.2 According to the author, neither the financial burden on the State's budget, nor the fact that many men are entitled to pensions of their own, can be used as arguments against the obligation to treat men and women equally, pursuant to article 26 of the Covenant. The author points out that the legislator could have established other, such as income-related, criteria to distinguish between those who are entitled to a full pension and those who are not. He further submits that the financial burden caused by the equal treatment of men and women under the Pension Act would be comparatively low, because of the small number of widowers who are entitled to such a pension.

Examination of the merits

. . .

7.2 The Committee has already had the opportunity to express the view [citations to views adopted by the Committee in 1987 have been omitted] that article 26 of the Covenant is applicable also to social security legislation. It reiterates that article 26 does not of itself contain any obligation with regard to the matters that may be provided for by legislation. Thus it does not, for example, require any State to enact pension legislation. However, when it is adopted, then such legislation must comply with article 26 of the Covenant.

7.3 The Committee reiterates its constant jurisprudence that the right to equality before the law and to the equal protection of the law without any discrimination does not make all differences of treatment discriminatory.

A differentiation based on reasonable and objective criteria does not amount to prohibited discrimination within the meaning of article 26.

7.4 In determining whether the Austrian Pension Act, as applied to the author, entailed a differentiation based on unreasonable or unobjective criteria, the Committee notes that the Austrian family law imposes equal rights and duties on both spouses, with regard to their income and mutual maintenance. The Pension Act, as amended on 22 October 1985, however, provides for full pension benefits to widowers only if they have no other source of income; the income requirement does not apply to widows. In the context of said Act, widowers will only be entitled to full pension benefits on equal footing with widows as of 1 January 1995. This in fact means that men and women, whose social circumstances are similar, are being treated differently, merely on the basis of sex. Such a differentiation is not reasonable, as is implicitly acknowledged by the State party when it points out that the ultimate goal of the legislation is to achieve full equality between men and women in 1995.

8. The Human Rights Committee, acting under article 5, paragraph 4, of the Optional Protocol, is of the view that the application of the Austrian Pension Act in respect of the author after 10 March 1988, the date of entry into force of the Optional Protocol for Austria, made him a victim of a violation of article 26 of the International Covenant on Civil and Political Rights, because he, as a widower, was denied full pension on equal footing with widows.

9. The Committee notes with appreciation that the State party has taken steps to remove the discriminatory provisions of the Pension Act as of 1995. Notwithstanding these steps, the Committee is of the View that the State party should offer Mr. Dietmar Pauger an appropriate remedy.

10. The Committee wishes to receive information, within 90 days, on any relevant measures taken by the State party in respect of the Committee's views.

TOONEN V. AUSTRALIA

Communication No. 488/1992, Human Rights Committee
Views of Committee, March 31, 1994.
1 Int. Hum. Rts. Reports 97 (No. 3, 1994)

[The author of this communication was an Australian citizen resident in the Australian state of Tasmania, and a leading member of the Tasmanian Gay Law Reform Group. He claimed that he was a victim of violations by Australia of Articles 2(1), 17 and 26 of the ICCPR. Article 2(1) provides that each state party will ensure to all individuals the 'rights recognized in the present Covenant, without distinction of any kind, such as . . . sex . . . political or other opinion . . . or other status.' Article 17 provides that no one shall be subjected 'to arbitrary or unlawful interference with his privacy. . . .' Article

26 provides that all persons 'are equal before the law and are entitled without any discrimination to the equal protection of the law.' Forbidden discriminations are similar to those in Article 2(1).

Toonen challenged in particular two provisions of the Tasmanian Criminal Code (Australia being internationally responsible for acts of a component state within its federal structure) which made criminal 'various forms of sexual conduct between men, including all forms of sexual contacts between consenting adult homosexual men in private.' The Tasmanian police had not charged anyone with violations of these statutes, such as 'intercourse against nature,' but there remained a threat of enforcement. Moreover, the author alleged that the criminalization of homosexuality had nourished prejudice and 'created the conditions for discrimination in employment, constant stigmatization, vilification, threats of physical violence and the violation of basic democratic rights.' Tasmania alone among Australian jurisdictions continued to have such laws in effect, and the Federal Government's position before the Human Rights Committee was critical of those laws.

Excerpts from the Committee's view follow, starting with some of Australia's observations.]

6.5 The state party does not accept the argument of the Tasmanian authorities that the retention of the challenged provisions is partly motivated by a concern to protect Tasmania from the spread of HIV/AIDS, and that the laws are justified on public health and moral grounds. This assessment in fact goes against the Australian Government's National HIV/AIDS Strategy, which emphasizes that laws criminalizing homosexual activity obstruct public health programmes promoting safer sex. The State party further disagrees with the Tasmanian authorities' contention that the laws are justified on moral grounds, noting that moral issues were not at issue when article 17 of the Covenant was drafted.

6.6 None the less, the State party cautions that the formulation of article 17 allows for *some* infringement of the right to privacy if there are reasonable grounds, and that domestic social mores may be relevant to the reasonableness of an interference with privacy. The State party observes that while laws penalizing homosexual activity existed in the past in other Australian states, they have since been repealed with the exception of Tasmania. Furthermore, discrimination on the basis of homosexuality or sexuality is unlawful in three of six Australian states and the two self-governing internal Australian territories. The Federal Government has declared sexual preference to be a ground of discrimination that may be invoked under ILO Convention No. 111 (Discrimination in Employment or Occupation Convention), and created a mechanism through which complaints about discrimination in employment on the basis of sexual preference may be considered by the Australian Human Rights and Equal Opportunity Commission.

6.7 On the basis of the above, the State party contends that there is now a general Australian acceptance that no individual should be disadvantaged on the basis of his or her sexual orientation. Given the legal and social situation in all of Australia except Tasmania, the State party acknowledges that a complete prohibition on sexual activity between men is unnecessary to sustain the moral fabric of Australian society. On balance, the State party 'does not seek to claim that the challenged laws are based on reasonable and objective criteria'.

. . .

Examination of the merits:

. . .

8.2 Inasmuch as article 17 is concerned, it is undisputed that adult consensual sexual activity in private is covered by the concept of 'privacy', and that Mr. Toonen is actually and currently affected by the continued existence of the Tasmanian laws. . . .

8.3 The prohibition against private homosexual behaviour is provided for by law, namely, Sections 122 and 123 of the Tasmanian Criminal Code. As to whether it may be deemed arbitrary, the Committee recalls that pursuant to its General Comment 16 [32] on article 17, the 'introduction of the concept of arbitrariness is intended to guarantee that even interference provided for by the law should be in accordance with the provisions, aims and objectives of the Covenant and should be, in any event, reasonable in the circumstances'. The Committee interprets the requirement of reasonableness to imply that any interference with privacy must be proportional to the end sought and be necessary in the circumstances of any given case.

. . .

8.5 As far as the public health argument of the Tasmanian authorities is concerned, the Committee notes that the criminalization of homosexual practices cannot be considered a reasonable means or proportionate measure to achieve the aim of preventing the spread of AIDS/HIV. . . .

8.6 The Committee cannot accept either that for the purposes of article 17 of the Covenant, moral issues are exclusively a matter of domestic concern, as this would open the door to withdrawing from the Committee's scrutiny a potentially large number of statutes interfering with privacy. It further notes that with the exception of Tasmania, all laws criminalizing homosexuality have been repealed throughout Australia and that, even in Tasmania, it is apparent that there is no consensus as to whether Sections 122 and 123 should not also be repealed. Considering further that these provisions are not currently enforced, which implies that they are not deemed essential to the protection of morals in Tasmania, the Committee concludes that the provisions do not meet the 'reasonableness' test in the circumstances of the case, and that they arbitrarily interfere with Mr. Toonen's right under article 17, paragraph 1.

8.7 The State party has sought the Committee's guidance as to whether sexual orientation may be considered an 'other status' for the purposes of article

26. The same issue could arise under article 2, paragraph 1, of the Covenant. The Committee confines itself to noting, however, that in its view the reference to 'sex' in articles 2, paragraph 1, and 26 is to be taken as including sexual orientation.

9. The Human Rights Committee, acting under article 5, paragraph 4, of the Optional Protocol to the International Covenant on Civil and Political Rights, is of the view that the facts before it reveal a violation of articles 17, paragraph 1, *juncto* 2, paragraph 1, of the Covenant.

10. Under article 2 (3) (a) of the Covenant, the author, victim of a violation of articles 17, paragraph 1, *juncto* 2, paragraph 1, of the Covenant, is entitled to a remedy. In the opinion of the Committee, an effective remedy would be the repeal of Sections 122 (a), (c) and 123 of the Tasmanian Criminal Code.

11. Since the Committee has found a violation of Mr. Toonen's rights under articles 17 (1) and 2 (1) of the Covenant requiring the repeal of the offending law, the Committee does not consider it necessary to consider whether there has also been a violation of article 26 of the Covenant.

12. The Committee would wish to receive, within 90 days of the date of the transmittal of its views, information from the State party on the measures taken to give effect to the views.

QUESTIONS

1. Suppose that a state party to the Optional Protocol makes criminal the sale or use of contraceptives. Under the Committee's view in *Toonen*, are you prepared to say that the statute violates the ICCPR? If you need more information, what kind of information? Suppose that only sale and not use were banned.

2. Should the Committee resolve the problems raised in Question (1) by issuing a revised General Comment interpreting Article 17 that deals specifically with these problems?

3. The United States is a party to the ICCPR (though not to the Optional Protocol). Is the United States in violation of the Covenant because the laws of certain states within the U.S. make sodomy between consenting adults criminal? What are the obligations of the United States under Article 2(1) of the Covenant? If the United States is arguably in violation of the Covenant, how, to what institution, might an individual adversely affected by the sodomy statute make that claim?

COMMENT ON EXHAUSTION OF REMEDIES

One recurrent problem before the ICCPR Committee involves determination of the admissibility of communications with respect to the requirement in

Article 5(2)(b) of the Optional Protocol that '[t]he individual has exhausted all available domestic remedies.' That article further provides: 'This shall not be the rule where the application of the remedies is unreasonably prolonged.'

In *Miango v. Zaire*, Communication No. 194/1985, UN Doc. A/43/40 (1988), Supp. No. 40 at 218, the author alleged the torture and killing by the military of the author's brother. The communication stated that in Zaire only a military tribunal could deal with cases involving soldiers, and the author despaired of any hope that the case of his brother's death would be properly investigated. The Committee, following its normal procedures, transmitted the communication to the state and requested information. It received no submission from Zaire, and then found the communication admissible. In expressing its view (decision of April 2, 1986) that Articles 6 and 7(1) of the ICCPR had been violated, the Committee observed:

> 9. In formulating its views, the Human Rights Committee also takes into account the failure of the State party to furnish any information and clarifications. It is implicit in article 4, paragraph 2, of the Optional Protocol that the State party has the duty to investigate in good faith all allegations of violations of the Covenant made against it and its authorities, and to furnish to the Committee the information available to it. The Committee notes with concern that, despite its repeated requests and reminders and despite the State party's obligation under article 4, paragraph 2, of the Optional Protocol, no explanations or statements clarifying the matter have been received from the State party in the present case. In the circumstances, due weight must be given to the author's allegations.

In *D.B.-B v. Zaire*, Communication No. 463/1991, UN Doc. A/47/40 (1994), Supp. No. 40 at 432, the author fled Zaire and was granted refugee status in Switzerland after allegedly witnessing a slaughter of some 100–150 students by the security police. He described several violations of his rights under the Covenant. The author wrote to the Ministry for Citizens' Rights and Freedoms to complain about what he had witnessed, but no action was taken in response to his complaint. The Committee (in its meeting of Nov. 8, 1991) found the communication inadmissible. It stated:

> 4.2 With regard to the requirement of the exhaustion of domestic remedies, the Committee observes that the author, by letter of 7 March 1991, filed a complaint to the Zairian Ministry for Citizens' Rights and Freedoms, and that he has not as yet received any reply. It is, however, a well established principle that a complainant must display reasonable diligence in the pursuit of available domestic remedies. In the instant case, the author has not shown the existence of circumstances which would prevent him from further pursuing the application of domestic remedies in the case.

Consider in this connection the following excerpts from a report by the Lawyers Committee for Human Rights, *Zaire: Repression as Policy* (1990):

In recent years, the Zairian judiciary has been reduced to a weak and compliant participant in a system of widespread abuse. The goal of judicial independence, which was never reached in Zaire, has been all but destroyed by executive interference, corruption and the requirements of MPR party discipline. . . .

. . . In combating human rights abuses, these problems are compounded by the inability of the Judicial Council to control the excesses of security forces and subject them to the existing body of laws.

According to a number of lawyers and judges interviewed, the executive dictates the outcome of political trials which come before the State Security Court. . . .

NOTE

The record of compliance by states with views rendered by the Committee under the Optional Protocol is patchy. In a document submitted to the 1993 World Conference on Human Rights in Vienna,[3] the Committee expressed its concern and indicated several paths toward dealing effectively with this problem.

The document stressed the significance of the follow-up procedures by the Special Rapporteur, p. 536, *supra*, which were 'not only compatible' with the Committee's mandate but 'essential if the Committee is expected to discharge' its responsibilities. Since the practice began, the Committee had received requested follow-up information in respect of 71 views, but no information had been received in respect of 37 views. Moreover, over half of the replies received were unsatisfactory—terse, vague, unresponsive, and so on.

The Committee endorsed a recommendation issuing from a recent meeting of chairs of human rights treaty bodies, to the effect that '[v]iews and recommendations expressed by the treaty bodies in relation to individual communications should be fully respected.' The Committee believed that calling further on states 'to accept such views as binding would be another, desirable, step.' It suggested that addition of a new paragraph to article 5 of the Optional Procool could be envisaged, to the effect that '[s]tates parties undertake to comply with the Committee's Views under the Optional Protocol.' The recommendations went further:

Furthermore, it could be envisaged to broaden the competence of the Special Rapporteur on the follow-up on Views to include a fact-finding mandate. Thus, the Special Rapporteur could be endowed with the authority to make inquiries *in situ*, as to what a given State party has done, or failed to do, to give effect to the Committee's Views.

[3] Follow-up on Views adopted under the Optional Protocol to the International Covenant on Civil and Political Rights, UN Doc. A/48/40 (1993), Part I, Annex X, at 222.

QUESTIONS

1. Compare the function of the Committee in deciding about communications with the role of a national court in deciding a dispute before it. What similarities, what differences? Could this function of the Committee be accurately labelled 'judicial' in character? If not, would you urge such changes in process and powers as would make the proceedings 'judicial'?

2. The four principal views of the Committee appearing above involve Colombia, Zambia, Austria and Australia. In which of these cases are the probabilities greatest that the views would be complied with? In what situations would you tend to assume non-compliance? What do your answers suggest about the role of the Optional Protocol in addressing and correcting serious human rights violations around the world?

3. Be attentive to the fact-finding processes at work in deciding how to resolve a communication. Facts are contested in most of the cases, and often the state is simply silent, or issues a flat denial of the charges, or advances a radically different version of the facts. The question arises of how the Committee should resolve this dilemma of fact-finding. The issue is present in the preceding communications involving Colombia, Zambia and (with respect to the exhaustion of local remedies and issue of admissibility) Zaire.

(a) Who seems to bear the burden of proof (persuasion)? Does that burden 'shift' to the state if certain conditions are met? What are the consequences of such a shift?

(b) Do presumptions arise when, say, a state is silent, or when it issues only a general denial in situations where the relevant facts or documents are in its exclusive possession?

(c) Do you think that the Committee should issue a General Comment on this problem of evidence and proof—that is, how it will handle situations such as those in the communications referred to above? What should the General Comment say?

4. A member of the Committee, after observing that the Optional Protocol says nothing about the legal significance of the Committee's Views, makes the following argument:[4]

> [A]rticle 2(3) of the Covenant provides that, when a violation of an individual right occurs, the state is under a legal obligation to give the victim an effective and enforceable remedy. This provision and the Optional Protocol tend to achieve the same goal, although at different levels, i.e. to provide for an international guarantee in the case of a violation: article 2 sets forth a legal obligation for the state, the Optional Protocol provides for machinery to establish the existence of a violation

[4] Fausto Pocar, Legal Value of the Human Rights Committee's Views, 1991–92 Canadian Human Rights Yearbook 119.

> . . . [A]nother point has to be made. Since in the preamble to the Optional Protocol, this international instrument is defined as a means of implementation of the Covenant, the States parties, it may be assumed, are under an obligation to cooperate with the Committee when a violation is brought before it. Such cooperation cannot be considered as confined to the procedure leading to the adoption of the views of the Committee, but must logically include the views themselves.

Do you agree? As a member of the Committee, would you support the issuance of a General Comment to the effect that the Committee's Views under the Optional Protocol constitute binding obligations that states must obey?

5. 'Each of the Committee's general functions—considering state reports, issuing General Comments, expressing its views about communications—leaves separately much to be desired in terms of the goals of effective implementation and development, let alone enforcement, of the ICCPR. But a look at these three functions as a system, in their interrelationships and totality, suggests that the Committee is serving such goals well.'

Comment.

5. EVALUATION

TORKEL OPSAHL, THE HUMAN RIGHTS COMMITTEE

in Philip Alston (ed.), The United Nations and Human Rights 369 (1992), at 39b.

. . .

. . . [T]he best way to describe the functions and role of the [ICCPR] Committee has been a matter of controversy. Is it an organ for promotion and co-operation, or for protection and supervision? Is it quasi-judicial or conciliatory? The Covenant, avoiding any such description of the Committee, states laconically that it 'shall carry out the functions hereinafter provided'. The Committee is careful in its manner of presenting itself so as not to provoke underlying disagreements. The field is, in other words, open for analysis and opinion, and above all for development through practice.

. . .[T]he Committee cannot and does not act *ex officio* in any way, and does not take initiatives outside those functions. It does not intervene in 'situations', however acute or frustrating, unless its functions under the Covenant and Protocol justify the action. What kinds of action are possible within these limits remains a somewhat controversial question.

It is sometimes argued that the Committee cannot do anything which is not explicitly stated in the Covenant and Optional Protocol. But this is misleading. There is an obvious need, confirmed by practice, to fill in many gaps created by the wording of the Covenant and the Optional Protocol. First, the question arises as to the Committee's 'implied' or 'inherent' powers. At the very least, the functions explicitly given to it may imply certain steps which are not expressly mentioned. On this basis, the Committee has often taken such action as has seemed useful for its work.

. . .

As shown above, the Committee has never fully exercised its mandate. The main reasons have been political, administrative, and financial constraints as well as internal disagreements combined with the consequences of the consensus principle. From the Committee's point of view a change of mandate is not foreseeable. The Committee's mandate follows from the Covenant and Optional Protocol, which are not likely to be amended during this century. Extending the Committee's functions through new instruments is a possibility, as shown in practice by the Second Optional Protocol.

The only development which could affect the Committee's general mandate, and which is not wholly unlikely, would be some modification in its present functions, including necessary co-ordination with other bodies. Since these functions are described in terms open to interpretation, the limits of the mandate could be determined in a broad and dynamic manner through the adoption of liberal positions by the Committee itself. This has happened, for example, with the adoption of a rule that the Committee could determine under Article 40 that a State Party had not discharged its obligations, and with the Committee's decision to publish its views and other decisions under the Optional Protocol. The practical implementation of its mandate can be much developed through the active and innovative use of guidelines and specific requests to States Parties. Resistance to the Committee's own interpretation of its mandate is more likely if this interpretation assumes additional or implied duties for States Parties than if it only concerns action that the Committee may take on its own or with their co-operation and consent. . . .

Most of the outside constraints have to be accepted by the Committee, like its own composition and mandate. But is the Committee doomed . . . to 'degenerate into an empty political forum, a familiar fate indeed of United Nations bodies' or will it be able to liberate itself to some extent from the United Nations?

The answer depends to a large degree on the members themselves. They risk not being renominated or re-elected if they antagonize their environment, especially the States Parties. The Committee as such has no way to confront or break away from the United Nations. But this does not mean that it must be silent about its needs and how it wishes to develop.

NOTE

One of the perplexing questions about every international human rights institution (as well as about many national institutions) concerns the effect that it has had or may have on norms and practice in given states, or more grandly on the human rights movement as a whole. That question has several components, including:

How does one go about assessing, let alone measuring, that effect?

What would constitute proof or at least reasonably persuasive evidence that the ICCPR Committee has influenced the course of events in a given country, perhaps (a) through the experience of that country in formulating its periodic report and through the discussion of that report with the Committee, (b) by the thrust of a General Comment bearing on issues in that country, or (c) by the 'views' of the Committee with respect to communications sent to it by nationals of that country?

How would one assess the influence of the Committee's work on other human rights organs, perhaps the Commission on Human Rights or the European Court of Human Rights? Or assess its influence on the interpretation of particular human rights norms by government officials, or by NGOs?

At the end of his comprehensive, scholarly account of the work of the Human Rights Committee during its initial fifteen years, Dominic McGoldrick writes in *The Human Rights Committee* (1991), at 504:

> It is very difficult to provide positive evidence that the existence of the Covenant and the work of the HRC is having any concrete and positive effect on the human rights position in the States parties. However, many of the State representatives that have appeared before the HRC have stated that the Covenant and the work of the HRC have played an important role at the national level. It would be immensely helpful if the HRC could catalogue and reproduce those claims together with any more specific evidence of wholesale or partial national reviews of the implementation of the Covenant and of account being taken of the Covenant and the HRC, for example, in legislative assemblies, executive decision making, judicial or administrative decisions.

NOTE AND QUESTION

In light of the preceding materials on the ICCPR Committee, and of the analysis of the UN Commission on Human Rights in Chapter 7, you should be in a position to compare and assess the composition, powers, functions and goals of each organ in relation to the other. That comparison should highlight the choices made in the architecture of these novel international institutions, raise the questions of why they were designed and why they function so differently, and lead us to consider possible revisions.

One threshold question must be considered. Should we analyze the ICCPR Committee solely in terms of its actual or potential role *within* the Covenant, *independently* of all other human rights treaties and organs? If so, we are concluding that each treaty and treaty organ can more usefully be understood and evaluated within its distinctive context. Or should an evaluation take account of the place of the Committee and of the ICCPR itself within a more comprehensive framework of (at least the universal parts of) the human rights movement? That larger framework would include the Charter and its organs, as well as all universal human rights treaties and their organs. Within it, we would inquire into the *complementary* nature of the functions that different organs under different treaties should fill, with the goal of thereby maximizing the effectiveness of the human rights movement as a whole. What consequences in terms of practical proposals for reform might flow from one or the other framework for analysis?

A comparison between the ICCPR Committee and UN Commission can inquire into what the differences between these two institutions are, how they came about, what they say about the institutions and the political frameworks within which they were designed and now function, whether the differences make sense today, and, if not, how the institutions should be changed. The characteristics of the two organs that can be usefully compared include the following:

- which states participate in or are affected by the work of each organ?
- how is membership in each organ determined?
- what powers does each organ have to investigate?
- what powers does each organ have to evaluate, judge, condemn?
- what are the voting techniques and patterns?
- what types of violations are examined?
- are there marked political influences on the agenda and how are they exercised?
- what political pressures, what sanctions, can or does each organ impose on states?
- in what ways, if at all, have the mandates and powers of each organ evolved?
- what role have NGOs in each organ?

Consider whether the answers that you bring to these questions would justify a characterization of the ICCPR Committee as the more 'legal' body, and of the UN Commission as the more 'political'. In responding, be clear about what meanings you are attributing to the two vital terms.

And consider the following excerpt from Dominic McGoldrick, *The Human Rights Committee* (1991), at 54:

> . . .[O]n a number of occasions members of the HRC have indicated their perceptions of the institutional character and nature of the HRC and it is instructive to note some of these.
>
> 2.21 Mr Uribe-Vargas described it as a body whose work was of a 'judicial nature'. Mr Mora-Rojas said that 'The Committee was quite different in nature from other bodies and, even though it was not a court or a tribunal, it did hear testimony and had evidence presented to it'. Mr Tomuschat has commented that, 'The Committee was . . . ruled by the Covenant and while it was true that members were not judges they had the task of applying the provisions laid down in the Covenant and therefore had to exercise judgement. It was the duty of the Committee to ensure that States parties fulfilled their obligations under the Covenant.' Mr Tomuschat has also said that 'The Committee was not an international court but was similar to one in certain respects, particularly in regard to its obligation to be guarded by exclusively legal criteria—which rightly distinguished it from a political body'. Mr Ermacora was concerned that the Committee should avoid giving the impression that it was 'a sort of advisory service, or had technical, assistance functions, whereas in fact its activities were based on legally binding instruments, with all the attendant consequences that that entailed'. Mr Aguilar commented that the Committee 'was not a judicial body' and 'its role was not to find fault'. Mr Bouziri commented that the 'Committee was not a court of law'. Mr Pocar commented that the Committee's function 'was not to judge and then either to condemn or congratulate States parties'. Mr Graefrath 'did not share the view that the work of the Committee could be compared to that of a court . . . Unlike a court the Committee was not required to make judgements, but simply to consider and comment on reports and to act as a conciliatory body in dealing with complaints and communications'. Mr Opsahl has described the HRC as the 'executive organ' of the Covenant. Emphasis is often put on the HRC's role as a promoter, monitor, or supervisory body with respect to improved human rights performances. However, some members have expressed doubts as to whether the HRC can properly be called a 'supervisory body' or a 'parent organ'. Finally, Mr Suy (UN Legal Advisor) believed the HRC to be 'neither a legislative nor a judicial body but that every expert body was sui generis', and Mr Herndl, former Under Secretary-General of the United Nations, recently described the HRC as 'the guardian of the Covenant'.

B. COMPARISONS: OTHER HUMAN RIGHTS TREATY REGIMES

Chapter 9 has examined the ICCPR Committee as a case study. We here make comparisons with the other UN human rights treaty bodies. Not every universal human rights treaty—for example, the Genocide Convention—has created a special organ to implement the treaty or monitor state conduct. In addition to the ICCPR Committee, there are now five other treaty bodies. They are:

- the Committee on Economic, Social and Cultural Rights (CESCR), established to monitor compliance with the ICESCR (and considered in Chapter 5);
- the Committee on the Elimination of Racial Discrimination (CERD), established under the International Convention on the Elimination of All Forms of Racial Discrimination;
- the Committee on the Elimination of Discrimination against Women (CEDAW), established under the Convention on the Elimination of All Forms of Discrimination against Women;
- the Committee against Torture (CAT), established under the Convention against Torture and Other Cruel, Inhuman or Degrading Treatment or Punishment; and
- the Committee on the Rights of the Child (CRC), established under the Convention on the Rights of the Child.

COMMENT ON THE REPORTING AND COMPLAINTS PROCEDURES FOR THE SIX HUMAN RIGHTS TREATY BODIES

Although their mandates differ somewhat, reflecting both the respective political climates in which they were adopted and their subject-matters, the five treaty bodies listed above function essentially in ways similar to the ICCPR Committee. The following descriptions of the reporting and complaints procedures under the treaties note the salient differences among them, suggest common problems before them, and sketch current proposals for reform within specific treaties or for the UN human rights treaty system as a whole.

States' Reports

The principal activity of all of the treaty bodies remains the consideration of states parties' reports. For the most part, states must report within two years of becoming a party to the treaty and subsequently at five-year intervals. On occasion, urgent *ad hoc* reports are sought by some of the committees. Initially, the reporting function was envisaged by most governments to be of a relatively *pro forma* nature, but it has since evolved into a much more demanding exercise.

Three examples illustrate how these committees initiated changes in procedures (some of which we have seen with respect to the ICCPR Committee) that have strengthened the reporting procedures. Perhaps the most significant is the practice, initiated by CERD in 1972, of 'inviting' states to send representatives to respond to questions by Committee members when the report by that state is being considered. This procedure was not foreseen in the treaty but it has since been adopted by all treaty bodies and is now an indispensable and fully accepted part of the process. A second example, initiated by the CESCR in 1991, is the practice of examining the situation in countries that have failed to report, even if no government-provided information is available and no representative is present. This practice removes the pre-existing reward to governments that ratified and never reported, thus avoiding committee scrutiny. The third example is the move from a procedure that was restricted to the examination of 'official' information (i.e. produced by the government or an intergovernmental organization) to one that takes full account of all sources and invites NGOs to provide written, and in some cases oral, information.

Another indication of the extent to which the reporting procedures have been broadened in scope appears in the following statement of the functions of reporting. It was adopted in relation to economic rights but has been generally accepted by other treaty bodies. The functions are:

> [1] . . . to ensure that a comprehensive review is undertaken with respect to national legislation, administrative rules and procedures, and practices in an effort to ensure the fullest possible conformity with the Convenant . . .
> [2] . . . to ensure that the State party monitors the actual situation with respect to each of the rights on a regular basis and is thus aware of the extent to which the various rights are, or are not, being enjoyed by all individuals within its territory or under its jurisdiction . . .
> [3] . . . the principal value of . . . an overview [of the situation in the state] is to provide the basis for the elaboration of clearly stated and carefully targeted policies, including the establishment of priorities which reflect the provisions of the Covenant. . . .
> [4] . . . to facilitate public scrutiny of government policies with respect to economic, social and cultural rights and to encourage the involvement of the various economic, social and cultural sectors of society in the formulation, implementation and review of the relevant policies . . .
> [5] . . . to provide a basis on which the State party itself, as well as the Committee, can effectively evaluate the extent to which progress has been made towards the realization of the obligations contained in the Covenant . . .
> [6] . . . to enable the State party itself to develop a better understanding of the problems and shortcomings [existing] . . .
> [7] . . . to develop a better understanding of the common problems faced by States. . . .[5]

[5] Committee on Economic, Social and Cultural Rights, General Comment No. 1 (1989), UN Doc. HRI/GEN/1/Rev.1 (1994), p. 43.

Whatever the achievements of the treaty system with respect to reporting, there remains scope for great improvement. Critics have pointed to the timidity of the committees in general, and especially to their excessive tolerance of the delinquencies of states. One critic has said of governments that:

> In large numbers they fail to produce timely reports, do not engage in reform activities in the course of producing reports, author inadequate reports, send uninformed representatives to the examination of reports by the treaty bodies, fail to respond to questions during the examinations, discourage greater media attention of the examination of reports, fail to disseminate reports and the results of the examinations within the State, elect non-independent/government employees to treaty body membership, fail to object to reservations, and fail to challenge reservations by additional means.[6]

Another issue about reporting has become pressing, the proliferation of obligations under the different treaty regimes. A state that is party to many or all of these six treaties is subject to distinct duties to report under each, whatever the overlap between, say, parts of its report to the ICCPR Committee and the report to CAT. Observers have questioned whether the system will eventually become—or indeed has already become—unsustainable because of the number of and overlap among these obligations. States are subject to the complex bureaucratic demands associated with a regular dialogue with as many as six separate committees—for example, ensuring normative and policy consistency among treaty bodies, thus risking dilution of already inadequate staffing and financial resources. Moreover, NGOs with sparse resources must deal with the difficulties of monitoring states' compliance with each of these treaties and seeking to give adequate publicity to the work of each.

A report commissioned by the General Assembly has suggested three 'long-term options for reducing reporting burdens . . . : (i) reducing the number of treaty bodies and hence the number of reports required; (ii) encouraging States to produce a single "global" report to be submitted to all relevant treaty bodies; and (iii) replacing the requirement of comprehensive periodic reports with specifically-tailored reports.'[7] Although there has been significant support, at least in the abstract, for reform, many observers have expressed concern about any attempt to overhaul the entire reporting system. Depending on perspective, they fear that proposals for such systematic reform might either be used to mask a significant downgrading of reporting or, alternatively, to strengthen the system of reporting to a politically unacceptable extent.

[6] Anne Bayefsky, Making the Human Rights Treaties Work, in Louis Henkin and J. L. Hargrove (eds.), Human Rights: An Agenda for the Next Century 222, 239 (1994).

[7] Philip Alston, Interim Report on Study on Enhancing the Long-Term Effectiveness of the United Nations Human Rights Treaty Regime, UN Doc. A/CONF.157/PC/62/Add.11/Rev.1 (1993), para. 27.

Complaints Procedures

Initially, an interstate complaints procedure was considered to be an important element in an effective monitoring system. Thus the first of the treaties adopted—CERD—provides for any state party to bring a complaint against another. By the following year, however, when the ICCPR was adopted, a comparable procedure was made dependent upon the specific, separate acceptance by a state of Article 41 of that treaty, setting forth that procedure. The CAT has a similar provision. These procedures are determinedly 'statist' in nature in that they make no provision for involvement by NGOs or individuals, the emphasis is upon achieving a 'friendly solution', and the procedure is undertaken in closed meetings. The ICCPR also provides for the establishment of an *ad hoc* Conciliation Commission, in the event that an amicable solution cannot be found.

Despite the availability of the interstate complaint procedure to all of the 143 parties to CERD (as of September 1995) and the acceptance of Article 41 of the ICCPR by 45 of its 131 states parties (whereas the CAT procedure under Article 21 has been accepted by only one state), no interstate complaint has ever been brought under any of the UN treaty-based procedures. Compare this record with that of the interstate procedure established under the European Human Rights Convention, p. 594, *infra*.

Analogous procedures for individual complaints under the Optional Protocol of the ICCPR are also provided for under the CAT and CERD. Although this chapter has demonstrated the significance of that procedure under the ICCPR, the similar procedures under the other two treaties have generated very few complaints. Over all their years, CAT has considered only 15 complaints while CERD has only dealt with five. Note the dramatic comparison with the statistics under the Optional Protocol, p. 536, *supra*.

One explanation for the discrepancy in the number of complaints received is that the CAT and CERD procedures have been accepted by only 35 and 21 states respectively. Another is that the ICCPR covers much of the same territory as the other two specialized treaties and that its procedures are much better known to NGOs and other potential complainants. This relative under-utilisation notwithstanding, proposals have been endorsed by the CESCR in 1994 and CEDAW in 1995 for optional protocols to establish individual complaints procedures. The proposals are not yet in final form, however, and governmental reactions remain ambivalent.

Investigation Procedures

The Convention against Torture is distinctive among the six treaties in providing for a more intrusive procedure under which one or more of its members may be asked to undertake a confidential inquiry and to report urgently to the CAT Committee in cases in which it 'receives reliable information which appears to it to contain well-founded indications that torture is being systematically practised in the territory of a State Party' (Art. 20). This procedure applies automatically to all states parties unless, at the time of ratification, they specifically exclude its application. With the agreement of the state concerned, an on-site visit may be undertaken, although all of the relevant proceedings are confidential. The Committee may, however, decide to include a summary account of the results in its annual report. This occurred in relation to Turkey in 1993, and other countries have apparently been under consideration subsequently. UN Doc. A/48/44/Add.1 (1993).

Both CERD and the CESCR have also undertaken on-site missions in order to pursue their dialogue with certain states (compare the proposal for country visits by ICCPR members, p. 520, *supra*). But these missions have been undertaken on the basis of the general monitoring provisions of the relevant treaties, rather than for specific investigations. A proposal has been made by CEDAW to include a specifically mandated on-site fact-finding procedure in relation to that Convention. In addition, an optional protocol to CAT is currently being drafted which would provide for a system of both scheduled and unscheduled visits by that Committee to places of detention or other places where torture or cruel, inhuman or degrading treatment or punishment may be occurring. UN Doc. E/CN.4/1995738.

QUESTIONS

1. Why do you think there has never been an interstate complaint brought under the UN treaty procedures, even in cases of gross violations by relatively powerless states?

2. What are the arguments for and against a consolidation of the treaty system that would result in only one or two treaty bodies dealing with all of the rights covered by the six treaties? Would you support or oppose such a proposal?

3. How many of the seven functions attributed to the reporting system that are noted above are satisfied by the reporting under the ICCPR?

4. From what you have seen of the workings of the Optional Protocol under the ICCPR, do you believe that the adoption of individual complaints procedures—of course by separate protocols requiring ratifications—by each treaty body is sufficiently important to justify the bureaucratic and political energies involved in such an undertaking?

5. Assume that the UN Commission on Human Rights is working on a draft of a proposed Convention on Religious Intolerance that must ultimately be approved by the General Assembly and submitted to states for ratification. The draft Convention provides that everyone has the right to freedom of conscience and religion; that no one shall be subject to coercion impairing freedom to hold a belief of one's choice; that freedom to manifest religious belief shall be limited

only by requirements of public order, health or morals; and that there shall be no discrimination on the basis of religious beliefs. States are obligated to take appropriate measures to eliminate discrimination and to combat intolerance on grounds of religion.

You are on the staff of a member of the Commission, who has asked you to consider what kind of organ (committee) this proposed Convention should create—how the committee is to be constituted, what kinds of members it should have, what functions and powers, and so on. Taking into account whatever special needs this Convention may have, indicate what kind of committee you would propose.

10

Regional Arrangements

Part C on international institutions and processes concludes with a look at the world's major regional human rights systems. After an introductory section comparing advantages and disadvantages of a regional rather than universal system—in fact, in the case of human rights, of a regional system together with a universal system, rather than *only* a universal system—this chapter turns to the regional arrangements in Europe, the Americas and Africa.

The materials describe the institutional structure and international processes of the regional systems, and invite comparisons among them as well as with the UN system. The European and Inter-American systems have distinctive institutions and processes; the African system has distinctive norms. Thus the regional arrangements add in important ways to knowledge derived from the UN and from UN-related treaties like the ICCPR about possible avenues toward control of human rights abuses.

Rather than attempt comprehensive analyses of the three regions, Chapter 10 concentrates on one important aspect of each. The remarkable feature of the European system is its productive and effective Court; the materials illustrate the work of the Court through several revealing decisions. In the Inter-American system, the Commission has been a very significant organ. We examine it at work in one field of increasing importance for Latin America and for global human rights, institutionalizing the practices of democratic government. The African system is the least developed institutionally; the materials concentrate on a distinctive aspect of its norms, a stress on duties as well as rights.

A. COMPARISON OF UNIVERSAL AND REGIONAL SYSTEMS

The relationship between 'universal' (meaning, in this context, United Nations-sponsored) and regional human rights arrangements is a complex one. In addition to the three major systems, there is a largely dormant Arab system and a proposal for the creation of an Asian regional system. Although Chapter VIII of the United Nations Charter makes provision for regional arrangements in relation to peace and security, it is silent as to human rights cooperation at that level. Nevertheless, the Council of Europe moved as early as 1950 to adopt the European Convention on Human Rights. It was not until 1969 that the analogous American Convention was adopted. In the meantime, at least until the mid-1960s, the UN remained at best ambivalent about such developments. Vasak has noted some of the reasons for its hesitation:

> For a long time, regionalism in the matter of human rights was not popular at the United Nations: there was often a tendency to regard it as the expression of a breakaway movement, calling the universality of human rights into question. However, the continual postponements of work on the International Human Rights Covenants led the UN to rehabilitate, and to be less suspicious (less jealous, some would say) towards, regionalism in human rights, especially after the adoption of the Covenants in 1966.[1]

It was not coincidental then that the UN General Assembly began to contemplate the active encouragement of regional mechanisms only in 1966, when the two basic Covenants were finally adopted by the General Assembly. In 1977, it formally endorsed a new approach by appealing 'to States in areas where regional arrangements in the field of human rights do not yet exist to consider agreements with a view to the establishment within their respective regions of suitable regional machinery for the promotion and protection of human rights' (GA res. 32/127 (1977)). Four years later, the African Charter on Human and Peoples' Rights was adopted. Throughout this period the Communist states of Eastern Europe were strongly opposed to regional arrangements, and the Asian and Pacific countries generally argued that their region was much too heterogeneous to permit the creation of a regional mechanism.

Although there have been no significant new regional or sub-regional systems created since 1981 (despite the General Assembly's annual adoption of a resolution calling upon the relevant regions to act), the current climate is somewhat more hospitable than was previously the case. The desire of Eastern and Central European governments to become full partners in a united Europe, and even to join the European Union, has prompted many of them to join the European human rights system. Moreover, the transformation of the Organisation on Security and Co-operation in Europe (OSCE) from an East–West debating forum (or shouting match) into a complex set of arrangements designed to promote respect for a broadly defined range of human rights, has added another important dimension to regional human rights cooperation.

The successful example of the OSCE in promoting political cooperation, combined with the dramatically increased importance of the role of regional

[1] In Karel Vasak and P. Alston (eds.), The International Dimensions of Human Rights 451 (Vol. 2, 1982).

trading blocs—exemplified by the Single European Act and the Maastricht Treaty in Europe, and the North American Free Trade Agreement among the U.S., Canada and Mexico—and efforts to develop the Asian–Pacific Economic Cooperation (APEC) initiative, have given renewed impetus to regional co-operation arrangements. It is to be expected that this will, over time, have potentially important consequences in relation to human rights.

* * * * *

The following readings develop some comparisons between universal and regional systems, both in general and with respect to human rights. Several themes suggested by these materials recur throughout the chapter: what are the relations between these systems—for example, are there hierarchical 'controls' between them; in what ways is it preferable from the perspective of observance of human rights to have a regional system complement a universal one; how do we explain the different institutional structures and differences in norms among these system; what can we learn from these differences about effective architecture for intergovernmental human rights organizations?

INIS CLAUDE, SWORDS INTO PLOWSHARES

(4th ed., 1984), at 102.

Regionalism is sometimes put forward as an alternative to globalism, a superior substitute for the principle of universality. Emphasis is placed upon the bigness and heterogeneity of the wide world, and the conclusion is drawn that only within limited segments of the globe can we find the cultural foundations of common loyalties, the objective similarity of national problems, and the potential awareness of common interests which are necessary for the effective functioning of multilateral institutions. The world is too diverse and unwieldy; the distances—physical, economic, cultural, administrative, and psychological—between peoples at opposite ends of the earth are too formidable to permit development of a working sense of common involvement and joint responsibility. Within a region, on the other hand, adaptation of international solutions to real problems can be intelligently carried out, and commitments by states to each other can be confined to manageable proportions and sanctioned by clearly evident bonds of mutuality.

. . .

The advocacy of regionalism can be, and often is, as doctrinaire and as heedless of concrete realities as the passion for all-encompassing organization. It should be stressed that the suitability of regionalism depends in the first place upon the nature of the problem to be dealt with. Some problems of the modern world are international in the largest sense, and can be effectively treated only by global agencies. Others are characteristically regional, and lend themselves to solution by correspondingly delimited bodies. Still others are regional in nature, but require for their solution the mobilization of extra-regional resources.

. . .

The nature of a problem is significant not only for the determination of the most appropriate means of solution, but also for the measurement of the range of its impact. A problem may be regional in location, and susceptible of regional management, and yet have such important implications for the whole world that it becomes a fit subject for the concern of a general organization. The world-at-large cannot be disinterested in such 'regional' matters as the demographic problem in Asia or racialism in Southern Africa. Thus, the question of the ramifications of a problem as well as that of its intrinsic quality affects the choice between regional and universal approaches.

. . .

However, . . . [t]he world does not in fact break easily along neatly perforated lines. Rational regional divisions are difficult to establish, boundaries determined for one purpose are not necessarily appropriate for other purposes, and the most carefully chosen dividing lines have a perverse way of changing or coming to require change, and of overlapping. It is true that brave universalist experiments tend to give way to sober regionalist afterthoughts, but it is equally true that carefully cut regional patterns tend to lose their shape through persistent stretching in the direction of universalism. In a sense, the adoption of the universal approach is the line of least resistance, since it obviates the difficulties of defining regions and keeping them defined.

. . . Intraregional affinities may be offset by historically rooted intraregional animosities, and geographical proximity may pose dangers which states wish to diminish by escaping into universalism, rather than collaborative possibilities which they wish to exploit in regional privacy. While global organization may be too large, in that it may ask states to be concerned with matters beyond the limited horizons of their interests, regional organization may be too small, in that it may represent a dangerous form of confinement for local rivalries. Global stretching, in short, may be no worse than regional cramping.

. . .

In a very general sense, it may be contended that regional organizations are particularly suitable for the cultivation of intensive cooperation among states, while global organizations have special advantages for dealing with conflict among states. If the goal is the development of linkages that bind states together in increasingly intimate collaboration and perhaps culminate in their integration, it would appear to be essential to restrict the enterprise to a few carefully selected states.

. . . On the other hand, the capacity of an organization to promote the control and resolution of conflicts may be enhanced by its inclusiveness. A global agency is inherently better equipped than a regional one to provide the mediatorial services of governments and individuals whose disinterested attitude toward any given pair of disputants is likely to be regarded as credible, and its potency as a mobilizer of pressure upon states engaged in conflict is a function of the broad scope of its membership and jurisdiction. The European Community may be treated as the model for international organization as a workshop for collaboration, and the United Nations as the model for international organization as an arena for conflict.

REGIONALISM AND THE UNITED NATIONS

The atmosphere [during the 1945 Conference that drafted the UN Charter in] San Francisco was affected by the necessity of making the bow to regionalism which was demanded by those states that had already made heavy political investments in such arrangements as the Inter-American system, the Commonwealth, and the Arab League. . . .

. . .

The interaction between theoretical preference for universalism and political pressures for regionalism at San Francisco produced an ambiguous compromise. The finished Charter conferred general approval upon existing and anticipated regional organizations, but contained provisions indicating the purpose of making them serve as adjuncts to the United Nations and subjecting them in considerable measure to the direction and control of the central organization. The Charter reflected the premise that the United Nations should be supreme. . . .

. . .

REGIONAL PROMOTION AND PROTECTION OF HUMAN RIGHTS

Twenty-Eighth Report of the Commission to Study the Organization of Peace (1980), at 15.

. . . [In response to a 1968 UN study, the] Inter-American Commission on Human Rights supported those members of the Ad Hoc Study Group who favored regional human rights commissions, noting four grounds: (1) the existence of geographic, historical, and cultural bonds among States of a particular region; (2) the fact that recommendations of a regional organization may meet with less resistence than those of a global body; (3) the likelihood that publicity about human rights will be wider and more effective; and (4) the fact that there is less possibility of 'general, compromise formulae,' which in global bodies are more likely to be based on 'considerations of a political nature.'

. . .

Opposition to the establishment of regional human rights commissions has been expressed on numerous occasions by the Eastern European States and other Members of the United Nations, on several grounds. First, they argue that human rights, being global in nature and belonging to everyone, should be defined in global instruments and implemented by global bodies. 'The African and the Asian should have the same human rights as the European or the American.' Second, regional bodies in the human rights field would, at best, duplicate the work of United Nations bodies and, at worst, develop contradictory policies and procedures. . . . Third, the Eastern European States in particular object that any cooperation between regional commissions and the United Nations would add to the financial burdens of the latter. Fourth, several Western European States contend that preoccupation with regional arrangements might deflect official and public attention from the two International Covenants and delay their ratification.

It may be argued that the global approach and the regional approach to promotion and protection of human rights are not necessarily incompatible; on the contrary, they are both useful and complementary. The two approaches can be reconciled on a functional basis: the normative content of all international instruments, both global and regional, should be similar in principle, reflecting the Universal Declaration of Human Rights, which was proclaimed 'as a common standard of achievement for all peoples and all nations.' The global instrument would contain the minimum normative standard, whereas the regional instrument might go further, add further rights, refine some rights, and take into account special differences within the region and between one region and another.

Thus what at first glance might seem to be a serious dichotomy—the global approach and the regional approach to human rights—has been resolved satisfactorily on a functional basis. . . .

Implementation procedures may well vary even more from region to region, as the Governments therein desire. Indeed, they may vary within a region.

. . .

It may also be argued that the regional approach involves certain possible risks. First, a regional or sub-regional commission might serve to insulate the area from outside influences and encourage it to ignore the global standards and institutions of the United Nations system. Second, institutions of one region or sub-region might become involved in competition or conflict with those of another area. Given a modicum of good will and statesmanship on the part of any newly-established regional institutions, however, these risks should be minimal.

The further question arises whether if human rights commissions were established in certain regions, they might interpret international standards too narrowly and thus adversely affect the work of global bodies in this field. It might be necessary in such a case to establish the right of global institutions to consider a particular matter *de novo*.

Another difficulty might arise with respect to cases involving a 'consistent pattern of gross and reliably attested violations,' which are subject to special procedures established by Resolution 1503 (XLVIII) of the Economic and Social Council (27 May 1970). It would delay the consideration of such questions if prior exhaustion of available regional remedies were required.

NOTE

Consider the following cautions of Vasak about conditions for the success of a regional human rights organization:[2]

> The experience of the European Convention of Human Rights . . . tends to show that the regional protection of human rights can achieve full success only if it constitutes an element in a policy of integration on the part of the States of a given region. Only at this price is it possible to permit the blow struck by regionalism in the matter of human rights against that necessary universalism which springs from the intrinsically identical nature of all human beings. The recent entry into force of the United Nations Covenants on Civil and Political Rights and on Economic, Social and Cultural Rights, which should be preserved as a legal expression of the universal character of the human being, should even lead us to be more exacting in the future in respect of regionalism than we were in the past when no universal system for the effective protection of human rights seemed feasible. In the last analysis, regional protection must come within the framework of regional organization in accordance with the Charter of the United Nations and become one aspect of the policy of integration. If, however, regional protection were but a *form of intergovernmental co-operation*, the parochial and perhaps even selfish attitudes of which it would also be the expression, would by no means justify the danger of such a serious blow to universalism.

Note the flexibility of conceptions of what a 'region' constitutes. For a range of purposes, such as caucuses among state representatives in the UN Commission on Human Rights, the UN divides the world into five geo-political regions: Asia, Africa, Eastern Europe, Latin America, and Western Europe and Others (including the United States). That classification need bear no relation whatsoever to appropriate definition of regions for purposes of a human rights regime. For example, the Pacific region (with or excluding Australia and New Zealand), South Asia, West Asia, Southeast Asia and possibly other groupings of states might all be considered appropriate units for the creation of a given type of joint human rights mechanism.

[2] In Karel Vasak and P. Alston (eds.), The International Dimensions of Human Rights 455 (Vol. 2, 1982).

QUESTIONS

1. Consider the observation that '[r]egional and sub-regional blocs and groupings, whatever their purpose, are by their very nature inward-looking and designed to serve specific ends. Like the states of which they are composed, these blocs and groupings are more concerned with the exploitation of immediate advantages than with long-range world plans . . . '?[3] Is this view as appropriate for a human rights regime as for, say, a regional trading bloc?

2. To date, there have been no major conflicts (as opposed to minor differences) in interpretation, or between formal decisions of the existing regional bodies and their UN counterparts, although the texts of the different regional treaties suggest on their face that serious conflicts with UN-related treaties could arise. In theory, such conflicts are to be avoided through the application of some basic guidelines or rules. How do you assess the following guidelines, and what alternatives might you propose to them:

   a. The standards in the Universal Declaration and in any other UN-related treaties accepted by the state or states concerned must be respected.

   b. Human rights standards forming part of general principles of international law must also be respected.

   c. Where standards conflict, the one most favorable to the individual concerned should prevail.

3. Consider how you might handle some further complications of having numbers of states subject to regional and universal human rights regimes. Suppose that an individual alleging injury through violation of both regional and universal human rights instruments can under the terms of each treaty (say, the European Convention and the ICCPR) bring a communication or complaint before a committee or court. Should choice be open? Could both paths be entered simultaneously? If the individual either loses or wins in one or the other forum, is she then free to follow the other route?[4]

4. Should regional organizations provide an opening for cultural relativism (see Chapter 4)—that is, for regionally specific norms that should be respected rather than superseded by the universal systems? How do the guidelines in Question (2) above bear on that possibility?

[3] M. Moscowitz, The Politics and Dynamics of Human Rights 48 (1968).

[4] Treaty provisions handle some of these issues. See, for example, Article 4(2) (a) of the First Optional Protocol to the ICCPR, providing that the ICCPR Committee shall not consider a communication from an individual if the same matter is 'being examined under another procedure of international investigation . . . '

ADDITIONAL READING

John Burton, *Regionalism, Functionalism and the United Nations*, 15 Australian Outlook 73 (1961); Bruce Russett, *International Regions and the International System: A Study in Political Ecology* (1967); R. Taylor, *International Organization in the Modern World: The Regional and the Global Process* (1993); Burns Weston, Lukes and Hnatt, *Regional Human Rights Regimes: A Comparison and Appraisal*, 20 Vanderbilt J. Transnat'l L. 585 (1987); Chapter VIII: Regional Arrangements, in Bruno Simma *et al.* (eds.), *The Charter of the United Nations: A Commentary*, 679 (1994).

B. THE EUROPEAN CONVENTION SYSTEM: THE PARAMOUNT ROLE OF THE COURT

1. INTRODUCTION

The European Convention for the Protection of Human Rights and Fundamental Freedoms was signed in 1950 and entered into force in 1953. It is of particular importance within the context of international human rights for several reasons: it was the first comprehensive treaty in the world in this field; it established the first international complaints procedure and the first international court for the determination of human rights matters; it remains the most developed of the three regional systems; and it has generated a more extensive jurisprudence than any other part of the international system. Our principal concern in this selective examination of the European Convention is with the evolving institutional architecture, particularly with the European Court of Human Rights as the dominant institution and with the manner in which it has performed the judicial function. Before turning to the Court, however, this section locates the Convention within a broader European framework of post-war initiatives and institutions.

COMMENT ON BACKGROUND TO THE CONVENTION

The impetus for the adoption of a European Convention came from three factors. In the first place, it was a regional response to the atrocities committed in Europe during the Second World War and an affirmation of the belief that governments respecting human rights are less likely to wage war on their neighbors. Secondly, both the Council of Europe, which was set up in 1949 (and under whose auspices the Convention was adopted) and the European Union (previously the European Community or Communities, the first of which was established in 1952) were partly based on the assumption that the best way to ensure that Germany would be a force for peace, in partnership with France, the United Kingdom and other Western European states, was through regional integration and the institutionalizion of common values. This strategy contrasted strongly with the punitive, reparations-based, approach embodied in the 1919 Treaty oı Versailles after the First World War.

Thus, the Preamble to the European Convention refers (perhaps somewhat optimistically at the time) to the 'European countries which are likeminded and have a common heritage of political traditions, ideals, freedom and the rule of law . . . '. But this statement also points to the third major impetus towards a Convention—the desire to bring the non-Communist countries of Europe together within a common ideological framework and to consolidate their unity in the face of the Communist threat. 'Genuine democracy' (to which the Statute of the Council of Europe commits its members) or the 'effective political democracy' to which the Preamble of the Convention refers, had to be clearly distinguished from the 'people's democracy' which was practiced and promoted by the Soviet Union and its allies.

The European Convention's transformation of abstract human rights ideals into a concrete legal framework followed a path which has characterized virtually all subsequent attempts. The initial enthusiasm was soon tempered by concerns over 'sovereignty' and a reluctance to take the concept of a state's accountability too far. Thus a call by the Congress of Europe in 1948 for the adoption of a Charter of Human Rights to be enforced by a Court of Justice 'with adequate sanctions for the implementation of this Charter,' an approach which, to a large extent, was reflected in the first draft of the Convention in July 1949, went further than Western European Governments were prepared to go. Instead, the final version of the Convention acknowledges in the Preamble that it constitutes only 'the first steps for the collective enforcement of certain of the Rights stated in the Universal Declaration.'

In recent years, major reforms (noted below) of some institutional provisions of the Convention have helped to move the system closer to that envisaged by the maximalists of the early 1950s. As with most systems for the protection of human rights, progress has required the gradual growth of popular expectations and an accumulation of experience in the functioning of the procedures that has served to assuage the worst fears of governments.

COMMENT ON RIGHTS RECOGNIZED BY THE CONVENTION

Although the initial moves to create a European Convention pre-dated the UN's adoption of the Universal Declaration, the text of the latter was available to those responsible for the final drafting of the Convention. After rejection of an early proposal to do no more than list the same rights in the Convention, the drafters defined rights in terms similar to the early version of the draft Covenant on Civil and Political Rights. Since there were numerous changes in the latter before it was eventually adopted by the General Assembly in 1966, the formulations used in the two treaties sometimes differ significantly. Several weighty provisions appear in only one or the other. For example, the European Convention contains no provision relating to self-determination or to the rights of members of minority groups (Articles 1 and 27 of the ICCPR). Each treaty limits freedoms of expression, association and religion in similar ways (criteria of public safety or national security, for example), but the European Convention consistently requires that a limitation be 'necessary in a democratic society' (Articles 8–11). The derogation clauses for periods of war or public emergency threatening the life of the nation (Article 4 of the ICCPR, Article 15 of the Convention) differ with respect to the list of non-derogable provisions.

The rights recognized in Articles 2–12 of the European Convention include the right to life; freedom from torture and inhuman or degrading treatment; freedom from slavery, servitude or forced labor; liberty and security of person, and detention only in accordance with procedures prescribed by law; the right to a fair and public hearing in determining civil rights and obligations or criminal charges; respect for privacy and family life; freedom of thought, conscience and religion; freedom of expression, peaceful assembly and association; and the right to marry and found a family. Article 14 guarantees freedom from discrimination in relation to enjoyment of the recognized rights.

Article 1 requires the Parties to 'secure [these rights] to everyone within their jurisdiction', while Article 13 requires the state to provide 'an effective remedy before a national authority' for everyone whose rights are violated. Compare the more demanding Article 2 of the ICCPR, which specifically refers to states' duty to adopt legislative and other measures to give effect to the recognized rights and to 'develop the possibilities of judicial remedy.'

When the Convention was adopted in 1950, there were several outstanding proposals on which final agreement could not be reached. It was therefore agreed to adopt Protocols containing additional provisions. Since 1952 eleven protocols have been adopted. While the majority are devoted to procedural matters, others have recognised the following additional rights: the right to property ('the peaceful enjoyment of [one's] possessions'), the right to education, and the obligation to hold free elections (Protocol 1 of 1952); freedom from imprisonment for civil debts, freedom of movement and residence, freedom to leave any country, freedom from exile, the right to enter the country of which one is a national, and no collective expulsion of aliens (Protocol 4 of 1963); abolition of the death penalty (Protocol 6 of 1983); the right of an alien not to be expelled without due process, the right to appeal in criminal cases, the right to compensation for a miscarriage of justice, immunity from double prosecution for the same offence, and equality of rights and responsibility of spouses (Protocol 7 of 1984). Acceptance of each of the Protocols is, however, optional.

By March 1995 thirty States were parties to the European Convention.[5] Twenty-eight States had ratified Protocol No. 1, 22 had ratified Protocol No. 4, 23 had ratified Protocol No. 6 and 17 had ratified Protocol No. 7.

At this point students should familiarize themselves with the provisions of the European Convention that are reproduced in the Document Annex, *infra*.

2. THE BROAD EUROPEAN INSTITUTIONAL CONTEXT

Before examining the procedures and institutions through which the European system protects rights, it is necessary to consider the broader European institutional context within which the Convention is situated. The Convention is the creation of the Council of Europe, which is only one of three major regional mechanisms dealing with human rights within Europe. The other two are the European Union and the Organization (formerly termed Conference) for Security and Co-operation in Europe. None of the three is concerned exclusively with human rights. The Council of Europe has the longest and most significant track record in this field.

COMMENT ON THREE EUROPEAN ORGANIZATIONS

The Council of Europe

The Council was established in 1949 by a group of ten states, primarily to promote democracy, the rule of law, and greater unity among the nations of Western Europe. It represented both a principled commitment of its members to these values and an ideological stance against Communism. Over the years its activities have included the promotion of cooperation in relation to social, cultural, sporting and a range of other matters. Until 1990, the Council's membership was essentially confined to Western European countries. Since then, post-Cold War developments have made a major impact upon the Council. Finland, whose neutrality had previously been seen to be not readily compatible with membership, joined. It was followed in rapid succession over the next three years by (in chronological order): Hungary, Poland, Bulgaria, Estonia, Lithuania, Slovenia, the Czech Republic, Slovakia, Romania and

[5] Four members of the Council of Europe, listed in note 6, *infra*—Andorra, Estonia, Latvia and Lithuania—had signed the Convention but not yet ratified.

Latvia. By March 1995 the Council of Europe had 34 Member States, an increase of 50 per cent in only seven years.[6]

The conditions for the admission of a state to the Council of Europe are laid down in Article 3 of its Statute. The state must be a genuine democracy that respects the rule of law and human rights and must 'collaborate sincerely and effectively' with the Council in these domains. In practice, such collaboration involves becoming a party to the European Convention on Human Rights. An applicant state must satisfy the Council's Committee of Ministers that its legal order conforms with the requirements of Article 3. The opinion of the Parliamentary Assembly is sought and the Assembly in turn will appoint an expert group to advise it.

The opinion of the experts is based upon an on-site visit. For example, a 1994 expert report on the situation in Russia concluded that the requirements were not met. The report noted 'important shortcomings with regard to the rights to liberty and security of person and to fair trial' as well as the absence of the rule of law in view of the fact that the 'activities of public authorities are mainly decided upon according to general policy choices, personal allegiance and the effective power structure'.[7]

The importance attached by the states of Central and Eastern Europe to membership of the Council reflects not only a commitment to human rights but a determination to gain 'respectability' within Europe and, perhaps most importantly, to qualify for certain membership 'benefits' as well as for possible admission to the European Union. Although the process of becoming a party to the Convention is not required to be completed prior to obtaining membership in the Council, it is generally assumed that the domestic legislative and other measures required to enable the state to ratify or accede will be completed within a period of two years.[8]

The European Union

The origins of the European Union lie in the Treaty of Paris of 1952 establishing the European Coal and Steel Community (ECSC) and subsequently in the two Treaties of Rome of 1957 creating the European Economic Community (EEC) and the European Atomic Energy Community. The entry into force on November 1, 1993 of the Treaty on Economic Union converted these communities into the European Union. Since January 1, 1995, the Union has consisted of fifteen members: France, Germany, Italy, Belgium, the

[6] As of March 1995 the Members were: Andorra, Austria, Belgium, Bulgaria, Cyprus, Czech Republic, Denmark, Estonia, Finland, France, Germany, Greece, Hungary, Iceland, Ireland, Italy, Latvia, Liechtenstein, Lithuania, Luxembourg, Malta, Netherlands, Norway, Poland, Portugal, Romania, San Marino, Slovakia, Slovenia, Spain, Sweden, Switzerland, Turkey, United Kingdom.

[7] Rudolf Bernhardt et al, Report on the Conformity of the Legal Order of the Russian Federation with Council of Europe Standards, reprinted in 15 Hum. Rts. L.J. 249, 287, (1994).

[8] This explains why, in March 1995 for example, there were 34 members of the Council of Europe but only 30 parties to the Convention.

Netherlands, Luxembourg (original EEC members in 1957), United Kingdom, Ireland, Denmark (since 1973), Greece (since 1981), Spain, Portugal (since 1986), Sweden, Finland and Austria (since 1995).

The impetus for the first step of creating the ECSC came essentially from a desire to ensure that the heavy industries of the Ruhr, which had underpinned Germany's military might in two World Wars, would be 'contained' within an intergovernmental structure bringing together West Germany and its former antagonists. The expansion into an EEC in 1957 was an attempt to promote closer economic integration within Europe for both federalist and economic reasons. While the adoption of a bill of rights had been proposed in the early 1950s, none of the subsequent treaties contained such a bill or a list of enumerated rights. The 1957 treaties were more concerned with protecting states' rights from Community encroachments than with the rights of individuals. The latter were seen to be appropriately protected at the national level.

Despite the absence of a bill of rights, the European Court of Justice (the judicial organ of the European Union) began in 1969 to evolve a specific doctrine of human rights, the original motivation for which probably owed more to a desire to protect the competences of the Community than to any concern to provide extended protection to individuals. Over the years during which the human rights doctrine has evolved, the Court has identified several different normative underpinnings for it. They include certain provisions of the Treaty of Rome, the constitutional traditions of the member states, and international treaties accepted by member states. For the most part the European Court of Justice has applied this concept of human rights to the actions of the Community itself, but not to actions of the member states.[9]

The European Convention on Human Rights (of the Council of Europe) has been accorded a priveliged position within the Community legal order, but on a *de facto* rather than a formal basis. In 1977 the European Parliament, the Council and the Commission issued a Joint Declaration emphasizing the importance they attached to fundamental rights, as derived in particular from the constitutions of the member states and the European Convention, and pledged to respect them in the exercise of their powers. In 1979 the Commission proposed that the Community should formally adhere to the European Convention. This suggestion, which has been revived in the 1990s, drew both technical and political objections (primarily from member states such as the United Kingdom, Ireland and Denmark which had not incorpo-

[9] It now appears to be generally accepted by commentators, however, that member states are required to respect the Community concept of human rights in certain circumstances: (1) whenever that concept is used to interpret provisions in the Treaties or Community legislation; (2) where a provision of Community law grants rights which are subject to derogation, in which case the latter must be consistent with the concept; and (3) when member states implement Community rules, in which context they are bound to respect that concept. See Trevor Hartley, The Foundations of European Community Law, 148 (3rd. ed., 1994); Joseph Weiler, Eurocracy and Distrust: Some Questions Concerning the Role of the European Court of Justice in the Protection of Fundamental Human Rights Within the Legal Order of the European Communities, 61 Wash. L. Rev. 1103 (1986).

rated the European Convention into their domestic law).[10] To date, no action has been taken to achieve this objective, nor to adopt a Community bill of rights *per se*.[11]

Note that the institutional arrangements relating to the European Convention are quite separate from those of the Union despite the use of similar appellations. The following table illustrates only a part of the immense potential for confusion in discussions about European institutions:

Council of Europe	*European Union*
Parliamentary Assembly	European Parliament
Committee of Ministers	Council of the European Union European Council Ministerial Councils
European Commission of Human Rights	Commission of the European Union
European Court of Human Rights	European Court of Justice
Committee of Independent Experts on the European Social Charter	Economic and Social Committee

Although arrangements exist to facilitate consultation and coordination between the two sets of institutions, they remain separate entities operating in very different settings despite the fact that the activities of each organization are very relevant to those of the other.

The Organization for Security and Co-operation in Europe (OSCE)

The Conference on Security and Co-operation in Europe (CSCE) opened in 1973 and concluded in August 1975 with the signing of the Final Act of Helsinki (known as the Helsinki Accord) by the 35 participating states (including all European states except Albania, plus Canada and the United States). The Soviet Union was motivated mainly by a desire to obtain formal recognition of its European frontiers, while the West took advantage of a period of East–West *détente* to obtain concessions primarily in relation to security matters. Human rights were of only secondary concern.

The CSCE process continued in the form of long-running diplomatic conferences throughout the late 1970s and 1980s designed to follow up and elaborate on the obligations contained in the Helsinki Accord. These subsequent agreements were reflected in the 'Concluding Documents' of various

[10] See J. McBride and Brown, The United Kingdom, the European Community and the European Convention on Human Rights 1 Yearbook of European Law 167 (1981).

[11] See Jean Paul Jacqué, The Convention and the European Communities, in R. St J Macdonald et al (eds.), The European System for the Protection of Human Rights 889 (1993).

follow-up meetings. In the human rights context the most important were those held in relation to what came to be known as 'the human dimension of the CSCE.' They include meetings which concluded in Vienna and Paris in 1989, Copenhagen in 1990, Moscow in 1991 and Geneva in 1992.

Several characteristics distinguish the work of the CSCE from that of other entities in the human rights field. Its standards are all formally non-binding (in the sense that they are solemn undertakings, but are not in treaty form and thus not ratified or acceded to by states). Secondly, its membership is far broader than that of the European Union or even the Council of Europe and by 1995 had grown to 53 states. Thirdly, until 1991 it had no more than a token institutional structure designed only to arrange its periodic meetings. It performed no operational tasks.

The non-binding diplomatic nature of the Helsinki Process led many observers to question its utility. Whatever contribution the process ultimately made to the demise of Communism, it clearly played an important role, especially in the second half of the 1980s and early 1990s, in legitimating human rights discourse within Eastern Europe, providing a focus for non-governmental activities at both the domestic and international levels, and developing standards in relation to democracy, the rule of law, 'human contacts,' national minorities and freedom of expression which went beyond those already in existence in other contexts such as the Council of Europe and the UN. To a large extent, it was its formally non-binding nature that enabled the CSCE standard-setting process to yield more detailed and innovative standards than those adopted by its counterparts.

Since 1991 the CSCE's institutional dimension has been developed through the establishment of a general secretariat in Prague, an Office for Democratic Institutions and Human Rights and an Office for Free Elections, both based in Warsaw, a Conflict Prevention Centre, based in Vienna, and a High Commissioner on National Minorities, based in the Hague. The work of the last of these is in some ways illustrative of the CSCE's overall approach. The position was created in July 1992 and a former Dutch Foreign Minister, Mr Max van der Stoel, took office in January 1993. He is assisted by only a couple of staff members. On occasion CSCE organs or the country concerned have requested him to undertake a mission, but the decision to do so is his own. His on-site visits are to the country itself as well as to other relevant countries, such as neighboring states in which members of the minority group in question also reside. Statements to the press are very limited and the High Commissioner reports confidentially to the CSCE Chairman-in-Office (see *infra*) and to the Foreign Minister of the country concerned. His recommendations address both short-term policy towards minorities and longer-term measures to encourage a continuing dialogue between the government and minority members. The relevant correspondence is eventually released by mutual consent.

In 1995 the CSCE was officially transformed into the Organisation for Security and Co-operation in Europe (OSCE). Its official organs include the

Parliamentary Assembly of the OSCE, the Council of Ministers for Foreign Affairs, the Committee of Senior Officials (CSO) which meets regularly, the 'Chairman-in-Office' which is a rotating post held by each Member State Foreign Minister in turn, and the biennial summit or Review Conference. This move to institutionalize the OSCE has been welcomed by most observers, although the level of bureaucratic and infrastructural support remains low and the institutions are physically scattered throughout Europe. One long-time observer has suggested that the institutionalization process will have certain negative consequences including a greater detachment of OSCE activities from the NGO community which was instrumental in many of the early successes of the process.[12]

One issue which the OCSE's evolution has made more pressing is its relationship with the Council of Europe's human rights activities. Until recently, it was sufficient to contrast the Council's traditional focus on protecting the rights of individuals through judicial and quasi-judicial mechanisms with the OSCE's concern with inter- and intra-state conflict and with the larger issues concerning democracy and the rule of law. In that respect the division of labour seemed reasonably clear. Nevertheless, as the Council of Europe has sought to adapt to the radical changes within Europe it has begun to develop a much greater focus on broader rule of law issues, has established a European Commission for Democracy through Law, based in Venice, and has adopted a Framework Convention for the Protection of National Minorities (see *infra*). In addition, the European Convention has always provided for an inter-state procedure and for the possibility of on-site fact-finding.[13]

An assessment by the Council of its relationship with the CSCE states that the two 'have reached a normal working relationship characterised by accepting and acknowledging each other's character, working methods, expertise and strong points',[14] but this belies an inevitable and continuing tension as the two entities seek to develop their respective comparative advantages. This process is certain on occasion to involve unwelcome incursions into the domain of the other and the question remains as to whether such competition will be beneficial or destructive. In the meantime, the proliferation of institutions, mechanisms and normative statements is unavoidably confusing for even the most seasoned observers.[15]

[12] William Korey, The Promises We Keep: Human Rights, The Helsinki Process and American Foreign Policy 429–38 (1993).

[13] See Jochen Frowein, Fact-Finding by the European Commission of Human Rights, in R. Lillich (ed.), Fact-Finding Before International Tribunals, 237 (1992).

[14] Council of Europe, Information Sheet No. 34, 134 (1995).

[15] The Dutch Advisory Committee on Human Rights and Foreign Policy concluded in a September 1994 Advisory Letter on CSCE Mechanisms that there is 'overwhelming evidence' that coordination between the CSCE on the one hand and the Council of Europe, the European Union and the UN on the other 'is still extremely deficient. The relationship between the CSCE and the Council of Europe continues to pose particular problems.' See Philip Alston, The Emerging European-Wide Human Rights Regime: Too Much of a Good Thing?, in V. Bornschier and P. Lengyel (eds.), Waves, Formations and Values in the World System, 237 (1992).

ADDITIONAL READING

European Union: J. Weiler and N. Lockhart, '*Taking Rights Seriously*': *The European Court and Its Fundamental Rights Jurisprudence*, 32 Common Market L. Rev. 51 (1995); Andrew Clapham, *Human Rights and the European Community: A Critical Overview* (1991); A. Cassese, A. Clapham and J. Weiler (eds.), *Human Rights and the European Community: Methods of Protection* (1991); *idem. Human Rights and the European Community: The Substantive Law* (1991).

The OSCE: Arie Bloed (ed.), *The Conference on Security and Cooperation in Europe: Analysis and Basic Documents 1972–1993* (1993); William Korey, *The Promises We Keep: Human Rights, The Helsinki Process, and American Foreign Policy* (1993); Thomas Buergenthal, *The CSCE Rights System*, 25 Geo. Wash. J. Int'l L. & Econ. 333 (1991); E. B. Schlager, *The Procedural Framework of the CSCE: From the Helsinki Consultation to the Paris Charter, 1972–1990*, 12 Hum. Rts. L. J. 22 (1991); A. Rosas and J. Helgesen (eds.), *Human Rights in a Changing East/West Perspective* (1990); S. Terstal *et al*, *The Functioning of the CSCE High Commissioner on National Minorities*, 20 New Community 502 (1994); Arie Bloed, *Monitoring the CSCE Human Dimension: In Search of its Effectiveness*, in A. Bloed *et al* (eds.), *Monitoring Human Rights in Europe: Comparing International Procedures and Mechanisms 45 (1993)*.

3. OTHER HUMAN RIGHTS CONVENTIONS ADOPTED BY THE COUNCIL OF EUROPE

COMMENT ON THREE CONVENTIONS

The European Social Charter

Although economic and social rights were reflected in the post-World War II constitutions of France, Germany and Italy, they were not included in the European Convention. One of the key drafters, Pierre-Henri Teitgen, explained this decision in 1949 on the grounds that it was first necessary 'to guarantee political democracy in the European Union and then to co-ordinate our economies, before undertaking the generalisation of social democracy.' These rights were subsequently recognized in the European Social Charter of 1961.

The Charter consists of four parts. The first lists nineteen social 'rights and principles' which the contracting parties accept as the aim of their policy. They include the rights to: 'just conditions of work,' 'fair remuneration sufficient for a decent standard of living,' freedom of association, special protection for children and young persons, 'benefit from all measures enabling [enjoyment of] the highest possible standard of health attainable,' social security and social welfare services. These rights are not legally binding *per se*. The legal obligations designed to ensure the effective exercise of those rights are contained in Part II, which expands upon the specific measures to be taken in relation to each of the rights. Part III reflects the principle of progressive implementation tailored to suit the circumstances of individual states. Each contracting party must agree to be bound by at least five of seven rights which are considered to be of central importance. It must also accept at least five of the other rights as listed in Part II.

Part IV provides for a monitoring system based on the submission of regular reports by contracting parties. The reports are examined by the Council of Europe's Committee of Independent Experts (CIE) whose assessments of compliance and non-compliance are then reviewed by the Parliamentary Assembly and a Governmental Committee (also established under Council auspices). Finally, on the basis of all these views, the Committee of Ministers may, by a two-thirds majority vote, make specific recommendations to the state concerned. In addition to being cumbersome and time-consuming, this system vests excessive power in the hands of the representatives of governments who have shown an unsurprising reluctance to take forceful action even when both the CIE and the Assembly consider it to be warranted.

A member of the CIE commented in 1992 that '[a]lthough having a potentially valuable role, the Charter has been until now one of the less prominent of human rights treaties. The Conclusions, or Annual Report, of the [CIE] might almost be classified in the rare books category.'[16] A 1991 Amending Protocol will introduce major reforms when it enters into force by giving the CIE exclusive competence to interpret and apply the Charter, significantly downgrading the role of the Governmental Committee, enabling the CIE to engage in an oral dialogue with governments, and enhancing the role of NGOs.

The Charter entered into force in 1965.[17] In 1987 an Additional Protocol was adopted that extends the protection of certain rights, particularly in relation to equal employment opportunity. It entered into force for the relevant states in September 1992.[18]

[16] D. J. Harris, Introductory Note, 31 Int. Leg. Mat. 155, 156 (1992).

[17] As of March 1995, twenty States were parties. Member States of the Council of Europe that have signed but not ratified are: the Czech Republic, Hungary, Liechtenstein, Poland, Romania, Slovakia and Switzerland. Non-signatories are: Andorra, Bulgaria, Estonia, Latvia, Lithuania, San Marino and Slovenia.

[18] The European Social Charter should be distinguished from the European Community's Charter of Fundamental Social Rights of Workers. The latter is neither a treaty nor an instrument of Community law, and applies only to 'workers' rather than to all members of society. The emphasis is upon implementation at the national rather than regional level. The 1991 (Maastricht) Treaty on European Union includes a 'Social Policy' Protocol to which eleven of the then twelve member states committed themselves (with the United Kingdom objecting). See Erik Lundberg, The Protection of Social Rights in the European Community, in K. Drzewicki, Krause and Rosas (eds.), Social Rights as Human Rights: A European Challenge 169 (1994).

The European Convention for the Prevention of Torture

In 1987 the Council of Europe adopted the European Convention for the Prevention of Torture and Inhuman or Degrading Treatment or Punishment. This step was taken after the adoption of the UN Convention Against Torture and Other Cruel, Inhuman or Degrading Treatment or Punishment in 1984, but the European Convention is concerned especially with prevention and is far more innovative and intrusive in its approach to supervision. As of March 1995 the European Convention had been ratified by 29 states.

The Convention establishes a Committee for the Prevention of Torture (CPT) which is composed of independent experts. Its function is 'to examine the treatment of persons deprived of their liberty with a view to strengthening, if necessary, the protection of such persons' from torture, inhuman or degrading treatment (Art. 1). The Convention is not concerned solely with prisoners but with any 'persons deprived of their liberty by a public authority.' Each state party is required to permit the Committee to visit any such place within the state's jurisdiction (Art.2), unless there are exceptional circumstances (which will rarely be the case). Most visits are routine and scheduled well in advance but there is also provision for *ad hoc* visits with little advance notice (Art. 7).

The Committee meets *in camera* and its visits and discussions are confidential as, in principle, are its reports. The latter, however, may be released, either at the request of the state concerned or if a state refuses to cooperate and the Committee decides by a two-thirds majority to make a public statement. This occurred in December 1992 when the Committee concluded after three visits to Turkey that the Government had failed to respond to its recommendations to strengthen legal safeguards against police torture, specifically in relation to the activities of certain police 'Anti-Terror Departments'.[19] Virtually all other states visited have voluntarily agreed to the release of the Committee's report, together with the government response thereto.

Framework Convention for the Protection of National Minorities

Despite the importance of national minorities within Europe and discussions about appropriate measures since 1949, the issue had proven too controversial and complex for the Council of Europe to adopt specific standards until November 1994, when the Framework Convention was adopted. In part, the impetus was the adoption of the 1992 UN Declaration on Rights of Members of Minorities, p. 1001, *infra*, and the development of non-binding standards and promotional activities in this field by the CSCE. The Council sought to avoid longstanding controversies by, among other things, confining the Convention to programmatic obligations that are not directly applicable and

[19] Public Statement on Turkey, Doc. CPT/Inf (93)1 (15 Dec. 1992).

that leave considerable discretion about implementation to the state concerned. International supervision is to be undertaken by the Committee of Ministers of the Council based upon periodic reports to be submitted by states parties.

* * * * *

Note that even within one region, deep differences appear in the institutional structures of conventions adopted by the same body, as one compares these three conventions among themselves and with the European Convention on Human Rights to which the materials now turn.

QUESTIONS

1. Compare the powers and functions of the Committee for the Prevention of Torture with those of treaty organs under the UN-related conventions described in Chapter 9, particularly with the Committee formed under the Convention against Torture(CAT), p. 560, *supra*. In what respects is the European Committee for the Prevention of Torture 'innovative,' as the text above states? Are its innovative characteristics likely to be easier to include in a treaty relating to a regional rather than universal arrangement?

2. Even based on the brief descriptions in the preceding Comment, can you draw on prior materials to suggest some relationships between the different institutional structures of the three conventions and their different subject matters?

4. THE EUROPEAN CONVENTION'S IMPLEMENTATION MACHINERY: ARTICLE 25 INDIVIDUAL PETITIONS

The European Convention on Human Rights provides for two procedures by which member states (referred to in the Convention as the High Contracting Parties) may be held accountable by the Convention organs for violations of the recognized rights: the individual petition procedure pursuant to Article 25, and the inter-state procedure under Article 24.

Various provisions of the Convention make clear that the primary responsibility for implementation rests with the member states themselves at the national level. The implementation machinery provided for in the Convention comes into play only when domestic remedies are considered to have been exhausted. Indeed the great majority of cases received by the Commission are deemed inadmissible, frequently on the grounds that domestic law provides an effective remedy for any violation that may have taken place. Recall the obligations of member states under Articles 1 and 13 of the Convention to 'secure to everyone' the Convention's rights and to provide 'an effective remedy before a national authority' for violations of those rights.

That remedy given by, say, a domestic court may be pursuant to provisions of domestic law that stand relatively independently of the Convention although are perhaps influenced by it—a code of criminal procedure or a constitutional provision on free speech that are consistent with the Convention, for example. Or a remedy may be given pursuant to the substantive provisions of the Convention itself after the Convention has been incorporated into domestic law automatically or by special legislation. The second possibility is considered with respect to application of the Convention by the judiciary in some European states in Chapter 11 at pp. 725–38, *infra*.

This preference for domestic resolution is also reinforced by the requirement to seek a friendly settlement wherever possible and by the procedures for full government consultation in the examination of complaints. The confidentiality of the Commission's own proceedings, the role accorded to the Committee of Ministers, and the provision for there to be a judge and Commission member from every state party again underscore the state-centred nature of many of the Convention's procedures.

COMMENT ON THE COMMISSION AND COURT

The procedures prescribed by the Convention are relatively complex. For reasons noted below, a far-reaching set of institutional and procedural reforms was adopted by the member states in Protocol No. 11, which is unlikely to enter into force before 1998. The Convention as amended by Protocol No. 11 is hereinafter referred to as the Revised Convention. In the following readings the complexities of the existing arrangements are not dealt with in detail, partly because of the pending changes pursuant to Protocol No. 11, and partly because our primary concern is with the functioning of the European Court, rather than the Commission or Committee of Ministers. Protocol No. 11 provides for the establishment of one institution, a full-time Court, to replace the present related functions of the Commission and Court and the present role of the Committee of Ministers.

As now constituted, the Commission consists of a number of members equal to that of the states parties ('High Contracting Parties') to the Convention. The Court, on the other hand, consists of one judge for every member state of the Council of Europe. In March 1995, there were 30 members of the Commission and 34 judges. The individuals are usually judges, university professors, or practising lawyers. Some have previously been senior government officials. Few women have been elected, and sitting judges and commissioners tend to be reelected.

Proceedings under the individual petitions procedure of Article 25 (Article 34 of the Revised Convention) begin with a complaint by an individual, group or NGO against a state that has expressly accepted this procedure. By March 1995 some 550 million people in 30 countries were eligible to lodge petitions under this procedure. Under the Revised Convention acceptance of this procedure will be mandatory for all member states.

To be declared admissible a petition must not be anonymous, manifestly ill-founded, or constitute an abuse of the right of petition. Domestic remedies must have been exhausted, it must be presented within six months of the final decision in the domestic forum and it must not concern a matter which is substantially the same as one which has already been examined by the Commission or submitted to another procedure of international investigation or settlement. Once the application is declared admissible (a procedure which often, but not always, involves an oral hearing), the Commission's task is to establish the facts and encourage efforts to reach a friendly settlement. Decisions on admissibility are published.

Where a friendly settlement is reached between the parties, and provided it considers the settlement to be based upon 'respect for human rights as defined in the Convention,' the Commission reports on this to the Committee of Ministers, which consists of the Foreign Ministers or their deputies of the member states of the Council of Europe. It is up to the government concerned to take the measures to which it has agreed. When the Convention was first drawn up most governments considered it imperative that they retain the final say in relation to all cases decided by the Commission. According such a role to a quintessentially political body has been strongly criticized in the following terms by a former Director of Human rights of the Council of Europe.

> [Bodies such as the Committee of Ministers] are composed of State or Government representatives whose task is to defend what are—rightly or wrongly— regarded as 'State interests'. These political bodies are therefore, almost inevitably, a playground or battlefield of the 'raison d'Etat'.
>
> Although nowadays we are increasingly aware of the fact that human rights are not only violated by State power, international systems for the protection of human rights are mainly designed to control the action of States in this field. When political bodies are entrusted with this task, are we not confronted with an inherent contradiction? These bodies and their members whose normal task is to defend State interests are supposed to protect fundamental rights of individuals or groups against State power. Within these bodies, States are at one and the same time judges and parties.[20]

Where no friendly settlement is reached the Commission determines whether there has been a breach of the Convention. Its opinions, which are

[20] Peter Leuprecht, The Protection of Human Rights by Political Bodies: The Example of the Committee of Ministers of the Council of Europe, in Manfred Nowak, Steurer and Tretter (eds.), Progress in the Spirit of Human Rights: Festschrift für Felix Ermacora 95, 96 (1988).

usually published, are not *per se* legally binding on the states parties. Within three months of the Commission's report the case can be referred to the Court by either the Commission or by a state concerned, but not (until Protocol No. 9 comes into force) by the complainant. If the case is not referred to the Court, the Committee of Ministers decides whether there has been a violation of the Convention and what measures must be taken by the state concerned.

The Court cannot determine which cases it hears but rather is at the mercy of the Commission or the state concerned. It does not undertake the examination of a case *de novo*. The subject-matter of each case brought before it is, in essence, determined by the Commission's decision on admissibility. In practice, most cases have been referred by the Commission. There are no established criteria to guide the Commission about referral. One observer has identified elements that might be influential in these decisions: where the case raises a point of interpretation that has not previously arisen; where the Commission is divided as to whether there has been a violation; or where a case is perceived to have particularly serious political implications.[21]

Although the Commission is not a party to cases before the Court, one of its members participates in the proceedings to assist the Court. Since 1983, an individual applicant may be represented before the Court by a lawyer but is still not formally a party to the case. Although there is no *amicus curiae* provision, the President of the Court may authorize interested parties to submit comments on specific issues and various NGOs have been invited to do so.

The functioning of the Court is explored in Section 8 below in relation to specific cases.

5. RESPONSES OF STATES TO FINDINGS OF VIOLATIONS

ANDREW DRZEMCZEWSKI AND MEYER-LADEWIG, PRINCIPAL CHARACTERISTICS OF THE NEW ECHR CONTROL MECHANISM, AS ESTABLISHED BY PROTOCOL NO.11

15 Hum. Rts. L. J. 81 (1994), at 82.

. . .

The Convention's achievements have been quite staggering, the case-law of the European Commission and Court of Human Rights exerting an ever deeper

[21] J. G. Merrills, The Development of International Law by the European Court of Human Rights 4 (2nd. ed., 1993).

influence on the laws and social realities of the State Parties. A few examples may be mentioned.

In Austria, where the Convention has the rank of constitutional law, the Code of Criminal Procedures has had to be modified as a result of case-law in Strasbourg; so too the system of legal aid fees for lawyers. In Belgium, amendments have been made to the Penal Code, its vagrancy legislation and its Civil Code to ensure equal rights to legitimate and illegitimate children. In Germany modifications that bring legislation better into line with the Convention's provisions have also been made, e.g. the Code of Criminal Procedure concerning pre-trial detention was amended. Various measures have been taken to expedite criminal and civil proceedings, and transsexuals have been given legal recognition.

In the Netherlands, where most of the Convention's self-executing substantive provisions are endowed with a hierarchically superior status to the Constitution itself, changes have been made in the Military Criminal Code and the law on detention of mental patients. In Ireland, court proceedings have been simplified and civil legal aid and advice schemes set up. Sweden has introduced rules concerning time-limits for expropriation permits and legislation enacted concerning the regulation of building permits. Switzerland has amended its Military Penal Code and completely reviewed its judicial organisation and criminal procedure as applied to the federal army, as well as its Civil Code as regards deprivation of liberty in reformatory centres.

In France, the law relating to the secrecy of telephone communications had to be altered, while in Italy a new Code of Criminal Procedure was enacted to change the law concerning regulation of detention on remand. In the U.K., despite the fact that the Convention has not been incorporated into domestic law (as is also the case in Ireland, and for the time-being in Iceland, Norway and Sweden), its constitutional impact cannot be doubted, with changes in domestic law being made in the areas of freedom of information, privacy, prison rules, mental health legislation and payments of compensation for administrative miscarriages of justice, among others.

But the effects of the Convention are not limited to the follow-up given to judgments of the Court, decisions of the Commission and findings of violation by the Committee of Ministers. The procedure for friendly settlements under the Convention has also produced significant results in this respect. Indeed, over two hundred instances can be cited where settlements have been reached either formally or informally, often with the Commission's or the Court's approval, subsequent to concessionary measures taken by the governments concerned.

More generally, national courts in the States Parties to the Convention increasingly turn to the Strasbourg case-law when deciding on a human rights issue, and apply the standards and principles developed by the Commission and Court. Many instances can also be cited of States modifying legislation and administrative practices prior to their ratification of the Convention, particularly in the case of those States which have recently joined the Organisation.

CHRISTIAN TOMUSCHAT, QUO VADIS, ARGENTORATUM? THE SUCCESS STORY OF THE EUROPEAN CONVENTION ON HUMAN RIGHTS—AND A FEW DARK STAINS

13 Hum. Rts. L. J. 401 (1992).

. . . [Despite the system's successes] symptoms of a deep-seated crisis cannot be overlooked. As any mechanism for the protection of human rights, the Strasbourg system must be measured by the concrete results which it produces in favour of the aggrieved individual. It is not the legal perfection of its normative structure that matters in the last analysis, but its actual impact on the real enjoyment of human rights. In that respect, substantial reasons barring euphoria exist. However numerous and bold the decisions of the Strasbourg bodies may be, it has recently emerged that especially their implementation lacks sufficiently effective safeguards.

[One of Tomuschat's principal concerns is the failure of the Committee of Ministers to discharge its obligations under the Convention effectively, both in relation to inter-state and individual applications. That deficiency will be remedied once Protocol No. 11 enters into force. Another major deficiency relates to enforcement of the judgments of the Court.]

a) Reparation in individual cases

As far as obligations to make specific payments are concerned, in the telephone tapping case of *Kruslin* France took more than one year and four months before it made the required payment to the applicant in respect of costs and expenses . . . and in the *Ezelin* case the period amounted to almost one year . . . . But the unfortunate top position is held by Italy. It appears that the *Colozza* case, in which Italy was ordered to pay to the victim's widow six million Lire by way of just satisfaction, [had still not been wound up seven years later].

. . .

b) General amendment of legislation

The most far-reaching challenge arises for a State when it has to modify its legislation following a judgment of the Court.

. . .

. . . It took Belgium roughly eight years to change its legislation on the status of illegitimate children after the judgment in the *Marckx* case had been delivered, whereas Ireland responded within one year to the Court's conclusion that certain aspects of the position of illegitimate children amounted to unlawful discrimination. The Federal Republic of Germany needed more than five years to comply with the judgment in the *Öztürk* case, according to which provision of an interpreter free of charge for the accused is mandatory also with regard to regulatory offences (*Ordnungswidrigkeiten*). A strange response was given by the Swiss authorities to the judgment of 18 December 1987 in *F. v. Switzerland*, which held that a prohibition to remarry, as foreseen in Article 150 of the Swiss Code, was contrary to Article 12. Switzerland informed the Committee about its intention to abolish the controversial provision in connection with a general reform of the Swiss law on divorce which would probably enter into force in 1995, i.e. no less than eight years after the pronouncement of the judgment. Lastly, it is worth mentioning the judgment in *Norris v. Ireland* of 26 October 1988 . . . [see p. 618, *infra*. Four years later the legal position had remained unchanged].

. . .

More positive examples are the swift reaction of the Netherlands to the finding in the *Benthem* case that the system of review of administrative decisions was not in keeping with the required safeguards of a judicial procedure, and similarly Sweden introduced mechanisms to ensure judicial review of administrative conduct after it had emerged that the existing machinery was characterized by large gaps. It is also with praiseworthy celerity that France responded to the *Huvig* and *Kruslin* judgments of 24 April 1990 which had passed a verdict of incompatibility with the Convention of the French system for telephone tapping in criminal proceedings. Finally, one of the guiding forces behind the adoption of the new Italian Code of Criminal Procedure, which entered into force on 24 October 1989, may have been the pressures engendered by the numerous applications complaining about excessive length of criminal proceedings. It remains to be seen what actual effects in practice this reform will produce. In this connection, one reads with mixed feelings the information that the Italian Court of Cassation, in an amazing *tour de force*, disposed in 1990 of no less than 44,811 of the appeals pending before it.

QUESTION

On what criteria or standards would you evaluate the effectiveness of the European Convention system? Should we examine the system in its own terms, primarily through judicial decisions and state responses thereto? Or should we compare the functions and powers of the organs in the Convention system with the different institutional structures and powers of other human rights systems, universal or regional, to make a broader judgment? Note the differences from the other systems examined in Chapters 7 and 9.

6. REFORMING THE CONVENTION SYSTEM

Between 1953 and 31 December 1994 the Council of Europe had established 78,583 provisional files (letters received), of which 26,041 were registered as applications. Of those 2,027 had been declared admissible. In 223 of those cases friendly settlements were achieved, 506 were referred to the Court (of which 290 resulted in a finding of violations, 103 in a finding of no violations, 62 are pending and 51 were settled or struck off the list). There were 551 cases that were referred to the Committee of Ministers (of which 121 were considered to involve violations, 64 no violations, 332 pending and 34 settled or leading to other findings).

Partly because of the expansion of the Council of Europe's membership and partly because of the Convention system's perceived effectiveness, it has become overburdened. In recent years the Commission has received between 5,000 and 10,000 complaints (known as 'applications') annually. In the years prior to 1990 the Commission built up a large backlog which began to be reduced only when it was permitted by Protocol No. 8 to meet in Committees of Three and Chambers. In a five month period between August 1994 and January 1995, for example, the Commission, in examining admissibility, dealt with 2,378 applications under Article 25. It declared 396 applications admissible, 947 inadmissible, struck 48 applications off its list of cases pending, and communicated 489 applications to governments.[22] In examining applications which had already been admitted, it referred 36 to the Court, adopted 19 reports based on friendly settlements (Article 29 (2)) and adopted 203 reports on the merits (Article 31).[23] The figures for the preceding six months were similar.

The existing system is unable to deal with the greatly increased number of cases. For example, the number of applications registered by the Commission (i.e. accepted for examination rather than being deemed immediately irrelevant) went from 404 in 1981, to 500 in 1983, 700 in 1986, 2,037 in 1993 and close to 4,000 in 1994. This tenfold increase in less than 15 years has placed severe strains on the entire system. With this increased volume has come growing complexity as the Convention jurisprudence has evolved.

As a result of these combined factors, the procedures have become unacceptably time-consuming. In 1993 it took, on average, five years and eight months for a case to be finally decided (four years and four months before the Commission and one year and three months before the Court). This is especially ironical in view of the Court's case law, which has often found states whose courts take equivalent periods of time to decide a case to be in breach of the right to a hearing 'within a reasonable time' under Article 6. Another

[22] The figures for any given period do not 'add up' since not all petitions are processed in the period in which they are received.

[23] Information taken from European Commission of Human Rights, Information Note, Nos. 121–123.

factor favoring reform is the rapid expansion in the number of states parties to the Convention, with further additions expected in the future.

Protocol No. 11, of May 1994, will enter into force one year after it has been ratified by all of the Convention's High Contracting Parties. At the time of its adoption it was generally predicted that the necessary ratifications would not be achieved before 1997, making 1998 the earliest likely date for the implementation of the Revised Convention. The essential features of the reform are described in the following excerpt.

ANDREW DRZEMCZEWSKI AND MEYER-LADEWIG, PRINCIPAL CHARACTERISTICS OF THE NEW ECHR CONTROL MECHANISM, AS ESTABLISHED BY PROTOCOL NO. 11

15 Hum. Rts. L. J. 81 (1994).

. . .

[The basic features of the reform]

With the signature of Protocol No. 11 by 27 of the 28 States Parties to the European Convention on Human Rights (ECHR) on 11 May 1994, the reform of the ECHR has now taken on a clear structure and has been provided with the necessary political impetus. Since then, Slovenia has already ratified Protocol No. 11, thereby becoming the first State to firmly commit itself in the path of reform.

A permanent European Court of Human Rights will replace the existing Commission and Court as well as the Committee of Ministers, in so far as the last-mentioned functions concerning individual and inter-State complaints are concerned. The competence of the Committee of Ministers will henceforth be limited to the supervision of the execution of the Court's judgments.

The Court will sit permanently, i.e., it will work in a similar way as does the Court of Justice of the European Communities in Luxembourg. The salaries of judges will be paid by the Council of Europe. With the permanent presence of judges in Strasbourg, the way in which the system functions will of necessity change fundamentally. The influence of civil servants in the Secretariat of the Commission and Registry of the Court will diminish. On the other hand, the participation of judges—in particular as rapporteur–judges—will, from the very beginning of the procedure, become more intense. An infrastructure, even more professional than now, together with a specific role provided to legal secretaries, should ensure an improvement in the quality of the Convention's control mechanism.

1. Composition of the Court

The Court will consist of a number of judges equal to that of State Parties to the Convention. It will sit in committees, Chambers and a Grand Chamber.

The judge elected in respect of the State concerned will sit as of right in Chambers and the Grand Chamber when a case concerning that State is examined by the Court. The plenary Court will only deal with organisational matters.

Committees will be composed of three judges. Chambers of seven judges and the Grand Chamber of seventeen judges. Committees are to be set-up by Chambers for fixed periods of time. They will exercise a filter function with respect to applications as does presently the Commission; they will be able to declare individual applications inadmissible. Admissibility criteria are to remain unchanged.

Chambers will also be set-up by the Court for fixed periods of time. Judges will be able to be members of more than one Chamber at the same time. Chambers will be able to determine the admissibility as well as the merits of applications.

The Grand Chamber will have jurisdiction with respect, in effect, to certain matters which the present plenary Court possesses. It will decide—in exceptionally important cases—on individual as well as inter-State applications referred to it. The President of the Court, the Court's Vice-Presidents, the Presidents of the Chambers and the judge elected in respect of the State against which an application has been lodged will be *ex officio* members of the Grand Chamber. The other members will be chosen by the Court. The Grand Chamber could be set-up for a specific case or for a fixed period of time.

2. Outline of the procedure

As under the present system, individual applications will be able to be lodged by any person or a State Party. . . . The application will . . . be examined by a committee of three judges, one of whom would be the judge rapporteur, which will then declare the application admissible or, in the alternative, reject it. The committee will be able, unanimously, to declare an application inadmissible; such a decision would be final. If a judge rapporteur were to consider that the application raises an issue of principle or that the application cannot be declared inadmissible, the application would be transmitted directly to the Chamber. This procedure corresponds closely to the present system. . . .

There will normally be a hearing before a Chamber. The parties will present their observations in writing. The Chamber will be able, at any stage of proceedings, to put itself at the disposal of the parties so as to facilitate a friendly settlement. . . .

. . .

Following the delivery of judgment, the parties to a case will be able to request that it be referred to the Grand Chamber. This procedure will be restricted only to exceptional cases, i.e. when a case raises a serious question concerning the interpretation or application of the Convention or its protocols, or if it raises an issue of general importance. A panel of the Grand Chamber will determine whether the request for a re-hearing can be accepted (Article 43).

A judgment of the Court will become final once the possibility of referral to the Grand Chamber is no longer feasable. Such a judgment will be definitive and, as is the case at present, binding in international law. The Committee of Ministers will maintain its role of supervising the execution of the Court's judgments.

. . .

QUESTIONS

1. Is there an inherent contradiction in a system for protecting human rights that accords a significant role (as on the Committee of Ministers) to governmental representatives? If so, how should we view the role(s) played by such representatives on the UN Commission on Human Rights?

2. How well insulated from such political/governmental pressures will the European system be, once the Protocol No. 11 changes are in place?

3. The Revised Convention continues to provide for the possibility of having a judge from every state party. This is justified on the grounds that it gives the judiciary in each state a sense of confidence in the Court and it assures the Court of expertise in the functioning of every national legal system. Would an international (UN) human rights court therefore need to have a comparable system? If this is considered not to be feasible (e.g. 190 judges), does it cast serious doubt on the viability of any such court? What are the implications for the 18-member ICCPR Human Rights Committee rendering its 'views' under the Optional Protocol, p. 535, *supra*?[24]

4. Protocol No. 11 provides that two members of the first instance Chamber (its President and its member from the state concerned) will sit again in the Grand Chamber if there is an appeal. The Court has held this practice to be an infringement of the right to a fair trial under Article 6 of the Convention when it occurs at the national level. Is there any justification for this apparent double standard?

5. What challenges and risks would you identify for the Court's functioning and jurisprudence after the recent expansion of the Convention system to include

[24] Consider the following provisions in the Statute of the International Court of Justice, the judicial organ of the UN that is briefly described at p. 637, *infra*. Article 3(1) provides that the Court shall consist of fifteen members, 'no two of whom may be nationals of the same state.' Article 9 provides that in electing members, the electors (states) should bear in mind 'that in the body as a whole the representation of the main forms of civilization and of the principal legal systems of the world should be assured.' Under Article 31, if in a case before the Court (only states can be parties in cases) the Bench includes a judge of the nationality of one party, the other party may choose a person to sit as judge. If there is no judge of the nationality of either party, each may proceed to choose a judge.

many states from Central and East Europe with very different legal and political backgrounds and comparatively embryonic systems of judicial independence?

6. One of the great strengths of the European system has been its provision of an effective remedy for individual violations. How do you react to the following question of Tomuschat (op. cit., *supra* p. 588, at 406)?

> Can the system of the European Convention . . . operate successfully only under generally favourable conditions, which make violations an exceptional occurrence, an accident-like event which can easily be remedied? Does it need special procedures for looking into 'situations', where any individual case brought to the attention of the monitoring bodies is representative of a general phenomenon with much larger dimensions . . . ?

7. THE INTER-STATE PROCEDURE: ARTICLE 24

Whereas the individual petition procedure is optional, states parties to the Convention are automatically subject under Article 24 of the Convention (Article 33 of the Revised Convention) to a procedure by which one or more states may allege breaches of the Convention by another state party. Unlike the traditional approach to such cases under the international law of state responsibility for injury to aliens, p. 75, *supra,* it is not necessary for an applicant state to allege that the rights of its own nationals have been violated, although this has generally been the case.

As of March 1995 eleven inter-State applications had been lodged.[25] They concern only six different situations and are as follows:

(a) *Greece vs. the United Kingdom* (2 cases, in 1956 and 1957) relating to Cyprus. The case concerned the declaration of a state of exception in Cyprus (then a British colony) and the introduction of emergency measures by the U.K. Government. A political settlement was reached in 1959 before the Committee of Ministers had formulated its views.

(b) *Austria vs. Italy* in 1960 related to the murder trial of six members of the German-speaking minority in the South Tyrol. The Committee of

[25] See generally P. van Dijk and G. J. H. van Hoof, Theory and Practice of the European Convention on Human Rights 33 (2nd ed., 1990).

Ministers informed the parties of the Commission's view that clemency should be shown and sought to achieve a broader resolution of the issues surrounding the case.

(c) *Denmark, the Netherlands, Norway and Sweden vs. Greece* in 1967 and the same group, minus the Netherlands, again in 1970. This case is considered below.

(d) *Ireland vs. the United Kingdom* (2 cases, in 1971 and 1972) relating to a declared state of emergency in Northern Ireland. The three Commission delegates heard 118 witnesses before the Commission concluded that, although measures for detention without trial were 'strictly required by the exigencies of the situation,' certain interrogation techniques used by the British forces did constitute torture and inhuman treatment. The Irish Government referred the case to the Court which found that while the techniques involved 'inhuman and degrading treatment' (and thus violated Art. 3), they did not amount to 'torture'.

(e) *Cyprus vs. Turkey* (3 cases in 1974, 1975 and 1977) after the Turkish intervention by armed forces in Cyprus. The Commission found in response to the last of these complaints that violations, including large-scale evictions, had occurred. The Committee of Ministers requested Turkey to put an end to them, and urged the parties to resume inter-communal talks.

(f) *Denmark, France, the Netherlands, Norway and Sweden vs. Turkey* in 1982, alleging violations, including torture, by the military government. Under the settlement approved by the Commission, the Turkish Government gave a number of vague undertakings such as a commitment to instruct 'the State Supervisory Council . . . to have special regard to the observance by all public authorities' of the Convention's prohibition against torture. In relation to the lifting of the state of emergency the settlement noted that 'special regard is given to the following declaration made by the Prime Minister of Turkey on 4 April 1985 in Washington D.C.: "I hope that we will be able to lift martial law from the remaining provinces within 18 months"'.[26] The settlement was widely criticised on the ground that it was not based on the respect for human rights required by the Convention.

The Commission's response to the applications brought against Greece in 1967 and 1970 has been described as a model 'for demonstrating both the possibilities and the political limitations of the international protection of human rights.'[27] The case also provides an excellent illustration of fact-finding by the

[26] Report of the Commission, 25 Int'l Legal Materials 308 (1986).
[27] Francis Jacobs, The European Convention on Human Rights 27 (1975).

Commission. It should be noted, however, that this 'model' is in no sense representative of the outcome of the other inter-state cases to date.

The background to the case involved the seizure of power by 'the Greek colonels' in a *coup d'état* in 1967. The military government proclaimed a state of emergency and notified various derogations under Article 15 of the Convention. The applications by Denmark, the Netherlands, Norway and Sweden were declared admissible in January 1968. The Commission had to consider whether there was (under the criteria set for states of emergency under Article 15) an 'emergency threatening the life of the nation,' and if so, whether the measures taken by the military government were 'strictly required by the exigencies of the situation.' The prohibition against torture as well as eleven other articles of the Convention were alleged to have been violated. The Commission initiated a fact-finding exercise on the basis of Articles 28 and 31 of the Convention. The story continues in the following reading.

A. H. ROBERTSON AND MERRILLS, HUMAN RIGHTS IN EUROPE: A STUDY OF THE EUROPEAN CONVENTION ON HUMAN RIGHTS

(3rd ed., 1993), at 278.

. . . [A]fter hearing about thirty witnesses in Strasbourg in November and December 1968, the sub-commission 'fixed 6 February 1969 as the opening date for its investigation in Greece'. In accordance with Article 28 of the Convention the Greek government was consulted about the arrangements, but not about the question whether the investigation should take place. At its request the date for the opening of the investigation was postponed until the beginning of March.

The sub-commission met in Athens on 9 March 1969 and began its investigation the following day. In the main, the Greek government cooperated and facilitated its work and the sub-commission expressed its appreciation of this. Between 10 and 20 March it heard thirty-four witnesses with regard to allegations of torture and twenty witnesses about the existence of a state of emergency. In addition it visited the police stations of the Security Police in Athens and Piraeus and delegated one of its members to visit the Hagia Paraskevi detention camp. When the sub-commission wanted a medical opinion on the condition of witnesses who alleged that they had been tortured, it summoned two forensic experts from the University of Geneva, who were provided with the necessary facilities. On the other hand, the government prevented the sub-commission from hearing thirteen witnesses whom it wished to examine in connection with allegations of torture and also prevented it from inspecting detention camps on the island of Leros and from visiting the Averoff prison in Athens. Since it considered the reasons for these refusals unjustified, the sub-commission terminated its visit and reported the facts to the plenary Commission.

Although some of the facilities it required were refused, the sub-commission succeeded in making a thorough investigation into the allegations of torture and the question of the existence of a state of emergency. This is demonstrated by the fact that, in addition to hearing many witnesses on the torture issue, it visited and photographed the notorious Bouboulinas station of the Security Police in Athens. Furthermore, during the hearings on the largely political question of the existence of a state of emergency, the witnesses included three former Prime Ministers, the Governor of the Bank of Greece, the chief of the armed forces and the Director General of Security at the Ministry of Public Order.

. . . [T]he full Commission, acting on the information obtained by its sub-commission, concluded that there was not a public emergency in Greece at the material time and, as a consequence, the Greek derogations were invalid. It also found that there was a practice of torture and ill-treatment by the Athens Security Police and, furthermore, that there had been violations of eight other articles of the Convention, together with Article 3 of Protocol No. 1. The Commission's conclusions were endorsed by the Committee of Ministers in April 1970 and thereby became decisions of the Committee under Article 32 of the Convention. In the meantime, however, while the Committee had been considering a recommendation from the Consultative Assembly on the situation in Greece, the Greek Minister for Foreign Affairs announced that his government had decided to denounce the Statute and withdraw from the Council of Europe, and also to denounce the European Convention.

NOTE

After the restoration of civilian government, Greece rejoined the Convention regime in 1974. Although the inter-state procedure has been used only in relation to six situations, recall that a comparable procedure involving the Human Rights Committee under Articles 41–43 of the ICCPR has never been invoked.

QUESTIONS

1. Some of the largest European countries, including the United Kingdom, Germany and Spain, have never lodged an inter-state complaint against another European government. Does this mean that this procedure is only likely to be invoked by small countries with limited political clout? If so, would that indicate a fundamental weakness of the procedure?

2. Under what circumstances might the lodging of an inter-state complaint be most productive and when might it be considered counter-productive?

3. Rather than maintaining this rarely used inter-state judicial procedure, would the Council of Europe be better served by a fact-finding system similar to that developed by the UN Commission on Human Rights?

8. THE EUROPEAN COURT IN ACTION: SOME ILLUSTRATIVE CASES

The role of the European Court is of particular importance for several reasons. Consider first the volume of cases. Until the early 1980s the Court dealt with an average of only five to six cases annually. By contrast, between 1989 and 1992 this figure rose to 63 cases. Put another way, the Court heard its first 100 cases between 1959 and 1985, its second hundred between 1985 and 1989. In the space of only six months in 1994, it delivered judgments in 24 cases.

Secondly, many cases that are now brought are more complex and raise more complicated jurisprudential issues than those which tended to come before the Court in its earlier years. Thirdly, because the Court is the longest standing international human rights court (the Inter-American Court, p. 650, *infra*, being the only other international body fulfilling a comparable role), it is inevitably seen as the model against which to measure other regional courts or a possible universal human rights court. Finally, the jurisprudence of the Court (as well as that of the Commission, although to a lesser extent) has been influential in the normative development of other parts of the international human rights system. Thus, the Inter-American Court and the ICCPR Human Rights Committee have frequently referred to judgments of the European Court. Its unacknowledged and perhaps unrecognized influence may have been even greater.

The frequency with which states have been before the Court differs considerably from country to country.[28] Italy holds the record (136 cases as of April 1995 in 82 of which the Court found violations). A large number related, however, to a single issue, delay in bringing prosecutions to trial. The next best customer has been the United Kingdom (73 cases, of which 35 have been held to involve violations), followed by France (29 of 62 cases involving violations), Austria (27 of 55), Sweden (21 of 32) and Belgium (20 of 34). Towards the other end of the scale 11 cases of 28 have involved violations by Germany and only two in six for Denmark and one in three for Norway. To some extent these figures reflect the extent to which adequate domestic remedies are available, which in turn is influenced by whether the Convention has been incorporated in domestic law. See pp. 725–38, *infra*.

[28] For a detailed guide see Donna Gomien, Judgments of the European Court of Human Rights: Reference Charts (Council of Europe, 1995).

Before examining some of the Court's case law consider the following description of its role.

J.G. MERRILLS, THE DEVELOPMENT OF INTERNATIONAL LAW BY THE EUROPEAN COURT OF HUMAN RIGHTS

(2nd ed., 1993), at 9.

. . . Since there is no aspect of national affairs which can be said to be without implications for one or other of the rights protected by the Convention, there is no matter of domestic law and policy which may not eventually reach the European Court.

In terms of the character of the Court's work this has a double significance. In the first place, it means that the Court is required to investigate and pronounce on many issues which have not hitherto been regarded as appropriate subjects for international adjudication, which in turn raises the question of how far the Court is entitled to go in monitoring the laws and practices of the Contracting States. This is essentially a question about the impact of human rights law on national sovereignty and the role of international adjudication in establishing and enforcing uniform standards. The other way in which the nature of the Court's work is significant is that as a tribunal dealing with human rights, the Court is required to decide difficult and important questions concerning the proper relationship between the individual and the State. The issue here is what it means to have a particular right and how the balance is to be struck between such competing interests as, for example, privacy and national security, or prompt trial and the limitation of public expenditure.

. . . As a commentator has said of the Strasbourg [European Convention] institutions, 'Conceived as regional international organs with limited jurisdiction and even more limited powers, they have gradually acquired the status and authority of constitutional tribunals'. As we shall see, this transformation of the Convention and its institutions, which is still in progress, is the key to understanding the wider significance of the Court's decisions.

. . .

Although the Court is careful to avoid trespassing on what it sees as the function of the national authorities, investigating 'the international responsibility of the State' almost always calls for scrutiny of certain aspects of domestic law. Thus another feature of the Court's work is that the very nature of the obligations with which the Court is concerned makes the adequacy of the Contracting States' domestic law a matter for investigation.

. . .

The Court's decisions are binding on the Contracting States, which under Article 53 of the Convention, 'undertake to abide by the decision of the Court in any case to which they are parties'. When the Court concludes that the Convention has been violated, it is therefore incumbent on the respondent to take whatever steps may be needed to put matters right.

. . .

In each decision the Court is not just spelling out the obligations of the State which happens to be involved in the particular case.

. . . The common law, advancing from precedent to precedent, has a counterpart, then, in the developing law of the European Convention.

Judgments have this wider significance because the Court consistently seeks to justify its decisions in terms which treat its existing case-law as authoritative. In other words, it follows judicial precedent.

. . .

NOTE

The following decisions explore some of the characteristic issues when international tribunals decide human rights issues that may deeply implicate the internal order of states. They involve two fields of work—national security issues and rights of homosexuals—among the vast range of matters that have come before the Court, a range characteristic of national courts such as the Supreme Court of the United States that decide constitutional issues. Indeed, the United States offers a useful comparison: relationships between the European Court and the (judiciaries of the) member states of the European Convention, on the one hand; relationships between the U.S. Supreme Court and the (judiciaries of the) component states of the United States federalism, on the other.

The differences between, say, the work of the UN Commission on Human Rights or even the ICCPR Human Rights Committee and the European Court are striking. Of course moral and political premises inform the work of the Court and sometimes become explicit. But its opinions take the forms of the law—a stress on the text of the Convention, the facts of the dispute, arguments over interpretation of the text, reflection on the institutional role of the Court in relation to national political orders, the ultimate decision applying the Convention in a decision binding the states parties. From this point of view, a study of the European Court's decisions best illustrates the promise of an international (regional) *legal* order governing national human rights issues.

The question inevitably arises of how transferable this experience of the European human rights system may be. You should bear in mind that question—whether, for example, an equally effective judicial system functioning with so high a record of compliance by states would be as likely within a universal human rights regime such as the ICCPR, or in a regional regime like the Americas or Africa.

The decisions below fall into two parts. (1) *Brogan v. United Kingdom* deals with one of the few situations where the European human rights system has been required to deal with serious and ongoing violent conflict. No such violence in Europe has been longer lasting than the conflict in Northern Ireland over whether it remains part of the U.K. *Brogan* is then not typical of the case law of the European Court; its context of terrorism and extended detention resembles the kinds of circumstances brought to the attention of universal human rights organs like the UN Commission on Human Rights, or of the Inter-American Commission and Court on Human Rights. Indeed, the problems in the *Brogan* case can be compared with the situation before the Inter-American Court of Human Rights in the *Velasquetz-Rodríguez* decision, p. 650, *infra*.

(2) *Norris v. Ireland*, the second principal decision below of the European Court, is more reminiscent of the kinds of issues apt to be decided by constitutional courts within liberal democracies—and indeed, all member states of the European system have democratically elected governments. The decision deals with the issue of the right to privacy in relation to homosexual relations. For purposes of comparison of the substantive law in the two decisions and of the European system with the United States federalism, the materials include the decision of the U.S. Supreme Court in *Bowers v. Hardwick* on a similar issue of privacy and homosexual relations.

The dominant jurisprudential and political theme in these cases is the doctrine (or theory, or principle) of margin of appreciation, critical to an understanding of the dilemmas before this international court and the ways in which it tries to come to terms with them. After the decisions, there follow excerpts from writings of scholars that probe the ways in which the Court has employed and developed this theme.

BROGAN v. UNITED KINGDOM

European Court of Human Rights, 1988
Ser. A, No. 145-B, 11 EHRR 117.

[In the 1970s and 1980s, terrorism in Northern Ireland caused thousands of deaths and tens of thousands of injuries. Parliamentary legislation in 1974 and 1976, subject to renewal each year, gave to the police special powers of arrest and detention. The 1976 Act was renewed annually until 1984, when it was replaced by the Prevention of Terrorism Act (1984 Act) proscribing both the Provisional Irish Republican Army (IRA) and the Irish National Liberation Army (INLA) as terrorist organizations. Annual reports on the 1984 Act were required to be submitted to Parliament before annual renewal. The authors of these reports had concluded that ongoing terrorism made special powers of arrest and extended detention indispensable.

Section 12 of the 1984 Act authorized a constable's arrest without warrant of a person 'who he has reasonable grounds for suspecting to be' a person involved with 'acts of terrorism' in Northern Ireland. It provided for detention after arrest for not over 48 hours, except that 'the Secretary of State may, in any particular case, extend the period of 48 hours by a period or periods specified by him' not to exceed five days. Police requests for extended detention were forwarded to the Secretary of State for Northern Ireland. The 1984 Act contained no criteria governing decisions to extend the original period of detention. From 1984–87, about two percent of the police requests were denied by the Secretary of State.

These provisions of the 1984 Act displaced the normal legislative rule that a person arrested without warrant and held for 24 hours must then be brought before a Magistrates Court as soon as practicable, in any event not more than 48 hours after arrest. The principal remedies of detained persons under the 1984 Act were an application for a writ of habeas corpus (not made by any of the four persons involved in this case) and a civil action claiming damages for false imprisonment.

Under ordinary law, police had no power to arrest and detain a person merely to make inquiries, and arrest without warrant required reasonable suspicion that a specific crime had been committed. The 1984 Act differed. No charge had to be preferred during the permitted period of detention under that act. The detention was not necessarily the first step in a criminal proceeding leading to judicial investigation of a charge against the detained person.

Four persons who later initiated proceedings against the U.K. before the European Commission of Human Rights were arrested and detained in 1984 and 1985 under the 1984 Act. Each was told by the police that there were reasonable grounds for suspecting his involvement in terrorism, and was cautioned that he need not say anything. The Secretary of State agreed in each case to a police request for extension of detention. None was brought before a judge or charged after their release, which occurred after periods of detention ranging from 4 days and 6 hours to 6 days and 16 hours.

Two typical cases were as follows: Brogan was questioned after his arrest about his suspected involvement in an attack on a police patrol leading to death and injuries, as well as about his suspected membership in the IRA. He maintained a total silence to his police questioners throughout his detention. Tracey was questioned about an armed robbery and about a conspiracy to murder members of the security forces. He too remained silent. Each was visited during the period of detention by his solicitor.

The four applicants argued before the Commission that the U.K. had violated provisions of Article 5 of the European Convention that are set forth below. In its report in 1987, the Commission concluded by different votes, and in some cases with respect to certain but not all applicants, that two provisions of Article 5, but not other provisions relied on by the applicants, had been violated. It referred the case to the European Court of Human Rights. Excerpts follow from the Court's Judgment and from several concurring and dissenting opinions.]

II. General approach

48. The government has adverted extensively to the existence of particularly difficult circumstances in Northern Ireland, notably the threat posed by organised terrorism.

The Court, having taken notice of the growth of terrorism in modern society, has already recognised the need, inherent in the Convention system, for a proper balance between the defence of the institutions of democracy in the common interest and the protection of individual rights.

The government informed the Secretary General of the Council of Europe on 22 August 1984 that it was withdrawing a notice of derogation under Article 15 which had relied on an emergency situation in Northern Ireland. . . .

Consequently, there is no call in the present proceedings to consider whether any derogation from the United Kingdom's obligations under the Convention might be permissible under Article 15 by reason of a terrorist campaign in Northern Ireland. Examination of the case must proceed on the basis that the Articles of the Convention in respect of which complaints have been made are fully applicable. This does not, however, preclude proper account being taken of the background circumstances of the case. . . .

III. Alleged breach of Article 5(1)

49. The applicants alleged breach of Article 5(1) of the Convention, which, in so far as relevant, provides:

> Everyone has the right to liberty and security of person. No one shall be deprived of his liberty save in the following cases and in accordance with a procedure prescribed by law:
>
> . . .
>
> (c) the lawful arrest or detention of a person effected for the purpose of bringing him before the competent legal authority on reasonable suspicion of having committed an offence . . .

. . .

52. Article 5(1)(c) . . . requires that the purpose of the arrest or detention should be to bring the person concerned before the competent legal authority.

. . .

The applicants . . . referred to the fact that they were neither charged nor brought before a court during their detention. No charge had necessarily to follow an arrest under section 12 of the 1984 Act and the requirement under the ordinary law to bring the person before a court had been made inapplicable to detention under this Act. In the applicants' contention, this was therefore a power of administrative detention exercised for the purpose of gathering information, as the use in practice of the special powers corroborated.

53. The court is not required to examine the impugned legislation *in abstracto*, but must confine itself to the circumstances of the case before it.

. . . [S]ub-paragraph (c) of Article 5(1) does not presuppose that the police should have obtained sufficient evidence to bring charges, either at the point of arrest or while the applicants were in custody.

Such evidence may have been unobtainable or, in view of the nature of the suspected offences, impossible to produce in court without endangering the lives of others. There is no reason to believe that the police investigation in this case was not in good faith or that the detention of the applicants was not intended to further that investigation by way of confirming or dispelling the concrete suspicions which, as the Court has found, grounded their arrest. . . .

Their arrest and detention must therefore be taken to have been effected for the purpose specified in paragraph (1)(c).

54. In conclusion, there has been no violation of Article 5(1).

IV. Alleged breach of Article 5(3)

55. . . . The applicants claimed, as a consequence of their arrest and detention under this legislation, to have been the victims of a violation of Article 5(3) which provides;

> Everyone arrested or detained in accordance with the provisions of paragraph (1)(c) of this Article shall be brought promptly before a judge or other officer authorised by law to exercise judicial power and shall be entitled to trial within a reasonable time or to release pending trial. Release may be conditioned by guarantees to appear for trial.

The applicants noted that . . . there was no plausible reason why a seven-day detention period was necessary, marking as it did such a radical departure from ordinary law. . . . Nor was there any justification for not entrusting such decisions to the judiciary of Northern Ireland.

56. The government . . . drew attention to the difficulty faced by the security forces in obtaining evidence which is both admissable and usable in consequence of training in anti-interrogation techniques adopted by those involved in terrorism. Time was also needed to undertake necessary scientific examinations, to correlate information from other detainees and to liaise with other security forces. The government claimed that the need for a power of extension of the period of detention was borne out by statistics. For instance, in 1987 extensions were granted in Northern Ireland in respect of 365 persons. Some 83 were detained in excess of five days and of this number 39 were charged with serious terrorist offences during the extended period.

As regards the suggestion that extensions of detention beyond the initial 48-hour period should be controlled or even authorised by a judge, the government pointed out the difficulty, in view of the acute sensitivity of some of the information on which the suspicion was based, of producing it in court. Not only would the court have to sit *in camera* but neither the detained person nor his legal advisers could be present or told any of the details. . . . If entrusted with the power to grant extensions of detention, the judges would be seen to be exercising an executive rather than a judicial function. It would add nothing to the safeguards against abuse which the present arrangements are designed to achieve and could lead to unanswerable criticism of the judiciary. In all the circumstances, the Secretary of State was better placed to take such decisions and to ensure a consistent approach. . . .

. . .

The assessment of 'promptness' has to be made in the light of the object and purpose of Article 5. The Court has regard to the importance of this Article in the Convention system: it enshrines a fundamental human right, namely the protection of the individual against arbitrary interferences by the State with his right to liberty. Judicial control of interferences by the executive with the individual's right to liberty is an essential feature of the guarantee embodied in Article 5(3), which is intended to minimise the risk of arbitrariness. Judicial control is implied by the rule of law, 'one of the fundamental principles of a democratic society . . . , which is expressly referred to in the Preamble to the Convention' and 'from which the whole Convention draws its inspiration'.

59. The obligation expressed in English by the word 'promptly' and in French by the word '*aussitôt*' is clearly distinguishable from the less strict requirement in the second part of paragraph 3 ('reasonable time'/'*délai raisonnable*') and even from that in paragraph 4 of Article 5 ('speedily'/'*à bref délai*'). . . . Thus confronted with versions of a law-making treaty which are equally authentic but not exactly the same, the Court must interpret them in a way that reconciles them as far as possible and is most appropriate in order to realise the aim and achieve the object of the treaty.

The use in the French text of the word '*aussitôt*', with its constraining connotation of immediacy, confirms that the degree of flexibility attaching to the notion of 'promptness' is limited, even if the attendant circumstances can never be ignored for the purposes of the assessment under paragraph 3. . . .

. . .

61. The investigation of terrorist offences undoubtedly presents the authorities with special problems, partial reference to which has already been made under Article 5(1). The Court takes full judicial notice of the factors adverted to by the government in this connection. It is also true that in Northern Ireland the referral of police requests for extended detention to the Secretary of State and the individual scrutiny of each police request by a Minister do provide a form of executive control. In addition, the need for the continuation of the special powers has been constantly monitored by Parliament and their operation regularly reviewed by independent personalities. The Court accepts that, subject to the existence of adequate safeguards, the context of terrorism in Northern Ireland has the effect of prolonging the period during

which the authorities may, without violating Article 5(3), keep a person suspected of serious terrorist offences in custody before bringing him before a judge or other judicial officer.

. . .

62. As indicated above, the scope for flexibility in interpreting and applying the notion of 'promptness' is very limited. In the Court's view, even the shortest of the four periods of detention namely the four days and six hours spent in police custody by Mr. McFadden, falls outside the strict contraints as to time permitted by the first part of Article 5(3). To attach such importance to the special features of this case as to justify so lengthy a period of detention without appearance before a judge or other judicial officer would be an unacceptably wide interpretation of the plain meaning of the word 'promptly'. An interpretation to this effect would import into Article 5(3) a serious weakening of a procedural guarantee to the detriment of the individual and would entail consequences impairing the very essence of the individual and would entail consequences impairing the very essence of the right protected by this provision. The Court thus has to conclude that none of the applicants was either brought 'promptly' before a judicial authority or released 'promptly' following his arrest. The undoubted fact that the arrest and detention of the applicants were inspired by the legitimate aim of protecting the community as a whole from terrorism is not on its own sufficient to ensure compliance with the specific requirements of Article 5(3).

There has thus been a breach of Article 5(3) in respect of all four applicants.

[The Court considered Article 50 of the Convention stating that in event of a violation, if the internal law of a respondent state 'allows only partial reparation,' the Court's decision shall 'afford just satisfaction to the injured party.' The applicants sought exemplary damages because of the 'conscious and flagrant' character of the breaches of the Convention, suggesting assessment on the basis of £2,000 per hour for each hour of wrongful detention. The Court reserved this matter for later decision, 'taking due account of the possibility of an agreement between the respondent State and the applicants.']

Joint Dissenting Opinion of Judges Thór Vilhjálmssón, Binschedler-Robert, Gŏlcüklü, Matscher and Valticos

. . .

The background to the instant case is a situation which no one would deny is exceptional. Terrorism in Northern Ireland has assumed alarming proportions and has claimed more than 2,000 victims who have died following actions of this kind. . . .

It is therefore necessary to weigh carefully, on the one hand, the rights of detainees and, on the other, those of the population as a whole, which is seriously threatened by terrorist activity.

. . .

While considering, therefore, that there was no breach of Article 5(3) in the instant case, we are anxious to stress that this view can be maintained only in so far as such exceptional conditions prevail in the country, and that the authorities should monitor the situation closely in order to return to the practices of ordinary law as soon as more normal conditions are restored, and even that, until then, an effort should be made to reduce as much as possible the length of time for which a person is detained before being brought before a judge.

. . .

Dissenting Opinion of Judges Walsh and Carrillo Salcedo in respect of Article 5(1)(c)

We believe that Article 5 of the European Convention on Human Rights does not afford to the State any margin of appreciation. If the concept of a margin of appreciation were to be read into Article 5, it would change the whole nature of this all-important provision which would then become subject to executive policy.

. . . In our opinion Article 5 does not permit the arrest and detention of persons for interrogation in the hope that something will turn up in the course of the interrogation which would justify the bringing of a charge.

In our view the arrests in the present cases were for the purpose of interrogation at a time when there was no evidential basis for the bringing of any charge against them. No such evidence ever emerged and eventually they had to be released. That the legislation in question is used for such a purpose is amply borne out by the fact that since 1974 15,173 persons have been arrested and detained in the United Kingdom pursuant to the legislation yet less than 25 per cent. of those persons, namely 3,342, have been charged with any criminal offence arising out of the interrogation including offences totally unconnected with the original arrest and detention. Still fewer of them have been convicted of any offence of a terrorist type.

The Convention embodies the presumption of innocence and thus enshrines a most fundamental human right, namely the protection of the individual against arbitrary interference by the State with his right to liberty. The circumstances of the arrest and detention in the present cases were not compatible with this right and accordingly we are of the opinion that Article 5(1) has been violated.

The undoubted fact that the arrest of the applicants was inspired by the legitimate aim of protecting the community as a whole from terrorism is in our opinion not sufficient to ensure compliance with requirements of Article 5(1)(c). . . .

Partly Dissenting Opinion of Sir Vincent Evans

. . .

The commission for its part has for more than 20 years taken the view that in normal cases a period of up to four days before the detained person is brought before a judge is compatible with the requirement of promptitude and that a somewhat longer period is justifiable in some circumstances. The Court has not hitherto cast doubt on the Commission's view in these respects. If anything, the Court's judgments in the *De Jong, Baljet* and *Van Den Brink* and other cases have tended by implication to confirm it.

Furthermore, the Court has consistently recognised that States must, in assessing the compatibility of their laws and practices with the requirements of the Convention, be permitted a 'margin of appreciation' and that inherent in the whole Convention is the search for a fair balance between the demands of the general interest of the community and the protection of the individual's fundamental rights. . . .

In my opinion the case law thus far developed constitutes a reasonable interpretation of Article 5(3) and in particular of the word 'promptly'.

4. The need to assess the issue of promptness according to the special features of the case and to strike a fair balance between the different rights and interests involved are considerations which are surely relevant in the special circumstances of the situation in Northern Ireland where more than 30,000 persons have been killed, maimed or injured as a direct result of terrorist activity in the last twenty years. . . .

. . . The need for the exceptional powers under section 12 to which such factors give rise is supported by the statistics quoted in . . . paragraph [56] of the judgment—that in 1987, for instance, of some 83 persons detained in excess of five days, 39 were charged with serious terrorist offences during the extended period.

Dissenting Opinion of Judge Martens

. . .

3. As the Court rightly recalls in paragraph 48 of its judgment, terrorism is a feature of modern life, which has attained its present extent and intensity only since the Convention was drafted. Terrorism—and particularly terrorism on the scale obtaining in Northern Ireland—is the very negation of the principles the Convention stands for and should therefore be combatted as vigorously as possible. It seems obvious that to suppress terrorism the exec-

utive needs extraordinary powers, just as it seems obvious that governments should to a large extent be free to choose the ways and means which they think most efficacious for combatting terrorism. Of course, in combatting terrorism the States Parties to the Convention have to respect the rights and freedoms secured therein to everyone. I subscribe to that and I am aware of the danger of measures being taken which, as the Court has put it, may undermine or even destroy democracy on the ground of defending it. But I think that this danger must not be exaggerated—especially with regard to States which have a long and firm tradition of democracy—and should not lead to the wings of national authorities being excessively clipped, for that would unduly benefit those who do not hesitate to trample on the rights and freedoms of others.

4. It goes without saying that a person against whom there is a reasonable suspicion of being involved in acts of terrorism should be free from torture or inhuman or degrading treatment. But it seems to me legitimate to ask whether he may not be detained, before being brought before a judge, for a somewhat longer period than is acceptable under ordinary criminal law. In this connection, I consider that the Court by saying in the second section of paragraph 58 of its judgment, that Article 5 'enshrines a fundamental human right' somewhat overestimates the importance of this provision in the Convention system. Undoubtedly, the right to liberty and security of person is an important right, but it does not belong to that small nucleus of rights from which no derogation is permitted. This means that there is room for weighing the general interest in an effective combatting of terrorism against the individual interests of those who are arrested on a reasonable suspicion of involvement in acts of terrorism. . . .

. . .

Striking a fair balance between the interests of the community that suffers from terrorism and those of the individual is particularly difficult and national authorities, who from long and painful experience have acquired a far better insight into the requirements of effectively combatting terrorism and of protecting their citizens than an international judge can ever hope to acquire from print, are in principle in a better position to do so than that judge!

It is in this context that three factors seem to me to be of importance:

(i) The first factor is the particular extent, vehemence and persistence of the terrorism that has raged since 1969 in Northern Ireland. . . .
(ii) The second factor is that we are undoubtedly dealing with a society which has been a democracy for a long time and as such is fully aware both of the importance of the individual right to liberty and of the inherent dangers of giving too wide a power of detention to the executive.
(iii) The third factor is that the United Kingdom legislature, apparently being aware of those dangers, has each time granted the extraordinary

> powers only for a limited period, *i.e.* one year on each occasion, and only after due inquiry into the continued need. . . .

In my opinion, these factors also make it highly desirable for an international judge to adopt an attitude of reserve.

Against this background I think that the Court can find that the United Kingdom, when enacting and maintaining section 12 of the 1984 Act, overstepped the margin of appreciation it is entitled to under Article 5(3) only if it considers that the arguments for maintaining the seven-day period are wholly unconvincing and cannot be reasonably defended. In my opinion that condition has not been satisfied.

. . .

NOTE

Following the *Brogan* decision, rather than revise the Prevention of Terrorism legislation or the practice of non-judicial supervision of extended detention, the U.K. government lodged a derogation under Article 15 of the Convention on grounds of 'public emergency'. Article 5 was again at issue before the European Court in *Brannigan and McBride v. United Kingdom*, Judgment of 26 May 1993, 17 EHRR 539. The Court held by 22 votes to 4 that the derogation by the U.K. was justified under Article 15 and that detention of suspected terrorists in Northern Ireland without judicial control, for 6 days and 14 hours and 4 days and 6 hours respectively, was justified by the public emergency.

In defining its standard of review under Article 15, the Court stated that it fell to each state party, with its responsibility for the life of its nation, to determine when that life was threatened by a public emergency and, if so, what measures were required to overcome the emergency.

> By reason of their direct and continuous contact with the pressing needs of the moment, the national authorities are in principle in a better position than the international judge to decide both on the presence of such an emergency and on the nature and scope of derogations necessary to avert it. Accordingly, in this matter a wide margin of appreciation should be left to the national authorities.
>
> Nonetheless, Contracting Parties do not enjoy an unlimited power of appreciation. It is for the Court to rule on whether *inter alia* the States have gone beyond the 'extent strictly required by the exigencies' of the crisis. The domestic margin of appreciation is thus accompanied by a European supervision. At the same time, in exercising its supervision, the Court must give appropriate weight to such relevant factors as the nature of the rights affected by the derogation, the circumstances leading to, and the duration of, the emergency situation.

'Making its own assessment' in the light of all material before it about terrorism in Northern Ireland and elsewhere in the U.K., 'the Court considers there can be no doubt that such a public emergency existed at the relevant time.' It referred to the U.K. Government as having direct responsibility 'for establishing the balance between the taking of effective measures to combat terrorism on the one hand, and respecting individual rights on the other.' In view of all considerations, 'it cannot be said that the Government have exceeded their margin of appreciation in deciding, in the prevailing circumstances, against judicial control.'

> Having regard to the nature of the terrorist threat in Northern Ireland, the limited scope of the derogation and the reasons advanced in support of it, as well as the existence of basic safeguards against abuse, the Court takes the view that the Government have not exceeded their margin of appreciation in considering that the derogation was strictly required by the exigencies of the situation.

QUESTIONS

1. 'The U.K. effectively gets to benefit in *Brogan* from a derogation under Article 15 of the Convention even though it had withdrawn its notice of derogation.'

a. Do you agree? If so, in what respects does the Court treat the case as if Article 15 were applicable, and is such treatment justified under the Convention? Is the statement better applied to the dissent of Judge Martens than to the Court's Judgment?

b. How do you distinguish the 'balancing' in several opinions of 'interests' like protection of state institutions and protection of individual rights, from judicial analysis of the case if Article 15 had been applicable? Do they amount to the same thing?

2. The Court's opinion does not mention the margin of appreciation. Three of the other opinions invoke that principle in different ways. What operative force does the principle seem to have? What premises about state sovereignty and the role of the Court seem to underlie its invocation?

3. Does the fact that Britain has long been a democracy with a respected Parliament appear to influence any of the opinions? Does it appear to affect the burden of proof of the complainants? Might opinions have been written differently if, on similar facts, less established democracies like Portugal or Spain were the respondent states?

4. In *Brannigan and McBride v. U.K.*, the Court's opinion gives the margin of appreciation a central position. Why the difference from *Brogan*? Do the

criteria in Article 15 for a justified derogation inevitably draw the Court into discussing matters raised by the margin of appreciation? What are those matters?

5. In 1993 in *W v. Switzerland*, 14 Hum. Rts L. J. 178 (1993), a businessman was held in pre-trial detention for over four years on charges relating to serious and complex corporate fraud. The Government justified the need for detention on the grounds of the danger of W absconding, the risk that he would collude with other defendants, and the need to prevent further offences. The Commission held, by 19 votes to 3, that Article 5(3) had been violated. A Chamber of the Court overturned this finding by 5 votes to 4. The dissenting judges strongly criticized the majority, with Judge Pettiti, in particular, referring to the undermining of the principle that 'liberty is the rule, detention the exception,' suggesting that the presumption of innocence had been abandoned, and criticizing the 'perverse effects' of such lengthy pre-trial detention. At what point in such types of cases should the Court defer to national-level findings as to the need for an exceptional period of detention? Should rejection of the position of the national executive and courts require a showing of bad faith or procedural error?

NOTE

In *Otto-Preminger-Institut v. Austria*,[29] the European Court of Human Rights decided in 1994 by 6 votes to 3 that the seizure and forfeiture of a blasphemous film did not violate the freedom of expression guaranteed by Article 10 of the European Convention. The applicant association had advertised the screening of the film, *Das Liebeskonzil*, based on an 1894 play, which,

> . . . portrays the God of the Jewish religion, the Christian religion and the Islamic religion as an apparently senile old man prostrating himself before the devil with whom he exchanges a deep kiss and calling the devil his friend . . . . Other scenes show the Virgin Mary permitting an obscene story to be read to her and the manifestation of a degree of erotic tension between the Virgin Mary and the devil. The adult Jesus Christ is portrayed as a low grade mental defective and in one scene is shown lasciviously attempting to fondle and kiss his mother's breasts, which she is shown as permitting.

The film was presented by the association as a 'satirical tragedy.' 'Trivial imagery and absurdities of the Christian creed are targeted in a caricatural mode and the relationship between religious beliefs and worldly mechanisms of oppression is investigated.'

The Innsbruck Regional Court in Austria ordered seizure and forfeiture of the film under Section 188 of the Austrian Penal Code for the criminal offense

[29] Judgment of 20 September 1994—No. 11/1993/406/485, reprinted in 15 Hum. Rts. L. J. 371 (1994).

of 'disparaging religious precepts.' The criminal proceedings against the association were eventually dropped.

Since there was no dispute that the seizure constituted an interference with the association's freedom of expression, the European Court considered whether the seizure was permissible under the conditions set by Article 10, para. 2, p. 615, *infra*.

The Court concluded that the interference had the 'legitimate aim' of protecting the rights of others to freedom of religion. Interpreting Article 9 of the Convention to include the right to respect for one's religious feelings, the Court found that such considerations outweighed the film's contribution to public debate. The Court reasoned:

> The respect for the religious feelings of believers as guaranteed in Article 9 can legitimately thought to have been violated by provocative portrayals of objects of religious veneration; and such portrayals can be regarded as malicious violation of the spirit of tolerance, which must also be a feature of democratic society. The Convention is to be read as a whole and therefore the interpretation and application of Article 10 in the present case must be in harmony with the logic of the Convention. (Para. 47). . . . [T]he Court accepts that the impugned measures pursued a legitimate aim under Article 10 para. 2, namely 'the protection of the rights of others.'

The Court stressed that freedom of expression applies not only to ideas that are favorably received, but also to those

> that shock, offend or disturb the State or any sector of the population. Such are the demands of that pluralism, tolerance and broadmindedness without which there is no 'democratic society.'

Nonetheless, people exercising their rights under Article 10 were subject to duties, among which could legitimately be included

> an obligation to avoid as far as possible expressions that are gratuitously offensive to others and thus an infringement of their rights, and which therefore do not contribute to any form of public debate capable of further progress in human affairs.

It determined that the seizure could be considered 'necessary in a democratic society.' There was no 'uniform conception of the significance of religion in society' throughout Europe, 'even within a single country.' A 'certain margin of appreciation is therefore to be left to the national authorities in assessing the existence and extent of the necessity of such interference.' It is 'for the national authorities, who are better placed than the international judge, to assess the need for such a measure in the light of the situation obtaining locally.' Given that the Tyrolean population was 87% Roman Catholic, the Court found that the Austrian authorities had acted within their margin of appreciation 'to ensure religious peace in that region and to prevent that some people should feel the object of attacks on their religious beliefs in an unwarranted and offensive manner.'

Three judges dissented. Given the precautions against offense to viewers taken by the association through a warning announcement, the showing of the film to a paid audience only, and the restriction of viewing to those over 17 years of age, the dissent found the seizure and forfeiture to be disproportionate to the aim pursued, and thus not necessary in a democratic society.

QUESTION

During the colonial period, the British colonial government in India enacted several laws as part of the Indian Penal Code that defined offences including 'defiling a place of worship,' 'acts insulting religion or religious beliefs,' 'disturbing a religious assembly,' 'trespassing on burial grounds,' and 'utterances wounding religious feelings.' Punishments were a maximum of two years imprisonment, a fine, or both.

These laws were amended or supplemented by the Government of Pakistan. Section 295-B of the Pakistan Penal Code, added in 1982, provided:

> Whoever willfully defiles, damages or desecrates a copy of the Holy Quran or an extract therefrom or uses it in any derogatory manner or for any unlawful purpose shall be punishable with imprisonment for life.

Section 295-C was enacted in 1986. It states:

> Whoever by words, either spoken or written, or by any visible representation, or by any imputation, innuendo, or insinuation, directly or indirectly, defiles the sacred name of the Holy Prophet Mohammed (peace be upon him) shall be punished with death, or imprisonment for life and shall also be liable to fine.

Compare these statutes with the Austrian criminal law described in the *Otto-Preminger-Institut* case and with the proceedings in that case. What are the salient differences? How would the Pakistani statutes be judged under the European Convention?

NOTE

The *Handyside* decision formed part of the background of jurisprudence of the European Court when it decided *Norris v. Ireland, infra*. There follow excerpts from *Handyside* that pertain only to issues bearing directly on the *Norris* decision.

HANDYSIDE CASE

European Court of Human Rights, 1976
Ser. A, No. 24, 1 EHRR 737.

[Handyside, a U.K. citizen who owned a publishing house, had advertised in 1971 and was en route to publishing 'The Little Red Schoolbook,' an English translation of a Danish book published in a number of continental countries which was meant to serve as a reference work for schoolchildren. It treated education and teaching in general, advancing in many instances unorthodox, counter-cultural perspectives. About ten percent of the book dealt with sexual matters, including 'reference' sections on masturbation, intercourse, contraceptives, homosexuality, pornography and venereal disease. Here too the advice offered was unorthodox: experiment, learn for yourself, don't fear disapproval. The publicity given the book drew adverse reactions from quarters such as schools, churches and parents' groups.

Before publication, government authorities acted under the Obscene Publications Acts of 1959 and 1964, seizing all found copies of the book. Handyside was convicted of a violation of the Acts before a Magistrate's court, which fined him and ordered that all books be destroyed. The appellate court affirmed, concluding that the book, seen as a whole, would tend to corrupt and deprave a significant portion of children who would read it, and that despite its virtues as an educational document, the book could not benefit from a statutory defense to the effect that on balance publication could be justified as being in the public good. A revised edition of the book was published later in 1971 and was not interfered with.

In 1972 Handyside filed an application against the United Kingdom before the European Commission of Human Rights, alleging several violations of the Convention but stressing Article 10. The Commission in effect concluded that there had been no breach of the Convention, and it referred the case to the European Court of Human Rights in 1976. The Court decided that no breach of Article 10 (or any other article) had been established. Article 10 provides:

> 1. Everyone has the right to freedom of expression. This right shall include freedom to hold opinions and to receive and impart information and ideas without interference by public authority and regardless of frontiers. This Article shall not prevent States from requiring the licensing of broadcasting, television or cinema enterprises.
>
> 2. The exercise of these freedoms, since it carries with it duties and responsibilities, may be subject to such formalities, conditions, restrictions or penalties as are prescribed by law and are necessary in a democratic society, in the interests of national security, territorial integrity or public safety, for the prevention of disorder or crime, for the protection of health or morals, for the protection of the reputation or rights of others, for preventing the disclosure of information received in confidence, or for maintaining the authority and impartiality of the judiciary.

After review of the book and the evidence before the English courts, the Court concluded that the English judges had a basis, in the exercise of the discretion left them by the Convention, for finding that the book would have a pernicious effect on the morals of the likely readers between ages 12 and 18. The excerpts below from the opinion treat matters relevant to a determination whether the U.K. restrictions on publication were 'necessary in a democratic society' within the meaning of Article 10(b).]

48. . . . These observations apply, notably, to Article 10 § 2. In particular, it is not possible to find in the domestic law of the various Contracting States a uniform European conception of morals. The view taken by their respective laws of the requirements of morals varies from time to time and from place to place, especially in our era which is characterised by a rapid and far-reaching evolution of opinions on the subject. By reason of their direct and continuous contact with the vital forces of their countries, State authorities are in principle in a better position than the international judge to give an opinion on the exact content of these requirements as well as on the 'necessity' of a 'restriction' or 'penalty' intended to meet them. . . . [I]t is for the national authorities to make the initial assessment of the reality of the pressing social need implied by the notion of 'necessity' in this context.

Consequently, Article 10 § 2 leaves to the Contracting States a margin of appreciation. This margin is given both to the domestic legislator ('prescribed by law') and to the bodies, judicial amongst others, that are called upon to interpret and apply the laws in force. . . .

49. Nevertheless, Article 10 § 2 does not give the Contracting States an unlimited power of appreciation. The Court, which, with the Commission, is responsible for ensuring the observance of those States' engagements (Article 19), is empowered to give the final ruling on whether a 'restriction' or 'penalty' is reconcilable with freedom of expression as protected by Article 10. The domestic margin of appreciation thus goes hand in hand with a European supervision. Such supervision concerns both the aim of the measure challenged and its 'necessity'. . . .

The Court's supervisory functions oblige it to pay the utmost attention to the principles characterising a 'democratic society'. Freedom of expression constitutes one of the essential foundations of such a society, one of the basic conditions for its progress and for the development of every man. Subject to paragraph 2 of Article 10, it is applicable not only to 'information' or 'ideas' that are favourably received or regarded as inoffensive or as a matter of indifference, but also to those that offend, shock or disturb the State or any sector of the population. Such are the demands of that pluralism, tolerance and broadmindedness without which there is no 'democratic society'. This means, amongst other things, that every 'formality', 'condition', 'restriction' or 'penalty' imposed in this sphere must be proportionate to the legitimate aim pursued.

From another standpoint, whoever exercises his freedom of expression undertakes 'duties and responsibilities' the scope of which depends on his situation and the technical means he uses. The Court cannot overlook such a person's 'duties' and 'responsibilities' when it enquires, as in this case, whether 'restrictions' or 'penalties' were conducive to the 'protection of morals' which made them 'necessary' in a 'democratic society'.

50. It follows from this that it is in no way the Court's task to take the place of the competent national courts but rather to review under Article 10 the decisions they delivered in the exercise of their power of appreciation.

. . .

NOTE

The judgment below of the European Court of Human Rights in *Norris v. Ireland* refers both to the *Handyside* case, *supra*, and in some detail several times to the *Dudgeon Case*, Ser. A, No. 45, 4 EHRR 149 (1981). In that case, Dudgeon, a homosexual resident of Northern Ireland, brought proceedings against the United Kingdom based on his complaint against laws of Northern Ireland that made certain homosexual acts between consenting adult males criminal offenses. The Court concluded that Dudgeon had suffered an unjustified interference with his right to respect for his private life and accordingly found a breach by the U.K. of Article 8 of the Convention, also at issue in *Norris*.

Consider the following three provisions in human rights instruments bearing on the right to privacy:

Article 8 of the European Convention

> 1. Everyone has the right to respect for his private and family life, his home and his correspondence.
> 2. There shall be no interference by a public authority with the exercise of this right except such as is in accordance with the law and is necessary in a democratic society in the interests of national security, public safety or the economic well-being of the country, for the prevention of disorder or crime, for the protection of health or morals, or for the protection of the rights and freedoms of others.

Article 12 of the Universal Declaration of Human Rights

> No one shall be subjected to arbitrary interference with his privacy, family, home or correspondence, nor to attacks upon his honour or reputation. Everyone has the right to the protection of the law against such interference or attacks.

Article 17 of the ICCPR

The ICCPR provision is similar to the Universal Declaration, except that it refers to 'arbitrary *or unlawful* interference' and to '*unlawful* attacks' (emphasis added). The ICCPR Human Rights Committee has issued a General Comment No. 16, see pp. 524 and 529, *supra*, on Article 17 without mentioning questions of sexuality.

As indicated below in the decision of the United States Supreme Court in *Bowers v. Hardwick*, the U.S. Constitution makes no reference to a right to 'privacy' as such.

We turn to two judicial decisions interpreting and applying the European Convention and the U.S. Constitution.

NORRIS v. IRELAND

European Court of Human Rights, 1989
Ser. A, No. 142, 13 EHRR 186.

[The applicant, an Irish national and member of the Irish Parliament, was a homosexual and chairman of the Irish Gay Rights Movement. In 1977, he instituted proceedings in the High Court of Ireland, seeking a declaration that certain laws prohibiting homosexual relations were invalid under the Irish Constitution. Those laws included (i) Section 62 of the Person Act 1861 to the effect that '[w]hosoever shall attempt to commit the said abominable crime [of buggery], or shall be guilty of any . . . indecent assault upon a male person,' is guilty of a misdemeanor and subject to a prison sentence not exceeding ten years, and (ii) Section 11 of the Criminal Law Amendment Act 1885 to the effect that any 'male person who, in public or in private, commits . . . any act of gross indecency with another male person' is guilty of a misdemeanor and subject to imprisonment not exceeding two years. The term 'gross indecency' was not statutorily defined and was to be given meaning by courts on the particular facts of each case. Later acts gave courts discretion to impose more lenient sentences.

At no time was the applicant charged with any offence in relation to his admitted homosexual activities, although he was continuously at risk of being so prosecuted on the basis of an indictment laid by the Director of Public Prosecutions. The Director made a statement in connection with this litigation to the effect that '[t]he Director has no *stated* prosecution policy on any branch of the criminal law. He has no unstated policy *not* to enforce any offence. Each case is treated on its merits.' Since the Office of the Director was created in 1984, no prosecution had been brought in respect of homosexual activities except where minors were involved or the acts were committed in public or without consent.

Mr. Norris offered evidence of the ways in which this legislation had interfered with his right to respect for his private life, including evidence (i) of deep depression on realizing that 'any overt expression of his sexuality would expose him to criminal prosecution,' and (ii) of fear of prosecution of him or of another man with whom he had a physical relationship.

The judge in the High Court found that '[o]ne of the effects of criminal sanctions against homosexual acts is to reinforce the misapprehension and general prejudice of the public and increase the anxiety and guilt feelings of homosexuals leading, on occasions, to depression . . . .' However, he dismissed the action on legal grounds, and his decision was upheld in 1983 by the Supreme Court of Ireland. That court concluded that the applicant had standing (*locus standi*) to bring an action for a declaration even though he had not been prosecuted, for the threat continued.

The Supreme Court rejected the applicant's argument that the Irish Constitution should be interpreted in the light of the European Convention on Human Rights, for the Convention was 'an international agreement [which] does not and cannot form part of [Ireland's] domestic law, nor affect in any way questions which arise thereunder.' Article 29(6) of the Irish Constitution declared: 'No international agreement shall be part of the domestic law of the State save as may be determined by the Oireachtas,' and the Oireachtas (legislature) had not taken action to enact the Convention as domestic legislation.

The Supreme Court found the laws complained of to be consistent with the Constitution, since no right of privacy encompassing consensual homosexual activity could be derived from the 'Christian and democratic nature of the Irish State.' It observed (i) that homosexuality 'has always been condemned in Christian teaching as being morally wrong' and has been regarded for centuries 'as an offence against nature and a very serious crime,' (ii) that '[e]xclusive homosexuality, whether the condition be congenital or acquired, can result in great distress and unhappiness for the individual and can lead to depression, despair and suicide,' and (iii) that male homosexual conduct resulted in many states in 'all forms of venereal disease,' which had become a 'significant public health problem' in England.

Mr. Norris started proceedings before the European Commission on Human Rights, claiming that the Irish laws constituted a continuing interference with his right to respect for private life under Article 8 of the European Convention. In 1987, by 6 votes to 5, the Commission expressed its opinion that there had been a violation of Article 8. The case was then referred to the European Court. Excerpts from its Judgment follow:]

30. . . . Article 25 requires that an individual applicant should be able to claim to be actually affected by the measure of which he complains. Article 25 may not be used to fund an action in the nature of an *actio popularis*; nor may it form the basis of a claim made *in abstracto* that a law contravenes the Convention.

. . .

32. In the Court's view, Mr. Norris is in substantially the same position as the applicant in the *Dudgeon* case, which concerned identical legislation then in force in Northern Ireland. As was held in that case, 'either [he] respects the law and refrains from engaging—even in private and with consenting male

partners—in prohibited sexual acts to which he is disposed by reason of his homosexual tendencies', or he commits such acts and thereby becomes liable to criminal prosecution.

. . .

34. On the basis of the foregoing considerations, the Court finds that the applicant can claim to be the victim of a violation of the Convention within the meaning of Article 25 (1) thereof.

. . .

38. The Court agrees with the Commission that, with regard to the interference with an Article 8 right, the present case is indistinguishable from the *Dudgeon* case . . . . [E]nforcement of the legislation is a matter for the Director of Public Prosecutions who may not fetter his discretion with regard to each individual case by making a general statement of his policy in advance. . . .

. . . As was held in that case, 'the maintenance in force of the impugned legislation constitutes a continuing interference with the applicant's right to respect for his private life . . . within the meaning of Article (1). In the personal circumstances of the applicant, the very existence of this legislation continuously and directly affects his private life . . .'

. . .

39. The interference found by the Court does not satisfy the conditions of paragraph (2) of Article 8 unless it is 'in accordance with the law,' has an aim which is legitimate under this paragraph and is 'necessary in a democratic society' for the aforesaid aim.

40. It is common ground that the first two conditions are satisfied. As the Commission pointed out in paragraph 58 of its report, the interference is plainly 'in accordance with the law' since it arises from the very existence of the impugned legislation. Neither was it contested that the interference has a legitimate aim, namely the protection of morals.

41. It remains to be determined whether the maintenance in force of the impugned legislation is 'necessary in a democratic society' for the aforesaid aim. According to the Court's case law, this will not be so unless, *inter alia*, the interference in question answers a pressing social need and in particular is proportionate to the legitimate aim pursued.

. . .

42. . . . It was not contended before the Commission that there is a large body of opinion in Ireland which is hostile or intolerant towards homosexual acts committed in private between consenting adults. Nor was it argued that Irish society had a special need to be protected from such activity. In these circumstances, the Commission concluded that the restriction imposed on the applicant under Irish law, by reason of its breadth and absolute character, is disproportionate to the aims sought to be achieved and therefore is

not necessary for one of the reasons laid down in Article 8(2) of the Convention.

. . .

44. . . . As early as 1976, the Court declared in its *Handyside* judgment of 7 December 1976 that, in investigating whether the protection of morals necessitated the various measures taken, it had to make an 'assessment of the reality of the pressing social need implied by the notion of "necessity" in this context' and stated that 'every "restriction" imposed in this sphere must be proportionate to the legitimate aim pursued.' It confirmed this approach in its *Dudgeon* judgment.

. . . [A]lthough of the three aforementioned judgments two related to Article 10 of the Convention, it sees no cause to apply different criteria in the context of Article 8.

. . .

46. As in the *Dudgeon* case, '. . . not only the nature of the aim of the restriction but also the nature of the activities involved will affect the scope of the margin of appreciation. The present case concerns a most intimate aspect of private life. Accordingly, there must exist particularly serious reasons before interferences on the part of public authorities can be legitimate for the purposes of paragraph (2) of Article 8.

Yet the Government has adduced no evidence which would point to the existence of factors justifying the retention of the impugned laws which are additional to or are of greater weight than those present in the aforementioned *Dudgeon* case. At paragraph 60 of its judgment of 22 October 1981 the Court noted that 'As compared with the era when [the] legislation was enacted, there is now a better understanding, and in consequence an increased tolerance, of homosexual behaviour to the extent that in the great majority of the member States of the Council of Europe it is no longer considered to be necessary or appropriate to treat homosexual practices of the kind now in question as in themselves a matter to which the sanctions of the criminal law should be applied; the Court cannot overlook the marked changes which have occurred in this regard in the domestic law of the member States.' It was clear that 'the authorities [had] refrained in recent years from enforcing the law in respect of private homosexual acts between consenting [adult] males . . . capable of valid consent.' There was no evidence to show that this '[had] been injurious to moral standards in Northern Ireland or that there [had] been any public demand for stricter enforcement of the law.'

Applying the same tests to the present case, the Court considers that, as regards Ireland, it cannot be maintained that there is a 'pressing social need' to make such acts criminal offences. On the specific issue of proportionality, the Court is of the opinion that 'such justifications as there are for retaining the law in force unamended are outweighed by the detrimental effects which the very existence of the legislative provisions in question can have on the life of a person of homosexual orientation like the applicant. Although members of the public who regard homosexuality as immoral may be shocked, offended or disturbed by the commission by others of private homosexual acts, this cannot on its own warrant the application of penal sanctions when it is consenting adults alone who are involved.'

47. The Court therefore finds that the reasons put forward as justifying the interference found are not sufficient to satisfy the requirements of paragraph (2) of Article 8. There is accordingly a breach of that Article.

48. Under Article 50 of the Convention:

> If the Court finds that a decision or a measure taken by a legal authority or any other authority of a High Contracting Party is completely or partially in conflict with the obligations arising from the . . . Convention, and if the internal law of the said Party allows only partial reparation to be made for the consequences of this decision or measure, the decision of the Court shall, if necessary, afford just satisfaction to the injured party.

The applicant seeks compensation for injury and reimbursement of legal costs and expenses.

49. The applicant requested the Court to fix such amount by way of damages as would acknowledge the extent to which he has suffered from the maintenance in force of the legislation.

50. . . . [I]t is inevitable that the Court's decision will have effects extending beyond the confines of this particular case, especially since the violation found stems directly from the contested provision and not from individual measures of implementation. It will be for Ireland to take the necessary measures in its domestic legal system to ensure the performance of its obligation under Article 53.

For this reason and notwithstanding the different situation in the present case as compared with the *Dudgeon* case, the Court is of the opinion that its finding of a breach of Article 8 constitutes adequate just satisfaction for the purposes of Article 50 of the Convention and therefore rejects this head of claim.

. . .

NOTE

Consider the following observation of Andrew Clapham in *Human Rights in the Private Sphere* (1993), at 64:

> . . . [T]he courts, even when faced with a popular legitimate law, such as the law of the Isle of Man on birching [a form of whipping or spanking with a birch rod, used under the law referred to for discipline of children],

may prefer to follow the Strasbourg lead. A parallel can be drawn between this case and the situation concerning homosexuality in Northern Ireland, where the Court of Human Rights, faced with another popular law (which prohibited sexual relations between men), found a breach of human rights: *Dudgeon v. United Kingdom*. It is in these types of situation involving unpopular minority interests that human rights theory is really tested. Both laws were relatively popular in the Isle of Man and Northern Ireland respectively and it was the European Court of Human Rights in Strasbourg that held the laws to violate human rights. It may well be that such a court can bring a detachment to bear on domestic laws that national courts may find hard. It is for this reason that even if the United Kingdom were to adopt the European Convention in the form of a Bill of Rights, the right of individual petition to Strasbourg should still be kept open, so that the European Court of Human Rights has the chance to examine cases arising in the United Kingdom context and give authoritative judgments on the scope of the rights guaranteed by the Convention.

BOWERS v. HARDWICK

Supreme Court of the United States, 1986
478 U.S. 186 106 S. Ct. 2841.

[A Georgia statute defined the crime of sodomy as 'any sex act involving the sex organs of one person and the mouth or anus of another.' Hardwick (who was under some surveillance by the police) was arrested in his home bedroom immediately after engaging in oral sex there with a consenting adult male. He was charged with the crime of sodomy and, before being tried, challenged the statute.

The Supreme Court, by a 5–4 vote, held that the statute was constitutional. Five opinions were written, including the opinion for the Court by Justice White, two separate concurrences and two dissents. The excerpts below from several opinions touch on a few of the complex themes relevant to this litigation and are intended to provide comparisons with the opinion of the European Court of Human Rights in *Norris*.]

Opinion of Justice White for the Court:

. . . The issue presented is whether the Federal Constitution confers a fundamental right upon homosexuals to engage in sodomy and hence invalidates the laws of the many States that still make such conduct illegal . . .

We first register our disagreement with the [opinion of the Court of Appeals below] that the [Supreme] Court's prior cases have construed the Constitution to confer a right of privacy that extends to homosexual sodomy . . . . [The opinion referred to prior decisions involving child rearing and education of children, to procreation and contraception, to marriage, and (*Roe v. Wade*) to abortion. Some of those decisions relied on a right to privacy not explicit in the

Constitution but found to be within the protection provided by constitutional provisions such as the Due Process Clause of the Fourteenth Amendment.]

. . . [W]e think it evident that none of the rights announced in those cases bears any resemblance to the claimed constitutional right of homosexuals to engage in acts of sodomy that is asserted in this case. No connection between family, marriage, or procreation on the one hand and homosexual activity on the other has been demonstrated . . . .

. . .

. . . Proscriptions against [consensual sodomy] have ancient roots . . . Sodomy was a criminal offense at common law and . . . [i]n 1868, when the Fourteenth Amendment was ratified, all but 5 of the 37 States in the Union had criminal sodomy laws. In fact, until 1961, all 50 states outlawed sodomy, and today, 24 States and the District of Columbia continue to provide criminal penalties for sodomy performed in private and between consenting adults. . . . Against this background, to claim that a right to engage in such conduct is [quoting from prior decisions in unrelated cases seeking to give content to concepts like due process or basic rights] 'deeply rooted in this Nation's history and tradition' or 'implicit in the concept of ordered liberty' is, at best, facetious.

Nor are we inclined to take a more expansive view of our authority to discover new fundamental rights imbedded in the Due Process Clause. The Court is most vulnerable and comes nearest to illegitimacy when it deals with judge-made constitutional law having little or no cognizable roots in the language or design of the Constitution. . . .

. . .

Even if the conduct at issue here is not a fundamental right, respondent asserts that there must be a rational basis for the law and that there is none in this case other than the presumed belief of a majority of the electorate in Georgia that homosexual sodomy is immoral and unacceptable. This is said to be an inadequate rationale to support the law. The law, however, is constantly based on notions of morality, and if all laws representing essentially moral choices are to be invalidated under the Due Process Clause, the courts will be very busy indeed. . . .

Chief Justice Burger, concurring:

. . . Decisions of individuals relating to homosexual conduct have been subject to state intervention throughout the history of Western civilization. Condemnation of those practices is firmly rooted in Judeao-Christian moral and ethical standards. . . . To hold that the act of homosexual sodomy is somehow protected as a fundamental right would be to cast aside millennia of moral teaching.

. . .

Justice Blackmun (joined by Justices Brennan, Marshall and Stevens), dissenting:

This case is [not] about a 'fundamental right to engage in homosexual sodomy,' as the Court purports to declare. . . . Rather, this case is about [quoting here and below from prior decisions of the Court] 'the most comprehensive of rights and the rights most valued by civilized men,' namely, 'the right to be let alone.'

. . . I believe we must analyze respondent Hardwick's claim in the light of the values that underlie the constitutional right to privacy. If that right means anything, it means that, before Georgia can prosecute its citizens for making choices about the most intimate aspects of their lives, it must do more than assert that the choice they have made is an 'abominable crime not fit to be named among Christians.'

. . .

. . . We protect those rights [to family, procreation] not because they contribute, in some direct and material way, to the general public welfare, but because they form so central a part of an individual's life. . . . We protect the decision whether to have a child because parenthood alters so dramatically an individual's self-definition, not because of demographic considerations or the Bible's command to be fruitful and multiply. And we protect the family because it contributes so powerfully to the happiness of individuals, not because of a preference for stereotypical households. . . .

Only the most willful blindness could obscure the fact that sexual intimacy is 'a sensitive, key relationship of human existence, central to family life, community welfare, and the development of human personality.' The fact that individuals define themselves in a significant way through their intimate sexual relationships with others suggests, in a Nation as diverse as ours, that there may be many 'right' ways of conducting those relationships, and that much of the richness of a relationship will come from the freedom an individual has to choose the form and nature of these intensely personal bonds.

. . .

The assertion [in the State Attorney General's brief] that 'traditional Judeo-Christian values proscribe' the conduct involved cannot provide an adequate justification [for the Georgia statute]. That certain, but by no means all, religious groups condemn the behavior at issue gives the State no license to impose their judgments on the entire citizenry. The legitimacy of secular legislation depends instead on whether the State can advance some justification for its law beyond its conformity to religious doctrine. Thus, far from buttressing [the Attorney General's] case, invocation of Leviticus, Romans, St. Thomas Aquinas, and sodomy's heretical status during the Middle Ages undermines his suggestion that [the statute] represents a legitimate use of secular coercive power. A State can no more punish private behavior because of religious intolerance than it can punish such behavior because of racial animus. . . .

. . . [The State] and the Court fail to see the difference between laws that protect public sensibilities and those that enforce private morality. Statutes banning public sexual activity are entirely consistent with protecting the individual's liberty interest in decisions concerning sexual relations: the same recognition that those decisions are intensely private which justifies protecting them from governmental interference can justify protecting individuals from unwilling exposure to the sexual activities of others. But the mere fact that intimate behavior may be punished when it takes place in public cannot dictate how States can regulate intimate behavior that occurs in intimate places. . . .

. . . This case involves no real interference with the rights of others, for the mere knowledge that other individuals do not adhere to one's value system cannot be a legal cognizable interest, let alone an interest that can justify invading the houses, hearts, and minds of citizens who choose to live their lives differently.

. . .

NOTE

Compare Laurence Tribe, *American Constitutional Law* (2nd ed. 1987), at 1428:

> . . . Therefore, in asking whether an alleged right forms part of a traditional liberty, it is crucial to define the liberty at a high enough level of generality to permit unconventional variants to claim protection along with mainstream versions of protected conduct. The proper question, as the dissent in *Hardwick* recognized, is not whether oral sex as such has long enjoyed a special place in the pantheon of constitutional rights, but whether private, consensual, adult sexual acts partake of traditionally revered liberties of intimate association and individual autonomy.
>
> . . . It should come as no surprise that, in the kind of society contemplated by our Constitution, government must offer greater justification to police the bedroom than it must to police the streets. Therefore, the relevant question is not what Michael Hardwick was doing in the privacy of his own bedroom, but what the State of Georgia was doing there.

* * * * *

The margin of appreciation has figured in a number of the preceding court decisions. One Judge of the European Court has said that this concept lies 'at the heart of virtually all major cases that come before the Court, whether the judgments refer to it expressly or not.'[30] There follow excerpts from an article by van Dijk and van Hoof on the subject, and a summary of views of some commentators about the meaning and relevance of this concept.

P. VAN DIJK AND G. J. H. VAN HOOF, THEORY AND PRACTICE OF THE EUROPEAN CONVENTION ON HUMAN RIGHTS

(2nd ed. 1990), at 585.

The 'margin of appreciation' doctrine is rooted in national case-law concerning judicial review of governmental action. The same doctrine also makes sense within the framework of the European Convention. There, however, the scope of its applicability should in our view be rather limited. . . . As far as the determination of facts is concerned, application of the 'margin of appreciation' doctrine by the Strasbourg organs would seem to be justified and self-evident. It should be borne in mind that national law and its interpretation are also factual data from the point of view of the Strasbourg organs. For the determination of those facts, i.e. for answering the question of whether and in what way precisely the facts took place and what are the exact contents and meaning of national law, the national authorities are in a better position than international organs. . . .

The situation is in our opinion quite different as far as the determination of questions of law is concerned. For the latter, *viz.* the assessment whether the facts as they have ultimately been established in the Strasbourg proceedings constitute a violation of the Convention, only the Strasbourg organs are competent in the last resort. According to Article 19 of the Convention it is the Court's and the Commission's task 'To ensure the observance of the engagements undertaken by the High Contracting Parties in the present Convention'. In performing this task they should not refer to the opinion of the national authorities of the State concerned. . . .

. . .

As far as the scope of the margin of appreciation is concerned, a number of different approaches of the doctrine can be discerned in the Strasbourg case-law. . . . For the sake of convenience these may be labeled the narrow approach, the reasonableness test, and the not-unreasonable test.

The demarcation between a full-fledged review and the narrow approach is not always clear. Despite reference to the existence of a 'certain measure of appreciation' on the part of the competent national authorities it is not uncommon for the Commission and the Court to subsequently conduct a

[30] R. Macdonald, The Margin of Appreciation in the Jurisprudence of the European Court of Human Rights, in Essays in Honor of Roberto Ago 187, 208 (Vol. III, 1987).

comprehensive independent inquiry into the question of whether or not the requirement of necessity has been satisfied, without apparently being guided very much by the views of the national authorities.

. . .

The trend of widening the national authorities' discretion is taken to the extreme in cases where the *reasonableness* test is turned into a *not-unreasonable* test. In the latter approach it is required of the national authorities not that the government acted reasonably, but that they did not act unreasonably. This is more than a mere play of words, since the not-unreasonable test would seem to imply a change with respect to the burden of proof. While the reasonableness test requires the government to prove that the national authorities have acted reasonably, the not-unreasonable test seems to burden the applicant (or the Strasbourg organs themselves?) with sustaining that they have not.

. . .

. . . [T]he structure of the Convention provides every reason to elaborate common standards, thereby reducing the area of application of the margin of appreciation doctrine. However, the case-law up to this point shows only a modest use of common or uniform European standards and where such standards are resorted to for deciding cases, they do not seem to rest on very firm ground. One authoritative commentator has observed in this respect with reference to the *Dudgeon* Case and the *Sunday Times* Case:

> Search as one may, through both judgments, there is no reference whatsoever to the relevant legislation of all or even most Contracting States, let alone comparative analysis to establish what degree of unity may or may not have existed.

. . . In the view of the Court [in the *Handyside* Case], there existed no such thing as 'a uniform European conception of morals'. Even assuming that this was a correct conclusion it would still be up to the Court—and the Commission—to develop such a 'European conception', since the term 'morals' is frequently mentioned in the Convention. That this term is interpreted in each individual case by reference to the national conceptions on this point is irreconcilable with a collectively guaranteed set of international norms like the Convention, at least in the long run.

. . . [T]he Strasbourg practice up to now offers the following picture. First, the case-law does not show a determined effort on the part of the Commission and the Court to systematically search for and elaborate such common standards. Secondly, conclusions as to the existence or the lack of common standards do not appear to rest on the firm basis of comparative legal analysis. Thirdly—and maybe consequently—there is not infrequently disagreement as to the existence or lack of common standards. Fourthly, in areas where there is agreement, conclusions as to the lack or existence of common standards are not always consistently applied. . . .

The preceding discussion of the margin of appreciation doctrine may be summed up as follows. In our view within the framework of the European Convention the margin of appreciation doctrine to some extent is to be compared to a spreading disease. Not only has the scope of its application gradually been broadened to the point where in principle none of the Convention's rights and freedoms is excluded, but also has the illness been intensified in that wider versions of the doctrine have been added. . . .

. . . The Commission and the Court appear to follow in many cases what might be called a *raison d'état* interpretation: when they weigh the full enjoyment of the rights and freedoms on the one hand and the interests advanced by the State for their restriction on the other hand. They appear to be inclined to pay more weight to the latter. In our opinion, however, the 'object and purpose' of the European Convention would seem to dictate a different attitude, which might be referred to as the 'constitutional' attitude. The latter implies that the full enjoyment of the rights and freedoms constitutes the point of departure and that the possibilities to restrict this enjoyment are exceptions which should be interpreted restrictively, while the organ examining the case should review these restrictions in an independent way for their conformity with the conditions laid down by the Convention.

. . . The use of the not-unreasonable test, has not been kept confined to the above-mentioned matters concerning economic policy. It has also been applied to cases concerning secret surveillance measures and has been advocated within the Court even with respect to the requirement of promptness contained in Article 5(3). In the *Brogan* Case the majority of the Court conducted a full-fledged review as to this latter requirement. In his dissenting opinion Judge Martens in fact proposed the use of the not-unreasonable test for answering the question of whether a seven-day period for detaining an accused before bringing him before a judicial authority was compatible with the requirement of promptness:

> the Court can find that the United Kingdom . . . overstepped the margin of appreciation it is entitled to . . . only if it considers that the arguments for maintaining the seven-day period are wholly unconvincing and cannot be reasonably defended.

The issues at stake in the *Brogan* Case he described as 'questions on which reasonable people may hold different views'. According to Judge Martens

> This means that the Court should respect the United Kingdom Government's choice and cannot but hold that they did not overstep their margin of appreciation.

. . .

The emphasis on the striking of a balance between individual and general interests in the above-mentioned cases as well as in many other cases in which the margin of appreciation is involved, derives from the principle listed in the *Silver* Case according to which the phrase 'necessary in a democratic society' means that, to be compatible with the Convention, the interference must, *inter alia*, correspond to a pressing social need and be proportionate to the legitimate aim pursued. This weighing of the various interests involved is indeed characteristic for the very nature of the Convention. . . . However, the question is not so much whether a weighing of interests has to take place, but rather which authority in the final analysis is empowered to do the weighing. It is this latter question which the debate on the margin of appreciation doctrine sharply puts in perspective. Those who argue that the doctrine should not—or only in its narrow version—be applied within the framework of the Convention in fact hold that the Strasbourg organs ultimately are the only competent bodies in this respect. Those, on the other hand, who favour application of the wider versions of the doctrine opt for a primary role of the competent authorities of the contracting States.

The former position is defended here. In our view, the arguments underlying that position were aptly summarized by the Court in the *Norris* Case. . . . Article 19, which reflects the structure of the European Convention, leaves room for no other conclusion than that in the final analysis only the Strasbourg organs are competent to conduct the weighing of interests involved in the Convention. Leaving this task to the national authorities is eventually going to undermine the Convention's entire structure. Granting the extreme latitude inherent in the above-mentioned wide version of the margin of appreciation doctrine—and exemplified by the Irish Government's claim in the *Norris* Case—would reduce the Strasbourg supervisory machinery to a mechanism for rubberstamping almost anything a Government wants.

. . . However, with respect to questions of law, too, it may be necessary to grant the national authorities a certain discretion, if the aspects of expediency, which are usually of a highly factual nature, clearly predominate over the aspects of lawfulness.

The latter occurs, for instance, when the issue is whether the grant of guardianship or the grant of a visiting right to one of the divorced parents is in the interest of the child, a question which may be important for judging whether a violation of the right to family life of Article 8 is to be considered justified on the ground of the second paragraph. More or less the same applies to the question of whether the measures taken are 'strictly required' in the sense of Article 15. Although this is a question of law—have the conditions of Article 15 been complied with?—still questions of expediency play a very important part as well, and the local situation will be conclusive in many cases.

In our view, however, the situation is generally different for the question of whether a restriction imposed on one of the rights and freedoms is 'necessary in a democratic society in the interest of . . . /for the protection of . . . '. It is true that here too the circumstances of the individual case and the questions of expediency connected therewith play a part, but still the relevant norm, in view of the very reference to 'democratic society', can be objectified and interpreted independently of the actual situation to a much greater extent, so that the review here leaves greater scope for an independent assessment by the examining international organ. . . .

. . .

VIEWS OF COMMENTATORS ON THE MARGIN OF APPRECIATION

Paul Mahoney[31]

Mahoney refers to concerns that the 'doctrine' of the margin of appreciation constitutes an 'abdication by the Court of its duty of adjudication or . . . an improper reading into the text of a pro-government restriction.' He notes fears that the Court has 'emptied many of the strict conditions laid down in the Convention of their strength' in the exercise of 'judicial self-restraint.' Such criticism is, however, 'misconceived' for it fails to recognize the 'distribution of powers' between the Convention and state. The margin of appreciation is indeed 'the natural product of that distribution of powers; it serves to delineate the dividing line.'

Like a national constitution safeguarding basic rights, the Convention 'reflects a desire to disable democratic discretion in certain respects,' but not all state discretion is removed. The Convention system, Mahoney argues, is grounded on the belief

> that political democracy is the best system of government for ensuring respect of . . . human rights. Any theory of interpretation or review by the Court must be compatible with that basic underpinning of political theory. In a pluralistic democratic society there will on many topics covered by the Convention be a spectrum of different but acceptable opinions. The Convention issue for decision before the Court will often be whether the choice of the national authorities remained within the permissible spectrum. To attach a label saying 'margin of appreciation' to that aspect of adjudication is not to add anything to the Convention.

Judicial review protects individuals against the 'excesses of majoritarian rule,' but in political democracies there is a 'deeply held view' that majority decision-making is 'the best guarantee of the survival of society.'

Mahoney grants that the doctrine of margin of appreciation runs risks, given the Court's duty under Article 19 to ensure that the member states

[31] Judicial Activism and Judicial Self-Restraint in the European Court of Human Rights: Two Sides of the Same Coin, 11 Hum. Rts. L. J. 57, 80 (1990).

observe their engagements. 'Too liberal an application of the doctrine in favour of the respondent State will in the long run undermine the Court's claim to be the authoritative institution for interpretation and application of the Convention.' The Court would do well to clear up the doctrine, to 'articulate clearer criteria' as to its scope. Such clarification should allay fears bred by the doctrine's current vagueness.

Franz Matscher[32]

Matscher argues that Convention institutions 'are not entitled to dictate uniformity' but 'are obliged to respect, within certain bounds, the cultural and ideological variety, and also the legal variety, which are characteristic of Europe.' He refers to remarks of Hallstein, a former President of the European Community Commission, to the effect that

> what we need is not a streamlined Europe; rather, the preservation of the marvellous richness and inexhaustible variety of our continent must remain a goal as important as integration. It is precisely this insight which forms the ideological background to the theory of the so-called 'margin of appreciation' of discretion, which is also a characteristic of the case-law of the Convention institutions.

Matscher notes that the Convention does not expressly allow member states any discretion in relation to the protected rights but 'merely defines the extent of those rights.' He emphasizes the number of general, undefined terms in the Convention that states must apply—terms like 'family life,' 'protection of morals,' necessary in a democratic society,' and so on. But the Court

> has adopted a more realistic attitude here and de facto has allowed the States a certain discretion in their legislative and executive functions. The Court merely examines whether the States have used the discretion allowed them in a reasonable fashion . . . .

He sums up the 'theory' of margin of appreciation as 'the expression of a realistic "judicial self-restraint."'

R. St. J. Macdonald[33]

Macdonald sees the task of the Court 'to reconcile' the political, economic, social and cultural diversity of the member states with the 'development of an effective and reasonably uniform standard of protection for Convention

[32] Methods of Interpretation of the Convention, in R. St. J. Macdonald, Matscher and Petzold (eds.), The European System for the Protection of Human Rights 63, 75 (1993).

[33] The Margin of Appreciation, in Macdonald, Matscher and Petzold (eds.), The European System for the Protection of Human Rights 83, 122 (1993).

rights.' The margin of appreciation, 'more a principle of justification than interpretation,' helps the Court to show 'the proper degree of respect' for member states' objectives while 'preventing unnecessary restrictions' on the Conventions's rights. It is this 'flexibility' that has enabled the Court 'to avoid any damaging dispute' with states over authority of the Court and Convention on one hand, and the states and their legislatures on the other.

The margin of justification has then a 'pragmatic' justification. Progress toward the goal of a European-wide uniform standard of protection must be gradual.

> [T]he gradual refining of an originally expansive margin of appreciation reflects the increasing legitimacy of the Convention organs in the European legal order. As that legitimacy has increased, an obvious way of reducing the amount of discretion to which national authorities should consider themselves entitled is to hold them to standards which are observed Europe-wide.

This pragmatism is not without its costs, for it 'prevents the emergence of a coherent vision of the Court's function. . . . But perhaps the Convention system is now sufficiently mature to be able to move beyond the margin of appreciation and grapple more openly with the questions of appropriateness which that device obscures.'

J. G. Merrills[34]

Merrills criticizes the views of a commentator on the *Handyside* decision who suggested that it might appropriately be the function of the Court to develop a 'European conception of morals' to displace the variations among the Convention's member states. 'To this the answer is surely no. The Court's function is not to decree uniformity wherever there are national differences, but to ensure that fundamental values are respected.' Where there are 'clear differences of view' among member states, there must be 'room for a significant margin of appreciation.'

> [I]n any democratic society there are situations in which the rights set out in the Convention may be limited. The question which then arises is how and by whom the need for a disputed limitation is to be judged. No one, of course, would wish to argue that the matter should be wholly in the hands of the respondent State. The difficulty, however, is that the limitations are permitted for reasons which on the face of it only the State can properly assess. There is a great deal to be said for the view that 'If it is accepted that the State has a valid interest in the prevention of corruption and in the preservation of the moral ethos of its society, then the State has a right to enact such laws as it may reasonably think necessary to achieve

[34] The Development of International Law by the European Court of Human Rights 146 (1988).

> these objects. . . .' The result [of reviewing a number of the Court's judgments] is not so much an inconsistency in the Court's jurisprudence, as a demonstration . . . that decisions about human rights are not a technical exercise in interpreting texts, but judgments about political morality.

VIEWS OF COMMENTATORS ABOUT DYNAMIC INTERPRETATION AND CONSENSUS

The preceding views of commentators about the margin of appreciation also considered the problem of consensus among European states on a question before the Court. In fact, these two issues are complexly intertwined, as the opinions in *Dudgeon* and *Norris* demonstrate. Both refer to the changing laws and morals of European states, and indicate the relevance of those changes for the Court's judgments. Consider two commentators' views.

Laurence Helfer[35]

Helfer asserts that the Commission and Court, far from 'being bound by the intention of the drafters . . . interpret the Convention as a modern document that responds to and progressively incorporates changing European social and legal developments.' When 'rights-enhancing' practices among member states 'achieve a certain measure of uniformity, a "European consensus", so to speak, the Court and Commission raise the standard of rights-protection to which all states must adhere.' In this way, these institutions have extended the Convention's protection to groups not within the drafters' intentions.

Helfer warns that the 'failure to articulate with precision the scope and function of the consensus inquiry poses a potentially grave threat to the tribunals' authority as the arbiters of European human rights.' They must develop 'a more comprehensive and rigorous methodology for applying the European consensus inquiry.' To the present, these institutions have relied on three factors as 'evidence of consensus': legal consensus expressed through statutes or regional agreements; expert consensus; and European public consensus. But the tests lack specificity. Helfer asks, for example, what percentage of member states must alter their laws 'before a right-enhancing norm will achieve consensus status.'

P. Van Dijk[36]

Van Dijk refers to a decision of the Court of Human Rights to the effect that the Convention's right to marry guaranteed by Article 12 refers to 'traditional'

[35] Consensus, Coherence and the European Convention on Human Rights, 26 Cornell Int. L. J. 133, 134 (1993).

[36] The Treatment of Homosexuals under the European Convention on Human Rights, in K. Waaldijk and Clapham (eds.), Homosexuality: A European Community Issue 179, 198 (1993).

heterosexual marriage as the basis of a family. Even conceding that such a link between marriage and procreation was intended by the Convention's drafters, he questions whether Article 12 should have been interpreted decades later to refer to such 'traditional' marriage in view of the dynamic interpretation of 'family.'

Van Dijk discusses a later case involving a transsexual, in which the Court held that developments in the member states did not evidence a departure from the traditional concept of marriage.

> It is submitted that by requiring such evidence of a general abandonment the Court sets a limitation to its powers to adopt a dynamic approach in its case-law . . . . After all, in the *Dudgeon* case and the *Norris* case there was also no general abandonment of the penalisation of homosexual acts between adults. Application of the 'European consensus inquiry' in its extreme would exclude the possibility of finding any national legislation in violation of the ECHR.

Van Dijk refers to a dissenting opinion in that decision, which argued that in cases involving family law and sexuality, the Court adapt its interpretation of the Convention to societal changes (quoting here from the dissenting opinion)

> only if almost all member States have adopted the new ideas. In my opinion this caution is in principle not consistent with the Court's mission to protect the individual against the collectivity and to do so by elaborating common standards. . . . [I]f a collectivity oppresses an individual because it does not want to recognize societal changes, the Court should take care not to yield too readily to arguments based on a country's cultural and historical particularities.

QUESTIONS

1. What are the salient differences between the opinions of the courts in *Norris* and *Bowers*? For example, what significance do the opinions accord to trends in legislation and mores, in one case, in European states and, in the other, in the component states of the United States federalism? Are the trends similar?

2. Do you believe that it was significant for the opinions that the European Convention is explicit about the right to 'respect' for 'private life' and the U.S. Constitution is not? Would, or could, the European Court have ruled the same way if there had been no Article 8 provision on privacy, so that the Court would have to 'discover' or 'infer' a right to privacy from other provisions of the Convention?

3. Given the views of the commentators above about dynamic interpretation and consensus in the context of the European system, what method do you think the Court should explicitly follow in deciding a case like *Norris*, or the birching case from the Isle of Man, or the blasphemy decision in *Otto-Preminger-Institut v. Austria*? Should it survey the legislation and practice thereunder (whether or not to prosecute, nature of punishment, and so on) in all member states? What should it make of the survey? Would all states have equal 'weight' in the decision? What relevance would a 'trend' have—for example, thirty years ago 90% of the states barred the practice at issue, but now only 60% bar it?

4. Given the several opinions and the different views of the commentators, what meaning(s) or methods or premises do you attach to invocation by a judge of the margin of appreciation? Does it appear to have the same meaning in the different decisions, or does that meaning change with context? Consider the following assertions:

a. 'The margin of appreciation is simply a fancy name given to an inevitable concern of an international tribunal. It tells the Court not to overstep its role and to be respectful of the sovereign states that created and submitted to the Court in the first instance. Else the Court and the Convention could get into deep trouble. It is a rule of prudence.'

b. 'The margin of appreciation is another name for cultural relativism. It respects difference, and cautions the Court not to compress the varied European states and cultures into one juridical and moral form.'

c. 'The margin of appreciation is less an application than a perversion of cultural relativism. Europe possesses an important unity vis-à-vis the rest of the world. We are not talking of a universal system embracing the United States and Yemen, Russia and Ecuador. The effect of the principle is to permit ongoing oppression of the culturally deviant within a country, the minorities that characteristically have had to resort to universal human rights norms to challenge their national laws and practice.'

d. 'The margin of appreciation is consistent with the principle of subsidiarity in European Community Law, a principle expressed in the Treaty on European Union whose Preamble declares that decisions will be taken "as closely as possible to the citizen in accordance with the principle of subsidiarity."'

COMMENT ON POWERS AND JURISDICTION OF INTERNATIONAL TRIBUNALS

Consider some comparisons and contrasts between the European Court of Human Rights and other international tribunals.

(1) *International arbitral tribunals.* Recall the discussion of arbitral tribunals on p. 73, *supra*, and the *Chattin* case, p. 75, *supra*, decided by the United

States–Mexican Claims Commission under the 1923 General Claims Convention between the two states. Frequently these tribunals were tripartite, one member appointed by each party to the dispute and the third pursuant to a stated procedure. In most instances, as in *Chattin*, the states themselves were the parties to the arbitration, with the complainant state frequently acting on behalf of (extending diplomatic protection to) an allegedly injured national. In a few instances, the individuals alleging injury by another state could themselves initiate proceedings before the arbitral tribunal. The governing agreement between the states generally required that the individual involved had exhausted the respondent state's local remedies. The tribunal or claims commission could provide whatever relief (generally, not invariably, the award of damages) that the governing instrument (a general arbitration treaty, an *ad hoc* agreement to create the tribunal for a given controversy, and so on) provided. As in the United States–Mexican General Claims Convention, the governing agreement generally spelled out the applicable law, often through terse references to principles of justice or international law.

(2) *International Military Tribunal*, p. 99, *supra*. This Tribunal, whose Charter was annexed to the London Agreement of 1945 among the Allied powers, was created for the trial of the Nazi war criminals at Nuremberg. The Charter spelled out the governing norms. The IMT's four judges were drawn from each of the four major Allied powers. The punishments imposed by the IMT on the individual defendants included the death penalty. (Chapter 15 examines the contemporary International Criminal Tribunal for the Former Yugoslavia, which again exercises a criminal jurisdiction over individual defendants. It was created pursuant to a decision of the UN Security Council.)

(3) *The International Court of Justice*. The opinion in the *Minority Schools in Albania* case of the Permanent Court of International Justice, the predecessor court under the League of Nations, appears at p. 89, *supra*. The contemporary ICJ is provided for in Article 92 of the UN Charter. It has both a contentious and advisory jurisdiction. Under Article 94, each member state of the UN agrees to comply with a decision of the ICJ against it. The Court's jurisdiction depends on states' consent, and under its Statute, only states may be parties before it. The Statute defines the governing law in very general terms (see Article 38 of the Statute, p. 27, *supra*). The Court may order the payment of damages by a respondent state as well as provide other forms of relief.

(4) The *Court of Justice* created under the Treaty Establishing the European Community (Common Market), p. 576, *supra*, is to 'ensure that in the interpretation and application of this Treaty the law is observed.' (Art. 164). Generally that application will in the first instance be by the courts of member states. The Court of Justice exercises different types of jurisdiction—for example, in cases brought against a state by another institution of the European Community (such as the Commission) (Art. 169), or brought by one member state against another (Art. 170). States are 'required to take the necessary measures to comply with the judgment' of the Court (Art. 171). Depending on its precise jurisdictional base, the Court may enter judgment

against a party, annul an administrative act, award money damages, or afford other relief.

Article 177 grants the Court jurisdiction to give 'preliminary rulings' on matters including the Treaty's interpretation, and the validity and interpretation of acts of Community institutions.[37]

> Where such a question is raised before any court or tribunal of a Member State, that court or tribunal may, if it considers that a decision on the question is necessary to enable it to give judgment, request the Court of Justice to give a ruling thereon.
>
> Where any such question is raised in a case pending before a court of tribunal of a Member State, against whose decisions there is no judicial remedy under national law, that court or tribunal shall bring the matter before the Court of Justice.

Article 177 is not an 'appeals' procedure leading to affirmance, reversal, remand with instructions, and so on. The Court does not 'decide' the case. Rather, the Court must 'interpret' the relevant Treaty provision, not 'apply' it. (In these respects, Article 177 is analogous to a form of proceedings before and jurisdiction of appellate courts in a number of civil law countries.) The state court suspends proceedings and itself (rather than a party to the state court proceeding) certifies questions of interpretation to the Court of Justice, which renders its opinion in the form of an abstract interpretation. The state court then continues the proceedings in light of that opinion. At the start, the primary purpose of Article 177 was to assure a uniform application of Community law by the courts of the member states. Over the decades, most of the Court's major rulings deciding basic principles about the Community grew out of Article 177.

Article 177 involves then mutual cooperation and a shared jurisdiction of both a state and Community court. Predictable problems have arisen with respect to the responsibility of state courts (the conditions under which that responsibility arises) to certify a question to the Court of Justice rather than resolve the question themselves, and with respect to the duty of the state court to decide the case consistently with the interpretation of the Court of Justice. Attitudes of judiciaries in the several member states have varied.

QUESTIONS

1. Compare the situation of international tribunals as described in the preceding Comment with procedures and powers of appellate courts in relation to lower courts in the domestic law of states. In many countries, a typical procedure would

[37] For analyses of the problems that have arisen under Article 177, see P.S.R.F. Mathijsen, A Guide to European Union Law 97–103 (6th ed. 1995); and Josephine Steiner, Textbook on EEC Law 285–305 (3rd ed. 1992).

involve the certification to the appellate court of a record of proceedings in the lower court; a decision by the appellate court to affirm or reverse the lower court, perhaps with a remand ordering the lower court to enter judgment for one or the other party (in which event, the case is terminated if the decision is by the highest appellate court); or a remand by the appellate court to the lower court with instructions to proceed in other stated ways (to hold a new trial, modify an injunction, order release of a prisoner, dismiss or reinstate the case, and so on).

a. Would you view it as desirable to have an international tribunal linked to, say, the highest court in a state judiciary in the same way that an appellate court in a state is linked to a lower court within that state's judicial system?

b. Would you view such an arrangement as politically feasible? In what respects does it go beyond existing arrangements?

2. In the light of the illustrations in the preceding Comment, what changes do you think desirable in the procedures, powers and jurisdiction of the European Court of Human Rights within the European human rights system? If you believe that some changes are desirable, why do you suppose that they are not now part of that system?

ADDITIONAL READING

R. St J. Macdonald, Matscher and Petzold (eds.), *The European System for the Protection of Human Rights* (1993); Franz Matscher and Petzold (eds.), *Protecting Human Rights: The European Dimension* (2nd ed., 1990); J. G. Merrills, *The Development of International Law by the European Court of Human Rights* (2nd ed., 1993); A. H. Robertson and Merrills, *Human Rights in Europe: A Study of the European Convention on Human Rights* (3rd ed., 1993); P. Van Dijk and Van Hoof, *Theory and Practice of the European Convention on Human Rights* (2nd ed., 1990); Andrew Clapham, *Human Rights in the Private Sphere* (1993); Council of Europe, *Collected Edition of the 'Travaux Préparatoires' of the European Convention on Human Rights* (8 vols., 1975–85); Andrew Drzemczewski, *European Human Rights Convention in Domestic Law* (1983); James Fawcett, *The Application of the European Convention on Human Rights* (2nd ed., 1987); Jochen Frowein, *The European Convention on Human Rights as the Public Order of Europe*, 1/2 *Collected Courses of the Academy of European Law* 267 (1992); David Kinley, *The European Convention on Human Rights: Compliance without Incorporation* (1993); Kees Waaldijk and Clapham (eds.), *Homosexuality: A European Community Issue: Essays on Lesbian and Gay Rights in European Law and Policy* (1993); Rudolf Bernhardt, *Reform of the Control Machinery under the European Convention on Human Rights: Protocol No. 11*, 89 Am. J. Int'l L, 145 (1995).

C. THE INTER-AMERICAN SYSTEM: PROMOTING DEMOCRACY

The aspect of the Inter-American system that Section C selects to illustrate the workings of that system involves building respect for and promoting some basic features of democratic government within the Western Hemisphere. At the same time, although less extensively than with respect to the European human rights system, the materials describe the rights protected by the American Convention on Human Rights and the complex institutional arrangements that result from the parallel existence of organs based on both the Charter of the Organization of American States (OAS) and the Convention. The section includes a leading decision of the Inter-American Court of Human Rights and a prominent report of the Inter-American Commission on Human Rights.

1. BACKGROUND AND PROTECTED RIGHTS

COMMENT ON DEVELOPMENT OF THE INTER-AMERICAN SYSTEM

In May 1948 the ninth Inter-American Conference, held in Bogotá, established the Organization of American States (OAS). Its predecessor organizations date back to the International Union of American Republics of 1890. The 1948 Charter entered into force in December 1951 and has since been amended by the Protocol of Buenos Aires of 1967, the Protocol of Cartagena de Indias of 1985, the Protocol of Washington of 1992, and the Protocol of Managua of 1993. The latter two have not entered into force. The purposes of the OAS are:

> to strengthen the peace and security of the continent; to promote and consolidate representative democracy, with due respect for the principle of nonintervention; to prevent possible causes of difficulties and to ensure the pacific settlement of disputes that may arise among the Member States; to provide for common action on the part of those States in the event of aggression; to seek the solution of political, juridical and economic problems that may arise among them; to promote by cooperative action, their economic, social and cultural development, and to achieve an effective limitation of conventional weapons that will make it possible to devote the largest amount of resources to the economic and social development of the Member States. (Annual Report of the Inter-American Commission on Human Rights 1994 (1995), at 347.)

Its principal organs are the General Assembly that meets annually and in additional special sessions if required, the Meeting of Consultation of Ministers of Foreign Affairs that considers urgent matters, the Permanent Council and the General Secretariat. The latter two organs are based in Washington D.C.

The Bogotá Conference of 1948 also adopted the American Declaration of the Rights and Duties of Man. The Inter-American system thus had a human rights declaration seven months before the UN had adopted the Universal Declaration and two and a half years before the European Convention was adopted. Nevertheless, the development of a regional treaty monitored by an effective supervisory machinery was to take considerably longer. The Inter-American Commission on Human Rights was created in 1959 and the American Convention on Human Rights was adopted in 1969. It entered into force in 1978.

The gradual evolution of the Inter-American system can only be understood against the background of a complex relationship between Latin America and the United States. The commitment to democracy and respect for human rights given in treaties by the former was both tempered and inspired by a desire to prevent intervention by the latter. A shared dislike of communism complicated matters further, in so far as it encouraged a certain reluctance on the part of the U.S. and others to insist on the restoration of democracy by dictatorships that purported to be resisting communist infiltrators.

For these and other reasons the development of the Inter-American system followed a different path from that of its European counterpart. Although the institutional structure is superficially very similar and the normative provisions are in most respects directly comparable, the conditions under which the two systems developed were radically different. Within the Council of Europe, military and other authoritarian governments have been rare and short-lived, while in Latin America they were close to being the norm until the 1980s.

The major challenges confronting the European system are epitomized by issues such as the length of pre-trial detention and the implications of the right to privacy. Cases involving states of emergency have been relatively few. The European Commission and Court have rarely had to deal with completely unresponsive or even antagonistic governments or national legal systems. By contrast, states of emergency have been common in Latin America, the domestic judiciary has often been extremely weak or corrupt, and large-scale practices involving torture, disappearances and executions have not been uncommon. Many of the governments with which the Inter-American Commission and Court have had to work have been ambivalent towards those institutions at best and outrightly hostile at worst.

In September 1995 there were 35 Member States of the OAS. Twenty-five of them had ratified the American Convention on Human Rights and 16 of those had recognized the jurisdiction of the Court.[38] Although the United States signed the Convention in 1978, it has yet to ratify. Cuba remains, technically, a member of the OAS, but the Government of Fidel Castro has been excluded from participation in its work since 31 January 1962. This has not prevented the OAS from scrutinizing the human rights record of Cuba.

COMMENT ON RIGHTS RECOGNIZED IN THE AMERICAN DECLARATION AND CONVENTION

You should now become familiar with the American Convention, excerpts from which appear in the Document Annex.

In terms of rights, the American Declaration on the Rights and Duties of Man is similar in content to the Universal Declaration, including the economic and social rights therein. What distinguishes it are ten articles setting out the duties of the citizen: the duty 'so to conduct himself in relation to others that each and every one may fully form and develop his personality;' to 'aid support, educate and protect his minor children,' to 'acquire at least an elementary education,' to vote in popular elections, to 'obey the law and other legitimate commands of the authorities,' to 'render whatever civil and military service his country may require for its defence and preservation,' to cooperate with the state with respect to social security and welfare; to pay taxes; and to work.

The process of drafting an Inter-American treaty began in 1959. The Inter-American Council of Jurists prepared a draft, based largely on the European Convention. This was considered in 1965, along with a draft by Chile, which emphasised democratic rights, and one by Uruguay which stressed economic and social rights. A revised, consolidated draft was then prepared by the Inter-American Commission on Human Rights. The result was the American Convention on Human Rights of 1969 (also known as the Pact of San José, Costa Rica) which contains 26 rights and freedoms, 21 of which are formulated in similar terms to the provisions of the ICCPR. Consider some comparisons:

1. Article 27 of the ICCPR, which recognizes the rights of members of minority groups, has no counterpart in the American Convention.

2. The five provisions which are in that Convention but not in the ICCPR are the right of reply (Art. 14), the right to property (Art. 21), freedom

[38] The States that both ratified the Convention and recognized the Court's jurisdiction were: Argentina, Bolivia, Chile, Colombia, Costa Rica, Ecuador, Guatemala, Honduras, Nicaragua, Panama, Paraguay, Peru, Surinam, Trinidad and Tobago, Uruguay and Venezuela. The following states ratified the Convention: Barbados, Brazil, Dominica, the Dominican Republic, El Salvador, Grenada, Haiti, Jamaica, and Mexico. In addition to the United States, the OAS member states that have not become parties to the Convention are Antigua and Barbuda, The Bahamas, Belize, Canada, Chile, Cuba, Guyana, Saint Kitts and Nevis, Saint Lucia, and Saint Vincent and the Grenadines.

from exile (Art. 22(5)), the right to asylum (Art. 22(7)), and prohibition of 'the collective expulsion of aliens' (Art. 21(9)).

3. The American Convention also recognizes several rights not contained in the European Convention, including the right of reply, the rights of the child, the right to a name and nationality and the right to asylum.

4. Some provisions in the American Convention express the same general idea as in other human rights treaties but give it a distinctive specification—for example, Article 4 on the right to life that provides in para. 1 that the right 'shall be protected by law and, in general, from the moment of conception.'

5. Article 23 on participation in government, which figures in the later materials, contains the same rights and requirements as the analogous Article 21 of the UDHR and Article 25 of the ICCPR.

When the Convention was adopted in 1969 it was decided not to have a separate treaty relating to economic, social and cultural rights but rather to include a general provision (Art. 26) in the following terms:

> The States Parties undertake to adopt measures, both internally and through international cooperation, especially those of an economic and technical nature, with a view to achieving progressively by legislation or other means, the full realization of the rights implicit in the economic, social, educational, scientific and cultural standards set forth in the Charter of the Organization of American States as amended by the Protocol of Buenos Aires.

The OAS Charter, as amended, sets up an Inter-American Council for Education, Science and Culture, as well as an Economic and Social Council, both of which are supposed to set standards, consider reports made by States, and make recommendations. That machinery has, however, achieved very little indeed in relation to economic, social and cultural rights, although the Inter-American Commission on Human Rights has carried out occasional reviews of States' reports. In 1988 the OAS adopted an Additional Protocol to the American Convention on Human Rights in the Area of Economic, Social and Cultural Rights (known as the Protocol of San Salvador). It has received four ratifications and has not yet entered into force. It obliges parties to adopt measures, 'to the extent allowed by their available resources and taking into account their degree of development,' for the progressive achievement of the rights listed.

The rights recognized in the Protocol are similar to those in the International Covenant on Economic, Social and Cultural Rights, although the formulations differ significantly. The Protocol does not recognize the rights to adequate clothing and housing or to an adequate standard of living (Art. 11 of the ICESCR), but it does include the right to a healthy environment, the right to special protection in old age, and the rights of persons with disabilities, none of which are explicitly recognized in the ICESCR. Although the system of supervision relies primarily on reporting, the Protocol provides for a right to petition when trade union rights or the right to education 'are violated by action directly attributable to a State Party' (Art. 19(6)).

NOTE

Compare the duties expressed in the American Declaration with the ICCPR, in which duties are only referred to in a preambular paraphrasing of Article 29(1) of the Universal Declaration ('Everyone has duties to the community in which alone the free and full development of his personality is possible'). The conception and nature of the duties expressed in the African Charter on Human and Peoples' Rights, p. 689, *infra*, is radically different.

2. INSTITUTIONAL ARRANGEMENTS

CECILIA MEDINA, THE INTER-AMERICAN COMMISSION ON HUMAN RIGHTS AND THE INTER-AMERICAN COURT OF HUMAN RIGHTS: REFLECTIONS ON A JOINT VENTURE

12 Hum. Rts. Q. 439 (1990), at 440.

[When the Inter-American Commission was established in 1959, the assumption was that it would confine itself to 'abstract investigations'. When complaints immediately flowed in, the Commission was compelled to respond in some way.]

. . .

A significant part of the Commission's work was addressing the problem of countries with gross, systematic violations of human rights, characterized by an absence or a lack of effective national mechanisms for the protection of human rights and a lack of cooperation on the part of the governments concerned. The main objective of the Commission was not to investigate isolated violations but to document the existence of these gross, systematic violations and to exercise pressure to improve the general condition of human rights in the country concerned. For this purpose, and by means of its regulatory powers, the Commission created a procedure to 'take cognizance' of individual complaints and use them as a source of information about gross, systematic violations of human rights in the territories of the OAS member states.

The Commission's competence to handle individual communications was formalized in 1965, after the OAS reviewed and was satisfied with the Commission's work. The OAS passed Resolution XXII, which allowed the Commission to 'examine' isolated human rights violations, with a particular focus on certain rights. This procedure, however, provided many obstacles for the Commission. Complaints could be handled only if domestic remedies had been exhausted, a requirement that prevented swift reactions to violations. Also, the procedure made the Commission more dependent on the governments for information. This resulted in the governments' either not answering the Commission's requests for information or answering with a blanket denial that did not contribute to a satisfactory solution of the problem.

Furthermore, once the Commission had given its opinion on the case, there was nothing else to be done; the Commission would declare that a government had violated the American Declaration of the Rights and Duties of Man and recommend the government take certain measures, knowing that this was unlikely to resolve the situation. The fact that some of the Commission's opinions could reach the political bodies of the OAS did not solve the problem, because the Commission's opinions on individual cases were never discussed at that level. Consequently, in order not to lose the flexibility it had, the Commission interpreted Resolution XXII as granting the Commission power to 'examine' communications concerning individual violations of certain rights specified in the resolution without diminishing its power to 'take cognizance' of communications concerning the rest of the human rights protected by the American Declaration. The Commission preserved this broader power for the purposes of identifying gross, systematic human rights violations.

The procedure to 'take cognizance' of communications evolved and became the general case procedure and was later used in examining the general human rights situation in a country. This procedure, maturing with the Commission's practice, had several positive characteristics in view of the Commission's purposes. First, it could be started without checking whether the communications met any admissibility requirements or even in the absence of any communication. All that was necessary was for news to reach the Commission that serious violations were taking place in the territory of an OAS member state. Second, the Commission assumed a very active role by requesting and gathering information by telegram and telephone from witnesses, newspapers, and experts, and also requesting consent to visit the country at the Commission's convenience. Third, the Commission could publicize its findings in order to put pressure upon the governments. Finally, the report resulting from the investigation could be sent to the political bodies of the OAS, thereby allowing for a political discussion of the problem which, at least theoretically, could be followed by political measures against the governments involved.

Since financial and human resources were limited, the Commission concentrated all its efforts on the examination of the general situation of human rights in each country. The examination of individual cases clearly took a secondary place. The Commission appeared to process them only because it had a duty to do so and not because of a conviction that its intervention would be helpful. After all, the special procedure for individual cases did not improve the victims' possibilities for redress, and the Commission could attempt to solve the cases through an examination of the general human rights situation in the country.

In short, the Commission was the sole guarantor of human rights in a continent plagued with gross, systematic violations, and the Commission was part of an international organization for which human rights were definitely not the first priority, and these facts made an imprint on the way the Commission looked upon its task. Apparently, the Commission viewed itself more as an international organ with a highly political task to perform than as a technical body whose main task was to participate in the first phase of a quasi-judicial supervision of the observance of human rights. The Commission's past made it ill-prepared to efficiently utilize the additional powers the Convention subsequently granted it.

A. The System Under the American Convention on Human Rights

The Convention vested the authority to supervise its observance in two organs: the Inter-American Commission, which pre-existed the Convention, and Inter-American Court of Human Rights, which was created by the Convention.

The Inter-American Commission is composed of seven members elected in a nongovernmental capacity by the OAS General Assembly and represents all the OAS member states. The entry into force of the Convention in 1978 invested the Commission with a dual role. It has retained its status as an organ of the OAS, thereby maintaining its powers to promote and protect human rights in the territories of all OAS member states. In addition, it is now an organ of the Convention, and in that capacity it supervises human rights in the territories of the states parties to the Convention.

The Commission's functions include: (1) promoting human rights in all OAS member states; (2) assisting in the drafting of human rights documents; (3) advising member states of the OAS; (4) preparing country reports, which usually include visits to the territories of these states; (5) mediating disputes over serious human rights problems; (6) handling individual complaints and initiating individual cases on its own motion, both with regard to states parties and states not parties to the Convention; and (7) participating in the handling of cases and advisory opinions before the Court.

The Inter-American Court consists of seven judges irrespective of the number of states that have recognized the jurisdiction of the Court. Although the Court is formally an organ of the Convention and not of the OAS, its judges may be nationals of any member state of the OAS whether or not they are parties to the Convention.

The Court has contentious and advisory jurisdiction. In exercising its contentious jurisdiction, the Court settles controversies about the interpretation and application of the provisions of the American Convention through a special procedure designed to handle individual or state complaints against states parties to the Convention. Under its advisory jurisdiction, the Court may interpret not only the Convention but also any other treaty concerning the protection of human rights in the American states. The Court may also give its opinion regarding the compatibility of the domestic laws of any OAS member state with the requirements of the Convention or any human rights treaties to which the Convention refers. . . . The advisory jurisdiction of the Court may be set in motion by any OAS member state, whether or not it is a party to the Convention, or by any OAS organ listed in Chapter X of the OAS Charter, which includes the Commission.

The procedure for handling individual or state complaints begins before the Commission. The procedure resembles those set forth in the European Convention and in the Additional Protocol to the International Covenant on Civil and Political Rights. It is a quasi-judicial mechanism which may be started by any person, group of persons, or nongovernmental entity legally recognized in one or more of the OAS member states, regardless of whether the complainant is the victim of a human rights violation. This right of individual petition is a mandatory provision in the Convention, binding on all states parties. Inter-state communications, however, are dependent upon an explicit recognition of the competence of the Commission to receive and examine them. In addition, the Commission may begin processing a case on its own motion.

. . .

The Commission has powers to request information from the government concerned and, with the consent of the government, to investigate the facts in the complaint at the location of the alleged violation. If the government does not cooperate in the proceedings by providing the requested information within the time limit set by the Commission, Article 42 of the Commission's Regulations allows the Commission to presume that the facts in the petition are true, 'as long as other evidence does not lead to a different conclusion.' Following this, the Commission need investigate the case no further.

. . .

The Court may consider a case that is brought either by the Commission or by a state party to the Convention. For the Commission to refer a case to the Court, the case must have been admitted for investigation and the Commission's draft report sent to the state party. In addition, the state must recognize the Court's general contentious jurisdiction or a limited jurisdiction specified by a time period or case. For a state party to be able to place a case before the Court, the only requirement is that both states must have recognized the Court's contentious jurisdiction.

. . .

If a state does not comply with the decision of the Court, the Court may inform and make recommendations to the OAS General Assembly. There is no reference in the Convention to any action that the General Assembly might take; the assembly, being a political body, may take any political action it deems necessary to persuade the state to comply with its international obligations.

As may be apparent, a petition to the Court by the Commission or state party to the Convention is meant to handle isolated violations of human rights committed by a state which otherwise respects the rule of law. This procedure functions efficiently when the states concerned act in goodwill and cooperate with the human rights supervisory organs. In the reality of the inter-American system, unfortunately, goodwill and cooperation on the part of the states are seldom seen. This being the case, the procedure is bound to be inadequate.

In addition to the problems posed by the Commission's status as part of a political organization and by the lack of cooperation among the states, financial limitations are also potentially troublesome. The OAS does not provide the Commission with the necessary means to carry out all its various activities. The Commission usually receives about 500 complaints a year, with each complaint frequently involving more than one victim. At times the number is much higher; in 1980, when members of the Commission visited Argentina, 5,000 complaints were received. Furthermore, in any one year, the Commission carries out two or three observations on location and monitors the general situation of human rights in at least six or seven countries. To perform all these functions the Commission holds, in principle, two ten-day sessions a year. To support these activities the Commission currently has only seven lawyers, including the Executive Secretary, four secretaries and one administrative official, and its budget is less than 2 percent of the OAS budget. Under these circumstances, the Commission inevitably makes a choice as to what it can accomplish and places a priority on tasks it perceives as most likely to increase the general respect for human rights. In this ordering, the handling of individual complaints does not rank very high.

. . .

NOTE

The Commission has been criticized by commentators for its handling of individual petitions. Critics, for example, have argued that the Commission does not follow its own regulations and that its decisions are politicized and sometimes confused. One report on the Commission concluded that all such criticisms were 'aspects of one problem: the Commission does not function well as a neutral, quasi-judicial adjudicative body for individual cases.' The report also said:[39]

[39] Committee on International Human Rights [of the Association of the Bar of the City of New York], The Inter-American Commission on Human Rights: A Promise Unfulfilled, 48 The Record 589, 600 (1993).

> Nevertheless, the inadequacies of the Commission's work on individual complaints cannot be explained fully by the intractability of the cases or the obstruction by governments. Those causes will not explain the Commission's reluctance, still persisting in some instances, to prepare and refer cases to the Court. Because the Commission's charter-based jurisdiction long ante-dated the emergence of the Inter-American court, the Commission became accustomed to resolving its cases without the aid of the Court. After the creation of the Court, the custom continued; the Commission feared, for example, that resort to the Court might prolong the already protracted resolution of complaints. Whatever the reason, it is a habit that never completely changed. This created tension between the Commission and the Court, which appears, according to Court opinions and other cases described below, not entirely to have dissipated.

The effective functioning of human rights machinery is heavily dependent on adequate staffing and financial resources. The Commission has an annual budget (1995 figures) of $1,734,000. Of that about $1,000,000 is spent on salaries and other fixed expenses, leaving only a little over $400,000 to cover meetings of the Commissioners, on-site visits, witness and expert expenses and the preparation of reports. According to one report, many States are 'offended by the attentions of the Commission [and] work to keep it starved for funds . . . .' Medina notes that the Commission usually meets for four weeks a year and has very few staff. Its European counterpart meets for 16 weeks and has a staff of over 40 lawyers.

Much of the Commission's work has been devoted to on-site country visits. The reports which have resulted from these visits have been widely praised as the best of their kind in terms of fact-finding. The materials in this Chapter 10 do not, however, explore this aspect of the Commission's work because of the attention given in Chapter 7 to issues raised by fact-finding in the work of the UN Commission on Human Rights.[40] One observer has suggested that:

> . . . With the return to democracy in many countries in the hemisphere, the time is ideal for participation in restoring and protecting human rights through military and police training programs, educational programs on human rights obligations for public servants and political leaders, and the availability of expert legal consultants on drafting and implementing laws for the protection of human rights.[41]

[40] See generally Cecilia Medina Quiroga, The Battle of Human Rights: Gross, Systematic Violations and the Inter-American System (1988).

[41] Dinah Shelton, Improving Human Rights Protections: Recommendations for Enhancing the Effectiveness of the Inter-American Commission and the Inter-American Court of Human Rights, 3 Am. U. J. Int. L. & Policy 323, 331 (1988).

3. THE COURT IN ACTION

In her article, *supra*, Medina identifies three stages that have characterized the relationship between the Commission and the Court. From 1979 to 1981, the Commission neither sought an advisory opinion nor submitted any contentious cases to the Court. This was seen by many observers as a regrettable omission, since states (the only other potential source of 'business' for the Court) were reluctant for political reasons to initiate a use of the Court that might become regular. From 1982 to 1985 the Commission requested an advisory opinion from the Court for the first time (although the Court had already delivered one such opinion, in response to a request from Peru).

The second stage also involved the Commission's failure to refer to the Court a decision it reached upholding a Costa Rican law requiring journalists to be licensed. Since the Commission's decision had not been unanimous, and the issue at hand was both controversial and important, the Costa Rican Government agreed to refer the case to the Court. The Court took the opportunity to berate the Commission for not acting on its 'special duty to consider the advisability of coming to the Court.'[42]

In the third stage, from 1986 onwards, a cooperative relationship began to evolve, commencing with the referral by the Commission of three contentious cases to the Court, each involving instances of disappearances in Honduras. The principal case, *Velásquez Rodríguez*, appears below. It was the first contentious case initiated by an individual that involved systemic state violence, and was one of three cases leading to decisions by the Court on the question of disappearances in Honduras. The cases placed in a judicial setting the same issues that occupied the UN Commission on Human Rights and its Working Group on Disappearances, p. 421, *supra*, as well as the Inter-American Commission in its function of investigating and reporting.

The Commission has since generated a small but steady flow of cases for the Court. For example, three cases (from Nicaragua, Venezuela and Argentina) were submitted in the first half of 1994.

VELASQUEZ RODRIGUEZ CASE

Inter-American Court of Human Rights, 1988
Ser. C, No. 4, 9 Hum. Rts L. J. 212 (1988).

[This case arose out of a period of political turbulence, violence and repression in Honduras. It originated in a petition against Honduras received by the Inter-American Commission on Human Rights in 1981. The thrust of the petition was that Angel Manfredo Velásquez Rodríguez (often referred to as

[42] Compulsory Membership in an Association Prescribed by Law for the Practice of Journalism (Arts. 13 and 19 of the American Convention on Human Rights), Advisory Opinion OC-5/85, Ser. A No. 5, 86, 145 (1985).

Velásquez) was arrested without warrant in 1981 by members of the National Office of Investigations (DNI) and the G-2 of the Armed Forces. The 'arrest' was a seizure by seven armed men dressed in civilian clothes who abducted him in an unlicensed car. The petition referred to eyewitnesses reporting his later detention, 'harsh interrogation and cruel torture.' Police and security forces continued to deny the arrest and detention. Velásquez had disappeared. The petition alleged that through this conduct, Honduras violated several articles of the American Convention on Human Rights.

In 1986, Velásquez was still missing, and the Commission concluded that the Government of Honduras 'had not offered convincing proof that would allow the Commission to determine that the allegations are not true.' Honduras had recognized the contentious jurisdiction of the Inter-American Court of Human Rights, to which the Commission referred the matter. The Court held closed and open hearings, called witnesses and requested the production of evidence and documents. The statement of facts below is taken from the Court's opinion and consists both of its independent findings and its affirmation of some findings of the Commission.

The Commission presented witnesses to testify whether 'between the years 1981 and 1984 (the period in which Manfredo Velásquez disappeared) there were numerous cases of persons who were kidnapped and who then disappeared, these actions being imputable to the Armed Forces of Honduras and enjoying the acquiescence of the Government of Honduras,' and whether in those years there were effective domestic remedies to protect such kidnapped persons. Several witnesses testified that they were kidnapped, imprisoned in clandestine jails and tortured by members of the Armed Forces. Explicit testimony described the severity of the torture—including beatings, electric shocks, hanging, burning, drugs and sexual abuse—to which witnesses had been subjected. Several witnesses indicated how they knew that their captors and torturers were connected with the military. The Court received testimony indicating that 'somewhere between 112 and 130 individuals were disappeared from 1981 to 1984.'

According to testimony, the kidnapping followed a pattern, such as use of automobiles with tinted glass, with false license plates and with disguised kidnappers. A witness who was President of the Committee for the Defense of Human Rights in Honduras testified about the existence of a unit in the Armed Forces that carried out the disappearance, giving details about its organization and commanding personnel. A former member of the Armed Forces testified that he had belonged to the battalion carrying out the kidnapping. He confirmed parts of the testimony of witnesses, claiming that he had been told of the kidnapping and later torture and killing of Velásquez, whose body was dismembered and buried in different places. All such testimony was denied by military officers and the Director of Honduran Intelligence.

The Commission also presented evidence showing that from 1981–1984 domestic judicial remedies in Honduras were inadequate to protect human rights. Courts were slow and judges were often ignored by police. Authorities denied detentions. Judges charged with executing the writs of habeas corpus were threatened and on several occasions imprisoned. Law professors and lawyers defending political prisoners were pressured not to act; one of the two lawyers to bring a writ of habeas corpus was arrested. In no case was the writ effective in relation to a disappeared person.

In view of threats against witnesses it had called, the Commission asked the Court to take provisional measures contemplated by the Convention. Soon thereafter, the Commission reported the death of a Honduran summoned by the Court to appear as a witness, killed 'on a public thoroughfare [in the capital city] by a group of armed men who . . . fled in a vehicle.' Four days later the Court was informed of two more assassinations, one victim being a man who had testified before the Court as a witness hostile to the Government. After a public hearing, the Court decided on 'additional provisional measures' requiring Honduras to report within two weeks (1) on measures that it adopted to protect persons connected with the case, (2) on its judicial investigations of threats against such persons, and (3) on its investigations of the assassinations.

The excerpts below from the Court's opinion refer to several articles of the American Convention. *Article 4* gives every person 'the right to have his life respected. . . . No one shall be arbitrarily deprived of his life.' *Article 5* provides that no one 'shall be subjected to torture or to cruel, inhuman, or degrading punishment or treatment.' *Article 7* gives every person 'the right to personal liberty and security,' prohibits 'arbitrary arrest or imprisonment,' and provides for such procedural rights as notification of charges, recourse of the detained person to a competent court, and trial within a reasonable time or release pending trial.]

VII

. . .

123. Because the Commission is accusing the Government of the disappearance of Manfredo Velásquez, it, in principle, should bear the burden of proving the facts underlying its petition.

124. The Commission's argument relies upon the proposition that the policy of disappearances, supported or tolerated by the Government, is designed to conceal and destroy evidence of disappearances. When the existence of such a policy or practice has been shown, the disappearance of a particular individual may be proved through circumstantial or indirect evidence or by logical inference. Otherwise, it would be impossible to prove that an individual has been disappeared.

. . .

126. The Court finds no reason to consider the Commission's argument inadmissible. If it can be shown that there was an official practice of disap-

pearances in Honduras, carried out by the Government or at least tolerated by it, and if the disappearance of Manfredo Velásquez can be linked to that practice, the Commission's allegations will have been proven to the Court's satisfaction, so long as the evidence presented on both points meets the standard of proof required in cases such as this.

127. The Court must determine what the standards of proof should be in the instant case. Neither the Convention, the Statute of the Court nor its Rules of Procedure speak to this matter. Nevertheless, international jurisprudence has recognized the power of the courts to weigh the evidence freely, although it has always avoided a rigid rule regarding the amount of proof necessary to support the judgment (Cf. Corfu Channel, Merits, Judgment I. C. J. Reports 1949; Military and Paramilitary Activities in and against Nicaragua (Nicaragua v. United States of America), Merits, Judgment, I. C. J. Reports 1986, paras. 29–30 and 59–60.)

. . .

129. The Court cannot ignore the special seriousness of finding that a State Party to the Convention has carried out or has tolerated a practice of disappearances in its territory. . . .

130. The practice of international and domestic courts shows that direct evidence, whether testimonial or documentary, is not the only type of evidence that may be legitimately considered in reaching a decision. Circumstantial evidence, indicia, and presumptions may be considered, so long as they lead to conclusions consistent with the facts.

131. Circumstantial or presumptive evidence is especially important in allegations of disappearances, because this type of repression is characterized by an attempt to suppress any information about the kidnapping or the whereabouts and fate of the victim.

. . .

134. The international protection of human rights should not be confused with criminal justice. States do not appear before the Court as defendants in a criminal action. The objective of international human rights law is not to punish those individuals who are guilty of violations, but rather to protect the victims and to provide for the reparation of damages resulting from the acts of the States responsible.

135. In contrast to domestic criminal law, in proceedings to determine human rights violations the State cannot rely on the defense that the complainant has failed to present evidence when it cannot be obtained without the State's cooperation.

136. The State controls the means to verify acts occurring within its territory. Although the Commission has investigatory powers, it cannot exercise them within a State's jurisdiction unless it has the cooperation of that State.

. . .

138. The manner in which the Government conducted its defense would have sufficed to prove many of the Commission's allegations by virtue of the principle that the silence of the accused or elusive or ambiguous answers on its part may be interpreted as an acknowledgment of the truth of the allegations, so long as the contrary is not indicated by the record or is not compelled as a matter of law. This result would not hold under criminal law, which does not apply in the instant case. . . .

. . .

IX

147. The Court now turns to the relevant facts that it finds to have been proven. They are as follows:

a. During the period 1981 to 1984, 100 to 150 persons disappeared in the Republic of Honduras, and many were never heard from again. . . .

b. Those disappearances followed a similar pattern. . . .

c. It was public and notorious knowledge in Honduras that the kidnappings were carried out by military personnel, police or persons acting under their orders. . . .

d. The disappearances were carried out in a systematic manner, regarding which the Court considers the following circumstances particularly relevant:

i. The victims were usually persons whom Honduran officials considered dangerous to State security. . . .

[Omitted paragraphs deal with arms used, details of the kidnappings and interrogations, denials by officials of any knowledge about the disappeared person, and the failure of any investigative committees to produce results.]

e. On September 12, 1981, between 4:30 and 5:00 p.m., several heavily armed men in civilian clothes driving a white Ford without license plates kidnapped Manfredo Velásquez from a parking lot in downtown Tegucigalpa. Today, nearly seven years later, he remains disappeared, which creates a reasonable presumption that he is dead. . . .

f. Persons connected with the Armed Forces or under its direction carried out that kidnapping. . . .

g. The kidnapping and disappearance of Manfredo Velásquez falls within the systematic practice of disappearances referred to by the facts deemed proved in paragraphs a–d.

. . .

X

149. Disappearances are not new in the history of human rights violations. However, their systematic and repeated nature and their use, not only for causing certain individuals to disappear, either briefly or permanently, but also as a means of creating a general state of anguish, insecurity and fear, is a recent phenomenon. Although this practice exists virtually worldwide, it has occurred with exceptional intensity in Latin America in the last few years.

150. The phenomenon of disappearances is a complex form of human rights violation that must be understood and confronted in an integral fashion.

151. The establishment of a Working Group on Enforced or Involuntary Disappearances of the United Nations Commission on Human Rights by Resolution 20(XXXVI) of February 29, 1980, is a clear demonstration of general censure and repudiation of the practice of disappearances. . . . The reports of the rapporteurs or special envoys of the Commission on Human Rights show concern that the practice of disappearances be stopped, the victims reappear and that those responsible be punished.

152. Within the inter-American system, the General Assembly of the Organization of American States (OAS) and the Commission have repeatedly referred to the practice of disappearances and have urged that disappearances be investigated and that the practice be stopped. . . .

153. International practice and doctrine have often categorized disappearances as a crime against humanity, although there is no treaty in force which is applicable to the States Parties to the Convention and which uses this terminology. . . .

. . .

155. The forced disappearance of human beings is a multiple and continuous violation of many rights under the Convention that the States Parties are obligated to respect and guarantee. The kidnapping of a person is an arbitrary deprivation of liberty, an infringement of a detainee's right to be taken without delay before a judge and to invoke the appropriate procedures to review the legality of the arrest, all in violation of Article 7 of the Convention. . . .

156. Moreover, prolonged isolation and deprivation of communication are in themselves cruel and inhuman treatment, harmful to the psychological and moral integrity of the person and a violation of the right of any detainee to respect for his inherent dignity as a human being. Such treatment, therefore, violates Article 5 of the Convention. . . .

157. The practice of disappearances often involves secret execution without trial, followed by concealment of the body to eliminate any material evidence of the crime and to ensure the impunity of those responsible. This is a flagrant violation of the right to life, recognized in Article 4 of the Convention. . . .

158. The practice of disappearances, in addition to directly violating many provisions of the Convention, such as those noted above, constitutes a radi-

cal breach of that treaty in that it implies a crass abandonment of the values which emanate from the concept of human dignity and of the most basic principles of the inter-American system and the Convention. . . .

. . .

[The part of the Court's opinion examining the obligation of a state not only to respect individual rights (such as by not 'disappearing' the government's opponents), but also to ensure free exercise of rights (such as by protecting those expressing political opinions against violence by private, non-governmental actors), appears at p. 946, *infra*.]

XII

189. Article 63(1) of the Convention provides:

> If the Court finds that there has been a violation of a right or freedom protected by this Convention, the Court shall rule that the injured party be ensured the enjoyment of his right or freedom that was violated. It shall also rule, if appropriate, that the consequences of the measure or situation that constituted the breach of such rights or freedom be remedied and that fair compensation be paid to the injured party.

Clearly, in the instant case the Court cannot order that the victim be guaranteed the enjoyment of the right or liberty violated. The Court, however, can rule that the consequences of the breach of the rights be remedied and rule that just compensation be paid.

190. During this proceeding, the Commission requested the payment of compensation, but did not offer evidence regarding the amount of damages or the manner of payment. Nor did the parties discuss these matters.

191. The Court believes that the parties can agree on the damages. If an agreement cannot be reached, the Court shall award an amount. The case shall, therefore, remain open for that purpose. The Court reserves the right to approve the agreement and, in the event no agreement is reached, to set the amount and order the manner of payment.

[In the concluding paragraphs, the Court unanimously declared that Honduras violated Articles 4, 5 and 7 of the Convention, all three read in conjunction with Article 1(1); and unanimously decided that Honduras was required to pair fair compensation to the victim's next-of-kin.]

NOTE

Note several aspects of this opinion. (1) The petition was brought against Honduras in 1981. The decision of the Court is dated 1988. (2) The Commission is active in the Court proceedings, arguing on behalf of the individual seeking relief. (3) The proceedings include varied participants such as witnesses and non-governmental organizations. The Court itself acts as a blend of trial and appellate court.

The *Velásquez Rodríguez* decision was followed by two related judgments of the Inter-American Court of Justice that also involved disappearances in Honduras alleged to be the Government's responsibility: the *Godinez* Judgment, (Ser. C) No. 5 (1989), and the *Fairen Garbi and Solis Corrales* Judgment, (Ser. C) No. 6 (1989). The *Godinez* case was substantially similar to *Velásquez Rodríguez* and the Court reached a similar decision. In the *Fairen Garbi* case, the Court concluded that Honduras was not responsible for the disappearances there relevant. As far as was known, neither of the disappeared persons were involved in activities considered dangerous by the government. The Court said that there was insufficient evidence 'to relate the disappearance . . . to the governmental *practice* of disappearances. There is no evidence that Honduran authorities had [the disappeared persons] under surveillance or suspicion of being dangerous persons, nor that [they] were arrested or kidnapped in Honduran territory.'

QUESTIONS

1. Recall that the Court found a failure on the part of Honduras to fulfill its duties assumed under Article 1(1) of the Convention—namely, failure of the state apparatus to act to investigate the disappearance. It found a legal duty of states to use means at its disposal 'to identify those responsible, impose the appropriate punishment and ensure the victim adequate compensation.' How do you respond to the following comment:[43]

> . . . [In no other international law case known to the author] has the state been ordered to prosecute, and it is unlikely that the states drafting the American Convention intended to cede the requisite sovereignty to allow the Court to order them to initiate a criminal prosecution.
>
> [It] is submitted that . . . currently under international law [the Court] is not competent to order the government of Honduras to initiate an investigation or prosecution. This interpretation also appears to be supported by the Court's general philosophy on the purpose of the international human rights protection system:
>
>> The international protection of human rights should not be confused with criminal justice. States do not appear before the Court as defendants in a criminal action. The objective of international human rights law is not to punish those individuals who are guilty of violations, but rather to protect the victims and to provide for the reparation of damages resulting from the acts of the States responsible.

[43] Christina Cerna, The Inter-American Court of Human Rights, in Mark Janis (ed.), International Courts for the Twenty-First Century 117, 147 (1992).

2. The facts are contested, and the method employed by the Court to resolve them represents an important contribution of this opinion to human rights law. How does the Court 'find' the facts about the disappearance? What kinds of evidence does it find convincing? Does it employ such traditional notions of the law of evidence in systems of national law as burdens of proof (burdens of persuasion) or presumptions? If so, what burdens and presumptions? It may help in answering these questions to consider:

a. What is the relevance to the Court's finding of Honduran responsibility of the Court's use of terms like (the Honduran) 'practice' or 'policy', or the characterization of disappearances as 'systemic'?

b. What is the significance of the Court's observation that the state 'controls the means to verify acts occurring within its territory'? Is the state then under a special obligation to make every effort to come forth with the facts? Is the Court threatening the state with an adverse finding if it fails to make that effort?

3. In relation to the European Convention the argument in favor of having a judge and a commissioner for every state party is twofold: to ensure confidence in the European organs on the part of each state, and to ensure expertise in relation to each state legal system. In relation to the Inter-American system the argument in favour of having only seven judges and commissioners is that their independence is thereby assured. Can these propositions be reconciled? If not, which approach would you opt for and why?

4. DEMOCRATIC GOVERNMENT IN THE INTER-AMERICAN SYSTEM

a. Background in Theory and State Practice

COMMENT ON RELATIONSHIPS BETWEEN THE HUMAN RIGHTS MOVEMENT AND DEMOCRATIC GOVERNMENT

As background for the discussion in Sub-section (b) *infra* of the work of the Inter-American Commission relating to democracy, these materials explore links between the basic human rights instruments and political democracy, as well as recent trends in both norms and state practice that bear on the question of democratic governance as a human right. The exploration is cursory, barely noting the different theories or ideologies or historical realizations of democracy that inhere in the human rights debates and that underlie arguments based on the human rights instruments. This Comment complements the discussions in the related Comment on the liberal political tradition at p. 187, *supra*, and the Comment on Constitutionalism at p. 710, *infra*.

It is indeed difficult to bypass a discussion of democracy in relation to human rights in the contemporary world. Many will argue that democracy—often an unspecified form or realization thereof that may consist of no more than the essential element of periodic, genuine, contested elections—is now becoming or has indeed become a global norm. Hence, an article in this subsection is entitled an 'emerging right to democratic governance.'[44] The conception of democracy in much of the ongoing debate concentrates on, and sometimes does not go beyond, the fundamental premise of 'rule by the people' (or related expressions of this premise such as popular sovereignty, or government as expressive of the will of the people) through one or another form of representative government that elections are meant to achieve. Needless to say, the degree to which the contemporary observer detects a significant trend toward democratic governance will depend on the conception of democracy that is employed and on the related essential components of democratic governance.

Other observers, at least in the context of universal human rights rather than a regional system like the European, raise questions about the content of this 'emerging right' in terms of ideals and components of democracy other than genuine elections. Contested conceptions today of the nature of democratic government range from (1) the classical liberal democracy of a century ago with limited powers and activity of government and a sharp division between the state and civil society, to (2) the different types of social democracies in which governments assume larger responsibility for social and economic structures and are attentive to forms of inequality of wealth, power and opportunity in a society. They range, to employ another example, from (1) a notion of moderate and sporadic political participation by the citizenry primarily through elections to choose between the options offered, to (2) a notion of ongoing and active citizen participation in governance and in the related institutions of civil society. The debate raises vital issues of the relationship, necessary or contingent, between political democracy and an economic system characterized by private property and a market subject to a varying degree of governmental regulation.

Consider as one illustration the argument of Brad Roth[45] that '[g]reat ideological issues remain to be resolved, even if particular ideological positions

[44] See Franck, p. 672, *infra*. See also Gregory Fox, The Right to Political Participation in International Law, 17 Yale J. Int. L. 539 (1992). For discussion of related themes, see James Crawford, Democracy in International Law (1994). See also Gregory Fox and Georg Nolte, Intolerant Democracies, 36 Harv. Int. L. J. 1 (1995).

[45] Evaluating Democratic Progress: A Normative Theoretical Perspective, 9 Ethics & Int. Affairs 55 (1995).

now belong solely to the past. Democracy still admits of radically contradictory interpretations . . . ' Roth's article distinguishes among three ends

> to which the establishment of competitive electoral mechanisms may to a greater or lesser extent be relevant: (1) the furtherance of broad popular empowerment with respect to the full range of social decisions that condition life in the society ('substantive democracy'); (2) the establishment of a government to which the populace may in some manner be said to have manifested consent ('popular sovereignty'); and (3) the establishment of a broadly recognized basis for, and thereby limitation on, the legitimate exercise of power ('constitutionalism').

Roth examines the relations among these different ends in the older democracies and in the states of the former Communist world and of the developing world to which forms of democratic government have spread since the late 1980s. Within his framework, he criticizes the internal trends in numbers of the 'new' democracies that have moved to plural political parties and periodic election.[46]

Constitutions of a particular character have become characteristic of the new democracies, a character that has been captured in the rubric 'constitutionalism' that distinguishes such instruments from, for example, the constitutions of authoritarian states. Constitutionalism itself then becomes complexly related to conceptions of democracy, a relationship stressed in Section A of Chapter 11, *infra*, that describes the horizontal spread among states of liberal constitutions characteristic of Western democracies since the time of decolonization but particularly over the last decade. Those readings raise questions both of the adaptability of Western-style constitutionalism to regions and cultures of radically different histories and structures, and of the aspects of constitutionalism that might be said to inhere in or be required by democratic government—for example, the separation of powers, particularly the independence of the judiciary; the related notion of the Rule of Law; or the protection against majoritarianism offered by constitutionally entrenched individual rights and by judicial review of legislative as well as executive action.

In broad terms, one can pose two questions—perhaps two sides of the same question—about the relationships between the human rights movement and democratic governance, questions that have become more vital and pertinent since the collapse of the Soviet empire with its radically different ideology and practice of government. (1) Do the fundamental norms of the human rights instruments point toward or indeed require democratic government? Will observance of those norms inevitably lead to such government, or are they consistent with other forms of government-say, a monarchy, or theocracy, or military junta, or ideology requiring one ruling and guiding party? (2) Does

[46] There is a vast and growing literature about the spread of constitutionalism and democracy to the former Communist states and to other states of the developing world. Some are noted in the section in Chapter 11 on constitutionalism. Two recent examples are: Larry Diamond, Promoting Democracy, 87 Foreign Policy 25 (1992); and James Johnson, Does Democracy 'Travel'? Some Thoughts on Democracy and its Cultural Context, 6 Ethics & Int. Affairs 41 (1992).

political democracy require the protection and observance of the fundamental norms of the human rights movement—for example, physical security of citizens, due process, equal protection, freedoms of speech and association? Can democratic government be realized without that full array of norms—say, by a simple emphasis on majority rule—or are democracy and human rights linked in some fundamental way?

Such questions rest on premises that must be clarified before one can attempt answers. Those premises involve both the conception and related definition of democracy and of fundamental human rights norms. To use a graphic example, fundamental human rights norms include the rights recognized in the International Covenant on Economic, Social and Cultural Rights. But much of the discussion about human rights in relation to democracy considers only the fundamental civil and political rights noted in the preceding paragraph.

* * * * *

The two readings in this sub-section develop and clarify some aspects of the preceding discussion. Steiner examines theories of democratic participation that inform or could inform the provisions of the basic human rights documents on political participation and democracy, and underscores the basic choices left open by those provisions. Written four years later, after a period which saw the ongoing collapse of the Soviet Union and a trend from authoritarian government to electoral democracy in a number of developing countries, Franck's article argues on the basis of both human rights texts and state practice for an emerging norm of democratic governance.

HENRY STEINER, POLITICAL PARTICIPATION AS A HUMAN RIGHT

1 Harv. Hum. Rts. Y'bk 77 (1988), at 78.

. . .

The article . . . examines the norms expressing this right [to political participation] that are included in two principal international instruments of universal scope: the Universal Declaration of Human Rights, and the International Covenant on Civil and Political Rights. Those norms grant citizens the right to take part, directly or through representatives, in the conduct of public affairs and government, and to vote at genuine periodic elections based on universal suffrage and the secret ballot. My discussion emphasizes these two different ways in which the international norms express citizens' right to political participation: the relatively vague and abstract right to take part in the conduct of public affairs or government, and the relatively specific right to vote in elections. It builds upon that distinction to explore different modes of political participation. . . .

. . .

II. The Texts: Their Making and Interpretation

. . .

What emerged from the periods of drafting and debates were norms that expressed an important ideal of political participation. But they gave little indication of the different ways of institutionalizing that ideal. It cannot have been by chance that their language was sufficiently confined—with respect to the 'elections' clause—and sufficiently abstract and porous—with respect to the 'take part' clause—to permit democratic and nondemocratic states to assert that they satisfied the norms' demands. More specific norms would have put at risk the goal of achieving broad support for the human rights instruments as a whole.

A. The Two Basic Provisions

Article 25 of the International Covenant, the principal treaty declaring a right to political participation, states:

> Every citizen shall have the right and the opportunity, without . . . unreasonable restrictions:
>
> (a) To take part in the conduct of public affairs, directly or through freely chosen representatives;
>
> (b) To vote and to be elected at genuine periodic elections which shall be by universal and equal suffrage and shall be held by secret ballot, guaranteeing the free expression of the will of the electors. . . .

Alone among the International Covenant's provisions, this article restricts a right to citizens. It distinguishes between direct and representative participation, but validates both. The article does not indicate how citizens are to 'take part in the conduct of public affairs,' other than by identifying periodic and 'genuine' elections as an ingredient of the right to participate.

. . .

The provisions of Article 25 are unusual among human rights norms in that they do more than declare a right. They articulate a political ideal inspiring that right. Though not invoking a particular political tradition such as democracy, Article 25 affirms that the popular vote is meant to guarantee 'the free expression of the will of the electors.'

The Universal Declaration, an inspiration and important model for the

International Covenant and similar in content to it, contains the analogous provision in its Article 21:

> (1) Everyone has the right to take part in the government of his country, directly or through freely chosen representatives. . .
>
> (3) The will of the people shall be the basis of the authority of government; this will shall be expressed in periodic and genuine elections which shall be by universal and equal suffrage and shall be held by secret vote or by equivalent free voting procedures.

This earlier instrument, influenced to a greater degree than the International Covenant by the tradition of liberal democracy, gives more emphasis to the role of the 'will of the people' as the 'basis of the authority of government.' But like the International Covenant, Article 21 neither employs the term 'democracy,' nor makes explicit whether plural political parties and contested elections must be permitted.

The influence of the Universal Declaration on the International Covenant extends to the internal structure of Article 25. It too begins with, and thereby gives an apparent priority and emphasis to, the 'take part' clause, and continues with the 'elections' clause. . . . Generally I shall use the two clauses to refer, respectively, to non-electoral and electoral (voting) participation in political processes.

. . .

Notwithstanding its mixed ancestry and ambiguities, the Covenant gives some internal guidance for an understanding of Article 25. In view of their essential role in most political activities, the expressive rights to free speech, press, assembly and association must in some way inform any theory of participation. Their prominence in the International Covenant reminds us that Article 25 should not be approached as an isolated provision, detached from the larger structure of rights in the Covenant. That larger structure here suggests that the 'take part' and 'elections' clauses assume some degree of public political debate and of citizens' participation in political groups expressing their beliefs or interests.

. . .

III. Treaty Provisions in Relation to Contemporary Debate About Participation in the Liberal Democracies

An attempt to formulate a universal norm about political participation confronts mutually antagonistic theories and practices among the liberal democracies, communist states, military dictatorships and a range of third world governments. Consensus over a norm limited to West European, North

American and Commonwealth democracies should then be easier to achieve. Basic obstacles to agreement about reasonably specific provisions fall away when, for example, none of the proposed signatories is a one-party state or imposes strict censorship.

Even among the democracies, however, a norm about political participation must leave significant matters open. To gain adoption, a regional norm embracing West European and North American countries would have to be understood as consistent with distinct theories of participation informing these countries' political systems. It would have to accommodate important variations in forms of government and in related practices institutionalizing political participation.

. . .

A. Political Participation in the Large: Competing Conceptions

All regimes, including over time the most repressive, permit or encourage or even require some institutionalized modes of political participation. Reasons additional to the need to be informed of popular discontent argue for a government's inviting political participation. The government may thereby enlist popular support and gain international as well as domestic legitimacy. It may reduce the risks accompanying efforts to rule exclusively by force.

In nondemocratic societies, such participation will reach beyond the ruling elite, however defined, to include broader elements of the population. It may stop shy even of controlled elections, and involve no more than informal consultations or legally approved ways to express grievances. It may be relatively self-generated *or* mobilized and manipulated, relatively genuine and effectual *or* ceremonial and insubstantial. Like the forms of popular political participation in all societies, it will represent some combination of institutionalized ideals, practical functions, public ritual, and legitimating myth.

Liberal democracies, then, are not distinctive in institutionalizing modes of political participation, but the modes which they stress are. The traditional negative rights and expressive rights of liberalism protect the environment of political pluralism. Political processes culminating in a popular and contested election are meant to yield a representative legislature, and a chief executive or government elected directly (presidential system) or elected indirectly by such legislature (parliamentary system).

Elections serve a variety of purposes for both the voters and the polity. So do the less common forms of citizens' political activities in liberal societies. No one ideal of political participation succeeds in rationalizing these diverse purposes. Therefore, no single scale can measure political participation comparatively. . . . Scales calibrated to different values would weigh different phenomena, or give varying weights to phenomena like elections. . . .

. . . [I]t is useful to concentrate on the divide between theories in which elections constitute a near-sufficient form of political participation for the great majority of the people, and theories urging more. Within this article's framework, that divide could be expressed in terms of the 'elections' clause and 'take part' clause of the international norms. While recognizing that elections are indispensable, the more demanding theories argue for a continuing rather than episodic experience of participation, for a broad conception of a right to 'take part' in the conduct of public affairs. They involve varied and flexible modes of non-electoral participation that supplement rather than substitute for voting. This dispute about the sufficiency of electoral democracy reflects deeper disagreement over the ends of participation. . . .

Permeating these writings is a distinction between notions of participation as an instrumental and as an inherent good. . . . Like most such contrasts, it wrenches apart elements in the psychology and practice of political participation that may usefully complement each other and comfortably coexist for most citizens. But understanding one or the other as a dominant justification for political participation helps to elucidate some underlying themes.

Emphasis on participation's instrumental character suggests that the function of political participation must be to influence public policy, to gain governmental recognition of individual or group interests. The value of participation lies in its instrumental efficacy. It is a means toward a goal rather than an end in itself. Elections are the paramount means for influencing governmental action. They give the electorate some degree of control over and thereby impose some degree of responsibility on those exercising power. Of all modes of political participation, elections enlist the largest number of citizens, including the largest number of those toward the bottom of the socio-economic ladder.

Other more demanding political activities, engaged in by relatively small percentages of the population, are oriented toward and culminate in elections: active membership in a political party, raising campaign funds, soliciting votes for a candidate. They form part of a complex electoral process which is itself part of the larger political process. Some activities which continue between election campaigns require greater devotion from their participants and longer-run planning—for example, the work of the interest groups which have become endemic to liberal societies. Lobbying and other strategies of those groups supplement the electoral process.

Within this prevalent justification for political participation—influencing public policy and governmental action—citizens' votes play a role as vital as it is confined. Theorists emphasize that, in modern mass society, elections cannot convey with any clarity to elected officials the preferences of citizens which they are meant to take into account. . . . Contested elections mean that the people have a choice, but political elites rather than the people decide what that choice is between.

. . .

Despite the indirect and limited control that it exercises, electoral participation does establish boundaries for governmental action. It rules out policies that meet wide opposition, at least if groups with a political voice are among that opposition. Elections thereby serve a vital protective function, one also served by courts in societies where judicial review vindicates constitutional norms. By periodically subjecting elected officials to the approval of the electorate, they help to arrest governmental violations of widely valued rights, or to counter threats to the interests of sufficiently important groups among the electorate. . . .

Other theorists, faulting a solely functional approach to political participation, hold that participation constitutes an inherent as well as an instrumental good. They do not claim that influencing governmental policies and checking governmental abuses are mythical or trivial consequences of the vote. Rather, they argue that elections and political activities closely related to the electoral process are indispensable, but themselves insufficient to realize the democratic ideal of the citizenry's continuing involvement in public life. Within the framework of the international norms on participation, these theorists could be said to elaborate the 'take part' clause.

Such theorists criticize exclusively instrumental views of participation because of their tendency to value only liberal premises of individualism, of the hegemony of private rights, and of limited government whose primary function is to honor those rights. A nearly exclusive reliance on elections heightens the sense of powerlessness of the many to act other than passively by reacting to choices formulated by others. Most citizens, at least those who even choose to vote, treat that act as meeting their full responsibility for participating in public affairs. Voting satisfies that responsibility in an undemanding and individualistic way without need for collegial discussion or group action. Reducing the participation of most citizens to the periodic vote denies them the benefits of a continuing experience of involvement in public life, of 'taking part' in the conduct of public affairs.

Recognizing the impossibility of extensive participation in the central government of modern states, theorists of direct or strong or participatory democracy emphasize other contexts. Inevitably they advocate decentralization of authority, and direct attention toward local governments of a size and functions more conducive to continuing and active public involvement. That participation can, and in many states now does, take various forms: citizen representation on governmental boards, public meetings and discussions, formally structured relationships between the managers of public enterprises and their consumers or the general citizenry, more extensive functions of city government responsive to citizens' needs.

In effect, these theorists urge the further development of established trends in the legislation of many liberal democracies increasing citizen participation in local government units, ranging from school or zoning boards and consumer protection agencies to environmental agencies. The character of the participation becomes vital. Active involvement in a neighborhood group addressing local concerns would be favored over a passive dues-paying membership in a political party or interest group.

Such proposals reach beyond the involvement of citizens in local government. They urge active participation in the formulation of policies of non-governmental institutions like churches or clubs. That is, they challenge the traditional distinction between public and private in the organization of social life. All institutions which exert significant social influence are in that sense political and public. The conception of political participation therefore expands to embrace activities in a broad range of institutional settings. . . .

Theorists of direct and strong democracy apply these arguments with particular force to greater participation of workers in union government and in decision-making in the workplace. . . .

The purposes of this extensive popular involvement in an enlarged domain of public life are at once greater self-government and self-realization. The two are interrelated. By taking responsibility for public life and committing their personal resources to it, citizens experience differently their relationship to society. Through increased participation in the institutions affecting their lives, they develop a sense of their worth and significance. That is, the benefits of participation are at once material and psychological. More citizens will feel empowered to act rather than only react. This heightened sense of responsibility and competence strengthens an ethic of civic virtue which points toward participation for reasons additional to the advancement of self-interest. The inherent good of political participation stems from this possibility of self-realization through development of the social self as a member of the polity.

B. Electoral Systems: Their Range and Effects

. . .

Elections may offer relatively accessible or relatively closed avenues, real or sham avenues, toward the people's participation in governance. The factors determining which of these avenues is opened by a given country's electoral system largely escape consideration or regulation by the international norms.

. . .

Article 25 of the International Covenant and Article 21 of the Universal Declaration require periodic and genuine elections. . . . [T]he international norms leave to the states decisions which carry important consequences for the quality of participation and the distribution of political power.

Consider the alternative systems for election of a legislature: proportional representation ('PR' system), under which the legislature is divided among parties according to the percentage vote received by each party list in the popular election; and single-member constituencies, in which the winner in an electoral district, the first past the post, takes all ('district' system). A choice between these systems or among their many variations will not be determined by the logic of a broadly shared theory of representation, participation or democracy. No such choice will be detached from practical politics.

In a given context, it will generally be clear which groups will benefit from a proposed change in an electoral system. It is not, for example, surprising that suggestions in the United Kingdom for a form of PR come from the smaller parties that have been underrepresented by the district system. On the other hand, PR's structural problems, namely the party fragmentation often caused by PR and the related politics of coalition governments that may give disproportionate strength to small parties, have led to proposals in countries like Israel for significant modifications.

Electoral systems, including the nominating or selection processes through which parties' candidates are chosen, thus both reflect and mold political power. Each system will favor some interests at the expense of others by defining the avenues and strategies through which political power must be gained. Electoral participation can then be arranged consistently with human rights norms to yield many different arrangements of power.

. . .

. . . [H]uman rights law offers no guidelines for the selection of an electoral system in a given political and socio-economic context, no theory of broad or fair electoral participation or access which might influence the contextual choice by a state among the many possibilities. By its terms and in view of the debates during its drafting period, the 'elections' clause remains neutral about these choices.

C. Elections as the Paramount Mode of Political Participation in Liberal Democracies

. . .

The question here considered is not then *why* Article 25 of the International Covenant specifies elections, but why it specifies *only* elections and leaves the 'take part' clause suggestive and unelaborated. An answer to that question lies in part in the conceptual structure of the human rights instruments and in ideological premises to liberal society. Those instruments impose on governments, not on individuals or nongovernmental institutions, most of the duties that correspond to the individual rights which they declare. Their requirements address principally state action: respect by government for individuals' integrity, governmental protection of individuals against lawless behavior, provision of fair trials, equal protection in the making and enforcement of law. This observation holds for the right to political participation as well. A government's correlative duty requires it to permit, foster or arrange for such participation.

In liberal democracies, most political participation stems from the initiatives of individuals or of institutions that are not formally part of government. Consider, for example, the International Covenant's declaration of rights to free speech and association, rights related to the electoral process. Under the traditional understandings of liberal democratic theory, the correlative duties of government do not obligate it to create the institutional frameworks for political debate and action, or to assure all groups of equal ability to propagate their views. Rather, those traditional understandings require governments to protect citizens in their political organizations and activities: forming political parties, mobilizing interest groups, soliciting campaign funds, petitioning and demonstrating, campaigning for votes, establishing associations to monitor local government, lobbying.

Such governmental duties of tolerance and equal protection for all political activities could be viewed as a minimum, essential elaboration of the 'take part' clause. They respond to the liberal commitment to guarantee citizens their political and legal equality. . . . Choices about types and degrees of participation may depend on citizens' economic resources and social status. But it is not government's responsibility to alleviate that dependence, to open paths to political participation which lack of funds or education or status would otherwise block.

Emphasis upon a right to vote, without elaboration of companion rights to other ways of taking part in public affairs, therefore fits traditional liberal conceptions of public and private spheres of competence. . . . Limiting governmental duty to what only governments can do, the construction of an electoral system, leaves the rest of the participatory framework to private initiative. Although Article 25 imposes no such limitation on governmental action, neither does it impose specific obligations other than the conduct of elections.

These understandings do not, however, constitute the whole of the tradition of liberal democracy. Changing ideologies and practices—those associated with the classical and contemporary theorists of direct, strong and participatory democracy—have blurred older lines between private and public spheres. Many states no longer treat the establishment of electoral machinery and the protection of citizens' political speech and association as the outer boundary of a government's duty to honor citizens' right to participate. Governmental functions in liberal democracies have expanded in ways that heighten popular participation. . . .

[Discussion of regulation of political parties, primaries, campaign funding and access to media; of devolution and consultative arrangements; and of worker representation and related matters is omitted.]

Such developing involvement of government in the institutionalization of political participation within and outside electoral processes stems from a theory which my prior discussion of liberal democracy has characterized as participatory and strong rather than representative and thin. . . . That theory is, however, more complex and contentious than one which aims only at justifying elections. Difficulties lie not only in its formulation but also in assessing compliance with an international norm embodying it. Suppose, for example, that such theory were understood to inform the 'take part' clause of Article 25, and thereby to give coherent direction to its elaboration. Problems would arise in reaching a judgment whether a government had satisfied its duty to enact legislation extending opportunities for citizens—for all citizens, powerful and powerless—to 'take part in the conduct of public affairs.'

As generally understood, the 'elections' clause poses no comparable difficulty. A government's violation of that clause, like torture or mock trials, is relatively easy to determine. Elections either take place or they don't; they are or are not contested; fact finders can make judgments about the integrity of the casting and counting of ballots; those elected do or do not take and hold office.[47] The expression 'genuine elections' in the human rights instruments is generally understood to cover these formal, even measurable criteria. By meeting those criteria elections are understood to comply with the human rights norms.

One can imagine the use of a more complex criterion than procedural correctness to determine whether an election satisfies Article 25—for example, the criterion whether it constitutes, in the article's language, the 'free expression of the will of the electors.' That phrase, if interpreted to require more than a non-coerced and secret ballot, could make relevant those considerations which a concentration on the formal integrity of an electoral process enables one to ignore: the distribution of political and economic power, its effect on differential access of classes and ethnic groups to the electoral process and political power, the obstacles to greater access, the power of elected officials relative to other centers of power in the society. What escapes analysis in the contemporary understanding of international norms on political participation is the quality and significance of electoral participation itself.

. . .

V. Political Participation as a Programmatic Right

. . .

. . . Do the human rights norms about participation express any ideal common to diverse nations? Are they more than indeterminate prescriptions which, in a fractious world, were fated or even meant to be understood in dramatically different ways? Do they serve any useful purpose?

The answers which I propose turn on the suggestion that the right to political participation be viewed as a programmatic right, one responsive to a shared ideal but to be realized progressively over time in different ways in dif-

[47] Farer, Human Rights and Human Wrongs: Is the Liberal Model Sufficient?, 7 Hum. Rts. Q. 189 (1985). Farer, a former Chairman of the Inter-American Commission on Human Rights, comments on a report of the Commission on Colombia, prepared during his tenure. He criticizes the Commission for not attempting 'to peer behind the form of party competition to determine whether . . . the right to participate in government was inhibited' by such factors as limited access to the media by left-wing parties, and an agreement between leaders of the two major parties 'to keep certain potentially popular programs off the political agenda.' *Id.* at 199.

Farer observes that nothing in the text of the governing convention (the American Convention on Human Rights) 'compels [or] precludes deep analysis of the political process as a condition for assessing compliance with the right to participate.' He points to the deep liberal assumption informing such inquiries of human rights groups 'that in a state where associational rights are reasonably well protected, the right of political participation is realized in all cases where formal political power coincides with electoral achievement. As long as the franchise is not arbitrarily restricted and the ballots are accurately counted, traditional liberal criteria are satisfied.' *Id.*

ferent contexts through invention and planning that will often have a programmatic character. . . .

At first look, political participation falls within the . . . immediately effective category of international human rights. It figures in the International Covenant as a right so fundamental that the realization of many others depends upon it. Nothing in Article 25 or in other text of the Covenant justifies a distinction between the clarity or immediacy of a state's duties under that article and, say, its duty to refrain from torture. Citizens of a party to the Covenant would have a valid claim under international law if their government had seized power and abolished elections.

Nonetheless, I argue that the right to political participation can be better understood as sharing the programmatic character of many economic and social rights. So understood, it nourishes a vital ideal and serves important purposes. I mean, to be sure, a programmatic right of a distinct character from the typical economic rights. Moreover, my characterization applies to the 'take part' clause of the international norms rather than to the 'elections' clause. . . .

. . .

The clearer if still open-textured character of the 'elections' clause, together with the programmatic character of the 'take part' clause, nourish the argument that Article 25 now expresses some 'positive law,' but also contains an aspirational or hortatory element which distinguishes it from most provisions of the International Covenant. The aspirations that it expresses are of course shaped by the different strands of political theory and the different national practices with which the world is familiar. But this rich historical deposit of ideas and practices cannot exhaust the ways of understanding or institutionalizing an ideal of political participation.

. . .

Although the practice of participation may be severely suppressed in a given state, the norm stands as an invitation to the disenfranchised or repressed to draw on the example of other societies where it is better appreciated. Its recognition in national and universal instruments legitimates inquiry and may spur protest. Dormant or imprisoned within a hostile environment—as lay dormant for many decades the ideal of equality expressed in the Equal Protection Clause of the Fourteenth Amendment of the United States Constitution—the norm retains its subversive potential.

. . .

. . . Fresh understandings and different institutionalizations of the right in different cultural and political contexts may reveal what an increasing number of states believe to be a necessary minimum of political participation for all states. That minimum should never require less of a government than provision for meaningful exercise of choice by citizens in some form of electoral process permitting active debate on a broad if not unlimited range of issues. But it could require much more.

THOMAS FRANCK, THE EMERGING RIGHT TO DEMOCRATIC GOVERNANCE

86 Am. J. Int. L. 46 (1992), at 53.

[Franck refers to two notions: governments derive their just powers from consent of the governed, and the international legitimacy of a state needs to be acknowledged by 'mankind.' These two notions form a 'radical vision' which, he contends, 'is rapidly becoming, in our time, a normative rule of the international system.' We see the 'emergence of a community expectation' that 'those who seek the validation of their empowerment patently govern with the consent of the governed. Democracy, thus, is on the way to becoming a global entitlement, one that will be increasingly promoted and protected by collective international process.'

This 'democratic entitlement' is gradually being transformed 'from moral prescription to international legal obligation.' In support of this claim, 'data will be marshaled from three related generations of rule making and implementation.']

. . . The oldest and most highly developed is that subset of democratic norms which emerged under the heading of 'self-determination.' The second subset—freedom of expression—developed as part of the exponential growth of human rights since the mid-1950s and focuses on maintaining an open market-place of ideas. The third and newest subset seeks to establish, define and monitor a right to free and open elections.

These three subsets somewhat overlap, both chronologically and normatively. Collectively, they do not necessarily penetrate every nook and cranny of democratic theory. For example, the three subsets do not yet address normatively the thorny issue of the right of a disaffected portion of an independent state to secede; nor, as we shall see, is it conceptually or strategically helpful—at least at this stage of its evolution—to treat the democratic entitlement as inextricably linked to the claim of minorities to secession. Still, these three increasingly normative subsets are large building stones, gradually reinforcing each other and assuming the shape of a coherent normative edifice. . . .

[The author interprets the right to self-determination—as expressed in the UN Charter, several General Assembly resolutions and declarations, and Article 1 of the ICCPR and ICESCR; and as developed by state practice—to refer to the right of citizens of all nations to determine their collective political status through democratic means. 'The right now entitles peoples in all states to free, fair and open participation in the democratic process of governance freely chosen by each state.'

Franck then refers to the right of free political expression, and (his third strand) 'the emerging normative entitlement of a participatory *electoral* process' which, despite its infancy, is 'rapidly evolving toward that determinacy which is essential to being perceived as legitimate.' He draws on Article 21 of the UDHR and Article 25 of the ICCPR, and discusses the evolution in the UN and in regional systems such as the European and the OAS of insistence on a state's duty to promote representative democracy, primarily through elections.

Franck notes that this 'evolution of textual determinacy with respect to the electoral entitlement is a relatively recent development,' and traces the recent trend toward election monitoring by the UN, by regional organizations, by states, and by NGOs. He stresses, however, that problems abound about the degree to which the international community can insist on standards that involve concrete forms of intervention, given the residual force of notions like domestic jurisdiction in Charter Article 2(7).]

. . . To proclaim a general right to free elections is less intrusive than monitoring any particular election in an independent state. Effective monitoring is even more intrusive than the mere observation of balloting. And collective action to compel states to adhere to a standard is the most intrusive of all. Thus, the conflict of principles needs to be recognized, made explicit, and reconciled to the general satisfaction of the large preponderance of states before the democratic entitlement's global legitimacy is demonstrated by real, as opposed to formulaic, coherence. That will require action to meet the practical concerns of states that still regard the nonintervention principle as of overriding importance to their national well-being.

Also unclear is the extent to which the various parts of the democratic entitlement can yet claim the legitimacy that derives from 'treating like cases alike.' Are virtually all states, for example, ready to have their elections monitored by a credible global process? This and other issues need to be examined in detail before the democratic entitlement can be said to have achieved universal coherence.

A bright line links the three components of the democratic entitlement. The rules, and the processes for realizing self-determination, freedom of expression and electoral rights, have much in common and evidently aim at achieving a coherent purpose: creating the opportunity for all persons to assume responsibility for shaping the kind of civil society in which they live and work. There is a large normative canon for promoting that objective: the UN Charter, the Universal Declaration of Human Rights, the International Covenant on Civil and Political Rights, the International Convention on the Elimination of All Forms of Racial Discrimination, the International Convention on the Suppression and Punishment of the Crime of *Apartheid*, the Declaration on the Elimination of All Forms of Intolerance and of Discrimination Based on Religion or Belief, and the Convention on the Elimination of All Forms of Discrimination against Women. These universally based rights are supplemented by regional instruments such as the European Convention for the Protection of Human Rights and Fundamental Freedoms, the American Convention on Human Rights, the African Charter on Human and Peoples' Rights, the Copenhagen Document and the Paris Charter.

Each of these instruments recognizes related specific entitlements as accruing to individual citizens. These constitute internationally mandated restraints on governments. As we have seen, they embody rights of free and equal participation in governance, a cluster within which electoral rights are a consistent and probably necessary segment. The result is a net of participatory entitlements. The various texts speak of similar goals and deploy, for the most part, a similar range of processes for monitoring compliance, several of which have already become common usage in connection with the democratic entitlement. One can convincingly argue that states which deny their citizens the right to free and open elections are violating a rule that is fast becoming an integral part of the elaborately woven human rights fabric. Thus, the democratic entitlement has acquired a degree of legitimacy by its association with a far broader panoply of laws pertaining to the rights of persons vis-à-vis their governments.

. . .

There are . . . no legal impediments to institutionalizing voluntary international election monitoring as one way to give effect to the emerging right of all peoples to free and open electoral democracy, but this is not to say that states as yet have a *duty* to submit their elections to international validation. . . .

[Franck then relates the democratic entitlement to the central concern of the UN (as evidenced by the reversal of Iraq's attack on Kuwait) in stopping aggression and maintaining peace.]

If that principle indeed stands at the apex of the global normative system, the democratic governance of states must be recognized as a necessary, although certainly not a sufficient, means to that end. Peace is the consequence of many circumstances: economic well-being, security, and the unimpeded movement of persons, ideas and goods. States' nonaggressiveness, however, depends fundamentally on domestic democracy. Although the argument is not entirely conclusive, historians have emphasized that, in the past 150 years, 'no liberal democracies have ever fought against each other.' It has been argued persuasively that 'a democratic society operating under a market economy has a strong predisposition towards peace.' This stands to reason: a society that makes its decisions democratically and openly will be reluctant to engage its members' lives and treasure in causes espoused by leaders deluded by fantasies of grievance or grandeur.

. . .

Thus, it appears with increasing clarity, in normative text and practice, that compliance with the norms prohibiting war making is inextricably linked to observance of human rights and the democratic entitlement. . . . A distinction needs to be noted here. As we have observed, some governments have argued that the international community's jurisdiction to intervene in the domestic affairs of states to secure compliance with the democratic entitlement is (or should be) limited to cases where its violation has given rise to breaches of the peace. Others have disagreed, claiming that the jurisdiction to intervene is also based on broader human rights law, which authorizes various intrusive forms of monitoring and even envisages sanctions against gross violators. One can prefer this latter view, while still agreeing that the democratic entitlement does have a connection to the United Nations' 'peace' role, that the *legitimacy* of any collective international intervention to support a democratic entitlement is augmented by the entitlement's intimate link to peace. The substance of that link, however, is not merely the role of democracy in *making* or *restoring* peace after conflict has arisen but also—indeed preeminently—its role in *maintaining* peace and *preventing* conflict.

. . .

The democratic entitlement's newness and recent rapid evolution make it understandable that important problems remain. We have considered these primarily under the rubric of coherence, indicating that this entitlement is not yet entirely coherent. The key to solving these residual problems is: (1) that the older democracies should be among the first to volunteer to be monitored in the hope that this will lead the way to near-universal voluntary compliance, thus gradually transforming a sovereign option into a customary legal obligation; (2) that a credible international monitoring service should be established with clearly defined parameters and procedures covering all aspects of voting, from the time an election is called until the newly elected take office; (3) that each nation's duty to be monitored should be linked to a commensurate right to nonintervention by states acting unilaterally; and (4) that legitimate governments should be assured of protection from overthrow by totalitarian forces through concerted systemic action after—and *only* after— the community has recognized that such an exigency has arisen. In the longer term, compliance with the democratic entitlement should also be linked to a right of representation in international organs, to international fiscal, trade and development benefits, and to the protection of UN and regional collective security measures.

Both textually and in practice, the international system is moving toward a clearly designated democratic entitlement, with national governance validated by international standards and systematic monitoring of compliance. The task is to perfect what has been so wondrously begun.

NOTE

Compare with Franck's article the views of Michael Reisman in *Sovereignty and Human Rights in Contemporary International Law*, p. 157, *supra*, about the relationship between (a) traditional notions of state sovereignty and (b) popular sovereignty or consent of the people as a condition to a state's international legitimacy.[48] The discussion of different meanings of self-determination in Chapter 14, develops ideas appearing in both those articles.

b. The Commission, Political Participation and Democracy

Until the late 1980s the Inter-American Commission, like its UN counterparts, had rarely grappled with issues relating directly to the right to political participation. The following statement, taken from the Inter-American Commission's Annual Report for 1971, p. 35, illustrates the pre-occupation at that time with military dictatorships rather than the finer points of democratic theory.

> . . . [W]ith each passing day, more and more people in this part of the world are denied the opportunity to take part in the affairs of the governments of the states in which they live. Terrorism and guerrilla action have led to *de facto* governments in a number of states, where the activities of political parties have been suspended . . . and elections have been postponed. Thus, fundamental human rights can no longer be exercised . . . [T]his Commission considers it to be its duty to point out that . . . cultural development is not furthered by the people's belief that they will be better off by turning away from politics, political parties, politicians, and political institutions and practices. . . .

By the late 1980s, however, the world had changed. Democratic governments had been established or restored in many Latin American countries, while in Europe the Berlin Wall had fallen and communism was in its death throes. These developments created a climate that was more conducive to the elaboration of international legal norms relating to democracy. A group of cases from Mexico on elections provided the Commission with an important opportunity on this issue; the Commission's Final Report on these cases appears below. The 1990–91 Annual Report of the Inter-American Commission that follows examines some premises to democratization. As background for these materials on the Commission, the section begins with a review of references to democracy in the basic Inter-American documents.

[48] These themes are explored in two related discussions: Theme Panel I: Theoretical Perspectives on the Transformation of Sovereignty, in American Society of International Law, Proceedings of the 88th Annual Meeting 1994, 1; and Panel: National Sovereignty Revisited: Perspectives on the Emerging Norm of Democracy in International Law, American Society of International Law, Proceedings of the 86th Annual Meeting 1992, 249.

COMMENT ON PROVISIONS ON DEMOCRACY IN INTER-AMERICAN DOCUMENTS

This Comment describes provisions that refer to democracy in the OAS Charter, a Protocol thereto, the American Declaration and the American Convention.[49] The Annual Report of the Inter-American Commission, p. 684, *infra*, refers to additional documents.

The Charter is distinctive among comparable instruments in the attention that it gives to democratic theory and practice. Its Preamble declares that 'representative democracy is an indispensable condition for the stability, peace and development of the region.' Article 2(b) states that one of the 'essential purposes' of the OAS is '[t]o promote and consolidate representative democracy. . . .' Article 3(d) affirms the 'principle' that the 'solidarity of the American States and the high aims which are sought through it require the political organization of those States on the basis of the effective exercise of representative democracy.'

A 1992 Protocol of Washington amended the Charter to provide in a new Article 9 that an OAS member 'whose democratically constituted government has been overthrown by force may be suspended from the exercise of the right to participate in the sessions of' any organs of the OAS, including the General Assembly and the IACHR.

The 1948 American Declaration of the Rights and Duties of Man, whose normative status was recognized by the OAS in the 1967 Protocol of Buenos Aires, declares in Article 20 that every person has a right to 'participate in the government of his country,' and spells out some requirements of voting and elections. Article 32 converts the right to vote into a 'duty of every person to vote in the popular elections of the country of which he is a national, when he is legally capable of doing so.' Article 28 states: 'The rights of man are limited by the rights of others, by the security of all, and by the just demands of the general welfare and the advancement of democracy.'

The American Convention contains strong affirmations of democracy. Its Preamble declares the intention of the states parties 'to consolidate in this hemisphere, within the framework of democratic institutions, a system of personal liberty and social justice based on respect for the essential rights of man.' Article 23, close in wording to Article 21 of the UDHR and Article 25 of the ICCPR, does not employ the term 'democracy' in its statement of the right to political participation. Citizens enjoy the right under para. 1:

> a. to take part in the conduct of public affairs, directly or through freely chosen representatives;

[49] This Comment draws on a section of a seminar paper at Harvard Law School by A. James Vázquez-Azpiri, The Determinacy of the Democratic Entitlement in the Inter-American System, May 3, 1995.

> b. to vote and to be elected in genuine periodic elections, which shall be by universal and equal suffrage and by secret ballot that guarantees the free expression of the will of the voters. . . .

This provision is made non-derogable 'in time of war, public danger, or other emergency' by Article 27. Contrast the analogous Article 25 of the ICCPR, which does not figure in the list of non-derogable provisions in that instrument that is set forth in ICCPR Article 4.

Article 29(c) of the Convention, a provision dealing with interpretation of the instrument, provides that no Convention provision shall be interpreted as precluding other rights that are 'derived from representative democracy as a form of government.'

FINAL REPORT ON CASES 9768, 9780 AND 9828 OF MEXICO

Annual Report of the Inter-American Commission on Human Rights 1989–90 (1990), at 98.

[These cases before the Commission involved electoral processes in two Mexican states in 1985–86. The petitioners belonged to the National Action Party (PAN). They claimed that members of the ruling Institutional Revolutionary Party (PRI) committed fraud in the elections, including forgeries of voter rolls, cancellation of polling place, stuffing of ballot boxes, and the use of government-controlled police and the military on election day. As a consequence, petitioners alleged violations of their free exercise of political rights set forth in Article 23 of the American Convention.

The Mexican Government argued on several related grounds that the Commission lacked competence to consider these cases. (1) Under federal and state constitutions in Mexico, rulings of domestic electoral bodies are 'final' or 'irrevocable' and thus cannot be subjected to international review. (2) If electoral processes were subjected to international jurisdiction, 'a State would cease to be sovereign.' (3) An adverse finding by the Commission would infringe upon the political autonomy of the Mexican state and violate the principle of the self-determination of peoples. (4) The American Convention did not limit 'the sovereign powers of the States to elect their political bodies,' and when it ratified the Convention, Mexico did not imagine that an international body could review elections of political bodies.

Excerpts from the Commission's Final Report follow:]

40. It is important to point out that Article 27, paragraph 2, of the American Convention, referring to the suspension of guarantees 'In case of war, public danger or any other emergency that threatens the independence or security of the State party . . . ,' does not authorize the suspension of political rights.

41. Hemispheric legal discourse has insisted, for its part, on the existence of

a direct relationship between the exercise of political rights thus defined and the concept of representative democracy as a form of the organization of the State, which at the same time presupposes the observance of other basic human rights. . . .

. . .

44. In short, the exercise of political rights is an essential element of representative democracy, which also presupposes the observance of other human rights. Furthermore, the protection of those civil and political rights, within the framework of representative democracy, also implies the existence of an institutional control of the acts of the branches of government, as well as the supremacy of the law.

45. Since popular will is the basis for the authority of government, according to the terms of the Universal Declaration, it is consistent with a method for naming public officials through elections. Both the Universal Declaration and the American Declaration, the International Covenant on Civil and Political Rights and the American Convention on Human Rights coincide in that elections must have certain specific characteristics: they must be 'authentic' ('genuine' in the American Declaration), 'periodic', 'universal' and be executed in a manner that preserves the freedom of expression of the will of the voter.

. . .

47. The act of electing representatives must be 'authentic,' in the sense stipulated by the American Convention, implying that there must be some consistency between the will of the voters and the result of the election. In the negative sense, the characteristic implies an absence of coercion which distorts the will of the citizens.

48. The different pronouncements which the Inter-American Commission on Human Rights has made on the subject, and which will be presented below, show that the authenticity of elections covers two different categories of phenomena: on one hand, those referring to the general conditions in which the electoral process is carried out and, on the other hand, phenomena linked to the legal and institutional system that organizes elections and which implements the activities linked to the electoral act, that is, everything related in an immediate and direct way to the casting of the vote.

49. As to the *general conditions* in which the electoral contest takes place, from the concrete situations considered by the Commission we can deduce that they must allow the different political groups to participate in the electoral process under equal conditions, that is, that they all have similar basic conditions for conducting their campaign. . . .

. . .

57. With respect to Chile, in its 1987–1988 Annual Report, the Inter-American Commission noted that the mature and reasoned exercise of the

right to vote during the 1988 plebiscite demanded a series of conditions in effect for a sufficiently long period before the aforementioned electoral act. Those conditions were the lifting of the states of exception, a sufficient number of registered voters, equitable access by the different political positions to communications media and the absence of any form of pressure on voters (page 306).

58. After verifying the existence of the first two conditions, the Commission analyzes the situation of the communications media to point out that:

> The presentation makes it possible to draw the conclusion that access to communications media, during the period covered by the present Annual Report and with reference to the plebiscite's campaign, has been characterized by a disproportionate presence of the government, which has used all the resources at its disposal to promote messages and images that favor its position in the next plebiscite. To that, numerous restrictions, legal and *de facto*, must be added, those affecting independent organs of expression and journalists and political leaders. Also, it must be pointed out that the authorization to broadcast political programs constitutes progress that, nevertheless, does not compensate the unequal access to communications media derived from the aforementioned circumstances (pages 307–308).

. . .

62. Another aspect linked to the authenticity of elections is the *organization of the electoral process* and the actual casting of votes. . . .

63. [In its 1978 report on El Salvador the Commission noted:] There is a generalized skepticism on the part of citizens with regard to the right to vote and participation in government. In particular, opposition political parties go as far as doubting the possibility of having pure and free elections, not only in the light of experiences during recent elections, but also with respect to the structure of the electoral system and the obstacles which parties face to organize in the interior of the country. Because of all this, the Commission is of the opinion that electoral rights are not effective in the present circumstances.

. . .

65. In the Seventh Report on the Situation of Human Rights in Cuba, in 1983, the Commission deems that one of the elements that determines the limited political participation of the population in important matters is the result of electoral mechanisms and control exercised over it by the Government and the Cuban Communist Party. After analyzing the principal characteristics of the Cuban electoral system, it points out as a 'counterproductive' element the preponderance of that political party, whose leaders intervene 'in a decisive manner in the operation of mechanisms to select candidates to occupy free elective offices' (pages 44, 45, and 48).

. . .

72. With respect to the exercise of the right to freedom of expression, the Commission has considered the manner in which the government uses its power both for disseminating messages in its favor as well as restricting the possibility of the opposition to broadcast its message. . . .

73. With respect to the freedom of assembly, the experience of the Commission has led it to examine the restrictions of this right resulting from states of exception or other legal restrictions (police permits, for example) or the use of indirect control such as the obligatory participation of public employees in demonstrations.

74. An element of special importance with respect to the general conditions in which electoral processes are conducted, are the activities of groups informally linked to one of the participating parties—usually the government party—who, through acts of violence, tend to intimidate those who oppose them. . . .

75. As to specific features of the organization of elections, the Commission has referred to the laws that regulate them with the aim of determining whether those laws guarantee both the adequate casting of the vote, as well as their correct tally, underscoring the powers vested in those bodies entrusted with implementing the activities of the electoral process and of monitoring both the implementation as well as the results. The institutional system, therefore, has been thoroughly examined by the Commission.

. . .

77. In this regard, the Commission has examined aspects of practical operations such as electoral rolls and registration requirements; the composition of polling stations; the composition of the electoral tribunal and its powers, and the existence of understandable ballots, devoid of any influence on voters.

. . .

81. The Commission considers that the act of ratifying the American Convention presupposes acceptance of the obligation of not only respecting the observance of rights and freedoms recognized in it, but also guaranteeing their existence and the exercise of all of them. . . . [The Commission then cites with approval passages from the *Velásquez Rodríguez* decision, p. 650, *supra*, expounding states' obligations under Articles 1 and 2 of the Convention to respect and guarantee the recognized rights and, pursuant thereto, 'to organize all the State apparatus and, in general, all the structures through which the exercise of public power is manifested, in such a manner that they are able to legally insure the free and full exercise of human rights.' The Commission applied those statements of the Court in the instant case to Mexico's obligations to guarantee rights under Article 23.]

. . .

84. . . . [When Mexico] contracted the obligations derived from the Convention, it also accepted that the Inter-American Commission exercise the

functions and attributions conferred by the Convention; no reservations or limitations were recorded in the instruments deposited when the Convention was ratified.

. . .

86. [In relation to the Government's characterization of the right to legitimate elections as a 'progressively achievable right'] several observations are in order. First of all, it should be pointed out that in order for this interpretive distinction between individual rights of immediate enforceability (the right to vote and to be elected) and collective rights to be developed progressively (the right to elections with particular characteristics) to have validity in the cases under consideration, it would have been necessary for Mexico, at one time or another, to have advanced this interpretation of the article and to have stated this distinction unequivocally.

87. No reference to such a distinction can be found in [any Mexican Government document]. . . .

88. From the normative point of view, the structure of Article 23.1.b makes reference to certain features that should be present in order for the right to be recognized to be valid in practice. Indeed, any mention of the right to vote and to be elected would be mere rhetoric if unaccompanied by a precisely described set of characteristics that the elections are required to meet. . . .

89. The [Mexican] comments also contain the argument that ' . . . the need for the elections to be legitimate imposes upon the State an obligation to act: To progressively develop, in accordance with circumstances and conditions in each country, the guarantee that the voters may freely express their will.' This argument would condition the existence of human rights on 'the circumstances and situation of each country' leaving the whole legal system in a precarious state.

90. With respect to the argument contained in the Mexican Government's comments which holds that any opinion issued by the Commission on an electoral process on the basis of individual complaints constitutes a violation of the principle of nonintervention, it should be stated here once again that the Mexican State, by virtue of having signed and ratified the Convention, has consented to allow certain aspects of its internal jurisdiction to be a subject of judgments on the part of the organs instituted to protect the rights and guarantees recognized by the Convention. If this is true whenever a State signs any international instrument, it is even truer when the instrument is a treaty that recognizes the inalienable rights of man, which, antedate and are paramount over those of the State.

91. Moreover, as stated in Article 18 of the OAS Charter, the principle of nonintervention is a rule of conduct that governs the acts of States or groups of States. . . . The Inter-American Juridical Committee, in its 'Draft Instrument' on cases of violations of the principle of nonintervention (1972), indicated that one of the basic criteria followed preparing it was that 'only States can be subjects of intervention.'

. . .

93. The principle of nonintervention is therefore linked to the right of peoples to self-determination and independence and is described as a principle to be practiced in suitable harmony with human rights and fundamental freedoms. This important interrelation of principles of international law is formalized as a rule of law in Article 16 of the OAS Charter, which reads as follows:

> Each State has the right to develop its cultural, political, and economic life freely and naturally. In this free development, the State shall respect the rights of individuals and the principles of universal morality.

94. According to this rule, the right of the State to develop its internal life freely has a counterpart in its obligation to respect the rights of individuals. And in inter-American law these rights are formally recognized in the American Convention on Human Rights. The correct interpretation of the principle of nonintervention is, therefore, one based on protecting the right of States to self-determination provided that right is exercised in a manner consistent with respect for the rights of individuals.

95. The above leads to the conclusion that the Commission, based on its regulatory instruments, is empowered to examine and evaluate the degree to which the internal legislation of the State party guarantees or protects the rights stipulated in the Convention and their adequate exercise and, obviously, among these, political rights. The IACHR is also empowered to verify, with respect to these rights, if the holding of periodic, authentic elections, with universal, equal, and secret suffrage takes place, within the framework of the necessary guarantees so that the results represent the popular will, including the possibility that the voters could, if necessary, effectively appeal against an electoral process that they consider fraudulent, defective, and irregular or that ignores the 'right to access, under general conditions of equality, to the public functions of their country.'

. . .

Issues in this case

99. The three denunciations hold that the elections held were not authentic because they did not adequately represent the popular [will]. As for the specific allegations, the Commission has decided to refrain from making any reference to the *de facto* situations alleged in these cases because the validity of some of the allegations would have to rest on a presence of the Commission during the electoral campaign and at the time of the voting. To this should be added the fact that the Commission did not engage in any exhaustive monitoring of the situation in Mexico, as [it] had done in certain cases in which it had issued judgments on electoral processes and even made inspection visits to some of the countries concerned. Accordingly, it neither accepts nor denies the veracity of the facts as alleged. This precludes the possibility that the Commission

comment on the origin of the mandate of the officials chosen in these elections. At present, this is also the intention of the claimants.

100. In relation with the internal remedies and guarantees in Mexico, the matter to be examined is whether Mexican law offers adequate means or a simple and quick remedy or of 'any other effective remedy before competent judges or independent and impartial courts' that protect those who petition against 'acts that violate their fundamental rights,' as is the case with political rights. The Commission has been able to perceive that no such remedy [exists] in Mexico.

101. In view of the aforementioned . . . the Commission deems it advisable to remind the Government of Mexico of its duty to adopt measures of internal law, in accordance with its constitutional procedures and the provisions of the Convention, whether legislative or of another character, necessary to make effective the rights and liberties which the Convention recognizes.

102. . . . [T]he Commission . . . has been informed that there is underway an active process of reform of the electoral laws. The Commission hopes that these reforms will lead to the adoption of standards that will adequately protect the exercise of political rights and create a rapid and effective process assuring the protection of the same. The Commission places itself at the disposal of the Government of Mexico to cooperate with it . . . [and] requests . . . information relating to the electoral reform process currently underway in accordance with Article 43 of the American Convention on Human Rights.

INTER-AMERICAN COMMISSION ON HUMAN RIGHTS, ANNUAL REPORT 1990–91

(1991), at 514.

[This Annual Report of the IACHR included a section examining 'areas in which steps need to be taken' toward full observance of the human rights recognized in the American Declaration and the American Convention. Part III of that section is entitled 'Human Rights, Political Rights and Representative Democracy in the Inter-American System.' Excerpts from Part III follow.]

Human rights, political rights and representative democracy are concepts whose interrelationship has been repeatedly reasserted by the inter-American community. There have been numerous declarations and international commitments made by the American states. The Inter-American Commission on Human Rights believed this was a fitting time to summarize the most important milestones in that regard: . . .

. . .

1. *Representative democracy and human rights in the inter-American system*

Later, the postulate of the relationship between representative democracy and human rights was further refined in the Declaration of Santiago, adopted in 1959 by the Fifth Meeting of Consultation of Ministers of Foreign Affairs [of the OAS]. . . .

1. The principle of the rule of law should be assured by the separation of powers, and by the control of the legality of governmental acts by competent organs of the state.

. . .

4. The governments of the American states should maintain a system of freedom for the individual and of social justice based on respect for fundamental human rights.

. . .

7. Freedom of the press, radio, and television, and, in general, freedom of information and expression, are essential conditions for the existence of a democratic regime.

. . .

For its part, the General Assembly of the Organization has on numerous occasions articulated the relationship between representative democracy and human rights, emphasizing the need for political rights to be exercised so as to elect government authorities. And so, it has recommended '. . . to the member states that have not yet done so that they reestablish or perfect the democratic system of government. . . .'

. . .

The last of the General Assembly resolutions that should be cited is AG/RES. 890 (XVII-0/87), wherein it decides:

> To reiterate to those governments that have not yet reinstated the representative democratic form of government that it is urgently necessary to implement the pertinent institutional machinery to restore such a system in the shortest possible time, through free and open elections held by secret ballot, since democracy is the best guarantee of the full exercise of human rights and is the firm foundation of the solidarity among the states of the hemisphere.

. . .

The Commission has noted that it is no coincidence that the laws drafted in this hemisphere have emphasized the existence of a direct relationship between the exercise of political rights so defined and the concept of democracy as a form of State organization, which in turn presupposes the observance of other fundamental human rights. . . .

Moreover, the observance of these rights and freedoms calls for a legal and institutional order wherein the law take precedence over the will of the governing and where certain institutions exercise control over others so as to preserve the integrity of the expression of the will of the people—a constitutional state or a state in which the rule of law prevails.

On numerous occasions the Commission has made reference to several aspects associated with the exercise of political rights in a representative democracy and their relationship to the other fundamental rights of the individual. For the sake of brevity, only those texts that best illustrate this point will be cited. And so, the Inter-American Commission has said the following in this regard:

. . .

> The right to political participation leaves room for a wide variety of forms of government; there are many constitutional alternatives as regards the degree of centralization of the powers of the state or the election and attributes of the organs responsible for the exercise of those powers. However, a democratic framework is an essential element for establishment of a political society where human values can be fully realized.
>
> The right to political participation makes possible the right to organize parties and political associations, which through open discussion and ideological struggle, can improve the social level and economic circumstances of the masses and prevent a monopoly on power by any one group or individual. At the same time it can be said that democracy is a unifying link among the nations of this hemisphere. (Annual Report of the IACHR, 1979–80, p. 151).

. . .

> In this context, governments have, in the face of political rights and the right to political participation, the obligation to permit and guarantee: the organization of all political parties and other associations, unless they are constituted to violate human rights; open debate of the principal themes of socio-economic development; the celebration of general and free elections with all the necessary guarantees so that the results represent the popular will.
>
> As demonstrated by historical experience, the denial of political rights or the alteration of the popular will may lead to a situation of violence. (Annual Report of the IACHR, 1980–81, pp. 122–123).
>
> The analysis of the human rights situation in the States to which the Commission has made reference in the foregoing chapter, as well as in others where the human rights situation has been considered by the Commission in recent years, enables the Commission to affirm that only by means of the effective exercise of representative democracy can the observance of human rights be fully guaranteed.
>
> It is not a question merely, of pointing out the organic relation that exists between representative democracy and human rights; which is manifest in the Charter of the OAS and other instruments of the inter-American system. The Commission's factual experience has been that serious human rights violations that have occurred or that are occurring in some countries of the Americas are primarily the result of the lack of political participation on the part of the citizenry, which is denied by the

> authorities in power. The resistence of these authorities as regards taking the action necessary to reestablish representative democracy has increased the tyranny, on the one hand, and led to serious social strife, on the other. The result has been that both the government and the more extreme opposition sectors have shown a preference for the use of violence as the sole means of resolving conflicts in the face of a lack of peaceful and rational options.
>
> This experience confirms, therefore, the authentic social peace and respect for human rights can only be found in a democratic system. It is the only system that allows for the harmonious interaction of different political tendencies and within which, by means of the inter institutional equilibrium it establishes, the necessary controls can be invoked to correct errors or abuses by the authorities. (Annual Report of the IACHR, 1985–1986, p. 191).

Finally, in its 1987 Report on the Situation of Human Rights in Paraguay, after analyzing the most relevant legal texts produced in the hemisphere, the Commission [stated]. . . .

. . .

> Exercise of political rights is, in turn, an essential factor in the democratic system of government, which is also characterized by the presence of an institutional system of checks on the exercise of power, the existence of ample freedom of expression, association and meeting: and acceptance of a pluralism that would prevent the use of political proscription as an instrument of power.
>
> This hemispheric vision of the exercise of political rights within the context of a democratic system of government is completed by the requisite development and promotion of economic, social and cultural rights. Without them, the exercise of political rights is severely limited and the very permanence of the democratic regime is seriously threatened.

In short, for the Inter-American Commission on Human Rights the exercise of political rights is an essential ingredient of a representative democracy and presupposes the observance of other human rights as well; the protection of those civil and political rights in a representative democracy also implies the existence of some institutional control over the actions of the powers of the State; it also presupposes the supremacy of the law. It might be well to determine the scope of the definition of the political rights contained in Article 23 of the American Convention on Human Rights, drawing upon earlier statements made by the Inter-American Commission on Human Rights.

. . .

QUESTIONS

1. In what respects, if any, does the Final Report on the Mexican cases elucidate the requirements of democratic government beyond the specific requirements of elections in Article 23(1)(b)?

2. Are the Commission's reasons for not taking a decision on the facts of the case (para. 99) persuasive? What alternative approaches might have been adopted?

3. Can you identify a coherent conception of sovereignty or domestic jurisdiction that the Commission develops to respond to Mexico's challenges to its competence? What content does the Commission give to the broad principle of non-interference in states' domestic affairs?

4. In the light of these cases, is there any aspect of a state's electoral system with which the Commission should not, from either a legal or policy perspective, concern itself? If so, which aspects, and why?

5. Does the 1990–91 Annual Report present a 'theory' of democracy that requires a state to comply with more requirements than periodic elections? If so, what are the elements of that theory, and are all such elements part of the human rights corpus?

6. Recall the suggestion in Steiner that the right to political participation be viewed in some respects as programmatic, thus subject to progressive realization. Would that suggestion have been appropriate for the Commission to incorporate into its 1990–91 Annual Report as a way of defining states' duties with respect to democratic governance? Would it have been appropriate for the Commission in its Final Report on the Mexican cases to have viewed Mexico as in compliance with the American Convention because it was in the process of gradual realization of the right to political participation? Note the Mexican argument in para. 86 of the Final Report.

ADDITIONAL READING

Cecilia Medina Quiroga, *The Battle of Human Rights: Gross, Systematic Violations and the Inter-American System* (1988); Thomas Buergenthal, Norris and Shelton, *Protecting Human Rights in the Americas: Selected Problems* (3rd ed., 1990); Scott Davidson, *The Inter-American Court of Human Rights* (1992); Claudio Grossman, *Proposals to Strengthen the Inter-American System of Protection of Human Rights*, 32 Germ. Y.B. Int'l L. 264 (1990); Dinah Shelton, *Improving Human Rights Protections: Recommendations for Enhancing the Effectiveness of the Inter-American Commission and the Inter-American Court of Human Rights*, 3 Am. U. J. Int'l L. & Policy 323 (1988); Thomas Buergenthal, *The Inter-American System for the*

Protection of Human Rights, in Theodor Meron (ed.), *Human Rights in International Law: Legal and Policy Issues* 439 (1984); Christina Cerna, *The Structure and Functioning of the Inter-American Court of Human Rights* (1979–1992), 63 Brit. Y.B. Int'l L. 135 (1992); Tom Farer, *The Grand Strategy of the United States in Latin America* (1988); Tom Farer, *Finding the Facts: The Procedures of the Inter-American Commission on Human Rights of the Organization of American States*, in Richard Lillich (ed.), *Fact-Finding Before International Tribunals* 275 (1991); Stephen Schnably, *The Santiago Commitment as a Call to Democracy in the United States: Evaluating the OAS Role in Haiti, Peru, and Guatemala*, 25 Inter-American Law Rev. 393 (1994); Scott Davidson, *Remedies for Violations of the American Convention on Human Rights*, 44 Int'l & Comp. L. Q. 405 (1995).

D. THE AFRICAN SYSTEM: RIGHTS AND DUTIES

The newest, the least developed or effective, the most distinctive and the most controversial of the regional human rights regimes involves African states. In 1981 the Assembly of Heads of States and Government of the Organization of African Unity adopted the African Charter on Human and Peoples' Rights. It entered into force in 1986. As of January 1995, 49 African states were parties.

We describe the African human rights system and Charter briefly both because it has had less than a decade of experience and because its sole implementing organ, the African Commission on Human and Peoples' Rights, has few powers and has been hesitant in exercising or creatively interpreting and developing them. For these reasons the African system has not yet yielded anywhere near the same amount of information and 'output' of recommendations or decisions—state reports and reactions thereto, communications (complaints) from individuals and state responses thereto, studies of 'situations' or investigations of particular violations—as have the other two regional regimes, let alone the UN system. Thus far, the states parties and the weak, hesitant Commission have taken few forceful or persuasive actions to attempt to curb serious human rights violations.

Section D, like the two previous sections, emphasizes one significant feature of the regional system under examination, within the framework of a general description of the system. In this instance, the feature is the distinctive attention that the African system gives to duties as well as rights. (Recall that the American Declaration on the Rights and Duties of Man, p. 642, *supra*, includes ten articles on individual duties. The American Convention on Human Rights does not give special attention to duties.) This distinctive emphasis is obvious on the face of the African Charter, even before we have seen any elaboration or application of the provisions on duties by the African Commission. Examination of these provisions also serves to develop themes about rights and their relation to duties that were introduced in Chapter 4's broad discussion of cultural relativism, which indeed included African materials. See pp. 184 and 240, *supra*.

The description and analysis of the African human rights system begin with a brief description of the Organization of African Unity. We continue with the Charter's norms, and conclude with an evaluation of the work of the African Commission. You should now become familiar with the provisions of the Charter set forth in the Documents Annex.

COMMENT ON THE ORGANIZATION OF AFRICAN UNITY

The OAU is the official regional body of all African states. It was inspired by the anti-colonial struggles of the late 1950s, and was primarily dedicated to the eradication of colonialism. The emergent African states created through it a political bloc to facilitate intra-African relations and to forge a regional approach to Africa's relationships with external powers. The OAU's Charter (47 U.N.T.S. 39, reprinted in 2 Int. Leg. Mat. 766) was adopted in 1963 by a conference of Heads of State and Government. Today, virtually all African states are members of the OAU.

Unlike the UN Charter, that of the OAU makes no provision for the enforcement of its principles. It emphasizes cooperation among member states and peaceful settlement of disputes, and includes among its purposes in Article II(1) promotion of the 'unity and solidarity' of African states, as well as defense of 'their sovereignty, their territorial integrity and independence.' This inviolability of territorial borders, expressed through the principle of non-interference in the internal affairs of member states, has been one of the OAU's central creeds.

This creed has contributed to the reluctance of member states to promote human rights aggressively. The most visible failure in this regard has been the reluctance of member states to criticize one another about human rights violations. A prominent case in point was the failure of most African states, the single exception being Tanzania, to denounce the abusive regime of Ugandan dictator Idi Amin.

The single prominent exception with respect to human rights policy has been the adoption by the OAU in 1981 of the African Charter on Human and Peoples' Rights. The formal, legal basis for that act is found in Article II(1)(b) of the OAU Charter, which requires member states to 'coordinate and intensify their collaboration and efforts to achieve a better life for the peoples of Africa,' and Article II(1)(e), which asks member states to 'promote international co-operation, having due regard to the Charter of the United Nations and the Universal Declaration of Human Rights.'

The Comment at p. 699, *infra*, describes some relationships between the African Commission on Human and Peoples' Rights and the OAU. In considering the significance of those relationships, bear in mind the political character of the OAU as sketched above.

COMMENT ON COMPARISONS BETWEEN THE AFRICAN CHARTER ON HUMAN AND PEOPLES' RIGHTS AND OTHER HUMAN RIGHTS INSTRUMENTS

Even the Charter's Preamble suggests some of the striking differences from other human rights instruments, universal and regional. The key theme is regional cultural distinctiveness, as in '[t]aking into consideration the virtues of [African states'] historical tradition and the values of African civilization. . . .'

Rights

Consider first Ch. 1 of Part I, dealing with 'human and peoples' rights.'

1. Compare the important opening provisions (Articles 1 and 2) of the Charter on the obligations of states with the analogous provisions in Article 2 of the ICCPR.

2. Several of the rights are expressed in ways that differ in wording from equivalent provisions in other instruments but that lack any distinctive qualification and amount in the large to a similar conception. See, for example, Article 4.

3. Many rights are expressed in significantly different ways from the equivalent provisions in, say, the ICCPR. Compare, for example, Article 7 of the Charter on criminal procedure with Articles 14 and 15 of the ICCPR; and Article 13 of the Charter on political participation with Article 25 of the ICCPR. Note the respects in which Article 25 is more specific.

4. The protection of the property right in Article 14 recalls Article 17 of the UDHR but finds no equivalent in the ICCPR. The European Convention as first drafted included no such provision, but its First Protocol extends protection to the property right.

5. Some provisions state familiar norms, but illustrate them or make them specific in ways that recall Africa's experience with the Western slave trade and with colonization. They bear out the phrase in the Preamble quoted above. See, for example, Articles 5, 19 and 20. Other provisions refer especially to abuses in Africa's own post-colonial history, such as Article 12(5) that recalls Uganda's expulsion of its citizens of Asian descent.

6. A number of provisions draw attention to the attempts of the states to reconcile humane treatment of individuals with their interests in territorial integrity and security. See for example, Article 23(2).

7. Compare the characteristic limitations on rights in the Charter with those in the ICCPR. See, for example, Articles 18(3) and 22(2) of the ICCPR.

Compare, say, Article 6 of the Charter providing the right to liberty 'except for reasons and conditions previously laid down by law,' Article 8 providing that freedom of conscience and religion are 'subject to law and order,' and Article 10 declaring the right to free association 'provided that [the individual] abides by the law.' See also Articles 11 and 12(2).

8. Note that the Charter has no provision for derogation of rights in situations of national emergency, equivalent to Article 4 of the ICCPR.

9.The Charter includes economic–social rights as in Articles 15 and 16, but does not qualify these rights with respect to their progressive realization and resource constraints as does Article 2 of the International Covenant on Economic, Social and Cultural Rights.

10. The Charter includes several collective or peoples' rights, sometimes referred to as 'third-generation' human rights, in provisions like those in Articles 23 and 24 dealing with peoples' rights 'to national and international peace and security' and 'to a generally satisfactory environment favourable to their development.' The Charter's title itself signals this important feature.

Duties

Consider now the distinctive Chapter 2 of Part I, on 'duties.' To be sure, references to 'duties' are not alien to human rights instruments. Note, for example, Article 29 of the UDHR, to the effect that '[e]veryone has duties to the community in which alone the free and full development of his personality is possible.' The preamble to the ICCPR observes that 'the individual, having duties to other individuals and to the community to which he belongs, is under a responsibility to strive for the promotion and observance of the rights recognized in the present Covenant.' Similarly the preamble to the Charter states that 'the enjoyment of rights and freedoms also implies the performance of duties on the part of everyone.'

Nonetheless, the Charter is the first human rights treaty to include an enumeration of, to give forceful attention to, individuals' duties. In this respect, it goes well beyond the conventional notion that duties are correlative to rights, such as the obvious duties of states that are correlative (corresponding) to individual rights—for example, states' duties not to torture, or to provide a structure for political participation. It goes beyond declaring similar correlative duties of individuals—for example, the duty not to invade another individual's right to personal security. The Charter differs in defining duties that are not simply the 'other side' of individual rights, and that run from individuals to the state as well as to other groups and individuals.

QUESTIONS

Assume in the following questions that the African state involved has ratified the Charter, but (typically) has not incorporated it legislatively into its domestic legal system, so that individuals cannot invoke it in civil or criminal proceedings in domestic courts. The individual referred to in each question has been convicted of a crime under state law, and files a communication (complaint) with the African Commission protesting the conviction. How should the Commission decide the issues identified below?

1. Article 10 declares the individual's right to free association 'provided that he abides by the law.' Suppose that the law provides that associations must be registered to be legal. The government refuses to register companies and other associations committed to advocacy of change in the state's political structure. The state arrests and convicts an individual member of an unregistered company with such a goal. What arguments would you make to assert that the state has committed a violation by refusing to register the company, and by the criminal conviction?

2. Article 6 provides for the right to liberty and personal security. 'No one may be deprived of his freedom except for reasons previously laid down by law.' The law provides that persons suspected of disloyalty to the state may be summarily arrested and detained for six months before being charged with a crime. Your client has been so arrested and detained. You allege that the arrest and detention are illegal under the Charter. The government responds that the reasons for deprivation of freedom were 'previously laid down by law,' and thus there has been no violation. How do you respond?

COMMENT ON DUTIES AND THEIR IMPLICATIONS IN THE AFRICAN CHARTER

Depending on their interpretation and possible application within the African human rights regime, the duties declared in the Charter could constitute part of the deep structure of the society contemplated by that instrument. They could inform basic relationships between the individual on the one hand, and society and state on the other. They could resolve in a particular way the tension between the individual and the collective. They could stand more in contradiction than in harmony with the Charter's preceding elaboration of rights.

Note some of the vital phrases in Articles 27–29. Article 27 refers to duties towards one's 'family and society, the State and other legally recognized communities and the international community.' Rights are to be exercised with 'due regard to the rights of others, collective security, morality and common interest.'

The reach of Article 29 is striking. Note such phrases as the 'harmonious development of the family,' 'cohesion and respect,' 'serve the national community,' 'not to compromise the security of the state,' 'strengthen social and national solidarity,' 'strengthen positive African cultural values in [one's] relations with other members of the society,' and 'contribute to the best of [one's] abilities . . . to the promotion and achievement of African unity.'

This stress on duties recalls the materials in Chapter 4 exploring differences between rights-oriented and duty-oriented visions of social or political life, and related issues of cultural relativism. Our present discussion of the significance of duties in the African Charter involves these same themes. For example, the materials in Chapter 4 at pp. 180–87, *supra*, suggested links between a culture's duty-orientation, and ideals and traditions of community and solidarity. This theme of solidarity appears in the African Charter's discussion of rights. Article 10(2) protects individuals against being compelled 'to join an association,' but '[s]ubject to the obligation of solidarity provided for in Article 29'. Article 25 provides that states must 'promote and ensure through teaching, education and publication' respect for the rights declared and to assure that such rights 'as well as corresponding obligations and duties are understood.'

That is, depending on their interpretation and application, duties and ideals of solidarity may impinge in clear and serious ways on the Charter's definitions of rights themselves. Moreover, the Charter imposes individual duties that run not only to the state but also to different types of groups of communities within (or perhaps transcending) that state.

Consider some of the problems raised by these provisions about individual duties:

1. Sometimes what appear to be conventional terms of reference may bear plural meanings that affect the nature of the duty. For example, what definition applies to the word 'family' in Article 27—the nuclear or extended family? In the African context, one might think of the extended family. Nonetheless, the only specific reference to family relationships in the three articles on duties deals with parents and children (Article 29(1)).

Or how are we to understand 'society' in Article 27—as equivalent specifically to the nation state, or to prevailing social and cultural structures within the state? It is striking that the article does not mention or seem to include ethnic groups, for they are frequently not 'legally' recognized.

2. As in the other two articles on duties, the requirements put on the individual by Article 28 raise the question of whether, and by whom, these duties are to be enforced. Who or what institution is to give meaning and application to them, or provide general guidance for their performance? To the present, the African Commission has taken no steps toward interpretation or general elaboration. The question remains open whether the three articles are to constitute in any sense 'binding' and enforceable obligations.

3. The duties are of such breadth and so ambiguous in their connotations that a regime of serious enforcement without some degree of prior elaboration

is difficult to imagine. Consider, for example, Article 28's provision that non-discrimination is not simply a duty of the state, but individuals also must not discriminate against other individuals. The article does not list any forbidden grounds for discrimination. Nor does it on its face distinguish between discrimination in the so-called private and public spheres—that is, discrimination in personal social relationships, and in employment or housing. Compare the legislation in numerous countries, including federal and state jurisdictions in the United States, that forbids defined non-governmental actors such as employers or lessors to discriminate on stated grounds such as race or religion.

4. Article 29 raises a host of such issues, none more salient than the question whether it imposes on individuals a duty to uphold extant, traditional structures ranging from the family to the government. The critical terms seem to be 'harmonious,' 'cohesion,' 'community,' 'security,' 'social and national solidarity,' 'territorial integrity,' 'positive African cultural values,' 'moral well-being of society,' and last but not least, 'African unity.' How are these injunctions to be reconciled with the rights earlier declared, other than by giving great weight to the limitation clauses such as the phrase in Article 8, 'subject to law and order'?

QUESTIONS

1. Consider the questions that could arise about whether the Charter protects the advocacy of a radical reformer who seeks significant changes in gender relationships or in the nature of the family. How, for example, would one reconcile the provisions of Article 29 with Article 18, to the effect that the state should 'ensure the elimination of every discrimination against women'? How would you argue for the supremacy of Article 18?

2. What effect might Articles 27–29 have on the Charter's provisions for speech and association, the very conditions of effective political participation? How would you relate the limitations clauses in those earlier provisions to these articles on duties? As a member of the African Commission, how would you answer these questions?

3. Recall the provisions on individual duties in ten articles of the American Declaration on the Rights and Duties of Man, p. 642, *supra*. Do you find them fundamentally similar to or different from those in the African Charter? Illustrate.

4. Note that the African Charter lacks a provision similar to Article 5(1) of the ICCPR. Would you advocate amendment of the Charter to include such a provision? If so, what effect do you think it would have in shaping the understanding of the Charter's duties?

5. 'One can parse the African Charter, looking as it were for difficulties and inherent violations of universal human rights norms, looking for ways to criticize the prevalence of duties. One can also view it as an instrument that will be sensibly understood and interpreted, even while it seeks to preserve some small degree of the cultural distinctiveness systematically denied Africa by the West as the West invaded, colonized and manipulated Africa for its strategic values, and now penetrates Africa by trade and by imposing media conditions on national or international aid. The Charter should be seen as an affirmation of a modest degree of cultural relativism or particularism, the very occasion and reason for regional as well as universal human rights systems. That particularity speaks to the African tradition and constitutes a fundamental value in the entire human rights movement, namely the preservation of difference.' Comment.

6. 'Even if the provisions on duties are in some way given application and enforcement by the Commission, or indeed by state courts or governments, I do not see how they derogate seriously from fundamental universal human rights. Could this assertion of cultural relativism justify, for example, torture, or a sham criminal trial, or raw governmental discrimination on religious or ethnic grounds? Clearly not. The challenge to the human rights movement of the Charter's stress on duties is much overstated.' Comment.

7. 'Whatever the African Charter says, African states are subject like all states to the universal human rights system expressed through the UDHR and the two basic Covenants. If there is a conflict, if this regional regime requires or permits state conduct that universal norms prohibit, or prohibits state conduct that universal norms require, those norms must prevail. Else the human rights movement collapses into a hopeless regional anarchy, and we encourage the development of additional regional systems that will simply flout universal imperatives by asserting their own distinctive norms.' Comment.

COMMENTATORS' EVALUATIONS OF THE CHARTER'S DUTIES

Note the following excerpts from writings of four authors about the substantive rights and duties set forth in the Charter.

Wolfgang Benedek, Peoples' Rights and Individuals' Duties as Special Features of the African Charter on Human and Peoples' Rights

in Philip Kunig et al., *Regional Protection of Human Rights by International Law* 59, 63 (1985).

The human rights approach to be found in traditional African societies is characterized by a permanent dialectical relationship between the individual and the group, which fits neither into the individualistic nor the collectivistic concept of human rights. . . . It is inherent in the African community-oriented approach towards human rights as discussed above that rights and duties form a common whole. However, different from the marxist concept of human rights of Eastern-European states this does not mean symmetrical or reciprocal rights and obligations between the individual and the state. . . . [T]here is no doubt that the concept of duties in the African Charter reflects an original approach which tries to accommodate both the values of traditional African societies and the needs of modern African states. . . . [T]here is a danger that states could try to use duties to derogate certain human rights. . . . Misuse of the duty provision is further facilitated by the very general and even vague wording of this chapter. . . . [I]t would be important to clarify and ensure that the rights of the African Charter can't be defeated by the provisions on duties of the Charter. . . . In conclusion, the duties in the African Charter appear to be more of a programmatic character. Some spell out a general philosophy and principles of behavior rather than operational legal concepts.

Richard Gittelman, The Banjul Charter on Human and Peoples' Rights: A Legal Analysis

in Claude Welch and Ronald Meltzer (eds.), *Human Rights and Development in Africa* 152 (1984), at 154.

The African [Charter] reference [to duties] imposes an obligation upon the individual not only toward other individuals but also toward the State of which he is a citizen. The notion of individual responsibility to the community is firmly ingrained in African tradition and is therefore consistent with the historical traditions and values of African civilization upon which the Charter relied. The inclusion of this far-reaching clause [on duties, in the Charter's preamble] has roots, however, in factors other than mere tradition and, to a large extent, explains the various tensions throughout the Charter. The Socialist States such as Mozambique and Ethiopia had a difficult time reconciling traditional human rights conventions with socialist philosophy. The notion of 'individual' in a socialist State differs markedly from the notion in a capitalist State. As a result, to ensure the eventual adoption of the Charter by all States, the drafters in Dakar stated that if the individual is to have rights 'recognized' by the State, he also must have obligations flowing back to the State.

Asmarom Legesse, Human Rights in African Political Culture

in Kenneth Thompson (ed.), *The Moral Imperatives of Human Rights* 123 (1980), at 124.

In the liberal democracies of the Western world the ultimate repository of rights is the human person. The individual is held in a virtually sacralized position. There is a perpetual, and in our view obsessive, concern with the dignity of the individual, his worth, personal autonomy, and property. . . . Nothing is more despicable to the Westerner than societies that force individuals to be lost in the 'faceless crowd.' No aspect of Western civilization makes an African more uncomfortable than the concept of the sacralized individual whose private wars against society are celebrated. If we turn the situation around and view it from an African perspective, the individual who is fighting private wars against his society is no hero. . . . [T]he very idea of advertising cultural values by personifying them in the lives of individuals is a very strange idiom to an African. . . . The heart of African culture is egalitarian and antiheroic in character. . . . Most African cultures, whether they are formally egalitarian or hierarchical, have mechanisms of distributive justice that ensure that individuals do not deviate so far from the norm that they can overwhelm the society. This is the factor that was widespread throughout precolonial Africa and served as the cornerstone of African morality.

Makau wa Mutua, The African Human Rights System in a Comparative Perspective

3 Rev. Afr. Comm'n Hum. & Peo. Rts. 5, 8 (1993).

[T]he Charter incorrectly assumes the existence of permanent and static 'African culture' that it has the task of preserving. It fails to recognize that cultural values are socially and historically constructed. Cultures, and African ones are no exception, express values in the context of continuous and permanent social and political struggles over resources and power relations in a given society. Cultural values are not inanimate artifacts or sacrosanct ideals, frozen in time. . . . The Charter's underlying theme, that of preserving the state and 'traditional' structures through the 'duties' that it imposes, requires, in essence, the domination and subjugation of the individual to the authoritarian state.

COMMENT ON INSTITUTIONAL IMPLEMENTATION: THE AFRICAN COMMISSION

Basic Functions of the Commission

The eleven members of the Commission, elected by secret ballot by the Assembly of Heads of State and Government from a list of persons nominated by parties to the Charter, are to serve (Article 31) 'in their personal capacity.' Article 45 defines the mandate or functions of the Commission to be (1) to 'promote Human and Peoples' Rights,' (2) to 'ensure the protection of human and peoples' rights' under conditions set by the Charter, (3) to 'interpret all the provisions of the Charter' when so requested by states or OAU institutions, and (4) to perform other tasks that may be committed to it by the Assembly. So the three dominant functions appear to be promotion, ensuring protection, and interpretation.

The Commission's task of 'promotion' includes (Article 45) undertaking 'studies and researches on African problems in the field of human and peoples' rights,' as well as organizing seminars and conferences, disseminating information, encouraging 'local institutions concerned with human and peoples' rights,' giving its views or making recommendations to Governments, and formulating principles and rules 'aimed at solving legal problems related to human and peoples' rights . . . upon which African Governments may base their legislation.' Article 46 states tersely that the Commission 'may resort to any appropriate method of investigation.'

Communications and Reports

Communications (complaints) and state reports are the most significant functions or processes involving the Commission that are identified in the Charter. Thus far, the procedures in the Charter involving communications by a state party concerning another state party have not been used.

Individuals and national and international institutions can also send communications to the Commission, as provided in Articles 55–59. These provisions recall, but differ significantly from, the First Optional Protocol of the ICCPR examined at pp. 535–52, *supra*. Note the report of the Commission referred to in Article 58 that indicates one possible consequence of a series of communications revealing 'serious or massive violations.' What follows from such a report is not stated in the Charter, but depends on the Assembly's discretion and inherent powers. All 'measures taken' involving a communication are to remain confidential (Article 59) until the Assembly decides otherwise.

The Charter refers tersely to states' reports. Under Article 62, each party 'shall undertake to submit every two years . . . a report on the legislative or other measures taken with a view to giving effect to the rights' under the Charter. Compare the more elaborate provisions in Article 40 of the ICCPR about the role of the ICCPR Committee in reviewing states' reports under that covenant.

The Commission's Performance of its Functions

Article 45 talks of promotion, protection and interpretation. In general, the 'Charter gives pre-eminence to the promotion of human rights and vests a wide range of responsibility on the Commission. In this regard, it has functions that are not directly vested in the European and American Commission.'[50] Several steps have been taken to implement the task of promotion—for example, resolutions by the Commission to the effect that states should include the teaching of human rights at all levels of the educational curricula, should integrate the Charter's provisions into national laws, and should establish committees on human rights.

Although there is some irony in the observation that the Commission, addressing a continent rife with state-imposed abuses, should have promotion as its primary function, that concentration of energy makes some sense in view of Africa's large uneducated population that is ignorant of its rights or lacks organization and capacity for mobilization to vindicate them. Creating a 'rights awareness' could understandably be considered to be a primary function.

Even with respect to promotion, however, the Commission in its early years has not acted energetically in such a way as to maximize the effect of its activities under the Charter. It has not aggressively promoted rights consciousness and has taken few steps to publicize its work with and develop its relationship with struggling African human rights NGOs. It has, however, organized several seminars jointly with NGOs, over 150 of which have observer status with the Commission. It has not explored, for example, massive violations in certain states by undertaking 'studies and researches on African problems in the field of human and peoples' rights,' or done the maximum to 'disseminate information' about the Charter and any such studies (Article 45).

Moreover, in the long run, promotion alone will not be sufficient. This human rights regime governs states that have committed rampant violations, and that lack experience in—and established and respected institutions for—curbing the abuse of governmental power. Such a regime must depend on the effectiveness of institutional pressures for the protection of individuals, in order to achieve long-run change. The African system—in part though not exclusively through the work of the Commission—must raise the costs to states of violations through one or another of the sanctions with which other human rights regimes are familiar.

The 'protection' now offered by the Commission consists primarily of following the Charter's procedures for communications. Those procedures have

[50] U. O. Umozurike, The African Commission on Human and Peoples' Rights, 1 Rev. Afr. Comm. Hum. & Peo. Rts. 5, 8 (1991).

been used in a very cautious way. For example, the provision of Article 59 about the confidentiality of 'measures taken' involving communications had long been rigorously followed, to the point of excluding the petitioners, NGOs, and even the state concerned from all proceedings, even though Article 46 appears to give the Commission some flexibility. In the words of Amnesty International:

> The Charter does not require that the entire procedure be confidential. Much of the Commission's activities in the consideration of communications under Article 55 could be routinely open to the public: the text of the communications, the replies of the governments, the presentation of evidence in open session and questioning of the parties. In addition, the African Commission could request the Assembly [of the OAU] to permit it to include information about 'measures taken' in its annual report.[51]

As of 1994, the Commission relaxed this notion of confidentiality. An annex to its Seventh Activity Report for the year 1993–94 included some information about communications that had been decided on the merits: the name of the author, the country complained against, the subject of the complaint, and the Commission's decision. The facts, however, are presented only in summary form.[52] In the sixteenth session of the Commission in 1994, procedures for communications were also changed. For example, two petitioners on communications made presentations to the Commission in closed meetings, but were free to divulge or publicize what happened.

> For the first time victims of human rights violations or their representatives came to defend their case before the Commission. Though [these] decisions are not yet made public, there is every indication that the Commission seriously and efficiently studies the cases and that the decisions taken were longer and more profound than ever. This shows a promising development in the jurisprudence of the Commission . . . .[53]

The procedures for periodic reports by states have been to some degree creatively developed by the Commission but remain underutilized. Article 45 does not include among the Commission's functions and powers the examination of such reports. Nonetheless, the Commission sought and received from the OAU the authority to study reports and make pertinent observations. It then adopted guidelines for states' submission of reports. But most reports that have been submitted (a large number of states are long past their deadlines for submission) generally refer tersely to formal laws while ignoring actual practice. They are not serious. State representatives often do not appear

[51] Amnesty International, Statement at the Eighth Session of the African Commission on Human and Peoples' Rights, Banjul, The Gambia, Oct. 8, 1990, 1–2.

[52] Wolfgang Benedek, Human Rights News: Africa, 12 Neth. Q. Hum. Rts. 469 (1994).

[53] Astrid Danielsen and Gerd Oberleitner, Human Rights News: Africa, 13 Neth. Q. Hum. Rts. 83 (1995).

at the public sessions at which reports are discussed. Some way must be found to make reporting more probing and to engage high state officials in the process, as well as to give greater publicity to the reports.

Consider the following comments in a description of the fifteenth Session of the Commission in 1994:[54]

> . . . Of the three State reports due to be examined at this session, not a single one could be considered, as no representatives of Benin and Cape Verde attended the session and only an incomplete delegation of Mozambique, not authorized to answer the Commission's questions, was present. . . . The anodyne, almost ritualistic appeal of the Commission [to states to submit their reports on time, or even belatedly] seems unlikely, in the absence of other pressures, to budge recalcitrant states. . . . States should not exercise an implicit veto power by refusing to submit reports or to attend meetings . . . . The Commission has the potential to examine reports, compare them with its own findings and other available information, and draw necessary conclusion even in states' absence—to prevent the reporting system from becoming a high-minded but totally ineffective means of promotion and protection.

QUESTIONS

1. 'Despite the limited powers granted by the Charter to the Commission, the Charter offers enough open-textured provisions for the Commission to proceed inventively through expansive interpretation of its terms. Articles 60 and 61 put this beyond doubt.' Do you agree? For example, what processes for investigations of a state or sanctions for violations that are now characteristic of the UN Commission on Human Rights might be adopted by the African Commission without amendment of the Charter?

2. Suppose that an individual in a state party to the Charter is injured—say, arbitrary arrest and imprisonment without trial pursuant to judgment of a military court, with harsh treatment before release. You are a member of a human rights NGO in that state and must advise the individual of possible processes under the Charter (a) to remedy the violations by providing relief to this individual victim, or (b) to attempt to prevent future similar violations. What is your advice?

3. 'What the African human rights system needs is a court. Then it could achieve the effectiveness of a regional system like the European that has relied so heavily and successfully on judicial review of state action.' Comment.

[54] Gerd Oberleitner and Claude Welch, Jr., Africa: 15th Session African Commission on Human and Peoples' Rights, 12 Neth. Q. Hum. Rts. 333 (1994).

4. How do you assess the African human rights system as a whole? Do you have ideas (realistically, taking account of political constraints) as to how it could be improved, either all at once or through implementing stages?

COMMENTATORS' EVALUATIONS OF THE EARLY YEARS OF THE COMMISSION'S WORK

There follow summaries of comments of several participants as published in: Fund for Peace, Proceedings of the Conference on the African Commission on Human and Peoples' Rights (1991).

Professor U. O. Umozurike, Chairman of the Commission (p. 10):

Responding to the more general reservation raised about the Commission's capacity for effectiveness given all the constraints under which it operates, Prof. Umozurike reminded participants of the timing and political context in which the Charter was drafted. In 1981, Africa . . . was a completely different place. The Charter was drafted so that it would be accepted by African states which means that it is not as strong a document in certain areas as it would be if it were drafted today. The Charter is, however, flexible, and, with the changing political environment, it can be reinterpreted in a more liberal fashion in the future. One vehicle for such liberal reinterpretation exists in the present Charter where the principles guiding its interpretation (article 60) make reference to other instruments of international human rights law . . . and the practices of the United Nations. . . .

Makau wa Mutua (p. 25):

Mr. Mutua . . . started by reporting that the Commission faces a crisis of legitimacy in Africa [stemming] first and foremost from the history of government in Africa. Africans have not seen government as an institution designed to advance their welfare. This was true of the colonial state and, sadly, the independent state as well. . . . Therefore, anything governmental is regarded with suspicion, including a region-wide human rights body created by governments. Mr. Mutua assured participants that he is enormously impressed with the commitment of the Commissioners as individuals, but stressed that the Commission must be seen not as a collection of individuals but as an institution with integrity. The institution suffers from the reputations of its founders—people like Mengistu, Barre, Doe, Mobutu and Moi. . . . Thus, when it comes to public perceptions, the burden of proof is with the

Commission. 'NGOs will only take the Commission seriously because of what it accomplishes—not attending seminars here and there but results!' . . . Human rights abuses in Africa must be denounced by Africans! Speaking out about human rights conditions would increase the Commission's visibility on the continent and bolster its credibility with African NGOs. . . . The Commission should [encourage] more NGOs to obtain observer status and attend its meetings. . . . [S]ome Commissioners are unaware even of human rights groups based in their own countries, let alone across their borders.

Ellen Johnson Sirleaf (p. 27):

Ms. Sirleaf was struck by three things about the Commission: it is generally unknown and invisible; it is regarded with suspicion by those who do know of it; and 'as seen from the eyes of a casual observer,' it is not performing. 'I don't know of any cases that you have resolved related to any of the major human rights problems recently affecting our continent.'

. . . [T]he Commission must disseminate information on its own existence and that of other human rights bodies and instruments. In order to become more effective and more visible, Ms. Sirleaf urged the Commission to identify select cases and 'engineer its involvement'—a theme seconded by numerous participants in later discussions. . . . Ms. Sirleaf urged the Commission to expand its membership to include women and young people. Since these two groups comprise a large percentage of abuse victims, she found their perspective essential to the effective functioning of the Commission. Ms. Sirleaf also repeated the plea to the Commission to familiarize itself with the range of African NGOs . . . . The Commission should contact these and lesser-known organizations and identify with them, thus legitimizing them in the eyes of their own governments.

NOTE

Under Charter Articles 41 and 44, the OAU is to provide staff and services necessary for the effective functioning of the Commission, and to provide for salaries of members within the OAU's regular budget.

In its Sixth Annual Activity Report 1992–1993, the Commission commented on its financial situation (p. 7):

> 20. Since the entry into force of the [Charter, the Commission] has suffered from a chronic lack of staff, resources and services necessary for the effective discharge of its functions. . . .
> 22. The Secretariat has only a skeleton staff and is not endowed with the basic facilities for the execution of certain basic functions such as producing documents on time.

ADDITIONAL READING

In addition to the readings cited in this chapter's text and footnotes, see Astrid Danielsen, *The State Reporting Procedure under the African Charter* (1994); Eileen MacCarthy-Arnolds et al., *Africa, Human Rights, and the Global System* (1994); Olusola Ojo and Amadu Sessay, *The OAU and Human Rights: Prospects for the 1980s and Beyond*, 8 Hum. Rts. Q. 89 (1994); and Makau wa Mutua, *The Banjul Charter and the African Cultural Fingerprint: An Evaluation of the Language of Duties*, 35 Va. J. Int. L. 339 (1995).

PART D

STATES AS PROTECTORS AND ENFORCERS OF HUMAN RIGHTS

Part D completes the basic structure of this coursebook. We first examined in Part B the processes for the creation of international human rights norms and the basic categories of civil–political and economic–social rights. Our attention in Part C turned to the relations between norms and institutions, particularly to the significance of international institutions and processes for the development and enforcement of norms.

Those two parts gave primary attention to the international dimensions of the human rights movement. Of course, states—the creators of the norms, the designers and members of the institutions, the participants in the processes, as well as the primary duty-bearers under international human rights law—figured in these earlier materials. They appeared frequently as the violators, the defendants, the entities being monitored, investigated and reported on by intergovernmental and non-governmental organizations.

Part D shifts focus. Here we observe primarily the internal processes of the states themselves, particularly the decisions and acts of governments. Our perspective is that of the state rather than the international community. For the most part, we examine state action—executive, legislative, judicial—looking toward the observance and protection of human rights rather than action violating rights. In short, we here imagine states as the first-line enforcers of the international human rights system that they have created.

Part D has two chapters, both of which involve the interpenetration of national legal–political orders and the international system. Chapter 11 examines ways in which states internally observe and protect human rights. Chapter 12 inquires into the ways in which a state acts internationally as an enforcer of international human rights by applying pressure (shy of military force) against other, violator states in the effort to limit or stop the violations.

In both chapters, as in the entire coursebook, the materials view the state primarily *in its relationships* to the international system. That is, these chapters do not examine states intensively—their politics, economic system, history, culture, ideology—*independently* of that system. They are not studies in national or comparative law. The chapters, atypically, draw their illustrations from only one political community of states: the United States and Europe. They concentrate on the United States so as to permit a more contextual examination of the problems of internalizing international human rights norms and the possibilities of bilateral enforcement of those norms against violator states.

11

Interpenetration of International and National Systems: Internal Protection of Human Rights by States

Human rights violations occur *within* a state principally in relations between a government and its own citizens, rather than on the high seas or in outer space outside the jurisdiction of any one state. Ultimately, effective protection must come from within the state. The international human rights system does not place delinquent states in political bankruptcy and through some form of receivership take over the administration of a country in order to assure the enjoyment of human rights—although the recent intervention in Somalia made a modest and confused step in that direction in a chaotic situation where a central organized government had all but ceased to exist. Rather the international system seeks to persuade or pressure states to fulfill their obligations through one or another method—either observing national law (constitutional or statutory) similar to the international norms, or making the international norms themselves part of the national legal and political order.

Such is the focus of Chapter 11, which falls into three sections. *Section A* describes the adoption by many states during the last half century of constitutions informed by liberal political (and often economic) premises and frequently declaring human rights. It notes the many relationships between such state constitutions and the human rights movement. *Section B* inquires how states 'internalize' *treaty* norms—that is, how they incorporate the provisions of human rights treaty within the state's legal and political order so that they can be implemented and enforced by state authorities. *Section C* then turns to the internalization of *customary* international law. It examines distinctive legislation with the United States, the Alien Tort Statute, to explore how courts have used customary human rights norms in order, in a sense, to 'enforce' them against foreign violators. That third section thus provides a bridge to the themes of Chapter 12.

A. THE SPREAD OF STATE CONSTITUTIONS IN THE LIBERAL MODEL

This section examines briefly one of the important phenomena in the fifty years of the human rights movement: the spread among many states of constitutions expressing the principles of political (and sometimes economic) liberalism, and often declaring human rights. The tradition and history of many states adopting such constitutions have often involved neither liberal principles nor espousal of human rights.

One can approach this phenomenon from two perspectives. First, the adoption of liberal-style constitutions (or at least constitutions with a significant liberal component) can be seen as a *horizontal* trend among states, a consequence of the influence and pressures exerted by powerful countries within the liberal constitutional tradition like the United States. For the states achieving independence upon decolonization and adopting constitutions in the liberal model, the primary influence may have been the internal constitutional and political orders of the former colonial powers.

A second perspective stresses the links between the spread of such types of constitutions and the achievements of the human rights movement in constructing an international system of norms, institutions and processes. One could describe these links and the related influence and pressure as *vertical* rather than horizontal in character, since they come from international law and institutions 'above' the state rather than from other states.

Both horizontal and vertical influences and pressures were simultaneously at work in this process of state drafting of constitutions. They were indeed complexly intertwined, for the very states exerting a 'horizontal' influence were simultaneously creators and members of the international system exerting pressure from above. International human rights figured ever more pervasively in the discourse of international relations and made demands on states that in many respects could best be met through constitutions assuring citizens of the state conduct required by the treaties. Those constitutions both 'borrowed' from other states and revealed, through their governmental structure and declarations of rights, the influence on them of the UDHR and the two basic Covenants.

COMMENT ON CONSTITUTIONS AND CONSTITUTIONALISM

The range of constitutions

Most states of the world (the United Kingdom being the most striking exception) have a formal, written constitution. By itself, that statement says very little. A 'constitution' refers simply to a basic state document that organizes a political system by, for example, setting forth the basic institutions of government. It need not follow any particular structure, impose or reflect any particular political or economic system or ideology, or prescribe any particular form of government. It may be democratic or authoritarian, oriented to private property and markets or to collective ownership and central direction, multi-party or one-party, attentive or not to individual rights, and so on.

Constitutions vary radically in their practical significance and symbolism as well as content. At one extreme, entire instruments or particular provisions may be meant to be hortatory and aspirational rather than to form part of the state's legal system. Those aspirations may represent genuine goals to be worked toward, or amount to sham and pretense, a shallow disguise of radically different and less admirable state purposes and methods. At the other extreme, constitutional provisions may be judicially enforceable against the government in actions brought by private parties, even to the extent of judicial review—that is, testing legislation against constitutional norms, and invalidating legislation found to violate them.

From a different perspective, a constitution may be broadly understood by the people to be insignificant, a document freely manipulated by government. Or it may be understood as an authoritative charter or even solemn covenant between the government and people. If the latter, it will tend to be more stable and resistant to change; the amendment process will be correspondingly complex, demanding and time-consuming.

Constitutionalism

Scholars frequently use the term 'constitutionalism' to describe a particular genus of constitutional system. This fluid term is put to many different uses; there is no consensus over exact content, although most of the scholarly discourse would agree on the core meaning. Constitutionalism in this Comment refers to a constitutional system that falls within the liberal tradition (see the Comment on liberalism at p. 187, *supra*) and possesses many characteristics of the democratic state. (See the Comment on democratic government at p. 658, *supra*). Often constitutionalism is implied when one speaks of the spread of constitutions among states as a part of the human rights movement.

Constitutionalism refers broadly to the following characteristics of a constitution, found among states to a greater or lesser degree, in different combinations and with different emphases. Often based on a conception of popular sovereignty, the constitution assures accountability to the people through a range of techniques and institutions, the most important being the requirement of periodic and genuine elections in a multi-party system. Consistent with the liberal tradition, the constitution would control and limit the powers of government in several different ways, including a scheme of checks and balances through a separation of powers that must include an independent judiciary. That judiciary becomes the essential guardian of the rule of law, in the sense that the executive (or 'government') must act within established legal frameworks and according to established processes. The legality and legitimacy of that action must be subject to judicial review.

Constitutionalism implies that the constitution is a 'real' rather than merely hortatory instrument, truly the fundamental law or charter—in the words of the United States Constitution, 'supreme law.' It thus possesses a distinctive solemnity and force. It may come to symbolize the nation itself.

Constitutionalism generally involves a declaration of the individual rights associated with the liberal tradition, in the manner of the Bill of Rights of (the first ten amendments to) the U.S. Constitution. These rights are indispensable to setting limits to governmental action, particularly when they are coupled with judicial review of the constitutionality of legislation. Judicial review itself could be seen as a desired but not essential element of the constitutional scheme.

The list of rights declared will vary among states. For example, in the constitutional history of the United States, the property right protected by the Fifth and Fourteenth Amendments to the Constitution has played a large and influential role. That has been much less the case in many states whose constitutions now fall within the liberal tradition.

Two polar illustrations: the P.R.C. and South Africa

Since the People's Republic of China was founded in 1949, it has had five constitutions, the last three effective in 1975, 1978 and 1982.

> Each of these constitutions has signaled a change in power or a change in the economic or political objectives of the Chinese leadership. . . . The judiciary had virtually no significant role in this constitutional evolution. . . . [E]ach constitution has been premised on the belief that rights are granted to citizens by the state. Rights are not derived from human personhood, nor are they enshrined or fixed within the constitution itself. This premise or belief allows the state to change both the quality and quantity of a citizen's rights whenever it is believed necessary because the existence of such rights does not limit the power of the state. Consequently, each constitution has either implicitly or explicitly permitted the state to define the nature of a citizen's rights by legislation. . . . [T]he U.S. constitutional principle of checks and balances between the traditional branches of government (executive, judicial, and legislative) is absent.[1]

The Preamble to the 1982 Constitution states four principles to which the state is committed, including the people's democratic dictatorship, Mao Zedong Thought and the leadership of the Communist Party. In several respects that Constitution implemented the changes in economic policy that have subsequently led to a limited market sector with individual ownership of

[1] Ralph Folsom, J. Minan and L.A. Otto, Law and Politics in the People's Republic of China 56 (1992).

property complementing collective property and central direction. Like its predecessors, it rejected or changed the content of some prior constitutional rights. For example, it eliminated the provision in the 1975 Constitution for 'speaking out freely . . . and writing big character posters [as] new forms of carrying on socialist revolution created by the masses of the people.'[2]

The second illustration is a polar one, of a state in transition from a racist regime to constitutionalism. Perhaps no change in regime and principles has been as sudden and dramatic as that in South Africa. After the collapse of apartheid came the Constitution of the Republic of South Africa, an interim document adopted in 1993 and effective the next year, a transitional constitution valid for two years from April 1994 and serving the purpose of facilitating the transition to democracy.

Under Section 4, the Constitution is 'the supreme law of the Republic and any law or act inconsistent with its provisions shall [unless otherwise provided in the Constitution] be of no force and effect.' Some illustrative provisions, each representing a dramatic break with tradition, follow:

1. Chapter 3, entitled 'Fundamental Rights,' declares many of the traditional human rights. Section 8 provides that 'every person' has the right to 'equal protection of the laws.' The forbidden grounds of discrimination include 'race, gender, sex, ethnic or social origin, colour, sexual orientation, age, disability, religion, conscience, belief, culture or langauge.' There is probably no more inclusive provision in any state's constitution—surely not in any international human rights instrument. Section 8 goes on to state that it 'shall not preclude measures designed to achieve the adequate protection and advancement of persons or groups or categories of persons disadvantaged by unfair discrimination, in order to enable their full and equal enjoyment of all rights and freedoms.'

2. Under Section 35, courts shall interpret the provisions of Chapter 3 to 'promote the values which underlie an open and democratic society based on freedom and equality and shall, where applicable, have regard to public international law applicable to the protection of the rights entrenched in this Chapter. . . .'

3. The Constitution creates an independent and impartial judiciary (Sec. 96). A Constitutional Court has jurisdiction over all matters relating to the interpretation and enforcement of provisions of the Constitution, including 'any alleged violation or threatened violation of any fundamental right,' and 'any inquiry into the constitutionality of any law, including an Act of Parliament, irrespective of whether such law was passed or made before or after the commencement of this Constitution.'

[2] Ibid. at 65.

NOTE

The variation in declared rights among state constitutions exists as well between any one constitution and the international instruments such as the UDHR and the two Covenants. Consider the following comments about the United States.

LOUIS HENKIN, CONSTITUTIONALISM AND HUMAN RIGHTS

in Henkin and Rosenthal (eds.), Constitutionalism and Rights 383 (1990), at 390.

There are other differences between U.S. constitutional rights and international human rights. Rights under the United States Constitution have to be understood in the larger conception of the United States as a 'liberal state.' Individualism was historically the hallmark of the American society, related perhaps to our vast spaces and the distant frontier. In the liberal state, rights of autonomy, liberty, and property are axiomatic as are freedom of enterprise, minimal government, and essential laissez-faire in economic as in other matters. As a liberal state the United States stood in contradiction to a welfare state, or to any other utilitarian conception that one may restrict the rights of the few to enhance the rights and welfare of the vast majority.

On the other hand, the international human rights instruments were designed for acceptance by states generally. Most states do not have an individualist tradition and do not value individualism. They are committed not to the liberal state but to the welfare state, not to minimal government but to activist, efficient, maximal government. The liberties and immunities recognized as human rights, then, are exceptional limitations on government. Inevitably, then, even when the international instruments and the U.S. Constitution recognize the same rights, even in the same terms, the content, scope, and significance of a constitutional right in the United States may be very different from what it is in another state operating under the international norm.

A related difference in the U.S. and international conceptions and scope of rights derives from the original philosophical conception of U.S. rights. American political society was conceived as a social compact—presumably an agreement among the people to form a polity as well as an agreement as to the kind of government to be created and the conditions that should govern it. Only the last part of that compact is in the Constitution and in our constitutional jurisprudence. Except for the amendment abolishing slavery, the Constitution protects individual rights only against invasion by government, against state action; it does not address rights between individuals. The concept of state action, too, has to be seen in the context of the United States as a liberal state. Restrictions on the state are congenial to that concept.

Restrictions on private persons are not, since every right of an individual against his/her neighbor is a restriction of some right of the neighbor.

Apart from what particular states might be moved to do, it was necessary to find other foundations—principally the power of Congress to regulate interstate commerce—to support power for Congress to forbid private violations of rights, for example, private discrimination on account of race or religion. Neither Congress nor any state is constitutionally required to adopt such legislation. On the other hand, the International Covenant on Civil and Political Rights requires states not only to respect rights but to ensure them, and a state party would seem to be obliged to protect life, liberty, or property, reputation or honor from violation by one's neighbor.

. . .

The biggest difference between the international bill of rights and ours involves what in the international lexicon are called economic and social rights. The Universal Declaration of Human Rights and the International Covenant on Economic, Social and Cultural Rights imply that only a welfare state is a human rights state. . . .

. . . [E]conomic and social rights are not constitutional rights in the United States. The Constitution still reflects a commitment to limited government, to the view that 'that government governs best which governs least.' The Constitution tells government what not to do; it imposes few positive obligations. While the international human rights idea requires a state to organize itself and its resources to ensure basic human needs to those of its inhabitants who cannot meet them otherwise, there is no constitutional right in the United States to food, housing, health care, education, or other basic human needs, and no constitutional obligation to provide them. . . .

. . .

NOTE

As rights migrate, they often take on the cultural characteristics of their new homes. Hence an abstract right, such as that to free speech, may assume a markedly different form in the constitutions of states of a particular religious or political character. Those differences can be expressed through the types of limitations to the right that may depart radically both from limitations in other states and in the international instruments. Note the following illustrations.

> *Pakistan Constitution (1973), Art. 19*: 'Every citizen shall have the right to freedom of speech and expression, and there shall be freedom of the press, subject to any reasonable restrictions imposed by law in the interest of the glory of Islam or the integrity, security or defence of Pakistan

or any part thereof, friendly relations with foreign States, public order, decency or morality. . . .'

Constitution of the Former U.S.S.R., Art. 50: Citizens have the rights of free speech, press and assembly '[i]n accordance with the interests of the working people and with a view to strengthening the socialist system. . . .'

QUESTION

Assume that a state's constitutional declaration of a right like free speech meets on its face the relevant international criteria set forth in the UDHR and ICCPR. What further inquiries about the state's legal and political system would you wish to make in order to assess the degree to which the state is observing and protecting this right?

NOTE

The three following readings depict some of the dilemmas of constitutionalism in those parts of the world where its spread in the last half century has been most marked: the new states of Africa and Asia following decolonization, and the states of Central–Eastern Europe after the collapse of the Soviet Union and communist rule. Those dilemmas recall the issues of universalism and cultural relativism debated in Chapter 4. The term constitutionalism as used in these readings sometimes appears synonymous with political democracy. Several themes emerge:

the difficulties of 'transplanting' constitutions in the liberal model from countries in which they grew over time organically (even if the process started by violent revolution) to states of a radically different ethnic, cultural, religious, economic and political character

the pull on the new constitutions by indigenous forces and traditions, including nationalism, ethnic divisions and religion

the question of how much cultural and economic baggage accompanies a constitution drawn from other states and from international instruments—for example, the degree to which market organization and a strong private property regime inhere in the organization of government power and declarations of rights in the new constitutions

the problems stemming from the adoption of a constitution that may for its effective functioning require a different foundation (an active civil society, for example) from that which the society can now provide the lesser value placed by parts of the population on governmental and constitutional structures than on the achievements or programs of a government in (depending on which part of the population is consulted) bringing about economic development or continuing to provide social welfare.

YASH GHAI, THE THEORY OF THE STATE IN THE THIRD WORLD AND THE PROBLEMATICS OF CONSTITUTIONALISM

in Douglas Greenberg, Katz, Oliviero and Wheatley (eds.), Constitutionalism and Democracy 186 (1993), at 187.

[The author states his purpose as trying to 'provide some explanation of why it has proven so hard to establish constitutionalism in many developing countries, but particularly in Africa.' He uses a dichotomy of Weber to illustrate his argument: legal–rational as opposed to patrimonial. Ghai relates the first term to the notion of the rule of law; the authority of the state is 'founded in law,' which provides the framework for state operations. No one is above the law. The impartiality of the state bureaucracy is ensured by fidelity to law, which requires independence of the judiciary.]

Patrimonialism is a different mode of domination. It is a form of personal rule, which does not tolerate opposition. Administration is based on the total power and discretion of the ruler. The bureaucracy is an extension of his household, and to which he delegates its powers. Officials owe their appointments to his trust and goodwill. There is no clear separation between the private and public spheres of the ruler. He is above the law, as are his officials, and dispenses justice; petitions to him for clemency and generosity substitute for legal writs. The ideological superstructure of such domination is the goodness, generosity, and concern of the ruler for his people. He is the 'father of his people,' 'the father of his nation.'

I argue that in many Third World countries the trend has been toward the patrimonial form of rule and away from the rational–legal. . . . The trend has fundamental implications for constitutionalism. . . .

It is unnecessary to argue that on the whole the record of most Third World governments on human rights is dismal. . . . The denial of human rights is part of the wider spectrum of undemocratic and authoritarian rule.

. . .

For a variety of reasons, countries emerging from colonial rule did so with constitutions closely modeled on those of the west. Whether they provided for parliamentary or presidential systems, they separated powers and personnel, diffused power through federal or regional devices, limited the scope of legislative and executive power by human rights, and separated religion under the state. A variety of devices protected minorities, and judicial review safeguarded the supremacy of the constitution. During the early years of decolonization, there was great faith in the ability of constitutions to settle the problems of new nations; later, considerable cynicism set in, and few, even those who participated in its preparation, believed that constitutional settlements made at independence would endure. Only a handful of countries still have the constitution they adopted on independence; yet, despite political vicissitudes in which military or other authoritarian regimes replaced democracy, aspirations for the return to the constitutional values of the independence period for the resolution of their problems animates the peoples of these countries.

Although it appears unfair to condemn them as utopian, one must question the feasibility of western models for most developing countries. Western constitutions (and constitutionalism) assume settled political and economic conditions and a broad consensus on social values. . . . The constitutions themselves created neither the conditions nor the consensus, although they consolidated and reinforced them. Constitutionalism represented the victory of particular groups and classes; but the victory itself was often the result of violence, exploitation, and repression.

In the developing countries the constitutions were expected to carry a much heavier burden. They had to foster a new nationalism, create a national unity out of diverse ethnic and religious communities, prevent oppression and promote equitable development, inculcate habits of tolerance and democracy, and ensure capacity for administration. These tasks are sometimes contradictory. Nationalism can easily be fostered on the basis of myths and symbols, but in a multiethnic society they are often divisive. Traditional sources of legitimacy may be inconsistent with modern values of equality. Economic development, closely checked and regulated during colonialism, also threatens order and ethnic harmony, as it results in the mobility of people and the intermingling of communities in contexts where there is severe competition for jobs and scarce resources. Democracy itself can sometimes evoke hostilities as unscrupulous leaders prey on parochialism, religion, and other similar distinctions.

The burden of constitutional tasks was, moreover, compounded *au fond* by the nature of Third World polities, different in crucial respects from the west. Although often attended by violence, the growth of the state in the west was more organic than it was in the Third World where it was an imposition. The state dominated the economy and was instrumental in shaping it; in the west the state reflected the economy. The political factor was consequently more important. Political power is harder to control because civil society is weak and fragmented, itself the result of colonial practices (which have proved congenial to new governments). The state in the west enjoyed relative autonomy from international forces, which facilitated indigenous control over society and enabled a degree of diffusion and institutionalization of power. The Third

World state owes not only its genesis to imperialism, but even contemporary international economics and politics condition its very nature and existence. Hardly in control of its destiny, such a society finds it hard to institutionalize power on the basis of general rule, any more than it can resist encroachments on rights and democracy engineered by the more powerful states and corporations. At the time many Third World states moved into independence, the tools of coercion were easily available, which made them careless of cultivating the consent of the ruled.

The ideology of constitutionalism has only the slenderest appeals to the rulers or the ruled in the Third World. Legitimacy comes from other sources, and some of these sources are antithetical to the rule of law. Closely associated with independence was the ideology of modernization and development, in whose name state structures were strengthened and their writ expanded. People appear to regard the promotion of development as the primary task of the government; and the governments, for their part, justify the aggregation and concentration of power (and dismiss debates on human rights) on the imperatives of development. Closely connected ideologies proclaim the supremacy of the party, reaching in some countries levels of deification, or, more frequently, the sanctification of the leader. . . .

Another important source of legitimacy, more significant in Asia than in Africa, is religion, of which the most influential has been Islam, the world view of many of its adherents being dominated by religion. Their religious belief underpins their self-identification and locates them within a community. The Islamic revolution in Iran inspired Muslims in many other parts of the world to establish a state based on Koranic principles. Islam makes no distinction between the temporal and the spiritual, and provides a complete system of philosophy and life. In many ways the resurgence of Islam is a reaction to unequal development as well as to the dominance of western influence in many Third World countries. As with other religious movements, it is in a significant measure a response to the tensions and uncertainties of modernization, the concentration and secularization of public power. But these lead to the theocratization of power, with less rather than more democracy.

If the rule of law as an ideology is unimportant, it will come as no surprise that it is also unimportant in its substantive aspect. In most developing countries, whether the economic system is a species of capitalism or socialism, general norms play a secondary role. Both kinds of economy are essentially administered economies, where the license is the king and discretion is the norm. . . . Capitalists seek the embrace of the governments, and concentration of state and capital is extensive and complex. Markets themselves are the creation of governments. . . . Multinationals accept that bargains have to be struck with the government and concessions negotiated. . . . [I]t is the character of the state and the nature of the process of accumulation in most Third World countries that undermine constitutionalism.

. . .

RADHIKA COOMARASWAMY, USES AND USURPATION OF CONSTITUTIONAL IDEOLOGY

in Douglas Greenberg, Katz, Oliviero and Wheatley (eds.), Constitutionalism and Democracy 159 (1993).

. . .

. . . [T]he term 'constitutionalism' will be used broadly, in both its ideological sense and imply a process and style of decision making specific to the genre of constitutions drafted in the Anglo-American tradition of jurisprudence. Although they vary in substance and although many of the provisions in South Asian Constitutions have been taken from socialist constitutions, . . . these constitutions have a similarity of tradition, style, and interpretation that reveals their origins in a liberal, social democratic political order. . . . In the South Asian context, constitutionalism has been enhanced by these traditions, which, precisely because of their source, share a similar crisis of confidence; they are not considered legitimate, and are not widely accepted as the only means of conducting political life. . . .

In today's world, with the exception of the United Kingdom, the source of legitimacy for liberal democratic values is a written constitution. In India and Sri Lanka, the source itself has been subject to repeal, reenactment, and an extraordinary number of amendments. This fact leads one commentator to state that the use of constitutions in these societies is 'instrumental' for those who actually capture state power. Therefore, constitutions and constitutionalism in South Asia cannot be seen as fundamental law but as a process in which the values of the constitution are mediated by realities of power and social antagonism.

The transfer of power in South Asian societies after the era of British imperialism saw the transfer on paper of the institutions of parliamentary democracy. Within a decade, Pakistani society experienced the strengthening of the executive presidency and the growth of military dictatorships. In Sri Lanka, three decades witnessed three constitutions, each drafted by a different government. Successive Indian governments in three decades of existence amended the constitution 50 times. Clearly the sense of a constitution as 'fundamental law' has yet to emerge as a settled consensus, accepted by all shades of political opinion. The process has begun but it may still take some time for the constitutions to become accepted social contracts.

Liberal scholarship in these societies has taken the constitution as granted and then spelled out the enormous evidence of situations and contexts that violate the basic tenets of any liberal constitution. If one adopts this line of reasoning, the picture is very clear; constitutions in South Asia are formal pieces of paper, whose basic provisions, such as fundamental rights, are rarely observed. However, what is also interesting but rarely analyzed are the ways and means in which liberal democratic values are transformed by the contexts of cultural nationalism and economic and social underdevelopment. This process has openly perverted liberal values. In certain creative instances, and in specific areas, experiments have resulted in what may be termed a 'genuine legitimacy' for liberal values.

. . .

Some major changes occurred in the 1970s and the 1980s. Liberal democratic values, no longer merely a 'passive inheritance' from the British, have become an active tool of political and social accountability. Examples of this change are the growth of social action litigation in India, the shift in the language and the discourse of political opposition against military dictatorship in Pakistan and in Sri Lanka, and a slowly growing judicial scrutiny in such areas as freedom of speech and criminal procedure. These changes cannot all be called revolutionary, but liberal values are not only longer valid as 'scraps of paper'; they have become active, albeit in association with other values, interests and struggles peculiar to the South Asian context.

The ideology of constitutionalism has also furthered political reform, both in government and as a mobilizing force, in the last four decades. In India, constitutionalism in the form of social action litigation has been most successful in economic and social areas where the state has not lived up to the standards that it has set for itself. This is especially true in India regarding legislation with respect to caste, class, and women. As we have seen, the ideology of constitutionalism has also been important as a langauge of protest against military regimes and authoritarian governments. . . . It is important to note, however, that constitutionalism as substantive ideology (as opposed to commitment to process and procedural style) does not have a monopoly on dissent. Marxism, nationalism, ethnic chauvinism, and religious fundamentalism are some of the alternative discourses competing for ideological legitimacy in many South Asian societies.

. . .

How then can constitutionalism acquire 'genuine legitimacy' in South Asia? One must recognize the inherent strength of a constitutional process—its potential for evolution and growth. None of the other ideologies prevalent in South Asia today is as committed to so specific and detailed a process of non-violent decision making. . . . The strength of the constitutionalism as an ideology is that it spells out the ways and means of resolving conflict, of electing representatives, of making and implementing public decisions, and of reconciling interests and rights in a systematic and open manner. No indigenous ideology in South Asia that has gained currency as a dominant political force has such a comprehensive project for consultation, compromise and conflict resolution. The political processes represented by the architects of indigenous ideology were either dynastic, religious, or tribal and not one of them was designed to cater to the needs of the modern nation-state.

To survive and grow as an important aspect of political life, constitutionalism must develop internal processes relevant to the actual struggles taking place in South Asian societies. The involvement of constitutional processes in elite concerns and in elite rights, such as property, lessens their legitimacy in the society at large. If, in the coming years, the processes of constitutionalism were to be attached to issues of poverty, political repression, social justice, regional backwardness, etc., it would be more likely that constitutionalism would enhance and enrich the democratic process in South Asian societies. Success will depend on a younger generation of lawyers, born in the post-colonial period, who have in some way managed to integrate values drawn from Asian civil society with the modern political demands for democracy and constitutionalism. The use of constitutionalism in the future cannot be restricted to that of a watchdog for an errant executive. Its uses must be extended to become active and dynamic, to protect values by the use of language, discourse, and doctrine that have some resonance in South Asian reality.

ANDRÁS SAJÓ AND VERA LOSONCI, RULE BY LAW IN EAST CENTRAL EUROPE: IS THE EMPEROR'S NEW SUIT A STRAITJACKET?

in Douglas Greenberg, Katz, Oliviero and Wheatley (eds.), Constitutionalism and Democracy 321 (1993), at 326.

Reformative and revolutionary forces considered the legal system one to be sued and conquered for their purposes. In Hungary, and to a certain extent in Poland, and in a different way in the Soviet Union, Communists were partly responsible for the beginning of reform or, at least, some of the Communists tried to control reform processes. In these countries, the transformation of the legal system was considered a proof of the good intentions of the reformists. It was an often-stated belief that a modernized legal system, which observed constitutionalism and promised a calculable legal environment for business, including a gradual transition to market economy, would grant legitimacy and credibility to the reform process both inside and outside the country. Furthermore, it was hoped that a system with more open rules would increase the efficiency of that system. The ruling elites hoped that once their system became constitutional, it would also be acceptable and, therefore, would continue to survive.

Given the nature of the system, the interests of the participants and the transition process, this proved to be a somewhat naive illusion. The results were contradictory, and certainly far from the expected with respect to the quality of the legal system. In the Soviet Union, Mikhail Gorbachev's rule partly destroyed the old inefficient but working network of illegal relations without replacing it with a system based on the rule of law. In Poland and Hungary, the spectacular changes in the public law contributed to democracy, but failed to instill respect for constitutionalism or to create the rule of law.

Both in the reform process and after the more or less revolutionary victories by the masses, the creation of a constitutional legal system with western types of liberal/formal solutions was a high priority. For the public this ideology was not self-legitimating. Religion, anticommunism, nationalism, and consumer values seem more important now to the masses. . . . There has only been a minimal common acceptance of the liberal tradition. Agreement exists on the importance of human rights, on the separation of powers (interpreted as denial of one-party rule), and on independent judiciary; but there is still no position of principle on welfare rights or on public interest priorities, in particular *vis-à-vis* private property. This level of acceptance obviously reflects the present political struggle for the control of the state, and more generally the slow emancipation of society from centralism.

Democratic legislation was thought to be a remedy for the malaise of the previous system. In this respect the east European transformation process follows the general pattern of revolutionary legal change: it is a denial of the previous solutions. The replacement of these solutions is, however, a matter of choice; there are many ways to part with the past. One group of possible choices is offered by the available western models (which are often little and selectively known or misunderstood, and are too sophisticated to be adapted). Another major choice consists of the pre-Communist legal solutions, which present linguistic barriers, and which are attractive because they can be legitimized as genuinely national solutions. The problem is that these solutions are often outdated; they did not meet criteria of democracy 50 or 60 years ago. Nor were these central European societies particularly modernized before World War II. One should keep in mind, however, that nationalism and religion were important sources of resistance, and important conservative voices among the emerging leaders in the society supported this solution.

If the new ruling classes and elites manage to convince the population that they are offering them rights through the new legal system, their legitimacy will increase. This is not an easy task, as economic hardship and political instability make it imperative to pass unpopular measures through law. . . .

. . .

One of the first steps of the social and political transformation taken in all former socialist societies was the amendment of the Communist constitutions. These amendments abolished the privileges of the Communist Party. Moreover, in the case of Hungary, they institutionalized the protection of human rights and separation of powers. Nevertheless, in every case the amendments were the result of an elite agreement with the former Communist leadership. They were not intended to be expressions of the 'will,' values, or demands of the masses who to varying degrees participated in the process resulting in the collapse of Communist rule. The amendments, openly intended to be revised, reflected the provisional arrangements; revisions became routine procedures as new conflicts emerged. Such confrontations surfaced over presidential powers in Poland, Bulgaria, and Hungary, over federalism in Czechoslovakia, and over the constitutional powers of the executive (government) in all the countries of the region.

From the point of view of a rule of law system or any system that takes law (i.e., the meaning of words) seriously, utilizing the constitution to rewrite political agreements is counterproductive to the goal of creating respect and belief in constitutionalism. If a constitution is easy to amend it loses its majestic special role. The Hungarian Constitution of 1989 was amended six times in the first year of its existence; the changes amount to one-third of the text. Issues of constitutional amendment aimed at settling conflicts absorb a considerable amount of time in parliamentary debates in Czechoslovakia and in Bulgaria. In that process law, including the text of the Constitution, easily becomes a matter of technicality. For instance, after the Hungarian Constitutional Court ruled unconstitutional an article of the Election law, the Parliament the next day amended the Constitution so as to make the article in question conform to the Election Law!

. . .

. . . [T]he emerging sentiments of nationalism and religion are not particularly tolerant of some of the formal 'impartial' criteria of law and constitutionalism. Other elements relate directly to the interests of the freely elected governments in maintaining the role of the state sector in the economy and perhaps in the welfare sector as well. Social groups for whom the reduction of the state sector means unemployment and poverty support these tendencies. The same social groups have less interest in and comprehension of individual rights and their protection through courts as their education, in a paternalistic tradition, has alienated them from law. Paternalism, in this context, means that they accept their role as clients and as more or less obedient servants of an uncontrolled state welfare system.

Political power factors also make the emergence of constitutionalism difficult. In all these societies at least part of the present ruling elite was involved in one way or another in the Communist system. Revision of the past through retroactive laws (political justice, revision of privatization decisions, reprivatizations, and indemnification of the economic and political victims of Communism) became crucial in one respect or another in all the countries of the region in 1990. Obviously, this is not the best education in constitutionalism. . . .

Forms of intolerance, in addition to anti-Communism, present problems for constitutionalism. In general, there is little tolerance for minorities, and protection of them is not high on the priority lists of political organizations capable of shaping constitutions. Political forces do not incline to undertake the commitments necessary to write a constitution for the future that would require a firm stance on constitutional values that are unpopular or still debatable. Constitution making is taking the form of a major power struggle in

Czechoslovakia in relation to the federalism issue and in Bulgaria. In Poland, in light of an increasing personal presidential power and an antiliberalism reflected in the pending antiabortion legislation, religious school education by ministerial decree, and nonprivatization, a new constitution may well not be a major contribution to constitutionalism.

The emerging elites and the lack of constitutional legal culture are not the only factors responsible for these conceptual insufficiencies. Communist rule and the lack of a capitalist society and a civic culture contributed to the passive acceptance and mistrust of law. On the other hand, as the emergence of civil society and a nonstate ownership system that would make personal independence a possibility is very slow, there is little actual interest or urge to have a rule of law system. Of course, the failure to develop a constitutional framework and an impartial legal system will make the formation of a constitutionalism-hungry society extremely painful if not impossible in the coming years. . . .

B. HUMAN RIGHTS TREATIES WITHIN STATES' LEGAL AND POLITICAL ORDERS

Constitutionalism as described in Section A represents states' efforts to create an internal order that is consistent to an important degree with fundamental aspects of liberal and democratic political theory and with the international human rights movement. Deeply influenced by the Western models and by the basic international human rights instruments, these new constitutions in many parts of the world are nonetheless emphatically *national* instruments, freshly created by each state.

Section B shifts attention to international instruments, specifically to human rights treaties. It examines the interpenetration of the international and national systems, the significance of treaties within as well as without states. The broad questions explored are: How do these treaties influence the national legal and political systems of states parties? Are they actually incorporated into a state legal system, or reproduced in state legislation, and, if so, with what effects on the different branches of government such as the judiciary? Or do they remain distinct from the state system, 'above' it as part of international law? The readings draw on the techniques and experiences of a number of European countries and of the United States to develop answers to these questions.

The materials then use the experience of the United States in its ratification of the International Covenant on Civil and Political Rights as a case study of problems in this field.

1. BASIC NOTIONS: ILLUSTRATIONS FROM DIFFERENT STATES

VIRGINIA LEARY, INTERNATIONAL LABOUR CONVENTIONS AND NATIONAL LAW

(1982), at 1.

The relationship between international law and national law, in the past, was largely an interesting theoretical problem, engaging legal scholars in a doctrinal debate; it has now become an important practical problem, primarily as a result of the increasing adoption of treaties whose scope is not inter-state relations but the relations of states with their own citizens. These treaties are concerned with 'the common interests of humanity' and have a quasi-legislative character. Their efficacy depends essentially on the incorporation of their provisions in national law. . . .

. . .

International law determines the validity of treaties in the international legal system, i.e., when and how a treaty becomes binding upon a state as regards other State Parties. It also determines the remedies available on the international plane for its breach. But it is the national legal system which determines the status or force of law which will be given to a treaty within that legal system, i.e., whether national judges and administrators will apply the norms of a treaty in a specific case. . . . When the treaty norms become domestic law, national judges and administrators apply them, and individuals in the ratifying states may receive rights as a result of the treaty provisions. Thus, developed municipal legal systems supplement the more limited enforcement system of international law.

While the international legal system does not reach *directly* into the national systems to enforce its norms it attempts to do so *indirectly*. States are required under international law to bring their domestic laws into conformity with their validly contracted international commitments. Failure to do so, however, results in an international delinquency but does not change the situation within the national legal systems where judges and administrators may continue to apply national law rather than international law in such cases. The work of the supervisory bodies of the International Labour Organization in influencing states to bring their law into conformity with ILO conventions is an example, however, of the increasing efficacy of the international legal system in this regard.

The status of treaties in national law is determined by two different constitutional techniques referred to in this study as 'legislative incorporation' and 'automatic incorporation'. In some states the provisions of ratified treaties do not become national law unless they have been enacted as legislation by the normal method. The legislative act creating the norms as domestic law is an act entirely distinct from the act of ratification of the treaty. The legislative bodies may refuse to enact legislation implementing the treaty. In this case the provisions of the treaty do not become national law. This method, referred to as 'legislative incorporation', is used, inter alia, in the United Kingdom, Commonwealth countries and Scandinavian countries. In other states, which have a different system, ratified treaties become domestic law by virtue of ratification. This method is referred to as 'automatic incorporation' and is the method adopted, inter alia, by France, Switzerland, the Netherlands, the United States and many Latin American countries and some African and Asian countries. It is important to note that in many of these states, proclamation or publication of the treaty may also be necessary for its force as national law. Even in such states, however, some treaty provisions require implementing legislation before they will be applied by the courts. Such provisions are categorized as 'non-self-executing.'

. . .

International law does not dictate that one or the other of the methods of legislative or automatic incorporation must be used. Either is satisfactory assuming that the norms of treaties effectively become part of national law. Conversely, neither method is ipso facto satisfactory under international law, if, in practice, the norms of ratified treaties are not applied by national judges and administrators. The method by which treaties become national law is a matter in principle to be determined by the constitutional law of the ratifying state and not a matter ordained by international law. The international community, lacking more effective means of enforcement, is often dependent on the constitutional system of particular states for the effective application of treaties intended for internal application.

Some national constitutions provide for automatic incorporation of treaty provisions. In other states, judicial decisions have determined that treaties are to be automatically incorporated. A correlation appears to exist between legislative consent to ratification and automatic incorporation. In states with the system of automatic incorporation, legislative consent by at least one house of the legislature is generally required before the executive may ratify treaties. In states with the system of legislative incorporation, ratification of treaties is frequently a purely executive act not requiring prior approbation of the legislature. In the United Kingdom, and other common law countries which have followed U.K. precedent in this regard, parliamentary consent to ratification is normally not required and express legislative enactment of treaty provisions is necessary before they become domestic law.

. . .

An individual may invoke the provisions of a treaty before national courts in automatic incorporation states in the absence of implementing legislation only when its provisions are considered to be self-executing and when he has standing to do so. . . . It suffices at this point to state that, in general, treaty provisions are considered by national courts and administrators as self-executing when they lend themselves to judicial or administrative application without further legislative implementation. . . .

. . .

RUDOLPH BERNHARDT, THE CONVENTION AND DOMESTIC LAW

in R. St. J. Macdonald, F. Matscher and H. Petzold. (eds.), The European System for the Protection of Human Rights 25 (1993), at 26.

. . .

One could imagine an express or implicit rule in national law that the Convention (or treaties in general) is superior to any provision in the domestic law, including the constitution of the State concerned. No such rule seems at present to exist, so we turn to the solutions which are adopted in the different Member States of the Convention.

A. The Convention as part of a Member State's constitution

The Convention can be expressly incorporated into a Member State's constitutional law. This solution has until now been adopted only in Austria, where the Convention formally has the rank of constitutional provisions. This has the consequence that no statute or secondary legislation is valid if it violates conventional norms. . . . It is obvious that even this solution cannot avoid difficult constitutional problems, if contradictions become visible between the Convention (as interpreted by the Convention organs) and other constitutional norms and principles. . . .

B. The Convention as superior to statutory law

In the great majority of Member States of the Council of Europe, treaties in general and the Convention in particular are part of the domestic legal order. To this extent, the Convention may have a higher rank than normal legislation, but below the constitution. Such an intermediate position of the Convention may follow from Article 55 of the French Constitution or from similar provisions in other constitutions. Article 55 provides: '[T]reaties or agreements duly ratified or approved shall, upon their publications, have an authority superior to that of laws. . . .'

If, under such rules, the Convention clearly prevails in conflicts between its requirements and provisions in national legislation, a satisfactory solution has undoubtedly been found, especially if the individual can invoke the Convention before domestic courts and institutions against formal legislative enactments. . . .

C. The Convention as having the rank of statutory law

In several States the Convention (like other treaties) is part of domestic law with the same rank as normal legislation. In these States, the Convention is approved by the legislature and introduced into the internal legal order similarly to normal statutes. . . .

In theory, the Convention has the same rank as other legislation in these States, and the *lex posterior rule* in principle applies, in the sense and with the consequence that the Convention supersedes older statutes, and later statutes prevail over Convention provisions. This theoretical position does not have greater practical importance, since other principles also come into play: the principle *lex specialis derogat leges generales* (under which the treaty rule is considered to be the more specific law), and the widely recognized principle that a State's statutes should be interpreted in harmony with the international obligations it has incurred. Thus, the formal equal rank of general statutes and the Convention leads to the elimination of conventional norms only in the rare cases where the later statute stands in clear and intended contradiction to the Convention. In actual practice, this possibility appears without real significance. The German Federal Constitutional Court has expressly held that priority must be given to the Convention even over statutes that have been enacted later, because it cannot be assumed that the legislature, without clearly stating so, wanted to deviate from Germany's obligations under public international law.

D. The Convention as lacking internal legal validity

Finally, a minority of the Member States of the Council of Europe still follow the principle that domestic law and treaty law are totally distinct legal areas, and treaties are not introduced into the internal legal order except in cases where an enabling law expressly declares a treaty to be part of the law of the country. This principle is still valid for the United Kingdom of Great Britain and Northern Ireland, for the Irish Republic and also for the majority of the Scandinavian States. . . .

For those States in which the Convention remains an inter-State treaty without incorporation into the internal legal order, some legal problems remain important. An initial question is whether States are not obliged to incorporate the Convention into domestic law, and whether States are not in breach of the Convention if they do not declare the Convention to be part of the law of the country. Some writers have in fact taken the position that States are obliged to make the Convention part of domestic law, but this is still a minority view among the scholars of international law. The European Court of Human Rights has repeatedly declared that there is no legal obligation to make the Convention part of domestic law, even if the incorporation of the Convention conforms better to the *telos* of this Convention than the separation of the two legal orders. . . .

There seems to be another encouraging tendency at least in some of the

States which have not incorporated the Convention in their domestic law. National courts in several of these States refer to the Convention and its interpretation by the Convention organs in their decisions, and they try to avoid conflicts between the Convention and national law. . . .

. . .

JORG POLAKIEWICZ AND VALÉRIE JACOB-FOLTZER, THE EUROPEAN HUMAN RIGHTS CONVENTION IN DOMESTIC LAW

12 Hum. Rts. L. J. 65 and 125 (1991), at 65.

. . . [T]he Convention for the Protection of Human Rights and Fundamental Freedoms . . . has gradually acquired the status of a 'constitutional instrument of European public order in the field of human rights'. . . .

It therefore seems appropriate to give a survey of the Convention's impact on the development of domestic law in the member States. Special emphasis has been given to the case-law of the European Court of Human Rights. . . .

. . . [This] study is limited to those 17 countries that formally incorporated the substantive provisions of the Convention in their internal law (i.e. Austria, Belgium, Cyprus, Finland, France, Germany, Greece, Italy, Liechtenstein, Luxembourg, Malta, the Netherlands, Portugal, San Marino, Spain, Switzerland and Turkey). It is obvious that here the influence of Strasbourg case-law is far more important than in those countries where the Convention has only the status of international law (i.e. Denmark, Iceland, Ireland, Norway, Sweden and the United Kingdom). The study will focus on the status of the Convention in domestic law, its rank in the hierarchy of norms, legislative reforms directly or indirectly prompted by judgments of the European Court and reference to Strasbourg case-law by municipal courts. . . .

. . . It is not surprising that in the Netherlands, a monistic country with a constitution giving precedence to self-executing treaty provisions over domestic law, the influence of the Convention in the field of human rights is paramount. . . .

THE NETHERLANDS

1. Status of the Convention in domestic law

The ECHR and First Additional Protocol were signed by the Netherlands on 4 November 1950 and 20 March 1952. After having been approved by the Law of 28 July 1954, both were ratified on 31 August 1954. At the same time the Government recognised the compulsory jurisdiction of the European Court of Human Rights. A declaration recognising the right of individual petition in accordance with Article 25 ECHR was deposited with the Secretary General of the Council of Europe on 5 July 1960. . . . The Netherlands also ratified the 2nd, 3rd, 4th, 5th, 6th and 8th Protocols.

The relationship between international law and domestic law is regulated in a monistic way by the Dutch Constitution of 1983. Art. 93 of the Constitution provides that provisions of treaties and decisions of international organisations, the contents of which may be binding on everyone, shall have this binding effect as from the time of publication. The words 'the contents of which may be binding on everyone' are generally understood to refer to the self-executing character which is required for their application by Dutch Courts. The rights contained in the ECHR are considered self-executing by the courts and are therefore directly applicable.

According to Art. 94 of the Dutch Constitution

> regulations which are in force in the Kingdom of the Netherlands shall not be applied if this application is not in conformity with provisions of treaties or decisions of international organisations which are binding upon everyone.

The Dutch Courts have therefore to give precedence to self-executing treaty provisions over domestic law that is not in conformity therewith, be it antecedent or posterior, statutory or constitutional law. In a recent judgment the Supreme Court even affirmed the priority of the self-executing provisions of the Convention and the Sixth Protocol over conflicting [provisions of other treaties], when it refused to allow an American serviceman to be handed over to the U.S. authorities on the grounds that he would face capital punishment. But the courts have no competence to nullify, repeal or amend the legislation in question. The provision remains in force, but will not be applied.

There is no Dutch case-law on the reopening of a case after a judgment of the European Court finding that the ECHR had been violated during the proceedings leading to the judgment. According to Dutch procedural law, only criminal and administrative cases may be reopened on the grounds that new evidence has come to light since the judgment was given and which could have led to a different finding if it had been known earlier. Given the high rank accorded to international law by the Dutch Constitution, it is not improbable that Dutch Courts may regard an adverse judgment of the European Court as constituting new evidence.

2. Legislation

There are several instances of legislative reform provoked or, at least, influenced by judgments of the European Court of Human Rights.

In the aftermath of the *Marckx* case, the Dutch legislature amended various provisions of the Civil Code in order to eliminate the distinction between legitimate and illegitimate parentage in the law of succession. It is noteworthy that the Law of 27 October 1982 was given retroactive effect from 13 June 1979, the date when the European Court rendered its judgment in the *Marckx* case.

The *Benthem* case, directly concerning the Netherlands, provoked an important reform of Dutch administrative procedure. To meet the Court's requirements with regard to Article 6 § 1 ECHR, the Provisional Act on disputes before the Crown, which provisionally regulates proceedings in matters of litigation where the Crown has thus far been competent, was adopted on 18 June 1987. . . .

3. Case-law

Until about 1980, the attitude of Dutch courts towards human rights treaties in general and towards the ECHR in particular did not reflect the favourable attitude of the Dutch legislature towards international law embodied in the Constitution. In spite of its direct applicability, the ECHR was only treated as a subsidiary source of law. When specific provisions of the Convention were invoked before Dutch courts, these almost always came to the conclusion that no violation had occurred. As Van Dijk has indicated, the courts reached this result by:

– applying a comparable provision of Dutch law and, if necessary, giving it a very broad scope, while ignoring the provision of the Convention, or
– denying the self-executing character of a provision of the Convention, or
– giving to the provision of the Convention a very restrictive scope, which made it not applicable to the case before the court, or
– giving a broad scope to the restriction allowed for in the Convention to solve the non-conformity, or
– interpreting the provision of the Convention and the applicable domestic law in such an 'embracing' way that a conflict between them is avoided.

. . .

This reticent attitude of Dutch courts towards the ECHR has changed quite dramatically during the 1980's. The statistical survey recently given by Van Dijk shows a considerable increase of references to the ECHR. The percentage of cases, however, in which the Supreme Court has found a violation of the Convention remains small (an average of 9%). When confronted with a conflict between a provision of the ECHR and a provision of Dutch law, the Supreme Court tends to circumvent it by giving to the latter an interpretation or scope different from its original meaning and from the anterior legal practice, or by inserting a new principle into Dutch law derived from the treaty provision. . . .

. . .

A last example is the judgment of 1 July 1983. The Insanity Act empowers the public prosecutor in certain cases to prevent a detainee from applying to a court for release from detention. Following the European Court's judgment in the *Winterwerp* case, the Supreme Court declared that the relevant provision was incompatible with Article 5 § 4 ECHR.

There are several examples where the Supreme Court has interpreted provisions of national legislation in the light of the provisions of the ECHR in order to avoid a possible conflict. In a judgment of 18 January 1980, the Supreme Court declared that Article 959 of the Civil Procedure Act concerning appeal against decisions of the local courts in matters of custody over infants, although originally intended to cover only cases of legitimate children, now had to be interpreted as applying equally to illegitimate children. The Court based its conclusions on the interpretation of Article 8 in connection with Article 14 ECHR given by the European Court of Human Rights in the *Marckx* case. The Court quashed the decision of the District Court rendered before the *Marckx* judgment was even delivered. It did not wait for the amendment of the national law in question, which was being prepared by the legislature. . . .

Many judgments concern the right of the accused to be tried within a reasonable time (Article 6 § 1 ECHR). Between 1980 and 1986, the Supreme Court dealt with 120 cases concerning this problem, in which 20 were found to contain a violation of the Convention. . . .

. . .

The other example concerned the conviction of three persons on the basis of evidence given by anonymous witnesses. Only one of the accused filed an application to the European Commission that found a violation of the Convention. The two others asked for their release pending the examination of the case before the European Court. In its judgment, the District Court ordered the release of the two, considering that it was almost certain that the European Court would not accept a conviction on the sole basis of anonymous witnesses. The European Court in fact concluded that there had been a violation of Article 6 § 3 (d) taken together with Article 6 § 1 ECHR.

. . .

CONCLUDING REMARKS

The foregoing survey offers a disparate but on the whole encouraging picture. The interpretation of the Convention given by the European Court of Human Rights has proved to be highly persuasive with regard to national jurisdictions and legislatures. With the exception of some rulings by the Austrian Constitutional Court, the Belgian Court of Cassation and the French *Conseil d'Etat* on the applicability of Article 6 § 1 ECHR to certain proceedings, it has never been openly defied by national courts. This persuasive authority may be attributed not only to the weight of the Court's arguments, but also to the possibility now existing in all member States that domestic court rulings could eventually be successfully challenged in Strasbourg.

. . .

Several factors can be distinguished which determine the impact of Strasbourg case-law in domestic law. Most important among them is certainly the status of the Convention in the hierarchy of internal norms. . . .

A restraining factor is the existence of parallel constitutional safeguards against human rights abuses supplanting those of the Convention. Especially if the existence of such guarantees is coupled with a strong tradition of judicial protection of political and civil rights, as in Italy and the Federal Republic of Germany, influences from Strasbourg are of minor importance. Nevertheless, the German example shows that even under these conditions the practice of the European Court of Human Rights can become more relevant if the national constitutional court is willing to modify its purely domestic view. . . .

. . .

. . . The role of municipal courts in assuring effective protection of the rights and freedoms contained in the Convention can hardly be overestimated. It is primarily before these courts that its guarantees are invoked. The possibility to file an application to the European Commission of Human Rights and to have the case eventually decided by the Court remains a time-consuming exercise which can only be envisaged as a last resort. It is therefore encouraging to see that, even in the absence of a formal procedure that regulates the relationship between the European Court and national jurisdictions, like the one in Article 177 of the EEC Treaty, we are witnessing the beginning of a dialogue between these different jurisdictions. In this respect, the practice of the European Court of Human Rights contributes substantially to the development of a truly European constitutional jurisprudence.

ANDREW CLAPHAM, HUMAN RIGHTS IN THE PRIVATE SPHERE

(1993), at 4.

[Although this book discusses the relevance of several treaties to the United Kingdom's internal legal order, the excerpts below refer only to the European Convention on Human Rights.]

Introduction

. . . [T]he Convention has a unique status in the United Kingdom. Of the twenty-four States which are bound by the Convention, eighteen have the Convention as part of their domestic law. This arises either constitutionally or by legislative enactment. Of the six States which do not have the Convention as part of domestic law, only the United Kingdom has no written constitution.

. . . [I]n the twenty-three States which have the Convention as part of domestic law or have a written constitution, the rights and values found in the Convention are often reproduced in their constitutions. This means that when questions of civil or human rights arise at the national level, they are decided not on the basis and case-law of the Convention, but usually by reference to the States' own constitutional and legal values. . . .

. . . [T]he role of the European Convention on Human Rights in the United Kingdom is in a state of flux. Different judges have different ideas as to its usefulness. . . .

. . . [T]he legal culture in the United Kingdom is very different from that of the vast majority of its European neighbours. Not only do the judges have a very different career structure and training, but also in the United Kingdom there is no tradition of written constitutional values or fundamental human rights with which the legislator may not interfere.

The Relevance of the Convention in the United Kingdom Courts

[The author stresses that the Convention, although not incorporated into domestic law, 'is surprisingly relevant in the domestic courts of the United Kingdom.' He notes several contexts in which the Convention may be relevant, including as an aid to statutory interpretation; as part of the British common law; as a factor to be considered by administrative bodies when exercising their discretion; and as a consequence of the case-law of the European Court of Human Rights. The excerpts below deal with these contexts.]

In 1974 the House of Lords first used the Convention as an aid to statutory interpretation. In *R.* v. *Miah* Lord Reid, who delivered the only opinion, relied on Article 11(2) of the Universal Declaration of Human Rights and Article 7 of the European Convention on Human Rights to demonstrate that it was 'hardly credible that any government department would promote, or that Parliament would pass, retrospective criminal legislation'.

This conclusion stemmed from the general principle that, so far as the language permits, Parliament is presumed to legislate in accordance with international law. Where the rule of international law is straightforward, as it was in this case—States may not create criminal offences retroactively—then the solution is relatively easy. But other human rights are more problematic; their ambit depends largely on how much recognition is given to the individual right by any one court or judge.

. . .

. . . [I]t is suggested that the concept of giving effect to the intentions of Parliament is an unhelpful one. A better justification for using the Convention as an aid to statutory interpretation is that all statutes ought to be interpreted 'so as to be in conformity with international law'. This is not, however, the attitude which was taken by the House of Lords Select Committee on a Bill of Rights in 1978; yet there is some evidence that judges are prepared to take this broad brush approach.

Lord Scarman, after stating that neither the European Convention nor the decision of the European Court of Human Rights in The *Sunday Times* case was part of the law of the United Kingdom, went on to justify his reference to the Convention:

> I do not doubt that, in considering how far we should extend the application of Contempt of Court, we must bear mind the impact of whatever decision we may be minded to make on the international obligations assumed by the United Kingdom under the European Convention. If the issue should ultimately be, as I think in this case it is, a question of legal policy, we must have regard to the Country's international obligation to observe the European Convention as interpreted by the European Court of Human Rights.

This seems to be a more 'honest' approach to the use of the Convention, as no reference is made to the implied intention of the framers of the secondary legislation under consideration. It could, however, be challenged on the grounds that it denies the 'transformation' tradition of English law. This states that treaties ratified by the executive are not part of the law until transformed by Parliament through legislation into domestic laws. It is suggested that this challenge fails. It fails due to the special nature of the European Convention on Human Rights and other relevant human rights instruments. The Convention declares principles; these principles can be legitimately used to interpret statutes where there is evidence of an intention by the legislature to give effect to those principles, either in the statute under consideration or in another statute. . . .

. . .

Before leaving the area of statutory interpretation mention should be made of the situation where there is a perceived clash between a statute and the Convention. Of course it is not clear at what point it is no longer possible to interpret a statute so as to be in conformity with the Convention, so that a judge is obliged to find the two in irreconcilable opposition. In the United Kingdom, when a statute is in opposition to a treaty, the statute must prevail. However, in the context of the European Convention on Human Rights Lord Denning felt able to depart from this orthodoxy. In *Birdie* v. *Secretary of State for Home Affairs* Lord Denning, MR, stated that 'if an Act of Parliament did not conform to the Convention I might be inclined to hold it invalid'; this surprising statement was repudiated in a later case in the same year, when Lord Denning returned to the orthodox view: that treaties do not become part of the law until made so by Parliament, and that 'if an Act of Parliament contained any provisions contrary to the Convention, the Act of Parliament must prevail'. He continued, 'But I hope that no Act ever will be contrary to the Convention. So the problem should not arise.'

The point did arise for Lord Denning in *Taylor* v. *Co-Op. Retail Services*.

After examining the *Case of Young, James and Webster*, decided by the European Court of Human Rights in Strasbourg, he concluded:

> Mr Taylor was subjected to a degree of compulsion which was contrary to the freedom guaranteed by the European Convention on Human Rights. He was dismissed by his employers because he refused to join a 'closed shop'. He cannot recover any compensation from his employers under English law because under the Acts of 1974 and 1976, his dismissal is to be regarded as fair. But those Acts themselves are inconsistent with the freedom guaranteed by the European Convention. The United Kingdom Government is responsible for passing those Acts and should pay him compensation. He can recover it by applying to the European Commission, and thence to the European Court of Human Rights.

. . .

[The following paragraphs consider the Convention as a part of the Common Law.]

. . . [T]he last word in this section has to be from the most recent case of *Derbyshire County Council* v. *Times Newspapers Ltd. and Others*. In this case the Court of Appeal had to define the extent of the Common Law tort of libel. The *Sunday Times* had published two articles questioning the propriety of certain investments made by the Council. The articles had headings such as 'Bizarre Deals of a Council Leader and Media Tycoon' and 'Revealed: Socialist Tycoon's Deals with a Labour Chief'. The judgment concedes that full use should be made of [Convention] Article 10 to resolve an uncertainty in municipal law. The court found that to allow a local authority to sue for libel was not necessary in a democratic society. It was an unjustifiable restriction on freedom of expression. The case really turns on proportionality, as the court considered that the other options open to the local authority—actions for criminal libel or malicious falsehood—might be legitimate restrictions on freedom of expression where the reputation of the local authority might be damaged so as to impair its function for the public good. It was the existence of alternative less intrusive measures which made the possibility of a civil libel action disproportionate and unnecessary.

The case is important because instead of the court using the Convention to bolster conclusions arrived at by other means the Convention was the starting-point for the judicial reasoning. . . .

Moreover the Court made extensive use of the Strasbourg case-law to accommodate the stress which the European Court and Commission of Human Rights have placed on freedom of the press and the legitimate discussion of persons with public functions in the public domain. This case probably represents the high point in the judicial consideration of the Convention as a tool for resolving uncertainty in the Common Law. . . .

. . .

The Relevance of the Strasbourg Proceedings for the United Kingdom Courts

. . .

Even if some of these cases suggest that the courts have been unwilling really to examine the case-law of the Commission and Court of Human Rights this looks set to change. At least in the recent *Derbyshire County Council* case considered [above] the logic and dynamic of the Convention's case-law was fully considered and cited at length by the Court of Appeal. The court weighed whether allowing the Council the right to bring libel proceedings would respond to a 'pressing social need' and was 'necessary in a democratic society' to the extent that was 'proportionate' to the goal: the protection of the 'reputation' of the Council.

Another case concerns the Isle of Man Court of Appeal. In *Teary (Sergeant of Police)* v. *O'Callaghan* this court recognized the significance of the *Tyrer* case, which found that birching as a judicial punishment was degrading punishment contrary to Article 3 of the Convention. O'Callaghan (16 years old) had been sentenced by magistrates to be whipped with the birch (four strokes). Despite the willingness of O'Callaghan to be birched, the Court of Appeal annulled the sentence, but pointed out that the court would take no consideration of the political consequences which might follow from allowing the sentence to stand, such as the United Kingdom being expelled from the Council of Europe, and that the sentence was nevertheless lawful. The decision to annul the sentence was based on the grounds that the courts should have regard to international obligations.

This case shows that the courts, even when faced with a popular legitimate law, such as the law of the Isle of Man on birching, may prefer to follow the Strasbourg lead. A parallel can be drawn between this case and the situation concerning homosexuality in Northern Ireland, where the Court of Human Rights, faced with another popular law (which prohibited sexual relations between men), found a breach of human rights: *Dudgeon* v. *United Kingdom*. It is in these types of situation involving unpopular minority interests that human rights theory is really tested. Both laws were relatively popular in the Isle of Man and Northern Ireland respectively and it was the European Court of Human Rights in Strasbourg that held the laws to violate human rights. It may well be that such a court can bring a detachment to bear on domestic laws that national courts may find hard. It is for this reason that even if the United Kingdom were to adopt the European Convention in the form of a Bill of Rights, the right of individual petition to Strasbourg should still be kept open, so that the European Court of Human Rights has the chance to examine cases arising in the United Kingdom context and give authoritative judgments on the scope of the rights guaranteed by the Convention.

. . .

COMMENT ON AUSTRALIAN RESPONSES TO THE TOONEN CASE

Recall the Toonen case, a communication to the ICCPR Committee that led to the Committee's 'views' reported at p. 545, *supra*. The case involved a law of Tasmania (a component state within the Australian federalism) making certain homosexual acts criminal; the Committee concluded that the statute violated provisions of the ICCPR on privacy. In this comment, we trace the consequences of its opinion for the Australian legal and political order.

The complaint in the Toonen case was sent to the ICCPR Committee on the same day that the First Optional Protocol entered into force for Australia, December 25, 1991. It thus immediately raised issues concerning the relationship between the ICCPR and Australian legal and political processes that had previously received little consideration from the public or media.

In Australia, as in the United Kingdom, ratification of a treaty has no formal effect on the internal legal order, although of course the state is bound internationally. For the Covenant (or part thereof) to become effective as internal law that could be invoked before Australian courts, Parliament would have to enact the Covenant (or the relevant part) as normal legislation.

At the time of ratifying the Covenant in 1980, the Australian Government had asserted that no new legislation was required since existing law was in full conformity with the requirements of the Covenant. As a result, individuals believing that the Covenant was violated by laws or acts of the Federal Government or an Australian state and wishing to litigate the relevant issue were obliged to do so on the basis of independent Australian laws which, in at least some instances, did not directly address the Covenant's provisions. Since Australia has no Bill of Rights and existing privacy legislation did not deal with the issues raised by the *Toonen* case, the ICCPR Committee correctly concluded in its views that there were no available domestic legal remedies.

Another important element of this case concerned the Australian system of federalism. Like its United States counterpart, that system allocates specific powers to the Commonwealth (the Federal level) and reserves the remaining powers to the states. Criminal law, at issue in this case, is a state matter unless it relates to a head of Federal power such as the Constitution's provisions on corporations or commerce. Under the 'external affairs power,' however, the Federal Government, which has exclusive power to enter into treaties, can undertake obligations which then bind the country as a whole, including the states.

It is generally accepted that the subject matter of a treaty need not relate to any enumerated head of Federal power. Australia's ratification of the ICCPR therefore provided the Commonwealth with the possibility of legislating in relation to any matter dealt with in the Covenant, including matters previously considered to be part of the exclusive domain of the states. Cf. *Missouri v. Holland*, p. 745, *infra*. Article 50 of the ICCPR reinforces this result by providing that the Covenant's provisions 'shall extend to all parts of federal States without any limitations or exceptions.'

Another complication related to federalism arose when the Australian Government's written submission to the Committee supported the position of the complainant. Although the submission also provided information about the position of the Tasmanian Government, the latter had no opportunity to present its own case.

The Committee's final views in the case drew a mixed and often heated response within Australia. Although the Federal Government welcomed the outcome, the Tasmanian Government and many of its political allies at the Federal level expressed grave concern. The following sampling of media reaction was tabled in the Federal Parliament:[3]

> Geoffrey Barker, The Age, *A Wrong Move for the Right Reasons*, 21 April 1994
>
> A sceptic might conclude that the Government is prepared to acquiesce in a significant surrender of Australia's sovereign independence to a remote international committee whose membership and methods fall short of what, in other contexts, would be called world's best practice.
>
> Crispin Hull, Canberra Times, *We need a Proper Bill of Rights*, 2 August 1994
>
> On the human-rights matters, it is folly for Australia to permit its citizens to appeal to UN committees half-full of nominees from countries where the rule of tyranny is paramount over the rule of law. They are totally unaccountable to the Australian people.
>
> . . .
>
> John Hyde, The Australian, *Let Public Discretion Decide Our Rights*, 6 May 1994
>
> Governments have no moral right to get their way against Australian public opinion by selectively referring domestic issues to foreign tribunals.
>
> Malcolm Fraser, The Australian *UN Poses biggest threat to our sovereignty*, 17 August 1994
>
> Human rights are as well safeguarded in Australia as in any country I could name. Such rights are obviously much better protected here than in most countries. Yet we give UN committees concerned with human rights the capacity to review and to take action that can lead to the overturning of Australian law.
>
> The Government is doing all this, of course, without telling the Australian people, without having asked their permission. It is not the Government's sovereignty that is being given away, it is the sovereignty that belongs to all Australians acting together.

[3] Commonwealth of Australia, Senate, Daily Hansard, 8 Dec. 1994, pp. 4298–9.

A Senator also tabled a list of the membership of the Human Rights Committee and of the other two UN treaty bodies to which Australians can complain. The Federal Government responded cautiously. It sought the views of the independent Federal Human Rights and Equal Opportunity Commission and the matter was referred to the Senate Legal and Constitutional Legislation Committee for a full report. The resulting *Human Rights (Sexual Conduct) Act 1994* provided only that:

> (1) Sexual conduct involving only consenting adults acting in private is not to be subject, by or under any law . . . , to any arbitrary interference with privacy within the meaning of Article 17 of the [ICCPR].
>
> (2) For the purposes of this section, an adult is a person who is 18 years old or more.

The legislation was criticized in some quarters on several grounds. (a) It did not repeal the offending Tasmanian legislation but rather provided a basis on which any prosecution could be defended. (b) Invalidation of any law regulating sexual conduct depends upon a determination by the courts that it constitutes 'arbitrary interference'. (c) The legislation mischaracterizes the right to privacy by protecting that right only when exercised 'in private'.[4] Gay and lesbian groups continued to challenge the law and policy domestically and vowed to take the matter back to the Human Rights Committee. Many commentators concluded that the adoption of a domestic Bill of Rights or the incorporation of the ICCPR into domestic legislation would be necessary in order to stem a rising tide of complaints submitted to the ICCPR that the Australian courts could not entertain.

QUESTION

'The upshot of the differences between, say, the treaty systems in the U.K. and in The Netherlands is that Dutch citizens have a far easier path toward vindicating their rights under the European Convention on Human Rights—at least until Parliament legislates the Convention as internal law. This seems wrong. The Convention should have equal effects in all states parties, including ease of remedies for victims of violations, and particularly human rights treaties should so provide.' Comment.

[4] Wayne Morgan, Identifying Evil for What It Is: Tasmania, Sexual Perversity and the United Nations, 19 Melb. U. L. Rev 740 (1994); and *idem.*, Protecting Rights or Just Passing the Buck? The Human Rights (Sexual Conduct) Bill 1994, 1 Aust. J. Hum. Rts 409 (1994).

COMMENT ON TREATIES IN THE UNITED STATES

Read the references to 'treaties' in the following provisions of the Constitution: Art. I, Sec. 10; Art. II, Sec. 2; Art. III, Sec. 2; and Art. VI. The term 'treaty' has a special constitutional significance in the United States. The following materials speak of *treaties* in this constitutional sense, as opposed to another form of international agreement (so-called executive agreements) into which the United States enters. The information below about the United States law of treaties complements the Comment on Treaties at p. 30, *supra*, which describes treaties from an international-law rather than national perspective. Like the other materials in Chapter 11, this Comment ignores the distinctive issues about treaties' relations to domestic legal orders that are posed by the federal (component States) characteristic of the United States.

The conclusion of a treaty binding on the United States normally involves three stages. (1) Negotiation of the treaty is usually conducted by an agent of the Executive, although members of the Senate have occasionally been brought into the process at an early stage as observers and advisers. (2) The President submits the treaty to the Senate for the advice and consent required by Art. II, Sec. 2. If the treaty fails to receive the required two-thirds vote of those present, no further action may be taken on it. If it receives that vote, the President may ratify it. (3) Ratification takes place by an exchange of instruments or, in the case of multilateral agreements, by deposit with a designated depositary. The President then proclaims the treaty, making it a matter of public notice and often effective as of that time.

Of course the United States has had to resolve the same issues as other countries about the internal status and effect of treaties. Constitutional decisions have brought reasonably clear answers to some basic questions. For example, treaties that have become part of the internal legal order have the same domestic effect as federal statutes. A treaty thus supersedes earlier inconsistent legislation. Just as a statute can be superseded by a later inconsistent statute, so can a treaty be superseded, although maxims of interpretation encourage a judicial effort to construe the later-in-time statute so as not to violate the treaty. If that effort fails, the legislative rule prevails internally, although as a matter of international law the United States would have broken its obligations to the other treaty party.

Perhaps the most vital question about a treaty effective as domestic law is its status *vis-à-vis* the Constitution. Will a treaty provision—perhaps one requiring a government to ban certain types of 'hate' speech—be given effect internally even if legislation to the same effect that was independent of any treaty commitment would be judged to be unconstitutional? The U.S. Supreme Court considered this question in the following decision.

REID v. COVERT

Supreme Court of the United States, 1957.
354 U.S. 1, 77 S. Ct. 1222.

[Mrs. Covert and Mrs. Smith killed their husbands, who were then performing military service in England and Japan, respectively. They were each tried by courts-martial convened under Article 2(11) of the Uniform Code of Military Justice, which provided:

> The following persons are subject to this code:
>
> (11) Subject to the provisions of any treaty or agreement to which the United States is or may be a party or to any accepted rule of international law, all persons serving with, employed by, or accompanying the armed forces without the continental limits of the United States. . . .

After conviction, each woman sought release on a writ of habeas corpus, which was granted in the case of Mrs. Covert and denied in the case of Mrs. Smith. On direct appeal the Court affirmed Mrs. Covert's case and reversed Mrs. Smith's. There were four opinions. Six members of the Court agreed that civilian dependents of members of the armed forces overseas could not constitutionally be tried by a court-martial in time of peace for capital offenses, even if committed abroad. Justice Black (in an opinion joined by the Chief Justice, Justice Douglas and Justice Brennan) concluded that military trial of civilians was inconsistent with the Constitution—particularly with those provisions of Article III, Section 2 and of the Fifth and Sixth Amendments which assure indictment by grand jury and trial by jury. Justice Frankfurter and Justice Harlan, in concurring opinions, limited their holdings to capital cases. There was a dissenting opinion.

The excerpts below from the opinion of Justice Black refer to 'executive agreements.' For present purposes, they can be considered to be the equivalent of treaties.]

At the time of Mrs. Covert's alleged offense, an executive agreement was in effect between the United States and Great Britain which permitted United States' military courts to exercise exclusive jurisdiction over offenses committed in Great Britain by American servicemen or their dependents.[5] For its

[5] Executive Agreement of July 27, 1942, 57 Stat. 1193. The arrangement now in effect in Great Britain and the other North Atlantic Treaty Organization nations, as well as in Japan, is the NATO Status of Forces Agreement, 4 U.S. Treaties and Other International Agreements 1792, T.I.A.S. 2846, which by its terms gives the foreign nation primary jurisdiction to try dependents accompanying American servicemen for offenses which are violations of the law of both the foreign nation and the United States. Art. VII, §§ 1(b), 3(a). The foreign nation has exclusive criminal jurisdiction over dependents for offenses which only violate its laws, Art. VII, § 2(b). However, the Agreement contains provisions which require that the foreign nations provide procedural safeguards for our nationals tried under the terms of the Agreement in their courts. Art. VII, § 9. Generally, see Note, 70 Harv. L. Rev. 1043.

Apart from those persons subject to the Status of Forces and comparable agreements and certain other restricted classes of Americans, a foreign nation has plenary criminal jurisdiction, of course, over all Americans—tourists, residents, businessmen, government employees and so forth—who commit offenses against its laws within its territory.

part, the United States agreed that these military courts would be willing and able to try and to punish all offenses against the laws of Great Britain by such persons. In all material respects, the same situation existed in Japan when Mrs. Smith killed her husband. Even though a court-martial does not give an accused trial by jury and other Bill of Rights protections, the Government contends that article 2(11) of UCMJ, insofar as it provides for the military trial of dependents accompanying the armed forces in Great Britain and Japan, can be sustained as legislation which is necessary and proper to carry out the United States' obligations under the international agreements made with those countries. The obvious and decisive answer to this, of course, is that no agreement with a foreign nation can confer power on the Congress, or on any other branch of Government, which is free from the restraints of the Constitution.

. . .

. . . There is nothing in [the language of Article VI of the Constitution] which intimates that treaties and laws enacted pursuant to them do not have to comply with the provisions of the Constitution. Nor is there anything in the debates which accompanied the drafting and ratification of the Constitution which even suggests such a result. These debates as well as the history that surrounds the adoption of the treaty provision in Article VI make it clear that the reason treaties were not limited to those made in 'pursuance' of the Constitution was so that agreements made by the United States under the Articles of Confederation, including the important peace treaties which concluded the Revolutionary War, would remain in effect. It would be manifestly contrary to the objectives of those who created the Constitution, as well as those who were responsible for the Bill of Rights—let alone alien to our entire constitutional history and tradition—to construe Article VI as permitting the United States to exercise power under an international agreement without observing constitutional prohibitions. In effect, such construction would permit amendment of that document in a manner not sanctioned by Article V. The prohibitions of the Constitution were designed to apply to all branches of the National Government and they cannot be nullified by the Executive or by the Executive and the Senate combined.

There is nothing new or unique about what we say here. This Court has regularly and uniformly recognized the supremacy of the Constitution over a treaty. For example, in Geofroy v. Riggs, 133 U.S. 258, 267, 10 S.Ct. 295, 297, 33 L.Ed. 642, it declared:

> 'The treaty power, as expressed in the constitution, is in terms unlimited except by those restraints which are found in that instrument against the action of the government or of its departments, and those arising from the nature of the government itself and of that of the States. It would not be contended that it extends so far as to authorize what the constitution forbids, or a change in the character of the government or in that of one of

the States, or a session of any portion of the territory of the latter, without its consent.

This Court has also repeatedly taken the position that an Act of Congress, which must comply with the Constitution, is on a full parity with a treaty, and that when a statute which is subsequent in time is inconsistent with a treaty, the statute to the extent of conflict renders the treaty null. It would be completely anomalous to say that a treaty need not comply with the Constitution when such an agreement can be overridden by a statute that must conform to that instrument.

There is nothing in State of Missouri v. Holland, 252 U.S. 416, 40 S.Ct. 382, 64 L.Ed. 641, which is contrary to the position taken here. There the Court carefully noted that the treaty involved was not inconsistent with any specific provision of the Constitution. The Court was concerned with the Tenth Amendment which reserves to the States or the people all power not delegated to the National Government. To the extent that the United States can validly make treaties, the people and the States have delegated their power to the National Government and the Tenth Amendment is no barrier.

In summary, we conclude that the Constitution in its entirety applied to the trials of Mrs. Smith and Mrs. Covert. Since their court-martial did not meet the requirements of Art. III, § 2, or the Fifth and Sixth Amendments we are compelled to determine if there is anything *within* the Constitution which authorizes the military trial of dependents accompanying the armed forces overseas. . . .

[The opinion concluded that the Constitution did not authorize such trials.]

NOTE

Although *Reid* v. *Covert* involved a capital case, later decisions such as *Kinsella* v. *United States ex rel. Singleton*, 361 U.S. 234, 80 S.Ct. 297 (1960), extended it to non-capital cases as well.

COMMENT ON SELF-EXECUTING TREATIES

A question that frequently arises in the United States, as in the European states earlier examined, is whether a treaty is 'self-executing,' in the sense that it creates rights and obligations for individuals that are enforceable in the courts without legislative implementation of the treaty. The concept of 'self-executing' is close to the concept of 'automatic incorporation' in the excerpts from Virginia Leary, p. 726, *supra*.

Each country here faces distinctive problems. In the United States, the answer to the question posed is bound up in constitutional text and in the allocation of powers over treaties among the Executive Branch, the Senate and the Congress as a whole. For example, note the status of 'supreme law' that is accorded the treaty under Art. VI of the Constitution (the Supremacy Clause), and the relationship of that clause to the self-executing character of treaties.

Consider the following excerpts from Section 111 of the *Restatement (Third), Foreign Relations Law of the United States* (1987):

> (3) Courts in the United States are bound to give effect to international law and to international agreements of the United States, except that a 'non-self-executing' agreement will not be given effect as law in the absence of necessary implementation.
>
> (4) An international agreement of the United States is 'non-self-executing' (a) if the agreement manifests an intention that it shall not become effective as domestic law without the enactment of implementing legislation, (b) if the Senate in giving consent to a treaty, or Congress by resolution, requires implementing legislation, or (c) if implementing legislation is constitutionally required.

Comment (h) to Section 111 provides:

> In the absence of special agreement, it is ordinarily for the United States to provide how it will carry out its international obligations. Accordingly, the intention of the United States determines whether an agreement is to be self-executing in the United States or should await implementation by legislation or by appropriate executive or administrative action. If the international agreement is silent as to its self-executing character and the intention of the United States is unclear, account must be taken of . . . any expression by the Senate or by Congress in dealing with the agreement. . . . Whether an agreement is to be given effect without further legislation is an issue that a court must decide when a party seeks to invoke the agreement as law. . . . Some provisions of an international agreement may be self-executing and others non-self-executing. If an international agreement or one of its provisions is non-self-executing, the United States is under an international obligation to adjust its laws and institutions as may be necessary to give effect to the agreement.

Certain types of treaties have traditionally been understood to be self-executing and have been applied by courts without any implementing legislation. Consider bilateral treaties giving (reciprocally) rights to nationals of each party to establish residence for certain purposes in the territory of the other party, establish corporations, conduct business there, and so on, frequently on national-treatment terms. Courts have long entertained actions by nationals of a treaty party seeking to enforce one or another of the rights provided for in the treaty.

Under U.S. law (as developed through constitutional decisions of the courts), certain types of treaties cannot be self-executing but require implementing legislation to have domestic effects. Note Section 111(4)(c) above of the Restatement. For example, a treaty obligating the United States to make certain conduct criminal and closely defining that conduct and its penalty would nonetheless require such legislation. A treaty obligating the United States to pay funds to another state may require an appropriation of funds by the Congress.

Generally it is not relevant from an international-law perspective whether a treaty is self-executing, since a state is obligated under international law to do whatever may be required under its internal law (such as legislative enactment) to fulfill its treaty commitments.

SEI FUJII v. STATE

Supreme Court of California, 1952.
38 Cal.2d 718, 242 P.2d 617.

[In this litigation, Fujii, a Japanese who was ineligible for citizenship under the United States naturalization laws then in effect, brought an action to determine whether an escheat of certain land that he had purchased had occurred under provisions of the California Alien Land Law. That Law (1 Deering's Gen.Laws, Act 261, as amended in 1945) provided in part:

> §1. All aliens eligible to citizenship under the laws of the United States may acquire, possess, enjoy, use, cultivate, occupy, transfer, transmit and inherit real property, or any interest therein, in this state, and have in whole or in part the beneficial use thereof, in the same manner and to the same extent as citizens of the United States except as otherwise provided by the laws of this state.
>
> §2. All aliens other than those mentioned in section one of this act may acquire, possess, enjoy, use, cultivate, occupy and transfer real property, or any interest therein, in this state, and have in whole or in part the beneficial use thereof, in the manner and to the extent, and for the purposes prescribed by any treaty now existing between the government of the United States and the nation or country of which such alien is a citizen or subject, and not otherwise.
>
> §7. Any real property hereafter acquired in fee in violation of the provisions of this act by any alien mentioned in section 2 of this act, . . . shall escheat as of the date of such acquiring, to, and become and remain the property of the state of California.

. . .

The Superior Court of Los Angeles County concluded that the property purchased by Fujii had escheated to the State. This decision was reversed by the District Court of Appeals, Second District. That court held that the Alien Land Law was unenforceable because contrary to the letter and spirit of the Charter of the United Nations, which as treaty was superior to state law. That decision was reviewed by the California Supreme Court. Excerpts from its opinion by Chief Justice Gibson appear below.]

It is first contended that the land law has been invalidated and superseded by the provisions of the United Nations Charter pledging the member nations to promote the observance of human rights and fundamental freedoms without distinction as to race. Plaintiff relies on statements in the preamble and in Articles 1, 55 and 56 of the Charter, 59 Stat. 1035.

It is not disputed that the charter is a treaty, and our federal Constitution provides that treaties made under the authority of the United States are part of the supreme law of the land and that the judges in every state are bound thereby. U.S. Const., art. VI. A treaty, however, does not automatically supersede local laws which are inconsistent with it unless the treaty provisions are self-executing. In the words of Chief Justice Marshall: A treaty is 'to be regarded in courts of justice as equivalent to an act of the Legislature, whenever it operates of itself, without the aid of any legislative provision. But when the terms of the stipulation import a contract—when either of the parties engages to perform a particular act, the treaty addresses itself to the political, not the judicial department; and the Legislature must execute the contract, before it can become a rule for the court.' Foster v. Neilson, 1829, 2 Pet. 253, 314, 7 L.Ed. 415.

In determining whether a treaty is self-executing courts look to the intent of the signatory parties as manifested by the language of the instrument, and, if the instrument is uncertain, recourse may be had to the circumstances surrounding its execution. . . . In order for a treaty provision to be operative without the aid of implementing legislation and to have the force and effect of a statute, it must appear that the framers of the treaty intended to prescribe a rule that, standing alone, would be enforceable in the courts. . . .

It is clear that the provisions of the preamble and of Article 1 of the charter which are claimed to be in conflict with the alien land law are not self-executing. They state general purposes and objectives of the United Nations Organization and do not purport to impose legal obligations on the individual member nations or to create rights in private persons. It is equally clear that none of the other provisions relied on by plaintiff is self-executing. . . . Although the member nations have obligated themselves to cooperate with the international organization in promoting respect for, and observance of, human rights, it is plain that it was contemplated that future legislative action by the several nations would be required to accomplish the declared objectives, and there is nothing to indicate that these provisions were intended to become rules of law for the courts of this country upon the ratification of the charter.

The language used in Articles 55 and 56 is not the type customarily employed in treaties which have been held to be self-executing and to create rights and duties in individuals. For example, the treaty involved in Clark v. Allen, 331 U.S. 503, 507–508, 67 S.Ct. 1431, 1434, 91 L.Ed. 1633, relating to the rights of a national of one country to inherit real property located in another country, specifically provided that 'such national shall be allowed a term of three years in which to sell the [property] . . . and withdraw the proceeds . . .' free from any discriminatory taxation. See, also, Hauenstein v. Lynham, 100 U.S. 483, 488–490, 25 L.Ed. 628. In Nielsen v. Johnson, 279 U.S. 47, 50, 49 S.Ct. 223, 73 L.Ed. 607, the provision treated as being self-executing was equally definite. There each of the signatory parties agreed that 'no higher or other duties, charges, or taxes of any kind, shall be levied' by one country on removal of property therefrom by citizens of the other country 'that are or shall be payable in each state, upon the same, when removed by a citizen or subject of such state respectively.' In other instances treaty provisions were enforced without implementing legislation where they prescribed in detail the rules governing rights and obligations of individuals or specifically provided that citizens of one nation shall have the same rights while in the other country as are enjoyed by that country's own citizens. . . .

It is significant to note that when the framers of the charter intended to make certain provisions effective without the aid of implementing legislation they employed language which is clear and definite and manifests that intention.[6] In Curran v. City of New York, 191 Misc. 229, 77 N.Y.S.2d 206, 212, these articles were treated as being self-executory.

. . .

The provisions in the charter pledging cooperation in promoting observance of fundamental freedoms lack the mandatory quality and definiteness which would indicate an intent to create justiciable rights in private persons immediately upon ratification. Instead, they are framed as a promise of future action by the member nations. Secretary of State Stettinius, Chairman of the United States delegation at the San Francisco Conference where the charter was drafted, stated in his report to President Truman that Article 56 'pledges the various countries to cooperate with the organization by joint and separate action in the achievement of the economic and social objectives of the organization without infringing upon their right to order their national affairs according to their own best ability, in their own way, and in accordance with their own political and economic institutions and processes.' The same view was repeatedly expressed by delegates of other nations in the debates attending the drafting of article 56.

The humane and enlightened objectives of the United Nations Charter are, of course, entitled to respectful consideration by the courts and Legislatures of every member nation, since that document expresses the universal desire of thinking men for peace and for equality of rights and opportunities. The charter represents a moral commitment of foremost importance, and we must not permit the spirit of our pledge to be compromised or disparaged in either our domestic or foreign affairs. We are satisfied, however, that the charter

[6] The opinion here refers to Charter Article 104 (giving the UN 'such legal capacity as may be necessary for the exercise of its functions' in member states) and Article 105 (giving the UN 'such privileges and immunities as are necessary').

provisions relied on by plaintiff were not intended to supersede existing domestic legislation, and we cannot hold that they operate to invalidate the alien land law. . . .

[The Court then upheld plaintiff's alternative allegation that the Alien Land Law was invalid since violative of the Equal Protection Clause of the Fourteenth Amendment.]

QUESTIONS

1. What advantages do you see in the U.K. system (legislation) and in the U.S. system (self-executing treaties) for giving treaty provisions internal effect? If you were drafting the U.S. Constitution freshly, which of the constitutional arrangements in the prior readings for giving treaties internal effect would you select?

2. What relation do you see between the conception of self-executing treaties in the U.S. and the provision of Article VI of the Constitution that treaties consistent with the Constitution form part of the 'supreme law' of the land?

3. 'The path of self-executing treaty can frustrate fundamental democratic principles. It would be satisfactory if the House of Representatives, the more popular and representative House in Congress, participated in giving consent to ratification, but only the Senate does. If two-thirds of that body will go along with treaty provisions that might bring about deep internal change in U.S. law, the treaty has the force of 'supreme law.' But there has been no full legislative process and debate, and that's not how laws should be made in the U.S.' Comment. Can you give realistic illustrations for the argument made? Are they apt to be common in treaty making?

2. RATIFICATION BY THE UNITED STATES OF THE ICCPR

In comparison with other democratic states, and even with many one-party and authoritarian states that are persistent and cruel violators of basic human rights, the United States has a modest record of ratification of human rights treaties. That comparison cuts both ways. One might say that the U.S. has a lesser commitment to and concern with developing international human rights than do many (say, European and Commonwealth) states of a roughly similar political and economic character. As the world's leading power, its lesser commitment necessarily weakens the human rights movement. *Or*, one might say that the United States does not engage in the hypocrisy of many states in ratifying and then ignoring treaties. If it ratifies, it means to comply, and hence will take a careful look to be certain that full compliance is possible.

One can be certain that neither of these 'pure' explanations captures the complexity of the arguments within the Executive Branch and the Senate about ratification of these treaties. This section examines in detail the debates over ratification of the ICCPR to illustrate that complexity.

As background to the materials on the ICCPR, the introductory readings below describe earlier attitudes within the United States about involvement in the international human rights system, and indicate the reasons why the United States effectively withdrew from participation in human rights treaties in the early 1950s. Recall the description at p. 119, *supra*, of the significant role played by the United States just a few years earlier in helping to launch the International Bill of Rights through the drafting of the Universal Declaration.

LOUIS SOHN AND THOMAS BUERGENTHAL, INTERNATIONAL PROTECTION OF HUMAN RIGHTS (1973) at 961.

1. In 1945, during the Senate Foreign Relations Committee's hearings on the U.N. Charter, a principal Department of State expert on the Charter, Dr. Leo Pasvolsky, was questioned extensively on the relationship between Article 2(7) and the human rights provisions of the Charter. The following is an excerpt from his testimony:

> Senator Millikin. I notice several reiterations of the thought of the Charter that the Organization shall not interfere with domestic affairs of any country. How can you get into these social questions and economic questions without conducting investigations and making inquiries in the various countries?
>
> Mr. Pasvolsky. Senator, the Charter provides that the Assembly shall have the right to initiate or make studies in all of these economic or social fields. . . .
>
> Senator Millikin. Might the activities of the Organization concern themselves with, for example, wage rates and working conditions in different countries?
>
> Mr. Pasvolsky. The question of what matters the Organization would be concerned with would depend upon whether or not they had international repercussions. This Organization is concerned with international problems. International problems may arise out of all sorts of circumstances. . . .
>
> Senator Millikin. Could such an Organization concern itself with various forms of discrimination which countries maintain for themselves, bloc currency, subsidies to merchant marine, and things of that kind?
>
> Mr Pasvolsky. I should think that the Organization would wish to discuss and consider them. It might even make recommendations on any

matters which affect international economic or social relations. The League of Nations did. The International Labor Office has done that. This new Organization being created will be doing a great deal of that. . . .

Senator Millikin. Would the investigation of racial discriminations be within the jurisdiction of this body?

Mr. Pasvolsky. Insofar, I imagine, as the Organization takes over the function of making studies and recommendations on human rights, it may wish to make studies in those fields and make pronouncements.

Senator Vandenberg. At that point I wish you would reemphasize what you read from the Commission Report specifically applying the exemption of domestic matters to the Social and Economic Council.

Mr. Pasvolsky. I will read that paragraph again.

Senator Vandenberg. Yes, please.

Mr. Pasvolsky. (reading):

'The members of Committee 3 of Commission II are in full agreement that nothing contained in chapter IX can be construed as giving authority to the Organization to intervene in the domestic affairs of Member states'. . . .

Senator Millikin. Is there any other international aspect to a labor problem or a racial problem or a religious problem that does not originate domestically? . . .

Mr. Pasvolsky. Well, Senator, I suppose we can say that there is no such thing as an international problem that is not related to national problems, because the word 'international' itself means that there are nations involved. What domestic jurisdiction relates to here, I should say, as it does in all of these matters, is that there are certain matters which are handled internally by nations which do not affect other nations or may not affect other nations. On the other hand, there are certainly many matters handled internally which do affect other nations and which by international law are considered to be of concern to other nations.

Senator Millikin. For example, let me ask you if this would be true. It is conceivable that there are racial questions on the southern shores of the Mediterranean that might have very explosive effects under some circumstances; but they originate locally, do they not, Doctor?

Mr. Pasvolsky. Yes.

Senator Millikin. And because they might have explosive effects, this Organization might concern itself with them; is that correct?

Mr. Pasvolsky. It might, if somebody brings them to the attention of the Organization.

Senator Millikin. And by the same token, am I correct in this, that in any racial matter, any of these matters we are talking about, that originates in one country domestically and that has the possibility of making international trouble, might be subject to the investigation and recommendations of the Organization?

Mr. Pasvolsky. I should think so, because the Organization is created for that.

2. A number of different versions of the proposal to amend the treatymaking power under the U.S. Constitution were considered by the Congress in the

course of the so-called 'Bricker Amendment' debate, which lasted roughly from 1952 to 1957.[7]

. . .

4. It is generally acknowledged that the defeat of the proposed constitutional amendment was due in large measure to the vigorous lobbying by the Eisenhower Administration and its concomitant undertaking, articulated in the above-quoted testimony by Secretary of State John Foster Dulles, not to adhere to human rights treaties. This undertaking was also embodied in a policy statement issued by Mr. Dulles in the form of a letter addressed to Mrs. Oswald B. Lord, the United States Representative on the United Nations Commission on Human Rights. 28 DSB 579-80 (1953); 13 M. M. Whiteman, Digest of International Law 667-68 (Washington, D.C., 1970). This letter read in part as follows:

> In the light of our national, and recently, international experience in the matter of human rights, the opening of a new session of the Commission on Human Rights appears an appropriate occasion for a fresh appraisal of the methods through which we may realize the human rights goals of the United Nations. These goals have a high place in the Charter as drafted at San Francisco and were articulated in greater detail in the Universal Declaration of Human Rights. . . .
>
> Since the establishment of these goals, much time and effort has been expended on the drafting of treaties, that is, Covenants on Human Rights, in which it was sought to frame, in mutually acceptable legal form, the obligations to be assumed by national states in regard to human rights. We have found that such drafts of Covenants as had a reasonable chance of acceptance in some respects established standards lower than those now observed in a number of countries.
>
> While the adoption of the Covenants would not compromise higher standards already in force, it seems wiser to press ahead in the United Nations for the achievement of the standards set forth in the Universal Declaration of Human Rights through ways other than the proposed Covenants on Human Rights. This is particularly important in view of the likelihood that the Covenants will not be as widely accepted by United Nations members as initially anticipated. Nor can we overlook the fact that the areas where human rights are being persistently and flagrantly violated are those where the Covenants would most likely be ignored.
>
> In these circumstances, there is a grave question whether the completion, signing and ratification of the Covenants at this time is the most desirable method of contributing to human betterment particularly in areas of greatest need. Furthermore, experience to date strongly suggests

[7] (Eds.) The Bricker Amendment, a series a proposals for constitutional amendments, would have significantly limited executive power over treaties and correspondingly increased the power of the Senate or the Congress as a whole. Different versions of the amendments would have curtailed the use of self-executing treaties by requiring treaties followed by legislation, and redrawn the boundary line between treaties and executive agreements so as to require larger Senate participation.

> that even if it be assumed that this is a proper area for treaty action, a wider general acceptance of human rights goals must be attained before it seems useful to codify standards of human rights as binding international legal obligations in the Covenants.
>
> With all these considerations in mind, the United States Government asks you to present to the Commission on Human Rights at its forthcoming session a statement of American goals and policies in this field; to point out the need for reexamining the approach of the Human Rights Covenants as the method for furthering at this time the objectives of the Universal Declaration of Human Rights; and to put forward other suggestions of method, based on American experience, for developing throughout the world a human rights conscience which will bring nearer the goals stated in the Charter. . . .
>
> . . . By reason of the considerations referred to above, the United States Government has reached the conclusion that we should not at this time become a party to any multilateral treaty such as those contemplated in the draft Covenants on Human Rights, and that we should now work toward the objectives of the Declaration by other means. While the Commission continues, under the General Assembly's instructions, with the drafting of the Covenants, you are, of course, expected to participate. This would be incumbent on the United States as a loyal Member of the United Nations.

NOTE

In the years following the decision to withdraw from participation in the two major Covenants, the United States did ratify a few human rights treaties, including the Slavery Convention, the Protocol Relating to the Status of Refugees, the Convention on the Political Rights of Women, and the four Geneva Conventions on the laws of war. But it was not until the Carter administration in the late 1970s that a President sought the Senate's consent for ratification of a number of major treaties (including the two Covenants).

In recent years, the record of the United States has substantially improved, for it has become a party not only to the ICCPR but also to the Convention on the Prevention and Punishment of the Crime of Genocide; the Convention against Torture and other Cruel, Inhuman or Degrading Treatment or Punishment; and the International Convention on the Elimination of All Forms of Racial Discrimination. But there has never been sustained debate in the Senate or broader political debate in the country about participation in three major and widely ratified treaties: the Convention on the Elimination of All Forms of Discrimination against Women, the American Convention on Human Rights, or indeed the International Covenant on Economic, Social and Cultural Rights.

The following materials deal with aspects of the ratification process of the International Covenant on Civil and Political Rights. Given the similarities between the provisions of that Covenant and the U.S. tradition of liberal constitutionalism and a Bill of Rights, no opponents of ratification then expressed doubt about the broad consistency between the principles of the Covenant and the United States Constitution. There were statements from civil liberties groups stressing significant if more limited ways in which the U.S., were it to become a party without making numerous legislative and policy changes, would be in violation of several ICCPR provisions.[8]

Debate in the Senate hearings over the ICCPR concentrated on relatively few provisions in and omissions from the Covenant that raised questions for some or many participants in this process, as well as on the question of whether the Covenant was to be self-executing. Throughout the ratification process much attention was given to the use of reservations, declarations and understandings to 'cure' the few problems that emerged. The materials below have the same focus.

COMMENT ON THE INTERNAL EFFECT OF THE ICCPR

One of the recurrent issues before the Executive Branch and the Senate in deciding whether the U.S. should become a party to the International Covenant on Civil and Political Rights was whether the Covenant, or salient parts of it, should be understood to be self-executing. Note that the ICCPR itself makes no reference to its self-executing or non-self-executing character but provides in Article 2(2):

> Where not already provided for by existing legislation or other measures, each State Party to the present Covenant undertakes to take the necessary steps, in accordance with its constitutional processes and with the provisions of the present Covenant, to adopt such legislative or other measures as may be necessary to give effect to the rights recognized in the present Covenant.

Article 2(3) bears out this obligation by stating the further undertaking to 'ensure that any persons whose rights . . . are violated shall have an effective remedy,' and to 'ensure that any person claiming such a remedy shall have his right thereto determined by competent judicial, administrative or legislative authorities . . . and to develop the possibilities of judicial remedy.' How the states fulfill these obligations lies within their discretion; they are not obligated to incorporate the treaty *as such* within their domestic legal order, whether through automatic incorporation (self-executing treaty) or legislative incorporation. Consider the following comments on the ICCPR:

> In practice, differences in the domestic status of the Covenant are substantial. [The author describes the differences among several European

[8] See, e.g., Human Rights Watch and American Civil Liberties Union, Human Rights Violations in the United States: A report on U.S. compliance with the International Covenant on Civil and Political Rights (1993).

states that were described in the materials, pp. 728–38, *supra*, examining the internal effect in those states of the European Human Rights Convention.] In its examination of individual communications and State reports, the [ICCPR Human Rights Committee] has confirmed that the States Parties may implement the Covenant domestically as they see fit. There is, however, a certain tendency to promote the *direct applicability* of the Covenant.[9]

In his book on the ICCPR Committee,[16] McGoldrick describes the consistent attention that the Committee has given in its examination of states' reports to the internal effect of the Covenant.

> States are requested to explain how their respective legal regimes would resolve the problems of conflict between the provisions of the Covenant and those of its Constitution and internal laws including the role of customary laws, traditional institutions, and tribal traditions. Details are sought of the account taken of the Covenant for the purpose of interpreting provisions of domestic legislation and as a standard for the administrative authorities in the exercise of discretionary powers.

COMMENT ON TREATY RESERVATIONS

At the urging of the Executive Branch, the ratification by the United States of the ICCPR was made subject by the Senate to a number of reservations, understandings and declarations. This Comment provides background on relevant aspects of the international law of treaties.

Article 2(1)(d) of the Vienna Convention on the Law of Treaties defines a reservation to mean 'a unilateral statement' made by a State when ratifying a treaty 'whereby it purports to exclude or to modify the legal effect of certain provisions of the treaty in their application to that State.' Article 19 provides that a State ratifying a treaty may make a reservation unless it is 'prohibited by the treaty' or 'is incompatible with the object and purpose of the treaty.' Section 313 of the *Restatement (Third), Foreign Relations Law of the United States* (1987), is to the same effect. Comment (g) to Section 313 refers to the terms *declaration* and *understanding*, both of which appear in the following materials:

> When signing or adhering to an international agreement, a state may make a unilateral declaration that does not purport to be a reservation. Whatever it is called, it constitutes a reservation in fact if it purports to exclude, limit, or modify the state's legal obligation. Sometimes, however, a declaration purports to be an 'understanding,' an interpretation of the agreement in a particular respect. Such an interpretive declaration is not

[9] Manfred Nowak. U.N. Covenant on Civil and Political Rights: CCPR Commentary 54 (1993).

[10] Dominic McGoldrick, The Human Rights Committee 270–71 (1991).

a reservation if it reflects the accepted view of the agreement. But another contracting party may challenge the expressed understanding, treating it as a reservation which it is not prepared to accept.

The issue of the effect of reservations to a multilateral human rights treaty was earlier addressed by the International Court of Justice in its 1951 advisory opinion on *Reservations to the Genocide Convention*,[11] which influenced the Vienna Convention's provisions above. The principal questions put to the I.C.J. by the UN General Assembly was whether a reserving state could be regarded as a party to the Genocide Convention if its reservation was objected to by one or more existing parties but not by others, and, if so, what effect the reservation then had between the reserving state and the accepting or rejecting parties.

In responding to that question[12], the Court addressed the 'traditional concept . . . that no reservation was valid unless it was accepted by all the contracting parties without exception. . . .' In the context of the Genocide Convention, the Court found it 'proper' to take into account circumstances leading to 'a more flexible application of this principle.' It emphasized the universal character and aspiration of multilateral human rights treaties. Widespread ratifications had 'already given rise to greater flexibility in the international practice' concerning them.

After concluding that the Genocide Convention (whose provisions were silent on the issue of reservations) permitted a state to enter a reservation, the Court considered 'what kind of reservations may be made and what kind of objections may be taken to them.' It underscored the special character of the Convention, which was 'manifestly adopted for a purely humanitarian and civilizing purpose.' In such a convention the contracting States do not have any interests of their own; they merely have, one and all, a common interest in the 'accomplishment of those high purposes which are the *raison d'être* of the convention.' In such circumstances, one cannot 'speak of individual advantages or disadvantages to States, or of the maintenance of a perfect contractual balance between rights and duties.' Permitting any one state party that objected to another state's reservation of any type to block adherence to the convention by that other state would frustrate the convention's goal of universal membership.

On the other hand, the Court could not accept the argument that 'any State entitled to become a party to the Genocide Convention may do so while making any reservation it chooses by virtue of its sovereignty.' It followed that 'it is the compatibility of a reservation with the object and purpose of the

[11] International Court of Justice, Reservations to the Convention on the Prevention and Punishment of the Crime of Genocide, Advisory Opinion, 1951 I.C.J. 15.

[12] The Court concluded (1) that a state whose reservation has been objected to by one or more parties but not by others can be regarded as a party to the Convention 'if the reservation is compatible with the object and purpose of the Convention;' and (2) that a state party objecting to a reservation that it views as incompatible with the Convention can consider the reserving state not to be a party.

Convention that must furnish the criterion for the attitude of a State in making the reservation on accession as well as for the appraisal by a State in objecting to the reservation.'

Subsequent to the ratification of the ICCPR by the United States that was made subject to certain reservations, the ICCPR Human Rights Committee adopted its General Comment No. 24 relating to reservations made upon accession to that convenant. The General Comment, which appears at p. 774, *infra*, develops some of the ideas in this Comment.

SENATE HEARINGS ON INTERNATIONAL HUMAN RIGHTS TREATIES

S. Comm. For. Rel., 96th Cong., 1st Sess. (1979).

[In 1977, President Carter signed four human rights treaties on behalf of the United States and soon thereafter submitted them to the Senate for its consent to ratification. The Carter Administration proposed that the Senate adopt a number of reservations, understandings and declarations as part of its consent. The following excerpts from the 1979 Senate hearings on the treaties concern only one of them, the International Covenant on Civil and Political Rights. Some references to the proposals for reservations, understandings and declarations appear in these excerpts. The treaties were never brought to vote in the Senate, and the matter effectively died until the Bush Administration revived in 1991 the question of U.S. participation in the International Covenant on Civil and Political Rights. It submitted the Covenant afresh to the Senate, together with modestly amended proposals for reservations, understandings and declarations that are set forth at p. 767, infra.]

Statement of Charles Yost, Former Ambassador to United Nations

. . .

There are, in my judgment, few failures or omissions on our part which have done more to undermine American credibility internationally than this one. Whenever an American delegate at an international conference, or an American Ambassador making representations on behalf of our Government, raises a question of human rights, as we have in these times many occasions to do, the response public or private, is very likely to be this: If you attach so much importance to human rights, why have you not even ratified the United Nations' conventions and covenants on this subject? Why have you not taken the steps necessary to enable you to sit upon and participate in the work of the United Nations Human Rights [Committee]?

Our refusal to join in the international implementation of the principles we so loudly and frequently proclaim cannot help but give the impression that we do not practice what we preach, that we have something to hide, that we are afraid to allow outsiders even to inquire whether we practice racial discrimination or violate other basic human rights. Yet we constantly take it upon ourselves to denounce the Soviet Union, Cuba, Vietnam, Argentina, Chile, and many other states for violating these rights. We are in most instances quite right to do so, but we seriously undermine our own case when we resist joining in the international endeavor to enforce these rights, which we ourselves had so much to do with launching.

Many are therefore inclined to believe that our whole human rights policy is merely a cold war exercise or a display of self-righteousness directed against governments we dislike. . . .

. . .

Moreover, the powers of the two international bodies applying the Convention on Racial Discrimination and the Covenant on Civil and Political Rights are so strictly limited that it is impossible for them to intrude in the domestic affairs of a party to these instruments without the consent of that party.

. . .

Prepared Statement of Senator Jesse Helms

. . .

. . . Article 17 [of the Universal Declaration of Human Rights] reads: 'Everyone has the right to own property alone as well as in association with others. No one shall be arbitrarily deprived of his property.'

President Truman insisted that the Universal Declaration contain article 17 guaranteeing the individual's right to property, before the United States would support the declaration. It has been the consistent position of the United States since that time—of Presidents Truman, Eisenhower, Kennedy, Johnson, Nixon and Ford—that before the United States would become a party to any implementing treaty of the declaration, the treaty must recognize each person's right to property as a basic human right protected under international law.

On October 5, 1977, President Carter reversed this American position of nearly three decades by signing, on behalf of the United States, the International Covenants on Economic, Social and Cultural Rights and on Civil and Political Rights.

On February 23 of last year, the President transmitted these covenants to the Senate for ratification. Because of the history of these covenants regarding article 17 of the declaration and personal property rights, their ratification by the Senate would for the first time legitimize the unlawful expropriation without compensation or arbitrary seizure of Americans' property overseas.

Furthermore, ratification by the Senate would again for the first time have the United States formally acquiesce to Socialist and Marxist governments' denial of basic individual economic rights. As presently written, these U.N. covenants require the United States to ignore basic constitutional rights of Americans as a matter of international law.

. . .

How is it that we are suddenly asked to abandon a right so central to the Constitution and Bill of Rights? Would we not hesitate to abandon its companion rights central to the existence of a free society? Rights such as freedom of speech, of the press, and of religion?

. . .

Prepared Statement of Robert Owen, Legal Adviser, Department of State

. . .

. . . But because these treaties do concern themselves with the relations between governments and individuals rather than solely with those between States, objections have been raised to them. It is feared by some that these treaties could be used to distort the constitutional legislative standards that shape our federal and our state governments' treatment of individuals within the United States. These criticisms deserve response.

Such objections tend to fall into three categories. First, it is said that the human rights treaties could serve to change our laws as they are, allowing individuals in courts of law to invoke the treaty terms where inconsistent with domestic law or even with the Constitution. The second type of objection is that the treaties could be used to alter the jurisdictional balance between our federal and state institutions. Since these first two objections will be addressed during these hearings by the Department of Justice, I will go into them only briefly. The third type of objection is that the relationship between a government and its citizens is not a proper subject for the treaty-making powers at all, but ought to be left entirely to domestic legislative processes. This last point I shall address in somewhat more detail.

As others have noted, the treaties do diverge from our domestic law in a relatively few instances. Critics fear that this divergence will cause changes in that domestic law outside the normal legislative process, or at least will subject the relations between the government and the individual to conflicting legal standards.

This fear is not well-founded, in our judgment, for two reasons. First, the President has recommended that to each of the four treaties there is appended a declaration that the treaties' substantive provisions are not self-executing.

. . .

. . . This does not mean that vast new implementing legislation is required, as the great majority of the treaty provisions are already implemented in our domestic law. It does mean that further changes in our laws will be brought about only through the normal legislative process. This understanding as to the non-self-executing nature of the substantive provisions of the treaties would not derogate from or diminish in any way our international obligations under the treaties; it touches only upon the role the treaty provisions will play in our domestic law.

A second reason why we need not fear a confusion of standards due to possible conflicts between the treaty provisions and domestic law rests in this Administration's recommended reservations and understandings. In the few instances where it was felt that a provision of the treaties could reasonably be interpreted to diverge from the requirements of our constitution or from federal or state law presently in force, the Administration has suggested that a reservation or understanding be made to that provision. In our view, these reservations do not detract from the object and the purpose of the treaties—that is, to see to it that minimum standards of human rights are observed throughout the world—and they permit us to accept the treaties in a form consonant with our domestic legal requirements.

. . .

The third objection that has been raised is that the subject matter of these treaties lies beyond the scope of the treaty-making power. The text of the Constitution, of course, gives no guidance as to what may or may not be the subject of a treaty. The Supreme Court has said a number of times that '[t]he treaty-making power of the United States is not limited by any express provision of the Constitution, and, though it does not extend "so far as to authorize what the Constitution forbids," it does extend to all proper subjects of negotiation between our government and other nations.' Although the Court has not elaborated upon what a 'proper subject of negotiation' might or might not be, it has come to be commonly accepted that the treaty power extends to any 'matter of international concern.' See *Restatement of Foreign Relations Law*. (§40, comment b at 117 (1965))

Although there have in the past been differences of opinion as to what is and is not a matter of 'international concern,' it seems clear today that no matter how widely or narrowly the boundaries of 'international concern' be drawn, a treaty concerning human rights falls squarely within them.

. . .

. . . The primary objective is the fostering of international commitments to erect and observe a minimum standard of rights for the individual as set forth by the treaties. This standard is met by our domestic system in practice, although not always in precisely the same way that the treaties envision. By ratification, we would commit ourselves to maintain the level of respect we already pay to the human rights of our people; we would commit ourselves not to backslide, and we would be subjecting this commitment and our human rights performance as a whole to international scrutiny.

Our main goal in suggesting the reservations that I have described is thus not to evade the minimum standards imposed by the treaties whenever they touch our system. The rationale behind the reservations is, rather, that we take our international legal obligations seriously, and therefore will commit ourselves to do by treaty only that which is constitutionally and legally permissible within our domestic law. In this respect, it should be noted, we are in good company. . . .

Another reason why the Administration has proposed a number of reservations, understandings and declarations is pragmatic. We believe these treaties to be important and necessary, and we are anxious to secure the advice and consent of the Senate to their ratification. It is our judgment that the prospects for securing that ratification would be significantly and perhaps decisively advanced if it were to be clear that, by adopting these treaties, the United States would not automatically be bringing about changes in its internal law without the legislative concurrence of the federal or state governments.

. . .

Statement of Jack Goldklang, Office of Legal Counsel, Department of Justice

. . .

. . . Because we already have on the books a very complicated legal system of local, State, and Federal law, we felt it would be better not to have an additional body of self-executing law. If we did, the courts then would be left with the job of trying to see whether differences in language which exist between the treaties and our own laws made a legal difference, and whether the rights that we have here in the United States have been changed.

If one reads through these treaties, one finds that in substance we have, by and large, the rights that are required, but one also sees that the language in the treaties is not necessarily the language that we are familiar with and comfortable with. Therefore, I think it is entirely appropriate that we recommend a declaration [that the treaty not be self-executing] and that the Senate adopt a declaration on this matter.

. . .

* * * * *

Senator Pell. Do you think by affixing reservations we may be making an error in that we would be permitting other nations also to affix reservations and reinterpret the covenants according to their own ideologies?

Mr. Owen. The reservations that we have recommended in some cases are absolutely essential in order to avoid conflicts with our own Constitution.

As to the other reservations, if the Senate should decide that they are not necessary, I think the administration would be willing to dispense with them. Then we would be, in effect, bringing about a more rigorous civil rights regime and there would be no possible criticism that we were not fulfilling the treaties as a whole.

. . .

Response by the Department of State to a Memorandum of the Lawyers Committee for International Human Rights

. . .

The provisions of the Covenants do not differ greatly from protections offered under U.S. law. If the United States ratifies these Covenants without reservations in the few instances where they do diverge from U.S. law, the nation assumes an international obligation to adopt legislation to rectify the differences. The Departments of State and Justice recommended reservations in these few instances so as to leave to the federal and state legislatures decisions as to whether such legislative changes should be undertaken. At a time of concern that the Executive Branch excessively encroaches upon legislative prerogatives, it was felt that these issues of domestic policy were best left in the hands of both Houses of Congress and the state legislatures.

. . .

Opposition to Ratification

Senator Pell. Where do you think the opposition has been to the passage of these treaties? Why is it we have had to delay for 20 years or more?

. . .

Mr. Farer. . . . I think that race relations have been one factor, if we will be perfectly frank. A lot of opposition came from representatives of States where law or practice were crudely discriminatory.

But I also think that like most other countries, particularly large countries, we tend to react instinctively with some belligerence to the idea that other countries and peoples can assess for themselves what we are doing, and the idea that they may fault the level of achievement that we have managed to reach.

. . .

Senator Pell. What would you think, Professor Sohn?

Mr. Sohn. I agree with the two other speakers that the fears have been exaggerated and that it is simply part of the general feeling that the United States knows better about various things and therefore should not be subject to other peoples' judgments. It reminds me of what happened in the United Kingdom when they finally ratified the European Convention on Human Rights. The Foreign Minister made a statement in the House of Commons saying of course we are willing to ratify it because nobody can find anything wrong with the British laws on human rights. Well, of course, 2 weeks later all of the cases relating to immigration from Kenya to the United Kingdom by people nominally British citizens and the restrictions on them by immigration authorities immediately were taken to the European Commission. The United Kingdom had to admit that its administrative procedures were not in accordance with the standards of the Convention.

I think on the one hand we always say to everybody else that our standards are higher than those of anyone else; but we will discover, if we are subject to international supervision, that there are some skeletons in our closet and they will be paraded in public, and we do not like that idea.

. . .

Statement of Phyllis Schlafly

. . .

I oppose Senate ratification of these international human rights treaties for the following reasons.

First, the treaties do not give Americans any rights whatsoever. They do not add a minuscule of benefit to the marvelous human rights proclaimed by the Declaration of Independence, guaranteed by the U.S. Constitution, and extended by our Federal and State laws.

Second, the treaties imperil or restrict existing rights of Americans by using treaty law to restrict or reduce U.S. constitutional rights, to change U.S. domestic Federal or State laws, and to upset the balance of power within our unique system of federalism.

Third, the treaties provide no tangible benefit to peoples in other lands and, even if they did, that would not justify sacrificing American rights.

. . .

This covenant sets up a Human Rights [Committee] of 18 members on which the United States would have at most one or perhaps no representative at all. It would have the competence to hear complaints against us, and who knows what they would do.

. . .

Statement of Oscar Garibaldi, Professor, Virginia Law School

. . .

The implementation systems of the Covenants are considerably weaker than the European or inter-American systems. They basically consist in the submission of reports, to be considered by the Human Rights Committee or the ECOSOC, with the addition of a voluntary fact-finding and conciliation procedure in the Covenant on Civil and Political Rights. I am very skeptical about the effectiveness of these procedures, if by this we mean their ability to influence the States Parties' conduct. . . .

In any event, these procedures will engender practices, which will significantly contribute to the interpretation of the treaties. In this sense, we should remember that the composition of the Human Rights Committee follows the ubiquitous principle of equitable distribution, which ensures the presence of members from totalitarian and authoritarian countries. As a result, the practice of the Committee already shows attempts to redefine the language of the Covenant on Civil and Political Rights, dilute its standards, and enforce it selectively.

. . .

Let me now summarize the reasons why I conditionally support . . . the Covenant on Civil and Political Rights only if the U.N. makes a number of reservations—not necessarily those proposed by the Administration—designed to make these treaties compatible with some basic principles of American law and foreign policy. . . . The reasons for my support of [the Covenant and one other of the four submitted treaties] are the following. First, although they are far from perfect, these treaties are largely the philosophical offering of classical liberalism; for the most part, they set limits to governmental power over individuals' lives, they prescribe not what governments should do, but what governments may not do. Second, the language of these treaties—indeed the language of human rights as a whole—is particularly susceptible to semantic infiltration, that is, the redefinition of a term to serve the purposes of totalitarian doublespeak. Because of the high emotional appeal of human rights language and its operational function in international law, we cannot afford to abandon these concepts to the enemies of freedom. But this long-term semantic struggle will be waged in the implementation organs set up by these treaties, which argues for a responsible American voice in these fora. . . .

. . .

QUESTIONS

1. In these excerpts from the Hearings, the Legal Adviser to the State Department noted the common acceptance of the view that the treaty power extended to any 'matter of international concern' (quoting from a Restatement). He argued that, whatever the boundaries of that conception, human rights treaties clearly fall within it. Suppose he were challenged on that assertion. With what arguments could he best support it?

2. Based on the prior readings, how would you identify the principal concerns that a President who believed it important to ratify a given human rights treaty should be aware of when seeking the Senate's consent? Can you think of ways in which the President might seek to alleviate or dispose of those concerns before submitting the treaty to the Senate?

LETTER FROM PRESIDENT BUSH TO SENATE FOREIGN RELATIONS COMMITTEE

Rep. of S. Comm. For. Rel. to Accompany Exec. E, 95–2 (1992), at 25.

August 8, 1991.

DEAR MR. CHAIRMAN: I am writing to urge the Senate to renew its consideration of the International Covenant on Civil and Political Rights with a view to providing advice and consent to ratification.

The end of the Cold War offers great opportunities for the forces of democracy and the rule of law throughout the world. I believe the United States has a special responsibility to assist those in other countries who are now working to make the transition to pluralist democracies. . . .

United States ratification of the Covenant on Civil and Political Rights at this moment in history would underscore our natural commitment to fostering democratic values through international law. The Covenant codifies the essential freedoms people must enjoy in a democratic society, such as the right to vote, freedom of peaceful assembly, equal protection of the law, the right to liberty and security, and freedom of opinion and expression. Subject to a few essential reservations and understandings, it is entirely consonant with the fundamental principles incorporated in our own Bill of Rights, U.S. ratification would also strengthen our ability to influence the development of appropriate human rights principles in the international community and provide an additional and effective tool in our efforts to improve respect for fundamental freedoms in many problem countries around the world.

. . .

PROPOSALS BY BUSH ADMINISTRATION OF RESERVATIONS TO INTERNATIONAL COVENANT ON CIVIL AND POLITICAL RIGHTS

Rep. of S. Comm. For. Rel. to Accompany Exec. E, 95-2 (1992) at 10.

[In 1978, the Carter Administration had proposed a list of reservations, understandings and declarations when it put four human rights treaties, including the ICCPR, to the Senate for its consent to ratification. In 1991, when the Bush Administration revived the ICCPR alone among the four treaties, it submitted a revised list to the Senate Foreign Relations Committee. The following excerpts from this 1991 submission include the Bush Administration's explanation of the reasons for proposing several of the reservations, understandings and declarations.]

GENERAL COMMENTS

. . .

In a few instances, however, it is necessary to subject U.S. ratification to reservations, understandings or declarations in order to ensure that the United States can fulfill its obligations under the Covenant in a manner consistent with the United States Constitution, including instances where the Constitution affords greater rights and liberties to individuals than does the Covenant. Additionally, a few provisions of the Covenant articulate legal rules which differ from U.S. law and which, upon careful consideration, the Administration declines to accept in preference to existing law. Specific proposals dealing with both situations are included below.

FORMAL RESERVATIONS

1. Free Speech (Article 20)

Although Article 19 of the Covenant specifically protects freedom of expression and opinion, Article 20 directly conflicts with the First Amendment by requiring the prohibition of certain forms of speech and expression which are protected under the First Amendment to the U.S. Constitution (i.e., propaganda for war and advocacy of national, racial or religious hatred that constitutes incitement to discrimination, hostility or violence). The United States cannot accept such an obligation.

Accordingly, the following reservation is recommended:

> Article 20 does not authorize or require legislation or other action by the United States that would restrict the right of free speech and association protected by the Constitution and laws of the United States.

. . .

2. Article 6 (capital punishment)

Article 6, paragraph 5 of the Covenant prohibits imposition of the death sentence for crimes committed by persons below 18 years of age and on pregnant women. In 1978, a broad reservation to this article was proposed in order to retain the right to impose capital punishment on any person duly convicted under existing or future laws permitting the imposition of capital punishment. The Administration is now prepared to accept the prohibition against execution of pregnant women. However, in light of the recent reaffirmation of U.S. policy towards capital punishment generally, and in particular the Supreme Court's decisions upholding state laws permitting the death penalty for crimes committed by juveniles aged 16 and 17, the prohibition against imposition of capital punishment for crimes committed by minors is not acceptable. Given the sharply differing view taken by many of our future treaty partners on the issue of the death penalty (including what constitutes 'serious crimes' under Article 6(2)), it is advisable to state our position clearly.

Accordingly, we recommend the following reservation to Article 6:

> The United States reserves the right, subject to its Constitutional constraints, to impose capital punishment on any person (other than a pregnant woman) duly convicted under existing or future laws permitting the imposition of capital punishment, including such punishment for crimes committed by persons below eighteen years of age.

. . .

4. Article 15(1) (post-offense reductions in penalty)

Article 15, paragraph 1, precludes the imposition of a heavier penalty for a criminal offense than was applicable at the time the offense was committed, and requires States Party to comply with any post-offense reductions in penalties: '[i]f, subsequent to the commission of the offense, provision is made by law for the imposition of the lighter penalty, the offender shall benefit thereby.' Current federal law, as well as the law of most states, does not require such relief and in fact contains a contrary presumption that the penalty in force at the time the offense is committed will be imposed, although post-sentence reductions are permitted (see 18 U.S.C. §3582 (c)(2) and the Federal Sentencing Guidelines) and are often granted in practice when there have been subsequent statutory changes. Upon consideration, there is no disposition to require a change in U.S. law to conform to the Covenant.

Accordingly, we recommend a reservation similar to the one proposed in 1978:

> Because U.S. law generally applies to an offender the penalty in force at the time the offense was committed, the United States does not adhere to the third clause of paragraph 1 of Article 15.

. . .

UNDERSTANDINGS

1. Article 2(1), 4(1) and 26 (non-discrimination)

The very broad anti-discrimination provisions contained in the above articles do not precisely comport with long-standing Supreme Court doctrine in the equal protection field. In particular, Articles 2(1) and 26 prohibit discrimination not only on the bases of 'race, colour, sex, language, religion, political or other opinion, national or social origin, property, birth' but also on any 'other status.' Current U.S. civil rights law is not so open-ended: discrimination is only prohibited for specific statuses, and there are exceptions which allow for discrimination. For example, under the Age Discrimination Act of 1975, age may be taken into account in certain circumstances. In addition, U.S. law permits additional distinctions, for example between citizens and non-citizens and between different categories of non-citizens, especially in the context of the immigration laws.

In interpreting the relevant Covenant provisions, the Human Rights Committee has observed that not all differentiation of treatment constitutes discrimination, if the criteria for such differentiation are reasonable and objective and if the aim is to achieve a purpose which is legitimate under the Covenant. In its General Comment on non-discrimination, for example, the Committee noted that the enjoyment of rights and freedoms on an equal footing does not mean identical treatment in every instance.

Notwithstanding the very extensive protections already provided under U.S. law and the Committee's interpretive approach to the issue, we recommend [an understanding that expresses the preceding concerns. Its text is here omitted.]

4. Article 14 (right to counsel, compelled witness, and double jeopardy)

In a few particular aspects, this Article could be read as going beyond existing U.S. domestic law. In particular, current Federal law does not entitle a defendant to counsel of his own choice when he is either indigent or financially able to retain counsel in some form; nor does federal law recognize a right to counsel with respect to offenses for which imprisonment is not imposed. With respect to the compelled attendance and examination of witnesses, a criminal defendant must show that the requested witness is necessary to his defense. Under the Constitution, double jeopardy attaches only to multiple prosecutions by the same sovereign and does not prohibit trial of the same defendant for the same crime in, for example, state and federal courts or in the courts of two states. See *Burton v. Maryland*, 395 U.S. 784 (1969).

To clarify our reading of the Covenant with respect to these issues, we recommend the following understanding, similar to the one proposed in 1978:

> The United States understands that subparagraphs 3(b) and (d) of Article 14 do not require the provision of a criminal defendant's counsel of choice when the defendant is provided with court-appointed counsel on grounds

> of indigence, when the defendant is financially able to retain alternative counsel, or when imprisonment is not imposed. The United States further understands that paragraph 3(e) does not prohibit a requirement that the defendant make a showing that any witness whose attendance he seeks to compel is necessary for his defense. The United States understands the prohibition upon double jeopardy in paragraph 7 to apply only when the judgment of acquittal has been rendered by a court of the same governmental unit, whether the Federal Government or a constituent unit, as is seeking a new trial for the same cause.

DECLARATIONS

1. Non-self-executing Treaty

For reasons of prudence, we recommend including a declaration that the substantive provisions of the Covenant are not self-executing. The intent is to clarify that the Covenant will not create a private cause of action in U.S. courts. As was the case with the Torture Convention, existing U.S. law generally complies with the Covenant; hence, implementing legislation is not contemplated.

We recommend the following declaration, virtually identical to the one proposed in 1978 as well as the one adopted by the Senate with respect to the Torture Convention:

> The United States declares that the provisions of Articles 1 through 27 of the Covenant are not self-executing.

. . .

3. Article 41 (state-to-state complaints)

Under Article 41, States Party to the Covenant may accept the competence of the Human Rights Committee to consider state-to-state complaints by means of a formal declaration to that effect. . . .

Accordingly, we recommend informing the Senate of our intent, subject to its approval, to make an appropriate declaration under Article 41 at the time of ratification, as follows:

> The United States declares that it accepts the competence of the Human Rights Committee to receive and consider communications under Article 41 in which a State Party claims another State Party is not fulfilling its obligations under the Covenant.

. . .

NOTE

The Senate consented to ratification subject to the described (and other) reservations, understandings and declarations. In the Senate debate preceding the approving vote, Senator Moynihan noted that '[o]thers have raised the legitimate concern that the number of reservations in the administration's package might imply to some that the United States does not take the obligations of the covenant seriously.' He stressed how few and selective the reservations were, in the context of the entire covenant, and argued that 'a wholly different interpretation' could be placed on them—namely, as an indication of the seriousness with which the United States approached its new obligations, unlike 'nations of the totalitarian block [that] ratified obligations without reservation—obligations that they had no intention of carrying out.' He observed that 'a Senator might well conclude that it is in the interests of the United States to ratify the covenant with this package of reservations even if that Senator disagrees strongly with a particular domestic practice which has prompted a reservation.' Efforts to change that domestic practice through legislation could continue. 138 Cong. Rec. S4781, April 2, 1992. On ratification, the United States became a party to the Covenant.

* * * * *

With respect to the reservation covering Article 20, note the ambiguous wording of para. 2 of that article, providing that '[a]ny advocacy of national, racial or religious hatred that constitutes incitement to discrimination, hostility or violence shall be prohibited by law.' The ICCPR Committee has not issued a General Comment on Article 20. Manfred Nowak, in his *U.N. Covenant on Civil and Political Rights: CCPR Commentary* (1993), observes (p. 365) that the 'legal formulation of this provision is not entirely clear.' The wording of para. (2)

> literally means that incitement to discrimination without violence must also be prohibited. . . . Particularly inexplicable is the insertion of the word 'discrimination'. . . . It is most difficult to conceive of an advocacy of national, racial or religious hatred that does not simultaneously incite discrimination. . . . Art. 20(2) as well may be sensibly interpreted only in light of its object and purpose, i.e., taking into consideration its *responsive character* with regard to the Nazi racial hatred campaigns. . . . Thus, despite its unclear formulation, States Parties are not obligated by Art. 20(2) to prohibit advocacy of hatred in private circles that instigates non-violent actions of racial or religious discrimination. What the delegates . . . had in mind was to . . . prevent the public incitement of racial hatred and violence within a State or against other States and peoples.

QUESTIONS

1. Recall that a number of NGOs participating in the Senate hearings on ratification of the ICCPR opposed the proposed (and later adopted) declaration to the effect that the substantive provisions of the Covenant would not be self-executing. Note the following comments in a report by the largest domestic civil-liberties organization and the largest U.S.-based international human rights organization in the United States: Human Rights Watch and American Civil Liberties Union, *Human Rights Violations in the United States: A report on U.S. compliance with the International Covenant on Civil and Political Rights* 2 (1993):

> . . . Americans would have been able to enforce the treaty in U.S. courts either if it had been declared to be self-executing or if implementing legislation had been enacted to create causes of action under the treaty. The Bush administration rejected both routes. The result was that ratification became an empty act for Americans: the endorsement of the most important treaty for the protection of civil rights yielded not a single additional enforceable right to citizens and residents of the United States.
>
> We issue this report to demonstrate the inaccuracy of the view that Americans do not need the protection of the ICCPR. As we show, the Bush administration was wrong in its assessment that the United States is already complying with all the treaty's obligations, even after the administration nullified some of the rights through its reservations, declarations and understandings. In the areas of racial and gender discrimination, prison conditions, immigrants rights, language discrimination, the death penalty, police brutality, freedom of expression and religious freedom, we show that the United States is now violating the treaty in important respects. As a result, the Clinton administration is under an immediate legal obligation to remedy these human rights violations at home, through specific steps that we outline.
>
> Moreover, to ensure that these remedies are sufficient, we believe the U.S. government is obligated to grant Americans the right to invoke the protections of the treaty in U.S. courts, at least through specific legislation enabling them to do so, but preferably through a formal declaration that the treaty is self-executing, and thus invocable in U.S. courts without further legislation. . . .

Do you agree with these observations about the need for a self-executing Covenant? What arguments would you make against this position?

2. What remedial path has a U.S. citizen who plausibly claims that the government has violated his rights under the ICCPR, but who lacks any plausible claim under U.S. law? What steps could she realistically take to pressure the U.S. government to accept her position?

3. Ratification by the U.S. of the ICCPR Optional Protocol does not seem to have been discussed. No such proposal was put to the Senate. (a) Why do you suppose this to have been the case? (b) As a member of the State Department, would you have argued for or against joining the Optional Protocol? (c) 'Ratification of the Optional Protocol would have been the correct solution, preferable to making the ICCPR self-executing.' Comment.

NOTE

A number of states parties to the ICCPR objected to one or more of the reservations of the United States. [13] Some illustrations follow.

The governments of several states—including Belgium, Denmark, Finland, France, Germany, Italy, Netherlands, Norway, Portugal, Spain and Sweden—objected to the reservation regarding Article 6, para. 5, prohibiting the imposition of the death sentence for crimes committed by persons below 18 years of age, and found that reservation incompatible with the ICCPR's provisions and with its object and purpose. Most of these states also objected to other reservations (or to understandings), particularly the one relating to Article 7. These states, however, stressed that (to take one illustration) the state's position on the relevant reservations (p. 56) 'does not constitute an obstacle to the entry into force of the Covenant between the Kingdom of Spain and the United States of America.'

In objecting to three reservations and three understandings, Sweden observed (p. 57) that under international treaty law, the name 'assigned to a statement' that excluded or modified the effect of certain treaty provisions

> does not determine its status as a reservation to the treaty. Thus, the Government considers that some of the understandings made by the United States in substance constitute reservations to the Covenant.
>
> A reservation by which a State modifies or excludes the application of the most fundamental provisions of the Covenant, or limits its responsibilities under that treaty by invoking general principles of national law, may cast doubts upon the commitment of the reserving State to the object and purpose of the Covenant. The reservations made by the United States of America include both reservations to essential and non-derogable provisions, and general references to national legislation. Reservations of this nature contribute to undermining the basis of international treaty law. All States parties share a common interest in the respect for the object and purpose of the treaty to which they have chosen to become parties.

[13] The objections are set forth in alphabetical order of the states involved, starting at p. 47 in CCPR/C/2/Rev. 4, 24 August 1994, entitled 'Reservations, Declarations, Notifications and Objections Relating to the International Covenant on Civil and Political Rights and the Optional Protocols Thereto.'

GENERAL COMMENT NO. 24 OF HUMAN RIGHTS COMMITTEE

CCPR/C/21/Rev.1/Add.6, 2 Nov. 1994.

[At its 52nd session in 1994, the ICCPR Committee adopted General Comment No. 24 entitled: 'General comment on issues relating to reservations made upon ratification or accession to the Covenant or the Optional Protocols thereto, or in relation to declarations under article 41 of the Covenant.' (Earlier General Comments (GCs) of the Committee appear at p. 522, *supra*). This GC was adopted after the ratification of the ICCPR by the United States described above, and preceded the Committee's consideration of the first periodic report submitted by the U.S. in 1995. It refers to the judicial decision and to the provisions of the Vienna Convention on the Law of Treaties described in the Comment on Treaty Reservations, p. 756, supra.

The GC notes that as of its date, 46 of the 127 states parties to the ICCPR had entered a total of 150 reservations, ranging from exclusion of the duty to provide particular rights, to insistence on the 'paramountcy of certain domestic legal provisions' and to limitation of the competence of the Committee. Those reservations 'tend to weaken respect' for obligations and 'may undermine the effective implementation of the Covenant.' The Committee felt compelled to act, partly under the necessity of clarifying for states parties just what obligations had been undertaken, a clarification that would require the Committee to determine 'the acceptability and effects' of a reservation or unilateral declaration.

The GC notes that the ICCPR itself makes no reference to reservations (as is true also for the First Optional Protocol; the Second Optional Protocol limits reservations), and that the matter of reservations is governed by international law. It finds in Article 19(3) of the Vienna Convention on the Law of Treaties 'relevant guidance.' Therefore, that article's 'object and purpose test . . . governs the matter of interpretation and acceptability of reservations.' The GC continues:]

8. Reservations that offend peremptory norms would not be compatible with the object and purpose of the Covenant. Although treaties that are mere exchanges of obligations between States allow them to reserve *inter se* application of rules of general international law, it is otherwise in human rights treaties, which are for the benefit of persons within their jurisdiction. Accordingly, provisions in the Covenant that represent customary international law (and *a fortiori* when they have the character of peremptory norms) may not be the subject of reservations. Accordingly, a State may not reserve the right to engage in slavery, to torture, to subject persons to cruel, inhuman or degrading treatment or punishment, to arbitrarily deprive persons of their lives, to arbitrarily arrest and detain persons, to deny freedom of thought, conscience and religion, to presume a person guilty unless he proves his innocence, to execute pregnant women or children, to permit the advocacy of

national, racial or religious hatred, to deny to persons of marriageable age the right to marry, or to deny to minorities the right to enjoy their own culture, profess their own religion, or use their own language. And while reservations to particular clauses of Article 14 may be acceptable, a general reservation to the right to a fair trial would not be.

9. Applying more generally the object and purpose test to the Covenant, the Committee notes that, for example, reservation to article 1 denying peoples the right to determine their own political status and to pursue their economic, social and cultural development, would be incompatible with the object and purpose of the Covenant. Equally, a reservation to the obligation to respect and ensure the rights, and to do so on a non-discriminatory basis (Article 2(1)) would not be acceptable. Nor may a State reserve an entitlement not to take the necessary steps at the domestic level to give effect to the rights of the Covenant (Article 2(2)).

10. The Committee has further examined whether categories of reservations may offend the 'object and purpose' test. In particular, it falls for consideration as to whether reservations to the non-derogable provisions of the Covenant are compatible with its object and purpose. While there is no hierarchy of importance of rights under the Covenant, the operation of certain rights may not be suspended, even in times of national emergency. This underlines the great importance of non-derogable rights. But not all rights of profound importance, such as articles 9 and 27 of the Covenant, have in fact been made non-derogable. One reason for certain rights being made non-derogable is because their suspension is irrelevant to the legitimate control of the state of national emergency (for example, no imprisonment for debt, in article 11). Another reason is that derogation may indeed be impossible (as, for example, freedom of conscience). At the same time, some provisions are non-derogable exactly because without them there would be no rule of law. . . .

11. The Covenant consists not just of the specified rights, but of important supportive guarantees. These guarantees provide the necessary framework for securing the rights in the Covenant and are thus essential to its object and purpose. Some operate at the national level and some at the international level. Reservations designed to remove these guarantees are thus not acceptable. Thus, a State could not make a reservation to article 2, paragraph 3, of the Covenant, indicating that it intends to provide no remedies for human rights violations. Guarantees such as these are an integral part of the structure of the Covenant and underpin its efficacy. . . .

The Committee's role under the Covenant, whether under article 40 or under the Optional Protocols, necessarily entails interpreting the provisions of the Covenant and the development of a jurisprudence. Accordingly, a reservation that rejects the Committee's competence to interpret the requirements of any provisions of the Covenant would also be contrary to the object and purpose of that treaty.

12. . . . Domestic laws may need to be altered properly to reflect the requirements of the Covenant; and mechanisms at the domestic level will be needed

to allow the Covenant rights to be enforceable at the local level. Reservations often reveal a tendency of States not to want to change a particular law. And sometimes that tendency is elevated to a general policy. Of particular concern are widely formulated reservations which essentially render ineffective all Covenant rights which would require any change in national law to ensure compliance with Covenant obligations. No real international rights or obligations have thus been accepted. And when there is an absence of provisions to ensure that Covenant rights may be sued on in domestic courts, and, further, a failure to allow individual complaints to be brought to the Committee under the first Optional Protocol, all the essential elements of the Covenant guarantees have been removed.

. . .

17. . . . [Human rights] treaties, and the Covenant specifically, are not a web of inter-State exchanges of mutual obligations. . . . Because the operation of the classic rules on reservations is so inadequate for the Covenant, States have often not seen any legal interest in or need to object to reservations. The absence of protest by States cannot imply that a reservation is either compatible or incompatible with the object and purpose of the Covenant. . . .

18. It necessarily falls to the Committee to determine whether a specific reservation is compatible with the object and purpose of the Covenant. . . . Because of the special character of a human rights treaty, the compatibility of a reservation with the object and purpose of the Covenant must be established objectively, by reference to legal principles, and the Committee is particularly well placed to perform this task. The normal consequence of an unacceptable reservation is not that the Covenant will not be in effect at all for a reserving party. Rather, such a reservation will generally be severable, in the sense that the Covenant will be operative for the reserving party without benefit of the reservation.

19. Reservations must be specific . . . . States should not enter so many reservations that they are in effect accepting a limited number of human rights obligations, and not the Covenant as such. So that reservations do not lead to a perpetual non-attainment of international human rights standards, reservations should not systematically reduce the obligations undertaken only to the presently existing in less demanding standards of domestic law. Nor should interpretative declarations or reservations seek to remove an autonomous meaning to Covenant obligations, by pronouncing them to be identical, or to be accepted only insofar as they are identical, with existing provisions of domestic law.

20. . . . It is desirable for a State entering a reservation to indicate in precise terms the domestic legislation or practices which it believes to be incompatible with the Covenant obligation reserved; and to explain the time period it requires to render its own laws and practices compatible with the Covenant, or why it is unable to render its own laws and practices compatible with the Covenant. . . .

QUESTIONS

1. 'The sequence from the *Reservations to the Genocide Convention* decision of the I.C.J. (p. 757, *supra*) to General Comment No. 24 is less surprising than it is inevitable. The judicial decision relaxes the relevant international law rules. All are welcome to the human rights treaties, so make entrance as easy as possible. What followed was foreseeable, including the reaction of the ICCPR Committee.' Comment.

2. Could General Comment No. 24 stand as an interpretation and elaboration of the Vienna Convention, covering all types of multilateral treaties, or do you believe that it has a special relevance to (a) the ICCPR, or (b) all human rights treaties in general?

3. What is the status of the U.S. ratification of the ICCPR from the perspective of General Comment No. 24? If the General Comment had preceded that ratification, should the U.S. now be considered a state party? What other factors might be relevant to that determination?

4. Why was there near unanimity among the states parties to the ICCPR that objected to the reservations by the United States about the reservation concerning Article 6, para. 5 (death sentence)? Did that reservation raise a special problem under the Covenant?

5. In the light of General Comment No. 24, if you were a Senator committed to U.S. ratification of the major human rights instruments, would you have voted for *any* reservation? Did any one of the reservations have a special justification?

COMMENT ON U.S. RATIFICATION OF THE CONVENTION ON ELIMINATION OF RACIAL DISCRIMINATION

In 1994, the United States ratified the International Convention on the Elimination of All Forms of Racial Discrimination (CERD) subject to three reservations, one understanding, and one declaration (stating that the Convention is not self-executing).

One reservation was addressed to Article 4 of CERD, which is the equivalent of Article 20 of the ICCPR. States Parties 'condemn all propaganda . . . based on ideas or theories of superiority of one race or group of persons of one colour or ethnic origin.' They undertake to declare a punishable offence 'all dissemination of ideas based on racial superiority or hatred, incitement to racial discrimination, as well as all acts of violence' against such a race or group. Article 1 defines 'racial discrimination' to mean any distinction based on 'race, colour, descent, or national or ethnic origin' that has the purpose or effect of impairing equal enjoyment of rights 'in the political, economic, social, cultural or any other field of public life.'

Another reservation concerned the extent to which CERD reaches beyond state discrimination to prohibit certain private or non-governmental conduct. Recall Article 1's definition of 'racial discrimination' and consider the following provisions:

> *Article 2. Para. 1:* 'States Parties condemn racial discrimination and undertake to pursue by all appropriate means . . . a policy of eliminating racial discrimination in all its forms . . . and, to this end: . . . (d) Each State Party shall prohibit and bring to an end, by all appropriate means, including legislation . . . , racial discrimination by any persons, group or organization; . . .'
>
> *Article 5:* States Parties undertake to guarantee the right of everyone to the enjoyment of stated rights, including '(e) Economic and social rights, in particular: . . . (iii) the right to housing; (f) The right of access to any place or service intended for use by the general public, such as transport, hotels, restaurants, cafes, theatres and parks.'

The reservation noted that the U.S. Constitution and laws reached 'significant areas of non-governmental activity' in protecting against discrimination. It continued:

> Individual privacy and freedom from governmental interference in private conduct, however, are also recognized as among the fundamental values which shape our free and democratic society. The United States understands that the identification of the rights protected under the Convention by reference in Article 1 to fields of 'public life' reflects a similar distinction between spheres of public conduct that are customarily the subject of governmental regulation, and spheres of private conduct that are not. To the extent, however, that the Convention calls for a broader regulation of private conduct, the United States does not accept any obligation under this Convention to enact legislation or take other measures under [designated provisions, including paragraph (1) and subparagraph 1(d) of Article 2, and Article 5] with respect to private conduct except as mandated by the Constitution and laws of the United States.[14]

QUESTIONS

1. What is your view of what the United States in its reservation sees as a possible contradiction: (a) the definition of discrimination in Article 1 which appears to refer only to fields of 'public life,' and (b) the provisions of Articles 2 and 5? What are the different meanings that you might attribute to the term 'public'?

[14] 140 Cong. Rec. S 7643, 103rd Cong. 2nd Sess., June 7, 1994.

2. In what respects could Articles 2 and 5 be read to go beyond existing U.S. regulation of discrimination by private (i.e. non-governmental) actors? Can you give concrete illustrations?

3. Why would you have supported or opposed this particular reservation?

C JUDICIAL ENFORCEMENT OF CUSTOMARY NORMS: THE ALIEN TORT STATUTE

The following decisions applying the Alien Tort Statute bridge in some sense Chapters 11 and 12. They fit within this chapter's scheme because the judiciary in the United States absorbs international human rights norms—unlike the prior sections, customary rather than treaty norms—into the domestic legal system. On the other hand, the decisions anticipate Chapter 12 because they involve civil claims based on alleged human rights violations that occurred outside the United States. In some sense, the United States, through its judiciary, here seeks to enforce human rights in foreign countries by affording remedies to foreign victims of violations committed by officials in those countries.

The decisions return to themes about customary international law that were first explored in Chapter 2. Recall *The Paquete Habana*, p. 60, *supra*, and its illustration of earlier, traditional approaches to identifying international customary norms. Compare the approach in the *Filartiga* decision that follows, in terms of (i) the court's method and (ii) the kinds of legal and other materials drawn on by the court as it decides whether a given norm has been established at customary international law and thus forms part of the law of the United States.

FILARTIGA v. PENA-IRALA

United States Court of Appeals, Second Circuit, 1980.
630 F.2d 876.

Irving R. Kaufman, Circuit Judge

. . .

Implementing the constitutional mandate for national control over foreign relations, the First Congress established original district court jurisdiction over 'all causes where an alien sues for a tort only [committed] in violation of the law of nations.' Judiciary Act of 1789, ch. 20, §9(b), 1 Stat. 73, 77 (1789), *codified at* 28 U.S.C. §1350. Construing this rarely-invoked provision, we hold that deliberate torture perpetrated under color of official authority violates universally accepted norms of the international law of human rights, regardless of the nationality of the parties. Thus, whenever an alleged torturer is found and served with process by an alien within our borders, § 1350 provides federal jurisdiction. Accordingly, we reverse the judgment of the district court dismissing the complaint for want of federal jurisdiction.

I

The appellants, plaintiffs below, are citizens of the Republic of Paraguay. Dr. Joel Filartiga, a physician, describes himself as a longstanding opponent of the government of President Alfredo Stroessner, which has held power in Paraguay since 1954. His daughter, Dolly Filartiga, arrived in the United States in 1978 under a visitor's visa, and has since applied for permanent political asylum. The Filartigas brought this action in the Eastern District of New York against Americao Norberto Pena-Irala (Pena), also a citizen of Paraguay, for wrongfully causing the death of Dr. Filartiga's seventeen-year old son, Joelito. Because the district court dismissed the action for want of subject matter jurisdiction, we must accept as true the allegations contained in the Filartigas' complaint and affidavits for purposes of this appeal.

The appellants contend that on March 29, 1976, Joelito Filartiga was kidnapped and tortured to death by Pena, who was then Inspector General of Police in Asuncion, Paraguay. . . . The Filartigas claim that Joelito was tortured and killed in retaliation for his father's political activities and beliefs.

Shortly thereafter, Dr. Filartiga commenced a criminal action in the Paraguayan courts against Pena and the police for the murder of his son. As a result, Dr. Filartiga's attorney was arrested and brought to police headquarters where, shackled to a wall, Pena threatened him with death. This attorney, it is alleged, has since been disbarred without just cause.

. . .

In July of 1978, Pena sold his house in Paraguay and entered the United States under a visitor's visa. He was accompanied by Juana Bautista Fernandez Villalba, who had lived with him in Paraguay. The couple remained in the United States beyond the term of their visas, and were living in Brooklyn, New York, when Dolly Filartiga, who was then living in Washington, D.C., learned of their presence. Acting on information provided by Dolly the Immigration and Naturalization Service arrested Pena and his companion, both of whom were subsequently ordered deported on April 5, 1979 following a hearing. They had then resided in the United States for more than nine months.

Almost immediately, Dolly caused Pena to be served with a summons and civil complaint at the Brooklyn Navy Yard, where he was being held pending deportation. The complaint alleged that Pena had wrongfully caused Joelito's death by torture and sought compensatory and punitive damages of $10,000,000. The Filartigas also sought to enjoin Pena's deportation to ensure his availability for testimony at trial. The cause of action is stated as arising under 'wrongful death statutes; the U.N. Charter; the Universal Declaration on Human Rights; the U.N. Declaration Against Torture; the American Declaration of the Rights and Duties of Man; and other pertinent declarations, documents and practices constituting the customary international law of human rights and the law of nations,' as well as 28 U.S.C. §1350, Article II, sec. 2 and the Supremacy Clause of the U.S. Constitution. Jurisdiction is claimed under the general federal question provision, 28 U.S.C. §331 and, principally on this appeal, under the Alien Tort Statute, 28 U.S.C. §1350.

Judge Nickerson stayed the order of deportation, and Pena immediately moved to dismiss the complaint on the grounds that subject matter jurisdiction was absent and for *forum non conveniens.* On the jurisdictional issue, there has been no suggestion that Pena claims diplomatic immunity from suit. . . . Pena, in support of his motion to dismiss on the ground of *forum non conveniens,* submitted the affidavit of his Paraguayan counsel, Jose Emilio Gorostiaga, who averred that Paraguayan law provides a full and adequate civil remedy for the wrong alleged. Dr. Filartiga has not commenced such an action, however, believing that further resort to the courts of his own country would be futile.

Judge Nickerson heard argument on the motion to dismiss on May 14, 1979, and on May 15 dismissed the complaint on jurisdictional grounds.[15]

. . .

The district court continued the stay of deportation for forty-eight hours while appellants applied for further stays. These applications were denied by a panel of this Court on May 22, 1979, and by the Supreme Court two days later. Shortly thereafter, Pena and his companion returned to Paraguay.

II

Appellants rest their principal argument in support of federal jurisdiction upon the Alien Tort Statute, 28 U.S.C. §1350, which provides: 'The district courts shall have original jurisdiction of any civil action by an alien for a tort only, committed in violation of the law of nations or a treaty of the United States.' Since appellants do not contend that their action arises directly under a treaty of the United States, a threshold question on the jurisdictional issue is whether the conduct alleged violates the law of nations. In light of the universal condemnation of torture in numerous international agreements, and the renunciation of torture as an instrument of official policy by virtually all of the nations of the world (in principle if not in practice), we find that an act of torture committed by a state official against one held in detention violates

[15] The court below accordingly did not consider the motion to dismiss on *forum non conveniens* grounds, which is not before us on this appeal.

established norms of the international law of human rights, and hence the law of nations.

The Supreme Court has enumerated the appropriate sources of international law. The law of nations 'may be ascertained by consulting the works of jurists, writing professedly on public law; or by the general usage and practice of nations; or by judicial decisions recognizing and enforcing that law.' United States v. Smith, 18 U.S. (5 Wheat.) 153, 160–61, 5 L.Ed. 57 (1820); Lopes v. Reederei Richard Schroder, 225 F.Supp. 292, 295 (E.D.Pa.1963). In *Smith*, a statute proscribing 'the crime of piracy [on the high seas] as defined by the law of nations,' 3 Stat. 510(a) (1819), was held sufficiently determinate in meaning to afford the basis for a death sentence. The *Smith* Court discovered among the works of Lord Bacon, Grotius, Bochard and other commentators a genuine consensus that rendered the crime 'sufficiently and constitutionally defined.' *Smith*, supra, 18 U.S. (5 Wheat.) at 162, 5 L.Ed. 57.

[The Court then discussed *The Paquete Habana*, p. 60 *supra*, stressing its lesson that 'courts must interpret international law not as it was in 1789, but as it has evolved and exists among the nations of the world today.']

The requirement that a rule command the 'general assent of civilized nations' to become binding upon them all is a stringent one. Were this not so, the courts of one nation might feel free to impose idiosyncratic legal rules upon others, in the name of applying international law. Thus, in Banco Nacional de Cuba v. Sabbatino, 376 U.S. 398, 84 S.Ct. 923, 11 L.Ed.2d 804 (1964), the Court declined to pass on the validity of the Cuban government's expropriation of a foreign-owned corporation's assets, noting the sharply conflicting views on the issue propounded by the capital-exporting, capital-importing, socialist and capitalist nations. Id. at 428-30, 84 S.Ct. at 940–41.

The case at bar presents us with a situation diametrically opposed to the conflicted state of law that confronted the *Sabbatino* Court. Indeed, to paraphrase that Court's statement, id. at 428, 84 S.Ct. at 940, there are few, if any, issues in international law today on which opinion seems to be so united as the limitations on a state's power to torture persons held in its custody.

The United Nations Charter (a treaty of the United States, see 59 Stat. 1033 (1945)) makes it clear that in this modern age a state's treatment of its own citizens is a matter of international concern. It provides:

> With a view to the creation of conditions of stability and well-being which are necessary for peaceful and friendly relations among nations . . . the United Nations shall promote . . . universal respect for, and observance of, human rights and fundamental freedoms for all without distinctions as to race, sex, language or religion.

Id. Art. 55. And further:

> All members pledge themselves to take joint and separate action in cooperation with the Organization for the achievement of the purposes set forth in Article 55.

Id. Art. 56.

While this broad mandate has been held not to be wholly self-executing, Hitai v. Immigration and Naturalization Service, 343 F.2d 466, 468 (2d Cir.1965), this observation alone does not end our inquiry. For although there is no universal agreement as to the precise extent of the 'human rights and fundamental freedoms' guaranteed to all by the Charter, there is at present no dissent from the view that the guaranties include, at a bare minimum, the right to be free from torture. This prohibition has become part of customary international law, as evidenced and defined by the Universal Declaration of Human Rights, General Assembly Resolution 217 (III)(A) (Dec. 10, 1948) which states, in the plainest of terms, 'no one shall be subjected to torture.' The General Assembly has declared that the Charter precepts embodied in this Universal Declaration 'constitute basic principles of international law.' G.A.Res. 2625 (XXV) (Oct. 24, 1970).

Particularly relevant is the Declaration on the Protection of All Persons from Being Subjected to Torture, General Assembly Resolution 3452, 30 U.N. GAOR Supp. (No. 34) 91, U.N.Doc. A/1034 (1975). . . . The Declaration goes on to provide that '[w]here it is proved that an act of torture or other cruel, inhuman or degrading treatment or punishment has been committed by or at the instigation of a public official, the victim shall be afforded redress and compensation, in accordance with national law.' This Declaration, like the Declaration of Human Rights before it, was adopted without dissent by the General Assembly. . . .

. . . [A] U.N. Declaration is, according to one authoritative definition, 'a formal and solemn instrument, suitable for rare occasions when principles of great and lasting importance are being enunciated.' 34 U.N. ESCOR, Supp. (No. 8) 15, U.N. Doc. E/cn.4/1/610 (1962) (memorandum of Office of Legal Affairs, U.N. Secretariat). Accordingly, it has been observed that the Universal Declaration of Human Rights 'no longer fits into the dichotomy of "binding treaty" against "nonbinding pronouncement," but is rather an authoritative statement of the international community.' *E. Schwelb, Human Rights and the International Community* 70 (1964). Thus, a Declaration creates an expectation of adherence, and 'insofar as the expectation is gradually justified by State practice, a declaration may by custom become recognized as laying down rules binding upon the States.' 34 U.N. ESCOR. supra. Indeed, several commentators have concluded that the Universal Declaration has become, *in toto*, a part of binding, customary international law. . . .

. . . The international consensus surrounding torture has found expression in numerous international treaties and accords. . . . The substance of these international agreements is reflected in modern municipal—i.e. national—law as well. Although torture was once a routine concomitant of criminal interrogations in many nations, during the modern and hopefully more enlightened era it has been universally renounced. According to one survey, torture is prohibited, expressly or implicitly, by the constitutions of over fifty-five nations, including both the United States and Paraguay. . . . We have been directed to no assertion by any contemporary state of a right to torture its own or another nation's citizens. Indeed, United States diplomatic contacts confirm the universal abhorrence with which torture is viewed:

> In exchanges between United States embassies and all foreign states with which the United States maintains relations, it has been the Department of State's general experience that no government has asserted a right to torture its own nationals. Where reports of torture elicit some credence, a state usually responds by denial or, less frequently, by asserting that the conduct was unauthorized or constituted rough treatment short of torture.[16]

Memorandum of the United States as *Amicus Curiae* at 16 n. 34.

Having examined the sources from which customary international law is derived—the usage of nations, judicial opinions and the works of jurists—we conclude that official torture is now prohibited by the law of nations. The prohibition is clear and unambiguous, and admits of no distinction between treatment of aliens and citizens. Accordingly, we must conclude that the dictum in Dreyfus v. von Finck, [534 F.2d 24 (2d Cir.1976), at 31] to the effect that 'violations of international law do not occur when the aggrieved parties are nationals of the acting state,' is clearly out of tune with the current usage and practice of international law. The treaties and accords cited above, as well as the express foreign policy of our own government,[17] all make it clear that international law confers fundamental rights upon all people vis-a-vis their own governments. While the ultimate scope of those rights will be a subject for continuing refinement and elaboration, we hold that the right to be free from torture is now among them. We therefore turn to the question whether the other requirements for jurisdiction are met.

III

. . .

It is not extraordinary for a court to adjudicate a tort claim arising outside of its territorial jurisdiction. A state or nation has a legitimate interest in the

[16] The fact that the prohibition of torture is often honored in the breach does not diminish its binding effect as a norm of international law. As one commentator has put it, 'The best evidence for the existence of international law is that every actual State recognizes that it does exist and that it is itself under an obligation to observe it. States often violate international law, just as individuals often violate municipal law; but no more than individuals do States defend their violations by claiming that they are above the law.' J. Brierly, The Outlook for International Law 4–5 (1944).

[17] E.g., 22 U.S.C. § 2304(a)(2) ('Except under circumstances specified in this section, no security assistance may be provided to any country the government of which engages in a consistent pattern of gross violations of internationally recognized human rights.'); 22 U.S.C. § 2151(a) ('The Congress finds that fundamental political, economic, and technological changes have resulted in the interdependence of nations. The Congress declares that the individual liberties, economic prosperity, and security of the people of the United States are best sustained and enhanced in a community of nations which respect individual civil and economic rights and freedoms').

orderly resolution of disputes among those within its borders, and where the *lex loci delicti commissi* is applied, it is an expression of comity to give effect to the laws of the state when the wrong occurred. . . .

. . . Here, where *in personam* jurisdiction has been obtained over the defendant, the parties agree that the acts alleged would violate Paraguayan law, and the policies of the forum are consistent with the foreign law,[18] state court jurisdiction would be proper. Indeed, appellees conceded as much at oral argument.

. . . [W]e proceed to consider whether the First Congress acted constitutionally in vesting jurisdiction over 'foreign suits,' . . . alleging torts committed in violation of the law of nations. A case properly 'aris[es] under the . . . laws of the United States' for Article III purposes if grounded upon statutes enacted by Congress or upon the common law of the United States. . . . The law of nations forms an integral part of the common law, and a review of the history surrounding the adoption of the Constitution demonstrates that it became a part of the common law *of the United States* upon the adoption of the Constitution. Therefore, the enactment of the Alien Tort Statute was authorized by Article III.

During the eighteenth century, it was taken for granted on both sides of the Atlantic that the law of nations forms a part of the common law. 1 Blackstone, Commentaries 263–64 (1st Ed. 1765–69); 4 id. at 67. . . .

. . .

As ratified, the judiciary article contained no express reference to cases arising under the law of nations. Indeed, the only express reference to that body of law is contained in Article I, sec. 8, cl. 10, which grants to the Congress the power to 'define and punish . . . offenses against the law of nations.' Appellees seize upon this circumstance and advance the proposition that the law of nations forms a part of the laws of the United States only to the extent that Congress has acted to define it. This extravagant claim is amply refuted by the numerous decisions applying rules of international law uncodified in any act of Congress. E.g., Ware v. Hylton, 3 U.S. (3 Dall.) 198, 1 L.Ed. 568 (1796); *The Paquete Habana*, supra, 175 U.S. 677, 20 S.Ct. 290, 44 L.Ed. 320; *Sabbatino,* supra, 376 U.S. 398, 84 S.Ct. 923, 11 L.Ed.2d 804 (1964). . . .

The Filartigas urge that 28 U.S.C. §1350 be treated as an exercise of Congress's power to define offenses against the law of nations. While such a reading is possible . . . we believe it is sufficient here to construe the Alien Tort Statute, not as granting new rights to aliens, but simply as opening the federal courts for adjudication of the rights already recognized by international law. The statute nonetheless does inform our analysis of Article III, for we

[18] Conduct of the type alleged here would be actionable under 42 U.S.C. § 1983 or, undoubtedly, the Constitution, if performed by a government official.

recognize that questions of jurisdiction 'must be considered part of an organic growth part of an evolutionary process,' and that the history of the judiciary article gives meaning to its pithy phrases. Romero v. International Terminal Operating Co., 358 U.S. 354, 360, 79 S.Ct. 468, 473, 3 L.Ed.2d 368 (1959). The Framers' overarching concern that control over international affairs be vested in the new national government to safeguard the standing of the United States among the nations of the world therefore reinforces the result we reach today.

. . . The paucity of suits successfully maintained under the section is readily attributable to the statute's requirement of alleging a '*violation* of the law of nations' (emphasis supplied) at the jurisdictional threshold. Courts have, accordingly, engaged in a more searching preliminary review of the merits than is required, for example, under the more flexible 'arising under' formulation. . . . Thus, the narrowing construction that the Alien Tort Statute has previously received reflects the fact that earlier cases did not involve such well-established, universally recognized norms of international law that are here at issue.

[Discussion omitted of prior decisions under the Alien Tort Statute involving different fact situations and distinct claims such as suits for fraud or for wilful negligence leading to an accident.]

Since federal jurisdiction may properly be exercised over the Filartigas' claim, the action must be remanded for further proceedings. Appellee Pena, however, advances several additional points that lie beyond the scope of our holding on jurisdiction. Both to emphasize the boundaries of our holding, and to clarify some of the issues reserved for the district court on remand, we will address these contentions briefly.

IV

Pena argues that the customary law of nations, as reflected in treaties and declarations that are not self-executing, should not be applied as rules of decision in this case. In doing so, he confuses the question of federal jurisdiction under the Alien Tort Statute, which requires consideration of the law of nations, with the issue of the choice of law to be applied, which will be addressed at a later stage in the proceedings. The two issues are distinct. Our holding on subject matter jurisdiction decides only whether Congress intended to confer judicial power, and whether it is authorized to do so by Article III. The choice of law inquiry is a much broader one, primarily concerned with fairness, see Home Insurance Co. v. Dick, 281 U.S. 397, 50 S.Ct. 338, 74 L.Ed. 926 (1930); consequently, it looks to wholly different considerations. See Lauritzen v. Larsen, 345 U.S. 571, 73 S.Ct. 921, 97 L.Ed. 1254 (1954). Should the district court decide that the *Lauritzen* analysis requires it to apply Paraguayan law, our courts will not have occasion to consider what law would govern a suit under the Alien Tort Statute where the challenged conduct is actionable under the law of the forum and the law of nations but not the law of the jurisdiction in which the tort occurred.[19]

Pena also argues that '[i]f the conduct complained of is alleged to be the act of the Paraguayan government, the suit is barred by the Act of State doctrine.' This argument was not advanced below, and is therefore not before us on this appeal. We note in passing, however, that we doubt whether action by a state official in violation of the Constitution and laws of the Republic of Paraguay, and wholly unratified by that nation's government, could properly be characterized as an act of state. See Banco Nacionale de Cuba v. Sabbatino, supra, 376 U.S. 398, 84 S.Ct. 923, 11 L.Ed.2d 804; Underhill v. Hernandez, 168 U.S. 250, 18 S.Ct. 83, 42 L.Ed. 456 (1897). Paraguay's renunciation of torture as a legitimate instrument of state policy, however, does not strip the tort of its character as an international law violation, if it in fact occurred under color of government authority. . . .

. . .

In the twentieth century the international community has come to recognize the common danger posed by the flagrant disregard of basic human rights and particularly the right to be free of torture. Spurred first by the Great War, and then the Second, civilized nations have banded together to prescribe acceptable norms of international behavior. From the ashes of the Second World War arose the United Nations Organization, amid hopes that an era of peace and cooperation had at last begun. Though many of these aspirations have remained elusive goals, that circumstance cannot diminish the true progress that has been made. In the modern age, humanitarian and practical considerations have combined to lead the nations of the world to recognize that respect for fundamental human rights is in their individual and collective interest. Among the rights universally proclaimed by all nations, as we have noted, is the right to be free of physical torture. Indeed, for purposes of civil liability, the torturer has become—like the pirate and slave trader before him—*hostis humani generis*, an enemy of all mankind. Our holding today, giving effect to a jurisdictional provision enacted by our First Congress, is a small but important step in the fulfillment of the ageless dream to free all people from brutal violence.

[19] In taking that broad range of factors into account, the district court may well decide that fairness requires it to apply Paraguayan law to the instant case. See Slater v. Mexican National Railway Co., 194 U.S. 120, 24 S.Ct. 581, 48 L.Ed. 900 (1904). Such a decision would not retroactively oust the federal court of subject matter jurisdiction, even though plaintiff's cause of action would no longer properly be 'created' by a law of the United States. See American Well Works Co. v. Layne & Bowler Co., 241 U.S. 257, 260, 36 S.Ct. 585, 586, 60 L.Ed. 987 (1916) (Holmes, J.). Once federal jurisdiction is established by a colorable claim under federal law at a preliminary stage of the proceeding, subsequent dismissal of that claim (here, the claim under the general international proscription of torture) does not deprive the court of jurisdiction previously established. . .

NOTE

Following remand, the defendant took no part in the action and a default judgment was entered. The district court, 577 F. Supp. 860 (E.D.N.Y. 1984), awarded punitive damages of $5,000,000 to each plaintiff, so that the total judgment amounted to $10,385,364. The judgment was never collected.

Like many cases under the Alien Tort Statute (ATS), *Filartiga* was not defended throughout, and the case ended in a default judgment. The appellate courts in these cases have frequently been required to rule on motions by defendants to dismiss for want of subject matter jurisdiction or for failure to state a cause of action. For purposes of their decisions on these matters, they have taken plaintiffs' allegations as true. That is, such trial and appellate court decisions do not involve fact finding.

Filartiga and the following decisions under the ATS take different positions about the statute's purpose and effect and the bases for jurisdiction. Bear in mind several possible positions: (1) The only question before a court (as in the *Filartiga* decision) is subject matter jurisdiction. This is all that the ATS purports to do. It does not create a cause of action, point toward resolution of the merits, and so on. (2) The ATS creates a federal cause of action. The court fills in the remedial provisions. (3) The court can assert general federal question jurisdiction under 28 U.S.C.§1331, for the civil action is one 'arising under the . . . laws . . . of the United States.'

QUESTIONS

1. Describe the court's argument in concluding that torture is a violation of customary international law, a part of the 'law of nations' under §1350. Compare and contrast the approach of the Supreme Court in *The Paquete Habana*, p. 60, *supra*. Specifically, (a) what use is made in *Filartiga* of UN resolutions and declarations and of human rights treaties? What relevance to this case have those treaties? (b) Does the court's conception of custom stress consensus over norms or over practice? What is the relation between the two? (c) How would you criticize the court's discussion of state practice that it uses to support its finding of an international custom? (d) What relevance to this decision has the classical notion of *opinio juris*?

2. Note the difficulty in drawing on the 'law of nations' to support civil cases like those under the ATS rather than criminal cases. (a) What relevance to this decision has, for example, the Nuremberg precedent, p. 102, *supra*? (b) In what instances does international law provide for civil remedies against perpetrators for violations of human rights (recall the provisions of universal and regional human rights treaties)? Are those examples helpful to plaintiffs in ATS cases? (c) What relevance to this issue has Article 8 of the UDHR or Article 2(3) of the ICCPR?

COMMENT ON THE TEL-OREN CASE, THE ACT-OF-STATE DOCTRINE, AND SOVEREIGN IMMUNITY

Tel-Oren v. Libyan Arab Republic

In 1978, in a few violent hours, members of the Palestine Liberation Organization (PLO) entered Israel for a terrorist raid, killing 24 people and wounding 77 more. With few exceptions, the victims were Israeli civilians. Plaintiffs in *Tel-Oren* v. *Libyan Arab Republic*, 726 F. 2d 774 (D.C. Cir. 1984), included most of the wounded and relatives of most of the killed. Plaintiffs alleged that the PLO had trained the terrorists and planned the raid, and that Libya had participated. Jurisdiction was claimed under 28 U.S.C. §1331 and under the Alien Tort Statute, 28 U.S.C. §1350. The court of appeals affirmed *per curiam* a dismissal by the district court for want of subject matter jurisdiction. Each of the three judges wrote a separate opinion. The description below of two of the opinions deals only with the claim under the Alien Tort Statute and only with the PLO as defendant.

In his opinion, *Judge Edwards* drew on the *Filartiga* decision, *supra*, stressing its observation that the law of nations was not stagnant but should be construed 'as it exists today among the nations of the world,' and its conclusion that §1350 opened federal courts for 'adjudication of rights already recognized by international law.' Plaintiffs were not required to point to a specific right to sue under the law of nations in order to establish jurisdiction under §1350. That section requires only that the action rest upon a violation of the law of nations. This permits countries 'to meet their international obligations as they will'—or as a treaty (no such treaty being relevant to this case) might specifically provide.

One could infer from *Filartiga*, said Judge Edwards, that persons could be subject to civil liability under §1350 either by (i) committing a crime that traditionally warranted universal jurisdiction (such as the criminal prosecution of a pirate or slave trader) or (ii) committing an offense comparably violating contemporary international law. Commentators had begun to identify 'a handful of heinous actions—each of which violates definable, universal and obligatory norms—and in the process are defining the limits of section 1350's reach.' Judge Edwards referred to a then current draft of the *Restatement of the Law of Foreign Relations* that included as violations of customary international law murder and torture, both of which were within plaintiffs' allegations.

Nonetheless, Judge Edwards affirmed the dismissal on grounds that the PLO was not a recognized state, and that (unlike *Filartiga*) the case did not therefore involve 'official torture' or persons acting under color of state law. He was unwilling to extend the notion of the law of nations (and hence §1350) to include conduct of non-state actors. Nor, alternatively, was he willing to view terrorism (by analogy to piracy) as a violation itself of the law of nations that, without any state's involvement, led to individual responsibility of those committing terrorist acts.

Judge Bork took a very different position. He argued that §1350 was no more than a grant of jurisdiction to federal courts, and criticized the opinion in *Filartiga* for its failure to inquire whether international law created a cause of action enforceable by private parties in municipal (national) courts. Recalling that there was no concept of international human rights in 1789, Judge Bork stressed that neither was there any recognition of a right of private parties to recover under customary international law. 'That problem is not avoided by observing that the law of nations evolves.' Contemporary customary international law should not be construed to create such a civil action, for international law is fundamentally a law among states in which individuals have no direct enforcement or other role.

In these types of cases, Judge Bork continued, the understanding of international law argued for by §1350 plaintiffs would require that 'our courts must sit in judgment of the conduct of foreign officials in their own countries with respect to their own citizens.' To recognize such an action would amount to 'judicial interference with nonjudicial functions, such as the conduct of foreign relations.' Adjudication 'would present grave separation of powers problems.' A different conclusion might be reached if §1350 'had been adopted by a modern Congress that made clear its desire that federal courts police the behavior of foreign individuals and governments.'

Drawing on Blackstone, Judge Bork concluded that the 1789 Congress enacting the Alien Tort Statute might have been thinking of the then principal offenses by individuals against the law of nations: violation of safe-conducts, infringement of rights of ambassadors, and piracy. This modest list of civil actions based on the law of nations continues in effect today. Nor should the contemporary human rights treaties be construed to create private causes of action, except where (as with respect to the European Human Rights Convention) they specifically do so.

The Act-of-State Doctrine

In his observations about potential judicial interference with the conduct of foreign relations, Judge Bork drew on a strand of Supreme Court decisions dealing with the so-called act-of-state doctrine. That doctrine has figured prominently in recent decisions under the Alien Tort Statute (together with other doctrines and principles such as sovereign immunity and the political question defense).

Bypassing the complexities in the act-of-state doctrine and its exceptions, this Comment sketches some of its characteristics that are relevant to litigation under §1350. An early statement of the doctrine appears in *Underhill* v. *Hernandez*, 168 U.S. 250 (1897). An American citizen in Venezuela was operating under contract a local water system. Over a period of several months, a

Venezuelan general effectively exercising governmental power denied him the necessary documents to permit him to leave, thus coercing him to continue his operations. That citizen later brought an action against the general in a U.S. federal court to recover damages for unlawful detention. The Supreme Court affirmed a judgment for the defendant, stating in part:

> Every sovereign state is bound to respect the independence of every other sovereign state, and the courts of one country will not sit in judgment on the acts of the government of another, done within its own territory. Redress of grievances by reason of such acts must be obtained through the means open to be availed of by sovereign powers as between themselves.

It agreed with the conclusion of the lower court that 'the acts of the defendant were the acts of the government of Venezuela, and as such are not properly the subject of adjudication in the courts of another government.'

The principal decision launching the act-of-state doctrine on its modern career is *Banco Nacional de Cuba* v. *Sabbatino*, 376 U.S. 398 (1964), a case involving a Cuban expropriation of sugar properties owned by U.S. residents. The question arose in the context of deciding whether a U.S. court would apply the act-of-state doctrine and therefore presume the expropriation to be legal, *or* would examine the legality of the Cuban expropriation and, if it were found illegal, would deny it effect.

It was the contention of the U.S. commercial interests involved in the litigation that the doctrine did not apply to acts of state violating international law. Such a violation was claimed to have occurred in this case, since the compensation offered by Cuba did not meet an alleged customary international law standard of prompt, adequate and effective compensation for takings by a state of alien-owned property. The Supreme Court, however, pointed out that this standard was disputed among states. 'There are few if any issues in international law today on which opinion seems to be so divided as the limitations on a state's power to expropriate the property of aliens.'

In finding the act-of-state-doctrine applicable (thus leading to the presumed legitimacy of the Cuban expropriation), Justice Harlan wrote for the Court:

> [The doctrine's] continuing vitality depends on its capacity to reflect the proper distribution of functions between the judicial and political branches of the Government on matters bearing upon foreign affairs. It should be apparent that the greater the degree of codification or consensus concerning a particular area of international law, the more appropriate it is for the judiciary to render decisions regarding it, since the courts can then focus on the application of an agreed principle to circumstances of fact rather than on the sensitive task of establishing a principle not inconsistent with the national interests or with international justice. It is also evident that some aspects of international law touch more sharply on national nerves than do others; the less important the implications of an issue are for our foreign relations, the weaker the justification for exclusivity in the political branches. . . . Therefore, rather than laying down or

> reaffirming an inflexible and all-encompassing rule in this case, we decide only that the Judicial Branch will not examine the validity of a taking of property within its own territory by a foreign sovereign government . . . in the absence of a treaty or other unambiguous agreement regarding controlling legal principles, even if the complaint alleges that the taking violates customary international law.

The *Sabbatino* decision, together with later decisions in this field, stress the relevance to the doctrine's application of several other factors, including whether the Executive Branch (generally the Department of State) informs a court that it opposes judicial examination of the legality of the foreign government's act on the ground of the act-of-state doctrine.

Sovereign Immunity

Recall Judge Edward's opinion in the *Tel-Oren* case dismissing the §1350 action because the PLO was not a recognized state and thus could not commit 'official torture' or action under color of state law. However, had the PLO been a recognized state, a judicial action would have been barred under the doctrine of sovereign immunity, as indeed was the case with the government of Libya, the second defendant in *Tel-Oren*. This question of sovereign immunity has frequently arisen in cases under the Alien Tort Statute, sometimes in a close relation to the act-of-state doctrine.

The Foreign Sovereign Immunities Act of 1976 (FSIA), codified principally at 22 U.S.C.A. §§1602–11, provides a comprehensive legislative framework for deciding on claims of foreign defendants for immunity. Foreign states, including 'an agency or instrumentality' thereof, are immune from judicial jurisdiction, subject to enumerated exceptions. Those exceptions include court actions growing out of a state's commercial activities occurring in the United States, and actions involving property expropriated by the foreign government in violation of international law.

Argentine Republic v. *Amerada Hess Shipping Corp.*, 488 U.S. 428 (1989), involved a §1350 action growing out of the Falklands (Malvinas) war between the U.K. and Argentina. It was based on the damage to plaintiff's ship by an attack of Argentine aircraft. Plaintiff claimed that the attack was in violation of international law. The Supreme Court refused to draw an exception to the rule of immunity for suits under the Alien Tort Statute because of Argentina's alleged violation of international law. It drew from the FSIA 'the plain implication that immunity is granted in those cases involving alleged violations of international law that do not come within one of the FSIA's exceptions.'

Saudi Arabia v. *Nelson*, 113 S. Ct. 1471 (1993), involved tort claims based on alleged human rights violations. The Supreme Court held that alleged acts of unlawful detention and torture by the defendant state do not fall within any of the FSIA's exceptions. It concluded that '[t]he conduct [complained of by plaintiff] boils down to abuse of the power of its police by the Saudi

Government, and however monstrous such abuse undoubtedly may be, a foreign state's exercise of the power of its police has long been understood . . . as peculiarly sovereign in nature.' Sovereign immunity was granted Saudi Arabia.

In *Siderman de Blake* v. *Republic of Argentina*, 965 F.2d 699 (9th Cir. 1992), the court of appeals agreed with the plaintiff's argument that official acts of torture attributed to Argentina constituted a violation of a *jus cogens* norm of the 'highest status within international law.' Nonetheless, taking its lesson from the *Amareda Hess* decision in which the Supreme Court was so specific, the court concluded that it was Congress that would have to make any further exceptions to sovereign immunity. 'The fact that there has been a violation of *jus cogens* does not confer jurisdiction under the FSIA.'

In view of the FSIA and the decisional law described above, plaintiffs in §1350 actions have sought to avoid the issue of sovereign immunity by suing not the state itself but individual perpetrators—as indeed occurred in *Filartiga*. Courts required to sort out the relevance of sovereign immunity when individual defendants are before it have taken different approaches. For example, in *Chuidian* v. *Philippine National Bank*, 912 F.2d 1095 (9th Cir. 1990), the court held that the individual defendant, who had been acting in his official capacity, had immunity as 'an agency or instrumentality of a foreign state.' It observed that 'a suit against an individual in his official capacity is the practical equivalent of a suit against the sovereign directly.' The court, however, conceded that sovereign immunity would not be granted to individuals acting beyond the scope of their official duties, or in their individual capacities.

That very situation was present in *In re Estate of Ferdinand Marcos*, 25 F.3d 1467 (9th Cir. 1994). The court held that the FSIA did not bar jurisdiction under §1350 over the estate of former President Ferdinand Marcos for alleged acts of torture and wrongful death, since those were not official acts perpetrated within the scope of his official authority in the Philippines but rather acts outside the scope of his authority as President. Quoting from a prior related case, the court said that '[o]ur courts have had no difficulty in distinguishing the legal acts of a deposed ruler from his acts for personal profit that lack a basis in law.' At the same time, the requirement of state action under the definition of official torture (recall Judge Edward's discussion of this issue in *Tel-Oren*) could still be met by an official acting under color of authority, though not within an official mandate. That official could violate international law for purposes of the Alien Tort Statute.

All these issues are raised by the *Suarez-Mason* decision that follows.

FORTI v. SUAREZ-MASON

United States District Court, Northern District California, 1987.
672 F. Supp. 1531.

MEMORANDUM DECISION AND ORDER
JENSEN, DISTRICT JUDGE.

. . .

Having now considered all of the parties' submissions in support of and opposition to defendant's motions, the Court hereby denies defendant's motion to dismiss and grants a temporary stay. An explanation of the Court's ruling follows.

I

FACTS

This is a civil action brought against a former Argentine general by two Argentine citizens currently residing in the United States. Plaintiffs Forti and Benchoam sue on their own behalf and on behalf of family members, seeking damages from defendant Suarez-Mason for actions which include, *inter alia*, torture, murder, and prolonged arbitrary detention, allegedly committed by military and police personnel under defendant's authority and control. . . .

. . .

Plaintiffs' action arises out of events alleged to have occurred in the mid-to late 1970s during Argentina's so-called 'dirty war' against suspected subversives. In 1975 the activities of terrorists representing the extremes of both ends of the political spectrum induced the constitutional government of President Peron to declare a 'state of siege' under Article 23 of the Argentine Constitution. President Peron also decreed that the Argentine Armed Forces should assume responsibility for suppressing terrorism. The country was accordingly divided into defense zones, each assigned to an army corps. In each zone the military was authorized to detain suspects and hold them in prison or in military installations pursuant to the terms of the 'state of siege.' Zone One—which included most of the Province of Buenos Aires and encompassed the national capital—was assigned to the First Army Corps. From January 1976 until January 1979 defendant Suarez-Mason was Commander of the First Army Corps.

On March 24, 1976 the commanding officers of the Armed Forces seized the government from President Peron. The ruling military junta continued the 'state of siege' and caused the enactment of legislation providing that civilians accused of crimes of subversion would be judged by military law. In the period from 1976 to 1979, tens of thousands of persons were detained without charges by the military, and it is estimated that more than 12,000 were 'disappeared,' never to be seen again. *See generally Nunca Mas: The Report of the Argentine National Commission on the Disappeared* (1986).

In January 1984 the constitutionally elected government of President Raul Alfonsin assumed power. The Alfonsin government commenced investigations of alleged human rights abuses by the military, and the criminal prosecution of certain former military authorities followed. The government vested the Supreme Council of the Armed Forces with jurisdiction over the prosecution of military commanders. Summoned by the Supreme Council in March 1984, defendant failed to appear and in fact fled the country. In January of 1987 Suarez-Mason was arrested in Foster City, California pursuant to a provisional arrest warrant at the request of the Republic of Argentina. While defendant was in custody awaiting an extradition hearing he was served with the Complaint herein.

[The court described the allegations in the Complaint about acts committed in the zone under the defendant's command. Those acts included (a) the torture, killing and continued disappearance of several family members of the plaintiffs, (b) harsh imprisonment without judicial process of one plaintiff for four years, (c) abduction and seizure of family members without charge, and (d) the seizure of personal property.]

Although the individual acts are alleged to have been committed by military and police officials, plaintiffs allege that these actors were all agents, employees, or representatives of defendant acting pursuant to a 'policy, pattern and practice' of the First Army Corps under defendant's command. Plaintiffs assert that defendant 'held the highest position of authority' in Buenos Aires Province; that defendant was responsible for maintaining the prisons and detention centers there, as well as the conduct of Army officers and agents; and that he 'authorized, approved, directed and ratified' the acts complained of.

. . . Although both plaintiffs retain their Argentine citizenship, both reside currently in Virginia. Plaintiffs predicate federal jurisdiction principally on the 'Alien Tort Statute,' 28 U.S.C. §1350, and alternatively on federal question jurisdiction, 28 U.S.C. §1331. Additionally, they assert jurisdiction for their common-law tort claims under principles of pendent and ancillary jurisdiction.

Based on these above allegations, plaintiffs seek compensatory and punitive damages for violations of customary international law and laws of the United States, Argentina, and California. They press eleven causes of action. Both allege claims for torture; prolonged arbitrary detention without trial; cruel, inhuman and degrading treatment; false imprisonment; assault and battery; intentional infliction of emotional distress; and conversion. Additionally Forti claims damages for 'causing the disappearance of individuals,' and Benchoam asserts claims for 'murder and summary execution,' wrongful death, and a survival action.

. . .

II

SUBJECT MATTER JURISDICTION

As a threshold matter, defendant argues that the Court lacks subject matter jurisdiction under 28 U.S.C. §1350, the 'Alien Tort Statute.' Defendant urges the Court to follow the interpretation of §1350 as a purely jurisdictional statute which requires that plaintiffs invoking it establish the existence of an independent, private right of action in international law. Defendant argues that the law of nations provides no tort cause of action for the acts of 'politically motivated terrorism' challenged by plaintiffs' Complaint. Alternatively, defendant argues that even if §1350 provides a cause of action for violations of the law of nations, not all of the torts alleged by plaintiffs qualify as violations of the law of nations. For the reasons set out below, the Court rejects defendant's construction of §1350 and finds that plaintiffs allege sufficient facts to establish subject matter jurisdiction under both the Alien Tort Statute and 28 U.S.C. §1331. . . .

A. The Alien Tort Statute

[The court agreed with the views of Judge Edwards in *Tel-Oren* v. *Libyan Arab Republic*, 726 F.2d (D.C. Cir. 1984), *supra*. Plaintiffs are not required to 'establish the existence of an independent express right of action' under the law of nations, for that law doesn't create or define civil actions. Hence such a requirement 'would effectively nullify' a vital portion of §1350. That section provides both for jurisdiction and for a federal cause of action arising by recognition of certain 'international common law torts.' As indicated in Judge Edward's opinion, such an international tort must be 'definable, obligatory (rather than hortatory) and universally condemned. . . . The requirement of international consensus is of paramount importance . . . .']

B. Analysis Under 28 U.S.C. §1350

In determining whether plaintiffs have stated cognizable claims under Section 1350, the Court has recourse to 'the works of jurists, writing professedly on public law; . . . the general usage and practice of nations; [and] judicial decisions recognizing and enforcing that law.' *United States* v. *Smith*, 18 U.S. (5 Wheat.) 153, 160–61, 5 L.Ed. 57 (1820). . . .

1. Official Torture

In Count One, plaintiffs both allege torture conducted by military and police personnel under defendant's command. The Court has no doubt that official torture constitutes a cognizable violation of the law of nations under §1350. . . . The claim would thus allege torture committed by *state officials* and so fall within the international tort first recognized in *Filartiga*.

2. Prolonged Arbitrary Detention

. . .

There is case law finding sufficient consensus to evince a customary international human rights norm against arbitrary detention. *Rodriguez-Fernandez* v. *Wilkinson*, 505 F.Supp. 787, 795–98 (D. Kan.1980) (citing international treaties, cases, and commentaries), *aff'd*, 654 F.2d 1382 (10th Cir.1981); . . .

The consensus in even clearer in the case of a state's *prolonged* arbitrary detention of its own citizens. *See, e.g., Restatement (Revised) of the Foreign Relations Law of the United States* §702 (Tent. Draft No. 6, 1985) (prolonged arbitrary detention by state constitutes international law violation). The norm is obligatory, and is readily definable in terms of the arbitrary character of the detention. The Court finds that plaintiffs have alleged international tort claims for prolonged arbitrary detention.

3. Summary Execution

. . .

The proscription of summary execution or murder by the state appears to be universal, is readily definable, and is of course obligatory. The Court emphasizes that plaintiff's allegations raise no issue as to whether or not the execution was within the lawful exercise of state power; rather, she alleges murder by state officials with neither authorization nor recourse to any process of law. Under these circumstances, the Court finds that plaintiff Benchoam has stated a cognizable claim under §1350 for the 1977 murder/summary execution of her brother by Argentine military personnel.

4. Causing Disappearance

. . .

Sadly, the practice of 'disappearing' individuals—i.e., abduction, secret detention, and torture, followed generally by either secret execution or release—during Argentina's 'dirty war' is now well documented in the official report of the Argentine National Commission on the Disappeared, *Nunca Mas*. Nor are such practices necessarily restricted to Argentina. With mounting publicity over the years, such conduct has begun to draw censure as a violation of the basic right to life. Plaintiff cites a 1978 United Nations resolution and a 1980 congressional resolution to this effect. U.N.G.A.Res. /173 (1978); H.R.Con.Res. 285, 96th Cong., 2d Sess. The Court notes, too, that the proposed Restatement of the Law of Foreign Relations lists 'the murder or causing the disappearance of individuals,' where practiced, encouraged, or condoned by the state, as a violation of international law. However, plaintiffs do not cite the Court to any case finding that causing the disappearance of an individual constitutes a violation of the law of nations.

Before this Court may adjudicate a tort claim under §1350, it must be satisfied that the legal standard it is to apply is one with universal acceptance and definition; on no other basis may the Court exercise jurisdiction over a claimed violation of the law of nations. Unfortunately, the Court cannot say, on the basis of the evidence submitted, that there yet exists the requisite degree of international concensus which demonstrates a customary international norm. Even if there were greater evidence of universality, there remain definitional problems. It is not clear precisely what conduct falls within the proposed norm, or how this proscription would differ from that of summary execution. . . . Yet there is no apparent international consensus as to the additional elements needed to make out a claim for causing the disappearance of an individual. For instance, plaintiffs have not shown that customary international law creates a presumption of causing disappearance upon a showing of prolonged absence after initial custody.

For these reasons the Court must dismiss Count Four for failure to state a claim upon which relief may be grounded.

. . .

5. Cruel, Inhuman and Degrading Treatment

Finally, in Count Five plaintiffs both allege a claim for 'cruel, inhuman and degrading treatment' based on the general allegations of the Complaint and consisting specifically of the alleged torture, murder, forcible disappearance and prolonged arbitrary detention.

This claim suffers the same defects as Count Four. Plaintiffs do not cite, and the Court is not aware of, such evidence of universal consensus regarding the right to be free from 'cruel, inhuman and degrading treatment as exists, for example, with respect to official torture.' Further, any such right poses problems of definability. . . . Accordingly, the Court dismisses Count Five of the Complaint for failure to state a claim upon which relief may be granted.

. . .

[The court also found jurisdiction under 28 U.S.C. §1331, which covers claims based on federal common law. It cited cases including *The Paquete Habana*, 175 U.S. 677, *supra* p. 60, for the proposition that federal common law incorporated international law.

The court, after describing decisions such as *Underhill* and *Sabbatino*, *supra* p. 791, next concluded that the act of state doctrine, which if applicable would block adjudication of the legality of defendant's conduct, did not apply. It said in part:]

. . . Defendant maintains that all of the challenged acts were taken pursuant to the 'state of siege' declared by the constitutional government and reaffirmed by the military junta. Thus, defendant argues, he was a government official acting under policies promulgated by the junta, and this Court cannot adjudicate the question of his liability without also passing on the question of the legality of the acts of the Argentine government. This, he concludes, is precisely the sort of case which the act of state doctrine removes from the courts' scrutiny.

. . .

Here, by contrast, [with prior discussed decisions], plaintiffs allege acts by a subordinate government official in violation not of economic rights, but of fundamental human rights lying at the very heart of the individual's existence. These are not the public official acts of a head of government, nor is it clear at this stage of the proceedings to what extent defendant's acts were 'ratified' by the *de facto* military government. . . .

Equally unpersuasive is defendant's argument that allegations of 'official' conduct sufficient to establish jurisdiction under 28 U.S.C. §1350 automatically implicate the act of state doctrine. . . . [F]or purposes of §1350 a plaintiff must allege 'official' (as opposed to private) action—but this is not necessarily the governmental and public action contemplated by the act of state doctrine. That is, a police chief who tortures, or orders to be tortured, prisoners in his custody fulfills the requirement that his action be 'official' simply by virtue of his position and the circumstances of the act; his conduct may be wholly unratified by his government and even proscribed by its constitution and criminal statutes. . . .

The procedural context of defendant's motion governs here. Inasmuch as this is a Rule 12(b) (6) motion, the Court must accept as true the allegations of the Complaint, and may not dismiss the Complaint unless plaintiffs can prove no set of facts to establish their claims. The Court cannot say at this stage of the proceedings that adjudication of plaintiffs' claims will necessarily entail considering the legality of the official acts of a foreign sovereign.

. . .

NOTE

Plaintiffs in the above case filed a motion for reconsideration of the court's order dismissing the claims for 'disappearance' and 'cruel, inhuman or degrading treatment.' In *Forti* v. *Suarez-Mason*, 694 F. Supp. 707 (N.D. Cal. 1988), Judge Jensen denied reconsideration of the second claim. 'Plaintiffs' submissions fail to establish that there is anything even remotely approaching universal consensus as to what constitutes "cruel, inhuman or degrading treatment."' He granted, however, plaintiffs' motion with respect to disappearance and reinstated that claim, stating in part:

> The legal scholars whose declarations have been submitted in connection with this Motion are in agreement that there is universal consensus as to the two essential elements of a claim for 'disappearance.' In Professor Franck's words:

> The international community has also reached a consensus on the definition of a 'disappearance.' It has two essential elements: (a) abduction by a state official or by persons acting under state approval or authority; and (b) refusal by the state to acknowledge the abduction and detention.

Plaintiffs cite numerous international legal authorities which support the assertion that 'disappearance' is a universally recognized wrong under the law of nations. For example, United Nations General Assembly Resolution 33/173 recognizes 'disappearance' as violative of many of the rights recognized in the Universal Declaration of Human Rights, G.A. Res. 217 A (III), adopted by the United Nations General Assembly, Dec. 10, 1948, U.N. Doc. A/810 (1948) [*hereinafter* Universal Declaration of Human Rights]. These rights include: (1) the right to life; (2) the right to liberty and security of the person; (3) the right to freedom from torture; (4) the right to freedom from arbitrary arrest and detention; and (5) the right to a fair and public trial. Id., articles 3, 5, 9, 10, 11. *See also* International Covenant on Political and Civil Rights, articles 6, 7, 9, 10, 14, 15, 17.

Other documents support this characterization of 'disappearance' as violative of universally recognized human rights. The United States Congress has denounced 'prolonged detention without charges and trial' along with other 'flagrant denial[s] of the right to life, liberty, or the security of person.' 22 U.S.C. §2304 (d)(I). The recently published Restatement (Third) of the Foreign Relations Law of the United States §702 includes 'disappearance' as a violation of the international law of human rights. The Organization of American States has also denounced 'disappearance' as 'an affront to the conscience of the hemisphere and . . . a crime against humanity.' Organization of American States, Inter-American Commission on Human Rights, General Assembly Resolution 666 (November 18, 1983).

. . .

QUESTIONS

1. What differences do you see between the positions on the ATS of the opinions in *Filartiga* and *Suarez-Mason*?

2. Do you agree with the court's resolution of the defense of act of state? Does this case present a stronger argument than *Filartiga* for such a defense?

3. Suppose that the former dictatorial president of a state that continues to be governed by an authoritarian elite supported by the military is sued for wrongful death damages under the Alien Tort Statute when he is in the United States (where he owns valuable real property). The plaintiff produces convincing evidence of the torture and murder of his parents, political opponents of the defen-

dant, on the orders of the defendant. Torture and killing (or forcing flight from the state through threats) were and continue to be the government's strategy for dealing with potentially threatening political opponents. The presidential successor to the defendant protests to the State Department this civil suit in the United States, which he views as an insult to the state's sovereignty? How should the court rule on the defenses of act of state and sovereign immunity? How, for example, might the defendant distinguish the *Filartiga* decision?

4. The plaintiff, a citizen of State X now in the U.S., brings an ATS action against a visiting official from X for categorically denying the plaintiff in X a permit to hold a parade in X's capital city. The parade's purpose was to publicize the program of the opposition party in X (to which plaintiff belongs). X is a tightly controlled state in which opposition parties are permitted to operate within severe limits. Elections are sham, the press is censored, the jails hold some political prisoners. Assume that there is no issue of sovereign or diplomatic immunity. What problems do you see in such an action? How does the action differ from *Filartiga* and *Suarez-Mason*?

5. How do you assess the significance of ATS actions and decisions, given that the collection of damages in most (not all) cases is unlikely? From the perspective of the human rights movement or the development of that movement within the United States, are ATS actions helpful? Why? Can you see any disadvantages in them?

TORTURE VICTIM PROTECTION ACT

106 Stat. 73 (1992), 28 U.S.C.A. §1350 Notes.

. . .

SEC. 2. ESTABLISHMENT OF CIVIL ACTION.

(a) LIABILITY.—An individual who, under actual or apparent authority, or color of law, of any foreign nation—

(1) subjects an individual to torture shall, in a civil action, be liable for damages to that individual; or

(2) subjects an individual to extrajudicial killing shall, in a civil action, be liable for damages to the individual's legal representative, or to any person who may be a claimant in an action for wrongful death.

(b) EXHAUSTION OF REMEDIES.—A court shall decline to hear a claim under this section if the claimant has not exhausted adequate and available remedies in the place in which the conduct giving rise to the claim occurred.

(c) STATUTE OF LIMITATIONS.—No action shall be maintained under this section unless it is commenced within 10 years after the cause of action arose.

SEC. 3. DEFINITIONS.

(a) EXTRAJUDICIAL KILLING.—For the purposes of this Act, the term 'extra-judicial killing' means a deliberated killing not authorized by a previous judgment pronounced by a regularly constituted court affording all the judicial guarantees which are recognized as indispensable by civilized peoples. Such term, however, does not include any such killing that, under international law, is lawfully carried out under the authority of a foreign nation.

(b) TORTURE.—For the purposes of this Act—

(1) the term 'torture' means any act, directed against an individual in the offender's custody or physical control, by which severe pain or suffering (other than pain or suffering arising only from or inherent in, or incidental to, lawful sanctions), whether physical, or mental, is intentionally inflicted on that individual for such purposes as obtaining from that individual or a third person information or a confession, punishing that individual for an act that individual or a third person has committed or is suspected of having committed, intimidating or coercing that individual or a third person, or for any reason based on discrimination of any kind; and

(2) mental pain or suffering refers to prolonged mental harm caused by or resulting from—

(A) the intentional infliction or threatened infliction of severe physical pain or suffering;

(B) the administration or application, or threatened administration or application, of mind altering substances or other procedures calculated to disrupt profoundly the senses or the personality;

(C) the threat of imminent death; or

(D) the threat that another individual will imminently be subjected to death, severe physical pain or suffering, or the administration or application of mind altering substances or other procedures calculated to disrupt profoundly the senses or personality.

SENATE REPORT ON THE TORTURE VICTIM PROTECTION ACT

S. Rep. 102–249, Committee on the Judiciary, 102nd Cong., 1st Sess., 1991.

. . . This legislation will carry out the intent of the Convention Against Torture and Other Cruel, Inhuman or Degrading Treatment or Punishment, which was ratified by the U.S. Senate on October 27, 1990. The convention obligates state parties to adopt measures to ensure that torturers within their territories are held legally accountable for their acts. This legislation will do precisely that—by making sure that torturers and death squads will no longer have a safe haven in the United States.

. . .

The TVPA would establish an unambiguous basis for a cause of action that has been successfully maintained under an existing law, section 1350 of title 28 of the U.S. Code, derived from the Judiciary Act of 1789 (the Alien Tort Claims Act) which permits Federal district courts to hear claims by aliens for torts committed 'in violation of the law of nations.' (28 U.S.C. 1350). Section 1350 has other important uses and should not be replaced. . . .

. . .

The TVPA would . . . enhance the remedy already available under section 1350 in an important respect: while the Alien Tort Claims Act provides a remedy to aliens only, the TVPA would extend a civil remedy also to U.S. citizens who may have been tortured abroad.

. . .

Congress clearly has authority to create a private right of action for torture and extrajudicial killings committed abroad. Under article III of the Constitution, the Federal judiciary has the power to adjudicate cases 'arising under' the 'law of the United States.' The Supreme Court has held that the law of the United States includes international law. In *Verlinden B.V.* v. *Central Bank of Nigeria* 461 U.S. 480, 481 (1983), the Supreme Court held that the 'arising under' clause allows Congress to confer jurisdiction on U.S. courts to recognize claims brought by a foreign plaintiff against a foreign defendant. Congress' ability to enact this legislation also drives from article I, section 8 of the Constitution, which authorizes Congress 'to define and punish . . . Offenses against the Laws of Nations.'

IV. ANALYSIS OF LEGISLATION

. . .

D. Who can be sued

First and foremost, only defendants over which a court in the United States has personal jurisdiction may be sued. In order for a Federal court to obtain personal jurisdiction over a defendant, the individual must have 'minimum contacts' with the forum state, for example through residency here or current travel. Thus, this legislation will not turn the U.S. courts into tribunals for torts having no connection to the United States whatsoever.

The legislation uses the term 'individual' to make crystal clear that foreign states or their entities cannot be sued under this bill under any circumstances: only individuals may be sued. Consequently, the TVPA is not meant to override the Foreign Sovereign Immunities Act (FSIA) of 1976, which renders foreign governments immune from suits in U.S. courts, except in certain instances.

The TVPA is not intended to override traditional diplomatic immunities which prevent the exercise of jurisdiction by U.S. courts over foreign diplomats. . . .

. . .

Similarly, the committee does not intend the 'act of state' doctrine to provide a shield from lawsuit for former officials. In *Banco Nacional de Cuba* v. *Sabbatino*, 376 U.S. 398 (1964), the Supreme Court held that the 'act of state' doctrine is meant to prevent U.S. courts from sitting in judgment of the official public acts of a sovereign foreign government. Since this doctrine applies only to 'public' acts, and no state commits torture as a matter of public policy, this doctrine cannot shield former officials from liability under this legislation.

E. Scope of liability

. . .

The legislation is limited to lawsuits against persons who ordered, abetted, or assisted in the torture. It will not permit a lawsuit against a former leader of a country merely because an isolated act of torture occurred somewhere in that country. However, a higher official need not have personally performed or ordered the abuses in order to be held liable. Under international law, responsibility for torture, summary execution, or disappearances extends beyond the person or persons who actually committed those acts—anyone with higher authority who authorized, tolerated or knowingly ignored those acts is liable for them. . . .

Finally, low-level officials cannot escape liability by claiming that they were acting under orders of superiors. Article 2(3) of the Torture Convention explicitly states that 'An order from a superior official or a public authority may not be invoked as a justification for torture.'

. . .

VII. MINORITY VIEWS OF MESSRS. SIMPSON AND GRASSLEY

. . .

The executive branch, through the Department of Justice, has expressed a most serious concern with S. 313, which we share. Senate bill 313 could create difficulties in the management of foreign policy. For example, under this bill, individual aliens could determine the timing and manner of the making of allegations in a U.S. court about a foreign country's alleged abuses of human rights.

There is no more complex and sensitive issue between countries than human rights. The risk that would be run if an alien could have a foreign country judged by a U.S. court is too great. Judges of U.S. courts would, in a sense, conduct some of our Nation's foreign policy. The executive branch is and should remain, we believe, left with substantial foreign policy control.

In addition the Justice Department properly notes that our passage of this bill could encourage hostile foreign countries to retaliate by trying to assert jurisdiction for acts committed in the United States by the U.S. Government against U.S. citizens. For example, if this bill's principles were adopted abroad, Saddam Hussein could try a United States citizen police officer who happened to be present in Iraq, in an Iraqi court, for alleged human rights abuses against any United States citizen that the policeman happened to arrest while performing his duties in the United States.

. . .

QUESTIONS

Compare the TVPA with the prior §1350 decisions. What are the differences with respect to coverage and parties in civil actions under the two statutes? What are the differences with respect to the bearing on the civil actions of international law?

COMMENT ON CRIMINAL PROSECUTION UNDER CONVENTION AGAINST TORTURE

Although the preceding Senate Report states that the TVPA 'will carry out the intent' of the Torture Convention, it does so only with respect to *civil liability*. Article 14 of that Convention provides that each State Party 'shall ensure in its legal system that the victim of an act of torture obtains redress and has an enforceable right to fair and adequate compensation. . . .'

Compare the provisions for *criminal prosecution* under Article 1 of the Convention, which applies (with respect to torture, as there defined) to pain and suffering that is 'inflicted by or at the instigation of or with the consent or acquiescence of a public official or other person acting in an official capacity.' Article 2 provides that States Parties 'shall take effective . . . measures to prevent acts of torture' in their territory. But the Convention reaches beyond this traditional territorial base for a state's criminal jurisdiction. Under Article 4, each State Party 'shall ensure that all acts of torture are offences under its criminal law,' and 'shall make these offences punishable by appropriate penalties.' Under Article 5(1), each Party 'shall take such measures as may be necessary to establish its jurisdiction over the offences referred to in article 4'

(a) when offenses are committed in the state's territory;
(b) when 'the alleged offender is a national of that State;' and
(c) when 'the victim is a national of that State if that State considers it appropriate.'

In addition, jurisdiction is to be established under Section 5(2) where the alleged offender is present in the state's territory and the state does not extradite him pursuant to the Convention to other indicated and involved states for criminal prosecution there.

In 1990, the Senate Committee on Foreign Relations reported favorably on the Convention and recommended that the Senate give its consent to ratification. With respect to Article 5, the Committee report states:[20]

> A major concern in drafting Article 5 . . . was whether the Convention should provide for possible prosecution by any State in which the alleged offender is found—so-called universal jurisdiction. The United States strongly supported the provision for universal jurisdiction, on the grounds that torture, like hijacking, sabotage, hostage-taking, and attacks on internationally protected persons, is an offense of special international concern, and should have similarly broad, universal recognition as a crime against humanity, with appropriate jurisdictional consequences. Provision for 'universal jurisdiction' was also deemed important in view of the fact that the government of the country where official torture actually occurs may seldom be relied on to take action. . . . [E]xisting federal and state law appears sufficient to establish jurisdiction when the offense has allegedly been committed in any territory under U.S. jurisdiction. . . . Implementing legislation is therefore needed only to establish Article 5(1)(b) jurisdiction over offenses committed by U.S. nationals outside the United States, and to establish Article 5(2) jurisdiction over foreign offenders committing torture abroad who are later found in territory under U.S. jurisdiction. . . . Similar legislation has already been enacted to implement comparable provisions of the Conventions on Hijacking, Sabotage, Hostages, and Protection of Diplomats.

In 1990, the Senate consented to ratification of the Convention, provided that the criminal legislation required by the Convention first be enacted by Congress. The Torture Convention Implementing Legislation was enacted in 1994, 18 U.S.C.A. §§2340–2340B. Section 2340A gives courts jurisdiction for torture (as defined) committed 'outside the United States' where the alleged offender is a U.S. national or 'is present in the United States, irrespective of the nationality of the victim or alleged offender.' The United States then ratified the Convention.

The different jurisdictional bases referred to in this Comment, including the notion of universal jurisdiction, are further developed in Chapter 15 in connection with an analysis of the jurisdictional provisions of the International Criminal Tribunal established to deal with crimes in the former Yugoslavia and Rwanda. See pp. 1022–25, *infra*.

[20] Sen. Comm. For. Relations, Report on Convention against Torture, 100th Cong., 2d Sess. (1990).

COMMENT ON RELEVANCE OF INTERNATIONAL LAW TO CONSTITUTIONAL LITIGATION OVER THE DEATH PENALTY

Courts in the United States have drawn on international law in a number of contexts involving alleged violations of human rights other than the core examples of torture, disappearances and summary execution that figured in the cases discussed above. They have referred explicitly to customary international law, and more generally to state practice and consensus, and to treaties including some not ratified by the U.S. The other contexts include treatment of prisoners, detention of refugees and Eighth Amendment jurisprudence about 'cruel and unusual punishment.' The last illustration concerns particularly the constitutionality of the death penalty. This Comment suggests the difficulties confronted in advancing arguments about or related to international law in such constitutional litigation.

Several of the major human rights treaties contain provisions limiting the use of the death penalty, although none prohibit it outright. For example, Article 6, para. 2 of the ICCPR permits the death sentence 'only for the most serious crimes,' while para. 5 prohibits the death sentence for pregnant women or for crimes committed by persons below the age of eighteen. Article 4 of the American Convention on Human Rights states similar injunctions, and prohibits capital punishment for political offenses.

In recent years, there has been a significant movement to abolish the death penalty altogether. Optional Protocol No. 2 to the ICCPR (effective 1991, 28 ratifications as of September 1995) does precisely that, subject to a limited exception for war-time crimes. Protocol No. 6 to the European Convention on Human Rights (effective 1985, 23 ratifications) (the Convention itself being silent with respect to the death penalty) and a Protocol to the American Convention on Human Rights to Abolish the Death Penalty (effective 1991, 3 ratifications) are to the same effect.

State practice reflects this trend toward abolition. Amnesty International reports that as of December 1994

> 54 countries have abolished the death penalty for all offenses, while 15 have done so for all but exceptional crimes such as war crimes. Twenty-seven countries can be considered abolitionists *de facto*. They retain the death penalty in law but have not carried out any executions for the past 10 years or more.[21]

The trend in the United States had been in the other direction. In April 1995, the laws of 38 states of the U.S. federalism made provision for the death penalty, and the number of executions increased twenty-fold in the past

[21] Amnesty International, The Death Penalty List of Abolitionist and Retentionist Countries (December 1994), AI Index: ACT 50/01/95.

decade. The size of death row grew from about 500 inmates in 1976 to nearly 3,000 in 1994.[22]

The U.S. Supreme Court considered in several cases the argument that imposition of the death penalty violates customary international law that is binding on the United States. In *Coker v. Georgia* 433 U.S. 584 (1977), the Court referred to that law to bolster its decision that imposition of the death penalty for the rape of an adult woman was 'cruel and unusual' within the meaning of the Eighth Amendment. Justice White relied on UN data, noting that it was 'thus not irrelevant here than out of 60 major nations in the world surveyed in 1965, only 3 retained the death penalty for rape where death did not ensue.'

In *Thompson v. Oklahoma*, 478 U.S. 815 (1988), the plurality in favor of finding unconstitutional capital punishment for those who committed a crime when under sixteen years of age relied on a comparative survey of those states that had either outlawed the death penalty or did not apply it to juveniles, and on human rights treaties proscribing such punishment. Justice Stevens wrote for the plurality, stating in part:

> The conclusion that it would offend civilized standards of decency to execute a person who was less than 16 years old at the time of his or her offense is consistent with the views that have been expressed by respected professional organizations, by other nations that share our Anglo-American heritage, and by the leading members of the Western European community. Thus, the American Bar Association and the American Law Institute have formally expressed their opposition to the death penalty for juveniles. Although the death penalty has not been entirely abolished in the United Kingdom or New Zealand (it has been abolished in Australia, except in the State of New South Wales, where it is available for treason and piracy), in neither of those countries may a juvenile be executed. The death penalty has been abolished in West Germany, France, Portugal, The Netherlands, and all of the Scandinavian countries, and is available only for exceptional crimes such as treason in Canada, Italy, Spain, and Switzerland. Juvenile executions are also prohibited in the Soviet Union.

Justice Scalia, dissenting, wrote that reliance on Amnesty International's account of practices in foreign states was 'totally inappropriate as a means of establishing the fundamental beliefs of this Nation.' He continued:

> That 40% of our States do not rule out capital punishment for 15-year-old felons is determinative of the question before us here, even if that position contradicts the uniform view of the rest of the world. We must never forget that it is a Constitution for the United States of America that we are expounding. The practices of other nations, particularly other democracies, can be relevant to determining whether a practice uniform among our people is not merely an historical accident, but rather so 'implicit in the

[22] Death Penalty Information Center, Facts about the Death Penalty (December 14, 1994).

> concept of ordered liberty' that it occupies a place not merely in our mores but, text permitting, in our Constitution as well. See *Palko* v. *Connecticut*, 302 U.S. 319, 325, 58 S.Ct. 149, 152, 82 L.Ed. 288 (1937) (Cardozo, J.). But where there is not first a settled consensus among our own people, the views of other nations, however enlightened the Justices of this Court may think them to be, cannot be imposed upon Americans through the Constitution. In the present case, therefore, the fact that a majority of foreign nations would not impose capital punishment upon persons under 16 at the time of the crime is of no more relevance than the fact that a majority of them would not impose capital punishment at all, or have standards of due process quite different from our own.

The minority view in *Thompson* became the majority in *Stanford v. Kentucky*, 492 U.S. 361 (1989), upholding the constitutionality of the imposition of capital punishment on those who had committed crimes when sixteen or seventeen years old. In determining what 'standards of decency' relevant to capital punishment had evolved, the Court looked to 'those of modern American society as a whole. We emphasize that it is *American* conceptions of decency that are dispositive.' It rejected the dissent's view that sentencing practices of other countries were relevant. The dissenting opinion of Justice Brennan presented comparative data about sentencing and observed:

> In addition to national laws, three leading human rights treaties ratified or signed by the United States explicitly prohibit juvenile death penalties [citing the ICCPR and the American Convention on Human Rights, then signed but not ratified, and the Geneva Convention Relative to the Protection of Civilian Persons in Time of War]. Within the world community, the imposition of the death penalty for juvenile crimes appears to be overwhelmingly disapproved.

QUESTIONS

1. When an opinion sets forth data about the number of states that have, for example, abolished capital punishment when a crime is committed by a juvenile, is its author making an international law argument? What type of argument? For a rule of customary international law?

2. How do you assess the following claims addressed to a state that is not party to any treaty abolishing capital punishment: (a) international law now bars the death sentence apart from serious crimes during war; (b) international law now bars the imposition of the death sentence on those who committed the relevant crimes when under 18.

3. If you believe claim (b) above to be strong, what is the effect of the U.S. reservation to Article 6 of the ICCPR, p. 768, *supra* that was attached to the ratification of that Covenant?

ADDITIONAL READING

There is a large legal literature on the Alien Tort Statute. Two recent articles on the implications of the current Ferdinand Marcos litigation that is referred to in a Comment above are: Joan Fitzpatrick, *The Future of the Alien Tort Claims Act of 1789: Lessons from In re Marcos Human Rights Litigation*, 67 St. John's L. Rev. 491 (1993) and Ralph Steinhardt, *Fulfilling the Promise of Filartiga: Litigating Human Rights Claims against the Estate of Ferdinand Marcos*, 20 Yale J. Int. L. 65 (1995).

12

Enforcement by States against Violator States

Chapter 11 represents the promise or ideal. States will take the necessary measures to assure internal compliance with international human rights. They will adopt constitutions implementing the international norms and internalize those norms directly. When they do not, when they are violators, we have seen the efforts of intergovernmental organizations and organs, universal and regional, to secure compliance. Their modes of enforcement are then *vertical*, in the sense that pressures are exerted and perhaps sanctions applied by international organs 'above' the state. From the perspective of international law, those organs exercise authority over all member states in accordance with the terms of the treaty.

Here, in Chapter 12, we consider *horizontal* modes of enforcement of human rights. Acting singly or as part of a consortium, states themselves may apply pressures against a violator state that are shy of the military force proscribed (without the authorization of the Security Council) by the UN Charter. Such forms of pressure include boycotts or embargoes. They also include conditioning bilateral or multilateral security assistance, development aid, or trade advantages such as a most-favored-nation clause, on the recipient or trading state's compliance with basic human rights norms. Within such a 'horizontal' (state-to-state) application of sanctions and pressures, the state imposing conditions ('conditionality') becomes part of a multi-layered system of enforcement of international human rights.

The debate within the United States about the relevance of human rights to foreign policy—including policy on security or development assistance and on trade—has been active and contentious since Congress started in the 1970s to write types of conditionality (imposing human rights conditions on aid or trade) into law. The contexts in which these issues have been argued include the central crises of American foreign policy since the early 1970s—for example, U.S. security aid to (or to opponents of) or trade with regimes in El Salvador, Nicaragua, Chile and South Africa. The economic importance of aid and trade has involved major sectors of the U.S. economy in the politics of conditionality—defense industries anxious to export their products, manufacturing and service-oriented firms fearful of losing foreign markets in retaliation for elimination by the U.S. of other states' advantages in trade, consumer groups fearful of losing cheaper imported products, and so on.

The chapter starts with a general discussion of the 'national interest' in relation to human rights enforcement, continues with general analyses of the U.S. experience in aid conditionality, and concludes with a case study of a contemporary issue: whether most-favored-nation treatment for imports into the United States from the People's Republic of China should be conditioned on that country's compliance with stated human rights norms.

A. NATIONAL INTEREST AND HUMAN RIGHTS

The clash of positions in the debates over conditionality evokes similar clashes throughout American history about the actual or necessary or desired character of foreign policy. The clash can be defined in various ways, including such broad notions as: idealism vs. realism, altruism vs. self-interest, moral ideals vs. national interest, principles vs. power.

The first term (the more 'interventionist') in each of these pairs could be characterized in most instances as tending to favor forms of conditionality in U.S. aid and trade legislation. The second term could be understood as tending to oppose human rights conditions, for they frustrate aid or trade policies that are believed to be in the national interest. Often the debate has had a more ambiguous and contextual character, in which questions like the effectiveness of aid–trade sanctions as compared with alternative paths to achieve human rights goals have been dominant.

Why indeed should a country like the United States take human rights considerations into account in its bilateral relations with other states? If it is in the 'national interest' to provide a given country with security (military) assistance or with development aid, does not making such aid dependent on the recipient's compliance with human rights norms impair that 'interest'? Is not the United States (or any other country following similar policies) surrendering its practical and ideological concerns in order to act as a global policeman, to enforce human rights norms at a cost to its own interests? Is not a purpose of the human rights movement to lift the problem of enforcement from the level of states to that of international organizations in which many states participate?

These are the issues debated in the following introductory readings.

HANS MORGENTHAU, HUMAN RIGHTS AND FOREIGN POLICY

reprinted in Kenneth Thompson (ed.), Moral Dimensions of American Foreign Policy 341 (1984), at 344.

. . .

That consideration brings me to the popular issue with which the problem of morality in foreign policy presents us today, and that is the issue of what is now called human rights. That is to say, to what extent is a nation entitled and obligated to impose its moral principles upon other nations? To what extent is it both morally just and intellectually tenable to apply principles we hold dear to other nations that, for a number of reasons, are impervious to them? It is obvious that the attempt to impose so-called human rights upon others or to punish others for not observing human rights assumes that human rights are of universal validity—that, in other words, all nations or all peoples living in different nations would embrace human rights if they knew they existed and that in any event they are as inalienable in their character as the Declaration of Independence declares them to be.

I'm not here entering into a discussion of the theological or strictly philosophic nature of human rights. I only want to make the point that whatever one's conception of that theological or philosophical nature, those human rights are filtered through the intermediary of historic and social circumstances, which will lead to different results in different times and under different circumstances. One need only look at the unique character of the American polity and at these very special, nowhere-else-to-be-found characteristics of our protection of human rights within the confines of America. You have only to look at the complete lack of respect for human rights in many nations . . . to realize how daring—or how ignorant if you will, which can also be daring—an attempt it is to impose upon the rest of the world the respect for human rights or in particular to punish other nations for not showing respect for human rights. What we are seeing here is an abstract principle we happen to hold dear, which we happen to have put to a considerable extent into practice, presented to the rest of mankind not for imitation but for acceptance.

It is quite wrong to assume that this has been the American tradition. It has not been the American tradition at all. Quite the contrary. I think it was John Quincy Adams who made the point forcefully that it was not for the United States to impose its own principles of government upon the rest of mankind, but, rather, to attract the rest of mankind through the example of the United States. And this has indeed been the persisting principle the United States has followed. . . . This has been the great difference between the early conception of America and its relations to the rest of the world on the one hand and what you might call the Wilsonian conception on the other.

For Wilson wanted to make the world safe for democracy. He wanted to transform the world through the will of the United States. The Founding Fathers wanted to present to the nations of the world an example of what man can do and called upon them to do it. So there is here a fundamental difference, both philosophic and political, between the present agitation in favor of human rights as a universal principle to be brought by the United States to the rest of the world and the dedication to human rights as an example to be offered to other nations—which is, I think, a better example of the American tradition than the Wilsonian one.

There are two other objections that must be made against the Wilsonian conception. One is the impossibility of enforcing the universal application of human rights. We can tell the Soviet Union, and we should from time to time tell the Soviet Union, that its treatment of minorities is incompatible with our conception of human rights. But once we have said this we will find that there is very little we can do to put this statement into practice. . . . There are other examples where private pressure—for example, the shaming of public high officials in the Soviet Union by private pressure—has had an obvious result. But it is inconceivable I would say on general grounds, and more particularly in view of the experiences we have had, to expect that the Soviet Union will yield to public pressure when public pressure becomes an instrument of foreign policy and will thereby admit its own weakness in this particular field and the priority of the other side as well. . . .

There is a second weakness of this approach, which is that the United States is a great power with manifold interests throughout the world, of which human rights is only one and not the most important one, and the United States is incapable of consistently following the path of the defense of human rights without maneuvering itself into a Quixotic position. This is obvious already in our discriminating treatment of, let me say, South Korea on the one hand and the Soviet Union on the other. Or you could mention mainland China on the one hand and the Soviet Union on the other. We dare to criticize and affront the Soviet Union because our relations in spite of being called détente, are not particularly friendly. We have a great interest in continuing the normalization of our relations with mainland China, and for this reason we are not going to hurt her feelings. On the other hand South Korea is an ally of the United States, it is attributed a considerable military importance, and so we are not going to do anything to harm those relations.

In other words, the principle of the defense of human rights cannot be consistently applied in foreign policy because it can and it must come in conflict with other interests that may be more important than the defense of human rights in a particular instance. And to say—as the undersecretary of state said the other day—that the defense of human rights must be woven into the fabric of American foreign policy is, of course, an attempt to conceal the actual impossibility of consistently pursuing the defense of human rights. And once you fail to defend human rights in a particular instance, you have given up the defense of human rights and you have accepted another principle to guide your actions. And this is indeed what has happened and is bound to happen if you are not a Don Quixote who foolishly but consistently follows a disastrous path of action.

. . .

SAMUEL HUNTINGTON, AMERICAN IDEALS VERSUS AMERICAN INSTITUTIONS

in G. John Ikenberry (ed.), American Foreign Policy 223 (1989), at 239.

. . .

In the eyes of most Americans not only should their foreign-policy institutions be structured and function so as to reflect liberal values, but American foreign policy should also be substantively directed to the promotion of those values in the external environment. This gives a distinctive cast to the American role in the world. In a famous phrase Viscount Palmerston once said that Britain did not have permanent friends or enemies, it only had permanent interests. Like Britain and other countries, the United States also has interests, defined in terms of power, wealth, and security, some of which are sufficiently enduring as to be thought of as permanent. As a founded society, however, the United States also has distinctive political principles and values that define its national identity. These principles provide a second set of goals and a second set of standards—in addition to those of national interest—by which to shape the goals and judge the success of American foreign policy.

This heritage, this transposition of the ideals-versus-institutions gap into foreign policy, again distinguishes the United States from other societies. Western European states clearly do not reject the relevance of morality and political ideology to the conduct of foreign policy. They do, however, see the goal of foreign policy as the advancement of the major and continuing security and economic interests of their state. Political principles provide limits and parameters to foreign policy but not to its goals. As a result European public debate over morality versus power in foreign policy has except in rare instances not played the role that it has in the United States.

. . .

The effort to use American foreign policy to promote American values abroad raises a central issue. There is a clear difference between political action to make American political practices conform to American political values and political action to make *foreign* political practices conform to American values. Americans can legitimately attempt to reduce the gap between American institutions and American values, but can they legitimately attempt to reduce the gap between other people's institutions and American values? The answer is not self-evident.

The argument for a negative response to this question can be made on at least four grounds. First, it is morally wrong for the United States to attempt to shape the institutions of other societies. Those institutions should reflect the values and behavior of the people in those societies. To intrude from outside is either imperialism or colonialism, each of which also violates American values. Second, it is difficult practically and in most cases impossible for the United States to influence significantly the institutional development of other societies. The task is simply beyond American knowledge, skill, and resources. To attempt to do so will often be counterproductive. Third, any effort to shape the domestic institutions of other societies needlessly irritates and antagonizes other governments and hence will complicate and often endanger the achievement of other more important foreign-policy goals, particularly in the areas of national security and economic well-being. Fourth, to influence the political development of other societies would require an enormous expansion of the military power and economic resources of the American government. This in turn would pose dangers to the operation of democratic government within the United States.

A yes answer to this question can, on the other hand, also be justified on four grounds. First, if other people's institutions pose direct threats to the viability of American institutions and values in the United States, an American effort to change those institutions would be justifiable in terms of self-defense. Whether or not foreign institutions do pose such a direct threat in any given circumstance is, however, not easily determined. Even in the case of Nazi Germany in 1940 there were widely differing opinions in the United States. After World War II opinion was also divided on whether Soviet institutions, as distinct from Soviet policies, threatened the United States.

Second, the direct-threat argument can be generalized to the proposition that authoritarian regimes in any form and on any continent pose a potential threat to the viability of liberal institutions and values in the United States. A liberal democratic system, it can be argued, can only be secure in a world system of similarly constituted states. In the past this argument did not play a central role because of the extent to which the United States was geographically isolated from differently constituted states. The world is, however, becoming smaller. Given the increasing interactions among societies and the emergence of transnational institutions operating in many societies, the pressures toward convergence among political systems are likely to become more intense. Interdependence may be incompatible with coexistence. In this case the world, like the United States in the nineteenth century or Western Europe in the twentieth century, will not be able to exist half-slave and half-free. Hence the survival of democratic institutions and values at home will depend upon their adoption abroad.

Third, American efforts to make other people's institutions conform to American values would be justified to the extent that the other people supported those values. Such support has historically been much more prevalent in Western Europe and Latin America than it has in Asia and Africa, but some support undoubtedly exists in almost every society for liberty, equality, democracy, and the rights of the individual. Americans could well feel justified in supporting and helping those individuals, groups, and institutions in other societies who share their belief in these values. At the same time it would also be appropriate for them to be aware that those values could be realized in other societies through institutions significantly different from those that exist in the United States.

Fourth, American efforts to make other people's institutions conform to American values could be justified on the grounds that those values are universally valid and universally applicable, whether or not most people in other societies believe in them. For Americans not to believe in the universal validity of American values could indeed lead to a moral relativism: liberty and democracy are not inherently better than any other political values; they just happen to be those that for historical and cultural reasons prevail in the United States. This relativistic position runs counter to the strong elements of moral absolutism and messianism that are part of American history and culture, and hence the argument for moral relativism may not wash in the United States for relativistic reasons. In addition the argument can be made that some element of belief in the universal validity of a set of political ideals is necessary to arouse the energy, support, and passion to defend those ideals and the institutions modeled on them in American society.

Historically Americans have generally believed in the universal validity of their values. At the end of World War II, when Americans forced Germany and Japan to be free, they did not stop to ask if liberty and democracy were what the German and Japanese people wanted. Americans implicitly assumed that their values were valid and applicable and that they would at the very least be morally negligent if they did not insist that Germany and Japan adopt political institutions reflecting those values. Belief in the universal validity of those values obviously reinforces and reflects those hypocritical elements of the American tradition that stress the United States's role as a redeemer nation and lead it to attempt to impose its values and often its institutions on other societies. . . .

. . .

QUESTIONS

1. The article by Morgenthau was written before the growth of the human rights movement. What changes in these excerpts, if any, would a half century of that movement require?

2. What criticism could be made of Huntington's discussion of 'American values' and of values that are 'universally valid and universally applicable,' as applied to notions of democracy and individual rights?

B. HUMAN RIGHTS CONDITIONALITY FOR U.S. SECURITY AND DEVELOPMENT ASSISTANCE

LOUIS HENKIN, THE AGE OF RIGHTS (1990), at 66.

Ch. 5. Human Rights and United States Foreign Policy

. . .

What is the human rights policy of the United States and what is the place of human rights in United States foreign policy generally? I suggest that the confusion of United States policy reflects not only, or principally, different policies at different times by different administrations, but, rather, more than one policy at any time—a Congressional policy and a different executive policy; one policy in respect of international human rights in some countries and another policy for other countries; one policy abroad and another at home.

EXECUTIVE AMBIVALENCE

. . .

Human rights in other countries was the particular preoccupation of 'liberal,' 'idealistic' elements which had come into the 'foreign policy establishment' during the war, and remained when the war was over. . . .

. . . [L]eading members of the traditional foreign policy establishment, notably the career foreign service, tended to find the new international human rights movement 'unsophisticated,' and at best a nuisance. They were inclined to consider human rights conditions in any other country that country's business, and active concern with those conditions by the United States, or by international institutions, to be meddlesome, officious, unprofessional, disturbing of 'friendly relations' and disruptive of sound diplomacy. During war, they had seen no reason to resist rhetorical declarations that served the needs of morale and psychological warfare, but they looked with growing concern when the wartime spirit and the influence of its amateur supporters continued in the postwar years. They were skeptical of international institutions generally, and resisted particularly their involvement in the internal affairs of states, such as human rights.

'Idealists' and 'realists' served the United States side-by-side, but they looked in different directions and saw United States interests differently. As the glow of victory and the 'spirit of the United Nations' waned, the influence of the human rights contingent receded and traditional diplomats again dominated. They concerned themselves with other important things: security, alignments, military bases, trade. But the human rights movement continued to command wide support from church and other 'do-good' bodies, and from particular ethnic constituencies, and therefore some support in Congress and even in the White House. It was a continuing activity of international organizations and therefore of those who represented the United States in those bodies, principally part-time citizen-diplomats at periodic meetings, and of the newfangled bureaus of the State Department. On the United Nations side-track, the United States joined and often led the human rights bandwagon. 'Realists' in the State Department remained skeptical but were not disposed to challenge that program as long as it remained on the plane of rhetoric and was not allowed to disturb the sensibilities of particular states or roil relations with them. . . .

. . . For the most part, however, Congress did not attend seriously to the condition of human rights in other countries during the first twenty-five years of the postwar era and generally acquiesced in what the executive branch did. Congress had little occasion for formal involvement in the development of United States human rights policy. . . . The Senate, whose advice and consent is constitutionally required to human rights treaties as to others, had few occasions to consider agreements that aimed at the condition of rights in other countries. . . .

An independent Congressional initiative to shape United States human rights policy developed in the early 1970s. Under influence of concerned liberal members of the House of Representatives, and responding to inadequacies in United Nations and other multilateral responses to human rights violations, Congress enacted a series of statutes declaring the promotion of respect for human rights to be a principal goal of United States foreign policy, and denying foreign aid, military assistance, and the sale of agricultural commodities to states guilty of gross violations of internationally recognized human rights. In addition, United States representatives were directed to act in international financial institutions so as to prevent or discourage loans to governments guilty of such violations. Congress also established a human rights bureau in the Department of State, and directed the department to report annually on the condition of human rights in every country in the world.

The Congressional program, it should be clear, was directed not at deviations from democratic governance as practiced by the United States (and by its European allies) but against 'consistent patterns of gross violations of internationally recognized human rights,' those that nations publicly decried and that none claimed the right to do or admitted doing. Congress specified clearly the violations at which it aimed—'torture or cruel, inhuman or degrading treatment or punishment, prolonged detention without charges, causing the disappearance of persons by the abduction and clandestine detention of those persons, or other flagrant denial of the right to life, liberty or the security of person.' Also, it should be clear, this general legislation was not aimed at Communism and the Communist states since they received neither arms nor aid from the United States, but at the non-Communist Third World. In addition, Congress addressed human rights in particular countries, e.g., denying various aid to Chile, Argentina, South Africa, Uganda and others, when the condition of human rights in those countries was particularly egregious. Later, Congress imposed various human rights conditions on assistance to particular countries in Central America. In 1986, Congress enacted the Comprehensive Anti-Apartheid Act.

The Congressional program was never popular with the executive branch (regardless of political party), particularly with those who reflected the dominant, traditional attitudes in the Department of State. That program limited executive autonomy in the conduct of foreign policy. It required embassies to collect information often critical of the countries in which they 'lived'; it required the Department of State to publish information often critical of countries with which the United States had friendly relations. It injected into foreign policy elements that foreign governments, and many in the State Department, thought not to be United States business. It sometimes disturbed alliances and alignments, base agreements or trade arrangements, and friendly relations generally.

Congress made some concessions to executive branch resistance. It gave the Aid Administrator authority to disregard the statutory limitation when assistance 'will directly benefit the needy people in such country.' It authorized security assistance to a country guilty of gross violations if the President certified that 'extraordinary circumstances exist warranting provision of such assistance,' or if the President finds that 'such a significant improvement in its human rights record has occurred as to warrant lifting the prohibition on furnishing such assistance in the national interest of the United States.' . . .

In the main, the tension between Congress and the executive branch reflected not partisan or political differences, but the different positions and perspectives of the two branches. Congress was closer to popular sentiment in the United States, which was responsive to the human condition in other countries, and wished to do something about it or at least to dissociate the United States from repressive regimes in general or in particular countries. The executive branch, more removed from constituent influence in the United States, was closer to official sentiment in other countries with which it had to deal; it was not indifferent to, but less swayed by, moral concerns, more attuned to international political and diplomatic needs and mores. . . .

. . .

I have described tensions within the executive branch and differences between the two branches. There have also been differences between Presidents and between presidential administrations, reflecting some partisan or ideological differences and some personal differences.

. . .

In sum, human rights legislation in the United States does not govern as other law does. Although gross violations of human rights are rampant in many countries, including some that are important beneficiaries of United States aid and arms trade, there have been virtually no cases in which military assistance or foreign aid was in fact cut off on human rights grounds. But the law is hardly a dead letter and Congress is not a toothless tiger. The existence of the law, the constitutional posture of Congress and President, establish a political context and generate a process that have important human rights consequences. Sometimes they deter Presidents from asking for aid for blatant violators, as in Guatemala. Sometimes they compel the President to press would-be beneficiary governments to act to improve the human rights condition in these countries, so that the President could certify to Congress at least significant improvement. Law and process also cause human rights policy to respond to political events, to United States relations with particular countries, to degrees of human rights violations. By a kind of *pas de deux* of President and Congress, by a combination of promise and threat, United States human rights policy has achieved a spectrum of influence on the condition of human rights around the world.

. . .

PROSPECT

The human rights policy of the United States is part of a larger foreign policy in the national interest broadly conceived. National interest is not a simple, single concern, and United States foreign policy has often struggled with competing national interests. But the ambiguities, ambivalences, and contradictions of United States human rights policy have been particularly glaring. . . . There are few voices now to insist that human rights elsewhere are not the business of the United States. President Carter proclaimed and made it so, and in different ways and with different emphasis, the Reagan administration confirmed that it is, originally only where Communism is or threatens, later elsewhere—in South Africa, and through pressures for democracy in Guatemala, Haiti, the Philippines.

No one in the United States suggests that the United States should end human rights violations in other countries by war and conquest; that would be a violation of international law and would bring more human suffering than it would cure. Some favor economic sanctions such as boycotts but most knowledgeable Americans recognize that such measures are generally ineffective and might damage other United States interests, though it may sometimes be necessary to make a political-moral statement as Congress did in enacting the Comprehensive Anti-Apartheid Act of 1986. But human rights advocates insist that the United States is fully entitled to withhold its foreign aid and deny arms to regimes that are guilty of consistent patterns of gross violations of human rights, as it has denied them to Communist countries. . . .

. . .

UNITED STATES LEGISLATION ON MILITARY AND ECONOMIC AID: HUMAN RIGHTS PROVISIONS

Human Rights and Security Assistance

Sec. 502B of the Foreign Assistance Act of 1961, as amended, 22 U.S.C.A. §2304.

(a) Observance of human rights as principal goal of foreign policy; Implementation requirements

(1) The United States shall, in accordance with its international obligations as set forth in the Charter of the United Nations and in keeping with the constitutional heritage and traditions of the United States, promote and encourage increased respect for human rights and fundamental freedoms throughout the world without distinction as to race, sex, language, or religion. Accordingly, a principal goal of the foreign policy of the United States shall be to promote the increased observance of internationally recognized human rights by all countries.

(2) Except under circumstances specified in this section, no security assistance may be provided to any country the government of which engages in a consistent pattern of gross violations of internationally recognized human rights. Security assistance may not be provided to the police, domestic intelligence, or similar law enforcement forces of a country . . . the government of which engages in a consistent pattern of gross violations of internationally recognized human rights unless the President certifies in writing . . . that extraordinary circumstances exist warranting provision of such assistance. . . .

(3) In furtherance of paragraphs (1) and (2), the President is directed to formulate and conduct international security assistance programs of the United States in a manner which will promote and advance human rights and avoid identification of the United States, through such programs, with governments which deny to their people internationally recognized human rights and fundamental freedoms, in violation of international law or in contravention of the policy of the United States as expressed in this section or otherwise.

(b) Report by Secretary of State on practices of proposed recipient countries; considerations

The Secretary of State shall transmit to the Congress, as part of the presentation materials for security assistance programs proposed for each fiscal year, a full and complete report, prepared with the assistance of the Assistant Secretary of State for Democracy, Human Rights, and Labor, with respect to practices regarding the observance of and respect for internationally recognized human rights in each country proposed as a recipient of security assistance. Wherever applicable, such report shall include information on practices regarding coercion in population control, including coerced abortion and involuntary sterilization. In determining whether a government falls within the provisions of subsection (a)(3) of this section and in the preparation of any report or statement required under this section, consideration shall be given to—

(1) the relevant findings of appropriate international organizations, including nongovernmental organizations, such as the International Committee of the Red Cross; and

(2) the extent of cooperation by such government in permitting an unimpeded investigation by any such organization of alleged violations of internationally recognized human rights.

(c) Congressional request for information; information required; 30-day period; failure to supply information; termination or restriction of assistance.

(1) Upon the request of the Senate or the House of Representatives by resolution of either such House, or upon the request of the Committee on Foreign Relations of the Senate or the Committee on Foreign Affairs of the House of Representatives, the Secretary of State shall, within thirty days after receipt of such request, transmit to both such committees a statement, prepared with the assistance of the Assistant Secretary of State for Democracy, Human Rights, and Labor, with respect to the country designated in such request, setting forth—

(A) all the available information about observance of and respect for human rights and fundamental freedom in that country, and a detailed description of practices by the recipient government with respect thereto;

(B) the steps the United States has taken to—

(i) promote respect for and observance of human rights in that country and discourage any practices which are inimical to internationally recognized human rights, and

(ii) publicly or privately call attention to, and disassociate the United States and any security assistance provided for such country from, such practices;

(C) whether, in the opinion of the Secretary of State, notwithstanding any such practices—

(i) extraordinary circumstances exist which necessitate a continuation of security assistance for such country, and, if so, a description of such circumstances and the extent to which such assistance should be continued (subject to such conditions as Congress may impose under this section), and

(ii) on all the facts it is in the national interest of the United States to provide such assistance; and

(D) such other information as such committee or such House may request.

. . .

(3) In the event a statement with respect to a country is requested pursuant to paragraph (1) of this subsection but is not transmitted in accordance therewith within thirty days after receipt of such request, no security assistance shall be delivered to such country except as may thereafter be specifically authorized by law from such country unless and until such statement is transmitted.

. . .

(d) Definitions

For the purposes of this section—

(1) the term 'gross violations of internationally recognized human rights' includes torture or cruel, inhuman, or degrading treatment or punishment, prolonged detention without charges and trial, causing the disappearance of persons by the abduction and clandestine detention of those persons, and other flagrant denial of the right to life, liberty, or the security of person. . . .

. . .

(e) Removal of prohibition on assistance

Notwithstanding any other provision of law, funds authorized to be appropriated under subchapter I of this chapter may be made available for the furnishing of assistance to any country with respect to which the President finds that such a significant improvement in its human rights record has occurred as to warrant lifting the prohibition on furnishing such assistance in the national interest of the United States.

. . .

* * * * *

Human Rights and Development Assistance

Sec. 116 of the Foreign Assistance Act of 1961, as amended 22 U.S.C.A. §2151n.

(a) Violations barring assistance; assistance for needy people

No assistance may be provided under subchapter I of this chapter to the government of any country which engages in a consistent pattern of gross violations of internationally recognized human rights, including torture or cruel, inhuman, or degrading treatment or punishment, prolonged detention without charges, causing the disappearance of persons, by the abduction and clandestine detention of those persons, or other flagrant denial of the right to life, liberty, and the security of person, unless such assistance will directly benefit the needy people in such country.

(b) Protection of children from exploitation

No assistance may be provided to any government failing to take appropriate and adequate measures, within their means, to protect children from exploitation, abuse or forced conscription into military or paramilitary services.

[Eds. This statute as amended contains two subsections '(b)'.]

(b) Information to Congressional committees for realization of assistance for needy people; concurrent resolution terminating assistance

In determining whether this standard is being met with regard to funds allocated under subchapter I of this chapter, the Committee on Foreign Relations of the Senate or the Committee on Foreign Affairs of the House of Representatives may require the Administrator primarily responsible for administering subchapter I of this chapter to submit in writing information demonstrating that such assistance will directly benefit the needy people in such country, together with a detailed explanation of the assistance to be provided (including the dollar amounts of such assistance) and an explanation of how such assistance will directly benefit the needy people in such country. . . .

(c) Factors considered

In determining whether or not a government falls within the provisions of subsection (a) of this section and in formulating development assistance programs under subchapter 1 of this chapter, the Administrator shall consider, in consultation with the Assistant Secretary for Democracy, Human Rights, and Labor—

(1) the extent of cooperation of such government in permitting an unimpeded investigation of alleged violations of internationally recognized human rights by appropriate international organizations, including the International Committee of the Red Cross, or groups or persons acting under the authority of the United Nations or of the Organization of American States; and

(2) specific actions which have been taken by the President or the Congress relating to multilateral or security assistance to a less developed country because of the human rights practices or policies of such country.

(d) Report to Speaker of House and Committee on Foreign Relations of the Senate

The Secretary of State shall transmit to the Speaker of the House of Representatives and the Committee on Foreign Relations of the Senate, by January 31 of each year, a full and complete report regarding—

(1) the status of internationally recognized human rights, within the meaning of subsection (a) of this section—

(A) in countries that receive assistance under subchapter I of this chapter, and

(B) in all other foreign countries which are members of the United Nations and which are not otherwise the subject of a human rights report under this chapter;

(2) wherever applicable, practices regarding coercion in population control, including coerced abortion and involuntary sterilization. . .

. . .

QUESTIONS

1. 'The U.S. legislation imposing human rights conditions on security and development aid is too restricted, selective. It arbitrarily carves out from the human rights corpus certain types of violations and ignores other vital ones. There seems to be no justification in the human rights movement for making these distinctions.'

a. What kinds of human rights violations are included in the statutory definitions? Is there any unity among them?

b. Would a bare reference to 'gross violations of internationally recognized human rights' have been preferable? If a selective list is preferable, do you agree with the present inclusions/exclusions?

2. 'I fail to see what long-run purposes this legislation serves. A country is asked to stop a given practice, like arbitrary detentions. It does so. All bad structures—military control, repression of speech and association, economic control by a small elite, and so on—stay in place. As soon as the practice is relaxed, as soon as it's on a downward curve, aid can be resumed. Of course the practice itself will be reinstated in its prior vigor when politically necessary. What has been accomplished?' Comment.

SECRETARY OF STATE CYRUS VANCE, SPEECH ON HUMAN RIGHTS AND FOREIGN POLICY

76 Dep't of State Bull. 505 (1977).

. . . Let me define what we mean by 'human rights.'

First, there is the right to be free from governmental violation of the integrity of the person. Such violations include torture; cruel, inhuman, or degrading treatment or punishment; and arbitrary arrest or imprisonment. And they include denial of fair public trial and invasion of the home.

Second, there is the right to the fulfillment of such vital needs as food, shelter, health care, and education. We recognize that the fulfillment of this right will depend, in part, upon the stage of a nation's economic development. But we also know that this right can be violated by a government's action or inaction—for example, through corrupt official processes which divert resources to an elite at the expense of the needy or through indifference to the plight of the poor.

Third, there is the right to enjoy civil and political liberties: freedom of thought, of religion, of assembly; freedom of speech; freedom of the press; freedom of movement both within and outside one's own country; freedom to take part in government.

Our policy is to promote all these rights. They are all recognized in the Universal Declaration of Human Rights, a basic document which the United States helped fashion and which the United Nations approved in 1948. There may be disagreement on the priorities these rights deserve. But I believe that, with work, all of these rights can become complementary and mutually reinforcing.

. . .

In pursuing a human rights policy, we must always keep in mind the limits of our power and of our wisdom. A sure formula for defeat of our goals would be a rigid, hubristic attempt to impose our values on others. A doctrinaire plan of action would be as damaging as indifference.

We must be realistic. Our country can only achieve our objectives if we shape what we do to the case at hand. . . .

. . .

A second set of questions concerns the prospects for effective action:

Will our action be useful in promoting the overall cause of human rights?

Will it actually improve the specific conditions at hand? Or will it be likely to make things worse instead?

Is the country involved receptive to our interest and efforts?

Will others work with us, including official and private international organizations dedicated to furthering human rights?

Finally, does our sense of values and decency demand that we speak out or take action anyway, even though there is only a remote chance of making our influence felt?

3. We will ask a third set of questions in order to maintain a sense of perspective:

Have we steered away from the self-righteous and strident, remembering that our own record is not unblemished?

Have we been sensitive to genuine security interests, realizing that outbreak of armed conflict or terrorism could in itself pose a serious threat to human rights?

Have we considered *all* the rights at stake? If, for instance, we reduce aid to a government which violates the political rights of its citizens, do we not risk penalizing the hungry and poor, who bear no responsibility for the abuses of their government?

If we are determined to act, the means available range from quiet diplomacy in its many forms, through public pronouncements, to withholding of assistance. Whenever possible, we will use positive steps of encouragement and inducement. Our strong support will go to countries that are working to improve the human condition. We will always try to act in concert with other countries, through international bodies.

. . .

Our policy is to be applied within our own society as well as abroad. We welcome constructive criticism at the same time as we offer it.

No one should suppose that we are working in a vacuum. We place great weight on joining with others in the cause of human rights.

The U.N. system is central to this cooperative endeavor. . . .

. . .

What results can we expect from all these efforts?

We may justifiably seek a rapid end to such gross violations as those cited in our law: 'torture or cruel, inhuman, or degrading treatment or punishment, (or) prolonged detention without charges. . . .'

. . .

The promotion of other human rights is a broader challenge. The results may be slower in coming but are no less worth pursuing. And we intend to let other countries know where we stand.

We recognize that many nations of the world are organized on authoritarian rather than democratic principles—some large and powerful, others struggling to raise the lives of their people above bare subsistence levels. We can nourish no illusions that a call to the banner of human rights will bring sudden transformations in authoritarian societies.

. . .

We seek these goals because they are right—and because we, too, will benefit. Our own well-being, and even our security, are enhanced in a world that shares common freedoms and in which prosperity and economic justice create the conditions for peace. And let us remember that we always risk paying a serious price when we become identified with repression.

Nations, like individuals, limit their potential when they limit their goals. The American people understand this. I am confident they will support foreign policies that reflect our traditional values. To offer less is to define America in ways we should not accept.

. . .

U.S. DEPARTMENT OF STATE, COUNTRY REPORTS ON HUMAN RIGHTS PRACTICES FOR 1994

(1995), at xi.

Preface

This report is submitted to the Congress by the Department of State in compliance with sections 116(d) and 502(b) of the Foreign Assistance Act of 1961 (FAA), as amended, and section 505(c) of the Trade Act of 1974, as amended.

. . .

. . .

In August 1993, the Secretary of State moved to strengthen further the human rights efforts of our embassies. All sections in each embassy were asked to contribute information and to corroborate reports of human rights violations, and new efforts were made to link mission programming to the advancement of human rights and democracy. This year, the Bureau of Human Rights and Humanitarian Affairs was reorganized and renamed as the Bureau of Democracy, Human Rights, and Labor, reflecting both a broader sweep and a more focused approach to the interlocking issues of human rights, worker rights, and democracy. The 1994 human rights reports reflect a year of dedicated effort by hundreds of State Department, Foreign Service, and other U.S. Government employees.

Our embassies, which prepared the initial drafts of the reports, gathered information throughout the year from a variety of sources across the political spectrum, including government officials, jurists, military sources, journalists, human rights monitors, academics, and labor activists. This information-gathering can be hazardous, and U.S. Foreign Service Officers regularly go to great lengths, under trying and sometimes dangerous conditions, to investigate reports of human rights abuse, monitor elections, and come to the aid of individuals at risk, such as political dissidents and human rights defenders whose rights are threatened by their governments.

After the embassies completed their drafts, the texts were sent to Washington for careful review by the Bureau of Democracy, Human Rights, and Labor, in cooperation with other State Department offices. . . . The guiding principle was to ensure that all relevant information was assessed as objectively, thoroughly, and fairly as possible.

The reports in this volume will be used as a resource for shaping policy, conducting diplomacy, and making assistance, training, and other resource allocations. They will also serve as a basis for the U.S. Government's cooperation with private groups to promote the observance of internationally recognized human rights.

NOTE

An article entitled *Tidings of Abuse Fall on Deaf Ears*, in the *New York Times*, Feb, 5, 1995, Sec. 4, p. 4, refers to a report by Amnesty International (AI) on human rights violations committed during the civil war in Afghanistan: rape of young girls, vicious treatment of prisoners, a 'human rights catastrophe' of 'appalling proportions.' Yet, notes AI, 'governments around the world are ignoring the tragedy.' The U.S. was no exception; reports distributed by AI to members of Congress were disregarded. 'There was no interest.' The article questions what impact NGO or other reports on human rights violations have. 'Has the public become inured' to the horrors of Rwanda, Chechnya, China and so on?

> With America turning inward, with foreign aid making up only one percent of the Federal budget and with the United States and Western Europe reluctant to intervene in Bosnia . . . is it possible that human rights reports have become a dead end?
>
> Even if no one else is paying attention to human rights reports, the accused governments are. Their efforts to soften a report before it is published, or failing that, to criticize it after it is published indicate how seriously such reports are taken.

STEPHEN COHEN, CONDITIONING U.S. SECURITY ASSISTANCE ON HUMAN RIGHTS PRACTICES

76 Am. J. Int. L. 246 (1982).

. . .

The international law of human rights imposes obligations on governments in the exercise of their domestic sovereignty. Under the Universal Declaration, the United Nations Covenants, the European and Inter-American Conventions, and the Helsinki accords, states have defined and agreed to respect certain basic freedoms of persons within their jurisdiction. These obligations suggest a corresponding duty of one government not to support another engaged in serious violations of internationally recognized human rights. In the world of states, the enforcement of international obligations depends on the recognition and implementation of such a duty. While its contours have yet to be articulated under international law, this duty has already been the subject of domestic American statutes that set forth rules for the Executive's conduct of foreign relations with repressive governments. . . .

Congress believed this legislation to be required by American self-interest, as well as supported by principles of international law. The gift of aid and the sale of arms, it is true, can promote important national security objectives of the United States. In addition to enhancing the defense capability of friendly countries, military ties give the United States influence with and access to other governments. Yet governments to which the United States sends military aid and arms are viewed, by their own people and by the world community, as in part sustained and even approved by us. When the government in question is repressive, the perception of American support for it can impose considerable costs for American interests. The government may be overthrown and the supply of American aid and arms deeply resented by its successors. The support of repressive governments also conflicts with traditional American values and damages the international reputation of the United States because of the appearance of complicity in repugnant practices.

The principal (although not exclusive) legislative enactment on human rights and military ties has been section 502B of the Foreign Assistance Act of 1961. . . .

. . .

In 1974, when Congress originally enacted section 502B, it also began to pass legislation to limit military aid and arms sales to designated countries. The country-specific legislation was contained in bills authorizing or appropriating funds for military aid, and it named, at various times, Argentina, Brazil, Chile, El Salvador, Guatemala, Paraguay, the Philippines, South Korea, Uruguay, and Zaire. The legislation usually mentioned only military aid and was effective for a single year, although in two cases it expressly prohibited arms sales as well and was of indefinite duration.

Congress has successfully attached country-specific legislation to bills authorizing or appropriating military aid for every fiscal year since 1975. For example, Chile was prohibited from receiving any military aid in fiscal 1975 and the amount requested for South Korea was substantially reduced. . . .

. . .

The opposition of the Foreign Service to section 502B was a logical consequence of its conception of its special role or of (what one student of the bureaucracy has labeled) its 'organizational essence.' The Foreign Service views its primary role or essence as the maintenance of smooth and cordial relations with other governments. It believes that military aid and arms sales are an indispensable means to achieving this goal. . . .

Keeping other governments happy becomes an end in itself. This phenomenon is often referred to as 'clientism' because the Foreign Service views other governments as 'clients' with whose interests it identifies, rather than as parties to be dealt with at arm's length according to the national interest of the United States.

. . .

Furthermore, costs are typically excluded from the calculation. The time horizon of the career bureaucracy is short and the possible long-run disadvantage of aiding repressive governments seems alien to its thinking. Ever after the examples of Iran and Nicaragua, there is rarely consideration that a particular regime may be overthrown and United States support for it resented by those who follow. Nor is the damage to U.S. standing in the world, or the conflict with both international legal obligations and traditional American values, weighed in the balance. In effect, the career bureaucracy develops its position by reckoning the national interest in terms of pleasing its 'clients,' the governments with which it conducts diplomatic relations.

The phenomenon of 'clientism' has a number of causes. A Foreign Service officer is typically required to develop personal relations and spend substantial periods of time with high officials of other governments. He tends, therefore, to sympathize and identify with their point of view. If the other government is accused of human rights abuses, he deals with officials who either deny the accusations or explain the excesses as regrettable, but necessary to stem 'terrorism' and avoid social chaos. He is much less likely to encounter the victims of repression and hear their point of view.

. . .

. . . [T]he career bureaucracy attempted to distort information about human rights conditions in particular countries. The extent of abusive practices was consistently underreported. For example, the Bureau for East Asia persisted in arguing that reports of Indonesian abuses in East Timor were grossly exaggerated and that few abuses had been committed by the Indonesian Army against the Timorese. In fact, an accumulation of reports from reputable sources indicates that a hundred thousand or more Timorese may have died at the hands of the Indonesian military. In the case of Argentina, the Latin American Bureau argued that, at most, hundreds of individuals had been summarily executed by security forces. As the evidence became incontrovertible that the number was actually 6,500 or more, the bureau shifted gears and argued that only Marxist terrorists were the victims. When it was documented that most of the victims were neither Marxists nor terrorists, the bureau maintained that the abuses were the work of local military commanders whom the ruling junta was struggling to control.

As it minimized or concealed negative aspects of a 'client's' human rights practices, the career bureaucracy exaggerated positive signs. Improvements were said to have occurred on the basis of insubstantial evidence or self-serving declarations of the government in power. . . .

. . . [T]he regional bureaus overstated the extent of U.S. interests at stake in particular cases and the damage that could possibly result from failure to approve proposed security assistance. The Latin American Bureau, for example, insisted (albeit unsuccessfully) that failure to provide military aid to Paraguay would seriously undermine a relationship with the

Stroessner Government that was important to fighting communism in Latin America.

. . .

Given the resistance of the career bureaucracy, concentrated in the regional bureaus, implementation of section 502B during the Carter administration depended on the newly created Bureau of Human Rights, headed by an outsider who was personally committed to the policy of the statute and staffed, to a significant degree, by persons from outside the career bureaucracy. The new bureau began to serve as a counterweight to the 'clientism' of the regional bureaus. . . .

. . .

Perhaps the most remarkable evidence of the Carter administration's conservative approach to section 502B was its policy never to determine formally, even in a classified decision, that a particular government was engaged in gross abuses. The primary reason for this policy was the belief that such a determination, even if classified, would inevitably be leaked to the press and become generally known. It was feared that each country named would then consider itself publicly insulted, with consequent damage to our bilateral relationship. In addition, there was concern that once such a finding was revealed, the freedom to alter it might be severely constrained by public political pressures. . . .

In practice, the Secretary of State had to resist pressures both from Congress and within his own Department to make such findings. Administration representatives repeatedly refused congressional requests for a list of governments considered to be engaged in gross abuses, stating that it was administration policy not to draw up such a list. Within the Department, the Secretary of State sought to avoid explaining the reason for decisions on security assistance either in writing or even informally. When he resolved a dispute, he often communicated simply whether the request was approved or disapproved and little more. Particularly in cases of disapproval, the Secretary strenuously avoided ever stating that it was required to section 502B since such a statement would have meant that the government in question was considered to be engaged 'in a consistent pattern of gross violations of human rights.'

. . .

. . . After the initial period, however, when a body of precedents was created, the amount of 'litigation' began to decrease as contesting bureaus inferred by the pattern of outcomes how the Secretary of State was interpreting section 502B. The major issues of interpretation are discussed below. The three key questions were:

(1) When was a foreign government considered to be engaged 'in a consistent pattern of gross violations of internationally recognized human rights'?

(2) What U.S. interests constituted 'extraordinary circumstances'?
(3) What was encompassed by the category 'security assistance'?

1. Gross Violations. The threshold issue under section 502B is whether a particular government 'engages in a consistent pattern of gross violations of internationally recognized human rights.' At the outset, it is necessary to explain that the issue never arose with respect to many governments that appear to have practiced gross abuses. The question whether section 502B might apply to them was never raised because they were never seriously considered for security assistance for a variety of reasons other than human rights. . . .

After these countries were eliminated, approximately 70 remained that were seriously considered for military aid and arms sales. A narrow reading of the 'gross violations' language was adopted, so that, in the end, only about 12 of the 70 were thought to fall within it. While in some respects the narrow reading was consistent with the congressional purpose, in others it appeared to subvert the intended meaning of the statute. The narrow reading was derived from the way that the four basic elements of the category were interpreted. First, there must be violations of 'internationally recognized human rights.' Second, the violations must be 'gross.' Third, the frequency of violations must result in a 'consistent pattern.' Fourth, the government itself must be engaged, which is to say, responsible for the violations.

. . .

The Carter administration followed Fraser's interpretation. In its decisions on security assistance, it was careful to go no further than required by the abuses specifically listed in subsection (d)(1). A government was considered to fall within it only when it practiced arbitrary imprisonment, torture, or summary execution of relatively large numbers of its own people. That a government was authoritarian, denied basic civil and political liberties, or failed to promote basic economic and social rights was not, by itself, enough to invoke the statutory prohibition on military ties.

The second element, that the violations must be 'gross,' was read to mean that they must be significant in their impact. For example, although arbitrary imprisonment is one of the listed violations, detention without charges for several days was not considered 'gross' because of the relatively brief period of confinement.

Third, the element of a 'consistent pattern' was held to mean that abuses had to be significant in number and recurrent. Isolated instances of torture or summary execution, while certainly gross abuses, would not trigger termination of security assistance under section 502B. For example, the reported imprisonment and torture of several dozen labor leaders in Tunisia was considered not to be sufficient by itself to constitute a 'consistent pattern.' . . .

. . .

2. Extraordinary Circumstances. The Carter administration always gave considerable weight to arguments that other U.S. interests might require continuation of security assistance, even when the government in question was thought to be a 'gross violator.' Thus, the charge that its pursuit of human rights was 'single-minded' and to the exclusion of other interests was far wide of the mark. If anything, the administration gave excessive credence to claims that some specific foreign policy objective would and could be promoted only if security assistance were provided, and often failed to subject such claims to rigorous analysis.

. . .

. . . In the end, human rights concerns resulted in the termination of security assistance to only eight countries, all in Latin America: Argentina, Bolivia, El Salvador, Guatemala, Haiti, Nicaragua, Paraguay, and Uruguay.

Extraordinary circumstances were found for all of the other countries considered to be gross violators. Thus, Indonesia (although technically not on the list) was a key member of ASEAN (the pro-Western association of Southeast Asian countries) and important to countering Soviet and Vietnamese influence in the region. Iran was judged critical because it shared a long border with the Soviet Union, was a major supplier of oil to the West, and defended our strategic interests in the Persian Gulf. Military ties with South Korea were deemed essential to deterring the threat of an invasion from the north. Military bases in the Philippines were judged critical to the United States and a security assistance relationship essential to keeping the bases. Finally, Zaire, the third largest country by area in Africa, was the source of nearly all the West's cobalt, a material crucial to the performance of high-performance jet engines.

. . .

IV. CONCLUSIONS

What are the lessons to be drawn for implementation by the Executive when Congress attempts to legislate foreign policy? The history of section 502B is a case study of executive frustration of congressionally mandated foreign policy and underlines the need, particularly with this kind of legislation, for clearer directives, less discretion, and more assiduous congressional oversight. While these observations emerge from experience in the human rights area, they are relevant to other congressional attempts to influence the Executive's conduct of foreign policy. They are especially apt when congressional objectives may require decisions that displease particular governments and that will therefore be resisted by the Foreign Service bureaucracy whose paramount interest is maintaining cordial relations. For example, conclusions about the effectiveness of legislation conditioning security assistance on human rights practices are likely to be highly relevant to legislation conditioning nuclear exports on practices with respect to nuclear proliferation.

Congress has the most decisive impact when its directives allow the Executive no discretion at all, as in the country-specific legislation stating precisely which governments are to be denied military aid and arms sales on human rights grounds. On the other hand, whenever the statute permits any exercise of discretion at all, the career Foreign Service is likely to use it to attempt to thwart congressional objectives.

. . .

There are, perhaps, some drawbacks to statutes that deny all discretion, such as the country-specific legislation. Once such a provision is enacted, the Executive lacks the flexibility to respond quickly to changed conditions. While Congress has a legitimate role to play in setting basic foreign policy goals, it may be less well equipped to make day-to-day decisions about how best to fulfill those goals in specific cases. Its members, by and large, are not specialists in foreign affairs, and they generally lack the detailed knowledge and expertise required to make careful assessments of human rights practices and United States interests or to balance the benefits and costs of providing military aid or selling arms to a particular government.

For these reasons, a general rule that sets forth basic goals may be preferable to country-specific legislation. Yet the history of section 502B suggests that the creation or tightening of a general rule will produce, by itself, little change in executive behavior. . . .

. . .

This examination of attempts by Congress to require the Executive to withhold military aid and arms sales on human rights grounds suggests another important issue: whether withholding is an effective instrument for enforcing adherence to international human rights law. A definitive answer to this question is far from easy to obtain. It may be difficult to determine whether a government has taken positive steps to improve human rights conditions. Changes that are merely 'cosmetic' must be distinguished from those which indicate real improvement. When positive changes do occur, it may be difficult to say whether U.S. actions or other factors were decisive. But the fact that positive changes are lacking does not mean that U.S. actions have been ineffective. Although a targeted government may not have altered current practices, it may be deterred from worse violations. Moreover, regimes other than the immediate target may be influenced by the risk of being denied United States security assistance if they engage in repression. While the impact of withholding security assistance on human rights practices is beyond the scope of this article, it is the logical next question for scholars interested in international human rights.

LAWYERS COMMITTEE FOR HUMAN RIGHTS, LINKING SECURITY ASSISTANCE AND HUMAN RIGHTS

1989.

. . .

That Section 502B [of the Foreign Assistance Act of 1961, as amended] has never been applied formally does not mean, however, that it has been useless or insignificant. The statute retains both symbolic and practical importance, and has given Congress political leverage to pressure the executive branch to support human rights.

Section 502B(c) has been used—although not frequently, and less often in recent years—by members of Congress (often responding to concerns raised by private human rights groups) to force an administration to file detailed reports justifying the provision of security assistance to certain countries. This mechanism may not have effected any specific changes in U.S. foreign policy, but it at least has forced successive administrations to confront human rights issues. The general Country Reports prepared pursuant to Section 502B(b) serve a similar function, and may have discouraged executive branch requests for security assistance to countries receiving unfavorable reports. Indeed, the sensitivity of certain governments to the reports has sometimes made further action unnecessary. Thus, in 1977 El Salvador, Guatemala, Brazil, and Uruguay responded to unfavorable Country Reports by refusing to accept any security assistance predicated on human rights conditions.

The statute has also been used to 'set the stage for follow up congressional action pertaining to specific countries. Congress can convene hearings and initiate inquiries to examine Section 502B implementation; these in turn highlight congressional interest in human rights, and can have a significant impact within the country in question. Section 502B thus provides a foundation on which subsequent human rights inquiries or legislation can build.

On a symbolic level, Section 502B remains important as the first legislation linking security assistance to human rights. Many have stressed that Section 502B should not be viewed simply as a traditional statute, but rather as a mechanism to heighten sensitivity and force an administration to consider human rights issues. As such, the statute lends 'legal and moral authority to the arguments of human rights proponents' and broadcasts the United States' interest in the protection of international human rights. This message is diluted when an administration publicly opposes the implementation of human rights measures or applies the law inconsistently. Nevertheless, Section 502B makes an administration politically accountable for its human rights policies by focusing public and congressional attention on the executive branch's treatment of human rights concerns.

. . .

Country-specific legislation has proven more effective than Section 502B because the legislative conditions can be tailored to the varying circumstances and capabilities of the target country, enabling members of Congress to address human rights issues on a measured, case-by-case basis.

A further advantage of country-specific legislation is that it may be structured to reward human rights progress by providing security assistance, rather than merely punish a country for violations by withholding already-appropriated aid. Although the characterization of country-specific legislation as a 'carrot' and Section 502B as a 'stick' is untenable—after all, country-specific legislation also may penalize human rights violators by withholding aid—the former *can* be structured as an incentive intended to reward and foster improved country practices.

...

One example of country-specific legislation is Section 728 of the International Security and Development Cooperation Act of 1981 (ISDCA), which applied to El Salvador. This provision required the President to certify within 30 days of its enactment, and every 180 days thereafter, that the Salvadoran Government had met four conditions: that it (1) make a 'concerted and significant effort to comply with internationally recognized human rights;' (2) achieve 'substantial control' over its armed forces to prevent serious human rights abuses; (3) continue economic and political reforms and (4) resolve the country's civil war through free elections.

Despite published reports that the Salvadoran security forces had killed thousands of civilians in 1981, President Reagan found, in a certification filed within 30 days of the legislation, that the four conditions had been met. The certification focused on the 'beleaguered government' and the violations of human rights by the 'insurgents,' and cited the disbanding of the paramilitary group *Orden*, decreasing violence (according to U.S. Embassy statistics), and a new code of conduct drafted for the military. These assertions were met with considerable skepticism in Congress.

...

... The final certification, in July 1983, was undoubtedly the most critical, as it acknowledged a rise in civilian deaths at the hands of security forces. Still, it stated that 'change is occurring,' whatever 'our disappointment' at its pace, and, according to human rights groups, understated the extent of political violence.

By the time of this final certification, both Congress and the Administration were dissatisfied with the certification process, although for very different reasons. Many in Congress shared the view of one member that 'the certification process has become more and more of a joke.' The Reagan Administration, meanwhile, felt that certification left it with an 'unacceptable choice' between abandoning an important strategic ally by denying it assistance, or exaggerating the extent of actual human rights progress. ...

...

Two of the aforementioned recommendations for improving the certification process—cutting only a portion of the security assistance and requiring concrete conditions to be met—are drawn from what has been called 'action-specific' legislation. Action-specific legislation 'avoids the vagueness of the certification process and provides Congress with a wider range of conditioning options than simply cutting off all military aid.'

The most prominent example of action-specific legislation is the Specter Amendment, enacted in November 1983. Dissatisfied with El Salvador's response to the conditions of Section 728 of the ISDCA, as discussed above, which included a provision requiring the government's 'good faith efforts' to investigate the December 1980 murders of four American church workers, Congress withheld thirty percent of $65 million of appropriated military assistance pending a verdict in the trial of National Guardsmen charged in the murders. The aid was delayed for six months until a verdict was obtained in May 1984. The condition was concrete, and did not involve the entire aid package. Although yielding a relatively small victory for human rights, the Specter Amendment proved more useful than had the general certification process.

While the Specter Amendment may have achieved its objective, it remains open to some criticism. First, it did not require the achievement of due process standards; thus, the Salvadoran Government had an incentive to find a scapegoat whose conviction would enable the resumption of $20 million in aid. Second, the condition did not 'extend accountability further up the military chain of command.' Third, contrary to the objectives of Section 502B, it did not promote human rights on a broad scale.

Similar criticisms apply to other examples of 'action-specific' legislation, such as the condition that Chile extradite two officials charged with the 1976 Washington D.C. slayings of Orlando Letelier and Ronni Moffitt, or the law withholding $5 million of aid authorized for FY 1985 until El Salvador took specific measures to investigate the January 1981 killings of two American labor advisers.

On balance, however, 'action-specific' legislation does appear to facilitate human rights improvements, without interfering drastically with the executive branch's ability to maintain the bilateral relationship. . . .

. . .

MAKAU MUTUA AND PETER ROSENBLUM, ZAIRE: REPRESSION AS POLICY

(1990), at 187.

United States Human Rights Policy Towards Zaire

The United States has been closely involved in Zairian politics since its independence. In 1960, the United States actively opposed the country's first elected leader, Patrice Lumumba. It supported the rise of Mobutu to the Presidency in 1965; and intervened with military support to protect the Mobutu government during the second invasion of Shaba in 1978. When the government of Zaire has been threatened with economic crisis and internal insurgency, the United States has stepped in to support it.

. . .

Throughout the last 25 years, the United States has provided significant bilateral economic and military assistance to Zaire. It has supported Zaire in international lending institutions such as the World Bank, the International Monetary Fund (IMF) and the African Development Bank. . . .

The United States is perceived by many Zairians as being responsible, at least in part, for Mr. Mobutu's continuation in power. Robert Remole, a former Political Counselor at the U.S. Embassy in Kinshasa and a retired 28-year veteran of the State Department, testified before the Subcommittee on Africa on this subject in 1980. 'I have heard this innumerable times,' he said, 'from educated Zairians, members of the Legislative Council, lawyers, other people who will be of importance when the Mobutu regime does finally collapse. 'Mr. Remole emphasized that blaming the United States was 'of course irrational, unreasonable,' but he argued that the policy of U.S. support for and identification with the government of Zaire 'bears great risk.' 'The risk,' he testified, 'is one that has been alluded to obliquely; that is, another Iran.'

U.S. support for the government of Zaire has created considerable resentment among Zairians. As one American-educated Zairian told the Lawyers Committee, 'When Mobutu goes, and he is going to go one day, the United States may find it extremely difficult to make friends in this country, because of its support for Mobutu.' He perceived the risk as very personal because he has agreed to work for an American-funded agency. 'Individuals like myself,' he told the Lawyers Committee, 'are targets of this anger.'

Nevertheless, though fully informed about President Mobutu's tactics and unpopularity, Congress and the Administration have repeatedly accepted the same justification for support to his government, namely, Zaire's strategic importance and its support for U.S. policies in the region. Professor Nzongola-Ntalaja, a Zairian academic living in the United States has observed that despite Zaire's dismal human rights record, it has earned the 'steadfast support' of the United States by faithfully promoting Western interests in Africa, especially in Angola and Chad.

In 1980, Lannon Walker, the Deputy Assistant Secretary of State for African Affairs during the Carter Administration, testified about Zaire. After noting its size and importance, he emphasized:

> And it is a fact that Zaire's moderate foreign policy orientation and close relations with the West stand in marked contrast to several countries in the area who favor more radical policies and have turned to the USSR, East Germany and Cuba for military and economic support.

Four years later, Elliot Abrams, Assistant Secretary of State for Human Rights and Humanitarian Affairs under the Reagan Administration, articulated a similar justification for U.S. support for the government of Zaire:

> As you are well aware, Zaire has long been a friend and key regional partner of the United States and the West. It has consistently worked with us . . . . Zairian minerals, notably copper and cobalt, are important to the West. Zaire's strategic relevance, which is due to its large size and population as well as its common borders with nine African countries, has never been more important than now, as we work towards political solutions in southern Africa.

In his testimony, Mr. Abrams claimed that 'Human rights conditions in Zaire had improved over the past 20 years,' and that 'progress has been made' in this area.

. . .

In 1982, as criticism of Zaire's human rights record mounted, the United States Congress reduced assistance to Zaire from $70 million to $20 million. In response, President Mobutu announced in May 1982 that he would no longer accept U.S. aid. The following month UN Ambassador Jeanne Kirkpatrick travelled to Kinshasa, in an effort to mend fences with President Mobutu. Following her visit, aid was delivered.

Two years later, President Mobutu was treated as an honored guest by the President of the United States. In an August 1984 visit to the White House, he received a 'warm welcome' from President Reagan. Mr. Reagan praised President Mobutu as 'a faithful friend for some 20 years.' He expressed admiration for Mr. Mobutu's courage in sending troops to Chad. Mr. Reagan did not comment on Zaire's dismal human rights record. In 1986, President Reagan renewed his praise for President Mobutu as 'a voice of good sense and goodwill' and an important U.S. ally in Africa.

President Mobutu has effectively used Ambassador Kirkpatrick's visit, his meetings with U.S. presidents and other symbolic gestures of support to sustain the perception that the United States condones his conduct, and will stand by his government.

. . .

Despite the continued intransigence of the government of Zaire, and the ongoing pattern of human rights abuses, the Bush Administration continues to press for continued U.S. aid to that country. The Administration is using virtually the same language it has in the past to justify its request for $56 million in military and economic aid for FY 1991. In its budget request for FY 1991, the State Department says:

> Zaire has been a staunch supporter of U.S. and Western policies for over two decades . . . . More recently, Zaire is playing a key role in mediating

a ceasefire in Angola. Zaire's size, population, economic importance, resources and location make it a focus of U.S. interests in the region. The U.S. has an interest in Zaire in having a stable and responsible government which influences the stability, as well as the foreign and domestic policies, of its nine bordering states.

. . .

JANET FLEISCHMAN, THE LIBERIAN TRAGEDY

2 Reconstruction 47 (1992).

The Liberian civil war—which killed twenty to thirty thousand people, turned one-third of the population into refugees, and destroyed the country's infrastructure— was a predictable explosion. By the late 1980s, the vicious and corrupt policies of Samuel Doe had brutalized his people, ignited ethnic conflict, and ruined the economy, bringing the country to the brink of violent eruption. Meanwhile, during most of the decade, the United States spent half a billion dollars in foreign aid for Liberia, making it the largest recipient of U.S. aid in sub-Saharan Africa. The U.S. propped up the Doe regime despite overwhelming evidence that he was cruel, unreliable and had no intention of keeping his promises to institute democracy. Liberia now lies tragically in ruins—a situation caused in large part by the U.S. refusal to condition aid on respect for human rights.

Africa's first republic and the only country on the continent never to have been colonized or attached to an empire, Liberia has had a 'special relationship' with the U.S. since 1847, when it was founded by freed African slaves. The new republic was immediately controlled by the settlers, who came to be know as Americo-Liberians. Liberia is the closest that the U.S. ever came to having a colony in Africa. U.S. influence has always been strong and apparent: the Liberian flag imitates that of the U.S., the U.S. dollar was used as legal tender, Liberia's constitution was written at Harvard Law School, and many Liberians have relatives in the U.S. Even the names of towns and streets reflect the U.S. connection—Monrovia, the capital, is named after President James Monroe.

The U.S. used this relationship to further its own interests. Liberia was the only country in West Africa where the U.S. had landing and refueling rights for military planes on twenty-four hour notice, and the Voice of America transmitter in Monrovia broadcast throughout West Africa. Under the regime of Samuel Doe, Liberia contributed to the U.S. Cold War strategy by expelling the Libyan Embassy and nine Soviet diplomats. The CIA and Defense Department also had facilities in Liberia for intelligence communications throughout Africa. In addition, private American investments were substantial. Firestone's rubber plantation in Liberia, for example, was one of the largest in the world.

. . .

After his campaign to silence the opposition, Doe stole the presidential elections in October 1985. Although the U.S. had provided considerable funding to ensure that the elections would be free and fair, the Reagan Administration ignored the widespread fraud and intimidation and quickly gave a stamp of approval to the new government. According to Blaine Harden's *Africa: Dispatches From a Fragile Continent*, a senior American diplomat commented.: 'It was one of those rare times when U.S. foreign policy could have made the difference. We funded the elections, we organized it, we supervised the voting, and then when Doe stole it, we didn't have the guts to tell him to get his ass out of the mansion.'

The Administration became an apologist for the Doe government. Chester Crocker, Assistant Secretary of State for African Affairs, refused to acknowledge that the elections were fraudulent. Testifying before Congress in December 1985, Crocker declared: 'There is now the beginning, however imperfect, of a democratic experience that Liberia and its friends can use as a benchmark for future elections, one on which they want to build.' He went even further: 'The prospects for national reconciliation were brightened by Doe's claim that he won only a narrow 51 percent election victory . . . ' Many informed Liberians were shocked by the Administration's efforts to legitimize the fraudulent elections, further damaging American credibility.

. . .

Contrary to the Reagan Administration's inaction, Congress made its opposition to Doe's policies clear. After the 1985 elections, both houses of Congress declared the elections fraudulent. In 1986, Congress drastically reduced U.S. aid. In 1985 U.S. military and economic aid totaled about $78 million. That amount was cut almost in half over the next two years, partly because of human rights abuses, but also due to blatant corruption in the Doe government and constraints on U.S. foreign aid imposed by America's own economic woes. The Reagan and Bush administrations, however, continued to rely on 'quiet diplomacy' and refused to engage in forceful public advocacy to curb human rights violations. While visiting Monrovia in January 1987, Secretary of State George Shultz praised Liberia's 'genuine progress' toward democracy.

Despite the introduction of a new constitution in January 1986 which guaranteed fundamental rights, many of the repressive features of martial law remained in effect, including Decree 88A. Freedom of expression and political activity continued to be tightly curtailed, arbitrary arrest and imprisonment were frequently used against opponents, and Doe's government refused to investigate or punish those responsible for the widespread killing that followed the 1985 coup attempt. Nevertheless, in a letter to Human Rights Watch in 1987, Assistant Secretary Crocker commented: 'We believe there has been a movement in a positive direction. If you take a moving picture, it shows a trend which we think is a good one. If you take a snapshot, then in that snapshot you can see problems. Problems are not absent, but the situation has improved.'

With the U.S. unwilling to use its considerable leverage to restrain Doe, the human rights situation in Liberia deteriorated.

. . .

Doe's death on September 10, 1990 did not end the war. The country disintegrated into warring factions.

. . .

QUESTIONS

1. 'I do not understand why the United States should be concerned with recipient states' reactions to decisions to cut off military or economic aid because of human rights violations. States cannot make serious objections to cut-offs. No one has a *right* to aid; it's not like, say, blocking trading relationships that are a normal part of the the world's business within a world regulatory framework. Consistent or inconsistent, effective or ineffective in trying to bring about change, decisions under the Foreign Assistance Act are this country's business alone.' Comment.

2. The materials make clear how radically inconsistent U.S. policy about aid and human rights appears to be.

a. Do you view inconsistency as such to be a substantial criticism of human rights conditionality? Is it possible for a great power like the United States to act consistently on these matters?

b. What factors could, in your view, properly influence considerations about suspensions of aid? What factors that have been historically relevant do you view as improper? Why?

3. What do you view as the primary justification for this human rights policy? If, for example, you believe that it should persuade or pressure other states to change their ways, would you abandon the policy unless you had proof that it had been effective in these ways? How would you go about collecting that proof? How, for example, would you solve the problems of causation in explaining why a country changed its practices (stopped systematic torture)?

4. Would you support an amendment to the foreign aid legislation eliminating the President's power to continue aid under the clauses dealing with 'extraordinary circumstances' and 'national interest'? Or do you wish to leave considerable discretion with the Executive? Why?

5. If a suspension of economic aid is being considered, what steps could realistically be taken to assure that the cut-off will not severely hurt the poor and needy in the target country, the very groups that may be the primary victims of human rights violations?

C. CASE STUDY: MOST FAVORED NATION TREATMENT AND THE PEOPLE'S REPUBLIC OF CHINA

COMMENT ON MOST FAVORED NATION TREATMENT

The following materials about the bearing of human rights considerations on trade with China draw on human rights conditions to trade that were stated in legislation, proposed legislation and a recent Executive Order. Neither of the statutory provisions previously considered—Section 502B and Section 116 of the Foreign Assistance Act of 1961, as amended, p. 822, *supra*—are here relevant, for the United States has provided neither security assistance nor development assistance to the PRC.

The so-called Jackson–Vanik Amendment to the Trade Act of 1974 (Sec. 402, 19 U.S.C. § 2432) at once authorizes the granting of trade benefits to designated categories of states and qualifies its authorization by stating:

> To assure the continued dedication of the United States to fundamental human rights . . . products from any nonmarket economy country shall not be eligible to receive nondiscriminatory treatment (most-favored-nation treatment) . . . during the period beginning with the date on which the President determines that such country—
>
> (1) denies its citizens the right or opportunity to emigrate;
>
> (2) imposes more than a nominal tax on emigration . . .; or
>
> (3) imposes more than a nominal tax . . . or other charge on any citizen as a consequence of the desire of such citizen to emigrate to the country of his choice. . . .

This provision is subject to Presidential waiver under stated conditions and procedures.

The most favored nation (MFN) treatment to which the amendment refers is a familiar clause and concept regulating many types of relationships among states. In general, a MFN clause (in, say, a treaty) grants a state benefiting from it (state 'X') the same treatment (say, with respect to the ability of X's citizens to do business in state 'Y', the other party to the treaty) as has been given by Y to some other 'most-favored' state. In other words, X and its citizens benefit from the most favored treatment that Y has granted to any other state on this matter.

For example, suppose that state 'Z' had previously entered into a treaty with Y giving Z's citizens extensive rights to do business in Y, rights that are as favorable as or more favorable than those given by Y to any other state. By virtue of the MFN clause in its treaty with Y, X is assured that its citizens doing business in Y will be treated as favorably as the nationals of Z.

In the case of the Jackson–Vanik Amendment, the MFN clause appears in domestic legislation (not a treaty) and affects tariffs, a field highly regulated both by multilateral arrangements such as the General Agreement on Tariffs and Trade (GATT) and by national legislation. Special categories of states may fall within special treaty or statutory regimes, such as the System of Generalized Preferences covering a range of exports from less developed countries. In the case of the PRC, which the U.S. views as not being a party to the GATT, MFN treatment plays a vital role. If such treatment were withdrawn by the U.S., statutory tariff rates that are often much higher would apply to imports from the PRC.

The Jackson–Vanik Amendment was enacted primarily in response to bars erected by the Soviet Union to emigration. Those bars particularly affected members of several groups anxious to leave, including principally Soviet Jews. After the (ongoing) transformation of the former Soviet Union, the question of MFN treatment has involved principally issues other than a state's denial of the right of its citizens to emigrate. That has surely been the case with respect to decisions about MFN treatment for China. Conditions other than those stated in the Jackson–Vanik Amendment have become relevant to tariffs on imports from the PRC, such as conditions set forth in proposed legislation and in the Presidential Executive Order at p. 851, *infra.*

The following materials look at the Chinese-American conflict, and at the related Congressional-Executive conflict, over human rights conditions to MFN treatment for tariffs on trade from China. A previous reading—Kausikan, *Asia's Different Standard*, p. 226, *supra*—is particularly relevant to the present discussion.

ROBERT DRINAN AND TERESA KUO, THE 1991 BATTLE FOR HUMAN RIGHTS IN CHINA

14. Hum. Rts. Q. 21 (1992), at 22.

. . .

While [President Bush] believes that the unconditional extension of MFN is the best way to advance US ideals of freedom and democracy in China, the majority of lawmakers disagree. . . .

. . . [T]he President recognized in 1991 that an overwhelming majority of the House of Representatives likely will not support unconditional MFN renewal. Therefore, from the initial stages of the 1991 MFN debate, the administration focused its efforts on garnering the thirty-four Senate votes needed to sustain a presidential veto of conditional MFN renewal.

. . .

The Jackson–Vanik Amendment allows the President to waive the emigration requirement only when it would 'substantially promote' the goals of the Amendment *and* the President has received assurances that the country's emigration practices will 'lead substantially' to such objectives. Since 1980 three presidents have used the waiver method to grant MFN status to China.

. . .

In 1991 both the Senate and the House took early action to disapprove the presidential waiver renewing China's MFN status. In addition to voicing its increasing dissatisfaction over China's human rights abuses, Congress also expressed growing concerns both over China's unfair trade practices toward the United States and with regard to the sales of nuclear weapons to developing nations. . . .

. . .

Examples of ongoing human rights violations in China include restrictions on emigration, a lack of fair trials and due process, deplorable prison conditions, suppression of political dissent, religious persecution, interference with foreign journalists, harrassment of Chinese nationals in the United States, repression of Tibetan nationals, and exporting of products of prison labor.

. . . Restrictive regulations adopted by Chinese authorities after June 1989 . . . have made emigration much more difficult for political dissidents. In addition to getting approval from both the police and their work units, post-Tiananmen passport applicants also must give detailed explanations of their whereabouts during the spring of 1989. Passports were not issued for applicants who gave unsatisfactory answers regarding their involvement in the pro-democracy movement or who lacked the requisite authorization from the appropriate authorities.

. . .

Despite the long list of human rights violations, the issue which captured the attention of many members of Congress last year was China's $10.4 billion trade [surplus] with the United States in 1990. The Treasury, along with the Central Intelligence Agency, attributed the trade imbalance to China's use of protectionist policies toward US imports. . . .

. . .

The bills which received the strongest support among congressional members and which observers saw as the most likely to override an expected presidential veto were the bills sponsored by Mitchell and Pelosi. Both bills allowed the extension of MFN trade status for another year, but with the condition that when it comes up for renewal in 1992, the President must certify, among other things, that China has:

- accounted for citizens detained, accused, or sentenced because of pro-democracy protests;
- released political prisoners taken from the pro-democracy protests;

– adhered to the joint Declaration on Hong Kong, which states terms under which Britain is to transfer the territory to China;
– stopped jamming Voice of America broadcasts; and
– made 'significant' progress in (1) ending the harassment of Chinese citizens in the United States, (2) granting access for humanitarian and human rights groups to prisoners, trials, and detention centers, and (3) taking action to stop human rights violations.

. . .

. . . However, five amendments were added to Pelosi's original bill, making the final bill resemble the more restrictive Mitchell bill. The amended conditions required that China abandon its practice of coercive abortions or sterilization, provide assurances that it is not assisting non-nuclear weapon states to develop nuclear devices, not contribute to the proliferation of missiles, prohibit the export of prison-made goods, and moderate its opposition to Taiwan's access to the General Agreement on Tariffs and Trade (GATT).

. . .

. . . A breakdown of the final votes shows that three major factors influenced the Senators to vote against the Mitchell bill: deference to the President, concerns about the overly restrictive amendments to the MFN bill, and the desire to protect US farmers from possible Chinese retaliation at MFN revocation.

. . .

In addition, pro-MFN lobby groups also made a big impact on the Senate vote last year. Unlike 1990, when the farmers mostly stayed out of the MFN debate, during 1991 the National Association of Wheat Growers was one of the most active lobbying groups for unconditional MFN extension. As the final Senate decision showed, the farmers' efforts influenced key Democratic votes.

. . .

In addition, the key players for unconditional MFN renewal from 1990—the US and Hong Kong companies doing business in China—were again involved in lobbying lawmakers during the summer of 1991. A leading pro-MFN group, the Business Coalition for US-China Trade, mobilized the employees of its 100 companies and trade associations to contact members of Congress.

None of these lobbying groups, however, had an interest in preserving unconditional MFN status quite as intense as that of the Chinese government. Despite the Beijing leaders' repeated rhetoric about not bending under international pressure, the Chinese government hired the public relations giant Hill and Knowlton in mid-July to lobby on its behalf. The Chinese government also hired former executive director of the National Republican Congressional

Committee, R. Marc Nuttle, and the law firm of Jones, Day, Reavis & Pogue to boost its image among influential figures in Washington.

On the other side of the MFN debate, overseas Chinese students, human rights watch groups, and labor organizations argued that MFN status either should be revoked completely or only should be renewed with conditions attached. Of these groups, the most active lobby in support of conditional MFN status renewal was the Independent Federation of Chinese Students and Scholars (IFCSS), which represents over 40,000 Chinese citizens from some 150 US universities. . . .

Although some have argued that MFN status should not be used as a tool to advance the cause of human rights in China, the United States has had a history of denying MFN status to other countries with similar human rights violations. For example, not only has it refused to grant MFN status to Vietnam for its human rights abuses, but the United States also has led Europe and Japan on a decade-long embargo against investing in that country. The United States also has steadfastly refused to grant MFN status to countries such as Afghanistan, Albania, Cambodia, Cuba, and North Korea, all run by tyrannical rulers. From 1982 to 1987, while Poland was under martial law, the United States refused to renew Poland's MFN status in order to protest the government's mass arrests and imprisonment of people connected with the Solidarity Movement. . . .

. . .

NOTE

In September 1992, Congress enacted the United States China Act of 1992, 138 Cong. Rec. H. 8841, 102nd Cong. 2d Sess. It provided that the President could not recommend from July 1993 the continuation of a waiver related to nondiscriminatory MFN treatment for China unless he reported that China had taken 'appropriate actions' to 'begin adhering' to the UDHR in China and Tibet; allowed emigration for reasons of political or religious persecution; accounted for and released political prisoners and prevented export to the U.S. of goods made by prison labor. Moreover, the President's report had to state that China had 'made overall significant progress' in other respects, including ending unfair trade practices against American business.

Even if the President could not report as required, nondiscriminatory MFN treatment was still to apply to goods manufactured by a business venture 'that is not a state-owned enterprise' of China, provided that such goods were not marketed or exported by a state-owned enterprise.

President Bush vetoed the Act. The House of Representatives overrode the veto by a vote of 345–74, but the Senate vote of 59–40 fell short of the required two-thirds majority for an override and hence sustained the veto.

ROGER SULLIVAN, DISCARDING THE CHINA CARD

86 For. Pol. 3 (1992), at 8.

. . .

. . . American policy treated [human rights] as a secondary issue during the Cold War because there seemed little to be done. For the most part, Chinese human rights violations were simply ignored. In fact, before the 1989–91 worldwide collapse of communism, few questioned two key assumptions guiding Washington's China policy: first, that communist regimes cannot be fundamentally reformed or overthrown; and, second, that the Chinese communist government was firmly in control and that it enjoyed the support of the Chinese people. Events in Eastern Europe and the Soviet Union have done much more to undercut these assumptions than has the massacre at Tiananmen.

. . .

American overtures to China in 1971 made sense for a number of reasons, none of which apply today. Achieving normal relations with the PRC enabled the United States to end military support for South Vietnam, which had become politically insupportable in America. It also put the United States in a position to exploit the Sino-Soviet split and reap the obvious benefits of blocking the advance of Soviet influence in Asia and drawing Soviet forces away from Europe.

. . .

Immediately after Tiananmen, it was not clear how the administration would respond.

. . .

. . . [T]he administration reverted to the old approach to China. The annual process of renewing non-discriminatory (so-called 'most-favored-nation,' or MFN) tariff treatment for China became the outlet for congressional frustration and the vehicle for engaging the administration in a general policy debate. Renewal had been routine every year since MFN treatment was first extended to China in 1980 because Congress was well aware that MFN treatment is not a special benefit but is considered standard for almost all trading partners. Congress even let Bush's decision to renew MFN treatment for China immediately after Tiananmen pass without a vote. But when Congress saw no change in administration China policy, despite the collapse of communism in Eastern Europe and the Soviet Union and the clear evidence that China was becoming more repressive, congressional opposition to China's MFN status mounted.

. . .

Nor would Chinese leaders compromise in any substantial way to secure continued MFN status. They are convinced that compromise and reform brought down the communist parties in Eastern Europe, and they have no intention of suffering the same fate. The Chinese leadership also does not believe compromise is necessary because it thinks the MFN threat is a bluff. . . .

. . .

. . . As long as China is ruled by a clique whose overriding concern is survival, which sees peaceful demonstrations as 'counterrevolutionary rebellion' aided by 'anti-China forces in the United States,' and which dismisses even such a moderate summary of human rights abuses as the State Department's 1990 human rights report on China as slander, flagrant interference in China's internal affairs, and 'expressions of out-and-out hegemonism and power politics,' foreign government attempts to talk to Beijing about human rights will prove fruitless and perhaps even counterproductive.

. . .

NOTE

In May 1993, President Clinton issued Executive Order 12850, Conditions for Renewal of Most Favored Nation Status for the People's Republic of China in 1994 (56 Fed. Reg. 31327, June 1, 1993). The Order continued the waiver for China for an additional twelve months to June 1994, so as to assure it of most favored nation treatment during this period, but made renewal after June 1994 subject to stated conditions.

Those conditions included a determination that further extension of the waiver would promote 'freedom of emigration objectives,' and that China was complying with the 1992 agreement on prison labor. Further determinations were required to the effect that China had made 'overall, significant progress' with respect to matters including: 'taking steps to begin adhering' to the UDHR, giving an accounting of and releasing political prisoners, ensuring humane treatment of prisoners, 'protecting Tibet's distinctive religious and cultural heritage,' and permitting international radio and television broadcasts into China.

U.S. DEPARTMENT OF STATE, COUNTRY REPORTS ON HUMAN RIGHTS PRACTICES FOR 1994

(1995), at 555.

CHINA

The People's Republic of China (PRC) is an authoritarian state in which the Chinese Communist Party (CCP) monopolizes decisionmaking authority.

Almost all top civilian, police, and military positions at the national and regional levels are held by party members. A 22-member Politburo and retired senior leaders hold ultimate power, but economic decentralization has increased the authority of regional officials. Socialism continues to provide the ideological underpinning, but Marxist ideology has given way to pragmatism in recent years. The party's authority rests primarily on the success of economic reform, its ability to maintain stability, and control of the security apparatus.

. . .

In 1994 there continued to be widespread and well-documented human rights abuses in China, in violation of internationally accepted norms, stemming both from the authorities' intolerance of dissent and the inadequacy of legal safeguards for freedom of speech, association, and religion. Abuses include arbitrary and lengthy incommunicado detention, torture, and mistreatment of prisoners. Despite a reduction during the year in the number of political detainees from the immediate post-Tiananmen period, hundreds, perhaps thousands, of other prisoners of conscience remain imprisoned or detained. The Government still has not provided a comprehensive, credible public accounting of all those missing or detained in connection with the suppression of the 1989 demonstrations. Chinese leaders moved swiftly to cut off organized expressions of protest or criticism and detained government critics, including those advocating greater worker rights. Citizens have no ability peacefully to change their government leaders or the form of government. Criminal defendants are denied basic legal safeguards such as due process or adequate defense. The regime continued severe restrictions on the freedoms of speech, press, assembly and association, and tightened controls on the exercise of these rights during 1994. Serious human rights abuses persisted in Tibet and other areas populated by ethnic minorities.

The human rights situation in 1994 was, however, marked by the same diversity that characterizes other aspects of Chinese life. In several instances, the Government acted to bring its behavior into conformity with internationally accepted human rights norms. These actions included releasing several prominent political and religious prisoners, granting passports to some critics of the regime and their relatives, and adopting a law, which became effective in January 1995, that allows citizens to recover damages from the Government for infringement of their rights. The Government continued to acknowledge the need to implement the rule of law and build the necessary legal and other institutions, but it has not yet significantly mitigated continuing repression of political dissent. In 1994 China also continued a human rights dialog with some foreign critics, and reaffirmed its adherence to the Universal Declaration of Human Rights. Chinese officials provided limited information about the status of several hundred specific cases of international concern.

. . .

SUBSEQUENT DEVELOPMENTS ABOUT CHINA–U.S. TRADE: NEWSPAPER ACCOUNTS

There follow excerpts from articles in the *New York Times* that indicate the considerations and forces at work with respect to U.S. action bearing on human rights violations in China.

China Rejects Call from Christopher for Rights Gains

by Elaine Sciolino, *New York Times*, March 13, 1994, p. A1

Chinese officials today flatly rejected Secretary of State Warren Christopher's demand that Beijing improve its human rights performance and warned that stripping China of its trade privileges would backfire.

. . .

This took place against a backdrop of heightened American demands for steady and substantive progress on human rights, and in large part they clashed with China's traditional resistance against being told what to do about its internal affairs and with the authorities' desire to maintain strict control as democracy advocates begin to become more assertive.

Mr. Christopher told Foreign Minister Qian Qichen this morning that China's actions toward its citizens, including the recent detention of a number of prominent dissidents, 'suggest that China attaches relatively little weight to core human rights issues,' the State Department spokesman, Michael McCurry, told reporters.

If that is so, the Secretary added, it 'certainly bodes in' for the chances of renewing the most-favored-nation trade status that gives China lower tariffs accorded to most American trading partners.

. . .

In an afternoon meeting, Prime Minister Li Peng seemed to reverse China's earlier pledges this year to try to make progress on human rights.

'China will never accept the U.S. human rights concept,' Mr. Li told Mr. Christopher, according to a Foreign Ministry spokesman, Wu Jianmin. 'History has already proven that it is futile to apply pressure on China.'

. . .

In taking a tougher line than at any time since the Clinton Administration took office, the Chinese appear willing to gamble that Washington has too much to lose by revoking the trade privileges, a policy they see as driven by anti-China lawmakers in Congress. Both Mr. Qian and Mr. Li warned Mr. Christopher today that the United States would suffer more than China if its trade privileges were revoked.

China is nearly self-sufficient economically, Mr. Li boasted, adding that if the United States were to revoke its trade status, 'it means the United States would lose its share of the big Chinese market.' He said Chinese imports were growing at a rate of 9 percent a year and would reach a trillion dollars by the turn of the century.

. . .

Gauging the Consequences of Spurning China

by Edward Gargan, *New York Times*, March 21, 1994, p. D1

As the Clinton Administration weighs whether to renew China's most favored nation status after the strained visit by Secretary of State Warren Christopher to Beijing, legions of American business leaders and Washington lobbyists for industry and trade associations are describing the consequences of withdrawing the designation in near-apocalyptic terms.

But some economists and legislators, pointing to the United States' $22.7 billion trade deficit with China—second only to its $59.32 billion deficit with Japan—suggest that much of the oratory about damage to American exports and investment in China is overstated and that higher duties could go some distance toward trimming the trade deficit.

In fact, both sides, while summoning considerable data, have shaped their analyses to suit their arguments. Any clear-cut assessment remains elusive.

'Our exports were $9 billion last year to China,' said Donald M. Anderson, the president of the United States–China Business Council, a group that promotes trade between the countries. 'No one really knows how much they would be hit, but they would be hit because the Chinese would unquestionably retaliate.'

Jagdish Bhagwati, a trade economist at Columbia University, does not agree. 'All this number-crunching is purely hypothetical,' he said. 'Would the Chinese retaliate by saying, "The U.S. wants to play hardball so we'll play hardball"? That is a vague, general fear, in my view. If they start playing that game, we are a superpower and we have lots of leverage. We have a whole range of punishments and rewards at our command.'

The threat of Chinese retaliation against American companies selling to China and what would be a surge in duties on Chinese products exported to this country have rattled industries with large stakes in China and worried trade experts about the broader relationship between the two countries should China's trade privileges be revoked.

By June, President Clinton must decide whether China has made 'significant, overall progress' toward improving its human rights record, a condition of an executive order that the President issued last year on renewing China's trade status. Virtually every country with which the United States trades holds most favored trading status, which results in significantly lower duties on their exports to this country than would otherwise be the case.

The Boeing Company, one of the most influential American manufacturers—and with orders from China for $3.9 billion in new airplanes—has led the charge against any change in China's trade status. Last Tuesday, at a panel discussion in Washington sponsored by the Council on Foreign Relations, Lawrence W. Clarkson, Boeing's corporate vice president for planning and international development, attacked proposals to link renewal of most favored nation status to China's record on human rights.

'Boeing currently has orders for 64 planes for delivery to China with a value of $3.9 billion,' Mr. Clarkson said, 'These orders, plus an additional $2 billion to $3 billion in near-term potential orders, would be at great risk if M.F.N. was not renewed. During the next 15 years, China will need 800 airplanes worth $40 billion, making it the world's third-largest aviation market.'

Like many executives who have urged renewal of China's trade privileges, Mr. Clarkson couched his argument in broader social and political terms as well. 'We believe strongly that this U.S. trade is contributing to the rapid decentralization and transformation of China's economy, helping millions of Chinese to obtain greater freedom to choose their work, their employer and their place of residence,' he said. 'As in Taiwan and South Korea, the development of a vigorous middle class will bring the development of social and political freedom for all Chinese more than any other factor.'

American industry leaders often say, as some business executives did to Secretary of State Christopher in Beijing on March 13, that 167,000 American jobs depend on trade with China. This figure comes from a Commerce Department estimate that every $1 billion in exports generates 19,000 jobs; thus, the roughly $9 billion in exports to China accounts for 167,000 jobs.

The calculation implies that most if not all of these jobs are in jeopardy if China responds to a cut-off of trade privileges by slashing imports from the United States. But, as economists like Mr. Bhagwati note, this inference is not warranted given the uncertainty of the Chinese response and the complexity and size of trade between the two countries.

Still, Lawrence Chimerine, the chief economist at the Economic Strategy Institute, a Washington based economic and trade policy lobby organization, said cases like Boeing suggest that American manufacturers could lose large chunks of the China market. 'I think Japan and Europe would get a lot of orders that we wouldn't get,' he said.

In 1993, according to the Commerce Department, the United States exported $8.76 billion worth of goods to China, including airplanes, fertilizer, wheat and automobiles. China, during the same period, sold $31.53 billion in goods to this country, products ranging from toys and shoes, to combs and clocks.

Peter T. Mangione, president of the Footwear Distributors and Retailers of America, said that ending China's access to low duties would mean higher shoe prices for Americans because China accounted for one of every two pairs of shoes sold here. 'Almost 90 percent of all shoes sold in America are imported,' Mr. Mangione said 'And China is by far the largest supplier of imported shoes; it supplies 60 percent of all imported shoes.'

Neal M. Soss, the chief economist at CS First Boston, has calculated the potential inflationary impact of revoking China's most favored nation status. If this trade privilege is denied, he said, the increase on duties on Chinese products imported into this country would fuel inflation by at least 0.15 percent annually. 'The slippery slope of protectionism is the reason,' he said.

But Representative Nancy Pelosi, a Democrat from San Francisco who has led the campaign in Congress against renewing China's trade status, argues that the size of America's trade deficit with China suggests that China needs American markets.

'China's economic growth depends on preferential access to the United States markets,' Ms. Pelosi said. 'The deficit is growing. Compound that with cheap labor and you see that they are flooding our markets with products made by people making $24 a month.'

. . .

U.S. Signals China It May End Annual Trade-Rights Battles

by Thomas Friedman, *New York Times*, March 24, 1994, p. A1

The Clinton Administration, in an effort to resolve the crisis in relations with China, has been quietly signaling Beijing that if it met Washington's minimum human rights demands, the United States would consider ending the annual threat of trade sanctions to change Chinese behavior.

Administration officials said today that there was already a consensus within President Clinton's economic team, which is now spreading to the White House and even parts of the State Department, that the annual rite of threatening China with a withdrawal of its beneficial trade status if it does not meet certain human rights conditions is outmoded and should be replaced.

That sentiment, which has been conveyed to China by the Administration, also seems to be taking hold in Capitol Hill, from which the pressure to link human rights and trade originally came.

. . .

Administration officials are increasingly arguing that the current policy is outmoded because, as the President has told aides and lawmakers in recent days, the last thing Mr. Clinton wants to do is withdraw China's trade benefits, which would costs thousands of American jobs and billions of dollars in contracts. So in some ways the policy rests on an empty threat.

It is also outmoded, other officials argue, because trade is now such an important instrument for opening up Chinese society, for promoting the rule of law and the freedom of movement there, and for encouraging the Chinese authorities to allow citizens to own everything from their own satellite dishes to foreign newspapers, that revoking it would be self-destructive for both sides.

Finally, American policy toward China is only effective if there is a united front, and that united front has broken down in recent years, as more and more Americans do business in China and refuse to see their economic interests imperiled by making them an annual weapon for pressing human rights concerns.

. . .

U.S. Is To Maintain Trade Privileges for China's Goods: A Policy Reversal

by Douglas Jehl, *New York Times*, May 27, 1994, p. A1

President Clinton renewed China's favorable trade status today with virtually no conditions, and said he would abandon his effort to use trade as a lever to force Beijing to make progress on human rights.

While Mr. Clinton said there had been some improvement in human rights, he acknowledged that China's actions had fallen far short of the steps he set forth a year ago as conditions for renewing the trade privileges. But he said that as punishment he would do little more than prohibit imports of Chinese guns and ammunition.

. . .

. . . Thus Mr. Clinton cast aside the executive order issued last year.

Human Rights Watch, which led the battle to block renewal of the trade benefits, voiced even greater disdain, accusing Mr. Clinton of 'effectively removing all pressure on China to improve its human rights practices.'

But the decision was applauded by business groups, including the United States Chamber of Congress, which likened the policy of tying human rights to trade to 'using a blunt instrument for brain surgery.'

The decision by Mr. Clinton leaves intact the Jackson–Vanik Amendment of two decades ago that ties trade benefits for Communist countries to their [*emigration*] policies. But since it first won the privilege in 1980, that barrier has never prevented China from winning renewal of its 'most favored nation' trade status, which allows it to send its exports to the United States at the lowest applicable tariffs.

Instead, Mr. Clinton's approach becomes roughly comparable to Washington's strategy toward other nations—using diplomatic pressure but not holding trade hostage in efforts to make progress on rights.

Today, Mr. Clinton mentioned China's importance in putting pressure on North Korea to abandon its nuclear weapons program as one of the reasons he believed it wise to tread softly. He also cited China's role as the most populous country, with 1.2 billion people, and the third-largest economy.

But the fundamental reason he offered for the shift amounted to a repudiation of a year's worth of American pressure tactics. He said there would be 'inevitably a reluctance' for China to act on rights issues 'if it looks like every step that is taken, is taken under the pressure of the United States, some outside power making them do it.'

. . .

Rights Issues Aside, Asia Deals Rise

by Key Brown, *New York Times*, Aug. 1, 1994, p. D1

After years of trying unsuccessfully to quiet criticism of human -rights records, Asian nations have found powerful allies: American businesses eager to invest in their growing markets and the Clinton Administration with its emphasis on economic interests in post-cold-war foreign policy.

The Clinton Administration's spring decision to continue China's favorable trade status gave some of the largest and fastest growing countries new confidence that their influence is now strong enough to have an effect on American policy. And from India and Indonesia to smaller nations like Singapore, Vietnam, Thailand and Malaysia, they are showing it.

'It seems to me the message is loud and clear: if American business gets strongly involved in a country, human rights takes second place or even less than that,' Ashutosh Varshney, an India expert at Harvard, said of the China decision.

. . .

The Administration's China decision was also closely watched by Indonesia, which is in the middle of the latest trade case brought against it by American labor unions. The case comes under the Guaranteed System of Preferences, which was created to help developing countries, and will be decided by the United States trade representative. Under the preferences program, Indonesia can export $650 million worth of goods duty free to the United States.

The A.F.L.-C.I.O. is protesting what it says is Indonesia's quashing of its labor movement. A ruling is due this month, but there is a growing feeling in Indonesia that the battle is already won.

'I think the Indonesians are resting a little easier that the decision won't go against them,' said Wayne J. Forrest, the executive director of the American Indonesian Chamber of Commerce, a group founded in 1949, whose members include Mobil, Texaco and several other large oil and mining companies as well as some small businesses.

A month after the Clinton Administration's May 26 decision to continue China's most-favored-nation status with virtually no conditions, despite its spotty record on human rights, the Indonesian Government shut three of the country's most influential magazines and broke up a protest in Jakarta over the closings. Amnesty International said 56 people were beaten and arrested. In mid-April, more than 100 workers and labor campaigners were arrested after strikes and demonstrations.

Meanwhile, abuses continue in the disputed territory of East Timor, a former Portuguese colony that was invaded and annexed by Indonesia in 1976, human rights groups say.

An Indonesian Embassy official in Washington, who insisted on anonymity, said Indonesia had followed China's lead and enlisted American oil and mining companies that do business in Indonesia to stress the country's economic importance to the United States. Nearly 800 big American companies wrote to President Clinton supporting China, a tactic crucial to China's success in the trade decision.

. . .

President Imposes Trade Sanctions on Chinese Goods

by David Sanger, *New York Times*, Feb. 5, 1995, p. A1

The Clinton Administration today imposed punitive tariffs on more than $1 billion of Chinese goods, the largest trade sanctions in American history, and warned of further action if the Communist Government continued to refuse to crack down on rampant piracy of American software, movies and music.

The decision to impose 100 percent punitive tariffs on goods ranging from silk blouses to cellular telephones was met almost immediately by an angry Chinese announcement of tariffs against American-made goods.

The trade confrontation comes at a particularly delicate moment in Chinese-American relations, with disagreements brewing on human rights and arms control as well, and with China's paramount leader, Deng Xiaoping, reported to be seriously ill.

. . .

Each country said its penalties would take effect on Feb. 26. . . .

Today's action was the culmination of a dispute between the United States and China that has bubbled along for nearly two years, a period in which the Administration had set aside serious concerns about human rights in China in the hope of bolstering trade, among other things.

. . .

[Eds: On February 26, 1995, the two countries signed an agreement purporting to end the dispute over intellectual property rights. Thus the announced trade sanctions that were to become effective on that date were avoided. See *New York Times*, Feb. 27, 1995, p. A1.]

China Warns of New Peril to U.S. Ties
by Patrick Tyler *New York Times*, Feb. 23, 1995, p. A9

China is threatening a new rupture in relations unless the United States backs away from a proposed United Nations resolution criticizing its human rights record.

The warning was delivered today by officials who summoned the American Ambassador to discuss the resolution, which has been raised and defeated repeatedly, with the United States as co-sponsor, at the annual meeting of the United Nations Human Rights Commission.

. . .

Mr. Clinton pledged to press human rights concerns when he decided last May to separate American trade policy toward China from its human rights record.

. . .

A State Department official in Washington said that because 'this year there did not seem to be any significant, concrete' improvement in China's human rights record, the Administration had concluded that it could not 'justify taking a different course of action' in Geneva.

The United States Embassy in Beijing opposed bringing the resolution this year, arguing that the United States would get nothing by embarrassing China and would stoke anti-American sentiment as China prepares for the death of its paramount leader, Deng Xiaoping.

Taking a hard line last fall, the Chinese Foreign Minister, Qian Qichen, warned in a meeting at the White House that China would end its high-level talks on human rights if Washington pressed the condemnatory resolution again in 1995.

. . . In addition to condemning China's human rights performance, the proposed resolution calls on China to cooperate with United Nations investigators and to 'take further measures to insure the observance of all human rights, including the rights of women, and to improve the impartial administration of justice.'

A year ago, the resolution failed by four votes. This year the Administration has undertaken a broad international campaign to win it support. New Commission members include countries with strong ties to the United States that could change the voting pattern. They are Egypt, Nicaragua, the Philippines, El Salvador and the Dominican Republic.

The United States has co-sponsored a human rights resolution on China every year since the 1989 military crackdown on the Tiananmen Square democracy movement, with one exception. The exception was in 1991, when the Bush Administration was seeking China's support for a Persian Gulf war resolution.

. . .

U.N. Rights Panel Declines to Censure China

New York Times, March 9, 1995, p. A5

Despite heavy lobbying by the United States, China narrowly escaped criticism of its human rights record today in a vote by the United Nations Human Rights Commission.

The commission voted 21 to 20 to reject a resolution expressing concern about human rights in China. Russia voted against the proposal.

The resolution would have obliged the commission to investigate human rights in China. That would have been considered an insult to China as a permanent member of the United Nations Security Council.

The proposed resolution cited reports of torture and other violations, but it also praised the Government for improving the lives of its people through economic and legal reforms.

China's representative, Jin Yongjian, said the sponsors had 'turned a blind eye to the achievements of China in the field of human rights.'

Even though the measure was defeated, human rights groups regarded the vote as a victory because it was the first time the commission had formally considered the China rights issue at all.

. . .

QUESTIONS

1. 'Trying to enforce human rights against other countries by threatening withdrawal of MFN status is a losing battle. Too many constituencies are involved, too many pressure groups both domestic and foreign have a keen interest in the outcome. Instead of maintaining a sharp and single moral purpose, the threat of withdrawal becomes enmeshed in a great power struggle involving many different economic interests and actors.' Comment. Is the situation notably different from threats of suspending military or economic aid?

2. Note the catalog of conditions in the proposed legislation and in the President's Executive Order. What coherence does it have? Why were these conditions chosen and other more significant issues overlooked?

3. If China had accepted all these conditions and made the necessary changes, would something substantial have been accomplished? Or is it not the purpose of this legislation to achieve something basic, structural? Does it mean only to secure a few precise steps, as with the focus of the original Jackson–Vanik Act on emigration?

4. 'All of this is conventional politics, good or bad drama, not serious. Human rights are not taken that seriously. When the U.S. really cares, as with respect to the announced trade sanctions in retaliation for trade piracy, it knows how to act.' Comment.

D. A COMPARISON: THE EUROPEAN COMMUNITY (UNION)

Among the major powers, the United States has the most developed legal scheme about restraints on aid and trade to serve human rights purposes. Multilateral institutions formed on regional or other bases have also pursued human rights goals through their aid policies, either by imposing conditions on their own lending to developing countries or by acting to coordinate the policies of their member states with respect to developmental aid and human rights.

Section D offers a comparison with the policies, laws and experience of the United States. The European Union (earlier known as the European Community, see pp. 574–77, *supra*) has given human rights an important place in its development cooperation policy; through that policy, it ensures coordination among policies of its member states and gives substantial aid to developing countries around the world. The following article by Marantis traces the recent evolution of its policy, and indicates some comparisons and many contrasts with the U.S. approach described in the preceding materials.

DEMETRIOS MARANTIS, HUMAN RIGHTS, DEMOCRACY, AND DEVELOPMENT: THE EUROPEAN COMMUNITY MODEL

7 Harv. Hum. Rts. J. 1 (1994).

. . .

Guided by the 28 November 1991 Resolution on Human Rights, Democracy and Development ('28 November Resolution'), the European Community[1] has recently placed human rights and democracy at the heart of its development cooperation policy.' The import of this Resolution is twofold. First, the Resolution confirms the relationship between human rights,

[1] As of November 1993, the European Community has become known as the European Union. Because the events discussed in this Article occurred prior to this date, the Article will refer to the European Community.

accountable government, and development. Second, it comprehensively defines this relationship to include both proactive and reactive measures that together tackle the problems associated with development cooperation. In so doing, the European Community has broken from past models of development aid which regarded human rights and democracy as marginal to development cooperation.

. . .

. . . [T]he European Community has attempted to . . . assist governments and local organizations in their own efforts to lay the foundations for democracy and human rights. Today's funding of an election enables tomorrow's construction of a health center sensitive to the needs of a particular constituent group. Today's support for a legal training seminar facilitates the codification of tomorrow's penal code. With a human rights and democracy orientation, the European Community is attempting to promote a development process that is both meaningful and accessible to expanding segments of donee populations.

. . .

. . . The European Community's development assistance policy is determined primarily by the Commission of the European Communities (Commission) and the Council of Ministers (Council).[2] The Commission makes policy suggestions to the Council, negotiates and implements aid agreements with donees, and ultimately determines whether to curtail or interrupt aid. The Council, composed of the development ministers from the twelve Member States, sets the guiding principles of European Community development policy and helps to ensure coordination between the policies of the Member States. . . .

The European Community provides a substantial amount of aid to the developing world.[3] The nations of the African, Caribbean, and Pacific group (ACP) receive almost half of the total sum. Since 1975, the European Community has adopted four successive Lomé Conventions to govern development cooperation links between the Community, its Member States, and the ACP group.[4] The Lomé IV Convention is currently in force and expires

[2] European Community development assistance policy is distinct from the policies of each of its Member States. However, the Member States recently have taken steps to coordinate their development cooperation policies through the European Community framework. *See* Treaty on European Union and Final Act, Feb. 7, 1992, art. 130x, 31 I.L.M. 247, 287.

[3] In 1991 alone, the Community advanced $3.8 billion, mainly in grant form, to developing countries and extended an additional $160 million of non-concessional flows through the European Investment Bank (EIB). DAC, 1992 Development Co-Operation Report 9, 102–3 (1992).

[4] ACP–EEC Convention, Feb. 28, 1975, 14 I.L.M. 596 (1975) (hereinafter Lomé I); Second ACP-EEC Convention, Oct. 31, 1979, 1277 U.N.T.S. 3 (hereinafter Lomé II); Third ACP-EEC Convention, Dec. 8, 1984, 24 I.L.M. 571 (1985) (hereinafter Lomé III); Fourth ACP-EEC Convention, Dec. 1, 1989, 29 I.L.M. 783 (1990) (hereinafter Lomé IV). The Lomé Conventions govern development issues such as trade cooperation, technical assistance, the stabilization of export earnings, and aid. . . .

in the year 2000. Currently, most European Community aid flows to the ACP through Lomé IV's National Indicative Programs, which are negotiated by the Commission and individual ACP states.[5] In addition, Lomé IV provides an institutional forum for ACP–EC development cooperation.

. . .

By the late 1980s, political conditions in Europe and the developing world had changed dramatically. Both the European Community and the developing world became more committed to the principles of human rights and democracy. At the same time, the economic conditions in the ACP states collapsed, making the ACP heavily dependent on foreign aid to meet basic foreign exchange needs, domestic requirements, and recurrent costs. This new political and economic climate undermined ACP resistance to European demands to incorporate human rights and democracy concerns into development cooperation.

These developments facilitated the inclusion of prominent references to human rights in Lomé IV.[6] The most significant reference is Article 5, which states:

> Cooperation shall be directed towards development centred on man, the main protagonist and beneficiary of development, which thus entails respect for and promotion of all human rights. Cooperation operations shall thus be conceived in accordance with the positive approach, where respect for human rights is recognized as a basic factor of real development and where cooperation is conceived as a contribution to the promotion of these rights. . . .
>
> Hence, the Parties reiterate their deep attachment to human dignity and human rights. . . . The rights in question are all human rights, the various categories thereof being indivisible and inter-related, each having its own legitimacy: non-discriminatory treatment; fundamental human rights; civil and political rights; economic, social and cultural rights.

. . . Furthermore, Lomé IV endorses decentralized aid distribution as a means of strengthening the role of grassroots organizations in preparing and implementing development programs.

[5] . . . Much of the financing for these projects is provided in grant form from the European Development Fund (EDF), to which each Member State contributes. The balance of the funding comes from risk capital from the EDF or loans from the European Investment Bank (EIB). . . .

[6] References to human rights appear throughout the Convention. The Preamble refers for the first time to the Universal Declaration of Human Rights and to both the Civil and Political and the Economic, Social and Cultural Covenants. Other provisions of the Convention advance a number of key economic, social, and cultural rights. These provisions include: environmental protection (tit. I), rural promotion (art. 46), enterprise and service sector development (tits. VIII–IX), cultural development (arts. 139–40), education and training (art. 151), scientific and technical cooperation (art. 152), the advancement of women (art. 153), and improved access to health care (art. 154). The Convention is supplemented by joint declarations on apartheid (Annex IV to the Final Act) and on ACP migrant workers and students in the Community. . . .

In spite of the Convention's symbolic emphasis on human rights, Lomé IV suffers from two major weaknesses. First, like its predecessors, Lomé IV pays scant attention to civil and political rights. Second, the Convention does not provide a legal basis for responding to human rights violations. Firm ACP opposition to the use of Article 5 as a basis for sanctions leaves consultation through the Convention's weak institutional network as the only means of confronting human rights violations.[7] These shortcomings have prevented Lomé IV from becoming a comprehensive human rights, democracy, and development program.

The 28 November Resolution officially inaugurated the European Community's current approach to development cooperation. . . .

The Resolution advances the Community's development policy beyond Article 5 of Lomé IV in five significant ways. First, the Resolution establishes a framework for European Community action in both positive and negative terms. Second, the Resolution places equal emphasis on economic, social, and cultural rights and civil and political rights and, for the first time, recognizes the importance of fostering 'democracy' and 'good governance' in development. Third, the Resolution applies human rights and democracy criteria to European Community development cooperation links around the world, not just with the ACP. Fourth, the Resolution builds on Lomé IV's endorsement of decentralized funding, explicitly adopting 'a decentralized approach to cooperation.' Fifth, the Resolution, as one of the Council and of the Member States meeting in the Council, represents the first time that the Member States have agreed in principle to coordinate certain aspects of their individual development policies.

. . .

A. Negative Measures

Curtailing or suspending development aid in reponse to human rights abuses can be instrumental in pressuring a wayward government to institute reform. Traditionally, donors have considered punitive strategies the primary means of linking development assistance to human rights and democracy. There are, however, many drawbacks to the exclusive use of reactive, or negative, measures.

First, negative measures do not address the root causes of human rights abuses. Rather, they punish governments and populations for the consequences of inadequacies in their legal, political, economic, and educational systems. Second, a reactive policy assumes that the recipient is both capable and willing to respond to external pressure. Countries with no capacity to act

[7] The ACP–EC Joint Assembly, for example, serves as a forum for negotiating responses to human rights violations. *See*, e.g., 1992 O.J. (C 31) 50 (criticizing the human rights situation in the Sudan). The Joint Assembly also sends special fact-finding missions to investigate human rights situations and submit recommendations. ACP–EC consultation, however, has proved problematic on the issue of human rights. . . .

and countries with low aid dependence are scarcely affected by donor threats to decrease or suspend aid.

Third, negative measures often antagonize donee states which view them as interferences with their sovereign right to determine internal policies. Recipient states have repeatedly challenged the international legal basis for conditioning development aid on adherence to norms of human rights and democracy. Fourth, negative measures often have devastating impacts on donee populations. Because negative measures are designed to influence governments, they are often applied without regard to their effects on local populations.

Fifth, reactive measures have not been guided by coherent implementation criteria. Conflicting foreign policy objectives, shifting funding priorities, and domestic pressure groups often thwart the consistent application of human rights conditions. The United States, for instance, has failed to implement seriously its policy of suspending aid to countries engaging in 'a consistent pattern of gross violations of internationally recognized human rights.'[8] Sixth, negative measures narrowly construe human rights as civil and political rights and ignore economic, social, and cultural rights. While violations of civil and political rights are often dramatic and easy to expose, they are no more serious than the denial of basic human needs.

. . .

B. Positive Measures

In contrast to negative measures, positive measures address the structural obstacles to sustained and equitable development. Positive measures incorporate the people and their needs into the development process, whether by helping to create a democractic environment conducive to popular participation in development or by building the political and social institutions necessary for a functioning civil society.

Whereas negative measures focus on the violator government, positive measures can target the actual needs of the donee population. Through flexible mechanisms, such as decentralized cooperation and policy dialogue, donors and recipients at the grassroots, non-governmental and government levels can work together to identify areas for funding and implement effective programs. Using decentralized cooperation, donors reinforce sustainable development efforts of the population through financial support to local groups and non-governmental organizations (NGOs). To complement these efforts, donors engage in joint consultation with donee states to devise strategies that satisfy the overall development needs of the recipient.

[8] Foreign Assistance Act of 1961, §502(b)75, Stat. 424 (codified as amended at 22 U.S.C. §2304(a)(2) (1993)) (prohibiting security assistance to gross violators). *See also* 22 U.S.C. §215(n) (1993) (linking other forms of foreign aid to human rights). Wide presidential discretion and interbranch conflicts have allowed foreign policy objectives to overshadow human rights concerns. . . .

This model of employing positive measures through decentralized cooperation and policy dialogue overcomes the six weaknesses of an exclusively reactive policy. . . .

Third, proactive development programs encroach less upon donees' sovereignty than do negative measures. Because positive measures aim to build the infrastructure necessary for sustainable development, donees and donors can jointly target priority areas for funding. . . .

Fifth, donors are less likely to behave inconsistently when pursuing promotional programs than when deciding whether to suspend aid. Proactive policies are more insulated from donors' foreign policy considerations because promotional programs tend to focus on immediate human needs and deliver funding through nongovernmental and grassroots projects rather than through governments.[9] As a result, donors are more apt to judge proactive programs by their usefulness as development tools rather than by their effects on foreign policy. Sixth, donors are likely to use promotional measures to foster a broader range of human rights. Unlike negative measures, which focus on egregious violations of civil and political rights, such as genocide and torture, positive measures can generally be used to advance both civil and political rights and economic, social, and cultural rights.

Despite their potential for addressing some of the traditional problems associated with negative measures, positive measures have a major limitation: they are an insufficient response to current patterns of gross human rights violations and interruptions of democratic processes. For this reason, effective development aid programs must encompass both strategies in a way that builds on the strengths of each approach while avoiding its weaknesses. . . .

. . .

The European Community has funded a variety of proactive initiatives to solidify democratic reforms, and reinforce civil society. Efforts to promote democracy dominate the European Community's development program. The Community has acted to safeguard the right of political participation through fair and open democratic transitions. Unfortunately, these programs have not always met with success. . . .

Because the problems of new democracies do not end with the holding of free and fair elections, the European Community has recognized the need for post-election projects to strengthen the political and legal institutions of fledgling democracies. In fact, the Community has pledged to expand its support for institution-building. By helping to make these new democratic institutions work, the Community hopes to demonstrate to the people of the developing world that 'democracy can answer some of the problems of development.'

. . .

[9] For instance, the European Community has continued programs to benefit the population while ceasing official cooperation assistance in countries like Haiti, Sudan, and Zaire.

At the grassroots level, the European Community has employed decentralized cooperation in order to involve donee populations in programs that address their basic needs. . . .

Decentralized cooperation at the intermediary level entails donor support for NGOs that link donors to the population. Through direct aid and co-financing schemes, the European Community provides necessary financial assistance for NGOs to implement development projects that empower community-based organizations, stimulate local initiatives and promote human rights awareness of democracy.

. . .

For the European Community, cooperation with NGOs offers tremendous advantages. These organizations have proved to be efficient and reliable development partners with unique abilities to carry out development activities at the grassroots level. . . .

. . .

With its preference for positive measures, the 28 November Resolution also recognizes the possible need for negative measures to respond to 'grave and persistent human [rights] violations or the serious interruption of democratic processes.' . . .

. . .

While the European Community has eschewed fixed procedures for responding to egregious human rights violations, it has established three general principles to frame its action. First, 'measures taken must be guided by objective and equitable criteria.' One Commission official explained this phrase to mean that the Community should be courageous enough to respond to violations irrespective of its political, economic, and security interests. In practice, however, the European Community has suspended aid only to economically weak states, such as Sudan, Zaire, Haiti, and Malawi, and not to human rights violators with greater geopolitical stature, like Indonesia.

In addition, objective and equitable criteria would demand a response free from the particular political and economic interests of the various Member States. Coordinating a united Member State response, however, remains difficult in light of these individual interests. Nevertheless, the Member States have agreed in principle to coordinate their individual policies, and the 28 November Resolution envisions some form of coordination at Council meetings and at the level of European Political Cooperation. Still, the proposed coordination procedures are too weak to structure a united response.

. . .

. . .[T]he European Community has developed four responses to human rights violations. At the lowest diplomatic level are confidential *démarches* to governments. Next are public statements that voice the European Community's human rights and democracy concerns. More serious measures include the deferral of signing National Indicative Programmes or financing agreements. The most severe negative measure is the interruption of cooperation agreements. Although hesitant to do so in the past, the Community has interrupted its cooperation programs with several nations, such as Sudan, Haiti, Zaire, and Malawi.

. . .

The European Community's third guiding principle stipulates that negative measures should 'avoid penalizing the population of the country in question and particularly its poorest sections.' This principle requires the European Community to redirect its development assistance away from the government and toward the poorest sections of the population through non-governmental networks. The Commission has established two preconditions to continued NGO financing: first, there must be no interference by the government; and second, the financed projects must directly aid those in need or promote some form of grassroots democracy.

This principle allows the European Community to interrupt official cooperation with the government while maintaining a humanitarian aid relationship with the population of the given country.[10] . . .

QUESTIONS

1. Assume that the goal of a human rights policy is not to 'punish' a state for violations but rather to induce its government to end abuses, move toward participatory government, and so on. How do you assess the different approaches of the United States and the European Union with respect to reaching this goal?

2. 'The approach of the European Union makes great sense if you're dealing with a state and government that has some genuine concern for the well being and development of its population. It makes less sense when serious violations are committed by governments that are heedless of the population's general welfare and intent on controlling all foreign funding directly so as to prevent the growth of independent centers of power and influence.' Comment.

[10] The Community continued to furnish humanitarian aid to the Sudan, Haiti, Zaire, and Malawi despite its interruption of governmental assistance. Until mid-September 1992, the Community financed 17 NGO projects totaling Ecu 3,914 million mainly in Haiti and Zaire that have supported NGO networks, workers' cooperatives, peasant farmer associations, integrated rural development projects, food security and health care programs, and the promotion of women. The Community recently granted Sudan Ecu 3 million in emergency aid to help the victims of the drought. . . .

E. CODES OF CONDUCT FOR FOREIGN OPERATIONS OF U.S. BUSINESS FIRMS

The following materials shift attention *from* governmental conditions and sanctions relating to aid, trade and investment to or with states engaged in serious human rights violations, *to* conditions imposed by U.S. businesses engaged in manufacturing or service operations abroad, whether directly (as through a subsidiary in a foreign state) or indirectly (as by contracting with an unrelated foreign firm to manufacture or assemble specified products in that state).

Should the U.S. business firm be concerned with human rights conditions in the foreign state, or should it remain attentive only to the goal of generating profits? Does it matter whether those foreign conditions (a) bear on torture in the prisons and repression of the press (that is, matters not of direct interest to the U.S. business), or (b) bear on repression of unions and other forms of association within the labor force and regulation of working conditions with which U.S. firms will be directly or indirectly involved? If U.S. business is concerned, what policies might firms devise? Should those policies be conceived, implemented and enforced solely within the U.S. business sector, or ought the government to play some regulatory role? Such are the issues here considered.

DIANE ORENTLICHER AND T. GELATT, PUBLIC LAW, PRIVATE ACTORS: THE IMPACT OF HUMAN RIGHTS ON BUSINESS INVESTORS IN CHINA

14 Northwestern J. of Int. L. & Bus. 66 (1993).

. . .

The focus of debate has been the annual renewal of China's most-favored-nation (MFN) trade status, but concerns about China's enduring human rights problems have pervaded virtually every aspect of U.S.–China trade and investment relations. However inconsistently enforced, sanctions ranging from a ban on military and high-technology sales to China to a prohibition on involvement by the Overseas Private Investment Corporation in Chinese projects have been used to promote human rights improvements.

But if the most visible debates have centered on U.S. trade and aid policies, a potentially more far-reaching debate is taking place in the boardrooms of corporate America. From Levi Strauss & Co. to Phillips–Van Heusen, from Sears, Roebuck and Co. to Reebok International Ltd., companies are asking how their role as investors can and should be shaped by human rights concerns in the PRC and other countries.

The answers that have emerged from these companies' deliberations reflect a pathbreaking reconception of corporate responsibility—one in which human rights occupy a central place. In March 1992, Sears, Roebuck and Co. announced that it would not import products produced by prison or other involuntary labor in China, and established a monitoring procedure to ensure compliance with its policy. In November 1990, Reebok International Ltd. condemned military repression in China and vowed that it 'will not operate under martial law conditions' or 'allow any military presence on its premises.' Two years later, Reebok adopted a human rights code of conduct governing workplace conditions in all of its overseas operations, including those in China. Phillips–Van Heusen currently threatens to terminate orders from suppliers that violate human rights principles enshrined in its ethical code. Adopting the most far-reaching policies, Levi Strauss & Co. and the Timberland Company apply human rights criteria in their selection of business partners, and avoid investing at all in countries where there are pervasive violations of basic human rights. In February 1993, the Timberland Company decided to end its sourcing from China. Two months later, Levi Strauss & Co. announced that it would end its relations with business partners in China, and would not initiate any direct investment there.

While these companies are in the vanguard of an emerging trend, their approach to human rights remains exceptional within the business community. . . .

. . . Are there basic principles that transnational companies should observe to ensure, at a minimum, that they do not become complicit in a host government's abrogation of universally-recognized human rights? Should such principles be enforced by Executive or congressional fiat, or should companies take primary responsibility for policing themselves? How can companies that wish to factor human rights considerations into their business decisions be assured that they will not pay a price in lost investment opportunities or reduced market share?

This article addresses these questions in light of relevant principles of international law and U.S. foreign policy. A central thesis of this article is that businesses that may or do invest in China bear a responsibility to ensure that their actions do not, however inadvertently, contribute to the systematic denial of human rights in the PRC. We believe, moreover, that international human rights law provides an objective basis for identifying those responsibilities.

We also believe that, in some circumstances, companies that invest in China can and should play a more proactive role in advancing respect for human rights. This view is based, above all, on the unique influence of major foreign investors in China. Today, after a temporary downturn in business activity in the year following the Tiananmen incident, the U.S. business presence in China is at an all-time high. With total committed investment close to six billion dollars, in 1993 the United States is China's second largest foreign investor. In the 1980s, U.S. companies became one of China's top providers of foreign investment, technology and management expertise. . . .

The importance of this large and growing U.S. presence to China's drive for economic and technical modernization cannot be overestimated. . . .

. . .

c. The Role of the Business Community

While views within the business community have not been monolithic, U.S. companies that invest in China have, on the whole, strongly opposed efforts to attach human rights conditions to renewal of China's MFN status. . . .

Demonstrating a sophisticated grasp of the U.S. political process, the PRC government exploited these apprehensions in the period preceding President Clinton's decision about renewal of China's MFN status. As the MFN debate approached, Chinese trade delegations went on a buying frenzy throughout the United States, spending more than $800 million for jetliners, $160 million for cars, and $200 million for oil exploration equipment. . . .

Throughout this process, China made it clear that it expected U.S. companies to lobby for continuation of its MFN status in return for its purchases. U.S. companies are 'regularly threatened with cancellation of orders or loss of future deals if China loses its preferred status,' according to business sources cited in *The Washington Post*. The message was not lost on U.S. companies, who mounted a campaign of unprecedented scope and intensity to secure unconditional renewal of China's MFN status in the period leading up to President Clinton's determination on this issue.

By letter dated May 12, 1993, some 370 companies and business associations, representing virtually every U.S. company active in China, stated their case to President Clinton:

> . . .We represent companies that exported products to China worth nearly $7.5 billion in 1992, and that employ an estimated 157,000 American workers producing those goods. We represent the aerospace industry which exported products to China worth over $2 billion in 1992, and which expects China to purchase approximately $40 billion in new aircraft over the next twenty years. We represent the farmers whose largest market for wheat is China. . . . America's economic stake in maintaining trade relations with China is high. Withdrawing or placing further conditions on MFN could terminate the large potential benefits of the trading relationship, lead the Chinese to engage in retaliatory actions that would harm U.S. exporters, farmers, laborers and consumers. . . .

But while emphasizing U.S. economic stakes, the signatories to this letter endorsed the human rights goals of the Clinton Administration's policy toward Beijing. Echoing arguments by spokesmen for business interests that had become increasingly common during previous debates about renewal of China's MFN status, they asserted:

> We in the business community . . . believe that our continued commercial interaction fuels positive elements for change in Chinese society. The expansion of trade and free market reforms has strengthened the pro-democratic forces in China. . . .

. . .

5. Code of Conduct Legislation

While MFN conditionality has dominated the U.S. human rights policy debate about China, a little-noticed legislative initiative has introduced a new, and potentially vital, plank in the policy options. Senator Edward Kennedy (D-Massachusetts) and Representative Jolene Unsoeld (D-Washington) have agreed to sponsor bills, in the Senate and House respectively, to establish a voluntary code of conduct governing the Chinese operations of U.S. companies. By directly focusing on the role of U.S. corporations in addressing human rights concerns in China, this legislation sharpens the broader debate about corporate responsibility vis-a-vis human rights violations in the PRC.

a. Background and Overview

The Kennedy and Unsoeld initiatives build upon a similar effort by then-Congressman John Miller (R.-Washington), who on June 21, 1991, introduced legislation that would have established a set of human rights principles governing the conduct of U.S. companies with investments and other business operations in the PRC. . . .

The proposed code-of-conduct bill does not seek to impose sanctions on China for failing to meet human rights standards, nor does it discourage U.S. businesses from investing in China. Instead, the bill asks companies with a significant presence in China to adhere to a set of basic human rights principles, on a 'best efforts' basis, in the course of their operations.

In this way, the proposed law seeks to assure that U.S. business activities in the PRC do not inadvertently encourage or themselves contribute to repressive practices, but instead make a constructive contribution to human rights. Under the proposed law, these goals would be promoted by encouraging U.S. nationals conducting industrial cooperation projects in China to adhere to nine principles that have the cumulative effect of (1) assuring that U.S. businesses operating in China extend to their foreign employees the same type of minimum human rights protections that they have long been required to provide to employees in the United States, such as protections against discrimination on the basis of religious beliefs, political views, gender and ethnic or national background; (2) assuring that the premises of U.S. business operations are not used in a fashion that violates fundamental rights (for example, the proposed code of conduct includes a pledge to discourage compulsory political indoctrination programs from taking place on the premises of U.S. nationals' industrial cooperation projects in the PRC); and (3) bringing the considerable—and indeed unique—influence of the U.S. business community to bear to promote an end to flagrant violations of human rights (for example, the proposed code of conduct urges U.S. nationals to use their access to Chinese officials informally to raise cases of individuals detained solely because of their nonviolent expression of political views).[11]

There are no penalties for failure to comply with the principles, and in this respect compliance with the code depends upon the voluntary efforts of U.S. companies. The bill itself is framed as a 'sense of Congress that any United States economic cooperation project in the People's Republic of China or Tibet should adhere to.' 'Adherence' is defined as 'agreeing to implement the principles set forth' in the bill, 'implementing those principles by taking good faith measures with respect to each such principle,' and 'reporting accurately to the Department of State on the measures taken to implement those principles.'

The bill imposes only two 'requirements' on U.S. companies: (1) the U.S. parent company of a PRC investment project must register with the Secretary of State and indicate whether it will implement the principles; and (2) the parent company must report on an annual basis to the Department of State describing the China project's adherence to the code. The Secretary of State is directed to review these reports to determine whether the project is adhering to the principles, and may request additional information to supplement company reports. The Secretary is further required to submit an annual report to Congress and the Secretariat of the Organization for Economic Coöperation and Development (OECD) describing the level of adherence to the principles by U.S. company projects in China.

. . .

. . .[W]hile some members of the U.S. business community have expressed support for the principles established in the legislation, many others have spoken out against it. The critics have raised two principal objections. The first, in essence, is that Congress should not dictate business practices to U.S. companies operating in China, and that the latter should not appear to be the 'lackeys' of U.S. policy. . . .

. . .

[11] Somewhat analogous codes have been developed to promote human rights in other countries. The best known of these are the Sullivan Principles for businesses operating in South Africa. First developed in 1977 and subsequently amplified, the Sullivan Principles were for many years adopted by corporations on a voluntary basis. In 1985, President Reagan issued an executive order that included a provision forbidding U.S. export assistance to any U.S. firm with 25 or more employees that had not adopted the principles enumerated in the Sullivan code. The Anti-Apartheid Act of 1986, which superseded President Reagan's executive order, incorporated the Sullivan Principles by, *inter alia*, requiring '[a]ny national of the United States that employs more than 25 persons in South Africa [to] take the necessary steps to insure that the Code of Conduct [based on the Sullivan Principles] is implemented with respect to the employment of those persons.' Comprehensive Anti-Apartheid Act of 1986 §207(a), 22 U.S.C. §5034(a) (1988). Another precedent is the MacBride Principles, which set forth employment standards for companies operating in Northern Ireland. A number of city and state governments have enacted laws supporting the MacBride Principles (by, for example, threatening to bar firms that do not adhere to the Principles from city contracts).

The general tenor of substantive objections by members of the business community is that the code would require U.S. companies to take action that may be 'impractical' or provocative, and that could jeopardize their position in China. This line of objection, which is more pronounced with respect to some provisions of the code than others, has often been backed by the claim that compliance with the bill's principles would require U.S. companies to violate Chinese law or policy.

A close examination of the proposed legislation suggests that these concerns are unwarranted. [T]he bill grants companies wide leeway to avoid taking action that could imperil their business relationships in China, urging only that they endeavor, on a 'best efforts' basis, to comply with and promote basic international standards in their Chinese operations.

. . .

II. RESPONSIBILITIES OF THE BUSINESS COMMUNITY

. . .

. . . Business leaders now rarely press the claim, once commonplace, that social policy and corporate practice occupy distinct spheres and that a rigid separation should be preserved.[12] . . .

Further, no company can afford to disregard the impact of massive human rights violations on the investment climate in a country where it may operate. The rule of law—the bedrock of human rights protection—is also essential to a stable and predictable environment for investment. One need only consider the devastating effect on the economies of the Latin American countries ruled by military dictatorships throughout the 1970s and, in many cases, into the 1980s to appreciate the correlation between massive human rights violations and investment risk.

. . .

Is there, for example, any principled reason to fault corporations for taking advantage of cheap labor in a developing country? Does the answer to this question depend on whether labor conditions fall below a minimum standard

[12] For a classic statement of the view that businesses should not be concerned with 'social responsibility,' see Milton Friedman, The Social Responsibility of Business is to Increase its Profits, N.Y. TIMES, Sept. 11, 1970 (Magazine), at 32. This claim has long been discredited, at least in its most sweeping form, in part because it is hopelessly circular. Our beliefs about what are proper concerns of the business community are themselves social constructs, and have evolved significantly over time in tandem with broader changes in the social and political environment. Further, to the extent that this argument asserts that it is the role of government to fashion and implement policy—in this case human rights policy—it also necessarily concedes to government the right to further specific policies by, *inter alia*, regulating the practices of U.S. corporations. Examples of such regulation, from legislation restricting companies' ability to discriminate or pollute at home, to laws prohibiting corrupt practices by corporations abroad, are too numerous to leave any room for doubting the legitimacy of government efforts to advance social policies in part by regulating corporate behavior.

of acceptability? Do transnational investors in a nation like China bear some measure of responsibility for the country's human rights problems on the ground that their investments help sustain a highly repressive government? On the other hand, in today's economy, can U.S. companies afford *not* to invest in the world's largest and fastest-growing market? Does their investment indeed serve human rights goals—as many companies claim—as well as their economic interests?

. . . In the context of China, business leaders have frequently asserted that the web of contacts between Chinese citizens and U.S. investors that develops in the course of business relationships promotes the transfer of liberal democratic values from this side of the Pacific to the East. Further, advocates of 'constructive engagement' also claim that transnational investment in repressive nations promotes greater integration of the host country in the international community, thereby enlarging its exposure to the shared values of civilized nations. It is sometimes further asserted that liberal political values are an inevitable concomitant of a liberal market economy, and that transnational efforts to foster development of such an economy in China through expanded trade and investment practices will therefore promote political liberalization as well.

A third and related claim is that U.S. investment in developing countries promotes economic growth, thereby fostering development of a middle class. Since, the argument continues, it is when this happens that citizens begin to assert demands for fundamental liberties, business investment spurs longer-term progress in respect of human rights. In the shorter term, foreign investment creates opportunities for employment that enhance the economic and social rights of the direct beneficiaries.

These are compelling arguments, and cannot be readily dismissed. But are the claims justified?

It depends. Whether a substantial U.S. business presence contributes to improved human rights conditions or helps bolster a repressive regime depends on the particular circumstances of each country, the conditions under which businesses operate, and the behavior of the businesses themselves. When, for example, the manager of a joint venture operation discharges a Chinese employee because of government pressure based on the individual's support for democracy, that manager becomes an agent for the Chinese government's denial of internationally-recognized human rights. When, instead, a potential investor insists as a precondition of investing on assurances that its employees' right to freedom of association will be fully protected, that investor's presence may in fact help foster improved human rights conditions. But here, too, an investor's ability to promote human rights may vary widely depending on both the conditions in a host country and on the nature of its investment. A company with direct investments in a country may, for example, have greater scope to promote human rights than a corporation that merely utilizes contractors there.

. . .

The claim that enhanced employment opportunities made possible by foreign investment in and of themselves advance human rights is initially appealing, but proves problematic upon closer scrutiny. To the extent that the transnationalization of investment has engendered a global chase for the cheapest labor markets, international investment practices inevitably drive down wage levels as developing countries compete for foreign investment. In this setting, it has become increasingly difficult to persuade governments of developing countries to respect internationally-recognized labor rights, particularly the right to receive a wage that meets the 'basic human needs' of workers.

In the longer term, this phenomenon has in many developing countries apparently retarded further expansion of the middle class, and instead has widened the economic gap between laborers and the management class. Against this background, it is increasingly difficult to assume that investment in and of itself will promote expansion of a middle class. . . .

. . .

Reebok International Ltd. adopted a human rights policy that responded to specific concerns raised by the human rights situation in China, and subsequently adopted a more comprehensive set of human rights principles governing workplace conditions in all of its overseas operations, including those in China.[13] . . .

. . .

III. UNIVERSAL HUMAN RIGHTS PRINCIPLES FOR TRANSNATIONAL COMPANIES

. . . [I]nternational law provides objective standards for determining the human rights responsibilities of transnational corporations.

. . . The [Treaty of Versailles] also established the International Labour Organisation (ILO), a tripartite organization comprising governments and representatives of employers and workers, which has established and monitored compliance with a broad array of labor standards. The ILO has promulgated some 170 international conventions elaborating labor standards, and members of the ILO are automatically bound to respect the core principles of freedom of association.

Basic workers' rights have also been incorporated in various international human rights instruments, such as the Universal Declaration of Human Rights, the International Covenant on Civil and Political Rights, and the International Covenant on Economic, Social and Cultural Rights, which collectively are considered the international equivalent of the U.S. Bill of Rights. Over the course of the past decade, moreover, the United States has adopted

[13] Those standards, adopted in December 1992, are set forth in Appendix B.

a range of laws that incorporate these international standards into U.S. trade policy.

. . .

In 1984 Congress enacted a law that incorporated workers rights criteria in the General System of Preferences (GSP) program initiated in 1974. That program authorizes the President to grant duty-free treatment to eligible imports from certain developing countries. The GSP law conditions duty-free treatment to otherwise eligible imports on whether the exporting country . . . 'has taken or is taking steps to afford to workers in that country (including any designated zone in that country) internationally recognized worker rights.' Drawing on ILO-established standards, the GSP law defines 'internationally recognized worker rights' to include:

(1) the right of association;
(2) the right to organize and bargain collectively;
(3) a prohibition on the use of any form of forced or compulsory labor;
(4) a minimum age for the employment of children; and
(5) acceptable conditions of work with respect to minimum wages, hours of work, and occupational safety and health.

As this and other legislation tying U.S. trade privileges to worker rights suggest, international law establishes a core set of universally-protected labor rights, and those rights can be fully protected only through international cooperation in enforcement. And as our previous analysis suggests, a key component of an effective international strategy for protecting workers' rights is adherence by multinational corporations to minimum standards. . . .

. . .

APPENDIX B

Reebok International Ltd.'s 'Human Rights Production Standards'

Reebok's devotion to human rights worldwide is a hallmark of our corporate culture. As a corporation in an ever-more global economy we will not be indifferent to the standards of our business partners around the world.

We believe that the incorporation of internationally recognized human rights standards into our business practice improves worker morale and results in a higher quality working environment and higher quality products.

In developing this policy, we have sought to use standards that are fair, that are appropriate to diverse cultures and that encourage workers to take pride in their work.

Non-Discrimination

Reebok will seek business partners that do not discriminate in hiring and employment practices on grounds of race, color, national origin, gender, religion, or political or other opinion.

Working hours/overtime

Reebok will seek business partners who do not require more than 60-hour work weeks on a regularly scheduled basis, except for appropriately compensated overtime in compliance with local laws, and we will favor business partners who use 48-hour work weeks as their maximum normal requirement.

Forced or Compulsory Labor

Reebok will not work with business partners that use forced or other compulsory labor, including labor that is required as a means of political coercion or as punishment for holding or for peacefully expressing political views. In the manufacture of its products, Reebok will not purchase materials that were produced by forced prison or other compulsory labor and will terminate business relationships with any sources found to utilize such labor.

Fair Wages

Reebok will seek business partners who share our commitment to the betterment of wage and benefits levels that address the basic needs of workers and their families so far as possible and appropriate in light of national practices and conditions. Reebok will not select business partners that pay less than the minimum wage required by local law or that pay less than prevailing local industry practices (whichever is higher).

Child Labor

Reebok will not work with business partners that use child labor. The term 'child' generally refers to a person who is less than 14 years of age, or younger than the age of completing compulsory education if that age is higher than 14. In countries where the law defines 'child' to include individuals who are older than 14, Reebok will apply that definition.

Freedom of Association

Reebok will seek business partners that share its commitment to the rights of employees to establish and join organizations of their own choosing. Reebok will seek to assure that no employee is penalized because of his or her non-violent exercise of this right. Reebok recognizes and respects the right of all employees to organize and bargain collectively.

Safe and Healthy Work Environment

Reebok will seek business partners that strive to assure employees a safe and healthy workplace and that do not expose workers to hazardous conditions.

AMERICAN SOCIETY OF INTERNATIONAL LAW, PANEL: HUMAN RIGHTS, BUSINESS AND INTERNATIONAL FINANCIAL INSTITUTIONS

Proceedings of 88th Annual Meeting, 1994, 271 (1995), 272.

John Keller (employee of Citicorp)

. . . [W]hat obligations do corporations have in the face both of host country human rights violations and of their need to pursue low-wage environments and rapidly expanding markets?

A wide spectrum of response is available.

At one end is the option chosen by most multinational corporations. I call this the 'don't ask don't tell' option. Let me caricature it here. Mr./Ms. Corporate Executive speaking: 'The human rights policies of a particular country are none of our business. We're here to produce product and make money for our shareholders, many of whom are pension funds and school districts. Our only obligation is to obey the laws of our host country and the laws of the United States of America, which are complicated enough to begin with. We might be paying low wages, but if we don't pay them somebody else will. And, by the way, if it weren't for 'exploiters,' people wouldn't have jobs in the first place. So go bother somebody else.'

. . .

. . . [S]hould businesses even care about human rights or any 'rights' other than those of their shareholders?

I believe that the answer is, clearly, Yes. And here I'd like to suggest an approach that tries to find some common ground between both sides of this very painful debate.

. . .

. . . [L]eadership must come from within the business community—U.S. companies need to make clear that their paramount fiduciary obligation to their shareholders resides in a unique context—the context of the democratic values on which this nation was founded and which are prized by their customers, employees and shareholders alike. I propose this is an open-ended starting point for a dialogue between the business and human rights communities—a dialogue that would explore appropriate Codes of Conduct for multinationals operating abroad. These codes do not need to tell a local government how to behave. There are countless ways, short of 'preaching' to local officials, that U.S. companies can affirm their values overseas. These codes could just state, for example, that a corporation will not purchase goods from a supplier that is known to use child labor or that operates with substandard wage and working conditions. Several U.S. corporations have already taken the lead in this area.

. . . The business community needs to broaden its definition of self-interest when it comes to human rights. I believe that there is more common ground between the objectives of business and the goals of the human rights community than is commonly acknowledged. Let me cite three examples:

> – In the long run, an educated and healthy workforce is necessary for continued economic development. It is logical for business to encourage local government policies that support these goals.
> – Employees whose basic nutritional and medical needs are met will be better and more productive workers. Company policies that assure adequate wages, workplace access to a good meal and—very importantly—superior local health care, will all support this objective.
> – Employees who do not have to live in fear of a battered-in door in the middle of the night, or of speaking freely to one another at appropriate times during the work day, will be more focused on the profitable task at hand. Company policies that refuse to cooperate with the presence of political or government 'monitors' in the workplace will help ensure this focus.

. . .

Bruce Landay (counsel to The Timberland Company)

. . .

The Timberland Policy drafters, who considered a firm policy to boycott countries where certain human rights and other standards are not met, ultimately opted for a more flexible human rights standard:

> While Timberland respects cultural differences, Timberland believes that basic human rights and nondiscrimination should cross all cultural barriers.
> Timberland will favor a partnership in a country where:
> – there is social and/or political commitment to the basic tenets of human rights, including free, democratic elections.
> – the government is proactive about eliminating discrimination, including discrimination based on race, color, national origin, gender, religion, disability, sexual orientation and political opinion.
> Timberland will not favor a partnership in a country where:
> – basic human rights are pervasively violated.

In addition, the 'Partnership Standards' section of the Timberland Policy contains a general policy statement on constructive engagement:

> We recognize that there will be situations in which a business partner meets our standards but operates within a country that does not. We will

> consider each such case very carefully, balancing the potential for reinforcing repressive countrywide conditions with the opportunity for meaningful change through constructive engagement.

. . .

. . . Were a corporation to shun all countries where governments are alleged to violate civil and human rights, perhaps as construed by organizations such as Amnesty International, the United States and many of its most important and developed allies might be off limits. In addition, the individuals whose rights are violated by their governments would in many cases be hurt by the loss of manufacturing work that the corporation provides.

. . . [T]he Timberland Policy contains 'Partnership Standards' regarding human rights to guide it in selecting contractors:

> While Timberland respects the necessity to balance our concern for human rights with respect for local culture and customs, Timberland believes that basic human rights and nondiscrimination should cross all cultural barriers.
>
> Timberland will favor Partners who:
>
> – have written policies regarding their position on human rights and discrimination.
>
> – are proactive in their commitment to the basic tenets of human rights and to the elimination of discrimination based on race, color, national origin, gender, religion, disability, sexual orientation and political opinion.
>
> – actively support community and cultural development projects.
>
> – take affirmative steps to eliminate repression.
>
> Timberland will not continue or initiate a partnership when:
>
> – the Partner discriminates against others based on race, color, national origin, gender, religion, disability, sexual orientation or political opinion.
>
> – the Partner is involved in any unfair or unethical trade practices;
>
> – the Partner has been found guilty of willful noncompliance with local, national or international human or civil rights requirements.

A key element in the Partnership Standards is the mechanism used for monitoring third-party conduct. The Timberland Policy provides that Timberland will require its contractors (1) to complete an annual compliance letter and (2) to consent to periodic, unannounced inspections of its facilities by Timberland, or such other measures as Timberland may deem necessary to confirm that the contractor has integrated the standards of the Timberland Policy into its operations. . . .

NOTE

As indicated in the article of Orentlicher and Gelatt, proposals have issued from the business community and from Congress about codes of conduct for

U.S. business operations abroad. Consider the following (*New York Times*, March 27, 1995, p. D1):

> In its latest attempt to win back the good will of human rights groups disappointed by President Clinton's decision last year to extend trade benefits to China without tying them to human rights concerns, the White House will announce on Monday a voluntary code of human rights principles for American companies operating abroad.
>
> . . .
>
> The chief problem is that America's allies, who are also its biggest economic competitors, have no such codes, and companies in those nations stand to sweep on business that the United States passes by. That issue was highlighted earlier this month when the Clinton Administration barred Conoco Inc. from entering a million-dollar development deal with Iran. Now French oil companies are negotiating with Iran to replace Conoco.
>
> It is unclear how many businesses will abide by the Administration's proposed code. Some executives reached last week said that if the wording is sufficiently weak or vague and appears unlikely to affect their competitiveness, there will probably be widespread adherence to what the Administration calls 'a model of behavior.'
>
> But already there are signs that the final draft has so many compromises that it will disappoint the human rights groups it was trying to satisfy in the first place.
>
> . . .
>
> People who have seen parts of the code say its strongest statements refer to bans on child labor and the use of prison laborers, actions that most large companies already say they oppose. The code will have deliberately vague language urging companies to respect the right of workers to organize and stops short of asking them to encourage such activities. The fear, several administration officials said, is that such activity could be viewed in many stations—including China—as an effort to undermine national laws that strictly regulate labor activities.
>
> . . . [T]he code will make no mention of paying a 'fair wage' to workers in developing countries and emerging markets, and it sidesteps the question of how companies should treat workers who are prosecuted for engaging in political dissent in countries like China. . . .
>
> President Clinton's plan quickly evolved into the development of what Secretary of State Warren Christopher described to Congress last month as 'the promulgation of a worldwide standard for the conduct of American business, not focused primarily on China, but focused on the world as a whole.'
>
> Many private companies already have codes of conduct for their operations abroad, including running-shoe manufacturers like Reebok and Nike, which have advertised their standards and used them as a selling point with American consumers. . . .

PART E
AN ILLUSTRATIVE STUDY

Drawing on the framework that Parts B–D have elaborated, this Part explores one topic of central importance to the human rights movement, women's rights. It has two functions. First, Part E examines one substantive field of human rights in a more systematic and comprehensive way than was possible in prior chapters developing other ideas. That field touches many others, from economic development to political participation, from sociology and history to religious and political culture.

Second, the topic of women's rights brings together within one illustrative study most of the ideas within the framework, both reviewing and elaborating on earlier chapters. The materials in this chapter involve relationships among norms, institutions and processes; the interpenetration of international and national systems; and broader themes pervading the human rights movement, like the claims of cultural relativism and the reach of international human rights to 'private' or non-governmental conduct.

Above all, the study of women's rights illustrates the increasing diffuseness and complexity of the human rights movement. We see a proliferation of instruments and institutions, world conferences and NGO initiatives, active proposals running in several directions, and growing conflicts about premises and goals within the women's movement itself. All such complexity is captured in what may be the human rights movement's most innovative and ambitious treaty, the Convention to Eliminate All Forms of Discrimination against Women.

13

Women's Rights

Of the several blind spots in the development of the human rights movement from 1945 to the present, none is as striking as that movement's failure to give to violations of women's (human) rights the attention, and in some respects the priority, that they require. It is not only that these problems adversely affect half of the world's population. They affect all of us, for a deep change in women's circumstances means corresponding change throughout social life.

Even in fields where the human rights movement has acted with vigor in setting standards, passing resolutions and at times imposing sanctions—for example, racial discrimination—it is too often clear that progress has been measured or slight, and that problems of the most serious character remain entrenched. Nonetheless, it is instructive to contrast the vigor of the movement in trying to 'eliminate' racial discrimination with its relative apathy over several decades in responding to gender discrimination—and to explore why this is so.

The subject of women's rights as international human rights offers distinctive perspectives on the human rights movement as a whole. The basic treaty in the field—the Convention on the Elimination of all Forms of Discrimination against Women (CEDAW, effective 1981, 144 ratifications as of September 1995)—has exceptional reach. At the same time, the problems that it addresses have exceptional depth and complexity.

The materials in this chapter suggest the complexly interwoven socio-economic, legal, political and cultural strands to the problem of women's subordination and women's rights. Indeed, it is difficult to know where to begin inquiry and analysis. Each starting point implicates others and by itself seems patently insufficient for yielding an adequate understanding of the problem, let alone solutions. When one focuses specifically on what appear to be women's issues, links between those issues and other aspects of social order (disorder) appear pervasive. All is interrelated. The problem is truly systemic.

The materials in Chapter 4 on cultural relativism used gender issues to illustrate the conflict between universalism and relativism, as well as questions about the reach of international norms to non-governmental practices. Those materials—particularly the article of An-Na'im and the readings on female circumcision, pp. 210 and 240, *supra*—could now be usefully reviewed.

A. BACKGROUND TO CEDAW: SOCIO-ECONOMIC CONTEXT

These introductory materials presents reports about women's circumstances in different parts of the world. They suggest the complex relationships among diverse phenomena that bear on women's rights. Several themes recur in the readings.

> Legal norms capture and reinforce deep cultural norms and community practices. They entrench ideas and help give them the sense of being natural, part of the order of things.
>
> Social forces and analysts insist that change is possible, so that what was seen as natural, if not inevitable, comes to be understood as socially constructed and thus contingent, open to change.
>
> Property rights and economic dependence interact with patterns of authority within family and workplace, and with vital issues like education and health.
>
> Major economic and political programs, like a development or privatization scheme or structural adjustment requirements, impose particular and severe costs on women that are not apparent on the face of the programs.
>
> The statistics created by bureaucracies or scholars structure and confine our imagination. We often view them as objective data, without awareness of what contentious concepts and information determine their formulation. What they record as well as what they do not record influence policies as well as perceptions.

INITIAL REPORT OF GUATEMALA SUBMITTED TO THE CEDAW COMMITTEE

CEDAW/C/Gua/1–2, 2 April 1991.

[In its introduction to this report submitted to the CEDAW Committee pursuant to the requirement of Article 18 of the Convention, Guatemala noted the difficulty of assembling the report, stressing that 'studies of this type are only a recent innovation.' The task of preparation 'has also been a positive exercise in thought, analysis and self-appraisal with respect to the position of women in Guatemala in 1983, and the changes made to date.' That work stimulated action to design 'strategies and targets . . . to improve the situation encountered in the short and medium term.' The following excerpts from the report deal with Articles 5 and 16 of the Convention.]

ARTICLE 5

46. Guatemala is a multi-ethnic, multi-cultural and multilingual country with traditional, cultural patterns that reinforce the subordination of women on the social, cultural, economic and political planes. Extended Guatemalan families in the country and nuclear families in the city are governed by a patriarchal system in which decisions are taken by men (husband, father or eldest son), who are considered the heads of the household, a role assumed by women only in their absence.

47. In Guatemalan society the man is expected to be the breadwinner, the legal representative, the repository of authority; the one who must 'correct' the children, while the mother is relegated to their care and upbringing, to household tasks, and to 'waiting on' or looking after her husband or partner. These roles often have to be performed in addition to engaging in some profitable activity which generates earnings that are always regarded as 'complementary'.

48. For their childhood, little boys and girls are guided towards work considered 'masculine' or 'feminine'; for example, boys play at working outside the home as carpenters, mechanics, farmers or pilots, and in all those jobs that are considered 'tough' or that require physical strength. Girls, on the other hand, are taught to interest themselves in cooking, weaving, sewing, washing, ironing, or cleaning the house and, especially, caring for the children and helping the mother, as a responsibility and duty more than just a game.

49. Care of the children is strictly considered the responsibility of the mother, grandmother, and/or sister; and in the event of divorce, separation or dissolution of the marriage, custody of the children is generally awarded to the mother.

50. The aforementioned patterns vary slightly with the socio-economic stratum, which generally also determines the social class to which the women belong and which in addition is related to their level of education and knowledge.

51. Notwithstanding what has been said, the woman is the chief social agent in the majority of spheres of action. An empirical profile of a Guatemalan woman may cover the following characteristics.

52. She is responsible for family health and hygiene and for the supervision of the formal and informal upbringing of the children in the home; she organizes and maintains living and sanitary conditions and a supply of water for domestic use. She produces nutritional supplements for the family, including animal proteins (cattle, sheep and goats) and sources of vitamins (fruit and vegetables); she is the one in charge of the purchase, preparation, stocking and distribution of food within the home. In addition, she manages the family income, ensuring that payment in kind and in cash is used in such a way as to maximise the material well-being of the family.

53. She takes responsibility for generating additional income or for producing consumer goods when her partner's income does not cover the minimum family requirements.

54. In the case of an irresponsible father, the entire responsibility for the support of the children devolves upon her, reflected in particular by a considerable increase in her hours of work.

55. Her work is poorly paid or not paid at all and is generally of low productivity owing to lack of access to capital.

56. It is falsely assumed that the man is the one who makes the principal economic contributions to the family, for which reason he is the owner and beneficiary of all payments and services.

57. The educational level of the woman is low, which reflects on the effectiveness of her efforts to maintain and improve the health, feeding, housing and other living conditions of her family.

58. In the paid work that she does, her salary is inferior to a man's and her instability in the sense of a job is greater.

59. The man has traditionally been considered the 'head of the household'.

ARTICLE 16

184. Family relations in Guatemala, as regards the guardianship, wardship, trusteeship and adoption of children, the ownership of property, its disposition and enjoyment, etc. are governed by the Guatemalan Civil Code (Decree-Law No. 106).

. . .

190. The woman's rights and responsibilities in marriage are as follows:

. . .

2. The husband owes his wife protection and assistance, and must provide her with all the means necessary to maintain the household, in accordance with his financial resources. *The woman has a special right and duty to nurture and care for her children during their minority, and to take charge of domestic affairs.*

. . .

5. The woman may be employed or ply a trade, occupation, public office or business, where she is able to do so without endangering the interests and the care of her children, or other needs of her household.

. . .

197. Married women are restricted in representing the marriage and in administration of marital assets, roles which are assigned by law to the husband, and this constitutes a relative incapacity.

198. Parental authority is a right which is virtually forbidden to women, since it is assigned to the father. *Women only come to exercise this right when the father is imprisoned or legally barred from such.*

. . .

201. The legal context allows the husband to object to the wife engaging in activities outside the home, thus barring her from the right and freedom to work. The legal context restricts her right to personal fulfilment in areas outside her function as mother and housewife and restricts her personal liberty.

. . .

203. A judicial declaration of paternity in cases of rape, rape of juveniles and abduction is dependent on the conduct of the mother, based on what the law terms 'notoriously disorderly conduct', an express form of discrimination against women and the product of conception resulting from forced intercourse.

. . .

209. Adultery defined as an 'offence against honour' protects the legal right of filiation and 'the interests of the family', but makes a clear distinction concerning the gravity of the act, depending on whether it involves the man or the woman, providing a tougher sentence for the woman; the proof and the procedure are different in the two cases, so that in practice it is only applied to women.

210. Offences 'against life' in which women are most affected are defined as abortion, which is defined as criminal conduct by which the death of the foetus is caused deliberately, within the mother's womb or by its premature expulsion. Medical abortion to avoid danger to the health or death of the mother, or due to deformities of the foetus, is not punishable. This is not envisaged when it is the result of rape.

211. With regard to the offence of rape, the punishment is graded according to the age of the victim and the relationship of authority which may exist between the victim and the offender. Reference is made to the 'honourable woman', requiring that the offender has used seduction, promise of marriage or deceit and the woman is a virgin; this emphasizes the value of 'honour', defining it as an offence against honour rather than against personal integrity, as would be correct.

212. Maltreatment of women and children and domestic violence are not defined as offences against the person and in practice are lumped together with injuries, coercion and threats, causing serious difficulties with regard to proof and other procedural problems.

. . .

WENDY PATTEN AND J. ANDREW WARD, EMPOWERING WOMEN TO STOP AIDS IN CÔTE D'IVOIRE AND UGANDA

6 Harv. Hum. Rts. J. 210 (1993).

The AIDS death toll continues to mount. The Global AIDS Policy Coalition estimates that as of 1992, 2.6 million people worldwide had developed AIDS. Africa, which contains only 9% of the global adult population, has absorbed roughly 67.7% of the adult AIDS cases, and 70% of cumulative adult AIDS deaths. Worldwide, these cases are occurring increasingly among women. More women developed AIDS in 1991 than in the entire preceding decade. In Africa, women are already bearing a disturbingly high share of the AIDS burden. Women constitute 48% of AIDS cases in Africa as compared to 13.7% in the United States and 17.2% in Europe. Côte d'Ivoire and Uganda, the countries examined in this Article, have particularly high levels of AIDS infection.

Violations of women's human rights make them especially vulnerable to AIDS in Africa. . . .

Although some international health experts are beginning to recognize the link between women's rights and HIV infection, the subject is conspicuously absent from HIV-prevention campaigns. Most public health practitioners see themselves as health engineers and consider human rights to be beyond the scope of their profession. Internationally, HIV prevention programs focus on disseminating information as a crucial step in ultimately changing the behavior of individuals. However, if people cannot change their behaviour because they lack the means to control their own destinies, information dissemination wastes time and resources.

The legal systems of both Côte d'Ivoire and Uganda disempower women, making it difficult for them to exercise control over their lives and to reduce their vulnerability to HIV infection. Côte d'Ivoire and Uganda, like most African countries, have a dual system of law: customary, or tribal law; and formal, or statutory law. In both countries, formal law preempts customary law. However, the principles of the formal law, largely inherited from the European colonial powers, are unknown or inaccessible to the vast majority of women, whose rights are defined by local custom. A central feature of both systems of law is the subordinate status of women.

This subordination is most apparent in marriage. Women's sexual availability is underscored by the tradition of bride price, whereby a man and his relatives pay the family of his prospective bride in order to marry her. This practice reinforces the notion that a husband has purchased his wife's sexual services, her labor, and her perpetual obedience and consent. While the majority of women have sexual relations only with their spouses, husbands are allowed greater sexual liberty. First, men often have open and socially accepted extramarital relationships. In the context of AIDS, men's multiple sexual partners increase their exposure to HIV, thereby endangering their wives. Second, the institution of polygamy allows men multiple wives. The addition of an HIV-positive co-wife to a stable polygamous union allows transmission of the virus to the husband and the other co-wives.

. . .

Although a woman may suspect that her husband has multiple sexual partners, her access to information about his HIV status is often blocked. Men are ashamed of the disease, maintaining a veil of secrecy around their HIV status. Since abstinence or condom use can raise suspicions, infected men often avoid such crucial protective measures. Even when women know that their husbands have multiple sexual partners, they rarely question them. When women do confront their husbands, they are often subjected to verbal and physical abuse, even rape.

. . .

Women's extreme reluctance to discuss sex reflects their relative powerlessness in the domestic sphere. Their husbands or partners exert almost exclusive control over decision making regarding sexual behaviour. Educating women to protect themselves from AIDS by using condoms is therefore ineffective since women find it extremely difficult to negotiate with men on this issue.

Traditional reproductive expectations also discourage women from insisting on condom use. In both Côte d'Ivoire and Uganda, it is a wife's duty to produce offspring, who, in turn, serve to insure vital family interests . . . Moreover, in both countries, women who wish to use birth control to protect themselves against AIDS have only limited access to contraceptives.

. . .

Both countries criminalize rape. In Uganda, however, rape, defilement (the term for statutory rape), and indecent assault are increasing. Most women are raped by men known to them, resulting in family intervention and settlements out of court. In Uganda, as elsewhere, women are reluctant to report rapes and to testify in humiliating court proceedings. When cases are reported to the police or to Resistance Councils, they are neither investigated effectively nor prosecuted seriously. Without an effective criminal penalty for rape, women remain vulnerable to AIDS.

. . .

In Côte d'Ivoire and Uganda, women's efforts to exercise control over their bodies are also frustrated by their lack of economic opportunity. Both economic dependence and social norms operate to coerce women to remain in relationships. Since a woman must submit to her husband's sexual demands if she remains in the home, her only other option is to leave, thereby abandoning her husband, her sole source of economic stability. Women whose husbands are infected with HIV must choose between the risk of AIDS and a life of extreme poverty and hardship.

Low education levels, lack of access to credit and appropriate technology, and employment discrimination all serve to limit women's economic opportunities. . . .

Property laws in both countries reflect and reinforce women's lack of economic rights, limiting their financial and social independence. In Côte d'Ivoire, the dominant marital property regime, community property, gives the husband full legal control over all family property including any property brought to the marriage by the wife. . . .

Discriminatory laws also limit a wife's right to inherit. . . .

. . .

COMMENT ON WOMEN'S SOCIAL AND ECONOMIC CONDITIONS

The status of women within the international human rights regime and the task of ensuring human rights for women are incomprehensible without taking into account the social and economic conditions that characterize women's lives around the world.

Later readings underscore the degree to which rights abuses are strongly correlated to victims' slight social and economic power, hence political power. Those who are most vulnerable to human rights abuses often lack the favour or protection of the state, as well as the power within their communities to protect and further their basic needs and interests.

According to virtually every indicator of social well-being and status—political participation, legal capacity, access to economic resources and employment, wage differentials, levels of educations and health care—women fare significantly and sometimes dramatically worse than men. The following information from the *Human Development Report* for 1993 of the United Nations Development Programme suggests the dimensions of the problem (at p. 25):

• *Literacy*—Women are much less likely than men to be literate. In South Asia, female literacy rates are only around 50% those of males. . . . in Nepal 35% . . . Sudan 27%. Women make up two-thirds of the world's illiterates.

• *Higher education*—Women in developing countries lag far behind men. In Sub-Saharan Africa, their enrolment rates for tertiary education are only a third of those of men. Even in industrial countries, women are very poorly represented in scientific and technical study. . . .

• *Employment*—In developing countries women have many fewer job opportunities; the employment participation rates of women are on average only 50% those of men (in South Asia 29% and in the Arab States only

16%). . . . Wage discrimination is also a feature of industrial countries; in Japan, women receive only 51% of male wages. Women who are not in paid employment are, of course, far from idle. Indeed, they tend to work much longer hours than men. . . .

• *Health*—Women tend on average to live longer than men. But in some Asian and North African countries, the discrimination against women—through neglect of their health or nutrition—is such that they have a shorter life expectancy. . . .

• *National statistics*—Women are often invisible in statistics. If women's unpaid housework were counted as productive output in national income accounts, global output would increase by 20–30%

Although employment outside the home provides women with increased income and often social status, employment remains a major source of discrimination. Women are doubly disadvantaged, occupying lower status and lower wage jobs in virtually every society while retaining the overwhelming burden of child care and household responsibilities. Alternatively, the labor women do outside caring for the family may remain unvalued and uncompensated (to a large extent, unvalued because uncompensated) where it does not form part of the cash economy. Consider the following (*Human Development Report* for 1993, at p. 45):

By 1990, women's share of the total economically active population in the industrial countries increased dramatically to 42%. In East Asia, it had risen to 43%, in Latin America and the Caribbean to 32%, and in North Africa and the Arab States to 13%.

But women are generally employed in a restricted range of jobs—in low-paid, low-productivity work . . . . In Africa, about 78% of economically active women work in agriculture (compared with 64% for men) . . .

Low status is reflected in low productivity and low pay, with women's earnings frequently only 50–80% those of men. . . .

In fact, there is considerable evidence that the well-being of women can be at risk rather than advanced through development including economic globalization and restructuring, in view of the types of jobs and wages that will be available for them. Women's tasks in the process of development tend to remain those involving long hours at basic labor like water collection, while men receive the necessary technical assistance.[1]

The structural adjustment programs that were intended to manage the debt crises endured by many developing countries during the 1980s further exacerbated the precarious situation of women. The orthodox formula for economic reform that has been promoted by institutions such as the World Bank and

[1] See generally Ester Boserup, Women and Economic Development (1970).

the International Monetary Fund involves some or all of the following elements: privatization of government services and corporations, economic deregulation and liberalization of trade, and reduction in the civil service and social services spending. As a consequence, these programs have led to sharp cutbacks in 'social safety net' spending in health, education and social services. The burden often fell disproportionately on women and children.[2]

In East and Central European states and the former Soviet Union, the burden of transition to market economies has similarly been borne disproportionately by women, despite their education levels that frequently exceed men's. In the new 'privatized' economies, job retraining programs have been targeted almost exclusively toward men. Women are regarded as more expensive and less attractive workers because of their responsibilities for the household and children. Employers may advertise explicitly for or tend to hire men. At the same time, the free or subsidized child care facilities which enabled women to work outside the home under the socialist regimes are disappearing or have been privatized, rendering it increasingly difficult for women to participate in the formal economy. Governments have indeed found it convenient to encourage women to leave paid employment and 'return home' as a means of dealing with the problem of unemployment.[3]

AMARTYA SEN, MORE THAN 100 MILLION WOMEN ARE MISSING

New York Review of Books, Dec. 20, 1990, at 61.

It is often said that women make up a majority of the world's population. They do not. This mistaken belief is based on generalizing from the contemporary situation in Europe and North America, where the ratio of women to men is typically around 1.05 or 1.06, or higher. In South Asia, West Asia, and China, the ratio of women to men can be as low as 0.94, or even lower, and it varies widely elsewhere in Asia, in Africa, and in Latin America. How can we understand and explain these differences, and react to them?

At birth, boys outnumber girls everywhere in the world, by much the same proportion—there are around 105 or 106 male children for every 100 female children. Just why the biology of reproduction leads to this result remains a subject of debate. But after conception, biology seems on the whole to favor women. . . . When given the same care as males, females tend to have better survival rates than males.

[2] Third Periodic Report of Ecuador to CEDAW Committee, CEDAW/C/ECU/3. 10 Jan. 1992; Lourdes, Beneria and Feldman (eds.), Unequal Burden: Economic Crises, Persistent Poverty, and Women's Work (1992); Akua Kuenyehia, The Impact of Structural Adjustment Programs on Women's International Human Rights: The Example of Ghana, in Rebecca Cook (ed.), Human Rights of Women: National and International Perspectives 422 (1944).

[3] Barbara Einhorn, Cinderella Goes to Market: Citizenship, Gender and Women's Movements in East Central Europe (1993); UN Centre for Social Development and Humanitarian Affairs, The Impact of Economic and Political Reform on the Status of Women in Eastern Europe (1992).

Women outnumber men substantially in Europe, the US, and Japan, where, despite the persistence of various types of bias against women (men having distinct advantages in higher education, job specialization, and promotion to senior executive positions, for example), women suffer little discrimination in basic nutrition and health care. . . .

The fate of women is quite different in most of Asia and North Africa. In these places the failure to give women medical care similar to what men get and to provide them with comparable food and social services results in fewer women surviving than would be the case if they had equal care. In India, for example, except in the period immediately following birth, the death rate is higher for women than for men fairly consistently in all age groups until the late thirties. This relates to higher rates of disease from which women suffer, and ultimately to the relative neglect of females, especially in health care and medical attention. Similar neglect of women vis-à-vis men can be seen also in many other parts of the world. The result is a lower proportion of women than would be the case if they had equal care—in most of Asia and North Africa, and to a lesser extent Latin America.

This pattern is not uniform in all parts of the third world, however. Sub-Saharan Africa, for example, ravaged as it is by extreme poverty, hunger, and famine, has a substantial excess rather than deficit of women, the ratio of women to men being around 1.02. The 'third world' in this matter is not a useful category, because it is so diverse.

. . .

To get an idea of the numbers of people involved in the different ratios of women to men, we can estimate the number of 'missing women' in a country, say, China or India, by calculating the number of extra women who would have been in China or India if these countries had the same ratio of women to men as obtain in areas of the world in which they receive similar care. If we could expect equal populations of the two sexes, the low ratio of 0.94 women to men in South Asia, West Asia, and China would indicate a 6 percent deficit of women; but since, in countries where men and women receive similar care, the ratio is about 1.05, the real shortfall is about 11 percent. In China alone this amounts to 50 million 'missing women,' taking 1.05 as the benchmark ratio. When that number is added to those in South Asia, West Asia, and North Africa, a great many more than 100 million women are 'missing.' These numbers tell us, quietly, a terrible story of inequality and neglect leading to the excess mortality of women.

To account for the neglect of women, two simplistic explanations have often been presented or, more often, implicitly assumed. One view emphasizes the cultural contrasts between East and West (or between the Occident and the Orient), claiming that Western civilization is less sexist than Eastern. That women outnumber men in Western countries may appear to lend support to this Kipling-like generalization. . . . The other simple argument looks instead at stages of economic development, seeing the unequal nutrition and health care provided for women as a feature of underdevelopment, a characteristic of poor economies awaiting economic advancement.

. . .

Despite their superficial plausibility, neither the alleged contrast between 'East' and 'West,' nor the simple hypothesis of female deprivation as a characteristic of economic 'underdevelopment' gives us anything like an adequate understanding of the geography of female deprivation in social wellbeing and survival. We have to examine the complex ways in which economic, social, and cultural factors can influence the regional differences.

It is certainly true that, for example, the status and power of women in the family differ greatly from one region to another, and there are good reasons to expect that these social features would be related to the economic role and independence of women. For example, employment outside the home and owning assets can both be important for women's economic independence and power; and these factors may have far-reaching effects on the divisions of benefits and chores within the family and can greatly influence what are implicitly accepted as women's 'entitlements.'

[Sen discusses decision-making within the family as the pursuit of cooperation 'in which solutions for the conflicting aspects of family life are implicitly agreed on.' Analysis of these 'cooperative conflicts' in different regions and cultures can 'provide a useful way of understanding the influences that affect the "deal" that women get in the division of benefits within the family.' Perceptions of who is doing 'productive' work or contributing to the family's welfare can be very influential, and such social perceptions are 'of pervasive importance in gender inequality,' particularly 'in sustaining female deprivation in many of the poorer countries.'

Division of a family's joint benefits are apt to be more favorable to women if (1) they earn outside income, (2) their work is recognized as productive, (3) they own some economic resources or hold economic rights, and (4) there is an understanding of ways in which women are deprived. 'Considerable empirical evidence' suggests that gainful employment such as working outside the home for a wage as opposed to unpaid housework 'can substantially enhance the deal that women get.' Not only access to funds but also women's status and standing in the family improve. Moreover, women bring home experience of the outside world, a form of education. Such factors can 'counter the relative neglect of girls as they grow up,' as women are seen as economic producers.

Sen discusses the different situation in China, where other explanatory factors may be important, such as the strong measures to control the size of families in the framework of a strong cultural preference for boys.]

In comparing different regions of Asia and Africa, if we try to relate the relative survival prospects of women to the 'gainful employment' of both sexes—

i.e., work outside the home, possibly for a wage—we do find a strong association. . . .

. . .

Analyses based on simple conflicts between East and West or on 'underdevelopment' clearly do not take us very far. The variables that appear important—for example, female employment or female literacy—combine both economic and cultural effects. To ascribe importance to the influence of gainful employment on women's prospects for survival may superficially look like another attempt at a simple economic explanation, but it would be a mistake to see it this way. The deeper question is why such outside employment is more prevalent in, say, sub-Saharan Africa than in North Africa, or in Southeast and Eastern Asia than in Western and Southern Asia. Here the cultural, including religious, backgrounds of the respective regions are surely important. Economic causes for women's deprivation have to be integrated with other—social and cultural—factors to give depth to the explanation.

Of course, gainful employment is not the only factor affecting women's chances of survival. Women's education and their economic rights—including property rights—may be crucial variables as well. Consider the state of Kerala in India. . . . It does not have a deficit of women—its ratio of women to men of more than 1.03 is closer to that of Europe (1.05) than those of China, West Asia, and India as a whole (0.94). The life expectancy of women at birth in Kerala, which had already reached sixty-eight years by the time of the last census in 1981 (and is estimated to be seventy-two years now), is considerably higher than men's sixty-four years at that time (and sixty-seven now). While women are generally able to find 'gainful employment' in Kerala—certainly much more so than in Punjab—the state is not exceptional in this regard. What is exceptional is Kerala's remarkably high literacy rate: not only is it much higher than elsewhere in India, it is also substantially higher than in China, especially for women.

. . .

Moreover, in parts of Kerala, property is usually inherited through the family's female line. These factors, as well as the generally high level of communal medicine, help to explain why women in Kerala do not suffer disadvantages in obtaining the means for survival. While it would be difficult to 'split up' the respective contributions made by each of these different influences, it would be a mistake not to include all these factors among the potentially interesting variables that deserve examination.

In view of the enormity of the problems of women's survival in large parts of Asia and Africa, it is surprising that these disadvantages have received such inadequate attention. . . . We confront here what is clearly one of the more momentous, and neglected problems facing the world today.

HILARY CHARLESWORTH, C. CHINKIN AND S. WRIGHT, FEMINIST APPROACHES TO INTERNATIONAL LAW

85 Am. J. Int. L. 613 (1991), at 614.

I. INTRODUCTION

. . .

International law has thus far largely resisted feminist analysis. The concerns of public international law do not, at first sight, have any particular impact on women: issues of sovereignty, territory, use of force and state responsibility, for example, appear gender free in their application to the abstract entities of states. Only where international law is considered directly relevant to individuals, as with human rights law, have some specifically feminist perspectives on international law begun to be developed.

In this article we question the immunity of international law to feminist analysis–why has gender not been an issue in this discipline?—and indicate the possibilities of feminist scholarship in international law. . . Our approach requires looking behind the abstract entities of states to the actual impact of rules on women within states. We argue that both the structures of international lawmaking and the content of the rules of international law privilege men; if women's interests are acknowledged at all, they are marginalized. International law is a thoroughly gendered system.

. . .

IV. TOWARD A FEMINIST ANALYSIS OF INTERNATIONAL LAW

The right to development. The right to development was formulated in legal terms only recently and its status in international law is still controversial. Its proponents present it as a collective or solidarity right that responds to the phenomenon of global interdependence, while its critics argue that it is an aspiration rather than a right. The 1986 United Nations Declaration on the Right to Development describes the content of the right as the entitlement 'to participate in, contribute to, and enjoy economic, social, cultural and political development in which all human rights and fundamental freedoms can be fully realized.' [4] Primary responsibility for the creation of conditions favorable to the right is placed on states:

> States have the right and the duty to formulate appropriate national development policies that aim at the constant improvement of the well-being of the entire population and of all individuals, on the basis of their active, free and meaningful participation in development and in the fair distribution of the benefits resulting therefrom.

[4] G.A. Res. 41/128, Art. 1(1) (Dec. 4, 1986).

The right is apparently designed to apply to all individuals within a state and is assumed to benefit women and men equally: the preamble to the declaration twice refers to the Charter exhortation to promote and encourage respect for human rights for all without distinction of any kind such as of race or sex. Moreover, Article 8 of the declaration obliges states to ensure equality of opportunity for all regarding access to basic resources and fair distribution of income. It provides that 'effective measures should be undertaken to ensure that women have an active role in the development process.'

Other provisions of the declaration, however, indicate that discrimination against women is not seen as a major obstacle to development or to the fair distribution of its benefits. For example, one aspect of the right to development is the obligation of states to take 'resolute steps' to eliminate 'massive and flagrant violations of the human rights of peoples and human beings'. The examples given of such violations include apartheid and racial discrimination but not sex discrimination.

Three theories about the causes of underdevelopment dominate its analysis: shortages of capital, technology, skilled labor and entrepreneurship; exploitation of the wealth of developing nations by richer nations; and economic dependence of developing nations on developed nations. The subordination of women to men does not enter this traditional calculus. . . .

. . .

. . . Women and children are more often the victims of poverty and malnutrition than men. Women should therefore have much to gain from an international right to development. Yet the position of many women in developing countries has deteriorated over the last two decades: their access to economic resources has been reduced, their health and educational status has declined, and their work burdens have increased. . . .

The distinction between the public and private spheres operates to make the work and needs of women invisible. Economic visibility depends on working in the public sphere and unpaid work in the home or community is categorized as 'unproductive, unoccupied, and economically inactive.' Marilyn Waring has recently argued that this division, which is institutionalized in developed nations, has been exported to the developing world, in part through the United Nations System of National Accounts (UNSNA).

The UNSNA, developed largely by Sir Richard Stone in the 1950s, enables experts to monitor the financial position of states and trends in their national development and to compare one nation's economy with that of another. It will thus influence the categorization of nations as developed or developing and the style and magnitude of the required international aid. The UNSNA measures the value of all goods and services that actually enter the market and of other nonmarket production such as government services provided free of charge. Some activities, however, are designated as outside the 'production boundary' and are not measured. Economic reality is constructed by the UNSNA's 'production boundaries' in such a way that reproduction, child care, domestic work and subsistence production are excluded from the measurement of economic productivity and growth. This view of women's work as nonwork was nicely summed up in 1985 in a report by the Secretary-General to the General Assembly, 'Overall socio-economic perspective of the world economy to the year 2000.' It said: 'Women's productive and reproductive roles tend to be compatible in rural areas of low-income countries, since family agriculture and cottage industries keep women close to the home, permit flexibility in working conditions *and require low investment of the mother's time.*'

The assignment of the work of women and men to different spheres, and the consequent categorization of women as 'nonproducers,' are detrimental to women in developing countries in many ways and make their rights to development considerably less attainable than men's. For example, the operation of the public/private distinction in international economic measurement excludes women from many aid programs because they are not considered to be workers or are regarded as less productive than men. . . .

Although the increased industrialization of the Third World has brought greater employment opportunities for women, this seeming improvement has not increased their economic independence or social standing and has had little impact on women's equality. Women are found in the lowest-paid and lowest-status jobs, without career paths; their working conditions are often discriminatory and insecure. Moreover, there is little difference in the position of women who live in developing nations with a socialist political order. The dominant model of development assumes that any paid employment is better than none and fails to take into account the potential for increasing the inequality of women and lowering their economic position.

. . .

B. CEDAW: PROVISIONS AND COMMITTEE

COMMENT ON PROTECTION OF WOMEN UNDER CONVENTIONS PRIOR TO CEDAW

Of course women benefit, and are meant to benefit, from provisions of the basic human rights instruments on state-inflicted torture or killing, denials of due process, freedom of speech and so on. The emphasis of this Comment is on issues of distinctive concern to women, on issues related to gender whether or not they so appear on the surface.

One fundamental protection of women stems from the assurance of equal protection. Such protection has been a dominant theme in the human rights movement, and prohibition of discrimination because of sex has been an essential ingredient of that theme. It is true that application of the equal protection principle in the field of women's rights raises particularly subtle and difficult problems, but the thrust of that principle is also direct and clear on basic issues like voting or admission to higher study.

Recall the stress on the equal protection norm in the following provisions in the UN Charter and the International Bill of Rights.

1. The Charter's preamble states the determination of the peoples of the United Nations to reaffirm faith 'in the equal rights of men and women.' Article 1(3) sets forth the organization's purpose of promoting respect for human rights 'for all without distinction as to race, sex, language, or religion.' Article 55(c) is to similar effect.

2. Article 2 of the Universal Declaration states that 'everyone' is entitled to the rights declared 'without distinction of any kind, such as race, colour, sex, language, religion, political or other opinion, national or social origin, property, birth or other status.' Under Article 16, men and women are 'entitled to equal rights as to marriage, during marriage and at its dissolution.'

3. Under Article 2 of the ICCPR, states undertake to ensure to all within their territory the rights recognized in the Covenant 'without distinction of any kind.' The list of prohibited distinctions is identical with that above in the UDHR. States further undertake in Article 3 to 'ensure the equal right of men and women' to enjoyment of all rights set forth in the Covenant. Under Article 23(4), states are to take 'appropriate steps to ensure equality of rights and responsibilities of spouses as to marriage, during marriage and at its dissolution.' Article 26 contains undertakings by states to prohibit discrimination on the same grounds as those identified in the UDHR.

4. Similar anti-discrimination provisions appear in the International Covenant on Economic, Social and Cultural Rights.

These instruments have been sharply criticized, in terms of both their substantive provisions bearing explicitly or implicitly on women and the means by which those provisions have been implemented. Some of the most trenchant analyses have been made by scholars and advocates writing from a feminist perspective—for present purposes, a perspective that examines instruments so as to identify their overt or covert bearing on women, particularly with respect to gender bias.

Consider, for example, the following analysis of Article 16 of the UDHR by Helen Holmes.[5] The article states that '[t]he family is the natural and fundamental group unit of society and is entitled to protection by society and the State.' Holmes notes the widely divergent definitions of the family in different cultures—nuclear, extended, matrilineal, patrilineal, and so on—and asserts that 'those with the power to implement Article 16 will implement it to conform to their own concepts of "family".' She notes the ongoing conflicts over the advantages and disadvantages of different types of families, including the prototypical ideal in the developed West of the nuclear family. She questions the wisdom or fairness, in the midst of such conflicts and changing perceptions, of institutionalizing a vision of the family as (quoting from the Article 16) the 'natural and fundamental group unit.'

What, moreover, is meant by (again quoting) 'protection by society and the State' of the family? What are such protection's implications, say, for divorce or for the desire of a child to leave a family because of abuse or other reasons? How will such protection be reconciled with the provisions about children in Article 25(2)? Holmes asks whether protection of the family would in fact amount to the defense of patriarchy and hierarchy.

Another critic, Laura Reanda, develops related perspectives on the older human rights corpus.[6] Writing in 1981, she observes that the main international human rights organs like the UN Commission on Human Rights or the ICCPR Human Rights Committee 'do not appear to deal specifically with violations of the human rights of women, except in a marginal way or within the framework of other human rights issues.' Through the UN Commission on the Status of Women and the CEDAW Committee, p. 911, *infra*, there has been a 'ghettoization' of questions relating to women and their relegation to structures endowed with less power and resources than the general human rights structures.

Reanda notes that the periodic reports submitted by states to the Human Rights Committee under Article 40 of the ICCPR rarely deal with the situation of women and, when they do so, generally stress laws on the books rather than custom or practice. Thus 'forms of oppression not specifically defined in the Covenant tend to be neglected.' She continues (at p. 15):

> . . . It is well known, for instance, that in many states parties to the Covenant, discriminatory practices against women persist, whether in law or in fact, such as segregation from public life, polygamy, the dowry system, bride-price, and genital mutilation. Another example is forced prostitution, which is considered a form of slavery by other United Nations organs but has never been raised in connection with the antislavery provision of the Covenant.
>
> While none of these institutions and practices are specifically mentioned by the Covenant, it is hard to see how women can be guaranteed the full enjoyment of the human rights and fundamental freedoms to which they are entitled under the Covenant so long as these and other forms of oppression continue. To bring these practices under the purview of the

[5] A Feminist Analysis of the Universal Declaration of Human Rights, in Carol Gould (ed), Beyond Domination: New Perspectives on Women and Philosophy 250 (1983).

[6] Human Rights and Women's Rights: The United Nations Approach, 3 Hum. Rts. Q. 11 (No. 2, 1981).

> Covenant, however, would require a concerted effort in terms of fact-gathering and interpretation, and there is no evidence that this is among the goals of the Committee. In fact, none of these practices has been mentioned in state reports or in the debates of the Committee. It would appear that the interpretation given to the Covenant so far is that the Covenant does not apply to these obvious violations of the human rights of women.

Felice Gaer[7] after describing some provisions that are distinctive to CEDAW rather than characteristic of most human rights conventions, observes that 'women's issues are *not* commonly dealt with in the mainstream international human rights debates and bodies of the UN or by the international human rights NGOs.' She notes the stress of IGOs and NGOs on violations of bodily security by state actors. 'Put simply, the human rights organizations do not rank the problem [of gender discrimination] very high on a hierarchical list of human rights violations.' Gaer suggests several reasons:

> (1) A sociological factor. Most of the delegates to UN human rights bodies are men; most are lawyers. They tend to look principally at due process questions and traditional formal procedures. The same is true of nongovernmental human rights organizations.
>
> (2) The *de facto* hierarchy of international human rights practice is to favor civil and political rights over economic and social. To the extent the latter are considered at all, it is from a development perspective, not in terms of gender or other discrimination.
>
> (3) There is a lack of consciousness of gender discrimination among the human rights organizations. In part, this is due to a lack of information regularly transmitted. Women become visible to many human rights groups only when they are 'victims' of traditional mainstream violations: torture, arbitrary arrest and imprisonment.
>
> (4) There is a widespread sense that gender discrimination issues are private and outside the responsibilities of government. Even issues like forced prostitution are seen as private matters not involving the government.

Gaer concludes by arguing that there is a 'great need for reconceptualization of the issues. . . . Once reconceptualized, the issues must be made visible. Documentation is a key factor. . . .'

NOTE

Section 702 of the *Restatement (Third), Foreign Relations Law of the United States*, p. 145, *supra*, sets forth categories of the contemporary *customary*

[7] Human Rights at the UN: Women's Rights Are Human Rights, Int. League for Hum. Rts., In Brief, No. 14 (Nov. 1989).

international law of human rights. It provides that a state violates international law if, 'as a matter of state policy, it practices, encourages, or condones' conduct that includes genocide, torture, systematic racial discrimination, or 'a consistent pattern of gross violations of internationally recognized human rights.' The section does not include gender discrimination among the seven identified categories. Note the *Restatement*'s comments following the text of §702.

> a. *Scope of customary law of human rights.* This section includes as customary law only those human rights whose status as customary law is generally accepted (as of 1987) and whose scope and content are generally agreed. The list is not necessarily complete, and is not closed: human rights not listed in this section may have achieved the status of customary law, and some rights might achieve that status in the future.
>
> . . .
>
> b. *Gender discrimination.* The United Nations Charter (Article 1(3)) and the Universal Declaration of Human Rights (Article 2) prohibit discrimination in respect of human rights on various grounds, including sex. Discrimination on the basis of sex in respect of recognized rights is prohibited by a number of international agreements, including the Covenant on Civil and Political Rights, the Covenant on Economic, Social and Cultural Rights, and more generally by the Convention on the Elimination of All Forms of Discrimination Against Women, which, as of 1987, had been ratified by 91 states and signed by a number of others. The United States had signed the Convention but had not yet ratified it. The domestic laws of a number of states, including those of the United States, mandate equality for, or prohibit discrimination against, women generally or in various respects. Gender-based discrimination is still practiced in many states in varying degrees, but freedom from gender discrimination as state policy, in many matters, may already be a principle of customary international law. . .

QUESTION

As an advocate of women's rights, in what respects and contexts would you find it helpful to argue that prohibition of gender discrimination has become part of the customary international law of human rights? If this argument were persuasive, would you nonetheless prefer to rely if possible on CEDAW? Why?

COMMENT ON CEDAW'S SUBSTANTIVE PROVISIONS

The Convention is among the many that elaborate in one particular field the norms and ideals that are generally and tersely stated in the Universal Declaration, and stated somewhat more amply in the ICCPR. Its preamble suggests how deeply the issues run and that the norms of this Convention must be placed in a broader transformative context. It recognizes 'that a change in the traditional role of men as well as the role of women in society and in the family is needed to achieve full equality between men and women.'

You should review the provisions of the Convention. To understand its structure and varied aspirations, as well as the radically different kinds of duties that are placed on states parties, it will be useful to characterize the obligations of states under CEDAW's varied provisions within the following scheme:[8]

(1) respect: the duty to treat persons equally, to respect their individual dignity and worth, hence not to interfere with or impair their declared rights—that is, the classical 'hands-off' duties of liberal states that are correlative to individual rights;

(2) protect, prevent: the duty to extend protection against violations of rights by the state as well as by non-state actors (individuals or organizations), hence the duty to create and administer an adequate system of police, law enforcement, and civil and criminal justice;

(3) provide: the duty to ensure, to assure individuals of defined minimum levels of welfare, to improve (and not only refrain from worsening) the situation of individuals up to such levels—that is, the type of duty elaborated in the International Covenant on Economic, Social and Cultural Rights, as in its Article 11 where states parties undertake to 'ensure the realization' of the right of everyone to an adequate standard of living; and

(4) promote: the (varied and often indeterminate) duty to take measures such as education to reduce violations of rights, to train people to help to gain recognition of their own rights, to transform (to one or another extent) existing attitudes inimical to realization of rights.

Article 1: Note three vital characteristics of the definition of 'discrimination against women.' (a) The article refers to *effect* as well as *purpose*, thus directing attention to the consequences of governmental measures as well as the intentions underlying them. (b) The definition is not limited to discrimination through 'state action' or action by persons acting under color of law,

[8] A similar scheme is set forth, and elaborated with respect to the right to food, in G.J.H. Van Hoof, The Legal Nature of Economic, Social and Cultural Rights, in Philip Alston and K. Tomasevski (eds.), The Right to Food 97 (1984). Excerpts appear at p. 279, *supra*. The terms and categories in the text above vary in several respects from the article.

as are the definitions of many rights such as the definition of torture under the Convention against Torture. (c) The definition's range is further expanded by the concluding phrase, 'or any other field.'

Article 2: The goals stated in this article are to be pursued 'without delay.' Consider the possible meanings of the terms 'equality' in clause (a) and 'any act of discrimination' in clause (c). Note the breadth of clauses (e) and (f) with respect to the private, non-governmental sectors of society, particularly in relation to the definition in Article 1. Note throughout the Convention the blurred lines between the private and public spheres of life, and the range of obligations on states to intervene in the private sector, to go beyond 'respect' to 'protect,' 'ensure,' and 'promote.'

Article 3: Note the grand goal set forth for states, to 'ensure the full development and advancement of women,' and consider whether the other human rights instruments examined contain a similar conception for any group, or for people in general.

Article 4: This 'affirmative action' clause, however qualified, appears as well in the Convention on the Elimination of all Forms of Racial Discrimination, but not in the ICCPR. Nonetheless, note General Comment 18 of the ICCPR Human Rights Committee, p. 529, *supra*, particularly pars. 7–9. Consider this Article 4 in relation to Article 2(e) and (f), and Article 11.

Article 5: The breadth and aspiration of this article are surely striking. Provisions such as Article 10(c) impose a similar obligation on states in defined contexts. Other human rights treaties lack a similar provision, although Article 2 of the Racial Discrimination Convention, p. 777, *supra*, comes close. Consider how a state in good faith might decide on 'appropriate measures' under this article, bearing in mind the injunction in Article 2 to proceed 'without delay' as well as the claims of the Convention's other provisions.

Article 6–16: These articles evidence how a treaty devoted to one set of problems—here, ending discrimination against women and achieving equality—makes possible discrete, disaggregated treatment of the different issues relevant to these problems. Clearly the variety and detail in these articles would have been out of place, indeed impossible, in a treaty of general scope like the ICCPR. Note the great range of verbs that are used throughout these articles to define states parties' duties, including: eliminate, provide, encourage, protect, introduce, accord, ensure.

> *Article 6* is typical of many provisions in requiring a state party to regulate specific non-governmental activity.
>
> *Articles 7–9*, to the contrary, deal with the traditional notion of state action, here barring discrimination by the state.
>
> *Article 10* concerns a particular field, education, and lists specific goals which, in their totality, take on a programmatic character. Note paragraph (h) on family planning and its relationship to three other provi-

sions: Articles 12(1), 14(2)(b) and 16(e). The Convention does not address as such the question of abortion.

Article 12 together with a number of other provisions indicate the degree to which CEDAW involves and interrelates the classical categories of civil–political rights and economic–social rights. It imposes a limited duty to provide free health care.

Article 14 disaggregates women's problems in regional and functional terms. It underscores strategies for realizing goals that permeate the entire Convention, such as mobilization through functional grass roots groups and participation in local decision making. CEDAW is not a convention in which solutions are to be provided only by the central authority of the state.

Article 16 orders the states to sweep away a large number of fundamental, traditional discriminations against and forms of subordination of women. Like several other articles, it could be understood as a complement to, one specification of, the broad goals stated in Article 5. Compare its provisions with the Report of Guatemala, p. 888, *supra*.

QUESTIONS

1. Consider CEDAW's stress (as in its title) on eliminating discrimination to achieve equality between men and women, as well as its means for realizing that equality. The phrase 'on the basis of equality of men and women' recurs in many articles. Compare the notion of equality in the ICCPR—say, in ICCPR Article 3 ('to ensure the equal rights of men and women to the enjoyment of all civil and political rights' in that covenant), or ICCPR Article 26 ('All persons are equal before the law and are entitled without any discrimination to the equal protection of the law'). Are the two treaties' conceptions of equality identical, similar, very different?

2. Note that CEDAW has no provision specifically addressing bodily security in the manner that other human rights instruments do by, for example, prohibiting arbitrary deprivation of life, torture, or arbitrary detention. What provisions of CEDAW would you rely on to assert a woman's right to bodily security?

3. Do the provisions of CEDAW on their face make any concession to cultural relativism, to cultural diversity in regional, ethnic, religious or other terms? Or do they insist throughout on a uniform universalism of norms and their applications?

4. Under Article 2, states parties agree to pursue the required policies, 'by all appropriate means and without delay.' Contrast the description of state obligations in Article 2 of the Covenant on Economic, Social and Cultural Rights:

'achieving progressively the full realization' of the recognized rights. Is this textual contrast accurate with respect to CEDAW? How do you understand the question of CEDAW's 'time frame' in comparison, say, with the ICCPR?

5. Anti-discrimination measures followed a certain chronology both in the United States and in the international human rights movement: prohibition first of types of racial discrimination and then of types of sex discrimination, while today the debate rages about what measures to take against discrimination on the basis of sexual orientation.

As far as the international movement is concerned, racism was considered an evil from the start. Because of its connections with the issues of apartheid and colonialism it fuelled the entire movement. Sex discrimination was also barred in the Universal Declaration of Human Rights, but had to wait decades longer to become a focus of attention and to be incorporated into the mainstream discussion of human rights. The normative consensus regarding elimination of gender discrimination started to develop much later. Today that consensus obtains on certain core issues, while on many others dispute continues at the practical, doctrinal and ideological level in political fora and within the women's movement. As yet we see no consensus at the universal level regarding the rights of sexual minorities or the right to sexual orientation.

Why this sequence? Is it significant, or merely an accident of history? Do the feared consequences of ending discrimination differ radically among these three fields? Is there an implicit hierarchy, some forms of discrimination being considered worse than others? What explanations can we offer?

NOTE

Two institutions within the universal human rights system are concerned exclusively with women's rights. The older of the two, the UN Commission on the Status of Women, is formally the body with primary responsibility for monitoring and encouraging implementation of international law on women's rights.[9] It was established by the General Assembly at the same time as the Human Rights Commission, but it has been a less effective and influential body.

The more significant and influential organ has been the Committee formed under CEDAW (referred to in this chapter as the CEDAW Committee). Hence the following materials examine only that Committee. You should now read Articles 17–21 of the Convention, and compare them with the equivalent provisions concerning another treaty organ, the ICCPR Human Rights Committee considered in Chapter 9.

[9] See Sandra Coliver, United Nations Commission on the Status of Women: Suggestions for Enhancing its Effectiveness, 9 Whittier L. Rev. 435 (1987).

ANDREW BYRNES, THE 'OTHER' HUMAN RIGHTS TREATY BODY: THE WORK OF THE COMMITTEE ON THE ELIMINATION OF DISCRIMINATION AGAINST WOMEN

14 Yale J. Int. L. 1 (1989), at 6.

. . .

II. Criteria for Assessing the Committee's Work

In assessing the work of CEDAW [used herein to refer to the Committee itself], one must recognize at the outset the limitations of such an international supervisory body. Although CEDAW is not a body of governmental representatives, it is nonetheless created by governments and its members are nominated and elected to positions on the Committee by States Parties. Governments are also in a position to influence the allocation of resources to the Committee. Thus, the scope of its activities and its effectiveness are ultimately constrained by the support which States Parties are willing to give it. By the same token, the nature of the Committee means that it has a level of access to governments which other groups may lack.

CEDAW's powers to promote implementation of the Convention are relatively limited. The Committee has no quasi-judicial powers enabling it to pronounce a State Party in violation of the Convention and to order an appropriate remedy. While it may offer suggestions to individual States, or to the States Parties generally, as to appropriate ways to pursue the Convention's goals, its major means for exerting pressure on States to comply with their obligations lies in its public review of individual country reports. Many governments care whether the supervisory committees make positive or adverse comments on their human rights performance. A positive appraisal in an international forum of a country's commitment and efforts can give impetus to further progress. An adverse assessment can embarrass a government at home and abroad, ideally providing it with some incentive to do more in the future. . . .

In both cases, the impact of the Committee's assessment depends on support from the States Parties as a whole. CEDAW needs the States Parties to ensure that it is given adequate resources to function efficiently, to support its role as a critic of individual States Parties which have not fulfilled their obligations, and to support the pursuit of the Convention's goals in other international fora. Thus the Committee must, as a tactical matter, conduct its work in a manner so as not to alienate a large number of States Parties.

. . .

Any assessment of the effectiveness of the work of CEDAW must also take into account factors internal and external to the review of national reports by the Committee. Among these factors are the following: whether the reporting procedure itself functions effectively and meaningfully, whether the work of the Committee has a significant and useful impact on other international bodies responsible for promoting rights guaranteed by the Convention, and perhaps most importantly, whether the process has had any impact on national laws, practices and conditions.

. . .

The ultimate criterion for success is whether the process contributes to a greater awareness and observance of the human rights of women in domestic fora. . . . The reporting process must have official and unofficial linkages back into the domestic forum if it is to have any significant impact on domestic policies. Developing these linkages represents the Committee's most important challenge, but it requires the cooperation of governmental organizations and independent NGOs to do so.

III. The Composition of the Committee

A. CEDAW'S Membership

Article 17(1) of the Convention provides that, in the election of the twenty-three members of the Committee, consideration be given 'to equitable geographical distribution and to the representation of the different forms of civilization as well as the principal legal systems.' The current composition of the Committee is largely in accordance with that injunction. At the same time, one of the most striking features of CEDAW's membership is that, with one exception, all its members have been women. CEDAW thus provides a stark contrast to the other human rights treaty bodies where, as of 1988, women comprised a grand total of seven out of sixty-four members.

. . .

. . . In contrast [to other treaty bodies], only about half of the experts who have served as CEDAW members are lawyers. Other members come from such areas as medicine, public health and hospital administration, political science, geography, trade union and labor relations, education, social work and engineering. This diversity of experience has been reflected in the Committee's questions and has been valuable to the Committee's work, particularly in the areas of economic and social rights and development.

. . .

Perhaps the major reason for the relatively critical stance that CEDAW as a collective body has been able to maintain is . . . the feminist background of most of CEDAW's members. Nearly all the members of CEDAW have been involved in some manner with feminist or other groups working to advance the position of women. In fact, in many cases their expertise on issues of particular concern to women and their commitment to the cause of women's equality have been decisive factors in their appointment to the official positions they hold as well as in their nomination to CEDAW.

As a result of their involvement in feminist activities, CEDAW members' commitment to women's rights is often deeper than their commitment to government institutions. This dimension of experts' backgrounds also means that they have access to networks and communities outside the governmental structure to which other officials, particularly male officials, may not; these contacts provide not only information but also a different perspective and a sense of solidarity and support for the work of the Committee.

. . .

V. Consideration of Reports by the Committee

A. 'A Constructive Dialogue'

The Committee has endorsed the concept of a 'constructive dialogue' with States Parties as the basis for its consideration of reports. The notion of a constructive dialogue is one that has developed in the work of other committees; it embodies a distinction between a procedure in which allegations of specific human rights violations can be made and considered, and one in which the supervisory body considers the overall progress made in the implementation of a convention by examining the reports of States Parties. This approach envisions the States Parties and the Committee as engaged in a joint enterprise to advance the goals of the Convention by cooperative endeavors involving the exchange of information, ideas and suggestions.

. . .

It is essential to the integrity of the review process that CEDAW members be prepared to adopt an adversary stance towards States Parties in appropriate cases. Of course, these questions and criticisms are those of individual members and not a formal collective pronouncement by CEDAW.[10]

. . .

The other dimension of a constructive dialogue, the exchange of ideas and experience between the Committee and States in order to assist States in the implementation of the Convention, has remained largely rhetorical. The Committee has not made a concerted effort to make detailed suggestions to individual States about specific measures and, in its general recommendations addressed to all States Parties, has only provided broad guidance on measures which the Committee would like States Parties to take.

[10] [Eds.] The Committee subsequently adopted the practice of issuing collective concluding remarks at the end of the examination of the country report. These remarks concern matters of procedure and reporting, as well as particular problems thought to demand instant attention or to be addressed in the state's next report. This development probably reflects both the high degree of consensus among members about the most serious problems facing women and the Committee's growing confidence about strategies that it should pursue. Compare the collective comments of the ICCPR Human Rights Committee, p. 517, *supra*.

B. General Features

A number of general features of the Committee's examination of reports can be identified; many of them parallel the experiences of the other supervisory committees. One feature of CEDAW's approach has been to press States Parties to provide information that shows the actual position of women in their societies and not just the formal legal status. To this end, it has consistently sought meaningful statistics and other empirical information about women's position in each country.

Another characteristic of CEDAW's consideration of reports has been an effort to overcome the common tendency of States Parties to paint an excessively favorable picture of the condition of women in their societies, making no reference to the difficulties experienced in their implementation of the Convention. However, the development of a more candid, self-critical approach to reporting by States Parties may only be possible when there is an adequate source of independent information such as that provided by NGOs, against which the Committee, the public and other governments can assess the accuracy of the State Party's report.

. . .

[The author signals two major problems facing CEDAW. (1) Many states submit reports late, sometimes years late. CEDAW's powers to ensure timely submission are very limited. (2) Reports are often 'grossly inadequate,' so that the Committee must seek (with mixed success) additional information. Much 'crucial material' for a probing discussion is missing. 'In the absence of a significant input from the specialized agencies, NGOs or other sources, the Committee has great difficulty in assessing the progress made by the State Party in implementing the Convention. . . .']

. . .

VI. Substance of Issues Concerning CEDAW

While the questions asked by CEDAW members have ranged over all the areas covered in the Convention, the Committee has not explicitly articulated any collective view of its understanding of the models of equality/nondiscrimination embodied in the Convention. Furthermore, although there is widespread agreement on the Committee about the existence of women's inequality and subordination, no unified theory about the nature and causes of that oppression has emerged.

The Convention itself embodies a number of different perspectives about the causes of women's oppression and the steps needed to over come it. It imposes an obligation to ensure that women enjoy formal equality under the law, and it recognizes that temporary affirmative action measures are necessary in many cases if guarantees of formal equality are to become reality. Various provisions of the Convention also embody a concern that women's reproductive lives should be under their own control, and that the State should ensure that women's choices are not coerced and do not prejudice them in their access to social and economic opportunities. The Convention also recognizes that there are experiences to which women are subjected which need to be eliminated (such as rape, sexual harassment and other forms of violence against and sexual exploitation of women which affect women asymmetrically), whether or not they can be fitted neatly within an equality model which requires that there be a direct male comparison available. In short, underlying the Convention is a view that women are entitled to all the rights and opportunities which men enjoy; in addition, their particular abilities and needs arising from biological differences between the sexes must also be recognized and accommodated, but without detracting from their entitlement to equal rights and opportunities with men.

. . .

If there is an overarching theme to the Committee's questioning it is probably article 5's obligation to take steps to discourage stereotyped attitudes about the roles of men and women.[11] The Committee has been particularly interested in the roles of the media and the educational system in perpetuating and changing stereotypes. It has been extremely critical of general policy statements or particular social arrangements which give primacy to motherhood, to the neglect of women's other roles and of men's responsibilities as fathers.

B. Information from NGOs

. . .

At this stage the Committee has not developed any procedure whereby it may formally request or receive reports from non-governmental organizations. Doubts have been expressed about the possibilities for the formal involvement of NGOs in the work of the Committee in view of the fact that there is no mention of them in the Convention and that a proposal to permit NGO participation was not taken up during the drafting process. Nonetheless, neither of these facts seems to pose insuperable barriers to such involvement, particularly in the light of the experience of other committees which have, by their practice, expanded their jurisdiction beyond what was arguably envisaged by those who originally drafted their enabling treaties.

. . .

[11] One issue which has been significantly absent from the Committee's agenda is that of sexual preference and discrimination against lesbians, an issue of major importance to many feminists from all parts of the world. Apart from the areas of child custody and adoption, the matter has hardly been mentioned. This reflects the largely heterosexist orientation of the Convention and, presumably, caution on the part of Committee members who would be somewhat apprehensive of the likely hostile reaction of many States Parties if the issue were raised with any frequency.

VIII. Suggestions and General Recommendations

A. The Power

Article 21 of the Convention provides that the Committee may 'make suggestions and general recommendations based on the examination of reports and information received from the States Parties.'

...

Article 21 provides that the Committee may make 'suggestions' and '*general* recommendations.' . . .

. . . [D]espite some division of opinion among its members, the Committee has taken the view that, in appropriate cases, it can base both suggestions and general recommendations on its examination of the report of an individual State Party. Nonetheless, a significant number of Committee members are clearly reluctant to adopt formal recommendations and suggestions directed to an individual State Party, lest this be misconstrued as an exercise of investigative and adjudicatory functions in relation to individual complaints for which the Convention makes no provision.

...

C. Development of a 'Jurisprudence' of the Convention

. . . While CEDAW does not have the formal power to interpret the Convention authoritatively, it necessarily interprets the Convention in the course of its work, even if only implicitly. In any event, the sort of interpretive exercise that might be undertaken, if worded appropriately, would not be viewed as an attempt to arrogate ultimate interpretive power to itself but rather as an appropriate exercise of the power to make suggestions and general recommendations. Both the Human Rights Committee and CERD offer models of how that task might be undertaken.

...

An area of particular importance to which CEDAW must direct its attention is the developing jurisprudence of the other committees, in particular that of the Human Rights Committee. The Human Rights Committee has now elaborated a large number of general comments interpreting the articles of the ICCPR and giving influential interpretations of major civil and political rights. With the exception of those comments concerning the articles of the ICCPR which deal specifically with non-discrimination, there is virtually no explicit mention of women in those general comments nor does there appear to be any recognition that gender is an important dimension in defining the substantive content of individual rights.

For example, the general comment dealing with privacy [see General Comment No. 16, pp. 524 and 529, *supra*] makes no mention of the importance that this right has assumed in women's struggle for control over their reproductive lives. Similarly, the general comments on the right to life [see General Comments No. 6 and 14, p. 526, *supra*] do not address the implications of that right for women's access to abortion. Although they do refer to the need to reduce infant mortality, they ignore the differential threats to life facing women and girls and the consequences for a State's obligation to ensure enjoyment of the right to life without discrimination. This neglect of women's experiences pervades the general comments, despite the fact that women are assured equal enjoyment of the rights guaranteed by the Covenant.

Thus, CEDAW could make a major contribution to the development of substantive human rights law by exploring the limitations of these mainstream interpretations of important civil and political rights and introducing women's different perspectives into that discourse. As the Convention guarantees equal enjoyment of essentially the same rights guaranteed by both Covenants, it is quite appropriate for CEDAW to undertake this task.

. . .

COMMENT ON THE CEDAW COMMITTEE AND SOCIO-ECONOMIC FACTORS IMPAIRING WOMEN'S RIGHTS

The earlier Comment at p. 894, *supra*, indicated the significance of economic and social factors in placing and maintaining women in their low status. For its part, the CEDAW Committee has placed great emphasis on overcoming stereotypes preventing women from fully participating in all fields of economic and social life. It continually stresses the need for states parties to adopt programs countering entrenched social and cultural attitudes about appropriate sex roles, and to encourage girls to enrol in education and job training that will give them access to the highest paying jobs. The Committee has also been concerned about the concomitant stereotypes that consign child care and the bulk of domestic duties to women, thus contributing to their disproportionate work load.

The earlier Comment noted that much women's work tends to be 'invisible' and accordingly undervalued or not valued at all. In its General Recommendation No. 17,[12] the Committee called on states parties to take steps to 'measure and value' domestic activities, such as by conducting time-use surveys and collecting gender-based statistics on time spent working in the home and on the labor market. States should quantify such work and include it in the gross national product, both to reveal the true contribution of women to the national economy and to provide the basis on which policies for the advancement of women could be formulated.

In recent sessions, the CEDAW Committee has been attentive to the

[12] Measurement and Quantification of the Unremunerated Domestic Activities of Women and their Recognition in the Gross National Product. General Recommendation No. 17 of CEDAW Committee, Tenth Sess. 1991, UN Doc. A/46/38, 1 Int. Hum. Rt. Rep. 24 (1994).

increasing importance of structural adjustment programs for the implementation of the Convention's objectives. Note the following comments in the third periodic report submitted by Ecuador to the Committee, CEDAW/C/ECU/3, 10 Jan. 1992:

> 207. In addition to external economic and political conditions, such as the country's external debt burden, adjustment policies that are impoverishing the economy have increased the deficit in the health services. The fact is that, under growing economic pressure, 37 of the poorest countries have reduced their spending on health and education (in Ecuador, by 50 and 25 per cent, respectively).
>
> 208. What can be stated with confidence is that if the position of women before the crisis was extremely difficult, it has now worsened both in quantitative and qualitative terms.

The Committee has expressed concern that structural adjustment programs, if they continue to ignore their effects on women, risk exacerbating historical disadvantages. It has taken the view that women and issues of concern to women must be integrally involved in any economic restructuring or development plan.

QUESTIONS

1. 'I suppose CEDAW had to pay the price somewhere. The provisions are bold, far-reaching, innovative. Compared with, say, the ICCPR Human Rights Committee, let alone the UN Commission on Human Rights, the CEDAW Committee is weak, isolated, a marginal rather than mainstream voice in the human rights movement. A trade-off.' Comment.

2. In view of your knowledge of other human rights organs within the UN or serving the human rights treaty regimes, what changes would you urge for the CEDAW Committee—including both changes that could be realized within the present treaty and those that require a protocol? Can you predict with some confidence what effect those changes would have in advancing the cause of women's rights?

COMMENT ON RESERVATIONS TO THE CONVENTION

The high number of reservations that have accompanied ratifications of CEDAW have become a regrettably notorious feature of the Convention, which is in this respect first among the human rights treaties. By way of contrast, only four states parties have entered reservations to the Convention on

Racial Discrimination. Moreover, many of the CEDAW reservations have eliminated or eroded fundamental provisions.

Unlike the ICCPR, which is silent on the issue, CEDAW addresses reservations in Article 28(2), which prohibits those incompatible with the 'object and purpose' of the Convention. See the Comment on reservations at p. 756, *supra*, describing the provisions in the Vienna Convention on the Law of Treaties and the advisory opinion of the International Court of Justice in the *Reservations to the Genocide Convention* case.

Tolerance of reservations has been urged on various grounds—for example, the desirability of securing widespread participation in treaties serving a 'purely humanitarian and civilizing purpose' (in the words of the *Genocide Convention* case), and hence the reluctance to view a ratification as invalid because of its reservations. A commentator emphasizes another ground:[13]

> Most states are apprehensive about the possible consequences of accepting a human rights treaty, not least because such treaties may have a dynamic force, and interpretation of their scope and impact is less certain than that of commercial treaties. . . . Reservations are seen to offer an assurance that the state can protect its interest to the fullest extent possible.

Other commentators have considered reservations to Article 2 to be 'manifestly incompatible' with the object and purpose of the Convention.[14] Several states parties have objected to these reservations on grounds that they threaten the integrity of the Convention and the human rights regime in general. See p. 921, *infra*.

Unlike, say, objections to the United States reservations to the ICCPR, p. 773, *supra*, these objections that purport to be applying Article 28(2) of CEDAW raise issues of religious intolerance and of cultural relativism. The net result, claims one commentator, has been the widespread view that international obligations assumed through the ratification of CEDAW are somehow 'separate and distinct' from and less binding than those of other human rights treaties.[15]

Consider the following suggestions of Rebecca Cook about criteria for distinguishing between reservations that are compatible and incompatible with the Convention:[16]

> The thesis of this article is that the object and purpose of the Women's Convention are that states parties shall move progressively towards elimination of all forms of discrimination against women and ensure equality between men and women. Further, states parties have an obligation to

[13] Rebecca Cook, Reservations to the Convention on the Elimination of All Forms of Discrimination against Women, 30 Va. J. Int. L. 643, 650 (1990).

[14] Belinda Clark, The Vienna Convention Reservations Regime and the Convention on Discrimination against Women, 85 Am. J. Int. L. 281 (1991).

[15] *Ibid.*

[16] *Op. cit.* Note 13, *supra*, at 648.

> provide the means to move progressively toward this result. Although the Women's Convention envisions that states parties shall move progressively towards elimination of all forms of discrimination against women and ensure equality between men and women, reservations to the Convention's substantive provisions pose a threat to the achievement of this goal. . . . Accordingly, reservations that contemplate the provision of means towards the pursuit of this goal will be regarded as compatible with 'the object and purpose of the treaty' as provided by article 28(2) of the Women's Convention and article 19 (c) of the Vienna Convention. Similarly, any reservation that contemplates enduring inconsistency between state law or practice and the obligations of the Women's Convention is incompatible with the treaty's object and purpose.

RESERVATIONS OF PARTIES TO CEDAW

[There follow a few of the reservations made by states to ratifications of CEDAW, and an example of objection of other states parties to those reservations.

DECLARATIONS AND RESERVATIONS

Bangladesh

'The Government of the People's Republic of Bangladesh does not consider as binding upon itself the provisions of articles 2, 13 (*a*) and 16 (1) (*c*) and (*f*) as they conflict with Shariah law based on Holy Koran and Sunna.'

Brazil

Reservation made upon signature and confirmed upon ratification:

'The Government of the Federative Republic of Brazil hereby expresses its reservations to article 15, paragraph 4, and to article 16, paragraph 1 (*a*), (*c*), (*g*) and (*h*), of the Convention on the Elimination of All Forms of Discrimination against Women.

'Furthermore, Brazil does not consider itself bound by article 29, paragraph 1, of the above-mentioned Convention.'

Egypt

Reservations made upon signature and confirmed upon ratification:

In respect of article 9:

Reservation to the text of article 9, paragraph 2, concerning the granting to women of equal rights with men with respect to the nationality of their children, without prejudice to the acquisition by a child born of a marriage of the nationality of his father. This is in order to prevent a child's acquisition of two nationalities where his parents are of different nationalities, since this may be prejudicial to his future. It is clear that the child's acquisition of his father's nationality is the procedure most suitable for the child and that this does not infringe upon the principle of equality between men and women, since it is customary for a woman to agree, upon marrying an alien, that her children shall be of the father's nationality.

In respect of article 16:

Reservation to the text of article 16 concerning the equality of men and women in all matters relating to marriage and family relations during the marriage and upon its dissolution, without prejudice to the Islamic Shariah's provisions whereby women are accorded rights equivalent to those of their spouses so as to ensure a just balance between them. This is out of respect for the sacrosanct nature of the firm religious beliefs which govern marital relations in Egypt and which may not be called in question and in view of the fact that one of the most important bases of these relations is an equivalency of rights and duties so as to ensure complementarity which guarantees true equality between the spouses. The provisions of the Shariah lay down that the husband shall pay bridal money to the wife and maintain her fully and shall also make a payment to her upon divorce, whereas the wife retains full rights over her property and is not obliged to spend anything on her keep. The Shariah therefore restricts the wife's rights to divorce by making it contingent on a judge's ruling, whereas no such restriction is laid down in the case of the husband.

Iraq

Reservations:

1. Approval of and accession to this Convention shall not mean that the Republic of Iraq is bound by the provisions of article 2, paragraphs (*f*) and (*g*), of article 9, paragraphs 1 and 2, nor of article 16 of the Convention, the reservation to this last-mentioned article shall be without prejudice to the provisions of the Islamic Shariah according women rights equivalent to the rights of their spouses so as to ensure a just balance between them. Iraq also enters a reservation to article 29, paragraph 1, of this Convention with regard to the principle of international arbitration in connection with the interpretation or application of this Convention.

OBJECTIONS

Germany, Federal Republic of

The Federal Republic of Germany considers that the reservations made by Egypt regarding article 2, article 9, paragraph 2, and article 16, by Bangladesh regarding article 2, article 13 (*a*) and article 16, paragraph 1 (*c*) and (*f*), by Brazil regarding article 15, paragraph 4, and article 16, paragraph 1 (*a*), (*c*), (*g*) and (*h*), by Jamaica regarding article 9, paragraph 2, by the Republic of Korea regarding article 9 and article 16, paragraph 1 (*c*), (*d*), (*f*) and (*g*), and by Mauritius regarding article 11, paragraph 1 (*b*) and (*d*), and article 16, paragraph 1 (*g*), are incompatible with the object and purpose of the Convention (article 28, paragraph 2) and therefore objects to them. In relation to the Federal Republic of Germany, they may not be invoked in support of a legal practice which does not pay due regard to the legal status afforded to women and children in the Federal Republic of Germany in conformity with the above-mentioned articles of the Convention.

This objection shall not preclude the entry into force of the Convention as between Egypt, Bangladesh, Brazil, Jamaica, the Republic of Korea, Mauritius and the Federal Republic of Germany.

QUESTIONS

1. Suppose that the substance of General Comment No. 24 of the ICCPR Committee (p. 774, *supra*), stating criteria for determining the consistency of a reservation with the ICCPR, was adopted with appropriate changes in a General Recommendation of the CEDAW Committee. What would be the status under such a General Recommendation of a broad reservation to Article 2 of CEDAW?

2. Suppose that State X submits with its ratification of CEDAW several reservations of the type noted on the preceding pages. Other states object on the basis of Article 28(2) of CEDAW. What arguments would you make on behalf of State X to justify the reservations?

3. How do you assess the criteria suggested by Cook? Can you give illustrations of two reservations to CEDAW that would be judged differently under those criteria?

COMMENT ON EFFORTS TOWARD U.S. RATIFICATION OF CEDAW

Ratification has been considered by the U.S. on several occasions and, in October 1994, almost occurred. The President had submitted the Convention to the Senate, and the Senate Committee on Foreign Relations recommended the Senate's consent to ratification. [17] It observed among other things that failure to ratify had limited U.S. leadership in the promotion of equality for women.

The Committee recommended ratification subject to a number of reservations, understandings and declarations, including the following:

[17] See S384–10, Exec. Rep. Sen. Comm. on For. Rel. Oct. 3, 1994., Various related documents and the text of all reservations are set forth in 89 Am. J. Int. L. 102 (1995).

> [T]he Constitution and laws of the United States establish extensive protections against discrimination, reaching all forms of governmental activity as well as significant areas of non-governmental activity. However, individual privacy and freedom from governmental interference in private conduct are also recognized as among the fundamental values of our free and democratic society. The United States understands that by its terms the Convention requires broad regulation of private conduct, in particular under Articles 2, 3 and 5. The United States does not accept any obligation under the Convention to enact legislation or to take any other action with respect to private conduct except as mandated by the Constitution and laws of the United States.

Compare the reservation attached by the United States to its ratification of the Convention on Racial Discrimination, p. 778, *supra*.

The Committee also proposed reservations to the right to equal pay understood as comparable worth, the right to paid maternity leave, and any obligation under Articles 5, 7, 8 and 13 of the Convention that might restrict constitutional rights to speech, expression and association.

Five senators (a Republican minority) on the Committee objected to ratification. Their statement of Minority Views recognized the 'unfortunate prevalence of violence and human rights abuses against women around the world' and shared the 'majority's strong support for eliminating discrimination against women.' Nonetheless, they were not 'persuaded' that CEDAW was 'a proper or effective means of pursuing that objective.' The Minority Views made the following points:

> (1) CEDAW may enable ratifying states to generate 'political capital,' but is 'unlikely to convince governments to make policy changes they would otherwise avoid.'
>
> (2) Countries like the U.S. 'must guard against treaties that overreach,' and must not promise 'more than we can deliver or we risk diluting the moral suasion that undergirds existing covenants.' Indeed, the fear exists that 'creating another set of unenforceable international standards will further dilute respect for international human rights.'
>
> (3) More than 30 states ratifying CEDAW have made significant reservations, sometimes 'so broad as to appear to be at variance with the object and purpose of the treaty itself.' The Minority Views drew illustrations from Islamic states. The statement questioned 'whether such behavior does not, in fact, 'cheapen the coin' of human rights treaties generally.' These reservations suggest that CEDAW 'may reach beyond the necessarily restrictive scope of an effective human rights treaty.'
>
> (4) 'Improvement in the status of women in countries such as India, China, and Sudan will ultimately be made in those countries, not in the United States Senate.'

> (5) Evolution of 'internationally accepted norms' on human rights 'is important and must be carefully encouraged. It must, however, take place within an international system of sovereign nations with differing cultural, religious and political systems. Pushing a normative agenda beyond that system's ability to incorporate it leads, we believe, to what is represented by this convention. . . .'

In the end, as the session of Congress came to an end, the Convention never reached the Senate floor.

QUESTIONS

1. Which objections to ratification of CEDAW that were set forth in the Minority Views seem particular to CEDAW, and which could refer generally to many human rights treaties?

2. The Minority Views state that CEDAW 'may reach beyond the necessarily restrictive scope of an effective human rights treaty.' The Views refer to the differing cultural, religious and political systems among states, and assert that '[p]ushing a normative agenda beyond [the international system's] ability to incorporate it leads . . . to what is represented by this convention. . . .' Does CEDAW raise special and more difficult problems for U.S. ratification than did the ICCPR, Genocide Convention, Torture Convention, or Racial Discrimination Convention? If so, why?

3. Assume that an appropriately modified General Comment No. 24 of the ICCPR Human Rights Committee, p. 774, *supra*, is adopted by the CEDAW Committee as a General Recommendation. What would be the status under such a General Recommendation of the proposed reservation set forth in the preceding Comment on the private–public issue in Articles 2, 3 and 5?

4. Note the ambiguity in the term 'private conduct' in the proposed reservation referred to in Question 3. Should that reservation have been drawn more narrowly—for example, by distinguishing between (a) the 'privacy' values to which it refers and (b) conduct within the 'private' sphere? How would you have drafted such a reservation, and what illustrations might you have used to clarify and justify your distinction?

C. PRESENT PROPOSALS

There follow a number of proposals about the implementation and reform of CEDAW and its Committee that range from the specific to the broad and programmatic. What these materials also indicate is the wealth of sources that now form part of an international legal and political process contributing to the development of women's rights as part of international human rights law. They include world conferences on human rights in general, on particular human rights problems, or on problems that have an important human rights dimension. In Chapter 13 as a whole, we see together with those conferences such other materials as NGO or scholars' proposals, reports of special rapporteurs, acts of treaty organs, demands of women's groups lobbying their own governments, statements of religious officials and movements, declarations and resolutions, and state reports.

NOTE

The shortcomings of provisions for the CEDAW Committee were underscored by the advocates of a stronger treaty regime. Theodor Meron noted several of them in a 1990 article.[18] His suggested reforms included a period longer than two weeks for the Committee's annual meeting; a change in venue for the Committee from Vienna to Geneva, where it would be less isolated and its work less fragmented; and the drafting and submission to states for ratification of an optional protocol permitting individual communications against states. Meron noted (at 215):

> . . . Both the geographical and the programmatical aspects of the Committee's work, including the possibility of a change of venue to Geneva, should be reexamined. As a result of this fragmentation, the struggle against sex discrimination has received inadequate attention and has not benefited from some salutary innovations in UN human rights procedures, such as the appointment of special rapporteurs. To reduce these shortcomings, the UN Commission on Human Rights, in coordination with another functional commission of ECOSOC, the Commission on the Status of Women, might consider appointing a 'thematic' rapporteur with a mandate similar, *mutatis mutandis*, to that of the rapporteur on religious intolerance or on torture. Acting under the powers flowing from the Charter prohibition of gender discrimination, the Commissions would authorize the rapporteur to investigate and report upon serious violations of sexual equality. The rapporteur would be permitted to receive information from governments and intergovernmental and nongovernmental organizations, to respond effectively to credible information of substantial violations and to recommend measures to prevent continuing violations. He or she could review the practice of all states members of the United Nations, including those not parties to the Convention. . . .

Since the article by Meron was published, prospects for an optional protocol to the Convention have improved. In 1994, the CEDAW Committee

[18] Enhancing the Effectiveness of the Prohibition of Discrimination against Women, 84 Am. J. Int. L. 213 (1990).

expressed its support for the right to petition through an optional protocol, suggesting that the Commission on the Status of Women request the Secretary General to arrange an expert group meeting on the subject.[19]

The expert group meeting organized by two NGOs and held in November 1994 prepared a draft optional protocol providing in part for an individual complaints procedure permitting a communication to the Committee from any 'individual, group or organization' that suffered because of the failure of a state party to fulfil its obligations under the Convention. In investigating the communication, the Committee may with the agreement of the state party involved visit its territory. Under Article 8, if the Committee concludes that a state party has failed to fulfill its obligations, it may recommend that the state take specific measures to remedy any violation. 'The State Party shall take all steps necessary to remedy any violation of the rights set forth in the Convention. . . . The State Party shall implement any recommendations made by the Committee, and shall ensure that adequate reparation or other appropriate remedy is provided.'

QUESTION

'An optional protocol to CEDAW permitting individual communications will not significantly advance the treaty's goals. As CEDAW's provisions evidence, the real problems blocking realization of goals like equality are structural, cultural, deep. These are exactly the kinds of problems that a communications procedure handling individual complaints against a state is not able to deal with effectively.' Comment.

COMMENT ON WORLD CONFERENCES ON WOMEN'S ISSUES

Another mechanism within the international system has been increasingly used to advance human rights as well as other objectives: large inter-governmental conferences organized around specific themes. These conferences provide an occasion for governments to discuss and perhaps ultimately agree on common strategies of action to resolve issues of global concern. Recent conferences have included the 1990 World Summit for Children, the 1992 Rio de Janeiro Conference on Environment and Development, the 1993 Second World Conference on Human Rights at Vienna, the 1994 Cairo Conference on Population and Development, and the 1995 Copenhagen World Summit for Social Development.

[19] Suggestion no. 5. Report of the Committee on the Elimination of Discrimination against Women, UN Doc. A/49/38.

These conferences are by no means identical in structure or flavor; the character of each is influenced by the particular purpose or issue at hand. Nonetheless, recent conferences have tended to follow a similar format. Apart from publicizing the member states' recognition of the importance of the subject under consideration, the main object of the conference is often to obtain agreement on a draft document, such as a declaration or charter, and on a related program of action.

Much of the difficult work in identifying the nature of the issue and proposing solutions to particular problems, is accomplished prior to the opening of the conference. An initial draft may be drawn up by a branch of the UN Secretariat. Expert group meetings are often convened around specific issues or areas of concern. Through a series of regional and global preparatory meetings attended by delegations of the member states, the draft document may be refined or significantly altered. By the time the conference opens, normally a significant portion of the text or program has been agreed upon. At this point it becomes a 'battle of the brackets', so called because of the practice of bracketing clauses in the text that are not yet agreed on—that is to say, an exercise in resolving the parts of the program that have remained contentious.

Throughout this process, sometimes at parallel meetings designed to facilitate it, NGOs engage in an attempt to influence the substance of the document, by providing relevant information to the Secretariat, lobbying Member States or publicizing issues of concern in the press. Reflecting NGOs' increasing influence and activity, it is becoming more common for member states to consult with them—at least ones from their own countries—prior to establishing their positions, and even to include NGO representation in their official delegations to the preparatory and world conferences.

Some of these conferences have proved to be significant milestones in the articulation and publicization of issues of concern to women. For example, a major focus and achievement of the 1993 World Conference on Human Rights was recognition of the central place of women in the human rights agenda. In Beijing in September 1995, another major conference on women's issues will take place. Delegations of the member states will meet while members of the NGO community participate in the NGO Forum, running roughly concurrently with the intergovernmental conference.

This Fourth World Conference on Women in Beijing is the most recent in a series of intergovernmental conferences devoted specifically to the status of women and the specific impact of social phenomena on women. Commencing in Mexico City in 1975, the beginning of the UN Decade on Women, and again 'mid-decade' in Copenhagen in 1980, and finally in Nairobi in 1985, the member states of the United Nations have met periodically to consider impediments to the advancement of women. The 'Forward-Looking Strategies' for the advancement of women that were adopted at the Nairobi World Conference were intended to 'review and appraise' the achievements of the United Nations Decade for Women. They provided a comprehensive statement of the problems women faced at that time and outlined the programs and strategies for countering them.

Given the increased emphasis in recent years on issues involving women at world conferences on general issues (Vienna) or special themes (children, environment, population, social development), one might well ask in what ways it remains useful or necessary to convene a conference devoted exclusively to women's issues. The following observations may give at least a partial answer.

Although issues such as violence against women have received increased attention in recent years, many if not most of the concerns identified in the Nairobi Forward-Looking Strategies remain unaddressed or unresolved in most countries. A main focus of the Beijing Conference will be implementation by member states of the Platform for Action, including commitments to the provision of the necessary resources and institutions.

In addition, events during the ten years since Nairobi have altered the agenda of pressing concerns for women. Wars, ethnic strife and political instability in many parts of the world have created an increasing flow of internally and externally displaced persons, many of whom are women at severe risk of violence and poverty. Phenomena such as debt restructuring, political transition (including, for example, the move to market economies) and economic globalization have revealed the significance of macroeconomic issues for all efforts to improve the position of women. Although such changes may adversely affect populations as a whole, typically the burden of transition and adjustment falls disproportionately on women (and children).

The Platform for Action at the Beijing Conference addresses eleven substantive areas of concern: poverty, education, health, violence, armed conflict, economic structures and policies, decision-making, mechanisms for the advancement of women, women's human rights, mass media and the environment.

VIENNA DECLARATION AND PROGRAMME OF ACTION

UN World Conference on Human Rights, Vienna, 1993.
14 Hum. Rts. L. J. 352 (1993).

The World Conference on Human Rights . . .

Solemnly adopts the Vienna Declaration and Programme of Action.

. . .

18. The human rights of women and of the girl–child are an inalienable, integral and indivisible part of universal human rights. The full and equal participation of women in political, civil, economic, social and cultural life, at the national, regional and international levels, and the eradication of all forms of

discrimination on grounds of sex are priority objectives of the international community. . . .

. . .

37. The equal status of women and the human rights of women should be integrated into the mainstream of United Nations system-wide activity. These issues should be regularly and systematically addressed throughout relevant United Nations bodies and mechanisms. In particular, steps should be taken to increase cooperation and promote further integration of objectives and goals between the Commission on the Status of Women, the Commission on Human Rights, the Committee for the Elimination of Discrimination against Women, the United Nations Development Fund for Women, the United Nations Development Programme and other United Nations agencies. In this context, cooperation and coordination should be strengthened between the Centre for Human Rights and the Division for the Advancement of Women.

38. In particular, the World Conference on Human Rights stresses the importance of working towards the elimination of violence against women in public and private life, the elimination of all forms of sexual harassment, exploitation and trafficking in women, the elimination of gender bias in the administration of justice and the eradication of any conflicts which may arise between the rights of women and the harmful effects of certain traditional or customary practices, cultural prejudices and religious extremism. . . .

39. . . . States are urged to withdraw reservations that are contrary to the object and purpose of the Convention or which are otherwise incompatible with international treaty law.

40. . . . The Commission on the Status of Women and the Committee on the Elimination of Discrimination against Women should quickly examine the possibility of introducing the right of petition through the preparation of an optional protocol to the Convention on the Elimination of All Forms of Discrimination against Women. The World Conference on Human Rights welcomes the decision of the Commission on Human Rights to consider the appointment of a special rapporteur on violence against women at its fiftieth session.

41. . . . [T]he World Conference on Human Rights reaffirms, on the basis of equality between women and men, a woman's right to accessible and adequate health care and the widest range of family planning services, as well as equal access to education at all levels.

42. Treaty monitoring bodies should include the status of women and the human rights of women in their deliberations and findings, making use of gender-specific data. States should be encouraged to supply information on the situation of women *de jure* and de facto in their reports to treaty monitoring bodies. The World Conference on Human Rights notes with satisfaction that the Commission on Human Rights adopted at its forty-ninth session resolution 1993/46 of 8 March 1993 stating that rapporteurs and working groups in the field of human rights should also be encouraged to do so. Steps should also be

taken by the Division for the Advancement of Women in cooperation with other United Nations bodies, specifically the Centre for Human Rights, to ensure that the human rights activities of the United Nations regularly address violations of women's human rights, including gender-specific abuses. Training for United Nations human rights and humanitarian relief personnel to assist them to recognize and deal with human rights abuses particular to women and to carry out their work without gender bias should be encouraged.

43. The World Conference on Human Rights urges Governments and regional and international organizations to facilitate the access of women to decision-making posts and their greater participation in the decision-making process. It encourages further steps within the United Nations Secretariat to appoint and promote women staff members in accordance with the Charter of the United Nations, and encourages other principal and subsidiary organs of the United Nations to guarantee the participation of women under conditions of equality.

REPORT OF THE INTERNATIONAL CONFERENCE ON POPULATION AND DEVELOPMENT, CAIRO 1994

UN Doc. A/Conf. 171/13, 18 Oct. 1994.

[These excerpts from the Report are taken from Chapter IV, 'Gender Equality, Equity and Empowerment of Women,' Section C, 'Male Responsibilities and Participation.']

4.24. Changes in both men's and women's knowledge, attitudes and behaviour are necessary conditions for achieving the harmonious partnership of men and women. Men play a key role in bringing about gender equality since, in most societies, men exercise preponderant power in nearly every sphere of life, ranging from personal decisions regarding the size of families to the policy and programme decisions taken at all levels of Government. It is essential to improve communication between men and women on issues of sexuality and reproductive health, and the understanding of their joint responsibilities, so that men and women are equal partners in public and private life.

. . .

4.26. The equal participation of women and men in all areas of family and household responsibilities, including family planning, child-rearing and housework, should be promoted and encouraged by Governments. This should be pursued by means of information, education, communication, employment legislation and by fostering an economically enabling environment, such as family leave for men and women so that they may have more choice regarding the balance of their domestic and public responsibilities.

4.27. Special efforts should be made to emphasize men's shared responsibility and promote their active involvement in responsible parenthood, sexual and reproductive behaviour, including family planning; prenatal, maternal and child health; prevention of sexually transmitted diseases, including HIV; prevention of unwanted and high-risk pregnancies; shared control and contribution to family income, children's education, health and nutrition; and recognition and promotion of the equal value of children of both sexes. Male responsibilities in family life must be included in the education of children from the earliest ages. Special emphasis should be placed on the prevention of violence against women and children.

. . .

4.29. National and community leaders should promote the full involvement of men in family life and the full integration of women in community life. Parents and schools should ensure that attitudes that are respectful of women and girls as equals are instilled in boys from the earliest possible age, along with an understanding of their shared responsibilities in all aspects of a safe, secure and harmonious family life. Relevant programmes to reach boys before they become sexually active are urgently needed.

QUESTION

What are the functions of world conferences like those at Vienna and Cairo? What practical consequences for women's rights and the women's movement might flow from the Vienna Declaration or Cairo Report?

D. VIOLENCE AGAINST WOMEN BY THE STATE AND IN THE HOME

It was the rare report of a human rights NGO not so many years ago that gave specific attention to human rights violations against women, although of course women figured with men as victims in reports dealing with themes like arbitrary detention, disappearances, or torture. That has changed as part of a larger change, including for example the appointment by the UN Commission of a special rapporteur on violence against women, and the introduction in the U.S. State Department's 1993 Country Reports on Human Rights Practices of a special section on women.

There follow two NGO reports exclusively on violence against women, in one case violence perpetrated directly by the state and in the other violence occurring in the home.

AMNESTY INTERNATIONAL, RAPE AND SEXUAL ABUSE: TORTURE AND ILL TREATMENT OF WOMEN IN DETENTION
(1992), at 1.

. . .

. . . In countries around the world, government agents use rape and sexual abuse to coerce, humiliate, punish and intimidate women. When a policeman or a soldier rapes a woman in his custody, that rape is no longer an act of private violence, but an act of torture or ill-treatment for which the state bears responsibility. . . .

Yet many governments persistently refuse to recognize that rape and sexual abuse by government agents are serious human rights violations. In country after country effective investigations into cases of rape do not take place nor are the perpetrators brought to justice. From the emergency zones of Peru, for instance, AI has received dozens of reports of members of the security forces raping women and girls. AI [Amnesty International] knows of no official investigations into such incidents since the state of emergency was first declared in October 1981 and the current government has not demonstrated the political will to institute such investigations. In 1986 a Peruvian prosecutor told an AI delegation in Ayacucho that rape was to be expected when troops were conducting counter-insurgency operations, and that prosecutions for such assaults were unlikely.

Even when public outrage forces officials into conducting investigations and prosecutions, the punishments imposed by the courts on government agents found guilty of rape are seldom commensurate with the enormity of the crime.

. . .

Although men are sometimes raped in custody by government agents, it is a form of torture primarily directed against women, and to which women are uniquely vulnerable. Women are also more likely to suffer sexual abuse and harassment short of rape, including fondling, verbal humiliation, excessive body searches, and other intentionally degrading treatment. . . .

When governments use military force to suppress armed insurgency movements, troops are often given extensive powers and are not held accountable to civilian legal authorities for their actions. In the course of counter-insurgency operations, government soldiers sometimes use rape and sexual abuse to try and extract information from women suspected of involvement with the armed opposition or even to punish women who simply live in areas known to be sympathetic to the insurgents. The indiscriminate use of torture and ill-treatment also helps create a permanent sense of fear and insecurity, against which the capacity for independent political action can be dulled or thwarted. The official failure to condemn or punish rape gives it an overt political sanction, which allows rape and other forms of torture and ill-treatment to become tools of military strategy.

. . .

Maria Nicolaidou was among 33 young men and women detained in Athens, Greece on 2 November 1991 after policemen found them sticking up political posters. The detainees were taken to a police station, where all 12 of the women were ordered to strip naked and were kept in an open room in full view of a number of policemen, who made obscene gestures and comments. Several of the women said they were beaten by police officers. . . .

In many countries policemen use sexual harassment and threats of rape as an interrogation tactic. The interrogators may be after something specific, like information or a signature on a confession, or they may simply want to frighten the victim and other local women. Rose Ann Maguire was arrested in July 1991 in Northern Ireland and held for five days in Castlereagh interrogation centre. During questioning sessions, she was reportedly sexually harassed, physically abused and threatened with death. . . .

Dozens of Palestinian women and children detained in the Israeli-Occupied Territories have reportedly been sexually abused or threatened in sexually explicit language during interrogation. Fatimah Salameh was arrested near Nablus in July, 1990. Her interrogators allegedly threatened to rape her with a chair leg and told her they would photograph her naked and show the pictures to her family. 'They called me a whore and said that a million men had slept with me,' she said. Fatimah Salameh agreed to confess to membership in an illegal organization and was sentenced to 14 months' imprisonment. . . .

In Turkey rape and sexual abuse are frequently used in attempts to extract possessions from both men and women during interrogation. . . .

Some women are raped or sexually abused because they happen to be the wives, mothers, daughters or sisters of men the authorities cannot capture. These women become substitutes for the men in their families and government agents torture and abuse them to punish and shame their male relatives or to coerce these men into surrendering.

. . .

Pregnant women who are tortured or held in inhumane conditions face the additional threat of suffering miscarriage or permanent injury. The special needs of pregnant women are recognized in international instruments such as the United Nations Standard Minimum Rules for the Treatment of Prisoners. Some governments not only ignore these needs, but take advantage of the vulnerability of pregnant women to inflict severe physical and emotional pain.

Wafa' Murtada was a 27-year-old civil engineer and nearly nine months pregnant when the Syrian authorities arrested her in September 1987. The authorities apparently suspected her husband, Yahya Murtada, of belonging to a banned opposition group and tried to extract the names of his associates from her through torture. Wafa' Murtada gave birth in prison and lost her child, apparently because of the torture. She was held without charge or trial until her recent release from Fara' Falastin detention centre in Damascus.

. . .

Some governments pursue policies that result in persistent human rights violations against people of particular ethnic or national origin. Indigenous peoples, who are often denied civil and political rights, have little recourse against the governments that allow these violations to occur. Those who work on their behalf have also been attacked. . . .

All government agents who encourage, condone or participate in the rape of women in their custody should be brought to justice. Yet many of the perpetrators go free because their victims are too terrified or ashamed to file a complaint. Some women try to obliterate the memory of the assault; others feel degraded and fear that they would be shunned or abandoned if they reveal what has been done to them. In some traditional societies raped women are thought to be tainted or defiled, and the economic and social pressures to conceal a rape can be considerable: if a married woman is raped, her husband may exercise his right to desert her; a single woman who has been raped may no longer be seen as fit for marriage. . . .

Many rape victims are threatened with additional violence if they complain to anyone about the attack. . . .

And some governments maintain legislation making it possible for the victims of rape to be charged with criminal offences. Under Pakistan's Hudood Ordinance, women convicted of extra-marital sexual relations—including rape and adultery—can be sentenced to be publicly whipped, imprisoned or stoned to death. In August 1989 two nurses were raped at gunpoint by three interns in a Karachi hospital. One of the victims tried to file a complaint and was herself charged with admitting to sexual intercourse. As a result of the charges she has lost her job and her marital engagement has been broken off. 'No one else can ever know how I feel inside,' she said. 'I may seem all right on the outside but inside I feel as if I no longer exist.'

. . .

AMERICAS WATCH, CRIMINAL INJUSTICE: VIOLENCE AGAINST WOMEN IN BRAZIL

(1991), at 12.

. . .

I. INTRODUCTION

In April of this year, Americas Watch, together with the Women's Rights Project of Human Rights Watch, travelled to Brazil to assess the response of the Brazilian government to the problem of domestic violence. This report contains the findings of that mission. It focuses on wife-murder, domestic battery and rape. It constitutes the first report of the newly formed Women's

Rights Project of Human Rights Watch which monitors violence against women and discrimination on the basis of sex throughout the world.

The crime of domestic violence is not unique to Brazil. According to recent United Nations reports, it exists in all regions, classes and cultures. Women all over the world and from all walks of life are at risk from violence in the home, usually at the hands of their husband or lover. Although the exact number of abused women will probably never be known, available information indicates unequivocally that domestic violence is a common and serious problem in developed and developing countries alike.

Although domestic violence is common and widespread, it has traditionally been perceived as a private, family problem, beyond the scope of state responsibility. Indeed, in the past husbands have had the legal right to punish or even kill their wives with impunity. Only gradually changing social attitudes and increased reporting have propelled the problem into the public eye. And as the nature and severity of violence in the home has become evident, so has the responsibility of governments to prosecute such abuse as they would any other violent crime.

. . .

Moreover, female victims still have little reason to expect that their abusers—once denounced—will ever be punished. A police chief in Rio de Janeiro told Americas Watch that to her knowledge, of more than 2,000 battery and sexual assault cases registered at her station in 1990, not a single one had ended in punishment of the accused. The São Luis women's police station in the northeastern state of Maranhão reported that of over 4,000 cases of physical and sexual assault registered with the station, only 300 were ever forwarded for processing and only two yielded punishment for the accused.

Brazil's criminal law is part of the problem. In the Brazilian Penal Code, rape is defined as a crime against custom rather than a crime against an individual person—society rather than the female victim is the offended party. Most other sex crimes are deemed crimes only if the victim is a 'virgin' or 'honest' woman. If a woman does not fit this 'customary' stereotype, she is likely to be accused of having consented to the crime and it is unlikely to be investigated. Moreover, pursuit of these cases by law depends on the initiative of the victim, not the state; if at anytime she desists from prosecution the case will be dropped. Of over 800 cases of rape reported to the São Paulo women's police stations from 1985 to 1989, less than 1/4 were ever investigated.

Marital rape, in particular, is severely under-reported and least likely to be prosecuted. While marital rape theoretically is included within the general prohibition against rape, in practice it is not commonly viewed by the courts as a crime. Under the Brazilian Civil Code, the refusal of sexual relations is cause for legal separation. According to several attorneys with whom Americas Watch spoke, when a husband uses violence to compel his wife to have sexual relations, it is viewed by the courts as enforcing the wife's conjugal obligations, not as rape. As a result, rape in the home, with the exception of incest, is almost never punished.

. . .

II. BACKGROUND

A. The Role of the Women's Movement

. . .

Women's increased economic and political power coupled with the development of autonomous and state-affiliated women's institutions, enabled the women's movement to press for fundamental changes in the state's response to gender-specific violence. In 1985, women's groups, together with the state council on women, persuaded São Paulo's opposition party mayor to establish a woman's police station, staffed entirely by women and dedicated solely to crimes of violence against women, excluding homicide, which was not viewed as a gender-specific crime. By late 1985, eight women's police stations (*Delegacias De Defesa Da Mulher*, hereafter *delegacias*) had opened in the state of São Paulo, and by 1990 there were 74 throughout the country.

The women's *delegacias* represented an integrated approach to the problem of domestic violence. They were designed to investigate gender-specific crimes, and to provide psychological and legal counseling. The female police officers (*delegadas*) were to receive training in all aspects of domestic violence, from its psychological impact to the legal remedies available to the victim. . . .

. . .

In 1984, women's rights advocates secured Brazil's adoption of the Convention on the Elimination of All Forms of Discrimination Against Women (CEDAW), although with several reservations.[20] In 1986, the women's movement held a national constitutional forum to draft a list of recommendations for consideration by the Constituent Assembly, which was elected in the 1986 congressional elections to take up the task of writing a new Brazilian Constitution. . . .

The new Constitution, enacted in 1988, reflects many of the national women's movement's demands. In particular, Article 226, Paragraph 8 provides that 'the state should assist the family, in the person of each of its members, and should create mechanisms so as to impede violence in the sphere of

[20] Brazil entered reservations regarding the sections of the Convention that guarantee the equal rights of men and women to choose their residence and domicile, to have equal rights to enter into marriage, during marriage and at its dissolution, and to have the same personal rights, including the right to choose a family name. Brazil also entered reservations to the Convention article according the same rights 'for both spouses in respect of ownership, acquisition, management, administration, enjoyment and disposition of property. . . .'

its relationships.' Similar provisions have been adopted in state constitutions throughout Brazil.

. . .

Yet, there is still a long way to go. Changes in political and economic power since the onset of the new republic have diminished the scope and effectiveness of initiatives launched in the euphoria of the early years. . . .

. . .

B. Domestic Violence: Statistical Evidence

. . . A marked difference emerged in the nature of violence suffered by women as opposed to men. For Brazilian men, murder and physical abuse primarily involve acquaintances or strangers and occur outside the home. For Brazilian women, the opposite is true. The 1988 census showed that men were abused by relatives (including spouses) only ten percent of the time, while women are related to their abuser in over half of the cases of reported physical violence.

. . .

. . . Examples of physical abuse cited in a 1989 study of violence against women in the state of Pernambuco included beating, tying up and spanking, burning the genitals and breasts with cigarettes, strangulation, inserting objects in the victim's vagina (such as bottles and pieces of wood) and throwing alcohol and fire on the victim. The study also noted repeated physical abuse of pregnant women in which the aggressors 'aimed for the womb, breasts and vagina.'

. . .

NOTE

An article entitled *Failed by Law, Women Seek Bodyguards*, the *New York Times*, April 3, 1992, p. A10, reported on the opening of a bodyguard agency in Milwaukee, Wisconsin to protect physically abused women inside and outside their homes against attacks by men such as estranged lovers. The bodyguards were working in connection with a local women's group, the Task Force on Battered Women, one of whose coordinators stated: 'Domestic violence is not a priority with the police or with the courts here or anywhere else in the country. So we're getting a professional bodyguarding agency to do what the system is not doing: protecting battered women.'

The article describes a scene among women being trained to fill out forms to file for restraining orders against those threatening them.

> The women sitting around the scarred table were black, white and Hispanic. They were middle-aged and barely out of their teens. Two were

pregnant. All said they were scared. 'I don't see what purpose this serves,' one woman said, looking over the restraining order form. 'If he's going to get you, he's going to get you.' 'That's what I'm afraid of,' someone else said. 'But we have to do something.'

CEDAW COMMITTEE, VIOLENCE AGAINST WOMEN

General Recommendation No. 19, Eleventh Sess. No. 1, 1992
UN Doc. A/47/38, 1 Int. Hum. Rt. Rep. 25 (No. 1, 1994).

. . .

Background

1. Gender-based violence is a form of discrimination that seriously inhibits women's ability to enjoy rights and freedoms on a basis of equality with men.

. . .

4. The Committee concluded that not all the reports of States parties adequately reflected the close connection between discrimination against women, gender-based violence, and violations of human rights and fundamental freedoms. The full implementation of the Convention required States to take positive measures to eliminate all forms of violence against women.
5. The Committee suggested to States parties that in reviewing their laws and policies, and in reporting under the Convention, they should have regard to the following comments of the Committee concerning gender-based violence.

General comments

6. The Convention in article 1 defines discrimination against women. The definition of discrimination includes gender-based violence, that is, violence that is directed against a woman because she is a woman or that affects women disproportionately. It includes acts that inflict physical, mental or sexual harm or suffering, threats of such acts, coercion and other deprivations of liberty. Gender-based violence may breach specific provisions of the Convention, regardless of whether those provisions expressly mention violence.
7. Gender-based violence, which impairs or nullifies the enjoyment by women of human rights and fundamental freedoms under general international law or under human rights conventions, is discrimination within the meaning of article 1 of the Convention. These rights and freedoms include:

(a) The right to life;
(b) The right not to be subject to torture or to cruel, inhuman or degrading treatment or punishment;
(c) The right to equal protection according to humanitarian norms in time of international or internal armed conflict;

(d) The right to liberty and security of person;
(e) The right to equal protection under the law;
(f) The right to equality in the family;
(g) The right to the highest standard attainable of physical and mental health;
(h) The right to just and favourable conditions of work.

8. The Convention applies to violence perpetrated by public authorities. Such acts of violence may breach that State's obligations under general international human rights law and under other conventions, in addition to breaching this Convention.
9. It is emphasized, however, that discrimination under the Convention is not restricted to action by or on behalf of Governments (see articles 2(e), 2(f) and 5). For example, under article 2 (e) the Convention calls on States parties to take all appropriate measures to eliminate discrimination against women by any person, organization or enterprise. Under general international law and specific human rights covenants, States may also be responsible for private acts if they fail to act with due diligence to prevent violations of rights or to investigate and punish acts of violence, and for providing compensation.

Comments on specific articles of the Convention

. . .

Articles 2(f), 5 and 10(c)

11. Traditional attitudes by which women are regarded as subordinate to men or as having stereotyped roles perpetuate widespread practices involving violence or coercion, such as family violence and abuse, forced marriage, dowry deaths, acid attacks and female circumcision. Such prejudices and practices may justify gender-based violence as a form of protection or control of women. The effect of such violence on the physical and mental integrity of women is to deprive them of the equal enjoyment, exercise and knowledge of human rights and fundamental freedoms. While this comment addresses mainly actual or threatened violence the underlying consequences of these forms of gender-based violence help to maintain women in subordinate roles and contribute to their low level of political participation and to their lower level of education, skills and work opportunities.
12. These attitudes also contribute to the propagation of pornography and the depiction and other commercial exploitation of women as sexual objects, rather than as individuals. This in turn contributes to gender-based violence.

. . .

Article 11

17. Equality in employment can be seriously impaired when women are subjected to gender-specific violence, such as sexual harassment in the workplace.

18. Sexual harassment includes such unwelcome sexually determined behaviour as physical contact and advances, sexually coloured remarks, showing pornography and sexual demands, whether by words or actions. Such conduct can be humiliating and may constitute a health and safety problem; it is discriminatory when the woman has reasonable grounds to believe that her objection would disadvantage her in connection with her employment, including recruitment or promotion, or when it creates a hostile working environment.

Article 12

19. States parties are required by article 12 to take measures to ensure equal access to health care. Violence against women puts their health and lives at risk.
20. In some States there are traditional practices perpetuated by culture and tradition that are harmful to the health of women and children. These practices include dietary restrictions for pregnant women, preference for male children and female circumcision or genital mutilation.

. . .

Article 16 (and article 5)

22. Compulsory sterilization or abortion adversely affects women's physical and mental health, and infringes the right of women to decide on the number and spacing of their children.
23. Family violence is one of the most insidious forms of violence against women. It is prevalent in all societies. . . .

Specific recommendations

24. In light of these comments, the Committee on the Elimination of Discrimination against Women recommends:

(a) States parties should take appropriate and effective measures to overcome all forms of gender-based violence, whether by public or private act;
(b) States parties should ensure that laws against family violence and abuse, rape, sexual assault and other gender-based violence give adequate protection to all women, and respect their integrity and dignity. Appropriate protective and support services should be provided for victims. Gender-sensitive training of judicial and law enforcement officers and other public officials is essential for the effective implementation of the Convention;
(c) States parties should encourage the compilation of statistics and research on the extent, causes and effects of violence, and on the effectiveness of measures to prevent and deal with violence;

(d) Effective measures should be taken to ensure that the media respect and promote respect for women;

(e) States parties in their report should identify the nature and extent of attitudes, customs and practices that perpetuate violence against women, and the kinds of violence that result. They should report the measures that they have undertaken to overcome violence, and the effect of those measures;

(f) Effective measures should be taken to overcome these attitudes and practices. States should introduce education and public information programmes to help eliminate prejudices which hinder women's equality;

. . .

(k) States parties should establish or support services for victims of family violence, rape, sex assault and other forms of gender-based violence, including refuges, specially trained health workers, rehabilitation and counselling;

. . .

(m) States parties should ensure that measures are taken to prevent coercion in regard to fertility and reproduction, and to ensure that women are not forced to seek unsafe medical procedures such as illegal abortion because of lack of appropriate services in regard to fertility control;

. . .

COMMENT ON INTERNATIONAL INSTRUMENTS ON VIOLENCE AGAINST WOMEN

In 1993, the General Assembly adopted a Declaration on the Elimination of Violence against Women.[21] The Preamble asserted that the Declaration would 'strengthen and complement' the process of effective implementation of CEDAW, and recognized that violence against women 'is a manifestation of historically unequal power relations between men and women, which have led to domination over and discrimination against women.' Violence is 'one of the crucial social mechanisms by which women are forced into a subordinate position.'

Violence against women covers (Article 1) 'gender-based violence . . . whether occurring in public or in private life.' Under Article 2, it includes (a) violence in the family, marital rape, female genital mutiliation and 'other traditional practices harmful to women,' (b) similar violence within the general community,

[21] UN Doc. A/48/629, reprinted in 33 Int. Leg. Mat. 1050 (1994).

including sexual harassment at work, and (c) violence 'perpetrated or condoned by the State, wherever it occurs.'

Article 4 provides that states 'should not invoke any custom, tradition or religious consideration to avoid their obligations with respect to' elimination of violence. The state is to adopt measures, especially in education (para. j), 'to modify the social and cultural patterns of conduct of men and women and to eliminate prejudices, customary practices and all other practices based on the idea of the inferiority or superiority of either of the sexes and on stereotyped roles for men and women.'

To date only one treaty directly addresses this issue. In June 1994, the Organization of American States adopted the Inter-American Convention on the Prevention, Punishment and Eradication of Violence Against Women (effective March 1995, with two ratifications).[22] Like the UN Declaration, its preambular paragraphs recognize that violence against women is 'a manifestation of the historically unequal power relations between women and men'. The substantive provisions and the definition of the Convention's reach echo provisions in the Declaration.

This Convention is, however, more explicit in recognizing the broad scope and nature of violence, encompassing (Article 2) 'physical, sexual and psychological violence.' Article 3 makes explicit a woman's rights 'to be free from violence in both the public and private sphere.'

States agree (Article 7) 'to pursue, by all appropriate means and without delay, policies to prevent, punish and eradicate such violence.' These duties include due diligence in investigating and imposing penalties for violence, and the establishment of effective legal procedures including protective measures for women.

NOTE

In 1994, the UN Economic and Social Council (Decision 1994/254) endorsed the resolution of the UN Commission on Human Rights (Resolution 1994/45) to appoint for a three-year term a special rapporteur on violence against women, its causes and consequences. The special rapporteur is empowered to engage in field missions either independently or jointly with other rapporteurs and working groups; to seek and receive information from governments, other treaty bodies, specialized agencies of the UN and NGOs; and to consult with the CEDAW Committee in the course of her investigations.

[22] 33 Int. Leg. Mat. 1534 (1994).

SPECIAL RAPPORTEUR ON VIOLENCE AGAINST WOMEN, PRELIMINARY REPORT

Commission on Human Rights, E/CN.4/1995/42, 22 Nov. 1994.

[The following excerpts from the Preliminary Report of Radhika Coomaraswamy are taken from sections entitled 'Historically unequal power relations,' and 'Cultural ideology.']

49. . . . [V]iolence against women is a manifestation of historically unequal power relations between men and women. Violence is part of a historical process and is not natural or born of biological determinism. The system of male dominance has historical roots and its functions and manifestations change over time. The oppression of women is therefore a question of politics, requiring an analysis of the institutions of the State and society, the conditioning and socialization of individuals, and the nature of economic and social exploitation. The use of force against women is only one aspect of this phenomenon, which relies on intimidation and fear to subordinate women.

50. Women are subject to certain universal forms of abuse . . . . It is argued that any attempt to universalize women's experience is to conceal other forms of oppression such as those based on race, class or nationality. This reservation must be noted and acknowledged. And yet it must be accepted that there are patterns of patriarchal domination which are universal, though this domination takes a number of different forms as a result of particular and different historical experiences.

. . .

54. The institution of the family is also an arena where historical power relations are often played out. On the one hand, the family can be the source of positive nurturing and caring values where individuals bond through mutual respect and love. On the other hand, it can be a social institution where labour is exploited, where male sexual power is violently expressed and where a certain type of socialization disempowers women. Female sexual identity is often created by the family environment. The negative images of the self which often inhibit women from realizing their full potential may be linked to familial expectation. The family is, therefore, the source of positive humane values, yet in some instances it is the site for violence against women and a socialization process which may result in justifying violence against women.

. . .

57. In the context of the historical power relations between men and women, women must also confront the problem that men control the knowledge systems of the world. Whether it be in the field of science, culture, religion or language, men control the accompanying discourse. Women have been excluded from the enterprise of creating symbolic systems or interpreting

historical experience. It is this lack of control over knowledge systems which allows them not only to be victims of violence, but to be part of a discourse which often legitimizes or trivializes violence against women. . . .

. . .

64. The ideologies which justify the use of violence against women base their discussion on a particular construction of sexual identity. The construction of masculinity often requires that manhood be equated with the ability to exert power over others, especially through the use of force. Masculinity gives man power to control the lives of those around him, especially women. The construction of femininity in these ideologies often requires women to be passive and submissive, to accept violence as part of a woman's estate. Such ideologies also link a woman's identity and self-esteem to her relationship to her father, husband or son. An independent woman is often denied expression in feminine terms. In addition, standards of beauty, defined by women, often require women to mutilate themselves or damage their health, whether with regard to foot binding, anorexia nervosa and bulimia. It is important to reinvent creatively these categories of masculinity and femininity, devoid of the use of force and ensuring the full development of human potential.

. . .

67. Certain customary practices and some aspects of tradition are often the cause of violence against women. Besides female genital mutilation, a whole host of practices violate female dignity. Foot binding, male preference, early marriage, virginity tests, dowry deaths, sati, female infanticide and malnutrition are among the many practices which violate a woman's human rights. Blind adherence to these practices and State inaction with regard to these customs and traditions have made possible large-scale violence against women. States are enacting new laws and regulations with regard to the development of a modern economy and modern technology and to developing practices which suit a modern democracy, yet it seems that in the area of women's rights change is slow to be accepted.

QUESTIONS

1. Consider paras. 6 and 7 of the General Recommendation, p. 938, *supra*. How many of the enumerated rights in para. 7 are stated in CEDAW? Do you view these paragraphs as an application, elaboration, or expansion of the provisions of CEDAW?

2. In what ways is the state involved in the acts and practices described in paras. 11, 12, 20 and 21? Does the General Recommendation suggest that those perpetrating the described acts—sexual harassment, family violence, female circumcision—are themselves violating the Convention?

3. The Preliminary Report of the Special Rapporteur states in para. 64 (on ideologies justifying violence against women) that it is 'important to reinvent creatively these categories of masculinity and femininity.' From this perspective, how would you draft a governmental program for sharply reducing violence against women? What function would the criminal law serve? Other types of laws?

E. THE PUBLIC–PRIVATE DIVIDE: DISCRIMINATION AND VIOLENCE BY NON-GOVERNMENTAL ACTORS

Earlier materials in this chapter sometimes described abuse of women not by state ('public') but by non-governmental ('private') actors and action. This theme—the relevance of the public–private divide—inevitably surfaces with respect to women's rights, as much as, if not more than, with respect to other human rights fields.

The distinction between these two terms or concepts is stated in different ways. In the preceding paragraph, it had to do with the nature or character of the actor (governmental or non-governmental). From other perspectives, the distinction concerns different spheres of life and action. The 'private' is frequently associated with the home, family, domestic life. The 'public' is identified with the interactions of a working life: salaried employment, business, professions, the give and take of the market, being 'out in the world'.

The following readings examine some of the issues presented by opposition of private and public:

the practical, political and ideological significance of the divide;

the shifting boundary line between the two as conceptions of their significance and content change;

the degree to which human rights instruments should require states parties to regulate non-governmental (private) conduct; and

the degree to which non-governmental actors should themselves become directly subject to duties under international human rights law and liable for their violation.

The readings start with what has become a classic judicial opinion in international human rights because of its clarification of a state's duty with respect to violence committed by non-governmental actors, and with questions by Meron about the effects of CEDAW on the realm of the private. The further readings by contemporary feminists suggest the importance of keeping clear the different meanings of private and public, as well as the need to distinguish between the conception of 'privacy' and certain definitions of the private–public divide.

VELASQUEZ RODRIGUEZ CASE

Inter-American Court of Human Rights, 1988
Ser. C, No. 4, 9 Hum. Rts L. J. 212 (1988).

[The facts relevant to this case, and portions of the opinion of the Inter-American Court of Human Rights, are set forth at p. 650, *supra*. In a nutshell, a petition against Honduras was received by the Inter-American Commission of Human Rights, alleging that one Velásquez was arrested without warranty by national security units of Honduras. Knowledge of his whereabouts was consistently denied by police and security forces. Velásquez had disappeared. Petitioners argued that through this conduct, Honduras had violated several articles of the American Convention on Human Rights. After hearings and conclusions, the Commission referred the matter to the Inter-American Court of Human Rights, whose contentious jurisdiction had been recognized by Honduras. The Court concluded that Honduras had violated the Convention.

In the excerpts below, the Court addresses the issue of just what the obligations of Honduras were under the Convention. Was Honduras obligated only to 'respect' individual rights and not directly violate them, as by torture or illegal arrest? Or was Honduras obligated to take steps, within reasonable limits, to protect people like Velásquez from seizure even by non-governmental, private persons? In an earlier portion of the opinion, the Court had found that the Honduran state was implicated in the arrest and disappearance, and that the acts of those arresting Velásquez could be imputed to the state. In the present excerpts, the Court reviews that information, and considers what might be the responsibility of Honduras even if the seizure and disappearance of Velásquez were caused by private persons unconnected with the government.]

161. Article 1(1) of the Convention provides:

> 1. The States Parties to this Convention undertake to respect the rights and freedoms recognized herein and to ensure to all persons subject to their jurisdiction the free and full exercise of those rights and freedoms. . . .

. . .

164. Article 1(1) is essential in determining whether a violation of the human rights recognized by the Convention can be imputed to a State Party. In effect, that article charges the States Parties with the fundamental duty to respect and guarantee the rights recognized in the Convention. Any impairment of those rights which can be attributed under the rules of international law to the action or omission of any public authority constitutes an act

imputable to the State, which assumes responsibility in the terms provided by the Convention itself.

165. The first obligation assumed by the States Parties under Article 1(1) is 'to respect the rights and freedoms' recognized by the Convention. . . .

166. The second obligation of the States Parties is to ['ensure'] the free and full exercise of the rights recognized by the Convention to every person subject to its jurisdiction. This obligation implies the duty of the States Parties to organize the governmental apparatus and, in general, all the structures through which public power is exercised, so that they are capable of juridically ensuring the free and full enjoyment of human rights. As a consequence of this obligation, the States must prevent, investigate and punish any violation of the rights recognized by the Convention and, moreover, if possible attempt to restore the right violated and provide compensation as warranted for damages resulting from the violation.

. . .

169. According to Article 1(1), any exercise of public power that violates the rights recognized by the Convention is illegal. . . .

170. This conclusion is independent of whether the organ or official has contravened provisions of internal law or overstepped the limits of his authority. Under international law a State is responsible for the acts of its agents undertaken in their official capacity and for their omissions, even when those agents act outside the sphere of their authority or violate internal law.

. . .

172. Thus, in principle, any violation of rights recognized by the Convention carried out by an act of public authority or by persons who use their position of authority is imputable to the State. However, this does not define all the circumstances in which a State is obligated to prevent, investigate and punish human rights violations, nor all the cases in which the State might be found responsible for an infringement of those rights. An illegal act which violates human rights and which is initially not directly imputable to a State (for example, because it is the act of a private person or because the person responsible has not been identified) can lead to international responsibility of the State, not because of the act itself, but because of the lack of due diligence to prevent the violation or to respond to it as required by the Convention.

. . .

174. The State has a legal duty to take reasonable steps to prevent human rights violations and to use the means at its disposal to carry out a serious investigation of violations committed within its jurisdiction, to identify those responsible, impose the appropriate punishment and ensure the victim adequate compensation.

175. This duty to prevent includes all those means of a legal, political,

administrative and cultural nature that promote the safeguard of human rights and ensure that any violations are considered and treated as illegal acts, which, as such, may lead to the punishment of those responsible and the obligation to indemnify the victims for damages. It is not possible to make a detailed list of all such measures, as they vary with the law and the conditions of each State Party. Of course, while the State is obligated to prevent human rights abuses, the existence of a particular violation does not, in itself, prove the failure to take preventive measures. . . .

. . .

177. In certain circumstances, it may be difficult to investigate acts that violate an individual's rights. The duty to investigate, like the duty to prevent, is not breached merely because the investigation does not produce a satisfactory result. Nevertheless, it must be undertaken in a serious manner. . . . Where the acts of private parties that violate the Convention are not seriously investigated, those parties are aided in a sense by the government, thereby making the State responsible on the international plane.

178. In the instant case, the evidence shows a complete inability of the procedures of the State of Honduras, which were theoretically adequate, to ensure the investigation of the disappearance of Manfredo Velásquez and the fulfillment of its duties to pay compensation and punish those responsible, as set out in Article 1(1) of the Convention.

179. As the Court has verified above, the failure of the judicial system to act upon the writs brought before various tribunals in the instant case has been proven. Not one writ of habeas corpus was processed. No judge had access to the places where Manfredo Velásquez might have been detained. The criminal complaint was dismissed.

180. Nor did the organs of the Executive Branch carry out a serious investigation to establish the fate of Manfredo Velásquez. There was no investigation of public allegations of a practice of disappearances nor a determination of whether Manfredo Velásquez had been a victim of that practice. The Commission's requests for information were ignored to the point that the Commission had to presume, under Article 42 of its Regulations, that the allegations were true. . . .

. . .

182. The Court is convinced, and has so found, that the disappearance of Manfredo Velásquez was carried out by agents who acted under cover of public authority. However, even had that fact not been proven, the failure of the State apparatus to act, which is clearly proven, is a failure on the part of Honduras to fulfill the duties it assumed under Article 1(1) of the Convention, which obligated it to guarantee Manfredo Velásquez the free and full exercise of his human rights.

. . .

THEODOR MERON, HUMAN RIGHTS LAW-MAKING IN THE UNITED NATIONS
(1986), at 60.

. . .

. . . [Article 1 of the] Convention on the Elimination of All Forms of Discrimination Against Women clearly extends the prohibition of discrimination to private life. . . .

The provisions of the Convention on the Elimination of All Forms of Discrimination Against Women apply to a broad range of activities: unintentional as well as intentional discrimination is prohibited (as indicated by the 'effect' clause); private as well as public actions are regulated (as indicated by the phrase 'any other field'). By proscribing practices which have the effect of discriminating against women, the Convention guards against the use of facially neutral criteria as a pretext for discrimination, for instance the use of height and weight requirements which are not related to the requirements of the job and which tend to exclude women as a group. The 'effects' standard avoids the difficulties inherent in proving specific discriminatory motive. The prohibition of unintentional discrimination is necessary to achieve systemic change, because policies undertaken without discriminatory motive may perpetuate inequalities established by prior acts of purposeful discrimination. . . .

The definition of discrimination against women does not prohibit certain distinctions *per se*, but only when they have the purpose or the effect of denying women the enjoyment of human rights and fundamental freedoms on a basis of equality with men. . . . [T]he Committee on the Elimination of Racial Discrimination has tended to regard all ethnic distinctions as suspect, unless made in the context of affirmative action, without engaging in a substantive inquiry as to whether they have in fact had an adverse effect on the enjoyment of 'rights'. If the parties to the Convention on the Elimination of All Forms of Discrimination Against Women and the Committee established under Art. 17 follow a similar approach, a further expansion in the reach of the Convention would result. . . .

It is not clear whether it was appropriate to extend the field of application of the Convention to encompass even private, interpersonal relations (except, of course, when the conduct which is challenged takes forms customarily regulated pursuant to the police power). It is certainly true that discrimination against women in personal and family life is rampant and may obviate equal opportunities which may be available in public life. There is danger, however, that state regulation of interpersonal conduct may violate the privacy and associational rights of the individual and conflict with the principles of freedom of opinion, expression, and belief. Such regulation may require invasive state action to determine compliance, including inquiry into political and religious beliefs. Attempts to regulate discrimination in interpersonal conduct may invite abuse of the discretion vested in the State by the broad language of Art. 1.

. . .

[Article 5 of the Convention] mandates regulation of social and cultural patterns of conduct regardless of whether the conduct is public or private. Coupled with the broad and vague language of the preambular sentence discussed above ('all appropriate measures'), para. (a) might permit States to curtail to an undefined extent privacy and associational interests and the freedom of opinion and expression. Moreover, since social and cultural behaviour may be patterned according to factors such as ethnicity or religion, state action authorized by para. (a) which is directed towards modifying the way in which a particular ethnic or religious group treats women may conflict with the principles forbidding discrimination on the basis of race or religion.

The danger of intrusive state action and possible violation of the rights of ethnic or religious groups might have been mitigated by limiting state action to educational measures. Social and cultural patterns of conduct could be regulated by the substantive provisions which govern actual practices in a particular field, for example employment practices, without loss of substantive rights under the Convention.

. . .

ELIZABETH SCHNEIDER, THE VIOLENCE OF PRIVACY

23 Conn. L. Rev. 973 (1992), at 974.

. . .

Historically, the dichotomy of 'public' and 'private' has been viewed as an important construct for understanding gender. The traditional notion of 'separate spheres' is premised on a dichotomy between the 'private' world of family and domestic life (the 'women's' sphere), and the 'public' world of marketplace (the 'men's' sphere).

. . .

Although a dichotomous view of the public sphere and the private sphere has some heuristic value, and considerable rhetorical power, the dichotomy is overdrawn. The notion of a sharp demarcation between public and private has been widely rejected by feminist and Critical Legal Studies scholars. There is no realm of personal and family life that exists totally separate from the reach of the state. The state defines both the family, the so-called private sphere, and the market, the so-called public sphere. 'Private' and 'public' exist on a continuum.

Thus, in the so-called private sphere of domestic and family life, which is purportedly immune from law, there is always the selective application of law. Significantly, this selective application of law invokes 'privacy' as a rationale for immunity in order to protect male domination. For example, when the police do not respond to a battered woman's call for assistance, or when a civil court refuses to evict her assailant, the woman is relegated to self-help, while the man who beats her receives the law's tacit encouragement and support. Indeed, we can see this pattern in recent legislative and prosecutorial efforts to control women's conduct during pregnancy in the form of 'fetal' protection laws. These laws are premised on the notion that women's child-bearing capacity, and pregnancy itself, subjects women to public regulation and control. Thus, pregnant battered women may find themselves facing criminal prosecution for drinking liquor, but the man who battered them is not prosecuted.

The rhetoric of privacy that has insulated the female world from the legal order sends an important ideological message to the rest of society. It devalues women and their functions and says that women are not important enough to merit legal regulation.

. . .

The tension between public and private also is seen in the issue of what legal processes are available to battered women, and the social meaning of those processes to battered women, in particular, and to society at large. Over the last several years, the range of legal remedies has expanded and there has been an explosion of statutory reforms. For example, there are civil remedies, known as restraining orders or orders of protection. These are court orders with flexible provisions that a battered woman can obtain to stop a man from beating her, prevent him from coming to the house, or evict him from the house. There are also criminal statutes that provide for the arrest of batterers, either for beating or for violation of protective orders. Although there remain serious problems in the enforcement and implementation of these orders, the fact that such formal legal processes exist is evidence of a developing understanding of the public dimension of the problem. By giving battered women remedies in court there is, at least theoretically, public scrutiny, public control and the possibility of public sanction.

. . .

CATHERINE MacKINNON, ON TORTURE: A FEMINIST PERSPECTIVE ON HUMAN RIGHTS

in Kathleen Mahoney and P. Mahoney (eds.), Human Rights in the Twenty-First Century 21 (1993).

Torture, with accompanying disappearance and murder, is widely recognized as a core violation of human rights. Inequality on the basis of sex is equally widely condemned, and sex equality affirmed as a core human rights value and legal guarantee, both nationally and internationally. My question is, given these two facts, why is torture on the basis of sex in the form of rape, domestic battering, and pornography not seen as a violation of human rights?

The purpose of this paper is to illustrate the way the realities of women's condition illuminate what must be included within the concept of human rights unless it is implicitly assumed that women are a lesser form of human life. My argument is that law and policy—national as well as international—and society must be held to human rights standards when torture, violence, assault and other forms of inequality are sex-based, just as much as when they are based on anything else.

Internationally, torture has a recognized profile. . . . Verbal abuse, humiliation, and making the victim feel worthless are part of the necessary degradation. Often torture victims are selected and specifically tortured as members of a social group. . . .

There is also international consensus on what torture does to a human being. It is recognized that people break and change under such extreme pressure, studied under the rubrics of brainwashing, post-traumatic stress and the Stockholm Syndrome. Long-term consequences to survivors are both mental and physical. Many victims disassociate to survive and find the response hard to reverse. What one learns about life from being tortured, and what one has to do to survive it when others do not, makes being alive unbearable for some. Suicide can result even years later.

It is also generally recognized that the usual purpose of torture is to control, intimidate, or eliminate those who insult or challenge or are seen to undermine a regime. Torture is thus imagined to be political, although it often seems that its political overlay is a facilitating pretext for the pure exercise of sadism, which is a politics in itself.

When these things happen, human rights are deemed violated. These recognitions delineate a standard for acknowledging atrocities.

With this framework in mind, consider the following accounts.

[The author recounts in detail terrifying, unspeakable and ongoing acts of savagery, sadism and humiliation against three women by men—one man a pimp, the other two husbands. The accounts, stressing sexual brutality and (by any standard) torture leaving permanent physical and psychic injuries, were based on statements by two of the women involved and, in one case, on newspaper sources.]

In the accounts by these women, *all the same things happen* as happen in the Amnesty International accounts of torture, except that they happen in homes in Nebraska or in pornography studios in Los Angeles rather than prison cells in Chile or detention centres in Turkey. But the social and legal responses to their experiences are not the same. Torture victims are not generally asked, 'How many like you are there?', as if it is not important and we do not have to worry about it if it happened only to you, or a few like you. Torture is not generally attributed to some sick individual and dismissed as exceptional. With torture, one doesn't hear that maybe the increase is just an increase in reporting, as if a constant level of such abuse would be acceptable. Billions of dollars are not made selling pictures of what is regarded as torture, nor is torture as such generally regarded as sexual entertainment. Never is a victim of torture asked, didn't you really want it?

A simple double standard is at work here. What fundamentally distinguishes torture, understood in human rights terms, from the events these women have described is that torture is done to men as well as to women. Torture is regarded as politically motivated not personal; the state is involved in it. I want to ask why the torture of women by men is not seen as torture, why it is not seen as politically motivated, and what is the involvement of the state in it?

Women comprise half the human race. . . .

. . .

Why isn't this political? The abuse is neither random nor individual. The fact that you may know your assailant does not mean your membership in a group selected for violation is irrelevant to your abuse. This abuse is systematic and group-based. It defines the quality of community life and the distribution of power in society. It would seem that something is not considered political if it is done to women by men, especially if it is considered to be sex. Then, it is not considered political or even violent because what is political is when men control and hurt and use other men, persons who are deserving of dignity, on some basis men have decided is deserving of dignity, like political ideology. So their suffering has the dignity of politics and is called torture. . .

A reason often given or implicit in considering atrocities to women not human rights violations, politically or legally, is that they do not involve acts by states. They happen between non-state actors, in civil society, unconscious and unorganized and unsystematic and undirected and unplanned. They do not happen by virtue of state policy. International instruments (and national constitutions) control only state action.

But the state is not all there is to power. To act as if it is produces an exceptionally inadequate definition for human rights when so much of the second class status of women, from sexual objectification to murder, is done by men to women prior to express state involvement. If 'the political' is to be defined in terms of men's experiences of being subjected to power, it makes some sense to centre its definition on the state. But if one is including the unjust power involved in the subjection of half of the human race by the other half—male dominance—it makes no sense whatsoever to define power exclusively in terms of what the state does. The state is only one instrumentality of it. To fail to see this is pure gender bias. Usually, this bias flies under the flag of privacy, so that those areas which are defined as inappropriate for state involvement, where the discourse of human rights is irrelevant, are those 'areas in which the majority of the world's women live out their days.' Moreover, the fact that there is no single state or organized group expressly dedicated to this pursuit does not mean that all states are not more or less dedicated to it on an operative level or that it is not a deep structure of social, political, and legal organization.

Actually, though, the state is deeply and actively complicit in all the abuses mentioned. . . .

. . .

. . . The abuse of women I described does not pretend to be official. But the cover-up, the legitimization, and the legalization of the abuse is. It is done with official impunity and legalized disregard. The abuse is systematic and known, the disregard is official and organized, and the effective governmental tolerance is a matter of policy.

Legally, the pattern is one of national and international guarantees of equality coexisting with massive rates of rape and battering and traffic in women through pornography effectively condoned by law. Some progressive international human rights bodies are beginning to inquire into some dimensions of these issues under equality rubrics—none into pornography, some into rape and battering. Still, rape is more likely to look like a potential human rights violation when it happens in official custody. A woman's human rights are deemed violated because the state was an instrumentality of the rape. The regular laws and their regular everyday administration are not seen as official state involvement in legalized sex inequality. . . .

. . .

Why are there no human rights standards for tortures of women as a sex? Why are these atrocities not seen as sex equality violations? . . .

. . .

. . . [T]o act consistent with equality guarantees is to move to end sex inequality through law. Wherever the law reinforces gender hierarchy, it violates legal equality guarantees in national constitutions and in international covenants as well.

. . . We need to look at unenforced laws: those that exist but virtually nothing is ever done with them, such as the law against battering, which can make violence women's only survival option. This violates women's human rights. We need to look at the wrong laws: the laws that don't fit the violation, have nothing to do with what really happens to women, such as the law of self-defense, the law of rape, and the law of obscenity in most places. They violate women's human rights. All these are affirmative state acts or positive omissions. To go deeper, we also need to look at those areas where there are no laws at all for the harms women experience in society because we are women, such as most of the harms of pornography. Women are human there, too. Women's human rights are being violated there, too.

If, when women are tortured and disappear because we are women, the law recognized that a human being had her human rights violated, the term 'rights' would begin to have something of the content to which we might aspire, and the term 'woman' would, to paraphrase Richard Rorty, begin to be 'a name for a way of being human'.

KAREN ENGLE, AFTER THE COLLAPSE OF THE PUBLIC/PRIVATE DISTINCTION: STRATEGIZING WOMEN'S RIGHTS

in Dorinda Dallmeyer (ed.), Reconceiving Reality: Women and International Law (1993), at 143.

. . .

Central to the critiques of international law have been analyses of the public/private distinction. They generally take one of two forms. Either women's rights advocates argue that public international law, and particularly human rights theory, is flawed because it is not really universal. That is, because international law excludes from its scope the private, or domestic sphere—presumably the space in which women operate—it cannot include them. Or advocates argue that international law does not really exclude the private, but rather uses the public/private divide as a convenient screen to avoid addressing women's issues.

Those who take the first approach, then, take for granted that public international law in its present form cannot enter what they see as the private sphere. For women to be included, they maintain, international law must be reconceptualized to include the private.

Those who take the second approach, on the other hand, assume that doctrinal tools are present in international law—particularly in human rights law—to accommodate women. The public/private dichotomy, they assert, is both irrational and inconsistently applied. According to this second approach, the human rights regime and states are disingenuous in their claim that they do not enter the private sphere. The private sphere is entered all the time, for example, through regulation of the family, or the ability to impute to states the acts of non-state actors in disappearance cases. Moreover, these critics argue, a state's failure to protect rights in the private sphere is not distinguishable from direct state action. Finally, they point out that the human rights regime applies a double standard when talking about women. The international legal regime would never argue, for example, that it could not intervene to ensure that states end certain forms of 'private' violence, such as cannibalism or slavery.

. . .

Concentrating too much on the public/private distinction excludes important parts of women's experiences. Not only does such a focus often omit those parts of women's lives that figure into the 'public,' however that gets defined, it also assumes that 'private' is bad for women. It fails to recognize that the 'private' is a place where many have tried to be (such as those involved in the market), and that it might ultimately afford protection to (at least some) women.

. . .

. . . [The critiques of the private–public distinction] make us think of the unregulated private as something that is necessarily bad for women. We rarely look at the ways in which privacy (even if only because it seems the best available paradigm) is seen by at least some women to offer them protection. A number of examples immediately come to mind, each of which centers on women's bodies and, not surprisingly, on women's sexuality. The language of privacy, and sketching out zones of privacy, many would argue, is our best shot at legally theorizing women's sexuality. In United States legal jurisprudence, the First Amendment has been used to a similar end as often seen in the debates about pornography.

Examples of where 'the private' is sometimes seen to have liberating potential for women are abortion (which is most obvious to us in the United States); battering; the protection of 'alternative' sexual lifestyles; prostitution; right to wear the veil as protection from sexual harassment; right to participate in *or be free of* clitoridectomies, sati, breast implants, the wearing of spike-heeled shoes. Failure to focus on these issues affecting women's relationships to our bodies obscures the ways that many women see their lives.

. . . [T]he critiques often prevent us from taking seriously women who claim not to want the regulation or protection of international law. Arguments about 'culture,' particularly those that attempt to use claims of cultural integrity or community to maintain practices that some women might find abhorrent, get transformed into arguments about the private. That is, women's rights advocates often treat these arguments as though they are yet another manifestation of the mainstream legal regime's exclusion of the private or women (or both) at all costs. As a result, advocates either ignore those women who defend practices they see as an important part of their culture, or assume that such women are replete with false consciousness.

. . .

. . .[T]he critiques often lead us to conspiracy theories. Our exclusion indicates that they're all out to get us. We point out that *they* don't include *us* in the mainstream, that *they* give *us* our own marginal institutions and then ensure that the institutions lack enforcement mechanisms, that *they* don't really care about *us* or take us seriously. We rarely ask, though, who *they* are, who *we* are, and why they're out to get us. We also fail to notice that others are singing a similar tune to our own. Those who argue for economic and social rights, for example, seem to feel just as isolated and outside as women's advocates do. In fact, sometimes it seems that international law, particularly human rights law, has been built by its own criticism. That is every time some group or cause feels outside the law, it pushes for inclusion, generally through a new official document. The vast proliferation of human rights documents, then, is as much a testament to exclusion as it is to inclusion.

. . .

QUESTIONS

1. How do you assess the 'danger of intrusive state action' to which Meron refers? Do such dangers as may appear to you stem from the degree to which (a) CEDAW regulates private or non-governmental actors and acts, (b) requires changes in cultural attitudes and customary practices, or both?

2. Is a state party to CEDAW required to take measures against a television series on a privately owned network that sympathetically portrays the women who figure in its story as financially dependent on men, as homekeepers, and as persons caring for the family without outside work or career ambitions? If so, under what provisions of CEDAW, and what measures?

3. 'Two radically different conceptions are presented in the readings of how violence against women by non-state actors leads to the conclusion that the relevant state has violated international human rights. (1) The state fails to act reasonably through the processes of the criminal law to protect against violence, to punish the violators, and to deter further violence. (2) There is no division of state and society or culture, no satisfactory way of distinguishing between the public and private with respect to the exercise of power against women. In this context, private power is public power. The state is deeply implicated in cultural attitudes that foster violence, and can meet its human rights duties only by changing those attitudes. It is not a matter of patching up criminal laws or enforcement here or there, but of transforming deep structures of beliefs and behavior.'

a. Is this an accurate description of the readings? If not, how would you formulate the different conceptions at work?

b. Are both conceptions, one or the other, or neither one, appropriate for a human rights treaty on women's rights?

4. 'CEDAW is more of a long-run political platform or program than a human rights treaty concerned with equal protection.' Do you agree? If so, what follows from it?

COMMENT ON CONFLICTS BETWEEN CEDAW AND RELIGIOUS AND CUSTOMARY LAWS AND PRACTICES

The potential for conflict between the objectives of the Convention and states' religious and customary laws and practices has become a salient concern. Readings in Chapter 4 by An-Na'im on *Human Rights in the Muslim World*, p. 210, *supra*, and about female circumcision, p. 240, *supra*, illustrate some of those laws and practices.

The concern about serious conflict has grown in recent years with the increasing power and prominence of fundamentalist religious groups, many of which actively oppose the transformative impetus of the human rights movement for women. A common feature of such groups is their commitment to the maintenance of traditional gender roles. The practices expressing those roles are often defended in terms of sex role 'complementarity' or 'equity,' in contrast with the preeminent CEDAW principle of equality. These divergent paths are illustrated in a less dramatic way by the 'benign neglect' shown so far by many states parties of basic provisions of the Convention—for example, of Articles 2(f) and 5(a) that require parties to take 'all appropriate measures' to modify customary practices constituting discrimination.

This conflict between women's rights as elaborated in the Convention and the customary or religious law in many states parties assumes a large significance since the control and influence over women of customary law may exceed that of modern secular law, particularly in non-Western states.[23] Such law and practice often govern crucial matters including marriage, divorce, child custody, maintenance, inheritance, succession, and the ownership and control of property. In addition, cultural or religious norms may influence the availability of family planning information and devices, as well as the scope and character of reproductive rights.

In states without a clear distinction between religious and secular law, women may be subject to discrimination in formal legal processes including criminal cases. The law, for example, may distinguish between the legal capacity of men and women as witnesses and as to matters of proof and corroboration in cases of sexual assault. Recall the reservations to CEDAW at p. 920, *supra*, particularly those of Bangladesh, Egypt and Iraq to the provisions of Article 2 insofar as it relates to issues governed by the Islamic Shari'a.

Even where states have not entered reservations to Articles 2(f) or 5(a), serious obstacles remain to practical implementation of the Convention.[24] A government may find it difficult to disregard the sentiments of politically powerful groups or segments of society that wish to maintain religious or customary

[23] A survey of these issues appears in Donna Sullivan, Gender Equality and Religious Freedom: Toward a Framework for Conflict Resolution, 24 N.Y.U.J. Int. L. & Politics 795 (1992).

[24] Abdullahi An-Na'im, State Responsibility under International Human Rights Law to Change Religious and Customary Beliefs, in Rebecca Cook (ed), Human Rights of Women: National and International Perspectives 167 (1994).

law. States may deny that certain practices, such as male land tenure or female genital mutilation, contravene CEDAW's provisions. Assuming that a state recognizes its obligations to change some customary norms and practices, its motivation to bring about that change will depend on that change's relation to other state objectives and on the depth of the socio-cultural roots of the practices. Indeed, the state might not possess the necessary influence or power to proceed. Authority may be divided among the central government and regional or ethnic leaders. The supervision and enforcement of some customary laws may rest not with the state but with another body, such as a religious court or officials.

Finally, secular remedies, even where available, may be of limited utility if women face the difficult choice of accepting customary practices or risking (in some cases certain) exclusion from the communities from which they also derive support.

QUESTIONS

1. How do you assess the argument of states parties entering reservations to CEDAW that principles of religious freedom are fundamental to the human rights movement and validate their decisions? Can such an argument be based on the provisions of CEDAW itself? Consider the precise phrasing and relevance of Article 18 of the UDHR and Article 18 of the ICCPR.

2. How do you assess the argument of states entering reservations that respect for diversity, for cultural relativism over a dogmatic and alien universalism, ought to prevail with respect to contested matters of customary and religious norms and practices?

3. In the light of the readings in this Section E, what can we determine about effective strategies for implementing Convention rights in relation to conflicting religious and customary laws and practices?

a. What financial and institutional resources should states parties be expected to commit to changing or eliminating social or customary norms and practices that are detrimental to women? What tactics and methods should be employed?

b. Should priorities be established, given the recognition that not all can be done at once and that progressive realization is inevitable? What matters should have top priority?

c. How could the state act to reduce or eliminate the risks faced by women from their own communities if they break from community practices that they and the CEDAW Committee view as denials of rights?

COMMENT ON THE REGULATORY REACH OF HUMAN RIGHTS TREATIES

This Comment calls attention to comparable provisions of several treaties that require states to develop programs and regulation in order to eliminate certain ideas and social practices and to substitute for them other ideas and practices.

(1) CEDAW is not the only human rights instrument with aspirations to a deep transformation of social life. It is simply more explicit and forceful about those aspirations. Articles 5 and 10(c) are the two preeminent provisions that state general transformative goals. Compare them with provisions of two associated instruments: (i) Article 24(d) of the CEDAW Committee's General Recommendation No. 19, p. 938, *supra*, and (ii) Article 4 of the UN Declaration on the Elimination of Violence against Women, p. 941, *supra*.

(2) The International Convention on the Elimination of All Forms of Racial Discrimination (CERD) also states broad transformative goals. Some of its provisions in Articles 1, 2 and 5 on state regulation of discrimination by private or non-governmental groups, appear in the Comment at p. 778, *supra*. Article 2(e) states:

> Each State Party undertakes to encourage, where appropriate, integrationist multi-racial organizations and movements and other means of eliminating barriers between races, and to discourage anything which tends to strengthen racial division.

Article 7 of CERD provides:

> States Parties undertake to adopt immediate and effective measures, particularly in the fields of teaching, education, culture and information, with a view to combating prejudices which lead to racial discrimination and to promoting understanding, tolerance and friendship among nations and racial or ethnical groups. . . .

(*3*) Recall that the United States, when it ratified the ICCPR and CERD, entered reservations on the following analogous provisions in the two treaties.

(a) *Article 20(2) of the ICCPR* provides that '[a]ny advocacy of national, racial or religious hatred that constitutes incitement to discrimination, hostility or violence shall be prohibited by law.' A commentator's interpretation of this provision appears at p. 771, *supra*.
(b) Under *Article 4 of CERD*, States Parties undertake to adopt measures 'designed to eradicate all incitement to, or acts of' racial discrimination; to declare a punishable offense 'all dissemination of ideas based on racial superiority or hatred, incitement to racial discrimination' or acts of violence; to prohibit organizations 'which promote and incite racial discrim-

ination'; and to 'recognize participation in such organizations or activities as an offence punishable by law . . . .'

CEDAW does not by its terms require states parties to prohibit and make criminal the advocacy or dissemination of ideas. It does refer in numerous articles, including Article 5, to the obligation of parties to 'take all appropriate measures,' including legislation, to effectuate the aim of the given article.

QUESTIONS

'It is a proper function of human rights treaties to require or prohibit defined conduct by the state, including requiring the state to regulate non-governmental conduct like discrimination in employment that is offensive to the treaty's principles. It is not a proper function to regulate citizens or impose penalties on them for the ideas they may hold and may wish to persuade others to hold.'

a. Do you agree or disagree? Why?

b. Compare from this perspective CERD and CEDAW. From the statement's perspective, is one or the other convention subject to criticism?

c. How might you explain two salient differences between CERD and CEDAW: a strong Article 5 in CEDAW, which unlike CERD has no provision making advocacy of particular ideas a criminal offense?

F. PARTICIPATION OF WOMEN IN POLITICAL INSTITUTIONS AND LIFE

Low political participation and a low level of representation of a group in political institutions surely contribute to marginalization on the political agenda of that group's issues and interests. Such has traditionally been the case for women.

Consider some statistical information. 'In some countries, women are still not allowed to vote. . . . [I]n 1980, they made up just over 10% of the world's parliamentarians and less than 4% of national cabinets. In 1993, only six countries had women as heads of government.'[25] The following percentages[26] refer to the seats in parliament (or equivalent legislative body) occupied by women in 1992: Finland 39%, Norway 38%, Netherlands 23%, Cuba 23%, China 21%, Tanzania 11%, U.S.A. 10%, U.K. 9%, Russia 9%, India 7%, Mexico 7%, France 6%, Peru 6%, Kenya 3%, Egypt 2%.

[25] United Nations Development Programme, Human Development Report of 1993, 25.

[26] United Nations Development Programme, Human Development Report for 1994, 144.

A recent UN study[27] reports that high levels of economic development do not appear to increase automatically the number of women in political life. For example, the number of women elected to the national legislative bodies in Japan remains well below 10%. The collapse of the Communist states in Eastern Europe has coincided with a dramatic decline in the number of women in elected positions—and in the region which formerly had the highest rate of representation by women. It is not then surprising that the CEDAW Committee looks favourably upon specific programs to improve this situation, such as targets for the numbers of women in the civil service and in legislative bodies.

The problem of representation of course affects international as well as national institutions, perhaps particularly within the international human rights system given its goals of political participation. Hence the readings begin with a criticism of the international legal order before turning to the organs of national governments.

HILARY CHARLESWORTH, C. CHINKIN AND S. WRIGHT, FEMINIST APPROACHES TO INTERNATIONAL LAW

85 Am. J. Int. L. 613 (1991), at 621.

III. THE MASCULINE WORLD OF INTERNATIONAL LAW

In this section we argue that the international legal order is virtually impervious to the voices of women and propose two related explanations for this: the organizational and normative structures of international law.

The Organizational Structure of International Law.

The structure of the international legal order reflects a male perspective and ensures its continued dominance. The primary subjects of international law are states and, increasingly, international organizations. In both states and international organizations the invisibility of women is striking. Power structures within governments are overwhelmingly masculine: women have significant positions of power in very few states, and in those where they do, their numbers are minuscule. Women are either unrepresented or underrepresented in the national and global decision-making processes.

. . .

International organizations are functional extensions of states that allow them to act collectively to achieve their objectives. Not surprisingly, their structures replicate those of states, restricting women to insignificant and subordinate roles. Thus, in the United Nations itself, where the achievement of

[27] Women in Politics and Decision-Making in the Late Twentieth Century (1992).

nearly universal membership is regarded as a major success of the international community, this universality does not apply to women.

Article 8 was included in the United Nations Charter to ensure the legitimacy of women as permanent staff members of international organizations. Article 8 states: 'The United Nations shall place no restrictions on the eligibility of men and women to participate in any capacity and under conditions of equality in its principal and subsidiary organs.' . . .

In reality, women's appointments within the United Nations have not attained even the limited promise of Article 8. The Group on Equal Rights for Women in the United Nations has observed that 'gender racism' is practiced in UN personnel policies 'every week, every month, every year.'

Women are excluded from all major decision making by international institutions on global policies and guidelines, despite the often disparate impact of those decisions on women. . . .

The silence and invisibility of women also characterizes those bodies with special functions regarding the creation and progressive development of international law. Only one women has sat as a judge on the International Court of Justice and no woman has ever been a member of the International Law Commission. . . .

Despite the common acceptance of human rights as an area in which attention can be directed toward women, they are still vastly underrepresented on UN human rights bodies. The one committee that has all women members, the Committee on the Elimination of Discrimination against Women (CEDAW Committee), the monitoring body for the Convention on the Elimination of All Forms of Discrimination against Women (Women's Convention), has been criticized for its 'disproportionate' representation of women by the United Nations Economic and Social Council (ECOSOC). When it considered the CEDAW Committee's sixth report, ECOSOC called upon the state parties to nominate both female and male experts for election to the committee. Thus, as regards the one committee dedicated to women's interests, where women *are* well represented, efforts have been made to decrease female participation, while the much more common dominance of men in other United Nations bodies goes unremarked. . . .

Why is it significant that all the major institutions of the international legal order are peopled by men? Long-term domination of all bodies wielding political power nationally and internationally means that issues traditionally of concern to men become seen as general human concerns, while 'women's concerns' are relegated to a special, limited category. Because men generally are not the victims of sex discrimination, domestic violence, and sexual degradation and violence, for example, these matters can be consigned to a separate sphere and tend to be ignored. The orthodox face of international law and politics would change dramatically if their institutions were truly human in composition: their horizons would widen to include issues previously regarded as domestic—in the two senses of the word. Balanced representation in international organizations of nations of differing economic structures and power has been a prominent theme in the United Nations since the era of decolonization in the 1960s. The importance of accommodating interests of developed, developing and socialist nations and of various regional and ideological groups is recognized in all aspects of the UN structure and work. This sensitivity should be extended much further to include the gender of chosen representatives.

. . .

CONSIDERATION BY CEDAW OF COUNTRY REPORTS: POLITICAL PARTICIPATION

UN Doc. A/48/38, 28 May 1993.

[The following excerpts bearing on political participation of women are taken from the CEDAW Committee's annual report to the General Assembly. In each case, the excerpts are from the summary record of the Committee's consideration of the part of the country reports that is directed to Article 7 of the Convention addressing political participation. The participants are, in the discussion, the representative of the reporting country and the members of the Committee.]

FRANCE

329. The representative said that French women had not obtained a share of power in 1945 when they had been given the right to vote, but rather in the 1970s when the process of dissociation between sexuality and procreation had been accomplished through the adoption of the contraception and abortion laws. She said that contraception and abortion were the true revolutions of the twentieth century, constituting not a power-sharing between women and men, but a transfer of power from men to women. Women alone could decide on maternity, they could determine whether they chose to live with the child's father, to be married and to recognize the father's rights. They alone currently had the power in the family under the law.

330. The representative said that the French mentality was still influenced by the Napoleonic Code, which had given women an inferior position in society. Considerable strides had been made in general, but women continued to suffer from that heritage. . . .

. . .

339. The representative said that women in France were still excluded from the political arena. That backlog was a heritage from the past as the French Revolution had not encompassed gender issues, and women had obtained the right to vote very late. In the French Parliament only 5 per cent of the deputies were women. The reason why there were more French women in the European Parliament than in the National Assembly was that voting was done on the

basis of lists and the European Parliament was not of such political concern. Women were still excluded from political participation in spite of such dynamic steps as appointing women to six out of 45 ministers' posts or a woman as prime minister.

340. Asked whether similar actions as those undertaken to combat sexual violence would initiate progress, the representative said that the political activity of women depended on the political determination of the parties. Candidates for elections were nominated by the party officials. One way of enabling more women to obtain political power would be to establish positive discrimination procedures; however, such measures were not popular with the French people.

. . .

KENYA

113. The Committee wished to know what the Women's Bureau was doing to increase awareness of the need to place women in higher positions, whether the KANU party [the official government party] had a special programme to encourage women to participate in politics at the local and national levels, and whether women had the same financial support as men for their electoral campaigns. With regard to women in politics, the representative explained that the awareness-creation of the democratization process among women had produced encouraging results during the elections on 29 December 1992. Initially, 80 women had showed interest in representative leadership. Eliminations at the party preliminaries had left 20 women to compete for leadership against their male counterparts for 188 parliamentary seats. Six of them had been elected to the Seventh Parliament, the highest number of women since independence. The only woman elected within KANU had been appointed as Assistant Minister in the Ministry of Culture and Social Services.

. . .

SOUTH KOREA

. . .

424. The Committee noted that in the last National Assembly, only six out of a total of 299 deputies were women, whereas in the election of 24 March 1992, the number of female parliamentarians had further dropped to three. Members asked what was preventing women from attaining high positions and what had been done to ensure their equal representation. The representative said that in the Thirteenth National Assembly for the term 1988–1991, all the parliamentarians had been elected in the national electoral districts. However, no female candidates in the local electoral districts had been successful. Since the composition of the members of parliament depended on the political parties and three of the four main parties had merged, the number of female parliamentarians representing the parties in the electoral districts had accordingly decreased.

425. Concerning public affairs, she said that the level of women in politics and in decision-making had not changed significantly. Women were poorly represented in decision-making positions in the Government. . . .

SWEDEN

508. Regarding legislation that would ensure a certain proportion of women in the lists of candidates for popular election, the representative stated that such legislation did not exist. However, many political parties had internal rules or practices on the nomination of women.
509. Asked about whether Sweden still promoted the 'Fifty–Fifty' campaign as seen in some international forums, the representative stated that the goal remained the same, although the Government could not impose a certain behaviour on political parties. The Government was only responsible for certain areas (boards of public bodies, committees, working groups etc.). In order to affect the political parties, public opinion needed to be created. Generally, women remained active in political life; all political parties had their own women's organizations.
510. The representative gave a positive answer to a question concerning the current target of attaining 30 per cent of women on public bodies. On committees and at the regional level, the number was slightly under 30. The next goal was 40 per cent by 1995. . . .
511. Asked about the efforts of employers' and employees' organizations to increase the number of women in decision-making bodies, the representative mentioned that general activity had risen in that field. She referred to training and projects, which were aimed at both encouraging women to seek higher positions and bringing about a change in attitude. Special reference was made to the Swedish Trade Union Confederation, which had created a large network and acted successfully as a pressure group contributing to an increased visibility of the problem.

NOTE

An article in the *New York Times*, Dec. 31, 1993, at A4, *Frenchwomen Say 'It's Time to Be a Bit Utopian,'* reports that a group of French women were working toward the goal of requiring that seats in the lower house of the French Parliament be shared equally by men and women. Several weeks earlier, Michel Rocard, head of France's Socialist opposition, said that he would head his party's ticket in coming elections for the European Parliament only if it nominated an equal number of men and women as candidates.

> Olivier Duhamel, a well-known political scientist, nonetheless warned that sexual parity in representation introduced the notion of quotas, with the risk that the old and young or ethnic and religious groups might eventually also claim a right to be equally represented in Parliament.

An article in *The Economist*, March 18, 1995, p. 51, noted that the socialist candidate in the French presidential elections that were about to take place (who ultimately lost) had said that if elected he would change the electoral system to give women a quota of about 30% in local-government elections chosen by proportional representation.

> The Socialist government tried to introduce a quota system (with the share at 25%) for the local elections in 1982. The idea was ruled unconstitutional on the ground that it discriminated between men and women. To get around that, Mr. Balladur has promised to put the constitutional changes required by his proposal to a referendum this autumn.

QUESTIONS

1. Would you support a proposed law (assume its constitutionality under a state legal system and its consistency with international human rights) requiring that a stated percentage of a legislative body consist of women? Would your answer depend on the percentage—say, 15% or 30% or 50%? Would it depend on the length of time of the law's effectiveness, on the level of government involved, on other factors?

2. Would your reaction be the same for a proposed law stating percentages for a judiciary, or for heads of cabinet departments?

3. What advantages and disadvantages do you see in such laws?

PART F
CURRENT TOPICS

Drawing on the framework of Chapters 1–12, Part F examines several additional human rights topics. There is indeed no shortage of topics that could be employed with advantage to complete the coursebook. Human rights has become a pervasive concern, as in children's rights, issues of refugees and asylum or humanitarian intervention. It offers a fresh perspective on many global issues, such as the bearing of human rights on population control, environment, or religious movements.

Among such rich possibilities, we have selected three topics that explore more briefly than did Chapter 13 issues of current importance for the human rights movement: self-determination and autonomy regimes, international crimes and criminal tribunals, and the relation between development and human rights.

14

Self-Determination and Autonomy Regimes

The vital concept or right or ideal or vision of self-determination continues to play a major role in political and legal debate. Its place of honor as Article 1 of the two major Covenants itself suggests its status as a 'super rule,' as a concept that stands apart from the normal discourse of rights and directly affects political power and organization within and among states. No one can deny its deep historical significance for several phases of the human rights movement, starting with decolonization and extending to the contemporary focus on democratization.

This chapter presents in Section A a number of readings that offer different perspectives on self-determination and on the 'peoples' that enjoy a right to it. These readings trace the history of the concept, and raise issues that are as contemporary as the ongoing conflicts in the former Yugoslavia, in India, in the Americas and elsewhere. Many of those conflicts have an ethnic character, involving minorities and indigenous peoples.

Section B, for which the readings in Section A provide a necessary background, examines in more detail trends in the human rights movement toward recognition of rights of ethnic minorities and indigenous peoples, and of their members. Those materials stress the character and significance of autonomy regimes—political systems or subsystems organized within a state for purposes of fostering political participation and self-government by ethnic minorities and indigenous peoples. Together with several other human rights norms, the notion of self-determination continues to exert a strong influence on the debate over and possible forms of realization of these autonomy regimes.

A. SELF-DETERMINATION: SOME HISTORY AND COMMENTARY

HURST HANNUM, RETHINKING SELF-DETERMINATION

34 Va. J. Int. L. 1, at 3 (1993).

I. *HISTORICAL DEVELOPMENT*

A. Nationalism, Woodrow Wilson, and the League of Nations

. . .

'As an agency of destruction the theory of nationalism proved one of the most potent that even modern society has known.' Along with the physically destructive power of the machine gun, airplane, and other weapons used on a large scale for the first time, nationalist fervor hastened the disintegration of the Austro-Hungarian and Ottoman empires prior to and during World War I. The territory of the former empires required new sovereigns, and the principle of self-determination as a means of drawing new 'nationstate' boundaries became the vehicle for legitimizing the victorious powers' re-division of Europe.

Although President Woodrow Wilson was the most public advocate of 'self-determination' as a guiding principle in the post-war period, neither he nor the other Allied leaders believed that the principle was absolute or universal. Indeed, in Wilson's celebrated 'Fourteen Points' speech to the United States Congress on January 8, 1918, the phrase 'self-determination' is conspicuous by its absence, even though the speech dealt with specific territorial settlements, including the creation of independent states out of the remnants of the Austro-Hungarian and Ottoman empires.

A month later, Wilson addressed the question of self-determination directly:

> National aspirations must be respected; peoples may now be dominated and governed only by their own consent. 'Self-determination' is not a mere phrase. It is an imperative principle of action, which statesmen will henceforth ignore at their peril.
>
> . . .
>
> . . . [A]ll well-defined national aspirations shall be accorded the utmost satisfaction that can be accorded them *without introducing new or perpetuating old elements of discord and antagonism that would be likely in time to break the peace of Europe and consequently of the world.*

. . .

Modern commentators often forget the relative nature of Wilson's concept of self-determination. They also have neglected, until very recently, the 'internal' aspect of self-determination promoted by Wilson and others: democracy. Indeed, this internal aspect, the conviction that the only legitimate basis for government is the consent of the governed, provided the ultimate justification for decolonization. 'Self-determination postulates the right of a people organized in an established territory to determine its collective political destiny in a democratic fashion and is therefore at the core of the democratic entitlement.'

. . .

B. The United Nations and Decolonization

In part because of the inconsistent manner in which it was applied following the First World War, the principle of self-determination was not recognized initially as a fundamental right under the United Nations regime created in 1945. In addition, there was great reluctance to revive a concept used to justify Hitler's attempts to reunify the German 'nation.'

The 'principle' of self-determination is mentioned only twice in the 1945 Charter of the United Nations, both times in the limiting context of developing 'friendly relations among nations' and in conjunction with the principle of 'equal rights . . . of peoples.' . . .

Neither self-determination nor minority rights is mentioned in the 1948 Universal Declaration of Human Rights, although the Declaration does contain a preambular reference to developing amicable international relations. Whatever its political significance, the principle of self-determination had not attained the status of a rule of international law by the time of the drafting of the United Nations Charter or in the early United Nations era.

Under the moral and political imperatives of decolonization, however, the vague 'principle' of self-determination soon evolved into the 'right' to self-determination. This evolution was most clearly demonstrated by the General Assembly's 1960 Declaration on the Granting of Independence to Colonial Countries and Peoples ('Declaration on Colonial Independence'). Premised, *inter alia*, on the need for stability, peace, and respect for human rights, the Declaration on Colonial Independence '[s]olemnly proclaims the necessity of bringing to a speedy and unconditional end colonialism in all its forms and manifestations.' It declares that '[a]ll peoples have the right to self-determination; by virtue of that right they freely determine their political status and freely pursue their economic, social and cultural development.' It also maintains that '[i]nadequacy of political, economic, social or educational preparedness should never serve as a pretext for delaying independence.' The final paragraph reaffirms 'the sovereign rights of all peoples and their territorial integrity.'

Paragraph 6 of the declaration sets forth a fundamental limiting principle, without which one almost never (at least in United Nations forums) finds a reference to self-determination: 'Any attempt aimed at the partial or total disruption of the national unity and the territorial integrity of a country is incompatible with the purposes and principles of the Charter of the United Nations.'

The thrust of the declaration is clear: all colonial territories have the right of independence. However, a closer reading reveals uncertainties arising from varying uses of the terms 'peoples,' 'territories,' and 'countries.' Although the title of the declaration refers only to 'colonial' countries and peoples, operative paragraph 2 refers expansively to the right of '[a]ll peoples' to self-determination. . . .

. . .

The questions raised by the Declaration on Colonial Independence were also addressed ten years later by the Declaration on Principles of International Law Concerning Friendly Relations and Co-operation Among States in Accordance with the Charter of the United Nations ('Declaration on Friendly Relations'). Adopted without a vote by the General Assembly after years of negotiation, the Declaration on Friendly Relations may be considered to state existing international law. Its provisions therefore possess unusual significance for a General Assembly resolution.

As its title suggests, the [1970] Declaration on Friendly Relations addresses a wide range of issues. The section concerned with equal rights and self-determination of peoples is worth quoting at length:

> By virtue of the principle of equal rights and self-determination of peoples enshrined in the Charter of the United Nations, all peoples have the right freely to determine, without external interference, their political status and to pursue their economic, social and cultural development, and every State has the duty to respect this right in accordance with the provisions of the Charter.
>
> Every State has the duty to promote, through joint and separate action, realization of the principle of equal rights and self-determination of peoples, in accordance with the provisions of the Charter, and to render assistance to the United Nations in carrying out the responsibilities entrusted to it by the Charter regarding the implementation of the principle, in order:
>
> (a) To promote friendly relations and co-operation among States; and
>
> (b) To bring a speedy end to colonialism, having due regard to the freely expressed will of the peoples concerned; . . .
>
> The establishment of a sovereign and independent State, the free association or integration with an independent State or the emergence into any other political status freely determined by a people constitute modes of implementing the right to self-determination by that people.
>
> Every State has the duty to refrain from any forcible action which deprives peoples referred to above in the elaboration of the present principle of their right to self-determination and freedom and independence. In their actions against, and resistance to, such forcible action in pursuit

> of the exercise of their right to self-determination, such peoples are entitled to seek and to receive support in accordance with the purposes and principles of the Charter.
>
> The territory of a colony or other Non-Self-Governing Territory has, under the Charter, a status separate and distinct from the territory of the State administering it; and such separate and distinct status under the Charter shall exist until the people of the colony or Non-Self-Governing Territory have exercised their right to self-determination in accordance with the Charter, and particularly its purposes and principles.
>
> Nothing in the foregoing paragraphs shall be construed as authorizing or encouraging any action which would dismember or impair, totally or in part, the territorial integrity or political unity of sovereign and independent States conducting themselves in compliance with the principle of equal rights and self-determination of peoples as described above and thus possessed of a government representing the whole people belonging to the territory without distinction as to race, creed or colour.
>
> Every State shall refrain from any action aimed at the partial or total disruption of the national unity and territorial integrity of any other State or country.

The resolution also provides:

> All States enjoy sovereign equality. . . . In particular, sovereign equality includes the following elements:
>
> . . .
>
> (d) The territorial integrity and political independence of the State are inviolable;
>
> (e) Each State has the right freely to choose and develop its political, social, economic and cultural systems.

. . . First, the Declaration on Friendly Relations offers no definition of 'peoples.' Neither of the two purposes it sets forth suggests that self-determination is intended to provide every ethnically distinct people with its own state. In fact, the particular mention of the 'distinct' status of 'a colony or other Non-Self-Governing Territory' suggests a limited scope for the right of self-determination. Similarly, the use in the same paragraph of the singular 'people' suggests that various minorities within a territory may not enjoy the same right of self-determination as that possessed by the people as a whole.

Following previous United Nations formulations of the principle of self-determination, the Declaration on Friendly Relations places the goal of territorial integrity or political unity as a principle superior to that of self-determination: 'Nothing in the foregoing paragraphs' shall be construed to authorize or encourage 'any action' which would limit this principle. However, this restriction applies only to those states which conduct themselves 'in compliance with the principle of equal rights and self-determination of peoples as described above and [are] *thus* possessed of a government representing the whole people belonging to the territory without distinction as to race, creed or colour.' The requirement of representativeness suggests internal democracy. However, such a requirement does not imply that the only government that can be deemed 'representative' is one that explicitly recognizes all of the various ethnic, religious, linguistic, and other communities within a state. Indeed, such a state might itself be considered to violate the requirement that it represent 'the whole people . . . *without distinction* as to race, creed or colour.'

A more persuasive interpretation, consistent with the concerns of most United Nations members when the declaration was adopted in 1970, is that a state will not be considered to be representative if it formally excludes a particular group from participation in the political process, based on that group's race, creed, or color (such as South Africa or Southern Rhodesia under the Smith regime). At the very least, a state with a democratic, non-discriminatory voting system whose political life is dominated by an ethnic majority would not be unrepresentative within the terms of the Declaration on Friendly Relations.

C. The International Covenants on Human Rights

. . .

The Indian reservation to article 1 [of the ICCPR] exemplifies the view of many countries that support a restricted interpretation of 'self-determination':

> With reference to article 1 [of both covenants]. . . the Government of the Republic of India declares that the words 'the right of self-determination' appearing in [those articles] apply only to the peoples under foreign domination and that these words do not apply to sovereign independent States or to a section of a people or nation—which is the essence of national integrity.

Three states, each a former colonial power, filed formal objections to the Indian reservation. . . .

The Federal Republic of Germany 'strongly object[ed]' to the Indian reservation, stating:

> The right of self-determination . . . applies to all peoples and not only to those under foreign domination. . . . The Federal Government cannot consider as valid any interpretation of the right of self-determination which is contrary to the clear language of the provisions in question. It moreover considers that any limitation of their applicability to all nations is incompatible with the object and purpose of the Covenants.

. . .

II. *THE MEANING OF SELF-DETERMINATION*

Although it is debatable whether the right of self-determination is *jus cogens*, self-determination has undoubtedly attained the status of a 'right' in international law. . . .

. . .

The various texts discussed above and the *travaux préparatoires* of the covenants do not establish that the right of self-determination, *defined as a unilateral right to independence*, was intended to apply outside the context of decolonization. As noted above, self-determination has meant at least decolonization since 1945. However, when addressing self-determination claims based on ethnicity or nationalist sentiment, one must recognize the shift from the territorially based *right* of self-determination developed by the United Nations in the context of decolonization to the ethnic–linguistic–national *principle* of self-determination advocated by Wilson and others in 1919. The difference is not only semantic. It reflects a fundamental limit on the definition that self-determination has acquired during the past four decades. . . .

Despite the apparently absolute formulation in various United Nations resolutions and the two international covenants on human rights, self-determination has never been considered an absolute right to be exercised irrespective of competing claims or rights, except in the limited context of 'classic' colonialism. . . .

. . .

This does not mean that other aspects of the right of self-determination are also so limited. The covenants' description of the right of self-determination as a right of 'all peoples' . . . cannot be ignored. Both the right of a people organized as a state to freedom from external domination and the right of the people of a state to a government that reflects their wishes are essential components of the right of self-determination. These rights have universal applicability, and the statement that '[n]o State has accepted the right of *all* peoples to self-determination' is correct only if one equates self-determination exclusively with secession or independence.

. . .

Self-determination is also relevant to the matrix of human rights law which has developed over the past four decades, including specific rights applicable to minorities and indigenous peoples. Defining self-determination as self-government is consistent with early United Nations formulations, including the Charter, and its implications are only now being fully considered in the context of autonomous and other domestic constitutional arrangements.

. . .

CHRISTIAN TOMUSCHAT, SELF-DETERMINATION IN A POST-COLONIAL WORLD

in Tomuschat (ed.), Modern Law of Self-Determination 1 (1993), at 2.

According to the applicable texts, self-determination could never be considered an exclusive right of colonial peoples. Even General Assembly resolution 1514 (XV) [the 1960 Declaration on Colonial Independence], in spite of its specific object of bringing about speedy decolonization, provides that self-determination is a right of all peoples. The two International Covenants on human rights use identical language. . . .

In spite of the clarity of the legal position, until the symbolic crumbling of the Berlin wall in 1989 self-determination was mainly invoked and relied upon by colonial peoples eager to gain full control of their own destiny. With the fall of communism in Eastern Europe, however, a new era seems to have begun. . . . What can be witnessed today, however, are mostly moves in the opposite direction. Ethnic groups—to use a neutral term—within many States invoke the right to self-determination, claiming political rights which range from internal autonomy to a right of secession pure and simple. Increasingly, self-determination, instead of uniting peoples, is becoming a divisive force.

It is not only the former communist States like the Soviet Union and Yugoslavia that are affected by this new development. In the Americas, the voice of indigenous peoples is growing ever louder. Yielding to their moral and political pressure, Canada has recently made important concessions to its indigenous communities. In Latin America, too, today's political structures, the result of 500 years of Spanish and Portuguese domination, are being challenged. Similarly, in spite of its preparedness to let the Baltic Republics, Belarus, the Ukraine as well as the southern republics leave the country, the Soviet Union, now reduced to Russia, will not be able to escape claims of the populations in Siberia, which were annexed to Russia through the Tsarist policies of conquering the large areas East of the Ural, to establish themselves as political entities with a great measure of self-government and possibly true independence. Quite visibly, self-determination is becoming a tool for attempts to revise historical developments that have extended not only over decades, but centuries. The big question is whether such processes can be directed and regulated by a principle of international law like self-determination. After all, international law, although requiring compliance by States which are its main addressees, is a creation by those same States which, through its rules, seek to guarantee stability in their mutual relations. If on the contrary international law should aspire critically to assess the *raison d'être* of States, including judgments to the effect that a State must disintegrate, it would become an instrument of revolutionary change, almost the opposite of a legal rule.

Hitherto, the international community has been extremely reluctant to accept that aspect of self-determination. The civil war in Nigeria, originating from an upheaval of the Ibo community in the Biafra region against the central government, was largely ignored by the United Nations. Similarly, the United Nations avoided to take a stand on the secession of East Pakistan from the State of Pakistan. Only after Bangladesh had emerged and consolidated itself, did the United Nations take note of that fact and admitted the new State. In the case of Cyprus, the proclamation of a Republic of Northern Cyprus was formally condemned by the Security Council and declared legally invalid. This reluctance *vis-à-vis* claims to break away from existing States is also reflected in the basic texts on self-determination. General Assembly resolution 2625 (XXV) [the 1970 Declaration on Friendly Relations] sounds a formal warning against claims imperilling territorial integrity (paragraph 6), and the comment of the Human Rights Committee on the scope of Article 1 of the International Convenant on Civil and Political Rights in absolutely silent on a right to secession.

. . .

Criteria Justifying Claims for Self-Determination Vis-à-Vis an Established State

In pursuing further this line of thought, the student encounters two questions, which follow one another in a logical sequence. The first question is whether it may be at all possible, against the background of present day international law, to define certain criteria for legitimizing claims directed against the political and territorial integrity of an existing State. One hundred years ago, even to raise such a question would have been considered preposterous or nonsensical. States were the only actors on the international stage. Their constitutional processes were considered a purely domestic matter. The veil of sovereign statehood could not be pierced. The proper role of international law was confined to regulating relationships among States. Its rules were not allowed to penetrate into matters subject to municipal jurisdiction, in particular relationships between a State and its citizens.

With the emergence of international human rights law, in particular, the traditional picture has changed dramatically. The consolidation of this new branch of international law amounts to a general recognition that States are not objectives in and by themselves and that, conversely, their finality is to discharge a task incumbent upon them in the service of their citizens. In other words, States are no more sacrosanct. Their existence is not exempt from challenge, even on a legal plane. Rather, they have a specific *raison d'être*. If they fundamentally fail to live up to their essential commitments they begin to lose their legitimacy and thus even their very existence can be called into question.

It is with a great measure of caution that one should approach answering the question as to what situations exhibit the degree of gravity required to give rise to a right of self-determination to the benefit of a group within an existing State. On one hand, it is obvious that any State is under a basic obligation to protect the life and the physical integrity of its citizens. Therefore, if a State machinery turns itself into an apparatus of terror which persecutes specific groups of the population, those groups cannot be held obligated to remain loyally under the jurisdiction of that State. Genocide is the ultimate of all international crimes. Any government that engages in genocide forfeits its right to expect and require obedience from the citizens it is targeting. If international law is to remain faithful to its own premises, it must give the actual victims a remedy enabling them to live in dignity. It is a matter of common knowledge that, in its elaboration on self-determination, the Friendly Relations Declaration of the General Assembly would appear to go even further when stating that the principle of national unity and territorial integrity may have to yield if the State concerned is not possessed of a government 'representing the whole people belonging to the territory without distinction as to race, creed or colour'. This formulation seems to be somewhat too loose if it is intended to sanction a right of secession. In fact, secession can be only a step of last resort and should not be granted lightly as a remedy.

. . .

MARTTI KOSKENNIEMI, NATIONAL SELF-DETERMINATION TODAY: PROBLEMS OF LEGAL THEORY AND PRACTICE

43 Int. & Comp. L.Q. 241 (1994), at 245.

In the first place, national self-determination acts as a *justification* of a State-centred international order. . . . As the 1970 Friendly Relations Declaration[1] puts it, all peoples have the right freely to determine, without external interference, their political status. A people choosing to live as a State has the perfect right to do so and in such case 'every State has the duty to respect this right'. Without a principle that entitles—or perhaps even requires—groups of people to start minding their own business within separately organised 'States', it is difficult to think how statehood and everything that we connect with it—political independence, territorial integrity and legal sovereignty—can be legitimate. As a background principle, self-determination expresses the political phenomenon of State patriotism and explains why we in general endow acts of foreign States with legal validity even when we do not agree with their content and why we think there is a strong argument against intervening in other States' political processes.

This idea of national self-determination as a patriotic concept, a justification of statehood, has not always been apparent. Much of the classical positivist writing has simply accepted States as the factual foundation of international law. The need to look behind States—into self-determination—however, becomes necessary when statehood itself is or becomes uncertain.

. . .

[1] [Eds.] see p. 974, *supra*.

. . . A Hungarian writer has claimed that national self-determination 'is the most important basic ideological principle of the 1990s'. But it is more: it is a legal-constitutional principle that claims to offer a principal (if not the only) basis on which political entities can be constituted, and among which international relations can again be conducted 'normally'. The centrality of the State to the political order becomes comprehensible only if we regard it as the formal, political shell for which nationhood provides the substance.

But of course there is another sense of national self-determination which far from supporting the formal structures of statehood provides a *challenge* to them. According to this sense, true self-determination is not expressed in the normal functioning of existing participating processes and in the duty of other States not to interfere but in the existence and free cultivation of an authentic communal feeling, a togetherness, a sense of being 'us' among the relevant group. If, in extreme cases, this may be possible only by leaving the State, then the necessity turns into a right.

. . .

National self-determination, then, has an ambiguous relationship with statehood as the basis of the international legal order. On the one hand, it supports statehood by providing a connecting explanation for why we should honour existing *de facto* boundaries and the acts of the State's power-holders as something other than gunman's orders. On the other hand, it explains that statehood *per se* embodies no particular virtue and that even as it is useful as a presumption about the authority of a particular territorial rule, that presumption may be overruled or its consequences modified in favour of a group or unit finding itself excluded from those positions of authority in which the substance of the rule is determined.

The extraordinary difficulties into which an attempt at a *consistent* application of the principle leads stem from the paradox that it both supports and challenges statehood and that it is impossible to establish a general preference between its patriotic and secessionist senses.

GREGORY FOX, SELF-DETERMINATION IN THE POST-COLD WAR ERA: A NEW INTERNAL FOCUS?

16 Mich. J. Int. L. 733 (1995), at 743.

. . .

The dissolutions of the Soviet Union and the Socialist Federal Republic of Yugoslavia, despite giving birth to a myriad of new states ultimately recognized by the international community, have made only uncertain contributions to a substate conception of the self. The strongest *opinio juris* that could have emerged from either break-up would have been statements of an entitlement to self-determination before *the fact* of independent statehood was clearly evident. Established conceptions of the self have followed this pattern. Resolution 1514 [the 1960 Declaration on Colonial Independence] was such a general statement regarding colonial territories, as was the General Assembly's recognition of SWAPO, the ANC, and the PLO as 'legitimate' representatives of peoples well before independence (or majoritarian elections) were presented to the Assembly as a *fait accompli*. Similarly, the OAU's decision to seat a POLISARIO delegation in November 1985 represented a statement that the people of the Western Sahara held *rights against* those actively denying their aspirations toward autonomy—notably Morocco. Recognition after statehood has been achieved, or after the state resisting independence finally acquiesces, does not necessarily affirm a prior right to seek independence. It may simply constitute a recognition by states or international organizations that according to the prevailing declarative theory, a new state has come into existence and must be dealt with as such.

International reaction to the break-up of the Soviet Union—and in particular the Baltic states, whose departure precipitated its dissolution—generally fell into this second weaker category. While the United States had, in principle, never recognized the incorporation of the Baltics into the Soviet Union, this was mostly a status of symbolic significance. In the aftermath of the August 1991 coup, President Bush announced not that preexisting relations with the Baltics would continue in some heightened fashion but that the United States was 'prepared immediately to *establish* diplomatic relations with their governments.' More importantly, prior to President Yeltsin's decree of August 24, 1991 recognizing the independence of Latvia and Estonia (Russia had recognized Lithuania in 1990), only Iceland had established diplomatic relations with a Baltic state, and that was with Lithuania. The European Community waited until August 27 to call for the establishment of relations. The United States extended recognition on September 2. The CSCE and the United Nations waited still longer—until after independence had been affirmed by the State Council of the Soviet Union on September 6—to admit the Baltics to membership.

Recognition of the former Yugoslav republics unfolded in a similar fashion. The unraveling of the Socialist Federal Republic of Yugoslavia (SFRY) began formally on September 27, 1990 when the Slovenian Parliament declared it would no longer recognize federal legislation as binding. Slovenes voted overwhelmingly for independence in a referendum on December 23, as did Croatia on May 19, 1991, Macedonia on September 9, and Bosnia (in a disputed vote boycotted by Bosnian Serbs) on October 14. Warfare had been raging since June 1990 when federal troops attacked the provisional Slovenian militia. By December 7, 1991, the cohesion of the federal structures had deteriorated to such an extent that the Badinter Commission of the European Community determined that the governmental organs of the SFRY 'no longer meet the criteria of participation and representativeness inherent in a federal State' and that, as a result, 'the Socialist Federal Republic of Yugoslavia is in the process of dissolution.'

Yet the response of the international community up through the time the Badinter Commission made its finding (and for a short time thereafter) was, by and large, to work at holding the old federal structures together. . . . The United States repeatedly opposed early recognition of the republics, a position it maintained through mid-December 1991. In December the Coordinating Bureau of the Non-Aligned Countries denounced 'all attempts aimed at undermining the sovereignty, territorial integrity and international legal personality of Yugoslavia.' Also in December—at a time when Germany had begun making clear its intention to offer early recognition—Secretary-General Boutros-Ghali urged forbearance until an overall peace settlement could be reached. Germany and Italy recognized Slovenia and Croatia on December 23; other states and international organizations followed over the next five months, culminating in the admission of Croatia, Slovenia, and Bosnia-Herzegovina to membership in the United Nations on May 22, 1992.

While this attenuated process might be read as affirming these states' right to secession, the repeated attempts to achieve peace by discouraging fragmentation do not suggest that was a motive in extending recognition. Formal recognition came at a time when any right to secede (if it indeed existed) had already been exercised. The establishment of relations is more plausibly attributable to states' realistic assessment that diplomatic relations with the republics could only be carried out through their newly established governments (most of which had been or would soon be elected). Alternatively, some states may have judged that the republics would be better protected against external aggression if they were Member States of the United Nations. In either case, there was virtually no prospect that the old federal system might be resuscitated. The former republics were, in fact, states and had to be dealt with as such. Given such practical necessities of the moment, coupled with earlier efforts to forestall recognition, the ultimate decision to treat the republics as sovereign implies little about their independence having been achieved pursuant to a legal right.

. . .

It may be that in attempting to support the cohesion of multiethnic states such as the Congo, Nigeria, the Soviet Union, or the SFRY, the international community is now fighting a losing battle against centrifugal nationalist forces. Yet the normative assumptions underlying that fight continue to discourage auto-defined conceptions of the self that are subjective in all but name.

. . .

FREDERIC KIRGIS, JR., THE DEGREES OF SELF-DETERMINATION IN THE UNITED NATIONS ERA

88 Am. J. Int. L. 304 (1994), at 306.

. . .

[The disclaimer in the 1970 Declaration on Friendly Relations, p. 974, *supra*] referred only to a government representing the whole people belonging to the territory without distinction as to race, creed or color. The disclaimer was reiterated in the Vienna Declaration emanating from the 1993 UN World Conference on Human Rights, with one significant change. The Vienna Declaration exempted only 'a Government representing the whole people belonging to the territory without distinction *of any kind*.' These disclaimers are a far cry from the General Assembly's formulation in 1960, when it said: 'Any attempt aimed at the partial or total disruption of the national unity and the territorial integrity of a country is incompatible with the purposes and principles of the Charter of the United Nations.'

Of course, these are nonbinding instruments. Nevertheless, they purport to, and probably do, reflect an *opinio juris*. In the human rights field, a strong showing of *opinio juris* may overcome a weak demonstration of state practice to establish a customary rule. . .

An arguable, limited right of secession is only one of the numerous faces of self-determination. Mentioned below are the more prominent ones that have appeared. Of course, some of them remain quite controversial because of disagreement either over what is meant by 'peoples' or over what is meant by self-determination itself. The many faces include:

(1) The established right to be free from colonial domination, with plenty of well-known examples in Africa, Asia and the Caribbean.

(2) The converse of that—a right to remain dependent, if it represents the will of the dependent people who occupy a defined territory, as in the case of the Island of Mayotte in the Comoros, or Puerto Rico.

(3) The right to dissolve a state, at least if done peacefully, and to form new states on the territory of the former one, as in the former Soviet Union and Czechoslovakia. The breakup of the former Yugoslavia except for Serbia and Montenegro might even be considered an example of this, after the initial skirmish in Slovenia ended and the Yugoslav federal forces ceased operating as such in Croatia and Bosnia–Hercegovina. The later fighting in Croatia and Bosnia–Hercegovina could be seen as efforts not so much to hold the old state of Yugoslavia together as to define the territories and ethnic composition of the new states, including possible new states within Bosnia–Hercegovina.

(4) The disputed right to secede, as in the case of Bangladesh and Eritrea.

(5) The right of divided states to reunite, as in Germany.

(6) The right of limited autonomy, short of secession, for groups defined territorially or by common ethnic, religious and linguistic bonds—as in autonomous areas within confederations.

(7) Rights of minority groups within a larger political entity, as recognized in Article 27 of the Covenant on Civil and Political Rights and in the General Assembly's 1992 Declaration on the Rights of Persons Belonging to National or Ethnic, Religious and Linguistic Minorities.

(8) The internal self-determination freedom to choose one's form of government, or even more sharply, the right to a democratic form of government, as in Haiti.

One must ask which of these is actually a right under international law, and which just an aspiration of some groups or putative governments. Clearly, the right to be free from alien colonial control is an established rule of international law. One cannot categorically say the same about the other manifestations. Their juridical status varies. Nevertheless, we can make some headway toward evaluating each claim by borrowing some of the tools of social scientists. That is, we can try to identify key variables that reduce the vast complexity of the problem to a manageable form and can serve as rough predictors for normative assessment of claims as they are made.

The key variables can be found in the General Assembly's 1970 Declaration of Principles of International Law concerning Friendly Relations and in the 1993 Vienna Declaration. . . . Dismemberment is at the end of a scale of claims, ranging from modest to extremely destabilizing, that includes all of those listed above. Moreover, as there are degrees of claim, there are degrees of representative government, with absolute dictatorship and all-inclusive democracy at the opposite extremes.

One can thus discern degrees of self-determination, with the legitimacy of each tied to the degree of representative government in the state. The relationship is inverse between the degree of representative government, on one hand, and the extent of destabilization that the international community will tolerate in a self-determination claim, on the other. If a government is at the high end of the scale of democracy, the only self-determination claims that will be given international credence are those with minimal destabilizing effect. If a government is extremely unrepresentative, much more destabilizing self-determination claims may well be recognized.

In this schema, a claim of right to secede from a representative democracy is not likely to be considered a legitimate exercise of the right of self-determination, but a claim of right by indigenous groups within the democracy to use their own languages and engage in their own noncoercive cultural practices is likely to be recognized—not always under the rubric of self-determination, but recognized nevertheless. Conversely, a claim of a right to secede from a repressive dictatorship *may* be regarded as legitimate. Not all secessionist claims are equally destabilizing. The degree to which a claimed right to secede will be destabilizing may depend on such things as the plausibility of the historical claim of the secessionist group to the territory it seeks to slice off.

. . .

To summarize: The right of self-determination may be seen as a variable right, depending on a combination of factors. The two most important of these seem to be the degree of destabilization in any given claim, taking into account all the circumstances surrounding it, and the degree to which the responding government represents the people belonging to the territory. If a government is quite unrepresentative, the international community may recognize even a seriously destabilizing self-determination claim as legitimate.

NOTE

Compare the following comments of Rodolfo Stavenhagen; [2]

> It does not help matters that 'self-determination' means different things to different persons. It is, as one international lawyer asserts, 'one of those unexceptionable goals that can be neither defined nor opposed'. Is it then, a goal, an aspiration, an objective? Or is it a principle, a right? And if the latter, is it only a moral and political right, or is it also a legal right? Is it enforceable? Should it be enforceable? Or is it none of these, or all of these at the same time, and more? . . . [S]elf-determination has become, indeed *is*, a social and political fact in the contemporary world, which we are challenged to understand and master for what it is: an *idée-force* of powerful magnitude, a philosophical stance, a moral value, a social movement, a potent ideology, that may also be expressed, in one of its many guises, as a legal right in international law. Whereas for some the 'self' in self-determination can only be the singular, individual human being for others the right of collective self-determination, that is, the claim of a group of people to choose the form of government under which they will live, must be treated as a myth in the Lévi-Straussian sense (that is, as a blueprint for living); not as an enforceable or enforced legal, political or moral right.

B. AUTONOMY REGIMES

We here turn to rights and protections of ethnic minorities and indigenous peoples and members of each of them. In particular, Section B examines political arrangements relevant to both kinds of groups that the materials refer to

[2] Self-Determination: Right or Demon?, IV Law and Society Trust (Issue No. 67, November 1993), at 12.

as autonomy regimes (defined below). Such regimes may be bargained for, created and governed by both ethnic minorities and indigenous peoples. The materials explore the degree to which international law has recognized or is moving toward recognition of a 'right' of either or both of these kinds of groups or collective entities to autonomy regimes.

Although there are important similarities between the categories of ethnic minorities and indigenous peoples, there are also significant differences. It has been important for strategic and other reasons for indigenous peoples to emphasize these differences and therefore to argue for a separate regulatory regime under international law. In fact, the UN has addressed the rights and protections of ethnic minorities and indigenous peoples in two separate contemporary standard-setting processes; one led to a 1992 declaration, while the other is now in mid-stream en route toward a declaration. The materials below follow this division.

As described at p. 96 *supra*, the special regimes for a number of minorities in Central and East Europe that grew out of World War I fell into disfavor because of their ineffectiveness and the malign uses to which some of their concepts were put. As a consequence, and in the belief that observance by states of individually based norms would solve the historical problems of oppression and brutality that many minorities had confronted, the UN Charter, the Universal Declaration on Human Rights and the International Covenant on Civil and Political Rights paid scant attention to minorities as such or (subject to the major exception of the self-determination clauses, and to the very conception of the state) to collective rights as such. Not until 1992, when the General Assembly adopted the Declaration on Minorities, p. 1001, *infra*, did the human rights movement produce a universal instrument dedicated to the problems and rights of minorities and their members.

In a general sense, the full range of the classical and individually based human rights norms becomes relevant to the rights and protections of minorities and indigenous peoples, from rights to personal security or due process or equal protection[3] to rights of speech or association and the right to political participation. Three ICCPR provisions, however, are particularly pertinent to the discussion in this section of autonomy regimes: Article 1 on self-determination, the subject of Section A; Article 25 on political participation, discussed at pp. 661–72, *supra*; and Article 27, discussed in the Comment below.

1. ETHNIC MINORITIES

The following materials concern minority groups that are typically ethnic, racial, religious, linguistic, or national origin in character. The text below uses the term 'ethnic' as a shorthand reference to all such minorities, whatever their

[3] See, for an earlier illustration in a minority regime after World War I, the advisory opinion of the Permanent Court of International Justice in Minority Schools in Albania, p. 89, *supra*.

distinctive characteristic. Hence that term embraces groups as diverse as Muslims of North African background in France, blacks and Jews in the United States, Gypsies in Hungary, Kurds in Iraq or Turkey, Russians in Georgia, Tamils in Sri Lanka, Copts in Egypt, and Turks in Germany. Frequently these ethnic minorities bear two or more of the defining characteristics, such as Basques in Spain, the Francophone population in Canada, or Palestinians in Israel.

To characterize a group within a state as a minority tells us little about its political, economic or cultural situation. Relations with a dominant majority or with other minority groups may be amicable or hostile. The minority may be well integrated, indeed moving towards a voluntary assimilation, or may be both rejected by dominant groups and intent on maintaining its own distinctive character. The differences in religion, culture and so on between majority and minority may be formal and inconsequential, or dramatic and involving basic world-views.

Two world wars, and many savage conflicts during the last half century of the human rights movement, have made evident the ways in which ethnic conflicts deeply affect international politics. Third-party countries become involved in such conflicts, particularly when their co-religionists or other groups with which they identify are involved—India in Sri Lanka, Turkey and Greece in Cyprus, several Muslim states in Bosnia, and so on. Conflicts spill across frontiers, as the violence in the Former Yugoslavia continues to threaten to do. Refugee flows involve other countries in serious ways. Universal or regional security systems observe and sometimes act, effectively or ineffectively. Hence steps to alleviate tensions between ethnic minorities and a state, or among such minorities, and to assure a necessary minimum of protection for minorities, would contribute not only to the well being of countless individuals but also to international peace and stability.

These materials are concerned with the degree to which, and the ways in which, minority issues have become 'internationalized'—that is, are now subject to regulation by international law and particularly by human rights norms. What indeed are 'minorities' within the discourse of international law? No authoritative instrument like a treaty or declaration imposes a definition. Hence the term remains to some degree politically disputed, a subject of debate among scholars, spokespersons for minorities or indigenous peoples, and governments. There appears to be a consensus over two broad parts of a definition, sometimes expressed in terms of objective and subjective criteria. *Objectively*, the group at issue must constitute a non-dominant minority of the population (usually a relatively small percentage of the population, even if a substantial number of people), and its members must share distinctive characteristics such as race, religion or language. Some of those characteristics will be natural, immutable; others (subject to cultural constraints) may be open to change. *Subjectively*, (most) members of this group must hold or evidence a sense of belonging to the group, and evidence the desire to continue as a distinctive group. (See discussions in the Reports of Asbjørn Eide, fn. 8, p. 1004, *infra.*)

One caveat should be noted. These materials are concerned only with the universal human rights system. Minority issues have figured as well in regional settings, particularly within the system of the Council of Europe. In some respects, the instruments growing out of Europe—particularly through the processes of the Conference on Security and Cooperation in Europe—are bolder than the analogoues instruments within the UN, in term of both their norms and institutional development. The problems are most intense in a global setting, where profound differences among groups, cultures and political systems burden the task of developing common standards.

ROSALYN HIGGINS, COMMENTS

in Catherine Brölmann, R. Lefeber & M. Zieck (eds.), Peoples and Minorities in International Law 29 (1993), at 30.

[Higgins contends that 'it is one of the great myths that the UN Charter provided for and required self-determination in the form in which it evolved.' The concept developed in ways 'quite *un*intended' by the Charter, whose Articles 1(2) and 55 referred in context only to 'rights of the peoples of one state to be protected from interference by other states or governments.' But ideas develop, and the growing identification of self-determination with independence from colonial rule 'has long since been regarded on all sides as legitimate and desirable.'

Higgins then inquires into what the concept has come to mean since the 1960 Declaration on Colonial Independence, p. 973, *supra*.]

. . . Can terms be invoked to mean whatever the user finds it convenient for them to mean? I am aware that I here approach the dangerous waters of linguistics and the current controversies on deconstructionism. My position is that I believe that legal ideas develop, and that that is proper. But that is not to say that they can mean simply whatever those using them want them to mean. The degree of change from a given normative understanding, that we should as lawyers tolerate, must depend in large part upon the policy implications.

. . .

. . . The concept of the self-determination of peoples requires us to answer not only what self-determination means, but the 'peoples' to whom it applies. Minority rights are the rights held by minorities. . . . But the right of self-determination is the right of *peoples*. The Political Covenant gives entirely discrete rights to minorities on the one hand (minority rights, as elaborated in Article 27) and to peoples (self-determination rights, as provided in Article 1). One cannot—though many today try, lawyers as well as politicians—assert that minorities are *peoples* and that therefore minorities are entitled to the right of self-determination. This is simply to ignore the fact that the Political Covenant provides for two discrete rights. It also, more insidiously, denies the right of self-determination to those to whom it was guaranteed—the peoples of a state *in their entirety*.

. . .

It follows that international law provides no right of secession, in the name of self-determination, to minorities. . . .

Quebec has no legal right to secede on the alleged ground that it is composed of a linguistic minority within a state in which the majority are of a different linguistic grouping. Francophone Quebecois *are*, of course, fully entitled to use their own language. . . . Croatia and Bosnia–Hercegovina had no automatic legal right to secede by invocation of a right of self-determination of ethnic or religious groups who formed a minority in the larger federal state of Yugoslavia. The perceived need of secession is understandable when minorities are denied their rights as minorities or where they cannot participate, as part of the entire peoples of a country, in the political and economic life of the country. But I am less sure . . . that even this entails a legal right to secession, in contra-distinction to a compelling political imperative. Where I certainly agree . . . is that, even if international law does not authorize secession, it will eventually recognise the reality once it has occurred and been made effective.

. . .

So I return to the importance of using concepts with some care. Looking at the ideas behind this current battle of the words, I am of course very aware that there are those who use the armoury of words in full knowledge of what they do. What, they ask, is wrong with secession/self-determination for every minority that wants it? Why shouldn't this be their right? And what is so wrong with the prospect of a world of two thousand states? I can only give my own answer. . . . There is, quite simply, no end to the disintegrative processes that are encouraged. The facts relating to Bosnia–Hercegovina bear witness. A majority of its inhabitants favours independence, including 750,000 Croats resident there as well as its 1.9 million Muslim Slavs. But most of its 1.4 million Serbs oppose the republic's independence. And if the regions where 1.4 million Serbs live are, through the use of force by Serbia, integrated into Greater Serbia, within it there will be disenfranchised Muslims—can they then, as a further minority, now in turn secede? And what of the few Croatians who may still live in those areas? And so on, *ad infinitum*.

Because I believe in diversity, and plurality, and tolerance, and mutual respect, I favour multilateralism and multinationalism. The use of force is appalling, indiscriminate barbarity unforgivable. But the move to uninational and unicultural states that constitutes postmodern tribalism is profoundly illiberal. The attempt to legitimate these tendencies by the misapplication of legal terms runs the risk of harming the very values that international law is meant to promote.

COMMENT ON AUTONOMY REGIMES

Forms and Types

Autonomy regimes refer to governmental systems or subsystems within a state that are directed or administered by a minority or its members. They take many forms. Three general types figure in this discussion, each subject to variations that range from strong to weak regimes. In each case, a distinguishing mark of the regime is that it depends on legal authorization—customary, statutory, or constitutional law. In most instances, the state's formal legal system defines the powers and scope of the regime. That is, autonomy regimes are instituted in law, and those governing or administering them exercise a form of governmental power.

The first type, personal law regimes, provides that members of a defined ethnic group (that could include both majority and minority groups) will be governed with respect to matters of personal law—marriage, divorce, adoption, perhaps inheritance, and so on—by a law distinctive to it, usually religious in character. Thus all members of a religious community—Jews, Muslims and members of different Christian communities in Israel; Hindus and Muslims in India—may be subject to a personal law applied by religious courts. Depending on the state, members of such groups may or may not be able to 'opt out' by selecting a nation-wide secular law.

A second type of regime has a territorial organization, and hence is plausible only when the ethnic minority at issue is regionally concentrated—as is true, for example, of the Tamil minority in Sri Lanka or the Kurdish minority in several states. This organization may take the form of component part of a federalism, or of a regional government to which powers have been devolved within a unitary state. The ethnic minority exercises one or another degree of political control over the territory and to that degree governs its own affairs. Self-government including regional elective government can extend to matters ranging from regulation of natural resources or the tax system to control of regional schools. Contemporary illustrations include Catalonia in Spain, or the states in India.

The third type, power-sharing regimes, assures that one or several ethnic groups will benefit from a particular form of participation in governance, in economic activity, or in other fields. Like personal law regimes, it does not demand that the ethnic minority be regionally concentrated. Power sharing can assume many forms. It may affect the composition of the national legislature—for example, through provision that members of an ethnic minority are entitled to elect a stated percentage of legislators through the use of separate voting rolls specific to the minority. It may require approval by a majority of the legislative representatives of a minority group before certain changes—say, a change in constitutional provisions giving the minority group stated protections—can be made. A certain percentage of the civil service, or of the army officer corps, or of cabinet positions may be reserved for members of the minority. Belgium and Lebanon illustrate relatively successful and failed power-sharing arrangements.

Article 27 of the ICCPR

Article 27 has become the preeminent human rights norm for discussion of rights and protections of minorities and their members. It provides:

> In those States in which ethnic, religious or linguistic minorities exist, persons belonging to such minorities shall not be denied the right, in community with the other members of their group, to enjoy their own culture, to profess and practise their own religion, or to use their own language.

Note several aspects of this provision: (1) It refers to 'minorities' rather than, as in ICCPR Article 1, to 'peoples.' (2) It protects 'persons belonging to' minorities rather than the minorities themselves. (3) Nonetheless, the right is to be exercised 'in community with the other members' of a minority.

In 1994, the ICCPR Committee on Human Rights adopted *General Comment No. 23 on Article 27 of the Covenant*.[4] A summary of several of its provisions follows:

> Para. 3.2: Enjoyment of rights under Article 27 'does not prejudice the sovereignty and territorial integrity of a State party.' Nonetheless, aspects of rights of individuals protected under that article, such as enjoyment of a particular culture, 'may consist of a way of life which is closely associated with territory,' particularly for members of indigenous communities.
>
> Para. 6.1: A state party is 'under an obligation to ensure that the existence and the exercise of [the right declared by Article 27] are protected against their denial or violation. Positive measures of protection are, therefore, required . . . also against the acts' of non-state actors.
>
> Para. 6.2: Although the rights protected are individual, they depend on the ability of the minority group to maintain its culture. 'Accordingly, positive measures by States may also be necessary to protect the identity of a minority' and the rights of its members. Such positive measures must respect the non-discrimination clauses of the Covenant with respect to treatment among minorities and between minorities and the rest of the population. 'However, as long as those measures are aimed at correcting conditions' impairing enjoyment of the guaranteed rights, they may 'constitute a legitimate differentiation . . . [if] based on reasonable and objective criteria.'

[4] CCPR/C/21/Rev. 1/Add,5. 26 April 1994. General Comments are adopted under Article 40(4) of the Covenant. See pp. 522–35, *supra*.

Para. 7: Cultural rights under Article 27 extend to ways of life 'associated with the use of land resources, especially in the case of indigenous peoples. . . . The enjoyment of those rights may require positive legal measures of protection and measures to ensure the effective participation of members of minority communities in decisions which affect them.'

Para. 8: None of the protected rights under Article 27 may be exercised 'to an extent inconsistent with the other provisions of the Covenant.'

Autonomy Regimes: Individual or Collective Rights

Most rights declared in the human rights instruments have an individual character. Many of them are at the same time germane, indeed essential, to the formation of groups. Thus the rights to advocacy and association are vital to the organization of interest groups. Moreover, certain rights such as those identified in Article 27 are inherently collective, such as the right of members of minorities to use their own language or practice their own religion. Indeed, Article 27 states that such rights may be exercised 'in community with the other members' of a linguistic, religious or other ethnic group.

Other types of rights that are characteristically expressed in individual terms, such as the right to equal protection in ICCPR Articles 2(1) and 26, also have a group aspect. The prohibited governmental conduct amounts to the disadvantaging of an individual because of that individual's group characteristic or identity. It is the individual's link to the group that provides the very occasion for discrimination. Moreover, the entire group benefits from the protection against discrimination accorded to any one of its members.

What then would be the nature of a 'right' to an autonomy regime? If we assume that such a right exists—say, to extensive regionally concentrated self-government, or to a form of power sharing—it would have a markedly different character even from the individual rights noted above that have some collective or group association. Autonomy rights are unmistakably collective, in the sense that they can be exercised *only* by the group—by its spokespersons or representatives, however they are selected—and cannot be reduced to or expressed through rights of its members. The group itself would have bargained for the arrangements that are confirmed in law for some form of autonomy regime. Officials or official institutions of the group—its clergy, a regional government, an ethnic political party—implement and administer autonomy schemes.

HENRY STEINER, IDEALS AND COUNTER-IDEALS IN THE STRUGGLE OVER AUTONOMY REGIMES FOR MINORITIES

66 Notre Dame L. Rev. 1539 (1991), at 1547.

[This article examines autonomy regimes, in the sense described in the preceding Comment, from the perspective of norms and ideals in the human rights movement. The excerpts below assess such regimes against those ideals as expressed primarily in the Universal Declaration and in the International Covenant on Civil and Political Rights.]

IV. Ideals and Counter-Ideals

. . . We now inquire whether [autonomy] schemes tend to realize vital ideals pervasive to the entire human rights movement, and paradoxically threaten at the same time to subvert those very ideals.

Let us first consider how autonomy schemes reinforce a basic human rights ideal. The Universal Declaration and Civil–Political Rights Covenant accept and, indeed, encourage many forms of diversity. They insist on respect for difference, an insistence expressed only in part through the particular attention given ethnic minorities in article 27. The value placed on the survival (and creation) of diversity in cultural, religious, political, and other terms permeates human rights law, which evidences throughout its hostility to imposed uniformity.

. . .

Other rights declared in basic human rights instruments complement the ideal of equal respect and confirm the value placed on diversity. Everyone has a right to adopt 'a religion or belief of his choice' and has freedom 'either individually or in community with others and in public or private' to manifest belief or religion in practice and teaching. Rights to 'peaceful assembly' and 'freedom of association with others,' in each case qualified by typical grounds for limitation like public order or national security, further commit the human rights movement to the protection of people's ongoing capacity to form, develop, and preserve different types of groups.

We have noted that such provisions of the Universal Declaration and Civil–Political Rights Covenant are expressed in terms of individuals' freedom and choice of action, but that their individual and collective characters are nonetheless inextricably linked. This must be so. Groups and communities, not isolated individuals, transmit culture from one generation to the next. They embody and give significance to cultural and social differences in a society. Hence we see the link between autonomy regimes and an ideal of maintaining diversity. Since those regimes protect, indeed entrench, diversity in group terms, they must constitute an effective means to realize this fundamental human rights ideal.

. . .

By valuing diverse cultural traditions, and by its related protection of groups, human rights law evidences what must be a basic assumption—namely that differences enrich more than endanger the world. They contribute to a fund of human experience on which all individuals and groups can draw in the ongoing processes of change and growth. Ethnic groups nourish that fund. The diversity that they supply endows a society with different histories, experiences, beliefs, ideals, arts, and cultures. The survival of distinctive characteristics of ethnic minorities guards against the trend toward homogenization that has accompanied Western development, technology, and material prosperity, and Western influence on the rest of the world. Autonomy regimes of ethnic minorities defend cultural survival rights in counteracting this trend. In given contexts, they may be useful or essential to the preservation of a culture.

. . .

. . . The ideal in the human rights movement of preserving difference cannot [however] so readily be bent to support the creation of autonomy regimes. To the contrary, a further elaboration of that ideal prompts a deep criticism of such regimes and their fragmenting effects.

Consider at the outset the relation of autonomy schemes to the norm of equal protection. We noted earlier that these schemes in some ways complement one purpose of a nondiscrimination norm of preserving differences. Institutionalized separateness, however, also violates the spirit and perhaps the letter of that norm.

A state must give all its citizens equal protection. Power-sharing schemes proceed on a contradictory premise. They are cast in ethnic terms (group X is assured of x% of the legislature or cabinet or judiciary or military, group Y of y%) and thus explicitly discriminate among groups on grounds like religion, language, race or national origin. Separate voting rolls or application forms for civil service positions drive home the lesson that socioeconomic life and career turn on ethnic bonds. Separate personal laws for each religious community, particularly if enforced within a mandatory assignment system permitting no escape by community members to a nationwide secular law, dramatize citizens' particular rather than shared characteristics.[5]

[5] A contrast can be drawn with school racial segregation in the United States. That system compulsorily separated blacks from whites, and amounted to a form of autonomy regime instituted and enforced by the state (by component states within the American federal system). Black citizens were given no choice, and the compulsory character of school segregation was important to the decisions of the Supreme Court finding it unconstitutional. Autonomy regimes considered in this Article differ. They generally stem from negotiations between the central government and minority groups, or negotiations directly between such groups. The resulting accommodations are then ratified in law. Nonetheless, even assuming that a rigorous standard of free and voluntary choice were met by the processes leading to agreement over a scheme, the fact remains that differences among ethnic groups expressed in autonomy regimes are institutionalized and enforced both by central government and by ethnic communities. The concept of equal protection/ nondiscrimination must at least be qualified to justify such regimes. Quite clearly, all are not treated the same, different treatment stemming from ethnic identity.

. . .

To a lesser or greater degree, autonomy schemes frustrate a major objective of the human rights movement of assuring that societies remain open to challenge and change. That movement institutionalizes no one ideal of social order. To the contrary, it explicitly allows for many faiths and ideologies while denying to any one among them the right or power to impose itself by force. It expresses a humanistic commitment to ongoing inquiry and diversity. . . . It denies governments the right to close avenues of reflection, criticism, advocacy, and innovation in order to impose any orthodoxy. To the extent that autonomy regimes protect historical differences but inhibit the creation, as it were, of fresh differences, they would convert the human rights movement's framework of protection of open inquiry and advocacy into the protection of static traditions. A state composed of segregated autonomy regimes would resemble more a museum of social and cultural antiquities than any human rights ideal.

. . .

Two deep characteristics of human rights law are here relevant. The first characteristic involves the relation between groups and individuals. Communities have a right to guard their own integrity, but only within the constraints imposed by human rights norms on governments' treatment of individuals, including the equal respect due to all members of a society. The central government of a state must respect individual conscience and choice. So then must ethnic communities within a state when they exercise (constitutionally based or statutorily delegated) governmental power over their members by applying a personal law or exercising a territorial authority or administering a power-sharing scheme in the central government.

In wielding governmental power, ethnic minorities may violate human rights norms in numerous ways—for example, by discriminating among their members on grounds forbidden by those norms, except to the extent that disadvantaged members can be understood to 'accept' the discriminatory treatment as part of their cultural tradition.[6] Within practical and cultural constraints, which may indeed be so deep and powerful as to block individuals from exercising any choice, all persons should be seen as empowered by

[6] I make this point boldly to set forth the argument clearly, but recognize that it raises many complex questions and, while vital, must be qualified. A group's treatment of its own members within an autonomy regime poses issues about cultural relativism and universalism in human rights norms, and about the degree to which cultural survival for many communities may be understood to require practices violating human rights instruments—practices like gender discrimination or government by a nonelected leadership. The notion of members' 'acceptance' of practices like discrimination or severe forms of punishment is itself problematic: what kind of acceptance, in what, if any, context of choice or even knowledge of alternative social arrangements? The notion of 'disadvantaged' members may itself be problematic, as a concept drawn from an alien political and moral framework. See the discussion of some of these issues in Minow, Putting Up and Putting Down: Tolerance Reconsidered, in Comparative Constitutional Federalism: Europe and America 77, 88–94 (M. Tushnet ed. 1990)

human rights norms to decide whether to remain on one side of a cultural boundary, to shift to another side, or to seek a life not committed to one or the other community.

A second characteristic of human rights law pointing toward open boundaries within a state concentrates on the polity as a whole rather than on individuals. Strong ethnic consciousness and concentration may stamp out the desire or even capacity for broader political or cultural associations. Formal legal barriers reinforce the natural tendencies of a group's members to look inward and assume only a particular identity, rather than to experience the tension between the particular and a more diffuse or broader identity as a citizen or human being.

As the sense of a common humanity weakens, alliances among people that shatter ethnic boundaries become more unlikely. Polyethnic political formations resting on generic interests like economic class or political ideology will confront decisive obstacles. Demonization of 'the other' may become part of an ethnic group's credo and stamp out the possibility of developing empathy for others. Such extreme manifestations of ethnic bonds will become the cardinal obstacle to a human rights consciousness, at least with respect to attitudes of members of the ethnic group toward 'the other.' Enforced ethnic separation both inhibits intercourse among groups, and creative development within the isolated communities themselves. It impoverishes cultures and peoples.

The distinctions between the ideals or pictures of social life that inhere in the human rights movement as a whole and those that point toward autonomy regimes can be stated with different degrees of contrast. Much turns on the strength and exclusivity of the autonomy regimes. For example, advocates of these regimes may argue that the inclusive ideals of the human rights movement are ultimately desirable for a state, but that autonomy is necessary for a transitional period. Such an argument may be based on circumstances of necessity, such as the incidence of violence described below, or on the view that some degree of autonomy is essential for maintenance of the group's culture against the onslaught of the modern state. On the other hand, those advocates may decisively reject human rights ideals as alien and evil since they risk the subversion from within or without of a religious or other tradition that binds a community together.

Autonomy regimes differ in other salient ways. The regime may involve relatively slight group differentiation and barely affect outsiders (a personal, religious law of marriage and divorce), or it may separate groups with respect to political matters of vital significance for all members of the polity (separate voting rolls, quotas, and so on). Within a federal state, an autonomy scheme for a geographically concentrated ethnic minority may grant that minority modest self-government and retain vital powers for the central government, or grant it extensive powers that border on self-rule. Those administering an autonomy regime may invite popular participation or may subject a population to decision-making powers of, say, religious officials.

For purposes of clarity and emphasis, the following list of ideals or pictures of social life bypasses these important distinctions, assumes strong and exclusive autonomy regimes, and hence states the differences in terms of stark contradictions.

Human Rights Ideals and Pictures of Social Life	*Strong Autonomy Regime Ideals and Pictures of Social Life*
Self and others, intercourse with strangers	Self vs. others, avoid strangers
Ethnic identity in tension with human identity and potential	Ethnic identity as total identity
Relevance of both particular and universal identity	Stress on the particular, what is exclusive rather than shared
Open, pluralism of spirit	Closed, separate lives within boundaries
Learn from own tradition and from others	Learning only within own tradition
Forward looking, potential for change in cultural tradition	History (myth) looking, stability-oriented
Attention to individual choice in affiliation with ethnic group	Affiliation as identity, a given that is not subject to individual choice

V. The Need for Autonomy Regimes

One way of testing the preceding observations is to inquire whether justifications as deep as those supporting, for example, rights to personal security or freedoms of conscience or expression, also support autonomy schemes for ethnic minorities. Can the same broad consensus among many political traditions and religions be identified with respect to 'rights' to separate regimes for different ethnic groups within the modern state? Or, on the other hand, are autonomy arrangements better understood as contingent, perhaps contextually necessary, ways to achieve desired ends such as peaceful coexistence rather than as ends that are valued in and of themselves?

The question raised is whether autonomy regimes express respect for difference or express despair over the possibility that diverse peoples can live with difference. Their widespread use in diverse countries and their potential effectiveness in realizing important goals indicate that these arrangements, if not ideal, are not anathema. They may rest on genuine acceptance by all affected communities. They may grow out of concrete and agonizing histories, frequently involving authoritarian and abusive rule over a minority that has been denied a fair share of power, resources, and opportunities. Often the product of negotiations between hostile communities seeking to contain discord that may have led to violence, they appear to be justified in such circumstances as the best available solution to otherwise unyielding problems.

Separation of ethnic communities through power-sharing arrangements, regional governments, and personal laws may then constitute a practical necessity, a 'least worst' solution that is surely preferable to ongoing violence and systemic oppression. The solutions that they bring to ethnic conflict may improve chances for pacific co-existence and provide the only realistic alternative not only to continuing oppression, but to a split of the contending ethnic groups into two states. Internal self-determination through autonomy schemes may blunt a minority's demands for external self-determination.

The distinction that I have drawn is a vital one: recognizing that autonomy regimes in given contexts may be preferable to other arrangements, or alternately characterizing autonomy regimes as a 'right' to be declared in international norms. The rhetoric of rights legitimates claims and mobilizes support for groups demanding autonomy. That rhetoric empowers and encourages. It goes beyond the welfare maximizing, contextual justification for autonomy regimes as 'least worst' solutions to pressing situations.

. . .

Despite these problems, the argument for recognizing in certain circumstances the justice or legitimacy of minority claims for some form of autonomy regime deserves support. The high incidence of ethnic conflicts and the particular dangers that they pose for the ethnic minorities require that the United Nations and regional institutions examine the possibility of international regulation. International attention to the systemic, structural aspects of ethnic conflict must be given a chance to show what it is capable of achieving, including achievement through the development of international law in this field. Given the reluctance of a government to support the development of international law principles favorable to autonomy schemes that may be relied on by dissident ethnic groups in its own state, or its reluctance to intervene in ethnic conflicts in foreign state and thereby risk others' intervention in its own internal disputes, the effort to achieve progress at the start through a United Nations declaration confronts dramatic obstacles.[7] But the effort must be made.

. . .

[7] It should be borne in mind that international regulation of ethnic minority claims for some form of autonomy would represent an important, but not a dramatically different, kind of 'intervention' into a state's affairs. The entire human rights movement amounts to such an 'intervention,' as demonstrated by the broad recognition of how deeply that movement has eroded the concept of domestic jurisdiction. All human rights influence the distribution and exercise of power in a country, from the prohibition of torture (limiting the means that a government can use to suppress opposition) to the vote (which, if freely exercised by all, would radically transform government and economy in many states).

NOTE

Consider the following observations of Claude Lévi-Strauss in *Race and History* (1958), at 12. Rather than view the diversity of cultures as a 'natural phenomenon,' people 'have tended rather to regard diversity as something abnormal or outrageous,' to the point of rejecting 'out of hand the cultural institutions . . . which are furthest removed from those with which we identify ourselves.' One sees 'crude reactions,' 'instinctive antipathy,' 'repugnance' toward ways of life or beliefs to which we are unaccustomed and which we term 'barbarous.'

Nonetheless, collaboration between cultures has been the key to achievement. The greater achievements stem not from 'isolated' cultures but from those which have 'combined their play' through such means as migration, borrowing, trade or warfare. 'For, if a culture were left to its own resources, it could never hope to be "superior".' Indeed, the greater the diversity between the cultures concerned, the 'more fruitful' such a 'coalition of cultures' will be.

Lévi-Strauss turns to a problem stemming from his preceding observations, that of cultural uniformity. He proposes some solutions to it, but emphasizes that he is describing a process that is inherently 'contradictory.' That is, progress requires collaboration among cultures, but 'in the course of their collaboration, the differences in their contributions will gradually be evened out, although collaboration was originally necessary and advantageous simply because of these differences.' To this contradiction he finds no clear solution. He stresses that

> man must, no doubt, guard against the blind particularism which would restrict the dignity of humankind to a single race, culture or society; but he must never forget, on the other hand, that no section of humanity has succeeded in finding universally applicable formulas; and that it is impossible to imagine mankind pursuing a single way of life for, in such a case, mankind would be ossified.

The currents toward both unification and diversity are essential. One must preserve diversity as such, diversity itself, the idea and substance of it, rather than any one historically realized form. One must look for 'stirrings of new life, foster latent potentialities,' and see each form of diversity as 'contributions to the fullness of all the others.'

* * * * *

A proposal in 1978 for commencing work in the UN Commission on Human Rights on a declaration on minorities and their members initiated a 14-year process that came to fruition in 1992 with the adoption of the declaration by the General Assembly. It is a supreme irony that the Government of Yugoslavia, a country in dissolution in 1992 when gripped by an intensifying violence among its ethnic groups, was the author of the original proposal.

As initially introduced, the draft concerned the 'rights of national, ethnic, linguistic and religious minorities.' A later draft referred to the 'rights of persons belonging to' such minorities. The tension between these two formulations of the project, with their significantly different implications, continued until the end. Successive drafts postponed a decision by indicating uncertainty through the use of brackets in phrases like, '[persons belonging to] minorities have the right to' establish associations, and so on. Ultimately, in the great majority of the articles, the brackets themselves rather than the bracketed phrases were deleted. That decision in the Commission led to the title of the declaration that introduces the following reading.

DECLARATION ON THE RIGHTS OF PERSONS BELONGING TO NATIONAL OR ETHNIC, RELIGIOUS OR LINGUISTIC MINORITIES

GA Res. 47/135, 18 December 1992.

The General Assembly

. . .

Inspired by the provisions of article 27 of the International Covenant on Civil and Political Rights concerning the rights of persons belonging to ethnic, religious or linguistic minorities,

Considering that the promotion and protection of the rights of persons belonging to national or ethnic, religious and linguistic minorities contribute to the political and social stability of States in which they live,

Emphasizing that the constant promotion and realization of the rights of persons belonging to national or ethnic, religious and linguistic minorities, as an integral part of the development of society as a whole and within a democratic framework based on the rule of law, would contribute to the strengthening of friendship and cooperation among peoples and States,

Considering that the United Nations has an important role to play regarding the protection of minorities,

. . .

Proclaims this Declaration on the Rights of Persons Belonging to National or Ethnic, Religious and Linguistic Minorities:

Article 1

1. States shall protect the existence and the national or ethnic, cultural, religious and linguistic identity of minorities within their respective territories and shall encourage conditions for the promotion of that identity.

2. States shall adopt appropriate legislative and other measures to achieve those ends.

Article 2

1. Persons belonging to national or ethnic, religious and linguistic minorities (hereinafter referred to as persons belonging to minorities) have the right to enjoy their own culture, to profess and practise their own religion, and to use their own language, in private and in public, freely and without interference or any form of discrimination.

2. Persons belonging to minorities have the right to participate effectively in cultural, religious, social, economic and public life.

3. Persons belonging to minorities have the right to participate effectively in decisions on the national and, where appropriate, regional level concerning the minority to which they belong or the regions in which they live, in a manner not incompatible with national legislation.

4. Persons belonging to minorities have the right to establish and maintain their own associations.

5. Persons belonging to minorities have the right to establish and maintain, without any discrimination, free and peaceful contacts with other members of their group and with persons belonging to other minorities, as well as contacts across frontiers with citizens of other States to whom they are related by national or ethnic, religious or linguistic ties.

Article 3

1. Persons belonging to minorities may exercise their rights, including those set forth in the present Declaration, individually as well as in community with other members of their group, without any discrimination.

2. No disadvantage shall result for any person belonging to a minority as the consequence of the exercise or non-exercise of the rights set forth in the present Declaration.

Article 4

1. States shall take measures where required to ensure that persons belonging to minorities may exercise fully and effectively all their human rights and fundamental freedoms without any discrimination and in full equality before the law.

2. States shall take measures to create favourable conditions to enable persons belonging to minorities to express their characteristics and to develop their culture, language, religion, traditions and customs, except where specific practices are in violation of national law and contrary to international standards.

3. States should take appropriate measures so that, wherever possible, persons belonging to minorities may have adequate opportunities to learn their mother tongue or to have instruction in their mother tongue.

4. States should, where appropriate, take measures in the field of education, in order to encourage knowledge of the history, traditions, language and culture of the minorities existing within their territory. Persons belonging to minorities should have adequate opportunities to gain knowledge of the society as a whole.

5. States should consider appropriate measures so that persons belonging to minorities may participate fully in the economic progress and development in their country.

Article 5

1. National policies and programmes shall be planned and implemented with due regard for the legitimate interests of persons belonging to minorities.

2. Programmes of cooperation and assistance among States should be planned and implemented with due regard for the legitimate interests of persons belonging to minorities.

. . .

Article 8

. . .

2. The exercise of the rights set forth in the present Declaration shall not prejudice the enjoyment by all persons of universally recognized human rights and fundamental freedoms.

3. Measures taken by States to ensure the effective enjoyment of the rights set forth in the present Declaration shall not *prima facie* be considered contrary to the principle of equality contained in the Universal Declaration of Human Rights.

4. Nothing in the present Declaration may be construed as permitting any activity contrary to the purposes and principles of the United Nations, including sovereign equality, territorial integrity and political independence of States.

. . .

NOTE

Consider the clauses in the Declaration on Minorities that refer to 'participation.' Article 2(2) provides that '[p]ersons belonging to minorities have the right to participate effectively in cultural, religious, social, economic and public life.' Article 2(3) gives such persons the right 'to participate effectively' in national or regional decisions concerning the minority, 'in a manner not incompatible with national legislation.' Under Article 4(5), states are to consider 'appropriate measures' so that such persons 'may participate fully in the economic progress and development in their country.'

In a report submitted to the UN Sub-commission (of the UN Commission) on Prevention of Discrimination and Protection of Minorities, Asbjørn Eide[8] states (para. 17) that members of minorities must be given 'opportunities for effective participation in the political organs of society,' but 'no single formula exists which is appropriate to all minority situations.' He suggests the exploration of various options, including the representation of minorities on advisory and decision-making bodies in fields like religion, education or cultural activities; self-administration ('functional autonomies') by the minority on matters essential to its particular identity such as development of language or religious rituals; local forms of self-government 'on a territorial and democratic basis;' and measures to assure representation of minorities in national legislatures 'even when their numerical strength is too small to have representation under normal conditions.'

In his comments on Article 2 of the Declaration on Minorities, Patrick Thornberry[9] notes that in the drafting process, the phrase 'public life' in para. 2 was preferred to 'political life;' some believed that it was a more comprehensive term. He suggests that 'effective participation' can involve the creation of ethnic associations as well as political parties, and that doctrine will 'probably move in the direction of greater decentralization, toward levels of government appropriate to continuing "effective" involvement. . . . No specific formula is mandated by the Declaration.' Thornberry emphasizes that the principle of participation 'has the advantage of turning the face of the minority to the general society, and represents an inclusive and not a separating concept.'

QUESTIONS

1. The preamble to the Declaration on Minorities states that it was 'inspired' by Article 27. Compare the Declaration and article.

   a. As a spokesperson of a minority, does the Declaration strengthen your hand (relative to your relying on Article 27 alone) in advancing claims of right for the minority or claims for protection of the minority by the state? For example, does it better support a claim for an autonomy regime that would assure the minority of, say, 20% of the seats in a national legislature, or of extensive self-government in the region where the minority is concentrated? Does it better support a claim for financial support by the state for a minority's religious schools or for cultural activities like theatre, art or music?

   b. Does the Declaration rely explicitly or implicitly on the right of peoples to self-determination or on the individual right to political participation in order to support its provisions? If so, how?

[8] Possible Ways and Means of Facilitating the Peaceful and Constructive Solution of Problems Involving Minorities: Recommendations. E/CN.4/Sub.2/1993/34/Add.4. 11 August 1993. Eide submitted several reports including valuable descriptions and analysis. See, *eg*, E/CN.4/ Sub.2/1993/ 34. 10 August 1993.

[9] The UN Declaration on Rights of Persons Belonging to Minorities, in Alan Phillips and A. Rosas (eds.), The UN Minority Rights Declaration 11, 41 (1993).

c. 'The Declaration was unnecessary. The problem lies in persuading states to observe the rights of individuals (including of course members of minorities under Article 27) that are stated so fully in the basic human rights instruments. If individual rights were honored, minorities would have no legitimate or defensible claim to further and special protection under international law.' Comment.

d. 'Individual rights take the form of the abstract and universal. Collective rights take the form of the concrete and local. To reach the goals of cultural survival and preservation of differences, collective rights are essential. Article 27 goes only half way toward meeting that goal, and is patently inadequate.' Comment.

2. 'It is said that the UN bodies drafting and adopting the Declaration, by repetition of the phrase "persons belonging to minorities," purged the Declaration of any recognition of collective rights or interests. Minorities are present only through their members, not as communities or collectivities in their own right. This is a mistaken view. The Declaration is instinct with the notion of group protection, group identity, group cultural survival, group association, group participation. Given the obvious political constraints on the drafting of the Declaration, it went as far in this direction as it could possibly go.' Comment, indicating what you believe the 'obvious political constraints' to be.

3. How do you react to Steiner's assessment of autonomy regimes for ethnic minorities? Would you support, say, a supplementary declaration or convention that would to some extent authorize such regimes? If, as Steiner suggests, 'rights language' is not an appropriate way to express arguments in favor of such regimes, what alternative language would you propose for an instrument that went beyond the provisions of the Declaration on Minorities?

4. Power-sharing arrangements are not necessarily restricted to ethnic groups. Recall the materials in Chapter 13 (p. 961) on political participation of women, and the provisions and proposals in several states for allocating to women stated percentages of seats in, say, a local or national legislative body. Would you view such proposals more or less favorably than the same form of power sharing for an ethnic minority?

5. 'Autonomy regimes (such as local government within a federalism, or a personal law regime governing marriage and divorce) should be accepted as legitimate within the Declaration on Minorities, or accepted by human rights law in general, only if they observe basic human rights, including popular participation in the formation of local government, freedoms of expression and association, and non-discrimination norms.'

a. The minority might deny this assertion and claim the right to follow its own traditions, practices, and norms within its field of self-government, even if they departed from some human rights standards. It might claim that the very purpose of an autonomy regime is to permit the minority to preserve its cultural identity that by definition differs from the state's majority culture. How do you assess such arguments?

b. Would you distinguish among different kinds of 'departures' by autonomy regimes from universal human rights norms, some being acceptable and others not? What would be the basis of the distinction?

c. 'To be considered legitimate and consistent with the human rights norms, an autonomy regime that internally practices, say, gender discrimination in family and political life, or that imposes punishment after criminal conviction that would be found to violate Article 7 of the ICCPR ("cruel, inhuman or degrading" punishment), should put such practices to a vote by the minority community.' Comment.

6. As a member of the UN Commission concerned about ethnic minorities, what steps might you urge the Commission to take in relation to the Declaration on Minorities—for example, steps related to implementation of the Declaration's provisions, or to monitoring of dangerous situations involving minorities, or to efforts to arrest growing ethnic violence?

ADDITIONAL READING

In addition to the sources noted in this section, see Francesco Capotorti, *Study on the Rights of Persons Belonging To Ethnic, Religious and Linguistic Minorities* (1979); Hurst Hannum, *Autonomy, Sovereignty, and Self-Determination: The Accommodation of Conflicting Rights* (1990); Donald Horowitz, *Ethnic Groups in Conflict* (1985); Patrick Thornberry, *International Law and the Rights of Minorities* (1991).

2. INDIGENOUS PEOPLES

COMMENT ON RIGHTS OF INDIGENOUS PEOPLES UNDER INTERNATIONAL LAW

As these materials use the term, *indigenous peoples* include the native tribes of the Americas, as well as non-dominant groups in other parts of the world that are culturally distinctive and distanced in significant ways from state governments and state political activity as well as from the modern economy—groups ranging from tribal peoples in India or Sri Lanka to Australian aboriginal communities. Although no authoritative text defines indigenous peoples, the term is usually thought to refer to communities or nations having an important historical continuity with societies that inhabited the same general territory and that predated colonization or invasion by other peoples. We are here concerned with such peoples who seek to preserve their ethnic and cultural identity, often through preservation of ancestral territory, and to continue as distinctive communities with their own social and legal institutions. These characteristics include both the 'objective' and 'subjective' criteria referred to above in the discussion of minorities.

The history of indigenous peoples has universally included exploitation by other groups, often in the form of harsh colonization, and generally involving one or another degree of dispossession from hereditary lands and destruction of native culture. Such peoples—referred to in different cultures or states in different ways, including Indians, nations, aboriginals, natives, tribes or bands—have long been plagued by the sense of inferiority imposed upon them by the dominant culture. They have rarely participated in a serious way in governance or decision-making in their states of residence on matters of vital significance for them. They have generally gained little from the economic development of those states. The centuries-long struggle in the Western Hemisphere between such peoples and the states in which they live forms part of a well-known history.

This Comment traces recent developments that have tended to 'internationalize' these struggles around the world.[10] In that sense, it is parallel to the discussion in Section (B)1 about ethnic minorities, ICCPR Article 27 and the Declaration on Minorities.

During and following the period of colonization of the Americas, international law justified in many respects the conduct (civilizing mission) of the European states, effectively legitimating colonization and subjugation while demeaning the culture, religious beliefs and achievements of the conquered or hostile native peoples. Civilization and Christianity stood against paganism and barbarism. Although these beliefs and doctrines were moderated, indigenous peoples did not start to become a significant protected category under international law until the start of the human rights movement. Progress was slow, and only in the last decade has the human rights movement acted vigorously to recognize the interests of indigenous peoples. In the process, international law has given increased recognition to indigenous peoples as distinctive communities meriting a special international-law regime distinct from both the general regime of individual rights and the regime for minorities and members thereof discussed in Section B(1). One commentator[11] describes the struggle of indigenous peoples for recognition of their right to self-determination, which would establish that they are

[10] Some information in this Comment is drawn from Robert Coulter, Commentary on the UN Draft Declaration on the Rights of Indigenous Peoples, Cult. Surviv. Q. 37 (Spring 1994); and from Stephan Marquardt, International Law and Indigenous Peoples, 3 Int. J. Group. Rts. 47 (1995).

[11] Russel Barsh, Indigenous Peoples in the 1990s: From Object to Subject of International Law?, 7 Harv. Hum. Rts. J. 33, 35 (1994).

> members of the international community who have legal personality under international law—'subjects' of international legal rights and duties rather than mere 'objects' of international concern. Although most United Nations Member States shrink fearfully from explicit references to 'peoples' and 'self-determination' in United Nations documents relating to indigenous peoples, the legal status . . . is changing incrementally, in deed if not always in word.

There now exist two binding universal instruments dealing specifically with indigenous peoples. A 1957 International Labor Organization (ILO) Convention No. 107, concerned with 'protection and integration' of indigenous 'populations' in independent states, was subjected to increasing criticism over the years because of its integrationist—most would say, assimilationist—approach to the issue of indigenous peoples. Many indigenous groups came to view the convention as paternalist in character, and as implicitly assuming the cultural inferiority of indigenous communities. Pressure grew for its revision. Cultural integrity, autonomy and survival displaced integration as a major goal.

ILO Convention No. 169, the Convention Concerning Indigenous and Tribal Peoples in Independent Countries, became effective in 1991 and, as of January 1995, had seven states parties. It partially revises Convention No. 107, which however remains effective to the extent that its parties do not ratify the new convention. There was much dispute over use of the term 'peoples,' as a result of which Article 1 states that use of that term 'shall not be construed as having any implication as regards the rights which may attach to the term under international law.'

ILO Convention No. 169 takes a different route in imposing obligations on states to protect the recognized rights of indigenous peoples and to protect their social, cultural, religious and spiritual values. Relative to the Draft Declaration referred to below, it takes an important step but a short one. The few duties imposed on governments include duties of consultation rather than, say, duties to obtain consent of indigenous peoples before enacting certain measures. The Convention does not indicate the means or forms of participation by indigenous peoples in national decision-making. It gives no effective rights of autonomy.

Over the last decade a Draft Declaration on the Rights of Indigenous Peoples has been prepared within the UN System. In 1995, it came before the UN Commission on Human Rights for consideration. Work on the Declaration by Indian and other native leaders from the Americas began in 1977. In 1982 the UN created a Working Group on Indigenous Populations of the UN Sub-Commission on Prevention of Discrimination and Protection of Minorities, a body subordinate to the UN Commission and, unlike that Commission, consisting of independent expert members.

The Working Group started drafting a declaration in 1985. It consisted of five Sub-Commission members from different regions. Under its influential chairperson Erica-Irene A. Daes, it became the pre-eminent forum for discussion of issues relating to indigenous peoples and for proposals for UN action. At its 1993 session (held in Geneva at the same time as the annual meeting of the Sub-Commission), over 600 persons attended as observers—principally indigenous leaders and representatives of about 125 peoples and organizations from all parts of the world, and experienced human rights NGOs and experts. Many indigenous groups held their own annual meetings in Geneva to develop proposals for the Declaration and draft language that they submitted to the Working Group. The annual August meetings of the Working Group were considered among the most broadly participatory in the entire UN system.

Note that this process did not formally involve state governments, although representatives of several governments attended the annual meetings as observers and commented like the NGOs on the developing draft. Many leading states did not so participate.

The Draft Declaration was agreed upon by the Members of the Working Group and was submitted to the UN Sub-Commission, which adopted the draft and submitted it in 1994 in the same form to the UN Commission on Human Rights for its consideration. The matter is likely to remain before the Commission, a body of governmental representatives, for several years. By Resolution 1995/32, 3 March 1995, the Commission created an open-ended inter-sessional working group 'with the sole purpose of elaborating a draft declaration,' considering the draft submitted to it by the Sub-Commission.

Recall the journey (starting in this instance with the Sub-Commission's working group) with many stops that is traced at p. 120, *supra*, for declarations or draft conventions within the UN system. At each stop there may be amendments: Commission to ECOSOC to the General Assembly for adoption by that body. Recall also that a number of declarations adopted by the General Assembly have been followed by the preparation within the UN of draft conventions on the same subject that, after adoption by the Assembly, have been submitted to states for ratification.

NOTE

Why should indigenous peoples not come within the category of minorities, so that Article 27 and the Declaration on Minorities would constitute the formal protection of such peoples by international law? Representatives of indigenous peoples have long resisted that classification. Consider the following observations:[12]

> . . . The fact that the rights of minorities and the rights of indigenous peoples are the object of separate international documents strongly suggests that under international law—at least *de lege ferenda*—, indigenous peoples are to be distinguished from minorities.

[12] Stephan Marquardt in n. 10 *supra*, *op. cit.*, at 70.

> From a legal point of view, this position can further be supported with reference to the different types of rights pertaining to minorities and indigenous peoples respectively, as illustrated by the applicable international instruments. . . .
>
> In contrast, the rights of indigenous peoples as expressed in the Draft Declaration are primarily collective by nature, although they may also have an individual component. These rights may in principle only be exercised by a 'people', i.e. by a collective entity, and not by individual persons. . . . The recognition of collective rights appears necessary for the protection of those characteristics forming a people's identity, i.e. those features pertaining to a people as a whole. The mere protection of individual rights would not be sufficient to protect such features. Accordingly, the Draft Declaration expressly provides for a number of collective rights of indigenous peoples. . . .
>
> Even if collective rights of minorities were to be recognized, it is submitted here that indigenous peoples are still to be distinguished from minorities under international law.
>
> Art. 27 of the ICCPR refers to ethnic, religious and linguistic minorities and the Declaration on Minorities makes an additional reference to national minorities. These qualifications indicate that minorities are characterized by elements which do not encompass all of the characteristics of indigenous peoples. A common ethnic origin, religion or language are in many cases the only elements which distinguish a minority group from the majority of the population in a given State. Many minorities are to greater or lesser degrees integrated into the population of the State in which they live and do not object to be treated as national citizens. As reflected in the rights protected under the Declaration on Minorities, minorities often seek only to practice their own religion, language or culture. . . .
>
> Most indigenous peoples, by contrast, constitute distinct entities within the States in which they live. The essential element which distinguishes them from minorities is their ancestral, 'pre-colonial' link to the territory, which is not the case for most minorities. Furthermore, indigenous peoples in most cases seek some form of political autonomy on the ground of their separate identity. These differences may be seen in the U.S.A. and Canada, where Indian (and Inuit) peoples all have a certain degree of autonomy in the form of limited self-governing powers. Most Indian and Inuit peoples also categorically reject classification as Canadian or US citizens and claim their own nationality.

Compare the following comments of Rosalyn Higgins[13], which immediately follow her argument that minorities should not be considered 'peoples' within ICCPR Article 1 and are not therefore entitled to self-determination, which is a group rather than individual right.

[13] Minority Rights: Discrepancies and Divergencies between the International Covenant and the Council of Europe System, in Rick Lawson and M. de Blois (eds.), The Dynamics of the Protection of Human Rights in Europe: Essays in Honour of Henry G. Schermers 195, 198 (Vol. III, 1994).

Two caveats may be entered. The first is that there is today an increasingly stated view that, exceptionally, minorities may be entitled to secede if their oppression is of a duration and magnitude that they are suffering gross violations with no prospect of relief or remedy. The second is that there is a developing tendency to grant certain rights, increasingly referred to as self-determination rights, to indigenous peoples. Indigenous peoples see their status as other than minorities in their own land. They are 'first peoples' and often have a culture and rights closely associated with the land. The UN Draft Declaration on Indigenous Rights provides that indigenous peoples have the right to autonomy in respect of a variety of functions traditionally reserved to State structures. A Convention on Indigenous Rights is being prepared but is still some way off: but the concept of autonomy for this special group appears to be gaining ground.

UN DRAFT DECLARATION ON THE RIGHTS OF INDIGENOUS PEOPLES

UN Doc. E/CN.4/Sub.2/1994/2/Add.1, 20 April 1994.

[The Preamble anticipates many themes of the Declaration's articles. A few of its distinctive clauses state that 'all peoples contribute to the diversity and richness of civilizations and cultures, which constitute the common heritage of humankind;' that doctrines advocating superiority of peoples or individuals on the basis of racial, religious, ethnic or cultural differences 'are racist, scientifically false, legally invalid, morally condemnable and socially unjust;' and that 'respect for indigenous knowledge, cultures and traditional practices contribute to sustainable and equitable development and proper management of the environment.' Excerpts from the Draft Declaration follow:]

. . .

Article 2

Indigenous individuals and peoples are free and equal to all other individuals and peoples in dignity and rights, and have the right to be free from any kind of adverse discrimination, in particular that based on their indigenous origin or identity.

Article 3

Indigenous peoples have the right of self-determination. By virtue of that right they freely determine their political status and freely pursue their economic, social and cultural development.

Article 4

Indigenous peoples have the right to maintain and strengthen their distinct political, economic, social and cultural characteristics, as well as their legal systems, while retaining their rights to participate fully, if they so choose, in the political, economic, social and cultural life of the State.

. . .

Article 6

Indigenous peoples have the collective right to live in freedom, peace and security as distinct peoples and to full guarantees against genocide or any other act of violence, including the removal of indigenous children from their families and communities under any pretext.

In addition, they have the individual rights to life, physical and mental integrity, liberty and security of person.

Article 7

Indigenous peoples have the collective and individual right not to be subjected to ethnocide and cultural genocide, including prevention of and redress for:

(a) Any action which has the aim or effect of depriving them of their integrity as distinct peoples, or of their cultural values or ethnic identities;
(b) Any action which has the aim or effect of dispossessing them of their lands, territories or resources;
(c) Any form of population transfer which has the aim or effect of violating or undermining any of their rights;
(d) Any form of assimilation or integration by other cultures or ways of life imposed on them by legislative, administrative or other measures;
(e) Any form of propaganda directed against them.

. . .

Article 9

Indigenous peoples and individuals have the right to belong to an indigenous community or nation, in accordance with the traditions and customs of the community or nation concerned. No disadvantage of any kind may arise from the exercise of such a right.

. . .

Article 12

Indigenous peoples have the right to practise and revitalize their cultural traditions and customs. This includes the right to maintain, protect and develop the past, present and future manifestations of their cultures, such as archaeological and historical sites, artifacts, designs, ceremonies, technologies and visual and performing arts and literature, as well as the right to the restitution of cultural, intellectual, religious and spiritual property taken without their free and informed consent or in violation of their laws, traditions and customs.

Article 13

Indigenous peoples have the right to manifest, practise, develop and teach their spiritual and religious traditions, customs and ceremonies; the right to maintain, protect, and have access in privacy to their religious and cultural sites; the right to the use and control of ceremonial objects; and the right to the repatriation of human remains.

States shall take effective measures, in conjunction with the indigenous peoples concerned, to ensure that indigenous sacred places, including burial sites, be preserved, respected and protected.

. . .

Article 15

Indigenous children have the right to all levels and forms of education of the State. All indigenous peoples also have this right and the right to establish and control their educational systems and institutions providing education in their own languages, in a manner appropriate to their cultural methods of teaching and learning.

Indigenous children living outside their communities have the right to be provided access to education in their own culture and language.

States shall take effective measures to provide appropriate resources for these purposes.

Article 16

Indigenous peoples have the right to have the dignity and diversity of their cultures, traditions, histories and aspirations appropriately reflected in all forms of education and public information.

States shall take effective measures, in consultation with the indigenous peoples concerned, to eliminate prejudice and discrimination and to promote tolerance, understanding and good relations among indigenous peoples and all segments of society.

. . .

Article 19

Indigenous peoples have the right to participate fully, if they so choose, at all levels of decision-making in matters which may affect their rights, lives and destinies through representatives chosen by themselves in accordance with their own procedures, as well as to maintain and develop their own indigenous decision-making institutions.

Article 20

Indigenous peoples have the right to participate fully, if they so choose, through procedures determined by them, in devising legislative or administrative measures that may affect them.

States shall obtain the free and informed consent of the peoples concerned before adopting and implementing such measures.

. . .

Article 25

Indigenous peoples have the right to maintain and strengthen their distinctive spiritual and material relationship with the lands, territories, waters and coastal seas and other resources which they have traditionally owned or otherwise occupied or used, and to uphold their responsibilities to future generations in this regard.

Article 26

Indigenous peoples have the right to own, develop, control and use the lands and territories, including the total environment of the lands, air, waters, coastal seas, sea-ice, flora and fauna and other resources which they have traditionally owned or otherwise occupied or used. This includes the right to the full recognition of their laws, traditions and customs, land-tenure systems and institutions for the development and management of resources, and the right to effective measures by States to prevent any interference with, alienation of or encroachment upon these rights.

Article 27

Indigenous peoples have the right to the restitution of the lands, territories and resources which they have traditionally owned or otherwise occupied or used, and which have been confiscated, occupied, used or damaged without their free and informed consent. Where this is not possible, they have the right to just and fair compensation. Unless otherwise freely agreed upon by the peoples concerned, compensation shall take the form of lands, territories and resources equal in quality, size and legal status.

Article 28

Indigenous peoples have the right to the conservation, restoration and protection of the total environment and the productive capacity of their lands, territories and resources, as well as to assistance for this purpose from States and through international cooperation. . . .

. . .

Article 30

Indigenous peoples have the right to determine and develop priorities and strategies for the development or use of their lands, territories and other resources, including the right to require that States obtain their free and informed consent prior to the approval of any project affecting their lands, territories and other resources, particularly in connection with the development, utilization or exploitation of mineral, water or other resources. Pursuant to agreement with the indigenous peoples concerned, just and fair compensation shall be provided for any such activities and measures taken to mitigate adverse environmental, economic, social, cultural or spiritual impact.

Article 31

Indigenous peoples, as a specific form of exercising their right to self-determination, have the right to autonomy or self-government in matters relating to their internal and local affairs, including culture, religion, education, information, media, health, housing, employment, social welfare, economic activities, land and resources management, environment and entry by non-members, as well as ways and means for financing these autonomous functions.

. . .

Article 33

Indigenous peoples have the right to promote, develop and maintain their institutional structures and their distinctive juridical customs, traditions, procedures and practices, in accordance with internationally recognized human rights standards.

Article 34

Indigenous peoples have the collective right to determine the responsibilities of individuals to their communities.

. . .

Article 36

Indigenous peoples have the right to the recognition, observance and enforcement of treaties, agreements and other constructive arrangements concluded with States or their successors, according to their original spirit and intent, and to have States honour and respect such treaties, agreements and other constructive arrangements. Conflicts and disputes which cannot otherwise be settled should be submitted to competent international bodies agreed to by all parties concerned.

Article 37

States shall take effective and appropriate measures, in consultation with the indigenous peoples concerned, to give full effect to the provisions of this Declaration. The rights recognized herein shall be adopted and included in national legislation in such a manner that indigenous peoples can avail themselves of such rights in practice.

Article 38

Indigenous peoples have the right to have access to adequate financial and technical assistance, from States and through international cooperation, to pursue freely their political, economic, social, cultural and spiritual development and for the enjoyment of the rights and freedoms recognized in this Declaration.

. . .

Article 42

The rights recognized herein constitute the minimum standards for the survival, dignity and well-being of the indigenous peoples of the world.

Article 43

All the rights and freedoms recognized herein are equally guaranteed to male and female indigenous individuals.

. . .

QUESTIONS

1. ILO Convention No. 107 of 1957 referred to 'indigenous *populations*.' The Draft Declaration was prepared by a Working Group on Indigenous *Populations*. Nonetheless the Draft Declaration uses the term 'indigenous *peoples*.' In what respects is the use of that term significant for the Draft Declaration? Which provisions would you emphasize?

2. Compare the Draft Declaration with the 1992 Declaration on Minorities with respect to the relative significance in each document of collective and individual rights.

a Which provisions of the Draft Declaration fall within the first category?

b. What significance do you attach to such collective rights? Do you believe that they were essential to realize the goals of the Draft Convention? Why would individual rights alone not have sufficed?

3. 'At this stage, the Draft Declaration is radically different from the 1992 Declaration on Minorities or Article 27, far bolder in its assertion of rights, stronger in the duties that it imposes on states. How much of this contrast will remain after the General Assembly adopts a declaration is another question.' How do you explain why the Draft Declaration is now bolder and stronger? Which of the Draft Declaration's provisions are apt to be subjected to the most critical review?

4. What meaning or meanings—secession and sovereign independence, democratic self-government, autonomy regimes, and so on—does the Draft Declaration (which has no definitive gloss) appear to attribute to the concept of self-determination? Recall the discussion at pp. 974–76, *supra*, of the relevance of the 1970 Declaration on Friendly Relations to the right of a people to secede in its exercise of self-determination. Should the 1970 Declaration be as applicable to indigenous peoples? Does the Draft Declaration take a stand on this issue?

5. Assume that the Draft Declaration is adopted in its present form by the General Assembly. What is the effect of Article 37? Assume, for example, that a state refuses to provide funds for education in their own culture and language of indigenous children living outside their communities. The relevant Indian nation, tribe or band believes that the state has violated Article 15. How would it develop its argument about the state's duty under international law, and where (before what fora) could it advance that argument?

LOVELACE v. CANADA

Communication No. R.6/24/1977, Human Rights Committee
Views of Committee, July 30, 1981
UN Doc. A/36/40, Supp. No. 40, reprinted in 2 Hum. Rts. L. J. 158 (1981).

[Sandra Lovelace, the author of this communication to the ICCPR Human Rights Committee under the Optional Protocol (see pp. 535–52, *supra*), was born in Canada where she had been registered as a Maliseet Indian. In 1970, Lovelace married a non-Indian. As a consequence, she lost her rights and status as an Indian under Section 12(1)(b) of the Canadian Indian Act. That Act provides that if an Indian woman who is a member of a band (such as the Maliseet) marries a person who is not a member, she 'ceases to be a member of that band.' She therefore loses her right to the use and benefits of the land allotted to the band, as well as her right to reside on an Indian reserve. Under the Act, an Indian man who marries a non-Indian woman does not lose Indian status.

Canada became a party to the ICCPR in 1976. Lovelace, who had been divorced and therefore sought to live as of right on a reserve, argued to the ICCPR that the Indian Act discriminated on grounds of sex and violated Articles 2(1), 3, 23(1) and (4), 26 and 27 of the Covenant. Her last claim was that 'the major loss to a person ceasing to be an Indian is the loss of the cultural benefits of living in an Indian community, the emotional ties to home, family, friends and neighbours, and the loss of identity.'

The Canadian Government stressed to the Committee the 'necessity of the Indian Act as an instrument designed to protect the Indian minority in accordance with article 27 of the covenant. . . . Traditionally, patrilineal family relationships were taken into account for determining legal claims.' For this and other reasons,

> legal enactments as from 1869 provided that an Indian woman who married a non-Indian man would lose her status as an Indian. These reasons were still valid. A change in the law could only be sought in consultation with the Indians themselves, who, however, were divided on the issue of equal rights. The Indian community should not be endangered by legislative changes.

Nonetheless, the Government noted its intention to put a reform bill on the issue to Parliament.

Lovelace denied that legal relationships within all Indian families were traditionally patrilineal in nature; her own cultural background was matrilineal. She denied that the Indian Act accurately expressed any prevailing pattern of rule or practice within Indian communities, and requested the Committee to recommend that Canada now amend the provisions in question.

The Committee decided to assess the post-1976 situation of Canada as involving a possible continuing violation of the Covenant in view of Lovelace's inability to reside as of right in an Indian reserve. Of the ICCPR provisions invoked by Lovelace, it considered Article 27 to be 'most directly applicable to this complaint.' Excerpts from the Views of the Committee under Article 5(4) of the Optional Protocol follow:]

14. The rights under article 27 of the Covenant have to be secured to 'persons belonging' to the minority. At present Sandra Lovelace does not qualify as an Indian under Canadian legislation . . . Protection under the Indian Act and protection under article 27 of the Covenant . . . have to be distinguished. . . . Since Sandra Lovelace is ethnically a Maliseet Indian and has only been absent from her home reserve for a few years during the existence of her marriage, she is, in the opinion of the Committee, entitled to be regarded as 'belonging' to this minority and to claim the benefits of article 27 of the Covenant. . . .

15. . . . [I]n the opinion of the Committee the rights of Sandra Lovelace to access to her native culture and language 'in community with the other members' of her group, has in fact been, and continues to be interfered with, because there is no place outside the Tobique Reserve where such a community exists. On the other hand, not every interference can be regarded as a denial of rights within the meaning of article 27. Restrictions on the right to

residence, by way of national legislation, cannot be ruled out under article 27 of the Covenant. . . . The Committee recognizes the need to define the category of persons entitled to live on a reserve, for such purposes as those explained by the Government regarding protection of its resources and preservation of the identity of its people. . . .

16. In this respect, the Committee is of the view that statutory restrictions affecting the right to residence on a reserve of a person belonging to the minority concerned, must have both a reasonable and objective justification and be consistent with the other provisions of the Covenant, read as a whole. Article 27 must be construed and applied in the light of the other provisions mentioned above, such as articles 12, 17 and 23 in so far as they may be relevant to the particular case, and also the provisions against discrimination, such as articles 2, 3 and 26, as the case may be. . . .

17. The case of Sandra Lovelace should be considered in the light of the fact that her marriage to a non-Indian has broken up. It is natural that in such a situation she wishes to return to the environment in which she was born, particularly as after the dissolution of her marriage her main cultural attachment again was to the Maliseet band. Whatever may be the merits of the Indian Act in other respects, it does not seem to the Committee that to deny Sandra Lovelace the right to reside on the reserve is reasonable, or necessary to preserve the identity of the tribe. The Committee therefore concludes that to prevent her recognition as belonging to the band is an unjustifiable denial of her rights under article 27 of the Covenant, read in the context of the other provisions referred to.

18. . . . The Committee's finding . . . makes it unnecessary to examine the general provisions against discrimination (articles 2, 3 and 26) in the context of the present case . . . .

19. Accordingly, the Human Rights Committee, acting under article 5(4) of the Optional Protocol to the International Covenant on Civil and Political Rights, is of the view that the facts of the present case, which establish that Sandra Lovelace has been denied the legal right to reside on the Tobique Reserve, disclose a breach by Canada of article 27 of the Covenant.

[Canada subsequently recognized the right of Lovelace to live on an Indian reserve, and amended the Indian Act.]

QUESTIONS

1. The ICCPR Committee assumes that, rather than debates whether, Article 27 is applicable to the case. What issue might have been debated?

2. In the actual case, it was disputed whether the Canadian Indian Act accurately reflected Indian customary or traditional rules or practices. Assume, however, that this issue was not disputed, that under prevailing Indian law as well as

under the Indian Act Lovelace had lost her previous rights. Assume, moreover, that the Committee had reached the issue of equal protection under ICCPR Articles 3 and 26.

a. How should it have resolved that issue?

b. In your proposed resolution, would you consider it relevant whether the group or community at issue were (i) a 'minority' within Article 27 and the Declaration on Members of Minorities (like, for example, Tamils in Sri Lanka, Sikhs in India, or Turkish Cypriots), or (ii) an indigenous people as in the *Lovelace* case? If so, why?

c. Are your answers to (a) and (b) affected by the fact that the issue in *Lovelace* involved gender discrimination rather than, say, a form of punishment pursuant to a tribal tradition of criminal justice that would be considered to violate the proscription of ICCPR Article 7 of 'cruel, inhuman or degrading' punishment?

3. Suppose that the Draft Declaration had been adopted by the General Assembly at the time of these proceedings in the form appearing above.

a. Should the ICCPR Committee have considered such a declaration relevant to its decision?

b. If so, in what ways do the provisions of such a declaration differ from the provisions of the ICCPR?

ADDITIONAL READING

Russel Barsh, *Indigenous Peoples in the 1990s: From Object to Subject of International Law?*, 7 Harv. Hum. Rts. J. 33 (1994); Julian Burger, *Report from the Frontier: The State of the World's Indigenous Peoples* (1987); Alexander Ewen (ed.), *Voice of Indigenous Peoples* (1994); Marie Leger (ed.), *Aboriginal Peoples: Towards Self-Government* (1994); Douglas Sanders, *Collective Rights*, 13 Hum. Rts. Q. 368 (1991).

15

International Crimes and Criminal Tribunals

This chapter explores a number of related themes: international crimes, the *ad hoc* International Criminal Tribunal for the Former Yugoslavia, proposals for a permanent international criminal court, and issues of amnesty for violators of international human rights norms. In several of these respects, the chapter builds on the discussion of the Nuremberg tribunal and proceedings at p. 98, *supra*.

Each of these themes has a current significance. Among other considerations before the Security Council, the massive brutality and serious human rights violations of the conflicts in the former Yugoslavia and related public pressure contributed toward persuading it to establish an *ad hoc* international criminal tribunal, the first such tribunal since the post-World War II tribunals constituted by the victorious powers in Nuremberg and Tokyo. Conventions that entered into force over the last few decades and the related evolution of international customary law have greatly expanded the number of crimes imposing an individual criminal responsibility that international law now defines. Recent work of the International Law Commission responding to requests of the General Assembly have led to drafts of documents on (i) crimes under international law against the peace and security of mankind, and (ii) the establishment of a permanent rather than *ad hoc* international criminal tribunal. Issues of punishment, impunity, amnesty and pardon of those involved in serious violations of human rights have become endemic to the resolution of conflicts whether internal or international in character. They regularly figure as components of the arrangements or agreements leading to new governments that pledge observance of human rights.

A. UNIVERSAL JURISDICTION AND INTERNATIONAL CRIMES

We here consider (1) the nature of crimes for which states have a 'universal jurisdiction' to define and prescribe punishment, even if a given state has no significant links to the conduct or persons involved, (2) crimes defined under international law, and (3) relationships between these first two categories.

As used in this chapter, 'international crimes' refer to crimes committed not by states as such but by individuals who bear a personal criminal responsibility. As the materials indicate, the meaning of the term 'international crime' is not self-evident. For example, is a crime 'international' in character simply by virtue of being within the scope of states' universal jurisdiction? Or within the subject-matter jurisdiction of an international tribunal? Or by virtue of having been defined by a treaty that obligated states parties to take the necessary measures (such as legislation) to make that crime applicable to individuals within its territory? Or by virtue of becoming part of customary international law?

Section A begins with a review of basic jurisdictional principles on which states prescribe (make law), and of crimes imposing individual responsibility that are defined by international law and are tried before state courts (perhaps under laws based on a 'universal jurisdiction') or before an international criminal tribunal. The *Eichmann* case in Israel then serves for an examination of these ideas in a distinctive context. The section concludes with a description of a Draft Code of Crimes against the Peace and Security of Mankind.

COMMENT ON JURISDICTION OF STATES TO PRESCRIBE CRIMINAL LAW

Certain distinctive characteristics of criminal legislation and of the adjudication of criminal cases affect the bases of judicial jurisdiction and give special importance to the jurisdiction of a state to prescribe (legislate) criminal law that reaches events (acts, effects) or persons outside its boundaries.

(1) Unlike civil litigation, where courts may entertain an action and render a default judgment when any of several bases for adjudicatory jurisdiction are present, custody of the person in criminal cases is ordinary essential. Some states' practice of conducting certain types of criminal trials *in absentia* is rare.

(2) In civil litigation, a court may invoke principles of choice of law (private international law, conflict of laws) to apply the law of a foreign state to certain aspects of the litigation when the persons or acts involved have links to other states—for example, applying the tort rules of the foreign state where the accident at issue occurred. But choice of law generally does not figure in criminal litigation. The court applies only the law of the state from which it derives its authority, almost invariably the one in which it sits, even if the prosecution is based on conduct that took place in other states. (Exceptions include military tribunals sitting in foreign states, and the reliance by states on the principle of universal jurisdiction to prescribe that is discussed below.)

If courts then are to apply their own law to their criminal prosecutions, the bases on which they can enact those laws become all the more critical. Jurisdiction of states to prescribe with respect to criminal law falls into certain conventional categories. Some of these categories are more broadly accepted internationally—more generally recognized as valid by international law—than others. This Comment, drawing on *Restatement (Third) The Foreign Relations Law of the United States* (1987), summarizes several categories that are relevant to the later materials.

Restatement §402 provides that a state has jurisdiction to prescribe law under several principles:

> (1) *Territorial principle.* A state can prescribe law with respect to 'conduct that, wholly or in substantial part, takes place within its territory.' This 'most common basis for the exercise of jurisdiction to prescribe' has 'generally been free from controversy.'
>
> (2) *Effects principle.* A state can prescribe law with respect to 'conduct outside its territory' that has 'substantial effect within its territory.' This principle is 'an aspect of jurisdiction based on territoriality, although it is sometimes viewed as a distinct category.' A classic illustration would involve shooting or sending libellous publications across a boundary.
>
> (3) *Nationality principle.* A state can prescribe with respect to 'activities, interests, status, or relations of its nationals outside as well as within its territory.' Typical criminal laws of the United States based on the nationality principle include a treason statute, statutes requiring citizens to register for military service, and judicial subpoenas addressed to nationals abroad. Many states provide that nationals may be prosecuted for all crimes, or serious crimes, regardless of where they are committed.
>
> (4) *Protective principle.* A state can prescribe law with respect to 'certain conduct outside its territory by persons not its nationals that is directed against the security of the state or against a limited class of other state interests.' That limited class concerns primarily 'offenses threatening the integrity of governmental functions that are generally recognized as crimes by developed legal systems' including espionage, counterfeiting currency, and perjûry before consular officials.
>
> (5) *Passive personality principle.* This principle asserts a state's jurisdiction to prescribe with respect to an act committed outside the state by a person not its national 'where the victim of the act was its national. The principle has not been generally accepted for ordinary torts or crimes, but it is increasingly accepted as applied to terrorist and other organized attacks on a state's nationals by reason of their nationality.'

Frequently a criminal statute will rest on more than one of these principles—for example, a combination of the effects and nationality principles (as when the conduct of a national abroad has substantial effects within a state) or a combination of the effects and protective principles.

The Restatement indicates a number of qualifications and competing considerations relevant to the exercise of jurisdiction to prescribe on the bases noted above. For example, §403 provides that even when such bases are present, a state 'may not exercise jurisdiction to prescribe law with respect to a person or activity having connections with another state when the exercise of such jurisdiction is unreasonable.' That section notes factors relevant to a conclusion of unreasonableness, including the strength of the connections to the prescribing state, the character of the activity to be regulated, the extent to which another state may have an interest in regulating the activity, the likelihood of conflict with another state, and the existence of justified expectations that might be hurt by the prescribing state.

There follows Restatement §404 on the principle of universal jurisdiction that is particularly relevant to the later materials.

RESTATEMENT (THIRD), FOREIGN RELATIONS LAW OF THE UNITED STATES

American Law Institute, 1987.

§404. Universal Jurisdiction to Define and Punish Certain Offenses
A state has jurisdiction to define and prescribe punishment for certain offenses recognized by the community of nations as of universal concern, such as piracy, slave trade, attacks on or hijacking of aircraft, genocide, war crimes, and perhaps certain acts of terrorism . . . .

COMMENT:

a. Expanding class of universal offenses. . . . [I]nternational law permits any state to apply its laws to punish certain offenses although the state has no links of territory with the offense, or of nationality with the offender (or even the victim). Universal jurisdiction over the specified offenses is a result of universal condemnation of those activities and general interest in cooperating to suppress them, as reflected in widely-accepted international agreements and resolutions of international organizations. These offenses are subject to universal jurisdiction as a matter of customary law. Universal jurisdiction for additional offenses is provided by international agreements, but it remains to be determined whether universal jurisdiction over a particular offense has become customary law for states not party to such an agreement. . . .

. . .

REPORTERS' NOTES

1. *Offenses subject to universal jurisdiction.* Piracy has sometimes been described as 'an offense against the law of nations'—an international crime. Since there is no international penal tribunal, the punishment of piracy is left to any state that seizes the offender. . . . Whether piracy is an international

crime, or is rather a matter of international concern as to which international law accepts the jurisdiction of all states, may not make an important difference.

. . .

That genocide and war crimes are subject to universal jurisdiction was accepted after the Second World War . . . .

The [Genocide] Convention provides for trial by the territorial state or by an international penal tribunal to be established, but no international penal tribunal with jurisdiction over the crime of genocide has been established. Universal jurisdiction to punish genocide is widely accepted as a principle of customary law. . . .

International agreements have provided for general jurisdiction for additional offenses, e.g., the Hague Convention for the Suppression of Unlawful Seizure of Aircraft . . . and the International Convention against the Taking of Hostages. . . . These agreements include an obligation on the parties to punish or extradite offenders, even when the offense was not committed within their territory or by a national . . . . Articles on State Responsibility, prepared for the International Law Commission, would include a provision that an international crime may result from 'a serious breach on a widespread scale of an international obligation of essential importance for safeguarding the human being, such as those prohibiting slavery, genocide and apartheid.' . . . An international crime is presumably subject to universal jurisdiction.

. . .

YORAM DINSTEIN, INTERNATIONAL CRIMINAL LAW

5 Israel Y'bk on Hum. Rts. 55 (1975).

I

The individual human being is manifestly the object of every legal system on this planet, and consequently also of international law. The ordinary subject of international law is the international corporate entity: first and foremost (though not exclusively) the State. Yet the corporate entity is not a tangible *res* that exists in reality, but an abstract notion, moulded through legal manipulation by and within the ambit of a superior legal system. When the veil is pierced, one can see that behind the legal personality of the State (or any other international corporate entity) there are natural persons: flesh-and-blood human beings. . . . In other words, the individual human being is not merely the object of international law, but indirectly also its subject, notwithstanding the fact that, ostensibly, the subject is the international corporate entity.

It is important, however, to stress that the individual human being is not merely the indirect subject of international law. Sometimes, he bears international rights and duties directly, without the interposition of the legal personality. Those who are interested in international human rights must also pay heed to the other side of the coin, namely, to international human duties.

. . . [I]ndividual responsibility means subjection to criminal sanctions. When an individual human being contravenes an international duty binding him directly, he commits an international offence and risks his life, liberty or property. Hence, international human duties are inextricably linked to the development of international criminal law.

II

In its origins, international criminal law goes back to customary international rules, which have long since prohibited piracy and war crimes. At the end of the nineteenth and the beginning of the twentieth century, other international offences (*delicta juris gentium*)—defined in international treaties—were gradually added to the list, and the pace of international legislation in this field has quickened as of the Second World War. It is possible nowadays to draw up a fairly long roster of international offences:

1. *Piracy*. The pirate is regarded by international law as the 'enemy of mankind (*hostis humani generis*),' whom every State is entitled to prosecute and punish. Piracy *jure gentium* is today delineated in Articles 15 to 17 of the Geneva Convention on the High Seas of 1958. . . .

The Geneva Convention expressly recognizes, in Article 19, the universal jurisdiction of every State in the world in respect of the offence of piracy.

2. *War Crimes*. War crimes constitute particularly grave offences against the laws of war. The most authoritative definition of war crimes appears in the Charter of the International Military Tribunal, annexed to the 1945 London Agreement for the Prosecution and Punishment of the Major War Criminals of the European Axis. [The definitions in the Charter of the IMT and the Judgment of the IMT appear at pp. 100 and 102, *supra*.]

. . .

There is no doubt that each belligerent is entitled to put on trial war criminals (including in particular war criminals from among enemy soldiers who fall into its hands), and the prevailing view is that jurisdiction in this case, as in piracy, is universal, *i.e.*, granted to all States (even neutrals).

3. *Crimes against Peace*. The London Charter provides, in Article 6(a), that war of aggression or war waged in violation of international treaties constitutes a crime against peace. The *Nuremberg* Judgment proclaims in unequivocal terms that this stipulation is declaratory in character, and that modern international law treats a war of aggression as a serious crime. Although the *Nuremberg* ruling encountered some doubts in 1946, today it is indisputable that war has not merely been proscribed by international law (subject to the exceptions of self-defence and collective security), but constitutes a crime; in fact not just a crime, but *the* crime against the international community. For jurisdictional purposes, war as a crime (against peace) is to be assimilated to war crimes.

Granted that at *Nuremberg* only the major war criminals were prosecuted, the provision of Article 6(a) of the London Charter is phrased in sweeping language, which appears to condemn as a criminal every soldier who fights in an aggressive war. Yet, in 1948, in the *German High Command Trial*, an American Military Tribunal determined (in the course of the 'subsequent proceedings' at Nuremberg) that officers below policy level must be acquitted of the charge of crimes against peace. In the *I. G. Farben Trial,* another American Military Tribunal, in the same year, held that a departure from the concept that only major war criminals—those persons in the political, military or industrial spheres who were responsible for the formulation and execution of policies—are to be convicted for crimes against peace would lead to incongruous conclusions: it would be necessary to indict the whole population, including the private soldier on the battlefield, the farmer who supplied the armed forces with foodstuffs and even the housewife who conserved essential commodities (say, canned beans) for the military industry.

4. *Crimes against Humanity*. Article 6(c) of the London Charter defines crimes against humanity as murder, extermination, enslavement, deportation and other inhumane acts committed against any civilian population before or during war, or persecutions on political, racial or religious grounds in connection with any crime within the jurisdiction of the International Military Tribunal. . . . The *Nuremberg* Judgment reaches the explicit conclusion that the Nazi persecutions against the Jews in Germany, prior to the outbreak of the Second World War in September 1939, were not connected with crimes against peace or war crimes, and therefore were not to be considered crimes against humanity. . . . As a result, though in theory it does not matter whether or not crimes against humanity—under Article 6(c)—are committed before or during war, in general practice these crimes must necessarily be committed in war-time.

The definition of crimes against humanity in paragraph (c) of Article 6 overlaps in many respects the definition of war crimes in paragraph (b): for example, murder and deportation of civilian population are specified in both places. It is noteworthy, however, that paragraph (b) applies only to the civilian population in occupied territories, whereas paragraph (c) relates to any civilian population, not excluding that of the State whose armed forces commit the crimes (*e.g.*, Jews in Germany). The outcome is that murder or deportation of the civilian population in occupied territories is both a war crime and a crime against humanity, while murder or deportation of any other civilian population in time of—and in connection with—war is only a crime against humanity. By contrast, crimes against humanity are circumscribed in their application to civilian population, unlike war crimes which can be—and usually are—directed against the armed forces too (by way of illustration, murder or maltreatment of prisoners-of-war or use of poisoned weapons in battle).

Insofar as crimes against humanity are concerned, it is obvious that universal jurisdiction over offenders is afforded to all States. . . .

5. *Genocide*. . . . In 1948, the General Assembly of the United Nations

adopted a Convention on the Prevention and Punishment of the Crime of Genocide. In its second Article, the Convention defines genocide as any of the following acts committed with intent to destroy—in whole or in part—a national, ethnical, racial or religious group, as such: (a) killing members of the group; (b) causing serious bodily or mental harm to members of the group; (c) inflicting on the group conditions of life calculated to bring about its physical destruction; (d) imposing measures intended to prevent births within the group; or (3) forcibly transferring children of the group to another group.

. . . [T]he essence of genocide is not the actual destruction of a group, but the intent to destroy it as such (in whole or in part). This has a dual consequence: first, if a group was destroyed through acts committed without an intent to bring about such destruction, there is no genocide; secondly and conversely, the murder of a single individual may be categorized as genocide if it constitutes a part of a series of acts designed to attain the destruction of the group to which the victim belonged. . . .

. . .

Though genocide may be looked upon as a crime against humanity *par excellence*, it should be pointed out that such is the case only when destruction of civilian population is carried out in connection with war. The definition of genocide transcends that of crimes against humanity in Article 6(c) of the London Charter both in its extent (for instance, as regards the forcible transfer of children from one group to another) and in the time of its application. Whereas crimes against humanity must be committed in connection with war, the first Article of the Genocide Convention expressly confirms that genocide may be committed either in time of war or in time of peace. . . .

In contradistinction to crimes against humanity, the Genocide Convention refrains from according criminal jurisdiction over offenders to every State. Article VI of the Convention provides that persons charged with genocide shall be tried by the courts of the State in whose territory the offence was committed (or by an international penal tribunal having appropriate jurisdiction, if and when established). This Article, needless to say, signifies a victory of the territoriality principle over the universality principle. A thorny problem arose in this context in the *Eichmann Trial*, inasmuch as the defendant was prosecuted and punished for genocide by a State (Israel) that could not rely on the territoriality principle. The Supreme Court of Israel resolved the problem in the following way:

> Article VI reflects a contractual obligation of the Parties to the Convention, to be applied from that moment on. That is to say, it binds them to prosecute cases of genocide that will take place within their boundaries in the future. But this undertaking has nothing to do with the universal *power* granted to every State to prosecute cases of this type that took place in the past, a power which is based on *customary* international law.

The idea is that the crime of genocide exists under both customary and conventional international law, and, whereas no universal jurisdiction pertains to it under conventional law (which is future-oriented), there is—and continues to be—a universal jurisdiction under customary law (which relates also to the past. . . .

6. *Grave Breaches of the Geneva Conventions for the Protection of Victims of War.* The Four Geneva Conventions, of 1949, for the Protection of Victims of War contain common Articles dealing with grave breaches of the Conventions that involve acts committed against protected persons or property. Grave breaches are defined as follows in Article 50 of the (First) Geneva Convention for the Amelioration of the Condition of the Wounded and Sick in Armed Forces in the Field, and in Article 51 of the (Second) Geneva Convention for the Amelioration of the Condition of Wounded, Sick and Shipwrecked Members of Armed Forces at Sea: wilful killing; torture or inhuman treatment, including biological experiments; wilfully causing great suffering or serious injury to body or health; extensive destruction and appropriation of property, not justified by military necessity and carried out unlawfully and wantonly. Article 130 of the (Third) Geneva Convention Relative to the Treatment of Prisoners of War repeats most of the constituent elements of the definition, but replaces the last part (on destruction and appropriation of property) by other grave breaches, namely, compelling a prisoner-of-war to serve in the forces of a hostile Power; wilfully depriving a prisoner-of-war of the rights of fair and regular trial. Finally, Article 147 of the (Fourth) Geneva Convention Relative to the Protection of Civilian Persons in Time of War reiterates the entire definition of the first two Conventions, but adds to it the following grave breaches: unlawful deportation or transfer or unlawful confinement of a protected person; compelling a protected person to serve in the forces of a hostile Power; wilfully depriving a protected person of the rights of fair and regular trial; taking hostages.[1]

. . .

All these grave breaches in effect constitute war crimes or crimes against humanity. The Four Geneva Conventions obligate all Contracting Parties to search for persons who commit grave breaches and bring them to trial regardless of their nationality (or alternatively to extradite them). The Conventions impose a duty on—and therefore also authorize—neutral States to prosecute the offenders. What this boils down to is establishing universal jurisdiction.

. . .

[1] [Eds.] The First Protocol to the Geneva Conventions—Protocol Relating to the Protection of Victims of International Armed Conflict, which entered into force in 1978—adds to the list of grave breaches. Its definitions of such breaches includes medical or scientific experimentation on protected persons; making the civilian population the object of attack; launching an indiscriminate attack affecting the civilian population in the knowledge that such attack will cause excessive loss of life or injury to civilians; transference by an occupying Power of part of its own civilian population into territory it occupies, or deportation of all or parts of the population of the occupied territory; and practices of racial discrimination.

9. *Enslavement and Slave Trade.* Under Article 6 of the 1956 Supplementary Convention on the Abolition of Slavery, the Slave Trade, and Institutions and Practices Similar to Slavery, the act of enslaving another person shall be a criminal offence under the laws of Contracting Parties, and offenders must be prosecuted and punished. A similar provision appears in Article 3 of the Convention in respect of conveying slaves from one country to another by whatever means of transport (including ships and aircraft), with a view to preventing and punishing slave trade. The 1958 Geneva Convention on the High Seas includes a specific clause very much like it in Article 13 relating to the transport of slaves in ships.

10. *Traffic in Persons for Prostitution.* The 1950 Convention for the Suppression of the Traffic in Persons and of the Exploitation of the Prostitution of Others—which supersedes several previous instruments relating to traffic in women and children (once dubbed 'white slavery' as distinct from the regular slavery of coloured people)—imposes on Contracting Parties the obligation to punish a number of acts pertaining to the exploitation of prostitution and the traffic in persons for such purposes.

11. *Narcotic Drugs.* According to Article 36 of the 1961 Single Convention on Narcotic Drugs, Contracting Parties are in duty bound to enjoin as punishable offences and to punish a whole series of activities in regard to the production, manufacture, extraction, distribution, delivery, sale and purchase of narcotic drugs.

. . .

15. *Aircraft Hijacking.* The 1970 Hague Convention for the Suppression of Unlawful Seizure of Aircraft provides in its first Article that any person who unlawfully—by force or threat thereof, or by any other form of intimidation—seizes or exercises control of a civil aircraft in flight, commits an offence. The Convention practically sets up universal jurisdiction over the offence, but this is not done in a clear-cut way.

. . .

III

These are the main international offences which are recognized—or attempts are made to have them recognized—in the practice of States, more particularly in international treaties.[2] What turns an act into an international offence? The answer is, its definition as a punishable offence in international (usually conventional) law. From time to time new offences are added to the rolls, in conformity with the changing needs of the international community, and it is impossible to determine a priori which acts are criminal under international law. One may patently identify certain features which characterize interna-

[2] [Eds.] See also the discussion at p. 805 *supra*, of the provision in the Convention against Torture on criminal prosecution of offenders.

tional offences—chiefly, the gravity of the acts under consideration and the fact that they harm fundamental interests of the whole international community—but the presence of these general traits by themselves does not denote that a specific act (such as an act of terrorism) is an international offence *de lege lata*. The practice of States is the conclusive determinant in the creation of international law (including international criminal law), and not the desirability of stamping out obnoxious patterns of human behaviour.

The very use of the general term 'international offences' is controversial and problematic. When international treaties define offences, they do not employ an unambiguous—or even a consistent—terminology. There are no less than five principal formulas, which the treaties tend to use in diminishing grades of clarity:

a. A categorical provision to the effect that the forbidden act constitutes an international crime involving the application of international criminal responsibility to individuals. Such limpid prose is used . . . in the London Charter. . . .

b. A less detailed provision confining itself to a statement that the forbidden act constitutes an international crime. The *locus classicus* is the Genocide Convention, which declares that genocide 'is a crime under international law'.

c. A vague provision according to which the forbidden act is a crime or an offence, without spelling out explicitly that it is necessarily an international crime or offence. This is the case, for example, in the Hague [Convention].

d. A provision which refrains from describing the forbidden act as a crime or an offence, and merely imposes a duty on Contracting Parties to prosecute and punish those who commit it. This is the method used in the Geneva Conventions for the Protection of Victims of War.

e. A provision which just prescribes that the forbidden act shall be a crime or an offence under the internal law of the Contracting Parties. Such a provision appears in the Slavery Convention. . . .

IV

Of course, at bottom all five formulas have an important common denominator: in all instances, it is international law that defines the acts under consideration as offences. But what is the legal significance of such a definition? It must be conceded that, at this phase in the evolution of international criminal law, the significance is primarily reflected in rather minor aspects:

a. International offences—and only international offences—may sometimes be subject to the universal jurisdiction of all States (or all Contracting States). That is to say, any State which lays its hands on the offender may put him on trial and punish him, irrespective of the locus of the offence and the nationality of the offender (or the victim). This, to be sure, is an exceptional jurisdiction, but one must realize that, although applicable exclusively to international offences, it does not automatically cover them all, and it is contingent on the treaties defining the offences. Thus, as we have seen whereas universal jurisdiction is recognized in regard to piracy—and, in less pellucid form, in respect of aircraft hijacking and sabotage—it is excluded from the Genocide Convention.

b. Contracting States are bound to collaborate in suppressing international offences. Consequently, they are required to exercise due diligence in attempting to forestall the forbiddent act, seize the offenders, try them—if criminal jurisdiction is available (and to the extent available)—and punish them. This is an important obligation, but it must be qualified in two substantial ways:

> (1) It is not enough that a State has undertaken, in an international treaty, to prosecute offenders: the question is whether the undertaking has in fact been incorporated in its domestic law. That is why the international treaties usually include a specific stipulation imposing on Contracting States a duty to enact the necessary internal legislation with a view to punishing offenders effectively. . . .
>
> (2) It is not enough that a State has enacted the necessary internal statutes: the question is whether it is sincerely interested in prosecuting the offenders. A special problem arises when the offenders are organs of the very same State. . . .

c. When Contracting Parties are under an obligation to extradite international offenders, international law does not permit the offenders to rely on the political nature of their acts, so as to evade extradition. International offences do not come within the purview of the rule of non-extradition of political criminals. The hitch is that treaties defining international offences do not always create an absolute duty of extradition, and, short of such a duty by treaty, the State holding the offender is not required to respond to the extradition request (regardless of the political or other nature of the offence).

. . .

e. In the absence of an express provision—in a treaty defining an international offence—in respect of the maximal penalty that can be meted out to offenders, the prosecuting State may determine the sentence as its discretion. In effect this means that any usual punishment (including the death sentence) can be imposed, depending on the circumstances. Often the treaty instructs that 'severe penalties' be imposed. . . .

V

. . . [T]he pivotal question is whether, over and above the duties devolving on States, there are also international obligations imposed directly on individuals. Many jurists maintain that this is by no means the case. In their opinion, the duty of trial and punishment—incurred by the Contracting State as such—is the sole international obligation created by the treaties, and the individual is only indirectly affected by this obligation as an object of the conduct of the State. Or . . . the crime created by the Convention is a crime 'under international law' (i.e., a crime defined by that law), not a crime against international law: the act can only be a crime against internal law, provided that the prohibitions of the Convention are incorporated in the domestic legislation. Even the fact that special rules relating to criminal jurisdiction may be applicable to offences defined by international law, does not impress the protagonists of this school of thought, for, to their mind, even if universal jurisdiction is established over the offence, that does not suffice to impose duties directly on the individual: it is merely a case of an extraordinary extension of the regular criminal jurisdiction of the State. The extension of jurisdiction has its palpable impact on the individual caught in its net, but it does not transform him from an object into a subject of international law.

It is not easy to contradict this approach—and to insist on the existence of international offences in the full sense of the term—as long as no permanent International Criminal Court has been established for the trial of international offenders. It is true that International Military Tribunals functioned successfully at *Nuremberg* and *Tokyo*, but those were judicial bodies that were set up ad hoc for the prosecution of the major war criminals of the Second World War, and when their duties were terminated they ceased to operate. It is symptomatic that, when Israel seized Eichmann, it did not possess an option of handing him over for trial by any extant international court.

The idea of founding a permanent international penal tribunal has been contemplated for a long time.

Faute de mieux, as long as no penal tribunal has been established on the international plane, the trial of persons charged with offences defined by international law must take place in the national courts of States (mainly, States Parties to international treaties). These courts may be regarded, for this purpose, as organs of the international community applying international criminal law and bringing it home to the individual, who is directly subjected to international obligations. . . .

While this situation lasts, international criminal law is admittedly beclouded by doubts: from the viewpoint of the offender—facing a regular judge in a domestic court—the criminal trial looks like ordinary municipal proceedings: and from the standpoint of the national judge, that judge does not apply international law unless it is incorporated into the national legal system and to the extent of its incorporation. If and when a permanent international criminal court comes into being, it will be possible to distinguish between real international offences (namely, international duties incurred directly by the individual, who is criminally liable for their infraction—over which the court will have jurisdiction—and national offences originating in international treaties (that is, international obligations imposed on the State, which is required to take measures to suppress the forbidden acts through the application of effective sanctions against their perpetrators). But as long as no such court exists, the distinction is not easy to draw.

. . .

The *Nuremberg* Judgment transcends in its importance the specific question of crimes connected with war. In many respects, the International Military Tribunal laid the groundwork of modern international criminal law when it proclaimed:

> That international law imposes duties and liabilities upon individuals as well as upon States has long been recognized. . . . individuals can be punished for violations of international law. Crimes against international law are committed by men, not by abstract entities, and only by punishing individuals who commit such crimes can the provisions of international law be enforced.

. . .

QUESTION

As indicated in editors' footnotes to the preceding article, the scope of 'international criminal law' has continued to expand since the time that the article was written. Note the contemporary effort to develop provisions on 'terrorism' described at p. 1039, *infra*.

Do you find any coherence among the the historical categories, a distinctive character of such 'international crimes' that binds them within a unified framework? Or does the historical list simply reflect the changing priorities of the international community over time, so that expansion will be less a matter of logic than of new and broadly shared perceptions of need together with an ability to build a consensus over the definition of a crime?

COMMENT ON THE EICHMANN TRIAL

Adolf Eichmann, operationally in charge of the mass murder of Jews in Germany and German-occupied countries, fled Germany after World War II. He was abducted from Argentina by Israelis, and brought to trial in Israel under the Nazi and Nazi Collaborators (Punishment) Law, enacted after Israel became a state. Section 1(a) of the Law provided:

A person who has committed one of the following offences—(1) did, during the period of the Nazi regime, in a hostile country, an act constituting a crime against the Jewish people; (2) did, during the period of the Nazi regime, in a hostile country, an act constituting a crime against humanity; (3) did, during the period of the Second World War, in a hostile country, an act constituting a war crime; is liable to the death penalty.

The Law defined 'crimes against the Jewish people' to consist principally of acts intended to bring about physical destruction. The other two crimes were defined similarly to the like charges at Nuremberg, p. 100, *supra*. The 15 counts against Eichmann involved all three crimes. The charges stressed Eichmann's active and significant participation in the 'final solution to the Jewish problem' developed and administered by Nazi officials. Eichmann was convicted in 1961 and later executed. There appear below summaries of portions of the opinions of the trial and appellate courts.

The Attorney-General of the Government of Israel v. Eichmann[3]

Eichmann argued that the prosecution violated international law by inflicting punishment (1) upon persons who were not Israeli citizens, (2) for acts done by them outside Israel and before its establishment, (3) in the course of duty, and (4) on behalf of a foreign country. In reply, the Court noted that, in event of a conflict between an Israeli statute and principles of international law, it would be bound to apply the statute. However, it then concluded that 'the law in question conforms to the best traditions of the law of nations. The power of the State of Israel to enact the law in question or Israel's "right to punish" is based . . . from the point of view of international law, on a dual foundation: The universal character of the crimes in question and their specific character as being designed to exterminate the Jewish people.'

Thus the Court relied primarily on the universality and protective principles to justify the Israeli statute and its application. It held such crimes to be offenses against the Law of Nations, much as was the traditional crime of piracy. It compared the conduct made criminal under the Israeli statute (particularly the 'crime against the Jewish people') and the crime of genocide, as defined in Article 1 of the Convention for the Prevention and Punishment of Genocide.

The Contracting Parties confirm that genocide, whether committed in time of peace or in time of war, is a crime under international law which they undertake to prevent and to punish.[4]

[3] District Court of Jerusalem, Judgment of December 11, 1961. This summary and the selective quotations are drawn from 56 Am. J. Int. L. 805 (1962) (unofficial translation).

[4] [Eds.] Article 6 of the Convention, the meaning and implications of which were viewed differently by the parties, states: 'Persons charged with genocide or any of the other acts enumerated in Article III shall be tried by a competent tribunal of the State in the territory of which the act was committed, or by such international penal tribunal as may have jurisdiction with respect to those Contracting Parties which shall have accepted its jurisdiction.' No such tribunal has been created.

The Court also stressed the relationship between the Law's definition of 'war crime' and the pattern of crimes defined in the Nuremberg Charter. It rejected arguments of Eichmann based upon the retroactive application of the legislation, and stated that 'all the reasons justifying the Nuremberg judgments justify *eo ipse* the retroactive legislation of the Israeli legislator.'

The Court then discussed another 'foundation' for the prosecution—the offence specifically aimed at the Jewish people.

> [This foundation] of penal jurisdiction conforms, according to [the] acknowledged terminology, to the protective principle . . . . The 'crime against the Jewish people,' as defined in the Law, constitutes in effect an attempt to exterminate the Jewish people . . . . If there is an effective link (and not necessarily an identity) between the State of Israel and the Jewish people, then a crime intended to exterminate the Jewish people has a very striking connection with the State of Israel. . . . The connection between the State of Israel and the Jewish people needs no explanation.

Eichmann v. The Attorney-General of the Government of Israel[5]

After stating that it fully concurred in the holding and reasoning of the district court, the Supreme Court proceeded to develop arguments in different directions. It stressed that Eichmann could not claim to have been unaware at the time of his conduct that he was violating deeply rooted and universal moral principles. Particularly in its relatively underdeveloped criminal side, international law could be analogized to the early common law, which would be similarly open to charges of retroactive law making. Because the international legal system lacked adjudicatory or executive institutions, it authorized for the time being national officials to punish individuals for violations of its principles, either directly under international law or by virtue of municipal legislation adopting those principles.

Moreover, in this case Israel was the most appropriate jurisdiction for trial, a *forum conveniens* where witnesses were readily available. It was relevant that there had been no requests for extradition of Eichmann to other states for trial, or indeed protests by other states against a trial in Israel.

The Court affirmed the holding of the district court that each charge could be sustained. It noted, however, much overlap among the charges, and that all could be grouped within the inclusive category of 'crimes against humanity.'

[5] Supreme Court sitting as Court of Criminal Appeals, May 29, 1962. This summary is based upon an English translation of the decision appearing in 36 Int'l. L. Rep. 14–17, 277 (1968).

QUESTIONS

1. Consider the alternatives to trial of Eichmann by the Israeli court. Would any international tribunal have been competent? What would have been involved in an effort to establish another *ad hoc* international criminal tribunal like Nuremberg, and would that effort have been likely to succeed? Would trial before the courts of another state have been preferable? Which state?

2. If you had been the author of the opinion, how would you have responded to Eichmann's charges that the prosecution violated international law ?

3. Which of the Israeli legislative bases for the trial of Eichmann do you view as most satisfactory? Does any provision of the Nazi and Nazi Collaborators Law raise troublesome issues with respect to its possible use as precedent for prosecutions in different contexts in other states?

NOTE

The work of the International Law Commission figures in the following article on international crimes, and in a later article on a permanent international criminal court. This Commission, constituted under its statute enacted by General Assembly Resolution 174(II) of November 21, 1947, as amended, has 25 members who are to be 'persons of recognized competence in international law.' Article 15 of the ILC Statute distinguishes between *development* ('the preparation of draft conventions on subjects which have not yet been regulated by international law or in regard to which the law has not yet been sufficiently developed in the practice of States') and *codification* ('the more precise formulation and systematization of rules of international law in fields where there already has been extensive state practice, precedent and doctrine'). The distinction is sometimes more easily stated that realized in practice. The ILC examines a subject sometimes at the request of the General Assembly, sometimes at its own initiative. Drafts adopted by the ILC are often submitted to the General Assembly for its consideration.

JOHN MURPHY, INTERNATIONAL CRIMES

in Oscar Schachter and C. Joyner (eds.), United Nations Legal Order 993 (Vol. 2, 1995), at 997.

Following the General Assembly's affirmation of the Nuremberg principles, the International Law Commission began work on a Draft Code of Offenses Against the Peace and Security of Mankind as well as on a Draft Statute for an International Criminal Court [see p. 1081, *infra*]. By 1954 a Draft Code . . .

had been developed by the ILC. . . . The Draft Code was in 1978 again placed on the agenda of the Sixth (Legal) Committee of the General Assembly and in 1982 the ILC resumed work on it. On July 12, 1991, the ILC adopted, on first reading, Draft Articles on the Draft Code of Crimes Against the Peace and Security of Mankind and transmitted them, through the Secretary-General, to governments, for their comments and observations. . . .

. . . Although the ILC adopted the Draft Articles on a first reading, many provisions remain controversial, and there are likely to be several critical comments forthcoming from the governments to which they have been submitted. Moreover, it has not yet been determined whether these articles will be adopted in binding legal form—perhaps as part of a treaty or as a model law—or in a non-binding mode along the lines of the Helsinki Accords. This issue is closely linked with the question of whether the Code should serve, in whole or in part, as the statute for an international criminal court. . . .

. . . [I]t should be noted that the Draft Articles are limited in their coverage to crimes that threaten the peace and security of mankind. Because of this bedrock test, Article 21 of the Draft Articles covers 'systematic or mass violations of human rights' but not human rights violations that do not rise to this level of magnitude, on the rationale that the latter are unlikely to cause disputes that would threaten the peace. The international nature of the crimes covered by the Draft Articles is highlighted by Article 2, which provides in pertinent part that the 'characterization of an act or omission as a crime against the peace and security of mankind is independent of internal law' . . . .

Part II of the Draft Articles sets forth specific crimes against the peace and security of mankind. Twelve such crimes are listed: aggression (Article 15); threat of aggression (Article 16); intervention (Article 17); colonial domination and other forms of alien domination (Article 18); genocide (Article 19); apartheid (Article 20); systematic or mass violations of human rights (Article 21); exceptionally serious war crimes (Article 22); the recruitment, use, financing and training of mercenaries (Article 23); international terrorism (Article 24); illicit traffic in narcotic drugs (Article 25); and willful and severe damage to the environment (Article 26). . . .

It should first be realized that the Draft Articles both codify and develop the law of international crimes. For example, some of the acts covered clearly constitute international crimes under current law, including aggression, threat of aggression (probably), genocide, international terrorism (at least as defined by the Commission) and illicit traffic in narcotic drugs. Other acts, although arguably already constituting international crimes and in some cases covered by international conventions, are more debatable. Many developed Western countries in particular would deny some are currently established as international crimes, for example, intervention; colonial domination and other forms of alien domination; apartheid; and the recruitment, use, financing and training of mercenaries. With respect to still other acts the Commission would seem to be proposing new categories of international crimes. Systematic or mass violations of human rights, for example, is a new crime against the peace and security of mankind. Some of the human rights violations covered—torture, slavery and deportation or forcible transfer of population—are already established as international crimes when committed on a nonsystematic or nonmass basis. Others, namely murder and persecution on social, political, racial, religious, or cultural grounds, are not generally regarded as international crimes absent the higher level of magnitude.

Similarly, although war crimes, including 'grave breaches,' have long been established as international crimes, the concept of 'exceptionally serious war crimes' as a crime against the peace and security of mankind is new.

Finally, although there are some provisions in treaties and conventions on the law of armed conflict regarding damage to the environment, the provision on willful and severe damage to the environment is an exercise in law development. Deliberate and widespread damage to the environment during the Gulf war, with global repercussions, has helped generate support for such a provision.

. . .

The historical record indicates that many, perhaps most, of these crimes against the peace and security of mankind will prove difficult to prosecute and punish. Aggression and genocide have long been established as among the most severe of international crimes, yet since the Nuremberg and Tokyo trials, there have been no prosecutions—much less punishments—for their commission. The crime of aggression is not even listed among the crimes covered by most national criminal codes and statutes.

. . .

It should also be noted that experts in criminal jurisprudence have suggested that, as currently worded, the definitions of the crimes covered by the Draft Articles are so ambiguous and vague as to make their prosecution in national courts difficult or perhaps even unconstitutional.

. . .

Despite efforts to do so, neither the United Nations nor its specialized agencies have been able to agree on a definition of 'international terrorism.' Partially because of this, the United Nations has also been unable to agree on a single convention on the legal control of terrorism. Rather, the United nations has adopted a piecemeal approach to the problem through the adoption of separate conventions aimed at suppressing aircraft hijacking, unlawful acts against the safety of civil aviation, or of airports serving international civil aviation, unlawful acts against internationally protected persons, including diplomatic agents, the taking of hostages, unlawful acts against the safety of maritime navigation and, most recently, the use of plastic explosives. . . .

Although these treaty provisions are often loosely described as 'antiterrorist,' the acts themselves that they cover are criminalized regardless of whether, in a particular case, they could be described as 'terrorism.' Whether the crimes covered by the antiterrorist conventions may be classified as 'international crimes' is debatable. At the least, they establish a legal framework for states parties to cooperate toward punishment of the perpetrators of these crimes. They also create a system of universal jurisdiction over these crimes for states parties. . . .

. . .

B. THE INTERNATIONAL CRIMINAL TRIBUNAL FOR THE FORMER YUGOSLAVIA

THEODOR MERON, THE CASE FOR WAR CRIMES TRIALS IN YUGOSLAVIA

72 Foreign Affairs 122 (No. 3 1993), at 123.

. . . Except in the case of a total defeat or subjugation—for example, Germany after World War II—prosecutions of enemy personnel accused of war crimes have been both rare and difficult. National prosecutions have also been rare because of nationalistic, patriotic or propagandistic considerations.

The Versailles Treaty after World War I illustrates the case of a defeated but not wholly occupied state. Germany was obligated to hand over to the allies for trial about 900 persons accused of violating the laws of war. But even a weak and defeated country such as Germany was able to effectively resist compliance. The allies eventually agreed to trials by German national courts of a significantly reduced number of Germans. The sentences were both few and clement. The Versailles model proved to be clearly disappointing.

On the other hand, after the four principal victorious and occupying powers established an international military tribunal (IMT) following World War II, several thousand Nazi war criminals were tried either by national courts under Allied Control Council Law No. 10 or by various states under national decrees. Nuremberg's IMT, before which about 20 major offenders were tried, and the national courts functioned reasonably well; the Allies had supreme authority over Germany and thus could often find and arrest the accused, obtain evidence and make arrangements for extradition.

Despite the revolutionary development of human rights in the U.N. era, no attempts have been made to bring to justice such gross perpetrators of crimes against humanity or genocide as Pol Pot, Idi Amin or Saddam Hussein, perhaps because the atrocities in Cambodia, Uganda and Iraq (against the Kurds) did not occur in the context of international wars. Internal strife and even civil wars are still largely outside the parameters of war crimes and the grave breaches provisions of the Geneva conventions.

The Persian Gulf War, as an international war, provided a classic environment for the vindication of the laws of war so grossly violated by Iraq by its plunder of Kuwait, its barbaric treatment of Kuwait's civilian population, its mistreatment of Kuwaiti and allied prisoners of war and during the sad chapter of the U.S. and other hostages. Although the Security Council had invoked the threat of prosecutions of Iraqi violators of international humanitarian law, the ceasefire resolution did not contain a single word regarding criminal responsibility. Instead, the U.N. resolution promulgated a system of war reparations and established numerous obligations for Iraq in areas ranging from disarmament to boundary demarcation.

This result is not surprising, for the U.N. coalition's war objectives were limited, and there was an obvious tension between negotiating a ceasefire with Saddam Hussein and demanding his arrest and trial as a war criminal. A historic opportunity was missed to breathe new life into the critically important concept of individual criminal responsibility for the laws of war violations. At the very least, the Security Council should have issued a warning that Saddam and other responsible Iraqis would be subject to arrest and prosecution under the grave breaches provisions of the Geneva conventions whenever they set foot abroad.

. . .

Warnings of war crimes trials have been unsuccessful deterrents in past wars and may prove no more effective in the case of the former Yugoslavia. The precedent and moral considerations for the establishment of the tribunal require action in any event. Furthermore, several factors may yet strengthen deterrence. First, modern media ensures that all actors in the former Yugoslavia know of the steps being taken to establish the tribunal. Second, the tribunal will probably be established while the war is still being waged. Even the worst war criminals involved in the present conflict know that their countries will eventually want to emerge from isolation and be reintegrated into the international community. Moreover, they themselves will want to travel abroad. Normalization of relations and travel would depend on compliance with warrants of arrest. A successful tribunal for Yugoslavia will enhance deterrence in future cases; failure may doom it.

. . .

The establishment of an ad hoc tribunal should not stand alone, however, as a sole or adequate solution. The world has failed to prosecute those responsible for egregious violations of international humanitarian law and human rights in Uganda, Iraq and Cambodia. To avoid charges of Eurocentrism this ad hoc tribunal for the former Yugoslavia should be a step toward the creation of a permanent criminal tribunal with general jurisdiction. The drafting of a treaty on a permanent tribunal, on which work has begun by the U.N. International Law Commission, should be expedited, providing an opportunity to supplement the substantive development of international law by an institutional process.

. . .

COMMENT ON SECURITY COUNCIL DECISIONS

The materials following this Comment consider the role of the Security Council in establishing the International Criminal Tribunal for the Former Yugoslavia. They indicate the degree to which the Council acted by taking decisions under Chapter VII of the UN Charter. This Comment provides some background about the historical use by the Council of Chapter VII.

The collective security system provided for in the UN Charter accords 'primary responsibility' for the maintenance of international peace and security to the Security Council (Art. 24). Member states are obligated to carry out the Council's decisions (Art. 25). Throughout the Cold War years, the system as a whole, and particularly the provisions for enforcement under Chapter VII, remained largely paralyzed. Apart from a limited range of decolonization matters, the Council was relatively inactive. Its decision in 1990 to respond to the invasion of Kuwait by Iraq marked the beginning of a new phase of Council activism.

The statistics tell the story as to the degree of involvement of the Council and the highly politicized selection of the few situations to which it was attentive. Between 1946 and 1989 the Council held 2,903 meetings and adopted 646 resolutions. Between 1990 and mid-1994 alone, it held 495 meetings (plus innumerable consultations) and adopted 288 resolutions. Until 1986, the Council declared breaches of the peace only in Korea in 1950 and the Falklands/Malvinas in 1982. Its determinations during that period that a state had committed acts of 'aggression' reached only to Israel and South Africa, and it recognized the existence of a 'threat to international peace and security' only in relation to seven situations: Palestine in 1948, the Congo in 1961, Southern Rhodesia in 1966, Bangladesh in 1971, Cyprus in 1974, South Africa in 1977, and Israel's attack on the PLO headquarters in Tunis in 1985.

Prior to 1990 the Council authorized the use of military force only three times (Korea, the Congo and Southern Rhodesia), and imposed sanctions twice (Southern Rhodesia 1966–79, and South Africa 1977–94, see. p. 365, *supra*). In contrast, between 1990 and 1995, the Council authorized the use of military force in five situations (Iraq, Somalia, the former Yugoslavia, Rwanda and Haiti), and imposed sanctions in eight situations (the preceding five, plus Liberia, Libya and Angola).

Peace-keeping operations show a similar trend. At the end of 1990, the UN was involved in eight operations with a total of 10,000 troops. Four years later there were 17 such operations (including the UN Protection Force in the former Yugoslavia established in 1992) involving more than 70,000 troops and costing over $3 billion a year.

Such heightened activity was both a cause for and a consequence of a considerable expansion in the Council's self-determined competence. In particular, the Council gave far greater attention to internal conflicts. Of 21 operations undertaken between 1988 and 1994, only eight related to inter-state wars; the Secretary-General characterized the balance of 13 conflicts as intra-state.

A large proportion of the situations addressed by the Council involved human rights violations, often systemic and brutal violations. Nonetheless, such aspects of these conflicts remained an incidental rather than primary focus for the Council. Chapter VII of the Charter does not refer to human rights violations as a condition for Council decisions ranging from sanctions to the use of armed force. Many commentators and a few states called on the Council to play a more active role in responding to such violations. This has not, however, occurred. The reasons of course involve the distinctive powers of the Council to make 'decisions' under Chapter VII. The General Assembly has no such power, but it does exercise a competence to debate, discuss, and recommend on a broader range of issues that include human rights concerns. Thus states fear a further erosion of their sovereignty, and seek to uphold the division of institutional competence between the Council and the Assembly. Many argue that the Council is subject to greater control by the West and thus would not act objectively if given a major human rights brief.

Nonetheless, the Council has recently taken up a number of situations that have involved significant human rights components. They include:

> Resolution 688 (1991) in which it characterized Iraq's repression of the Kurds, leading to cross-border refugee flows, as a threat to international peace and security and insisted that Iraq 'allow immediate access by international humanitarian organizations;'
>
> Resolution 794 (1992) in which it authorized multinational forces to establish a secure environment for humanitarian relief operations in Somalia;
>
> various similar resolutions in relation to the former Yugoslavia; and
>
> Resolution 940 (1994) authorizing the use of 'all necessary means to facilitate the departure from Haiti of the military leadership . . . , the prompt return of the legitimately elected president and the restoration of the legal authorities of the Government of Haiti'.

Despite various calls on it to recognize a right to humanitarian intervention, the Council has not used such terminology. Indeed, its insistence that unimpeded access be given to the delivery of humanitarian relief supplies has met with only limited success in situations such as the former Yugoslavia and Rwanda.

The establishment of the office of the UN High Commissioner for Human Rights in 1994 has led to suggestions that the High Commissioner act as an early warning mechanism to alert the Council to actual or potential situations involving gross violations of human rights. The same suggestion has been made in relation to the human rights treaty bodies, such as the ICCPR Human Rights Committee. The uncertain attitude of the Council toward such suggestions, together with the reluctance of the Secretary-General to develop the role given him under Article 99 of the Charter, indicate the significant political obstacles to such developments.

ADDITIONAL READING

N. D. White, *The United Nations and the Maintenance of International Peace and Security* (1990); Helmut Freudenschuss, *Between Unilateralism and Collective Security: Authorizations to Use Force by the Security Council*, 5 Eur. J. Int. L. 492 (1994); Andrew Hurrell, *Collective Security and International Order Revisited*, 11 Int. Rel. 37 (1992); Boutros Boutros-Ghali, *An Agenda for Peace* (2nd ed. 1995); Philip Alston, *The Security Council and Human Rights: Lessons to Be Learned from the Iraq–Kuwait Crisis and its Aftermath*, 13 Aust. Y. B. Int. L. 107 (1992); Sean Murphy, *The Security Council, Legitimacy and the Concept of Collective Security after the Cold War*, 32 Colum. J. Transnat'l L. 201 (1994); Martti Koskenniemi, *The Police in the Temple: A Dialectical View of the United Nations*, 6 Eur. J. Int. L. (forthcoming 1995).

SECURITY COUNCIL RESOLUTIONS ON ESTABLISHMENT OF AN INTERNATIONAL TRIBUNAL FOR THE FORMER YUGOSLAVIA

Reprinted in 14 Hum. Rts. L. J. 197 (1993).

Resolution 808, 22 Feb. 1993

. . .

Recalling paragraph 10 of its resolution 764 (1992) of 13 July 1992, in which it reaffirmed that all parties are bound to comply with the obligations under international humanitarian law and in particular the Geneva Conventions of 12 August 1949, and that persons who commit or order the commission of grave breaches of the Conventions are individually responsible in respect of such breaches,

. . .

Recalling further its resolution 780 (1992) of 6 October 1992, in which it requested the Secretary-General to establish, as a matter of urgency, an impartial Commission of Experts. . . with a view to providing the Secretary-General with its conclusions on the evidence of grave breaches of the Geneva Conventions and other violations of international humanitarian law committed in the territory of the former Yugoslavia,

. . .

Expressing once again its grave alarm at continuing reports of widespread violations of international humanitarian law occurring within the territory of the former Yugoslavia, including reports of mass killings and the continuance of the practice of 'ethnic cleansing',

Determining that this situation constitutes a threat to international peace and security,

Determined to put an end to such crimes and to take effective measures to bring to justice the persons who are responsible for them,

Convinced that in the particular circumstances of the former Yugoslavia the establishment of an international tribunal would enable this aim to be achieved and would contribute to the restoration and maintenance of peace.

. . .

1. *Decides* that an international tribunal shall be established for the prosecution of persons responsible for serious violations of international humanitarian law committed in the territory of the former Yugoslavia since 1991;

2. *Requests* the Secretary-General to submit for consideration by the Council. . . a report on all aspects of this matter, including specific proposals and where appropriate options for the effective and expeditious implementation of the decision contained in paragraph 1 above, taking into account suggestions put forward in this regard by Member States;

. . .

Resolution 827, 25 May 1993

. . .

Acting under Chapter VII of the Charter of the United Nations,

1. Approves the report of the Secretary-General;

2. Decides hereby to establish an international tribunal for the sole purpose of prosecuting persons responsible for serious violations of international humanitarian law committed in the territory of the former Yugoslavia between 1 January 1991 and a date to be determined by the Security Council upon the restoration of peace and to this end to adopt the Statute of the International Tribunal annexed to the above-mentioned report;

. . .

4. Decides that all Sates shall cooperate fully with the International Tribunal and its organs in accordance with the present resolution and the Statute of the International Tribunal and that consequently all States shall take any measures necessary under their domestic law to implement the provisions of the present resolution and the Statute, including the obligation of

States to comply with requests for assistance or orders issued by a Trial Chamber under Article 29 of the Statute;

. . .

7. Decides also that the work of the International Tribunal shall be carried out without prejudice to the right of the victims to seek, through appropriate means, compensation for damages incurred as a result of violations of international humanitarian law;

. . .

REPORT OF THE SECRETARY-GENERAL UNDER SECURITY COUNCIL RESOLUTION 808

Doc. S/2504, 3 May 1993, reprinted in 14 Hum. Rts. L. J. 198 (1993).

. . .

I. THE LEGAL BASIS FOR THE ESTABLISHMENT OF THE INTERNATIONAL TRIBUNAL

18. Security Council resolution 808 (1993) states that an international tribunal shall be established for the prosecution of persons responsible for serious violations of international humanitarian law committed in the territory of the former Yugoslavia since 1991. It does not, however, indicate how such an international tribunal is to be established or on what legal basis.

19. The approach which, in the normal course of events, would be followed in establishing an international tribunal would be the conclusion of a treaty by which the States parties would establish a tribunal and approve its statute. This treaty would be drawn up and adopted by an appropriate international body (e.g., the General Assembly or a specially convened conference), following which it would be opened for signature and ratification. Such an approach would have the advantage of allowing for a detailed examination and elaboration of all the issues pertaining to the establishment of the international tribunal. It also would allow the States participating in the negotiation and conclusion of the treaty fully to exercise their sovereign will, in particular whether they wish to become parties to the treaty or not.

20. As has been pointed out in many of the comments received, the treaty approach incurs the disadvantage of requiring considerable time to establish an instrument and then to achieve the required number of ratifications for entry into force. Even then, there could be no guarantee that ratifications will be received from those States which should be parties to the treaty if it is to be truly effective.

21. A number of suggestions have been put forward to the effect that the General Assembly, as the most representative organ of the United Nations, should have a role in the establishment of the international tribunal in addi-

tion to its role in the administrative and budgetary aspects of the question. The involvement of the General Assembly in the drafting or the review of the statute of the International Tribunal would not be reconcilable with the urgency expressed by the Security Council in resolution 808 (1993). The Secretary-General believes that there are other ways of involving the authority and prestige of the General Assembly in the establishment of the International Tribunal.

22. In the light of the disadvantages of the treaty approach in this particular case and of the need indicated in resolution 808 (1993) for an effective and expeditious implementation of the decision to establish an international tribunal, the Secretary-General believes that the International Tribunal should be established by a decision of the Security Council on the basis of Chapter VII of the Charter of the United Nations. Such a decision would constitute a measure to maintain or restore international peace and security, following the requisite determination of the existence of a threat to the peace, breach of the peace or act of aggression.

23. This approach would have the advantage of being expeditious and of being immediately effective as all States would be under a binding obligation to take whatever action is required to carry out a decision taken as an enforcement measure under Chapter VII.

24. In the particular case of the former Yugoslavia, the Secretary-General believes that the establishment of the International Tribunal by means of a Chapter VII decision would be legally justified, both in terms of the object and purpose of the decision, as indicated in the preceding paragraphs, and of past Security Council practice.

25. As indicated in paragraph 10 above, the Security Council has already determined that the situation posed by continuing reports of widespread violations of international humanitarian law occurring in the former Yugoslavia constitutes a threat to international peace and security. The Council has also decided under Chapter VII of the Charter that all parties and others, concerned in the former Yugoslavia, and all military forces in Bosnia and Herzegovina, shall comply with the provision of resolution 771 (1992), failing which it would need to take further measures under the Charter. Furthermore, the Council has repeatedly reaffirmed that all parties in the former Yugoslavia are bound to comply with the obligations under international humanitarian law and in particular the Geneva Conventions of 12 August 1949, and that persons who commit or order the commission of grave breaches of the Conventions are individually responsible in respect of such breaches.

26. Finally, the Security Council stated in resolution 808 (1993) that it was convinced that in the particular circumstances of the former Yugoslavia, the establishment of an international tribunal would bring about the achievement of the aim of putting an end to such crimes and of taking effective measures to bring to justice the persons responsible for them, and would contribute to the restoration and maintenance of peace.

27. The Security Council has on various occasions adopted decisions under

Chapter VII aimed at restoring and maintaining international peace and security, which have involved the establishment of subsidiary organs for a variety of purposes. Reference may be made in this regard to Security Council resolution 687 (1991) and subsequent resolutions relating to the situation between Iraq and Kuwait.

28. In this particular case, the Security Council would be establishing, as an enforcement measure under Chapter VII, a subsidiary organ within the terms of Article 29 of the Charter, but one of a judicial nature. This organ would, of course, have to perform its functions independently of political considerations, it would not be subject to the authority or control of the Security Council with regard to the performance of its judicial functions. As an enforcement measure under Chapter VII, however, the life span of the international tribunal would be linked to the restoration and maintenance of international peace and security in the territory of the former Yugoslavia, and Security Council decisions related thereto.

29. It should be pointed out that, in assigning to the International tribunal the task of prosecuting persons responsible for serious violations of international humanitarian law, the Security Council would not be creating or purporting to 'legislate' that law. Rather, the International Tribunal would have the task of applying existing international humanitarian law.

30. On the basis of the foregoing considerations, the Secretary-General proposes that the Security Council, acting under Chapter VII of the Charter establish the International Tribunal. . . .

II. COMPETENCE OF THE INTERNATIONAL TRIBUNAL

. . .

33. According to paragraph 1 of resolution 808 (1993), the international tribunal shall prosecute persons responsible for serious violations of international humanitarian law committed in the territory of the former Yugoslavia since 1991. This body of law exists in the form of both conventional law and customary law. While there is international customary law which is not laid down in conventions, some of the major conventional humanitarian law has become part of customary international law.

34. In the view of the Secretary-General, the application of the principle *nullum crimen sine lege* requires that the international tribunal should apply rules of international humanitarian law which are beyond any doubt part of customary law so that the problem of adherence of some but not all States to specific conventions does not arise. This would appear to be particularly important in the context of an international tribunal prosecuting persons responsible for serious violations of international humanitarian law.

35. The part of conventional international humanitarian law which has beyond doubt become part of international customary law is the law applicable in armed conflict as embodied in: the Geneva Conventions of 12 August 1949 for the Protection of War Victims; the Hague Convention (IV)

Respecting the Laws and Customs of War on Land and the Regulations annexed thereto of 18 October 1907; the Convention on the Prevention and Punishment of the Crime of Genocide of 9 December 1948; and the Charter of the International Military Tribunal of 8 August 1945.

36. Suggestions have been made that the international tribunal should apply domestic law in so far as it incorporates customary international humanitarian law. While international humanitarian law as outlined above provides a sufficient basis for subject-matter jurisdiction, there is one related issue which would require reference to domestic practice, namely, penalties (see para. 111 below).

Grave breaches of the 1949 Geneva Conventions

37. The Geneva Conventions constitute rules of international humanitarian law and provide the core of the customary law applicable in international armed conflicts. . . .

38. Each Convention contains a provision listing the particularly serious violations that qualify as 'grave breaches' or war crimes. Persons committing or ordering grave breaches are subject to trial and punishment. The lists of grave breaches contained in the Geneva Conventions are reproduced in the article which follows.

39. The Security Council has reaffirmed on several occasions that persons who commit or order the commission of grave breaches of the 1949 Geneva Conventions in the territory of the former Yugoslavia are individually responsible for such breaches as serious violations of international humanitarian law.

40. The corresponding article of the statute would read: [Article 2 appears in the Statute, *infra*.]

Violations of the laws or customs of war

41. The 1907 Hague Convention (IV) Respecting the Law and Customs of War on Land and the Regulations annexed thereto comprise a second important area of conventional humanitarian international law which has become part of the body of international customary law.

. . .

44. These rules of customary law, as interpreted and applied by the Nürnberg Tribunal, provide the basis for the corresponding article of the statute which would read as follows: [Article 3 appears in the Statute, *infra*.]

Genocide

45. The 1948 Convention on the Prevention and Punishment of the Crime of Genocide confirms that genocide, whether committed in time of peace or in time of war, is a crime under international law for which individuals shall be tried and punished. The Convention is today considered part of international customary law as evidenced by the International Court of Justice in its

Advisory Opinion on Reservations to the Convention on the Prevention and Punishment of the Crime of Genocide, 1951.

46. The relevant provisions of the Genocide Convention are reproduced in the corresponding article of the statute, which would read as follows: [Article 4 appears in the Statute, *infra.*]

Crimes against humanity

47. Crimes against humanity were first recognized in the Charter and Judgement of the Nürnberg Tribunal, as well as in Law No. 10 of the Control Council for Germany. Crimes against humanity are aimed at any civilian population and are prohibited regardless of whether they are committed in an armed conflict, international or internal in character.

48. . . . In the conflict in the territory of the former Yugoslavia, such inhumane acts have taken the form of so-called 'ethnic cleansing' and widespread and systematic rape and other forms of sexual assault, including enforced prostitution.

49. The corresponding article of the statute would read as follows: [Article 5 appears in the Statute, *infra.*]

STATUTE OF THE INTERNATIONAL TRIBUNAL FOR THE FORMER YUGOSLAVIA

Reprinted in 14 Hum. Rts. L. J. 211 (1993).

Having been established by the Security Council acting under Chapter VII of the Charter of the United Nations, the International Tribunal for the Prosecution of Persons Responsible for Serious Violations of International Humanitarian Law Committed in the Territory of the Former Yugoslavia since 1991 (hereinafter referred to as 'the International Tribunal') shall function in accordance with the provisions of the present Statute.

Article 1

Competence of the International Tribunal

The International Tribunal shall have the power to prosecute persons responsible for serious violations of international humanitarian law committed in the territory of the former Yugoslavia since 1991 in accordance with the provision of the present Statute.

Article 2

Grave breaches of the Geneva Conventions of 1949

The International Tribunal shall have the power to prosecute persons committing or ordering to be committed grave breaches of the Geneva Conventions of 12 August 1949, namely the following acts against persons or property protected under the provisions of the relevant Geneva Convention:

(a) wilful killing;
(b) torture or inhuman treatment, including biological experiments;
(c) wilfully causing great suffering or serious injury to body or health;
(d) extensive destruction and appropriation of property, not justified by military necessity and carried out unlawfully and wantonly;
(e) compelling a prisoner of war or a civilian to serve in the forces of a hostile power;
(f) wilfully depriving a prisoner of war or a civilian of the rights of fair and regular trial;
(g) unlawful deportation or transfer or unlawful confinement of a civilian;
(h) taking civilians as hostages.

Article 3
Violations of the laws or customs of war

The International Tribunal shall have the power to prosecute persons violating the laws or customs of war. Such violations shall include, but not be limited to:

(a) employment of poisonous weapons or other weapons calculated to cause unnecessary suffering;
(b) wanton destruction of cities, towns or villages, or devastation not justified by military necessity;
(c) attack, or bombardment, by whatever means, of undefended towns, villages, dwellings, or buildings;
(d) seizure of, destruction or wilful damage done to institutions dedicated to religion, charity and education, the arts and sciences, historic monuments and works of art and science;
(e) plunder of public or private property.

Article 4
Genocide

1. The International Tribunal shall have the power to prosecute persons committing genocide as defined in paragraph 2 of this article or of committing any of the other acts enumerated in paragraph 3 of this article.

2. Genocide means any of the following acts committed with intent to destroy, in whole or in part, a national, ethnical, racial or religious group, as such:

(a) killing members of the group;
(b) causing serious bodily or mental harm to members of the group;
(c) deliberately inflicting on the group conditions of life calculated to bring about its physical destruction in whole or in part;
(d) imposing measures intended to prevent births within the group;

(e) forcibly transferring children of the group to another group.

3. The following acts shall be punishable:

(a) genocide;
(b) conspiracy to commit genocide;
(c) direct and public incitement to commit genocide;
(d) attempt to commit genocide;
(e) complicity in genocide.

Article 5
Crimes against humanity

The International Tribunal shall have the power to prosecute persons responsible for the following crimes when committed in armed conflict, whether international or internal in character, and directed against any civilian population:

(a) murder;
(b extermination;
(c) enslavement;
(d) deportation;
(e) imprisonment;
(f) torture;
(g) rape;
(h) persecutions on political, racial and religious grounds;
(i) other inhumane acts.

Article 6
Personal jurisdiction

The International Tribunal shall have jurisdiction over natural persons pursuant to the provisions of the present Statute.

Article 7
Individual criminal responsibility

1. A person who planned, instigated, ordered, committed or otherwise aided and abetted in the planning, preparation or execution of a crime referred to in articles 2 to 5 of the present Statute, shall be individually responsible for the crime.

2. The official position of any accused person, whether as Head of State or Government or as responsible Government official, shall not relieve such person of criminal responsibility nor mitigate punishment.

3. The fact that any of the acts referred to in articles 2 to 5 of the present Statute was committed by a subordinate does not relieve his superior of crim-

inal responsibility if he know or had reason to know that the subordinate was about to commit such acts or had done so and the superior failed to take the necessary and reasonable measures to prevent such acts or to punish the perpetrators thereof.

4. The fact that an accused person acted pursuant to an order of a Government or of a superior shall not relieve him of criminal responsibility, but may be considered in mitigation of punishment if the International Tribunal determines that justice so requires.

Article 8

Territorial and temporal jurisdiction

The territorial jurisdiction of the International Tribunal shall extend to the territory of the former Socialist Federal Republic of Yugoslavia, including its land surface, airspace and territorial waters. The temporal jurisdiction of the International Tribunal shall extend to a period beginning on 1 January 1991.

Article 9

Concurrent jurisdiction

1. The International Tribunal and national courts shall have concurrent jurisdiction to prosecute persons for serious violations of international humanitarian law committed in the territory of the former Yugoslavia since 1 January 1991.

2. The International Tribunal shall have primacy over national courts. At any stage of the procedure, the International Tribunal may formally request national courts to defer to the competence of the International Tribunal in accordance with the present Statute and the Rules of Procedure and Evidence of the International Tribunal.

Article 10

Non-bis-in-idem

1. No person shall be tried before a national court for acts constituting serious violations of international humanitarian law under the present Statute, for which he or she has already been tried by the International Tribunal.

2. A person who has been tried by a national court for acts constituting serious violations of international humanitarian law may be subsequently tried by the International Tribunal only if:

(a) the act for which he or she was tried was characterized as an ordinary crime; or
(b) the national court proceedings were not impartial or independent, were designed to shield the accused from international criminal responsibility, or the case was not diligently prosecuted.

3. In considering the penalty to be imposed on a person convicted of a crime under the present Statute, the International Tribunal shall take into account

the extent to which any penalty imposed by a national court on the same person for the same act has already been served.

. . .

Article 18

Investigation and preparation of indictment

1. The Prosecutor shall initiate investigations ex-officio or on the basis of information obtained from any source, particularly from Governments, United Nations organs, intergovernmental and non-governmental organizations. The Prosecutor shall assess the information received or obtained and decide whether there is sufficient basis to proceed.

2. The Prosecutor shall have the power to question suspects, victims and witnesses, to collect evidence and to conduct on-site investigations. In carrying out these tasks, the Prosecutor may, as appropriate, seek the assistance of the State authorities concerned.

3. If questioned, the suspect shall be entitled to be assisted by counsel of his own choice, including the right to have legal assistance assigned to him without payment by him in any case if he does not have sufficient means to pay for it, as well as to necessary translation into and from a language he speaks and understands.

4. Upon a determination that a prima facie case exists, the Prosecutor shall prepare an indictment containing a concise statement of the facts and the crime or crimes with which the accused is charged under the Statute. The indictment shall be transmitted to a judge of the Trial Chamber.

Article 19

Review of the indictment

1. The judge of the Trial Chamber to whom the indictment has been transmitted shall review it. If satisfied that a prima facie case has been established by the Prosecutor, he shall confirm the indictment. If not so satisfied, the indictment shall be dismissed.

2. Upon confirmation of an indictment, the judge may, at the request of the Prosecutor, issue such orders and warrants for the arrest, detention, surrender or transfer of persons, and any orders as may be required for the conduct of the trial.

Article 20

Commencement and conduct of trial proceedings

1. The Trial Chambers shall ensure that a trial is fair and expeditious and that proceedings are conducted in accordance with the rules of procedure and evidence, with full respect for the rights of the accused and due regard for the protection of victims and witnesses.

2. A person against whom an indictment has been confirmed shall, pursuant

to an order or an arrest warrant of the International Tribunal, be taken into custody, immediately informed of the charges against him and transferred to the International Tribunal.

3. The Trial Chamber shall read the indictment, satisfy itself that the rights of the accused are respected, confirm that the accused understands the indictment, and instruct the accused to enter a plea. The Trial Chamber shall then set the date for trial.

4. The hearings shall be public unless the Trial Chamber decides to close the proceedings in accordance with its rules of procedure and evidence.

Article 21

Rights of the accused

1. All persons shall be equal before the International Tribunal.

2. In the determination of charges against him, the accused shall be entitled to a fair and public hearing, subject to article 22 of the Statute.

3. The accused shall be presumed innocent until proved guilty according to the provisions of the present Statute.

4. In the determination of any charge against the accused pursuant to the present Statute, the accused shall be entitled to the following minimum guarantees, in full equality:

(a) to be informed promptly and in detail in a language which he understands of the nature and cause of the charge against him;
(b) to have adequate time and facilities for the preparation of his defence and to communicate with counsel of his own choosing;
(c) to be tried without undue delay;
(d) to be tried in his presence, and to defend himself in person or through legal assistance of his own choosing; to be informed, if he does not have legal assistance, of this right; and to have legal assistance assigned to him, in any case where the interests of justice so require, and without payment by him in any such case if he does not have sufficient means to pay for it;
(e) to examine, or have examined, the witnesses against him and to obtain the attendance and examination of witnesses on his behalf under the same conditions as witnesses against him;
(f) to have the free assistance of an interpreter if he cannot understand or speak the language used in the International Tribunal;
(g) not to be compelled to testify against himself or to confess guilt.

Article 22

Protection of victims and witnesses

The International Tribunal shall provide in its rules of procedure and evidence for the protection of victims and witnesses. Such protection measures shall include, but shall not be limited to, the conduct of *in camera* proceedings and the protection of the victim's identity.

Article 23
Judgement

1. The Trial Chambers shall pronounce judgements and impose sentences and penalties on persons convicted of serious violations of international humanitarian law.

2. The judgement shall be rendered by a majority of the judges of the Trial Chamber, and shall be delivered by the Trial Chamber in public. It shall be accompanied by a reasoned opinion in writing, to which separate or dissenting opinions may be appended.

Article 24
Penalties

1. The penalty imposed by the Trial Chamber shall be limited to imprisonment. In determining the terms of imprisonment, the Trial Chambers shall have recourse to the general practice regarding prison sentences in the courts of the former Yugoslavia.

2. In imposing the sentences, the Trial Chambers should take into account such factors as the gravity of the offence and the individual circumstances of the convicted person.

3. In addition to imprisonment, the Trial Chambers may order the return of any property and proceeds acquired by criminal conduct, including by means of duress, to their rightful owners.

Article 25
Appellate proceedings

1. The Appeals Chamber shall hear appeals from persons convicted by the Trial Chambers or from the Prosecutor on the following grounds:

(a) an error on a question of law invalidating the decision; or
(b) an error of fact which has occasioned a miscarriage of justice.

2. The Appeals Chamber may affirm, reverse or revise the decisions taken by the Trial Chambers.

. . .

Article 27
Enforcement of sentences

Imprisonment shall be served in a State designated by the International Tribunal from a list of States which have indicated to the Security Council their willingness to accept convicted persons. Such imprisonment shall be in accordance with the applicable law of the State concerned, subject to the supervision of the International Tribunal.

. . .

Article 29

Cooperation and judicial assistance

1. States shall cooperate with the International Tribunal in the investigation and prosecution of persons accused of committing serious violations of international humanitarian law.

2. States shall comply without undue delay with any request for assistance or an order issued by a Trial Chamber, including, but not limited to:

(a) the identification and location of persons;
(b) the taking of testimony and the production of evidence;
(c) the service of documents;
(d) the arrest or detention of persons;
(e) the surrender or the transfer of the accused to the International Tribunal.

. . .

Article 32

Expenses of the International Tribunal

The expenses of the International Tribunal shall be borne by the regular budget of the United Nations in accordance with Article 17 of the Charter of the United Nations.

. . .

NOTE

By Resolution 955 (1994), the Security Council established an International Tribunal for Rwanda 'to prosecute persons responsible for genocide and other serious violations of international humanitarian law' committed in that country in 1994. This tribunal is closely related to the International Tribunal for the Former Yugoslavia, using the same Prosecutor and appellate judges, but having separate trial judges. The Statutes of the two tribunals are in most respects identical but differ in a few important ways. For example, the internal rather than international nature of the conflict in Rwanda required different references in that tribunal's Statute to treaties and provisions thereof on war crimes.

RULES OF PROCEDURE AND EVIDENCE OF THE INTERNATIONAL TRIBUNAL FOR THE FORMER YUGOSLAVIA

As amended, IT/32/Rev.3, 30 January 1995.

. . .

Rule 8

Request for Information

Where it appears to the Prosecutor that a crime within the jurisdiction of the Tribunal is or has been the subject of investigations or criminal proceedings instituted in the courts of any State, he may request the State to forward to him all relevant information in that respect, and the State shall transmit to him such information forthwith in accordance with Article 29 of the Statute.

Rule 9

Prosecutor's Request for Deferral

Where it appears to the Prosecutor that in any such investigations or criminal proceedings instituted in the courts of any State:

(i) the act being investigated or which is the subject of those proceedings is characterized as an ordinary crime;
(ii) there is a lack of impartiality or independence, or the investigations or proceedings are designed to shield the accused from international criminal responsibility, or the case is not diligently prosecuted; or
(iii) what is in issue is closely related to, or otherwise involves, significant factual or legal questions which may have implications for investigations or prosecutions before the Tribunal,

the Prosecutor may propose to the Trial Chamber designated by the President that a formal request be made that such court defer to the competence of the Tribunal.

Rule 10

Formal Request for Deferral

(A) If it appears to the Trial Chamber seised of a proposal for deferral that, on any of the grounds specified in Rule 9, deferral is appropriate, the Trial Chamber may issue a formal request to the State concerned that its court defer to the competence of the Tribunal.
(B) A request for deferral shall include a request that the results of the investigation and a copy of the court's records and the judgement, if already delivered, be forwarded to the Tribunal.
(C) Where deferral to the Tribunal has been requested by a Trial Chamber, any subsequent proceedings shall be held before the other Trial Chamber.

Rule 11

Non-compliance with a Request for Deferral

If, within sixty days after a request for deferral has been notified by the Registrar to the State under whose jurisdiction the investigations or criminal proceedings have been instituted, the State fails to file a response which satisfies the Trial Chamber that the State has taken or is taking adequate steps to comply with the order, the Trial Chamber may request the President to report the matter to the Security Council.

Rule 12

Determinations of Courts of any State

Subject to Article 10(2) of the Statute, determinations of courts of any State are not binding on the Tribunal.

Rule 13

Non Bis in Idem

When the President receives reliable information to show that criminal proceedings have been instituted against a person before a court of any State for a crime for which that person has already been tried by the Tribunal, a Trial Chamber shall, following *mutatis mutandis* the procedure provided in Rule 10, issue a reasoned order requesting that court permanently to discontinue its proceedings. If that court fails to do so, the President may report the matter to the Security Council.

. . .

Rule 54

General Rule

At the request of either party or *proprio motu*, a Judge or a Trial Chamber may issue such orders, summonses, subpoenas and warrants as may be necessary for the purposes of an investigation or for the preparation or conduct of the trial.

Rule 55

Execution of Arrest Warrants

(A) A warrant of arrest shall be signed by a Judge and shall bear the seal of the Tribunal. It shall be accompanied by a copy of the indictment, and a statement of the rights of the accused. . . .

(B) A warrant for the arrest of the accused and an order for his surrender to the Tribunal shall be transmitted by the Registrar to the national authorities of the State in whose territory or under whose jurisdiction or control the accused resides, or was last known to be. . . .

. . .

Rule 56

Cooperation of States

The State to which a warrant of arrest is transmitted shall act promptly and with all due diligence to ensure proper and effective execution thereof, in accordance with Article 29 of the Statute.

Rule 57

Procedure after Arrest

Upon the arrest of the accused, the State concerned shall detain him, and shall promptly notify the Registrar. The transfer of the accused to the seat of the Tribunal shall be arranged between the State authorities concerned, the authorities of the host country and the Registrar.

Rule 58

National Extradition Provisions

The obligations laid down in Article 29 of the Statute shall prevail over any legal impediment to the surrender or transfer of the accused to the Tribunal which may exist under the national law or extradition treaties of the State concerned.

Rule 59

Failure to Execute a Warrant

(A) Where the State to which a warrant of arrest has been transmitted has been unable to execute the warrant, it shall report forthwith its inability to the Registrar, and the reasons therefore.

(B) If, within a reasonable time after the warrant of arrest has been transmitted to the State, no report is made on action taken, this shall be deemed a failure to execute the warrant of arrest and the Tribunal, through the President, may notify the Security Council accordingly.

. . .

Rule 61

Procedure in Case of Failure to Execute a Warrant

. . .

(D) The Trial Chamber shall also issue an international arrest warrant in respect of the accused which shall be transmitted to all States.

. . .

Rule 68

Disclosure of Exculpatory Evidence

The Prosecutor shall, as soon as practicable, disclose to the defence the existence of evidence known to the Prosecutor which in any way tends to suggest the innocence or mitigate the guilt of the accused or may affect the credibility of prosecution evidence.

Rule 69

Protection of Victims and Witnesses

(A) In exceptional circumstances, the Prosecutor may apply to a Trial Chamber to order the non-disclosure of the identity of a victim or witness who may be in danger or at risk until such person is brought under the protection of the Tribunal.

(B) Subject to Rule 75, the identity of the victim or witness shall be disclosed in sufficient time prior to the trial to allow adequate time for preparation of the defence.

. . .

Rule 75

Measures for the Protection of Victims and Witnesses

(A) A Judge or a Chamber may, *proprio motu* or at the request of either party, or of the victim or witness concerned, order appropriate measures for the privacy and protection of victims and witnesses, provided that the measures are consistent with the rights of the accused.

(B) A Chamber may hold an *in camera* proceeding to determine whether to order:

(i) measures to prevent disclosure to the public or the media of the identity or whereabouts of a victim or a witness, or of persons related to or associated with him . . . ;
(ii) closed sessions, in accordance with Rule 79;
(iii) appropriate measures to facilitate the testimony of vulnerable victims and witnesses, such as one-way closed circuit television.

. . .

Rule 77

Contempt of the Tribunal

(A) Subject to the provisions of Sub-rule 90(E), a witness who refuses or fails contumaciously to answer a question relevant to the issue before a Chamber may be found in contempt of the Tribunal. The Chamber may impose a fine not exceeding US$10,000 or a term of imprisonment not exceeding six months.

. . .

Rule 95

Evidence Obtained by Means Contrary to Internationally Protected Human Rights

No evidence shall be admissible if obtained by methods which cast substantial doubt on its reliability or if its admission is antithetical to, and would seriously damage, the integrity of the proceedings.

Rule 96

Evidence in Cases of Sexual Assault

In cases of sexual assault:

(i) no corroboration of the victim's testimony shall be required;
(ii) consent shall not be allowed as a defence if the victim
  (a) has been subjected to or threatened with or has had reason to fear violence, duress, detention or psychological oppression, or
  (b) reasonably believed that if the victim did not submit, another might be so subjected, threatened or put in fear;
(iii) before evidence of the victim's consent is admitted, the accused shall satisfy the Trial Chamber *in camera* that the evidence is relevant and credible;
(iv) prior sexual conduct of the victim shall not be admitted in evidence.

. . .

Rule 103

Place of Imprisonment

(A) Imprisonment shall be served in a State designated by the Tribunal from a list of States which have indicated their willingness to accept convicted persons.

. . .

Rule 105

Restitution of Property

(A) After a judgement of conviction containing a specific finding as provided in Sub-rule 88(B), the Trial Chamber shall, at the request of the Prosecutor, or may, at its own initiative, hold a special hearing to determine the matter of the restitution of the property or the proceeds thereof, and may in the meantime order such provisional measures for the preservation and protection of the property or proceeds as it considers appropriate.

. . .

Rule 106

Compensation to Victims

(A) The Registrar shall transmit to the competent authorities of the States concerned the judgement finding the accused guilty of a crime which has caused injury to a victim.

(B) Pursuant to the relevant national legislation, a victim or persons claiming through him may bring an action in a national court or other competent body to obtain compensation.

(C) For the purposes of a claim made under Sub-rule (B) the judgement of the Tribunal shall be final and binding as to the criminal responsibility of the convicted person for such injury.

. . .

JAMES O'BRIEN, THE INTERNATIONAL TRIBUNAL FOR VIOLATIONS OF INTERNATIONAL HUMANITARIAN LAW IN THE FORMER YUGOSLAVIA

87 Am. J. Int. L. 639 (1993).

. . .

1. Establishment of the Tribunal

A confluence of political opportunity, disappointed expectations of peace, and publicity have allowed the Security Council to respond to atrocities in the former Yugoslavia. The breakdown of the bipolar world since 1989 has enabled major powers to find common, not opposed, interest in punishing violations of international law. The end of the Cold War may also have raised hopes of a comfortable peace, at least in Europe, perhaps because it threatened these expectations, the fighting in the former Yugoslavia sparked a strong international response. Daily media reports of atrocities—including the abuse of women, inhumane detention facilities, indiscriminate targeting of defenseless civilians, forced expulsions and deportations, and the obstruction of relief convoys—have maintained a high level of public outrage. There is cause to doubt that the available national judicial systems (which would usually be responsible for enforcing international humanitarian law) can be either impartial or effective in punishing those responsible for violations. Yet the victims of atrocities in the former Yugoslavia deserve justice, as do all those elsewhere who will be protected by strengthened international humanitarian law.

. . .

[One] part of the explanation for the Security Council's response lies in the exhaustion of alternative remedies. On becoming seized of the issue in 1991, the Council tried to facilitate negotiations and efforts to reduce the human toll, and it has continued that effort. Other organizations, such as the UN Human Rights Commission, the Conference on Security and Co-operation in Europe (CSCE), the European Community, and the International Conference on the Former Yugoslavia, have all tried, without success, to end the conflict or at least reduce the fighting.

By the summer of 1992, widespread and credible reports of atrocities in the former Yugoslavia prompted the Council to take four steps in succession: condemnation; publication; investigation; and, by establishing the tribunal, punishment. As a first step, the Council publicly *condemned* atrocities as violations of international humanitarian law. . . .

As the reported atrocities grew throughout the month of July 1992, the Council took a second step by *publicizing* them. In Resolution 771, on August 12, 1992, the Council called upon states and other bodies to submit 'substantiated information' to the Secretary-General, who would report to the Council 'recommending additional measures that might be appropriate.' . . .

Without waiting for the submission of a single report from a state or for the Secretary-General's recommendations for further steps, the Council, on October 6, took the third step of *investigating* violations through a 'Commission of Experts,' established by Resolution 780. It was hoped that the formation of the 'War Crimes Commission,' the first since World War II, would deter abuses. It was tasked with preparing cases for possible prosecution in national or international courts, and drawing up an internationally accepted record of atrocities and their perpetrators.

. . .

Finally, on February 22, 1993, Security Council Resolution 808 announced the fourth step: *punishment* through due process of law, by creating an international tribunal to prosecute those violating international humanitarian law in the former Yugoslavia. Pursuant to Resolution 808, the Secretary-General submitted his report, which included a draft statute for the tribunal, on May 6, 1993, taking into account the views of thirty-one states and several organizations. The Council established the tribunal on May 25, 1993, in Resolution 827, which unanimously adopted the statute developed by the Secretary-General, without change.

The Council has the authority under chapter VII to take measures necessary to maintain international peace and security. The Council may suspend commerce between states or authorize the start of hostilities; it would be odd if it could not take the lesser, surgical step of ameliorating a threat to international peace and security by providing for the prosecution of individuals who violate well-established international law. Moreover, the threat to international peace and security at issue—violations of international humanitarian law—is best addressed by a judicial remedy. . . .

Furthermore, provision for consent is built into the tribunal's statute. The UN General Assembly must approve the tribunal's budget, and thus approve its mandate and, in effect, its statute. The tribunal's authority to issue mandatory orders is limited to the transfer of indicted individuals and to other forms of judicial assistance. . . .

Perhaps most importantly, the Yugoslav tribunal is accountable to the defendants, who are not prohibited from challenging its authority (as they were under the Nuremberg Charter). The tribunal may—and probably should—decide that it lacks the power to consider the legitimacy of the Security Council's actions to establish it, because this would require that it construe the powers of the Security Council. . . .

II. Structure and Powers of the Tribunal

. . .

Structure of the Tribunal. Articles 11–15 of the statute provide that the tribunal will have eleven judges, three in each of two trial chambers and five in a single appellate chamber. There are no provisions for alternate judges or for the appellate court to sit in panels. These and related topics will presumably be addressed in the tribunal's rules of procedure, which will be adopted by the tribunal. Article 16 establishes the office of the prosecutor to conduct investigations and prepare indictments; the prosecutor will select his or her own staff. . . . Article 31 provides that the seat of the tribunal will be at The Hague. . . . Regardless of where the tribunal's headquarters are located, it may conduct proceedings wherever required to carry out its functions.

Subject matter jurisdiction. The Secretary-General emphasized that the law to be applied must be well established. Pursuant to Articles 1–5 of its statute, the tribunal will have jurisdiction over war crimes and crimes against humanity, including genocide, committed as of January 1, 1991, almost seven months before Slovenia became the first former Yugoslav republic to declare independence. The Charters of the Nuremberg and Tokyo Tribunals recognized a third crime, crimes against peace. Crimes against peace are appropriately omitted from the Yugoslav statute. Their inclusion would almost inevitably require the tribunal to investigate the causes of the conflict itself (and the justifications issued by the combatants), which would involve the tribunal squarely in the political issues surrounding the conflict.

Articles 2 and 3 provide the tribunal with jurisdiction over violations of the international law of armed conflict. These articles should provide the basis for most of the tribunal's work, as virtually any act that might constitute a crime against humanity or genocide would also be a war crime. Without question, rape—one crime that may fall into all three categories—is a war crime.

In explaining the scope of Articles 2 and 3, the Secretary-General referred only to the 1907 Hague Conventions, the Nuremberg judgment, and the 1949 Geneva Conventions. The elastic 'but not limited to' language of Article 3 ensures that all relevant, well-established international law falls within the tribunal's jurisdiction. The United States, the United Kingdom and France explained, when voting for Resolution 827, that the laws and customs of war under Article 3 included all law of armed conflict in force in the territory of the former Yugoslavia, in addition to customary international law. The tribunal may thus apply the provisions of the Geneva Conventions on nongrave breaches, including common Article 3 concerning internal armed conflict, and the Additional Protocols to the Conventions.

. . .

Common Article 3 and Additional Protocol II may prove to be even more important. The Hague Regulations, the Nuremberg judgment and Additional Protocol I apply to *international* armed conflicts. The conflict in the former Yugoslavia has internal and international elements. There was only one country until Slovenia declared independence on June 25, 1991. Thereafter, some fighting involved intranational conflict (particularly Bosnian against Bosnian) and some international elements, such as the Yugoslav National Army (JNA); the JNA was officially withdrawn from Bosnia-Hercegovina in the spring of 1992, ostensibly leaving only internal combatants there. It is possible (but not correct, as I note below) that the tribunal might determine the conflict to be internal. . . .

The tribunal should conclude that the conflict in the former Yugoslavia is covered by the law applicable to international armed conflict, at least after June 25, 1991, when the former republics of Yugoslavia began declaring their independence. Consistent with the provisions of the Geneva Conventions, the parties to the conflict have agreed to apply the law of international armed conflict. Moreover, some internal conflicts fall under the law of international armed conflict. Most importantly, the conflict is clearly international: three nations have fought, primarily in the territory of two of them (thus far), with a number of fronts and partisan or proxy groups participating on behalf of each. Once this determination is made, it should not matter that some combatants are citizens of the same nation-state. . . .

Articles 4 and 5 give the tribunal jurisdiction over crimes against humanity, including genocide. Although these crimes overlap considerably with war crimes, they may reach some acts that are not war crimes, for example, acts by persons not covered by the laws of armed conflict, acts outside an armed conflict and a state's acts against its own civilians. . . .

Genocide—the attempt to destroy a group in whole or in part—is a special case of crimes against humanity. Article 4 of the statute adopts verbatim the definition in the 1948 Genocide Convention of genocide . . . Because genocide requires proof of the intention to destroy a group, not only criminal intention against an individual (as is true of war crimes) or on the basis of group membership (as may be the case with crimes against humanity), it may be the most difficult charge to prove, particularly against lower-level persons whose connection to a policy may be hard to trace.

Article 5 of the statute defines crimes against humanity. . . .

Article 5 also requires that crimes against humanity be committed against a 'civilian population.' Thus, at a minimum, crimes against humanity must be systematic or organized, not simply episodic and/or scattered attacks on individuals. . . .

The most difficult question concerns the requirement that crimes be 'committed in armed conflict.' The legacy of the post-World War II tribunals is unclear on this point. Both the Nuremberg and Tokyo Charters restricted the vitality of crimes against humanity by requiring that they be committed 'in execution of, or in connection with' another crime in the Charters of the tribunals. . . . The United Nations has not endorsed a definition of crimes against humanity that severs such crimes from armed conflict. In making recommendations for the Yugoslav tribunal, however, most states that dealt with the topic did not connect crimes against humanity to another crime or an armed conflict.

On its face, the Yugoslav statute requires only a connection between crimes against humanity and armed conflict, which is not itself a crime under the statute; it thus marks a modest advance over the Nuremberg Charter by expressly removing the requirement of connection to another crime under international law. The question remains, however, of how to interpret 'committed in armed conflict.' It would make sense to interpret the phrase as 'during' the conflict, without requiring any proof of substantive connection between the crime and the armed conflict. This reading comports with the intention of the Secretary-General and the Security Council, which did not require substantive links between crimes against humanity and armed conflict. . . .

. . .

Individual criminal responsibility. . . .

Paragraphs 3 and 4 of Article 7 pose some of the more interesting questions facing the tribunal: the responsibility of superiors for acts of subordinates and of persons following orders. The tribunal should aim at higher officials who have guided or at least benefited from the atrocities that anger the world. It is not clear how useful Articles 7(1) and 4 will be in this regard. In the former Yugoslavia, the military situation appears (on the basis of information available at the time of writing) to be chaotic, with front-line forces operating under opaque (and perhaps nonexistent) lines of authority, spotty communications and rare direct orders.

If orders or other direct involvement cannot be found, prosecutions of political and military leaders will rest on the doctrine of command responsibility, which requires commanders to repress or punish violations of international humanitarian law by their subordinates. Article 7(3) provides that a commander is not 'relieved' of his individual responsibility if he or she did not prevent or punish violations of which he or she knew or had reason to know. Although the article does not indicate the source of the commander's 'responsibility,' the Secretary-General says, and the relevant statements by Security Council members agree, that the responsibility is to be defined in accordance with well-established international law:

> This imputed responsibility or criminal negligence is engaged if the person in superior authority knew or had reason to know that his subordinates were about to commit or had committed crimes and yet failed to take the necessary and reasonable steps to prevent or repress the commission of such crimes or to punish those who had committed them.

Here the Secretary-General accurately states international law. Yet to apply this standard to the facts of the conflict in the former Yugoslavia will present formidable and interesting issues, if persons of political and military authority appear before the tribunal.

. . .

In the former Yugoslavia, much of the fighting is done by small paramilitary bands, making it difficult to determine a chain of political or military authority. The application of the doctrine of command responsibility to persons with oblique or attenuated authority will depend heavily on specific facts and may prove very difficult. Nevertheless, a few legal doctrines derived from the case law provide a tentative start. Evidence of a pattern (similarity of atrocities, coordination among those who commit them) can support the inference that persons with military or political authority (even if not over those committing the atrocities) acquiesced in them. Further, those with occupational or executive responsibility for an area are responsible for repressing and punishing violations, even if committed by forces not under their operational or administrative control. The tribunal should be reluctant to doubt the authority of persons who present themselves (or their superiors or subordinates) as powerful, for example, in peace negotiations. In addition, persons cannot create ignorance or impotence by their own actions: the duty to repress or punish violations cannot be discharged by delegation, by forfeiting authority even in the face of a deteriorating command-and-control capacity, or by permitting troops not under one's control to take over an area. . . .

. . .

III. Conclusion

States are also accountable for the tribunal's success. They should provide it with adequate resources and their cooperation and persuade other states to cooperate fully as well. To the extent that states are concerned that the tribunal may encroach on their own responsibilities to enforce international humanitarian law, they should be encouraged to conduct thorough and impartial proceedings of their own. The early difficulties of the War Crimes Commission, which tried to operate without active state interest, demonstrate the importance of continual state involvement and teach the simple lesson that, in the end, states are responsible for these international endeavors. All states have their self-interest at stake: viable international humanitarian law will one day protect their own citizens, as well as deter violations that may quickly threaten international peace and security.

It may be more important that states create a political climate in which the tribunal will have a chance to succeed. States should speak out clearly and resolutely against violations of international humanitarian law. In addition, they should explain repeatedly that the tribunal lacks the mandate and ability to consider the responsibility of racial, ethnic or religious groups for the conflict in the former Yugoslavia. It is, instead, a means to determine *individual* responsibility, and both states and the tribunal itself must be careful to keep it separate from efforts to allocate responsibility for, or rewards from, the broader conflict. Simply put, the tribunal is aimed at bad soldiers (and those who should control them), not bad sides or groups. If the tribunal is merged into the broader political discourse of responsibility and resolution of the conflict, states and groups that perceive themselves as having little to gain from such a body may withhold their cooperation. It may also be easier to regard the tribunal as another element of the group conflict, to be settled (and perhaps traded away) between and among groups, without permitting recourse to a full and fair determination of individual responsibility. For the principle of individual accountability underlying international humanitarian law, this could be the most cynical result of all.

. . .

NOTE

The systematic incidence of mass rape in the former Yugoslavia has drawn particular attention from observers and scholars, both in terms of the legal characterization of rape under international humanitarian law[6], and in terms of its strategic use as an instrument of fear, shame and ethnic cleansing. The crime of rape is referred to once in the Tribunal's Statute, as an instance of crimes against humanity under Article 5.

Consider the following observations about the different motivations and consequences of rape in wartime, the excerpts below being directed to the former Yugoslavia:[7]

> Rape has long been mischaracterized and dismissed by military and political leaders . . . as a private crime, a sexual act, the ignoble conduct of the occasional soldier, or, worse still, it has been accepted precisely because it is so commonplace.
>
> . . .

[6] Theodor Meron, Rape as a Crime under International Humanitarian Law, 87 Am. J. Int. L. 424 (1993).

[7] Dorothy Thomas and Regan Ralph, Rape in War: Challenging the Tradition of Impunity, SAIS Review 81, 84 (Winter–Spring 1994). See generally 5 Hastings Women's L. J. (Summer 1994, No. 2).

In fact, rape is neither incidental nor private. It routinely serves a strategic function in war and acts as an integral tool for achieving particular military objectives. In the former Yugoslavia, rape and other grave abuses committed by Serbian forces are intended to drive the non-Serbian population into flight.

. . . The attention to rape's strategic function, however, has attached much significance to 'mass rape' and 'rape as genocide.' This emphasis on rape's scale as to what makes it an abuse demanding redress distorts the nature of rape in war by failing to reflect both the experience of individual women and the various functions of wartime rape.

Rape rises to the level of a war crime or a grave breach of the Geneva Conventions regardless of whether it occurs on a demonstrably massive scale or is associated with an overarching policy. Individual rapes that function as torture or cruel and inhuman treatment themselves constitute grave breaches of the Geneva Conventions. . . . When rape does occur on a mass scale or as a matter of orchestrated policy, this added dimension of the crime is recognized by designating and prosecuting rape as a crime against humanity.

. . .

Although rape is a sex-specific type of abuse, it generally functions like other forms of torture to intimidate and punish individual women. In some instances, however, it also can serve a striking sex-specific function, when, for example, it is committed with the intent of impregnating its victims. A Bosnian rape victim told Human Rights Watch, '. . . They wanted to humiliate us. They would say directly, looking into your eyes, that they wanted to make a baby.'

Soldiers are motivated to rape precisely because rape serves the strategic interests delineated above. But the fact that it is predominantly men raping women reveals that rape in war, like all rape, reflects a gender-based motivation, namely, the assertion by men of their power over women. . . . It is therefore difficult to distinguish the gender elements of a rapist's motivation from the specific political function served by the rape.

COMMENT ON ISSUES BEFORE THE TRIBUNAL

The establishment of this International Criminal Tribunal (ICT) for the Former Yugoslavia was a truly historic event, holding considerable promise and inescapable risk. This Comment notes several aspects of the ICT and its work.

1. Observers have read different motivations into the Security Council's decision to establish the Tribunal. Some understand the ICT to be an essential response by the Council to the public outcry after exposure by the media of the outrages in the conflict—a minimum response, an effort to do 'something' that could prove to be significant and that was politically manageable (unlike the failures in efforts at negotiation or in discussion of types of inter-

vention). Others understand the Tribunal to be a slave to conscience for the West, a way of dealing with ethnic cleansing and the accompanying brutality without effectively dealing with it. Whatever the motivations—and they were surely complex—the fact remains that a tribunal has been created, and the arguments for or against its establishment are now irrelevant. Success in its work—that is, some minimum of convictions after the fair trials by independent and impartial judges that the Statute requires—seems imperative, for failure at this stage would be far more serious than an earlier decision not to establish the tribunal at all.

2. The ICT is in a radically different situation from a court in a state observing fundamental principles of the Rule of Law in the sense that the state's executive and legislative branches comply with and execute court judgments. The Security Council has created an independent organ, as must be the case. Nonetheless, the ICT remains fundamentally dependent on an uncertain and changing political context; it lacks the relative autonomy of a court in a state with a strong tradition of an independent judiciary. The Tribunal depends for funds on a UN General Assembly whose members hold different views about it and who may judge its work differently. It must receive support from states and from the Security Council with respect to such basic matters as enforcing its orders related to gathering evidence, or obtaining custody of indicted persons. There is no equivalent to a 'national tradition' for the Tribunal to draw on.

3. Beyond its fundamental mission of bringing a sense of justice and reconciliation to the combatants and civilians in the area, the ICT possesses an exceptional opportunity to develop international law in the field of individual criminal responsibility in an authoritative way. Its distinguished Prosecutor (and his staff) and the many distinguished judges will confront in their presentations of cases and in their decisions numerous vexing issues, some of ancient lineage and some bred by the developments over the last half century in international humanitarian law including the crimes defined at Nuremberg. For example, prosecutions will likely involve the following issues: the nature of the 'intent' required for genocide; the relation between ethnic cleansing and genocide; the characterization of rape within the crimes stated in the Statute; the ambiguities in the concept of 'grave breaches' under the Geneva Conventions; the problems in imposing criminal liability for failure to exercise command responsibility; the problems of siege warfare and indiscriminate fire.

4. The Prosecutor, Judge Richard Goldstone, has publicly stated his intention to give priority to indicting the most culpable—that is, those who gave the orders, who were in charge. Here, however, the potentially conflicting demands of the ICT and of those attempting to negotiate an end to the conflict may be at their point of greatest tension with each other. The Prosecutor characterizes the Tribunal as an independent organ pursuing its professional and objective task as defined by the Security Council without attention to larger political considerations. Nonetheless, the political context may become very

relevant for the essential institutional supporters of the Tribunal: states individually, the General Assembly and the Security Council.

QUESTIONS

1. The determination of the Security Council stated in Resolution 808 (restated in Resolution 827) that 'this situation constitutes a threat to international peace and security,' as well as the stated conviction of the Council that 'in the particular circumstances of the former Yugoslavia the establishment of an international tribunal would . . . contribute to the restoration and maintenance of peace,' derive from the conditions for making a decision under Chapter VII (see Art. 39). What arguments would you make to support or to challenge these statements?

2. Observers have described the aims of the ICT in ways that evoke traditional notions of the aims of the criminal law generally, but that also address specific characteristics of this conflict. Consider:

a. Deterrence. Whom is the Tribunal attempting to deter, the present leaders in this conflict, or those who might instigate and commit crimes in future conflicts? Should different strategies be at work to achieve one or the other goal? Who indeed is deterrable in an ethnic conflict stirring such deep hatreds and cruel actions—only the leaders, or also the foot soldiers who commit many of the atrocities? What are the conditions for effective deterrence in this present conflict, and have they been met? Is a court the most effective instrument of deterrence, or does the Tribunal necessarily play this role because of failure of other means of addressing the conflict?

b. Punishment–retribution. How can the Tribunal best serve this function? Is symbolic justice through the conviction and imprisonment of a small number of people (in relation to the number of people committing the international crimes defined by the Statute) sufficient to create a broad sense of justice among the conflict's victims? What other means (shy of forceful intervention) are available to help build this sense of catharsis and justice?

c. Reconciliation, long-run peace and stability. Can reconciliation and a 'true' lasting peace be achieved partly through the work of the Tribunal? What role are convictions and imprisonments likely to play in this process of reconciliation relative to, for example, the kind of settlement about territorial control and residence that emerges, provisions for compensation of victims, and so on?

3. The preceding Comment refers to the 'considerable promise and inescapable risk' for the Tribunal and its work. How do you understand the promise, and the risk?

4. The Comment also states that the priority in seeking indictments that the Prosecutor has described may pose a situation of conflicting demands of the Tribunal and of those working toward an end of the conflict through negotiations. Why should this be the case?

5. Note the strict rules in the Tribunal's Statute and Rules of Procedure and Evidence that protect the rights of the accused by assuring a fair trial. Suppose that you were a member of the office of the Prosecutor sent on an investigative mission to find evidence supporting an indictment and prosecution of particular individuals who were (i) leaders of battalions that were alleged to have committed rape and other atrocities against the civilian population and (ii) persons in charge of detention camps where torture was routinely practiced. The relevant events took place in 1993. What kinds of problems would you anticipate in gathering adequate evidence for a trial?

6. What do you believe are the appropriate substantive conditions (as opposed to the formal legal conditions governing decisions by the Security Council) for the establishment of an *ad hoc* international criminal tribunal for a given conflict? For example, should it be relevant that the conflict has an international rather than internal character, or that the conflict as a whole reaches a certain level of systematic atrocity?

7. Suppose that negotiations among the parties end the conflict (whatever the resulting settlement). Do the relevant resolutions of the Security Council establishing the ICT remain in effect so as to permit ongoing indictments and prosecutions?

NOTE

The conflict in the former Yugoslavia has been ugly almost beyond belief. The reports of systematic murder of civilians, systematic and gross cruelty including rape and mutilation, and systematic deportations to achieve 'ethnic cleansing' are legion. Although all parties to the conflict are reported to have committed serious crimes under international law as described in the Tribunal's Statute, the large majority of violations have been attributed to the Serbian/Bosnian Serb participants.

Such charges grow out of the massive documentation, much of it of high quality, achieved through the remarkable work of the UN-established Commission on Experts under Professor Cherif Bassiouni as its chairman, the detailed reports of the Special Representative (Tadeusz Mazowiecki) appointed by the UN Commission on Human Rights, the many investigative reports of international human rights NGOs and numerous press accounts. All such sources have provided the Tribunal with a base on which to proceed with its own detailed investigations to accumulate evidence of the reliability and specificity necessary to support indictments and prosecutions. Here as elsewhere, the Tribunal has been hampered in its work by the contingent and inadequate funding provided by the UN.

As of July 1995, a number of indictments had been issued (approved, as required by the Statute, by a judge of a Trial Chamber upon application by the Prosecutor), and a number of requests by the Prosecutor for deferral to the Tribunal by state authorities had been approved by a Trial Chamber.[8]

KENNETH ANDERSON, NUREMBERG SENSIBILITY: TELFORD TAYLOR'S MEMOIR OF THE NUREMBERG TRIALS

7 Harv. Hum. Rts. J. 281 (1994).

[In this review of Telford Taylor's *Anatomy of the Nuremberg Trials* (1992), Anderson compares Nuremberg with the International Criminal Tribunal for the Former Yugoslavia.]

> The leitmotiv of Goering's defense at the Nuremberg trials returned time and time again to this theme: 'The victor will always be the judge, and the vanquished will always be the accused.'
>
> Albert Camus, *The Rebel*

> That four great nations, flushed with victory and stung with injury, stay the hand of vengeance and voluntarily submit their captive enemies to the judgment of the law is one of the most significant tributes that Power has ever paid to reason.
>
> Justice Robert H. Jackson

Justice Robert H. Jackson's opening statement at the Nuremberg trial has justly been characterized as one of the greatest orations in modern juristic literature. Yet behind its rhetorical power lies a fervent anxiety: a desire to silence the skeptical voices whispering that the Nuremberg trials were just the tarted-up revenge to which Camus alludes.

Nuremberg was indeed the profoundest tribute of power to reason, as Jackson claimed, but of what did the tribute consist? Jackson said it was to 'stay the hand of vengeance,' so to turn vengeance into something else in the instant before the hand falls. Vengeance, therefore, becomes justice by virtue of a *pause*: a pause in which the partial becomes impartial. The action that had seemed inevitable becomes suddenly, for a moment, contingent. A pause, a moment, a space of time, a gap in the historical action: this was the tribute that constituted Nuremberg. It resolves the skeptical anxiety that the trial was only about vengeance.

. . .

[8] For illustrative documents, see the application of Richard Goldstone, Prosecutor, for an indictment against Dragan Nikolic, and the Decision of the Trial Chamber granting the Prosecutor's application for a formal request for deferral by Germany to the competence of the Tribunal in the matter of Dusko Tadic, 15 Hum. Rts. L. J. 480 and 486 (1994).

Taylor shows that American legalist sensibilities . . . are, in fact, deeply rooted. Deepest of all is the belief that the courtroom is the right place for dealing with war crimes. The American lawyers viewed the war effort and the war itself as properly bringing the rule of law to bear on an ever-widening range of human conduct, subjecting mere politics to it . . . .The Nuremberg trial was needed to convert the contingent fact of Allied might into the eternal fact of Allied right.

. . .

Indeed, as Taylor makes clear, Jackson aimed to establish a judicial precedent on the criminality of aggressive war. Jackson 'made crimes against peace—the criminalization of initiating aggressive war—the foremost feature of the Nuremberg trial . . .' (p. 635). He saw the fundamental crime as a crime against peace, the crowning, pathbreaking charge that the Americans and British divided between them. For Jackson, war crimes were minor matters to be left to the French and Soviet prosecutors.

As today's Security Council establishes the Yugoslavia tribunal, Nuremberg remains relevant principally for matters of war crimes, crimes against humanity, and genocide—the 'minor' matters at Nuremberg. Jackson's aspiration for Nuremberg has faltered precisely on account of the existence of the Security Council. To be sure, Nuremberg did establish a rule of law concerning aggressive war; it is codified in the United Nations Charter. In that sense, at least, war is no longer merely the foreign policy option of an independent sovereign. The rule of law has replaced political decision. However, this rule of law is not justiciable; it is not, as at Nuremberg, a matter of judicial determination. Instead, the determination of aggressive war and threats to peace and security is committed to the hands of the Security Council itself. Great power politics is now dressed up in the rhetoric of the rule of law and sanctified by the U.N. Charter.

The Yugoslavia tribunal's charter, therefore, is noteworthy for what it does not reproduce from the vastly more sweeping Nuremberg charter. The division between *jus ad bellum* and *jus in bello* is fully restored in the Yugoslavia tribunal charter, with the Security Council having sole discretion as to the former, and the tribunal adjudicating only the latter. . . .

Of course, the Yugoslavia tribunal's authority is limited by a more practical matter: it has no one to try. Still, my own experience—private conversations with senior military officers and judge advocates of the NATO armies over the past two years—suggests that this tribunal has more than just practical difficulties: there is a deep malaise. Senior European military officers and diplomats have told me that they see no point in scheduling a trial if no one is willing to commit to a military victory. . . .

. . . Many of these officers argue that there is no justice but 'victor's justice.' . . . A military victory is not simply a practical prerequisite to a trial, they seem to say, but a moral necessity.

This assertion captures my own point about the sensibility of Nuremberg:

to reduce the world to a courtroom, to legal memoranda and pleadings and paperwork, is possible only once an army sits atop its vanquished enemy. Otherwise, the enormity of the crimes left unaddressed out in the hills of Bosnia so dwarf those raised before the tribunal that it mocks justice. A trial, Nuremberg taught, puts the symbolic seal of justice on what armies have rectified with force. These officers imply that to hold a trial without having 'fixed things' in the field is, symbolically, as much or more an act of ratification as condemnation.

In other words, to hold a war crimes trial in the former Yugoslavia today would be like holding Nuremberg after acquiescing in the German annexation of Poland, the Ukraine, and the rest of the eastern lands.

. . .

. . . The Yugoslavia tribunal invokes the law as a rhetorical device to make the U.N. appear to do more than it has, but everyone knows that those who would conduct the trial have not paid the price.

. . .The Yugoslavia tribunal seeks to make history; but the true lesson of Nuremberg is that the power of law to reduce history to its terms can only be accomplished where history—victory and surrender—has been made elsewhere.

QUESTION

'Anderson's argument should be stood on its head. To have rejected the idea of an international criminal tribunal for the conflict in the former Yugoslavia would have been to acquiesce in the argument of those who understand Nuremberg only as "victor's law." It was essential to make the point through the ICT that international standards (at least of *jus in bello* if not *jus ad bellum*) must be vindicated independently of whether there was or would ever be a military victory to roll back the illegal and cruelly-achieved gains. International law is more than an afterthought. This Tribunal indeed provides the golden opportunity for that law to demonstrate its autonomy from international military force as a means of protecting human rights.' Comment.

* * * * *

The following reading raises questions about amnesty in the Yugoslavian conflict that are examined in a broader setting in Section D of this chapter.

PAUL SZASZ, THE INTERNATIONAL CONFERENCE ON THE FORMER YUGOSLAVIA AND THE WAR CRIMES ISSUE

American Society of International Law, Proceedings of the 87th Annual Meeting 20 (1993).

. . .

It is clear that effective peace negotiations cannot be conducted in an atmosphere of mutual or unilateral accusations of criminal conduct. The Co-Chairmen [of a negotiating group] cannot seriously say: 'Please sign this agreement, Mr. K, and as soon as you do so and fighting has ceased we expect you to answer criminal charges in our war crimes court.' Therefore, they must keep their distance from the establishment or operation of such a tribunal, from the prosecution of alleged offenses within it, and even from the investigation of charges. . . .

The question has been raised: How can the Co-Chairmen talk to or negotiate with accused war criminals? (As if there were a choice!) . . . [P]eace negotiations must evidently be conducted with the persons who control the warring forces. It is no use seeking to talk only to persons uncompromised by the struggle, for most likely they could not cause its termination . . .

. . .

This brings me logically to the final question that is constantly raised in connection with our subject today: Will not an overall peace settlement negotiated within the framework of the Conference necessarily include an amnesty provision under which all accused war criminals escape responsibility for their acts? . . .

The answer is really quite simple. The parties can, individually or collectively, grant each other and themselves immunity only from prosecution under their own laws and within their own courts. War crimes, and crimes against humanity, such as the ones that would be prosecuted in an international war crimes tribunal, are, however, not primarily national crimes—they are crimes against the world community. . . . [N]ational organs cannot, whether by domestic actions or by regional treaties, exculpate their citizens from crimes committed against the world community.

. . .

If the world community, through the Security Council, decides that particular offenses should be prosecuted in a war crimes tribunal, then it is only the world community—presumably, but not necessarily, acting through the same organ—that can grant amnesty or respite from prosecution. . . .

There is another objection often raised in connection with the proposal to establish an ad hoc tribunal to deal with war crimes in former Yugoslavia:

Why start just with this conflict? Nothing was done about Cambodia, or the Iraqi invasions of Iran and Kuwait and the associated outrages and inhumanities, or Idi Amin in Uganda (and others I might list). What justifies a different attitude now? One answer is that, in some ways, what is being reported from the Balkans is indeed in many ways worse—in both actions and expressed motivations—than anything that has occurred since World War II. Even without entering into such a process of weighing, however, it should be said that we must make a start somewhere! If past inaction necessitates or merely excuses similar neglect now and in the future, then there is little hope for any improvement in the lot of humanity.

. . .

NOTE

On 20 March 1993 the Republic of Bosnia and Herzegovina filed a complaint against the Federal Republic of Yugoslavia (Serbia and Montenegro) in the International Court of Justice alleging violations of the Convention on the Prevention and Punishment of Genocide. In response to a request for provisional measures, the Court decided as follows:

> A. (1) Unanimously,
>
> The Government of the Federal Republic of Yugoslavia (Serbia and Montenegro) should immediately, in pursuance of its undertaking in the Convention on the Prevention and Punishment of the Crime of Genocide of 9 December 1948, take all measures within its power to prevent commission of the crime of genocide;
>
> (2) By 13 votes to 1,
>
> The Government of the Federal Republic of Yugoslavia (Serbia and Montenegro) should in particular ensure that any military, paramilitary or irregular armed units which may be directed or supported by it, as well as any organizations and persons which may be subject to its control, direction or influence, do not commit any acts of genocide, of conspiracy to commit genocide, of direct and public incitement to commit genocide, or of complicity in genocide, whether directed against the Muslim population of Bosnia and Herzegovina or against any other national, ethnical, racial or religious group;
>
> B. Unanimously,
>
> The Government of the Federal Republic of Yugoslavia (Serbia and Montenegro) and the Government of the Republic of Bosnia and Herzegovina should not take any action and should ensure that no action is taken which may aggravate or extend the existing dispute over the prevention or punishment of the crime of genocide, or render it more difficult of solution.

In response to a second request for interim measures, the Court in its Order of 13 September 1993 reaffirmed its earlier order but was not prepared to go further at that stage. The case remains pending.

C. PROPOSALS FOR A PERMANENT INTERNATIONAL CRIMINAL COURT

BERNHARD GRAEFRATH, UNIVERSAL CRIMINAL JURISDICTION AND AN INTERNATIONAL CRIMINAL COURT

1 Eur. J. Int. L. 67 (1990), at 73.

. . .

For a long time, the question of international implementation of criminal law was approached from the viewpoint of the need to prevent possible interference with state sovereignty and not from that of the need for coordinated struggle and cooperation in the fight against international crimes. Thus, states either cited the sovereignty principle as justification for objecting to the extension of universal criminal jurisdiction or as justification for rejecting the establishment of an international criminal court. This situation continues to exist today, though in a different fashion; there is increasing recognition that national security is at present achievable only by way of international cooperation. In this context, however, one cannot underestimate the importance of the fact that states are the essential structural elements of today's international legal order, that they represent the effective political organizational form of peoples and that they have particular protective functions which they actually exercise. However compelling the precept of cooperation may be, all states want to insure that other states will not be permitted to use criminal law to interfere with their sovereignty or to achieve goals incompatible with the interests of the international community and peoples' right to self-determination.

. . .

[S]tates use the sovereignty principle to justify their objections to the competence of an international criminal court. The underlying fear here is that criminal jurisdiction over crimes committed on one's own territory, where the victim is a citizen or national interests are at stake, will be at the mercy of an international system of criminal justice controlled by others. . . . [T]he Soviet Union and other socialist countries. . . have repeatedly rejected the creation of an international criminal court as a supra-national institution. Following the example of Nuremberg, they have always advocated the creation of *ad hoc* courts whose competence could be based on the existence of joint national criminal jurisdiction, decisively opposing attempts to create an international criminal court which would have the competence to act side by side, or in place of national criminal jurisdiction, decisively opposing attempts to create an international criminal court which would have the competence to act side by side, or in place of national criminal jurisdiction. They have regarded it as impossible for states to hand over their own citizens to an international court for punishment or to refrain from criminal prosecution of offences committed on their territory.

Rejection of, and skepticism about, an international criminal court is not in any way a typically socialist attitude or a position confined to the socialist states. Britain in particular, but the US too, has opposed creation of an international criminal court. Vehemently opposed to the proposal set forth by the International Law Commission, Fitzmaurice stated that the ILC had failed to establish that states regarded creation of any such institution as at all desirable. Obviously he was right. The debate on the Genocide Convention clearly showed that the majority of states did not favor an international criminal court. To date, no viable majority in the UN has been found to favour the creation of such a court.

Many states advocate the creation of an international criminal court, claiming that an international court is necessary if the Draft Code of Crimes against the Peace and Security of Mankind is to be implemented effectively. These states assert that the establishment of an international court is the only way to guarantee objective, impartial jurisdiction, which is particularly important and, at the same time, particularly hard to achieve. In addition these states often point out that the creation of an international criminal court is the only way to avoid differing punishment by individual states. But while these questions are certainly of central importance to the international criminal legal regime, it is not convincing to argue that the only solution to these problems is to transfer criminal jurisdiction from states to an international criminal court.

. . .

. . . [I]t seems to me important . . . to get away from the sterile thoroughly unproductive approach of setting up an alternative between universal criminal jurisdiction of national courts and criminal jurisdiction of an international criminal court. This approach, which led I. Sinclair to declare, 'it is implied that there are two alternatives: universal jurisdiction or an international criminal jurisdiction,' has been the position assumed by many states in the past. In fact, even the special rapporteur gave vent to this approach in his explanation of Article 4 of the International Law Commission's draft code, stating: 'The problem of competent jurisdiction was most serious, since it involved a choice between creating an international jurisdiction and extending the competence of national courts to cover such crimes.'

There is no real reason, however, to see the system of universal criminal jurisdiction and the power of an international criminal court as mutually exclusive. . . .

In fact, most drafts for an international criminal court are reconcilable with this position; most do not provide the international criminal court with court exclusive criminal jurisdiction for particular crimes. The notion that states would be prepared to delegate their sovereignty over crimes committed on their territory, against them, or by their citizens to an international criminal court is so far from reality that it has hardly been seriously defended.

. . .

NOTE

For several years, the International Law Commission (ILC) has been working on draft provisions for a permanent international criminal court. In 1994, it adopted a Draft Statute for an International Criminal Court, the subject of the following article.

JAMES CRAWFORD, THE ILC ADOPTS A STATUTE FOR AN INTERNATIONAL CRIMINAL COURT

89 Am. J. Int. L. 404 (1995), at 408.

. . . The Draft Statute does not correspond to any single criminal justice model, adopting elements from different national traditions. Thus, there is an essentially accusatorial procedure, with an independent prosecutor as distinct from an investigating magistrate. But decision making is by a multimember panel of judges, with no jury, and as in the civilian tradition dissents are suppressed.

. . .

The Court envisaged by the Draft Statute is to be:

(1) a permanent court in the sense that it will be available to act as required, but one that will not have a large infrastructure or permanent staff; in particular, it is intended that the judges will perform other roles (e.g., as national judges) unless called on;
(2) a court created by treaty under the control of the states parties, but in a close relationship with the United Nations;
(3) a court of defined jurisdiction over grave crimes of an international character under existing international law and treaties;
(4) a court the basis of whose jurisdiction is—with the significant exception of genocide—dependent on the acceptance of states; and
(5) a court whose operation is integrated with the existing system of international criminal assistance and which is not intended to displace that

> system in cases where it is functioning. As the preamble states, the Court 'is intended to exercise jurisdiction only over the most serious crimes of concern to the international community as a whole . . . [and] to be complementary to national criminal justice systems in cases where [their] trial procedures may not be available or may be ineffective.'

. . .

The Court is to have a defined jurisdiction over crimes of an international character or concern, and the Draft Statute thus lists specifically the crimes that are to be within the jurisdiction, avoiding merely generic definitions.

In the 1993 draft, three categories of crimes were envisaged: (1) crimes under general international law, (2) crimes under a list of treaties in force (the Genocide Convention, the four Geneva Conventions of 1949 and the first Protocol of 1977, and the various 'terrorism' conventions directed at hijacking, hostage taking, etc.), and (3) a further category of crimes under national law giving effect to what were described as 'suppression conventions.' This third category was meant to cover conventions such as the United Nations Convention against Illicit Traffic in Narcotic Drugs and Psychotropic Substances of December 20, 1988, which in terms envisage that the crimes punishable in accordance with the convention are crimes *under national law*. Those conventions cover a wide range of conduct, far wider than could justifiably be brought within the scope of an international criminal jurisdiction. Despite these characteristics, no list of suppression conventions was drawn up, comparable to the list of treaties within the second category.

The Draft Statute as adopted avoids the earlier generic definitions both of crimes under general international law and of suppression conventions. In the interest of simplicity and clarity, the number of categories is reduced to two, both consisting of enumerated crimes, and (with the exception of genocide) the prerequisites for the exercise of jurisdiction over such crimes are the same. As to crimes under general international law, the earlier definition had been criticized by states on the grounds of its uncertainty. The Court should not, it was said, have a general power to 'define' new crimes under general international law. Accordingly, the 1994 Draft Statute (Article 20) confers jurisdiction over four named crimes under general international law: '(a) the crime of genocide; (b) the crime of aggression; (c) serious violations of the laws and customs applicable in armed conflict; (d) crimes against humanity.' The content of these crimes is to be found in general international law. The Statute itself is conceived of as primarily procedural and adjectival, with the substantive law the Court applies deriving from other sources.

. . .

With the exception of aggression, the list of crimes under general international law in Article 20 reflects that in the Statute of the Tribunal for Former Yugoslavia, although discussion of their definition is contained in the commentaries rather than in the Draft Statute itself. Genocide is included only in Article 20 (a), since the Commission took the view that the definition of genocide in the Genocide Convention of 1948 reflects that under general international law. In other respects also, genocide is given special treatment, since it is the only crime within the 'inherent' jurisdiction of the Court. . . .

. . .

As noted already, the earlier distinction between crimes under listed treaties and crimes under suppression conventions was abandoned. In effect, the two categories are merged, combining the more restrictive elements of each. Thus, as was the case with crimes under listed treaties in the 1993 draft, there is a *numerus clausus* of treaties. If additional treaty crimes are created, the Statute will have to be amended to include them. On the other hand, the 1993 draft required for crimes under 'suppression conventions' that the crime in question reach a threshold level of seriousness, and this requirement has been extended to *all* treaty crimes under the 1994 draft. . . .

As to the crimes encompassed by the list of treaties, illicit traffic in narcotic drugs and psychotropic substances contrary to Article 3 (1) of the 1988 Drugs Convention is added, and also the crime of torture as defined by Article 4 of the Convention against Torture and Other Cruel, Inhuman and Degrading Treatment or Punishment of December 10, 1984. The omission of the latter Convention from the 1993 Statute had been widely criticized.

The same requirements of state consent or acceptance are laid down for all crimes referred to in Article 20 (other than genocide). . . .

. . .

Rather than focusing further on the detail of the ILC's draft, it may be more profitable to reflect on the underlying factors that will determine whether the 1994 Draft Statute, or any other version of an international criminal court, stands a chance of acceptance. . . .

. . .

. . .[T]he importance of the political changes in the United Nations following the end of the Cold War cannot be overestimated. Quite simply, things are now possible that were not when the likelihood of a veto lay ever present over the Security Council, and when consensus could often be achieved only at the price of inaction. But the very activism of the Security Council is one of the reasons for the change of mood about a criminal court. Many commentators, and not a few governments, are concerned by the rather ad hoc character of the Security Council's actions in establishing tribunals for the former Yugoslavia and Rwanda.

There are both practical and constitutional issues here. Practically, the establishment of a new tribunal and the appointment of its personnel are a difficult, expensive and time-consuming process. Particular difficulty was experienced with the office of the prosecutor.

Moreover, the establishment of a tribunal for a particular conflict carries with it the risk that the tribunal will be seen in a sense as *part* of the conflict. The creation of a special tribunal raises expectations that something will happen, and may divert attention from the resolution of conflict to the targeting and punishment of wrongdoers. . . .

A further fundamental issue is that of the constitutional security of a set of international criminal tribunals established ad hoc for particular conflicts. It is certainly arguable that the Security Council's chapter VII powers extend to the creation of criminal tribunals where necessary as part of an approach to resolving a conflict. But the creation of such tribunals by executive resolution in the exercise of emergency powers is less than satisfactory. The principle of legality is of particular significance in criminal cases. It connotes that criminal courts be established on a secure constitutional base; that the law to be applied be sufficiently defined in advance; that court personnel, the judges especially, have security of tenure, and that they be independent of executive interference in the performance of their judicial functions. These conditions may be met de facto in the case of the tribunals so far established—and it must be stressed that those involved in the two tribunals have been working with vigor and integrity to discharge their mandate. But, quite apart from underlying issues of *vires*, the tribunals could be abolished by Security Council resolution at any time, and meanwhile the financial and personal resources available to them have to be negotiated and cannot be described as secure.

The normal way in which institutions, and especially judicial institutions, are created is by legislation. At the international level this requires a treaty, and the implementation of the treaty within the national legal orders of states parties in turn requires compliance with national constitutional procedures. This is the secure and, in the long run, the only appropriate way to create an international criminal court. With the ILC's adoption of the Draft Statute, and the relatively favorable reception it received from a majority of states in the General Assembly, that idea is a step closer.

D. PUNISHMENT, AMNESTY, TRUTH COMMISSIONS

Sections B and C examined the role of international criminal tribunals for the prosecution of individuals accused of committing crimes under international law. Our attention in Section D remains on violent conflicts. Those here described have principally an internal character. The issues to be explored arise when serious human rights violations are committed by a controlling state regime or succession of regimes. At a certain stage, whether because of a strengthening internal opposition, international pressures, or distinctive international circumstances such as a war, negotiations may start that look to political change through a popularly elected regime that pledges to observe human rights. Or the older (often military) regime in power may suddenly collapse.

The question then arises how the new regime should act toward those suspected or accused of serious human rights violations in the prior period. Should there be trial and punishment of individuals, and, if so, of all violators or only of leaders? Should there be a general amnesty? Should other paths be followed?

Section D thus departs in several respects from the prior sections. The tribunals here at issue are national rather than international. The prosecutions are for crimes under national law. Of course no issue arises of the state's jurisdiction to prescribe, for the criminal conduct took place on its very territory. Those crimes—often of a systemic and vicious character involving mass murder, disappearances, rapes, and torture—of course violate rights declared by the international law of human rights. Much of the offensive conduct also constitutes crimes under international law that impose individual responsibility—for example, the obligation of parties to the Convention against Torture to enact legislation making torture by stated persons criminal. The issues discussed in this Section D concentrate on individual criminal liability, generally of persons who acted as state officials or under color of state authority, but the discussion also reaches matters that are chargeable directly to the state such as victims' claims for compensation.

The materials concentrate on Latin America, in which questions of amnesty have been particularly prominent, almost endemic to changes of government following the rule of an oppressive regime. The states selected to illustrate the major issues fall into three patterns. Argentina employed a combination of prosecutions, exoneration and pardons. Uruguay, through legislation supported by a plebescite, determined on a general amnesty without prosecutions and without a serious effort to uncover the truth of what had occurred. Chile decided against prosecutions but its President appointed a National Commission for Truth and Reconciliation to investigate and report in detail on what had occurred. The materials also include a description of El Salvador's experience with a UN Commission on the Truth for El Salvador. Several of these states made provision for compensation to victims.

The protracted negotiations in 1993 and 1994 involving President Aristide, the military rulers of Haiti, the United States and international organizations, and looking toward the departure of those rulers and the return of the President, are a good place to start. The negotiations frequently had to deal with the military leaders' demands for amnesty for them and their subordinates. The following account indicates the many complexities (only starting with the legal and the political ones) in such situations, and suggests the kinds of problems presented in Section D's illustrations of amnesty issues in Central and South American states.

HUMAN RIGHTS WATCH WORLD REPORT

1994 (events of 1993), at 107

. . .

In 1993, as previously, the issue of army accountability was a recurring stumbling block in negotiations to restore President Aristide and democracy in Haiti. While the Clinton administration and the U.N. promised large amounts of economic and military assistance to entice the military, and to a lesser extent President Aristide, to pursue negotiations, the carrot-and-stick approach foundered on the issue of accountability. Aristide was under consistent pressure from U.N. Special Envoy Caputo and from Amb. Lawrence Pezzullo, special envoy for President Clinton, to make concessions on the Haitian army's accountability for its crimes.

Before coming to a settlement, General Cédras required guarantees that President Aristide's opponents would be immune from prosecution and protected from acts of vengeance for participating in the military coup, and that U.N. observers would play a protective role. Cédras demanded amnesty and protection for himself, his family and other members of the high command. The U.S. and U.N. supported these conditions and put Aristide in the position of making or breaking the settlement.

In June, as the *de facto* leaders in Haiti were faced with increasingly harsh sanctions. Cédras agreed to negotiate with Aristide. . . .

The ill-fated Governors Island Accord was signed on July 3. It called for the resignation of General Cédras shortly before the return of President Aristide to Haiti. . . . The agreement also called for President Aristide to issue an amnesty in accordance with the Haitian Constitution, which allows amnesty for political crimes but not for common crimes. Aristide interpreted this constitutional norm as allowing an amnesty for the crime of overturning the constitutional order, but not for the murders, disappearances and torture that had taken place since the coup.

. . .

. . . General Cédras set new conditions for his resignation by demanding that the Haitian parliament pass legislation on an amnesty for crimes committed in connection with the coup. (President Aristide had already issued a decree in early October in accordance with the Governors Island process providing amnesty only for crimes against the state, not for crimes against human rights.) Although Cédras claimed he merely wanted Aristide's decree reinforced by amnesty legislation, it was understood that he sought a broader amnesty that would cover human rights crimes, or common crimes such as murder and torture; such an amnesty would violate the Haitian Constitution. In response to this new demand, the Clinton administration failed to state clearly that it supported the scope of Aristide's decreed amnesty or to oppose Cédras's demand for total impunity.

. . .

1995 (events of 1994), at 101

The Clinton administration made a sharp reversal in Haiti policy midway through the year, transforming its failed approach of accommodating the military regime into a face-off that resulted in the September intervention. Throughout the year, however, the administration was consistent in failing to promote accountability for human rights violations or to insist on safeguards to prevent their recurrence.

During the first half of 1994, U.S. officials actively promoted a blanket amnesty for human rights violations committed since the coup, in addition to the amnesty for crimes associated with the coup itself already decreed by President Aristide. Even after the U.S.-led occupation of Haiti, the administration consistently failed to oppose a broad amnesty that would deny victims of human rights crimes their internationally guaranteed right to a legal remedy.

. . .

Consistent with its pursuit of a power-sharing arrangement, the administration downplayed human rights abuses committed by the Haitian armed forces and its supporters, choosing not to condemn publicly serious abuses or to attribute responsibility for them to the military regime. . . .

. . .

On July 31, the U.N. Security Council passed Resolution 940, which invoked Chapter VII of the U.N. Charter and allowed the U.S. to form a multinational force 'to use all necessary means to facilitate the departure from Haiti of the Military leadership.' . . .

On September 13, the State Department rectified its past indifference by issuing a strong condemnation of human rights abuses in Haiti. Two days later, President Clinton addressed the nation with an emotional description of the regime's brutality as a principal rationale for invading Haiti. . . .

In a last-ditch effort to avoid an unpopoular hostile intervention, President Clinton authorized former President Jimmy Carter, Senator Sam Nunn, and General (Retired) Colin Powell to negotiate with the Haitian army's high command. On September 18, with war planes en route to Haiti, the Carter delegation produced an accord, signed by de facto president Emile Jonaissant, under which the top three coup leaders would step down by October 15, the Parliament would pass a general amnesty, and international sanctions would be lifted. . . .

. . .

On October 15, President Aristide returned to Haiti and was welcomed by jubilant crowds. After naming a new prime minister and government, he faced the enormous tasks of repairing the ravaged national economy, establishing a permanent, civilian police force answerable to civilian authority, and rehabilitating a crippled judiciary. He was also responsible for establishing a climate favorable to holding parliamentary and local government elections, tentatively scheduled for early in 1995. Most importantly, with financial and technical assistance of the international community, he would have to break the cycle of violence that has plagued Haiti for decades by assuring accountability for thousands of crimes committed under the coup regime.

. . .

VIEWS OF COMMENTATORS

Consider the following comments about issues of impunity and amnesty:

Amnesty International[9]

Peace settlements should include provisions for investigating past human rights abuses and ensuring that perpetrators are brought to justice. . . .

[P]rovisions for investigating past violations should be ensured in order to determine individual and collective responsibility and to provide a full account of the truth to the victims, their relatives and society. Investigations must be undertaken by impartial institutions, independent of the security forces. In the context of peace-keeping arrangements, investigations into the past should preferably include international participation. The results of the investigation must be made public. Where there has been an endemic pattern of human rights abuses, a public inquiry commission should be established to investigate the entire pattern of abuses and the reasons why they occurred. Such an inquiry should be able to examine the institutions and agencies responsible and make recommendations regarding the accountability of personnel; legislative, institutional, and procedural reforms; and human rights education and training for officials who continue in office.

Amnesty International considers that those . . . accused of human rights crimes should be tried, and their trials should include a clear verdict of guilt or innocence. Although Amnesty International takes no position on *post-conviction* pardons, where it is deemed that such a measure would be in the best interests of national reconciliation, the organization does oppose amnesty laws which prevent the emergence of the truth of individual cases as well as the pattern of abuses in the society, or which prevent the full completion of the judicial process.

[9] Peace-Keeping and Human Rights, AI Doc. IOR 40/01/94 (1994), at 38.

UN Special Rapporteur[10]

691. . . . [T]he Special Rapporteur wishes to emphasize that 'under no circumstances . . . shall blanket immunity from prosecution be granted to any person allegedly involved in extra-legal, summary or arbitrary executions' (principle 19 of the Principles on the Effective Prevention and Investigation of Extra-legal, Summary or Arbitrary Executions). Even if, in exceptional cases, Governments may decide that perpetrators should benefit from measures that would exempt them from or limit the extent of their punishment, their obligation to bring them to justice and hold them formally accountable remains, as does the obligation to carry out prompt, thorough and impartial investigations, grant compensation to the victims or their families and adopt effective preventive measures for the future. The Special Rapporteur appeals to all Governments concerned to revise any legislation as may be in force exempting those involved in violations of the right to life from prosecution.

. . .

699. The link between the effective investigation of human rights violations of the right to life and the prevention of their recurrence in the future cannot be over-emphasized. . . .

Diane Orentlicher[11]

. . .

There is . . . a paradox in the development of international law prescribing the role of punishment in securing fundamental rights: on the one hand, legal norms requiring states to punish those who commit atrocious crimes have been clarified and strengthened in recent years. At the same time, however, international bodies responsible for enunciating the norms have been reticent to insist on enforcement.

What, then, accounts for this discrepancy? At one level, the answer is simple. It is a lesser intrusion on sovereignty for an international body to enunciate a norm than to insist on its enforcement—particularly when the norm asserts duties in an area traditionally left to the broad discretion of states.

At a deeper level, ambivalence about the desirability of enforcing states' general duty to punish human rights crimes has at times inhibited efforts to secure prosecutions. This ambivalence is most apparent in the now-common circumstance in which a democratically elected government has replaced one responsible for massive human rights violations. In these situations, issues of

[10] UN Special Rapporteur on Extra-Judicial, Summary or Arbitrary Executions, UN Doc. E/CN.4/1994/7, at 157.

[11] Addressing Gross Human Rights Abuses: Punishment and Victim Compensation, in Louis Henkin and J. L. Hargrove (eds.), Human Rights: An Agenda for the Next Century 425, 431 (1994).

accountability frequently pose a daunting dilemma. On the one hand, the balance of power between the *ancien régime* and the new government is often precarious, and prosecuting depredations of the outgoing regime may seem to place an already fragile democracy at greater risk—particularly when the outgoing regime was dominated by military sectors that retain a monopoly on the use of force. On the other hand, impunity for atrocious crimes of the recent past undermines the law's authority just when a society is poised to reassert the supremacy of law. A Faustian pact with a brutal regime—a pact to allow impunity in exchange for the end of dictatorship—raises the specter of perpetuating the very lawlessness that is meant to be ended.

This dilemma has proved agonizing for the many nations that have confronted it in recent years—the cases of Argentina, Chile, El Salvador, the Philippines, and Uruguay are a few examples—and also seems to play a part in the reticence of some international human rights bodies to press for prosecutions during periods of transition from dictatorship to democracy. . . .

NOTE

Does international law impose on states affirmative duties to punish human rights violators, so that a state's grant of an amnesty or pardon[12] would be considered a violation of a treaty or of customary international law? Recall the discussion of a state's duty to prosecute under the Convention against Torture, p. 805, *supra*. The Report on Uruguay of the Inter-American Commission on Human Rights, p. 1098, *infra*, examines this issue with respect to the American Convention on Human Rights. Consider the following reading.

DIANE ORENTLICHER, SETTLING ACCOUNTS: THE DUTY TO PROSECUTE HUMAN RIGHTS VIOLATIONS OF A PRIOR REGIME

100 Yale L. J. 2537 (1991), at 2568.

[In her discussion of the International Covenant on Civil and Political Rights, Orentlicher notes that, unlike the Convention on Torture but like the European and American human rights conventions, the International Covenant does not explicitly require states to punish violations of rights set forth in it. Her discussion continues:]

[12] In general, *amnesties* foreclose prosecutions for stated crimes (often by reference to crimes or conduct that took place before a stated date), whereas *pardons* release convicted human rights offenders from serving their sentences (or the remainders thereof if they are prisoners at the time of pardon). Nonetheless, usage often views these terms as interchangeable, so that persons not yet tried are 'pardoned' and prisoners serving sentences are granted an 'amnesty.'

. . . Authoritative interpretations make clear, however, that these treaties require States Parties generally to investigate serious violations of physical integrity—in particular, torture, extra-legal executions, and forced disappearances—and to bring to justice those who are responsible. The rationale behind these duties is straightforward: prosecution and punishment are the most effective—and therefore only adequate—means of ensuring a narrow class of rights that merit special protection.

The duties derive from States Parties' affirmative obligation to ensure rights set forth in these conventions. Adherents to all three treaties pledge not only to respect enumerated rights, but also to ensure that persons subject to their jurisdiction enjoy the full exercise of those rights. The International Covenant and American Convention further require States Parties to adopt legislation or other measures necessary to give effect to the rights and freedoms recognized in the treaties, and all three conventions require Parties to ensure that individuals whose rights are violated have an effective remedy before a competent body, even if the violation was committed by someone acting in an official capacity. To underscore the importance of ensuring the right to life through legal guarantees, each convention provides that the right to life shall be protected by law.

The possibility of requiring States Parties to punish violations was never seriously considered by the drafters of the International Covenant. The United Nations Commission on Human Rights, which drafted the Covenant, debated extensively the nature of a State Party's duty under Article 2(3) to provide an 'effective remedy' for violations of the Covenant, and in this context briefly considered explicitly requiring States Parties to punish violators. . . .

. . .

But if the drafters did not seriously consider requiring States Parties to prosecute violations of the Covenant, nothing in the drafting history is inconsistent with such a duty. The text could, moreover, reasonably be interpreted to require States Parties to ensure at least some rights through use of criminal sanctions.

The Human Rights Committee established to monitor compliance with the Covenant has, in fact, repeatedly asserted that States Parties must investigate summary executions, torture, and unresolved disappearances; bring to justice those who are responsible; and provide compensation to victims. These pronouncements stand in notable contrast to the Committee's more characteristic practice of granting states substantial leeway to determine how they will implement the Covenant and to fashion an appropriate remedy when found in breach of their treaty obligations.

In 'general comments' interpreting Article 7 of the Covenant, which prohibits torture and cruel, inhuman or degrading treatment or punishment, the Committee has asserted:

> [I]t is not sufficient for the implementation of [article 7] to prohibit [torture or other cruel, inhuman or degrading] treatment or punishment or to

> make it a crime. Most States have penal provisions which are applicable to cases or torture or similar practices. Because such cases nevertheless occur, it follows from article 7, read together with article 2 of the Covenant, that States must ensure an effective protection through some machinery of control. Complaints about ill-treatment must be investigated effectively by competent authorities. Those found guilty must be held responsible, and the alleged victims must themselves have effective remedies at their disposal, including the right to obtain compensation.

As this comment makes clear, a state's duty to investigate allegations of torture and hold the wrongdoers responsible exists over and above its duty to provide victims an effective civil remedy.

. . .

Subsequent decisions on individual communications have reinforced the Committee's 'general comment' on Article 7. For example, in *Muteba v. Zaire*, the Committee found that the government of Zaire had committed torture in violation of Article 7, and concluded that the government was 'under an obligation to . . . conduct an inquiry into the circumstances of [the victim's] torture, to punish those found guilty of torture and to take steps to ensure that similar violations do not occur in the future.'

. . .

Under [the author's preceding interpretations of] the Committee's jurisprudence, prosecution leading to an appropriate sanction is generally required when a disappearance, an extra-legal execution, or torture is credibly alleged. Some departures from this norm might be justified, however. The clearest justification would be evidentiary constraints, provided they were not the product of government malfeasance. A measure of flexibility is implied, moreover, in the Committee's use of such phrases as 'bring to justice' and 'hold responsible.' While these terms seem to contemplate appropriately severe criminal penalties, in some instances administrative disciplinary procedures might satisfy a state's duty. Clearly, however, a State Party's routine failure to punish extra-legal killings, torture, and disappearances would constitute a breach of its duties.

While the Committee has made clear that punishment plays a necessary part in States Parties' fulfillment of certain duties under the Covenant, it has stopped short of recognizing a right by individuals to ensure that a particular person is prosecuted. In *H.C.M.A. v. the Netherlands*, the Committee rejected the complainant's claim that Article 14(1) of the Covenant, which guarantees the right to a fair trial, establishes a right 'to see another person criminally prosecuted.'

. . .

RONALD DWORKIN, INTRODUCTION TO *NUNCA MAS*

in Nunca Más (Report of the Argentine National Commission of the Disappeared) (1986), at xi.

Nunca Más, which was published in Argentina in 1984, is a report from Hell. The work of a special commission appointed by President Alfonsín, it describes in detail almost unbearable to read the system of licensed sadism the military rulers of Argentina created in their country from 1976 to 1979, when more than twelve thousand citizens were 'sucked' off the streets, tortured for months, and then killed. . . .

. . .

[Raúl Alfonsin was elected President after the military junta that had ruled Argentina fell.] He immediately appointed a commission of distinguished citizens, under the chairmanship of a prominent writer, Ernesto Sábato, called the Commission on the Disappeared, with full powers to investigate and report. The commission's interviews were methodical and painstaking; its members visited and explored the detention sites the witnesses they interviewed had mentioned, cross-checked the stories of each, created charts and flow charts of events, and confirmed the most pessimistic speculations about the fate of the thousands of people who had been bundled into the Ford Falcons with no license plates.

Nunca Más is the report of the Sábato Commission. . . .

The original point of the 'dirty war'—to create a climate of fear in which subversion would be impossible—was superseded, for the officers who actually carried it out, by an even more repellent purpose; the perverse exhilaration of absolute, uncontrolled dominion over others, which became an end in itself, a way of life. Nothing can seem out of bounds in a room where people are deliberately made to suffer excruciating pain. Every instinct of dignity was violated there: nuns and pregnant women were tortured with special glee, husbands and wives and children tortured in each other's presence, and babies taken from their mothers for military families who wanted children. . . .

Alfonsín had made two commitments to the nation: to investigate the disappearances, and to prosecute those responsible. The Sábato Commission was charged only with the former: it was not a judicial body, and its report, *Nunca Más*, made no judgments of individual responsibility. The new government had to decide how to proceed with criminal prosecution, and it faced a variety of problems, both legal and political. Alfonsín was anxious that the process vindicate not only justice but the rule of law. The accused were to be treated with every courtesy, and the strictest standards of evidence and procedure were to be applied.

Above all, the trials were to respect an important jurisprudential distinction: they were not to be trials in the style of Nuremberg, prosecutions by a conqueror imposing a new code of rules on a defeated regime, but acts of a constitutional government prosecuting past officials for acts that were criminal when they were performed. No new retrospective criminal laws were needed because the outrages the Sábato Commission had documented were plainly illegal under the law of Argentina as it stood during the military rule. The military had not enacted special laws permitting kidnapping, or torture, or determination without trial, or theft, or murder. It had . . . awarded itself an amnesty just before it handed over power, but that self-amnesty was unconstitutional, according to Alfonsín's legal advisers; the new Congress had already repealed it, and the Supreme Court had held this repeal valid.

Two further crucial decisions were necessary, however, and both were politically sensitive. First, who should be prosecuted? Alfonsín, at the same time as he appointed the Sábato Commission, had in his capacity as commander in chief ordered the arrest and trial of the military men at the top: the nine commanders who formed the three ruling juntas from 1976 to 1982. But should the government also prosecute the staff and junior officers who supervised the abductions and detention centers and the torture, or the thousands of ordinary soldiers who participated in these crimes? Argentine law provided a defense for military subordinates who were merely following orders. But how should this defense be interpreted? Should it protect soldiers who followed orders that were, in fact, illegitimate? Should it protect those who, following orders, committed obvious atrocities?

Second, in which courts should those who were prosecuted be tried? It was at least arguable that, as many Argentine lawyers believed, the law required that military men be tried only in military courts for crimes committed in connection with their duties. . . .

The government was subject to intense political pressures on both sides of these two issues. The human rights community, and particularly the Mothers of the Plaza, were outraged at the possibility that the army could be left to judge itself, or that those who had actually butchered and tortured their fellow citizens might escape condemnation altogether. But Argentina needed to bury its past as well as to condemn it, and many citizens felt that years of trials would undermine the fresh sense of community Alfonsín's victory had produced. And any general program of prosecution, reaching far down the command structure, might anger the military and make it regard the new government as its enemy, which would be unwise in a nation where military coups had become almost a ritual.

The new government formally declared its intentions in a comprehensive statute, Law 23.049 of February 14, 1984, drafted mainly by one of Alfonsín's advisers, Carlos Nino, a legal philosopher from the faculty of the Law School of the University of Buenos Aires. The statute resolved the issue of jurisdiction in this way: all prosecutions of the military for alleged crimes committed under cover of a war against subversives, including both those brought by the public prosecutor and those initiated by private citizens, were to be tried in the first instance by the Supreme Council of the Armed Forces under its 'summary proceedings' jurisdiction. But the decisions of the Supreme Council were subject to automatic review by the civilian Federal Chamber of Appeal, which could consider new evidence if it thought this necessary . . . .

The law also resolved the issue of criminal responsibility. It provided, in Article 11, that, in the absence of any evidence to the contrary, any member of the military who acted 'without decision-making capacity' would be presumed justifiably to have regarded all the orders he received as legitimate orders, except that this presumption would not hold if the acts he committed were 'atrocious' or 'aberrant.' That provision created, in effect, three categories of defendants: high-ranking officers, who were not entitled to the defense that they were merely following orders; junior officers and servicemen, who could use that defense because they were deemed entitled to treat the orders they had received as legitimate; and those of any rank who committed atrocities and were denied the defense for that reason. It was widely understood that abduction, for example, was not atrocious, so that the junior officers who formed the abduction squads would not be guilty under these standards, but that torture, rape, murder, and robbery were atrocities, so that those who could be proved to have committed those offenses would not be excused just because they had been ordered to do so.

. . .

The trial of the nine commanders in the civilian court [after the Supreme Tribunal of the military refused to participate in the trial of the nine] began on April 22, 1985, and held the nation enthralled for five months. . . .

The commanders' trial ended and the court gave its verdicts on December 9, 1985. Though the military rules under which the civilian trial had been conducted allowed for capital punishment, the prosecutor had asked only for life imprisonment for five of the nine defendants, fifteen-year sentences for two of them, and twelve- and ten-year sentences for the remaining two. Only two of the defendants were in fact given life sentences. . . .

Some Argentines were disappointed by these results. Many believed that all the defendants should have been convicted and jailed for life. But the court's policy of making distinctions among the defendants, acquitting four and giving lighter sentences to some of the others, was valuable in several ways. It showed the court's political independence from the government and from the prosecution, which had demanded much heavier sentences, and in that way reinforced the character of the trial as an exercise in due process of law rather than political vengeance. And it avoided any suggestion that there cannot be degrees of guilt in crimes against humanity, that those who have committed some outrages have nothing to fear, by way of future punishment, if they commit more.

. . .

But the most powerful evidence of complicity throughout the structure of command of each of the armed services was simply the sustained pattern of abduction, torture and murder, a pattern that could not possibly be explained as the work of a few aberrant officers. The court explicitly relied on *Nunca Más* as evidence of this pattern, though it added that the evidence it had heard directly was sufficient to establish the pattern of outrage on which its verdict was based, even without that report. . . .

. . .

It has become clear, moreover, that some officials of the present government do want a general amnesty for junior servicemen. . . . The arguments . . . in favor of some general amnesty are powerful and in no sense illegitimate. It is desperately important, not only for Argentina but for Latin America generally, that the Alfonsín government succeed. It is one of the few governments in the region firmly committed to constitutional democracy in our own sense, one of the few we can respect unreservedly. But it is vulnerable to a variety of economic and political forces.

Argentina's economy remains fragile, even though the austere new economic plan Alfonsín instituted last summer brought down the rate of inflation and secured some monetary stability. . . . of political stability. The congressional elections last autumn were marred by bombings attributed to the right, and Alfonsín thought it necessary to declare a state of emergency for a short period to meet what he called a serious threat of disturbances. . . . Nor is the new government fully confident of political stability.

In these circumstances, the government has good reason to avoid divisive policies that might alienate the armed forces. Nevertheless, we must hope that the Alfonsín government will take that risk and prosecute anyone it can prove tortured or killed civilians, even under orders, though it may turn out that only a relatively small number can be convicted. The world needs a taboo against torture. . . . [It is] almost everywhere used, and the discrepancy is partly the result of a widespread opinion that it is justifiable sometimes, that it is defensible when carefully aimed only at extracting information needed to save lives from terrorism, for example.

The Argentine nightmare shows one of the several fallacies of this view. Torture cannot be surgically limited only to what is necessary for some discrete goal, because once the taboo is violated the basis of all the other constraints of civilization, which is sympathy for suffering, is destroyed. The Mothers of the Plaza de Mayo, and the others who call for prosecution of all torturers and murderers in the military ranks, are right—not because they are entitled to vengeance, but because the best guarantee against tyranny, everywhere but especially in countries like Argentina where tyranny has often seemed acceptable to the majority, is a heightened public sense of why it is repulsive. Trials that explore and enforce the idea that torture can have no defense may encourage that sense. Allowing known torturers to remain in positions of authority, unchallenged and uncondemned, can only weaken it.

ARYEH NEIER, WHAT SHOULD BE DONE ABOUT THE GUILTY?

New York Review of Books, Feb. 1, 1990, at 32.

[The author discusses the trial of the nine military commanders in Argentina making up the first three juntas that ruled Argentina following the military coup. He praises the meticulous investigation, the scrupulous protection of the rights of the accused, and the dignified manner in which the trials were carried out, all of which 'inculcated respect for the rule of law.']

Nevertheless, something went seriously wrong. Some lower-ranking military officers who were also facing prosecution conspired to mount small military revolts during Alfonsín's final years in office. Several army posts were briefly taken over, and indicted officers were protected from arrest and trial. As a consequence of such actions, much of what was accomplished by Alfonsín was rescinded or reversed before he left office. That reversal has now proceeded much further under his successor Carlos Saul Menem . . . .

. . .

It is worth considering why the Argentine prosecution of crimes against human rights started so promisingly and why it ended so badly. It is often said that Alfonsín was able to order the prosecutions because, at the moment he took office, the armed forces had been so weakened and discredited by the Falklands war that they had no power to resist. But this is only a partial explanation. Alfonsín acted as soon as he assumed office, at a moment when his popularity was at its highest, and when heads of state from around the world had just visited Buenos Aires for his inauguration to celebrate the restoration of democracy. More significant, I believe, were Alfonsín's decision to establish the Sábato commission, in order to document and publicize the disappearances, and the fact that the commission performed this task admirably and quickly. Though the prosecutions were announced earlier, they did not seriously get underway until after the Sábato commission issued its report.

The decision to prosecute high military commanders who were members of the junta seems to me to have been correct. Indeed, if the prosecutions had been limited to these officers, the military uprisings against the Alfonsín government might never have taken place. Punishment of officers at the higher level who were responsible for crimes would likely have been accepted by most of the Argentine population, and many in the military. But when middle-level officers were indicated as well, local revolts among the military began to take place. The prosecution of middle-level officers who had presided over acts of torture and killing was fully justified; but it was also deeply resented by many officers who claimed they were only 'following orders,' and by their military comrades. In order to appease them, Alfonsín pushed through the congress a 'due obedience' law, exonerating most such officers.

In hindsight, this appears to have been a mistake. By giving in, Alfonsín probably encouraged the rebel officers to make further demands and to launch further revolts, preparing the way for the pardons issued by President Menem in October that covered all but five convicted commanders—one is still facing trial—and for Menem's announcement that he intends to pardon those commanders soon. In retrospect, Alfonsín might have done better to limit prosecutions to the top commanders, while continuing to publicize the crimes committed by lower ranking officers.

. . .

Of course, concerns for truth and justice are closely linked. In Argentina, the disclosure of the truth by the Sábato commission helped to create a climate of opinion that supported the government's prosecutions. In most cases, a government such as Argentina's tries to suppress the truth about its abuses while they are being committed. Whereas many people may be generally aware that killing, torture, and disappearances are taking place, they know little of their extent and details. When the prosecutions were brought, the information disclosed at the trial both confirmed the findings of the Sábato commission and gave a more precise identification of torturers and killers. Indeed, because a huge number of Argentine citizens watched the trial of members of the junta day by day, the trial did more than the Sábato commission itself to circulate information.

. . .

INTER-AMERICAN COMMISSION ON HUMAN RIGHTS, ANNUAL REPORT 1992–93

(1993), at 154.

Report No. 29/92, Uruguay, Oct. 2, 1992

[The Report on Uruguay grew out of petitions lodged with the Commission alleging that Uruguay had violated rights declared in the American Convention on Human Rights.]

. . .

2. The petitions denounced the legal effects of Law No. 15,848 (hereinafter 'the Law') and its application by the judiciary, which they allege violated rights upheld in the American Convention (hereinafter 'the Convention'): the right to judicial protection (Art. 25) and the right to a fair trial (Art. 8), among others.

3. The first article of that law states that: 'It is hereby recognized that as a consequence of the logic of the events stemming from the agreement between

the political parties and the Armed Forces in August 1984 and in order to complete the transition to full constitutional order, any State action to seek punishment of crimes committed prior to March 1, 1985, by military and police personnel for political motives, in the performance of their functions or on orders from commanding officers who served during the *de facto* period, has hereby expired.

. . .

5. Article 4 states that: 'The foregoing notwithstanding, the judge hearing the case shall remit to the Executive Branch all testimony offered in the complaint as of the date of enactment of this law, regarding measures involving individuals alleged to have been detained in military or police operations and who have since disappeared, as well as minors alleged to have been abducted under similar circumstances. The Executive Branch shall immediately order investigations to ascertain the facts. Within 120 days of the date of the communication received from the court, the Executive Branch shall advise the plaintiffs of the findings of these investigations and provide them with the information compiled.'

6. Since the Executive Branch entrusted the investigation to military judges, doubts were raised as to the seriousness and impartiality of the investigative proceeding, and as to whether the duty to provide the essential judicial guarantees has been observed (Articles 8 and 27 of the Convention).

7. The law was declared constitutional by the Uruguayan Supreme Court and was approved by a national referendum called for that purpose pursuant to the provisions of Article 79 of the Uruguayan Constitution.

Observations of the Government on the Report Adopted in Accordance with Article 50

21. Essentially, the Government contends that the Commission has failed to consider the 'democratic juridical–political context' inasmuch as it has not taken into account the domestic legitimacy of the law and has failed to consider important aspects of the present political situation, as well as the higher ethical ends of the Caducity Law [terminating prosecutions for acts committed prior to March 1, 1985].

What follows is a summary of the principal arguments in the Government's reply.

22. The Government avers that the amnesty question should be viewed in the political context of the reconciliation, as part of a legislative program for national pacification that covered all actors involved in past human rights violations, i.e., 'political crimes and related common and military crimes,' that the Caducity Law was adopted for 'the sake of legal symmetry and for very justified and serious reasons of the utmost political importance,' with 'unqualified adherence to its constitutional system and its international

commitments.' Uruguay emphasizes the fact that this law, approved by the necessary parliamentary majority, was also 'the subject of a plebiscite' by the electorate: that it cannot accept the Commission's finding that while the domestic legitimacy of the law is not within the Commission's purview, the legal effects denounced by the petitioners are; 'the express will of the Uruguayan people to close a painful chapter in their history in order to put an end, as is their sovereign right, to division among Uruguayans, is not subject to international condemnation.'

. . .

27. The Government further contended that the Commission had failed to note that the Caducity Law does not prevent the injured party from seeking damages in a Civil court; hence, the recommendation made to the Government that the victims be awarded just compensation for past human rights violations was out of order.

The Opinion and Conclusions of the Commission

. . .

A. As to the interpretation of the Convention

33. Article 29 of the Convention stipulates the following:

> No provision of this Convention shall be interpreted as:
>
> a. permitting any State Party, group, or person to suppress the enjoyment or exercise of the rights and freedoms recognized in this Convention or to restrict them to a greater extent than is provided for herein;
>
> b. restricting the enjoyment or exercise of any right or freedom recognized by virtue of the laws of any State Party or by virtue of another Convention to which one of the said States is a party;
>
> c. precluding other rights or guarantees that are inherent in the human personality or derived from representative democracy as a form of government; or
>
> . . .

34. The Commission notes that any interpretation of the Convention must be rendered in accordance with this provision.

B. As to the right to a fair trial

35. The law in question has the intended effect of dismissing all criminal proceedings involving past human rights violations. With that, the law eliminates any judicial possibility of a serious and impartial investigation designed to establish the crimes denounced and to identify their authors, accomplices, and accessories after the fact.

36. The Commission must also consider the fact that in Uruguay, no national investigatory commission was ever set up nor was there any official report on the very grave human rights violations committed during the previous *de facto* government.

37. It is fitting, in this regard, to cite the Commission's general position on the subject, as set forth in its Annual Report of 1985–1986:

> . . . [One] of the few matters that the Commission feels obliged to give its opinion in this regard is the need to investigate the human rights violations committed prior to the establishment of the democratic government. Every society has the inalienable right to know the truth about past events, as well as the motive and circumstances in which aberrant crimes came to be committed, in order to prevent a repetition of such acts in the future. Moreover, the family members of the victims are entitled to information as to what happened to their relatives. Such access to the truth presupposes freedom of speech, which of course should be exercised responsibly; the establishment of *investigating committees* whose membership and authority must be determined in accordance with the internal legislation of each country, or the provision of the necessary resources so that the *judiciary* itself may undertake whatever investigations may be necessary. (Emphasis added.)

38. The Commission must also weigh the nature and gravity of the events with which the law concerns itself; alleged disappearances of persons and the abduction of minors, among others, have been widely condemned as a particularly grave violation of human rights. The social imperative of their clarification and investigation cannot be equated with that of a mere common crime . . . .

. . .

D. With respect to the obligation to investigate

[The Commission quotes with approval a lengthy passage from the *Velásquez Rodríguez* case, p. 946, *supra*.]

51. When it enacted this law, Uruguay ceased to comply fully with the obligation stipulated in Article 1.1 and violated the petitioners' rights upheld in the Convention.

52. As for the interpretation of Article 1.1, Article 8.1 and Article 25.1, and the possible restrictions on those rights as set forth in Articles 30 and 32, the Commission respects but nevertheless disagrees with the Uruguayan Government's interpretation of those provisions.

53. As for compensatory damages, the Commission points out that while it is true that the text of the law did not affect the possibility of filing a suit for such damages, the ability to establish the crime in a civil court has been considerably curtailed since vital testimony from the moral and material authors, military and police personnel of the State, cannot be adduced or used.

The Commission also noted that four years after the fact, the State invoked the caducity exception, even though at the time the crimes were committed a dictatorial government was in power whose judiciary lacked any independence, especially in matters of this nature. In the past year, the Commission has noted, with satisfaction, a number of important damages agreements that the Uruguayan State and certain victims of past human rights violations have reached, including three petitioners in these cases. Nevertheless, the Commission must make clear that the purpose of these petitions is to object to the denial of justice (Articles 8 and 25 in relation to Article 1 of the Convention) with enactment and application of the 1986 Law, and not to the violations of the rights to life (Article 4), humane treatment (Article 5) and liberty (Article 7), among others, which triggered the right to a fair trial and the right to judicial protection, but that occurred before the Convention entered into force for Uruguay on April 19, 1985, and therefore were not a subject of these complaints.

54. The Commission has carefully weighed the political and ethical dimensions of the measure adopted by the Uruguayan Government and reached a conclusion different from that of the Government as to whether, with the law, the Government's highest mission according to the obligations of the American Convention, which is to defend and promote human rights, is being served.

Given the foregoing considerations, the

INTER-AMERICAN COMMISSION ON HUMAN RIGHTS,

1. Concludes that Law 15,848 of December 22, 1986, is incompatible with . . . Articles 1, 8 and 25 of the American Convention on Human Rights.

2. Recommends to the Government of Uruguay that it give the applicant victims or their rightful claimants just compensation for the violations to which the preceding paragraph refers.

3. Recommends to the Government of Uruguay that it adopt the measures necessary to clarify the facts and identify those responsible for the human rights violations that occurred during the *de facto* period.

THOMAS BUERGENTHAL, THE UNITED NATIONS TRUTH COMMISSION FOR EL SALVADOR

27 Vanderbilt J. Trans. L. 497 (1994), at 519.

[In 1992–93, the author served as one of three Commissioners (all foreign nationals) of the UN Commission on the Truth for El Salvador, a Commission charged with the investigation of acts of violence in El Salvador between 1980 and 1991. That period cost some 75,000 lives and was characterized by an extreme cruelty involving massacres, torture and other offenses against the civilian population.

The Commission was established in 1992 pursuant to the Salvadorean Peace

Accords negotiated under the auspices of the UN between the government and the principal opposition group, the FMLN. Beyond investigation and reporting of the relevant facts, the Commission was to take into account 'the need to create confidence in the positive changes' to be effected by the peace process and 'to assist the transition to national reconciliation.'

In its Report, the Commission recommended changes designed to transform El Salvador into a state with a democratic form of government that would observe human rights.]

The Commission could have recommended that the individuals identified as responsible for the serious acts of violence described in the *Report* be tried by Salvadoran courts. However, such a recommendation would have made sense only if the Commissioners believed that the justice system of that country was capable of doing justice, which we did not. Although the Peace Accords had ended the armed conflict and called for substantial reforms in the justice system, very few of these changes had been implemented or were likely to be implemented in the near future. This meant that the same judges who were in office during the war, including some accused by the Truth Commission of covering up various offenses, would be the ones to adjudicate these charges. As stated in the *Report*:

> These considerations confront the Commission with a serious dilemma. The question is not whether the guilty should be punished but whether justice can be done. Public morality demands that those responsible for the crimes described here be punished. However, El Salvador has no system for the administration of justice which meets the minimum requirements of objectivity and impartiality so that justice can be rendered reliably.

Taking this reality into account, the Commission decided not to call for trials, nor for that matter to recommend amnesties. The former made no sense until the full implementation of the Peace Accords. The latter seemed worthwhile only, if at all, after a national consensus that an amnesty would promote the goal of reconciliation in El Salvador. Ultimately, the decision whether to grant amnesty was one for the people of El Salvador to make after an appropriate dialogue on the subject.

At the same time, it was clear that the Commission's findings required some recommendations for immediate action. In the Commission's view, those identified as responsible for serious acts of violence had to be removed from the offices that had enabled them to commit these acts. To this end, the Commission made a series of recommendations. First, it called for the dismissal from the armed forces of those active military officers who had committed or covered up serious acts of violence. Second, the Commission recommended the dismissal from their positions of those civilian government officials and members of the judiciary who committed or covered up serious acts of violence or failed to investigate them. Third, the Commission declared that '[u]nder no circumstances would it be advisable to allow persons who committed acts of violence such as those which the Commission has investigated to participate in the running of the State.' Therefore, we recommended that appropriate legislation be adopted to ensure that all individuals found by the Commission to have been implicated in serious acts of violence—whether active or retired military officers, civilian officials, FMLN members or military commanders, judges, or civilians—should be disqualified from holding any public office for a period of no less than ten years. The Commission added that these persons should also be 'disqualified permanently from any activity related to public security or national defence.' Finally, the Commission recommended that all Supreme Court judges should resign from office to permit the designation of new judges untainted by illegal conduct.

The Commission made a series of recommendations designed to promote national reconciliation. It proposed, *inter alia*, the construction of a national monument listing the names of all victims of the Salvadoran conflict; a national holiday honoring them; and the creation of a Forum for Truth and Reconciliation, comprising representatives of all sectors of Salvadoran society, to address the conclusions and recommendations of the Truth Commission with a view to promoting their implementation. The Commission also recommended the establishment of a fund to compensate all victims of serious acts of violence. Calling on foreign governments to assist El Salvador with these payments, the Commission urged that at least one percent of all international assistance received from abroad be earmarked for such compensation.

. . .

A few days after the publication of the *Report*, the government of President Cristiani and the national legislature controlled by his party granted an across-the-board amnesty to all individuals charged with serious acts of violence. This measure did not, however, nullify the Commission's work or have a serious effect on it. The amnesty merely prevented those identified by the Commission as responsible for acts of violence from being tried in Salvadoran courts and resulted in the release from prison of a few others who had been convicted earlier in that country on similar charges. . . . [T]he amnesty did not affect the Commission's recommendations or override those calling for the dismissal from their positions of individuals named in the *Report*. . . .

[The remaining excerpts from the article deal with the problem of 'naming names' in the Report.]

. . . [T]he Parties to the Peace Accords wanted 'the complete truth be made known.' For that purpose, they empowered the Commission to investigate the 'serious acts of violence that have occurred since 1980 and whose impact on society urgently demands that the public should know the truth.' How could we make known 'the complete truth' about a murder or massacre, for example, without identifying the killers if we knew their identity?

. . .

The attitude of the government began to change dramatically as it became known that the Commission had gathered incriminating evidence against high-ranking government officials. . . . [T]he Truth Commission was going to present evidence and make public its Report, which posed a . . . serious threat. Moreover, some of the officers, particularly General Ponce, had been instrumental in convincing his military colleagues to go along with the Peace Accords. . . . Together, [members of an elite and powerful group in the military known as the *tandona*] had the power to make life very difficult for President Cristiani and to impede, if not scuttle, his efforts to proceed with the implementation of the Peace Accords.

The power and influence of the *tandona*, rumors that 'naming names' would lead to a military coup in El Salvador, and claims by many well-intentioned individuals in El Salvador and outside the country that the publication of names by the Truth Commission would make national reconciliation very difficult—that it would be like pouring gasoline on a smoldering fire—prompted the government to mount a fierce diplomatic campaign to force us to omit names from the Report. President Cristiani led the campaign by urging various Latin American leaders, the United States, and the UN Secretary-General to use their power and influence to prevent the publication of names. He also sent a ministerial delegation to meet with us in New York for the same purpose. . . . The government also attempted to convince the FMLN to agree with its position. The FMLN and the government together, as the Parties to the Peace Accords, presumably had it in their power to amend the Commission's mandate and to require us not to publish any names. Some in the FMLN leadership were quite sympathetic to this effort and implied as much in conversations with us; a majority was opposed. Eventually, after a lengthy and apparently acrimonious debate within the FMLN high command, the FMLN informed the Commission that the Peace Accords required the publication of names.

The diplomatic campaign mounted by the Salvadoran government against the publication of names by the Truth Commission made it necessary for the Commissioners to explain our position to government leaders in the United States, Europe, and Latin America who were being lobbied by the Salvadorans. . . . The Commission's stature also established the requisite credibility to explain why we believed that our mandate required us to name names and why in our judgment this action would promote rather than impede national reconciliation in El Salvador. . . . In the *Report*, we explained our decision as follows:

> . . . [T]he Commission was not asked to write an academic report on El Salvador, it was asked to investigate and describe exceptionally important acts of violence and to recommend measures to prevent the repetition of such acts. This task cannot be performed in the abstract, suppressing

> information (for example, the names of person responsible for such acts) where there is reliable testimony available, especially when the persons identified [continue to] occupy senior positions and perform official functions directly related to the violations or the cover-up of violations. Not to name names would be to reinforce the very impunity to which the Parties instructed the Commission to put an end.

. . . What was often not generally known and what engendered acrimonious debates was who committed the acts and who ordered them to be committed. The search for the 'intellectual authors' of these offenses was a national obsession. Many of the intellectual authors were still holding influential positions. To refrain from exposing them would have amounted to yet another cover-up. In our view, national reconciliation would be harmed rather than helped by a Commission report that told only part of the truth. If there had been an effective justice system in El Salvador at the time of the publication of our Report, it could have used the Report as a basis for an independent investigation of those guilty of the violations. In these circumstances, it might have made some sense for the Commission not to publish the names and, instead, to transmit the relevant information to the police or courts for appropriate action. But one reason for establishing the Commission was that the Parties to the Peace Accords knew, and the Truth Commission had ample evidence to confirm, that the Salvadoran justice system was corrupt, ineffective, and incapable of rendering impartial judgments in so-called 'political' cases.

. . .

JOSE ZALAQUETT, BALANCING ETHICAL IMPERATIVES AND POLITICAL CONSTRAINTS: THE DILEMMA OF NEW DEMOCRACIES CONFRONTING PAST HUMAN RIGHTS VIOLATIONS

43 Hastings L. J. 1425 (1992), at 1432.

. . .

With the inauguration of President Aylwin in March of 1990, Chile became the last of the Southern Cone countries ruled by similarly minded military dictatorships to achieve a restoration of democratic government. . . .

Unlike Argentina, the transition to democracy in Chile took place after the new government negotiated rules to the game with a united, undefeated military that continued to enjoy considerable, albeit minority, political support. Despite this difference, the Argentinean case was telling for Chileans. It proved the importance of a systematic effort to reveal the truth. It also showed the extent to which a government can lose authority when it raises expectations it cannot fulfill.

Uruguay also provided an example for Chile in that the transition there was made under conditions similar to those present in Chile. But in Uruguay the government took too cautious an approach, avoiding not only trials for past state crimes but also any significant official disclosure about past violations. . . .

Taking these lessons into account, the Aylwin government decided to follow a course it could sustain. It adopted the guiding principle that reparation and prevention must be the overall objectives of the policies regarding past human rights violations. Righting a wrong and resolving not to do it again is, at its core, the same philosophy that underpins Judeo-Christian beliefs about atonement, penance, forgiveness, and reconciliation.

At a societal level, the equivalent of penance is criminal justice. Yet the Chilean government's assessment of the situation led it to conclude that priority ought to be given to disclosure of the truth. This disclosure was deemed an inescapable imperative. Justice would not be foregone, but pursued to the extent possible given the existing political restraints. Forms of justice other than prosecuting the crimes of the past, such as vindicating the victims and compensating their families, could be achieved more fully. The underlying assumption, which I share, was that if Chile gave truth and justice equal priority, the result might well have been that neither could be achieved. Fearing that official efforts to establish the truth would be the first step toward widespread prosecutions, the military would have determinedly opposed such efforts.

. . .

To establish the truth, President Aylwin appointed the National Commission for Truth and Reconciliation, a panel of eight people from across the political spectrum. . . .

. . .

In February of 1991, the commission presented a 1,800-page unanimous report to the president including its decisions in each of the individual cases, several chapters of historical, political and legal references, and a detailed section containing recommendations for both reparations and preventions. It proposed pensions for the families of the dead and 'disappeared,' measures designed to commemorate the events and honor the victims, and other forms of relief. It further proposed legal, institutional, and educational reforms aimed at enhancing the promotion and protection of human rights. In a televised address to the nation, President Aylwin presented the commission's findings and, as the head of State, atoned for the crimes committed by its agents.

. . .

For me, as for many of my friends and colleagues in the human rights movement in Chile, an eighteen-year period is coming to a close. It began the moment we started to organize the defense of human rights in Chile after the coup d'etat of September 1973, and has continued through our present efforts to overcome this dark chapter in our country's history.

Of the many lessons learned, I will refer to one. Back in the hazardous days of late 1973, all of my friends and colleagues in the human rights movement had to face danger on a daily basis. None of them ever claimed to have been endowed with innate bravery. They realized that courage was just another name for learning how to live with your fears. Eighteen years later, we all have come to realize that under changed circumstances, a less striking form of courage is called for. It is the courage to forgo easy righteousness, to learn how to live with real-life restrictions, but to seek nevertheless to advance one's most cherished values day by day to the extent possible. Relentlessly. Responsibly.

QUESTIONS

1. Recall the discussion by Szasz, p. 1077, *supra*, about amnesty as part of a peace settlement in the Former Yugoslavia. Suppose that leaders of one or several of the ethnic groups who have been or might be indicted demand an amnesty with respect to prosecutions before the ICT as a condition to an agreement ending hostilities. Should the Security Council grant their request through one or another type of decision that effectively bars prosecutions? If an amnesty were granted, are there other paths that the UN could follow that might achieve some of the same goals as convictions and prison sentences?

2. 'All instrumentalist justifications for prosecution of those violating international criminal law are speculative. We simply don't know how much or whom convictions or the threat thereof will deter, or whether convictions will achieve reconcilation and ongoing stability. We know as little about the long-run effects of amnesties. Much depends on particular and changing contexts, on factors and emotions and mass psychology that we cannot predict. We do know, however, what justice requires. We should act on that knowledge, on principle, not on the basis of some dubious cost-benefit reckoning about how to achieve certain goals. From the perspective of justice, there can be no retreat from the necessity of serious punishment following serious crime.' Comment.

3. What do you view as the advantages and disadvantages of truth commissions as alternatives to prosecutions or amnesties?

4. Consider the decision of the Inter-American Commission on Human Rights on the Uruguayan law:

a. Are you persuaded that the provisions of the American Convention relied on by the Commission should be interpreted to invalidate the law? Does the Commission's interpretation show any awareness of the dilemma in which democratic governments may find themselves in relation to the amnesty issue?

b. Why should not the plebescite have settled the issue, as an expression of the will of the people? Would no expression of popular will have satisfied the Commission?

5. How, if at all, can Zalaquett's views based on the Chilean experience be reconciled with those expressed in categorical terms by Amnesty International and the UN Special Rapporteur?

6. Buergenthal suggests that the decision to 'name names' in the Report on El Salvador was justified in part by the knowledge that the Salvadorean justice system was incapable of honest investigation and impartial judgment. Are there nonetheless problems in a report that includes names of violators? What form of investigation and what methods or standards of proof and judgment would you recommend for a truth commission that intended to publish such names?

16

Development and Human Rights

The spacious title of this chapter embraces a multitude of subjects. Like the two other chapters in Part F, it selects a few among them to illustrate the kinds of problems that have been discussed under the broad rubric of 'development' and human rights. The chapter notes but does not examine other contemporary concerns about a 'right to development,' such as the complex relationship between development and environment, the distinction between the development of a state and of its people, and the policies of international financial organizations like the International Monetary Fund that are meant to assist in the process of development.

Ever since the first UN World Conference on Human Rights, held in Teheran in 1968, the relationship between human rights and development has occupied a prominent place in the international discourse of rights. Since 1977 the debate has been pursued with increasing vigor under the rubric of 'the right to development'. That debate brings together several important themes raised in earlier chapters. They include the legal foundations of the classical human rights and the basis for recognition of new rights, the priority to be accorded to the different sets of rights, the links between human rights and democratic governance, the extent to which the international community bears some responsibility for assisting states whose resources are inadequate to ensure the human rights of their own citizens, and the relationship between individual and collective rights (including 'peoples' rights'). In addition, an examination of the concept of the right to development and its implications in the 1990s cannot avoid consideration of the effects of the globalization of the economy and the consequences of the near-universal embrace of the market economy.

The list of internationally recognized human rights is by no means immutable. Just as the British sociologist T.H. Marshall characterized the eighteenth century as the century of civil rights, the nineteenth as that of political rights and the twentieth as that of social rights,[1] so too have some commentators over the past two decades put forward claims for the recognition of the new rights, in particular a category known as the 'third generation of solidarity rights'. By analogy with the slogan of the French Revolution these rights have been said to correspond to the theme of *fraternité*, while first gen-

[1] T.H. Marshall, Citizenship and Social Class 14 (1950).

eration civil and political rights correspond with *liberté* and second generation economic and social rights with *egalité*. Karel Vasak's list of solidarity rights includes 'the right to development, the right to peace, the right to environment, the right to the ownership of the common heritage of mankind, and the right to communication.'[2]

Far more significant has been the impact of the right to development. First recognized by the UN Commission on Human Rights in 1977, it was enshrined by the General Assembly in the 1986 Declaration on the Right to Development. It has never ceased to be controversial among governments as well as among academic and other commentators. Efforts to secure its recognition raised the question of the process by which new human rights should be recognized. Some commentators emphasized the need for the catalogue of rights to keep pace with new developments and to be responsive to new challenges. Others argued that the third-generation approach of rights of peoples was a flawed way of seeking to meet such needs.

> Not only is proliferation of rights considered to be dangerous, but also the use of the term 'generation' implies, the detractors say, that the rights belonging to earlier generations are outdated. It is also frequently said that the rights of the new generation are too vague to be justiciable and are no more than slogans . . . .[3]

In 1984 Philip Alston identified a large number of rights whose recognition had been proposed by various sources.[4] They ranged from the right to sleep and the right to social transparency to the right to co-existence with nature and the right to tourism. He examined whether the United Nations could or should adopt substantive criteria to be satisfied before a new human right could be recognized. His conclusion was that such an approach was unworkable within an inter-governmental setting. Instead, he proposed that procedural requirements should be met. These included formal acknowledgement of an intention to accord recognition, followed by a detailed analytical study by the Secretary-General, comments by governments, examination by a specialist committee, and formal endorsement by both the UN Commission on Human Rights and the General Assembly.

Two years later the General Assembly responded by adopting the following guidelines that states were to 'bear in mind' in developing human rights instruments.

> (a) Be consistent with the existing body of international human rights law;
> (b) Be of fundamental character and derive from the inherent dignity and worth of the human person;

[2] K. Vasak, For the Third Generation of Human Rights: The Rights of Solidarity, Inaugural Lecture, Tenth Study Session, International Institute of Human Rights, July 1979, at 3.

[3] Stephen Marks, Emerging Human Rights: A New Generation for the 1980s?, 33 Rutgers Law Rev. 435, 451 (1981).

[4] Alston, Conjuring Up New Human Rights: A Proposal for Quality Control, 78 Am J. Int. L. 607 (1984).

(c) Be sufficiently precise to give rise to identifiable and practicable rights and obligations;
(d) Provide, where appropriate, realistic and effective implementation machinery, including reporting systems;
(e) Attract broad international support. . . .[5]

A. THE RIGHT TO DEVELOPMENT AS A NEW HUMAN RIGHT

In the readings below, Alston explains the context in which the debate on a right to development unfolded within the United Nations; Abi-Saab and Bedjaoui then present the case for the existence of such a right in international law; in contrast, Jack Donnelly concludes that the right is not only without foundation but is dangerous as well.

PHILIP ALSTON, REVITALISING UNITED NATIONS WORK ON HUMAN RIGHTS AND DEVELOPMENT

18 Melb. U. L. Rev. 216 (1991), at 218.

. . . The concept [of the right to development] was first mooted in 1972. It was another five years before serious and sustained debate began in the Commission on Human Rights. But by 1981, the topic had become so entrenched that the debate was 'institutionalized' through the establishment of a separate Working Group of Governmental Experts on the Right to Development . . . Five years after the Working Group's creation, the General Assembly adopted the Declaration on the Right to Development and left the Commission to work out what procedures or institutional arrangements ought to be put in place to follow up on the Declaration.

. . . [In 1993 the Commission established another Working Group to clarify the concept and to suggest ways of making it operational. As of July 1995 the Group had achieved no significant breakthroughs.]

But while the chronology is simple, the political currents influencing the evolution of the concept are much less so. Briefly stated, the emergence of a numerically dominant group of developing countries, as a result of the wave of decolonization that peaked in the late 1960s, led to the elevation of economic development goals to the top of the international agenda. Given the level of resentment over the negative consequences of the colonial experience and the reticence of the former colonial powers to recognize continuing obligations towards the peoples concerned, the assumption that reparations were

[5] G.A. Res. 41/121 (1986).

payable was never far below the surface. In terms of the U.N.'s human rights debate, these concerns translated into demands that greater attention be paid to economic and social rights (cultural rights being largely neglected in this setting), that colonialism and neo-colonialism be recognized as gross violations of international law and that some forms of development co-operation should be seen as entitlements rather than as acts of welfare or charity.

The Eastern Europeans provided significant and enthusiastic political support for all of these demands, but they did so on the basis of several rather convenient understandings: that the pursuit of centrally-planned socialism was the best and perhaps the only effective guarantee of economic and social rights; that they themselves had never been involved in colonialism or the denial of self-determination to any peoples; and that large-scale aid transfers were owed only by the former colonizers to their victims and not by the industrialized countries in general. Thus understood, the issues involved in the right to development enjoyed the enthusiastic support of the Eastern Europeans and their allies in their collective struggle against the capitalist West.

The West, for its part, presented a less than united front, but in general there was significant support for the proposition that economic and social rights had been accorded insufficient. attention by the United Nations. Moreover, the obligation to co-operate to promote Third World development was accepted by many Western states in general terms, although by no means in the form of a legally binding obligation to provide specific transfers of capital, technology or other goods and services. Even the United States, which from 1981 onwards was to become an implacable opponent of the right to development, was, under President Carter, open to many of the goals and even some of the means contained in the demands for the establishment of a new international economic order.

But the West, supported by a reasonable number of Third World states, also had other concerns, at least in terms of human rights doctrine in the United Nations context, if not in terms of their own practice. Issues of equity and distribution, which manifested themselves in the development debate of the 1970s, under the guise of the 'basic needs strategy' . . . translated into support for a particular vision of economic and social rights which did not necessarily coincide with that of some of the Third World proponents of those rights. Similarly, the view that respect for civil and political rights was indispensable for the achievement of human development, as appropriately defined, was firmly held by virtually all Western states.

. . .

. . . [N]either the North nor the South (in so far as such general descriptive terms are analytically valid) were prepared to accept the logical conclusions which could reasonably be drawn from their general negotiating positions in the debate. The North, for its part, was anxious to insist that the development process should be predicated upon full respect for human rights and that economic and social rights should be taken seriously, but was not prepared to accept that these positions might have direct implications for its own policies towards Third World countries, especially in terms of aid and trade. The South, on the other hand, was anxious to demand concessions from the North and to constrain it in respect of various of its policy options, but was unprepared to accept any constraints on its own freedom of action. While these positions, even in their extreme forms, were hardly surprising, they also helped to ensure a relatively unproductive debate within the Commission on Human Rights.

GEORGES ABI-SAAB, THE LEGAL FORMULATION OF A RIGHT TO DEVELOPMENT

in Academy of International Law, The Right to Development at the International Level 159 (1980), at 163.

[Abi-Saab begins by noting that, for the right to development to be considered a legal right, it must be possible to identify the active and passive subjects of the right and its content. But those elements depend on the legal basis of the right, which in turn depends on whether the right is an individual or collective one.]

It is possible to think of different legal bases of the right to development as a collective right. The first possibility . . . is to consider the right to development as the aggregate of the social, economic and cultural rights not of each individual, but of all the individuals constituting a collectivity. In other words, it is the sum total of a double aggregation of the rights and of the individuals. This version . . . has the merit of shedding light on the link between the rights of the individual and the right of the collectivity; a link which is crucial. . . .

Another way . . . is to approach it directly from a collective perspective . . . by considering it either as the economic dimension of the right of self-determination, or alternatively as a parallel right to self-determination, partaking of the same nature and belonging to the same category of collective rights.

What can be the content of a right to development? And is it really possible to give a legal rendering of development whether in terms of municipal or international law? This is a very difficult task, but not an impossible one. . . .

. . . [W]hen we speak of global development, most emphasis is put on the development of the less developed countries (LDCs), because of the increasing awareness that this is the major and most urgent problem and the *conditio sine qua non* for the harmonious development of the international community as a whole.

What are these policy measures? . . . [The author then considers the evolution of international development law through the UN Development Decade Strategies, the sessions of the UN Conference on Trade and Development

(UNCTAD) and the program for the achievement of a New International Economic Order (NIEO), endorsed by the UN General Assembly in 1973–75.]

Briefly, these measures can be classified into four or five clusters each addressing itself to a major problem in North–South economic relations: the question of stabilizing commodity prices at an equitable level; the question of changing the international division of labour in order to give a greater opportunity for industrialization to the less developed countries and a greater share in international trade to their industrial or manufactured products; the question of transfer of technology; the question of redirecting the monetary and financial flows in a manner more compatible with the imperatives of development, etc.

The measures are either of a purely economic content (such as the integrated commodity programme, the generalized preference scheme, the fixing of targets for public aid) or of a political-normative nature which aims at the strengthening of the bargaining power of less developed countries (or rather at re-equilibriating to some extent the great disparity in power between them and the developed countries within the international economic system) such as the reaffirmation and extension of the principle of permanent sovereignty over natural wealth and resources, the legitimatization of producers' associations, the claim for more effective participation in decision-making in monetary and financial institutions, etc.

. . .

Whether this bundle of policy measures we now call the NIEO is in fact conducive to global development is another question, however; a question which is subject to much theoretical controversy. Indeed, the NIEO has been criticized both from the Right and from the Left: from the Right as being an incoherent amalgam of 'uneconomic' policies which would bring about chaos and disorder by destroying the market mechanisms; from the Left the NIEO is attacked by the supporters of the strategy of 'self-reliance' and 'de-linking', as consolidating the patterns of dependence of the South and its control by the North, and for ignoring the vital issue of a more equitable internal distribution of income and thus working for the exclusive benefit of the oppressive élites, who are the objective allies and the guardians of the interests of the North in the South.

But the crucial fact remains that now, as in 1974, as in 1964, this bundle of policy measures is the platform that has been officially, consistently and staunchly put forward by the countries of the Third World, and which is gradually (though only partially and in a piecemeal fashion) commanding a widening acceptance on the part of the Western industrialized countries. . . In other words, the NIEO reflects the long maturing and increasingly shared conception of a large majority of the international community of what constitutes the necessary and sufficient means of achieving equitable global development.

Thus, if we envisage the question from the perspective of law-formation, i.e., of the prospects of a legal right to development, it is this generally shared concept—whatever one's own evaluation of the NIEO—which constitutes the emerging social value within the body-politic of the international community, and which alone stands a chance of eventually being transformed or hardened into law. In other words, only social—i.e., widely shared—values, not individual desiderata, can become legal rules. For law is neither pure ideas, ideals or ideology, nor sheer social power or brute force. It is ideas backed by a sufficient amount of social power . . . .

. . .

. . .[I]t is very unlikely that the right to development will emerge exclusively through the spontaneous process of custom-formation, or out of one legal instrument which covers all its facets and provides it with the necessary legal sanction. Rather, the whole will gradually emerge from its parts.

But if negotiations on the parts produce different types of legal instruments with varying legal nature and effect, the right to development is emerging from them through the process of custom-formation. For it is through the cumulative effect of these policy measures, legal instruments and underlying principles that the right to development is slowly but surely taking shape and asserting itself.

As far as the beneficiaries or active subjects are concerned, the first answer that comes to mind is that they are those societies possessing certain characteristics which lead the international community to consider them wanting in terms of development and to classify them as 'developing' or 'less developed' countries (LDC). . . .

. . .

. . . Up to now, we have used societies, communities, countries and States as interchangeable, which they are not. In fact, here as with self-determination, the common denominator of these different ways of describing the beneficiary collectivity is the 'people' they designate, which constitutes the socially relevant entity or group in this context. . . . Suffice it to say here that the distinction between 'people' and 'State', though in theory it is as important in relation to the right to development as to the right of self-determination, in practice it is not . . . .

. . .

. . . [T]he passive subject of the right to development can only be the international community as such. But as the international community does not have at its disposal the means (organs, resources) of directly fulfilling its obligations under the right to development, it can only discharge them through a category of its members, that of the 'developed' States . . . .

. . .

. . . [S]atisfaction of the collective right is a necessary condition, a condition-precedent or a prerequisite for the materialization of the individual rights.

Thus without self-determination it is impossible to imagine a total realization of the civil and political rights of the individuals constituting the collectivity in question. Such rights can be granted and exercised at lower levels, such as villages and municipalities, but they cannot reach their full scope and logical conclusion if the community is subject to colonial or alien rule . . . .

The same with the right to development, which is a necessary precondition for the satisfaction of the social and economic rights of the individuals. And here, even more than in the case of self-determination, the causal link between the two levels is particularly strong; for without a tolerable degree of development, the society will not be materially in a position to grant and guarantee these rights to its members, i.e., of providing the positive services and securing the minimum economic standards which are required by these rights.

. . .

MOHAMMED BEDJAOUI, THE RIGHT TO DEVELOPMENT

in Bedjaoui (ed.), International Law: Achievements and Prospects 1177 (1991), at 1182.

. . .

III. ELEMENTS FOR THE DEFINITION AND IDENTIFICATION OF THE 'RIGHT TO DEVELOPMENT'

14. The right to development is a fundamental right, the precondition of liberty, progress, justice and creativity. It is the alpha and omega of human rights, the first and last human right, the beginning and the end, the means and the goal of human rights, in short it is the *core right* from which all the others stem. . . .

. . .

15. . . . In reality the international dimension of the right to development is nothing other than *the right to an equitable share in the economic and social well-being of the world.* It reflects an essential demand of our time since four fifths of the world's population no longer accept that the remaining fifth should continue to build its wealth on their poverty.

. . .

IV. BASIS OF THE RIGHT TO DEVELOPMENT

19. The most essential human rights have, in a sense, a meta-juridical foundation. For example, the right to life is independent both of international law and of the municipal laws of States. It pre-exists law. In this sense it is a 'primary' or 'first' law, that is to say a law commanding all the others. . . . Thus

the right to development imposes itself with the force of a self-evident principle and its natural foundation is as a corollary of the right to life. . . .

20. . . . Certain lawyers consider the right to development to be a legal concept and principle enshrined in the Charter and find its basis both in the preamble, in Article 1, paragraph 3, and, above all, in Article 55. These are commendable efforts at textual exegesis to which the remarkable flexibility of this exemplary instrument lends itself so generously. But the results of these efforts are necessarily limited since in 1945, and in the context of that time, 'the ideology of development for all' was still vague and its legal expression in the Charter necessarily still timid.

22. The 'right to development' flows from this right to self-determination and has the *same nature*. There is little sense in recognizing self-determination as a superior and inviolable principle if one does not recognize *at the same time* a 'right to development' for the peoples that have achieved self-determination. This right to development can only be an 'inherent' and 'built-in' right forming an inseparable part of the right to self-determination.

23. . . . [This makes the right to development] much more a right of the State or of the people, than a right of the individual, and it seems to me that it is better that way.

. . .

26. The present writer considers that international solidarity means taking into account the interdependence of nations. One may identify three stages in this search for the foundation of the right to development based on international solidarity:

(i) interdependence, the result of the global nature of the world economy;
(ii) the universal duty of every State to develop the world economy, which makes development an international problem *par excellence*;
(iii) preservation of the human species as the basis of the right to development.

. . .

V. CONTENT OF THE RIGHT TO DEVELOPMENT

34. . . . [This right] has several aspects, the most important and comprehensive of which is the right of each people freely to choose its economic and social system without outside interference or constraint of any kind, and to determine, with equal freedom, its own model of development. . . .

35. Where the elementary bases of independent national control over economic planning are non-existent because they have been expropriated in a more or less covert way by foreign powers, it is impossible to speak of the sovereign equality of States or of State sovereignty without lapsing into fiction. It is thus clear, and even blindingly obvious, that not all States can be equally

sovereign. Furthermore, one can demonstrate with perfect clarity the fact that the right to development is an imperative of sovereignty. There is thus a *necessary relationship* between authentic sovereignty and the right to development, between true sovereignty over the wealth of a country and that country's right to development. For any inequality in development gives rise to the effects of domination and dependence which compromise sovereignty.

. . .

44. . . . [A] people may in no case be deprived of its own means of subsistence. . . .

. . .

47. This right [to develop] which is due from all parties to the State which possesses the right has, in the present writer's view, two aspects which might be stated thus:

to each his due;
to each according to his needs.

48. The first point means that the State seeking its own development is entitled to demand that all the other States, the international community and international economic agents collectively *do not take away from it what belongs to it, or do not deprive it of what is or 'must be' its due in international trade*. In the name of this right to development, the State being considered may claim a *'fair price'* for its raw materials and for whatever it offers in its trade with the more developed countries.

The claim of such a State goes something like this: 'Before giving me charity or offering me your aid, give me my due. Perhaps I shall then have no need of your aid. Perhaps charity is no more than the screen behind which you expropriate what is due to me. Such charity does not deserve to be so called; it is my own property you are handing back to me in this way and, what is more, not all of it'.

49. This second meaning of the right to development which is due from the international community seems much more complex. It implies that the State is *entitled if not to the satisfaction of its needs at least to receive a fair share of what belongs to all, and therefore to that State also*.

This raises two questions:

(a) What things belong to the international community and may be claimed by any State in the context of the right to development?
(b) What is a 'fair' share which may be due to such State by virtue of the maxim 'to each according to his needs'?

50. As far as the first question is concerned, the satisfaction of the needs of a people should be perceived as a right and not as an act of charity. It is a

right which should be made effective by *norms and institutions*. The relation between the donor and the recipient States is seen in terms of responsibility and reciprocal rights over goods that are considered as belonging to all. There is no place in such an analysis for charity, the 'act of mercy', considered as being a factor of inequality from which the donor expects tokens of submissiveness or political flexibility on the part of the receiving State. The concept of charity thus gives place to that of justice. *The need*, taken as a criterion of equity, gives greater precision to the concept of '*equitable distribution*' which would otherwise be too vague.

51. To each according to his needs means that the '*fair price*' cannot be tied to the *market price*. The desirable objective of an equitable relationship between the producer and the consumer could be attained only through price manipulation, by means of other compensatory measures, to the advantage of the less developed country. This would signify a transfer of resources to the poorest countries. The fair price would be calculated according to the general principle of responsibility towards the most needy countries. It would be conceived as being a price which allows for amortization of the costs of production, including social costs, to ensure the well-being and minimal development of the developing countries.

52. What belongs to the international community and is 'the common heritage of mankind' should be shared among all States in accordance with the maxim '*to each according to his needs*'. . . .

There can be no denying that this innovative concept, the common heritage of mankind, is capable of giving world-wide solidarity a wealth of practical expression. It might prove especially productive for the future of world relations and be applied not only to the resources of the sea-bed and of space (those of the moon and celestial bodies), as is already the case, but also to the land, the air, the climate, the environment, inert or living matter and the animal and vegetable genetic heritage, the wealth and variety of which it is vital to preserve for future generations.

. . .

VI. DEGREE OF NORMATIVITY OF THE RIGHT TO DEVELOPMENT

53 Learned opinion is divided in its view of the legal validity of the right to development. Many writers consider that while it is undoubtedly an inalienable and imperative right, this is only in the moral, rather than in the legal, sphere. The present writer has, on the contrary, maintained that the right to development is, by its nature, so incontrovertible that it *should* be regarded as belonging to *jus cogens*.

. . .

55. It is clear, however, that a right which is not opposable by the possessor of the right against the person from whom the right is due is not a right

in the full legal sense. This constitutes *the challenge which the right to development throws down to contemporary international law* and the whole of the challenge which the underdevelopment of four fifths of the globe places, in political terms, before the rulers of the world. . . .

. . .

JACK DONNELLY, IN SEARCH OF THE UNICORN: THE JURISPRUDENCE AND POLITICS OF THE RIGHT TO DEVELOPMENT

15 Calif. Western Int. L. J. 473, at 482 (1985).

III. LEGAL SOURCES OF THE RIGHT TO DEVELOPMENT

. . .

. . . If the right to development means the right of peoples freely to pursue their development, then it can be plausibly argued to be implied by the Covenants' right to self-determination. However, such a right to development is without interest; it is already firmly established as the right to self-determination.

A substantially broader right to development, however, cannot be extracted from this right to self-determination. The right to self-determination recognized in the Covenants does not imply a right to live in a developing society; it is explicitly only a right to *pursue* development. Neither does it imply an *individual* right to development; self-determination, again explicitly, is a right of peoples only. In no sense does it imply a right to be developed. Thus the claim that the right to development is simply the realization of the right to self-determination is not based on the Covenants' understanding of self-determination.

It might also be argued that because development is necessary for self-determination, development is itself a human right. Such an argument, however, is fallacious. Since we will come across this form of argument again, let us look briefly at this 'instrumental fallacy.' Suppose that A holds mineral rights in certain oil-bearing properties. Suppose further that in order to enjoy these rights fully, she requires $500,000 to begin pumping the oil. Clearly A does not have a right to $500,000 just because she needs it to enjoy her rights. . . . The same reasoning applies to the link between development and the right to self-determination. Even assuming that development is necessary for, rather than a consequence of, full enjoyment of the right to self-determination, it simply does not follow that peoples have a right to development.

Allowing such an argument to prevail would result in a proliferation of bizarre or misguided rights. . . .

. . .

The second promising implicit source of a right to development is Article 28 of the Universal Declaration: 'Everyone is entitled to a social and international order in which the rights and freedoms set forth in this declaration can be fully realized.'

. . .

Unlike most other provisions of the Declaration, however, Article 28 was not confirmed by a parallel provision in the Covenants, which were drafted precisely to give the force of treaty law to the standards of the Universal Declaration. This absence of a passage equivalent to Article 28 is most plausibly interpreted as representing its tacit renunciation. . . .

More substantively, one might question whether 'development' falls under the notion of a social and international order referred to in Article 28. 'Development' suggest a process or result; the process of development or the condition of being developed. 'Order,' by contrast, implies a set of principles, rules, practices or institutions; neither a process nor a result but a structure. Article 28, therefore, is most plausibly interpreted as prohibiting *structures* that deny opportunities or resources for the realization of civil, political, economic, social or cultural human rights. . . .

. . .

Suppose, though, that Article 28 *were* to be taken to imply a human right to development. What would that right look like? It would be an *individual* right, and only an individual right; a right of persons, not peoples, and certainly not States. It would be a right to the enjoyment of traditional human rights, not a substantively new right. It would be as much a civil and political as an economic and social right—Article 28 refers to *all* human rights—and would be held equally against one's national government and the international community. . .

Continuing our review of alleged legal sources, we find that other provisions of the International Bill of Human Rights are unquestionably inadequate as a source of a legal right to development. For example, Article 22 of the Universal Declaration provides for the 'realization, through national effort and international cooperation and in accordance with the organization and resources of each [S]tate' of basic economic, social and cultural rights. This implies at most a right to a fair share of available resources, not a right to development. . . .

. . .

V. SUBJECTS OF THE RIGHT TO DEVELOPMENT

. . .

If human rights derive from the inherent dignity of the human person, collective human rights are logically possible only if we see social membership as an inherent part of human personality, *and* if we argue that as part of a nation or people, persons hold human rights substantively different from, and in no way reducible to, individual human rights. This last proposition is extremely controversial. . . .

The very concept of human rights, as it has heretofore been understood, rests on a view of the individual person as separate from, and endowed with inalienable rights held primarily in relation to, society, and especially the state. Furthermore, within the area defined by these rights, the individual is superior to society in the sense that ordinarily, in cases of conflict between individual human rights and social goals or interests, individual rights must prevail. The idea of collective *human* rights represents a major, and at best confusing, conceptual deviation.

I do not want to challenge the idea of collective rights *per se* or even the notion of peoples' rights; groups, including nations, can and do hold a variety of rights. But these are not *human* rights as that term is ordinarily understood. . . .

. . .

A further problem with collective human rights is determining who is to exercise the right; the right-holder is not a physical person, and thus an institutional 'person' must exercise it. In the case of a right held by a people, or by society as a whole, the most plausible 'person' to exercise the right is, unfortunately, the state. Again this represents a radical reconceptualization of human rights—and an especially dangerous one.

. . .

VI. THE SUBSTANCE OF THE RIGHT TO DEVELOPMENT

. . .

The political dangers of subordinating human rights are especially great when development or the NIEO appear as *preconditions* for the realization of human rights, or when the right to development itself is presented as a precondition to the enjoyment of traditional human rights. Since satisfaction of these preconditions is a long way off, such arguments readily lend themselves to claims that Third World regimes ought not be held to international human rights standards. As preconditions multiply, we reach a situation where it appears as if the world would need to be 'transformed into an earthly paradise' before human rights could be realized.

The right to development and its preconditions thus lend themselves to use as an excuse not to act on human rights now. This, I would argue, is the central contemporary political fact about the right to development. . . .

The right to development, therefore, in its substance as in other aspects, is subject to both relatively defensible and entirely undefensible interpretations.

The problem is that the defensible interpretation, as an individual right to full personal development in all areas of human life, is rather innocuous, while the indefensible interpretations are quite dangerous. In other words, the dangers posed by the right to development far outweigh its benefits.

. . .

DECLARATION ON THE RIGHT TO DEVELOPMENT

General Assembly Res. 41/128 (1986).

The General Assembly,

Bearing in mind the purposes and principles of the Charter of the United Nations relating to the achievement of international co-operation in solving international problems of an economic, social, cultural or humanitarian nature, and in promoting and encouraging respect for human rights and fundamental freedoms for all without distinction as to race, sex, language or religion,

Recognizing that development is a comprehensive economic, social, cultural and political process, which aims at the constant improvement of the well-being of the entire population and of all individuals on the basis of their active, free and meaningful participation in development and in the fair distribution of benefits resulting therefrom,

. . .

Article 1

1. The right to development is an inalienable human right by virtue of which every human person and all peoples are entitled to participate in, contribute to, and enjoy economic, social, cultural and political development, in which all human rights and fundamental freedoms can be fully realized.

2. The human right to development also implies the full realization of the right of peoples to self-determination, which includes, subject to the relevant provisions of both International Covenants on Human Rights, the exercise of their inalienable right to full sovereignty over all their natural wealth and resources.

Article 2

1. The human person is the central subject of development and should be the active participant and beneficiary of the right to development.

2. All human beings have a responsibility for development, individually and collectively, taking into account the need for full respect for their human rights and fundamental freedoms as well as their duties to the community, which alone can ensure the free and complete fulfilment of the human being, and they should therefore promote and protect an appropriate political, social and economic order for development.

3. States have the right and the duty to formulate appropriate national development policies that aim at the constant improvement of the well-being of the entire population and of all individuals, on the basis of their active, free and meaningful participation in development and in the fair distribution of the benefits resulting therefrom.

Article 3

1. States have the primary responsibility for the creation of national and international conditions favourable to the realization of the right to development.

. . .

Article 4

1. States have the duty to take steps, individually and collectively, to formulate international development policies with a view to facilitating the full realization of the right to development.

2. Sustained action is required to promote more rapid development of developing countries. As a complement to the efforts of developing countries, effective international co-operation is essential in providing these countries with appropriate means and facilities to foster their comprehensive development.

. . .

Article 6

1. All States should co-operate with a view to promoting, encouraging and strengthening universal respect for and observance of all human rights and fundamental freedoms for all without any distinction as to race, sex, language or religion.

2. All human rights and fundamental freedoms are indivisible and interdependent; equal attention and urgent consideration should be given to the implementation, promotion and protection of civil, political, economic, social and cultural rights.

3. States should take steps to eliminate obstacles to development resulting from failure to observe civil and political rights, as well as economic, social and cultural rights.

. . .

Article 8

1. States should undertake, at the national level, all necessary measures for the realization of the right to development and shall ensure, *inter alia*, equality of opportunity for all in their access to basic resources, education, health services, food, housing, employment and the fair distribution in income. Effective measures should be undertaken to ensure that women have an active role in the development process. Appropriate economic and social reforms should be carried out with a view to eradicating all social injustices.

2. States should encourage popular participation in all spheres as an important factor in development and in the full realization of all human rights.

Article 9

1. All the aspects of the right to development set forth in the present Declaration are indivisible and interdependent and each of them should be considered in the context of the whole.

. . .

QUESTIONS

1. Judge Bedjaoui, the President of the International Court of Justice, characterizes the right to development as the 'core right' from which all others stem, and as a pre-condition of liberty and justice. Other rights, such as that to self-determination, have also been labelled 'core rights.'

a. What do you think the notion means? An essential condition (but there may be many essential conditions)? The only essential condition? The logical source of all other rights? How would you flesh out any such argument?

b. Could free speech be a 'core right'? Freedom from torture?

2. 'Scholars search for a foundation in sacred human rights texts for a new right, as if a right declared by the UN depends for its legitimacy on a clear pedigree from the UDHR or ICCPR. Why must this be so? Rights change as times change, and if times demand amendments to older rights or the legislation of new ones, so be it. Let a General Assembly vote for a declaration together with state support thereof suffice.' Comment.

3. Based on the Declaration on the Right to Development, what do you take to be the components of this right? What are the key ideas? Are they unified within one conception?

4. Do you perceive actual or potential contradictions between this right and other rights in the human rights corpus? For example, does the Declaration indicate how to resolve what might be competing claims for economic growth and for preservation of the environment—or how to resolve what might be competing claims for aggregate economic growth that is independent of distributional concerns, and for growth consistent with maintenance of a certain level of welfare support?

5. Compare the right of peoples to self-determination, considered in Chapter 14(A), and the right to development. What similarities and differences?

6. 'This "Declaration" may be significant as a rallying point, as powerful and official rhetoric expressing a hope of developing states and their populations. It has nothing to do with "rights." Whose rights are they—an individual's, a people's, a state's? Against whom are they asserted, and through what agency

if we talk of peoples' rights? Like most rights of the so-called third generation such as the so-called right to peace, the Declaration is better understood as aspiration, desire, need and claim. Within that understanding, it makes sense as a political and moral statement.' Comment.

7. To what extent does the Declaration bear out the hopes expressed by Abi-Saab and Bedjaoui, and the concerns expressed by Donnelly?

B. THE INTERDEPENDENCE OF THE TWO SETS OF RIGHTS

The question of the interdependence of civil and political rights on the one hand, and economic, social and cultural rights on the other, figured in the examination in Chapter 5 of the second set of rights, and in Chapter 13 of the prospects for advancement and equality for women. During the Cold War, the nature of this relationship was a constant bone of contention, both between East and West and between North and South. In order to preserve the formal consensus reflected in the International Bill of Rights, all governments continued to pay lip-service to the notion that the two sets of rights are 'interdependent, indivisible and interrelated.' UN human rights organs have regularly endorsed such assertions. Although this consensus may be both necessary and desirable in intergovernmental settings, it means that the complex nature of the theoretical and practical relationships among different rights is not open to exploration in those settings.

Compare in the following readings the position put to the Vienna World Conference on Human Rights by the Committee on Economic, Social and Cultural Rights with the viewpoint expressed by Amartya Sen.

STATEMENT TO THE WORLD CONFERENCE ON HUMAN RIGHTS ON BEHALF OF THE COMMITTEE ON ECONOMIC, SOCIAL AND CULTURAL RIGHTS

UN Doc. E/1993/22, Annex III.

. . .

3. . . . [F]ull realization of human rights can never be achieved as a mere by-product, or fortuitous consequence, of some other developments, no matter how positive. For that reason, suggestions that the full realization of economic, social and cultural rights will be a direct consequence of, or will flow

automatically from, the enjoyment of civil and political rights are misplaced. Such optimism is neither compatible with the basic principles of human rights nor is it supported by empirical evidence. The reality is that every society must work in a deliberate and carefully structured way to ensure the enjoyment by all of its members of their economic, social and cultural rights. . . .

4. Just as carefully targeted policies and unremitting vigilance are necessary to ensure that respect for civil and political rights will follow from, for example, the holding of free and fair elections or from the introduction or restoration of an essentially democratic system of government, so too is it essential that specific policies and programmes be devised and implemented by any Government which aims to ensure the respect of the economic, social and cultural rights of its citizens and of others for whom it is responsible.

. . .

12. [Some] Governments in the industrialized countries tend to assume that the existence of a genuinely democratic system and the generation of relatively high levels of per capita income are sufficient evidence of comprehensive respect for human rights. Yet, the Committee's experience shows that such conditions are perfectly capable of coexisting with significant areas of neglect of the basic economic, social and cultural rights of large numbers of their citizens. High infant mortality rates, a significant incidence of hunger or malnutrition, mass unemployment, large-scale homelessness and high drop-out rates from educational institutions are all indicators, at least prima facie, of violations of economic, social and cultural rights and hence of human rights.

AMARTYA SEN, FREEDOMS AND NEEDS

The New Republic 31 (Jan. 10 and 17, 1994), at 32.

. . .

. . . Do needs and rights represent a basic contradiction? Do the former really undermine the latter? I would argue that this is altogether the wrong way to understand, first, the force of economic needs and, second, the salience of political rights. The real issues that have to be addressed lie elsewhere, and they involve taking note of extensive interconnections between the enjoyment of political rights and the appreciation of economic needs. Political rights can have a major role in providing incentives and information toward the solution of economic privation. But the connections between rights and needs are not merely instrumental they are also constitutive. For our conceptualization of economic needs depends on open public debates and discussions, and the guaranteeing of those debates and those discussions requires an insistence on political rights.

. . .

Those who are skeptical of the relevance of political rights to poor countries would not necessarily deny the basic importance of political rights. Some of them would not even deny my contention that the nastiness of the violation of liberty can go well beyond other forms of disadvantage. Instead their arguments turn on the impact of political rights on the fulfillment of economic needs, and they take this impact to be firmly negative and overwhelmingly important.

The belief abounds that political rights correlate negatively with economic growth. Indeed, something of a 'general theory' of this relationship between political liberty and economic prosperity has been articulated recently by that unlikely theorist Lee Kuan Yew, the former prime minister of Singapore; and the praise of the supposed advantages of 'the hard state' in promoting economic development goes back a long way in the development literature. . . .

. . .

It is true that some relatively authoritarian states (such as Lee's Singapore, South Korea under military rule and more recently China) have had faster rates of economic growth than some less authoritarian states (such as India, Costa Rica and Jamaica). But the overall picture is much more complex than such isolated observations might suggest. Systematic statistical studies give little support to the view of a general conflict between civil rights and economic performance. In fact, scholars such as Partha Dasgupta, Abbas Pourgerami and Surjit Bhalla have offered substantial evidence to suggest that political and civil rights have a *positive* impact on economic progress. Other scholars find divergent patterns, while still others argue, in the words of John Helliwell, that on the basis of the information so far obtained 'an optimistic interpretation of the overall results would thus be that democracy which aparently has a value independent of its economic effects, is estimated to be available at little cost in terms of subsequent lower growth.'

There is not much comfort in all these findings for the 'Lee Kuan Yew hypothesis' that there exists an essential conflict between political rights and economic performance. The general thesis in praise of the tough state suffers not only from causal empiricism based on a few selected examples, but also from a lack of conceptual discrimination. Political and civil rights come in various types, and authoritarian intrusions take many forms. It would be a mistake, for example, to equate North Korea with South Korea in the infringement of political rights, even though both have violated many such rights; the complete suppression of opposition parties in the North can hardly be taken to be no more repressive than the roughness with which opposition parties have been treated in the South. Some authoritarian regimes, both of the 'left' and of the 'right', such as Zaire or Sudan or Ethiopia or the Khmer Rouge's Cambodia, have been enormously more hostile to political rights than many other regimes that are also identified, rightly, as authoritarian.

The processes that led to the economic success of, say, South Korea are now reasonably well understood; a variety of factors played a part, including the use of international markets, an openness to competition, a high level of literacy, successful land reforms and the provision of selective incentives to encourage growth and exports. There is *nothing* to indicate that these economic and social policies were inconsistent with greater democracy, that they had to be sustained by the elements of authoritarianism actually present in South Korea. The danger of taking *post hoc* to be *propter hoc* is as real in the making of such political and strategic judgments as it is in any empirical reasoning.

Thus the fundamental importance of political rights is not refuted by some allegedly negative effect of these rights on economic performance. In fact, the instrumental connections may even give a very positive role to political rights in the context of deprivations of a drastic and elementary kind: whether, and how, a government responds to intense needs and sufferings may well depend on how much pressure is put on it, and whether or not pressure is put on it will depend on the exercise of political rights (such as voting, criticizing, protesting and so on).

Consider the matter of famine. I have tried to argue elsewhere that the avoidance of such economic disasters as famines is made much easier by the existence, and the exercise, of various liberties and political rights, including the liberty of free expression. . . . But famines have never afflicted any country that is independent, that goes to elections regularly, that has opposition parties to voice criticisms, that permits newspapers to report freely and to question the wisdom of government policies without extensive censorship.

. . .

Why might we expect a general connection between democracy and the nonoccurrence of famines? The answer is not hard to seek. Famines kill millions of people in different countries in the world, but they do not kill the rulers. The kings and the presidents, the bureaucrats and the bosses, the military leaders and the commanders never starve. And if there are no elections, no opposition parties, no forums for uncensored public criticism, then those in authority do not have to suffer the political consequences of their failure to prevent famine. Democracy, by contrast, would spread the penalty of famine to the ruling groups and the political leadership.

There is, moreover, the issue of information. A free press, and more generally the practice of democracy, contributes greatly to bringing out the information that can have an enormous impact on policies for famine prevention, such as facts about the early effects of droughts and floods, and about the nature and the results of unemployment. . . . Indeed, I would argue that a free press and an active political opposition constitute the best 'early warning system' that a country threatened by famine can possess.

The connection between political rights and economic needs can be illustrated in the specific context of famine prevention by considering the massive Chinese famines of 1958–61. Even before the recent economic reforms, China had been much more successful than India in economic development. The average life expectancy, for example, rose in China much more than it did in India, and well before the reforms of 1979 it had already reached something like the high figure—nearly seventy years at birth—that is quoted now. And yet China was unable to prevent famine. It is now estimated that the Chinese famines of 1958–61 killed close to 30 million people—ten times more than even the gigantic 1943 famine in British India.

The so-called 'Great Leap Forward,' initiated in the late 1950s, was a massive failure, but the Chinese government refused to admit it, and continued dogmatically to pursue much the same disastrous policies for three more years. It is hard to imagine that this could have happened in a country that goes to the polls regularly and has an independent press. . . .

The lack of a free system of news distribution even misled the government itself. It believed its own propaganda and the rosy reports of local party officials competing for credit in Beijing. Indeed, there is evidence that just as the famine was moving toward its peak, the Chinese authorities mistakenly believed that they had 100 million more metric tons of grain than they actually did. . . .

. . .

. . . [o]f course, there is the danger of exaggerating the effectiveness of democracy. Political rights and liberties are permissive advantages, and their effectiveness depends on how they are exercised. Democracies have been particularly successful in preventing disasters that are easy to understand, in which sympathy can take an especially immediate form. Many other problems are not quite so accessible. Thus India's success in eradicating famine is not matched by a similar success in eliminating non-extreme hunger, or in curing persistent illiteracy or in relieving inequalities in gender relations. While the plight of famine victims is easy to politicize, these other deprivations call for deeper analysis, and for greater and more effective use of mass communication and political participation—in sum, for a fuller practice of democracy.

. . .

It is important to acknowledge, however, the special difficulty of making a democracy take adequate notice of some types of deprivation, particularly the needs of minorities. One factor of some importance is the extent to which a minority group in a particular society can build on sympathy rather than alienation. When a minority forms a highly distinct and particularist group, it can be harder for it to receive the sympathy of the majority, and then the protective role of democracy may be particularly constrained.

. . .

. . . The comprehension and the conceptualization of economic needs themselves may require the exercise of political and civil rights, open discussion and public exchange.

Human lives suffer from miseries and deprivations of various kinds, some more amenable to alleviation than others. The totality of the human predicament would be an undiscriminating basis for the social analysis of needs. There are many things that we might have good reason to value if they were feasible, maybe even immortality; yet we do not see them as needs. Our conception of needs relates to our analysis of the nature of deprivations, and also to our understanding of what can be done about them. Political rights, including freedom of expression and discussion, are not only pivotal in inducing political responses to economic needs, they are also central to the conceptualization of economic needs themselves.

QUESTIONS

1. Can the appropriate relationship between the two sets of rights be determined on the basis of either instrumentalist arguments or empirical studies, or is the question rather a matter of political axiom and faith?

2. If Sen's position is accepted, what are the implications for economic and social rights within the international human rights framework? Does it follow that the emphasis in developing countries under authoritarian rule should clearly and consistently be placed upon civil and political rights?

3. How do you relate the Committee's stress on forming programs on economic and social rights (as opposed to emphasis on averting disasters through a free press and public debate) to Sen's argument?

C. IS THERE AN OBLIGATION TO ASSIST?

Many observers believe that the single most important objective of the proponents of the right to development is to establish an obligation on the part of wealthier countries to provide financial and other types of assistance to poorer countries. This issue has also been prominent in relation to the obligations contained in the International Covenant on Economic, Social and Cultural Rights (ICESCR).

In the next reading, Alston and Quinn consider the implications of the provision in that Covenant relating to 'international assistance and co-operation.' The argument that there is an obligation to provide assistance is sometimes linked to human rights in general and sometimes directly to the right to development. It also arises in the general literature on economic and social development. In the readings that follow by Bedjaoui, and by Griffin and Khan, the authors put forward proposals designed to ensure automatic assistance to poor countries.

The furthest that any of the international human rights bodies has gone in linking respect for human rights to the role of international cooperation is a General Comment adopted by the UN Committee on Economic, Social and Cultural Rights. The key parts of that statement are provided below.

PHILIP ALSTON AND GERARD QUINN, THE NATURE AND SCOPE OF STATES PARTIES' OBLIGATIONS UNDER THE ICESCR

9 Hum. Rts. Q. 156 (1987), at 186.

. . .

The Covenant contains three provisions that could be interpreted as giving rise to an obligation on the part of the richer states parties to provide assistance to poorer states parties in situations in which the latter are prevented by a lack of resources from fulfilling their obligations under the Covenant to their citizens. The first is the phrase ['individually and through international assistance and co-operation, especially economic and technical'], which appears in Article 2(1). The second is the provision in Article 11(1) according to which states parties agree to 'take appropriate steps to ensure the realization of this right [to an adequate standard of living], recognizing to this effect the essential importance of international co-operation based on free consent.' Similarly, in Article 11(2) states parties agree to take, 'individually and through international co-operation,' relevant measures concerning the right to be free from hunger.

Almost inevitably, dramatically diverging interpretations of the significance of these provisions have been put forward. On the one hand they have been said to give rise to quite specific international obligations on the part of industrialized countries and to provide the foundations for the existence of a right to development. On the other hand, the Carter administration, in seeking the advice and consent of the U.S. Senate to U.S. ratification of the Covenant, proposed a reservation to the effect that: 'It is also understood that paragraph 1 of article 2, as well as article 11 . . . import no legally binding obligation to provide aid to foreign countries.'

. . .

The principle of international cooperation is recognized in Articles 55 and 56 of the United Nations Charter. . . .

. . .

During the preparatory work [for the Covenant] it was conceded by virtually all delegations that the developing states would require some forms of international assistance if they were to be able to promote effectively the realization of economic and social rights. . . .

. . . [T]he U.S. representative considered it 'quite essential for the article . . . to indicate the necessity of international co-operation in the matter.' The consensus, however, did not extend much, if at all, beyond that general proposition.

Those arguing in favor of imposing a strong obligation on the developed countries invoked a wide range of justifications. Perhaps the least controversial was the argument based on interdependence. . . .

On occasion, however, this argument was closely linked to the view that international cooperation was owed to the formerly colonized states in reparation for 'the systematic plundering of their wealth under colonialism.' As another representative put it, 'nations that were or had been colonized did not go begging, but called for the restoration of their rights and property.' . . . It was also argued that the absence of a provision relating to international cooperation would render the undertakings of developing countries 'purely academic' because they would be unable to afford to implement them.

The only formal suggestion of the existence of a binding obligation came from the Chilean representative who observed 'that international assistance to under-developed countries had in a sense become mandatory as a result of commitments assumed by States in the United Nations.'

The arguments against that proposition took a variety of forms and came from a significant range of states. France argued simply that 'multilateral assistance could not be mandatory' and an almost identical argument was made by the Soviet Union. In the view of the representative of Greece 'developing countries like her own had no right to demand financial assistance through such an instrument; they could ask for it, but not claim it.'

Other representatives opposed the reference to international cooperation on the grounds that it would enable states seeking to evade their obligations to invoke the inadequacy of international development assistance as an excuse. . .

. . .

Thus, on the basis of the preparatory work it is difficult, if not impossible, to sustain the argument that the commitment to international cooperation contained in the Covenant can accurately be characterized as a legally binding obligation upon any particular state to provide any particular form of assistance. It would, however, be unjustified to go further and suggest that the relevant commitment is meaningless. In the context of a given right it may, according to the circumstances, be possible to identify obligations to cooperate internationally that would appear to be mandatory on the basis of the undertaking contained in Article 2(1) of the Covenant.

NOTE

Consider the following comments of Mohammed Bedjaoui:[6]

> . . . [W]e are advocating that 'the world food stocks' essential to life, that is to say principally *grain* stocks, be declared to be 'the common heritage of mankind' so as to guarantee every people the vital minimum of a bowl of rice or loaf of bread in order to eradicate the hunger which kills fifty million human beings a year. We are not suggesting this out of moral idealism but out of a concern to avoid a dangerous impasse in international relations. . . .
>
> . . .
>
> Why, for example, should not the twentieth century also be equal to the spirituality of the seventh century when the Koran announced to all mankind that 'all wealth, all things, belong to God' and thus to all members of the human community and that consequently the 'Zakat', the act of charity, should be seen rather as a compulsory institutionalized act, a manifestation of human solidarity, making it every man's duty to give away one tenth of his wealth each year?
>
> Is the twentieth century incapable of matching the principles of solidarity stated by the lawyer Emeric de Vattel in 1758 when he affirmed that each nation must contribute, by every means in its power, to the happiness and perfection of the others?
>
> . . .
>
> How can we make the world food stocks, which are essential to survival, the common heritage of mankind? How should this '*new world food order*' be brought into being? It is a matter for politicians, economists, lawyers and financiers to think about. An immediate and provisional first step might be the creation of a *universal agency*, provided with an operational administration, which could be called the 'International Fund for Food Stocks' (IFFS). It would have a budget with funds provided from a tax levied in each State on a number of manufactured products of high added value made of raw materials from the countries of the third world, or (and) by a one per cent tax on military budgets.

[6] The Right to Development, in Bedjaoui (ed.). International Law: Achievements and Prospects 1177, 1196 (1991).

KEITH GRIFFIN AND A.R. KHAN, GLOBALIZATION AND THE DEVELOPING WORLD (1992), at 90.

. . .

The most important issue linking human development and international capital revolves around the role of foreign aid. Foreign aid includes the highly concessional loans and outright grants to developing countries from the OECD and other rich countries, channelled both bilaterally and through multilateral development agencies such as the World Bank and the Regional Development Banks. It is often taken for granted that foreign assistance actually assists developing countries. Often, alas, this is not the case. The Committee for Development Planning reflects an emerging consensus when it states that

> far too much aid serves no developmental purpose but is used instead to promote the exports of the donor country, to encourage the use of (imported) capital-intensive methods of production or to strengthen the police and armed forces of the recipient country.

But while agreeing that much aid in the past has been wasted, there is evidence that when donor and recipient act responsibly, foreign aid can indeed be of benefit.

The first thing that needs to be done is to depoliticize aid by bringing it under the control of a supranational authority operating under clearly defined and agreed principles. These principles should include both the mobilization and allocation of aid funds. It may be too much to expect that the leading donor countries would agree to channel all foreign assistance through a supranational authority, but it should be possible to reach agreement to channel most foreign aid through such an authority while leaving individual countries free to supplement multilateral assistance with bilateral programmes if they wish to do so. This would not be a radical departure from present practice although it would change the balance decisively in favour of multilateral assistance.

Agreement among donors might be facilitated if it were understood that all multilateral aid would be allocated to countries representing the poorest 60 per cent of the world's population. This implies that only countries with a per capita income of about 700 US dollars or less would be eligible for assistance. Having determined which countries are eligible, the next step is to agree on how the available funds would be distributed among the recipients. We suggest that the criteria for determining the amount of aid to be allocated to eligible countries reflect (a) the severity of poverty as measured by the shortfall of real per capita income from the agreed threshold of 700 US dollars; (b) the degree of commitment to human development as demonstrated by recent success and current programmes; and (c) the size of the population.

The desired total amount of foreign aid available for distribution might be set as an agreed proportion of the combined GNP of all potential recipients. The burden of financing this total should be distributed among the donor countries progressively so that a richer country contributes a higher proportion of its per capita income than a less rich country. This would make the total volume of aid predictable and the distribution of its burden among the contributors equitable.

. . .

COMMITTEE ON ECONOMIC, SOCIAL AND CULTURAL RIGHTS, GENERAL COMMENT NO. 2

UN Doc. E/1990/23, Annex III.

International technical assistance measures (Article 22 of the Covenant)

1. Article 22 of the Covenant establishes a mechanism by which the Economic and Social Council may bring to the attention of relevant United Nations bodies any matters arising out of reports submitted under the Covenant 'which may assist such bodies in deciding, each within its field of competence, on the advisability of international measures likely to contribute to the effective progressive implementation of the . . . Covenant'. While the primary responsibility under Article 22 is vested in the Council, it is clearly appropriate for the Committee on Economic, Social and Cultural Rights to play an active role in advising and assisting the Council in this regard.

2. Recommendations in accordance with Article 22 may be made to any 'organs of the United Nations, their subsidiary organs and specialized agencies concerned with furnishing technical assistance'. The Committee considers that this provision should be interpreted so as to include virtually all United Nations organs and agencies involved in any aspect of international development co-operation. It would therefore be appropriate for recommendations in accordance with Article 22 to be addressed, *inter alia*, to the Secretary-General, subsidiary organs of the Council such as the Commission on Human Rights, the Commission on Social Development and the Commission on the Status of Women, other bodies such as UNDP [the UN Development Programme], UNICEF [the UN Children's Fund] . . ., agencies such as the World Bank and IMF [International Monetary Fund], and any of the other [UN] specialized agencies . . . .

. . .

6. . . . United Nations agencies involved in the promotion of economic, social and cultural rights should do their utmost to ensure that their activities are fully consistent with the enjoyment of civil and political rights. In negative

terms this means that the international agencies should scrupulously avoid involvement in projects which, for example, involve the use of forced labour in contravention of international standards, or promote or reinforce discrimination against individuals or groups contrary to the provisions of the Covenant, or involve large-scale evictions or displacement of persons without the provision of all appropriate protection and compensation. In positive terms, it means that, wherever possible, the agencies should act as advocates of projects and approaches which contribute not only to economic growth or other broadly-defined objectives, but also to enhanced enjoyment of the full range of human rights.

7. . . .[D]evelopment co-operation activities do not automatically contribute to the promotion of respect for economic, social and cultural rights. Many activities undertaken in the name of 'development' have subsequently been recognized as ill-conceived and even counter-productive in human rights terms. In order to reduce the incidence of such problems, the whole range of issues dealt with in the Covenant should, wherever possible and appropriate, be given specific and careful consideration.

8. Despite the importance of seeking to integrate human rights concerns into development activities, it is true that proposals for such integration can too easily remain at a level of generality. Thus, in an effort to encourage the operationalization of the principle contained in Article 22 of the Covenant, the Committee wishes to draw attention to the following specific measures. . . .

(*a*) As a matter of principle, the appropriate United Nations organs and agencies should specifically recognize the intimate relationship which should be established between development activities and efforts to promote respect for human rights in general, and economic, social and cultural rights in particular. . . .

(*b*) Consideration should be given by United Nations agencies to the proposal, made by the Secretary-General in a report of 1979, that a 'human rights impact statement' be required to be prepared in connection with all major development co-operation activities;

(*d*) Every effort should be made, at each phase of a development project, to ensure that the rights contained in the Covenants are duly taken into account. This would apply, for example, in the initial assessment of the priority needs of a particular country, in the identification of particular projects, in project design, in the implementation of the project, and in its final evaluation.

9. A matter which has been of particular concern to the Committee in the examination of the reports of States parties is the adverse impact of the debt burden and of the relevant adjustment measures on the enjoyment of economic, social and cultural rights in many countries. The Committee recognizes that adjustment programmes will often be unavoidable and that these will frequently involve a major element of austerity. Under such circumstances,

however, endeavours to protect the most basic economic, social and cultural rights become more, rather than less, urgent. States parties to the Covenant, as well as the relevant United Nations agencies, should thus make a particular effort to ensure that such protection is, to the maximum extent possible, built-in to programmes and policies designed to promote adjustment. Such an approach, which is sometimes referred to as 'adjustment with a human face' or as promoting 'the human dimension of development' requires that the goal of protecting the rights of the poor and vulnerable should become a basic objective of economic adjustment. Similarly, international measures to deal with the debt crisis should take full account of the need to protect economic, social and cultural rights through, *inter alia*, international co-operation. In many situations, this might point to the need for major debt relief initiatives.

. . .

NOTE

Para. 9 of the Committee's General Comment discusses adjustment measures that are meant to alleviate the debt burdens of developing states. This problem has become a major issue of contention between international financial institutions and states benefiting from their assistance. The term 'structural adjustment programs' describes policies advocated by institutions such as the International Monetary Fund and by individual governments in order to enable beneficiary states to achieve an improved balance of payments. In the case of the IMF, such states must agree to implement those policies in order to qualify for future assistance. The policies generally involve a move toward greater market ordering relative to government planning, and one or another degree of fiscal austerity through heightened taxation and reduction of governmental subsidies and spending (including reduction in government employment and provision of welfare).[7]

QUESTIONS

1. Is there any foundation in the UN Charter, the UDHR, the ICESCR or other international instruments with which you are familiar for the following views put to the U.S. Senate Committee on Foreign Relations in 1979 by an opponent of ratification by the U.S. of the ICESCR? Phyllis Schlafly argued that the ICESCR 'would obligate us to take steps by all measures, including legislation, to distribute food all over the world and to finance a rising standard of living' for other nations?

[7] See Sigrun Skogly, Structural Adjustment and Development: Human Rights—An Agenda for Change, 15 Hum. Rts. Q. 751 (1993).

2. Is Bedjaoui's proposal justified by reference either to the ICESCR or the right to development? How do you assess it with respect to its effect on human rights?

3. Consider the relation between developmental aid on the one hand, and humanitarian assistance in crisis situations on the other.

a. Is there a correlative obligation on any state claiming an expansive right to development and to the financial assistance necessary for development to accept assistance of a humanitarian nature (as in cases of civil conflict) for which there is a clearly demonstrated need?

b. Conversely, can a provider country, which claims that there is an obligation upon a developing state to accept urgently needed humanitarian assistance, maintain with logical consistency that it is under no obligation to provide developmental aid?

4. Are the views in the General Comment adopted by the Committee on Economic, Social and Cultural Rights adequate to meet the concerns of developing countries that have figured in the debate over the right to development? If not, what more should the General Comment have stated, and on the basis of what authority?

D. THE SPREAD OF THE MARKET ECONOMY AND GLOBALIZATION: CONSEQUENCES

Developments since the fall of communism in Central and Eastern Europe and the spread of one or another form of a (regulated) market economy have brought about some important consequences for the enjoyment of human rights. Civil and political rights have been greatly strengthened in many countries. Nonetheless, related contemporary phenomena—including privatization, deregulation, the expanded provision of incentives to entrepreneurial behavior, and structural adjustment programs and related pressures from international financial institutions and developed countries—have had mixed, and sometimes seriously adverse, effects on the enjoyment of economic and social rights.

In the first reading below, Yash Ghai explores the consequences of recent developments for the debate in Asia over human rights. His ideas further develop issues introduced in Section C of Chapter 4, on cultural relativism and Asian critiques of universal human rights. In the next reading, Steven Lukes responds to another scholar's analysis of why the Solidarity-led Government, elected in Poland to replace the Communist regime, opted for a 'market utopian approach'.

YASH GHAI, HUMAN RIGHTS AND GOVERNANCE: THE ASIA DEBATE

15 Aust. Y. B. Int'l L. 1994, 1 (1995), at 2.

. . .

. . . Much effort has gone into the establishment of markets in Eastern Europe and formerly communist States in Asia; indeed a veritable and diverse profession has developed around this enterprise. In some respects these efforts represent an extension of privatisation and structural adjustment policies that had been adopted by or imposed on various States in Africa, Asia and Latin America, the effect of which was to diminish the economic role of the State. The marketisation of domestic economies was paralleled by the globalisation of capital, markets and services, as national barriers to investments and trade were removed (under pressure from world economic institutions). The thrust of globalisation came largely from the West, and served to increase its economic influence and hegemony over the rest of the world. But it was presented, instead, principally in terms of the virtues of entrepreneurship, economic rationality and an efficient legal regime marked by security, predictability and transparency. The links of these developments to human rights were sought to be established through claims of increased choices for investors, consumers and workers, the restrictions on State power and the emergence or strengthening of civil society. So democracy, marketisation of economies, the promotion of human rights and the emergence of civil society were declared to be all of a piece.

The result of the approach of the West was to bring out clearly the implications of the human rights work steadily . . . developed through the United Nations and its agencies. It brought to the fore the responsibility of the international community for the protection of human rights everywhere, and thereby highlighted the ways in which national sovereignty has been qualified by the UN Charter and the human rights conventions. . . . International relations themselves are increasingly mediated through human rights discourse and practice. Aid conditionalities, unthinkable a decade ago, have become common place. . . . However, not all governments have taken kindly to the internationalisation of human rights and democracy, although many of them are unable to carry their opposition to it to international fora because of their fragile political and economic systems and the dependence on external donors. Several Asian countries, particularly in Southeast and East Asia, have offered a spirited rebuttal of this internationalisation. They resent the interposition of the international community (and particularly the hegemony of particular countries within it) in their relations with their citizens, in an ever increasing number of areas. They also resent the leverage it has given these other countries (and international institutions) over their policies, and see the new approach as attempts to undermine their moral authority, disrupt their political stability, and retard their economic progress. Rather than, as in some other regions, withdraw from the debate, they have sought to provide an alternative framework for the discourse of human rights and democracy.

STEVEN LUKES, IS THERE AN ALTERNATIVE TO MARKET UTOPIANISM?

in Christopher Bryant and E. Mokrzycki (eds.), The New Great Transformation? Change and Continuity in East–Central Europe 218 (1994).

In the previous chapter Maurice Glasman asks three related but unfashionable questions[8] . . .

First, he asks why the West German post-war settlement, with its distinctive form of welfare state and system of co-determination, successfully combining liberal-democratic institutions and considerable trade-union power, has been so comprehensively ignored in the post-communist transitions in Eastern and Central Europe, among actors, advisers, and observers alike, both within and outside the region.

It is an excellent question, though many will think that there are several answers and that they are rather obvious. They will point to various factors that, they will claim, make the comparison between the two transitions dubiously relevant. There is, first, the enormous change that has intervened in the relationship between the international and the national political economies, above all in respect of the impact of international financial markets on domestic economic policies. There is an ever-diminishing scope for the pursuit of national social and economic policies, in the face of the ever-present threat of capital flight, as the French socialists discovered in the early 1980s. Second, there are the massive changes in the class structure and the division of labour, both nationally and internationally, the fragmentation and feminization of many areas of work, the growth of the service sector, and the secular decline in the manual working class, leading to the declining role of trade unions. Third is the difference in the climate of expectations: post-war Europe was widely held to be ripe for social reconstruction, after the ravages of the Depression and the War, while post-communist Eastern Europe is seen to be in need of social deconstruction and the dismantling of obstructions to economic development. The social interest groups and institutions, if not the personnel, of the Nazi regime perished with its defeat, whereas communism has left behind it a legacy of patterns of interest, dependency and state dependency which at best constitute a constraint and at worst a blocking mechanism upon

[8] [Eds.] This article by Lukes is a comment on an article in the same book: Maurice Glasman, The Great Deformation: Polanyi, Poland and the Terrors of Planned Spontaneity.

the privatization and marketization of the economy, and overcoming them becomes a central priority. Fourth is the different international climate: as the cold war set in, the securing of social consensus and stability in West Germany had a high priority for the Western allies, whereas such threats to consensus and stability as exist in various areas of the East, while arguably greater, are endogenous and widely seen as intractable.

At least some of these apparent objections are put into question by Glasman's argument itself, for it does not, as I read it, depend on the claim that the cases are analogous, and that the Poles could now implement a West German-style social market and welfare state. His case is rather that after the war the West Germans and their Western allies created (without designing it) a functioning capitalist order that built upon 'existing institutions and patterns of co-operation' which centred around the workplace and drew upon indigenous religious and working-class traditions. By contrast, and ironically, the Solidarity government in Poland sought to impose an alien and abstract model, at the behest of international financial institutions and under the influence of advisers with no local knowledge. His case, in short, is that the response of the West to the collapse of communism has been both narrow-minded and short-sighted, echoing and magnifying the reaction within, and failing to take into account the social context or 'embeddedness' of market relations, and thus failing to foresee the likely consequences of allowing it to be damaged further by too headlong a rush towards the implementation of a market utopia.

Here arises Glasman's second question: why has the market utopian 'model' . . . seemed so exclusively appealing to all the significant actors in the post-communist states of Eastern Europe, and of Poland in particular. Here he may be exaggerating. Contrary voices . . . have certainly been heard, even if they have not prevailed. Nevertheless he is right in broad terms: most of the debate has been over the speed of the proposed reforms, over the dosage of the medicine rather than over the availability of other cures or even an alternative medicine.

Here too many will think that several convincing answers are at hand. They will point to the discrediting of the alternatives—'real existing socialism' in the East and social democracy in the West. They will probably suggest . . . that realism was only delayed by long-lasting illusions about the latter . . . a misinformed and inappropriate longing for 'Sweden', as opposed to Sweden. And they will doubtless add that Sweden has in any case run out of steam, in terms of both efficiency and legitimacy. They will point to all the real failures of real socialism—not just the barriers to minimally efficient production and the massive surveillance system, but also the corruption and graft in the provision of welfare and the systematic destruction of self-esteem, trust and initiative. All these interconnected evils seem to call for a radically alternative system based on an alternative principle and indeed an end to the very idea of realizing social solidarity and social justice by design. From this standpoint, the aftermath of 1989 symbolizes the defeat of an indefensible system, not the failure to defend some alternative defensible one. For a long time, in the last years of communist rule, and above all in East Germany in the phoney pre-election period, there was a hankering by many intellectuals for some supposed 'Third Way'. But they could never make clear what this was and came to be seen as timid trimmers whose life-experiences, professional qualifications and interests tied them to the old regime. Moreover, one could indeed argue that the very concept of the 'Third Way' embodied an ideological distortion, suggesting that capitalism was one way (when it in fact is indeterminately many ways) and that real socialism is a second way (when it is no way at all).

Yet here too Glasman's argument puts a question-mark beside these objections. For . . . he argues that the trouble with the market utopian model is that it is a utopia and a model—indeed that it is a utopia because it is a model—which abstracts from the historical realities of 'capitalism' as an entire socio-economic formation that can take innumerable forms. In particular it abstracts from the social and moral contexts in which markets can function compatibly with rights guaranteeing basic freedoms, social justice securing fairness for all but especially the weakest, and democracy, including economic democracy. He makes the bold claim that the Solidarity programme of 1981 set out the essentials of a non-utopian conception that defines a historical alternative (capitalist) path that could have been taken but was not. His claim, in short, is that the programme was both feasible and legitimate since it drew upon still living Catholic and working-class traditions.

This leads to the third question which Glasman's chapter raises: was this path indeed feasible? Could one even imagine some future Polish government stumbling back onto it, thereby affording brighter prospects for the Polish transition—suggesting, indeed, that there might *be* a transition, rather than stagnation or disintegration?

. . . Here we return to the answers I listed above to Glasman's first question. To what extent did international economic and above all financial constraints set limits to the implementation of the Solidarity programme? To what extent did that programme take account of the changes in the class structure and the division of labour (albeit far less developed in Poland than in Western Europe)? To what extent were the constraints facing the first Polish governments ideological, confining them to a narrower range of options than was objectively available, and inducing them to perpetrate unnecessary social destruction in the name of economic progress? If this last suggestion is right, to what extent are such constraints conjunctural—the product of a particular time and place, and to what extent can they be overcome by developing theoretical and practical alternatives? Such questions are of the greatest theoretical interest and current political relevance.

QUESTIONS

1. The common theme of both preceding analyses is that the process of globalization, in so far as it constrains domestic freedom of action in the economic as well as other spheres, might frustrate efforts to develop an international consensus on human rights and in particular might prejudice the protection of economic and social rights in the growing number of countries subject to structural adjustment pressures.

a. 'Economic and social rights will necessarily bow to the demands of the market and economic growth. At least in this field, international human rights law has no option but to adapt itself to changing realities.' Comment.

b. If there is adaptation, what changes might be required? How do you assess para. 9 of the General Comment of the Committee on Economic, Social and Cultural Rights, p. 1138, *supra*?

2. 'To some extent, human rights have become for many developed countries a tool of foreign economic policy to pry open foreign markets for trade and investment and thereby permit competition by multinational enterprises in internal markets.' Comment.

3. The Director-General of the ILO, in a statement on 'the decline of the Nation-State,' notes that '[f]inancial regulation has left States dependent upon the world financial markets' and that

> the multinationalization of enterprises removes decision-making power further and further away from the workplace. . . . In their discussions with the State and with employers, workers often have the impression of dealing with agents who are neither very credible nor motivated to enter into firm commitments because they are so dependent upon an international economic environment over which they have less and less control. . . . Some political leaders are concluding that the concept of the State as a community with a single destiny has seen its day.[9]

If we accept the validity of the process of 'globalization' reflected in this analysis, should we conclude that domestic jurisdiction is a concept doomed to extinction? Or will there be a backlash against such trends, as suggested by this comment:

> [T]he very immensity of the global village will serve in the long run to alienate its citizens. . . . [E]stranged citizens will see the

[9] Defending Values, Promoting Change, Social Justice in a Global Economy: An ILO Agenda 15 (1994).

> global community, and the national government that imposes international rules, as alien from them, and from what they care about.[10]

ADDITIONAL READING

Philip Alston, *A Third Generation of Solidarity Rights: Progressive Development or Obfuscation of International Human Rights Law?*, 29 Neths. Int. L. Rev. 307 (1982); Kéba M'Baye, *Introduction*, in Mohammed Bedjaoui (ed.), *International Law: Achievements and Prospects* 1043 (1991), Adrien Nastase, *The Right to Peace*, *ibid.*, at 1219; Hector Gros Espiell, *The Right of Development as a Human Right*, 16 Texas Int. L. J. 189 (1981); Richard Kiwanuka, *Developing Rights: The UN Declaration on the Right to Development*, 35 Neths. Int. L. Rev. 257 (1988); Ian Brownlie, *The Human Right to Development* (1989); David Forsythe (ed.), *Human Right and Development: International Views* (1989); and Russell Barsh, *The Right to Development as a Human Right: Results of the Global Consultation*, 13 Hum. Rts. Q. 322 (1991).

[10] B. Friedman, Federalism's Future in the Global Village, 47 Vanderbilt L. Rev. 1441, 1479 (1994).

Annex on
DOCUMENTS

This annex sets forth documents that are referred to in more than one section of the coursebook. Documents that pertain to only one section appear in it.

The documents have been edited to delete provisions that are unnecessary for an understanding of the coursebook's materials. Hence you should not rely on any of them as a full and official version. The Annex on Citations provides the official citation to each document and in many cases a reference to a more readily available source in which the complete document appears.

In the Table of Contents below, there appear within parentheses after the official title of a document the acronym or abbreviated name(s) by which the text or the readings sometimes refer to it.

CONTENTS

CHARTER OF THE UNITED NATIONS

We The Peoples of The United Nations Determined

to save succeeding generations from the scourge of war, which twice in our lifetime has brought untold sorrow to mankind, and

to reaffirm faith in fundamental human rights, in the dignity and worth of the human person, in the equal rights of men and women and of nations large and small, and

to establish conditions under which justice and respect for the obligations arising from treaties and other sources of international law can be maintained, and

to promote social progress and better standards of life in larger freedom,

And for These Ends

to practice tolerance and live together in peace with one another as good neighbours, and

to unite our strength to maintain international peace and security, and

to ensure, by the acceptance of principles and the institution of methods, that armed force shall not be used, save in the common interest, and

to employ international machinery for the promotion of the economic and social advancement of all peoples,

Have Resolved to Combine Our Efforts to Accomplish These Aims

. . .

CHAPTER I. PURPOSES AND PRINCIPLES

Article 1

The Purposes of the United Nations are:

1. To maintain international peace and security, and to that end: to take effective collective measures for the prevention and removal of threats to the peace, and for the suppression of acts of aggression or other breaches of the peace, and to bring about by peaceful means, and in conformity with the principles of justice and international law, adjustment or settlement of international disputes or situations which might lead to a breach of the peace;

2. To develop friendly relations among nations based on respect for the principle of equal rights and self-determination of peoples, and to take other appropriate measures to strengthen universal peace;

3. To achieve international co-operation in solving international problems of an economic, social, cultural, or humanitarian character, and in promoting and encouraging respect for human rights and for fundamental freedoms for all without distinction as to race, sex, language, or religion; and

4. To be a centre for harmonizing the actions of nations in the attainment of these common ends.

Article 2

The Organization and its Members, in pursuit of the Purposes stated in Article 1, shall act in accordance with the following Principles.

1. The Organization is based on the principle of the sovereign equality of all its Members.

2. All Members, in order to ensure to all of them the rights and benefits resulting from membership, shall fulfill in good faith the obligations assumed by them in accordance with the present Charter.

3. All Members shall settle their international disputes by peaceful means in such a manner that international peace and security, and justice, are not endangered.

4. All Members shall refrain in their international relations from the threat or use of force against the territorial integrity or political independence of any state, or in any other manner inconsistent with the Purposes of the United Nations.

5. All Members shall give the United Nations every assistance in any action it takes in accordance with the present Charter, and shall refrain from giving assistance to any state against which the United Nations is taking preventive or enforcement action.

. . .

7. Nothing contained in the present Charter shall authorize the United Nations to intervene in matters which are essentially within the domestic jurisdiction of any state or shall require the Members to submit such matters to settlement under the present Charter; but this principle shall not prejudice the application of enforcement measures under Chapter VII.

CHAPTER II. MEMBERSHIP

. . .

Article 4

1. Membership in the United Nations is open to all other peace-loving states which accept the obligations contained in the present Charter and, in the judgment of the Organization, are able and willing to carry out these obligations.

. . .

CHAPTER III. ORGANS

Article 7

1. There are established as the principal organs of the United Nations: a General Assembly, a Security Council, an Economic and Social Council, a Trusteeship Council, an International Court of Justice, and a Secretariat.

2. Such subsidiary organs as may be found necessary may be established in accordance with the present Charter.

Article 8

The United Nations shall place no restrictions on the eligibility of men and women to participate in any capacity and under conditions of equality in its principal and subsidiary organs.

CHAPTER IV. THE GENERAL ASSEMBLY

Article 9

1. The General Assembly shall consist of all the Members of the United Nations.

. . .

Article 10

The General Assembly may discuss any questions or any matters within the scope of the present Charter or relating to the powers and functions of any organs provided for in the present Charter, and, except as provided in Article 12, may make recommendations to the Members of the United Nations or to the Security Council or to both on any such questions or matters.

. . .

Article 13

1. The General Assembly shall initiate studies and make recommendations for the purpose of:

a. promoting international co-operation in the political field and encouraging the progressive development of international law and its codification;

b. promoting international co-operation in the economic, social, cultural, educational, and health fields, and assisting in the realization of human rights and fundamental freedoms for all without distinction as to race, sex, language, or religion.

. . .

CHAPTER V. THE SECURITY COUNCIL

Article 23

1. The Security Council shall consist of fifteen Members of the United Nations. The Republic of China, France, the Union of Soviet Socialist Republics, the United Kingdom of Great Britain and Northern Ireland, and the United States of America shall be permanent members of the Security Council. The General Assembly shall elect ten other Members of the United Nations to be non-permanent members of the Security Council, due regard being specially paid, in the first instance to the contribution of Members of the United Nations to the maintenance of international peace and security and to the other purposes of the Organization, and also to equitable geographical distribution.

2. The non-permanent members of the Security Council shall be elected for a term of two years. In the first election of the non-permanent members after the increase of the membership of the Security Council from eleven to fifteen, two of the four additional members shall be chosen for a term of one year. A retiring member shall not be eligible for immediate re-election.

. . .

Article 24

1. In order to ensure prompt and effective action by the United Nations, its Members confer on the Security Council primary responsibility for the maintenance of international peace and security, and agree that in carrying out its duties under this responsibility the Security Council acts on their behalf.

2. In discharging these duties the Security Council shall act in accordance with the Purposes and Principles of the United Nations. The specific powers granted to the Security Council for the discharge of these duties are laid down in Chapters VI, VII, VIII, and XII.

. . .

Article 25

The members of the United Nations agree to accept and carry out the decisions of the Security Council in accordance with the present Charter.

. . .

Article 27

1. Each member of the Security Council shall have one vote.

2. Decisions of the Security Council on procedural matters shall be made by an affirmative vote of nine members.

3. Decisions of the Security Council on all other matters shall be made by an affirmative vote of nine members including the concurring votes of the permanent members; provided that, in decisions under Chapter VI, and under paragraph 3 of Article 52, a party to a dispute shall abstain from voting.

CHAPTER VI. PACIFIC SETTLEMENT OF DISPUTES

Article 33

1. The parties to any dispute, the continuance of which is likely to endanger the maintenance of international peace and security, shall, first of all, seek a solution by negotiation, enquiry, mediation, conciliation, arbitration, judicial settlement, resort to regional agencies or arrangements, or other peaceful means of their own choice.

2. The Security Council shall, when it deems necessary, call upon the parties to settle their dispute by such means.

Article 34

The Security Council may investigate any dispute, or any situation which might lead to international friction or give rise to a dispute, in order to determine whether the continuance of the dispute or situation is likely to endanger the maintenance of international peace and security.

. . .

Article 36

1. The Security Council may, at any stage of a dispute of the nature referred to in Article 33 or of a situation of like nature, recommend appropriate procedures or methods of adjustment.

2. The Security Council should take into consideration any procedures for the settlement of the dispute which have already been adopted by the parties.

3. In making recommendations under this Article the Security Council should also take into consideration that legal disputes should as a general rule be referred by the parties to the International Court of Justice in accordance with the provisions of the Statute of the Court.

. . .

CHAPTER VII. ACTION WITH RESPECT TO THREATS TO THE PEACE, BREACHES OF THE PEACE, AND ACTS OF AGGRESSION

Article 39

The Security Council shall determine the existence of any threat to the peace, breach of the peace, or act of aggression and shall make recommendations, or decide what measures shall be taken in accordance with Articles 41 and 42, to maintain or restore international peace and security.

. . .

Article 41

The Security Council may decide what measures not involving the use of armed force are to be employed to give effect to its decisions, and it may call upon the Members of the United Nations to apply such measures. These may include complete or partial interruption of economic relations and of rail, sea, air, postal, telegraphic, radio, and other means of communication, and the severance of diplomatic relations.

Article 42

Should the Security Council consider that measures provided for in Article 41 would be inadequate or have proved to be inadequate, it may take such action by air, sea, or land forces as may be necessary to maintain or restore international peace and security. Such action may include demonstrations, blockade, and other operations by air, sea, or land forces of Members of the United Nations.

Article 43

1. All Members of the United Nations, in order to contribute to the maintenance of international peace and security, undertake to make available to the Security Council, on its call and in accordance with a special agreement or agreements, armed forces, assistance, and facilities, including rights of passage, necessary for the purpose of maintaining international peace and security.

2. Such agreement or agreements shall govern the numbers and types of forces, their degree of readiness and general location, and the nature of the facilities and assistance to be provided.

3. The agreement or agreements shall be negotiated as soon as possible on the initiative of the Security Council. They shall be concluded between the Security Council and Members or between the Security Council and groups of members and shall be subject to ratification by the signatory states in accordance with their respective constitutional processes.

Article 44

When the Security Council has decided to use force it shall, before calling upon a Member not represented on it to provide armed forces in fulfilment of the obligations assumed under Article 43, invite that Member, if the Member so desires, to participate in the decisions of the Security Council concerning the employment of contingents of that Member's armed forces.

. . .

Article 48

1. The action required to carry out the decisions of the Security Council for the maintenance of international peace and security shall be taken by all the Members of the United Nations or by some of them, as the Security Council may determine.

2. Such decisions shall be carried out by the Members of the United Nations directly and through their action in the appropriate international agencies of which they are members.

. . .

Article 51

Nothing in the present Charter shall impair the inherent right of individual or collective self-defence if an armed attack occurs against a Member of the United Nations, until the Security Council has taken measures necessary to maintain international peace and security. Measures taken by Members in the exercise of this right of self-defence shall be immediately reported to the Security Council and shall not in any way affect the authority and responsibility of the Security Council under the present Charter to take at any time such action as it deems necessary in order to maintain or restore international peace and security.

CHAPTER VIII. REGIONAL ARRANGEMENTS

Article 52

1. Nothing in the present Charter precludes the existence of regional arrangements or agencies for dealing with such matters relating to the maintenance of international peace and security as are appropriate for regional action, provided that such arrangements or agencies and their activities are consistent with the Purposes and Principles of the United Nations.

. . .

CHAPTER IX. INTERNATIONAL ECONOMIC AND SOCIAL CO-OPERATION

Article 55

With a view to the creation of conditions of stability and well-being which are necessary for peaceful and friendly relations among nations based on respect for the principle of equal rights and self-determination of peoples, the United Nations shall promote:

a. higher standards of living, full employment, and conditions of economic and social progress and development.

b. solutions of international economic, social, health, and related problems; and international cultural and educational co-operation; and

c. universal respect for, and observance of, human rights and fundamental freedoms for all without distinction as to race, sex, language, or religion.

Article 56

All Members pledge themselves to take joint and separate action in co-operation with the Organization for the achievement of the purposes set forth in Article 55.

. . .

CHAPTER X. THE ECONOMIC AND SOCIAL COUNCIL

Article 61

1. The Economic and Social Council shall consist of fifty-four Members of the United Nations elected by the General Assembly.

. . .

Article 62

1. The Economic and Social Council may make or initiate studies and reports with respect to international economic, social, cultural, educational, health, and related matters and may make recommendations with respect to any such matters of the General Assembly, to the Members of the United Nations, and to the specialized agencies concerned.
2. It may make recommendations for the purpose of promoting respect for, and observance of, human rights and fundamental freedoms for all.
3. It may prepare draft conventions for submission to the General Assembly, with respect to matters falling within its competence.
4. It may call, in accordance with the rules prescribed by the United Nations, international conferences on matters falling within its competence.

. . .

Article 68

The Economic and Social Council shall set up commissions in economic and social fields and for the promotion of human rights, and such other commissions as may be required for the performance of its functions.

. . .

CHAPTER XIV. THE INTERNATIONAL COURT OF JUSTICE

Article 92

The International Court of Justice shall be the principal judicial organ of the United Nations. It shall function in accordance with the annexed Statute, which is based upon the Statute of the Permanent Court of International Justice and forms an integral part of the present Charter.

Article 94

1. Each Member of the United Nations undertakes to comply with the decision of the International Court of Justice in any case to which it is a party.

2. If any party to a case fails to perform the obligations incumbent upon it under a judgment rendered by the Court, the other party may have recourse to the Security Council, which may, if it deems necessary, make recommendations or decide upon measures to be taken to give effect to the judgment.

. . .

CHAPTER XV. THE SECRETARIAT

Article 97

The Secretariat shall comprise a Secretary-General and such staff as the Organization may require. The Secretary-General shall be appointed by the General Assembly upon the recommendation of the Security Council. He shall be the chief administrative officer of the Organization.

. . .

Article 99

The Secretary-General may bring to the attention of the Security Council any matter which in his opinion may threaten the maintenance of international peace and security.

Article 100

1. In the performance of their duties the Secretary-General and the staff shall not seek or receive instructions from any government or from any other authority external to the Organization. They shall refrain from any action which might reflect on their position as international officials responsible only to the Organization.

2. Each Member of the United Nations undertakes to respect the exclusively international character of the responsibilities of the Secretary-General and the staff and not to seek to influence them in the discharge of their responsibilities.

. . .

CHAPTER XVI. MISCELLANEOUS PROVISIONS

. . .

Article 103

In the event of a conflict between the obligations of the Members of the United Nations under the present Charter and their obligations under any other international agreement, their obligations under the present Charter shall prevail.

. . .

CHAPTER XVIII. AMENDMENTS

Article 108

Amendments to the present Charter shall come into force for all Members of the United Nations when they have been adopted by a vote of two-thirds of the members of the General Assembly and ratified in accordance with their respective constitutional processes by two-thirds of the Members of the United Nations, including all the permanent members of the Security Council.

UNIVERSAL DECLARATION OF HUMAN RIGHTS

PREAMBLE

Whereas recognition of the inherent dignity and of the equal and inalienable rights of all members of the human family is the foundation of freedom, justice and peace in the world,

Whereas disregard and contempt for human rights have resulted in barbarous acts which have outraged the conscience of mankind, and the advent of a world in which human beings shall enjoy freedom of speech and belief and freedom from fear and want has been proclaimed as the highest aspiration of the common people,

Whereas it is essential, if man is not to be compelled to have recourse, as a last resort, to rebellion against tyranny and oppression, that human rights should be protected by the rule of law,

Whereas it is essential to promote the development of friendly relations between nations,

Whereas the peoples of the United Nations have in the Charter reaffirmed their faith in fundamental human rights, in the dignity and worth of the human person and in the equal rights of men and women and have determined to promote social progress and better standards of life in larger freedom,

Whereas Member States have pledged themselves to achieve, in co-operation with the United Nations, the promotion of universal respect for and observance of human rights and fundamental freedoms,

Whereas a common understanding of these rights and freedoms is of the greatest importance for the full realization of this pledge,

Now, therefore,
The General Assembly,

Proclaims this Universal Declaration of Human Rights as a common standard of achievement for all peoples and all nations, to the end that every individual and every organ of society, keeping this Declaration constantly in mind, shall strive by teaching and education to promote respect for these rights and freedoms and by progressive measures, national and international, to secure their universal and effective recognition and observance, both among the peoples of Member States themselves and among the peoples of territories under their jurisdiction.

Article 1

All human beings are born free and equal in dignity and rights. They are endowed with reason and conscience and should act towards one another in a spirit of brotherhood.

Article 2

Everyone is entitled to all the rights and freedoms set forth in this Declaration, without distinction of any kind, such as race, colour, sex, language, religion, political or other opinion, national or social origin, property, birth or other status.

Furthermore, no distinction shall be made on the basis of the political, jurisdictional or international status of the country or territory to which a person belongs, whether it be independent, trust, non-self-governing or under any other limitation of sovereignty.

Article 3

Everyone has the right to life, liberty and security of person.

Article 4

No one shall be held in slavery or servitude; slavery and the slave trade shall be prohibited in all their forms.

Article 5

No one shall be subjected to torture or to cruel, inhuman or degrading treatment or punishment.

Article 6

Everyone has the right to recognition everywhere as a person before the law.

Article 7

All are equal before the law and are entitled without any discrimination to equal protection of the law. All are entitled to equal protection against any discrimination in violation of this Declaration and against any incitement to such discrimination.

Article 8

Everyone has the right to an effective remedy by the competent national tribunals for acts violating the fundamental rights granted him by the constitution or by law.

Article 9

No one shall be subjected to arbitrary arrest, detention or exile.

Article 10

Everyone is entitled in full equality to a fair and public hearing by an independent and impartial tribunal, in the determination of his rights and obligations and of any criminal charge against him.

Article 11

1. Everyone charged with a penal offence has the right to be presumed innocent until proved guilty according to law in a public trial at which he has had all the guarantees necessary for his defence.
2. No one shall be held guilty of any penal offence on account of any act or omission which did not constitute a penal offence, under national or international law, at the time when it was committed. Nor shall a heavier penalty be imposed than the one that was applicable at the time the penal offence was committed.

Article 12

No one shall be subjected to arbitrary interference with his privacy, family, home or correspondence, nor to attacks upon his honour and reputation. Everyone has the right to the protection of the law against such interference or attacks.

Article 13

1. Everyone has the right to freedom of movement and residence within the borders of each State.
2. Everyone has the right to leave any country, including his own, and to return to his country.

Article 14

1. Everyone has the right to seek and to enjoy in other countries asylum from persecution.
2. This right may not be invoked in the case of prosecutions genuinely arising from non-political crimes or from acts contrary to the purposes and principles of the United Nations.

Article 15

1. Everyone has the right to a nationality.
2. No one shall be arbitrarily deprived of his nationality nor denied the right to change his nationality.

Article 16

1. Men and women of full age, without any limitation due to race, nationality or religion, have the right to marry and to found a family. They are entitled to equal rights as to marriage, during marriage and at its dissolution.
2. Marriage shall be entered into only with the free and full consent of the intending spouses.
3. The family is the natural and fundamental group unit of society and is entitled to protection by society and the State.

Article 17

1. Everyone has the right to own property alone as well as in association with others.
2. No one shall be arbitrarily deprived of his property.

Article 18

Everyone has the right to freedom of thought, conscience and religion; this right includes freedom to change his religion or belief, and freedom, either alone or in community with others and in public or private, to manifest his religion or belief in teaching, practice, worship and observance.

Article 19

Everyone has the right to freedom of opinion and expression; this right includes freedom to hold opinions without interference and to seek, receive and impart information and ideas through any media and regardless of frontiers.

Article 20

1. Everyone has the right to freedom of peaceful assembly and association.
2. No one may be compelled to belong to an association.

Article 21

1. Everyone has the right to take part in the government of his country, directly or through freely chosen representatives.
2. Everyone has the right to equal access to public service in his country.
3. The will of the people shall be the basis of the authority of government; this will shall be expressed in periodic and genuine elections which shall be by universal and equal suffrage and shall be held by secret vote or by equivalent free voting procedures.

Article 22

Everyone, as a member of society, has the right to social security and is entitled to realization, through national effort and international co-operation and in accordance with the organization and resources of each State, of the economic, social and cultural rights indispensable for his dignity and the free development of his personality.

Article 23

1. Everyone has the right to work, to free choice of employment, to just and favourable conditions of work and to protection against unemployment.
2. Everyone, without any discrimination, has the right to equal pay for equal work.
3. Everyone who works has the right to just and favourable remuneration ensuring for himself and his family an existence worthy of human dignity, and supplemented, if necessary, by other means of social protection.
4. Everyone has the right to form and to join trade unions for the protection of his interests.

Article 24

Everyone has the right to rest and leisure, including reasonable limitation of working hours and periodic holidays with pay.

Article 25

1. Everyone has the right to a standard of living adequate for the health and well-being of himself and of his family, including food, clothing, housing and medical care and necessary social services, and the right to security in the event of unemployment, sickness, disability, widowhood, old age or other lack of livelihood in circumstances beyond his control.

2. Motherhood and childhood are entitled to special care and assistance. All children, whether born in or out of wedlock, shall enjoy the same social protection.

Article 26

1. Everyone has the right to education. Education shall be free, at least in the elementary and fundamental stages. Elementary education shall be compulsory. Technical and professional education shall be made generally available and higher education shall be equally accessible to all on the basis of merit.
2. Education shall be directed to the full development of the human personality and to the strengthening of respect for human rights and fundamental freedoms. It shall promote understanding, tolerance and friendship among all nations, racial or religious groups, and shall further the activities of the United Nations for the maintenance of peace.
3. Parents have a prior right to choose the kind of education that shall be given to their children.

Article 27

1. Everyone has the right freely to participate in the cultural life of the community, to enjoy the arts and to share in scientific advancement and its benefits.
2. Everyone has the right to the protection of the moral and material interests resulting from any scientific, literary or artistic production of which he is the author.

Article 28

Everyone is entitled to a social and international order in which the rights and freedoms set forth in this Declaration can be fully realized.

Article 29

1. Everyone has duties to the community in which alone the free and full development of his personality is possible.
2. In the exercise of his rights and freedoms, everyone shall be subject only to such limitations as are determined by law solely for the purpose of securing due recognition and respect for the rights and freedoms of others and of meeting the just requirements of morality, public order and the general welfare in a democratic society.
3. These rights and freedoms may in no case be exercised contrary to the purposes and principles of the United Nations.

Article 30

Nothing in this Declaration may be interpreted as implying for any State, group or person any right to engage in any activity or to perform any act aimed at the destruction of any of the rights and freedoms set forth herein.

INTERNATIONAL COVENANT ON CIVIL AND POLITICAL RIGHTS

PREAMBLE

The States Parties to the present Covenant,

Considering that, in accordance with the principles proclaimed in the Charter of the United Nations, recognition of the inherent dignity and of the equal and inalienable rights of all members of the human family is the foundation of freedom, justice and peace in the world,

Recognizing that these rights derive from the inherent dignity of the human person,

Recognizing that, in accordance with the Universal Declaration of Human rights, the ideal of free human beings enjoying civil and political freedom and freedom from fear and want can only be achieved if conditions are created whereby everyone may enjoy his civil and political rights, as well as his economic, social and cultural rights,

Considering the obligation of States under the Charter of the United Nations to promote universal respect for, and observance of, human rights and freedoms,

Realizing that the individual, having duties to other individuals and to the community to which he belongs, is under a responsibility to strive for the promotion and observance of the rights recognized in the present Covenant,

Agree upon the following articles:

PART I

Article 1

1. All peoples have the right of self-determination. By virtue of that right they freely determine their political status and freely pursue their economic, social and cultural development.
2. All peoples may, for their own ends, freely dispose of their natural wealth and resources without prejudice to any obligations arising out of international economic co-operation, based upon the principle of mutual benefit, and international law. In no case may a people be deprived of its own means of subsistence.
3. The States Parties to the present Covenant . . . shall promote the realization of the right of self-determination, and shall respect that right, in conformity with the provisions of the Charter of the United Nations.

PART II

Article 2

1. Each State Party to the present Covenant undertakes to respect and to ensure to all individuals within its territory and subject to its jurisdiction the rights

recognized in the present Covenant, without distinction of any kind, such as race, colour, sex, language, religion, political or other opinion, national or social origin, property, birth or other status.

2. Where not already provided for by existing legislative or other measures, each State Party to the present Covenant undertakes to take the necessary steps, in accordance with its constitutional processes and with the provisions of the present Covenant, to adopt such legislative or other measures as may be necessary to give effect to the rights recognized in the present Covenant.

3. Each State Party to the present Covenant undertakes:

(*a*) To ensure that any person whose rights or freedoms as herein recognized are violated shall have an effective remedy, notwithstanding that the violation has been committed by persons acting in an official capacity;

(*b*) To ensure that any person claiming such a remedy shall have his right thereto determined by competent judicial, administrative or legislative authorities, or by any other competent authority provided for by the legal system of the State, and to develop the possibilities of judicial remedy;

(*c*) To ensure that the competent authorities shall enforce such remedies when granted.

Article 3

The State Parties to the present Covenant undertake to ensure the equal right of men and women to the enjoyment of all civil and political rights set forth in the present Covenant.

Article 4

1. In time of public emergency which threatens the life of the nation and the existence of which is officially proclaimed, the States Parties to the present Covenant may take measures derogating from their obligations under the present Covenant to the extent strictly required by the exigencies of the situation, provided that such measures are not inconsistent with their other obligations under international law and do not involve discrimination solely on the ground of race, colour, sex, language, religion or social origin.

2. No derogation from articles 6, 7, 8 (paragraphs 1 and 2), 11, 15, 16 and 18 may be made under this provision.

3. Any State Party to the present Covenant availing itself of the right of derogation shall immediately inform the other States Parties to the present Covenant, through the intermediary of the Secretary-General of the United Nations, of the provisions from which it has derogated and of the reasons by which it was actuated. A further communication shall be made, through the same intermediary, on the date on which it terminates such derogation.

Article 5

1. Nothing in the present Covenant may be interpreted as implying for any State, group or person any right to engage in any activity or perform any act aimed at the destruction of any of the rights and freedom recognized herein or at their limitation to a greater extent than is provided for in the present Covenant.

2. There shall be no restriction upon or derogation from any of the fundamental human rights recognized or existing in any State Party to the present Covenant pursuant to law, conventions, regulations or custom on the pretext that the present Covenant does not recognize such rights or that it recognizes them to a lesser extent.

PART III

Article 6

1. Every human being has the inherent right to life. This right shall be protected by law. No one shall be arbitrarily deprived of his life.

2. In countries which have not abolished the death penalty, sentence of death may be imposed only for the most serious crimes in accordance with the law in force at the time of the commission of the crime and not contrary to the provisions of the present Covenant and to the Convention on the Prevention and Punishment of the Crime of Genocide. This penalty can only be carried out pursuant to a final judgement rendered by a competent court. – due process

3. When deprivation of life constitutes the crime of genocide, it is understood that nothing in this article shall authorize any State Party to the present Covenant to derogate in any way from any obligation assumed under the provisions of the Convention on the Prevention and Punishment of the Crime of Genocide.

4. Anyone sentenced to death shall have the right to seek pardon or commutation of the sentence. Amnesty, pardon or commutation of the sentence of death may be granted in all cases.

5. Sentence of death shall not be imposed for crimes committed by persons below eighteen years of age and shall not be carried out on pregnant women.

6. Nothing in this article shall be invoked to delay or to prevent the abolition of capital punishement by any State Party to the present Covenant.

Article 7

No one shall be subjected to torture or to cruel, inhuman or degrading treatment or punishment. In particular, no one shall be subjected without his free consent to medical or scientific experimentation.

Article 8

1. No one shall be held in slavery; slavery and the slave-trade in all their forms shall be prohibited.

2. No one shall be held in servitude.

. . .

Article 9

1. Everyone has the right to liberty and security of person. No one shall be subjected to arbitrary arrest or detention. No one shall be deprived of his liberty except on such grounds and in accordance with such procedure as are established by law.

2. Anyone who is arrested shall be informed, at the time of arrest, of the reasons for his arrest and shall be promptly informed of any charges against him.

3. Anyone arrested or detained on a criminal charge shall be brought promptly

before a judge or other officer authorized by law to exercise judicial power and shall be entitled to trial within a reasonable time or to release. It shall not be the general rule that persons awaiting trial shall be detained in custody, but release may be subject to guarantees to appear for trial, at any other stage of the judicial proceedings, and, should occasion arise, for execution of the judgement.

4. Anyone who is deprived of his liberty by arrest or detention shall be entitled to take proceedings before a court, in order that the court may decide without delay on the lawfulness of his detention and order his release if the detention is not lawful.

5. Anyone who has been victim of unlawful arrest or detention shall have an enforceable right to compensation.

Article 10

1. All persons deprived of their liberty shall be treated with humanity and with respect for the inherent dignity of the human person.

2. (*a*) Accused persons shall, save in exceptional circumstances, be segregated from convicted persons and shall be subject to separate treatment appropriate to their status as unconvicted persons;

(*b*) Accused juvenile persons shall be separated from adults and brought as speedily as possible for adjudication.

3. The penitentiary system shall comprise treatment of prisoners the essential aim of which shall be their reformation and social rehabilitation. Juvenile offenders shall be segregated from adults and be accorded treatment appropriate to their age and legal status.

Article 11

No one shall be imprisoned merely on the ground of inability to fulfil a contractual obligation.

Article 12

1. Everyone lawfully within the territory of a State shall, within that territory, have the right to liberty of movement and freedom to choose his residence.

2. Everyone shall be free to leave any country, including his own.

3. The above-mentioned rights shall not be subject to any restrictions except those which are provided by law, are necessary to protect national security, public order (*ordre public*), public health or morals or the rights and freedoms of others, and are consistent with the other rights recognized in the present Covenant.

4. No one shall be arbitrarily deprived of the right to enter his own country.

Article 13

An alien lawfully in the territory of a State Party to the present Covenant may be expelled therefrom only in pursuance of a decision reached in accordance with law and shall, except where compelling reasons of national security otherwise require, be allowed to submit the reasons against his expulsion and to have his case reviewed by, and be represented for the purpose before, the competent authority or a person or persons especially designated by the competent authority.

Article 14

1. All persons shall be equal before the courts and tribunals. In the determination of any criminal charge against him, or of his rights and obligations in a suit at law, everyone shall be entitled to a fair and public hearing by a competent, independent and impartial tribunal established by law. The press and the public may be excluded from all or part of a trial for reasons of morals, public order (*ordre public*) or national security in a democratic society, or when the interest of the private lives of the parties so requires, or to the extent strictly necessary in the opinion of the court in special circumstances where publicity would prejudice the interests of justice; but any judgement rendered in a criminal case or in a suit at law shall be made public except where the interest or juvenile persons otherwise requires or the proceedings concern matrimonial disputes of the guardianship of children.

2. Everyone charged with a criminal offence shall have the right to be presumed innocent until proved guilty according to law.

3. In the determination of any criminal charge against him, everyone shall be entitled to the following minimum guarantees, in full equality:

(*a*) To be informed promptly and in detail in a language which he understands of the nature and cause of the charge against him;

(*b*) To have adequate time and facilities for the preparation of his defence and to communicate with counsel of his own choosing;

(*c*) To be tried without undue delay;

(*d*) To be tried in his presence, and to defend himself in person or through legal assistance of his own choosing; to be informed, if he does not have legal assistance, of this right; and to have legal assistance assigned to him, in any case where the interests of justice so require, and without payment by him in any such case if he does not have sufficient means to pay for it;

(*e*) To examine, or have examined, the witnesses against him and to obtain the attendance and examination of witnesses on his behalf under the same conditions as witnesses against him;

(*f*) To have the free assistance of an interpreter if he cannot understand or speak the language used in court;

(*g*) Not to be compelled to testify against himself or to confess guilt.

4. In the case of juvenile persons, the procedure shall be such as will take account of their age and the desirability of promoting their rehabilitation.

5. Everyone convicted of a crime shall have the right to his conviction and sentence being reviewed by a higher tribunal according to law.

6. When a person has by a final decision been convicted of a criminal offence and when subsequently his conviction has been reversed or he has been pardoned on the ground that a new or newly discovered fact shows conclusively that there has been a miscarriage of justice, the person who has suffered punishment as a result of such conviction shall be compensated according to law, unless it is proved that the non-disclosure of the unknown fact in time is wholly or partly attributable to him.

7. No one shall be liable to be tried or punished again for an offence for which he has already been finally convicted or acquitted in accordance with the law and penal procedure of each country.

Article 15

1. No one shall be held guilty of any criminal offence on account of any act of omission which did not constitute a criminal offence, under national or international law, at the time when it was committed. Nor shall a heavier penalty be imposed than the one that was applicable at the time when the criminal offence was committed. If, subsequent to the commission of the offence, provision is made by law for the imposition of the lighter penalty, the offender shall benefit thereby.

2. Nothing in this article shall prejudice the trial and punishment of any person for any act or omission which, at the time when it was committed, was criminal according to the general principles of law recognized by the community of nations.

Article 16

Everyone shall have the right to recognition everywhere as a person before the law.

Article 17

1. No one shall be subjected to arbitrary or unlawful interference with his privacy, family, home or correspondence, nor to unlawful attacks on his honour and reputation.

2. Everyone has the right to the protection of the law against such interference or attacks.

Article 18

1. Everyone shall have the right to freedom of thought, conscience and religion. This right shall include freedom to have or to adopt a religion or belief of his choice, and freedom, either individually or in community with others and in public or private, to manifest his religion or belief in worship, observance, practice and teaching.

2. No one shall be subject to coercion which would impair his freedom to have or to adopt a religion or belief of his choice.

3. Freedom to manifest one's religion or beliefs may be subject only to such limitations as are prescribed by law and are necessary to protect public safety, order, health, or morals or the fundamental rights and freedoms of others.

4. The States Parties to the present Covenant undertake to have respect for the liberty of parents and, when applicable, legal guardians to ensure the religious and moral education of their children in conformity with their own convictions.

Article 19

1. Everyone shall have the right to hold opinions without interference.

2. Everyone shall have the right to freedom of expression; this right shall include freedom to seek, receive and impart information and ideas of all kinds, regardless of frontiers, either orally, in writing or in print, in the form of art, or through any other media of his choice.

3. The exercise of the rights provided for in paragraph 2 of this article carries with its special duties and responsibilities. It may therefore be subject to certain restrictions, but these shall only be such as are provided by law and are necessary:

(*a*) For respect of the rights or reputations of others;

(*b*) For the protection of national security or of public order (*ordre public*), or of public health or morals.

Article 20

1. Any propaganda for war shall be prohibited by law.
2. Any advocacy of national, racial or religious hatred that constitutes incitement to discrimination, hostility or violence shall be prohibited by law.

Article 21

The right of peaceful assembly shall be recognized. No restrictions may be placed on the exercise of this right other than those imposed in conformity with the law and which are necessary in a democratic society in the interests of national security or public safety, public order (*ordre public*), the protection of public health or morals or the protection of the rights and freedoms of others.

Article 22

1. Everyone shall have the right to freedom of association with others, including the right to form and join trade unions for the protection of his interests.
2. No restrictions may be placed on the exercise of this right other than those which are prescribed by law and which are necessary in a democratic society in the interests of national security or public safety, public order (*ordre public*), the protection of public health or morals or the protection of the rights and freedoms of others. This article shall not prevent the imposition of lawful restrictions on members of the armed forces and of the police in their exercise of this right.

. . .

Article 23

1. The family is the natural and fundamental group unit of society and is entitled to protection by society and the State.
2. The right of men and women of marriageable age to marry and to found a family shall be recognized.
3. No marriage shall be entered into without the free and full consent of the intending spouses.
4. States Parties to the present Covenant shall take appropriate steps to ensure equality of rights and responsibilities of spouses as to marriage, during marriage and at its dissolution. In the case of dissolution, provision shall be made for the necessary protection of any children.

Article 24

1. Every child shall have, without any discrimination as to race, colour, sex, language, religion, national or social origin, property or birth, the right to such measures of protection as are required by his status as a minor, on the part of his family, society and the State.
2. Every child shall be registered immediately after birth and shall have a name.
3. Every child has the right to acquire a nationality.

Article 25

Every citizen shall have the right and the opportunity, without any of the distinctions mentioned in article 2 and without unreasonable restrictions:

(*a*) To take part in the conduct of public affairs, directly or through freely chosen representatives;

(*b*) To vote and to be elected at genuine periodic elections which shall be by universal and equal suffrage and shall be held by secret ballot, guaranteeing the free expression of the will of the electors;

(*c*) To have access, on general terms of equality, to public service in his country.

Article 26

All persons are equal before the law and are entitled without any discrimination to the equal protection of the law. In this respect, the law shall prohibit any discrimination and guarantee to all persons equal and effective protection against discrimination on any ground such as race, colour, sex, language, religion, political or other opinion, national or social origin, property, birth or other status.

Article 27

In those States in which ethnic, religious or linguistic minorities exist, persons belonging to such minorities shall not be denied the right, in community with the other members of their group, to enjoy their own culture, to profess and practice their own religion, or to use their own language.

PART IV

Article 28

1. There shall be established a Human Rights Committee (hereafter referred to in the present Covenant as the Committee). It shall consist of eighteen members and shall carry out the functions hereinafter provided.
2. The Committee shall be composed of nationals of the States Parties to the present Covenant who shall be persons of high moral character and recognized competence in the field of human rights, consideration being given to the usefulness of the participation of some persons having legal experience.
3. The members of the Committee shall be elected and shall serve in their personal capacity.

. . .

Article 31

1. The Committee may not include more than one national of the same State.
2. In the election of the Committee, consideration shall be given to equitable geographical distribution of membership and to the representation of the different forms of civilization and of the principal legal systems.

. . .

Article 38

Every member of the Committee shall, before taking up his duties, make a solemn declaration in open committee that he will perform his functions impartially and conscientiously.

. . .

Article 40

1. The States Parties to the present Covenant undertake to submit reports on the measures they have adopted which give effect to the rights recognized herein and on the progress made in the enjoyment of those rights:

(*a*) Within one year of the entry into force of the present Covenant for the States Parties concerned;

(*b*) Thereafter whenever the Committee so requests.

2. All reports shall be submitted to the Secretary-General of the United Nations, who shall transmit them to the Committee for consideration. Reports shall indicate the factors and difficulties, if any, affecting the implementation of the present Covenant.

3. The Secretary-General of the United Nations may, after consultation with the Committee, transmit to the specialized agencies concerned copies of such parts of the reports as may fall within their field of competence.

4. The Committee shall study the reports submitted by the States Parties to the present Covenant. It shall transmit its reports, and such general comments as it may consider appropriate, to the States Parties. The Committee may also transmit to the Economic and Social Council these comments along with the copies of the reports it has received from States Parties to the present Covenant.

5. The States Parties to the present Covenant may submit to the Committee observations on any comments that may be made in accordance with paragraph 4 of this article.

Article 41

1. A State Party to the present Covenant may at any time declare under this article that it recognizes the competence of the Committee to receive and consider communications to the effect that a State Party claims that another State Party is not fulfilling its obligations under the present Covenant. Communications under this article may be received and considered only if submitted by a State Party which has made a declaration recognizing in regard to itself the competence of the Committee. No communication shall be received by the Committee if it concerns a State Party which has not made such a declaration. Communications received under this article shall be dealt with in accordance with the following procedure:

[Article 41 spells out a procedure involving efforts toward resolution, referral of the matter to the Committee, and a report by the Committee to the States Parties concerned that is confined 'to a brief statement of the facts; the written submissions and record of the oral submissions made by the States Parties concerned, shall be attached to the report.' Article 42 provides that if the matter is not resolved to the satisfaction of the States Parties concerned the Committee may, with the consent of those parties, appoint an *ad hoc* Conciliation Commission. If no amicable solution is reached, the Commission submits a report to the Chairman of the Committee. The report includes the Commission's finding on all relevant questions of fact, and its views on possibilities of an amicable solution.]

. . .

Article 44

The provisions for the implementation of the present Covenant shall apply without prejudice to the procedures prescribed in the field of human rights by or under the constituent instruments and the conventions of the United Nations and of the specialized agencies and shall not prevent the States Parties to the present Covenant from having recourse to other procedures for settling a dispute in accordance with general or special international agreements in force between them.

Article 45

The Committee shall submit to the General Assembly of the United Nations, through the Economic and Social Council, an annual report on its activities.

PART V

Article 46

Nothing in the present Covenant shall be interpreted as impairing the provisions of the Charter of the United Nations and of the constitutions of the specialized agencies which define the respective responsibilities of the various organs of the United Nations and of the specialized agencies in regard to the matters dealt with in the present Covenant.

Article 47

Nothing in the present Covenant shall be interpreted as impairing the inherent right of all peoples to enjoy and utilize fully and freely their natural wealth and resources.

PART VI

. . .

Article 50

The provisions of the present Covenant shall extend to all parts of federal States without any limitations or exceptions.

Article 51

1. Any State Party to the present Covenant may propose an amendment and file it with the Secretary-General of the United Nations. The Secretary-General of the United Nations shall thereupon communicate any proposed amendments to the States Parties to the present Covenant with a request that they notify him whether they favour a conference of States Parties for the purpose of considering and voting

upon the proposals. In the event that at least one third of the States Parties favours such a conference, the Secretary-General shall convene the conference under the auspices of the United Nations. Any amendment adopted by a majority of the States Parties present and voting at the conference shall be submitted to the General Assembly of the United Nations for approval.

2. Amendments shall come into force when they have been approved by the General Assembly of the United Nations and accepted by a two-thirds majority of the States Parties to the present Covenant in accordance with their respective constitutional processes.

3. When amendments come into force, they shall be binding on those States Parties which have accepted them, other States Parties still being bound by the provisions of the present Covenant and any earlier amendment which they have accepted.

. . .

PROTOCOLS TO THE INTERNATIONAL COVENANT ON CIVIL AND POLITICAL RIGHTS

(First) Optional Protocol

The States Parties to the present Protocol,

Considering that in order further to achieve the purpose of the International Covenant on Civil and Political Rights (hereinafter referred to as the Covenant) and the implemenation of its provisions it would be appropriate to enable the Human Rights Committee set up in part IV of the Covenant (hereinafter referred to as the Committee) to receive and consider, as provided in the present Protocol, communications from individuals claiming to be victims of violations of any of the rights set forth in the Covenant,

Have agreed as follows:

Article 1

A State Party to the Covenant that becomes a Party to the present Protocol recognizes the competence of the Committee to receive and consider communications from individuals subject to its jurisdiction who claim to be victims of a violation by that State Party of any of the rights set forth in the Covenant. No communication shall be received by the Committee if it concerns a State Party to the Covenant which is not a Party to the present Protocol.

Article 2

Subject to the provisions of article 1, individuals who claim that any of their rights enumerated in the Covenant have been violated and who have exhausted all available domestic remedies may submit a written communication to the Committee for consideration.

Article 3

The Committee shall consider inadmissible any communication under the present Protocol which is anonymous, or which it considers to be an abuse of the right of submission of such communications or to be incompatible with the provisions of the Covenant.

Article 4

1. Subject to the provisions of article 3, the Committee shall bring any communications submitted to it under the present Protocol to the attention of the State Party to the present Protocol alleged to be violating any provision of the Covenant.

2. Within six months, the receiving State shall submit to the Committee written explanations or statements clarifying the matter and the remedy, if any, that may have been taken by that State.

Article 5

1. The Committee shall consider communications received under the present Protocol in the light of all written information made available to it by the individual and by the State Party concerned.

2. The Committee shall not consider any communication from an individual unless it has ascertained that:

(*a*) The same matter is not being examined under another procedure of international investigation or settlement;

(*b*) The individual has exhausted all available domestic remedies. This shall not be the rule where the application of the remedies is unreasonably prolonged.

3. The Committee shall hold closed meetings when examining communications under the present Protocol.

4. The Committee shall forward its views to the State Party concerned and to the individual.

Article 6

The Committee shall include in its annual report under article 45 of the Covenant a summary of its activities under the present Protocol.

. . .

Second Optional Protocol

The States Parties to the present Protocol,

Believing that abolition of the death penalty contributes to enhancement of human dignity and progressive development of human rights,

Recalling article 3 of the Universal Declaration of Human Rights, adopted on 10 December 1948, and article 6 of the International Covenant on Civil and Political Rights, adopted on 16 December 1966,

Noting that article 6 of the International Covenant on Civil and Political Rights refers to abolition of the death penalty in terms that strongly suggest that abolition is desirable,

Convinced that all measures of abolition of the death penalty should be considered as progress in the enjoyment of the right to life,

Desirous to undertake hereby an international commitment to abolish the death penalty,

Have agreed as follows:

Article 1

1. No one within the jurisdiction of a State Party to the present Protocol shall be executed.

2. Each State Party shall take all necessary measures to abolish the death penalty within its jurisdiction.

Article 2

1. No reservation is admissible to the present Protocol, except for a reservation made at the time of ratification or accession that provides for the application of the death penalty in time of war pursuant to a conviction for a most serious crime of a military nature committed during wartime.

. . .

Article 3

The States Parties to the present Protocol shall include in the reports they submit to the Human Rights Committee, in accordance with article 40 of the Covenant, information on the measures that they have adopted to give effect to the present Protocol.

. . .

INTERNATIONAL COVENANT ON ECONOMIC, SOCIAL AND CULTURAL RIGHTS

PREAMBLE

The States Parties to the present Covenant,

. . .

Recognizing that these rights derive from the inherent dignity of the human person,

Recognizing that, in accordance with the Universal Declaration of Human Rights, the ideal of free human beings enjoying freedom from fear and want can only be achieved if conditions are created whereby everyone may enjoy his economic, social and cultural rights, as well as his civil and political rights,

. . .

Realizing that the individual, having duties to other individuals and to the community to which he belongs, is under a responsibility to strive for the promotion and observance of the rights recognized in the present Covenant,

Agree upon the following articles:

PART I

Article 1

1. All peoples have the right of self-determination. By virtue of that right they freely determine their political status and freely pursue their economic, social and cultural development.
2. All peoples may, for their own ends, freely dispose of their natural wealth and resources without prejudice to any obligations arising out of international economic co-operation, based upon the principle of mutual benefit, and international law. In no case may a people by deprived of its own means of subsistence.

. . .

PART II

Article 2

1. Each State Party to the present Covenant undertakes to take steps, individually and through international assistance and co-operation, especially economic and technical, to the maximum of its available resources, with a view to achieving progressively the full realization of the rights recognized in the present Covenant by all appropriate means, including particularly the adoption of legislative measures.
2. The States Parties to the present Covenant undertake to guarantee that the rights enunciated in the present Covenant will be exercised without discrimination of any kind as to race, colour, sex, language, religion, political or other opinion, national or social origin, property, birth or other status.

3. Developing countries, with due regard to human rights and their national economy, may determine to what extent they would guarantee the economic rights recognized in the present Covenant to non-nationals.

Article 3

The States Parties to the present Covenant undertake to ensure the equal right of men and women to the enjoyment of all economic, social and cultural rights set forth in the present Covenant.

Article 4

The States Parties to the present Covenant recognize that, in the enjoyment of those rights provided by the State in conformity with the present Covenant, the State may subject such rights only to such limitations as are determined by law only in so far as this may be compatible with the nature of these rights and solely for the purpose of promoting the general welfare in a democratic society.

Article 5

1. Nothing in the present Covenant may be interpreted as implying for any State, group or person any right to engage in any activity or to perform any act aimed at the destruction of any of the rights or freedoms recognized herein, or at their limitation to a greater extent than is provided for in the present Covenant.

2. No restriction upon or derogation from any of the fundamental human rights recognized or existing in any country in virtue of law, conventions, regulations or custom shall be admitted on the pretext that the present Covenant does not recognize such rights or that it recognizes them to a lesser extent.

PART III

Article 6

1. The States Parties to the present Covenant recognize the right to work, which includes the right of everyone to the opportunity to gain his living by work which he freely chooses or accepts, and will take appropriate steps to safeguard this right.

2. The steps to be taken by a State Party to the present Covenant to achieve the full realization of this right shall include technical and vocational guidance and training programmes, policies and techniques to achieve steady economic, social and cultural development and full and productive employment under conditions safeguarding fundamental political and economic freedoms to the individual.

Article 7

The States Parties to the present Covenant recognize the right of everyone to the enjoyment of just and favourable conditions of work which ensure, in particular:

(*a*) Remuneration which provides all workers, as a minimum, with:

(i) Fair wages and equal remuneration for work of equal value without distinction of any kind, in particular women being guaranteed conditions of work not inferior to those enjoyed by men, with equal pay for equal work;

(ii) A decent living for themselves and their families in accordance with the provisions of the present Covenant;

(*b*) Safe and healthy working conditions;

(*c*) Equal opportunity for everyone to be promoted in his employment to an appropriate higher level, subject to no considerations other than those of seniority and competence;

(*d*) Rest, leisure and reasonable limitation of working hours and periodic holidays with pay, as well as remuneration for public holidays.

Article 8

1. The States Parties to the present Covenant undertake to ensure:

(*a*) The right of everyone to form trade unions and join the trade union of his choice, subject only to the rules of the organization concerned, for the promotion and protection of his economic and social interests. No restrictions may be placed on the exercise of this right other than those prescribed by law and which are necessary in a democratic society in the interests of national security or public order or for the protection of the rights and freedoms of others;

(*b*) The right of trade unions to establish national federations or confederations and the right of the latter to form or join international trade-union organizations;

(*c*) The right of trade unions to function freely subject to no limitations other than those prescribed by law and which are necessary in a democratic society in the interests of national security or public order or for the protection of the rights and freedom of others;

(*d*) The right to strike, provided that it is exercised in conformity with the laws of the particular country.

2. This article shall not prevent the imposition of lawful restrictions on the exercise of these rights by members of the armed forces or of the police or of the administration of the State.

. . .

Article 9

The States Parties to the present Covenant recognize the right of everyone to social security, including social insurance.

Article 10

The States Parties to the present Covenant recognize that:

1. The widest possible protection and assistance should be accorded to the family, which is the natural and fundamental group unit of society, particularly for its establishment and while it is responsible for the care and education of dependent children. Marriage must be entered into with the free consent of the intending spouses.

2. Special protection should be accorded to mothers during a reasonable period before and after childbirth. During such period working mothers should be accorded paid leave or leave with adequate social security benefits.

3. Special measures of protection and assistance should be taken on behalf of all children and young persons without any discrimination for reasons of parentage or other conditions. Children and young persons should be protected from economic and social exploitation. Their employment in work harmful to their morals or health or dangerous to life or likely to hamper their normal development should be punishable by law. States should also set age limits below which the paid employment of child labour should be prohibited and punishable by law.

Article 11

1. The States Parties to the present Covenant recognize the right of everyone to an adequate standard of living for himself and his family, including adequate food, clothing and housing, and to the continuous improvement of living conditions. The States Parties will take appropriate steps to ensure the realization of this right, recognizing to this effect the essential importance of international co-operation based on free consent.

2. The States Parties to the present Covenant, recognizing the fundamental right of everyone to be free from hunger, shall take, individually and through international co-operation, the measures, including specific programmes, which are needed:

(*a*) To improve methods of production, conservation and distribution of food by making full use of technical and scientific knowledge, by disseminating knowledge of the principles of nutrition and by developing or reforming agrarian systems in such a way as to achieve the most efficient development and utilization of natural resources;

(*b*) Taking into account the problems of both food-importing and food-exporting countries, to ensure an equitable distribution of world food supplies in relation to need.

Article 12

1. The States Parties to the present Covenant recognize the right of everyone to the enjoyment of the highest attainable standard of physical and mental health.

2. The steps to be taken by the States Parties to the present Covenant to achieve the full realization of this right shall include those necessary for:

(*a*) The provision for the reduction of the stillbirth-rate and of infant mortality and for the healthy development of the child;

(*b*) The improvement of all aspects of environmental and industrial hygiene;

(*c*) The prevention, treatment and control of epidemic, endemic, occupational and other diseases;

(*d*) The creation of conditions which would assure to all medical service and medical attention in the event of sickness.

Article 13

1. The States Parties to the present Covenant recognize the right of everyone to education. They agree that education shall be directed to the full development of the human personality and the sense of its dignity, and shall strengthen the respect for

human rights and fundamental freedoms. They further agree that education shall enable all persons to participate effectively in a free society, promote understanding, tolerance and friendship among all nations and all racial, ethnic or religious groups, and further the activities of the United Nations for the maintenance of peace.

2. The States Parties to the present Covenant recognize that, with a view to achieving the full realization of this right:

(*a*) Primary education shall be compulsory and available free to all;

(*b*) Secondary education in its different forms, including technical and vocational secondary education, shall be made generally available and accessible to all by every appropriate means, and in particular by the progressive introduction of free education;

(*c*) Higher education shall be made equally accessible to all, on the basis of capacity, by every appropriate means, and in particular by the progressive introduction of free education;

(*d*) Fundamental education shall be encouraged or intensified as far as possible for those persons who have not received or completed the whole period of their primary education;

(*e*) The development of a system of schools at all levels shall be actively pursued, an adequate fellowship system shall be established, and the material conditions of teaching staff shall be continuously improved.

3. The States Parties to the present Covenant undertake to have respect for the liberty of parents and, when applicable, legal guardians to choose for their children schools, other than those established by the public authorities, which conform to such minimum education standards as may be laid down or approved by the State and to ensure the religious and moral education of their children in conformity with their own convictions.

4. No part of this article shall be construed so as to interfere with the liberty of individuals and bodies to establish and direct educational institutions, subject always to the observance of the principles set forth in paragraph 1 of this article and to the requirement that the education given in such institutions shall conform to such minimum standards as may be laid down by the State.

Article 14

Each State Party to the present Covenant which, at the time of becoming a Party, has not been able to secure in its metropolitan territory or other territories under its jurisdiction compulsory primary education, free of charge, undertakes, within two years, to work out and adopt a detailed plan of action for the progressive implementation, within a reasonable number of years, to be fixed in the plan, of the principle of compulsory education free of charge for all.

Article 15

1. The States Parties to the present Covenant recognize the right of everyone:

(*a*) To take part in cultural life;

(*b*) To enjoy the benefits of scientific progress and its applications;

(*c*) To benefit from the protection of the moral and material interests resulting from any scientific, literary or artistic production of which he is the author.

2. The steps to be taken by the States Parties to the present Covenant to achieve the full realization of this right shall include those necessary for the conservation, the development and the diffusion of science and culture.

3. The States Parties to the present Covenant undertake to respect the freedom indispensable for scientific research and creative activity.

4. The States Parties to the present Covenant recognize the benefits to be derived from the encouragement and development of international contacts and co-operation in the scientific and cultural fields.

PART IV

Article 16

1. The States Parties to the present Covenant undertake to submit in conformity with this part of the Covenant reports on the measures which they have adopted and the progress made in achieving the observance of the rights recognized herein.

. . .

Article 22

The Economic and Social Council may bring to the attention of other organs of the United Nations, their subsidiary organs and specialized agencies concerned with furnishing technical assistance any matters arising out of the reports referred to in this part of the present Covenant which may assist such bodies in deciding, each within its field of competence, on the advisability of international measures likely to contribute to the effective progressive implementation of the present Covenant.

Article 23

The States Parties to the present Covenant agree that international action for the achievement of the rights recognized in the present Covenant includes such methods as the conclusion of conventions, the adoption of recommendations, the furnishing of technical assistance and the holding of regional meetings and technical meetings for the purpose of consultation and study organized in conjunction with the Governments concerned.

Article 24

Nothing in the present Covenant shall be interpreted as impairing the provisions of the Charter of the United Nations and of the constitutions of the specialized agencies which define the respective responsibilities of the various organs of the United Nations and of the specialized agencies in regard to the matters dealt with in the present Covenant.

Article 25

Nothing in the present Covenant shall be interpreted as impairing the inherent right of all peoples to enjoy and utilize fully and freely their natural wealth and resources.

PART V

. . .

Article 28

The provisions of the present Covenant shall extend to all parts of federal States without any limitations or exceptions.

. . .

CONVENTION ON THE ELIMINATION OF ALL FORMS OF DISCRIMINATION AGAINST WOMEN

The States Parties to the present Convention,

Noting that the Charter of the United Nations reaffirms faith in fundamental human rights, in the dignity and worth of the human person and in the equal rights of men and women,

Noting that the Universal Declaration of Human Rights affirms the principle of the inadmissibility of discrimination . . .,

Noting that the States Parties to the International Covenants on Human Rights have the obligation to ensure the equal rights of men and women to enjoy all economic, social, cultural, civil and political rights,

. . .

Concerned, however, that despite these various instruments extensive discrimination against women continues to exist,

Recalling that discrimination against women violates the principles of equality of rights and respect for human dignity, is an obstacle to the participation of women, on equal terms with men, in the political, social, economic and cultural life of their countries, hampers the growth of the prosperity of society and the family and makes more difficult the full development of the potentialities of women in the service of their countries and of humanity,

. . .

Convinced that the full and complete development of a country, the welfare of the world and the cause of peace require the maximum participation of women on equal terms with men in all fields,

Bearing in mind the great contribution of women to the welfare of the family and to the development of society, so far not fully recognized, the social significance of maternity and the role of both parents in the family and in the upbringing of children, and aware that the role of women in procreation should not be a basis for discrimination but that the upbringing of children requires a sharing of responsibility between men and women and society as a whole,

Aware that a change in the traditional role of men as well as the role of women in society and in the family is needed to achieve full equality between men and women,

. . .

Have agreed on the following:

PART I

Article 1

For the purposes of the present Convention, the term 'discrimination against women' shall mean any distinction, exclusion or restriction made on the basis of sex which has the effect or purpose of impairing or nullifying the recognition, enjoyment or exercise by women, irrespective of their marital status, on a basis of equality of men and women, of human rights and fundamental freedoms in the political, economic, social, cultural, civil or any other field.

Article 2

States Parties condemn discrimination against women in all its forms, agree to pursue by all appropriate means and without delay a policy of eliminating discrimination against women and, to this end, undertake:

(*a*) To embody the principle of the equality of men and women in their national institutions or other appropriate legislation if not yet incorporated therein and to ensure, through law and other appropriate means, the practical realization of this principle;

(*b*) To adopt appropriate legislative and other measures, including sanctions where appropriate, prohibiting all discrimination against women;

(*c*) To establish legal protection of the rights of women on an equal basis with men and to ensure through competent national tribunals and other public institutions the effective protection of women against any act of discrimination;

(*d*) To refrain from engaging in any act or practice of discrimination against women and to ensure that public authorities and institutions shall act in conformity with this obligation;

(*e*) To take all appropriate measures to eliminate discrimination against women by any person, organization or enterprise;

(*f*) To take all appropriate measures, including legislation, to modify or abolish existing laws, regulations, customs and practices which constitute discrimination against women;

(*g*) To repeal all national penal provisions which constitute discrimination against women.

Article 3

States Parties shall take in all fields, in particular in the political, social, economic and cultural fields, all appropriate measures, including legislation, to ensure the full development and advancement of women, for the purpose of guaranteeing them the exercise and enjoyment of human rights and fundamental freedoms on a basis of equality with men.

Article 4

1. Adoption by States Parties of temporary special measures aimed at accelerating *de facto* equality between men and women shall not be considered discrimination as defined in the present Convention, but shall in no way entail as a consequence the maintenance of unequal or separate standards; these measures shall be discontinued when the objectives of equality of opportunity and treatment have been achieved.

2. Adoption by States Parties of special measures, including those measures contained in the present Convention, aimed at protecting maternity shall not be considered discriminatory.

Article 5

States Parties shall take all appropriate measures:

(*a*) To modify the social and cultural patterns of conduct of men and women, with a view to achieving the elimination of prejudices and customary and all other practices which are based on the idea of the inferiority or the superiority of either of the sexes or on stereotyped roles for men and women;

(*b*) To ensure that family education includes a proper understanding of maternity as a social function and the recognition of the common responsibility of men and women in the upbringing and development of their children, it being understood that the interest of the children is the primordial consideration in all cases.

Article 6

States Parties shall take all appropriate measures, including legislation, to suppress all forms of traffic in women and exploitation of prostitution of women.

PART II

Article 7

States Parties shall take all appropriate measures to eliminate discrimination against women in the political and public life of the country and, in particular, shall ensure to women, on equal terms with men, the right:

(*a*) To vote in all elections and public referenda and to be eligible for election to all publicly elected bodies;

(*b*) To participate in the formulation of government policy and the implementation thereof and to hold public office and perform all public functions at all levels of government;

(*c*) To participate in non-governmental organizations and associations concerned with the public and political life of the country.

Article 8

States Parties shall take all appropriate measures to ensure to women, on equal terms with men and without any discrimination, the opportunity to represent their

Governments at the international level and to participate in the work of international organizations.

Article 9

1. States Parties shall grant women equal rights with men to acquire, change or retain their nationality. They shall ensure in particular that neither marriage to an alien nor change of nationality by the husband during marriage shall automatically change the nationality of the wife, render her stateless or force upon her the nationality of the husband.

2. States Parties shall grant women equal rights with men with respect to the nationality of their children.

PART III

Article 10

States Parties shall take all appropriate measures to eliminate discrimination against women in order to ensure to them equal rights with men in the field of education and in particular to ensure, on a basis of equality of men and women:

(*a*) The same conditions for career and vocational guidance, for access to studies and for the achievement of diplomas in educational establishments of all categories in rural as well as in urban areas; this equality shall be ensured in pre-school, general, technical, professional and higher technical education, as well as in all types of vocational training;

(*b*) Access to the same curricula, the same examinations, teaching staff with qualifications of the same standard and school premises and equipment of the same quality;

(*c*) The elimination of any stereotyped concept of the roles of men and women at all levels and in all forms of education by encouraging coeducation and other types of education which will help to achieve this aim and, in particular, by the revision of textbooks and school programmes and the adaptation of teaching methods;

(*d*) The same opportunities to benefit from scholarships and other study grants;

(*e*) The same opportunities for access to programmes of continuing education, including adult and functional literacy programmes, particulary those aimed at reducing, at the earliest possible time, any gap in education existing between men and women;

(*f*) The reduction of female student drop-out rates and the organization of programmes for girls and women who have left school prematurely;

(*g*) The same opportunities to participate actively in sports and physical education;

(*h*) Access to specific educational information to help to ensure the health and well-being of families, including information and advice on family planning.

Article 11

1. States Parties shall take all appropriate measures to eliminate discrimination against women in the field of employment in order to ensure, on a basis of equality of men and women, the same rights, in particular:

(*a*) The right to work as an inalienable right of all human beings;

(*b*) The right to the same employment opportunities, including the application of the same criteria for selection in matters of employment;

(*c*) The right to free choice of profession and employment, the right to promotion, job security and all benefits and conditions of service and the right to receive vocational training and retraining, including apprenticeships, advanced vocational training and recurrent training;

(*d*) The right to equal remuneration, including benefits, and to equal treatment in respect of work of equal value, as well as equality of treatment in the evaluation of the quality of work;

(*e*) The right to social security, particularly in cases of retirement, unemployment, sickness, invalidity and old age and other incapacity to work, as well as the right to paid leave;

(*f*) The right to protection of health and to safety in working conditions, including the safeguarding of the function of reproduction.

2. In order to prevent discrimination against women on the grounds of marriage or maternity and to ensure their effective right to work, States Parties shall take appropriate measures:

(*a*) To prohibit, subject to the imposition of sanctions, dismissal on the grounds of pregnancy or of maternity leave and discrimination in dismissals on the basis of marital status;

(*b*) To introduce maternity leave with pay or with comparable social benefits without loss of former employment, seniority or social allowances;

(*c*) To encourage the provision of the necessary supporting social services to enable parents to combine family obligations with work responsibilities and participation in public life, in particular through promoting the establishment and development of a network of child-care facilities;

(*d*) To provide special protection to women during pregnancy in types of work proved to be harmful to them.

3. Protective legislation relating to matters covered in this article shall be reviewed periodically in the light of scientific and technological knowledge and shall be revised, repealed or extended as necessary.

Article 12

1. States Parties shall take all appropriate measures to eliminate discrimination against women in the field of health care in order to ensure, on a basis of equality of men and women, access to health care services, including those related to family planning.

2. Notwithstanding the provisions of paragraph 1 of this article, States Parties shall ensure to women appropriate services in connection with pregnancy, confinement and the post-natal period, granting free services where necessary, as well as adequate nutrition during pregnancy and lactation.

Article 13

States Parties shall take all appropriate measures to eliminate discrimination against women in other areas of economic and social life in order to ensure, on a basis of equality of men and women, the same rights, in particular:

(*a*) The right to family benefits;

(*b*) The right to bank loans, mortgages and other forms of financial credit;

(*c*) The right to participate in recreational activities, sports and all aspects of cultural life.

Article 14

1. States Parties shall take into account the particular problems faced by rural women and the significant roles which rural women play in the economic survival of their families, including their work in the non-monetized sectors of the economy, and shall take all appropriate measures to ensure the application of the provisions of the present Convention to women in rural areas.

2. States Parties shall take all appropriate measures to eliminate discrimination against women in rural areas in order to ensure, on a basis of equality of men and women, that they participate in and benefit from rural development and, in particular, shall ensure to such women the right:

(*a*) To participate in the elaboration and implementation of development planning at all levels;

(*b*) To have access to adequate health care facilities, including information, counselling and services in family planning;

(*c*) To benefit directly from social security programmes;

(*d*) To obtain all types of training and education, formal and non-formal, including that relating to functional literacy, as well as, *inter alia*, the benefit of all community and extension services, in order to increase their technical proficiency;

(*e*) To organize self-help groups and co-operatives in order to obtain equal access to economic opportunities through employment or self-employment;

(*f*) To participate in all community activities;

(*g*) To have access to agricultural credit and loans, marketing facilities, appropriate technology and equal treatment in land and agrarian reform as well as in land resettlement schemes;

(*h*) To enjoy adequate living conditions, particularly in relation to housing, sanitation, electricity and water supply, transport and communications.

PART IV

Article 15

1. States Parties shall accord to women equality with men before the law.

2. States Parties shall accord to women, in civil matters, a legal capacity identical

to that of men and the same opportunities to exercise that capacity. In particular, they shall give women equal rights to conclude contracts and to administer property and shall treat them equally in all stages of procedure in courts and tribunals.

3. States Parties agree that all contracts and all other private instruments of any kind with a legal effect which is directed at restricting the legal capacity of women shall be deemed null and void.

4. States Parties shall accord to men and women the same rights with regard to the law relating to the movement of persons and the freedom to choose their residence and domicile.

Article 16

1. States Parties shall take all appropriate measures to eliminate discrimination against women in all matters relating to marriage and family relations and in particular shall ensure, on a basis of equality of men and women:

(*a*) The same right to enter into marriage;

(*b*) The same right freely to choose a spouse and to enter into marriage only with their free and full consent;

(*c*) The same rights and responsibilities during marriage and at its dissolution;

(*d*) The same rights and responsibilities as parents, irrespective of their marital status, in matters relating to their children; in all cases the interests of the children shall be paramount;

(*e*) The same rights to decide freely and responsibly on the number and spacing of their children and to have access to the information, education and means to enable them to exercise these rights;

(*f*) The same rights and responsibilities with regard to guardianship, wardship, trusteeship and adoption of children, or similar institutions where these concepts exist in national legislation; in all cases the interests of the children shall be paramount;

(*g*) The same personal rights as husband and wife, including the right to choose a family name, a profession and an occupation;

(*h*) The same rights for both spouses in respect of the ownership, acquisition, management, administration, enjoyment and disposition of property, whether free of charge or for a valuable consideration.

2. The betrothal and the marriage of a child shall have no legal effect, and all necessary action, including legislation, shall be taken to specify a minimum age for marriage and to make the registration of marriages in an official registry compulsory.

PART V

Article 17

1. For the purpose of considering the progress made in the implementation of the present Convention, there shall be established a Committee on the Elimination of Discrimination against Women (hereinafter referred to as the Committee) consisting, at the time of entry into force of the Convention, of eighteen and, after ratifi-

cation of or accession to the Convention by the thirty-fifth State Party, of twenty-three experts of high moral standing and competence in the field covered by the Convention. The experts shall be elected by States Parties from among their nationals and shall serve in their personal capacity, consideration being given to equitable geographical distribution and to the representation of the different forms of civilization as well as the principal legal systems.

. . .

Article 18

1. States Parties undertake to submit to the Secretary-General of the United Nations, for consideration by the Committee, a report on the legislative, judicial, administrative or other measures which they have adopted to give effect to the provisions of the present Convention and on the progress made in this respect:

(*a*) Within one year after the entry into force for the State concerned;

(*b*) Thereafter at least every four years and further whenever the Committee so requests.

2. Reports may indicate factors and difficulties affecting the degree of fulfilment of obligations under the present Convention.

. . .

Article 20

1. The Committee shall normally meet for a period of not more than two weeks annually in order to consider the reports submitted in accordance with article 18 of the present Convention.

2. The meetings of the Committee shall normally be held at United Nations Headquarters or at any other convenient place as determined by the Committee.

Article 21

1. The Committee shall, through the Economic and Social Council, report annually to the General Assembly of the United Nations on its activities and may make suggestions and general recommendations based on the examination of reports and information received from the States Parties. Such suggestions and general recommendations shall be included in the report of the Committee together with comments, if any, from States Parties.

. . .

Article 22

The specialized agencies shall be entitled to be represented at the consideration of the implementation of such provisions of the present Convention as fall within the scope of their activities. The Committee may invite the specialized agencies to submit reports on the implementation of the Convention in areas falling within the scope of their activities.

PART VI

Article 23

Nothing in the present Convention shall affect any provisions that are more conducive to the achievement of equality between men and women which may be contained:

(*a*) In the legislation of a State Party; or

(*b*) In any other international convention, treaty or agreement in force for that State.

Article 24

States Parties undertake to adopt all necessary measures at the national level aimed at achieving the full realization of the rights recognized in the present Convention.

. . .

Article 28

1. The Secretary-General of the United Nations shall receive and circulate to all States the text of reservations made by States at the time of ratification or accession.

2. A reservation incompatible with the object and purpose of the present Convention shall not be permitted.

. . .

Article 29

1. Any dispute between two or more States Parties concerning the interpretation or application of the present Convention which is not settled by negotiation shall, at the request of one of them, be submitted to arbitration. If within six months from the date of the request for arbitration the parties are unable to agree on the organization of the arbitration, any one of those parties may refer the dispute to the International Court of Justice by request in conformity with the Statute of the Court.

. . .

EUROPEAN CONVENTION FOR THE PROTECTION OF HUMAN RIGHTS AND FUNDAMENTAL FREEDOMS

The governments signatory hereto, being members of the Council of Europe,

Considering the Universal Declaration of Human Rights proclaimed by the General Assembly of the United Nations on 10th December 1948;

Considering that this Declaration aims at securing the universal and effective recognition and observance of the Rights therein declared;

Considering that the aim of the Council of Europe is the achievement of greater unity between its members and that one of the methods by which that aim is to be pursued is the maintenance and further realisation of human rights and fundamental freedoms;

Reaffirming their profound belief in those fundamental freedoms which are the foundation of justice and peace in the world and are best maintained on the one hand by an effective political democracy and on the other by a common understanding and observance of the human rights upon which they depend;

Being resolved, as the governments of European countries which are like-minded and have a common heritage of political traditions, ideals, freedom and the rule of law, to take the first steps for the collective enforcement of certain of the rights stated in the Universal Declaration,

Have agreed as follows:

Article 1

The High Contracting Parties shall secure to everyone within their jurisdiction the rights and freedoms defined in Section I of this Convention.

SECTION I

Article 2

1. Everyone's right to life shall be protected by law. No one shall be deprived of his life intentionally save in the execution of a sentence of a court following his conviction of a crime for which this penalty is provided by law.

2. Deprivation of life shall not be regarded as inflicted in contravention of this article when it results from the use of force which is no more than absolutely necessary:

(*a*) in defence of any person from unlawful violence;

(*b*) in order to effect a lawful arrest or to prevent the escape of a person lawfully detained;

(*c*) in action lawfully taken for the purpose of quelling a riot or insurrection.

Article 3

No one shall be subjected to torture or to inhuman or degrading treatment or punishment.

Article 4

1. No one shall be held in slavery or servitude.

. . .

Article 5

1. Everyone has the right to liberty and security of person. No one shall be deprived of his liberty save in the following cases and in accordance with a procedure prescribed by law:

(*a*) the lawful detention of a person after conviction by a competent court;

(*b*) the lawful arrest or detention of a person for non-compliance with the lawful order of a court or in order to secure the fulfilment of any obligation prescribed by law;

(*c*) the lawful arrest or detention of a person effected for the purpose of bringing him before the competent legal authority on reasonable suspicion of having committed an offence or when it is reasonably considered necessary to prevent his committing an offence or fleeing after having done so;

. . .

2. Everyone who is arrested shall be informed promptly, in a language which he understands, of the reasons for his arrest and of any charge against him.

3. Everyone arrested or detained in accordance with the provisions of paragraph 1(*c*) of this Article shall be brought promptly before a judge or other officer authorised by law to exercise judicial power and shall be entitled to trial within a reasonable time or to release pending trial. Release may be conditioned by guarantees to appear for trial.

4. Everyone who is deprived of his liberty by arrest or detention shall be entitled to take proceedings by which the lawfulness of his detention shall be decided speedily by a court and his release ordered if the detention is not lawful.

5. Everyone who has been the victim of arrest or detention in contravention of the provisions of this article shall have an enforceable right to compensation.

Article 6

1. In the determination of his civil rights and obligations or of any criminal charge against him, everyone is entitled to a fair and public hearing within a reasonable time by an independent and impartial tribunal established by law. Judgment shall be pronounced publicly but the press and public may be excluded from all or part of the trial in the interests of morals, public order or national security in a democratic society, where the interests of juveniles or the protection of the private life of the parties so require, or to the extent strictly necessary in the opinion of the court in special circumstances where publicity would prejudice the interests of justice.

2. Everyone charged with a criminal offence shall be presumed innocent until proved guilty according to law.

3. Everyone charged with a criminal offence has the following minimum rights:

(*a*) to be informed promptly, in a language which he understands and in detail, of the nature and cause of the accusation against him;

(*b*) to have adequate time and facilities for the preparation of his defence;

(*c*) to defend himself in person or through legal assistance of his own choosing or, if he has not sufficient means to pay for legal assistance, to be given it free when the interests of justice so require;

(*d*) to examine or have examined witnesses against him and to obtain the attendance and examination of witnesses on his behalf under the same conditions as witnesses against him;

(*e*) to have the free assistance of an interpreter if he cannot understand or speak the language used in court.

Article 7

1. No one shall be held guilty of any criminal offence on account of any act or omission which did not constitute a criminal offence under national or international law at the time when it was committed. Nor shall a heavier penalty be imposed than the one that was applicable at the time the criminal offence was committed.

2. This article shall not prejudice the trial and punishment of any person for any act or omission which, at the time when it was committed, was criminal according to the general principles of law recognised by civilised nations.

Article 8

1. Everyone has the right to respect for his private and family life, his home and his correspondence.

2. There shall be no interference by a public authority with the exercise of this right except such as is in accordance with the law and is necessary in a democratic society in the interests of national security, public safety or the economic well-being of the country, for the prevention of disorder or crime, for the protection of health or morals, or for the protection of the rights and freedoms of others.

Article 9

1. Everyone has the right to freedom of thought, conscience and religion; this right includes freedom to change his religion or belief and freedom, either alone or in community with others and in public or private, to manifest his religion or belief, in worship, teaching, practice and observance.

2. Freedom to manifest one's religion or beliefs shall be subject only to such limitations as are prescribed by law and are necessary in a democratic society in the interests of public safety, for the protection of public order, health or morals, or for the protection of the rights and freedoms of others.

Article 10

1. Everyone has the right to freedom of expression. This right shall include freedom to hold opinions and to receive and impart information and ideas without

interference by public authority and regardless of frontiers. This article shall not prevent States from requiring the licensing of broadcasting, television or cinema enterprises.

2. The exercise of these freedoms, since it carries with it duties and responsibilities, may be subject to such formalities, conditions, restrictions or penalties as are prescribed by law and are necessary in a democratic society, in the interests of national security, territorial integrity or public safety, for the prevention of disorder or crime, for the protection of health or morals, for the protection of the reputation or rights of others, for preventing the disclosure of information received in confidence, or for maintaining the authority and impartiality of the judiciary.

Article 11

1. Everyone has the right to freedom of peaceful assembly and to freedom of association with others, including the right to form and to join trade unions for the protection of his interests.

2. No restrictions shall be placed on the exercise of these rights other than such as are prescribed by law and are necessary in a democratic society in the interests of national security or public safety, for the prevention of disorder or crime, for the protection of health or morals or for the protection of the rights and freedoms of others. This article shall not prevent the imposition of lawful restrictions on the exercise of these rights by members of the armed forces, of the police or of the administration of the State.

Article 12

Men and Women of marriageable age have the right to marry and to found a family, according to the national laws governing the exercise of this right.

Article 13

Everyone whose rights and freedoms as set forth in this Convention are violated shall have an effective remedy before a national authority notwithstanding that the violation has been committed by persons acting in an official capacity.

Article 14

The enjoyment of the rights and freedoms set forth in this Convention shall be secured without discrimination on any ground such as sex, race, colour, language, religion, political or other opinion, national or social origin, association with a national minority, property, birth or other status.

Article 15

1. In time of war or other public emergency threatening the life of the nation any High Contracting Party may take measures derogating from its obligations under this Convention to the extent strictly required by the exigencies of the situation, provided that such measures are not inconsistent with its other obligations under international law.

2. No derogation from Article 2, except in respect of deaths resulting from lawful

acts of war, or from Articles 3, 4 (paragraph 1) and 7 shall be made under this provision.

3. Any High Contracting Party availing itself of this right of derogation shall keep the Secretary General of the Council of Europe fully informed of the measures which it has taken and the reasons therefor. It shall also inform the Secretary General of the Council of Europe when such measures have ceased to operate and the provisions of the Convention are again being fully executed.

Article 16

Nothing in Articles 10, 11 and 14 shall be regarded as preventing the High Contracting Parties from imposing restrictions on the political activity of aliens.

Article 17

Nothing in this Convention may be interpreted as implying for any State, group or person any right to engage in any activity or perform any act aimed at the destruction of any of the rights and freedoms set forth herein or at their limitation to a greater extent than is provided for in the Convention.

Article 18

The restrictions permitted under this Convention to the said rights and freedoms shall not be applied for any purpose other than those for which they have been prescribed.

SECTION II

Article 19

To ensure the observance of the engagements undertaken by the High Contracting Parties in the present Convention, there shall be set up:

(1) A European Commission of Human Rights, hereinafter referred to as 'the Commission';

(2) A European Court of Human Rights, hereinafter referred to as 'the Court'.

SECTION III

Article 20

1. The Commission shall consist of a number of members equal to that of the High Contracting Parties. No two members of the Commission may be nationals of the same State.

. . .

Article 23

The members of the Commission shall sit on the Commission in their individual capacity. During their term of office they shall not hold any position which is incompatible with their independence and impartiality as members of the Commission or the demands of this office.

Article 24

Any High Contracting Party may refer to the Commission, through the Secretary General of the Council of Europe, any alleged breach of the provisions of the Convention by another High Contracting Party.

Article 25

1. The Commission may receive petitions addressed to the Secretary General of the Council of Europe from any person, non-governmental organisation or group of individuals claiming to be the victim of a violation by one of the High Contracting Parties of the rights set forth in this Convention, provided that the High Contracting Party against which the complaint has been lodged has declared that it recognises the competence of the Commission to receive such petitions. Those of the High Contracting Parties who have made such a declaration undertake not to hinder in any way the effective exercise of this right.

2. Such declarations may be made for a specific period.

. . .

Article 26

The Commission may only deal with the matter after all domestic remedies have been exhausted, according to the generally recognised rules of international law, and within a period of six months from the date on which the final decision was taken.

Article 27

1 The Commission shall not deal with any petition submitted under Article 25 which:

(*a*) is anonymous, or

(*b*) is substantially the same as a matter which has already been examined by the Commission or has already been submitted to another procedure of international investigation or settlement and if it contains no relevant new information.

. . .

Article 28

1. In the event of the Commission accepting a petition referred to it:

(*a*) it shall, with a view to ascertaining the facts, undertake together with the representatives of the parties an examination of the petition and, if need be, an investigation, for the effective conduct of which the States concerned shall furnish all necessary facilities, after an exchange of views with the Commission;

(*b*) it shall at the same time place itself at the disposal of the parties concerned with a view to securing a friendly settlement of the matter on the basis of respect for Human Rights as defined in this Convention.

2. If the Commission succeeds in effecting a friendly settlement, it shall draw up a Report which shall be sent to the States concerned, to the Committee of Ministers and to the Secretary General of the Council of Europe for publication. This Report shall be confined to a brief statement of the facts and of the solution reached.

. . .

Article 31

1. If the examination of a petition has not been completed in accordance with Article 28 (paragraph 2), 29 or 30, the Commission shall draw up a Report on the facts and state its opinion as to whether the facts found disclose a breach by the State concerned of its obligations under the Convention. The individual opinions of members of the Commission on this point may be stated in the Report.

2. The Report shall be transmitted to the Committee of Ministers. It shall also be transmitted to the States concerned, who shall not be at liberty to publish it.

3. In transmitting the Report to the Committee of Ministers the Commission may make such proposals as it thinks fit.

Article 32

1. If the question is not referred to the Court in accordance with Article 48 of this Convention within a period of three months from the date of the transmission of the Report to the Committee of Ministers, the Committee of Ministers shall decide by a majority of two-thirds of the members entitled to sit on the Committee whether there has been a violation of the Convention.

2. In the affirmative case the Committee of Ministers shall prescribe a period during which the High Contracting Party concerned must take the measures required by the decision of the Committee of Ministers.

3. If the High Contracting Party concerned has not taken satisfactory measures within the prescribed period the Committee of Ministers shall decide by the majority provided for in paragraph 1 above what effect shall be given to its original decision and shall publish the report.

4. The High Contracting Parties undertake to regard as binding on them any decision which the Committee of Ministers may take in application of the preceding paragraphs.

. . .

SECTION IV

Article 38

The European Court of Human Rights shall consist of a number of judges equal to that of the Members of the Council of Europe. No two judges may be nationals of the same State.

. . .

Article 44

Only the High Contracting Parties and the Commission shall have the right to bring a case before the Court.

Article 45

The jurisdiction of the Court shall extend to all cases concerning the interpretation and application of the present Convention which the High Contracting Parties or the Commission shall refer to it in accordance with Article 48.

Article 46

1. Any of the High Contracting Parties may at any time declare that it recognises as compulsory *ipso facto* and without special agreement the jurisdiction of the Court in all matters concerning the interpretation and application of the present Convention.

2. The declarations referred to above may be made unconditionally or on condition of reciprocity on the part of several or certain other High Contracting Parties or for a specified period.

. . .

Article 48

The following may bring a case before the Court, provided that the High Contracting Party concerned, if there is only one, or the High Contracting Parties concerned, if there is more than one, are subject to the compulsory jurisdiction of the Court or, failing that, with the consent of the High Contracting Party concerned, if there is only one, or of the High Contracting Parties concerned if there is more than one:

(*a*) the Commission;

(*b*) a High Contracting Party whose national is alleged to be a victim;

(*c*) a High Contracting Party which referred the case to the Commission;

(*d*) a High Contracting Party against which the complaint has been lodged.

Article 49

In the event of dispute as to whether the Court has jurisdiction, the matter shall be settled by the decision of the Court.

Article 50

If the Court finds that a decision or a measure taken by a legal authority or any other authority of a High Contracting Party is completely or partially in conflict with the obligations arising from the present Convention, and if the internal law of the said Party allows only partial reparation to be made for the consequences of this decision or measure, the decision of the Court shall, if necessary, afford just satisfaction to the injured party.

. . .

Article 52

The judgment of the Court shall be final.

Article 53

The High Contracting Parties undertake to abide by the decision of the Court in any case to which they are parties.

Articles 54

The judgment of the Court shall be transmitted to the Committee of Ministers which shall supervise its execution.

. . .

PROTOCOLS TO THE EUROPEAN CONVENTION FOR THE PROTECTION OF HUMAN RIGHTS AND FUNDAMENTAL FREEDOMS

PROTOCOL NO. 1

. . .

Article 1

Every natural or legal person is entitled to the peaceful enjoyment of his possessions. No one shall be deprived of his possessions except in the public interest and subject to the conditions provided for by law and by the general principles of international law.

The preceding provisions shall not, however, in any way impair the right of a State to enforce such laws as it deems necessary to control the use of property in accordance with the general interest or to secure the payment of taxes or other contributions or penalties.

Article 2

No person shall be denied the right to education. In the exercise of any functions which it assumes in relation to education and to teaching, the State shall respect the right of parents to ensure such education and teaching in conformity with their own religious and philosophical convictions.

Article 3

The High Contracting Parties undertake to hold free elections at reasonable intervals by secret ballot, under conditions which will ensure the free expression of the opinion of the people in the choice of the legislature.

. . .

PROTOCOL NO. 4

Article 1

No one shall be deprived of his liberty merely on the ground of inability to fulfil a contractual obligation.

Article 2

1. Everyone lawfully within the territory of a State shall, within that territory, have the right to liberty of movement and freedom to choose his residence.

2. Everyone shall be free to leave any country, including his own.

3. No restrictions shall be placed on the exercise of these rights other than such as are in accordance with law and are necessary in a democratic society in the interests

of national security or public safety, for the maintenance of *ordre public*, for the prevention of crime, for the protection of health or morals, or for the protection of the rights and freedoms of others.

4. The rights set forth in paragraph 1 may also be subject, in particular areas, to restrictions imposed in accordance with law and justified by the public interest in a democratic society.

Article 3

1. No one shall be expelled, by means either of an individual or of a collective measure, from the territory of the State of which he is a national.

2. No one shall be deprived of the right to enter the territory of the State of which he is a national.

Article 4

Collective expulsion of aliens is prohibited.

. . .

PROTOCOL NO. 6

Article 1

The death penalty shall be abolished. No one shall be condemned to such penalty or executed.

Article 2

A State may make provision in its law for the death penalty in respect of acts committed in time of war or of imminent threat of war; such penalty shall be applied only in the instances laid down in the law and in accordance with its provisions. The State shall communicate to the Secretary General of the Council of Europe the relevant provisions of that law.

Article 3

No derogation from the provisions of this Protocol shall be made under Article 15 of the Convention.

. . .

PROTOCOL NO. 7

Article 1

1. An alien lawfully resident in the territory of a State shall not be expelled therefrom except in pursuance of a decision reached in accordance with law and shall be allowed:

(*a*) to submit reasons against his expulsion,

(*b*) to have his case reviewed, and

(*c*) to be represented for these purposes before the competent authority or a person or persons designated by that authority.

2. An alien may be expelled before the exercise of his rights under paragraph 1(*a*), (*b*) and (*c*) of this Article, when such expulsion is necessary in the interests of public order or is grounded on reasons of national security.

Article 2

1. Everyone convicted of a criminal offence by a tribunal shall have the right to have his conviction or sentence reviewed by a higher tribunal. The exercise of this right, including the grounds on which it may be exercised, shall be governed by law.

2. This right may be subject to exception in regard to offences of a minor character, as prescribed by law, or in cases in which the person concerned was tried in the first instance by the highest tribunal or was convicted following an appeal against acquittal.

Article 3

When a person has by a final decision been convicted of a criminal offence and when subsequently his conviction has been reversed, or he has been pardoned, on the ground that a new or newly discovered fact shows conclusively that there has been a miscarriage of justice, the person who has suffered punishment as a result of such conviction shall be compensated according to the law or the practice of the State concerned, unless it is proved that the non-disclosure of the unknown fact in time is wholly or partly attributable to him.

Article 4

1. No one shall be liable to be tried or punished again in criminal proceedings under the jurisdiction of the same State for an offence for which he has already been finally acquitted or convicted in accordance with the law and penal procedure of that State.

2. The provisions of the preceding paragraph shall not prevent the reopening of the case in accordance with the law and penal procedure of the State concerned, if there is evidence of new or newly discovered facts, or if there has been a fundamental defect in the previous proceedings, which could affect the outcome of the case.

3. No derogation from this Article shall be made under Article 15 of the Convention.

Article 5

1. Spouses shall enjoy equality of rights and responsibilities of a private law character between them, and in their relations with their children, as to marriage, during marriage and in the event of its dissolution. This Article shall not prevent States from taking such measures as are necessary in the interests of the children.

. . .

AMERICAN CONVENTION ON HUMAN RIGHTS

PREAMBLE

The American states signatory to the present Convention,

Reaffirming their intention to consolidate in this hemisphere, within the framework of democratic institutions, a system of personal liberty and social justice based on respect for the essential rights of man;

Recognizing that the essential rights of man are not derived from one's being a national of a certain state, but are based upon attributes of the human personality, and that they therefore justify international protection in the form of a convention reinforcing or complementing the protection provided by the domestic law of the American state;

Considering that these principles have been set forth in the Charter of the Organization of American States, in the American Declaration of the Rights and Duties of Man, and in the Universal Declaration of Human Rights, and that they have been reaffirmed and refined in other international instruments, worldwide as well as regional in scope;

. . .

Have agreed upon the following:

Part I State Obligations and Rights Protected

Chapter I. General Obligations

Article 1

1. The States Parties to this Convention undertake to respect the rights and freedoms recognized herein and to ensure to all persons subject to their jurisdiction the free and full exercise of those rights and freedoms, without any discrimination for reasons of race, color, sex, language, religion, political or other opinion, national or social origin, economic status, birth, or any other social condition.
2. For the purposes of this Convention, 'person' means every human being.

Article 2

Where the exercise of any of the rights or freedoms referred to in Article 1 is not already ensured by legislative or other provisions, the States Parties undertake to adopt, in accordance with their constitutional processes and the provisions of this Convention, such legislative or other measures as may be necessary to give effect to those rights or freedoms.

Chapter II. Civil and Political Rights

Article 3.

Every person has the right to recognition as a person before the law.

Article 4

1. Every person has the right to have his life respected. This right shall be protected by law and, in general, from the moment of conception. No one shall be arbitrarily deprived of his life.

2. In countries that have not abolished the death penalty, it may be imposed only for the most serious crimes and pursuant to a final judgment rendered by a competent court and in accordance with a law establishing such punishment, enacted prior to the commission of the crime. The application of such punishment shall not be extended to crimes to which it does not presently apply.

3. The death penalty shall not be reestablished in states that have abolished it.

4. In no case shall capital punishment be inflicted for political offenses or related common crimes.

5. Capital punishment shall not be imposed upon persons who, at the time the crime was committed, were under 18 years of age or over 70 years of age; nor shall it be applied to pregnant women.

6. Every person condemned to death shall have the right to apply for amnesty, pardon, or commutation of sentence, which may be granted in all cases. Capital punishment shall not be imposed while such a petition is pending decision by the competent authority.

Article 5

1. Every person has the right to have his physical, mental, and moral integrity respected.

2. No one shall be subjected to torture or to cruel, inhuman, or degrading punishment or treatment. All persons deprived of their liberty shall be treated with respect for the inherent dignity of the human person.

3. Punishment shall not be extended to any person other than the criminal.

4. Accused persons shall, save in exceptional circumstances, be segregated from convicted persons, and shall be subject to separate treatment appropriate to their status as unconvicted persons.

5. Minors while subject to criminal proceedings shall be separated from adults and brought before specialized tribunals, as speedily as possible, so that they may be treated in accordance with their status as minors.

6. Punishments consisting of deprivation of liberty shall have as an essential aim the reform and social readaptation of the prisoners.

Article 6

1. No one shall be subject to slavery or to involuntary servitude, which are prohibited in all their forms, as are the slave trade and traffic in women.

. . .

Article 7

1. Every person has the right to personal liberty and security.

2. No one shall be deprived of his physical liberty except for the reasons and under the conditions established beforehand by the constitution of the State Party concerned or by a law established, pursuant thereto.

3. No one shall be subject to arbitrary arrest or imprisonment.

4. Anyone who is detained shall be informed of the reasons for his detention and shall be promptly notified of the charge or charges against him.

5. Any person detained shall be brought promptly before a judge or other officer authorized by law to exercise judicial power and shall be entitled to trial within a reasonable time or to be released without prejudice to the continuation of the proceedings. His release may be subject to guarantees to assure his appearance for trial.

6. Anyone who is deprived of his liberty shall be entitled to recourse to a competent court, in order that the court may decide without delay on the lawfulness of his arrest or detention and order his release if the arrest or detention is unlawful. In States Parties whose laws provide that anyone who believes himself to be threatened with deprivation of his liberty is entitled to recourse to a competent court in order that it may decide on the lawfulness of such threat, this remedy may not be restricted or abolished. The interested party or another person in his behalf is entitled to seek these remedies.

7. No one shall be detained for debt. This principle shall not limit the orders of a competent judicial authority issued for nonfulfillment of duties of support.

Article 8

1. Every person has the right to a hearing, with due guarantees and within a reasonable time, by a competent, independent, and impartial tribunal, previously established by law, in the substantiation of any accusation of a criminal nature made against him or for the determination of his rights and obligations of a civil, labor, fiscal, or any other nature.

2. Every person accused of a criminal offense has the right to be presumed innocent so long as his guilt has not been proven according to law. During the proceedings, every person is entitled, with full equality, to the following minimum guarantees:

(*a*) the right of the accused to be assisted without charge by a translator or interpreter, if he does not understand or does not speak the language of the tribunal or court;

(*b*) prior notification in detail to the accused of the charges against him;

(*c*) adequate time and means for the preparation of his defense;

(*d*) the right of the accused to defend himself personally or to be assisted by legal counsel of his own choosing, and to communicate freely and privately with his counsel;

(*e*) the inalienable right to be assisted by counsel provided by the state, paid or not as the domestic law provides, if the accused does not defend himself personally or engage his own counsel within the time period established by law;

(*f*) the right of the defense to examine witnesses present in the court and to obtain the appearance, as witnesses, of experts or other persons who may throw light on the facts;

(*g*) the right not to be compelled to be a witness against himself or to plead guilty; and

(*h*) the right to appeal the judgment to a higher court.

3. A confession of guilt by the accused shall be valid only if it is made without coercion of any kind.

4. An accused person acquitted by a nonappealable judgment shall not be subjected to a new trial for the same cause.

5. Criminal proceedings shall be public, except insofar as may be necessary to protect the interests of justice.

Article 9

No one shall be convicted of any act or omission that did not constitute a criminal offense, under the applicable law, at the time it was committed. A heavier penalty shall not be imposed than the one that was applicable at the time the criminal offense was committed. If subsequent to the commission of the offense the law provides for the imposition of a lighter punishment, the guilty person shall benefit therefrom.

Article 10

Every person has the right to be compensated in accordance with the law in the event he has been sentenced by a final judgment through a miscarriage of justice.

Article 11

1. Everyone has the right to have his honor respected and his dignity recognized.

2. No one may be the object of arbitrary or abusive interference with his private life, his family, his home, or his correspondence, or of unlawful attacks on his honor or reputation.

3. Everyone had the right to the protection of the law against such interference or attacks.

Article 12

1. Everyone has the right to freedom of conscience and of religion. This right includes freedom to maintain or to change one's religion or beliefs, and freedom to profess or disseminate one's religion or beliefs, either individually or together with others, in public or in private.

2. No one shall be subject to restrictions that might impair his freedom to maintain or to change his religion or beliefs.

3. Freedom to manifest one's religion and beliefs may be subject only to the limitations prescribed by law that are necessary to protect public safety, order, health, or morals, or the rights or freedoms of others.

4. Parents or guardians, as the case may be, have the right to provide for the religious and moral education of their children or wards that is in accord with their own convictions.

Article 13

1. Everyone has the right to freedom of thought and expression. This right includes freedom to seek, receive, and impart information and ideas of all kinds, regardless of frontiers, either orally, in writing, in print, in the form of art, or through any other medium of one's choice.

2. The exercise of the right provided for in the foregoing paragraph shall not be subject to prior censorship but shall be subject to subsequent imposition of liability, which shall be expressly established by law to the extent necessary to ensure:

(*a*) respect for the rights or reputations of others; or

(*b*) the protection of national security, public order, or public health or morals.

3. The right of expression may not be restricted by indirect methods or means, such as the abuse of government or private controls over newsprint, radio broadcasting frequencies, or equipment used in the dissemination of information, or by any other means tending to impede the communication and circulation of ideas and opinions.

4. Notwithstanding the provisions of paragraph 2 above, public entertainments may be subject by law to prior censorship for the sole purpose of regulating access to them for the moral protection of childhood and adolescence.

5. Any propaganda for war and any advocacy of national, racial, or religious hatred that constitute incitements to lawless violence or to any other similar illegal action against any person or group of persons on any grounds including those of race, color, religion, language, or national origin shall be considered as offenses punishable by law.

. . .

Article 15

The right of peaceful assembly, without arms, is recognized. No restrictions may be placed on the exercise of this right other than those imposed in conformity with the law and necessary in a democratic society in the interest of national security, public safety or public order, or to protect public health or morals or the rights or freedoms of others.

Article 16

1. Everyone has the right to associate freely for ideological, religious, political, economic, labor, social, cultural, sports, or other purposes.

2. The exercise of this right shall be subject only to such restrictions established by law as may be necessary in a democratic society, in the interest of national security, public safety or public order, or to protect public health or morals or the rights and freedoms of others.

3. The provisions of this article do not bar the imposition of legal restrictions, including even deprivation of the exercise of the right of association, on members of the armed forces and the police.

Article 17

1. The family is the natural and fundamental group unit of society and is entitled to protection by society and the state.

2. The right of men and women of marriageable age to marry and to raise a family shall be recognized, if they meet the conditions required by domestic laws, insofar as such conditions do not affect the principle of nondiscrimination established in this Convention.

3. No marriage shall be entered into without the free and full consent of the intending spouse.

4. The States Parties shall take appropriate steps to ensure the equality of rights and the adequate balancing of responsibilities of the spouses as to marriage, during marriage, and in the event of its dissolution. In case of dissolution, provision shall be made for the necessary protection of any children solely on the basis of their own best interests.

5. The law shall recognize equal rights for children born out of wedlock and those born in wedlock.

. . .

Article 19

Every minor child has the right to the measures of protection required by his condition as a minor on the part of his family, society, and the state.

Article 20

1. Every person has the right to a nationality.
2. Every person has the right to the nationality of the state in whose territory he was born if he does not have the right to any other nationality.
3. No one shall be arbitrarily deprived of his nationality or of the right to change it.

Article 21

1. Everyone has the right to the use and enjoyment of his property. The law may subordinate such use and enjoyment to the interest of society.
2. No one shall be deprived of his property except upon payment of just compensation, for reasons of public utility or social interest, and in the cases and according to the forms established by law.
3. Usury and any other form of exploitation of man by man shall be prohibited by law.

Article 22

1. Every person lawfully in the territory of a State Party has the right to move about in it, and to reside in it subject to the provisions of the law.
2. Every person has the right to leave any country freely, including his own.
3. The exercise of the foregoing rights may be restricted only pursuant to a law to the extent necessary in a democratic society to prevent crime or to protect national security, public safety, public order, public morals, public health, or the rights or freedoms of others.
4. The exercise of the rights recognized in paragraph 1 may also be restricted by law in designated zones for reasons of public interest.
5. No one can be expelled from the territory of the state of which he is a national or be deprived of the right to enter it.
6. An alien lawfully in the territory of a State Party to this Convention may be expelled from it only pursuant to a decision reached in accordance with law.
7. Every person has the right to seek and be granted asylum in a foreign territory, in accordance with the legislation of the state and international conventions, in the event he is being pursued for political offenses or related common crimes.
8. In no case may an alien be deported or returned to a country, regardless of whether or not it is his country of origin, if in that country his right to life or personal freedom is in danger of being violated because of his race, nationality, religion, social status, or political opinions.
9. The collective explusion of aliens is prohibited.

Article 23

1. Every citizen shall enjoy the following rights and opportunities:

(*a*) to take part in the conduct of public affairs, directly or through freely chosen representatives;

(*b*) to vote and to be elected in genuine periodic elections, which shall be by universal and equal suffrage and by secret ballot that guarantees the free expression of the will of the voters; and

(*c*) to have access, under general conditions of equality, to the public service of his country.

2. The law may regulate the exercise of the rights and opportunities referred to in the preceding paragraph only on the basis of age, nationality, residence, language, education, civil and mental capacity, or sentencing by a competent court in criminal proceedings.

Article 24

All persons are equal before the law. Consequently, they are entitled, without discrimination, to equal protection of the law.

Article 25

1. Everyone has the right to simple and prompt recourse, or any other effective recourse, to a competent court or tribunal for protection against acts that violate his fundamental rights recognized by the constitution or laws of the state concerned or by this Convention, even though such violation may have been committed by persons acting in the course of their official duties.

2. The States Parties undertake:

(*a*) to ensure that any person claiming such remedy shall have his rights determined by the competent authority provided for by the legal system of the state;

(*b*) to develop the possibilities of judicial remedy; and

(*c*) to ensure that the competent authorities shall enforce such remedies when granted.

Chapter III. Economic, Social and Cultural Rights

Article 26

The States Parties undertake to adopt measures both internally and through international co-operation, especially those of an economic and technical nature, with a view to achieving progressively, by legislation or other appropriate means, the full realization of the rights implicit in the economic, social, educational, scientific, and cultural standards set forth in the Charter of the Organization of American States as amended by the Protocol of Buenos Aires.

Chapter IV. Suspension of Guarantees, Interpretation, and Application

Article 27

1. In time of war, public danger, or other emergency that threatens the independence or security of a State Party, it may take measures derogating from its obligations under the present Convention to the extent and for the period of time strictly required by the exigencies of the situation, provided that such measures are not inconsistent with its other obligations under international law and do not involve discrimination on the ground of race, color, sex, language, religion, or social origin.

2. The foregoing provision does not authorize any suspension of the following articles: Article 3 (Right to Juridical Personality), Article 4 (Right to Life), Article 5 (Right to Humane Treatment), Article 6 (Freedom from Slavery), Article 9 (Freedom from *Ex Post Facto* Laws), Article 12 (Freedom of Conscience and Religion), Article 17 (Rights of the Family), Article 18 (Right to a Name), Article 19 (Rights of the Child), Article 20 (Right to Nationality), and Article 23 (Right to Participate in Government), or of the judicial guarantees essential for the protection of such rights.

3. Any State Party availing itself of the right of suspension shall immediately inform the other States Parties, through the Secretary General of the Organization of American States, of the provisions the application of which it has suspended, the reasons that gave rise to the suspension, and the date set for the termination of such suspension.

Article 28

1. Where a State Party is constituted as a federal state, the national government of such State Party shall implement all the provisions of the Convention over whose subject matter it exercises legislative and judicial jurisdiction.

2. With respect to the provisions over whose subject matter the constituent units of the federal state have jurisdiction, the national government shall immediately take suitable measures, in accordance with its constitution and its laws, to the end that the competent authorities of the constituent units may adopt appropriate provisions for the fulfillment of this Convention.

. . .

Chapter V. Personal Responsibilities

Article 32

1. Every person has responsibilities to his family, his community, and mankind.

2. The rights of each person are limited by the rights of others, by the security of all, and by the just demands of the general welfare, in a democratic society.

Part II. Means of Protection

Chapter VI. Competent Organs

Article 33

The following organs shall have competence with respect to matters relating to the fulfillment of the commitments made by the States Parties to this Convention:

(*a*) the Inter-American Commission on Human Rights, referred to as 'The Commission'; and

(*b*) the Inter-American Court of Human Rights, referred to as 'The Court.'

Chapter VII. Inter-American Commission on Human Rights

Section I. Organization

Article 34

The Inter-American Commission on Human Rights shall be composed of seven members, who shall be persons of high moral character and recognized competence in the field of human rights.

Article 35

The Commission shall represent all the member countries of the Organization of American States.

Article 36

1. The members of the Commission shall be elected in a personal capacity by the General Assembly of the Organization from a list of candidates proposed by the governments of the member states.

. . .

Section 2. Functions

Article 41

The main function of the Commission shall be to promote respect for and defense of human rights. In the exercise of its mandate, it shall have the following functions and powers:

(*a*) to develop an awareness of human rights among the peoples of America;

(*b*) to make recommendations to the governments of the member states, when it considers such action advisable, for the adoption of progressive measures in favour of human rights within the framework of their domestic law and constitutional provisions as well as appropriate measures to further the observance of those rights;

(*c*) to prepare such studies or reports as it considers advisable in the performance of its duties;

(*d*) to request the governments of the member states to supply it with information on the measures adopted by them in matters of human rights;

(*e*) to respond, through the General Secretariat of the Organization of American States, to inquiries made by the member states on matters related to human rights and, within the limits of its possibilities, to provide those states with the advisory services they request;

(*f*) to take action on petitions and other communications pursuant to its authority under the provisions of Articles 44 through 51 of this Convention; and

(*g*) to submit an annual report to the General Assembly of the Organization of American States.

. . .

Section 3. Competence

Article 44

Any person or group of persons, or any nongovernmental entity legally recognized in one or more member states of the Organization, may lodge petitions with the Commission containing denunciations or complaints of violation of this Convention by a State Party.

Article 45

1. Any State Party may, when it deposits its instrument of ratification of or adherence to this Convention, or at any later time, declare that it recognizes the competence of the Commission to receive and examine communications in which a State Party alleges that another State Party has committed a violation of a human right set forth in this Convention.

2. Communications presented by virtue of this article may be admitted and examined only if the are presented by a State Party that has made a declaration recognizing the aforementioned competence of the Commission. The Commission shall not admit any communication against a State Party that has not made such a declaration.

. . .

Section 4. Procedure

[Article 48 sets forth procedures for the Commission after it receives a petition or communication alleging violation of any protected rights. The Commission is to 'place itself at the disposal of the parties concerned with a view to reaching a friendly settlement of the matter on the basis of respect for the human rights recognized in this Convention.' Article 49 concerns situations where such a settlement has been reached.]

Article 50

1. If a settlement is not reached, the Commission shall, within the time limit established by its Statute, draw up a report setting forth the facts and stating its conclusions. . . .

2. The report shall be transmitted to the states concerned, which shall not be at liberty to publish it.

3. In transmitting the report, the Committee may make such proposals and recommendations as it sees fit.

Article 51

1. If, within a period of three months from the date of the transmittal of the report of the Commission to the states concerned, the matter has not either been settled or submitted by the Commission or by the state concerned to the Court and its jurisdiction accepted, the Commission may, by the vote of an absolute majority of its members, set forth its opinion and conclusions concerning the question submitted for its consideration.

2. Where appropriate, the Commission shall make pertinent recommendations and shall prescribe a period within which the state is to take the measures that are incumbent upon it to remedy the situation examined.

3. When the prescribed period has expired, the Commission shall decide by the vote of an absolute majority of its members whether the state has taken adequate measures and whether to publish its report.

Chapter VIII. Inter-American Court of Human Rights

Section 1. Organization

Article 52

1. The Court shall consist of seven judges, nationals of the member states of the Organization, elected in an individual capacity from among jurists of the highest moral authority and of recognized competence in the field of human rights, who possess the qualifications required for the exercise of the highest judicial functions in conformity with the law of the state of which they are nationals or of the state that proposes them as candidates.

2. No two judges may be nationals of the same state.

. . .

Section 2. Jurisdiction and Functions

Article 61

1. Only the States Parties and the Commission shall have the right to submit a case to the Court.

. . .

Article 62

1. A State Party may, upon depositing its instrument of ratification or adherence to this Convention, or at any subsequent time, declare that it recognizes as binding, *ipso facto*, and not requiring special agreement, the jurisdiction of the Court on all matters relating to the interpretation or application of this Convention.

2. Such declaration may be made unconditionally, on the condition of reciprocity, for a specified period, or for specific cases. . . .

3. The jurisdiction of the Court shall comprise all cases concerning the interpretation and application of the provisions of this Convention that are submitted to it, provided that the States Parties to the case recognize or have recognized such jurisdiction, whether by special declaration pursuant to the preceding paragraphs, or by a special agreement.

Article 63

1. If the Court finds that there has been a violation of a right or freedom protected by this Convention, the Court shall rule that the injured party be ensured the enjoyment of his right or freedom that was violated. It shall also rule, if appropriate, that the consequences of the measure or situation that constituted the breach of such right or freedom be remedied and that fair compensation be paid to the injured party.

2. In cases of extreme gravity and urgency, and when necessary to avoid irreparable damage to persons, the Court shall adopt such provisional measures as it deems pertinent in matters it has under consideration. With respect to a case not yet submitted to the Court, it may act at the request of the Commission.

Article 64

1. The member states of the Organization may consult the Court regarding the interpretation of this Convention or of other treaties concerning the protection of human rights in the American states. . . .

2. The Court, at the request of a member state of the Organization, may provide that state with opinions regarding the compatibility of any of its domestic laws with the aforesaid international instruments.

. . .

Section 3. Procedure

. . .

Article 67

The judgment of the Court shall be final and not subject to appeal. In case of disagreement as to the meaning or scope of the judgment, the Court shall interpret it at the request of any of the parties, provided the request is made within ninety days from the date of notification of the judgment.

Article 68

1. The States Parties to the Convention undertake to comply with the judgment of the Court in any case to which they are parties.

. . .

AFRICAN CHARTER ON HUMAN AND PEOPLES' RIGHTS

PREAMBLE

. . .

The African States members of the Organization of African Unity, parties to the present convention, entitled 'African Charter on Human and Peoples' Rights',

. . .

Considering the Charter of the Organization of African Unity, which stipulates that 'freedom, equality, justice and dignity are essential objectives for the achievement of the legitimate aspirations of the African peoples';

Reaffirming the pledge they solemnly made in Article 2 of the said Charter to eradicate all forms of colonialism from Africa, to coordinate and intensify their cooperation and efforts to achieve a better life for the peoples of Africa and to promote international cooperation having due regard to the Charter of the United Nations and the Universal Declaration of Human Rights;

Taking into consideration the virtues of their historical tradition and the values of African civilization which should inspire and characterize their reflection on the concept of human and peoples' rights;

. . .

Considering that the enjoyment of rights and freedoms also implies the performance of duties on the part of everyone;

Convinced that it is henceforth essential to pay a particular attention to the right to development and that civil and political rights cannot be dissociated from economic, social and cultural rights in their conception as well as universality and that the satisfaction of economic, social and cultural rights is a guarantee for the enjoyment of civil and political rights;

Conscious of their duty to achieve the total liberation of Africa, the peoples of which are still struggling for their dignity and genuine independence, and undertaking to eliminate colonialism, neo-colonialism, apartheid, zionism and to dismantle aggressive foreign military bases and all forms of discrimination, particularly those based on race, ethnic group, color, sex, language, religion or political opinions;

. . .

Firmly convinced of their duty to promote and protect human and peoples' rights and freedoms taking into account the importance traditionally attached to these rights and freedoms in Africa;

Have agreed as follows:

PART I. RIGHTS AND DUTIES

CHAPTER I. HUMAN AND PEOPLES'RIGHTS

Article 1

The Member States of the Organization of African Unity parties to the present Charter shall recognize the rights, duties and freedoms enshrined in this Charter and shall undertake to adopt legislative or other measures to give effect to them.

Article 2

Every individual shall be entitled to the enjoyment of the rights and freedoms recognized and guaranteed in the present Charter without distinction of any kind such as race, ethnic group, color, sex, language, religion, political or any other opinion, national and social origin, fortune, birth or other status.

Article 3

1. Every individual shall be equal before the law.
2. Every individual shall be entitled to equal protection of the law.

Article 4

Human beings are inviolable. Every human being shall be entitled to respect for his life and the integrity of his person. No one may be arbitrarily deprived of this right.

Article 5

Every individual shall have the right to the respect of the dignity inherent in a human being and to the recognition of his legal status. All forms of exploitation and degradation of man particularly slavery, slave trade, torture, cruel, inhuman or degrading punishment and treatment shall be prohibited.

Article 6

Every individual shall have the right to liberty and to the security of his person. No one may be deprived of his freedom except for reasons and conditions previously laid down by law. In particular, no one may be arbitrarily arrested or detained.

Article 7

1. Every individual shall have the right to have his cause heard. This comprises:

(*a*) the right to an appeal to competent national organs against acts of violating his fundamental rights as recognized and guaranteed by conventions, laws, regulations and customs in force;

(*b*) the right to be presumed innocent until proved guilty by a competent court or tribunal;

(*c*) the right to defence, including the right to be defended by counsel of his choice;

(*d*) the right to be tried within a reasonable time by an impartial court or tribunal.

2. No one may be condemned for an act or omission which did not constitute a legally punishable offence for which no provision was made at the time it was committed. Punishment is personal and can be imposed only on the offender.

Article 8

Freedom of conscience, the profession and free practice of religion shall be guaranteed. No one may, subject to law and order, be submitted to measures restricting the exercise of these freedoms.

Article 9

1. Every individual shall have the right to receive information.
2. Every individual shall have the right to express and disseminate his opinions within the law.

Article 10

1. Every individual shall have the right to free association provided that he abides by the law.
2. Subject to the obligation of solidarity provided for in Article 29 no one may be compelled to join an association.

Article 11

Every individual shall have the right to assemble freely with others. The exercise of this right shall be subject only to necessary restrictions provided for by law in particular those enacted in the interest of national security, the safety, health, ethics and rights and freedoms of others.

Article 12

1. Every individual shall have the right to freedom of movement and residence within the borders of a State provided he abides by the law.
2. Every individual shall have the right to leave any country including his own, and to return to his country. This right may only be subject to restrictions provided for by law for the protection of national security, law and order, public health or morality.
3. Every individual shall have the right, when persecuted, to seek and obtain asylum in other countries in accordance with laws of those countries and international conventions.
4. A non-national legally admitted in a territory of a State Party to the present Charter, may only be expelled from it by virtue of a decision taken in accordance with the law.
5. The mass expulsion of non-nationals shall be prohibited. Mass expulsion shall be that which is aimed at national, racial, ethnic or religious groups.

Article 13

1. Every citizen shall have the right to participate freely in the government of his country, either directly or through freely chosen representatives in accordance with the provisions of the law.
2. Every citizen shall have the right of equal access to the public service of his country.

3. Every individual shall have the right of access to public property and services in strict equality of all persons before the law.

Article 14

The right to property shall be guaranteed. It may only be encroached upon in the interest of public need or in the general interest of the community and in accordance with the provisions of appropriate laws.

Article 15

Every individual shall have the right to work under equitable and satisfactory conditions, and shall receive equal pay for equal work.

Article 16

1. Every individual shall have the right to enjoy the best attainable state of physical and mental health.
2. States parties to the present Charter shall take the necessary measures to protect the health of their people and to ensure that they receive medical attention when they are sick.

Article 17

1. Every individual shall have the right to education.
2. Every individual may freely, take part in the cultural life of his community.
3. The promotion and protection of morals and traditional values recognized by the community shall be the duty of the State.

Article 18

1. The family shall be the natural unit and basis of society. It shall be protected by the State which shall take care of its physical health and moral.
2. The State shall have the duty to assist the family which is the custodian of morals and traditional values recognized by the community.
3. The State shall ensure the elimination of every discrimination against women and also ensure the protection of the rights of the woman and the child as stipulated in international declarations and conventions.
4. The aged and the disabled shall also have the right to special measures of protection in keeping with their physical or moral needs.

Article 19

All peoples shall be equal; they shall enjoy the same respect and shall have the same rights. Nothing shall justify the domination of a people by another.

Article 20

1. All peoples shall have the right to existence. They shall have the unquestionable and inalienable right to self-determination. They shall freely determine their political status and shall pursue their economic and social development according to the policy they have freely chosen.
2. Colonized or oppressed peoples shall have the right to free themselves from the

bonds of domination by resorting to any means recognized by the international community.

3. All peoples shall have the right to the assistance of the States parties to the present Charter in their liberation struggle against foreign domination, be it political, economic or cultural.

Article 21

1. All peoples shall freely dispose of their wealth and natural resources. This right shall be exercised in the exclusive interest of the people. In no case shall a people be deprived of it.

2. In case of spoliation the dispossessed people shall have the right to the lawful recovery of its property as well as to an adequate compensation.

3. The free disposal of wealth and natural resources shall be exercised without prejudice to the obligation of promoting international economic cooperation based on mutual respect, equitable exchange and the principles of international law.

4. States parties to the present Charter shall individually and collectively exercise the right to free disposal of their wealth and natural resources with a view to strengthening African unity and solidarity.

5. States parties to the present Charter shall undertake to eliminate all forms of foreign economic exploitation particularly that practiced by international monopolies so as to enable their peoples to fully benefit from the advantages derived from their national resources.

Article 22

1. All peoples shall have the right to their economic, social and cultural development with due regard to their freedom and identity and in the equal enjoyment of the common heritage of mankind.

2. States shall have the duty, individually or collectively, to ensure the exercise of the right to development.

Article 23

1. All peoples shall have the right to national and international peace and security. The principles of solidarity and friendly relations implicitly affirmed by the Charter of the United Nations and reaffirmed by that of the Organization of African Unity shall govern relations between States.

2. For the purpose of strengthening peace, solidarity and friendly relations, States parties to the present Charter shall ensure that:

(*a*) any individual enjoying the right of asylum under Article 12 of the present Charter shall not engage in subversive activities against his country of origin or any other State party to the present Charter;

(*b*) their territories shall not be used as bases for subversive or terrorist activities against the people of any other State party to the present Charter.

Article 24

All peoples shall have the right to a general satisfactory environment favorable to their development.

Article 25

States parties to the present Charter shall have the duty to promote and ensure through teaching, education and publication, the respect of the rights and freedoms contained in the present Charter and to see to it that these freedoms and rights as well as corresponding obligations and duties are understood.

Article 26

States parties to the present Charter shall have the duty to guarantee the independence of the Courts and shall allow the establishment and improvement of appropriate national institutions entrusted with the promotion and protection of the rights and freedoms guaranteed by the present Charter.

CHAPTER II. DUTIES

Article 27

1. Every individual shall have duties towards his family and society, the State and other legally recognized communities and the international community.

2. The rights and freedoms of each individual shall be exercised with due regard to the rights of others, collective security, morality and common interest.

Article 28

Every individual shall have the duty to respect and consider his fellow beings without discrimination, and to maintain relations aimed at promoting, safeguarding and reinforcing mutual respect and tolerance.

Article 29

The individual shall also have the duty:

1. To preserve the harmonious development of the family and to work for the cohesion and respect of the family; to respect his parents at all times, to maintain them in case of need;

2. To serve his national community by placing his physical and intellectual abilities at its service;

3. Not to compromise the security of the State whose national or resident he is;

4. To preserve and strengthen social and national solidarity, particularly when the latter is threatened;

5. To preserve and strengthen the national independence and the territorial integrity of his country and to contribute to its defence in accordance with the law;

6. To work to the best of his abilities and competence, and to pay taxes imposed by law in the interest of the society;

7. To preserve and strengthen positive African cultural values in his relations with other members of the society, in the spirit of tolerance, dialogue and consultation and, in general, to contribute to the promotion of the moral well being of society;

8. To contribute to the best of his abilities, at all times and at all levels, to the promotion and achievement of African unity.

. . .

VIENNA CONVENTION ON THE LAW OF TREATIES

. . .

Article 2

1. For the purposes of the present Convention:

(*a*) 'Treaty' means an international agreement concluded between States in written form and governed by international law, whether embodied in a single instrument or in two or more related instruments and whatever its particular designation;

(*b*) 'Ratification', 'acceptance', 'approval' and 'accession' means in each case the international act so named whereby a State establishes on the international plane its consent to be bound by a treaty;

. . .

(*d*) 'Reservation' means a unilateral statement, however phrased or named, made by a State, when signing, ratifying, accepting, approving or acceding to a treaty, whereby it purports to exclude or to modify the legal effect of certain provisions of the treaty in their application to that State;

. . .

Article 18

A State is obliged to refrain from acts which would defeat the object and purpose of a treaty when:

(*a*) It has signed the treaty or has exchanged instruments constituting the treaty subject to ratification, acceptance or approval, until it shall have made its intention clear not to become a party to the treaty; or

(*b*) It has expressed its consent to be bound by the treaty, pending the entry into force of the treaty and provided that such entry into force is not unduly delayed.

Article 19.

A State may, when signing, ratifying, accepting, approving or acceding to a treaty, formulate a reservation unless:

(*a*) The reservation is prohibited by the treaty;

(*b*) The treaty provides that only specified reservations, which do not include the reservation in question, may be made; or

(*c*) In cases not falling under sub-paragraphs (*a*) and (*b*), the reservation is incompatible with the object and purpose of the treaty.

Article 20

1. A reservation expressly authorized by a treaty does not require any subsequent acceptance by the other contracting States unless the treaty so provides.

2. When it appears from the limited number of the negotiating States and the object and purpose of a treaty that the application of the treaty in its entirety between all the

parties is an essential condition of the consent of each one to be bound by the treaty, a reservation requires acceptance by all the parties;

. . .

4. In cases not falling under the preceding paragraphs and unless the treaty otherwise provides:

(*a*) Acceptance by another contracting State of a reservation constitutes the reserving State a party to the treaty in relation to that other State if or when the treaty is in force for those States;

(*b*) An objection by another contracting State to a reservation does not preclude the entry into force of the treaty as between the objecting and reserving States unless a contrary intention is definitely expressed by the objecting State;

. . .

Article 21

1. A reservation established with regard to another party in accordance with articles 19, 20 and 23:

(*a*) Modifies for the reserving State in its relations with that other party the provisions of the treaty to which the reservation relates to the extent of the reservation; and

(*b*) Modifies those provisions to the same extent for that other party in its relations with the reserving State.

2. The reservation does not modify the provisions of the treaty for the other parties to the treaty *inter se*.

3. When a State objecting to a reservation has not opposed the entry into force of the treaty between itself and the reserving State, the provisions to which the reservation relates do not apply as between the two States to the extent of the reservation.

. . .

Article 26

Every treaty in force is binding upon the parties to it and must be performed by them in good faith.

Article 27

A party may not invoke the provisions of its internal law as justification for its failure to perform a treaty. . . .

. . .

Article 31

1. A treaty shall be interpreted in good faith in accordance with the ordinary meaning to be given to the terms of the treaty in their context and in the light of its object and purpose.

. . .

3. There shall be taken into account, together with the context:

(*a*) Any subsequent agreement between the parties regarding the interpretation of the treaty or the application of its provisions;

(*b*) Any subsequent practice in the application of the treaty which establishes the agreement of the parties regarding its interpretation;

(*c*) Any relevant rules of international law applicable in the relations between the parties.

4. A special meaning shall be given to a term if it is established that the parties so intended.

Article 32

Recourse may be had to supplementary means of interpretation, including the preparatory work of the treaty and the circumstances of its conclusion, in order to confirm the meaning resulting from the application of article 31, or to determine the meaning when the interpretation according to article 31:

(*a*) Leaves the meaning ambiguous or obscure; or

(*b*) Leads to a result which is manifestly absurd or unreasonable.

. . .

Article 53.

A treaty is void if, at the time of its conclusion, it conflicts with a peremptory norm of general international law. For the purposes of the present Convention, a peremptory norm of general international law is a norm accepted and recognized by the international community of States as a whole as a norm from which no derogation is permitted and which can be modified only by a subsequent norm of general international law having the same character.

CONSTITUTION OF THE UNITED STATES

We the People of the United States, in Order to form a more perfect Union, establish Justice, insure domestic Tranquility, provide for the common defence, promote the general Welfare, and secure the Blessings of Liberty to ourselves and our Posterity, do ordain and establish this Constitution for the United States of America.

Article I

SECTION 1. All legislative Powers herein granted shall be vested in a Congress of the United States, which shall consist of a Senate and House of Representatives.

. . .

SECTION 7. . . . Every Bill which shall have passed the House of Representatives and the Senate, shall, before it become a Law, be presented to the President of the United States; If he approve he shall sign it, but if not he shall return it, with his Objections to that House in which it shall have originated, who shall enter the Objections at large on their Journal, and proceed to reconsider it. If after such Reconsideration two thirds of that House shall agree to pass the Bill, it shall be sent, together with the Objections, to the other House, by which it shall likewise be reconsidered, and if approved by two thirds of that House, it shall become a Law. . . .

. . .

SECTION 8. The Congress shall have Power To lay and collect Taxes, Duties, Imposts and Excises, to pay the Debts and provide for the common Defence and general Welfare of the United States. . . .

To regulate Commerce with foreign Nations, and among the several States, and with the Indian Tribes;

. . .

To define and punish Piracies and Felonies committed on the high Seas, and Offences against the Law of Nations;

To declare War, grant Letters of Marque and Reprisal, and make Rules concerning Captures on Land and Water;

. . .

To make all Laws which shall be necessary and proper for carrying into Execution the foregoing Powers, and all other Powers vested by this Constitution in the Government of the United States, or in any Department or Officer thereof.

SECTION 9. . . .

The privilege of the Writ of Habeas Corpus shall not be suspended, unless when in Cases of Rebellion or Invasion the public Safety may require it.

No Bill of Attainder or ex post facto Law shall be passed.

. . .

No money shall be drawn from the Treasury, but in Consequence of Appropriations made by Law. . . .

. . .

Article II

SECTION 1. The executive Power shall be vested in a President of the United States of America. . . .

. . .

SECTION 2. The President shall be Commander in Chief of the Army and Navy of the United States, and of the Militia of the several States, when called into the actual Service of the United States . . . .

He shall have Power, by and with the Advice and Consent of the Senate, to make Treaties, provided two thirds of the Senators present concur; and he shall nominate, and by and with the Advice and Consent of the Senate, shall appoint Ambassadors, other public Ministers and Consuls, Judges of the supreme Court, and all other Officers of the United States, whose Appointments are not herein otherwise provided for, and which shall be established by Law. . .

. . .

SECTION 3. . . . [H]e shall receive Ambassadors and other public Ministers; he shall take Care that the Laws be faithfully executed, and shall Commission all the Officers of the United States.

. . .

Article III

SECTION 1. The judicial Power of the United States, shall be vested in one supreme Court, and in such inferior Courts as the Congress may from time to time ordain and establish. . . .

SECTION 2. The judicial Power shall extend to all Cases, in Law and Equity, arising under this Constitution, the Laws of the United States, and Treaties made, or which shall be made, under their Authority;—to all Cases affecting Ambassadors, other public Ministers and Consuls;—to all Cases of admiralty and maritime Jurisdiction;—to Controversies to which the United States shall be a Party;—to Controversies between two or more States;—between a State and Citizens of another State;—between Citizens of different States;—between Citizens of the same State claiming Lands under Grants of different States, and between a State, or the Citizens thereof, and foreign States, Citizens or Subjects.

. . .

The trial of all Crimes, except in Cases of Impeachment, shall be by Jury; and such Trial shall be held in the State where the said Crimes shall have been committed; but when not committed within any State, the Trial shall be at such Place or Places as the Congress may by Law have directed.

. . .

Article V

The Congress, whenever two thirds of both Houses shall deem it necessary, shall propose Amendments to this Constitution, or, on the Application of the Legislatures of two thirds of the several States, shall call a Convention for proposing Amendments, which, in either Case, shall be valid to all Intents and Purposes, as part of this Constitution, when ratified by the Legislatures of three fourths of the several States, or by Conventions in three fourths thereof, as the one or the other Mode of Ratification may be proposed by the Congress. . . .

Article VI

This Constitution, and the Laws of the United States which shall be made in Pursuance thereof; and all Treaties made, or which shall be made, under the Authority of the United States, shall be the supreme Law of the Land; and the Judges in every State shall be bound thereby, any Thing in the Constitution or Laws of any State to the Contrary notwithstanding.

. . .

Articles in Addition to, and Amendment of, the Constitution of the United States of America, Proposed by Congress and Ratified by the Several States, Pursuant to the Fifth Article of the Original Constitution

Amendment I

Congress shall make no law respecting an establishment of religion, or prohibiting the free exercise thereof; or abridging the freedom of speech, or of the press; or the right of the people peaceably to assemble, and to petition the Government for a redress of grievances.

. . .

Amendment IV

The right of the people to be secured in their persons, houses, papers, and effects, against unreasonable searches and seizures, shall not be violated, and no Warrants shall issue, but upon probable cause, supported by Oath or affirmation, and particularly describing the place to be searched, and the persons or things to be seized.

Amendment V

No person shall be held to answer for a capital, or otherwise infamous crime, unless on a presentment or indictment of a Grand Jury, except in cases arising in the land or naval forces, or in the Militia, when in actual service in time of War or public danger; nor shall any person be subject for the same offence to be twice put in jeopardy of life or limb; nor shall be compelled in any criminal case to be a witness against himself, nor be deprived of life, liberty, or property, without due process of law; nor shall private property be taken for public use, without just compensation.

Amendment VI

In all criminal prosecutions, the accused shall enjoy the right to a speedy and public trial, by an impartial jury of the State and district wherein the crime shall have been committed, which district shall have been previously ascertained by law, and to be informed of the nature and cause of the accusation; to be confronted with the witnesses against him; to have compulsory process for obtaining witnesses in his favor, and to have the Assistance of Counsel for his defence.

Amendment VII

In suits at common law, where the value in controversy shall exceed twenty dollars, the right of trial by jury shall be preserved, and no fact tried by jury, shall be otherwise re-examined in any Court of the United States, than according to the rules of the common law.

Amendment VIII

Excessive bail shall not be required, nor excessive fines imposed, nor cruel and unusual punishments inflicted.

Amendment IX

The enumeration in the Constitution, of certain rights, shall not be construed to deny or disparage others retained by the people.

Amendment X

The powers not delegated to the United States by the Constitution, nor prohibited by it to the States, are reserved to the States respectively, or to the people.

. . .

Amendment XIV

SECTION 1. All persons born or naturalized in the United States, and subject to the jurisdiction thereof, are citizens of the United States and of the State wherein they reside. No State shall make or enforce any law which shall abridge the privileges or immunities of citizens of the United States; nor shall any State deprive any person of life, liberty, or property, without due process of law, nor deny to any person within its jurisdiction the equal protection of the laws.

. . .

SECTION 5. The Congress shall have power to enforce, by appropriate legislation, the provisions of this article.

Amendment XV

SECTION 1. The right of citizens of the United States to vote shall not be denied or abridged by the United States or by any State on account of race, color, or previous condition of servitude.

SECTION 2. The Congress shall have power to enforce this article by appropriate legislation.

. . .

Amendment XIX

The right of citizens of the United States to vote shall not be denied or abridged by the United States or by any State on account of sex.

Congress shall have power to enforce this article by appropriate legislation.

Annex on
CITATIONS

Unless otherwise indicated, details of the number of parties to a treaty are current as of 1 September 1995. The principal abbreviations used are as follows: HRLJ for *Human Rights Law Journal*, ILM for *International Legal Materials*, UNTS for *United Nations Treaty Series* and ETS for *European Treaty Series*.

Human Rights Document Collections

Brownlie, I., ed., *Basic Instruments on Human Rights*, 4th ed., Oxford: Clarendon Press, 1995.

Hamalengwa, M., C. Flinterman and E. Dankwa, *The International Law of Human Rights in Africa: Basic Documents and Annotated Bibliography*, Dordrecht: Martinus Nijhoff, 1988.

Human Rights: A Compilation of International Instruments, 2 vols., Geneva: United Nations Centre for Human Rights, 1994.

Human Rights in International Law: Basic Texts, Strasbourg: Council of Europe Press, 1992.

International Labour Conventions and Recommendations 1919-1991, 2 vols., Geneva: International Labour Office, 1992.

Roberts, A. and R. Guelff, *Documents on the Laws of War*, Oxford: Clarendon Press, 1982.

Twenty-Five Human Rights Documents, 2nd ed., NY: Columbia University, Center for the Study of Human Rights, 1994.

United Nations Treaties

Charter of the United Nations, adopted 26 June 1945, entered into force 24 Oct. 1945, as amended by G.A. Res. 1991 (XVIII) 17 Dec. 1963, entered into force 31 Aug. 1965 (557 UNTS 143); 2101 of 20 Dec. 1965, entered into force 12 June 1968 (638 UNTS 308); and 2847 (XXVI) of 20 Dec. 1971, entered into force 24 Sept. 1973 (892 UNTS 119). 185 Member States.

Convention Against Torture and Other Cruel, Inhuman or Degrading Treatment or Punishment, adopted 10 Dec. 1984, entered into force 26 June 1987, G.A. Res. 39/46, 39 UN GAOR, Supp. (No. 51), UN Doc. A/39/51, at 197 (1984), reprinted in 23 ILM 1027 (1984), minor changes reprinted in 24 ILM 535 (1985), 5 HRLJ 350 (1984). 90 states parties.

Convention on the Elimination of All Forms of Discrimination against Women, adopted 18 Dec. 1979, entered into force 3 Sept. 1981, G.A. Res. 34/180, 34 UN GAOR, Supp. (No. 46), UN Doc. A/34/46, at 193 (1979), reprinted in 19 ILM 33 (1980). 144 states parties.

Convention on the Prevention and Punishment of the Crime of Genocide, adopted 9 Dec. 1948, entered into force 12 Jan. 1951, 78 UNTS 277. 116 states parties as at 1 Jan. 1995.

Convention on the Rights of the Child, adopted 20 Nov. 1989, entered into force 2 Sept. 1990, G.A. Res. 44/25, 44 UN GAOR, Supp. (No. 49), UN Doc. A/44/49, at 166 (1989), reprinted in 28 ILM 1448 (1989). 176 states parties.

Convention Relating to the Status of Refugees, signed 28 July 1951, entered into force 22 April 1954, 189 UNTS 150, 123 states parties as at 1 Jan. 1995.

Protocol Relating to the Status of Refugees, opened for signature 31 Jan. 1967, entered into force 4 Oct. 1967, 606 UNTS 267, reprinted in 6 ILM 78 (1967). 124 states parties as at 1 Jan. 1995.

International Convention for the Elimination of All Forms of Racial Discrimination, adopted 21 Dec. 1965, entered into force 4 Jan. 1969, 660 UNTS 195, reprinted in 5 ILM 352 (1966). 143 states parties.

International Convention on the Protection of the Rights of All Migrant Workers and Members of Their Families, adopted 18 Dec. 1990, not in force, G.A. Res. 45/158, reprinted in 30 ILM 1517 (1991). 5 states parties.

International Convention on the Suppression and Punishment of the Crime of Apartheid, adopted 30 Nov. 1973, entered into force 18 July 1976, G.A. Res. 3068 (XXVIII), reprinted in 13 ILM 50 (1974). 99 states parties as at 1 Jan. 1995.

International Covenant on Civil and Political Rights, adopted 16 Dec. 1966, entered into force 23 March 1976, G.A. Res. 2200A (XXI), UN Doc. A/6316 (1966), 999 UNTS 171, reprinted in 6 ILM 368 (1967). 131 states parties.

Optional Protocol to the International Covenant on Civil and Political Rights, adopted 16 Dec. 1966, entered into force 23 March 1976, 999 UNTS 171, reprinted in 6 ILM 383 (1967). 84 states parties.

Second Optional Protocol to the International Covenant on Civil and Political Rights, adopted 15 Dec. 1989, entered into force 11 July 1991, G.A. Res. 44/128, reprinted in 29 ILM 1464 (1990). 28 states parties.

International Covenant on Economic, Social and Cultural Rights, adopted 16 Dec. 1966, entered into force 3 Jan. 1976, G.A. Res. 2200A (XXI), UN Doc. A/6316 (1966), 993 UNTS 3, reprinted in 6 ILM 360 (1967). 132 states parties.

Supplementary Convention on the Abolition of Slavery, the Slave Trade, and Institutions and Practices Similar to Slavery, adopted 7 Sept. 1956, entered into force 30 April 1957, E.S.C. Res. 608 (XXI), 226 UNTS 3. 113 states parties as at 1 Jan. 1995.

Other Universal Treaties

Geneva Convention for the Amelioration of the Condition of the Wounded and Sick in Armed Forces in the Field, adopted 12 Aug. 1949, entered into force 21 Oct. 1950, 75 UNTS 31. 185 states parties.

Geneva Convention for the Amelioration of the Condition of Wounded, Sick and Shipwrecked Members of Armed Forces at Sea, adopted 12 Aug. 1949, entered into force 21 Oct. 1950, 75 UNTS 85. 185 states parties.

Geneva Convention Relative to the Treatment of Prisoners of War, adopted 12 Aug. 1949, entered into force 21 Oct. 1950, 75 UNTS 135. 185 states parties.

Geneva Convention Relative to the Protection of Civilian Persons in Time of War, adopted 12 Aug. 1948, entered into force 21 Oct. 1950, 75 UNTS 287. 185 states parties.

Protocol I Additional to the Geneva Conventions of August 12, 1949, and relating to the Protection of Victims of International Armed Conflicts, adopted 8 June 1977, entered into force 7 Dec. 1978, UN Doc. A/32/144 Annex I, 1125 UNTS no. 17512, reprinted in 16 ILM 1391 (1977). 138 states parties.

Protocol II Additional to the Geneva Conventions of August 12, 1949, and relating to the Protection of Victims of Non-International Armed Conflicts, adopted 8 June 1977, entered into force 7 Dec. 1978, UN Doc. A/32/144 Annex II, 1125 UNTS no. 17513, reprinted in 16 ILM 1442 (1977). 129 states parties.

Vienna Convention on the Law of Treaties, adopted 23 May 1969, entered into force 27 Jan. 1980, U.N. Doc. A/CONF. 39/26, reprinted in 8 ILM 679 (1969). 76 states parties as at 1 Jan. 1995.

Other United Nations Instruments

Declaration on the Elimination of All Forms of Intolerance and of Discrimination Based on Religion or Belief, adopted 25 Nov. 1981, G.A. Res. 36/55, 36 UN GAOR, Supp. (No. 51), UN Doc. A/36/51, at 171 (1981).

Declaration on the Elimination of Violence Against Women, adopted 20 December 1993, G.A. Res. 48/104, UN Doc. A/48/29, reprinted in 33 ILM 1049 (1994).

Declaration on the Granting of Independence to Colonial Countries and Peoples, G.A. Res. 1514 (XV), 15 UN GAOR, Supp. (No. 16), UN Doc. A/4684, at 66 (1960).

Declaration on Principles of International Law concerning Friendly Relations and Co-operation among States in accordance with the Charter of the United Nations, G.A. Res. 2625 (XXV), 25 UN GAOR, Supp. (No. 28), UN Doc. A/8028, at 121 (1970), reprinted in 9 ILM 1292 (1970).

Declaration on the Protection of All Persons from Being Subjected to Torture and Other Cruel, Inhuman or Degrading Treatment or Punishment, adopted 9 Dec. 1975, G.A. Res. 3452 (XXX), 30 UN GAOR, Supp. (No. 34), UN Doc. A/10034, at 91 (1976).

Declaration on the Protection of All Persons from Enforced Disappearance, adopted 18 Dec. 1992, G.A. Res. 47/133, 32 ILM 903 (1993).

Declaration on the Right to Development, adopted 4 Dec. 1986, G.A. Res. 41/28.

Declaration on the Rights of Disabled Persons, adopted 9 Dec. 1975, G.A. Res. 3447 (XXX), 30 UN GAOR Supp. (No. 34), UN Doc. A/10034, at 88 (1975).

Declaration on the Rights of Persons Belonging to National or Ethnic, Religious or Linguistic Minorities, adopted 18 Dec. 1992, G.A. Res. 47/135, reprinted in 32 ILM 911 (1993), 14 HRLJ 54 (1993).

Declaration on the Rights of the Child, adopted 20 Nov. 1959, G.A. Res. 1386 (XIV), 14 UN GAOR, Supp. (No. 16) UN Doc A/4354, at 19 (1959).

Standard Minimum Rules for the Treatment of Prisoners, adopted 31 July 1957, E.S.C. Res. 663C (XXIV), 24 UN ESCOR, Supp. (No. 1) at 11 (1957), extended 13 May 1977, E.S.C. Res. 2076 (LXIII), 62 UN ESCOR, Supp. (No. 1) at 35 (1977).

Universal Declaration of Human Rights, adopted 10 Dec. 1948, G.A. Res. 217A (III), UN Doc. A/810, at 71 (1948).

United Nations World Conference on Human Rights, Vienna Declaration and Programme of Action, adopted 25 June 1993, reprinted in 32 ILM 1661 (1993), 14 HRLJ 352 (1993).

Instruments adopted by United Nations Agencies

Convention Concerning Forced or Compulsory Labour (I.L.O. No. 29), adopted 28 June 1930, entered into force 1 May 1932, 39 UNTS 55. 136 states parties.

Convention Concerning Indigenous and Tribal Peoples in Independent Countries (I.L.O. No. 169), adopted 27 June 1989, entered into force 5 Sept. 1991, reprinted in 28 ILM 1382 (1989). 8 states parties.

Convention Concerning the Abolition of Forced Labour (I.L.O. No. 105), adopted 25 June 1957, entered into force 17 Jan. 1959, 320 UNTS 291. 115 states parties.

Convention Concerning the Protection and Integration of Indigenous and Other Tribal and Semi-Tribal Populations in Independent Countries (I.L.O. No. 107), adopted 26 June 1957, entered into force 2 June 1959, 328 UNTS 247. 27 states parties.

UNESCO Convention against Discrimination in Education, adopted 14 Dec. 1960, entered into force 22 May 1962, 429 UNTS 93. 85 states parties.

Regional Instruments

(a)The Council of Europe

European Convention for the Prevention of Torture and Inhuman or Degrading Treatment or Punishment, signed 26 Nov. 1987, entered into force 1 Feb. 1989, Doc. No. H(87)4 1987, ETS 126, reprinted in 27 ILM 1152 (1988); 9 HRLJ 359 (1988). 30 states parties.

European Convention for the Protection of Human Rights and Fundamental Freedoms, signed 4 Nov. 1950, entered into force 3 Sept. 1953, 213 UNTS 221, ETS 5. 31 states parties.

Protocol No. 1 to the European Convention for the Protection of Human Rights and Fundamental Freedoms, adopted 20 March 1952, entered into force 18 May 1954, ETS 9. 28 states parties.

Protocol No. 2 to the European Convention for the Protection of Human Rights and Fundamental Freedoms, adopted 6 May 1963, entered into force 21 Sept. 1970, ETS 44. 31 states parties.

Protocol No. 3 to the European Convention for the Protection of Human Rights and Fundamental Freedoms, adopted 6 May 1963, entered into force 21 Sept. 1970, ETS 45. 31 states parties.

Protocol No. 4 to the European Convention for the Protection of Human Rights and Fundamental Freedoms, adopted 16 Sept. 1963, entered into force 2 May 1968, ETS 46. 23 states parties.

Protocol No. 5 to the European Convention for the Protection of Human Rights and Fundamental Freedoms, adopted 20 Jan. 1966, entered into force 20 Dec. 1971, ETS 55. 31 states parties.

Protocol No. 6 to the European Convention for the Protection of Human Rights and Fundamental Freedoms, adopted 28 April 1983, entered into force 1 March 1985, ETS 114, reprinted in 22 ILM 539 (1983); 6 HRLJ 77 (1985). 23 states parties.

Protocol No. 7 to the European Convention for the Protection of Human Rights and Fundamental Freedoms, adopted 22 Nov. 1984, entered into force 1 Nov. 1988, ETS 117, reprinted in 24 ILM 435 (1985); 6 HRLJ 80 (1985). 17 states parties.

Protocol No. 8 to the European Convention for the Protection of Human Rights and Fundamental Freedoms, adopted 19 March 1985, entered into force 1 Jan. 1990, ETS 118, reprinted in 25 ILM 387 (1986); 6 HRLJ 88 (1985). 31 states parties.

Protocol No. 9 to the European Convention for the Protection of Human Rights and Fundamental Freedoms, adopted 6 Nov. 1990, not in force, ETS 40 reprinted in 30 ILM 693 (1991). 19 states parties.

Protocol No. 10 to the European Convention for the Protection of Human Rights and Fundamental Freedoms, adopted 25 March 1992, entered into force 1 Oct 1994, ETS 146, reprinted in 13 HRLJ 182 (1992). 21 states parties.

Protocol No. 11 to the European Convention for the Protection of Human Rights and Fundamental Freedoms, adopted 11 May 1994, not in force, ETS 155, reprinted in 33 ILM 960 (1994), 15 HRLJ 86 (1994). 15 states parties.

European Social Charter, signed 18 Oct. 1961, entered into force 26 Feb. 1965, 529 UNTS 89; ETS 35: Protocol amending Charter, adopted 21 Oct. 1991, not in force, ETS 142, reprinted in 31 ILM 155 (1992). 20 states parties.

Additional Protocol to the European Social Charter, adopted 5 May 1988, not in force, ETS 128. 5 states parties.

Protocol Amending the European Social Charter, adopted 21 October 1991, not in force, ETS 142. 10 states parties.

Framework Convention for the Protection of National Minorities, adopted 10 Nov. 1994, opened for signature 1 Feb. 1995, not in force, ETS 157, reprinted in 34 ILM 351 (1995). 3 states parties.

Statute of the Council of Europe, adopted 5 May 1949, entered into force 3 Aug. 1949, ETS 1. 36 states parties.

(b)The Organization of American States

American Convention on Human Rights (Pact of San Jose), signed 22 Nov. 1969, entered into force 18 July 1978, OASTS 36, O.A.S. Off. Rec. OEA/Ser.L/V/II.23, doc.21, rev.6 (1979), reprinted in 9 ILM 673 (1970). 25 states parties.

Additional Protocol to the American Convention on Human Rights in the Area of Economic, Social and Cultural Rights (Protocol of San Salvador), adopted 17 Nov. 1988, not in force, OASTS 69, reprinted in 28 ILM 156 (1989), corrections at 28 ILM 573 and 1341 (1989). 4 states parties.

Protocol to the American Convention on Human Rights to Abolish the Death Penalty, adopted 8 June 1990, entered into force 28 Aug. 1991, OASTS 73, reprinted in 29 ILM 1447 (1990). 3 states parties.

American Declaration of the Rights and Duties of Man, signed 2 May 1948, OEA/Ser.L./V/II.71, at 17 (1988).

Charter of the Organization of American States, signed 1948, entered into force 13 Dec. 1951, amended 1967, 1985, 14 Dec. 1992, 10 June 1993. Integrated text of the Charter as amended by the Protocols of Buenos Aires and Cartagena de Indias; The Protocol of Amendment of Washington; and the Protocol of Amendment of Managua, reprinted in 33 ILM 981 (1994). 35 member states.

Inter-American Convention on the Forced Disappearance of Persons, signed 9 June 1994, not in force, reprinted in 33 ILM 1529 (1994). 0 states parties.

Inter-American Convention on the Prevention, Punishment and Eradication of Violence Against Women, signed 9 June 1994, entered into force 3 March 1995, reprinted in 33 ILM 1534 (1994). 3 states parties.

Inter-American Convention to Prevent and Punish Torture, signed 9 Dec. 1985, entered into force 28 Feb. 1987, OASTS 67, GA Doc. OEA/Ser.P, AG/doc.2023/85 rev.1 (1986) pp 46-54, reprinted in 25 ILM 519 (1986). 13 states parties.

(c) The Organization of African Unity

African Charter on Human and Peoples' Rights adopted 27 June 1981, entered into force 21 Oct. 1986, O.A.U. Doc. CAB/LEG/67/3 Rev. 5, reprinted in 21 ILM 58 (1982); 7 HRLJ 403 (1986). 49 states parties.

African Charter on the Rights and Welfare of the Child, adopted July 1990, not in force, O.A.U. Doc. CAB/LEG/TSG/Rev. 1.

Charter of the Organization of African Unity, adopted 25 May 1963, 47 UNTS 39, reprinted in 2 ILM 766 (1963). 51 states parties.

Convention Governing the Specific Aspects of the Refugee Problems in Africa, adopted 10 Sept. 1969, entered into force 20 June 1974, 1001 UNTS 45, reprinted in 8 ILM 1288 (1969). 42 states parties.

(d) The Organization on Security and Co-operation in Europe

Document of the Copenhagen Meeting of the Conference on the Human Dimension of the Conference of Security and Co-operation in Europe, adopted 29 June 1990, reprinted in 29 ILM 1305 (1990); 11 HRLJ 232 (1990).

Document of the Moscow Meeting of the Conference on the Human Dimension of the Conference on Security and Co-operation in Europe, adopted 3 Oct. 1991, reprinted in 30 ILM 1670 (1991).

Final Act of the Conference on Security and Co-operation in Europe, adopted 1 Aug. 1975, reprinted in 14 ILM 1292 (1975).

Human Rights in the Concluding Document of the Vienna Conference on Security and Co-operation in Europe, adopted 15 Jan. 1989, reprinted in 10 HRLJ 270 (1989).

Annex on BIBLIOGRAPHY

This bibliography draws on but modifies and updates information in Jack Tobin and Jennifer Green, *Guide to Human Rights Research* (1994), prepared by the staff of and published by the Harvard Law School Human Rights Program. The following list, restricted to materials in English, is suggestive and not meant to be complete. It does not include compilations of human rights documents, which appear in the accompanying Annex on Citations.

A. Current Human Rights Developments

Newspapers such as the *New York Times* and periodicals such as *The Economist* with good international coverage are invaluable for keeping abreast of events bearing on human rights. Listed below are some periodicals that systematically provide updates on formal developments such as ratification of human rights treaties, significant action by human rights bodies, or relevant court decisions or new statutes.

American Society of International Law. *International Legal Materials*. Washington, DC: Society of International Law, 1962– .

Human Rights Internet Reporter. Ottawa: Human Rights Internet, 1976– .

Human Rights Law Journal. Kehl am Rhein; Arlington, VA: N.P. Engel, 1980– .

Human Rights Monitor. Geneva: International Service for Human Rights, 1988–.

Human Rights Quarterly. Baltimore: Johns Hopkins University Press, 1981– .

Human Rights Tribune. Ottawa: Human Rights Internet, 1992– .

Human Rights Reports. Nottingham: Human Rights Law Centre, University of Nottingham 1994– .

The Review. Geneva: International Commission of Jurists, 1969– .

The Women's Watch. Minnneapolis, MN: International Women's Rights Action Watch, Humphrey Institute of Public Affairs, 1987– .

B. General Scholarly Journals that Include Articles on Human Rights

American Journal of International Law. Washington, D.C.: American Society of International Law, 1907– .

European Journal of International Law. Florence, Italy: European University Institute, 1990– .

International and Comparative Law Quarterly. London: British Institute of International and Comparative Law, 1952– .

C. Principal Scholarly Journals on Human Rights

Australian Journal of Human Rights. Sydney, NSW: Human Rights Centre, 1994–.

Canadian Human Rights Yearbook. Toronto: Carswell, 1983–.

Columbia Human Rights Law Review. New York: Columbia University School of Law, 1972– .

Harvard Human Rights Journal. Cambridge, MA: Harvard Law School, 1988– .

Human Rights Law Journal. Kehl am Rhein: N.P. Engel, 1980– .

Human Rights Quarterly. Baltimore, MD: Johns Hopkins University Press, 1979– .

Interntional Human Rights Reports. Nottingham, U.K: Human Rights Law Centre, 1994– .

Israel Yearbook on Human Rights. Tel Aviv: Tel Aviv University Faculty of Law, 1971– .

Netherlands Quarterly of Human Rights. Utrecht: Studie-en Informatiecentrum Mensenrechten, 1983– .

New York Law School Journal of Human Rights. New York: New York Law School, 1983– .

The Review. Geneva: International Commission of Jurists, 1969– .

South African Journal on Human Rights. Johannesburg: Centre for Applied Legal Studies, University of Witwatersrand, 1985– .

D. General Treatises on International Law with Sections on Human Rights

Akehurst, M. *A Modern Introduction to International Law*. 6th ed. London: HarperCollins Academic, 1987.

Bernhardt, R., ed. *Encyclopedia of Public International Law*. Amsterdam; New York: North Holland Publishing Co., 1981–. (See especially Vols. 5, 7 and 8.)

Brownlie, I. *Principles of Public International Law*. 4th ed. New York: Oxford University Press, 1990.

Dugard, J. *International Law: A South African Perspective*. Cape Town: Juta, 1994.

Henkin, L. *International Law: Politics, Values and Functions*. Dordrecht: M. Nijhoff, 1995.

Jennings, R. and A. Watts. *Oppenheim's International Law*. 9th ed. Harlow, Essex, England: Longman Group UK Limited, 1992.

Schachter, O. *International Law in Theory and Practice*. Dordrecht; Boston; London: M. Nijhoff, 1991.

Shaw, M. *International Law*. 3rd ed. Cambridge: Grotius, 1991.

Shearer, I. *Starke's International Law*. 11th ed. London: Butterworth's, 1994.

Simma, B. et al (eds), *The Charter of the United Nations: A Commentary*. Oxford: Oxford University Press, 1994.

E. General Readings on Human Rights

Alston, P., ed. *The United Nations and Human Rights: A Critical Appraisal*. Oxford: Clarendon Press, New York: Oxford University Press, 1992.

Alston, P., ed. *Promoting Human Rights through Bills of Rights: Comparative Perspectives*. Oxford: Clarendon Press, 1995.

Alston, P., *The Convention on the Rights of the Child: A Commentary*, 2 vols, Geneva, United Nations and UNICEF, 1995.

Alston, P. and K. Tomasevski, eds. *The Right to Food*. Boston: M. Nijhoff; Utrecht: Stichting Studie- en Informatiecentrum Mensenrechten, 1984.

An-Na'im, A., ed. *Human Rights in Cross-Cultural Perspectives: A Quest for Consensus*. Philadelphia: University of Pennsylvania Press, 1992.

An-Na'im, A. *Toward an Islamic Reformation: Civil Liberties, Human Rights, and International Law*. Syracuse, NY: Syracuse University Press, 1990.

An-Na'im, A. and F. Deng, eds. *Human Rights in Africa: Cross-Cultural Perspectives*. Washington, DC: The Brookings Institution, 1990.

Bassiouni, M., ed. *International Criminal Law*, 3 vols. Dobbs Ferry, NY: Transnational Publishers, 1986. (Vol. 1: Crimes; Vol. 2: Procedure; Vol. 3: Enforcement.)

Bassiouni, M. *Crimes Against Humanity in Internationmal Criminal Law*. Dordrecht, Boston: M. Nijhoff, 1992.

Beatty, D., ed. *Human Rights and Judicial Review: A Comparative Perspective*. Dordrecht: M. Nijhoff, 1994.

Brownlie, I. *Treaties and Indigenous Peoples*. Oxford: Clarendon Press; New York: Oxford University Press, 1992.

Buergenthal, T. *International Human Rights in a Nutshell*. 2nd ed. St. Paul, MN: West Publishing Co., 1995.

Buergenthal, T. et al. *Protecting Human Rights in the Americas*. 3rd ed. Kehl am Rhein: N.P. Engel, 1990.

Cassese, A. *International Law in a Divided World*. Oxford: Clarendon Press; New York: Oxford University Press, 1986.

Cassese, A. *Human Rights in a Changing World*. Philadelphia: Temple University Press, 1990.

Cassese, A., ed. *The International Fight Against Torture*. Baden-Baden: Nomos Verlagsgesellschaft, 1991.

Cassese, A., et al. *Human Rights and The European Community*, 3 vols. Baden-Baden: Nomos Verlagsgesellschaft, 1991.

Centre for Human Rights. *United Nations Action in the Field of Human Rights*. New York: United Nations, 1995.

Clapham, A. *Human Rights in the Private Sphere*. Oxford: Clarendon Press, 1993.

Claude, R.P., and B.H. Weston, eds. *Human Rights in the World Community*. 2nd ed. Philadelphia: University of Pennsylvania Press, 1992.

Cook, R., ed. *Human Rights of Women: National and International Perspectives*. Philadelphia: University of Pennsylvania Press, 1994.

Craven, M. *The International Covenant on Economic, Social and Cultural Rights: A Perspective on Development*. Oxford: Clarendon Press, 1995.

Crawford, J. ed. *The Rights of Peoples*. Oxford: Clarendon Press; New York: Oxford University Press, 1992.

Degener, T. and Y. Koster-Dreese, eds. *Human Rights and Disabled Persons: Essays and Relevant Human Rights Instruments*. Dordrecht; Boston: Federatie Nederlandse Gehandicaptenraad; M. Nijhoff, 1995.

Delissen, A. et al., eds. *Humanitarian Law of Armed Conflict: Challenges Ahead. Essays in Honor of Frits Kalshoven*. Boston; Dordrecht: M. Nijhoff, 1991.

Diemer, A. et al. *Philosophical Foundations of Human Rights*. Paris: United Nations Educational, Scientific and Cultural Organization, 1986.

Dijk, P. van and G.J.H. van Hoof. *Theory and Practice of the European Convention on Human Rights*. 2nd ed. Deventer; Boston: Kluwer Law and Taxation Publishers, 1990.

Donnelly, J. *The Concept of Human Rights*. New York: St. Martin's Press, 1985.

Donnelly, J. *International Human Rights*. Boulder, CO: Westview Press, 1993.

Donnelly, J. *Universal Human Rights in Theory and Practice*. Ithaca, NY: Cornell University Press, 1989.

Eide, A., C. Krause and A. Rosas, *Economic, Social and Cultural Rights: A Textbook*. Dordrecht: M. Nijhoff, 1995.

Fitzpatrick, J. *Human Rights in Crisis: The International System for Protecting Human Rights During States of Emergency*. Philadelphia: University of Penn - sylvania Press, 1994.

Gomien, D., ed. *Broadening the Frontiers of Human Rights: Essays in Honour of Asbjørn Eide*. Oslo: Scandinavian University Press; New York: Oxford University Press, 1993.

Goodwin-Gill, G. *The Refugee in International Law*. Oxford: Clarendon Press; New York: Oxford University Press, 1983.

Hannum, H. *Autonomy, Sovereignty and Self-Determination: The Accommodation of Conflicting Rights*. Philadelphia: University of Pennsylvania Press, 1990.

Hannum, H., ed. *Guide to International Human Rights Practice*. 2nd ed. Philadelphia: University of Pennsylvania Press, 1992.

Harris, D. *The Euorpean Social Charter*. Charlottesville, VA: University Press of Virginia, 1984.

Hathaway, J. *The Law of Refugee Status*. Toronto: Butterworths; Austin, TX: Butterworth Legal Publishers, 1991.

Heinze, Eric. *Sexual Orientation: A Human Right*. Dordrecht; Boston: M. Nijhoff, 1995.

Henkin, L. *The Age of Rights*. New York: Columbia University Press, 1990.

Henkin, L., ed. *The International Bill of Rights: The Covenant on Civil and Political Rights*. New York: Columbia University Press, 1981.

Henkin, L. and J. Hargrove, eds. *Human Rights: An Agenda for the Next Century*. Washington, DC: American Society of International Law, 1994.

Higgins, R. *Problems and Process: International Law and How We Use It*. Oxford: Clarendon Press, 1994.

International Dimensions of Humanitarian Law. Geneva: Henry Dunant Institute; Dordrecht; Boston: M. Nijhoff, 1988.

Jabine, T. and R. Claude, eds. *Human Rights and Statistics: Getting the Record Straight*. Philadelphia: University of Pennsylvania Press, 1992.

Janis, M., R. Kay and A. Bradley. *European Human Rigths*. Oxford University Press, 1995.

Laqueur, W. and B. Rubin, eds. *The Human Rights Reader*. 2nd ed. New York: New American Library, 1989.

Lauterpacht, H. *An International Bill of the Rights of Man*. New York: Columbia University Press, 1945.

Lawson, E., ed. *Encyclopedia of Human Rights*. New York: Taylor & Francis, 1991.

Macdonald, R.St.J. et al., eds. *The European System for the Protection of Human Rights*. Dordrecht: M. Nijhoff, 1993.

Mayer, A. *Islam and Human Rights: Tradition and Politics*. Boulder, CO: Westview Press, 1991.

McDougall, M., H. Lasswell and L. Chen. *Human Rights and World Public Order*. New Haven, CT: Yale University Press, 1980.

McGoldrick, D. *The Human Rights Committee*. Oxford: Clarendon Press; New York: Oxford University Press, 1991.

Medina Quiroga, C. *The Battle of Human Rights: Gross, Systematic Violations and the Inter-American System*. Dordrecht: M. Nijhoff, 1988.

Meron, T. *Human Rights in Internal Strife: Their International Protection*. Cambridge, U.K.: Grotius, 1987.

Meron, T. *Human Rights and Humanitarian Norms as Customary Law*. Oxford: Clarendon Press; New York: Oxford University Press, 1989.

Meron, T., ed. *Human Rights in International Law: Legal and Policy Issues*. 2 vols. Oxford: Clarendon Press, 1984.

Meron, T. *Human Rights Law-Making in the United Nations*. New York: Oxford University Press, 1986.

Merrills, J. *The Development of International Law by the European Court of Human Rights*. 2nd ed. New York: Manchester University Press, 1993.

Morris, V. and M. Schaft. *An Insider's Guide to the International Criminal Tribunal for the Former Yugoslavia: A Documentary History and Analysis*. Irvington-on-Hudson, NY: Transnational Publishers, 1995.

Mullerson, R. *International Law, Rights and Politics: Developments in Eastern Europe and the CIS*. London; New York: LSE/Routledge, 1994.

Nardin, T., and D. Mapel, eds. *Traditions of International Ethics*. Cambridge; New York: Cambridge University Press, 1992.

Nowak, M. *Commentary on the U.N. Covenant on Civil and Political Rights*. Kehl am Rhein: N.P. Engel, 1993.

Rhoodie, E. *Discrimination Against Women: A Global Survey of the Economic, Educational, Social and Political Status of Women*. Jefferson, NC: McFarland, 1989.

Robertson, A., and J. Merrills. *Human Rights in the World*. 3rd ed. New York: Manchester University Press, 1989.

Robertson, A., and J. Merrills. *Human Rights in Europe: A Study of the European Convention on Human Rights*. 3rd ed. Manchester: Manchester University Press, 1993.

Rodley, N. *The Treatment of Prisoners Under International Law*. Paris: United Nations Educational, Scientific and Cultural Organization; Oxford: Clarendon Press; New York: Oxford University Press, 1987.

Roht-Arriaza, N. *Impunity and Human Rights in International Law and Practice*. New York: Oxford University Press, 1995.

Shue, H. *Basic Rights*. Princeton, NJ: Princeton University Press, 1980.

Sieghart, P. *The International Law of Human Rights*. Oxford: Clarendon Press, 1983.

Sieghart, P. *The Lawful Rights of Mankind: An Introduction to the International Code of Human Rights*. Oxford; New York: Oxford University Press, 1985.

Sohn, L. *Rights in Conflict: The United Nations and South Africa*. Irvington, NY: Transnational Publishers, 1994.

Sohn, L., ed. *Guide to Interpretation of the International Covenant on Economic, Social, and Cultural Rights*. Irvington, NY: Transnational Publishers, 1993.

Sohn, L. and T. Buergenthal, eds. *The Movement of Persons Across Borders*. Washington, DC: American Society of International Law, 1992.

Steiner, H., ed. *Ethnic Conflict and the U.N. Human Rights System*. Cambridge, MA: Harvard Law School Human Rights Program, 1995.

Thornberry, P. *International Law and the Rights of Minorities*. Oxford: Clarendon Press; New York; Oxford University Press, 1991.

Tolley, H. *The U.N. Commission on Human Rights*. Boulder, CO: Westview Press, 1987.

Tolley, H. *The International Commission of Jurists: Global Advocates for Human Rights*. Philadelphia: University of Pennsylvania Press, 1994.

Van Dyke, V. *Human Rights, Ethnicity and Discrimination*. Westport, CT: Greenwood Press, 1985.

Vasak, K. and P. Alston, eds. *The International Dimensions of Human Rights*. 2 vols. Westport, CT: Greenwood Press; Paris: United Nations Educational, Scientific and Cultural Organization, 1982.

Vincent, R. J. *Human Rights and International Relations*. New York: Cambridge University Press, 1986.

Index

Note: This index is not exhaustive and does not include author's names. It should be used in conjunction with the Table of Contents.